HANDBOOK OF ENVIRONMENTAL FLUID DYNAMICS

OVERVIEW AND FUNDAMENTALS

VOLUME ONE

EDITED BY
H. J. S. FERNANDO

CRC Press
Taylor & Francis Group
Boca Raton London New York

CRC Press is an imprint of the
Taylor & Francis Group, an **informa** business

CRC Press
Taylor & Francis Group
6000 Broken Sound Parkway NW, Suite 300
Boca Raton, FL 33487-2742

First issued in paperback 2019

ISBN-13: 978-0-367-44587-4 (pbk)
ISBN-13: 978-1-4398-1669-1 (hbk)

Visit the Taylor & Francis Web site at
http://www.taylorandfrancis.com

and the CRC Press Web site at
http://www.crcpress.com

Library of Congress Cataloging-in-Publication Data

Handbook of environmental fluid dynamics / edited by Harindra Joseph Fernando.
v. cm.
Includes bibliographical references and index.
Contents: v. 1. Overview and fundamentals -- v. 2. Systems, pollution, modeling, and measurements.
ISBN 978-1-4398-1669-1
1. Fluid dynamics--Handbooks, manuals, etc. 2. Hydrogeology--Handbooks, manuals, etc. 3. Water quality--Handbooks, manuals, etc. 4. Fluids--Migration--Handbooks, manuals, etc. 5. Earth--Research--Handbooks, manuals, etc. I. Fernando, Harindra Joseph.

QC809.F5H33 2013
620.1'064--dc23 2012015977

HANDBOOK OF
ENVIRONMENTAL FLUID DYNAMICS

OVERVIEW AND FUNDAMENTALS

VOLUME ONE

In memory of

Owen M. Phillips

(December 30, 1930–October 13, 2010)

Contents

PART II Focus Areas for Study of Natural Systems

PART III Fundamental Flow Phenomena and Turbulence

Preface

The *Handbook of Environmental Fluid Dynamics* is a culmination of years of work by the contributing authors, publisher, and the volume editor and represents a state-of-the-science exposition of selected subtopics within environmental fluid dynamics (EFD). When I was first contacted by Luna Han (senior editor of Professional & Trade Books, Taylor & Francis Group LLC) in October 2008, regarding the editorship of this volume, I agreed to do so only with a brief reflection of its potential usefulness, perhaps not fully fathoming the task ahead. The original proposal included some 45 papers, and it quickly grew to about 85 topics with a moratorium declared thereafter. Some appealing topics had to be excluded, partly because of space limitations and last minute cancellations, with the hope of including them in a future edition. This two-volume set represents 81 chapters written by a cadre of 135 authors from 20 countries. Without the dedication and commitment of these authors, notwithstanding their regular professional obligations, this volume could not have come to light, and my sincere appreciation goes to all of them.

During the course of the preparation of this volume, two of our accepted contributors, Gerhard H. Jirka and Daewon W. Byun, passed away unexpectedly, and we fondly remember them. Gerhard's chapter, "Jets and Plumes," (Volume One, Chapter 25) was written by his students and collaborators. Daewon wrote his chapter, "Air Pollution Modeling and Its Applications" (Volume Two, Chapter 26), by himself. May their souls rest in peace!

The discipline of EFD descended from the revered geophysical fluid dynamics (GFD), the latter being the platform for fluid dynamics study of natural motions akin to climate, weather, oceanic circulation, and motions of the earth's interior. Conversely, EFD focuses on natural motions influenced by or directly affecting human (anthropogenic) activities. Applications of EFD abound in wind and ocean engineering as well as areas of human health and welfare. Air and water pollution are examples of the latter, which to a large extent sparked the early interest in EFD.

Environmental pollution has been recognized for centuries, as have the environmental regulations that date back to King Edward I of England, who decreed a ban on sea coal burning in 1306 because of its acrid odor. It was not until the 1940s that the pernicious human health effects of air and water pollution were realized beyond their aesthetic and human welfare woes. The appearance of Los Angeles smog or *gas attacks* in the 1940s and London *black fog* in December 1952 that killed more than 4000 clearly indicated the association of air pollution episodes with atmospheric inversions and complex topography. The roles of local pollutant emissions, flow, turbulent mixing, topography, and land use were soon realized in determining the *environmental quality* of urban areas. Thus was born EFD, which has now ramified into a plethora of other problems of the current epoch—the *anthropocene*. The public awareness of human-inflicted permanent damage to the environment started to develop, as exemplified by the 1962 novel by Rachel Carson, *Silent Spring*, and the policymakers became increasingly cognizant of environmental degradation that followed the industrial revolution. The U.S. EPA was established in 1970, and the advent of the Clean Air (1970) and Clean Water (1977) Acts prompted focused, funded research on how air and water flows interact with biogeochemophysical processes, providing a boost to EFD. While the early emphasis was on pollutant emissions, which could be effectively regulated, communities are now faced with additional existential challenges, for example, the hazards of climate change (floods, heat waves, and drainage), rapid urbanization (heat island), and scarcity of water and energy. The quest for economic development without further environmental and resource degradation led, in 1987, to the concept of *sustainability*.

Anthropogenic forcing occurs mainly at scales smaller than the urban scales, and thus are the problems of concern in EFD and for this two-volume handbook. Volume One starts with topics of general relevance (Part I), emphasizing the close relevance of EFD research in societal, policy, infrastructure, quality of life, security, and legal arenas. Some established or rapidly establishing focus areas akin to EFD are discussed next (Part II). Submesoscale flow processes and phenomena that form the building blocks of environmental motions are included in Part III, paying attention to turbulent motions and their role in heat, momentum, and species transport. In Volume Two, engineered structures and anthropogenic activities that affect natural flows and vice versa are discussed in Part I, and in its sequel, Part II, which deals with the most conspicuous anthropogenic effect, environmental pollution. Numerical methodologies that underpin research, predictive modeling, and cyber-infrastructure developments are included in Part III. All quantitative predictions on environmental motions need to be validated by laboratory experiments and field observations, which also help identify new phenomena and processes. These are addressed in Parts IV and V, respectively.

In closing, I would like to acknowledge those who contributed immensely to the preparation of this handbook. Luna Han patiently guided its development despite incessant delays. Jennifer McCulley carried out major administrative chores, assisted in the later stages by Marie Villarreal and Shenal Fernando. The back cover photo was kindly provided by a friend and skillful photographer, Jeff Topping. As always, all anxieties associated with the preparation of this handbook were magnanimously accommodated by my wife, Ravini.

It is my hope that this handbook will be a useful reference to students and professionals in EFD, and the efforts of those who labored to produce it will be handsomely rewarded through readership and dissemination of knowledge.

Harindra Joseph Shermal Fernando
Wayne and Diana Murdy Professor of
Engineering & Geosciences
University of Notre Dame
Notre Dame, Indiana

Editor

Harindra Joseph Shermal Fernando received his BSc (1979) in mechanical engineering from the University of Sri Lanka and his MS (1982) and PhD (1983) in geophysical fluid dynamics from the Johns Hopkins University. He received postdoctoral training in environmental engineering sciences at California Institute of Technology (1983–1984). From 1984 to 2009, he was affiliated with the Department of Mechanical and Aerospace Engineering at Arizona State University, ASU (assistant professor 1984–1987; associate professor 1988–1992; professor 1992–2009). In 1994, Fernando was appointed the founding director of the Center for Environmental Fluid Dynamics, a position he held till 2009, while holding a co-appointment with the School of Sustainability. In January 2010, he joined University of Notre Dame as the Wayne and Diana Murdy Endowed Professor of Engineering and Geosciences, with the primary affiliation in the Department of Civil and Environmental Engineering and Earth Sciences and a concurrent appointment in the Department of Aerospace and Mechanical Engineering.

Among awards and honors he received are the UNESCO Team Gold Medal (1979), Presidential Young Investigator Award (NSF, 1986), ASU Alumni Distinguished Research Award (1997), Rieger Foundation Distinguished Scholar Award in Environmental Sciences (2001), William Mong Lectureship from the University of Hong Kong (2004), and Lifetime Achievement Award from the Sri Lanka Foundation of the USA (2007). He is a fellow of the American Society of Mechanical Engineers, American Physical Society, and American Meteorological Society. He was elected to the European Academy in 2009. In 2007, he was featured in the *New York Times*, *International Herald Tribune*, and other international news media for his work on hydrodynamics of beach defenses. In closing the year 2008, the *Arizona Republic News* honored him by including in "Tempe Five Who Matter"—one of the five residents who have made a notable difference in the life of the city—in recognition of his work on the Phoenix Urban Heat Island.

He has served on numerous national and international committees and panels, including the Sumatra Tsunami Survey Panel (NSF, 2005), Louisiana Coastal Area Science and Technology Board (2006–2011), and the American Geophysical Union Committee on Natural Disasters (2006). He serves on the editorial boards of *Applied Mechanics Reviews* (associate editor), *Theoretical and Computational Fluid Dynamics*, IAHR *Journal of Hydro-environment Research* (associate editor), and *Nonlinear Processes in Geophysics* (editor). He is the editor in chief of *Environmental Fluid Dynamics*. He has published more than 225 papers spanning nearly 50 different international peer-review journals covering basic fluid dynamics, experimental methods, oceanography, atmospheric sciences, environmental sciences and engineering, air pollution, alternative energy sources, acoustics, heat transfer and hydraulics, river hydrodynamics, and fluids engineering. He is also the editor of the books entitled *National Security and Human Health Implications of Climate Change* (2012) and *Double Diffusive Convection* (1994).

Contributors

Malcolm J. Andrews
XCP-4, Methods and Algorithms
Los Alamos National Laboratory
Los Alamos, New Mexico

Jaywant H. Arakeri
Department of Mechanical Engineering
Indian Institute of Science
Bangalore, India

Peter G. Baines
Department of Civil and Environmental Engineering
University of Melbourne
Melbourne, Victoria, Australia

S. Balachandar
Department of Mechanical & Aerospace Engineering
University of Florida
Gainesville, Florida

Tobias Bleninger
Department of Environmental Engineering
Federal University of Paraná
Curitiba, Brazil

O. Bokhove
University of Leeds
Leeds, United Kingdom

and

Department of Applied Mathematics
University of Twente
Enschede, the Netherlands

Lydia Bourouiba
Department of Mathematics
Massachusetts Institute of Technology
Cambridge, Massachusetts

John W.M. Bush
Department of Mathematics
Massachusetts Institute of Technology
Cambridge, Massachusetts

R. Chant
Institute of Marine and Coastal Sciences
Rutgers—The State University of New Jersey
New Brunswick, New Jersey

K.W. Choi
Department of Civil Engineering
University of Hong Kong
Pokfulam, Hong Kong

A.B. Clarke
School of Earth and Space Exploration
Arizona State University
Tempe, Arizona

Benoit Cushman-Roisin
Thayer School of Engineering
Dartmouth College
Hanover, New Hampshire

Steven F. Daly
Engineer Research and Development Center
United States Army Corps of Engineers
Hanover, New Hampshire

Peter A. Davies
Department of Civil Engineering
University of Dundee
Dundee, United Kingdom

Robert L. Doneker
MixZon, Inc.
Portland, Oregon

Robert E. Ecke
Center for Nonlinear Studies
Los Alamos National Laboratory
Los Alamos, New Mexico

Robert Ettema
Department of Civil and Architectural Engineering
University of Wyoming
Laramie, Wyoming

Harindra Joseph Shermal Fernando
Department of Civil and Environmental Engineering and Earth Sciences
University of Notre Dame
Notre Dame, Indiana

J.H. Fink
Research & Strategic Partnerships
Portland State University
Portland, Oregon

J.J. Finnigan
Marine and Atmospheric Research
Commonwealth Scientific and Industrial Research Organisation
Canberra, Australian Capital Territory, Australia

Jørgen Fredsøe
Department of Mechanical Engineering
Technical University of Denmark
Lyngby, Denmark

Mohamed Gad-el-Hak
Department of Mechanical Engineering
Virginia Commonwealth University
Richmond, Virginia

M.S. Ghidaoui
Department of Civil and Environmental Engineering
Hong Kong University of Science and Technology
Kowloon, Hong Kong

Fernando F. Ginstein
Los Alamos National Laboratory
Los Alamos, New Mexico

Yakun Guo
School of Engineering
University of Aberdeen
Aberdeen, United Kingdom

D. Haidvogel
Institute of Marine and Coastal Sciences
Rutgers—The State University of New Jersey
New Brunswick, New Jersey

G.J.F. van Heijst
Fluid Dynamics Laboratory
Eindhoven University of Technology
Eindhoven, the Netherlands

Miki Hondzo
Department of Civil Engineering
University of Minnesota
Minneapolis, Minnesota

Huei-Ping Huang
School for Engineering of Matter, Transport and Energy
Arizona State University
Tempe, Arizona

Julian Hunt
University of Cambridge
Cambridge, United Kingdom

and

Department of Earth Sciences
University College London
London, United Kingdom

Herbert E. Huppert
Department of Applied Mathematics and Theoretical Physics
University of Cambridge
Cambridge, United Kingdom

Pablo Huq
College of Earth, Ocean and Environment
University of Delaware
Newark, Delaware

Richard M. Iverson
Cascades Volcano Observatory
United States Geological Survey
Vancouver, Washington

G.N. Ivey
School of Environmental Systems Engineering
and
Oceans Institute
University of Western Australia
Perth, Western Australia, Australia

Bryan W. Karney
Department of Civil Engineering
University of Toronto
Toronto, Ontario, Canada

Ben Kneller
Department of Geology and Petroleum Geology
University of Aberdeen
Aberdeen, United Kingdom

Satoru Komori
Department of Mechanical Engineering and Science
Kyoto University
Kyoto, Japan

Ruby Krishnamurti
Geophysical Fluid Dynamics Institute
Florida State University
Tallahassee, Florida

Edward Kwicklis
Earth and Environmental Science Division
Los Alamos National Laboratory
Los Alamos, New Mexico

M.Y. Lam
Department of Civil and Environmental Engineering
Hong Kong University of Science and Technology
Kowloon, Hong Kong

Joseph H.W. Lee
Department of Civil Engineering
Hong Kong University of Science and Technology
Clear Water Bay, Hong Kong

Sang-Mi Lee
South Coast Air Quality Management District
Diamond Bar, California

Rodman Linn
Earth and Environmental Science Division
Los Alamos National Laboratory
Los Alamos, New Mexico

Jim McElwaine
Department of Applied Mathematics and Theoretical Physics
University of Cambridge
Cambridge, United Kingdom

Chiang C. Mei
Department of Civil and Environmental Engineering
Massachusetts Institute of Technology
Cambridge, Massachusetts

Eckart Meiburg
Department of Mechanical Engineering
University of California, Santa Barbara
Santa Barbara, California

Iehisa Nezu
Department of Civil and Earth Resources Engineering
Kyoto University
Kyoto, Japan

Taka-aki Okamoto
Department of Civil and Earth Resources Engineering
Kyoto University
Kyoto, Japan

Eric R. Pardyjak
Department of Mechanical Engineering
University of Utah
Salt Lake City, Utah

E.G. Patton
Mesoscale and Microscale Meteorology (MMM)
National Center for Atmospheric Research
Boulder, Colorado

Patrick J. Paul
Snell & Wilmer L.L.P.
Phoenix, Arizona

Hillel Rubin
Faculty of Civil & Environmental Engineering
Technion—Israel Institute of Technology
Haifa, Israel

Martin Schmid
Surface Waters—Research and Management
Eawag—Swiss Federal Institute of Aquatic Science and Technology
Kastanienbaum, Switzerland

Scott A. Socolofsky
Zachry Department of Civil Engineering
Texas A&M University
College Station, Texas

Howard A. Stone
Department of Mechanical and Aerospace Engineering
Princeton University
Princeton, New Jersey

Bruce R. Sutherland
Department of Physics
and
Department of Earth and Atmospheric Sciences
University of Alberta
Edmonton, Alberta, Canada

D.M. Tartakovsky
Department of Mechanical and Aerospace Engineering
University of California, La Jolla
La Jolla, California

Jessica L. Thompson
Department of Human Dimensions of Natural Resources
Colorado State University
Fort Collins, Colorado

A.R. Thornton
Department of Applied Mathematics
and
Department of Mechanical Engineering
University of Twente
Enschede, the Netherlands

Marius Ungarish
Department of Computer Science
Technion—Israel Institute of Technology
Haifa, Israel

S. Karan Venayagamoorthy
Department of Civil and Environmental Engineering
Colorado State University
Fort Collins, Colorado

S.I. Voropayev
Department of Civil and Environmental Engineering and Earth Sciences
University of Notre Dame
Notre Dame, Indiana

and

P.P. Shirshov Institute of Oceanology
Russian Academy of Sciences
Moscow, Russia

C.L. Winter
Department of Hydrology and Water Resources
University of Arizona
Tucson, Arizona

Ken T.M. Wong
Department of Civil Engineering
University of Hong Kong
Pokfulam, Hong Kong

M. Grae Worster
Department of Applied Mathematics and Theoretical Physics
University of Cambridge
Cambridge, United Kingdom

Alfred Wüest
Surface Waters—Research and Management
Eawag—Swiss Federal Institute of Aquatic Science and Technology
Kastanienbaum, Switzerland

I

Overview and General Topics

1

Environmental Fluid Dynamics: A Brief Introduction

Harindra Joseph
Shermal Fernando
University of Notre Dame

1.1 Introduction

The planet Earth is just one of the 10^{22} planets in the universe, yet it is singularly unique, as it is the only planet known to maintain the precise conditions to sustain life. First, the earth is covered by an *atmosphere*, which is held in place by sufficiently large gravitational attraction due to its large mass (6×10^{24} kg) and radius (average 6371 km). Second, the atmosphere contains essential gases such as oxygen (~21% by volume), nitrogen (~78%), and carbon dioxide (~0.03%), in requisite amounts for breathing, biogeochemical cycling, and for maintaining a narrow temperature range conducive for life. Third, the average distance from the earth to the sun (1.49×10^{11} m) is such that the temperatures on the earth are neither extremely hot nor cold so that the water can be in liquid state, and the nearly circular (variation <3.5%) orbit of the earth around the sun ensures smaller seasonal temperature swings. Fourth, the rate of rotation and thermal and gravitational forcing of the earth are just right for the atmosphere, oceans, and the earth's interior to be in a *quasi stable* or an *equilibrium* state. This *state* of the planet, known as the *climate*, prima facie is in a delicate balance dictated by the ultimate forces of universe.

Five major components are believed to determine the climate: the *atmosphere* (~100 km blanket of gasses around the earth), *hydrosphere* (encompassing all water bodies from aquifers to clouds, e.g., lakes, streams, and oceans), *lithosphere* (upper part of the solid earth, containing sediments, rocks, and soil), *biosphere* (regions that sustain life, covering the lithosphere, portions of the hydrosphere, and the lower atmosphere), and *cryosphere* (ice, snow, glaciers, and permafrost; it holds ~2% of the water supply, covering 5.7% of the earth's surface). Common constituents, transformations, interactions, and feedbacks intertwine these spheres. For example, melting of the cryosphere would raise the sea level by tens of meters, shut off thermohaline circulation of oceans, and drastically change weather patterns and heat distribution in the atmosphere. Oceans, the major component of the hydrosphere, contain ~97% of the world's water supply. They cover 75% of the earth surface with an average depth of 3.7 km, maximum depth of 11 km (Mariana trench), and a mass of 1.4×10^{21} kg (Gill 1982). Oceans are also the dominant reservoirs of carbon and hence are central to the understanding of climate variability (Sarmiento and Gruber 2002). The intersection of the biosphere with other major climatic spheres, with lithosphere as the platform, is called the *ecosphere* or simply put—our *Environment* (Figure 1.1). The physical and biological (humans, plants, animals, and microorganisms) aspects of the ecosphere are studied in *Ecology* (Odum 1971).

In 2008, the world crossed an important milestone, in that most people now live in urban rather than rural areas (Science 2011). Concentrated human (anthropogenic) activities exert stresses on the environment to the extent of causing appreciable change of climate, and humans have now entered the epoch

Handbook of Environmental Fluid Dynamics, Volume One, edited by Harindra Joseph Shermal Fernando. © 2013 CRC Press/Taylor & Francis Group, LLC. ISBN: 978-1-4398-1669-1.

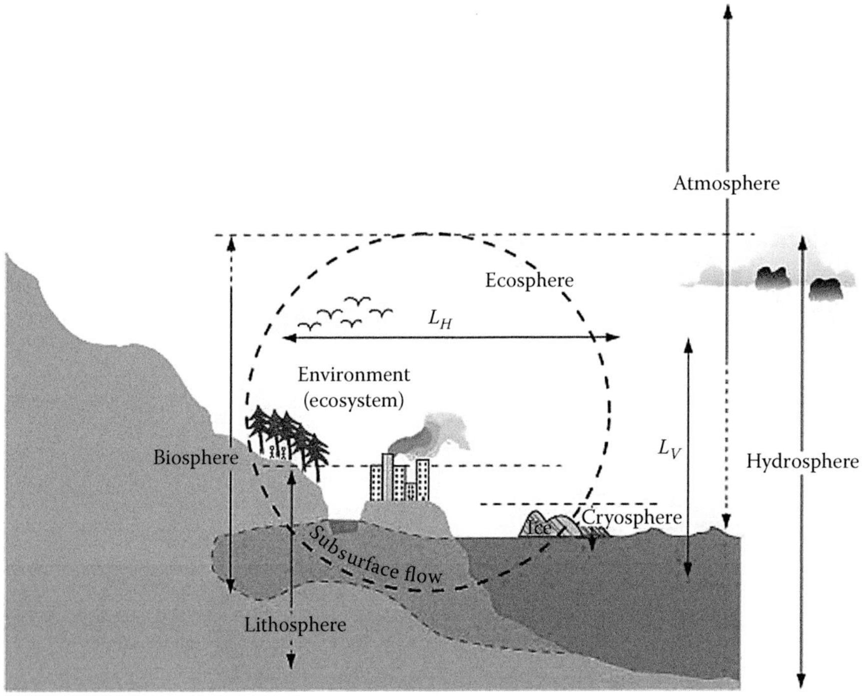

FIGURE 1.1 The *components* of the earth's climate system and spatial scales of an ecosystem. The influence of the polisphere propagates upscale via flow and dispersion, which, in turn, penetrates down the scales. Quantification of urban influence on itself and global scales remains a challenge.

of *anthropocene*. An additional climatic sphere, the *polisphere*, therefore needs to be introduced to represent anthropogenic stressors of urban systems (Fernando 2008). In environmental studies, appropriate spatial and temporal scales are selected, based on a *patch* of ecological interest (*ecosystem*), thus simplifying the task of delving into the vast global environment in piecemeal manner. Fluid motions on the earth, a communication medium between all spheres, span over 10 decades of scales, with strong interactions among the scales.

An ecosystem with horizontal and vertical spatial scales of L_H and L_V, respectively, is illustrated in Figure 1.1. Although a given spatial scale is associated with a myriad of time scales, it is customary to link it with a dominant energy-bearing time scale T. Tables 1.1 and 1.2 attempt to identify such scales for oceanic and atmospheric flows. The study of natural flows

TABLE 1.1 Atmospheric Scales

Motion Scale	Space Scale L_H	Timescale T
Global/climate	~10^4 km	Centuries/months
Synoptic (regional)	10^3–10^4 km	Days to months
Meso	~10^2 km	Hours to days
Urban	10–100 km	Minutes to days
Neighborhood	1–10 km	Hours to days
City (CBD)	1 km	Minutes to hours
Atmospheric boundary layer	1–100 m	Minutes to hours
Street Canyon scale	10–100 m	Seconds to hours
Personal cloud	~1 m	Seconds to minutes
Kolmogorov	~1 mm	Seconds
Sub Kolmogorov	100 nm–1 mm	<Seconds

TABLE 1.2 Oceanic Scales

Motion Scale	Space Scale L_H	Timescale T
Climate/global	10^3–10^4 km	Centuries/months
Regional	100–1000 km	Months to years
Basin	1000 km	Weeks to years
Meso	~10 km	Days to weeks
Fine	1–100 m	Minutes to hours
Small	<1 m	Seconds to minutes
Kolmogorov	1 mm	Seconds

based on equations of motion broadly falls within the area of *geophysical fluid dynamics* (GFD), which encompasses the realms of (physical) oceanography, limnology, meteorology, and solid-earth geophysics. Following the pioneering work of Laplace on tidal forcing, GFD has traditionally dealt with motions directly affected by earth's rotation, and hence with the low Rossby number regime. Meso- and larger scales are in this domain, with strong nonlinear interactions among themselves through processes such as barotropic and baroclinic instabilities and with smaller scales via shear and convective instabilities. An example is the link between global climate change and the reduction of the diurnal temperature range (DTR) of the atmospheric boundary layer (Karl et al. 1993). Conversely, changes to the local boundary layer, for example, due to land use change, can have climatic impacts (Fernando et al. 2012a).

Fluid motions akin to anthropogenic activities are studied in *environmental fluid dynamics* (EFD), which are usually (but not always) unaffected by earth's rotation. In other words, penetration of human influence to environmental motions is at the nub of EFD

FIGURE 1.2 A natural ecosystem, characterized by clean air, water, terrain, flora, and fauna. They are increasingly intruded by factitious elements. The challenge is to maintain ecosystem services with minimum impacts on the environment—known as *sustainable development*.

(Figure 1.2), thus demarcating it from GFD, albeit nebulously. The pragmatic connotations of EFD to humans and ecosystems permeate into *quality of life* (e.g., safety, health, and comfort), engineering designs, environmental planning and management, and hence policymaking. EFD problems tend to be multidisciplinary, and thus, many subtopics have emerged within EFD: urban fluid dynamics (Fernando et al. 2001), geological fluid dynamics (Huppert 2000), and ecological fluid dynamics (e.g., Goodman and Robinson 2008; Hondzo and Wuest 2009), to name a few.

A conspicuous application of EFD is flow and pollution dispersion in cities (Fernando et al. 2001, 2010). While external pressure gradients and thermal circulation driven by diurnal solar forcing are responsible for the approaching background mesoscale flow, which is studied in meteorology, flow adjustments and enhanced turbulence within building canyons, flow introduced by differential heating of building surfaces, heat and turbulence in automobile wakes, effects of HVAC systems, influence of irrigation and plant canopies, dust entrainment, and distribution of pollution from industrial sources, all of which have anthropogenic bent, are in the EFD domain. On the other hand, such topics as stable boundary layer cut across meteorology, oceanography, and EFD (Fernando and Weil 2010). In all, the divisions are more subjective, and vast overlaps of GFD and its off springs are evident—which reflects from the coverage of this volume.

1.2 Examples

Some applications of EFD are illustrated in Figure 1.3 and discussed in the following. In the urban context, historically cities have emerged surrounding water resources and hence are mostly

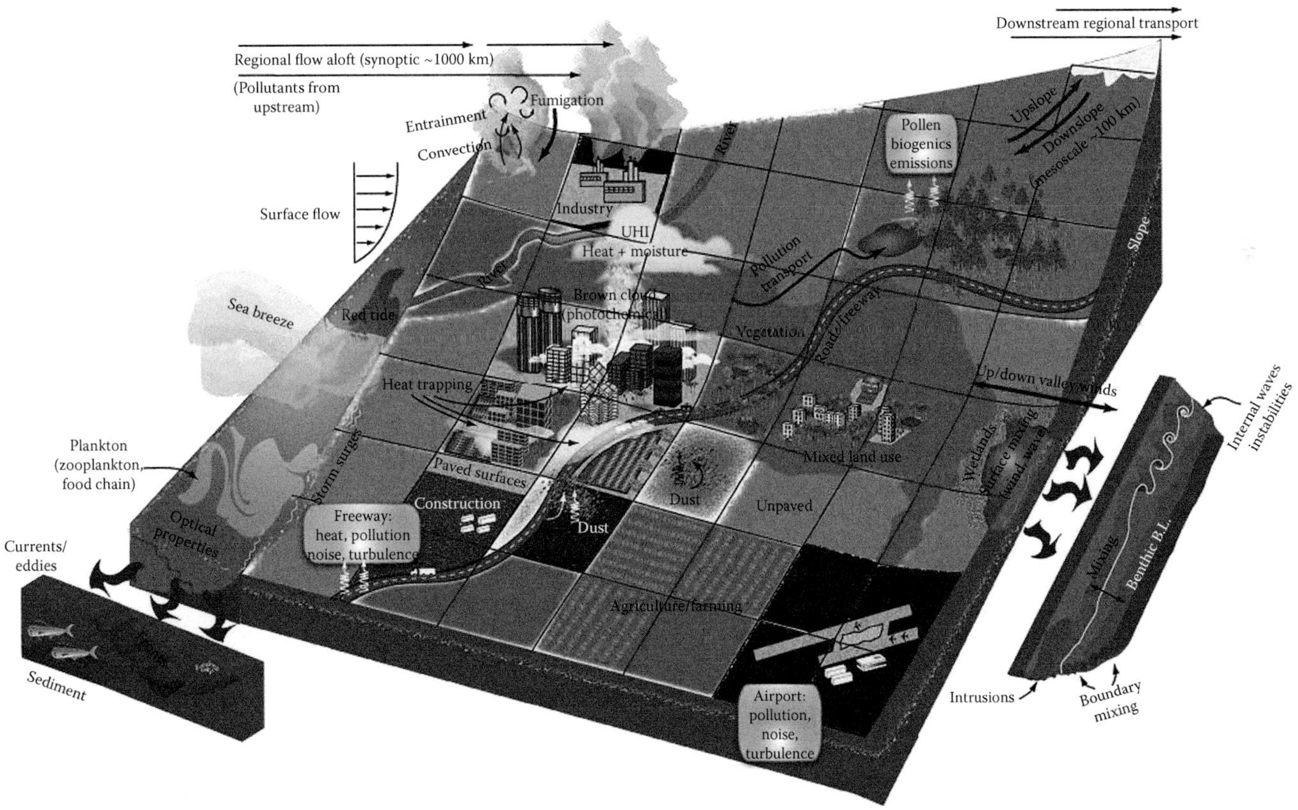

FIGURE 1.3 **(See color insert.)** A schematic of anthropogenically modified environs, and pertinent fluid motions. Coasts, water bodies, cities, and terrain add Sisyphean complexity to modeling and prediction. (Adapted from Fernando, H.J.S. et al., *Phys. Fluids*, 22, 051301, 2010.)

located in complex terrain (Fernando 2010). Water and air quality modeling for complex topography is challenging, which become even more problematic for coastal areas because of ocean–land–atmosphere interactions. More than 50% of urban areas are coastal, and currently only about 15% of the U.S. coastlines have been developed compared with 5% along the interior waterways; the former is expected to grow to 25% by 2025. Tides, salinity intrusions, wetlands, mangroves, estuaries, bays, beach erosion, coral cover, scour, shoaling, waves, storm surges, tsunamis on the ocean side, and sea/land breezes and urban heat island (UHI) on the atmosphere side are some overarching flow issues that impact the livability of coastal cities. These features are tied to economic activities such as mineral and oil exploration, effluent outfalls, harbors, navigation, fisheries, and recreation as well as urban security (Fernando et al. 2005; Paola et al. 2011). Engineering of rivers with levees, canals, and diversions has bearing on natural processes such as river meandering, sedimentation, wetlands, wildlife, and land formation.

Lakes provide water for recreation, agriculture, and human consumption, and quality of water therein depends on physical and biogeochemical processes (Imberger and Patterson 1990). Flow processes in limnology include seiching in lake basins, strait flows, small-scale mixing in the presence of weak turbulence (where molecular diffusion becomes important), sinking and trapping of microorganisms, and selective withdrawal. The quality and optical properties of coastal waters and lakes are determined by boundary and thermocline mixing, waves and turbulence, air–sea interactions, run off, subsurface processes, vegetation, and nutrients. Fundamental studies of flow–structure interaction, biological and nutrient cycling, two-phase flows, gravity and turbidity currents, debris flow, sediment transport, morphodynamics, and turbulence cut across aforementioned application areas.

The nonlinear response of anthropogenic forcing on the environment leads to the so-called *emergent* phenomena; examples include UHI in the atmosphere and dead zones in water bodies. The former arises due to the high heat capacity of urban built elements, which retain heat absorbed during the day for a period longer than that of (rural) surroundings. The result is a warmer city at night compared to the suburbs. The UHI is abetted by the lack of moisture within the built environment, whence incoming solar radiation is mostly used to raise the air temperature rather than consuming as latent heat. If irrigation is present, evaporation may cause the city to be cooler during daytime (*urban oasis island*). The UHI increases the evaporation, and this moisture can be transported downstream to trigger thunderstorms and precipitation. The heat budget of the upper urban lithosphere is not well established and is an area of active research. Studies of natural hazards (e.g., hurricanes, storm surges, tsunamis, and pyroclastic flows) as well as their urban impacts are also emerging areas that underpin risk assessment.

The built environment and anthropogenic forcing strongly influence the urban atmospheric boundary layer (Fernando et al. 2010). Processes therein determine the local air pollutant dispersion from industries (e.g., stacks, metalworking, and mining), transportation (e.g., roadways, airports, boating), and agricultural and construction sites. On larger scales, cities are pollution islands (area sources), contributing to background concentrations of air pollutants and greenhouse gases (Crutzen 2004). Rising convective plumes loft pollutants to regional flow, which transports pollutants downstream. Vertical mixing and downward diffusion of pollutants (known as *fumigation*) occur when the regional flow is perturbed by turbulence induced by the surface roughness and UHI of an urban area located downstream. The turbulence in a city is related to its morphometry (Britter and Hanna 2003). Pollution plumes of megacities are a topic of considerable international interest, alongside climate change (Molina et al. 2010). Aviation, vehicles, and industry generate noise pollution, which is strongly dependent on the local mean flow, shear, and stratification (Ovenden et al. 2009).

The change of hydrologic cycle due to global change impacts water availability and quality. Thus, sustainable water management techniques are being actively pursued, most notably novel designs for hydraulic dams, aquifer recharge, water treatment, and flood control. Similarly, understanding of circulation, eddies, turbulence, instabilities, mixing, stratification, nutrients, and biological production (Klein and Lapeyre 2008) are processes linked to ecosystem management. Excessive primary production (eutrophication) that leads to anoxia (oxygen deficiency) and release of toxins that affects aquatic life (Smith 1998) remain crucial issues.

Embedded in urban areas are scales of considerable practical interest. The neighborhood scales (Table 1.1) coincide with ecological patches of heightened air pollutants, which are selected as the grid size of mesoscale air quality computations. The central business district (CBD) scale is of interest to security agencies and fast responders, as CBDs are a target of choice for chemical, biological, and radiological (CBR) releases. Enhanced spatial inhomogeneity of CBDs requires the use of high-resolution computational fluid dynamics (CFD) models for predictions. Street canyons, courtyards, buildings, parks, vehicles, tress, and heat sources all contribute to this heterogeneity. A myriad of processes, from mesoscale approach flow to transport of sub-Kolmogorov scale aerosols to human lungs, are of interest in epidemiological studies.

Air, water, and noise pollution are the main considerations in setting environmental protection regulations. The U.S. Environmental Protection Agency (EPA) has set National Ambient Air Quality Standards (NAAQS) for pollutants considered harmful to public health (primary standards) as well as public welfare (secondary standard) such as those damaging to animals and plants, visibility, and buildings. Six principal pollutants identified by the EPA, called *criteria pollutants*, include ozone, carbon monoxide, lead, nitrogen and sulfur dioxides, and particulate matter (of aerodynamic diameter less than $10\,\mu m$ (PM_{10}) and $2.5\,\mu m$ ($PM_{2.5}$)). The PM_{10} is contributed by the turbulent entrainment of dust, whereas combustion sources are mainly responsible for $PM_{2.5}$. In addition, currently a group of 188 air pollutants have been identified, which are suspected of being health hazards when present even in small amounts (e.g., benzene). EPA has recommended water quality criteria for the protection of aquatic life and human health in surface water, considering the impacts of about 150 pollutants. Spatial and temporal distribution of pollutants in environmental flows, their chemical transformation, and human exposure form the basis of environmental epidemiology and toxicology.

1.3 Principles

1.3.1 Equations of Motion and Dimensionless Variables

The conservation equations of mass, momentum, energy, and species form the quantitative basis of EFD. Consider a motion field with typical (horizontal/vertical) velocity (u_H, u_V), time (T_H, T_V), length (L_H, L_V), pressure (p_0), and buoyancy (b_0) scales, with a time scale T_b for buoyancy fluctuations. The normalized continuity equation becomes (Fernando and Voropayev 2010)

$$\frac{\partial u_\alpha}{\partial x_\alpha}+\frac{\partial w}{\partial z}\left(\frac{u_V L_H}{u_H L_V}\right)=0, \quad \alpha=1,2, \qquad (1.1)^n$$

where the nondimensional variables are $u_\alpha^* = u_\alpha/u_H$, $w^* = w/u_V$, $x^* = x/L_H$, and $z^* = z/L_V$, z being the vertical coordinate antiparallel to gravitational acceleration $-g\delta_{i3}$. The superscript n on equation number indicates normalized variables, and (*) has been dropped. When the motions are three dimensional, $u_V L_H/u_H L_V \sim 1$. Similarly, the momentum equations, to the Boussinesq approximation, can be written as

$$\left(\frac{L_H}{u_H T_H}\right)\frac{\partial u_\alpha}{\partial t}+u_\beta\frac{\partial u_\alpha}{\partial x_\beta}+w\frac{\partial u_\alpha}{\partial z}+\left(\frac{f L_H}{u_H}\right)\varepsilon_{\alpha jk}\ell_j u_k$$
$$=-\left(\frac{p_0}{\rho_0 u_H^2}\right)\frac{\partial p}{\partial x_\alpha}+\left(\frac{\nu}{L_H u_H}\right)\left[\frac{\partial^2 u_\alpha}{\partial x_\beta \partial x_\beta}+\left(\frac{L_H}{L_V}\right)^2\frac{\partial^2 u_\alpha}{\partial z^2}\right] \qquad (1.2)^n$$

$$\left(\frac{L_V}{L_H}\right)^2\left\{\frac{L_H}{u_H T_V}\frac{\partial w}{\partial t}+u_\beta\frac{\partial w}{\partial x_\beta}+w\frac{\partial w}{\partial z}\right\}=-\left(\frac{p_0}{\rho_0 u_H^2}\right)\frac{\partial p}{\partial z}$$
$$+\left(\frac{b_0 L_V}{u_H^2}\right)\tilde{b}+\left(\frac{\nu}{L_H u_H}\right)\left[\left(\frac{L_V}{L_H}\right)^2\frac{\partial^2 w}{\partial x_\beta \partial x_\beta}+\frac{\partial^2 w}{\partial z^2}\right], \qquad (1.3)^n$$

where

$\alpha, \beta = 1, 2$

$f\ell_j$ is the Coriolis parameter

ρ_0 is a reference density

$\tilde{b} = -g(\rho - \rho_0)/\rho_0$ is the buoyancy, which can be written in terms of its mean $\bar{b}$ and perturbation b values as $\tilde{b} = \bar{b}(z) + b$

ρ is the density

The buoyancy conservation becomes

$$\left(\frac{b_0}{u_V N^2 T_b}\right)\frac{\partial b}{\partial t}+\left(\frac{b_0}{L_V N^2}\right)u_\beta\frac{\partial b}{\partial x_\beta}+\left(\frac{b_0}{L_V N^2}\right)w\frac{\partial b}{\partial z}+w$$
$$=\left(\frac{\nu}{u_H L_H}\right)\left(\frac{b_0}{L_V N^2}\right)\left(\frac{\kappa}{\nu}\right)\left[\frac{\partial^2 b}{\partial x_\beta \partial x_\beta}+\left(\frac{L_H}{L_V}\right)^2\frac{\partial^2 b}{\partial z^2}\right], \qquad (1.4)^n$$

where

$N^2 = d\bar{b}/dz$ is the background buoyancy (Brünt–Väisälä) frequency

κ is the molecular diffusivity of the stratifying solute

1.3.2 Dimensionless Numbers and Scales

Several familiar nondimensional parameters appear in (1.2)–(1.4): the *Reynolds number*; $Re = U_H L_H/\nu$; *aspect ratio* $A = L_H/L_V$; *Rossby number* $Ro = U_H/fL_H$; *Euler number* $Eu = p_0/\rho_0 U_H^2$; *Schmidt number* $Sc = \nu/\kappa$; *Bulk Richardson number* $Ri_B = \Delta b_0 L_V/U_H^2$ (equivalently, an *internal Froude number* $Fr_i = U_H/\sqrt{\Delta b_0 L_V}$ with $Ri_B = Fr_i^{-2}$); and the time scale ratios $L_H/U_H T_H$ and $L_H/U_H T_V = U_V/L_V T_V$, which is a *Strouhal number* $St = \omega L_H/U_H$, where $\omega = T_H^{-1}$ is a frequency scale. The Ri_B in general differs from the gradient Richardson number $Ri_g = N^2/(\partial u_\alpha/\partial z)^2$ encountered in hydrodynamic stability studies, but they can be related. If the Boussinesq approximation is not invoked, the gravity term in (1.3) is associated with the Froude number $Fr = U_H/\sqrt{gL_V}$. Each dimensionless number signifies the ratio of two forcing terms in the equations of motion.

From (1.2) and (1.3), it appears that at large Re and when

$$L_V \gg L_H Re^{-1/2}, \qquad (1.5)$$

the viscous forces can be neglected compared with the inertial forces (i.e., convective acceleration $\underset{\sim}{u}\cdot\nabla\underset{\sim}{u}$). The scale $L_V = L_H Re^{-1/2}$ is the commonly encountered viscous boundary layer scale, and (1.5) is satisfied outside the boundary layer (c.f., the concept of Reynolds number similarity in Section 1.4). At a distance sufficiently away from the boundary, the condition $L_V \sim L_H$ is satisfied, whence $Re \gg 1$ implies that the viscous effects can be neglected compared with the inertial effects. Note that $\underset{\sim}{u}\cdot\nabla\underset{\sim}{u}$ is the only explicit nonlinear term in the equation of motion and hence key to generating additional (smaller) scales and emergent phenomena. When $Re \gg 1$, the inertial forces need to be balanced by other forces. For example, if the inertial and pressure gradient forces balance each other while $Eu \sim O(1)$, $Ro \gg 1$ and $Ri_B \ll 1$, the flow generates *three-dimensional turbulence*. If the inertial, buoyancy, and pressure force balance is established at high Reynolds number, then the flow becomes *buoyancy-dominated* or *stratified turbulence* [$Eu \sim O(1)$, $Ri_B \sim O(1)$, $Ro \gg 1$]. In addition, if $Ro \sim 1$ the flow becomes a rotating and stratified turbulent flow (Riley and Lelong 2000).

Turbulent flows consist of a range of length-time scales. Although the condition $Re = u\ell/\nu \gg 1$ based on energy bearing scales u and ℓ is satisfied globally, there can be smaller scales where the viscous and inertial effects become of the same order. Viscous influence starts appearing at the scale $\ell Re^{-1/2}$, which is the well-known *Taylor microscale* λ of turbulence, as evidenced from the earlier discussion. It is generally believed that scales much smaller than ℓ are isotropic, and major viscous influence is realized at a still smaller length scale ℓ_K. This, together with corresponding velocity and time scales, u_k and t_k, respectively, satisfies $Re_K = u_K\ell_K/\nu = O(1)$ and $t_K = \ell_k/u_k$. By hypothesizing that ε (the kinetic energy dissipation) and ν are the governing parameters, the dissipation (Kolmogorov) scales of turbulence can be written as $l_K \sim (\nu^3/\varepsilon)^{1/4}$, $u_K \sim (\nu\varepsilon)^{1/4}$, and $t_K = (\nu/\varepsilon)^{1/2}$. As such, ε is believed to occur at Kolmogorov scales, but recent high Reynolds number observations adumbrate a spectrum roll off at the length scale $\sim\lambda$ (Figure 1.4). This disparity needs further investigation.

In large-scale geophysical problems, the Coriolis terms (fU_H) dominate the inertial terms U_H^2/L_H or $Ro \ll 1$. If the Reynolds number is high, a balance may occur between Coriolis and viscous terms in wall regions, $fL_H/U_H \sim Re^{-1}(L_H/L_V)^2$ or $L_V \sim (\nu/f)^{1/2} = L_{EK}$, which holds within the Ekman layer or wall boundary layer. It differs from the inertial boundary layer discussed earlier (at $Ro \ll 1$) for which $L_V \sim L_H Re^{-1/2}$. Note that $L_V/L_H \sim \left(\nu/fL_H^2\right)^{1/2} = E_K^{1/2}$, where $E_K = \nu/fL_H^2$ is the *Ekman number*, and at distances larger than $L_H E_K^{1/2}$ from the boundary, the viscous influence can be neglected. For the mid-latitude atmosphere, $L_{EK} \sim (0.1/10^{-4})^{½} \approx$ 30 cm and for oceans $L_{EK} \sim$ 10 cm based on the molecular viscosity. Because of the presence of turbulence, the effective momentum (eddy) diffusivity of natural flows K_e is larger than ν, whence it has been suggested $L_{EK} \approx \pi\,(2K_e/f)^{1/2}$ (Kaimal and Finnigan 1994).

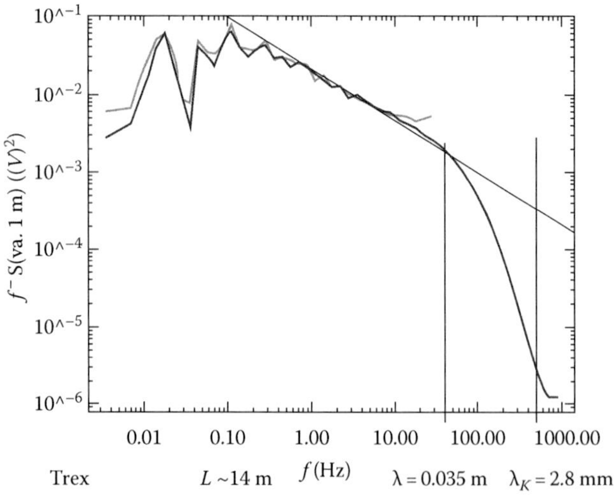

FIGURE 1.4 Variance preserving frequency (f) spectra of turbulent velocities in the atmosphere. Note the separation between Taylor (λ) and Kolmogorov (l_K) scales and the roll off of spectra at λ following Kolmogorov $f^{-5/3}$ regime. The full range of scales was obtained via hot-film anemometry and a limited range using sonic anemometry. (Courtesy of Greg Poulos, National Center for Atmospheric Research, Boulder, CO.)

1.3.3 Waves

When nonlinearities are negligible and a restoring force (e.g., gravity, buoyancy) is in place countering flow disturbances, repeating or nearly repeating distortions of physical properties may occur in space and time that typify wave motions. When the amplitude A is small, wave motion can be described by a set of linear equations (linear waves). In most cases the fluid (e.g., viscous) friction is small so as to maintain the repeatability. Any property P (e.g., pressure, surface displacement, etc.) of liner waves can be described by

$$L(P) = 0, \quad P = Ae^{i\chi}, \tag{1.6}$$

where χ is the "phase function," an indicator of the *repeatability* of waves. The wave number $\tilde{k}$ and the angular frequency n of waves are defined as $\tilde{k} = \nabla\chi = \partial\chi/\partial x_i$, $n = -\partial\chi/\partial t$, whence $\chi = \tilde{k}.\tilde{x} - nt$. Linear waves do not cause fluid medium to move, when averaged over a single cycle, and thus there is no mass transport. Nevertheless, waves can be efficient transporters of momentum. Waves frequented in natural flows include surface gravity waves, capillary waves, and internal gravity waves on the smaller scales, and inertial, inertio-gravity, and planetary waves on the larger scales (at low Ro). Because of the restoring force, exchange of energy present in different forms (e.g., potential, kinetic, or electromagnetic energy) may occur, thus sustaining oscillatory motions. For surface waves, this restoring force is either the gravity (surface gravity waves) or surface tension (capillary waves); for internal waves, it is the buoyancy ($L_H \sim L_V$; $L_H/U_H T_V \sim b_0 L_H/U_H^2 \gg 1$ or $T_V \sim U_H/b_0$); for inertial waves, the Coriolis force ($T_V \sim f^{-1}$); and for magneto-hydrodynamic waves, the Lorentz force.

Linear waves are a mature field of study, but most practical problems are in the realm of weakly nonlinear and nonlinear waves (Phillips 1977; Mei 2003). These include wave–current interactions, Langmuir circulation, transitioning of stratified/rotating turbulence to internal/inertial waves, differential momentum, heat and mass transfer, as well as sediment transport under waves.

1.3.4 Instabilities

Environmental motions are typically unstable in that perturbations to basic flow states grow in space and time, and after passing through a series of intermediate transition states, they break down into turbulence. A stability analysis would determine whether a perturbation to the basic flow would decay (stable), exponentially amplify (unstable), or remain the same (neutral). If A is sufficiently small, the governing equations can be linearized (linear stability analysis), and periodic solutions of the form (1.6) are sought. If the wave number or frequency becomes imaginary, the disturbance would grow exponentially, signifying instability. The flow is stable when all possible disturbances do not grow. The locus that demarcates stable and unstable states is the marginal stability curve on which the disturbances are neutral. In the marginal stability state, a small disturbance may evolve into a periodic state, for example, cellular patterns observed in Rayleigh Bernard convection. Sometimes the disturbance growth is slower than exponential, simultaneously exhibiting oscillatory behavior with a specific frequency/wave number, indicating that the restoring forces are greater than what caused the oscillations. Under these conditions, oscillations overshoot past the amplitude of previous excursion; this is referred to as *overstability* (Turner 1973).

Linear stability theory is also a mature topic in fluid mechanics, and both linear and weakly nonlinear stability theories are extensively used in EFD (Chandrasekhar 1961; Drazin and Reid 2004). Some examples include instabilities in stratified shear and vortex flows, two-phase flows, mushy zones, double-diffusive convection, morphological instabilities, and river meandering, many of which are addressed throughout this handbook. In addition to their amenability to elegant mathematical treatment, the aesthetic splendor of instabilities in nature has attracted the attention of scientists and artists alike.

1.3.5 Turbulence

Three-dimensional *active turbulence* can be defined as random motions where inertial-vortex forces are balanced by pressure gradient forces, while the influence of body (buoyancy, Coriolis, and electromagnetic) and viscous forces are negligible (Gibson 1991). When the body and inertial forces are on the same order, the turbulence (at those scales) is said to be *fossilized.* In general, buoyancy influence is first felt at larger length scales l of turbulence, say when $l > L_o$, where L_o is a threshold. When $L_o \gg L_K$, turbulence is active in $L_K < l < L_o$ and fossil at larger ($>L_o$) scales, leading to *active-fossil* turbulence. When $L_K \approx L_o$, no active three-dimensional turbulence can be present and the flow is *fully fossilized.* The hydrodynamic state of a stratified turbulent flow is dependent on a variety of factors, including the characteristics of stratification (N), the length (ℓ) and *rms* velocity ($\sqrt{u_i'^2}$) scales of turbulence, as well as the forcing mechanisms present; here the primed quantities denote turbulent fluctuations. Some features of stratified turbulence can be illustrated in terms of governing Equations 1.1 through 1.4 as follows.

If turbulence is imposed on a stratified fluid at time $t = 0$, for "active" turbulence to exist, the inertial terms should be of the leading order, and the flow is expected to follow

$$L_H \sim L_V \sim u_H t, \quad u_H \sim u_V, \quad \frac{b_0}{L_V N^2} \sim O(1). \tag{1.7}$$

This initial growth occurs until the buoyancy forces impede the growth of turbulence, when the conditions $b_0 L_V / u_H^2 \sim (L_V/L_H)^2$ or $N^2 L_V^2 / u_V^2 \sim N^2 L_H^2 / u_H^2 \sim O(1)$ are satisfied. Therefore, the vertical scale of turbulence at the onset of fossilization is $L_V \sim u_V/N$, which can also be derived using simple energy arguments (Fernando 1991). Two length scales of stratified turbulence can be identified using (1.7): $L_E = \sqrt{b'^2}/N^2 \sim b_0/N^2$ and $L_b = \sqrt{w'^2}/N \sim u_V/N$, the former (*Ellison*) scale representing the vertical scale of evolving turbulence and the latter (*buoyancy* scale) signifying the onset of buoyancy effects. The buoyancy influence sets in at a time scale $t_N \sim L_H/u_H \sim N^{-1}$, typically $t_N \approx (2-4)N^{-1}$. The Coriolis forces become important when $fL_H/u_H \sim O(1)$ or at a time $t_c \sim f^{-1}$. Since the time scale ratio t_C/t_N is large for oceans (~10) and atmosphere (~10^2), during the evolution of environmental turbulence, stratification effects come into play first.

At the onset of fossilization, L_b can be recast in terms of the well-known parameterization $\varepsilon \sim (\overline{w'^2})^{3/2}/L_V$ as $L_b \sim (\varepsilon/N^3)^{1/2}$, where $L_R = (\varepsilon/N^3)^{1/2}$ is the *Ozmidov* length scale. Since ε can be measured in oceans with a higher reliability than $\overline{w'^2}$, it is common to use L_R in oceanographic studies. Because of its space-time patchiness, stratified turbulence is intermittent, and hence, the flow can be undersampled during measurements. This is particularly acute in oceanic measurements, because repeated profiles conducive for averaging cannot be realized by multiple casts owing to platform (ship) drift and intermittency of turbulence (Gregg 1987). Although L_V is constrained by L_b, there are no immediate constraints on horizontal growth $\left(L_H \sim \int u_H dt \right)$, causing $(L_V/L_H) < 1$ characterized by large-scale anisotropy at large times. This anisotropy is felt even at smaller (dissipation) scales when the buoyancy scale $L_b \sim L_R \sim (\varepsilon/N^3)^{1/2}$ becomes on the order of the Kolmogorov scale ℓ_K or when the (Gibson) parameter $G = \varepsilon/\nu N^2$ drops below a critical value ~1000. As L_V/L_H becomes smaller, turbulence is confined to horizontal sheets or thin layers with strong shear in between.

Of great practical importance is the efficiency of mixing in stratified flows, defined in terms of the *flux Richardson number*

$$R_f = \frac{-\overline{w'b'}}{-\left(\overline{u'w'}(\partial U/\partial z) + \overline{v'w'}(\partial V/\partial z)\right)}, \tag{1.8}$$

where (U,V) are the horizontal mean velocities. Numerous experimental and theoretical studies exist on R_f (Peltier and Caulfield 2003; Ivey et al. 2008), given its direct relationship to "eddy diffusivities" that are critical for subgrid closure in environmental models. Direct measurements of R_f have been reported in the atmosphere that show disparities with previous laboratory measurements, which need to be addressed in future studies.

1.4 Methods of Analysis

1.4.1 Numerical Prediction

The prediction of environmental flows is fraught by complexities of initial and boundary conditions, buoyancy and Coriolis effects, turbulence, anthropogenic forcing, and a vast range of space-time scales of processes involved that cover more than 10 decades of scales. Extant computing technologies cannot resolve all natural processes within a single model nor are they being expected to be resolved in the foreseeable future. Therefore, analyses and predictions are conducted at a given scale or a limited range of scales by reducing the governing equations to a manageable

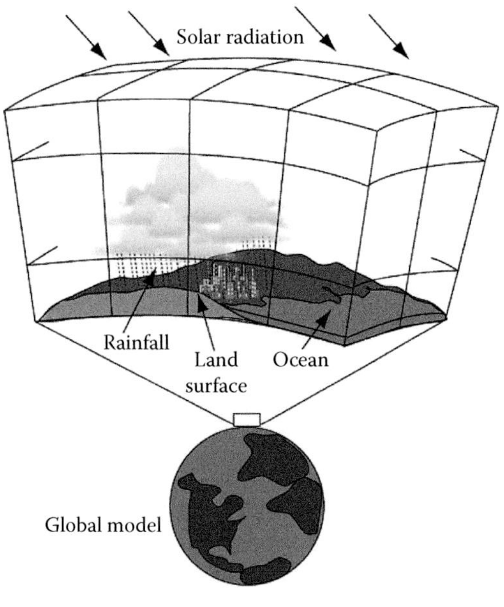

FIGURE 1.5 Depiction of an earth systems model. The processes (e.g., cities, mountain ranges) of scales smaller than horizontal resolution ~100 km need to be parameterized. The primitive equations of motions with subgrid closure are solved with boundary/initial conditions and global forcing. The atmosphere and oceans are treated separately.

mathematical set. As illustrated in Figure 1.5, the atmosphere-ocean general circulation models (AOGCM) cover the entire earth, and the primitive equations are solved with external inputs and bottom and top boundary conditions. Their bane is the low resolution (~100 km), requiring mathematical representation (*parameterization*) of salient unresolved processes within grid boxes via eddy coefficients of momentum (k_m) and species (k_c), viz.

$$\overline{u_i'u_j'} = 2k_m\left(\overline{e}_{ij} - \frac{1}{3}\theta\delta_{ij}\right) \quad (i \neq j) \tag{1.9}$$

and

$$\overline{u_i'c'} = k_c \frac{dC}{dx_i}, \tag{1.10}$$

where

$\theta = \partial u_i/\partial x_i$ is the dilatation

$\overline{e}_{ij}$ is the mean rate of strain

Naturally, the eddy coefficients are sensitively dependent on processes being modeled. In nested simulations, the output from AOGCMs is fed into mesoscale (~1–10 km resolution) models, which in-turn provides input to CFD models (resolution ~1–10 m). For the latter, variants of Reynolds Averaged Navier Stokes Simulations (RANS), Large-Eddy Simulation (LES), and Detached Eddy Simulation (DES) are used in practice. The most scientifically useful would be the Direct Numerical Simulation (DNS), where all turbulence scales down to Kolmogorov scales are resolved, but computer demands for such simulations are insurmountable. A challenge would be to develop eddy diffusivities of general utility to account for multiple processes. The *conditional diffusivities*, in which a particular process appears in a grid cell when certain criteria are satisfied, have been attempted; for example, see the work of Large et al. (1994). The addition of *k* values of prevailing processes has been used so far, and this approach appears to be too simplistic.

The indispensible role of eddy coefficients in modeling has prompted many fundamental studies, focusing on mixing processes and associated fluxes. An example is the parameterization of heat (k_T) and momentum (k_m) diffusivities for mesoscale models. A grid box near the ground, for instance, encompasses many processes—for example, turbulence due to roughness and instabilities, internal wave generation, radiation and breaking, land-air heat fluxes, and evapotranspiration. Biogeochemical processes are also of interest in integrated models, which sharply increases model complexity. Examples include anthropogenic activities of cities, biological transformation in natural waters, and evolution of particulate matter (PM) in the lower atmosphere due to entrainment of crustal material and chemical reactions.

Multiscale simulations consist of grid nesting and exchanging information between different scales, both down the scales (*downscaling*; Salathé et al. 2008) and from small to larger scales (*upscaling*). Downscaling is the most convenient and provides an opportunity for large-scale simulations to run independently of smaller scale models. *Upscaling* is tedious and couples simulations of multiple scales. Only limited progress has been made in upscaling studies, although their importance is widely appreciated. Scale interactions are of particular interest in cases where slow changes at larger scales may cause amplified response at smaller scales, sometimes leading to emergent phenomena (Section 1.2) as well as environmental disasters.

Experimentally capturing physical phenomena across a swath of space-time scales requires remote sensors or sensor networks of appropriate resolution and footprint. Dense, fast sensor networks with cyber infrastructure capable of transferring and processing information are being sought after for decision support and environmental management. On the global scale, for example, the World Meteorological Network's "Global Weather Watch" receives and archives atmospheric information. The Global Ocean Observing System (GOOS) under development is its oceanic counterpart. A number of networks operate on regional scales, for example, AmeriFlux for fluxes of momentum, CO_2, and energy. There are also networks with extensive spatial coverage with a myriad of observables. The FLUXNET, a global network of micrometeorological tower sites that measures fine-scale turbulence and mean quantities, is of this ilk so are the multispectral satellite instruments (e.g., ASTER, MODIS) that capture environmental information at meter scales. In model–data intercomparisons, individual data stations are usually taken as a footprint for a grid box, but within a given grid box the variability is extensive, and methods to handle such heterogeneity are required.

1.4.2 Process Modeling

Laboratory and numerical models are widely employed to gain fundamental insights on natural processes, concentrating on a single or at most a few processes. The translation of results from such models to improve operational predictability, nevertheless, occurs rather slowly. This is, in part, due to the facts that only a few research groups are working on operational model development/improvement and the laborious nature of model validation required after new implementations are made. In numerical modeling, the process-based approach has been extended to develop *reduced* or *parsimonious* models that include a selected set of processes, and such models have been useful for identifying new phenomena arising from nonlinear interactions between different processes. The literature on process modeling is voluminous, and this topic has been discussed extensively in reviews and monographs. A simple example is given in Section 1.5 on ocean bottom boundary layer, which, in laboratory or theoretical studies, is treated as equivalent to flow above an oscillating bottom of a semi-infinite volume of water.

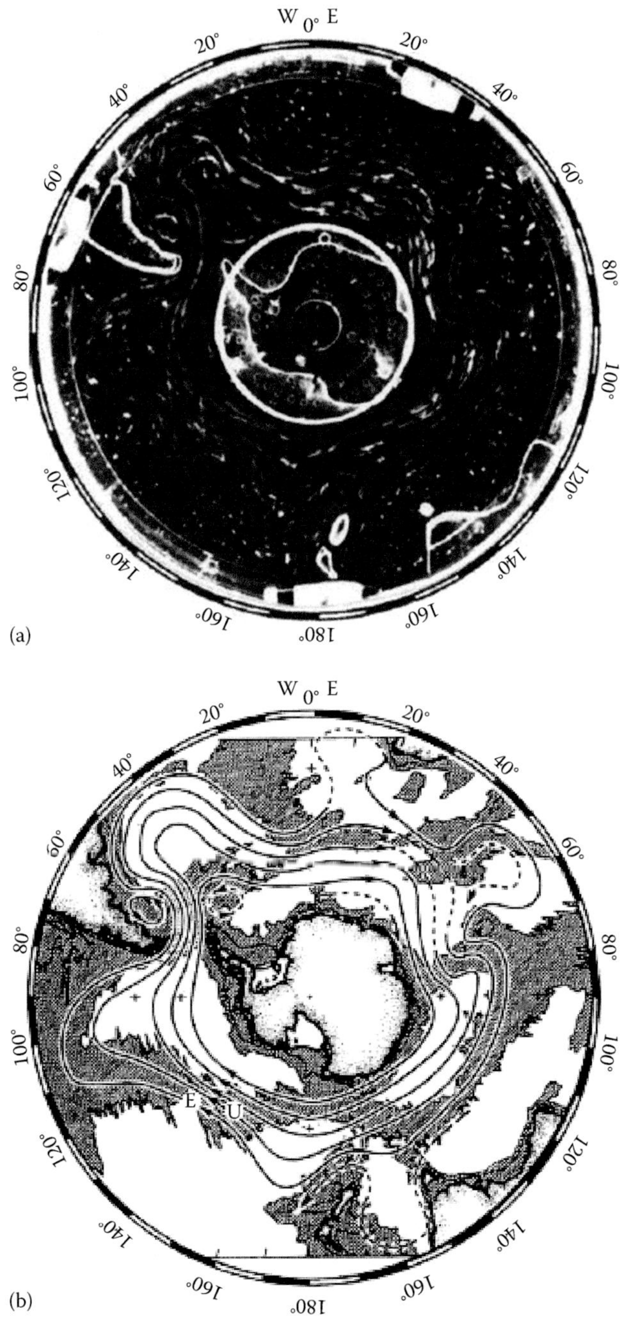

FIGURE 1.6 (a) Laboratory modeling of Antarctic Circumpolar Current in a rotating dishpan, with matching of a selected number of dimensionless parameters. (b) Streamlines from the experiments, which agree well with ocean observations. (From Boyer, D.L. et al., *J. Geophys. Res.*, 98(C2), 2587, 1993. With permission.)

Laboratory modeling has been used in two ways. The first uses traditional model testing based on dynamic similitude, where the matching of geometric and dynamic parameters is attempted to the extent possible between the model and actual (Boyer and Davies 2000). The most difficult to match are the aspect ratio (horizontal/vertical scales) and the Reynolds number, both of which are very large in geophysical flows. Similarity and physical arguments permit to reduce the number of relevant dynamic variables to a few and model environmental situations with some success (e.g., Figure 1.6). On the other hand, idealized flow configurations can be used without matching of nondimensional numbers, allowing the development of parameterizations of general utility and collection of benchmark data for model evaluation (Figure 1.7). Care must be taken to ensure that the model and actual flows are in the same hydrodynamic regime (e.g., laminar versus turbulent). Recent advances in flow

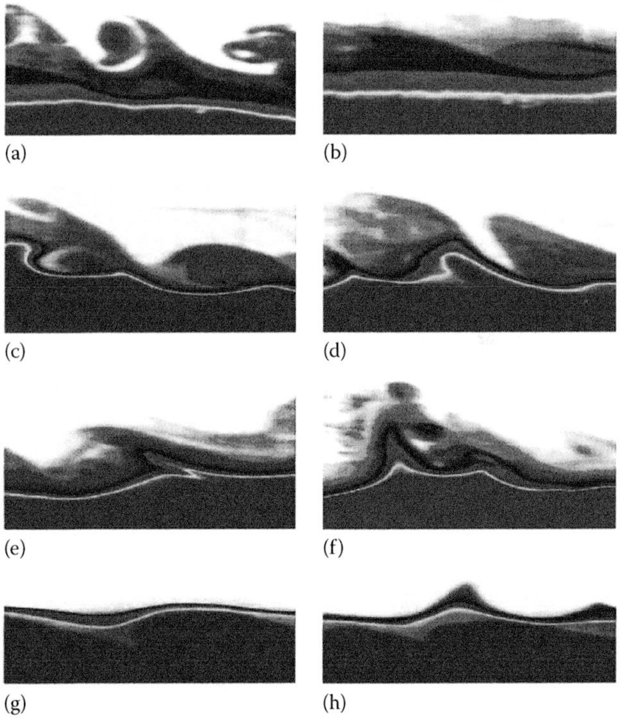

FIGURE 1.7 **(See color insert.)** Laboratory modeling of turbulent mixing in a two-layer stably stratified fluid, with fluid layers separated by a sheared density interface. A generic flow configuration was used here to obtain results of general validity. To (h) represent cases with different Bulk Richardson number ⟦Ri⟧_B (Section 1.3.2) with L_V replaced by the depth of the upper turbulent layer. (a) ⟦Ri⟧_B = 1.8, (b) 3.2, (c) 4.5, (d) 5.5, (e) shortly after (d), (f) 5.8, (g) 9.2 and (h) shortly after (d). (From Strang, E.J. and Fernando, H.J.S., *J. Fluid Mech.*, 428, 349, 2001. With permission.)

diagnostic techniques have greatly improved the quality, resolution, and volume of laboratory data.

1.4.3 Dimensional Arguments and Similarity Theories

Complexities of natural flows behoove a suite of mathematical tools for understanding and analysis of natural flows. In typical process studies, basic building blocks of the flow are identified and studied in depth. Perhaps the simplest analysis tool is the *dimensional analysis* (Bolster et al. 2011) and its elegant extensions such as the *generalized dimensional analysis* (Long 1963) and *similarity theories* (Baranblatt 1996). If a dependent parameter in dimensional domain is a function of n independent parameters ($a_1, a_2, \ldots, a_n$), the dimensional analysis axiomatizes that physical laws relating these quantities is independent on human-derived units and hence can be written as a functional relationship between dimensionless parameters. The Buckingham π Theorem states that the number of nondimensional (π) numbers possible from the total dimensional set is $(n - m + 1)$, m being the number of basic units involved, viz.,

$$\pi = \pi(\pi_1, \pi_2, \ldots, \pi_{n-m}), \tag{1.11}$$

where, generally in deriving π and π_i's, a_i's are involved at least once. The actual functional form of $\pi(\pi_i)$ needs to be derived using alternative methods, for example, laboratory/numerical experiments or physical arguments.

While classical dimensional analysis is a powerful tool, more detailed results are possible using extensions of it. Similarity solutions are one of them, where for evolving flows the results are expressed in terms of two dimensionless parameters, $\pi = \pi(\pi_1)$. Similarity theory allows obtaining of spatial distribution of a time evolving solution via similarity transformations. A simple example would be the turbulent wake of a cylinder of diameter D, where any property such as the velocity $u(x)$ at a downstream distance x along the centerline is dependent on the free-stream velocity U, D, and x, where the kinematic viscosity ν has been neglected (see the following). The ensuing similarity solution is simple, $u(x)/U = f(x/D)$, where $f, f_1, \ldots$ are functions.

When two or more independent π numbers are involved, similarity solutions are not guaranteed. In certain cases, however, it is possible to seek self-similar solutions at the extremes of π_i values. Consider

$$\pi = \pi(\pi_1, \pi_2), \tag{1.12}$$

which may be converted to a self-similar form of a single independent parameter in the limit of $\pi_2 \to 0$ or ∞ as

$$\pi = \pi(\pi_1)[\text{self similarity of the first kind}] \tag{1.13}$$

$$\pi = \pi_2^{\varpi}\left(\frac{\pi_1}{\pi_2^{\vartheta}}\right)[\text{self similarity of the second kind}], \tag{1.14}$$

where ϖ and ϑ are constant coefficients. An example for (1.13) would be the neglect of ν in the wake problem described earlier. When ν is included, $u(x)/U = f_1(x/D, Re)$, where $Re = UD/\nu$ is the Reynolds number, and thus for $Re \to \infty$ the dependence on it, and hence on ν, can be neglected. This procedure is referred to as the *Reynolds number similarity*. Although (1.13) and (1.14) are nominally applicable at extreme limits, they do have applications to intermediate stages of a phenomenon, when suitably interpreted. For example, if the wake example discussed earlier is complicated by the presence of stable stratification with buoyancy frequency N, then an extra π ($=U/ND$) term comes into play and the self-similarity is lost, viz.,

$$\frac{u}{U} = f_2\left(\frac{x}{D}, \frac{U}{ND}\right), \tag{1.15}$$

but it is possible to seek similarity solutions in the limit $x/D \to \infty$ as

$$\frac{u}{U} = \left(\frac{x}{D}\right)^m f_3\left(\frac{x}{D(U/ND)^n}\right) \quad \text{or} \quad \frac{u}{u_0} = f_3\left(\frac{x}{\ell_0}\right), \tag{1.16}$$

where (1.16) looks as if the flow has a new similarity velocity $u_0 = U(x/M)^m$ and length $\ell_0 = D(U/ND)^n$ scales, with diminished significance of D. Obviously, these new scales are valid when $x/D \to \infty$ but not exactly at those extremes (where the velocity tends to be either 0 or ∞, depending on m). Pragmatically, $x/D \to \infty$ solution is valid for $x \gg D$, that is at (intermediate) values of x much greater than D and less than $x \to \infty$. At these distances, the influence of D is decadent, leading to a new length scale. This exemplifies an *intermediate asymptotic* solution.

The scaling analysis presented in Section 1.3.1 is also helpful in delineating dominant dynamical balances and hence for experimental data analysis and obtaining exponents for (1.14). Such is a more rigorous form of dimensional analysis, allowing one to simplify the number of dimensional parameters based on physical insight and anisotropic scaling (Long 1963). The final solution, however, involves dimensionless (sometimes universal) constants, which need to be determined using numerical and experimental means. A combination of all these techniques is a powerful resource for quantification of processes while illuminating underlying physics.

1.4.4 Field Studies

The prediction of environmental motions are predicated by difficulties with subgrid closure, complexity of initial and boundary conditions, numerical issues, and sensitivities of numerical solution to boundary conditions (Lorenz 1976). As such, model validations with field data as well as model verifications are essential components of model development (Beven 2002). Field experiments are tedious because of the vast domains involved, spatial and temporal inhomogeneities within the domain, instrument limitations and complexities, logistics, need for a group of skilled personal with diverse expertise, setup and operational costs, efforts required for compliance with environmental regulations, as well as natural variability (Acts of God). The advent of remote sensors and satellites has added immensely to environmental research and monitoring, and if their scanning rates are faster than the variability of the processes of interest, useful four-dimensional information is possible, including delineation of individual processes and their evolution.

1.5 Illustrations

1.5.1 Environmental Simulations

In principle, at present, nested simulations downscale from global to 1 m scale. A repertoire of models exclusively dedicated for different scales are available, and Figure 1.8a and b show the results of a nested simulation where NCEP's eta model (synoptic), MM5 mesoscale model, and a CFD model with $k-\varepsilon$ closure (Baik et al. 2003) were used to predict flow in a downtown area at 2 m resolution. The computational domain is a commercialized area of Houston with buildings and street canyons. It shows velocity vectors at a specific point in time, and the model picks up typical flow features such as wakes, flow separation, and convergence/divergence zones. Obviously, simulations of this type need to be validated with well-designed field campaigns.

While nested simulations may include meter scale features, typical air quality calculations are based on mesoscale grids of size 1–12 km (Lee et al. 2007). Each grid represents a reasonable *footprint* of an urban, rural, or a mixed domain, known as the *neighborhood scale*. This is the scale over which health advisories are issued. Air quality predictions even on such scales are computationally daunting, mainly confined to research or for planning purposes. Although available photochemical models predict observed air quality distribution trends satisfactorily, quantitative predictions of pollutant concentrations, their distribution, and the timing of pollution episodes exhibit significant errors. Alternatives such as stochastic tools (e.g., neural networks) or empirical correlations are used for convenience, which gives area-specific predictions with acceptable reliability (Fernando et al. 2012b). In all, the current state of environmental forecasting leaves much to be desired, calling for improvements in model designs (architecture), numerical algorithms, and parameterizations. These as well as new paradigms of modeling ought to be given top priority in developing a vision for improved environmental forecasting (Jakob 2010).

1.5.2 Process Modeling

A case of simple process modeling is discussed in the following, by considering a deep water layer subjected to a reversing square wave shear stress at the bottom $z = 0$. The aim is to mimic the bottom boundary layer of a natural water body subjected to reversing tides. With respect to a coordinate system moving with the bottom, the flow aloft is seen as oscillating while exerting a reversing shear stress on the bottom. In the theoretical analysis, the problem can be posed as

$$\nu\frac{\partial^2 u}{\partial z^2} = \frac{\partial u}{\partial t},$$

with boundary conditions

$$-\mu\frac{\partial u}{\partial z} = T(t) \text{ at } z = 0, \quad T(t) = \begin{cases} \rho_0 u_*^2 & \text{for } 2ka < t < (2k+1)a \\ -\rho_0 u_*^2 & \text{for } (2k+1)a < t < (2k+2)a \end{cases}, \tag{1.17a,b}$$

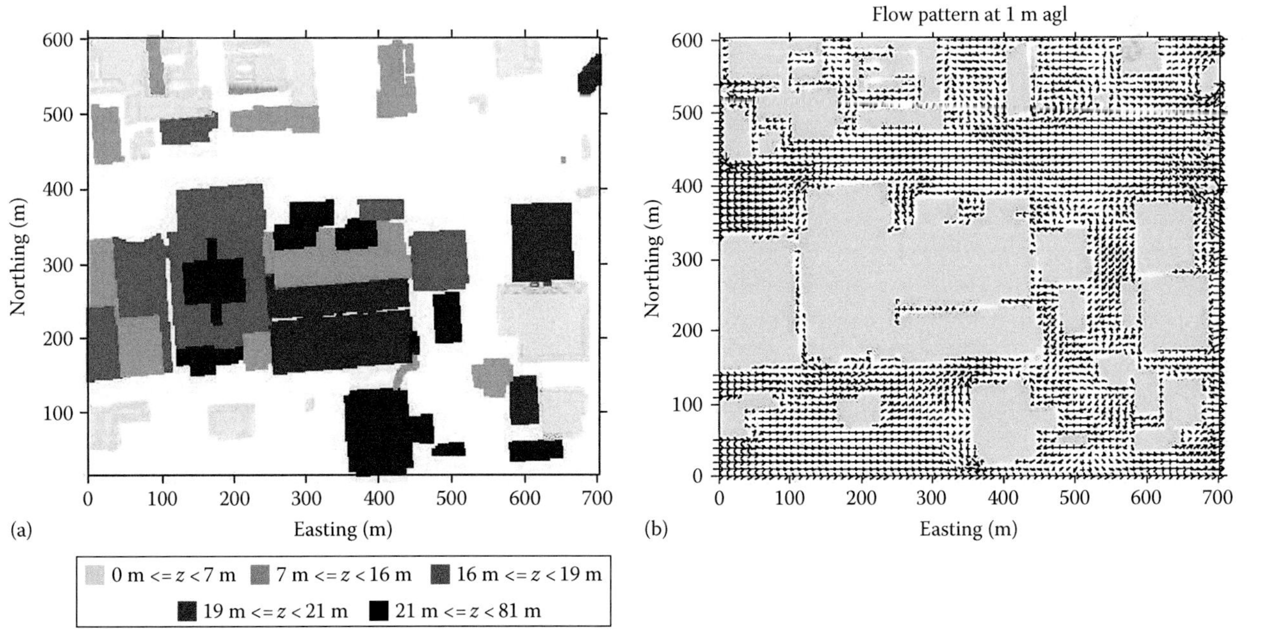

FIGURE 1.8 (a) The building geometry in the computational domain. (Courtesy of S.-M. Lee.) (b) Simulated flow field at 1 m above ground level. Filled polygons represent buildings.

where

$k > 0$ is an integer
u is the x-direction velocity
$\mu(=\nu\rho_0)$ is the dynamic viscosity
ρ_0 is the density
z is directed upward

The other initial and boundary conditions are $u(z \to \infty, t) \to 0$, and $u(z,0) = 0$. The analytical solution for the velocity field becomes, using Laplace transforms,

$$u(z,t) = \frac{4u_*^2}{\pi^{3/2}}\sqrt{\frac{a}{\nu}}\sum_{k=0}^{\infty}\frac{1}{(2k+1)^{3/2}}\exp\left\{-\left[\frac{(2k+1)\pi}{2a\nu}\right]^{1/2} z\right\}$$
$$\times\cos\left\{\frac{(2k+1)\pi t}{a} - \sqrt{\frac{(2k+1)\pi z^2}{2a\nu}} - \frac{3\pi}{4}\right\}$$
$$+\frac{u_*^2}{\pi\sqrt{\nu}}\int_0^{\infty}\left[\frac{1}{\xi^{3/2}}\tanh\left(\frac{a\xi}{2}\right)\exp(-\xi t)\cos\left(\frac{\xi}{\nu}\right)^{1/2} z\right]d\xi. \tag{1.18}$$

The bottom of the water column therefore should oscillate as

$$u(0,t) = \frac{4u_*^2}{\pi^{3/2}}\sqrt{\frac{a}{\nu}}\sum_{k=0}^{\infty}\frac{1}{(2k+1)^{3/2}}\cos\left\{\frac{(2k+1)\pi t}{a} - \frac{3\pi}{4}\right\}. \tag{1.19}$$

Although an elegant closed-form solution is possible, in nature the flow is turbulent and hence (1.19) is hardly valid.

As an alternative, consider conventional dimensional analysis with governing dimensional variables a, ν, z, t, and u_*. With fundamental dimensions being length and time, three independent dimensionless parameters are possible:

$$\frac{u}{u_*} = f_4\left(\frac{t}{a}, \frac{z}{\sqrt{a\nu}}, \frac{u_*^2 a}{\nu}\right). \tag{1.20}$$

Note that all dimensionless parameters in (1.20) also appear in (1.18), but in (1.20) the functional form f_2 is unknown and no similarity solutions are viable. Nevertheless, simplifications are possible using the generalized dimensionless analysis. By introducing the dimensionless variables $u' = u/u_0$, $t' = t/t_0$, and $z' = z/z_0$, where $u_0 = u_*^2\sqrt{a/\nu}$, $t_0 = a$, and $z_0 = \sqrt{a\nu}$ obtained using the scaling analysis of (1.17a,b), we get

$$\frac{\partial^2 u'}{\partial z'^2} = \frac{\partial u'}{\partial t'},$$

with

$$-\frac{\partial u'}{\partial z'} = T'(t') \quad \text{at } z' = 0; \quad \text{and}$$

$$T'(t) = \begin{cases} 1 & \text{for } 2k < t' < (2k+1) \\ -1 & \text{for } (2k+1) < t' < (2k+2) \end{cases}. \tag{1.21a,b}$$

The plausible solution for (1.21a,b)would be $u' = u'(z',t')$ or

$$\frac{u}{u_*^2\sqrt{a/\nu}} = f_5\left(\frac{t}{a}, \frac{z}{\sqrt{a\nu}}\right), \tag{1.22}$$

which contains one less variable than (1.20). Still the problem is not self-similar, but the generalized dimensional analysis helped simplify the functional form.

Natural flows are turbulent, and hence Reynolds number similarity can be invoked for the interior of the flow, away from the boundaries. This means that, as $z/\sqrt{a\nu} \to \infty$, the self-similarity of the second kind can be used for (1.22), viz.,

$$f_5 = \left(\frac{z}{\sqrt{a\nu}}\right)^{-1} f_6\left(\frac{t}{a}\right), \tag{1.23}$$

and hence

$$u = \frac{u_*^2 a}{z} f_6\left(\frac{t}{a}\right), \tag{1.24}$$

which is a self-similar solution for the velocity field. A similar analysis can be performed for the boundary layer thickness and *rms* turbulent velocities. While this idealized case can be readily verified in the laboratory and provide useful guidance for data analysis, in natural flows the shear stress variation is more irregular and additional effects such as background currents and Coriolis forces complicate the flow.

Figure 1.9a shows a typical oceanic case where a rotating tidal current is evident, which considerably differs from that discussed previously. Nonetheless, expressions of the form (1.24) are of some utility in presenting data of the oceanic bottom boundary layer. Figure 1.9b shows vertical velocity profiles taken in the same area. Lacking an alternative, the profiles near the bottom have been approximated by log-linear power laws conventionally used for steady boundary layers. The interior of the fluid sometimes shows a decrease in velocity with height, as in (1.24), but the variability from one profile to another is significant. The presence of background currents and other factors needed to be considered in interpreting field observations. Idealized laboratory simulations and analysis of fundamental processes, however, are immensely useful in interpreting field observations as well as inferring prevailing processes at a given field site.

1.6 Major Challenges

Improved understanding and quantification of the dynamics of environmental (air and water) flows and their interaction with anthropogenic, biological, cryological, chemical, and geological components of the earth system are at the heart of EFD. In a broader perspective, EFD parses the challenges of environmental prediction, management, and planning on a firm scientific footing while developing tools to successfully address them. EFD tends to focus on the small-scale end of environmental

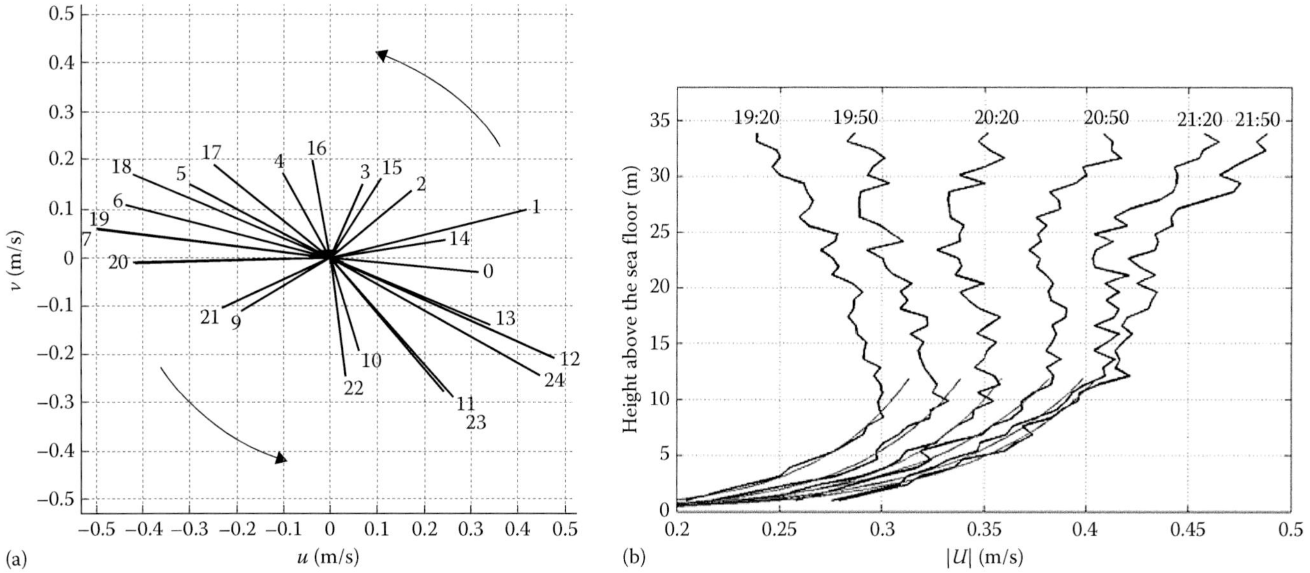

FIGURE 1.9 (a) A hodograph of rotating tidal flow measured at a mean water depth of 35.5 m (total water depth 38 m) on the shelf of the East China Sea (Yellow Sea); the numbers at the arrows are times of 1 h averaged current vectors after 15:00 of March 26, 2005. (From Lozovatsky, I. et al., *Cont. Shelf Res.*, 28, 318, 2008. With permission.) (b) Profiles of velocity magnitude at the location of (a). The oft-used log-layer region is indicated by the dotted lines. While (1.24) is applicable for an idealized oscillating tidal flows, the presence of a mix of processes in nature precludes its direct usage in the field.

motions, particularly the meso- and smaller scales, the latter being unresolved in most predictive models. The EFD research employs a suite of approaches—analytical, numerical, laboratory experimental, and field observational—with the aim of achieving better forecasts for environmental motions. Studies focusing on one or a few processes have had the most success, as evident from such discoveries as double diffusion, lee waves, solitary waves, and a host of instabilities via process studies. The progress in tackling nonlinear phenomena has been slow, although some important new (emergent) behaviors resulting from nonlinear processes have been delineated (e.g., lee waves). To this end, a reductionist approach is used, incorporating a selected number of processes.

Some grand challenges loom large for EFD. The burgeoning world's population has led to large (mega) cities and aggravated anthropogenic stresses on the environment. Urban areas are the sources of heat emissions, air pollution, drag (due to buildings), greenhouse gases, and hence anthropogenic climate change. Cities in turn bear the brunt of climate change, and their very sustainability may depend on local response to global change. Key elements for planning and management of *green, smart, and secure cities* are the local meteorology, air and water quality, human comfort, and water availability, which are common topics in EFD research. Further, *risk assessment* tools used in environmental-hazard mitigation studies hinge on flow predictions.

EFD also plays a key role in hydraulic engineering, for example, in the design of dams, wells, waterways, cooling systems, storm water and flood control systems, irrigation canals, and desalinization plants. Some specific issues of concern are dispersion, turbulent mixing, outfalls, wetlands, sediment transport, remediation, chemical reactions, dead zones (e.g., eutrophication and hypoxia), and underwater acoustics. Conversely, on the atmospheric side, issues of engineering interest are the plume stacks, urban designs, indoor–outdoor air exchange, HVAC systems, and sound pollution.

The emphasis on *nonfossil energy resources* (solar, hydro, wind, and nuclear power) is expected to continue. Small hydrokinetic and wind turbines are becoming popular, and ways of locating them for optimal efficiency and minimum environmental impacts are being sought out by commercial developers. Nuclear power plant operations have a strong EFD presence, including cooling water recirculation and emergency planning based on transport and diffusion.

On the technological front, smart buildings, transportation infrastructure, aircraft, ship and spacecraft operations, as well as emergency hazard alert systems rely on acquisition, processing, and mining of real-time environmental data as well as utilization of historic data in conjunction with predictive models. To this end, the deployment of *cyberinfrastructure* is a frontier issue of environmental research. Benefits of environmental resources (*ecosystem services*) sensitively depend on water availability, air and water quality, local microclimate, and health of ecosystems.

In all, environmental fluid dynamics suffuses a host of grand challenges of modern times, unfolding exciting research opportunities ranging from molecular mixing in reservoirs to pedestrian comfort in urban canyons to cyberinfrastructured (smart) cities with hazard warning capabilities. With reverberating societal calls for environmental *sustainability*, understanding and prediction of environmental motions are becoming ever so more important. This yearning combined with emergence of powerful analytical, laboratory, field experimental, remote sensing, numerical, and computing tools promise a bright future for environmental fluid dynamics.

References

Baik, J.J., Kim, J.J., and Fernando, H.J.S. 2003. A CFD model for simulating urban flow and dispersion. *J. Appl. Meteorol.* **42**(11): 1636–1648.

Baranblatt, G.I. 1996. *Scaling, Self Similarity and Intermediate Asymptotics*, Cambridge University Press, Cambridge, U.K.

Beven, K. 2002. Towards a coherent philosophy for modelling the environment. *Proc. R. Soc. Lond.* **458**(2026): 2465–2484.

Bolster, D., Hersberger, R.E., and Donnelley, R.J. 2011. Dynamics similarity, the dimensionless science. *Phys. Today* **64**(9): 42–47.

Boyer, D.L., Chen, R.-R., and Tao, L. 1993. Physical model of bathymetric effects on the Antarctic circumpolar current. *J. Geophys. Res.* **98**(C2): 2587–2608.

Boyer, D.L. and Davies, P.A. 2000. Laboratory studies of orographic effects in rotating and stratified flows. *Annu. Rev. Fluid Mech.* **32**(1): 165–202.

Britter, S. and Hanna, S. 2003. Flow and dispersion in urban areas. *Annu. Rev. Fluid Mech.* **35**: 469–496.

Chandrasekhar, S. 1961. *Hydrodynamic and Hydromagnetic Stability*, Courier Dover Publications, Mineola, NY.

Crutzen, P.J. 2004. New directions: The growing urban heat and pollution "island" effect—Impact on chemistry and climate. *Atmos. Environ.* **38**: 3539.

Drazin, P.G. and Reid, W.H. 2004. *Hydrodynamic Stability*, Cambridge University Press, Cambridge, U.K.

Fernando, H.J.S. 1991. Turbulent mixing in stratified fluids. *Annu. Rev. Fluid Mech.* **23**: 455–493.

Fernando, H.J.S. 2008. Polimetrics: The quantitative study of urban systems (and its applications to atmospheric and hydro environments). *J. Environ. Fluid Mech.* **8**(5–6): 397–409.

Fernando, H.J.S. 2010. Fluid dynamics of urban atmospheres in complex terrain. *Annu. Rev. Fluid Mech.* **42**: 365–389.

Fernando, H.J.S. and Voropayev, S.A. 2010. Shallow flows in the atmosphere and oceans: Geophysical and engineering applications. *J. Hydro-Environ. Res.* **4**: 83–87.

Fernando, H.J.S. and Weil, J. 2010. An essay: Whither the SBL?—A shift in the research agenda. *Bull. Am. Meteorol. Soc. (BAMS)*, **91**(11): 1475–1484.

Fernando, H.J.S., Klaic, Z., and McCulley, J. (Eds.). 2012a. *National Security and Human Health Implications of Climate Change*. NATO ARW Series, Springer Publishing, New York.

Fernando, H.J.S., Lee, S.-M., Anderson, J., Princevac, M., Pardyjak, E., and Grossman-Clarke, S. 2001. Urban fluid mechanics: Air circulation and contaminant dispersion in cities. *J. Environ. Fluid Mech.* **1**: 107–164.

Fernando, H.J.S., Mammarella, M.C., Grandoni, G., Fedele, P., DiMarco, R., Dimitrova, R., and Hyde, P. 2012b. Forecasting PM_{10} in metropolitan areas: Efficacy of neural networks. *Environ. Pollut.* **163**: 62–67.

Fernando, H.J.S., Mendis, G., McCulley, J., and Perera, K. 2005. Coral poaching worsens tsunami destruction. *EOS* **86**: 301–304.

Fernando, H.J.S., Zajic, D., Di Sabatino, S., Dimitrova, R., Hedquist, B., and Dallman, A. 2010. Flow, turbulence and pollutant dispersion in urban atmospheres. *Phys. Fluids* **22**: 051301–051319.

Gibson, C.H. 1991. Laboratory, numerical, and oceanic fossil turbulence in rotating and stratified flows. *J. Geophys. Res.* **96**: 12549–12566.

Gill, A.E. 1982. *Atmosphere-Ocean Dynamics*, Academic Press, New York.

Goodman, L. and Robinson, A. 2008. On the theory of advective effects on biological dynamics in the sea. III. The role of turbulence in biological–physical interactions. *Proc. R. Soc. A* **464**: 555–572.

Gregg, M.C. 1987. Diapycnal mixing in the thermocline: A review. *J. Geophys. Res.* **92**(C5): 5249–5286.

Hondzo, M. and Wüest, A. 2009. Do microscopic organisms feel turbulent flows? *Environ. Sci. Technol.* **43**(3): 764–768.

Huppert, H.E. 2000. Geological fluid mechanics. In: G.K. Batchelor, H.K. Moffatt, and M.G. Worster (Eds.) *Perspectives in Fluid Dynamics: A Collective Introduction to Current Research*, Cambridge University Press, Cambridge, U.K., pp. 447–506.

Imberger, J. and Patterson, J.C. 1990. Physical limnology. *Adv. Appl. Mech.* **27**: 303–475.

Ivey, G.N., Winters, K.B., and Koseff, J.R. 2008. Density stratification, turbulence, but how much mixing? *Annu. Rev. Fluid Mech.* **40**: 169–184.

Jakob, C. 2010. Accelerating progress in global atmospheric model development through improved parameterizations—Challenges, opportunities and strategies. *Bull. Am. Meteorol. Soc.* **91**: 869–875.

Kaimal, J.C. and Finnigan, J.J. 1994. *Atmospheric Boundary Layer Flows: Their Structure and Measurement*, Academic Press, New York, 289pp.

Karl, T.R., Jones, P.D., Knight, R.W., Kukla, G., Plummer, N., Razvayev, V., Gallo, K.P., Lindseay, J., Charlson, R.J., and Peterson, T.C. 1993. Asymmetric trends in surface temperature. *Bull. Am. Meteorol. Soc.* **74**: 1007–1023.

Klein, P. and Lapeyre, G. 2008. The oceanic vertical pump induced by mesoscale and submesoscale turbulence. *Annu. Rev. Mar. Sci.* **1**: 351–375.

Large, W., McWilliams, J.C., and Doney, S.C. 1994. Oceanic vertical mixing: A review and a model with a nonlocal boundary layer parameterization. *Rev. Geophys.* **32**(4): 363–403.

Lee, S.-M., Fernando, H.J.S., and Clarke, S. 2007. Modeling of ozone distribution in the State of Arizona in support of 8-hour non-attainment area boundary designations. *Environ. Model. Assess.* **12**: 63–74.

Long, R.R. 1963. The use of the governing equations in dimensional analysis. *J. Atmos. Sci.* **20**(3): 209–212.

Lorenz, E.N. 1976. Nondeterministic theories of climatic change. *Quat. Res.* **6**(4): 495–506.

Lozovatsky, I., Liu, Z., Wei, H., and Fernando, H.J.S. 2008. Tides and mixing in the northwestern East China Sea Part I: Rotating and reversing tidal flows. *Cont. Shelf Res.* **28**: 318–337.

Mei, C.C. 2003. *The Applied Dynamics of Ocean Surface Waves*, World Scientific, Singapore.

Molina, L.T. et al. 2010. An overview of the MILAGRO 2006 campaign: Mexico City emissions and their transport and transformation. *Atmos. Chem. Phys. Discuss.* **10**: 7819–7983.

Odum, E.P. 1971. *Fundamentals of Ecology*, 3rd edn., Saunders, New York.

Ovenden, N., Shaffer, S., and Fernando, H.J.S. 2009. Determining the impact of meteorological conditions on noise propagation away from freeway corridors. *J. Acoust. Soc. Am.* **126**(1): 25–35.

Paola, C., Twilley, R.R., Edmonds, D.A., Kim, W., Mohrig, D., Parker, G., Viparelli, E., and Voller, V.R. 2011. Natural processes in delta restoration: Application to the Mississippi delta. *Annu. Rev. Mar. Sci.* **3**: 67–91.

Peltier, W.R. and Caulfield, C.P. 2003. Mixing efficiency in stratified shear flows. *Annu. Rev. Fluid Mech.* **35**: 135–167.

Phillips, O.M. 1977. *Dynamics of Upper Ocean*, Cambridge University Press, Cambridge, U.K.

Riley, J.J. and Lelong, M.-P. 2000. Fluid motions in the presence of strong stable stratification. *Annu. Rev. Fluid Mech.* **32**: 613–657.

Salathé, E.P., Jr., Steed, R., Mass, C.F., and Zhan, P.H. 2008. A high-resolution climate model for the U.S. Pacific Northwest: Mesoscale feedbacks and local responses to climate change. *J. Climate* **21**: 5708–5726.

Sarmiento, J.L. and Gruber, N.L. 2002. Sinks for anthropogenic carbon. *Phys. Today* **55**(8): 30–36.

Science. 2011. Doom or vroom? Special section on population. *Science* **333**: 540–594.

Smith, V.H. 1998. Cultural eutrophication of inland, estuarine and coastal waters. In: M.L. Groffman and P.M. Pace (Eds.) *Success, Limitations, and Frontiers in Ecosystems Sciences*, Springer-Verlag, New York.

Strang, E.J. and Fernando, H.J.S. 2001. Turbulence and mixing in stratified shear flows. *J. Fluid Mech.* **428**: 349–386.

Turner, J.S. 1973. *Buoyancy Effects in Fluids*, Cambridge University Press, London, U.K.

2

Research in EFD and Its Policy Implications

Julian Hunt
University of Cambridge
and
University College London

2.1 Introduction

A powerful motive for engaging in research and teaching in environmental fluid dynamics (EFD) is to understand and explain the fluid motions in the natural environment and how they relate to other processes. Our subject was transformed in the 1950s and 1960s with the first photographs from space showing the wholeness of the Earth's unique fluid environment. However, at about this time, new research warned us that this environment was being progressively threatened by biologically significant levels of man-made contamination spreading everywhere from the top of the atmosphere to the bottom of the oceans (Carson 1962).

The scientific discipline of EFD is closely linked to meteorology, oceanography, hydrology, and environmental engineering but more distantly related to other disciplines such as environmental health, chemistry, geology, etc. As a branch of applied science and engineering, EFD's aim is to moderate and control the hazardous impacts of the environment while at the same time utilizing its potential for society's well-being. Teachers and researchers in EFD need to understand how environmental policies of government and industries are reliant on knowledge about the basic and the applied aspects of EFD. Indeed, in recognizing its importance, policy makers now want to influence the direction and funding of EFD research. Although only a minority of researchers have close connections with practitioners and policy makers, they are aware that their results have a significant impact. However, it is also in their interests to know how research programs are affected by the changing priorities of current and future directions of policy. The purpose of this chapter is to show how research in EFD contributes to policy and at the same time how developments in policy and new practical problems continue to be a source of interesting new research problems.

Most scientific and technological studies of the natural environment are specialized; the same is true with EFD, which has tended to develop in the separate disciplines of civil engineering, oceanography, ocean engineering, meteorology, geological, and geotechnical engineering (see other chapters in this book, and articles in the *Annual Review of Fluid Mechanics* including Hunt 1991). However, as the holistic nature of environmental problems have begin to be understood, research programs have become more integrated, particularly by considering the dynamical and evolutionary nature of natural science and also their interactions with human society (e.g., Whitehead 1925; Smuts 1926). Lovelock (1968) explained how life as well as the chemical and physical form of the rocks, atmosphere, and oceans emerged from complex interactions over thousands and millions of years. The methodology of engineering and biological systems (which is further discussed in Section 2.4) provides a general framework for studying such complex environmental problems as well as making decisions about practical environmental projects, some of which have to be planned over decades and centuries (e.g., Parry and Carter 1998).

EFD studies are usually focused on one component of a complex system involving several disciplines. These may or may not lead on to environmental policies and practical solutions. Practical problems usually involve studying fluid and other processes and

Handbook of Environmental Fluid Dynamics, Volume One, edited by Harindra Joseph Shermal Fernando. © 2013 CRC Press/Taylor & Francis Group, LLC. ISBN: 978-1-4398-1669-1.

over a wide range of length and time scales. Take, for example, improving regional water policies, or dealing with global, long terms such as climate change or nuclear wastes. Both require integrated solutions involving different aspects of science and technology as well as social science. These interactions between the environment and the society have increased in magnitude and complexity, which is partly why the directions that policies should take are unclear. Nevertheless, there is now general agreement about how the main long-term objective of environmental policy should be to attain a future state when these interactions reach a long-term, "sustainable" equilibrium (Brundtland 1987).

A schematic "systems" diagram in Figure 2.1 illustrates the way in which environmental science broadly relates to the policies and projects of government, industry, and communities. Their main purposes are usually to minimize adverse environmental impacts (I) and maximize the utilization (U) of environmental potential. Even in thirteenth century London (Brimblecombe 1989; Hunt 2005b), it was realized that the utilization (U) of the atmosphere or a river as a refuse dump produced serious side effects on the environment (SE), with a significant impact (I) on society through worsening human health and comfort. This led to societal response and policies (SRP), which were among the first environmental regulation in the United Kingdom. Nevertheless, the French painter Monet visiting London in nineteenth century marveled at how pollution could also make for a beautiful dawn or sunset! (Hunt 2005a) This was an early example of how societies everywhere have found that environmental impacts and utilization activities have to be moderated as the feedback effects of societal activities on the environment become more obtrusive and greater societal policy responses are required. The net result is that the environment changes, denoted by ΔE. As discussed later in Section 2.3, societies also act deliberately (SE) to change their local environment for greater human comfort and economic productivity (U) (e.g., Crate and Nuttall 2009).

Research in EFD began with the study of pure fluids moving in open regions of the atmosphere, oceans, and rivers, in restricted spaces (such as buildings and engineering systems), and in porous media (e.g., Rouse 1963; Hunt 1991; Figure 2.2a). In the earliest conceptual models and measurements of such flows, their properties were usually averaged over the whole flow region or a cross-section. Models developed for these integral properties are still the basis of most EFD computations used in practice and also for testing data and more complex models (the fundamental implications were explained by T. B. Benjamin; see Hunt 2005b).

In the twentieth century, EFD modeling and experimental studies have been extended to three-dimensional flows, complex fluids (e.g., in porous media; Dagan 1989), to fluids transporting diffuse matter, heat and particles, and to flows where the boundaries are flexible. In some cases, dynamically evolving flows have to be considered where the flow boundaries become indeterminate, and solid boundaries disintegrate as structures collapse in floods and storms. The descriptions and analysis of these kinds of transition have become more objective and succinct through analyzing the topology of the boundaries and the mathematically "singular points" of the changing flow patterns (Tobak and Peake 1982).

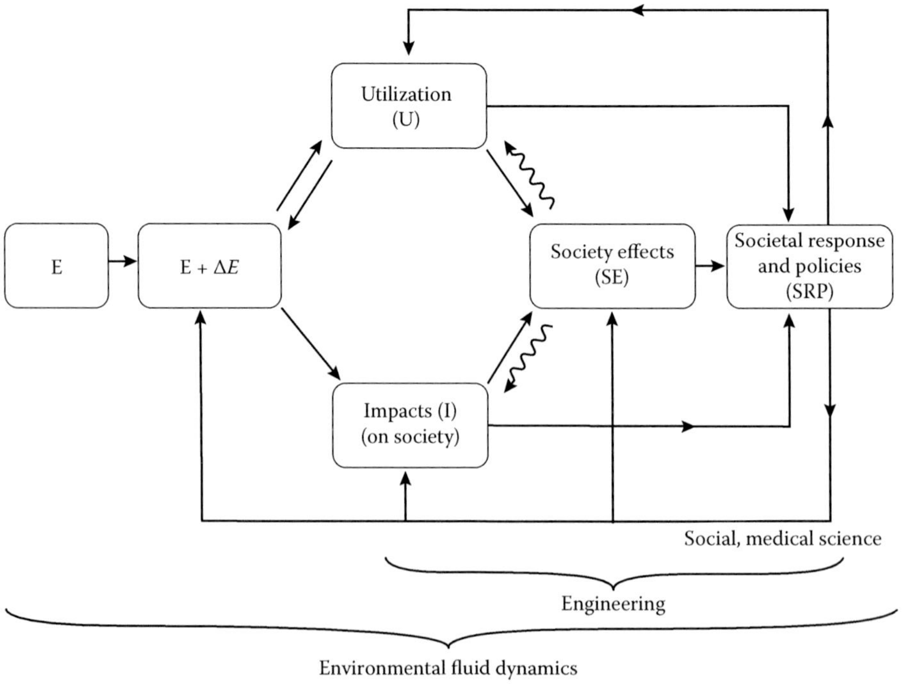

FIGURE 2.1 Interactions between the environment (E) and human activities: utilization (U), impacts on society (I), societal effects on the environment (SE), and societal policy response (SRP). Note that SE effects affect the utilization (U) and impacts (I) denoted by →. There are direct effects of U and I on SRP. Note that technology (T) is involved in all these interactions. Other disciplines that study these interactions are indicated.

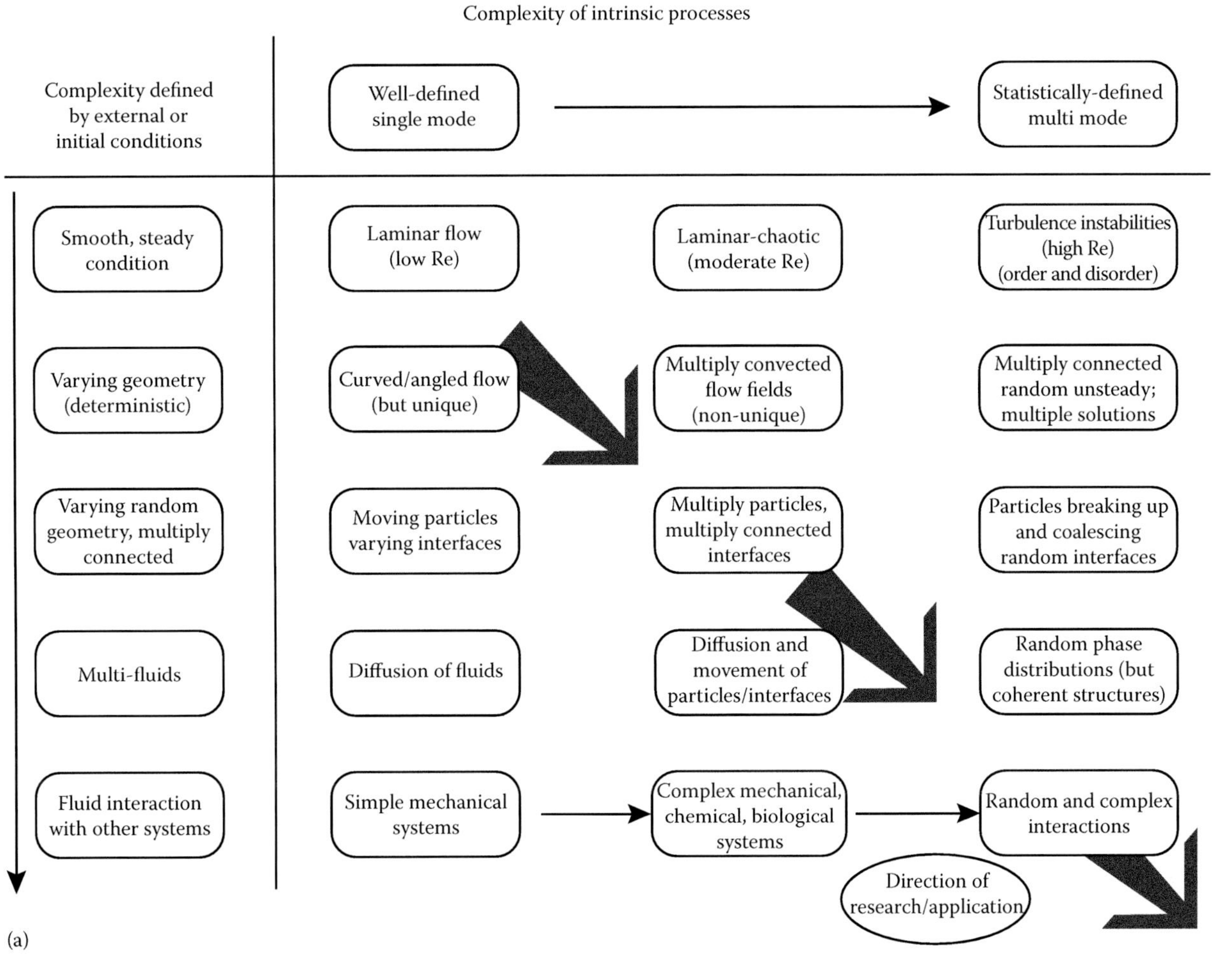

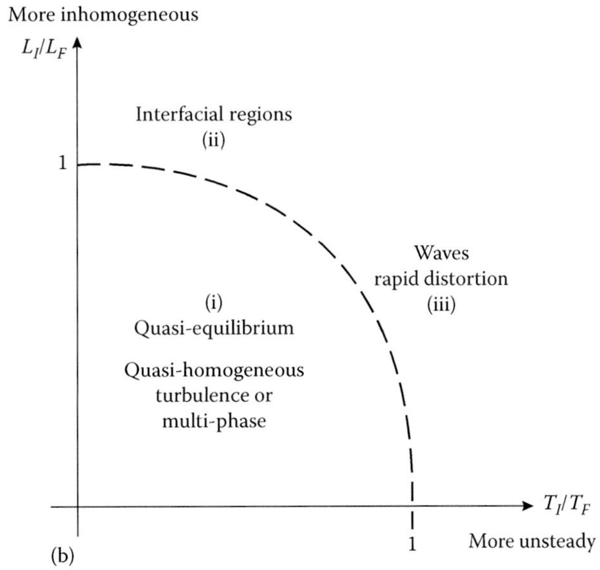

FIGURE 2.2 Overviews of environmental fluid dynamics research and modeling: (a) Table of EFD research in relation to the intrinsic and externally imposed complexity of flows and related processes, and the directions of new research. (b) Map of modeling of turbulent flow processes in terms of inhomogeneity and unsteadiness: zone (i) Statistical models based on quasi-equilibrium and weak inhomogeneity, (ii) ad hoc or integral models for particular types of flow, (iii) approximate (analytically based) statistical or real-time models, (ii) and (iii) Approximate real-time simulation models (e.g., with different types of filtering), T_I, L_I are the intrinsic time and length scales of the flow processes (e.g., for eddies or physical/chemical processes), T_F, L_F are the scales for the large-scale flow, as determined by boundary or initial conditions (and evolution of the flow).

New research techniques have developed for heterogeneous flows by a combination of micro modeling from processes on the smallest scales, such as molecular processes at solid–liquid surfaces (Poesio et al. 2006), to models that average certain properties of multiphase fluid–solid mixture over defined multiple elements of space and time or within certain boundaries (Gayev and Hunt 2007).

When EFD research began (Rouse 1963), a majority of the fundamental and applied investigations of impact and utilization studies were performed with laboratory and field experiments (e.g., Turner 1973; Lawson 2001). Computational modeling has now taken over as the preferred method for studying the impact and utilization of flows, especially those that are very large and very complex (such as coastal defenses, regional wind energy, or pollution of aquifers, or flood waves through groups of buildings). However, laboratory and field experiments, often in conjunction with computation and analysis, continue to be essential to test the basic processes that are represented approximately in the models, such as sand movement (Bagnold 1941), flowthrough groups of plants and buildings (Gayev and Hunt 2007), small-scale processes of turbulence (Tsinober 2001), liquid gas interfaces (Brutsaert and Jirka 1984) and particle interaction, and bubble or droplets in complex flow processes (Magnaudet and Eames 2003). With the higher resolution over space and time of modern instrumentation, the detailed structure of flows and fluid particle interactions can now be measured in three dimensions so that, for example, their conditional statistics can be defined and analyzed during extreme events, such as within eddy structures (Westerweel et al. 2009) or during events when particles break up or coalesce. As these areas of research into inhomogeneous, stochastic, and multicomponent flow processes have progressed, it has become clearer how different conceptual and practical models are relevant for different aspects of these flows (Hunt and Savill 2005; see Figure 2.2b).

For the foreseeable future, electronic computers will not be large and fast enough to describe the full range of fluid processes in most practical environmental problems. However, numerical methods for computational modeling have developed systematic methods for approximating the aspects of the processes that cannot be computed, such as the smallest eddies or biochemical reactions. Strategic mathematical and physical questions about modeling methodologies also have to be considered such as whether the answers are unique, or have extremely sensitive and chaotic tendencies. Their interpretation and practical use also depend on which properties of these models are conserved and the limits of their behavior (Hunt 2006). Lions et al. (1995), in analyzing the equations of global climate models, concluded that when there are large areas of polar ice covering the planet no mathematical solution might even exist—which indicated, as we know now, that models are very sensitive to assumptions about land and sea ice. Policy-makers and politicians, quite reasonably, ask about these delicate questions and enquire whether predictions are credible (e.g., Lawson 2008).

Theoretical or experimental studies of idealized situations are also very useful for investigating the properties of realistic complex models, although they are usually limited in time and space and have properties that only approximate realistic phenomena. They provide insights—as in Lorenz's famous investigation of chaos (see Palmer and Hagedorn 2006)—and sometimes useful formulae and also help explain how models of different components of a complex environmental system relate to each other. Simplified technical, and also economic, concepts and calculations are invaluable for explaining and debating the environmental implications of projects and policies, for example, to public inquires and to the media (e.g., Pasquill and Smith 1983; Stern 2007).

Although computational models become more detailed and accurately developed, and are applied more extensively, they always need to be explained using both the latest conceptual advances and new data. This is particularly necessary where models are used in policy making on controversial issues like climate change, environmental risks, and exploitation of environmental resources.

2.2 Classifying Interactions with the Environment

Policies and most research questions related to EFD tend to be focused on the interaction between societal activities and the environment. These can broadly be classified as impacts (I), utilization (U), and feedback effects (Figure 2.1). Studies of these interactions involve both natural and artificial fluid processes and also have to be related to other sciences and technologies. An instructive way to classify these complex interactions is by considering how natural flows are perturbed in different types of EFD problems. This classification indicates where different subdisciplines within EFD are applied and where new problems of basic fluid dynamics are emerging. Furthermore, it also demonstrates the complex way in which EFD has to be integrated in multidata disciplinary models, such as those for estimating impacts of natural hazards or climate change (e.g., Parry and Carter 1998; Hunt 2009).

2.2.1 Diffusive and Distributed Perturbations

In the natural environment (whose state is denoted by E), liquid, gaseous, and particulate flows all play a vital role in the distributions of heat, of various kinds of matter whether inert or changing chemically and biologically. Particles greatly influence heating and cooling by radiation. Human activities utilize these properties of flows to transport and disperse artificial and natural substances and facilitate their dispersion, especially those that are noxious (e.g., Brimblecombe 1989; Jacobson 2002). These diffusive or non-dynamical perturbations can be characterized as a change in the concentrations $\Delta C_s^{(n)}$ of the nth substance (or the temperature ΔT_s). Such substances might be directly introduced at some local source regions or might be generated by subsequent reactions in the environment, as it changes by ΔE. The atmosphere is significantly affected by gaseous and particulate emissions and by artificial or enhanced natural sources of combustion (e.g., from energy systems or forest fires). Similarly, liquids carrying particles or dissolved substances are discharged into or extracted from oceans, rivers, and porous media (Dagan 1989). Both gaseous and liquid sources can also affect the local temperature near the source ΔT_s either directly, as with cooling water from power stations along rivers, or indirectly, as when concentrations of particles in the atmosphere or ocean affect radiation fluxes and therefore the temperature perturbation ΔT_s. Temperatures in urban smog and dust storms decrease by a few degrees, but temperatures are raised in the surface layers of algae-rich rives and oceans.

As solid particles and fluids are dispersed into the fluid environment with significant concentrations, they affect physical and chemical processes, such as acid rain produced by sulfur dioxide emissions into the atmosphere. Ash particles from volcanoes can affect clouds, precipitation, and global weather for about 2 years after their eruptions. In water bodies, particles also have significant effects on biological processes. The result of such emissions and secondary processes is to change the concentrations of fluid substances and particles (each denoted here by index m). The concentrations of particles and fluids are also affected by their absorption and reaction with the boundary of the fluid region, for example, at the ground or at the atmosphere–ocean interface.

The effects of dispersive, nondynamic perturbations on the environment can be expressed schematically as

$$\Delta C_s^{(n)} + E \Rightarrow \Delta C_E^{(m)} + E + \Delta E \tag{2.1}$$

Here, $\Rightarrow$ denotes how environmental flows transport and disperse the substances or heat. When there are no reactions, no extra substances are produced, so that $m = n$.

As ΔC_s changes to ΔC_E, physical, chemical and biological processes in the environment can selectively transform (i.e., $m \neq n$) *and* concentrate substances, such as radioactive particles depositing in local areas following Chernobyl (Hunt 1991, Wells and Chamberlain 1969) or marine life concentrating chemicals in the Minamata disaster (Ui 1992). For these weak kinds of perturbations, the environmental flow processes are not significantly changed; although other aspects of the natural environment (ΔE) can be strongly affected, such as lake acidity and fish stocks down wind of atmospheric emissions of sulfur, or coastal ecology following oil spills. As Rachel Carson pointed out, devastating biological effects can spread over the whole globe. Specific studies of transport and diffusion processes in the different EFD "spheres" using different modeling systems also need to be explained in general terms. This contributes to a broad understanding of the similarities of the processes, which can be helpful to policy makers in developing consistent environmental policies.

2.2.2 Dynamic Perturbations

The second category of interactions involves changing the natural fluid motions (E_F) in the atmospheric and hydrospheric environments including geophysical motions. These perturbations, which are denoted by ΔE_F, can significantly affect the environment over large areas or they may only be significant locally where their impact (I) and utilization (U) are of interest. The main ways in which environmental flows are disturbed are as follows:

1. Changing the location and the physical properties of the boundaries of the flow region ΔB; first by directly changing their natural forms such as by remodeling terrain or river banks or by introducing engineered structures, whether rigid or moving, and by changes in surface vegetation; or secondly by changing boundaries indirectly through altering natural processes at the boundaries, which, for example, affect movements of deserts, or shore lines or riverbanks or ice formation. Internal boundary changes are also significant, as when the flow is affected by artificial or natural objects in the interior of the region (such as ships or particles).
2. Changing the velocity fields ΔV_B by moving the boundaries B (if it is solid or resistive) or by diverting or by injecting flows at the boundary, for example, gas entrainment in liquid flow, extraction of flows from porous media, or fluid exchange at a boundary between a flow and a porous region (Breugem et al. 2006). Dramatic natural examples are the eruptions of volcanic plumes (U.S. Geological Survey 2000). An equally serious event occurs when sudden movements of the ocean floor generate massive tsunami waves that travel across ocean basins at speeds of more than 100 m/s onto shorelines where they surge inland over about 1 km or more causing massive destruction and loss of life (Klettner et al. 2010).
3. Changing the density ρ at the boundary B or in the flow domain D by $\Delta\rho_{IB}$ or $\Delta\rho_{ID}$, respectively, through various physical processes, described later. This indirectly affects the velocity V in the interior domain D. (Note that the pressure at the boundary is not independent and is equivalent to the joint effect of ΔV_B and/or $\Delta\rho_{IB}$.)

Examples of these indirect processes in the atmosphere occur where the temperature or heat and water vapor fluxes vary at the ground or at the atmosphere–ocean boundary, in oceans and rivers where fluxes of heat and salinity vary (Turner 1973), and at solid boundaries at interfaces with other dynamic regions such as the atmosphere and ice (Kampf and Backhaus 1999). Where gases and particles are formed or are released into the atmosphere, they may directly affect the density (as in an avalanche) or by condensation (as in clouds) (Prupacher and Klett 1997), or they act indirectly by absorbing thermal radiation and then changing the temperature and density (as in the "greenhouse" effect, Houghton 2002). In liquid flows where gases can be dissolved or come out of solution, for example, in porous media, the dynamical effects on these liquid flows are at least as large as the effect of clouds on atmospheric flows.

These interactions can be summarized schematically as

$$E_F + \Delta B + \Delta V_B + \Delta\rho \Rightarrow E_F + \Delta E_F$$

where ΔE_F denotes the changes to the flow and other processes in the environment caused by the perturbation to the boundaries and velocity and density on the boundary and in the interior domain, namely ΔB, ΔV_B, $\Delta\rho$ (where $\Delta\rho = \Delta\rho_B + \Delta\rho_D$).

2.2.3 Effects and Causes of the Perturbations

Nondynamic and dynamic perturbations to the fluid environment caused by society (*SE*) may be deliberate or accidental. These usually result from actions to reduce or increase the impacts of the environment on society (*I*) and/or by utilizing (*U*) environmental flows. Society's responses and policies (*SRP*) for these three effects take place over time scales depending on the time scales of the environmental processes and on how society learns about the problems and is able to adapt (Figure 2.1).

The relation between urban communities and the environment is a good example of the whole interaction process. Their different effects noted in the previous section can be summarized symbolically as

$$I + U + SE + SRP \Rightarrow \Delta C_s + \Delta B + \Delta V_B + \Delta\rho \Rightarrow \Delta E$$
$$\Rightarrow \Delta SE + \Delta U + \Delta I + \Delta SRP$$

All four human interactions, *I*, *U*, *SE*, and *SRP*, affect the fluid flow and other processes. In general, there is a feedback effect between the societal interactions and the environment, leading to further changes, denoted by ΔSE, ΔU, etc.

In urban areas, the buildings and the natural environment (e.g., green spaces) are generally designed to reduce the flow-induced impacts (I) such as winds, high temperatures, floods, or blowing sand, while environmental flows are utilized (U) to disperse pollutants (in air, rivers, and sub soil), to ventilate and cool buildings, to extract water from rivers, subsurface reservoirs, etc., or to provide power (using wind/water turbines, Lighthill 1979) and water transport. There are many societal impacts (SE) on the regional and global environment, notably production of waste substances, for example, greenhouse gases (which lead to global warming and the local environment flows, i.e., ΔE, ΔE_F), waste heat (which can raise temperatures), and depletion of resources with direct and indirect effects on the fluid environment, for example, extraction of groundwater, or importing of agricultural/forestry products, which can adversely affect the environment in the originating regions. However, some by-products of human communities can restore the environment that had earlier been damaged, such as introducing vegetation to promote biodiversity, to stimulate rainfall (e.g., Abu-Zeid and Hamdy 2008), or reduce extremes in temperature.

Since such multiple interactions often result from intensive or large-scale activities involving society and the environment, these have to be considered carefully in any new project that is proposed, such as power plants (whether hydroelectric, wind, or nuclear), or significant developments in urbanization, agriculture, and forestry that can change the climate over entire regions, as several studies for the World Bank have shown.

In dealing with such planning problems, technical knowledge of EFD and other disciplines is needed both for the specific processes and for understanding their interactions (Figure 2.1), an aspect of EFD that should have higher priority in future.

2.3 Policies and EFD Research

To demonstrate how EFD research has contributed to the formulation of environmental policies and regulations, a number of different themes are considered here where societal activities have interacted strongly with the environment in the past and will continue to do so in the foreseeable future. The interactions and the changes in policy response and technology are plotted on a connectivity (or "phase plane") map using the notation introduced in the previous sections; see Figure 2.3a through c. In these and many others cases of EFD interactions, there have been changing cycles of utilizations and/or impact reduction (ΔU), societal impacts on the environment (ΔSE), and then followed by changes in societal response and policy making (ΔSRP). In most cases, this stimulates changes to *U*, *I*, *SE* activities, and another cycle develops, but sometimes the *SRP* phase leads to considering alternative paths of development outside the cycle. Each phase of the cycle raises a new question for EFD research, which the EFD researchers need to understand if their contributions are to be relevant.

2.3.1 Fossil Fuel Power Plants and the Environment

In the 1950s, electricity generating power plants throughout the world used mostly coal as a fuel. They did not remove smoke particles or polluting gases from their chimneys. Furthermore, these pollutants did not disperse far from the chimney because their heights barely exceeded those of nearby buildings. (Brimblecombe 1989). Because of the incomplete combustion in coal steam boilers, the thermal efficiency of most power plants was less than 30%. Because of their low temperature, the smoke plumes were scarcely buoyant and hardly rose above the tops of the chimneys. In urban areas, with many domestic and industrial chimneys, the levels of particulate and gas pollution at street level were high enough to cause disease and deaths. In London in 1952, thousands of people died over a few days. The particle concentrations were high enough to cause "smog" through increased condensation in clouds and in surface fog. At mid-day, the visibility was only a few yards and it was dark enough that lights were needed on the streets and in offices.

Societal response in Europe and North America led governments to introduce regulations within about 3 years of these events; first to limit pollutants emitted from power plants (first particles and then sulfur) and then by introducing taller chimney stacks to reduce the maximum concentration of the pollutants near the ground. At the same time, meteorological agencies in several countries introduced forecasts of urban air pollution largely based on the likelihood of inversion conditions. They provided warnings to hospitals one or two days in advance of when to expect large numbers of patients suffering from breathing difficulties (Pasquill and Smith 1992; an operational connection between EFD modeling and advice on health that is now widespread around the world, e.g., www.cerc.co.uk/yourair). Over 10–20 years, these policy responses that began in the 1950s were effective both in terms of health, fewer deaths and hospital admissions, and in terms of the environment; with cleaner air, and lower soot deposits on building surfaces, etc.

With more comprehensive and detailed monitoring of pollutants and atmospheric conditions, fundamental research in EFD was stimulated, which contributed better modeling and knowledge of buoyant plumes, atmospheric dispersion, and impacts of pollution on terrain (e.g., Snyder et al. 1984, Hunt 1985). Better regulations also resulted from medical and biological research on establishing health, agricultural, and fish limits for concentrations of pollutants. When it became clearer that fluctuations in concentrations over limited periods also affected the responses of humans and plants, this influenced EFD research into how turbulence affects concentration fluctuations (e.g., Durbin 1980; Yeung 2002).

However, 20 years later it became apparent in North America that regulations designed to meet local health standards were inappropriate for preventing another kind of longer term and larger scale environmental damage caused by the deposition of low concentrations of atmospheric sulfur to lakes, water courses, and forests. A combination of measurements and atmospheric dispersion modeling showed that the pollution was being carried by atmospheric winds hundreds of kilometers across regional and international boundaries (e.g., Scriven and Fisher 1975). The social response became international, with governments being pressured by those of neighboring countries to remove sulfur from their industrial emissions. Research in EFD and related sciences contributed to the national and international policy decisions (Jacobson 2002). In order to estimate how far SO_2 could

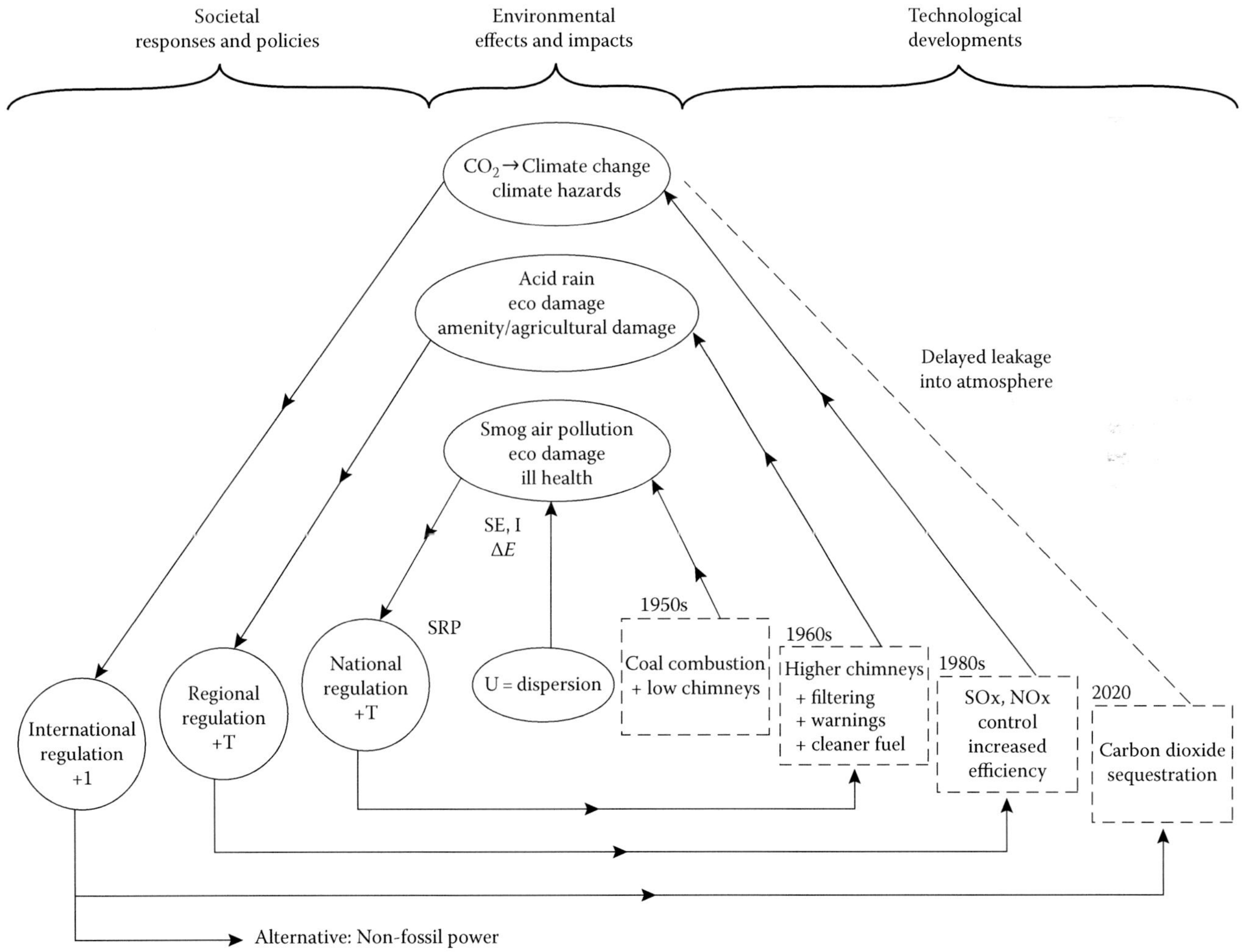

FIGURE 2.3 Themes of environmental developments showing how cycles of impacts (I), utilization (U), and societal effects on the environment (SE) lead to changing societal policy responses (SPR), technology (T), and new problems for EFD research. (a) Developments associated with changes in fossil fuel power plant and their effects.

(continued)

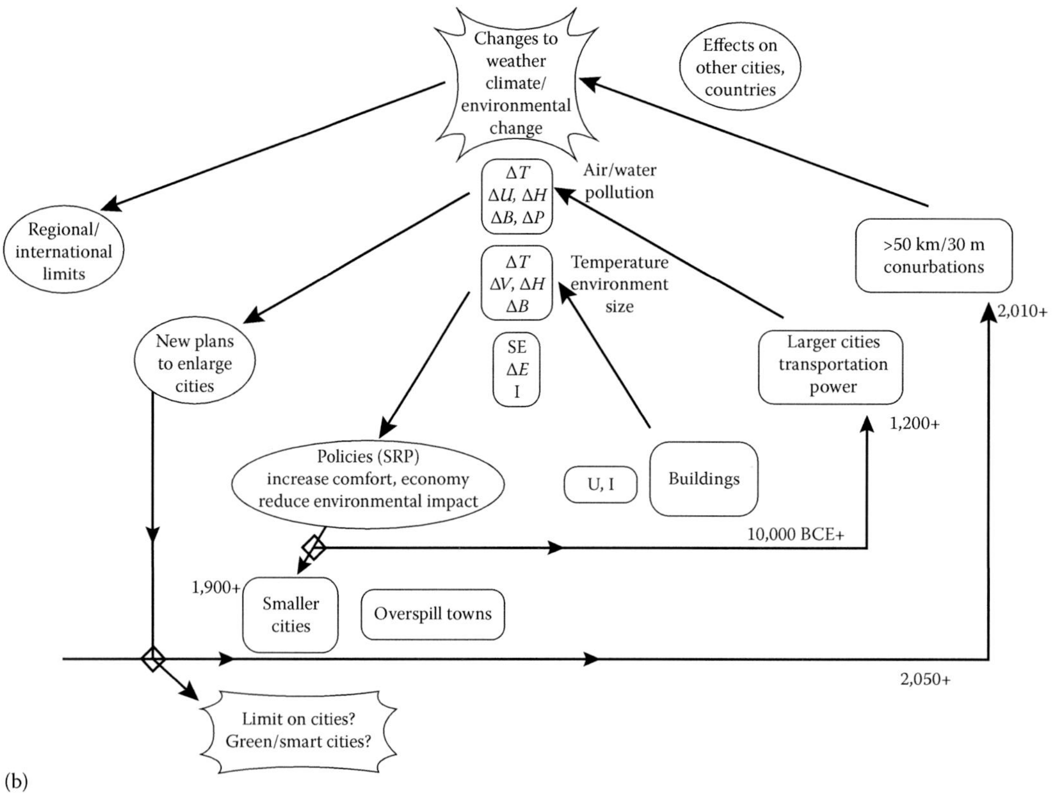

(b)

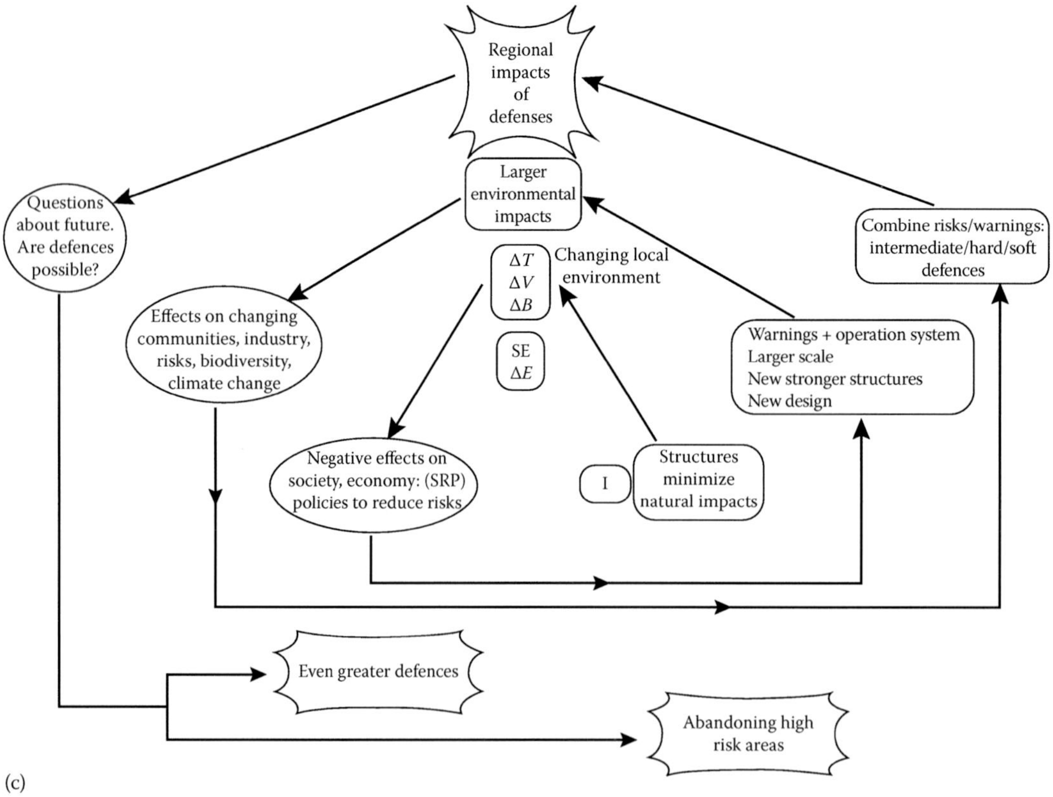

(c)

FIGURE 2.3 (continued) (b) Developments in the infrastructure and environment of urban areas associated with growth and transformation. (c) Stages in how communities have minimized the impacts of extreme environmental flow forces.

be transported, it was necessary estimate to estimate the mixing process and chemical transformations in the atmosphere, which are strongly affected by water vapor, water droplets, and particles. The concentration of air-borne sulfur decreased with downwind distance as the sulfur was deposited by the turbulent transport to seas, lakes, and land surfaces particularly over mountains where rainfall and deposition processes are enhanced (Carruthers and Choularton 1984).

In these scientific developments, research in one field of EFD problems benefited others. Models of buoyant plumes were derived from earlier military studies of how fog was entrained and dissipated in heated plumes situated along airfields for increasing visibility (Morton et al. 1956). Research on atmospheric deposition benefited from earlier studies of deposition of radioactive particles emitted from nuclear facilities (Pasquill and Smith 1983).

The first cycle of policy making to deal with urban particulate pollution followed within a few years from the evident impact on people's heath of unusually dirty urban atmospheres (i.e., I followed from SE; see Figure 2.1). By contrast, government policies to reduce sulfur emissions were not implemented in some countries for about 20 years after EFD studies had demonstrated the likelihood of significant levels of sulfate pollutants transported over hundreds of kilometers, and ecological studies had demonstrated their effects on the acidification of rivers and lakes, with consequences and damage to the biosphere. In this case, there was initially only limited societal and political response (SRP) to these effects caused by society on the environment (SE). But when their impacts (I) on society began to be understood and people realized the loss of amenity and threats to forests and agriculture, the public and then governments became concerned and new regulations were introduced (Mellanby 1988). New technology, which was stimulated by regulations and in the U.S. by charging companies for their sulfur emissions led to improved, helped in providing improved methods for removing sulfur dioxide from the emitted gases. These measures incentivized using coal with a lower sulfur content.

Another aspect of this second cycle was the impact on the environment associated with other gases being produced by high temperature combustion, especially nitrogen oxide. The internal combustion engine, which was an insignificant problem in the first cycle, now became the major source of pollution in urban areas through its emissions of unburnt hydrocarbons as well as NOx. These generated other chemicals, notably ozone and other oxides of nitrogen, in the atmosphere leading to the characteristic brown/yellow smog found in all urban areas with high levels of sunlight. The former type of simple regulations developed for lower temperature emissions from industrial and domestic chimneys had to be amended, to allow for these highly nonlinear interactions between these different types of gaseous emissions. The concentrations of these gases depended more sensitively on the spatial distribution of different sources. These considerations stimulated new approaches in EFD research using stochastic models and detailed numerical simulations to calculate fluctuating concentrations and mixing between reactive species in turbulent flows (Komori et al. 1991; Yeung 2002). As with long-range transport of acid rain, policies to regulate these reactive emissions had to be based on regional planning. This was successfully achieved in the Los Angeles valley where industrial emissions were moved to optimize the regional air quality.

The third cycle of the interactions between emissions and the environment emerged in the 1980s when it was realized that the increase in combustion-related emissions around the world, from natural burning, power plants, heating systems for buildings and internal combustion engines, is increasing the concentration of carbon dioxide (and other green house gases that absorb long wavelength radiation). The physical equilibrium and the basic mechanisms in the atmosphere are being changed, leading to significant changes in the atmosphere and oceanic environments, the earth's surface, and the biosphere (Houghton 2002; IPCC 2007). Measurements showed that the global average value of CO_2 in the atmosphere (below the tropopause) was steadily increasing, approximately in proportion to the emissions. This societal effect (SE) had first been proposed in the nineteenth century by Arrhenius who predicted that the trapping of long wave radiation by CO_2, even though its concentration is 0.1%–1% of all other gases, would lead to perceptible increases in the temperature across the world, that is, "global warming."

It is now estimated that the global average near-surface temperature would level out at about 2°C above preindustrial levels if immediate action was taken to reduce the emissions. Otherwise, it will probably level out at about 4°C–5°C after about 2100 when the world's population levels out, fossil fuels reserves are depleted, and technology increases the efficiency of combustion (IPCC 2007).

Detailed studies based on EFD, meteorology, oceanography, chemistry, and biology have shown that local environments would be transformed, with gradual melting of glaciers and polar ice caps, deforestation, rising sea levels, and flooding, etc. (IPCC 2007). These changes to the global environment will not be as large as those in the last ice age about 10,000 years ago, but they will, as happened then, have substantial impacts (I) on societies and cause mass migration in some areas of the world.

Despite the relative size of these predicted changes to the environment and because they will take place in the future, the public and politicians have taken many years (more than 20) to understand and respond, especially in the industrial countries at moderate to high latitudes where the effects are less. This has also been because of the uncertainty of scientific modeling, which, until 1995, did not even agree with the observed global temperature trends. It is also because the societal impacts on the environment (SE) are only just becoming generally evident, such as less snow on mountains, acidification of oceans, and rising sea level. The impacts on society (I) of the changing environment are more significant in parts of the world far from the industrial countries that are actually producing most of the emissions. However, even in these industrial countries, people are beginning to realize that they will experience significant impacts (I) on their economies and well-being, particularly with large changes or extreme events in the local climates (such as variation in rainfall pattern, droughts, and heat waves) and inundations from sea level rise.

In the previous cycles, the societal policy responses (SRP) to impacts were driven by their immediate and obvious effects. Because human-induced climate change has a longer term impact, SRP must be based on approximate predictions of damaging impacts in the future. These are made more credible through relating them to detailed studies of these environmental effects (SE) and also of their impacts (I). Some of these are detectable now and are beginning to affect societies quite significantly (Parry and Carter 1998). These considerations have influenced how EFD research is being applied and further developed.

EFD contributed to improving both the complex global climate models (GCM) through improved parameterization of processes. Many of these cannot be computed directly, such as the very inhomogeneous turbulence and radiation mechanism in clouds, turbulent transfer processes at air–sea interfaces, and coupling of flows over and within forests and in the surface layers of the ocean (Thorpe 2005; Palmer and Hagedorn 2006). Studies have shown (Davey and Hunt 2007) how the uncertainty in the modeling of certain EFD interactions has a more significant effect on the prediction than others, for example, rates of entrainment at the edges of turbulent flows. EFD research is also necessary to interpret the results and assess the sensitivity of certain processes in GCMs. Some of the most critical processes, for example, the generation of waves by wind and spray, and heat mass transfer processes across the atmosphere ocean interface and other parameterizations, are still based on fitting field data to idealized but controversial EFD models.

A challenge for EFD research is to understand and model fluid processes affected by interactions with many nonfluid processes, from biological flows to climate dynamics. In the latter case, large-scale EFD is contributing to multiscale interactions in the studies of regional climate patterns affected by global climate change. Decision makers are now worrying about possible changes to the El-Niño Southern oscillation, which dominates climatic variability, over years to decades, at lower latitudes, and affects the regularity of the monsoon in Asia and Africa. It is an urgent challenge for research to improve predictions about possible future scenarios and also provide qualitative explanations for decision makers and the public (Couper-Johnston 2000).

EFD research is already contributing to the next cycle of societal responses (SRP) that avoids emitting GHG emissions into the atmosphere. One of these policies being considered is by pumping CO_2 emissions in a highly compressed gaseous form into layers of porous rock, such as those that formerly contained oil and natural gas. How long the gases will stay in these rock formations is still uncertain, as is the time scale to be filled up—perhaps not more than 50–100 years (Bickle et al. 2007)?

Earlier EFD studies in the 1980s considered the practicality of depositing compressed solid CO_2 at the bottom of the ocean. But for those considering this proposal, as has happened on other occasions, the value of EFD research for policy was to advise against risky environmental solutions. Power companies are now replacing some fossil-fuelled power generators with alternative nonfossil methods. These, too, have a variety of complex impacts on the environment, which are raising other problems for EFD research, such as effects of wind energy farms on local climate, interactions of waves and wave-power generation, and dealing with radioactive wastes from nuclear power stations.

Similar cycles of environment interaction and technological developments stimulated by research in the EFD (as illustrated in Figure 2.3a) have occurred in other types of utilization of the environment, such as dispersing waste products. Just as with air pollution, when local problems of water pollution were tackled by treating effluents and designing discharges appropriately, other aspects become apparent such as long-term effects of dispersed chemicals, the concentration of waste products by different species, and the impact of dissolved nitrogen from rivers on the ecological health of ocean basins.

Ecological processes have also been affected by the acidification and rising temperatures of ocean waters brought about by the atmospheric problem of increasing GHG concentrations, another example of the linkages between different types of global environmental damage.

2.3.2 Developments in Infrastructure and the Environment of Urban Areas

Another example of how EFD is involved in cycles of environmental interaction is in the development of urban areas. In these environments, as shown in Figure 2.3b, atmospheric flows are deflected by being deflected by obstacles and buildings of the urban infrastructure and are affected by the buoyancy forces produced by their thermal imprint, while in the hydrosphere urban areas affect natural flows in rivers and subsurface water and introduce artificial flows in canals, pipes, leaking pipes, etc. Urban areas of course have other large effects (SE) on the local environment as well as on the air and water flows (see the text given earlier). They also affect their surrounding regions beyond their boundaries. Because these environmental impacts on society (I) are both positive and negative, they lead to complex responses of society (SRP). EFD research greatly contributes to understanding and making policies in all these aspects. (Cermak 1975, Fernando 2008).

From the earliest times, communities used artificial structures and the planting of crops and trees (or in forests, their removal) to moderate the winds, temperature excesses, and blowing of sand or snow (Finnigan 2000). They also had to regulate the levels of water and its movement along rivers and coasts. The social and economic advantages of cities outweighed some of the environmental dangers of these cities, such as ill health associated with dense population and the inadequate dispersion of air and water pollutants. As cities grew, the environment was further worsened as fossil fuel combustion and industry developed. However, starting in the mid-nineteenth century as higher chimneys were introduced and the water infrastructure of drains, supply, and water purification was established, there were considerable improvements in health and the environment.

EFD research (e.g., Rouse 1963) began and still continues with studies to improve urban water systems from aqueducts and drains (from Roman times) to pumps and pipework (based on the research of D. Bernoulli, H. Pitot, L. Prandtl, and T. von Karman; see Prandtl 1956). Outstanding examples of EFD and biology coming together, in this case for the benefit of human health, are the systems for cleaning contaminated water through filtration beds of varying porosity (Hunt 2005c). In the twentieth century, EFD has played a crucial role in ensuring the safety of tall structures that enable people to work and live at very high densities exceeding 1000 people per hectare. The construction of tall buildings first depended on structural and electrical engineering advances (e.g., Mumford 1961). Later, as they and the other structures grew higher and became lighter, the mean and fluctuating forces became comparable with gravitational forces in the static structure.

As in other areas of technology, here, too, science and engineering progressed by responding to system failures. Some of the ambitious stone structures in the middle ages were completed after some parts collapsed during construction, even twice as in the case of Beauvais Cathedral in France. Similarly ambitious lightweight, stressed structures in the past 70 years (bridges, cooling towers, skyscrapers) oscillated and collapsed because fluctuating aerodynamic forces were underestimated. EFD and micro-meteorological research was combined (Davenport 1961), to show how these forces depend sensitively on how the spectrum of atmospheric turbulence relates to the natural frequencies of the structure and its movements. Later EFD research showed how this relationship is affected by turbulent eddies of different scales being distorted as they interacted with structures and their wakes (Hunt et al. 1990). Equally important has been the research into the flows around different shapes of bluff body (Braza and Hourigan 2009). This has improved the reliability of wind-loading calculations on very tall structures as their design has changed from the porous construction of the tallest building in 1890 (the 350 m Eiffel Tower) to the nonporous 850 m Abu Dhabi Tower in 2010. In the latter case, engineers explained that the tapering design ensured decorrelation of vortex shedding and acceptable levels of vibration (Gaster 1969).

Equally challenging EFD problems associated with urban infrastructure emerged in the hydrosphere as flood defenses and harbors were enlarged and spread along coasts and rivers, and outward into deep water, As structures grew in size and depth, and affected the natural environment, they became subject to greater natural forces, from waves, wind, and sediment movement. Recent EFD research has shown how many of the largest of these fluid-structural interactions can now be studied by numerical simulations; for example, the way that large boulders protect shore structures from waves and turbulent erosions (Breugem et al. 2006) or the way that sediment movement and coastal erosion vary as global climate change affects storms and sea level rise. Research is helping establish policies to deal with these long-term problems that might cost 1%–3% of the gross national product of the countries that are most affected (Stern 2007; Government of Netherlands 2008, Pielke 2007).

The planning and associated technical problem of urban areas also changed as societies' responses and policies (SRP) led to demands for more green spaces, reduced urban heat island temperatures, less pollution, and cleaner industries and transport. Effective integrated policies to meet these demands are being explored through EFD and urban environmental modeling (Fernando et al. 2001; Hunt et al. 2010b). However, these challenges led some planners and politicians to promote an alternative response (see Figure 2.3b) in which urban populations would migrate to new kinds of smaller city such as garden cities, dormitories, over spill towns or science cities, etc. This choice was a bifurcation in the decision system. In fact, both options have been applied simultaneously around the world.

Regrettably there has also been a third option, namely for urban areas to regress to earlier standards of environment; an option forced on several mega-cities by population growth and economic difficulties; green areas have been built on, drainage and power are insufficient, housing becomes more concentrated and squalid, and there are greater risks of serious impacts from environmental hazards. However, some cities in China having had this experience then reversed the trend by constructing districts with very high rise buildings packed very close together. Inevitably, this leads to other environmental problems that are discussed later. (Lu et al. 2009).

Whichever of these paths are taken, EFD research is assisting decision makers to monitor the local environment, to plan for future changes in the environment, and to communicate with local people and organizations.

Energy and water supply is an example of a strategic problem, because expanding urban areas, with diameter L, will be limited by their power requirements (in proportion to L^6) for pumping the water that they need from distant sources, unless new more efficient methods of pumping, transport, use, and purification of water can be found through EFD research. About one third of the electrical power of Southern California is used for this purpose.

Another contribution has been in the design of the wind and temperature environment of urban areas, through studies of the airflow around the buildings and in the spaces between them. EFD sometimes explained why some innovations should not be repeated, such as tall buildings designed with large passageways underneath causing high winds at pedestrian level, or designed with parallelopiped cross-sections, which caused sickening and dangerous torsional oscillations (such as was tried in Boston). Aerodynamic and applied psychological research on people's reactions in various wind conditions led to warnings and advice to planners about when winds are dangerous to pedestrians (Hunt et al. 1976; Lawson 2001). Research on dispersion, mixing, and chemical transformation of pollution and on the effects of turbulence produced by vehicles (Britter and Hanna 2003) has also provided predictions of the highly variable concentrations along city streets. This local information can now be regularly communicated to the individuals who are particularly sensitive to these contaminants. Progressively, as more aspects of the local environment are being measured and predicted, they will be communicated more widely in the future (Hunt 2009).

The third cycle of urban environmental interactions is also posing new EFD research problems as urban areas become so large (with their diameters L exceeding 50 km or more) that they can significantly affect the atmosphere and hydrosphere over distances greater than L. This is because as observations and computational models demonstrate, the depth of the surface air flow varies over distances of the order of the Rossby deformation radius L_R (50–100 km), upwind and either side of the city (Hunt et al. 2004). Clouds and rainfall can be affected far downwind in wake regions extending up to several hundred kilometers (Cheng and Chan 2012) depending on the humidity and temperature in the atmosphere. Similar effects can occur as arrays of wind farms grow in size, as the average surface drag is significantly increased over distances of the area of L_R.

There are important policy implications that might lead to international decisions about urban planning and forestry, but EFD research to advise policies at this scale is only just beginning.

2.3.3 Minimizing the Impacts of Extreme Fluid Forces

The third aspect of how EFD is involved in society's interactions with the environment is in dealing with the impacts (I) of extreme flow conditions and with societal responses and policies (SRP). Although these physical and societal events vary greatly around the world (Srinivasan and Hunt 2011) both in the atmosphere and hydrosphere, many of them exhibit some common characteristic features, as is to be expected in such complex systems (e.g., Woo 1999; Hunt 2006).

Extreme environmental flow speeds V_{mx} occur in high winds (which can exceed 100 m/s in tornadoes), or air-particle flows in volcanic eruptions and avalanches (also about 100 m/s) (Huppert 2006), or ocean waves and tsunamis (up to about 10 m/s) (Grisler 2009). They exist within flow zones whose lengths can vary from tens of kilometers to a few meters, and which travel at speeds V_{fz} that may be much greater than V_{mx} (e.g., a factor of 10 as in ocean wave events) or much less (e.g., 1/10 as circulating flows like atmospheric tornados and hurricanes).

The impacts of such events can even destroy whole communities, including their economy and local environment. In some areas, such as low-lying coasts with maritime activities, such as ocean fishing or oil drilling, tropical cyclones these events occur frequently enough that they dominate people's lives in these regions and industries. EFD research contributes to the assessment by government, industries, and communities of the likelihood of these hazard events and on appropriate policies for impact reduction that also need to take into account social and economic implications.

As the world's population has grown and as technology, economies and communities have been transformed, the ways that societies deal with extreme hazard have also changed, as Figure 2.3c illustrates.

When communities were first established, they used empirical experience to design their structures and position their buildings and also modify their activities to minimize impacts of hazards associated with extreme flows. These empirical practices are benchmarks against which any new scientifically based understanding and new practice should be judged. For example, bridges had to be constructed to withstand high and deep currents and also to ensure that debris carried in the stream did not block the flow. Bridges often collapsed in such extreme conditions, as they still do today. Using the ideas of ship design, "streamlined" piers were introduced to reduce their drag. Sometimes (as in Toulouse's "Pont Neuf") the solid piers had openings high above the normal river level through which excess flows could pass.

This is another example of the advantages of using openings in the design of buildings and structures to reduce damage from winds or water flows. Another well-known advantage of open structures is that they have additional sheltering effect downwind (Castro 1971). Designers are discussing how such traditional knowledge about open structures may become more relevant as the frequency of extreme wind and water hazards increases with climate change (Stelling—private communication).

Engineering design can learn from nature: hedgerows and tree shelter belts survive extreme winds by losing their leaves. Now engineers are contemplating how structures can survive by dynamic adjustment, for example, allowing lower storey walls to break with high floods, or even buildings to float (the Noah's Ark solution). These examples show the value of topological thinking for analyzing flows and for innovative designs (Tobak and Peake 1982).

Another aspect of traditional knowledge now recast with the aid of modern science, including EFD, has been in positioning buildings and structures in relation to where environmental flows are least damaging. This is particularly important in mountainous terrain, to avoid danger of avalanches, mudslides, and near coasts to avoid very high storms sages, waves, and tsunamis. In cities on mountainous terrain, such as Hong Kong, EFD and solid mechanics have led to improved designs of retaining walls and drains on steep hill slides, to avoid mudslides and flooding from tropical precipitation. In the future, geophysically forced events are likely to become more frequent as the earth's crust adjusts to a warmer atmosphere and to reducing thickness of ice in cold regions of the world (McGuire 2010)—another challenge for EFD. Since coupling between electromagnetic and flow disturbances in the upper atmosphere can be triggered by surface movements, temperature, fluctuations and gases from the earth crust, these kinds of geophysically forced hazards and extreme flow events are likely to be predicted several days or more ahead in the future, which could greatly reduce their impacts on communities (Hunt and Kopec 2010).

The third aspect of EFD knowledge in dealing with extreme flow events involves the prediction of their occurrence (whether statistical or deterministic) and the nature of their flow fields. This is necessary both for the optimum design and the positioning of structures, communities, agriculture, etc., before, during, and after these events. The flow patterns within extreme flow events are now better understood thanks to modern measurement and communications systems (with radar, satellites, electromagnetic instruments in the atmosphere, and autosubs and buoys, infrasound in oceans) and also advances in EFD, for example, in downdrafts, gusts (Bradbury et al. 1994), rotating storms, and cyclones (e.g., Lighthill 1998). The data are fed into models and advanced warning systems designed especially for the different communities that are affected (Hunt and Kopec 2010).

The cycles of societal impacts and technical responses for extreme flow events (shown in Figure 2.3c) have changed as communities have grown and as infrastructure and economic systems have become liable to damage from hazards, especially in environmentally sensitive or critical areas. Although people's lives have become less risky through science and technology, societies now expect that hazards and their impacts should be predicted more accurately, and that dangers and any inconveniences should be averted, that is, a world where $I = 0$! Examples of extreme flow-hazard events show that there is progress toward this objective, but that there are many outstanding problems for EFD research to address.

The most extreme motions on the sea are waves, which depend on the structure as well as the strength of the winds and are now regularly forecast (Janssen 2004), but the highest waves (up to 30–40 m high) are often underestimated (Sajjadi et al. 1999). In some cases, they are isolated giant waves, which are large and steep enough to overturn large cargo ships exceeding 100 m length and 10^8 kg. The first such "solitary" waves on water surfaces were recognized as a hazard on canals in the 1840s and were first observed by Scott Russell, who applied the physical concept to the design of ships' hulls. These waves are likely to be the key to resolving the fundamental EFD question of how large amplitude isolated ocean waves can form spontaneously in a field of moderate/low amplitude waves, as recent laboratory experiments events have begun to demonstrate (Dysthe et al. 2008).

Tracking of large tsunami waves has now become more accurate using monitoring from satellites and fixed instruments on the ocean floor. However, the mechanisms for the initiation of these devastating waves by earth movements are still not well understood and cannot be predicted. More detailed EFD computation, analysis, and laboratory experimentation of these waves are necessary to estimate their motions onto beaches and coastal settlements. This is where they have their devastating impact, such as occurred in the Asian tsunami in December 2004, when 200,000 people were killed (Hunt 2005a). With new EFD research into these extreme phenomena (e.g., Klettner et al. 2010), all aspects of disaster mitigation are becoming more effective with better warnings and planning of the infrastructure, design, and emergency responses.

Preventing extreme flows from damaging flexible structures is vital to secure the safety of pipes and other elongated structures used for oil drilling and also lightweight industrial chimneys. This has particularly stimulated research on flow round dynamic buff bodies (e.g., Williamson and Govardhan 2004) and forces acting on them. New shapes of structures or additions to structures emerged in relation to the materials used and functions of the structures. In the 1960s, EFD research (Scruton 1965) led to vortex spoilers to reduce vibrational forces, and this enabled steel chimneys to be constructed without stay wires. A similar approach was tried for oil drilling structures in the ocean, where tubes 1000 m long had a length/diameter ratio of about 3000 or more. Crosscurrents and wave-induced motions led to large vortex-induced vibrations. In some marine applications, these are suppressed by attaching splitter plates (Bearman 1984) on the downstream side of the cylinder. Another solution, which worked even better in the laboratory, but not so well for very long pipes in the field, used bubbles from an air stream on the downstream side of the cylinder for damping the vortices (Abbassian and Moros 1996).

For the practical designer today, the great contribution of EFD computational and laboratory research has been that numerical methods are now sufficiently advanced that with large enough computer resources, vortex shedding flows around bluff bodies at high Reynolds number can be calculated to within a few percent. But more accuracy is needed when vortices interact in the presence of groups of cylinder and in more complex flows (Braza and Hourigan 2009). Improved accuracy is also needed to calculate the external flow environment around obstacles and their effects on people, sediment movement, heat transfer, and aerodynamic noise.

Extreme environmental flows tend to be particularly hazardous in areas where less affluent social communities tend to be concentrated. Often they are located on hillsides prone to landslides or in low-lying flood plains and coastal areas. During extreme waves or floods, the flows become even more complex and dangerous as structures fail, and trees and parts of buildings are transported by the wind and currents. Sometimes trees and other large objects can obstruct and deviate the whole flow (such as occurred in Aceh, Sumatra following the 2004 tsunami). Computational methods are being developed to help in the design and planning of these communities.

As governmental organizations and environmental research are now looking to the future, not only will urban areas become very large with diameters of 50 km or more, and populations of 30–40 million, but each will have distinct structural and environmental characteristics. In addition, as the climate and environment change in some of these areas, extreme winds, temperatures, and floods are likely to worsen. Earthquakes, desertification, and mudslides may add to the impacts. New developments in EFD research will have an increasingly valuable role in exploring the physical mechanisms that link these extreme events and also the means for detecting and predicting them (Hunt et al. 2010a).

2.4 Trends in the Connections between EFD Research and Policy

The examples of natural and human interactions in the environment described in the preceding sections have shown how our increased knowledge of EFD has influenced policies and practice; first in helping to identify and describe environmental problems and then in contributing to their solutions. The identification phase has been particularly prolonged for the problem of climate change. EFD research in new directions is also stimulated when impacts of environmental hazards become more severe, or, as in the case of air pollution, change their nature. Research in other scientific fields has enabled certain aspects of flows and flow processes to be better understood and defined, such as molecular dynamics and nanotechnology, which affects gas–solid–liquid boundary conditions, or biochemistry of living organisms and radiation physics, both of which significantly affect interactions in the flow produced by varying body forces.

Technological developments in measurement and computation are not only revolutionizing research methods but also our concepts of flow processes. These may have potential policy implications—provided the results are explained clearly to policy makers. One illustration comes from the refined computations and PIV measurements of 3-D velocity fluctuations, which have revealed for the first time the internal structures of eddies and shear layers in turbulent flows at high Reynolds number, showing how they are made up of intensive thin shear layers and very energetic microscale vortices (Ishihara et al. 2009; Hunt et al. 2010c). This intermittency of eddy structures has practical and regulatory implications. It means that when aircraft or blades of wind turbines move through atmospheric or wake turbulence, they experience step-like fluctuating forces, which are not accounted for in most present designs or regulations. New regulations are likely to follow in a few years. This is an example of how the time period of the cycle between research and policy decisions and then more research is shrinking.

Part of this cycle of course involves evaluating and modifying or discarding possible solutions—an urgent task for policy makers as they consider which combinations of methods to adopt in order to reduce green house emissions, while also attempting to reduce the societal impacts of environmental changes, from global warming to local water shortage. Evaluating the effectiveness of solutions often leads to new research questions.

An important general question for dealing with environmental problems is how to study integrated solutions and the interactions between the different aspects. For example, the enormous problem of providing engineering and societal solutions in coastal areas to deal with sea level rise, increasing storminess, and salt penetration can be solved either separately or through integrated systems. In Rotterdam harbor, concrete dykes are part of the defenses against sea level rise, for which plans are now ready beyond the year 2100 when the level might reach 2.0 m (Government of Netherlands 2008). But these structures can also assist policies to increase renewable energy by placing wind turbines on the dykes. This greatly reduces the cost of the foundation for the turbine (usually about 40% of the total), but there is a research question of determining how the dykes affect the mean wind and turbulence, which has to be considered in the design. For low-lying countries, integrated studies investigate the combined effects of wave dynamics, beach scour, cliff erosion, etc. This helps answer the key questions about what kinds of hard or soft (i.e., erodible and mobile) defenses are possible and appropriate (e.g., Bolndean 2001). The research topics in this case are also influenced by the considerable social impacts of these measures, which inevitably involve displacement of whole communities and changes to their way of life (Adger 2006).

This chapter has pointed out how the fruitful developments of EFD research, its application, and its policy implementation depend on specialists in different fields communicating effectively, especially for multidisciplinary projects. Explanations should point out clearly the inevitable constraints and risks resulting from research and its applications. They should be understandable to specialists in other fields and, where relevant, also to policy makers. As Poincare (1905) convincingly argued, although it is not easy, this is an essential part of any research investigation; first, because it actually helps the researcher to understand where his research is leading, and second, because it encourages useful inputs from others. Indeed nonspecialists also often point out other approaches or obvious points that can be missed.

Since environmental policies inevitably require choices between different options, this approach is particularly valuable, for example, in understanding why practical solutions in one environmental setting maybe inappropriate in another. As civil engineers know well, natural coastal defenses with sand-earth dykes are sensitive to the size and type of the particles as well as the impacts of winds, waves, and subsurface flows. This is why the design of such defenses can differ greatly between one country (e.g., England) and another (the Netherlands) only 100 km apart.

When explanations are made to specialists in related subjects or to nonspecialists, they have to be based on qualitative descriptions and order of magnitude estimates. EFD researchers interested in practical applications know that providing this kind of overview is just as important as improving the detailed understanding and accuracy of modeling of specific aspects of environmental flows. Of course, the two goals are complementary, especially for developing hypotheses for research and for explaining the results of detailed study.

There are many advantages in considering scientifically the general characteristic features of integrated systems, for example, about their response to external changes, their internal instabilities, and how they might evolve, like biological systems. Adopting the systems approach to the interactions between research and policy leads to new questions, such as how to control or optimize these systems, and their internal connections between subsystems. Progress in this interdisciplinary approach by engineers, mathematicians, and biological and social scientists show, for example, when systems might or might not have similar general properties and how systems evolve and respond to different types of input and feedback from the external systems (e.g., the public or community; see, e.g., Hitchins 2007).

In many examples where the overall system behavior is to be predicted, it is necessary to know which aspects are relevant (e.g., estimating the time to heat a saucepan of water requires an average description of the eddy motions but not their time history). This basic point, which needs to be better understood by politicians and commentators, is why climate predictions are in fact more reliable than weather forecasts. These are analogous to predicting the eddies in the saucepan, which as with the weather, can only be done for a short period ahead. Developing and evaluating models of conceptual systems is a fruitful way to invent and explore possible future environmental scenarios (Finnigan 2005; Hunt et al. 2010b). This approach is complementary to planning and policy making based simply on many computations for each kind of scenario. EFD knowledge and future developments in research are integral to both kinds of approach.

In looking forward, exchanges between researchers and policy makers should become more effective for two main reasons. First, as scientists and decision makers everywhere have access to more data from in situ and remote sensing measurements and visualizations, and from computer models and projections, there is now a greater common understanding about environmental changes, hazards, and interconnected issues. This speeds up the sharing of ideas about scientific questions. The second reason is society's realization that environmental issues are intimately connected to social and economic concerns. Because these have become globalized through international politics, media communications, travel and trade, this has forced science also to work faster and in a more global framework. In addition, these developments are enabling EFD to contribute to more integrated decision making, with the systems approach helping to provide a framework for dealing with unfamiliar problems.

As the examples in this chapter have shown, environmental research and policies have to be both national and, for the largest scale and most long lasting problems, international. Scientists can contribute to the international policies based on their research by publications and also by participating through academic and nongovernmental organizations, which are represented in official bodies such as IPCC and other UN and regional bodies. Although the work of these bodies is not widely known among either researchers or governments and politicians, they make recommendations based on the latest research to governments and industry on environmental regulations, data gathering, and operational decisions (e.g., forecasts and warnings) and directions for future research; topics have ranged from long-range transport of air pollution and causes of climate change, the ecology of oceans, transport of volcanic ash, fisheries and oceans, and advice and warnings about hazards and long-term developments. Their recommendations are taken seriously by governments and society generally. As environmental issues have become more global and urgent, these environmental organizations will have to collaborate and focus more on multidisciplinary and integrated solutions and use input from all aspects of EFD research.

References

Abbassian, F. and T. Moros. 1996. The use of air-bubble spoilers to suppress vortex-induced vibrations of risers. *SPE Prod. Facil.* 11: 35–40.

Abu-Zeid, M. and A. Hamdy. 2008. Coping with water scarcity in the Arab world. *Third International Conference on Water Resources and Arid Environment and First Arab Water Forum*, Riyadh, Saudi Arabia. www.icwrae-psipw.org

Adger, W. N. 2006. Vulnerability. *Global Environ. Change* 16: 268–281.

Bagnold, R. A. 1941. *The Physics of Blown Sand and Desert Dunes.* Chapman & Hall, London, U.K.

Bearman, P. W. 1984. Vortex shedding from oscillating bluff bodies. *Annu. Rev. Fluid Mech.* 16: 195–222.

Bickle, M., A. Chadwick, H. E. Huppert, M. A. Hallworth, and S. Lyle. 2007. Modelling carbon dioxide accumulation at Sleipner: Implications for underground carbon storage. *Earth Planet. Sci. Lett.* 255: 164–176.

Bolndean, P. 2001. Mechanics of coastal forms. *Annu. Rev. Fluid Mech.* 33: 339–370.

Bradbury, W. M. S., D. M. Deaves, J. C. R. Hunt, R. Kershaw, K. Nakamura, M. E. Hardman, and P. W. Bearman. 1994. The importance of convective gusts. *Meteorol. Appl.* 1: 365–378.

Braza, M. and K. Hourigan. 2009. Unsteady separated flows and their control. *IUTAM Symposium*, Corfu, Greece. Springer, Berlin, Germany.

Breugem, W. P., B. J. Boersma, and R. E. Uittenbogaard. 2006. The influence of wall permeability on turbulent channel flow. *J. Fluid Mech.* 562: 35–72, doi:10.1017/S0022112006000887.

Brimblecombe, P. 1989. *The Big Smoke: History of Air Pollution since Medieval Times.* Routledge, Chapman & Hall, Ltd., London, U.K.

Britter, R. E. and S. R. Hanna. 2003. Flow and dispersion in urban areas. *Annu. Rev. Fluid Mech.* 35: 469–496.

Brundtland, G. 1987. *Our Common Future.* Oxford University Press, Oxford, U.K.

Brutsaert, W. and G. H. Jirka. 1984. *Gas Transfer at Water Surfaces.* Reidel Pub. Co., Dordrecht, the Netherlands.

Carruthers, D. J. and T. W. Choularton. 1984. Acid deposition in rain over hills. *Atmos. Environ.* 18: 1905–1908.

Carson, R. 1962. *Silent Spring.* Houghton-Mifflin, Boston, MA.

Castro, I. P. 1971. Wake characteristics of 2-dimensional perforated plates normal to an air stream. *J. Fluid Mech.* 46: 599.

Cermak, J. E. March 1975. Applications of fluid mechanics to wind engineering. *J. Fluid Eng.* March: 9–38.

Cheng, C. K. M. and J. C. L. Chan. 2012. Impacts of land use changes and synoptic forcing on the seasonal climate over the Pearl River Delta of China. *Atm. Env.* (in press).

Couper-Johnston, R. 2000. *El-Nino.* Hodder & Stoughton, London, U.K.

Crate, S. and M. Nuttall. 2009. *Anthropology of Climate Change.* Left Coast Press, Walnut Creek, CA.

Dagan, G. 1989. *Flow and Transport in Formations.* Springer, Berlin, Germany.

Davenport, A. G. 1961. The application of statistical concepts to the wind loading of structures. *Proc. Inst. Civ. Eng.* 19: 449–472.

Davey, M. K. and J. C. R. Hunt. 2007. The importance of thin (layer)models—A workshop on critical regions in geophysical flows. *Math. Today* April: 53–57.

Durbin, P. A. 1980. A stochastic model of the two particle dispersion and concentration fluctuations in homogeneous turbulence. *J. Fluid Mech.* 100: 279–302.

Dysthe, K., H. E. Krogstad, and P. Muller. 2008. Oceanic rogue waves. *Annu. Rev. Fluid Mech.* 40: 287–310.

Fernando, H. J. S. 2008. Polimetrics: The quantitative study of urban systems and its application to atmospheric and hydroenvironments. *J. Environ. Fluid Mech.* 8: 397–409.

Fernando, H. et al. 2001. Urban fluid mechanics. *J. Environ. Fluid Mech.* 1: 107–164.

Finnigan, J. 2000. Turbulence in plant canopies. *Annu. Rev. Fluid Mech.* 32: 519–571.

Finnigan, J. 2005. The science of complex systems. *Australas. Sci.* June: 32–35.

Gaster, M. 1969. Vortex shedding from slender cones at low Reynolds numbers. *J. Fluid Mech.* 38: 565.

Gayev, Ye. A. and J. C. R. Hunt, Eds. 2007. *Flow and Transport Processes with Complex Obstructions*. Springer, Berlin, Germany.

Government of Netherlands. 2008. Delta commission report, www.deltacommisie.com, the Netherlands.

Grisler, G. R. 2009. Tsunami simulation. *Annu. Rev. Fluid Mech.* 40: 104–128.

Hitchins, D. 2007. *Systems Engineering: A 21st Century System Methodology*. Wiley, Chichester, U.K.

Houghton, J. 2002. *Physics of Atmospheres*, 3rd edn. Cambridge University Press, Cambridge, U.K.

Hunt, J. C. R. 1984. Turbulence structure and turbulent diffusion near gas-liquid interfaces. *Proceedings of Symposium on 'Gas Transfer at Water Surfaces'* (Eds.: W. Brutsaert and G. H. Jirka), pp. 67–82, D. Reidel, Dordrecht, the Netherlands.

Hunt, J. C. R. 1985. Turbulent diffusion from sources in complex flows. *Annu. Rev. Fluid Mech.* 17: 447–485.

Hunt, J. C. R., H. Kawai, S. R. Ramsay, G. Pedrizetti, and R. J. Perkins. 1990. A review of velocity and pressure fluctuations in turbulent flows around bluff bodies. *J. Wind Eng. and Ind. Aero.* 35: 49–85.

Hunt, J. C. R. 1991. Industrial and environmental fluid mechanics. *Annu. Rev. Fluid Mech.* 23: 1–41.

Hunt, J. C. R. 2005a. *London's Environment: Prospects for a Sustainable World City*. Imperial College Press, London, U.K. 336pp.

Hunt, J. C. R. 2005b. Non-linear and wave theory contributions of T. Brooke-Benjamin (1929–1995). *Annu. Rev. Fluid Mech.* 38: 1–25.

Hunt, J. C. R. 2005c. Tsunami waves and coastal flooding. *Math. Today* October pp. 144–146.

Hunt, J. C. R. 2006. Communicating big themes in applied mathematics. In *Mathematical Modelling-Education, Engineering and Economics, Proceedings of ICTMA12*, London, U.K., July 2005, Horwood, Chichester, U.K.

Hunt, J. C. R. 2009. Integrated policies for environmental resilience and sustainability. *Proc. Inst. Civ. Eng. Eng. Sus.* 162: 155–167, doi: 10.1680/ensu.

Hunt J. C. R., Y. Timoshkina, P. J. Baudain, and S. R. Bishop. 2012a. System dynamics applied to operation and public decisions. *European Review*, 20(3): 324–342b.

Hunt, J. C. R., S. Bohnenstengel, S. E. Belcher, and Y. Timoshkina. 2012b. Implications of climate change for cities world wide. *Proc. Inst. Civ. Eng. Urban Environ.* (in press).

Hunt, J. C. R., I. Eames, J. Westerweel, P. A. Davidson, S. Voropayev, J. Fernando, and M. Braza. 2010. Thin shear layers-the key to turbulence structure? *J. Hydro-Environ. Res.* 4: 75–82.

Hunt, J. C. R. and G. Kopec. 2010. Tsunamis and geophysical warnings. *Math. Today* June, pp. 116–118.

Hunt, J. C. R., A. Orr, J. W. Rottman, and R. Capon. 2004. Coriolis effects in mesoscale flows with sharp changes in surface conditions. *Q. J. R. Meterol. Soc.* 130: 2703–2731.

Hunt, J. C. R., E. C. Poulton, and J. C. Mumford. 1976. The effects of wind on people: New criteria based on wind-tunnel tests. *Build. Environ.* 11: 15–28.

Hunt, J. C. R. and A. M. Savill. 2005. Guidelines and criteria for the use of turbulence models in complex flows. In: *Turbulence Modelling* (Eds.: G. F. Hewitt and J. C. Vassilicos), Cambridge University Press, Cambridge, U.K.

Huppert, H. E. 2006. Gravity currents: A personal perspective. *J. Fluid Mech.* 554: 299–322.

Intergovernmental Panel on Climate Change. Fourth assessment report. 2007. See www.ipcc.ch for further details. Accessed January 12, 2007.

Ishihara, T., T. Gotoh, and Y. Kaneda. 2009. Study of high-Reynolds number isotropic turbulence by direct numerical simulation. *Annu. Rev. Fluid Mech.* 41: 165–178.

Jacobson, M. Z. 2002. *Atmospheric Pollution: History Science and Regulation*. Cambridge University Press, Cambridge, U.K.

Janssen, P. A. E. M. 2004. *The Interaction of Ocean Waves and Wind*. Cambridge University Press, Cambridge, U.K.

Kampf, J. and J. O. Backhaus. 1999. Ice-ocean interactions during shallow convection under conditions of steady wind: Three dimensional numerical studies. *Deep Sea Res.* 46: 1335.

Klettner, C. A., S. Balasubramanian, J. C. R. Hunt, H. J. S. Fernando, S. I. Voropayev, and I. Eames. 2010. Evolution of tsunami waves of depression and elevation. *Proceedings of the European Conference on Mathematics for Industry, July 2008* (Eds.: A. Fitt et al.), Springer, Berlin, Germany.

Komori, S., J. C. R. Hunt, T. Kanzaki, and Y. Murakami. 1991. The effects of turbulent mixing on the correlation between two species and on concentration fluctuations in non-premixed reacting flows. *J. Fluid Mech.* 228: 629–659.

Lawson, T. V. 2001. *Building Aerodynamics*. Imperial College Press, London, U.K.

Lawson, N. 2008. *An Appeal to Reason: A Cool Look at Global Warming*. Duckworth, London, U.K.

Lighthill, M. J. 1979. Two-dimensional analyses related to wave energy extraction by submerged resonant ducts. *J. Fluid Mech.* 91: 253–317.

Lighthill, M. J. 1998. Fluid mechanics of tropical cyclones. *Theor. Comp. Fluid Dyn.* 10: 3–21.

Lions, J. L., R. Temam, and S. Wang. 1995. Mathematics of the coupled atmosphere-ocean model. *J. Math. Pure App.* 74: 105–163.

Lovelock, J. E. 1968. *GAIA—A New Look at Life on Earth*. Oxford University Press, Oxford, U.K.

Lu, X., K. Chow, T. Yao, A. Lau, and J. Fung. 2009. Effects of urbanization on the land sea breeze circulation over the Pearl River Delta region in winter. *Int. J. Climatol.* 30: 1089–1104.

Magnaudet, J. and I. Eames. 2003. The motion of high Reynolds number bubbles in inhomogeneous flows. *Annu. Rev. Fluid Mech.* 32: 659–708.

McGuire, W. 2010. Climate forcing of geological and geomorphological hazards. *Phil. Trans. R. Soc. A* 368: 2311–2315.

Mellanby, K. (Ed.). 1988. *Air Pollution, Acid Rain and the Environment*. Elsevier, London, U.K.

Morton, K. W., G. I. Taylor, and J. S. Turner. 1956. Turbulent gravitational convection from maintained and instantaneous sources. *Proc. R. Soc. A* 234: 1–23.

Mumford, L. 1961. *The City in History*. Harcourt, New York.

Palmer, T. and R. Hagedorn. 2006. *Predictability of Weather and Climate*. Cambridge University Press, Cambridge, U.K.

Parry, M. and T. Carter. 1998. *Climate Impact and Adaptation Assessment: A Guide to the IPCC Approach*. Earth Scan, London, U.K.

Pasquill, F. and F. B. Smith. 1983. *Atmospheric Diffusion*. Ellis Horwood, Chichester, U.K.

Pielke, R., Jr. 2007. Climate change and coastal development. *Phil. Trans. R. Soc. A* 365: 2717–2730.

Poesio, R., G. Ooms, A. Tencote, and J. C. R. Hunt. 2006. Interaction and collision and "bridges" of particles in a linear shear flow near a wall. *Journal of Fluid Mechanics* 555: 113–130.

Poincare, H. 1905. *Science & Method*. Nelson, London, U.K.

Prandtl, L. 1956. *Essentials of Fluid Dynamics*. Blackie, Edinburgh, U.K.

Prupacher, H. R. and J. Klett. 1997. *Microphysics of Clouds and Precipitation*. Kluwer NL, Amsterdam, the Netherlands.

Rouse, H. 1963. *History of Hydraulics*. Dover, New York.

Sajjadi, S. J., J. C. R. Hunt, and N. H. Thomas. 1999. *Introduction in Proceedings of Wind over Waves*. Clarendon Press, Oxford, U.K.

Scriven, R. A. and B. E. A. Fisher. 1975. Long range transport of airborne material and its removal by deposition and washout. *Atmos. Environ.* 9: 49–68.

Scruton, C. 1965. Proceedings of a symposium on wind effects on buildings and structures. *Nat. Phys. Lab.* 2: 798–832.

Smuts, J. C. 1926. *Holism and Evolution*. Macmillan, London, U.K.

Snyder, W. H., R. S. Thompson, R. E. Eskridge, R. E. Lawson, I. P. Castro, J. T. Lee, J. C. R. Hunt, and Y. Ogawa. 1984. The structure of strongly stratified flow over hills: Dividing streamline concepts. *J. Fluid Mech.* 152: 249–288.

Srinivasan, J. and J. C. R. Hunt. 2011. Climate change in Asia. *Current Science,* 101(10): 1269–1270.

Stern, N. 2007. *The Economics of Climate Change*. The Stern Review, Cambridge University Press, Cambridge, U.K.

Thorpe, S. A. 2005. *The Turbulent Ocean*. Cambridge University Press, Cambridge, U.K.

Tobak, M. and D. J. Peake. 1982. Topology of three-dimensional separated flows. *Annu. Rev. Fluid Mech.* 14: 61–85.

Tsinober, A. 2001. *An Informal Introduction to Turbulence*. Springer, Berlin, Germany.

Turner, J. S. 1973. *Buoyancy Effects in Fluids*. Cambridge University Press, Cambridge, U.K.

Ui, J. 1992. *Industrial Pollution in Japan*. U. N. University Press, Tokyo, Japan.

U.S. Geological Survey. 2000. *Review of Volcano Hazards Programme*. National Academic Press, Washington, DC.

Wells, A. C. and A. C. Chamberlain. 1969. Particle motion and eddy diffusivity. *Atmos. Environ.* 3: 494–496.

Westerweel, J., C. Fukushima, J. H. Pedersen, and J. C. R. Hunt. 2009. Momentum and scalar transport at the turbulent/non-turbulent interface of a jet. *J. Fluid Mech.* 631: 199–230.

Whitehead, A. N. 1925. *Science and the Modern World*. Cambridge University Press, Cambridge, U.K.

Williamson, C. H. and R. Govardhan. 2004. Vortex induced vibrations. *Annu. Rev. Fluid Mech.* 36: 413–455.

Woo, G. 1999. *Mathematics of Catastrophes*. Imperial College Press, London, U.K.

Yeung, P. K. 2002. Lagrangian investigations of turbulence. *Annu. Rev. Fluid Mech.* 34: 115–142.

3

Interdisciplinary Dynamics in EFD Research

Eric R. Pardyjak
University of Utah

Jessica L. Thompson
Colorado State University

Environmental fluid dynamics (EFD) deals with the study of complex systems that integrate natural, chemical, biological, and physical processes with anthropogenically altered systems. From climate change to wastewater treatment, policy-relevant decisions are often based on EFD research. Hence, EFD is multidisciplinary and its applications and solutions have an inherent human element. This is a sensitive topic in applied scientific research with associated difficult questions, which must be addressed by interdisciplinary teams. Typical questions include: How do we facilitate effective interdisciplinary research (IDR)? How can a team of scientists wrestle with the necessary social and political choices in order to collaboratively produce new scientific knowledge? Who gets credit? How are data and resources shared? What is the appropriate level of trust? How does one negotiate new knowledge and the role it may play in environmental decision making? This chapter provides guidance toward meeting scientific and policy-related goals for researchers interacting in a multidisciplinary framework with an emphasis on communication among researchers of various backgrounds, as well as communication between researchers and policy makers.

In this chapter, we use the definitions proposed by the National Research Council (NRC, 2001) to define multidisciplinary and interdisciplinary. Namely, *multidisciplinary* indicates a "collaborative approach involving many disciplines," and *interdisciplinary* "implies integration of multidisciplinary knowledge" (NRC, 2001, p. 8).

In order for EFD research to effectively inform current environmental policy decisions, engineers must first be able to communicate with collaborative research partners, spanning multiple disciplines, and second, they must be able to communicate with local policy and decision makers, who have different disciplinary lenses for understanding environmental issues (NAS, 2005). Successful multidisciplinary and IDR begins with clearly defined and functional lines of communication among team members. Likewise, effective communication is necessary to integrate natural and human systems research into environmental policy design and planning efforts. Participants in an interdisciplinary team are bound together in interdependent relationships, and the integrity of their work depends upon developing a shared vision for the project, mutual learning, and satisfying collaboration experience.

Aims and Goals of Coupled Natural–Human Systems Research Projects: Biocomplexity in the Environment (BE) was launched in 1999 and was one of NSF's priority areas. It is no longer run as Biocomplexity; today, related work is typically funded through the Dynamics of Coupled Natural and Human Systems competition. The idea behind the Biocomplexity program was to provide multiyear funding to promote new approaches to investigating the interactivity of biota and the environment. BE was one of the first interdisciplinary funding opportunities that was explicitly designed to foster research and education on the complex interdependencies among the elements of specific environmental systems and interactions of different types of systems. All kinds of organisms and environments, from microbes to humans, fit within the BE framework. Biocomplexity projects were set in diverse environments ranging from frozen polar regions and volcanic vents to temperate forests and agricultural lands as well as the neighborhoods and industries of urban centers. The key connector of BE activities was complexity. Research on the individual components of environmental systems provides limited information about the behavior of the systems themselves, so BE projects were designed to investigate the dynamic interactions of systems, often at multiple scales.

Handbook of Environmental Fluid Dynamics, Volume One, edited by Harindra Joseph Shermal Fernando. © 2013 CRC Press/Taylor & Francis Group, LLC. ISBN: 978-1-4398-1669-1.

In this chapter, we provide guidance for successful interdisciplinary EFD research projects using as an example a coupled natural–human (CNH) system project that includes various aspects of EFD research situated within a larger ecological, social, and political context. Note that reviews of CNH system research in general can be found in a number of publications including Liu et al. (2007), the special issue on "Biocomplexity in Coupled Human-Natural Systems" published in Geoforum (Walsh and McGinnis, 2008), as well as in Vajjhala et al. (2007). In this chapter, we highlight the role of interdisciplinary collaboration in the context of EFD-focused research. Such collaborations are necessary to investigate the dynamic interrelationships among human and natural systems; however, such collaborations are challenging and require collective communication competence among all of the disciplinary experts on the team (Thompson, 2009).

3.1 Introduction: Interdisciplinary Research

IDR has a long, rich history that disciplinary experts are still struggling to learn from due to the evolving complexity of the issues interdisciplinary teams are asked to tackle. IDR typically targets current and complex issues facing today's citizens. Historically, IDR was linked with government agencies and industrial advancement. One of the first IDR grants awarded was in 1930 to a team of chemists, engineers, physicists, and meteorologists investigating the dynamics of steam boiler explosions (Wofle, 1981). Since then, IDR laboratories have launched various thematic research programs across the nation. The National Science Foundation (NSF) was founded in the late 1950s, and federal support for applied multidisciplinary research increased significantly in that decade. Currently, the NSF sponsors several billion dollars in IDR, including a growing program on CNH systems. CNH research projects tend to expand systemic knowledge of complex, interrelated systems beyond the standard research programs in engineering, the sciences, or social sciences (Liu et al., 2007). In addition to CNH, a number of other large-scale programs have focused on multidisciplinary research from NSF including Human and Social Dynamics (HSD), Long-Term Ecological Research (LTER), National Ecological Observation Network (NEON), and Consortium of Universities for the Advancement of Hydrologic Science (CUAHSI) (Vajjhala et al., 2007).

In the past few decades, as external funding has grown and more opportunities for IDR emerged, studies of the interdisciplinary process itself have followed. Beyond discussions of training the next generation of scientific problem solvers, there has been substantial research and theoretical development regarding the functionality of interdisciplinarity and research team dynamics (e.g., Klein, 1990, 1993, 1996, 2003; Nissani, 1995, 1997; Weingart and Stehr, 2000). Some studies have focused on *challenges* to interdisciplinary teamwork (e.g., Bauer, 1990; Kostoff, 2002; Pellmar and Eisenberg, 2000; Salter and Hearn, 1993; Turner and Carpenter, 1999; Wear, 1999) and *opportunities* for interdisciplinary success (e.g., Benda et al., 2002; Cassell, 1986; Gillespie and Birnbaum, 1980; Pickett et al., 1999; Turner and Carpenter, 1999). In our experience, understanding common challenges and opportunities for success is necessary to effectively frame IDR research and engage disciplinary partners in complex interdisciplinary collaborations.

3.2 Elements of Successful Interdisciplinary Projects

In 2007, Thompson completed a 4 year ethnographic investigation of a complex IDR team. The team was participating in an NSF-funded CNH project tasked with investigating the dynamics of urban air quality, greenhouse gases, and natural as well as human impacts on climate change that required the expertise of multiple disciplines (Pataki et al., 2009). The interdisciplinary team members designed a research agenda that included measuring the concentrations of emissions and pollutant gases, tracing their origins, and integrating that data into a computer simulation model while hosting a series of public outreach workshops to solicit the input from local stakeholders. Thompson's responsibility on the team was to assist in developing and facilitating the outreach workshops by translating scientific findings for the team members and stakeholders. Much of her work was focused on taking notes and facilitating team meetings, coordinating scientific poster sessions, and assisting in systems model building activities. However, during the course of this project she realized, along with the team, that there was a larger story to be told about how IDR teams operate.

The Team included 5 principal investigators (PIs), 14 coinvestigators, and 9 graduate research assistants. Participants represented 12 disciplines including biology, chemical engineering, civil engineering, communication, ecology, geography, hydrology, material science engineering, mechanical engineering, meteorology, psychology, and urban planning. The researchers ranged in age, interdisciplinary team experience, and academic seniority. For example, one coinvestigator was a full professor and retired near the beginning of the project, and it was another coinvestigator's first year at this university and his first experience working on a formal and funded IDR project. The researchers took on the responsibility of addressing a complex ecological issue but also realized that they needed to negotiate knowledge, science, power, and interpersonal relationships in order to make scientific progress.

The lead PI was in the College of Life Sciences, a second PI was in the College of Engineering, a third was in the physical sciences, a fourth in the social sciences, and the fifth was in the College of Humanities. Three of the PIs were senior, tenured faculty members, and two were full-time research scientists. Each PI was in charge of one of the team's subgroups. These subgroups were organized around specific aspects of the project, including measuring greenhouse gas emissions (measurements subgroup), modeling the urban ecosystem (modeling subgroup), preparing a series of outreach workshops (outreach subgroup), and devising emissions management policies (emissions management subgroup); and the steering committee subgroup oversaw data integration. In an attempt to protect the identities of the participants in this case study, team members will be referred to by their role or position in the team.

As the project evolved, Thompson, a trained social scientist, took ethnographic field notes focused on the team's communicative and collaborative interactions (methods detailed in Lindlof and Taylor, 2002). For example, she was asked to coordinate poster sessions for a series of outreach workshops. In coordinating that event, she met with 12 team members and helped to translate their research process and data into public-friendly poster presentations. Through this process, she gained extensive access to many team members. They invited her into their offices and laboratories, and she used that invitation as an opportunity to ask questions about their expertise, research, and experiences working on the project. These informal, ethnographic interviews informed a deeper investigation of the dynamics of this IDR team. Specifically, she set out to understand: how can a set of disciplinary experts truly integrate their research to produce emergent knowledge? The answer is in the relational dynamics of the collaborative process, which begins with spending time together.

1. *Spending time together*: The first requirement for team building and developing communication competence is to spend time together. Team members who are open and willing to learn from each other also appear willing to spend time together. One's willingness to spend time is often related to external variables such as the resources available and the amount and types of outside resistance members felt. In many interdisciplinary projects, team members report that they feel they do not have enough time to discuss, connect, and explore ideas with other team members. Even team members who were initially excited about the project and participated regularly in early meetings later disclosed in interviews that they did not have time to commit because of all of their other academic duties.
2. *Practicing trust*: The team members openly discussed their desires for trust and the consequences of a lack of trust. Such conversations about trust helped to build the team's collective ability to communicate. Building trust takes a lot of time, but several theorists have suggested (e.g., Meyerson et al., 1996) that swift trust can help move motivated teams through a transition from temporary trust to a more sustainable, lasting trust necessary for sharing data, analysis, and integrated insights. This team, like other research teams, was working with a common task and finite life span. With tight deadlines, there is not much time for relationship building, so members import trust from previous experiences in an effort to be successful in this situation. Likewise, the more a team practices trust, the more they are able to build—thus, practice becomes the norm and relationships build as shared knowledge about the science builds.
3. *Discussing language differences*: People from different scholarly backgrounds assign qualitatively different meanings to the same or similar terms. Thus, a major obstacle in IDR is that it is difficult to find a common language because of our entrenched disciplinary specializations and corresponding languages. Many scholars have explored the challenges of language differences in collaborative research projects (i.e., Benda et al., 2002; Glantz and Orlovsky, 1986) and conclude that differences in definitions of jargon often lead to miscommunication and misunderstandings. Therefore, discussing such differences openly is necessary for the team to fully engage in collaborative problem solving.
4. *A sense of presence*: Senge et al. (2005) articulate the idea of presence in a team context, which involves "deep listening [and] being open beyond one's preconceptions and historical ways of making sense" (p. 13). Presence is one's ability to engage in collaboration while being open to experiencing a shift in one's own thinking. This is often manifested in an expressed motivation to learn, listen, and see the world differently. Sometimes the enthusiasm for mutual learning of this magnitude is contagious and can ripple through the team, creating a positive feedback, building enthusiasm and synergy as the team continues to work and learn together.
5. *Reflexive communication*: Reflexivity and reflexive talk are terms used to describe the idea that humans, as participants in a complex social system, are able to observe, reflect, and ultimately affect change within their system. In an IDR team, this process is magnified when team members collectively realize the impact of mutual learning and awareness of how that learning, in combination with their own behavior, impacts the group dynamics. Reflexive talk is an opportunity for a team to step back and ask the following questions: How are we learning together? What is working? What needs to change? This type of dialogue helps team members to avoid "groupthinking" and identify processes that facilitate collaboration and understanding across disciplinary divides.
6. *A sense of humor*: Finally, humor allows us to take advantage of the inconsistencies and incongruencies of a human existence. In teams, humor helps to relieve stress, support group ideals, integrate ideas in creative ways, and show support for common values (Bolman and Deal, 1991). Every team can benefit from the power of laughter, when natural and appropriate. Joking in this team helped to build a sense of community and cohesiveness and ease tension.

3.3 Tools for Facilitating Successful Interdisciplinary Projects

The following activities are critical to facilitating interdisciplinary projects but oftentimes are not performed, as they are not part of the typical scientific training process:

1. Build in trust-building time.
2. Host explicit discussions about language differences.
3. Schedule social time.
4. Confront communication challenges early.
5. Allot time to develop a shared conceptual model of the project.

First, interdisciplinary teams should write project proposals and team agendas to include time for building trust and engaging in reflexive talk. Time is one of the most sacred assets for any researcher, and team members may become frustrated if meeting time is not being optimized. However, it is suggested that time be allotted for team members to recognize and discuss the need for trust and barriers to trust within their team. This time should also facilitate reflexive communication about the team's relational evolution and development of collaborative capacity.

Second, team managers should make an effort to supplement talk about tasks with explicit discussions about language differences. A common challenge in multidisciplinary teams is overcoming jargon and terminology differences; such barriers can be reframed as learning opportunities for the entire team. Managers should be cognizant of concepts and terms that are not accessible to all members of the team and make room during the meeting time to dissect, discuss, and learn from such differences.

Third, teams might also consider scheduling social time for members to build relationships and share laughter. We recommend that large teams set aside time for members to network and mingle among themselves. Researchers may be much more motivated to participate in collaborative work if some type of social benefit accompanies the professional rewards for participating.

Fourth, negative humor, debating expertise, communicating boredom, and power struggles may be an inherent part of interdisciplinary team dynamics; efforts should be made to confront such challenges before nasty habits become group norms. IDR teams could benefit from the use of a professional facilitator. A professional facilitator can help the team to manage negative aspects of group dynamics. Team members may not know how to confront such conflicts or challenges. Instead of avoiding collaborative research because of the possibility of unpleasant experiences, teams should include funding for a facilitator to help the team reflect upon and navigate through such challenges.

Finally, IDR teams need a map. They need to know how their expertise is connected to the research goal, as well as to each other. Taking several hours early in the project to map out the team's system can save hours of arguing later. When it comes to linking social, ecological, and engineered systems, there is a lot of room for miscommunication and misunderstanding; however, a visual depiction or qualitative systems map of the project can facilitate communication that transcends any one team member's disciplinary expertise. There are several ways to develop such maps; the only rule is that all team members contribute to the process so that they have a sense of shared ownership of the final model. Once the concepts are mapped out, it is equally helpful to have team members mark or locate themselves on the map. Ultimately, if implemented, these suggestions can help avoid pitfalls that are common to IDR research, potentially lead to more productive research projects, or aid in attaining a more efficient level of communication more rapidly.

References

Bauer, H. H. (1990). Barriers against interdisciplinarity: Implications for studies of science, technology, and society (STS). *Science, Technology, & Human Values*, 15, 105–119.

Benda, L. E., Poff, N. L., Tague, C., Palmer, M. A., Pizzuto, J., Cooper, S., Stanley, E., and Moglen, G. (2002). How to avoid train wrecks when using science in environmental problem solving. *BioScience*, 52(12), 1127–1136.

Bolman, L. G. and Deal, T. E. (1991). *Reframing Organizations: Artistry, Choice, and Leadership*. Oxford, U.K.: Jossey-Bass.

Cassell, E. J. (1986). How does interdisciplinary work get done? In D. E. Chubin, A. L. Porter, F. A. Rossini, and T. Connolly (Eds.), *Interdisciplinary Analysis and Research* (pp. 339–345). Mt. Airy, MD: Lomond Publications, Inc.

Gillespie, D. F. and Birnbaum, P. H. (1980). Status concordance, coordination, and success in interdisciplinary research teams. *Human Relations*, 33(1), 41–56.

Glantz, M. and Orlovsky, N. (1986). Desertification: Anatomy of a complex environmental process. In K. Dahlberg and J. Bennett (Eds.), *Natural Resources and People: Conceptual Issues in Interdisciplinary Research* (pp. 213–229). Boulder, CO: Westview Press.

Klein, J. T. (1990). *Interdisciplinarity: History, Theory, and Practice*. Detroit, MI: Wayne State University Press.

Klein, J. T. (1993). Blurring, cracking, and crossing: Permeation and the fracturing of discipline. In E. Messer-Davidow, D. R. Shumway, and D. J. Sylvan (Eds.), *Knowledges: Historical and Critical Studies in Disciplinarity*. Charlottesville, VA: University Press of Virginia.

Klein, J. T. (1996). *Crossing Boundaries: Knowledge, Disciplinarities, and Interdisciplinarities*. Charlottesville, VA: University of Virginia Press.

Klein, J. T. (2003). Thinking about interdisciplinarity: A primer for practice. *Colorado School of Mines Quarterly*, 103(1), 101–114.

Kostoff, R. N. (2002). Overcoming specialization. *BioScience*, 52(10), 937–941.

Lindlof, T. R. and Taylor, B. C. (2002). *Qualitative Communication Research Methods* (2nd edn.). Thousand Oaks, CA: Sage.

Liu, J., Dietz, T., Carpenter, S. R., Alberti, M., Folke, C., Moran, E., Pell, A. N. et al. (2007). Complexity of coupled human and natural systems. *Science*, 317, 1513–1516.

Meyerson, D., Weick, K. E., and Kramer, R. M. (1996). Swift trust and temporary groups. In R. M. Kramer and T. R. Tyler (Eds.), *Trust in Organizations: Frontiers of Theory and Research* (pp. 166–195). Thousand Oaks, CA: Sage.

National Academy of Sciences (2005). *Facilitating Interdisciplinary Research*. Washington, DC: National Academies Press.

National Research Council (2001). *Grand Challenges in Environmental Sciences*. Washington, DC: National Academies Press.

Nissani, M. (1995). Fruits, salads, and smoothies: A working definition of interdisciplinarity. *Journal of Educational Thought*, 29, 119–126.

Nissani, M. (1997). Ten cheers for interdisciplinarity: The case for interdisciplinary knowledge research. *Social Science Journal*, 34(2), 201–216.

Pataki, D. E., Emmi, P. C., Forster, C. B., Mills, J. I., Pardyjak, E. R., Peterson, T. R., Thompson, J. D., and Dudley-Murphy, E. (2009). An integrated approach to improving fossil fuel emissions scenarios with urban ecosystem studies. *Ecological Complexity*, 6, 1–14.

Pellmar, T. C. and Eisenberg, L. (Eds.) (2000). *Bridging Disciplines in the Brain, Behavioral, and Clinical Sciences*. Washington, DC: National Academies Press.

Pickett, S. T. A., Burch, W. R., and Grove, J. M. (1999). Interdisciplinary research: Maintaining the constructive impulse in a culture of criticism. *Ecosystems*, 2, 302–307.

Salter, L. and Hearn, A. (Eds.) (1993). *Outside the Lines: Issues and Problems in Interdisciplinary Research*. Montreal, Quebec, Canada: McGill-Queens Press.

Senge, P. M., Scharmer, C. O., Jaworski, J., and Flowers, B. S. (2005). *Presence: An Exploration of Profound Change in People, Organizations, and Society*. New York: Doubleday.

Thompson, J. L. (2007). *Interdisciplinary Research Team Dynamics: A Systems Approach to Understanding Communication and Collaboration in Complex Teams*. Saarbrucken, Germany: VDM Verlag.

Thompson, J. (2009). Building collective communication competence in interdisciplinary research teams. *Journal of Applied Communication Research*, 37(3), 278–297.

Turner, M. G. and Carpenter, S. R. (1999). Tips and traps in interdisciplinary research. *Ecosystems*, 2, 275–276.

Vajjhala, S., Krupnick, A., McCormick, E., Grove, M., McDowell, P., Redman, C., Shabman, L., and Small, M. (2007). Rising to the challenge: Integrating social science into NSF environmental observatories, resources for the future report. (http://www.rff.org/Documents/NSFFinalReport.pdf).

Walsh, S. J. and McGinnis, D. (2008). Biocomplexity in coupled human-natural systems: The study of population and environment interactions, *Geoforum*, 39, 773–775.

Wear, D. N. (1999). Challenges to interdisciplinary discourse. *Ecosystems*, 2, 299–301.

Weingart, P. and Stehr, N. (Eds.) (2000). *Practising Interdisciplinarity*. Toronto, Ontario, Canada: University of Toronto Press.

Wofle, D. L. (1981). Interdisciplinary research as a form of research. *Journal of the Society of Research Administrators*, 13, 5–7.

4

Climate Change and Its Effects on Environmental Fluid Systems

Huei-Ping Huang
Arizona State University

4.1 Introduction

Climate change due to anthropogenic causes can occur on planetary to continental scales that are greater than the typical scales of environmental flows that directly impact human life. Because fluid flows at these scales are generally turbulent, the changes in planetary-scale circulation can exert a cascading influence on regional and small-scale flows through nonlinear interaction. Currently, how the large-scale circulation will change in response to anthropogenic greenhouse-gas (GHG) forcing is a subject of intensive research, exemplified by the efforts that culminated in the publication of the Intergovernmental Panel on Climate Change report series (IPCC 2007). Although an upward trend in global surface air temperature due to anthropogenic GHG emission has been established by observational and modeling studies, uncertainty remains in the projected changes in the large-scale circulation pattern, especially its coupling to the hydrological cycle (e.g., Held and Soden 2006). Even less understood is how this large-scale change will impact regional and small-scale flows and what roles the principles of environmental fluid dynamics may play in interpreting and predicting such multiscale flow interactions. Making progress on this subject is not only a major scientific challenge but also an important endeavor with potential benefits for the society.

In Section 4.2, we will first outline how the anthropogenically induced global climate change affects large-scale and, in turn, regional and small-scale flows. While many of the detailed mechanisms in this chain of influences remain unknown, a consideration of the relevant spatial and temporal scales and the relevant processes corresponding to these scales will give us a head start. In Section 4.3, we will discuss some key features of large-scale climate change as predicted by IPCC-grade global climate models driven by GHG forcing. This sets the tone for the discussion in Section 4.4 on how these projected large-scale changes in flow pattern, temperature, stratification, etc., may be used to help predict regional environmental flows. This discussion will center on the emerging idea of climate downscaling, supplemented by the more speculative concept of upscaling. The remaining challenges in understanding and predicting the multiscale interaction from global to urban scales will be discussed in Section 4.5.

4.2 Principles

4.2.1 Relevant Spatial and Temporal Scales

The Earth's atmosphere is a multiscale system governed by the principles of fluid dynamics. The spatial scales of interest to environmental fluid dynamicists stretch from that of a building (~10 m) to that of a city (~10^4 m) or even a state (>10^5 m). At the large-scale end of that spectrum, the interaction between a

Handbook of Environmental Fluid Dynamics, Volume One, edited by Harindra Joseph Shermal Fernando. © 2013 CRC Press/Taylor & Francis Group, LLC.
ISBN: 978-1-4398-1669-1.

regional flow and the global- or continental-scale (~10^6–10^7 m) circulation becomes prominent. Computationally, in a simplified one-way interaction framework, it is customary to use a high-resolution model to treat the regional flow within a "box," using an imposed boundary condition deduced from the large-scale background flow. If the phenomenon of interest is local and occurs on a very short time scale (e.g., less than a day), the influence of the large-scale boundary condition is often negligible. On a longer time scale, $t > \tau$, where τ is the time for the flow in the box to adjust to the boundary condition, the influence of large-scale flow on regional flow becomes critical. This situation will be encountered when we are concerned with the prediction of seasonal-mean or annual-mean climate for a metropolitan area or a state.

One may alternatively view the aforementioned τ as the Lagrangian time scale during which the air within the regional "box" at the initial time is replaced by air of remote origin. Clearly, beyond that time scale one can no longer regard the local flow as isolated or devoid of the large-scale influence. Observations of tracer transport generally indicate that τ is on the order of one to a few weeks for a mesoscale flow domain with the size of a state. A recently observed dust storm that originated in northern China traversed the Pacific Ocean and spread to the continental United States within weeks illustrated how quickly air masses are exchanged around the globe (Thulasiraman et al. 2002). Since the air within a regional box is refreshed many times during a season or a year, regional "climate" is predominantly the local manifestation of the interaction between large-scale circulation and the detailed local orography and surface characteristics (land surface type, etc.). A useful analogy of this view of "large-scale control" is the classical fluid dynamical experiment of a flow passing a cylinder under an increasing Reynolds number. Fixing the dimension of the cylinder, L, and the viscosity of the fluid, ν, the local flow regime (defined by the detailed flow pattern at equilibrium in the vicinity of the cylinder) is determined by the strength of the impending uniform flow (i.e., the "large-scale flow"), U, from afar. Note that even though the impending large-scale flow is structureless, through its interaction with the cylinder (the "local topography") it can create detailed, coherent, local structures. Moreover, in the vicinity of the boundary in Reynolds-number space that separates one flow regime from another, increasing U (i.e., increasing the Reynolds number) slightly can lead to a large change in the resulted local flow pattern. In other words, given the right circumstances, climate sensitivity can *amplify* as it cascades from large to small scales. This underscores the importance of performing detailed downscaling simulations discussed in Section 4.4 for assessing regional climate change.

4.2.2 Multiscale Interaction

From the preceding discussion, any regional environmental prediction (e.g., for the inter-state pollutant transport between two major metropolitan areas) for the future must rely on an accurate prediction of the changes in the large-scale circulation.

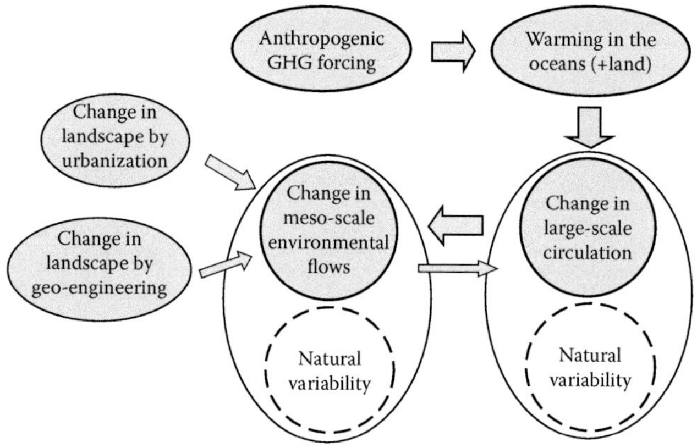

FIGURE 4.1 A schematic diagram showing the relation among the driving forces of climate change and the resulting chain of influences. The anthropogenically induced climate changes and their influences on mesoscale environmental flows are summarized by shaded ovals and arrows. These changes occur amid the background natural variability, which is not negligible at both global- and mesoscales.

The bold ovals and arrows in Figure 4.1 summarize the primary interactions that carry the influence of anthropogenic GHG-induced climate change from global to regional scales. The GHG forcing that drives climate change is of planetary scale, because GHGs such as CO_2 are well mixed (the observed increase in CO_2 concentration is highly uniform globally; IPCC 2007). The direct thermodynamic response to this GHG forcing is an approximately uniform (except for land-sea contrast) increase in the surface air temperature of the Earth (IPCC 2007). The first-order atmospheric response to this change in surface condition is however less uniform but exhibits distinctive spatial structures that prevail even in the absence of land-sea contrast (see Section 4.3). This nonuniform change in large-scale flow forms the basis for further climate downscaling to the mesoscale and beyond. In Nature, the interaction between large- and small-scale flows is always two-way and occurring simultaneously. Computationally, however, the limit in model resolution often forces us to treat the downscale and upscale influences separately, thereby a separate arrow is added to Figure 4.1 that directs the feedback from regional to large scales. Otherwise, the fluid dynamical models used to simulate global-scale and mesoscale atmospheric flows are remarkably similar. This is because all the terms that dominate global-scale flows, for instance Coriolis force, remain nonnegligible at the mesoscale. The major difference is quantitative: subgrid-scale processes are more heavily parameterized in the coarse-resolution global models.

Two other components in Figure 4.1 also play some roles in shaping the regional changes in environmental flows. Future urban development will transform land surface coverage from vegetation to concrete and built structures, affecting mesoscale flows both thermodynamically (by changing surface reflectivity/emissivity and soil moisture) and mechanically (by changing surface roughness). This process is already emerging to produce detectable effects, especially urban heat islands

(Fernando 2008), but may become even more prominent in the future (e.g., Georgescu et al. 2009). Second, geo-engineering such as constructions of large-scale wind farms may also impact regional or even global-scale circulation (e.g., Keith et al. 2004), although systematic studies are so far lacking in quantifying this effect.

4.3 Methods of Analysis

4.3.1 Climate Modeling and Its Justifications

To quantitatively project future changes in large-scale atmospheric circulation under increased GHG forcing, the only tool that has been extensively developed is the so-called global climate model. It consists of a coarse-resolution geophysical fluid dynamical (Navier–Stokes plus thermodynamics) code for the atmosphere, coupled to its counterpart for the oceans. The large-scale atmospheric and ocean models are very similar, except that the former requires an equation for water vapor and the related moist processes (precipitation, etc.), while the latter includes an extra equation for salinity. The two components are further coupled to land surface and sea ice processes that are modeled in a highly parameterized manner (Washington and Parkinson 2005). Note that this system has multiple scales not only in space but also in time. A winter storm has a life span of a week; the "memory" in soil moisture can last beyond a month, while the overturning circulations in the deep ocean have time scales of years to decades. The importance of the slower components emerges when a longer time scale is concerned; therefore, world oceans play a crucial role in regulating the response of Earth's climate system to anthropogenic GHG forcing that gradually rises on a centennial time scale. (As a useful mathematical analogy, recall that when an internal mode has the same frequency as the external forcing, resonance occurs that leads to a large-amplitude response in that mode.)

A typical daily weather forecast is similar to running a climate model for a short term (approximately a week), using the observed atmosphere-ocean state as the initial condition. Since slow processes associated with the ocean, land, and sea ice are unimportant on a short time scale, they are usually fixed, leaving the atmosphere as the only prognostic component. A weather prediction requires the accuracy to resolve the specific location and time for a specific weather event (e.g., severe rainfall produced by a winter storm) to occur. By that requirement, the predictability limit (quantified, for instance, by the time when the predicted anomaly correlation of 500 hPa geopotential height drops below 0.7) is merely a week (e.g., Lorenz 1982). This by no means contradicts the claim of usefulness of long-term prediction by climate models. The objective of the latter is to obtain an estimate of the time mean of a meteorological variable over a region (e.g., the annual mean of precipitation over Central Arizona), disregarding the detailed day-to-day or location-to-location variation of that variable. This large-scale, low-frequency variability in climate is largely driven by the variability and trend in the slowly varying ocean. Medium-range climate prediction with a lead time beyond a week is useful because the memory in the ocean (and in other components in land surface and the cryosphere) lasts far beyond that time scale: The sea surface temperature anomaly that carries a long-term memory can help drive a persistent anomaly in global atmospheric circulation pattern, which in turn influences the climate anomalies over land on seasonal and longer time scales. Thus, a useful climate prediction must invoke atmosphere-ocean coupling. It is with this reason that the oceanic response to GHG forcing is indicated as of primary importance in the top right oval in Figure 4.1.

In the IPCC-type of global climate model prediction, the future evolution of the GHG forcing (concentration of CO_2 and aerosol) is imposed based on various emission scenarios (detailed in IPCC 2007). Assuming that the emission scenarios are sound, the long-term projection is of value if a climate model correctly produces (1) the slow response in the ocean to the GHG forcing and (2) the response in large-scale atmospheric circulation to the predicted oceanic warming. The two are not always separable (e.g., changes in cloud type and/or cloud fraction in the atmosphere will modify local sea surface temperature), but there is evidence that long-term variability and trend in the troposphere largely reflects the large-scale atmospheric response to the variability and trend in SST (e.g., Stephenson and Held 1993 in the context of global warming; Huang et al. 2005; Compo and Sardeshmukh 2009; Deser and Phillips 2009 in the context of inter-decadal variability).

Note that the emission scenarios used to drive the climate models only provide the injection of anthropogenically produced GHG into the atmosphere. The total amount of CO_2 in the atmosphere also depends on natural carbon sources and sinks in the biosphere and especially in the ocean. The efficiencies of these sources/sinks can vary with a changing climate (e.g., Khatiwala et al. 2009), an aspect that is not yet fully resolved in climate models. Inaccuracies in the emission scenarios and the lack of an interactive scheme to predict the carbon cycle may further contribute to errors in the climate model prediction.

The cautions from the preceding discussion are balanced by optimism based on evidences of continued improvement of climate models in short-term climate prediction and in reproducing today's climate. It is reassuring that climate models now have verifiable skills to make seasonal-to-interannual prediction of climate anomalies during El Nino events (e.g., Gueremy et al. 2005). A recent study by Reichler and Kim (2008) also showed that the ability for climate models to reproduce present-day climate has been constantly improving from the Climate Model Intercomparison Project (CMIP) Phase 1 (CMIP1) to CMIP2, and then CMIP3, the latest formed the basis for the IPCC 2007 Report. The results from these efforts of model verification reflect an encouraging degree of realism in the atmosphere-ocean coupling process and the large-scale atmospheric dynamics in climate models. One then hopes that the atmosphere-ocean system in a climate model will respond correctly to an imposed long-term trend in GHG forcing.

4.3.2 Projected Large-Scale Climate Change due to GHG Forcing

To illustrate what information climate models can produce, Figure 4.2 shows an example of the projected changes in (a) 850 hPa geopotential height (as an indicator of pressure), (b) zonally averaged zonal velocity [*u*], and (c) zonally averaged temperature, [*T*], from the Geophysical Fluid Dynamics Laboratory (GFDL) climate model version CM2.1 simulation carried out for IPCC fourth Assessment (AR4) (based on data archived at PCMDI, Lawrence Livermore National Lab, www-pcmdi.llnl.gov). The simulation was forced by the standard A1B scenario of projected future increase in GHG (to doubling of CO_2 concentration by the end of the twenty-first century) and aerosol concentration. The climate change in Figure 4.2 is defined as the difference between future (2060–2080) and present (1980–2000) climate states. The change in the 850 hPa height field in Figure 4.2a shows a pair of ridges in the subtropics and troughs elsewhere; the structure of the trend is strongly "zonally symmetric." Outside the polar regions, the projected change in surface temperature is, to a large extent, uniform over the ocean with a typical 2°C–3°C increase by 2080, while land warms slightly more (not shown). The GFDL model also simulated a strongly "zonally symmetric" and "hemispherically symmetric" (northern hemispheric pattern being the mirror image of its southern hemispheric counterpart) change in the zonal velocity; therefore, the zonal mean pictures in Figure 4.2b and c are representative of the changes for the whole latitudinal belt.

In Figure 4.2b, in each hemisphere a pair of positive and negative changes in [*u*] straddle the climatological zonal wind maximum of present-day climate (the latter not shown). These changes and the corresponding climatological zonal jets all have

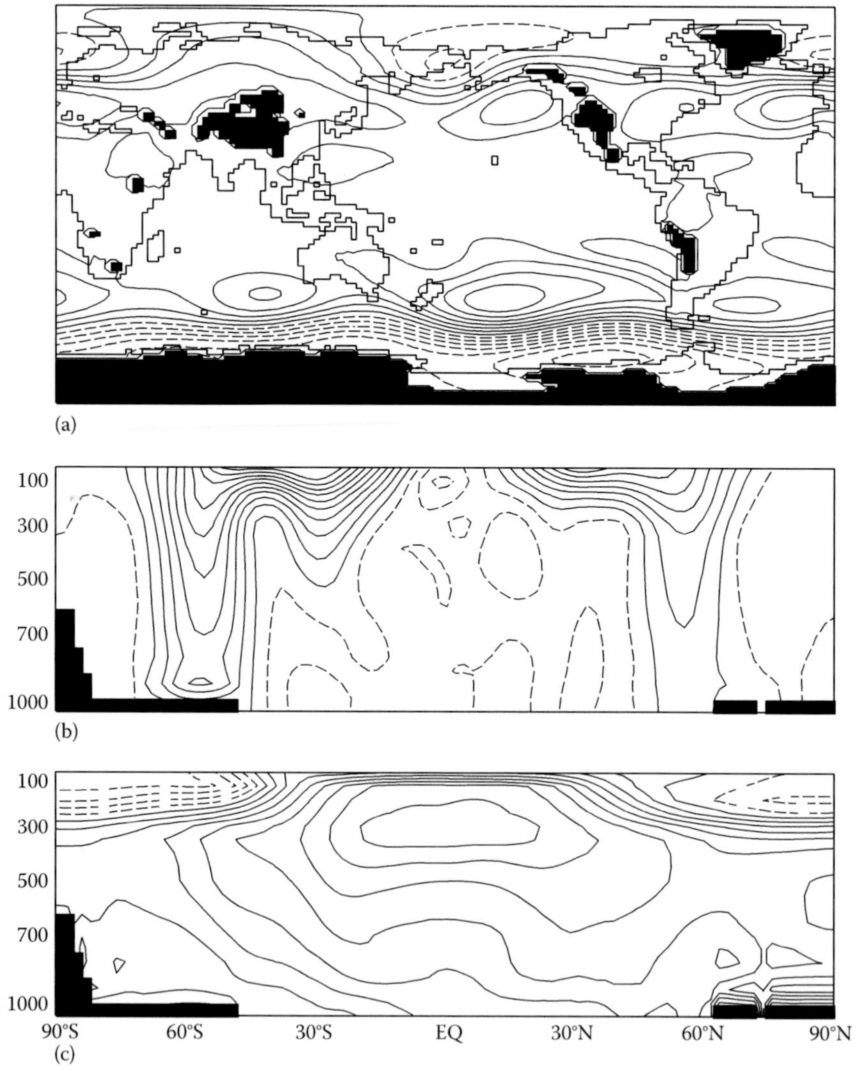

FIGURE 4.2 The projected large-scale climate change as simulated by the GFDL climate model used in IPCC AR4. The change is defined as future (2060–2080) minus present (1980–2000) climate state. (a) 850 hPa geopotential height field. Areas where topography rises above 850 hPa are masked in black. (b) Zonally averaged zonal (eastward) velocity as a function of latitude and height. The vertical coordinate is in pressure (hPa). (c) Zonally averaged temperature. Contour intervals are 5 m for (a), 0.5 m/s for (b), and 0.5°C for (c). Zero and negative contours are dashed.

a deep vertical extent in the troposphere and a significant surface component. Together, they indicate a poleward shift of the *eddy-driven* zonal jets. Accompanying this change is a poleward shift of the baroclinic eddies (the "storm tracks"), as has been discussed elsewhere (e.g., Yin 2005; Wu et al. 2010). This change is relevant to regional environmental assessment. The regions located at the boundary of the subtropics and midlatitude (e.g., the southwest United States) are expected to see less frequent occurrences of winter storms, potentially affecting water resources in those regions (Seager et al. 2007). Poleward of this latitudinal belt (e.g., New England and southern Canada), winter storms are projected to occur more frequently in the future (IPCC 2007, Chapter 10).

The projected temperature change in Figure 4.2c exhibits a nontrivial structure of a maximum warming not at the surface but near the tropopause. This structure is most pronounced in the tropics but is visible even in the Arctic region, where there is also a second maximum at the surface related to the reduction of sea ice coverage in late twenty-first century. The upper tropospheric maximum in the projected trend in temperature may be understood by a simple thermodynamic argument: Consider two air parcels at the surface with a small temperature difference. If we lift them following a typical moist adiabatic process (the parcel conserves entropy until condensation occurs, thereafter latent heat release augments the buoyancy of the parcel while liquid water falls out of the parcel immediately as it forms), the temperature difference between them will widen when they reach tropopause. (One can immediately appreciate this fact by tracing two adjacent moist adiabatic curves in a thermodynamic chart.) This interpretation requires pervasive happening of moist convection in the atmosphere for it to maintain a nearly moist adiabatic profile. While this is not unexpected for the tropics, there is evidence that moist convection occurs frequently even in the Arctic region in summertime (e.g., Korty and Schneider 2007), which renders our argument plausible. (Although Figure 4.2c shows the annual average, the upper tropospheric maximum in the temperature trend over the Arctic region is contributed mainly by the trend in summertime; see Frierson 2006.)

In addition to the principal prognostic variables of three-dimensional temperature, velocity, pressure, and moisture, global climate models used in the IPCC Report predict many other "parameterized" quantities such as precipitation. For example, the precipitation produced by a climate model with a 100 km horizontal resolution (typical of a state-of-the-art global climate model) is estimated by an empirical formula that links the large-scale velocity, temperature, and moisture at that resolution to the domain-averaged precipitation rate for a 100 km × 100 km grid box. Clearly, the prediction of these parameterized variables depends on model resolution. Despite this caveat, some interesting predictions, notably subtropical drying over semi-arid regions (IPCC 2007; Seager et al. 2007), have been made using the current generation of global climate models. How these predictions might change, especially over regions with complex terrains, with an improved resolution remains to be seen.

4.3.3 Learning from an Aquaplanet Model

While the details of land-sea contrast and topography help modify the response of large-scale circulation to GHG forcing, we will show that many first-order features in that response can be reproduced by running an "aquaplanet model" forced by a uniform surface warming. Here, the imposed surface warming serves as a proxy of the GHG forcing, an assumption justified by the importance of the control of tropospheric circulation anomalies by SST (see Section 4.3.1). The aquaplanet model is a global, three-dimensional atmospheric model with the full suite of parameterized physics (moist convection, cloud formation, radiation, etc.; Neale and Hoskins 2001) as in a full climate model, but with land-sea contrast and topography removed. The entire planet is covered by ocean; the model is driven by an imposed SST. Since the observed climatological SST is close to zonally symmetric except over the tropical Pacific warm pool and cold tongue regions, we use an SST that is a function of latitude only. In this case, the statistics of atmospheric circulation is zonally symmetric. To the extent that the aquaplanet model reproduces the projected trend by a full climate model, the simpler setting of the former helps us extract the essence of the large-scale atmospheric response to global warming. For this discussion, the aquaplanet model was modified by the author from a T42 (triangular-42 spectral truncation), 18-vertical-level version of the NCAR CCM3 global atmospheric model. As a further simplification, the simulations are performed under a perpetual equinoctial condition. Each simulation has a duration of 15 years.

To deduce the change in large-scale circulation, a pair of aquaplanet model simulations are performed. The "Control run" is forced by a standard, zonally symmetric SST that peaks at the equator and decreases toward higher latitudes (solid curve in Figure 4.3c). For simplicity, the SST is set to 0°C poleward of 60°N and 60°S to suppress the effect of sea ice. This restricts our focus to the tropics and midlatitude. A "Warming run" is then performed with the SST raised by a globally uniform +2.5°C (dashed curve in Figure 4.3c). The difference between the two runs is taken as the change in the tropospheric circulation in response to GHG-induced surface warming. Figure 4.3a and b are the latitude-height plots of the simulated zonal mean zonal velocity and the mean meridional circulation expressed as Stokes stream function (positive value corresponds to "clockwise" rotation in the figure). They capture the basic features of the eastward upper tropospheric zonal jets and the Hadley and Ferrel cells. (The polar cell is very weak in the simulated mean meridional circulation because we artificially flatten the SST poleward of 60°N and 60°S.) To highlight the effect of the uniform surface warming, the main contour plots (in black) show the time mean fields from the Control run, while the zero contours from the +2.5°C run are superimposed as the bold gray lines. They reveal a poleward expansion of both the Hadley and Ferrel cells. The zonal jets in the zonal velocity field also shift poleward, consistent with the results from the full climate model prediction discussed before. The poleward expansion of the Hadley cell means that the subtropical descent that maintains the arid or semi-arid

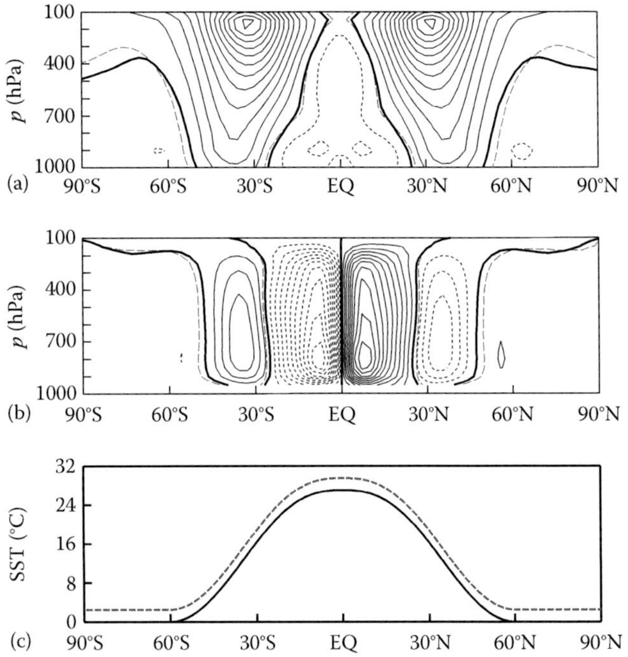

FIGURE 4.3 The thin black contours in (a) and (b) are the simulated climatological state from the Control run with the aquaplanet model forced by the SST shown as the solid curve in (c). (a) Zonally averaged zonal velocity; (b) mean meridional overturning circulation in terms of Stokes stream function. Contour intervals are 5 m/s for (a) and 2×10^{10} kg/s for (b), with negative contours dashed. In (b), positive values correspond to clockwise rotation for the flow. The bold gray dashed contours superimposed in (a) and (b) are the zero contours for the same fields but from the +2.5°C ("global warming") simulation forced with the increased SST as the dashed curve shown in (c).

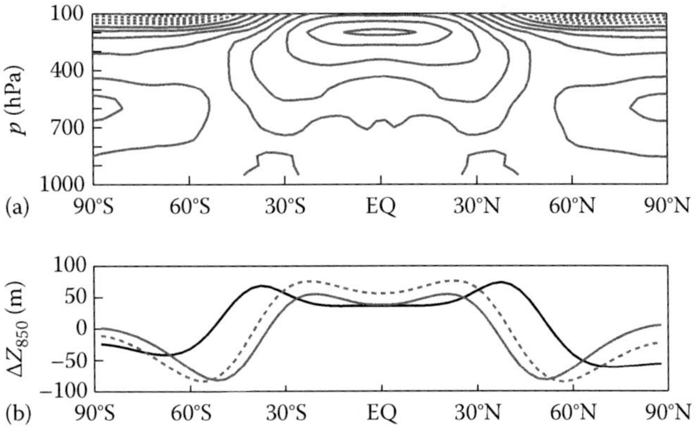

FIGURE 4.4 The "climate change" deduced from the aquaplanet model simulations as defined by the +2.5°C run minus Control run. (a) The change in zonally averaged temperature. Contour interval is 5°C, with negative contours dashed. (b) The black curve is the change in zonally averaged 850 hPa geopotential height, with scale shown at left. The gray solid and dashed curves are the climatological means for the Control (solid) and +2.5°C (dashed) runs. For plotting purpose, they are rescaled as $Z' = (Z_{850} - 1210\,\text{m})/2$, where Z_{850} is the original 850 hPa geopotential height.

climate at that latitude will shift poleward, thereby pushing the dry zone poleward. This, combined with previously discussed poleward shift of storm tracks, leads to overall future drying over the poleward edge of the subtropics (e.g., the Southwest United States, Seager et al. 2007).

Figure 4.4a shows the change in zonal mean temperature as a response to a uniform surface warming. It is deduced from the difference between the +2.5°C run and Control run of the aquaplanet model simulations. The simple model reproduces the upper tropospheric maximum in the trend in temperature simulated by the full climate models. The maximum warming at the surface in the arctic region is absent in this simulation, because we have held the SST poleward of 60°N and 60°S at 0°C and suppressed the effect of sea ice. Figure 4.4b shows the meridional profile of the zonal mean 850 hPa height from the aquaplanet model simulations. The gray lines are the climatological means for the Control (solid) and +2.5°C (dashed) runs, and the black line is the trend defined as +2.5°C minus Control run. The subtropical ridges in the trend are similar to those produced by the full climate model in Figure 4.2a. They reflect the poleward shift of the subtropical descents as discussed before. The aquaplanet model simulations reaffirm our argument that the first-order picture of the changes in large-scale tropospheric circulation reflects mainly the atmospheric response to a uniform surface warming, the latter being the direct thermodynamic response to GHG forcing.

The projected changes in temperature and velocity also imply changes in the stability properties of the large-scale circulation. Of particular interest are static stability that affects vertical convection, and baroclinic stability that affects the generation of midlatitude eddies (weather storms) that transport heat poleward by "slantwise convection." The baroclinic eddies also transport heat upward such that future changes in their characters may also affect the static stability (e.g., Juckes 2000). Figure 4.5a shows the zonal mean climatology of the static stability parameter, N^2, where N is Brunt–Vaisala frequency, from the Control run, while Figure 4.5b shows the trend in N^2 deduced from +2.5°C minus Control run. We find an increase in static stability in the warming world in mid-troposphere in the subtropics and part of midlatitude. The increased static stability in the subtropics is consistent with the poleward expansion of Hadley cell, in the sense that it allows an angular momentum conserving, upper tropospheric meridional flow to reach deeper into the subtropics without having the resulted upper-level zonal jet (or the vertical shear) violating the baroclinic stability criterion (Lu et al. 2007).

The few examples in Sections 4.3.1 and 4.3.2 provide a glimpse of the information that one expects to obtain from climate model predictions for the large-scale atmosphere and oceans. The horizontal resolution of the current generation of IPCC-grade climate models is close to 100 km (vertical resolution ranges from ~100 m for the boundary layer to ~1 km for mid-troposphere), implying that the large-scale prediction is meaningful only for the averaged climate condition of a region with $L > O$ (100 km). For instance, the prediction will not be particularly useful for determining the future climate change for a metropolitan area such as Phoenix, Arizona, which is covered by just one or two grid boxes of a global climate model.

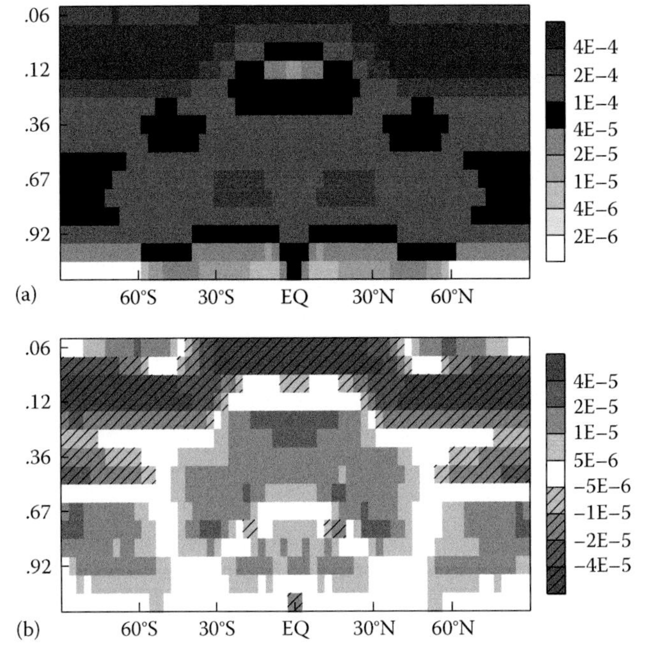

FIGURE 4.5 (a) The zonally averaged static stability parameter, N^2, where N is Brunt–Vaisala frequency, as a function of latitude and height, from the Control run of the aquaplanet model simulation. The numbers shown at left for the vertical axis are the normalized pressure, $\sigma = p/p_s$, where p_s is the surface pressure; $\sigma = 1$ is surface and $\sigma = 0.1$ is approximately the tropopause. (b) The change in N^2 is defined as the +2.5°C run minus Control run of the aquaplanet model simulations. Negative values (with a decrease in static stability under global warming) are hatched. The unit for the gray scale is s^{-2} for both panels.

4.4 Applications

4.4.1 Climate Downscaling

To make the climate model prediction useful for regional environmental assessment, the information of large-scale climate change has to be conveyed to smaller scales through additional high-resolution fluid dynamical simulations for the region. Directly running a global climate model at the desirable horizontal resolution, for example, 1 km, for local environmental prediction for the future is not yet feasible even with the most powerful computers. To illustrate the situation, note that from the First IPCC Assessment in 1991 to the Fourth Assessment in 2007, the typical horizontal resolution of global climate models used in those assessments has improved only by a factor of 5, from 500 to 100 km (IPCC 2007, Chapter 1). Extrapolating this trend, it would be another 16 years before the resolution is refined to ~20 km for the majority of the climate models. For environmental predictions for a limited region, an alternative is to run a high-resolution model only for that region but incorporate the large-scale flow from the global climate model prediction as the boundary conditions. This approach, with the large-scale boundary conditions injecting a one-way influence on the small-scale flows, is commonly called climate "downscaling" (e.g., Leung et al. 2004; Salate et al. 2008).

To predict regional environmental changes under the influence of global climate change, we can use a pair of downscaling simulations with their boundary conditions constrained by the observed present-day climate and the projected future climate state from IPCC-type of climate model predictions. The difference between the two reflects the influence of climate change on the regional environmental flow. To illustrate the benefit of climate downscaling, consider the idealized situation, shown in Figure 4.6, of a grid box in a coarse-resolution climate model within which lies an unresolved mesoscale mountain range (the elongated oval). Assuming that by running the climate model with GHG forcing, the prevailing low-level velocity is projected to change from the black arrow for today's climate to the gray arrow for the future. Invoking climate downscaling, we will further perform a pair of high-resolution atmospheric model simulations for the box that fully resolve the mountain range; one with the black arrow and another with the gray arrow imposed as the large-scale mean flow. Since the gray arrow now intersects with the mountain range, topographic lifting will help increase the probability of rainfall along the mountain range in the future, as indicated in the figure. This effect will be absent without downscaling. In Figure 4.6, the climate sensitivity (e.g., defined by the increase or decrease in precipitation with respect to double CO_2) is nonuniform within the large-scale grid box and may amplify regionally (a large increase in rainfall may occur over a small area along the mountain range). Having this information will benefit local environmental assessment and planning.

Since climate downscaling is a one-way street, the statistics of the small-scale environmental flow within the regional domain are largely controlled by the imposed large-scale boundary conditions. If there is a bias in the climate model simulation, it will propagate to the smaller scales through downscaling. In fact, just like climate sensitivity, climate model errors might also amplify as they interact with small-scale topography and

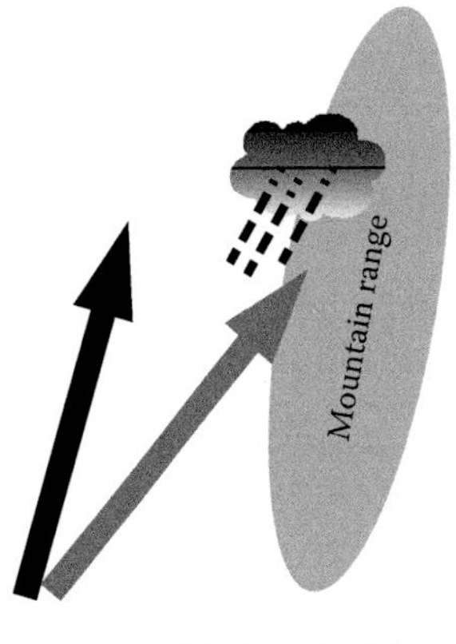

FIGURE 4.6 An illustration of the benefit of climate downscaling. The gray square represents a grid box of a coarse-resolution global climate model used for long-term climate projection. The black and gray arrows indicate the low-level air velocity in present-day and future climate as predicted by the global climate model. The oval is a mesoscale mountain range unresolved by the climate model. See the text for detail.

surface irregularities through downscaling. Despite increasing efforts in improving climate downscaling, this aspect is not yet well understood and should be kept in the mind of practitioners as a caveat.

4.4.2 Upscaling and Two-Way Interaction

While the downscaling approach described in Section 4.4.1 is useful, it does not fully represent the nonlinear interaction between the resolved and unresolved flows in a climate model. To illustrate what is missing, let us reconsider the idealized case in Figure 4.6. After the procedure of downscaling is complete, extra rainfall is produced along the mesoscale mountain range. At the same time, by conservation of water, one should expect drier air to exit the grid box through its northeast (upper right in the figure) boundary, leading to a reduced probability for rainfall to occur in the adjacent grid box in that direction. To account for this effect, an additional sink of water vapor determined from downscaling should be imposed to the large-scale flow in the global climate model. We will broadly call this procedure "upscaling."

It is clear that upscaling must be performed with downscaling, because we need the latter to determine the rectified effect of subgrid-scale processes on the large-scale flow. Once the large-scale flow is modified by upscaling, one might consider yet another iteration of downscaling and upscaling, and so on, until the upscaling influence diminishes. How this process may be implemented in a system of a global climate model and a regional mesoscale model remains to be investigated. Given that the idea of upscaling has barely been explored, even results from the "one iteration" (downscaling, then upscaling) case will be of great interest to practitioners of regional climate prediction.

As another reason to develop the technique for upscaling, recall that in addition to global-scale GHG forcing, regional climate change may also be driven by local urbanization and geo-engineering (e.g., building of wind farms). The downscaling/upscaling scheme will help merging the effects of large-scale GHG emission and local changes in landscape into a unified framework for predicting local environmental changes.

4.5 Major Challenges

4.5.1 Uncertainty in the Prediction of Large-Scale Climate Change

Our discussion of large-scale circulation changes in Section 4.3 has relied exclusively on climate model predictions. While observational records now consist of several decades of three-dimensional velocity and temperature data and slightly longer records for surface temperature and pressure, extracting anthropogenic climate change signal from these relatively short records is nontrivial, because in the late twentieth and early twenty-first century the GHG forcing is not yet strong enough to overwhelm natural variability on decadal to interdecadal time scales. Given this condition, it is also challenging to conclusively validate the model-predicted climate change over that short period of time. Some attempts in this vein have produced mixed results. For example, while climate models have predicted a strong maximum of warming in the upper troposphere (Figures 4.2c and 4.4a) that exceeds the surface warming, this feature is not yet clearly observed in the trend of tropospheric temperature (e.g., see a survey, and an attempt to reconcile the discrepancy, by Allen and Sherwood 2008). It is understood that this by no means indicates a deficiency of the climate models. Rather, it reflects the limitation in extracting signal from a short record with non-trivial internal variability. Developing alternative ways (e.g., based on the soundness of individual physical processes in the model) to measure the realism of climate model simulations with anthropogenic forcing will be an important task for fluid dynamicists working on climate change.

4.5.2 Validation of the Two-Way Downscaling and Upscaling Procedure

The technical challenges in climate downscaling and upscaling discussed in Section 4.4 involve the development of not only the mathematical formula for downscaling/upscaling but also an objective way to validate the scheme. Surveying related ideas, we find that the downscaling + upscaling scheme is close in spirit to the so-called "super-parameterization" scheme for convection (Randall et al. 2003), in which a two-dimensional cloud-resolving model is run along the grid lines of a coarse-resolution atmospheric model to produce the detail of cloud and moist convective processes, which in turn provide the diabatic heating for the large-scale flow. However, in our downscaling + upscaling framework, the downscaling part is not merely a way to produce large-scale forcing, but it fulfills our primary purpose of regional environmental prediction. Both downscaling/upscaling and super-parameterization are generalizations of conventional subgrid-scale parameterization schemes from one-way influence to two-way feedbacks. As a plausible way to measure the integrity of such schemes, note that a physically consistent parameterization scheme for the subgrid-scale processes should preserve the sum of the resolved and unresolved components of a given term in the Navier-Stokes equations (e.g., the topographic drag) as the model resolution improves (e.g., Brown 2004). Testing this property on the downscaling/upscaling scheme will be of both theoretical and practical interests. Interestingly, there is evidence that this conservative property is not preserved even by the conventional subgrid-scale parameterization schemes for the surface friction and orographic drag (Brown 2004).

4.5.3 Incorporation of Changes in Landscape due to Urbanization

We have so far focused on the effects of climate change due to anthropogenic GHG forcing. This is by no means the only kind of climate change that human activity may produce. For example, permanent changes in surface landscape by urbanization may also affect regional environmental flows. To simulate this effect, one needs a comprehensive scenario to describe future changes in the surface types (e.g., conversion from vegetation coverage to pavement that alters the radiative property and soil moisture of the surface) and future expansion of urban areas with concentrated built structures (that affects surface roughness). For local environmental fluid systems at city scales, these effects may emerge to become locally comparable to the large-scale GHG effect in future climate. Investigations in this direction are ongoing. A recent attempt to incorporate these effects into an assessment of the future climate for Phoenix and Central Arizona can be found in Georgescu et al. (2009).

References

Allen, R. J. and S. C. Sherwood, 2008. Warming maximum in the tropical upper troposphere deduced from thermal winds, *Nat. Geosci.*, **1**, 399–403.

Brown, A. R., 2004. Resolution dependence of orographic torques, *Q. J. R. Meteorol. Soc.*, **130**, 3029–3046.

Compo, G. P. and P. D. Sardeshmukh, 2009. Oceanic influences on recent continental warming, *Clim. Dyn.*, **32**, 333–342.

Deser, C. and A. S. Phillips, 2009. Atmospheric circulation trends, 1950–2000: The relative roles of sea surface temperature forcing and direct atmospheric radiative forcing, *J. Clim*, **22**, 396–413.

Fernando, H. J. S., 2008. Polimetrics: The quantitative study of urban systems (and its applications to atmospheric and hydro environments), *J. Environ. Fluid Mech.*, **8**, 397–409.

Frierson, D. M. W., 2006. Robust increases in midlatitude static stability in simulations of global warming, *Geophys. Res. Lett.*, **33**, L24816.

Georgescu, M., G. Miguez-Macho, L. T. Steyaert, and C. P. Weaver, 2009. Climatic effects of 30 years of landscape change over the greater Phoenix, Arizona, region: 1. Surface energy budget changes, *J. Geophys. Res.*, **114**, D05110.

Gueremy, J.-F., M. Deque, A. Braun, and J.-P. Piedelievre, 2005. Actual and potential skill of seasonal predictions using the CNRM contribution to DEMETER: Coupled versus uncoupled model, *Tellus A*, **57**, 308–319.

Held, I. M. and B. J. Soden, 2006. Robust responses of the hydrological cycle to global warming, *J. Clim.*, **19**, 5686–5699.

Huang, H.-P., R. Seager, and Y. Kushnir, 2005. The 1976/77 transition in precipitation over the Americas and the influence of tropical sea surface temperature, *Clim. Dyn.*, **24**, 721–740.

IPCC, 2007. Climate change 2007: The physical bases. *Contribution of Working Group I to the Fourth Assessment Report of the Intergovernmental Panel on Climate Change*, S. Solomon, D. Qing et al. (eds.), Cambridge University Press, Cambridge, U.K., 996p.

Juckes, M. N., 2000. The static stability of the midlatitude troposphere: The relevance of moisture, *J. Atmos. Sci.*, **57**, 3050–3057.

Keith, D. W., J. F. DeCarolis, D. C. Denkenberger, D. H. Lenschow, S. L. Malyshev, S. Pacala, and P. J. Rasch, 2004. The influence of large-scale wind power on global climate, *Proc. Natl. Acad. Sci. USA*, **101**, 16115–16120.

Khatiwala, S., F. Primeau, and T. Hall, 2009 Reconstruction of the history of anthropogenic CO_2 concentrations in the ocean, *Nature*, **462**, 346–349.

Korty, R. L. and T. Schneider, 2007. A climatology of the tropospheric thermal stratification using saturation potential vorticity, *J. Clim.*, **20**, 5977–5991.

Leung, L. R., Y. Qian, X. Bian, W. M. Washington, J. Han, and J. Road, 2004. Mid-century ensemble regional climate change scenarios for the western United States, *Clim. Change*, **62**, 75–113.

Lorenz, E. N., 1982. Atmospheric predictability experiments with a large numerical model, *Tellus*, **34**, 505–513.

Lu, J., G. A. Vecchi, and T. Reichler, 2007. Expansion of the Hadley cell under global warming, *Geophys. Res. Lett.*, **34**, L06805.

Neale, R. B. and B. J. Hoskins, 2001. A standard test for AGCMs including their physical parameterizations: I: The proposal, *Atmos. Sci. Lett.*, **1**, 101–107, doi:10.1006/asle.2000.0019.

Randall, D., M. Khairoutdinov, A. Arakawa, and W. Grabowski, 2003. Breaking the cloud parameterization deadlock, *Bull. Am. Meteorol. Soc.*, **84**, 1547–1564.

Reichler, T. and J. Kim, 2008. How well do coupled models simulate today's climate? *Bull. Am. Meteorol. Soc.*, **89**, 303–311.

Salate, E. P., Jr., R. Steed, C. F. Mass, and P. H. Zhan, 2008. A high-resolution climate model for the U. S. Pacific Northwest: Mesoscale feedbacks and local responses to climate change, *J. Clim.*, **21**, 5708–5726.

Seager, R., M. Ting, I. Held, Y. Kushnir, J. Lu, G. Vecchi, H.-P. Huang et al., 2007. Model projections of an imminent transition to a more arid climate in southwestern North America, *Science*, **316**, 1181–1184.

Stephenson, D. B. and I. M. Held, 1993. GCM response of northern winter stationary waves and storm tracks to increasing amounts of carbon dioxide, *J. Clim.*, **6**, 1859–1870.

Thulasiraman, S., N. T. O'Neill, A. Royer, B. N. Holben, D. L. Westphal, and L. J. B. McArthur, 2002. Sunphotometric observations of the 2001 Asian dust storm over Canada and the U.S., *Geophys. Res. Lett.*, **29**, 1255, doi:10.1029/2001GL014188.

Washington, W. M. and C. L. Parkinson, 2005. *An Introduction to Three-Dimensional Climate Modeling*, 2nd edn., University Science Books, Sausalito, CA, 355p.

Wu, Y., M. Ting, R. Seager, H.-P. Huang, and M. A. Cane, 2010. Changes in storm tracks and energy transports in a warmer climate simulated by the GFDL CM2.1 model, *Clim. Dyn.*, **37**, 53–72, doi:10.1007/s00382-010-0776-4.

Yin, J. H., 2005. A consistent poleward shift of the storm tracks in simulations of 21st century climate, *Geophys. Res. Lett.*, **32**, L18701.

5 Sustainability Implications

Bryan W. Karney
University of Toronto

S. Karan Venayagamoorthy
Colorado State University

5.1 Introduction

This chapter asks what contribution a better understanding of flow behavior can make for achieving a more sustainable engineered system. The answer, to largely misquote Tolkien's Bilbo Baggins, is "Lots, and none at all." No sustainable system could dare to seriously mistreat fluid behavior. Who would deny that the controlled movement of water and air—and all the varied and complex processes of transport, flow, dispersion, mixing, and exchange that connect them, as nicely summarized by H. Fernando in the Introduction to this volume—is a prerequisite to more sustainable systems? Yet it is clearly possible to have great mechanical or fluid efficiency in a completely unsustainable system. One might have, as but one example, an impressive tidal energy system that still does irreparable damage to an ecologically sensitive salt water estuary. Knowing and efficiently manipulating the fluid flow equations is clearly a necessary, but not a sufficient, condition for achieving sustainability.

Yet perhaps we are moving too fast. The word "sustainability" can mean many things. To those with a heartfelt concern for our planetary home, the word now unmistakably bears both positive and negative implications, being conceptually inviting and yet distressingly vague or even abused. The essential implication of sustainability that we adopt here is that something of value should continue for an unlimited number of generations, serving a well-conceived and well-thought-out purpose. Yet the negative aspect of the word is undeniable too, for once a word becomes imbued with a certain "good feeling" or intrinsic worth, it can be exploited, used by the unscrupulous to sanitize or whitewash activities or products that are, on reflection, often highly unsustainable. In this context, a classic line from the *Princess Bride* is worth recalling: "Life is pain, Highness. Anyone who says differently is selling something." Effective decision making needs more than good intentions or simplistic word associations: it needs painstaking analysis, reliable information, and wise and sometimes courageous choices between difficult alternatives.

We resist here adopting the smoke screen that, strictly speaking, sustainability is conceptually meaningless. It certainly seems inevitable that our sun will eventually burn out and, billions of years later, the universe itself will suffer a heat death. Yet our conviction is that humans are right now causing irretrievable, and often largely unnecessary, harm to a still remarkable planet, and thus, it behooves us all to do what we can to become *more* sustainable. The hope is that more sustainable engineering, based at least partly on a deep understanding of flow, will allow humans to continue to a larger fraction of the billions of years that lie between right now and our ultimate astronomical demise. The challenge is to turn the probable survival of our species, and the carrying capacity of the planet, into an "indefinite horizon" problem, not one of imminent catastrophe.

Even if it is not always easy to argue for strict sustainability, who would actually argue for its opposite? Unsustainability can only mean a premature curtailment or untimely death, of something ending before its time. It is a denial of validity and worth, of actions or outcomes that have undesirable consequences. If intelligence means anything, it certainly should mean that unsustainable actions—the industrial-scale production of persistent toxic wastes, the accidental creation of a huge oil slick that currently assaults Gulf of Mexico waters, the extinction of countless species, or the persistent violation of natural limits—must become much less likely in the future than they have been in the past. The problem, of course, is unsustainability is seldom

Handbook of Environmental Fluid Dynamics, Volume One, edited by Harindra Joseph Shermal Fernando. © 2013 CRC Press/Taylor & Francis Group, LLC. ISBN: 978-1-4398-1669-1.

something that we can simply give up—rather, serious consequences are a by-product of other actions, of the search for oil, or even the quest for solutions to human hunger, goals that are often understandable, sometimes defendable, and occasionally even noble. If we really believe we can and should do better, we must better evaluate the consequences of human actions; to this end, a better grasp of fluid behavior is surely a crucial and powerful asset.

5.1.1 Required Shift: From an Interior to an Exterior Perspective

One of the greatest challenges of sustainability is that it ultimately demands not only on getting the details of processes and fluid behavior right but also fitting these outcomes appropriately within their context. This requires a careful balancing of the global or overall requirements with all the details of the parts. As Fernando (2012) says, "understanding and modeling of non-linear interactions between scales remain a bane in environmental prediction." Sustainability is inevitably a compound endeavor, invariably multi-objective in scope and interdisciplinary in evaluation. Indeed, one benefit (and challenge) of the quest for sustainability is that we will be forced to put back together what would perhaps have been wiser to have never separated.

The process of detailed analysis alone is always a vexing one requiring many difficult compromises and tradeoffs. To focus on one part of the world—whether it is a specific location or in a defined domain of knowledge like fluid mechanics—is, by necessity, to pay less attention to other areas and other ways of knowing. This specialized view is essential if we are to make progress; we simply do not have a God-like capacity to know all things all at once. But such a choice introduces a distortion and fragmentation that does not exist in the original system (Vanderburg, 2005).

In fluid problems we often isolate a part of the flow field of immediate interest (the system or control volume) from the remainder that we call the environment. We look at a fish farm in the ocean, or waves breaking on a beach, or polluted air moving over complex terrain. We write equations for the interior space and couple them to the environment through boundary conditions (coupling the control system to the environment in space) and initial conditions (coupling it in time). This approach is practically essential and allows the analysis to proceed so that we can then solve or approximate the solutions of the equations and estimate the system's response. Eventually we must also judge whether the answers provided are humanly or physically acceptable, and we often change conditions, assumptions, or policies in order to better reflect our wishes and our preference for certain outcomes over others.

There are two immediate dimensions to the assessment of the system's performance that are raised by this approach. The first is whether our approximation itself is meaningful, raising the complex questions about the solution, its existence, uniqueness, regularity, stability, consistency, and the convergence of any numerical approximation to the solution of the governing equations that we believe better describe the flow. Few of these "internal" evaluations are trivial, and many can be overwhelmingly complex, demanding concentration, time, skill, and in-depth concentration from the researcher. Such preoccupations are certainly valid but almost invariably distract one from other, also valid, but more diffuse demands. Even assuming we are satisfied with the validity of our numerical answers—that we have a valid approximation to the original prediction problem—we still have to assess its overall merits. Are we humanly satisfied with the system's response? If not, what can we change about the design or configuration, the distance to the shore, or the shape of the breakwater, which might improve it? Is the natural system (the environment) going to be influenced by this application in ways that we can predict? Will it be influenced in other ways that are more difficult to quantify? Thus, we must also enter a challenging "outer loop" of design and assessment, of changing the assumptions or configurations of the system, and of adjusting what is under our control and questing through new outcomes for a more acceptable behavior. Naturally, this all takes time, money, and effort.

Yet, of course, all of this, though difficult, is routine and more or less comfortable for those accustomed to performing simulations or running models. Indeed, a successful project is often judged as one that is able to generate, at this scale, acceptable solutions having desired outcomes.

5.1.2 Harder Questions

The challenge, of course, is that there are two other dimensions of the problem that are often obscured or shortchanged by these traditional (though still challenging) preoccupations. Our focus is traditionally with the interior of the system in question, of what lies within our model itself—with assumed values, variations, interactions, and equations of state, and their immediate consequences on the environment. However, the next two dimensions we are now going to consider are largely more distant from the control system, and thus not in the modeling domain or its immediate vicinity that we typically interrogate. These additional dimensions are within the natural world, which gives rise to the question of loading, and within the human world. The human world more explicitly considers the decision space, which formally isolates one set of considerations from another, but for which no such clear boundaries actually exist. We must briefly lay out this territory, for it has a profound influence on issues of sustainability.

First is the question of loading. The "environment" external to the control system is typically capable of a wide range of behaviors, many of which have profound consequences to the system under consideration. Thus, for example, in ocean studies tides can be ignored, taken at average values, or at high or low water depths, themselves under either Spring or Neap conditions. Winds, currents, and waves vary widely in direction and intensity. Temperatures vary, as does the salinity, radiant environment and a host of other factors including ocean pH (tied also to the atmosphere) or water level (tied to world temperature and ice melt). How do we determine which are important and to which combination the modeled system should be subjected? Do we use statistical means, a stochastic simulation, or simply assume deterministic assumptions attempt to assign "worst-case" loads? What do we do if a combination of loading is predicted to be particularly problematic to the physical system, and yet relatively unlikely to occur? The challenge is to make good decisions that fit with human and natural values. It is not enough in most cases to simply argue, a posteriori after something unforeseen has happened, that *the system* was fine, or would have been if only the loading had cooperated. As humans we tend to push boundaries, assuming that we can drill oil wells at an ocean depth of 5000 ft, by compensating for what is known to change at more shallow depths, but miss the reality that our predictive and control capabilities are being overextended.

How should one properly come to grips with such issues of loading? A priori assumptions are dangerous. Collecting data is daunting, expensive, and never fully completed in a dynamic and complex world; we can, after all, always know more. Exhaustive simulations of "what if" scenarios not only are impossible to perform but also would usually further complicate the decision by blurring it in a confusion of contradictory outcomes. The temptation is to evade this task and to seek to enlarge the model to contain more of the world and to thereby make fewer assumptions. So we take a bigger slice of the ocean and couple it to the atmosphere in a few crucial ways, making a more complex simulation, and the model costs more to run, naturally requiring a larger computer (and grant!) and yet more researcher patience, skill, and cleverness. But is even this enough and, if so, how do we know? What about larger scale variations and complexities? Should these be included too? Perhaps the seasons and state of the sun are relevant, and the orbit of the moon that changes the tides, and a full climate model, and perhaps the off-gasing of volcanoes to boot? The more one considers the complexity and ambiguity of this path, the more appealing it is to simply assume the load and hope for the best. The only real alternative to being exhaustive is to ask more penetrating questions and to understand what part of the decision space is indeed influenced by the likely variation of the loads.

But these important issues of loading, so relevant to sustainability, do not exhaust the external challenges. Decisions on one system are seldom made in isolation: one outcome influences not only the attributes of the physical systems but also the attitudes and wishes of those who are making the decisions. To be sustainable requires that people care and continue to make more sustainable choices as the world unfolds. The human world changes in a dynamic way too, responding to goals, perceptions, attitudes, and experiences that include the perceived success or failure of previous attempts.

While not specifically a fluid example, it is one that provides a context for many choices that are made about fluid systems. This is an illustration of the so-called rebound effect, and why the decision space can itself be so vexing if one really cares about sustainability. The "rebound effect" arises when a decision is made to adopt, say, a more energy-efficient appliance or mode of transportation (an excellent review is found in Sorrell [2010]). Is the move toward sustainability directly related to improvements in efficiency? In fact, this is often a false assumption. The reason, on reflection, is obvious: if an energy user, say, saves a certain amount of money, say, on a reduced fuel bill, then the money saved can now be used to acquire other energy-intensive products or engage in energy-using activities. So, say, an expected 30% reduction in impacts, due to better efficiency or from adopting a conservation measure, may be reduced to a smaller fraction of this, perhaps to 15% or less. Under extreme conditions, the human system might "kick back" so that the new impacts actually exceed the old, meaning the new activities are less sustainable than the ones they replaced.

These issues of both the range of loading to which the system is subject, and the degree to which the system induces changes its human context, are large and vexing ones and will not be resolved here. Rather than dwelling indefinitely on what we have not yet achieved, it is perhaps time to shift to some of the areas where real progress has been made. We consider an example briefly here, a system that nicely shows what can be achieved, and the kind of interrelations that can begin to be explored. No single example can be exhaustive, but what is presented, like many of the other chapters in this volume, is meant to be suggestive that the best is yet to come. The quest for more sustainable systems will surely come from a greater ability to understand the consequences of our actions and their sensitivity to key environmental and human variables.

5.2 Example: Aquaculture Waste Plumes in Coastal Ecosystems

The rapid expansion of marine aquaculture is not only a potential solution to the problem of overfishing and fisheries depletion worldwide but also a major threat to ocean ecosystems. For example, wastes* and nutrient levels from fish pens alter the character of the surrounding ecosystem and potentially pose a threat to human health, escaped flora and fauna affect the genetic variation of the local population, and pathogen transmission from farmed to wild fish threatens the health and

* Wastes can be classified as solid (effects typically confined to the "near-field"), dissolved such as nitrogen (effects can be substantial in the "far-field"), chemical (if used in the facility), and biological (effects from escapes and pathogen transmission). The work discussed here focuses on dissolved wastes.

long-run viability of wild populations. One of the most widely cited but poorly quantified impacts of open fish pen aquaculture is its release of nutrients and other wastes to the surrounding environment (Naylor and Burke, 2005). In the United States there is considerable pressure on state and federal agencies to regulate the growth and mitigate the impacts of aquaculture operations in coastal waters (Naylor, 2006). There remains much uncertainty regarding the environmental impacts of aquaculture operations and the appropriate (and reliable) methods for regulating these impacts. The siting of the pens close to coastlines needs to be carefully investigated because evidence suggests that the dispersal of these wastes may not be "Gaussian" (monotonically decreasing from the source in all directions), and "dilution may not be the solution." Therefore, a better understanding of the effects of aquaculture on the marine environment is required for this industry to develop in an environmentally sustainable manner.

5.2.1 Numerical Modeling of Aquaculture Waste Plumes

5.2.1.1 Overview

In a recent study on aquaculture waste plumes, Venayagamoorthy et al. (2011) modeled the dispersal of "dissolved wastes" (considered here as passive scalars) such as nitrogen and phosphorus from aquaculture pens using high-resolution, two-dimensional, depth-averaged numerical simulations under different time-varying flows in an idealized coastal embayment. Their study was motivated by the need for a "science-based" tool to determine the "after the fact" impacts and "before the fact" alternatives for pen locations. The purpose of their study was not to perform a parameter study to investigate the quantitative effects of a single parameter on the dispersion, but rather to demonstrate the pronounced variability of the contaminant plume and how it is highly sensitive to different environmental forcing scenarios.

Here, we briefly describe this study to highlight the different dispersal patterns that may occur under various forcing scenarios (flows, tides, earth's rotation, and local sources) in a more idealized model bathymetry and to show how such a study can provide a science-based assessment tool in the sustainable development of aquaculture farms.

5.2.1.2 Simulation Setup

The hydrodynamics code SUNTANS (Stanford Unstructured Nonhydrostatic Terrain-following Adaptive Navier–Stokes Simulator), which employs unstructured grids to compute flows in the coastal ocean at high resolution was used as the numerical modeling tool for this study (see Fringer et al., 2006 for more details). The domain used for this study is a model coastal embayment with a width of 5 km as shown in Figure 5.1. The bathymetry of the domain consists of a shallow embayment incised by a deep channel as shown in Figure 5.2. Two sets of six 20 m diameter fish pens are used in this study, as shown in Figure 5.1, and the fish pens are all located close to the western edge of the embayment.

At the alongshore boundaries of the domain shown in Figure 5.1, a velocity field of the form was imposed:

$$u_y = U_m + U_t \sin(\omega t) \tag{5.1}$$

where

$U_m \leq 0$ is the amplitude of the mean current (where flow is in the north–south direction)

U_t is the amplitude of the sinusoidal tidal component of the flow field with forcing frequency ω

u_y is the alongshore component of the velocity field

A drag law formulation is employed to account for the flow reduction inside the pens and the resulting decrease in momentum downstream of the pens. This drag formulation is given by a quadratic drag-law on the right-hand side of the x- and y-momentum equations of the form

$$F_{D,x} = -\frac{pC_D\left(u_x^2 + u_y^2\right)^{1/2}}{D}u_x$$
$$F_{D,y} = -\frac{pC_D\left(u_x^2 + u_y^2\right)^{1/2}}{D}u_y \tag{5.2}$$

where

$C_D = 1$ is the drag coefficient

u_x and u_y are the easterly and northerly components of the velocity

$p = 1$ inside of the pens while $p = 0$ outside of the pens

An important nondimensional parameter for the simple model flow problem defined by Equation 5.1 is the ratio of the tidal to mean flow given by

$$\eta = \frac{U_t}{U_m} \tag{5.3}$$

which compares the amplitude of the oscillatory flow to the amplitude of the mean current and is an important parameter that determines the shape of the contaminant plume (Purnama and Kay, 1999). A second important parameter is the nondimensional tidal excursion length scale given by

$$K = \frac{2U_t}{\omega D} \tag{5.4}$$

where

U_t is the amplitude of the tidal current

ω is the forcing frequency

D is the pen diameter

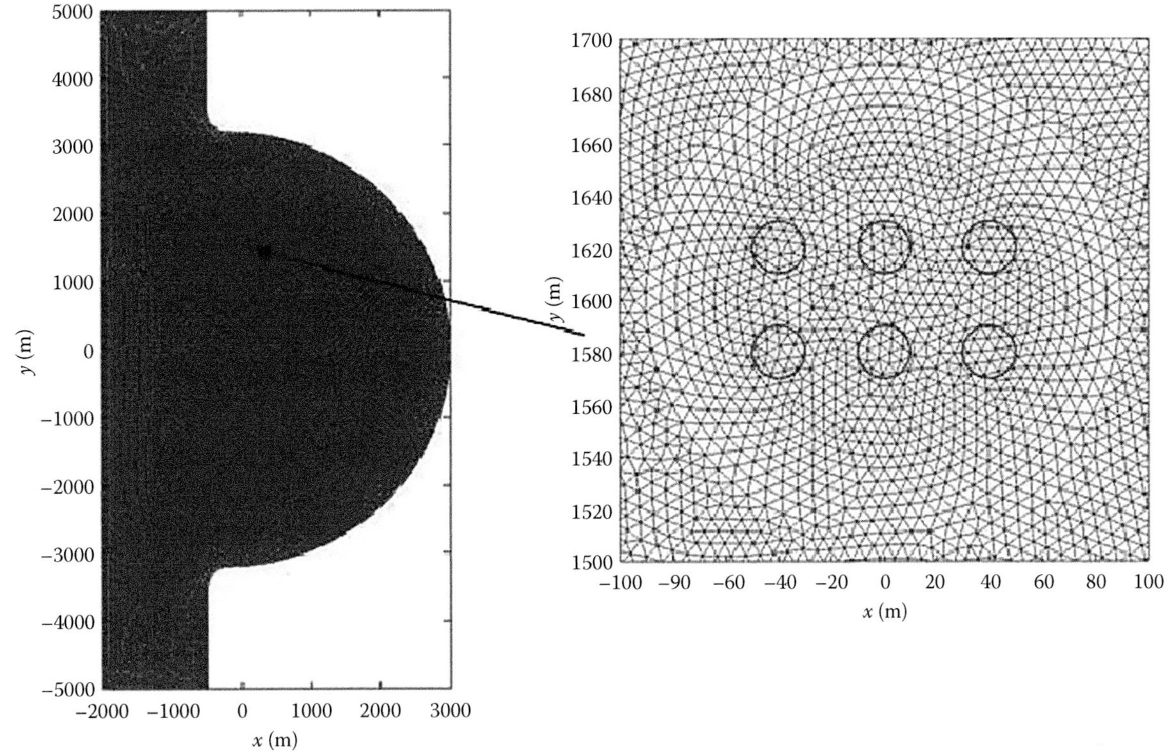

FIGURE 5.1 Unstructured computational mesh showing the embayment used in the simulations for this study. A velocity field described by Equation 5.1 is imposed at the northern boundary of the domain with a M_2 tidal frequency of 1.4×10^{-4} rad s^{-1}. The image on the right shows a zoomed view highlighting the grid refinement around an array of six 20 m diameter fish pens.

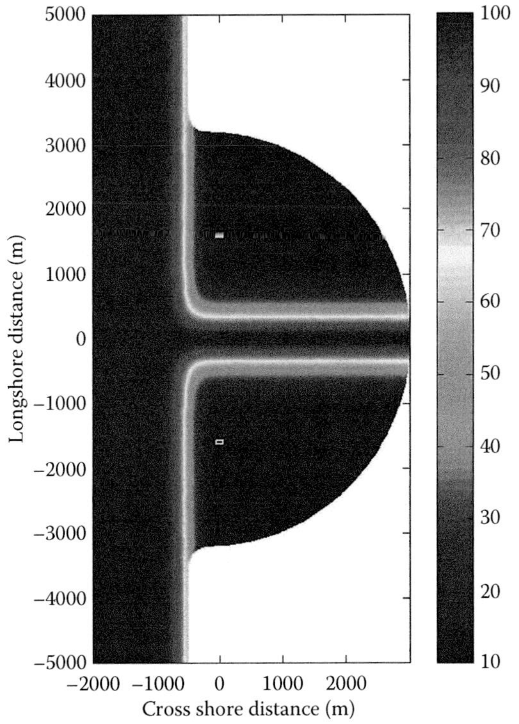

FIGURE 5.2 **(See color insert.)** Idealized depth contours of the model embayment depicting a shallow shelf incised by a deep channel. Locations of model offshore fish pens are depicted by the white boxes. The depth is indicated by the color bar in meters.

This represents the ratio of the tidal excursion to the pen diameter and is analogous to the Keulegan–Carpenter number used in wave-structure interaction studies. For the bulk of the simulations performed in this study, an alongshore velocity magnitude of $U_m = 0.1$ m s^{-1} and the tidal velocity magnitude were varied to yield different (field-scale) values of η and K. Detailed information on the simulation setup is given in Venayagamoorthy et al. (2011).

5.2.1.3 Results and Discussion

The passive scalar concentration field from a simulation run with two sets of fish farm pens located along the edge of the embayment is shown in Figure 5.3A as a time sequence over a duration of six tidal periods. For this "offshore base case" (as shown in Table 5.1), the drag induced by each of the pens on the flow field has been parameterized using Equation 5.2 and the effect of the earth's rotation has been included. Using a Coriolis parameter of $f = 8.7 \times 10^{-5}$ rad s^{-1}, and the embayment width as the length scale, the Rossby number for this case is $R = U_m/fL = 0.26$, which indicates that the Coriolis force will influence the flow dynamics. The flow field is driven by a tidal flow from north to south combined with a southerly flowing mean current. The relevant oscillating flow parameters are $\eta = 1$ and $K = 71$, based on a tidal velocity amplitude of $U_t = 0.1$ m s^{-1} and the M_2 tidal period of 12.42 h. The formation of the downstream vortex shedding is evident early on as shown in Figure 5.3Aa. As the flow reverses, the plume contracts as the flow retards due to reversal in the tidal component of the velocity field. The plume, while predominantly transported downstream (southward) due to the rather strong mean current, also tends to spread eastward into the bay (see Figure 5.3Ac through f).

The location of the fish pens is a key factor in regulating the concentration distribution of the waste plumes. A whole range of scenarios are possible and would be prohibitively expensive to completely simulate computationally. To get an indication of the effect of the pen location on the plume dispersion, a "near-shore base case" was simulated. Here, the southern facing farm in Figure 5.1 was moved into the bay and closer to the channel incision. All other conditions remain unchanged relative to the "offshore base case" shown in Table 5.1. The concentration distributions of the plume as shown in Figure 5.3B indicate much higher concentrations closer to the coast compared to the "off-shore base case."

Figure 5.4 shows the concentration distribution for all the offshore cases outlined in Table 5.1 (i.e., cases 1, 3, 4, and 5) at time $t = 4T$ to highlight the differences in the plume behavior between these cases. The effect of the earth's rotation on the dispersion of the plume was investigated by simply switching off the Coriolis term in the Navier–Stokes equation in the numerical code. Figure 5.4B shows the concentration field at time $t = 4T$ together with the concentration field for "offshore base case" for the same time (Figure 5.4A). For this scenario, it is seen that the plume does not spread deep into the embayment. The absence of the Coriolis force implies that the flow field is not deflected into the embayment especially during slack tides when the effects of rotation are expected to be strong. Hence, when the tide turns, the plume is flushed downstream (southward) with much higher concentrations.

Enclosed coastal embayments may often have rivers discharging into them. A small inflow river discharge with a

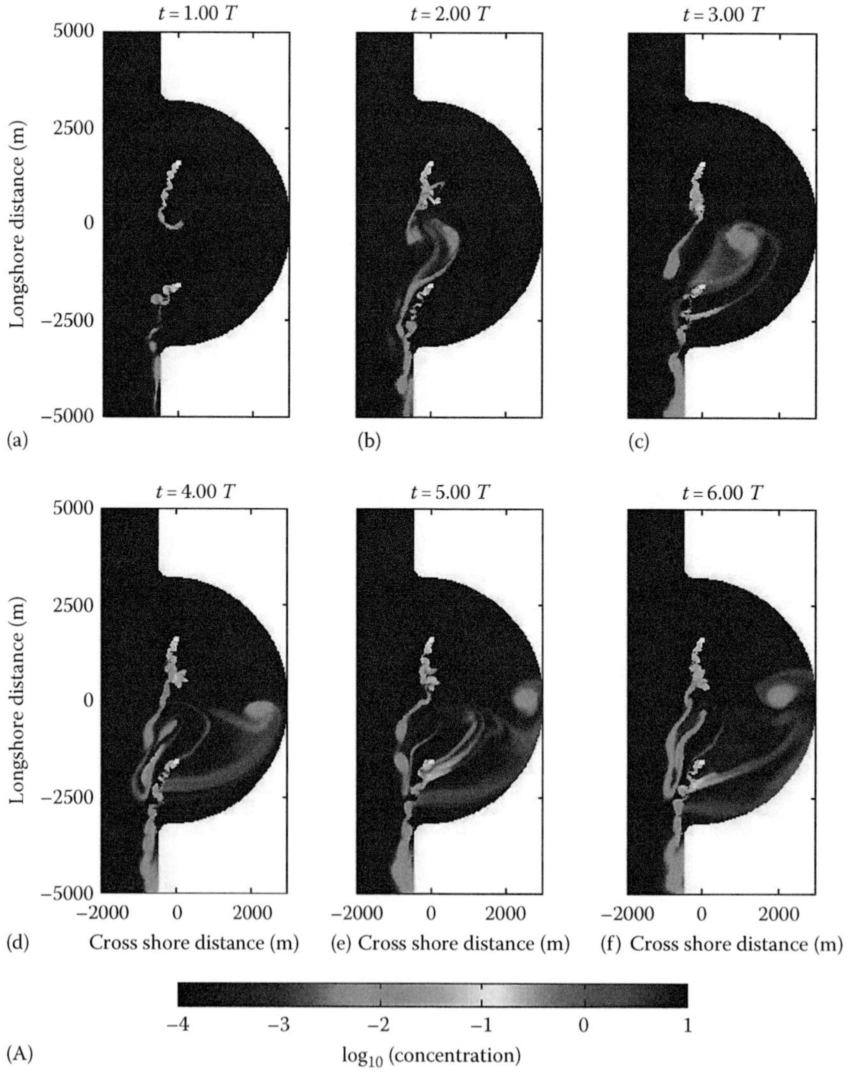

FIGURE 5.3 **(See color insert.)** "Birds-eye-view" of the time history of the normalized concentration in the vicinity of two sets of six 20 m diameter pens (depicted as white boxes) releasing a passive scalar at the coastal embayment for (A) the "offshore base case" (upper panel) and

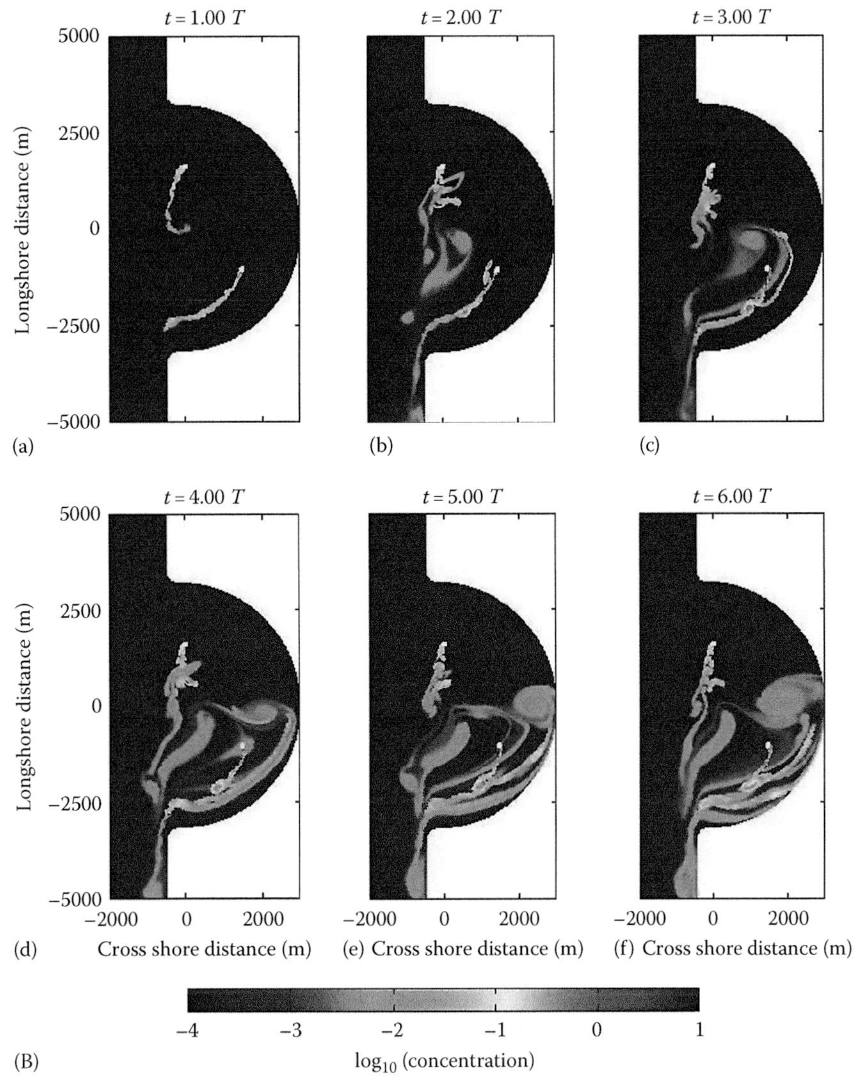

FIGURE 5.3 (continued) (B) "nearshore base case" (lower panel) shown in Table 5.1.

TABLE 5.1 Summary of Five Simulation Cases Discussed in Section 5.2

Case #	Domain Type	Case Name	η	K	Other
1	Embayment	Offshore base case	1	71	
2	Embayment	Nearshore base case	1	71	
3	Embayment	Offshore with no rotation	1	71	$R = \infty$
4	Embayment	Offshore with river inflow	1	71	$U_r/U_m = 0.5$
5	Embayment	Offshore with no pen drag	1	71	$C_D = 0$

The last column for cases 3, 4 and 5 indicates the parameter that was changed relative to the base cases 1 and 2, respectively.

velocity of 0.05 m s^{-1} was placed at the head of the channel incision at the central embayment coast. The resulting plume concentration at $t = 4T$ is shown in Figure 5.4C. With all other parameters being equal with the "offshore base case," it is evident (as would be expected) that the plume spread into the embayment is impeded by the river inflow. However, in contrast to the case with no rotation (Figure 5.4B), the time sequence of the plume distributions (not shown here) clearly indicates that the plume does not hug the embayment coast except further downstream resulting in a highly skewed distribution. This is clearly a desirable situation for fish farm operations since the waste plume is unlikely to reach the coastline due to the flushing action of the freshwater inflow.

The effect of the drag induced by the fish pens causes flow separation and vortex shedding, thus enhancing the local mixing. In an effort to understand the influence of blockage introduced by the pens, an "offshore case" without accounting for the drag from the pens was simulated. For this case (case number 5 in Table 5.1), it is clear from the concentration distribution shown in Figure 5.4D that the absence of drag results in less mixing compared to the "offshore case" shown in Figure 5.4A, even though the overall plume distribution looks very similar. This result highlights the need to correctly account for the drag induced by the pens in predictive numerical models for water quality applications. An overprediction of the drag results in a well-mixed plume, and on the other hand an underprediction implies poor mixing conditions.

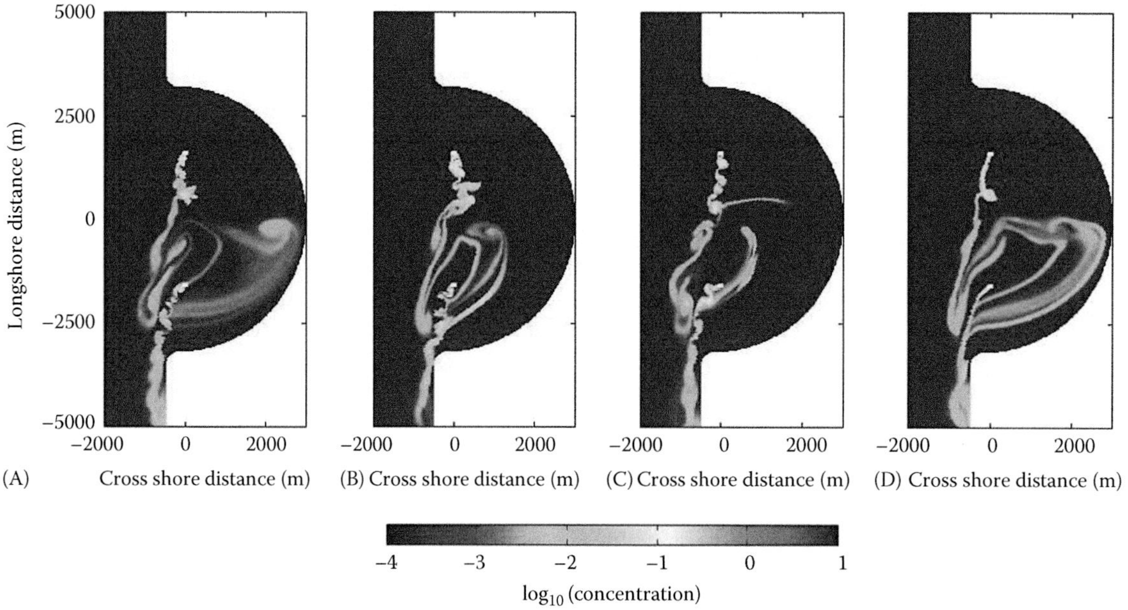

FIGURE 5.4 "Birds-eye-view" of the normalized concentration in the vicinity of two sets of six 20 m diameter pens (depicted by white boxes) releasing a passive scalar at the edge of the coastal embayment at time $t = 4T$ for (A) offshore base case (case 1), (B) offshore with no rotation case (case 3), (C) offshore with river inflow case (case 4), and (D) offshore with no pen drag case (case 5), respectively.

5.2.1.4 Summary

The results highlight the complex and different dispersion patterns that occur under oscillatory flow conditions coupled with mean flow. In particular, the results from this study demonstrate the following key points:

- The importance of identifying the correct (and dominant) hydrodynamic conditions at a particular fish farm site because these conditions dramatically influence how far a waste plume from a fish farm site would spread.
- Accounting for the drag induced by the pens is important to accurately predict the level of mixing in numerical models for water quality applications.
- The large-scale effect of the earth's rotation does indeed influence the dispersion of contaminant plumes and should be accounted for in numerical models.
- The local runoff from rivers and other tributaries should be taken into account in modeling the plume dispersion in near-coastal environments.

This study shows how environmental flow models can be used in the design of water quality regulations and the monitoring of wastes in order to ensure an environmentally sustainable aquaculture industry.

5.3 Major Challenges

The case study provided here is obviously specific, but the implications are not. They raise the broader issues, either directly or indirectly, of human use of natural space and of the consequences of those uses, whether beneficial or detrimental, to both humans and the natural system. But who is to judge the relative merits of the cases? If, as is so often the case, the advocate of the action is a human beneficiary, who will speak on behalf of the system that is deteriorated, which is so often without the same legal standing?

Paul Colinvaux, in the final chapter to his remarkable 1978 book *Why Big Fierce Animals Are Rare* argues that humans have fundamentally learned to occupy the ecological niche of a progressively larger range of organisms until we have learned to span the whole planet and the space of every other organism. He reasons that each technological extension has only further enlarged our reach, and we have not yet learned any restraint or humility. Thus, the danger is that greater understanding merely leads to greater exploitation, which leads to great human expectations and reduced sustainability. So, will a study such as the one presented earlier merely mean we push the ocean limits harder, so that they impinge (for at least a short time) less pressingly on humans to curb their expectations? Will we take such studies as a de facto sanction that fish farming is (despite the explicit challenges) manageable, controllable, and even beneficial, since certain of its impacts can be predicted? It is interesting that a recent comprehensive review of Global Biodiversity in *Science* (Butchart et al., May 28, 2010, pp. 1164–1168) concluded that our human commitments to biodiversity have been ineffective and that biodiversity has not only decreased in many contexts but also the loss shows no sign of slowing.

Donnelly and Boyle (2006) considered how well equipped engineers are for making calls relating to sustainability. Their conclusion is so relevant that it is worth quoting in detail:

The engineering profession is caught in a significant Catch-22 situation when it comes to delivering sustainable development.

Engineers have been told that they have a "tremendous responsibility in the implementation of development" because of their "unique capacity to address the priorities" and yet

- Current sustainability assessment tools are inadequate for assessing or delivering sustainable solutions.
- Engineers generally work at site and project levels but sustainability first needs to be addressed at the regional, system level that the site will tie into.
- Traditionally, engineers are problem solvers, but developing innovative, sustainable solutions requires engineers to become problem framers. Engineers are constrained in this as they are frequently unable to address the truly significant sustainability issues within the scope of a project.

It is widely accepted that complex decisions resulting in more sustainable outcomes require integration between different professional domains as well as those making and those affected by these decisions (stakeholders, actors). In practice, a lack of adequate knowledge sharing across domains leads to consideration only of tried and tested practices and an overreliance on monetary evaluations to determine appropriate solutions. Furthermore, stakeholder and actor engagement is often inadequate due to the failure of decision makers to connect to the reality of the problem being addressed from the point of view of those affected by it or by wider considerations.

Such challenges are obviously valid with respect to environmental fluid mechanics as to require little additional comment. Yet it is interesting that one of the greatest values of our modern skill with fluid mechanics is our ability to visualize outcomes so that the researcher can obtain a sense or feel of the system's response. Indeed, the word for theory has a root which is quite akin to the word for theatre, which both relate to the idea of a place for seeing things. We can see the fluid motion and make sense of it. We can see the impacts, but the long-term unfolding of those consequences that are implied by the term for sustainability is much more difficult to unpack. Again, one of the issues is one of scale; the time scales required to understand turbulence, wave motion, economic consequences, or ecological changes cover orders of magnitudes with all the consequent challenges association with that massive range.

5.4 Conclusions

Environmental sustainability of ecosystems is a global scale concern and a theme that is at the forefront of multidisciplinary research spanning from ecology to engineering to policy. Environmental fluid dynamics, which involves the application of established fluid mechanics principles to fluid flows in the natural and engineering environment, is an important component in addressing the complex issues that are directly or indirectly related to environmental sustainability. Some examples of how the fascinating subject of environmental fluid dynamics is directly related to environmental sustainability issues include a range of subjects: water quality and nutrient transport in estuaries, lakes, reservoirs, rivers, and coastal ocean; air quality from atmospheric pollution; and the implications of climate change on ecosystems. The intent of this chapter is to provide the reader with a somewhat philosophical overview (by no means exhaustive) of the implications of environmental fluid dynamics on the key issue of sustainability. The main goals are to highlight the importance of environmental fluid dynamics in addressing the issue of sustainability from a science-based viewpoint as well as to draw attention to some of the key larger challenges that need to be addressed in order to effectively address sustainability.

More specifically, a relevant example environmental problem that is linked to the theme of sustainability is discussed. This example focuses on numerical modeling study of aquaculture waste transport in a fictitious coastal embayment. Numerous other examples such as dispersion in urban canopies, sustainability of coral reefs, etc., could be cited but this one example should suffice to convey the message. Yet the overall challenges are huge, and only an ability to deal with both large-scale issue of system behavior along with the details of process and behavior will ultimately be able to address overall issues of sustainability.

Eric Zencey (2010) has recently provided an excellent primer on "Theses on Sustainability." The fourth of these theses is so to the point that is worth quoting in full:

Nature is malleable and has enormous resilience, a resilience that gives healthy ecosystems a dynamic equilibrium. But the resilience of nature has limits and to transgress them is to act unsustainably. Thus, the most diffuse usage [of the term sustainable], "sensibly far-sighted," is the usage that contains and properly reflects the strict ecological definition of the term: a thing is ecologically sustainable if it does not destroy the ecological preconditions for its own existence.

Certainly, in this quest for ecological sustainability, the role of the understanding of the fluid system can and must play a crucial role. But we must learn to not simply use our knowledge to more thoroughly exploit the world, but to live more gently within its intricate and, thus far, life-sustaining richness.

References

Colinvaux, P. (1978). *Why Big Fierce Animals Are Rare*. Princeton University Press, Princeton, NJ.

Fernando, H. (2012). *Environmental Fluid Dynamics: An Introduction*.

Fringer, O. B., M. G. Gerritsen, and R. L. Street. (2006). An unstructured-grid, finite-volume, nonhydrostatic, parallel coastal ocean simulator. *Ocean Modelling*, **14**: 139–173.

Hurley, L., R. Ashley, and S. Mounce. (2008). Addressing practical problems in sustainability assessment frameworks. *Engineering Sustainability*, **161**: 23–30.

Naylor, R. (2006). Environmental safeguards for open ocean aquaculture. *Issues in Science and Technology*, **22**: 53–58.

Naylor, R. and M. Burke. (2005). Aquaculture and ocean resources: Raising tigers of the sea. *Annual Review of Environment and Resources*, **30**: 185–218.

Purnama, A. and A. Kay. (1999). Effluent discharge into tidal water: Optimal or economic strategy? *Environmetrics*, **10**: 601–624.

Sorrell, S. (2010). A review of the rebound effect in energy efficiency programs, in *Current Affairs: Perspectives on Electricity Policy for Ontario*. D. Reeve, Donald N. Newees, and Bryan W. Karney, (eds.), University of Toronto Press, Toronto, Ontario, Canada, pp. 195–244.

Vanderburg, W. H. (2005). *Living in the Labyrinth of Technology*. University of Toronto Press, Toronto, Ontario, Canada.

Venayagamoorthy, S. K., H. Ku, O. B. Fringer, A. Chiu, R. L. Naylor, and J. R. Koseff. (2011). Numerical modeling of aquaculture dissolved waste transport in a coastal embayment. *Environmental Fluid Mechanics*, **11**: 329–352.

Zencey, E. (2010). Theses on sustainability: A primer. *Orion Magazine*, May/June 2010.

6

Air Quality and Management

Sang-Mi Lee
South Coast Air Quality Management District

6.1 Introduction

A pollutant emitted into the atmosphere is subject to various dynamic and chemical processes such as transport, mixing, recirculation, trapping, photochemical production, titration, coagulation, and deposition. These processes are closely related to natural environments as well as anthropogenic activities. The Los Angeles air basin is an example; worst air quality often occurs in an urban area over complex terrain, in which comprehensive flow patterns play a critical role in the buildup of pollutants. At the same time, climatological factors such as high actinic flux promote the formation of secondary pollutants, which are, like ozone and fine particulate matters, not directly emitted from sources (primary pollutants) but chemically produced in the air from those primary pollutants. Inhomogenous surfaces, including land/water contrast and urban/nonurban differences, add further complexity to the nature of the problems. Despite multidecade-long efforts to improve the understanding of such processes and eventually to protect public health and welfare from adverse air quality, the attainment of air quality to the level defined in National Ambient Air Quality Standards (NAAQS) remains elusive, especially for O_3 and $PM_{2.5}$. Many of the major urban centers in the United States have been classified as nonattainment of O_3 and PM standards with limited improvement to be expected in the near future. Still, better understanding of physical and chemical processes involved in air pollution is critical to achieve better air quality. In this context, we discuss fundamentals of transport, dispersion, and mixing processes as well as chemical reactions involved in the most commonly cited pollutants (Section 6.2), and typical methodologies to elucidate such fundamentals (Section 6.3). The Phoenix and Los Angeles air basins, two of the most populated and polluted urban centers in the United States, were taken as examples to illustrate such theories (Section 6.4). Finally, regulatory efforts to reduce adverse effects of air pollution on human health and welfare are briefly described in Section 6.5.

6.2 Principles

6.2.1 Multiscale Flows in the Atmosphere

Atmospheric motions are characterized by a series of different scales ranging from the whole globe to a narrow street in an urban center with their corresponding time scales of multi-years to a fraction of a second. These scales of motions are generally classified into three broad categories of global, regional (or synoptic), and local scales. The global scale phenomena include Hadley circulation, Walker circulation, El Nino and Southern Oscillation (ENSO), and La Nina. The Hadley circulation is governed by global energy balance of the Earth that has net surplus of thermal energy in the equator and net deficit in the polar areas, which derive ascending and descending motion, respectively, at the equator and at the poles. However, the rotation of the earth makes such a large cell unsustainable and splits into three cells—Hadley, Ferrel, and polar cell. The Hadley cell is located between the equator and approximately 30°, the Ferrel, also called as mid-latitude cell is in approximately 30°–60° region, and the polar cell is in the higher latitude area. Among them, the Hadley and the polar cells are thermally induced circulations with rising motion at a warm temperature (lower latitude) and

Handbook of Environmental Fluid Dynamics, Volume One, edited by Harindra Joseph Shermal Fernando. © 2013 CRC Press/Taylor & Francis Group, LLC. ISBN: 978-1-4398-1669-1.

sinking at a cold area (higher latitude). On the contrary, the mid-latitude Ferrel cell has a thermally indirect circulation forced by the other two cells—a downward branch near 30° mechanically induced by the strong descending motions of the direct thermal cell and surface divergence, while an upward branch near 60° triggered by upward motion and surface convergence.

While the Hadley circulation is on a meridional plane (North–South and vertical direction), the Walker circulation occurs on a zonal plane (East–West and vertical direction). It is driven by pressure gradient force due to the heat capacity difference between land and water. The circulation occurs in tropical atmosphere where coriolis force is zero so that air flows directly from a high pressure to a low pressure. A typical Walker circulation consists of two cells: the Pacific and Indonesian. The easterly trade wind is a part of the low-level component of the Walker circulation over the Pacific. Typically, the trade winds bring warm moist air toward the Indonesian region. Here, moving over normally very warm sea, the moist air ascends. The air then travels eastward before sinking over the eastern Pacific Ocean. The rising air is associated with a region of low air pressure and precipitation. High pressure and dry conditions accompany the sinking air. Opposite to the Pacific cell that has clockwise rotation in a zonal plane, the Indonesian cell is composed of rising air parcel over Indonesia and sinking over Somali coast forming an anticlockwise circulation.

Southern Oscillation is surface pressure oscillation between the East Pacific and the Indian Ocean. Typically, Southern Oscillation Index (SOI), a surface pressure difference between Tahiti (southeastern Pacific) and Darwin (western Pacific), is used to illustrate the southern oscillation, which has a close correlation with El Nino and La Nina. Negative SOI indicates an El Nino condition, while positive SOI signifies La Nina. El Nino is a global-scale phenomenon with a wide range of natural and socioeconomic effects. During normal conditions, fresh ice melting water from the Antarctic sunken into the Southern sea upwells along the western coast of Southern America building cold sea surface temperature (SST) in the east Pacific and warm in the west. However, during an El Nino condition, the cold water upwelling in the eastern Pacific weakens and the easterly trade wind in the equatorial region slows down, and consequently significant SST anomalies occur across the Pacific: warm anomaly in the east and cold anomaly in the west. The anomaly is a difference between climatological mean and in situ value. Under this condition, the two distinctive cells of the Walker circulation split into three cells and the precipitation pattern associated with its shifts—increased rainfall in the central and the east Pacific near the ascending branches of the Walker circulation and drought in Indonesia and Australia, which are associated with sinking motions. On the contrary, La Nina, which enhances the Walker circulation cell over the Pacific, is associated with a strong easterly trade wind at the Equator and a strong rising motion over the western pacific leading to enhanced precipitation in Indonesia and Australia.

Such global-scale motions are closely related with long-range transports of air pollutants. They also provide climatological background that are conducive to pollutant buildup. For example, westerly flow of the Ferrel cell transports dust particles originated from Asian deserts to the North American continent or even further. Similarly, the easterly trade wind in the tropical region attributes to the westward transport of Saharan dusts to the North American continent. Synoptic phenomena are closely linked with global motions. Synoptic flow is defined as a weather system with a size ranging from several hundred to several thousand kilometers, a scale of a typical mid-latitude migratory high- or low-pressure system. In the case of the Los Angeles air basin, being located near the end of a polar jet stream, it permanently experiences a strong descending motion, in other words a high-pressure system. The air mass converges near the exit of the jet stream due to the deceleration of the flow and then descends to a lower altitude. This sinking motion creates a semipermanent high pressure over the area. Consequently, Southern California experiences the highest number of sunny days and the largest amount of actinic flux among metropolitan cities in the United States, which results in being among the worst air quality in the nation (Lee et al. 2009).

In the synoptic scale, differential thermal forcing on inhomogenous surfaces is one of the major governing mechanisms. Land–sea contrast and its associated heat capacity difference cause seasonal monsoon as well as a diurnal land–sea circulation. A typical local thermal circulation spans approximately 10–100 km. The land/sea and mountain/valley breeze often enhance each other when they are lined up to the same direction in a coastal environment with complex terrain. In the Los Angeles air basin, westerly momentum initiated by onshore pressure gradient (from ocean to land) increases, as the sea breeze reaches to the sloped surface of mountains located in the inland of the basin during the daytime. Similarly, the easterly downslope flow accelerates as the land breeze develops during the nighttime. The local circulation signifies horizontal and vertical distribution of pollutants. Details associated with such circulation are further discussed in Section 6.4.

6.2.2 Atmospheric Boundary Layer

6.2.2.1 Vertical Structure

The lowest part of the atmosphere that is in direct contact with the earth surface is called Atmospheric Boundary Layer (ABL), which is important specifically to delineate the amount of vertical ventilation of pollutants. The bottom portion of the ABL is distinguished as the Surface Layer (SL), which spans from the ground up to a height that a surface turbulent flux remains within 10% of its variation in the vertical direction. The SL is further divided into canopy layer, roughness sublayer, and inertial sublayer. The canopy layer, the layer below a mean canopy height, has specific importance in an urban environment due to complex phenomena captured within urban street canyons. This layer is subject to anthropogenic emissions, pollutants trapping and ventilation, reflection, refraction and absorption of solar and terrestrial radiation, changes in moisture budget, etc.

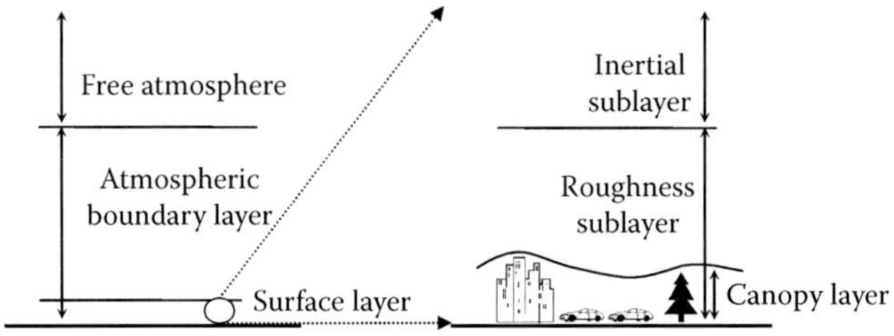

FIGURE 6.1 Schematic of the atmospheric boundary layer structure.

The roughness sublayer, including the canopy layer at the lowest part, extends up to two to five times of an average roughness height. Within this sublayer, atmospheric motion is observed to be three dimensional influenced by length scales of individual roughness elements. This indicates that the traditional surface layer scaling cannot be applied, which casts a great challenge in simulating urban environment (Britter and Hanna 2003). Note that most of micrometeorological theories, such as Monin–Obukhov similarity, assume the ABL to be one dimensional, allowing variation only in the vertical direction. The inertial sublayer is the rest of the surface layer where typical surface layer parameterizations such as Monin–Obukhov similarity apply (Figure 6.1).

6.2.2.2 Convective Period

The ABL has a distinctive diurnal variation. During the daytime, the heated ground due to the incoming shortwave radiation warms up the air parcel adjacent to the surface initiating an ascending motion. The rising air parcel often travels beyond its neutrally buoyant height due to inertial acceleration. Then, the overshooting parcel, cooler than the ambient air, begins to descend back, again, to further below its neutrally buoyant height. This oscillation with its center being at the neutrally buoyant height continues until the parcel loses the inertia and subsides to the neutral level. The frequency associated with this oscillation is called Brunt–Vaisala frequency,

$$N^2 = \frac{gd\theta}{\theta dz}, \tag{6.1}$$

where
θ is potential temperature
z is height

A layer formed as a result of such convections, called the convective boundary layer (CBL) or mixed layer (ML), typically extends approximately 1 km above ground level (agl). The CBL exerts a critical role in pollutant buildup since it determines the degree of vertical ventilation, as evident from the fact that heavy air pollution is often associated with a limited vertical mixing. The bottom and the top portions of the ML have distinctively different structures. The former is strongly unstable with super-adiabatic lapse rate attributed to vigorous radiational heating. The latter called *capping inversion* is a thin strip with strong stratification due to the vertical entrainment from the free atmosphere (FA).

When a flow encounters sudden or gradual changes in surface roughness, temperatures, wetness, or elevation, an internal boundary layer develops. Specifically an internal boundary layer associated with a cold air mass moving over a warm ground is called the thermal internal boundary layer (TIBL). This is commonly observed in a coastal area where a sea breeze pushes cold marine air mass over dry warm land. The depth of the TIBL increases with the ratio of the square root of the distance from the discontinuity and eventually merges with the convective boundary layer. However, the shallow boundary layer near the coast limits vertical convection of pollutants.

In photochemical perspective, the sea breeze/upslope flow transports O_3 precursors away from the coastal urban center into downstream suburban areas. Abundant NO_x from the urban area and the limited vertical mixing due to the shallow boundary layer suppress O_3 production in the city (*VOC* limited O_3 production regime: see the next section for further discussions). As the urban plume travels toward suburban areas where NO_x/VOC ratio is lower than the urban center, additional NO_x transported from the city promotes O_3 production, which results in the rapid increase of O_3 (NO_x limited O_3 production). A return flow aloft that is isolated from the ground level emissions maintains the high O_3 and carries it toward the urban center and the ocean (Solomon et al. 2000). A schematic to demonstrate the earlier-discussed processes is presented in Figure 6.2a.

6.2.2.3 Nocturnal Condition

As soon as the ground heating turns off at the sunset, a stable layer called surface boundary layer (SBL) starts to form near the ground, and lifts the air mass belonging to the CBL to the top of the stable layer. The layer aloft with neutral stability is called the residual layer (RL). In general, the exchange between the layers is minimal due to the stable stratification in the SBL and the FA. Thus, the pollutants contained within the RL remain inactive during the nocturnal period. However, in the following morning as the convection starts to grow and intercepts the RL, the pollutants aloft entrain down to the ground and contribute to the rapid increase of photooxidants in the morning and the early afternoon hours (McKendry et al. 1997).

Still, the stably stratified nocturnal boundary layer can be disturbed or broken down, thus transporting O_3 and NO_x vertically during the nighttime. As the stratification within the SBL enhances, the SBL decouples from the above RL and the flow near the interface becomes free from the ground friction, glides near the interface with high speed, and forms a jet-like flow called low level jet (LLJ) or nocturnal jet. As the jet further accelerates, the vertical wind shear in the vicinity of the jet increases enough to trigger Kelvin–Helmholtz (K–H) type instability and turbulent mixing, which subsequently removes the strong shear and stabilizes the flow. Still, the ongoing radiative cooling rebuilds the stable stratification within the SBL leading to the decoupling and the LLJ. Then,

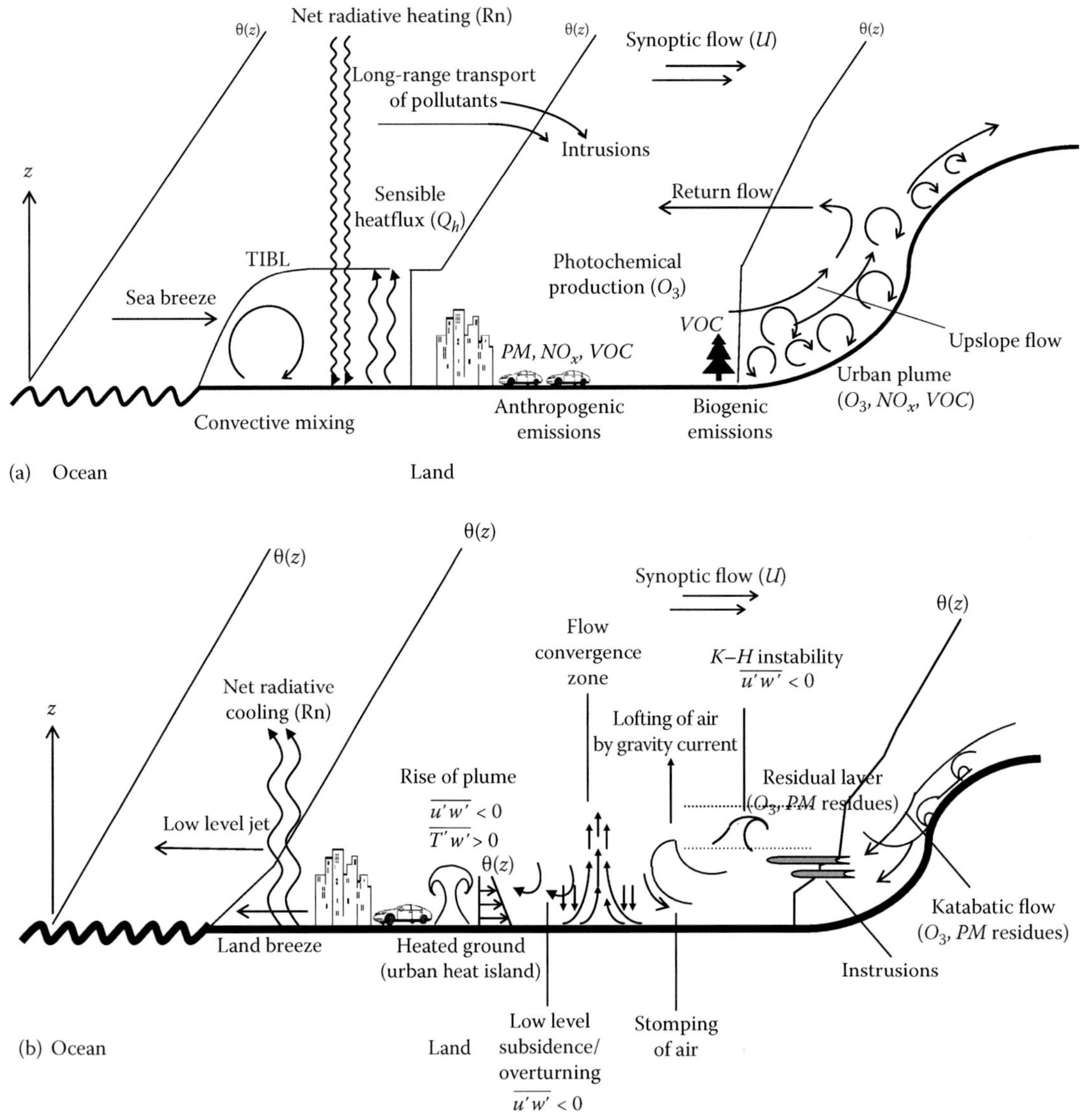

FIGURE 6.2 Transport and mixing processes associated with air pollution over heterogeneous surface with complex terrain for (a) convective period and (b) stable period. (Adapted from Lee, S.-M. et al., *Environ. Fluid Mech.*, 3, 331, 2003.)

the shear-generated turbulence recurs. This repeated building up of shear and the destruction by turbulence attributes to the intermittent nature of the nocturnal turbulence (Nappo 1991). In addition, as shown in Figure 6.2b, the unstable boundary layer that develops near the ground due to heat stored in the upper soil layer creates a plume-like motion (urban heat island effect), which lasts for several minutes. Near a mountain slope, a cold and dense air attributed to radiational cooling moves down the slope to warm area at lower altitude. This gravity current causes a vigorous mixing upon its arrival that can last a few hours. Regarding photooxidants, the SBL is still subject to vehicle emissions (NO_x) that titrate O_3, while the RL is chemically inactive due to the isolation from the ground emissions, thus maintains higher O_3. The intrusion of katabatic flow at a neutrally buoyant height creates O_3 rich layer within the low O_3 stable background.

6.2.3 Photochemical Reactions

Unlike other criteria pollutants, O_3 has a controversial role in the atmosphere. In the stratosphere, it is considered as "good" ozone, while it is "bad" ozone to cause adverse effects in human health and ecosystem in the troposphere. In an atmospheric column, approximately 90% of O_3 is confined in the stratosphere (10–50 km above the earth's surface) and the rest exists in the troposphere. The stratospheric ozone absorbs most of the harmful high-frequency ultraviolet (UV) light preventing it from reaching the ground level. On the contrary, tropospheric ozone is a pollutant considered as being among the most difficult to regulate. The health effects associated with ozone exposure include respiratory health problems ranging from decreased lung function and aggravated asthma to increased emergency department visits, hospital admissions, and premature death.

The environmental effects associated with seasonal exposure to ground-level ozone include adverse effects on sensitive vegetation, forests, and ecosystems. O_3 production is rather simple in the stratosphere. It is

$$O_2 + hv \rightarrow O + O, \quad (6.2)$$

$$O + O_2 + M \rightarrow O_3 + M, \quad (6.3)$$

where M represents a third body that absorbs the excess energy of the initial product and stabilizes it. The UV light of which wavelength is shorter than 242 nm causes the direct photo-dissociation of oxygen molecules (Reaction 6.2). At the same time, the O_3 molecule produced by these reactions absorbs UV light (240–320 nm) to decompose back to O_2 and O,

$$O_3 + hv \rightarrow O_2 + O, \quad (6.4)$$

or directly react with oxygen atom (O),

$$O_3 + O \rightarrow O_2 + O_2. \quad (6.5)$$

The earlier-discussed ozone cycle, called Chapman cycle, is to keep stratospheric ozone in an equilibrium status in the absence of free radical catalysts. However, measurements indicated other ozone destruction passages,

$$X + O_3 \rightarrow XO + O_2 \quad (6.6)$$

$$XO + O \rightarrow X + O_2. \quad (6.7)$$

Note that X, a free radical catalyst such as H, OH, NO, Cl or Br, is not consumed in the process but is regenerated and put back to the cycle. Chlorofluorocarbons ($CFCs$) such as $CFCl_3$ (CFC-11) and CF_2Cl_2 (CFC-12), the ban of whose use was agreed on Montreal protocol in 1987, release a chlorine atom (Cl) via photolysis reactions using 185–210 nm wavelength UV,

$$CFCl_3 + hv \rightarrow CFCl_2 + Cl \quad (6.8)$$

$$CF_2Cl_2 + hv \rightarrow CF_2Cl + Cl. \quad (6.9)$$

Then, the chlorine atom (Cl), highly reactive toward O_3, develops a rapid O_3 destruction cycle through Reactions (6.6) and (6.7). In addition, ice crystals in the polar region act as a catalyst to promote the ozone destruction. During the winter time in the Antarctic, the baroclinic instability due to a strong temperature gradient between cold air over the ice continent and warm air over the ocean creates a polar vortex and subsequently polar stratospheric clouds, which consist of ice crystals and nitric acid. Then, the ice crystals of which surface the $CFCs$ are attached to break the bonds in the CFC molecules and release chlorine compounds that consume the ozone molecules. The reactions involving ice crystals require much less energy than the typical break of the $CFCs$ through Reactions (6.8) and (6.9), as explained by the fact that the ozone depletion is more rapid in the polar region than other latitudes.

The ozone chemistry in the troposphere differs since most of the short wavelength UV light required in the dissociation of oxygen molecules is depleted in the stratosphere via reaction (6.2). Instead, it involves nitrogen oxides, a major constituent of vehicle emissions, through photodissociation using UV wavelength less than 424 nm,

$$NO_2 + hv \rightarrow NO + O \quad (6.10)$$

and reaction (6.3). Nitrogen monoxide is also involved in the destruction of O_3,

$$O_3 + NO \rightarrow NO_2 + O_2. \quad (6.11)$$

In the absence of hydroperoxyl (HO_2) and peroxyl radicals (RO_2), the ozone involved in Reactions (6.2), (6.10), and (6.11) establishes an equilibrium state with no net production or destruction (Figure 6.3a). However, HO_2 and RO_2 advance NO oxidation without the consumption of O_3 (Figure 6.3b),

$$NO + RO_2 \rightarrow NO_2 + RO \quad (6.12)$$

$$NO + HO_2 \rightarrow NO_2 + HO. \quad (6.13)$$

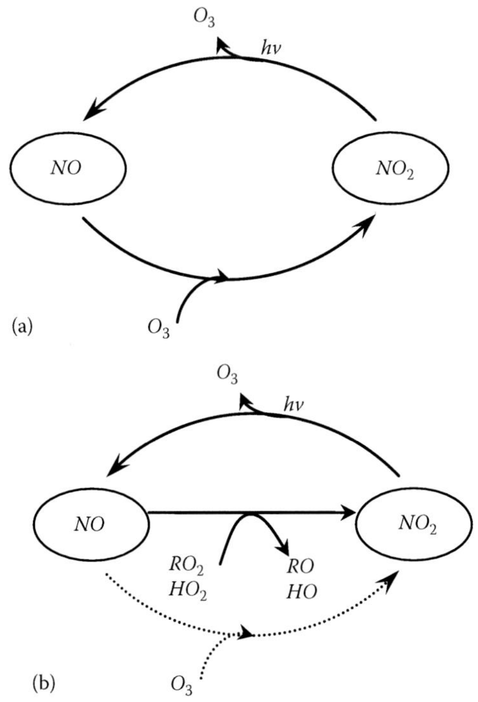

FIGURE 6.3 Schematics of tropospheric O_3 photochemical reactions (a) closed system with NO_x and O_3 and (b) with additional VOC introduced in the system.

Usually, these reactions are much faster than the oxidation of nitrogen monoxide in reaction (6.11). The hydroperoxyl radicals can be formed by the following reactions,

$$H_2O + O \rightarrow 2OH \tag{6.14}$$

$$CO + OH \rightarrow CO_2 + H \tag{6.15}$$

$$H + O_2 + M \rightarrow HO_2 + M. \tag{6.16}$$

Peroxyl radical is formed either by hydrocarbon (RH) based reactions or by non-hydrocarbon based reactions. The former path is

$$RH + OH \rightarrow R + H_2O \tag{6.17}$$

$$M + R + O_2 \rightarrow RO_2 + M, \tag{6.18}$$

and the latter example is, using formaldehyde ($HCHO$),

$$HCHO + OH \rightarrow HCO + H_2O \tag{6.19}$$

$$HCO + O_2 \rightarrow HO_2 + CO. \tag{6.20}$$

The hydroxyl radicals (OH), generated by (6.13), go back to the oxidated form (HO_2) through Reactions (6.17)–(6.20), reenter the NO_x oxidations in (6.13), and thus further promote O_3 accumulation. The alkoxy radical, RO, generated by RO_2 reactions can react with O_2, decompose, or undergo an isomerization, depending on the structure of the radical (Finlayson-Pitts and Pitts 1999).

In all, evidently, O_3 formation is highly dependent on NO_x concentration. Under a low NO_x condition, the O_3 production rate is primarily determined by NO_x concentration. This is due to the fact that hydroperoxyl radicals leave the HO_x cycle via,

$$HO_2 + HO_2 \rightarrow H_2O_2 + O_2. \tag{6.21}$$

Note that H_2O_2 is a stable final product that does not participate to further reactions. On the other hand, O_3 production is proportional to CO and inversely proportional to NO_x concentration in a NO_x abundant situation, where HNO_3 is a primary sink of hydroxyl radicals through,

$$OH + NO_2 \rightarrow HNO_3, \tag{6.22}$$

in that NO_2 derives the removal of free radicals that promote O_3 accumulation. At the same time, CO, through Reactions (6.15) and (6.16), generates hydroxyl radicals that promote O_3 accumulation. The O_3 dependency on NO_x and VOC is well demonstrated in a diagram called empirical kinetics modeling approach (EKMA), ozone isopleths as a function of NO_x and VOC concentrations (Figure 6.4). In the lower right side of the diagram marked by "A," O_3 decreases as NO_x decreases, while O_3 does not respond to VOC changes. This condition is referred as a NO_x limited regime, that is, NO_x availability determines O_3 formation. On the other hand, the regime near the upper left corner, marked by "B," is considered as VOC limited since the VOC increase enhances O_3 formation while additional NO_x further titrates O_3. The EKMA is particularly important to establish a control strategy for O_3 reduction. For example, an air mass of which VOC/NO_x ratio is close to "A", VOC reduction will not lower O_3, but NO_x reduction will be efficient to reduce O_3. On the other hand, an air mass that has VOC/NO_x ration similar to "B" will benefit from VOC reduction, but disbenefit from NO_x reduction. Therefore, the ratio of NO_x and VOC should be kept in a way to avoid following an O_3 isopleth or shifting to higher O_3.

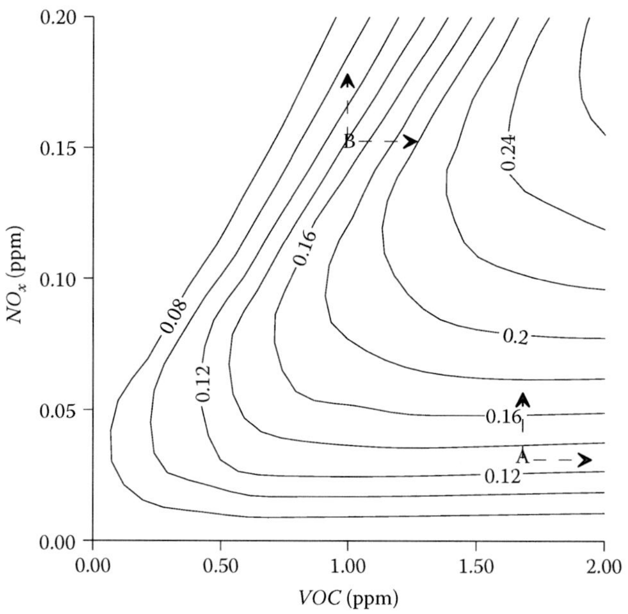

FIGURE 6.4 A typical O_3 isopleth as a function of NO_x and VOC concentration.

6.3 Methods and Analysis

Field experiments, physical modeling, and numerical simulations are used to investigate dynamical and chemical characteristics of air quality. The advantages and disadvantages of the three approaches are as follows.

Field measurements: A major advantage of in situ measurements is that no assumption or simplification is involved. A location of interest is instrumented with appropriate meteorological sensors and samplers for gaseous pollutants and particulate matters. The instrumentation can be operated either locally on the spot or in a remote sensing mode. It also can be set up to transmit data on a real-time basis or store data for later access (so-called off-line processing). Although the field measurements provide actual data for a site in question, they are site-specific values that require an extrapolation in order to apply to different locations, which even for a same phenomenon, often includes questionable assumptions. Besides, field measurements are expensive, usually nonrepeatable, and include nonlinear interactions of multiple processes. These interactions make delineation of various influences nearly impossible. To this extent, a physical modeling is suggested as an alternative to overcome such limitations of field measurements.

Physical modeling: Unlike the field measurements, physical modeling is conducted under carefully selected/controlled conditions. The most common types of physical modeling are wind tunnels, water channels, and smoke/environmental chambers. The former two are intended to investigate flow, turbulence, and dispersion, and the latter is designed to research chemical reactions that take place in the atmosphere. Opposite to the field measurements, physical modeling is relatively inexpensive, repeatable, and possible to isolate individual processes attributed to the phenomenon of interest. Laboratory setups for wind tunnels/water channels range from simplified building models to elaborate replicas of a whole city composed of high-rise buildings, parks, and vegetation. Usually, models are equipped with heating/cooling capabilities for the simulation of heat island and thermally driven up/down slope winds. Simple models provide in-depth fundamentals, while complex models may offer, similar to field measurements, explanations specified to the modeling configuration. Laboratory data are available for a larger span of space and time, which are beneficial to the development and evaluation of a numerical model. However, special attention is required to scale the laboratory data to an actual atmospheric event, especially if multiple effects are involved (e.g., pressure gradient and heating). Environmental chambers are primarily designed to investigate photochemical reactions of pollutants, gas-to-particle conversion, and the growth of particles under controlled light and temperature.

Numerical simulations: A numerical model discretizes a set of governing equations and integrates them with respect to time and space. Equations of fluid motion, radiational energy transfer, thermodynamics, and mass conservation are fundamental constituents of a meteorological model. A mesoscale model such as Pennsylvania State University/National Center for Atmospheric Research Mesoscale Model 5 (MM5) and Weather Research and Forecast model (WRF) are applied to a broad range of phenomena, such as regional climate, monsoons, cyclones, mesoscale fronts, land–sea breezes, and mountain-valley circulations. As for an air quality model, there are different types designed for specific situations. A 3-D comprehensive photochemical model, such as Community Multiscale Air Quality model (CMAQ) and Comprehensive Air Quality model with extensions (CAMx), uses the same approach of the meteorological model in the sense of discretizing and integrating over space and time, except that it has a conservation equation and chemical reactions. This type of models are to simulate transport, mixing, production, and removal of photooxidants over relatively larger domain—comparable to the synoptic to the local scale discussed in the previous section. However, this type of prognostic approach requires extensive computation, which makes it difficult to provide solutions in a timely manner. Therefore, a rather simple code with highly parameterized equations has been sought after for a quick response. This category includes Quick Urban and Industrial Complex Dispersion Modeling System (QUIC), a model to calculate concentrations within urban morphology based on empirical relationships and AERMOD, a dispersion model based on an assumption that pollutant concentrations from a plume released from a source would have a normal distribution across the plume centerline. Still, regardless of the type of model, every model has pros and cons and needs to be used with caution. And more importantly, model results need to be interpreted with in-depth scientific knowledge.

6.4 Applications

6.4.1 Horizontal Distribution of Photooxidants

As many urban centers in the world suffer from degraded air quality, the Los Angeles air basin has among the worst air quality in the United States. Under the Clean Air Act (CAA), the basin is designated as a "severe-17" nonattainment for 8-h ozone, nonattainment for $PM_{2.5}$, and maintenance area for CO and NO_2. The Los Angeles air basin includes Los Angeles metropolitan area, its surrounding suburbs, mountain ranges, and costal islands (Figure 6.5). A unique meteorological condition as well as emissions attributes to the poor air quality. During the summer months, strong subsidence induced by the semipermanent high-pressure system residing over the northeastern Pacific brings an elevated inversion to the basin. In that, a cool and moist marine layer near the surface is capped by a warm and dry air mass aloft, which, often, inhibits the growth of the mixing layer. In addition, light winds within the basin combined with the complex terrain that separates the basin from surrounding semiarid areas further limit pollutant dispersion. Furthermore, the region experiences more days of sunlight than any other major urban area in the United States except Phoenix, Arizona.

During the daytime, well-developed sea breeze transports pollutants such as NO_x and CO emitted from the urban center to the east and north of the Basin. Typically, the sea breeze entering into the Los Angeles downtown splits into two branches with one moving to the east (toward RIVR and SNBO in Figure 6.5) and the other to the north (toward SCLR in Figure 6.5). The greater Los Angeles (LA) metropolitan area has high mountain areas located in the east and the north of the basin, while the urban center is located near the coast. The northwest and southeast running coastal line attributes to the southwesterly momentum, as well. The upslope component along the sloped surface of the high terrain accelerates the transport, while, at the same time, the mountains function as a physical wall to block the transport to further downstream. In terms of air quality, while the downtown center is high in primary pollutants such as CO and NO_x, secondary pollutants like O_3 and $PM_{2.5}$ are high in the downstream as they are chemically produced during the transport (Figure 6.6). Like most urban centers, the O_3 formation in the LA urban center is limited by VOC, meaning an additional NO_x leads to the destruction of O_3 rather than production. Likewise, the suburban is a NO_x limited regime in which high NO_x transported from the urban results in high O_3 and often creates local hot spots. Evidently, in the LA basin, peak O_3 concentrations occur in San Bernardino/Riverside and Santa Clarita (Figure 6.5).

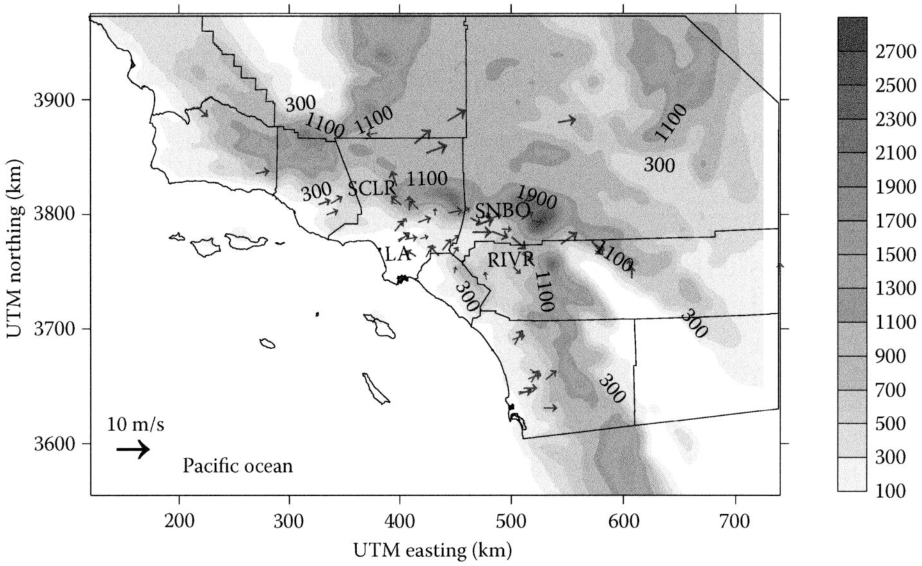

FIGURE 6.5 Near-surface wind field measured at 1700 PST on July 16, 2005. Los Angeles urban center and its downstream suburban areas are marked by LA, RIVR, SNBO, and SCLR, which represent Los Angeles, Riverside, San Bernardino, and Santa Clarita, respectively.

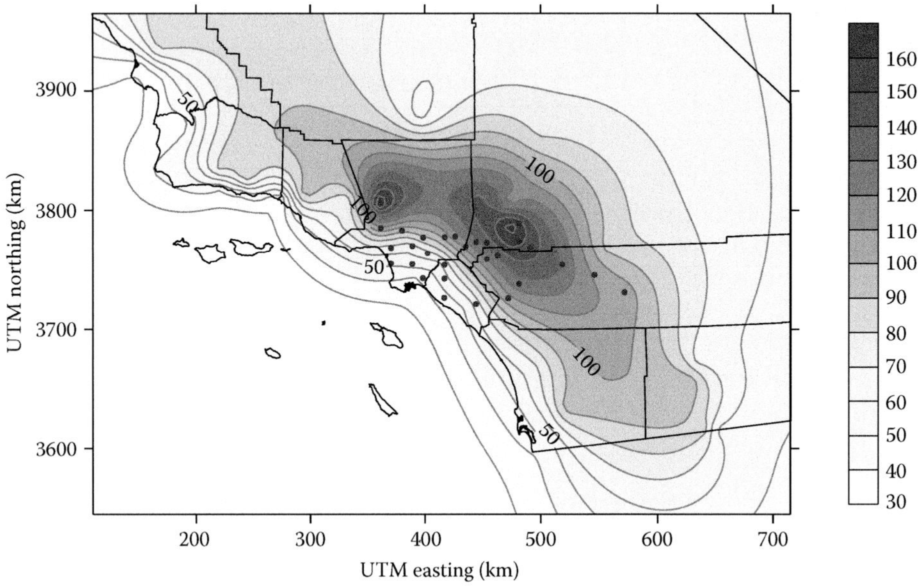

FIGURE 6.6 Maximum 1-h ozone concentrations observed on July 16, 2005. The locations of monitoring sites are marked by dots.

Vertically, the basin experiences a multilayered structure due to the complex topography and the landuse contrast (Lu and Turco 1996). The westerly sea breeze reaching the mountains in the inland merges into the upslope flow and gains the upward momentum. Then, it intrudes into a neutrally buoyant level and eventually returns toward the ocean. This typically occurs at approximately 1000 m above ground level, which is right above the CBL. This return flow carries the high O_3-laden air mass back into the coastal area. During the summer months when photochemistry is active and the land/sea breeze is strong, it is often observed that O_3 is higher aloft within the return flow layer than within the ABL where O_3 is subject to a vigorous titration by fresh NO_x emissions. At the same time, the TIBL inhibits vertical mixing and often leads to a shallow ML (only a few hundred meter deep) in the coastal areas. Above the return flow, another flow reversal exists due to the synoptic influence. A strong north Pacific high resides over the basin semipermanently especially during the smog season, which typically runs from the beginning of May through late October in the area. The high pressure brings the westerly synoptic flow in the upper atmosphere. The eastward momentum of the synoptic flow occasionally attributes to the recirculation of local pollutants transported back to the basin.

6.4.2 Vertical Cross Sections

As an attempt to elucidate the chemical aspect of an urban plume, we limit discussions to a relatively simple circumstance. The Phoenix metropolitan area, the eighth largest and the second fastest growing city in the United States, is nonattainment for 8 h ozone and serious nonattainment for PM_{10} under the federal law. Unlike the LA basin where a synoptic condition provides a critical background to conduce pollutant buildup, the synoptic flow in the Phoenix air basin is weak for most of the year; thus, a terrain-induced diurnal circulation (upslope and downslope flow) is dominant. The residence times of air masses within the basin are long due to weak wind, thus favoring recirculation of pollution and entraining down to the ground level in the city. The vertical cross sections presented in Figure 6.7 well elucidate both photochemical and fluid dynamical features of an urban environment located over complex terrain. The nitrogen oxides discussed in Figure 6.7 are surrogated as follows:

$$NO_x = NO + NO_2$$

$$NO_y = NO_x + NO_z$$

$$NO_z = NO_y - NO_x$$

NO_y, the sum of all reactive oxidized nitrogen species, is, for example, composed primarily of NO, NO_2, peroxyacetyl Nitrate (PAN), HNO_3, and aerosol NO_3^-. NO_z is usually considered as a "reservoir" or "termination product" of NO_x photooxidation process. The NO_z concentration presented here is defined as the sum of PAN, $HONO$, Peroxy-nitric acid (PNA), NO_3, N_2O_5, and HNO_3, while the VOC is the sum of paraffins, ethene olefins, toluene, xylene, isoprene, formaldehyde, and high molecular weight aldehydes.

A vertical cross section of ozone is consistent to the horizontal distribution shown in Figure 6.6, despite the geographical difference: a low-ozone dome located in the urban core and a high-ozone plume near ground in the far downstream (Figure 6.7a). Expected from O_3 and NO_x chemistry, the distribution of NO_x is out-out-of phase with that of O_3: high NO_x confined over the urban center and low NO_x everywhere else including the urban downstream (Figure 6.7b). Interestingly, NO_x was quickly depleted with height and downwind distance, compared with CO, which was dispersed along the downwind to form a ground-based plume shape (Figure 6.7c). Note that a major freeway runs in the east–west direction parallel to the cross section so that the CO plume is attributed to freeway emissions as well as the transport from the urban areas. The

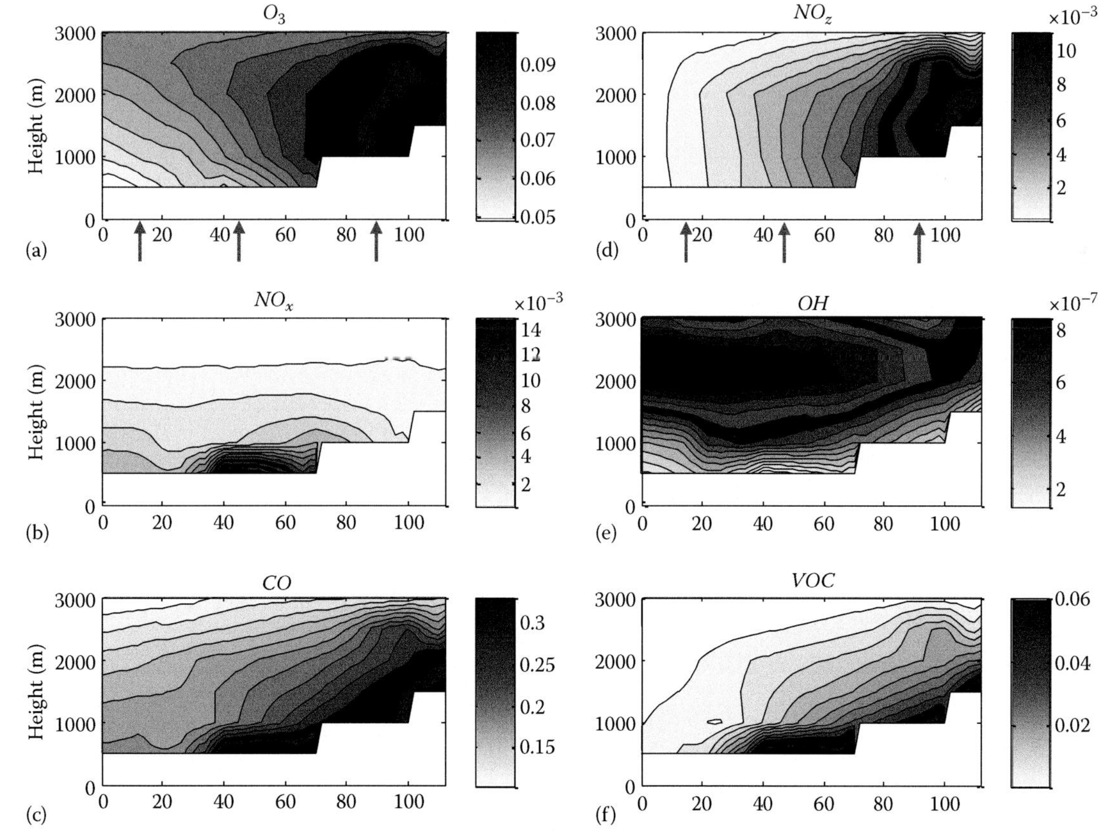

FIGURE 6.7 Vertical cross section of (a) O_3, (b) NO_x, (c) CO, (d) NO_2, (e) OH, and (f) VOC in Phoenix metropolitan area on 1600 LST, July 23, 1996 simulated by Community Multiscale Air Quality (CMAQ) model. The concentrations are in ppm unit. The arrows at the bottom of figures represent the locations of upstream, central downtown, and downstream of the metropolitan area. Terrain is depicted by white space at the bottom of each figure. The ordinate is mean sea level height in meter, and the abscissa is modeling grid points with 2 km distance.

NO_z distribution in Figure 6.7d, to a degree, explains the difference in NO_x and CO plumes. Strictly speaking, NO_z is not exactly the final product of nitrogen oxides, especially under the high heat in the Phoenix area. Instead, NO_z can be considered as a reservoir in the case of a rapid photolysis of nitrous acid (*HONO*) back to *NO* in the presence of solar radiation or a thermal decomposition of *PAN* to produce NO_2, both of which possibly occur under the climate of the Phoenix area. However, NO_z still gave a reasonable indication that the air mass in the downstream was chemically aged and had most of available NO_x transformed to NO_z. The location of high NO_z was almost identical to that of an elevated ozone concentration, which is consistent to that in Trainer et al. (1993) who found a linear relationship between O_3 and NO_z in photochemically aged air sampled at rural locations in the eastern United States and Canada. *OH* concentration was simulated to have maximum near the interface between the ABL and the FA over the urban center and minimum in the urban surface (Figure 6.7e). The low *OH* in the urban appeared to prevent the oxidization of *VOCs* to produce ozone, which, in part, contributed to the low-ozone dome in the urban center in addition to the NO_x titration. The *VOC* cross section in Figure 6.7f shows a high concentration near ground. Note that the mountainous area in the east is covered by coniferous trees that emit a significant amount of *VOC*. Considering that *CO* and NO_x are, respectively, inert and highly reactive, the dispersion of *VOC*, shaped in between the *CO* and NO_x plume, turned out to be influenced by both a direct emission chemical reactions. Needless to say, the ML height was critical to the dispersion of pollutants. The pollutants trapped within the ML were well depicted by the steep concentration gradient near 2500–3000 m above mean sea level. Note that the ABL heights in the urban and the downstream mountainous area were 1800 and 1500 m above ground level, respectively.

6.5 Major Challenges

6.5.1 Air Quality Standards

There are two different types of standards defined under the U.S. Clean Air Act Amendments (CAAA) of 1970: one is regulating ambient concentrations and the other targets for emission rates from sources. The former is through NAAQS that define maximum allowable atmospheric concentrations for six criteria pollutants: O_3, *CO*, NO_2, SO_2, PM_{10} and lead (Table 6.1). Regardless of the type or the amount of emissions, a localized area or a region is expected to have its ambient concentrations equal to or lower than the defined thresholds. Otherwise, the area is classified as nonattainment and requires various control measures to be implemented to achieve an attainment status. For example, an area classified as "serious" or as having higher levels of nonattainment for O_3 must achieve a 15% reduction of VOC in 6 years and continue to reduce 3% per year until the attainment under the federal law. There are five levels of nonattainment status for O_3 starting from "marginal" as the least degree of violation to "moderate," "serious," "severe," and the highest level of violation to be "extreme" category.

TABLE 6.1 National Ambient Air Quality Standard (NAAQS)

Pollutant	Standard	Averaging Time
Carbon monoxide	9 ppm (10 mg/m^3)	8 h[a]
	35 ppm (40 mg/m^3)	1 h[a]
Lead	0.15 µg/m^3	Rolling 3 month average
Nitrogen dioxide	53 ppb (100 µg/m^3)	Annual mean
	100 ppb(188 µg/m^3)	1 h[b]
Particulate matter (PM_{10})	150 µg/m^3	24 h[c]
Particulate matter ($PM_{2.5}$)	15 µg/m^3	Annual mean[d]
	35 µg/m^3	24 h[e]
Ozone	0.075 ppm	8 h[f]
Sulfur dioxide	0.14 ppm	24 h[g]
	0.5 ppm	3 h[a]

[a] Not to be exceeded more than once per year.
[b] To attain the standard, the 3 year average of the 98th percentile of 1 h concentrations must not exceed 100 ppb.
[c] Not to be exceeded more than once per year on average over 3 years.
[d] To attain this standard, the 3 year average of the weighted annual mean $PM_{2.5}$ concentrations must not exceed 15 µg/m3.
[e] To attain this standard, the 3 year average of the 98th percentile of 24 h concentrations must not exceed 35 µg/m3.
[f] To attain this standard, the 3 year average of the fourth-highest daily maximum 8 h average ozone concentrations measured over each year must not exceed 0.075 ppm.
[g] To attain the standard, the 3 year average of the 99th percentile of 1 h daily maximum concentrations must not exceed 75 ppb.

The other type of standards is New Source Performance Standards (NSPS), which limit maximum emission rates from a specific stationary source. The NSPS is, based on actual field test results collected from a number of sources, tailored to the type of pollutant and the size and the nature of the industry. This includes specific process, fuel type, and control equipments involved in the emission source. Thus, there are numerous entries listed under the NSPS aiming at very specific types and processes of sources. Selected examples of NSPS are given in Table 6.2. In terms of unit, the NSPS uses mass per time or mass per production (g/min or kg of pollutant per ton of product produced), while the NAAQS uses concentration (µg/m^3 or ppm). Overall, it is important to keep in mind that the standards are not fixed values but are constantly evolving toward being more stringent in accordance with advancements in science. For example, when U.S. EPA first issued standards for particulate matter in 1971, it was total suspended particles (*TSP*) that did not distinguish the size of particles. Afterward, the linkage of inhalable fine particles to health problems was delineated, and the focus has been shifted to regulating fine particles. This resulted in revoking *TSP* standards and introducing PM_{10} (1987). As of 2006, the NAAQS defines $PM_{2.5}$ for both annual and 24 h averages (refer Table 6.1). Similarly, O_3 standard used to be 80 ppb for 8 h in 1997 and was amended to be 75 ppb as of 2008.

TABLE 6.2 Selected Examples of National New Source Performance Standard (NSPS)

1. Petroleum refineries (construction, modification, or reconstruction begun after May 14, 2007)
   A. Fluid catalytic cracking unit (FCCU) or fluid coking unit (FCU)
      a. Particulate matters
         i. New FCCU: 0.5 lb per 1000 lb coke burn-off
         ii. Modified/reconstructed FCCU or any FCU: 1 lb per 1000 lb coke burn-off
      b. NO_x
         i. 80 ppmv, dry basis corrected to 0% excess air, on a 7-day rolling average basis
      c. SO_2
         i. 50 ppmv dry basis corrected to 0% excess air, on a 7-day rolling average basis and 25 ppmv dry basis corrected to 0% excess air, on a 365-day rolling average basis
   B. Sulfur recovery plant (SRP) with oxidation control system or a reduction control system followed by incineration
      a. SO_2
         i. 250 ppmv dry basis corrected to 0% excess air for an SRP capacity greater than 20 long tons per day (LTD)
         ii. 2500 ppmv dry basis corrected to 0% excess air for a SRP capacity less than 20 LTD
   C. Sulfur recovery plant (SRP) with a reduction control system not followed by incineration
      a. Reduced sulfur compound
         i. 300 ppmv dry basis corrected to 0% excess air for an SRP capacity greater than 20 LTD
         ii. 3000 ppmv dry basis corrected to 0% excess air for an SRP capacity less than 20 LTD
      b. H_2S
         i. 10 ppmv dry basis corrected to 0% excess air for an SRP capacity greater than 20 long tons per day (LTD)
         ii. 100 ppmv dry basis corrected to 0% excess air for an SRP capacity less than 20 LTD
2. Solid waste incinerator
   Particulate matters: 0.18 g/dscm[a] (0.08 gr/dscf)[b] corrected to 12% CO_2
3. Municipal solid waste landfill
   The landfill has a design capacity greater than or equal to 2.5 million Mg and 2.5 million m^3
   Non-methane organic compound: 50 Mg per year for a facility with design capacity above 2.5 million Mg or 2.5 million m^3
4. Portland cement plants
   A. From kiln
      a. Particulate matters: 0.15 kg/ton of feed (dry basis) to the kiln
      b. 20% opacity
   B. From a clinker cooler
      a. Particulate matters: 0.05 kg/ton of feed (dry basis) to the kiln
      b. 10% opacity
   C. From any affected facility: 10% opacity
5. Sewage treatment plants
   A. Particulate matter: 0.65 g/kg dry sludge input
   B. 20% opacity

[a] Dry standard cubic meter.
[b] Dry standard cubic feet.

6.5.2 Managements and Regulations

Even if an area is classified as nonattainment, there is a need to introduce a new source in the area. Among widely used regulatory tools to accommodate such necessity is the emissions offset program. A fundamental idea of the program is to offset the new emissions by simultaneous reductions in emissions from existing sources. Normally, it requires the offset ratio to be higher than 1 meaning that less than 1 ton of new emission will be produced per 1 ton reductions from existing sources. Thus, the total amount of emissions from the limited area reduces over time. Another commonly adapted regulatory method is emissions trading program. SO_2 and NO_x trading, under the federal policy, are widely implemented in various industrial sectors across the nation. Among them are SO_2 trading from electricity generators established under Title IV of the 1990 Clean Air Act Amendments (CAAA). The cap-and-trade approach quickly emerges to be the centerpiece of the green house gas control policy. The cap-and-trade approach is, in short, setting a legal limit on the quantity of pollutants to emit each year (cap) and allowing facilities to "trade" the permissions—or permits—to pollute among themselves. The program usually accommodates an allowance bank, such that aggregated emissions at a given year must be equal to or less than the quantity of allowances allocated for the year plus the surplus accrued from previous years. However, despite the aggregated emission reductions achievable through the trading and offset program, a major critic is that it does not have an adequate mechanism to protect sensitive areas or hot spots since it cannot control the spatial distribution of the allowances.

6.5.3 Public Outreach

When all the policies and regulations are in place, a practical concern is an efficient approach to reach out to general public and to advise a proper action to avoid unnecessary exposure to pollution. To this extent, U.S. EPA introduced an index called air quality index (AQI) to explain the degree of air pollution, its associated health effects, and actions to take (U.S. EPA, 1999). The AQI is calculated, as follows, for five major air pollutants: O_3, $PM_{2.5}$, PM_{10}, CO, and NO_2,

$$I_p = \frac{I_{Hi} - I_{Lo}}{BP_{Hi} - BP_{Lo}}(C_p - BP_p) + I_{Lo}, \tag{6.23}$$

where

I_p is the index value for pollutant p
C_p is the truncated concentration of pollutant p
BP_{Hi} is the breakpoint that is greater than or equal to C_p
BP_{Lo} is the breakpoint that is less than or equal to C_p
I_{Hi} is the AQI value corresponding to BP_{Hi}
I_{Lo} is the AQI value corresponding to BP_{Lo}

The AQI categories and breakpoints for each pollutant are given in Table 6.3. Similar to the AQI, Environment Canada developed

TABLE 6.3 Proposed Breakpoints for O_3, $PM_{2.5}$, PM_{10}, CO, SO_2, and NO_2 Subindices

Category	AQI	O_3 8 h (ppm)	O_3 1 h[a] (ppm)	PM_{10} 24 h (μg/m³)	$PM_{2.5}$ 24 h (μg/m³)	CO 8 h (ppm)	SO_2 24 h (ppm)	NO_2 24 h (ppm)
Good	0–50	0.000–0.064	—	0–54	0.0–15.4	0.0–4.4	0.000–0.034	[b]
Moderate	51–100	0.065–0.084	—	55–154	15.5–40.4	4.5–9.4	0.035–0.144	[b]
Unhealthy for sensitive groups	101–150	0.085–0.104	0.125–0.164	155–254	40.5–65.4	9.5–12.4	0.145–0.224	[b]
Unhealthy	151–200	0.105–0.124	0.165–0.204	255–354	65.5–150.4	12.5–15.4	0.225–0.304	[b]
Very unhealthy	201–300	0.125–0.374	0.205–0.404	355–424	150.5–250.4	15.5–30.4	0.305–0.604	0.65–1.24
Hazardous	301–400	[c]	0.405–0.504	425–504	250.5–350.4	30.5–40.4	0.605–0.804	1.25–1.64
Hazardous	401–500	[c]	0.505–0.604	505–604	350.5–500.4	40.5–50.4	0.805–1.004	1.65–2.04

[a] Areas are required to report the AQI based on 8 h O_3 values. However, there are areas where an AQI based on 1 h O_3 values would be more protective. In these cases, the index for both the 8 h and the 1 h O_3 values may be calculated and the maximum AQI reported.

[b] NO_2 has no short-term NAAQS and can generate an AQI only above a value of 200.

[c] 8 h O_3 values do not define AQI higher than 301, in which 1 h O_3 based AQI is used instead.

Air Quality Health Index (AQHI) based on epidemiological research, which estimates the short-term relative risks posed by a combination of multiple pollutants that are ground level O_3, $PM_{2.5}$, and NO_2, as follows:

$$\mathrm{AQHI} = K\left\{\left(e^{A[O_3]} - 1\right) + \left(e^{B[PM_{2.5}]} - 1\right) + \left(e^{C[NO_2]} - 1\right)\right\} \quad (6.24)$$

where empirical constants K, A, B, and C are, respectively, 96.153846, 0.000537, 1.000487, and 0.000871. The AQHI is measured on a scale ranging from 1 to 10 or higher with its value grouped into health risk categories as follows: 1–3 for low, 4–6 for moderate, 7–10 for high, and 10 and higher for very high health risk.

References

Britter, R.E. and S.R. Hanna. 2003. Flow and dispersion in urban areas. *Annu. Rev. Fluid Mech.* 35: 469–496.

Finlayson-Pitts, B.J. and J.N. Pitts. 1999. *Chemistry of the Upper and Lower Atmosphere: Theory, Experiments and Applications*. Academic Press, San Diego, CA.

Lee, S.-M., H.J.S. Fernando, M. Princevac et al. 2003. Transport and diffusion of ozone in the nocturnal and morning planetary boundary layer of the Phoenix valley. *Environ. Fluid Mech.* 3: 331–362.

Lee, S.-M., M. Princevac, S. Mitsutomi, and J. Cassmassi. 2009. MM5 simulations for air quality modeling: An application to a coastal area with complex terrain. *Atmos. Environ.* 43: 447–457.

Lu, R. and R.P. Turco. 1996. Ozone distributions over the Los Angeles Basin: Three-dimensional simulations with the smog model. *Atmos. Environ.* 30: 4155–4176.

McKendry, I.G., D.G. Steyna, J. Lundgrena et al. 1997. Elevated ozone layers and vertical down-mixing over the Lower Fraser Valley, BC. *Atmos. Environ.* 31: 2135–2146.

Nappo, C.J. 1991. Sporadic breakdowns of stability in the PBL over simple and complex terrain. *Boundary-Layer Meteorol.* 54: 69–87.

Solomon, P., E. Cowling, G. Hidy, and C. Furiness. 2000. Comparison of scientific findings from major ozone filed studies in North America and Europe. *Atmos. Environ.* 34: 1885–1920.

Trainer, M., D.D. Parrish, M.P. Buhr et al. 1993. Correlation of ozone with NO_y in photochemically aged air. *J. Geophys. Res.* 98(D2): 2917–2925.

Tropea, C., A.L. Yarin, and J.F. Foss. 2007. *Handbook of Experimental Fluid Mechanics*. Springer, Berlin, Germany.

U.S. Environmental Protection Agency. 1999. 40 CFR Part 58 Air Quality Index Reporting; Final Rule [FRL-6409-7] RIN 2060–AH92.

7

Forecasting and Management of Coastal Water Quality

Joseph H.W. Lee
Hong Kong University of Science and Technology

Ken T.M. Wong
University of Hong Kong

K.W. Choi
University of Hong Kong

7.1 Introduction

More than 70% of the world's megacities (greater than 8 million inhabitants) are located in coastal areas. By the year 2025, coastal populations are expected to account for 75% of the total world population (UNESCO, 2001). Coastal waters support a multitude of important social and economic activities that include navigation, amenities, fishing and mariculture, energy production, water supply and tourism. Rapid urbanization and industrialization place increasing pressure on the integrity of coastal and marine ecosystems; this concern is especially relevant in many fast-developing economies. Sound integrated environmental and resource management is needed for the sustainable development of coastal cities.

The sea is a valuable and unique resource that supports a huge variety of wildlife ranging from microscopic plankton, seagrass, and corals to large mammals such as dolphins and porpoises. To the uninitiated, the vastness of the sea and the ocean may give the impression that coastal water quality is hardly a problem of concern. Nevertheless, the reality is that the waters adjacent to many coastal cities are often heavily exploited for "beneficial uses"—as receiving water for municipal and industrial effluent discharges, cooling water discharges from power stations and liquefied natural gas plants, brine disposal from desalination, land creation by reclamation, source of sand borrowing, and dumping ground for contaminated mud. Aside from pollution discharges that would lead to deterioration of water quality, coastal infrastructure developments such as bridge crossing, dredging, harbor reclamation, offshore wind farms, and coastal protection works can also impact on the environment in various ways. For all of the aforementioned activities, the protection of the water environment is typically effected via an Environmental Impact Assessment (EIA) process whereby the impact of any development is legally required to be assessed quantitatively against "water quality objectives" or standards. Modern EIA relies heavily on the use of advanced 3D hydrodynamic and water quality models based on a large body of environmental fluid mechanics research developed since the 1960s. Reviews on environmental hydraulics related to waste disposal in the ocean can be found in Fischer et al. (1979) and Jirka and Lee (1994).

7.1.1 Water Quality Management

The scientific management of our coastal waters is essential for building a quality living environment for the next generation of coastal cities. While avoidance or reduction is the best means against pollution, there is always a practical limit. On the other hand, knowledge of the fate and transport of pollutants and the effectiveness of different mitigation options can significantly help reduce the pollution impact. Coastal water quality modeling has been plagued by several challenges: (1) pollution sources are often located in shallow water in close proximity to sensitive receivers such as bathing beaches, marine parks, and fish farms—with consequent demands on the accuracy of model predictions; (2) water quality data are typically sparse and costly, with associated uncertainties in calibration and validation of the fluid-mechanics-based models; (3) the water quality prediction deals with complex nonlinear interdisciplinary phenomenon

Handbook of Environmental Fluid Dynamics, Volume One, edited by Harindra Joseph Shermal Fernando. © 2013 CRC Press/Taylor & Francis Group, LLC.
ISBN: 978-1-4398-1669-1.

that are notoriously difficult to model; as an example, harmful algal blooms resulting from excessive nutrient inputs from sewage discharges may be governed by wind-driven physical–biological interactions rather than on the pollution load (e.g., Harrison et al., 2008); (4) model predictions are often presented in voluminous reports and numerous colorful contour plots; the massive data output is not easily digested even by informed professionals, with consequent significant gaps in public engagement.

Numerical hydrodynamic and water quality modeling is a useful tool for environmental planning and management. Robust 3D models are readily available and provide an effective means for understanding the cause–effect and control mechanism, hypothesis testing, and visualization of the pollution impact. Most importantly, judicious data assimilation using 3D models is an indispensable aid for unraveling the complex myriad of spatial and temporal variations of coastal water quality. Effective use of 3D modeling tools reinforces the basic principles of the EIA process: avoidance, reduction, and mitigation.

In this chapter, the basic principles of hydrodynamic and water quality modeling will first be outlined. This will be followed by an overview of Project WATERMAN—a water quality forecast and management system developed for Hong Kong. The key elements of the modeling system including daily beach water quality forecasts and a virtual reality GIS-based EIA system will be described. The 3D hydrodynamic model allows seamless near-far field coupling of effluent discharges. A simple algal bloom risk forecast and disaster mitigation system will be described. Finally, some remarks on future challenges and trends are given.

7.2 Basic Principles of Hydrodynamic Transport Modeling

Wastewater effluents find its way into the marine environment either as passive sources (e.g., diffuse pollution from stormwater runoff) or as active point sources (e.g., sewage outfalls from municipal wastewater treatment plants). Conceptually, the sources can be modeled as surface or submerged buoyant jet discharges—i.e., a jet flow possessing both momentum and buoyancy. Close to the discharge point, the interaction of the discharge momentum and buoyancy with the tidal flow results in rapid initial mixing and dilution of the sewage effluent. The pollutant concentration can be reduced 100 times or more by the time the mixed effluent rises to the surface, or to a trapped level in the presence of ambient stratification. Within the initial dilution zone, the pollutant concentration may exceed the target water quality criterion. As the time of plume rise is rapid (in the order of a couple of minutes), physical processes dominate and biochemical decay processes are not important in the near field (length scale in the order 10–100 m). The environmental impact of an outfall discharge is usually measured in terms of a target surface pollutant concentration (e.g., ammonia nitrogen) or an initial dilution (Jirka and Lee, 1994; Lee and Choi, 2008; Choi et al., 2009). In principle, the near field flow and concentration fields induced by the buoyant jets can be computed by solving the full Reynolds-Averaged Navier Stokes (RANS) equations. In practical engineering application, however, the initial dilution can usually be reliably predicted by well-validated plume models (e.g., Lee and Cheung, 1990; Lee et al., 2000; Jirka, 2008). The computation of near field mixing using CFD models may be required or justified in situations when the buoyant jets discharge in close proximity to solid boundaries (Lee and Chu, 2003).

Beyond the initial dilution zone, the mixed effluent is transported by the tidal current while undergoing horizontal and vertical turbulent mixing; the concentration of any soluble and/or suspended substance may decrease due to sedimentation or water quality transformations. The environmental impact in this "far field" is usually assessed using a 3D hydrodynamic transport model based on the shallow water approximation. Various concepts relevant to EIA are embodied in the hydrodynamic transport. First, advection plays an important role—a pollution source is transported by the horizontal flow field to impact surrounding waters. For example, massive fish kills may result from a toxic algal bloom transported into a fish farm on a flood tide under favorable tidal and wind conditions (e.g., Lee and Qu, 2004). Second, the overall water quality in a semi-enclosed coastal bay is governed by the tidal exchanges between the bay and the outer sea; this cleansing mechanism can be accurately determined in terms of a tidal flushing rate (or conversely, the retention time) (Choi and Lee, 2004). Third, the horizontal and vertical mixing result in far field dispersion and influence the pollution level at a sensitive receiver (e.g., nearby beach). For example, vertical mixing has been shown to play a key role in determining the formation of red tides in the southern waters in Hong Kong (Wong et al., 2009). Knowledge of environmental fluid mechanics can be gainfully employed for water quality control. For example, by proper outfall design, the sewage effluent can be engineered to be submerged beneath the surface; the impact to nearby beaches is thus minimized (Choi et al., 2009).

Hydrodynamic modeling is the cornerstone of water quality management. The basic governing equations of hydrodynamic transport are the shallow water equations:

Continuity equation

$$\frac{\partial U}{\partial x}+\frac{\partial V}{\partial y}+\frac{\partial W}{\partial z}=0 \tag{7.1}$$

x-Momentum

$$\frac{\partial HU}{\partial t}+\mathbf{V}\cdot\nabla(HU)-f(HV)=-\frac{H\partial P}{\rho\partial x}+\frac{\partial}{\partial z}\left[A_V H\frac{\partial U}{\partial z}\right]+F_x$$

y-Momentum

$$\frac{\partial HV}{\partial t}+\mathbf{V}\cdot\nabla(HV)+f(HU)=-\frac{H\partial P}{\rho\partial y}+\frac{\partial}{\partial z}\left[A_V H\frac{\partial V}{\partial z}\right]+F_y$$

The continuity equation is the condition for mass conservation while the two horizontal momentum equations (x,y) represent momentum balance between fluid acceleration, pressure gradient, turbulent shear stress, body force, and Coriolis forces. In most applications, the horizontal scale is much greater than the water depth (shallow water approximation); the vertical acceleration can hence be neglected, resulting in the hydrostatic approximation:

$$\frac{\partial P}{\partial z} = -\rho g$$

The fluid (water) is assumed to be incompressible; the sea water density depends on the temperature and salinity via an equation of state:

$$\rho = \rho(T,S)$$

In the previous equations, U, V, W are the turbulent-mean velocity components in the x, y, z directions; ρ is the water density; H is the water depth; P is the hydrostatic pressure; T and S are the water temperature and salinity, respectively; A_V is the vertical eddy viscosity; f is the Coriolis parameter. The horizontal turbulent shear terms F_x and F_y are defined by

$$F_x \equiv \frac{\partial}{\partial x}(H\tau_{xx}) + \frac{\partial}{\partial y}(H\tau_{xy})$$

$$F_y \equiv \frac{\partial}{\partial x}(H\tau_{yx}) + \frac{\partial}{\partial y}(H\tau_{yy})$$

where $\tau_{xx} = 2A_H\partial U/\partial x$, $\tau_{xy} = \tau_{yx} = A_H(\partial U/\partial y + \partial V/\partial x)$, $\tau_{yx} = 2A_H\partial V/\partial y$, with A_H being the horizontal eddy viscosity. In coastal applications, the turbulence mixing coefficients A_H and A_V have to be defined. The horizontal eddy viscosity A_H is related to the mean flow and also the grid size via a subgrid modeling scheme (Smagorinsky, 1963, Hamrick, 1992). For the vertical eddy viscosity A_V, a number of turbulence closure models have been developed. Most of these models are based on the eddy viscosity concept of Kolmogorov and Prandtl (Rodi, 1980; Burchard, 2002). The eddy viscosity is related to a characteristic length scale (mixing length):

$$A_V \propto L\sqrt{k}$$

where
- L is the mixing length
- k is the turbulent kinetic energy

In addition, stratification can also reduce the vertical mixing. This is characterized by the gradient Richardson number:

$$Ri = \frac{-g(\partial\rho/\partial z)}{\rho(\partial u/\partial z)^2}$$

For example, in the Mellor and Yamada 2.5 turbulence model (Mellor and Yamada, 1974; Mellor and Yamada, 1982), A_V is given by

$$A_V = 0.4\frac{(1+8Ri)qL}{(1+36Ri)(1+6Ri)}$$

where the turbulence intensity q and turbulence length scale L are determined by solving corresponding transport equations similar to salinity, temperature, and other dissolved constituents. Various turbulence models including k-L model, k-ε model, and Large Eddy Simulation (LES) models have been developed for various applications. The reader is referred to the review by Rodi (1984) and Burchard (2002).

7.2.1 Transport Modeling

The mass conservation of a dissolved constituent (e.g., salinity, temperature, bacteria, suspended sediment) can be expressed by a governing advective diffusion equation. The mass transport equation typically includes the rate kinetics as an additional source/sink term:

$$\frac{\partial C}{\partial t} = \underbrace{\boxed{-\vec{v}\cdot\vec{\nabla}C + E\nabla^2 C}}_{\textit{hydrodynamics}} + \underbrace{\boxed{\text{rate kinetics}}}_{\textit{water quality}}$$

For a conservative substance, the rate kinetic terms would be zero. In a general CFD code (e.g., EFDC; Hamrick, 1992), the transport equation for salinity and temperature are solved together with the continuity and momentum equations to give the 3D flow field.

7.2.2 Eulerian versus Lagrangian Models

Given the hydrodynamic flow field, the fate and transport of a pollutant and other environmental state variables can be simulated. There are two general approaches in simulating the hydrodynamic mass transport, namely the Lagrangian and Eulerian approaches. In the Lagrangian approach, the subjects (e.g., pollutant, microorganism) are simulated as individuals or patches. Their movement follows the hydrodynamic flow field and their locations are tracked—for example, hydrodynamic tracking of an algal bloom (see next section). Superimposed on the particle tracks are the biochemical processes which affects the growth and mortality of the particles and are simulated as increase/decrease in the size or number of particles (e.g., Wong and Lee, 2007).

While the Lagrangian approach mimics the reality better, it is also more computationally demanding since it involves tracking of the location and properties of tens of thousands of particles. Moreover, it is necessary to translate a prediction back to its Eulerian equivalent (e.g., counting the number of particles in a grid cell) for model interpretation. On the other hand, while the Eulerian approach is more computationally efficient, it can be subject to artificial numerical diffusion. Moreover, definition and tracking of ecological status are impossible.

Depending on the complexity of the problem, the rate kinetics may vary from a conservative substance (e.g., salinity) to a first-order decay (e.g., bacteria mortality or radioactive decay). For more complicated problems, a fully coupled water quality model involving tens to hundreds of state variables (e.g., in eutrophication modeling) may be required. For many problems, valuable insights can be gained by approximate models that capture the essential mechanisms at work. These simplified models provide a useful conceptual framework for coastal management, and can give surprisingly good results when combined with judicious use of hydrodynamic models. As an example, in many semi-enclosed and weakly flushed tidal inlets, the mass transport can be lumped into a tidal flushing rate which can be accurately determined using 3D hydrodynamic models (Choi and Lee, 2004). The flushing rate can then be used in a simplified box model to predict the water quality changes and the carrying capacity of fish farms for mariculture management. The use of a vertical structure model also proved to be most valuable to help interpret the complex algal bloom dynamics in Hong Kong waters (Wong and Lee, 2007; Wong et al., 2007).

7.3 WATERMAN: Water Quality Forecast and Management System

The Hong Kong Special Administrative Region (HKSAR) of China is a coastal city with a land area of 1100 km^2 and a population of over 7 million. It is located at the mouth of the Pearl River Estuary in Southern China, with coastal waters of around 1800 km^2. Hong Kong waters is divided into 10 water control zones (WCZ), based mainly on the hydrodynamic characteristics and pollution status; each WCZ has its own set of beneficial uses and water quality objectives. A systematic and extensive water quality monitoring program has been in operation by the Hong Kong Environmental Protection Department since 1986. The routine monitoring data are used to regularly evaluate the health of Hong Kong waters, to verify the performance of major sewage outfalls, and to assess their impact on the marine environment. Currently, there are 76 water monitoring stations in open waters where water quality is sampled once a month; three samples are typically taken at the surface, middle, and bottom. In addition, a phytoplankton monitoring program is in place to detect the development of red tides and potentially harmful algal species. Weekly water samples are taken from six representative fish culture zone stations and biweekly from five offshore stations year round for phytoplankton analysis. The monitoring data have been used to evaluate the effectiveness of major sewage infrastructure schemes (Xu et al., 2011).

Project WATERMAN is an interdisciplinary project to develop an internet and GIS-based real-time coastal water quality forecast and management system for Hong Kong (http://waterman.hku.hk). The project was funded by the Hong Kong Jockey Club Charities Trust and developed in collaboration with the Agriculture Fisheries and Conservation Department, Drainage Services Department, Environmental Protection Department, Hong Kong Observatory, and the Leisure and Cultural Services Departments of the Hong Kong Government. The objective of the WATERMAN system is to develop a modeling system to assimilate the monitoring data for more effective coastal management and public engagement.

The core of the WATERMAN system is a 3D hydrodynamic and transport model, the GIS database, and a 3D visualization system (Figure 7.1). The hydrodynamic model consists of a territorial model covering the Hong Kong and four regional submodels (Figure 7.2). The 3D GIS-based EIA system is characterized by seamless coupling of near and far field models, and enables simulation of environmental impact over the entire range of environmental conditions. It provides daily beach quality

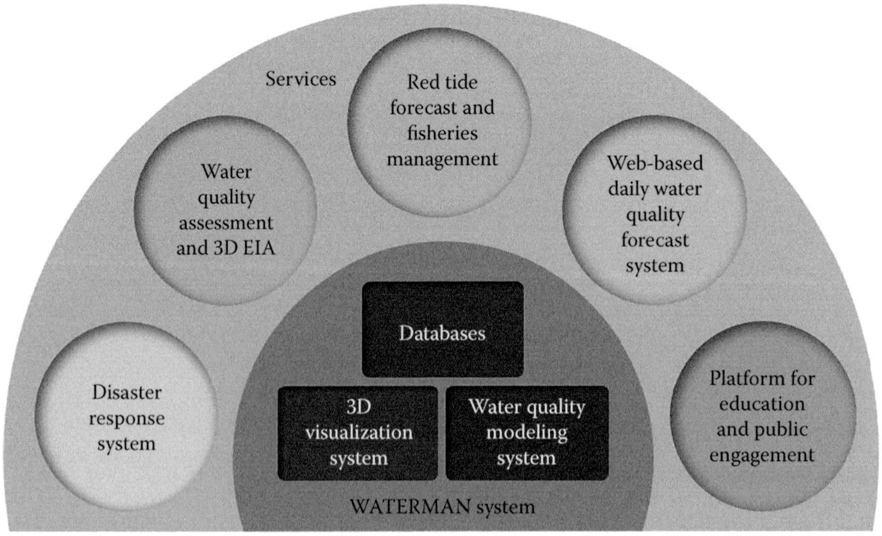

FIGURE 7.1 The WATERMAN water quality forecast and management system.

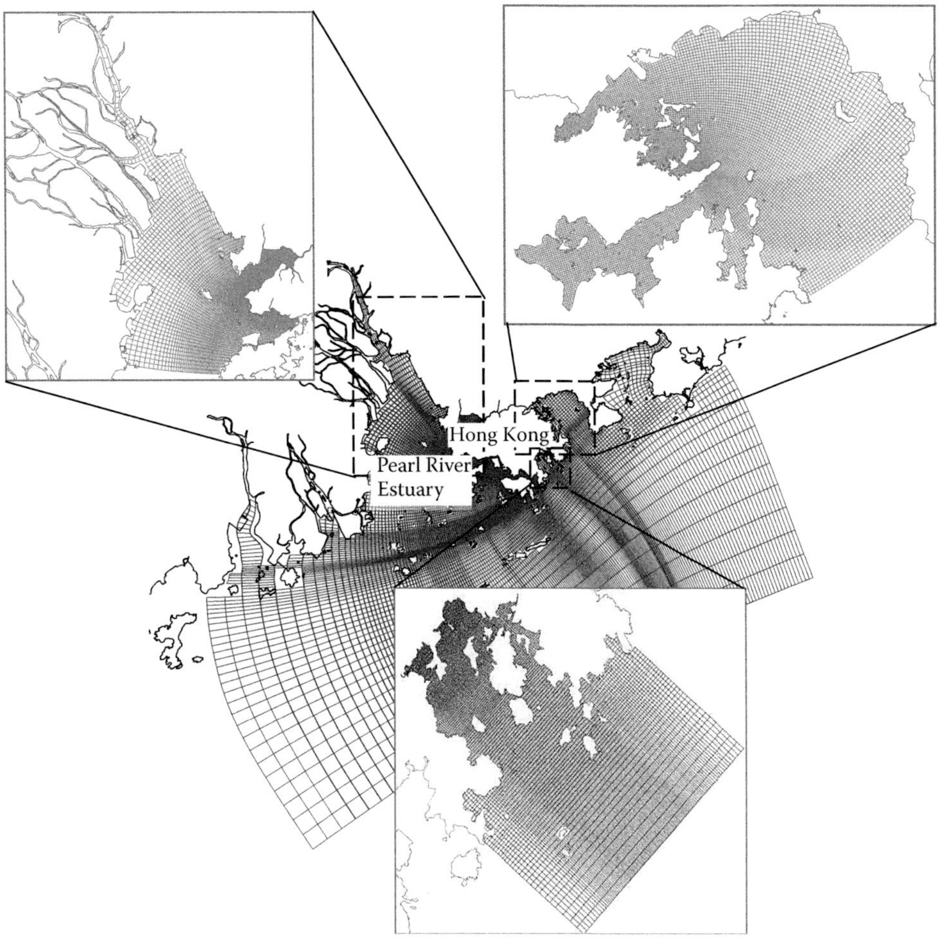

FIGURE 7.2 Hydrodynamic model grids for Hong Kong waters.

forecasts and project-specific virtual reality visualization and data interrogation of EIA scenarios over the internet. The hydrodynamics model is also linked to a red tide forecast and fisheries management module (Figure 7.1). The modeling system is integrated with a water quality database that provides input to models that forecast real-time beach water quality and red tide risks.

7.3.1 3D Environmental Impact Assessment

7.3.1.1 Near and Far Field Coupling

The determination of mixing zones of a polluting discharge and accurate impact assessment requires the proper coupling between the near and far field models. The coupling of plume models and 3D far field circulation models has been a difficult problem that is not fully resolved; the coupling issues have been discussed by Zhang and Adams (1999). In the WATERMAN system, the near field mixing can be simulated either by the Lagrangian virtual reality model VISJET (Lee et al., 2000; Lee and Chu, 2003) or the embedded version in the 3D hydrodynamic model. VISJET is based on the well-proven JETLAG model that predicts the near field mixing of an arbitrarily inclined round buoyant jet in a stratified crossflow, with a 3D trajectory (Lee and Cheung, 1990). The model tracks the evolution of the average properties of a plume element by conservation of horizontal and vertical momentum, conservation of mass accounting for the shear and vortex entrainments, and conservation of solute or tracer mass/heat. For a given set of ambient conditions (vertical profile of ambient horizontal velocity and density), the jet trajectory, jet velocity and radius, tracer concentration and dilution can be reliably predicted (Choi et al., 2009).

The plume model and the far field circulation model can be effectively coupled using the Distributed Entrainment Sink Approach (DESA) (Choi and Lee, 2007). With this method, the effect of the active plume mixing on the surrounding flow is modeled by a distribution of entrainment sinks along the plume trajectory and an equivalent diluted source flow at the predicted terminal height of rise. A two-way dynamic link can be established at grid cell level between the near and far field models. The coupling captures the key physical mechanisms of (1) turbulent jet entrainment and (2) 3D hydrodynamics with hydrostatic pressure approximation in the intermediate field. At any time instant, the ambient current, salinity, temperature, water density, and tracer concentration are provided by the far field solution at the grid cells upstream of the outfall discharge. Based on these ambient conditions, the embedded JETLAG computes the evolution of the average properties of the plume elements along the jet trajectory: jet velocity and width, tracer concentration, location, density, and the total fluid mass entrained into the plume element as a result of shear and vortex entrainment, ΔM_k. Based on the location of the center of the plume elements, the entrainment sinks, Q_e, for each far field model grid cell can then be computed by summing the entrainment flows corresponding to all the plume elements within that cell:

$$Q_e = \sum \left(\frac{\Delta M_k}{\rho_a \Delta t} \right)$$

The diluted source flow at terminal height of rise, Q_d, is then given by

$$Q_d = Q_o + \sum Q_e$$

and applied as a source term to the corresponding far field model grid cell, where Q_o is the effluent discharge flow. A source term representing the discharged pollution loading (tracer mass flux) is also introduced at the terminal height of rise (i.e., the near field mixing is represented as fluid mass source in the continuity equation and tracer mass source in the tracer transport equation for the far field flow model). In the WATERMAN system, the 3D flow model is based on the Environmental Fluid Dynamics Code (EFDC). It is a curvilinear finite difference model that solves the 3D shallow water equations using the Mellor and Yamada (1982) scheme for turbulent closure (Hamrick, 1992). The pressure is assumed to be hydrostatic and consists of barotropic (induced by external free surface gravity) and baroclinic (induced by the horizontal density gradient) components. The model uses a staggered grid for discretization and a sigma grid (terrain-following) coordinate in vertical direction. The modal splitting technique is used to solve the discretized equations, and the simulation is separated into the external and internal modes. The free surface elevation and the vertically integrated horizontal transports are solved first (external mode), then the 3D velocity, salinity, and thermodynamic properties are computed (internal mode). The dynamic linkage is illustrated in Figure 7.3; more details can be found in Choi and Lee (2007).

The major advantage of the DESA method is that the dynamic coupling at grid cell level (along the entire time-dependent jet trajectory) ensures the near field mixing is fully embedded into the 3D far field model in an essentially grid-independent manner. Unsteady tidal situations can be readily modeled, and there is no need to specify parameters such as the size of mixing zone or layer thickness. Full mass conservation is always ensured. The DESA method has been tested against the basic experimental data of a variety of complex flows including buoyant plumes in stagnant, flowing, and stratified environments (Choi and Lee, 2007).

Figure 7.4 shows a VISJET simulation of jet trajectories and dilution of the near field jet mixing of rosette risers mounted on a submerged ocean outfall for the dry and wet season; on each riser an eight-jet group discharges the effluent in the form of a rosette. It can be seen the sewage plumes are trapped in the presence of ambient stratification. Figure 7.5 shows a comparison of the predicted and observed buoyant surface layer in the intermediate field of a rosette jet group (Lai et al., 2007).

The impact of the Hong Kong Harbour Area Treatment Scheme (HATS) outfall at Stonecutters Island has been simulated using DESA on a full 3D model of the entire Hong Kong waters. A boundary-fitted curvilinear horizontal grid covering the entire Hong Kong waters is used. It has 13,888 active cells in the horizontal plan, with 10 vertical σ-layers and a grid size of 200–300 m near the outfall (Figure 7.6). The location of the outfall in relation to nearby beaches in Tsuen Wan (about 10 km away), the bottom terrain, and sewage outfall are all depicted in the GIS-based virtual reality simulation (Figure 7.6). Typical wet and dry season conditions with constant freshwater runoff from the Pearl River are considered. Figure 7.7 shows the effects of the HATS discharge upon the nearby bathing beaches during the flood tide. Under hydrodynamic transport, a streak of higher *Escherichia coli* concentration is formed along from Tsuen Wan region to the southeast coast of Hong Kong Island. With disinfection, *E. coli* concentration in this high concentration streak is reduced by an order of magnitude. Figure 7.8 shows the comparison between the predicted and measured *E. coli* concentrations at Gemini Beach during a period when sewage disinfection was in operation, from January 15 to March 1, 2010. It can be seen that the time variation in *E. coli* concentration can be reasonably captured. In particular, over the period between February 9 and February 20, the disinfection was temporarily switched off and the model correctly predicted a significant increase in *E. coli* concentration (by 1 order of magnitude).

7.3.1.2 Daily Beach Water Quality Forecast

Predictions of tidal currents can be obtained from the hydrodynamic model either in real-time mode or in offline mode for predefined scenarios (e.g., in environmental impact assessment). The deterministic model also serves as the engine for daily forecast of beach water quality. Hong Kong has a beautiful coastline that extends over 700 km. There are 41 gazetted beaches that are used by millions of users during the March–October bathing season. It is well known that swimming in polluted waters is hazardous to health. Epidemiological studies have shown that bacteria level is a good indicator of beach water quality. Beach water quality advisory is currently based on past water quality measurements collected roughly at weekly intervals. To supplement the existing monitoring-based beach grading system, a daily beach water quality forecast system using a combination of statistical, data-driven, and 3D hydrodynamic forecast models have been developed. The model is based on field and laboratory studies of bacterial decay in Hong Kong coastal waters (Chan, 2010). Figure 7.9 shows the comparison of an actual forecast of diurnal variation of *E. coli* with field data at Gemini Beach. It is seen that the water quality is significantly better at low tide than

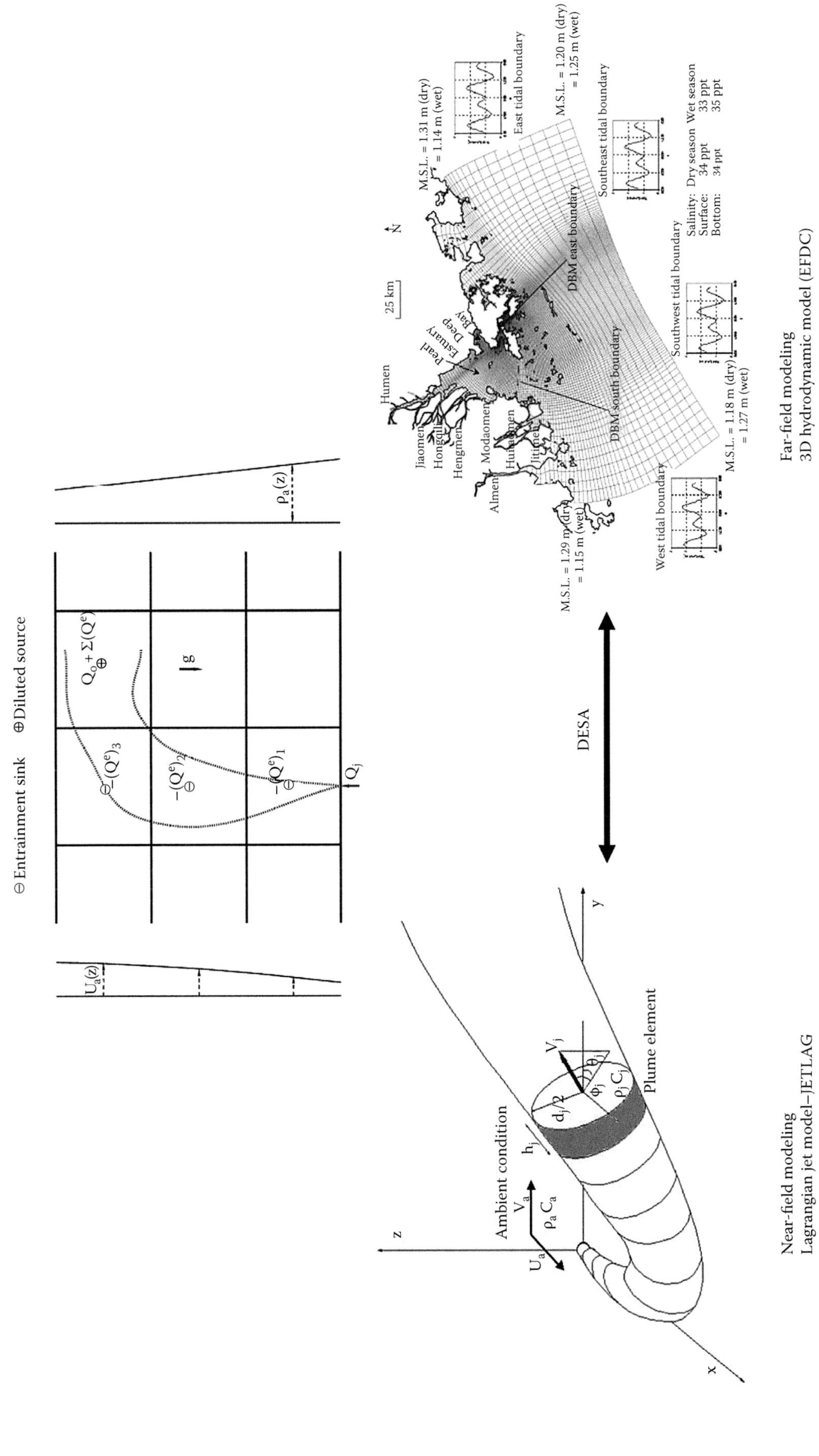

FIGURE 7.3 Coupling of near field plume model with 3D Pearl River Estuary hydrodynamic model.

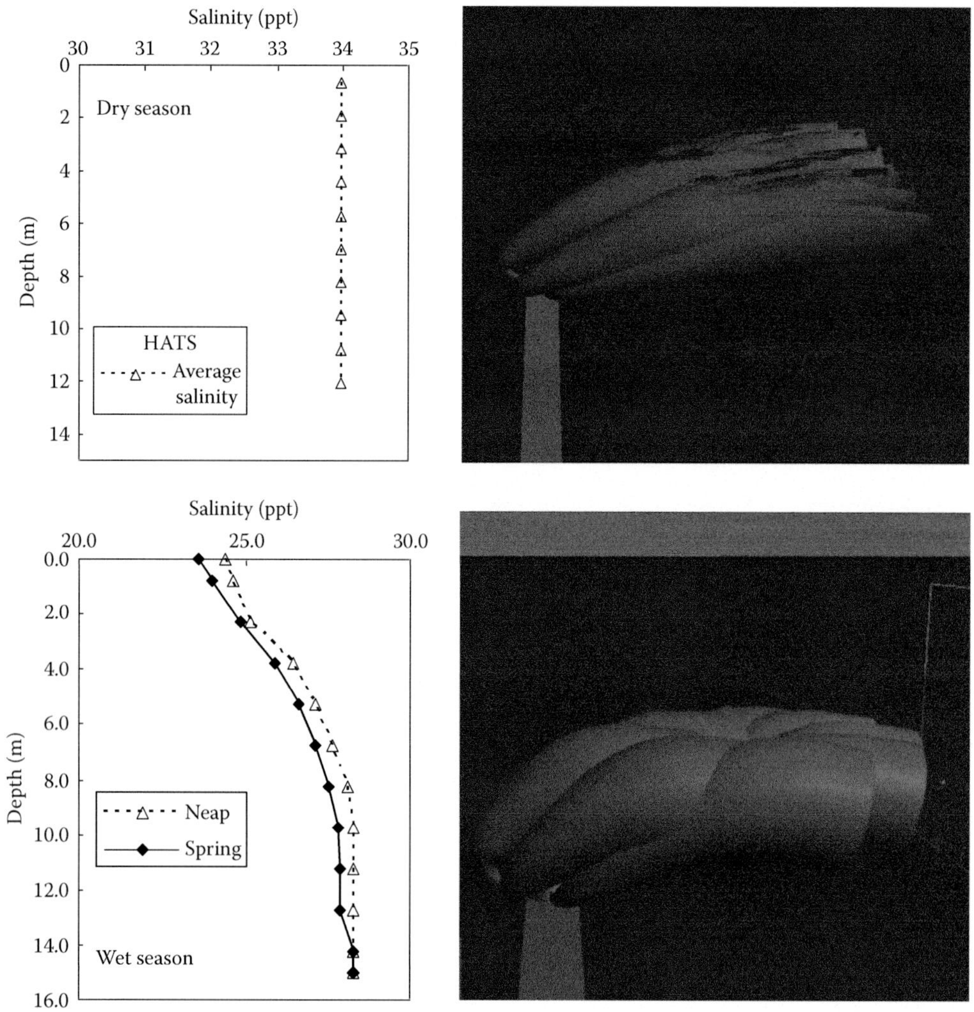

FIGURE 7.4 VISJET simulation of buoyant plumes from eight-jet rosette risers from the sewage outfall of the Hong Kong Harbour Area Treatment Scheme (HATS).

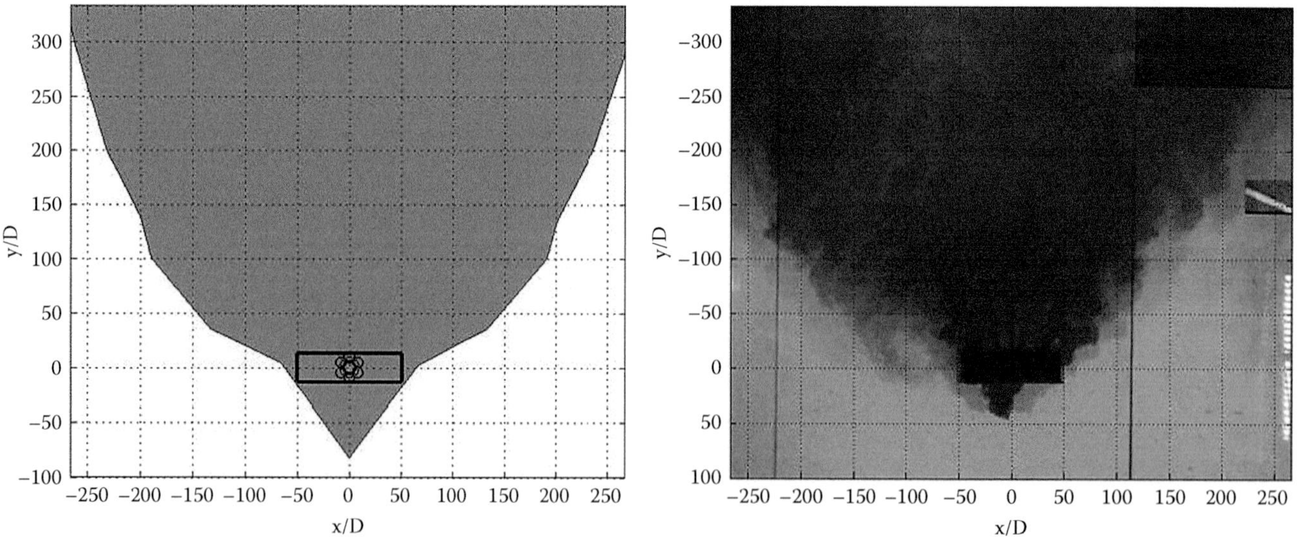

FIGURE 7.5 Comparison of the predicted buoyant surface layer in the near field of a rosette jet group with laboratory experiments (six jets, Fr = 15, Ua = 3 cm s^{-1}, H = 0.3 m, and D = 0.3 cm).

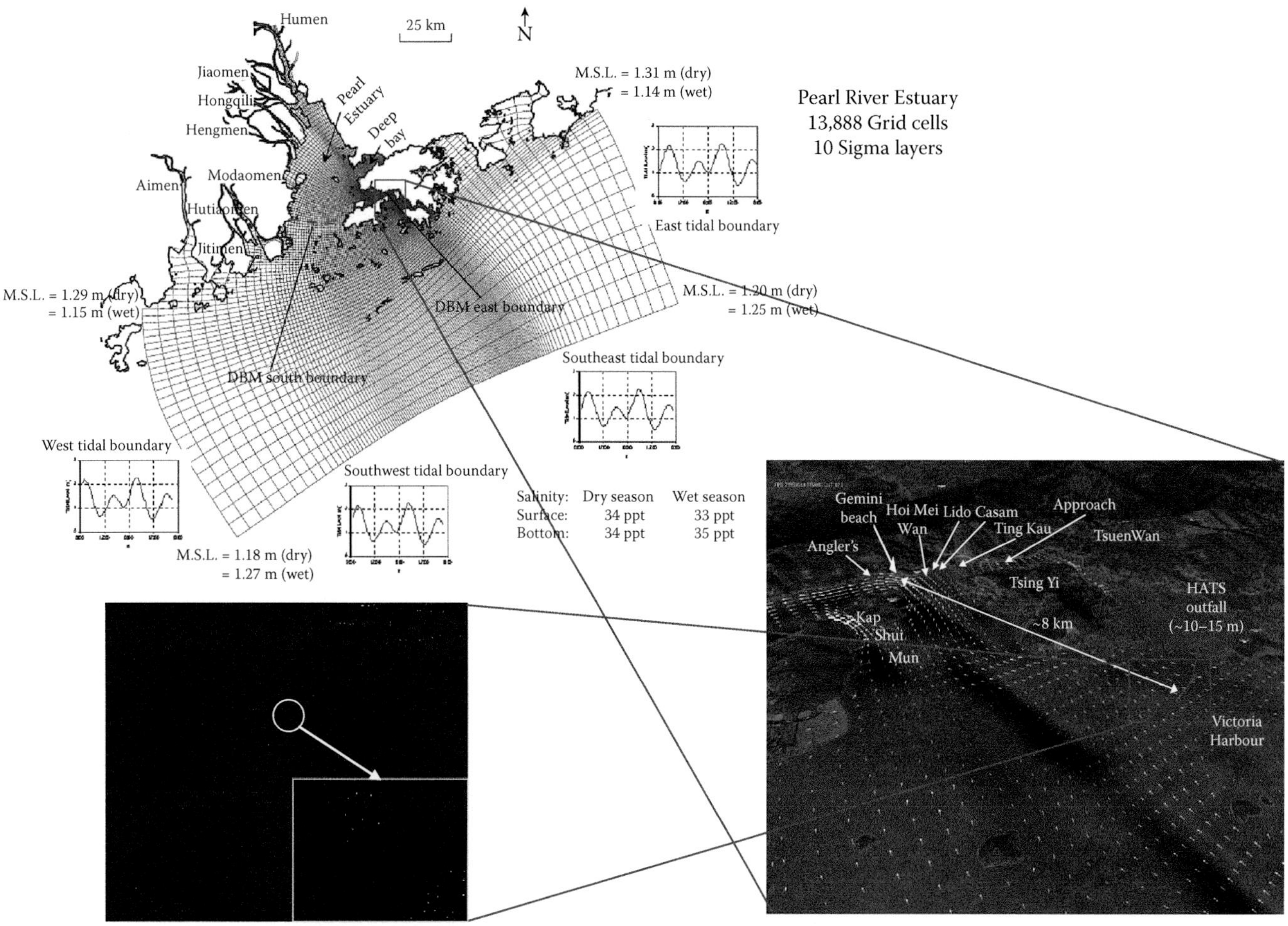

FIGURE 7.6 The Pearl River Estuary hydrodynamic model in the WATERMAN system.

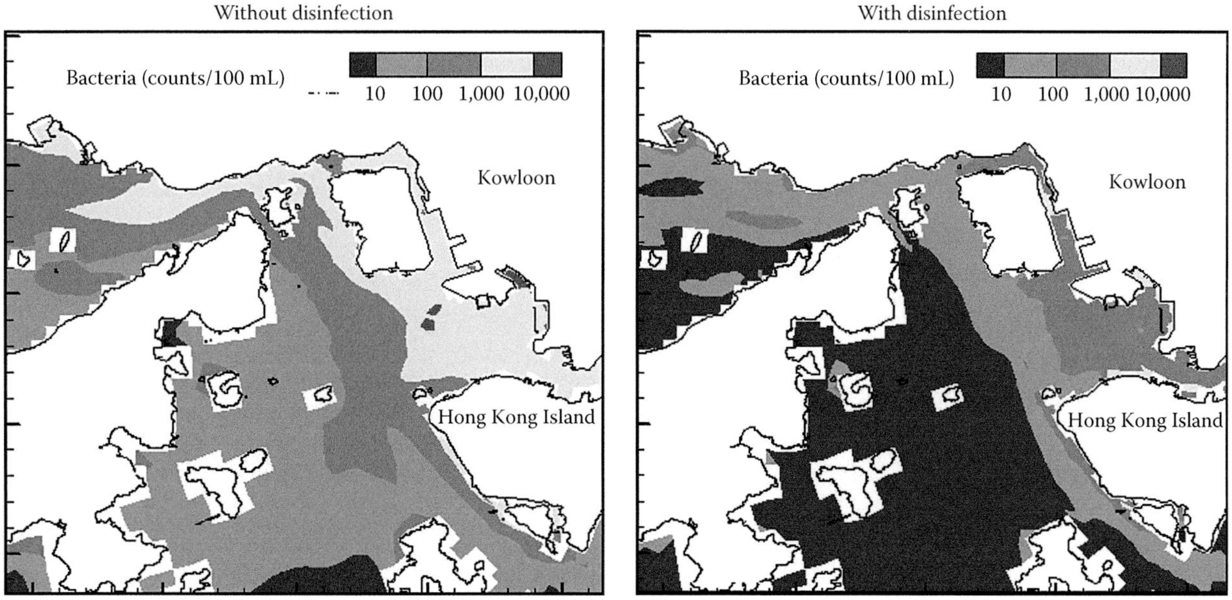

FIGURE 7.7 **(See color insert.)** Predicted *Escherichia coli* concentrations in the vicinity of the HATS outfall.

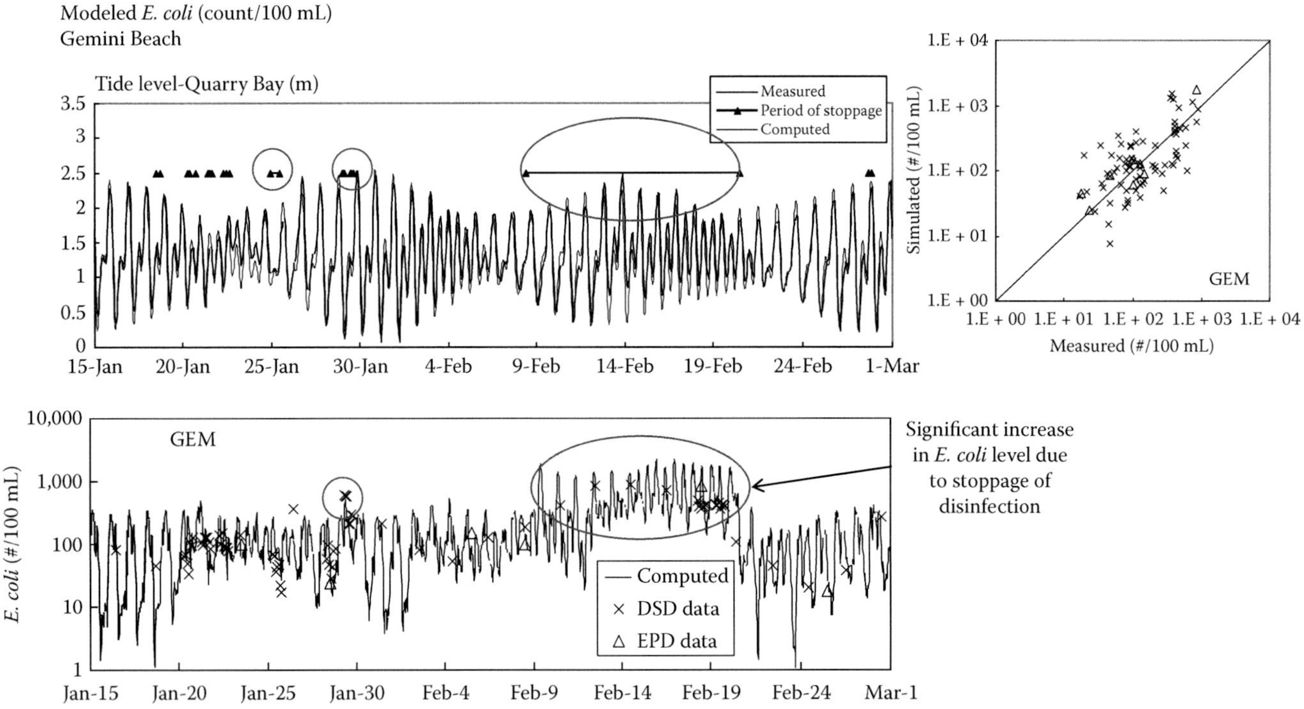

FIGURE 7.8 Comparison of predicted and measured *Escherichia coli* concentration at Gemini Beach near the HATS outfall; the impact of discontinuation of disinfection in mid-February can be seen.

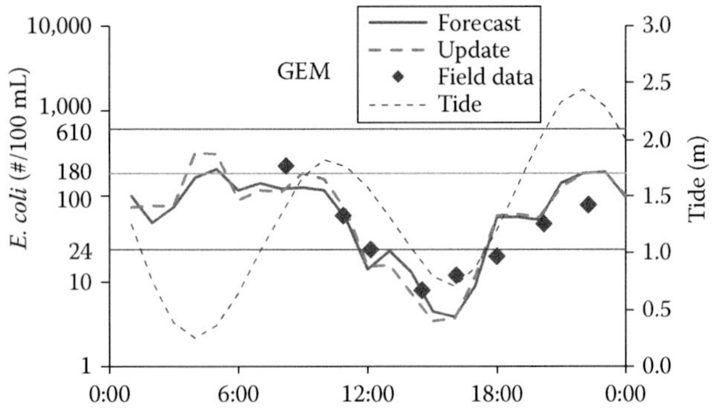

FIGURE 7.9 Predicted diurnal *Escherichia coli* level variation at Gemini Beach (October 27, 2011—semi-diurnal tide, sunny day).

at high tide. The 3D deterministic model is most useful for modeling significant point sources and complements statistical models of beach forecasts (Thoe, 2010).

7.3.2 Fisheries Management and Red Tide Forecast System

Live fish is an important food resource for coastal cities like Hong Kong. However, the deterioration of marine water quality has posed challenges to local fish farming. In particular, eutrophication, harmful algal blooms, and the occurrence of hypoxic (low dissolved oxygen) condition are recurrent threats to fish farming. The WATERMAN fisheries management system has two main tasks: (1) prediction and tracking the occurrence of harmful algal blooms and (2) estimating the carrying capacity for sustainable mariculture. Despite efforts to control water pollution, currently there are 20–30 red tide outbreaks every year in Hong Kong. To provide timely red tide warnings and mitigations, a real-time red tide forecast and early warning system has been developed to supplement the existing red tide reporting and monitoring network of the AFCD.

7.3.2.1 Red Tide Forecast and Early Warning

Red tide is an environmental problem of global importance. It is caused by the rapid growth and/or accumulation of microscopic phytoplankton (algae) and results in water discoloration. It is a complex hydro-biological phenomenon driven by a myriad of biological and environmental factors including light and nutrient availability, hydrodynamic and mass transport processes. In the subtropical Hong Kong coastal waters, algal blooms often occur suddenly and subside in a relatively short time frame (in the order of days). Red tide prediction has been a difficult challenge for decades.

Based on online telemetry monitoring for over 4 years with over 20 red tides successfully observed (Lee et al., 2005), a two-layer model and vertical stability theory have been developed to explain and predict the red tide occurrence in Hong Kong (Riley et al., 1949; Wong et al., 2007). In this model, algal bloom occurrence is linked quantitatively to water column stability in a predictive framework for the first time.

In the two-layer model, the water column is postulated to consist of a photic layer of depth l and an aphotic layer below where no algal growth can occur. The phytoplankton dynamics is governed by vertical turbulent diffusion, sinking, and phytoplankton growth/mortality. The governing equation can be written as

$$\frac{\partial C}{\partial t} = E\frac{\partial^2 C}{\partial z^2} - v\frac{\partial C}{\partial z} + \eta C \tag{7.2}$$

where

- $C(z,t)$ is the algal concentration
- E is the vertical turbulent diffusivity
- v is the algal settling velocity
- η equals to the growth rate μ in the upper euphotic layer or the morality rate d in the lower aphotic layer.

In discretized form, the vertical concentration profile $C(z)$ can be expressed as

$$C(z) = (C_1, C_2, C_3, \ldots)^T$$

with spacing Δz apart. This results in an eigenvalue problem

$$\mathbf{A}\phi_1 = k_1\phi_1$$

with the unique property that its asymptotic solution will be dominated by the largest real simple eigenvalue k (Di Toro, 1974; Schnoor and Di Toro, 1980). As a result, the variables z and t are separable, and $C(z,t)$ is given by

$$C(z,t) = f(z)e^{kt} \tag{7.3}$$

It can be seen that regardless of the initial condition, the algal concentration profile always approaches the same form; the algal population maintains the same profile and grows as a whole with the population growth rate given by k. The condition for which k is always positive can be used to infer the hydrodynamic criterion for red tides (Wong et al., 2007).

By substituting back the concentration profile Equation 7.3 into the governing Equation 7.2, we have

$$E\frac{\partial^2 f}{\partial z^2} - v\frac{\partial f}{\partial z} + (\eta - k)f = 0 \tag{7.4}$$

Assuming constant values for the coefficients, an analytical solution for bloom stability can be obtained (Wong et al., 2007). More generally, a numerical solution of the bloom stability can be obtained by imposing zero mass flux boundary conditions at the free surface and sea bottom ($z = -H$). The likelihood of bloom occurrence depends on the vertical diffusivity, algal settling speed, and algal growth rate. Figure 7.10 shows the bloom stability criterion obtained from extensive numerical experiments under different initial conditions and a wide range of model parameters typical of Hong Kong coastal waters (water depth $H = 5$–$20\,m$; $l = 2$–$8\,m$; $\mu = 0.5$–$2\ day^{-1}$; $d = 0.1$–$100\ day^{-1}$; $v = 0$–$5\,m\ day^{-1}$; $E = 0.0002$–$0.001\ m^2s^{-1}$). The numerical results show

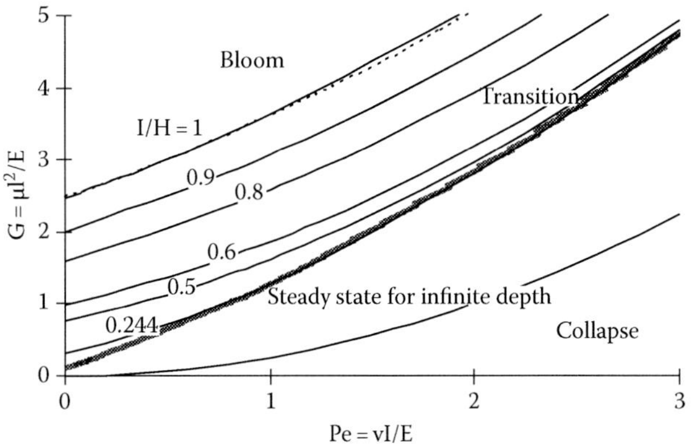

FIGURE 7.10 Stability diagram for red tide occurrence. The stability threshold is obtained for different euphotic zone thickness l/H.

that bloom stability is governed by two key parameters: a dimensionless growth number $G = \mu l^2/E$ and a euphotic layer Peclet number $Pe = vl/E$. The numerical result depends mainly on the dimensionless euphotic zone depth l/H. As l/H increases, the stability bound is shifted upward. If the upper envelope ($l/H = 1$) of the family of eigen conditions is adopted as a conservative estimate of the likelihood of algal blooms, it can be shown that the algal bloom stability can be expressed in terms of a critical turbulence below which k is always positive (Wong et al., 2007):

$$E < E_c \approx \frac{4\mu l^2}{\pi^2}$$

When this condition is fulfilled, diffusive loss to bottom layer is always less than production in the euphotic zone. The above criterion gives a first-order necessary condition for bloom formation. The criterion is validated by our algal bloom data as well as independent observations (Figure 7.11). Together with necessary condition imposed by nutrient criterion (Hodgkiss and Ho, 1997; Wong and Lee, 2007), a decision tree for the condition of red tide can be constructed (Figure 7.12). A daily red

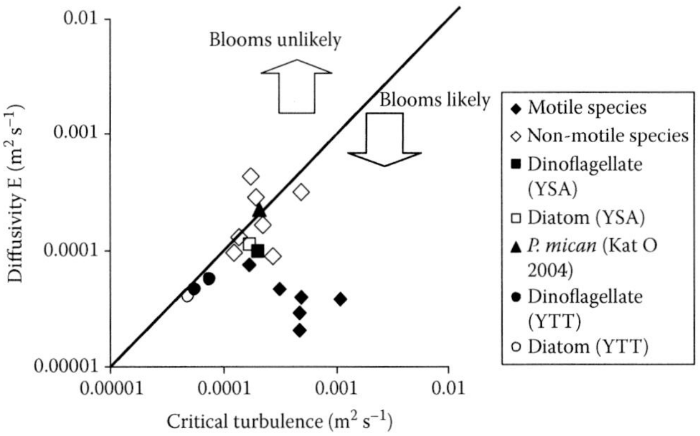

FIGURE 7.11 Comparison of predicted critical turbulent diffusivity for bloom occurrence with estimated vertical diffusivity for observed algal blooms in the field monitoring.

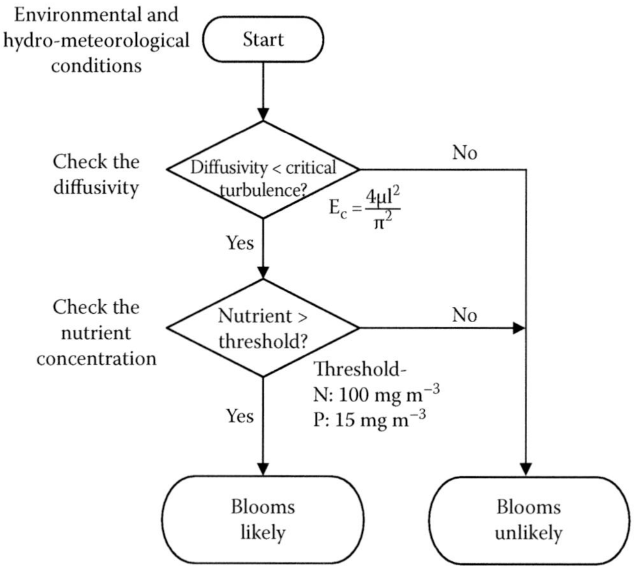

FIGURE 7.12 Decision model for prediction of likelihood of algal bloom.

tide forecast can then be hoisted based on field measurement in nutrient concentration, light attenuation (e.g., secchi depth), estimated algal growth rate, diffusivity based on salinity and temperature profiles, and hydrodynamic calculations. It is worth noting that, due to more dynamic changes in the hydrodynamic and meteorological condition compared to the slow process of nutrient input and regeneration, hydrodynamic stability is found to be the major triggering factor.

7.3.2.2 Forecast of Red Tide Risks

The prognostic forecast (Figure 7.12) can be cast in the form of a red tide likelihood risk map (Wong et al., 2009). A daily red tide occurrence probability is calculated at each of the water control zones. The red tide occurrence probability is defined in terms of the water column stability and nutrient limitation factors. A hydrodynamic stability risk factor is defined as the ratio between the critical turbulence E_c and the estimated vertical diffusivity E to quantify the hydrodynamic controlling effect:

$$\text{Hydrodynamic stability factor (R)} = \frac{\text{Critical turbulence } (E_c)}{\text{Estimated diffusivity (E)}} = \frac{4\mu l^2}{E\pi^2}$$

Based on the historical water quality data in Hong Kong, a probabilistic model of red tide risk can be developed. From the historical data, a sigmoid function relationship is observed between the red tide risk and the increase in water column stability and nutrient concentration (Figure 7.13). Consistent with the traditional water quality model framework, the interaction between stability (including the effect of light and temperature) and nutrient is expressed using the multiplication rule; the limiting nutrient is determined from the Liebig's law of minimum:

$$P(\text{bloom}) = P(R) \cdot \text{Min}[P(N), P(P)]$$

where

$$P(R) = \frac{1}{1 + (R/\mu_R)^{-a}}; \quad P(N) = \frac{1}{1 + (N/\mu_N)^{-b}}; \quad P(P) = \frac{1}{1 + (P/\mu_P)^{-c}}$$

When R is large ($P(R) \approx 1$), nutrient takes the primary controlling effect; when the nutrient concentrations are high (P(N) and $P(P) \approx 1$), hydrodynamics plays a primary role. From the historical data, the parameters μ_R, μ_N, μ_P, a, b, c are determined to be $\mu_R = 0.32$, $\mu_N = 40$, $\mu_P = 5$, a = 1.65, b = 1.4, c = 1.4. A high risk is defined for P(bloom) > 70%, and low risk when P(bloom) < 30%.

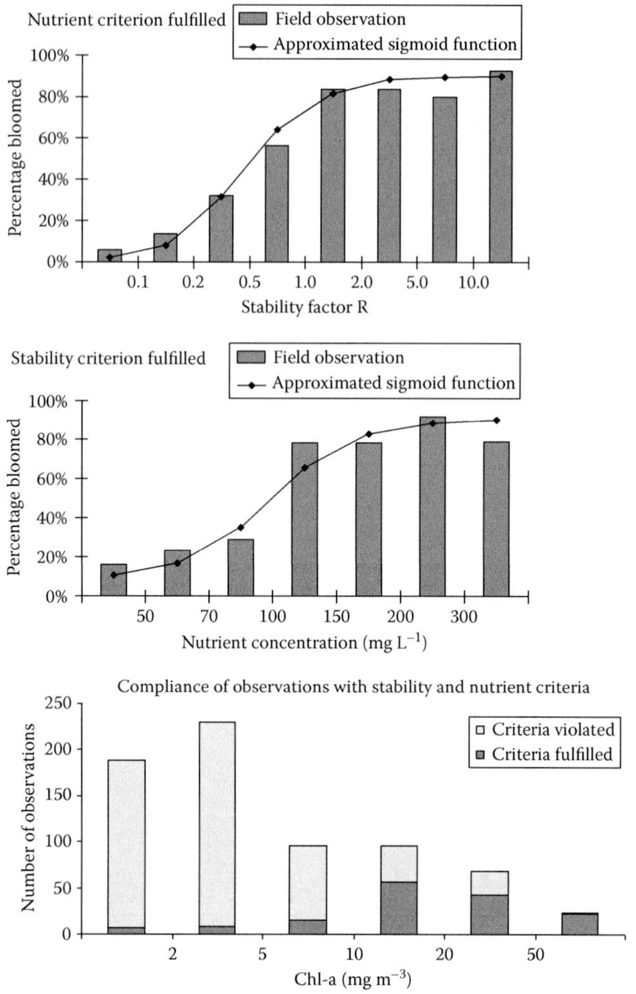

FIGURE 7.13 Statistical analysis of EPD field monitoring data using algal bloom risk prediction model.

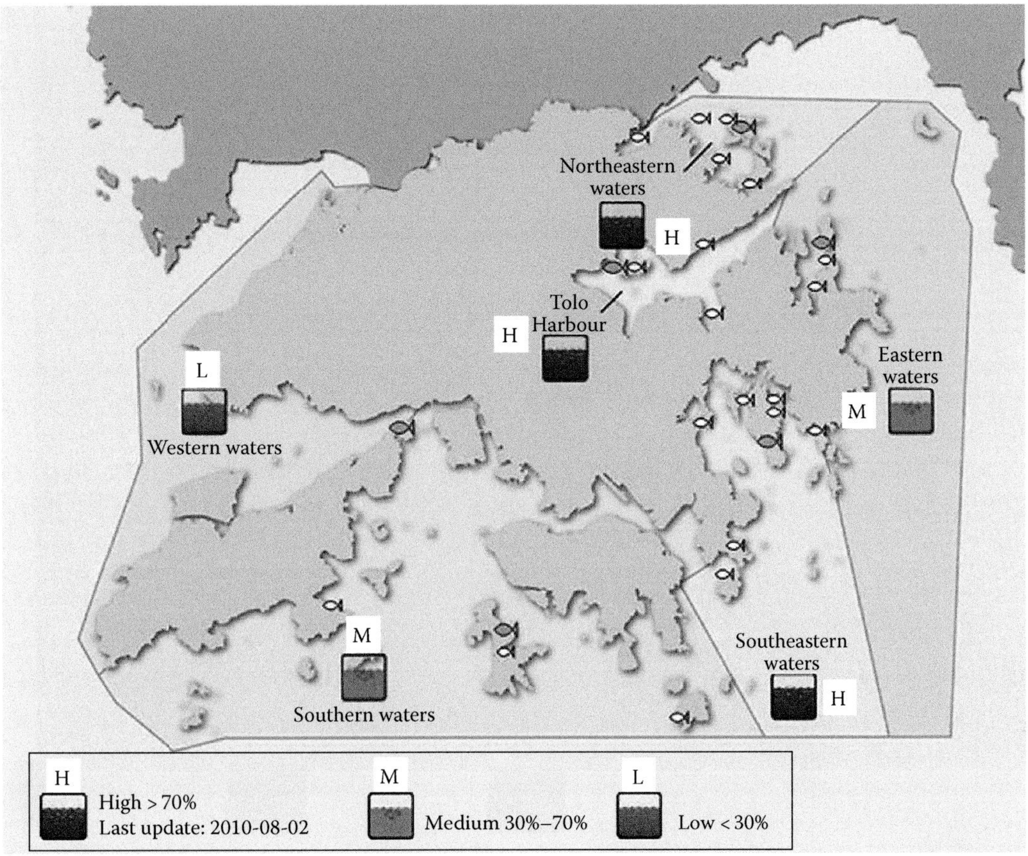

FIGURE 7.14 Algal bloom risk map forecast in Project WATERMAN on August 1, 2010.

Figure 7.14 shows an example of algal bloom risk map prediction in early August 2010. In this case, high risk is predicted for the large area in the eastern waters including Mirs Bay, Tolo Harbour, and Port Shelter. An alert was sent to the Agriculture, Fisheries, and Conservation Department. On August 3, a red tide formed by the dinoflagellates *Scrippsiella trochoidea* was reported in Sha Tau Kok and soon developed into a massive bloom. Field surveys were immediately carried out to investigate the conditions of the red tide. Based on the observed red tide, hydrodynamic particle tracking was carried out to predict the affected area (Figure 7.15). According to the tracking forecast, the red tide will persist in the area for a long time (it lasted until late August). Moreover, according to our field water samples, very high algal concentration (up to 10^5cells/mL) was observable, and low dissolved oxygen levels were observed in the bottom waters. Alerts were therefore sent to local fish farmers. Fish farmers in O Pui Tong Fish Culture Zone relocated the fish rafts to the outer waters not affected by the red tide (Figure 7.14). For fish farms which were difficult to relocate (when the fish farming scale is too big), fish farmers were advised to aerate their fish farms at night. Fortunately, probably due to the precautions taken, no fish kill was reported in this massive red tide incident.

Since the operation of the system in 2009, over 30 algal blooms have been successfully predicted and tracked by the system. In particular, timely response actions were taken during two massive algal blooms in August and October 2010; no significant mariculture loss was recorded.

7.3.2.3 Sustainable Mariculture Fish Loading (Carrying Capacity)

Carrying capacity is the maximum allowable fish culture production without breaching the water quality objective. The carrying capacity for all the fish culture zones in Hong Kong can be determined. Hydrodynamic models are used to determine the tidal flushing rates of each fish culture zone, which is the primary factor governing the carrying capacity. A diagenetic water quality model is then used to model the water quality by accounting for eutrophication kinetics and sediment–water interaction (Lee et al., 2003). The optimal fish farm loading capacity can then be determined. The main target is to prevent the occurrence of hypoxia and reduce the number of red tide occurrences.

7.4 Concluding Remarks and Future Challenges

Predicting coastal water quality is no easy task—especially in view of the uncertainty with model rate parameters, inflows, pollution loads, and boundary conditions. Extensive research

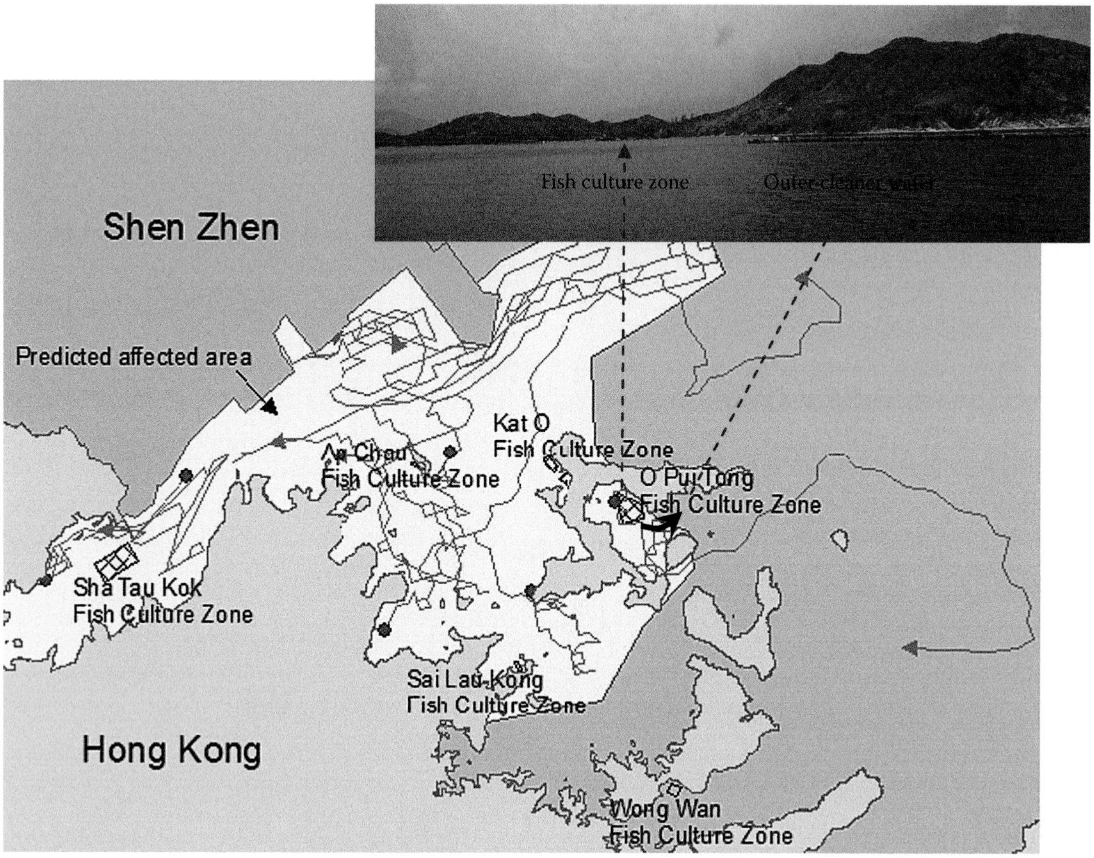

FIGURE 7.15 Predicted red tide track and affected area, and the temporary fish farm relocation at O Pui Tong Fish Culture Zone.

and practice over the past decades have confirmed the validity of 3D hydrodynamic and water quality models. The effective use of hydrodynamic and water quality models can assist with impact assessment and water quality control, provide water quality forecasts or prognosis to engage citizens in the use and protection of the marine environment, and to inform better planning of water quality monitoring programs (e.g., data sampling interval). A major promise of environmental fluid mechanics lies in the predictive ability of the models in emergency situations: when an environmental disaster strikes—whether it is a toxic cyanide spill or a massive red tide, the cause and response required is often unknown, and the needed data are usually not available. The use of real-time forecasts can provide useful hints; even a 1 day lead time prediction of what is likely to happen can make a huge difference toward disaster mitigation. On the other hand, the acceptance and adoption of scientifically proven water quality forecasts is hampered by an intrinsic social resistance toward innovation, and also depends on institutional and political considerations.

The use of advanced 3D models for daily forecasting requires effective data assimilation and coordination among various organizations and parties. In comparison to hydro-meteorological data, water quality data are typically sparse; data may be only available at weekly to monthly intervals at a number of selected stations. This can lead to large data gaps and uncertainty in the environmental forcing or boundary conditions. The acquisition of timely data remains a challenge for real-time forecasting. In addition, due to the time needed for analytical chemical tests and data processing, there is always a delay between field sampling and data availability. The use of data-driven methods such as Artificial Neural Networks (ANN) to predict some of the needed model inputs from real-time hydro-meteorological data (e.g., river flow and salinity from rainfall) has proved to be useful in water quality forecasts (e.g., Dong, 2010; Thoe, 2010). Moreover, to cater for model and observational errors, data assimilation methods such as Kalman Filtering can help reduce the prediction uncertainty and provide best estimates of the water quality state variables. For example, Mao et al. (2009) used the Extended Kalman Filter (EKF) to capture the dynamic changes in the algal growth rates in an algal dynamics model and provide an optimal estimate of the chlorophyll-*a* concentrations. The model uncertainties can also be generated using Ensemble-Kalman filtering—i.e., obtaining the envelope of model errors via simulation of an ensemble of solution trajectories.

The emergence and increasing availability of smart sensors would likely influence model-data integration in the future—remotely controlled automatic nutrient analyzers, underwater robots for water quality sampling, and unmanned helicopters for aerial monitoring are all attractive avenues to enhance model-data integration. In addition, there is increasing attention paid to the modeling of micro-pollutants such as endocrine-disrupting compounds (EDC) and pharmaceuticals. These pollutants have been detected in sewage effluents, stormwater, and sediments; they are usually present in the marine environment at very low concentrations, but can pose significant health risks. The use of hydrodynamic transport models to study the fate and transport of such emerging chemicals is receiving active interest.

Acknowledgments

This work is supported by the Hong Kong Jockey Club Charities Trust and by in part by a grant from the University Grants Committee of the Hong Kong Special Administrative Region, China (Project No. AoE/P-04/04) to the Area of Excellence (AoE) in Marine Environment Research and Innovative Technology (MERIT).

References

Burchard, H., 2002. *Applied Turbulence Modelling in Marine Waters*. Springer, Hong Kong, 215pp.

Chan, Y.M., 2010. Field and laboratory studies of *E. coli* decay rate at a coastal beach with reference to storm events. MPhil thesis. The University of Hong Kong, Hong Kong.

Choi, K.W. and Lee, J.H.W., 2004. Numerical determination of flushing time for stratified water bodies. *Journal of Marine Systems* 50: 263–281.

Choi, K.W. and Lee, J.H.W., 2007. Distributed entrainment sink approach for modelling mixing and transport in intermediate field. *Journal of Hydraulic Engineering, ASCE* 133(7): 804–815.

Choi, K.W., Lee, J.H.W., Kwok, K.P.W.H., and Leung, K.M.Y., 2009. Integrated stochastic environmental risk assessment of the Harbour Area Treatment Scheme (HATS) in Hong Kong. *Environmental Science and Technology* 43: 3705–3711.

Di Toro, D.M., 1974. Vertical interactions in phytoplankton populations - an asymptotic eigenvalue analysis. *Proceedings of the 17th Conference on Great Lakes Research, International Association for Great Lakes Research*, Ann Arbor, MI, pp. 17–27.

Dong, Y.H., 2010. Analysis of stratification and algal bloom risk in Mirs Bay. MPhil thesis. The University of Hong Kong, Hong Kong.

Fischer, H.B., List, E.J., Koh, R.C.Y., Imberger, J., and Brook, N.H., 1979. *Mixing in Inland and Coastal Waters*. Academic Press, Inc., New York, 483pp.

Hamrick, J.M., 1992. A three-dimensional environmental fluid dynamics computer code: Theoretical and computational aspects. The College of William and Mary, Virginia Institute of Marine Science, Special Report 317, 63pp.

Harrison, P.J., Yin, K.D., Lee, J.H.W., Gan, J.P., and Liu, H.B., 2008. Physical-biological coupling in the Pearl River Estuary. *Continental Shelf Research* 28: 1405–1415.

Hodgkiss, I.J. and Ho, K.C., 1997. Are changes in N:P ratios in coastal waters the key to increased red tide blooms. *Hydrobiologia* 352: 141–147.

Jirka, G.H., 2008. Improved discharge configurations for brine effluents from desalination plants. *Journal of Hydraulic Engineering, ASCE* 134(1): 116–129.

Jirka, G.H. and Lee, J.H.W., 1994. Waste disposal in the ocean. In: *Water Quality and Its Control* (IAHR Hydraulic Structures Design Manual, Vol. 5), M. Hino (Ed.), Balkema, Rotterdam, the Netherlands, pp. 193–242.

Lai, A.C.H., Yu, D., and Lee, J.H.W., 2007. Near and intermediate field mixing of a rosette jet group. *Proceedings of the 5th International Symposium on Environmental Hydraulics (ISEH V)*, Arizona State University, Tempe, AZ, December 4–7, 2007 (CDROM).

Lee, J.H.W., and Cheung, V., 1990. Generalized Lagrangian model for buoyant jets in current. *Journal of Environmental Engineering, ASCE* 116(6): 1085–1105.

Lee, J.H.W., Cheung, V., Wang, W.P., and Cheung, S.K.B., 2000. Lagrangian modeling and visualization of rosette outfall plumes. *Proceedings of Hydroinformatics 2000*, Iowa, July 23–27, 2000 (CDROM).

Lee, J.H.W. and Choi, K.W., 2008. Real-time hydro-environmental modeling and visualization system for public engagement. *Environmental Fluid Mechanics* 8: 411–421.

Lee, J.H.W., Choi, K.W., and Arega, F., 2003. Environmental management of marine fish culture in Hong Kong. *Marine Pollution Bulletin* 47: 202–210.

Lee, J.H.W. and Chu, V., 2003. *Turbulent Jets and Plumes—A Lagrangian Approach*. Kluwer Academic Publishers, Dordrecht, the Netherlands, 390pp.

Lee, J.H.W., Hodgkiss, I.J., Wong, K.T.M., and Lam, I.H.Y., 2005. Real time observations of coastal algal blooms by an early warning system. *Estuarine, Coastal and Shelf Science* 65: 172–190.

Lee, J.H.W. and Qu, B., 2004. Hydrodynamic tracking of the massive spring 1998 red tide in Hong Kong. *Journal of Environmental Engineering, ASCE* 130: 535–550.

Mao, J.Q., Lee, J.H.W., and Choi, K.W., 2009. The extended Kalman filter for forecast of algal bloom dynamics. *Water Research* 43(17): 4214–4224.

Mellor, G.L. and Yamada, T., 1974. A hierarchy of turbulence closure models for planetary boundary layers. *Journal of the Atmospheric Sciences* 31: 1791–1806.

Mellor, G.L. and Yamada, T., 1982. Development of a turbulence closure model for geophysical fluid problems. *Reviews of Geophysics and Space Physics* 20(4): 851–875.

Riley, G.A., Stommel, H., and Bumpus, D.F., 1949. Quantitative ecology of the plankton of the western North Atlantic. *Bulletin of the Bingham Oceanographic Collection*, Yale University 12: 1–169.

Rodi, W., 1980. *Turbulence Models and Their Application in Hydraulics*, IAHR Monograph, International Association for Hydraulic Research, Delft, the Netherlands.

Schnoor, J.L. and Di Toro, D.M., 1980. Differential phytoplankton sinking- and growthrates: An eigenvalue analysis. *Ecological Modelling* 9: 233–245.

Smagorinsky, J., 1963. General circulation experiments with the primitive equations, i. the basic experiment. *Month Weather Review* 91: 99–164.

Thoe, W., 2010. A daily forecasting system of marine beach water quality in Hong Kong. PhD thesis. The University of Hong Kong, Hong Kong.

UNESCO, 2001. Ensuring the sustainable development of oceans and coasts: A call to action. Report on the *Global Conference on Oceans and Coasts at Rio+10*, UNESCO, Paris, France, December 3–7, 2001.

Wong, K.T.M. and Lee, J.H.W., 2007. Simulation of harmful algal blooms using a deterministic Lagrangian particle separation-based method. *Journal of Hydro-Environment Research* 1(1): 43–55.

Wong, K.T.M., Lee, J.H.W., and Harrison, P.J., 2009. Forecasting of environmental risk maps of coastal algal blooms. *Harmful Algae* 8: 407–420.

Wong, K.T.M., Lee, J.H.W., and Hodgkiss, I.J., 2007. A simple model for forecast of coastal algal blooms. *Estuarine, Coastal and Shelf Science* 74: 175–196.

Xu, J., Lee, J.H.W., Yin, K., Liu, H.B., and Harrison, P.J., 2011. Environmental response to sewage treatment strategies: Hong Kong's experience in long term water quality monitoring. *Marine Pollution Bulletin* 62: 2275–2287.

Zhang, X.Y. and Adams, E.E., 1999. Prediction of near field plume characteristics using far field circulation model. *Journal of Hydraulic Engineering, ASCE* 125(3): 233–241.

8

Soil and Aquifer Management

Hillel Rubin
Technion—Israel Institute of Technology

8.1 Introduction

The objective of this chapter is to provide a survey of major concepts associated with soil and aquifer management. Within the list of references of this chapter, the interested reader may find comprehensive amounts of information about a variety of issues associated with the interaction between soil and groundwater pollution (e.g., Rubin et al. 1998). Major emphasis is given to the penetration and contamination of soils and groundwater by various types of contaminants. Also about such issues, basic concepts and approaches are represented. In some cases of particular importance, more details are given. Representation of some research work is also made in cases that concern fundamental issues of soil and groundwater contamination and remediation.

Geological formations that have the capacity to store and allow conveyance of groundwater are called "aquifers." Aquifers consist of permeable formations that may store water. In general, we identify two major types of aquifers:

1. Free surface (phreatic) aquifers (Figure 8.1)
2. Confined aquifers (Figure 8.2).

A free surface aquifer is characterized by an impermeable bottom, and water is accumulated on top of this bottom up to a level called "water table." The elevation of the water table is usually subject to seasonal and annual changes due to natural effects (precipitation) and man-made effects (pumping and/or injection). The free surface aquifer is subject to natural recharge (accretion) by rain water or snow melt percolating through the unsaturated (vadose) zone that is located on top of the aquifer.

A confined aquifer is characterized by impermeable bottom and impermeable top. The natural recharge of a confined aquifer is carried out via the vadose zone of an area where the aquifer top is permeable, namely, the confined aquifer starts as a free surface aquifer. There are various combinations of aquifers in different sites. Very often underneath the impermeable bottom of the phreatic aquifer, there are permeable layers saturated with water interbedded by layers of impermeable formations. In such cases, the series of confined aquifers is represented by the permeable layers existing underneath the phreatic aquifer. A system of phreatic aquifer and a single confined aquifer is typical of the coastal zone of Israel, as shown in Figure 8.3.

In Figure 8.3, the phreatic aquifer is called "The Coastal Plain Aquifer" and the confined aquifer is called "The Mountain Aquifer," as its recharge comes from the mountain area that is quite far from the sea shore. In cases of impermeable layers interbedded by permeable layers comprising a vertical series of confined aquifers, very often the impermeable formations allow some contacts between the different vertical aquifers. Then the aquifers are called "leaky aquifers."

The presence of groundwater in aquifers is associated with the global water cycle that starts with surface water, like oceans, seas, and lakes, which evaporates into the atmosphere and later returns to soil surface as precipitations. Part of the precipitations evaporates, another part leads to surface runoff, and

Handbook of Environmental Fluid Dynamics, Volume One, edited by Harindra Joseph Shermal Fernando. © 2013 CRC Press/Taylor & Francis Group, LLC. ISBN: 978-1-4398-1669-1.

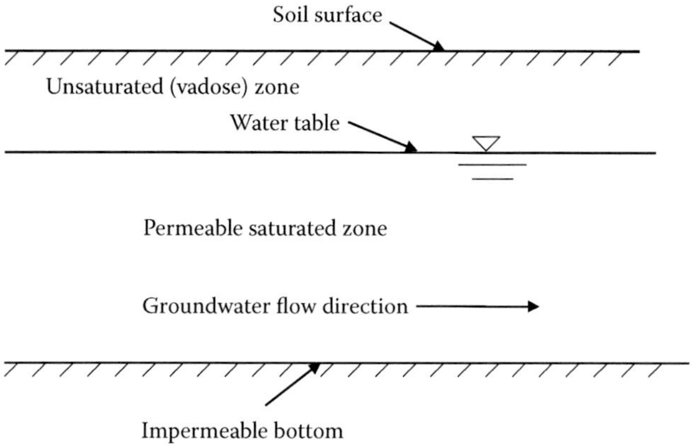

FIGURE 8.1 Schematic description of a free surface (phreatic) aquifer.

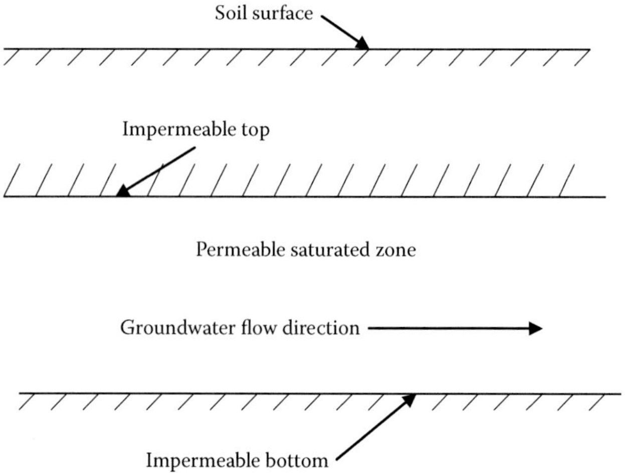

FIGURE 8.2 Schematic description of a confined aquifer.

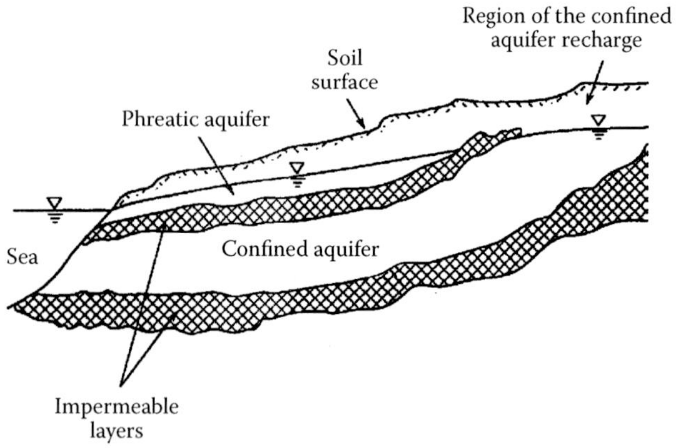

FIGURE 8.3 Schematic description of the two aquifer systems in the coastal zone of Israel.

another part infiltrates vertically through the vadose (unsaturated) zone into aquifers. In the aquifers the flow is mainly in the horizontal direction toward rivers, lakes, seas, and oceans. Proper management of aquifers is associated with soil management, as rainwater infiltrates into aquifers via the unsaturated soil, namely, the vadose zone. And also contaminants released into the environment may reach the aquifers via this zone. Further, contaminants arriving at groundwater are subject to advection and dispersion, which lead to migration and spreading of the contaminant to parts of the aquifer that are free of contaminants. Polluted groundwater arriving at surface water bodies like rivers, lakes, seas, and oceans convey contaminants into such waters. Groundwater conveying volatile contaminants release vapors of such materials into the gaseous phase of the vadose zone and later into the atmosphere. The qualitative phenomena described in this paragraph indicate the connection between groundwater pollution and potential spreading of contaminants into all components of the environment. However, most important is the intimate contact between soil and aquifer.

Groundwater represents the major source of drinking water in most parts of the world. And aquifers are major reservoirs of water resources. In arid and semi-arid areas, for many years the availability of sufficient quantities of water for the people and agricultural needs has represented a major national goal. In the first half of the twentieth century, big projects of dams have been carried out to create water reservoirs and transfer water from regions rich with water resources to regions with insufficient water resources. Such a project presently conveys water from northern parts of California to southern California. In Israel, the National Water Carrier delivers water from the northern part of the country to its southern part, which is a desert called "Negev." During summer time, all quantities of water conveyed by the National Water Carrier are used by various types of consumers. In winter time, parts of these quantities are injected into aquifers for later use during summer time.

During the last few decades of the twentieth century, scientists of different disciplines connected with hydrogeology and environmental fluid mechanics directed the authorities and the public opinion to be aware about the continuous deterioration of groundwater quality in most countries over the globe. It should be noted that the presence of minute quantities of some contaminants (micropollutants) like halogenated organic compounds and heavy metals, in groundwater may comprise a major threat to the environment, human health, and natural habitats. Very common is the negative effect on groundwater quality by the increasing chloride concentrations that originate from irrigation and enhanced by using uncontrolled treated wastewater for such purposes. Also nitrate concentrations are subject to continuous increase mainly due to fertilization.

8.2 Principles

8.2.1 Groundwater Flow

The science and engineering of groundwater hydrology started in the twentieth century due to the increasing interest and using groundwater as a reliable source of good-quality drinking water. During that time major interest of management of aquifers was concentrated in groundwater flow and efficient pumping of groundwater by systems of wells. Such management concerns the ability of the soil to infiltrate rainwater and allow natural recharge of the aquifers. This approach concerns the flow of a single phase (water) through porous media, mainly through the aquifer (saturated zone). Aquifer management of this type is also somewhat connected with the ability of various types of precipitations to infiltrate through the unsaturated (vadose) zone and increase the storage of groundwater in aquifers. In the following paragraphs we review the basic approach for calculating and analyzing single-phase flow (groundwater) through the saturated zone (aquifers).

The groundwater is subject to flow in the aquifer according to the equation called Darcy's law:

$$\mathbf{q} = \mathbf{K} \cdot \mathbf{J} \tag{8.1}$$

where
- $\mathbf{q}$ is called Darcy velocity, or specific discharge vector
- $\mathbf{K}$ is the hydraulic conductivity tensor
- $\mathbf{J}$ is the hydraulic gradient vector

The Darcy velocity is defined as the discharge per unit area of the porous medium; it is connected with the average interstitial flow velocity vector of groundwater through the pores of the porous medium $\mathbf{V}$ by

$$\mathbf{q} = \phi \mathbf{V} \tag{8.2}$$

where ϕ is the porosity of the porous medium. It is important to note here that dissolved contaminants move in the aquifer with the advection velocity $\mathbf{V}$. Equation 8.1 represents the conservation of momentum in flow through porous media. This equation is usually introduced into the principle of mass conservation in order to obtain the differential equation of single-phase flows in aquifers. This approach allows carrying out calculations of groundwater mass balances in aquifers. We will discuss this issue later within this section. The hydraulic conductivity depends on the properties of the porous medium and also on the ratio of the viscosity to the density of the fluid, namely, the fluid kinematic viscosity. Therefore, it is common to apply the permeability tensor, $\mathbf{k}$, as a property of only the porous medium in order to express the hydraulic conductivity. Such a reference is required in cases of multiphase flow, but of course it can also be applied to single-phase flows. The relationship between the hydraulic conductivity and the permeability of the porous medium is given as

$$\mathbf{K} = \frac{\rho g}{\mu} \mathbf{k} \tag{8.3}$$

where
- ρ is the density of the fluid
- g is the gravitational acceleration
- μ is the fluid viscosity

The permeability and the hydraulic conductivity are usually considered as symmetric second-rank tensors (e.g., Bear 1972, p. 137). However, in cases of isotropic porous media, they can be represented by scalar quantities. The hydraulic gradient $\mathbf{J}$ is the negative gradient of the piezometric head, h, which is defined as

$$h = \frac{P}{\rho g} + z \tag{8.4}$$

where
- P is the pressure
- z is an elevation above an arbitrary datum

Therefore, Darcy's law for groundwater flow through an isotropic porous medium is given by

$$\mathbf{q} = -K \nabla h \tag{8.5}$$

where K is the hydraulic conductivity (a scalar quantity). If the hydraulic conductivity and the porosity are approximately constant in the considered flow domain, then it is possible to assume that the specific discharge and the average interstitial velocity of the groundwater originate from potential functions, Φ_q and Φ_v, respectively, which are defined as

$$\mathbf{q} = \nabla \Phi_q \quad \text{where} \quad \Phi_q = -Kh \tag{8.6a}$$

$$\mathbf{V} = \nabla \Phi_v \quad \text{where} \quad \Phi_v = -\frac{Kh}{\phi} \tag{8.6b}$$

These expressions allow using the potential flow theory for analyzing complicated phenomena of flow through porous media (e.g., Polubarinova-Kochina 1962). Equations 8.1 through 8.6 are useful for calculating characteristics of single-phase flows in porous media. As shown later, flow of a single phase (groundwater) in the aquifer is usually associated with transport of dissolved species, whose effects on groundwater quality may be detrimental. On the other hand, the presence of non-aqueous-phase-liquids (NAPLs), like petroleum distillates and chlorinated hydrocarbons, in aquifers may be associated with multiphase flow, which leads to transport of dissolved organic contaminants whose concentrations in parts per billion (ppb) may be very risky to human health and the environment. Quantifying issues of dissolved contaminants in groundwater is discussed later in this chapter.

Reference to Darcy's velocity $\mathbf{q}$ is usually sufficient for calculating water production from aquifers. On the other hand, as the rate of dissolved species advection in aquifers is equal to $\mathbf{V}$, in cases dealing with contaminant transport in groundwater the reference to the average interstitial velocity is needed, as shown later in this chapter.

Due to the limited thickness of common aquifers and the much larger horizontal extent of interest in the groundwater flow for water extraction from the porous medium, usually Dupuit approximation is applied to calculating flow in aquifers. According to Dupuit approximation the aquifer flow is in the

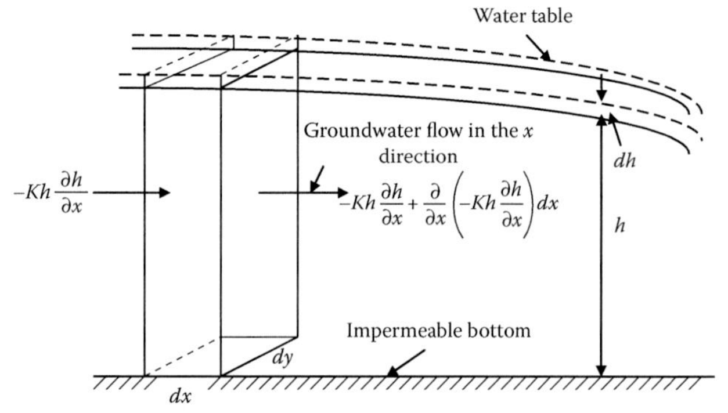

FIGURE 8.4 Applying Dupuit's approximation to formulate the flow in a phreatic aquifer.

horizontal direction, and lines of constant piezometric head are approximated as vertical lines. If the impermeable bottom of the phreatic aquifer is almost horizontal, then it is convenient to consider it as the datum for calculating the piezometric head. Under such conditions, the piezometric head is represented by the elevation of the water table above the impermeable bottom of the aquifer, as shown in Figure 8.4.

Figure 8.4 shows how Dupuit approximation is introduced into the calculation of the water mass balance in an elementary volume of the aquifer. The height of this elementary volume is h and its horizontal cross-sectional area is represented by the product of dx by dy. The water flows in the x and y directions. However, in order to simplify the schematic description of Figure 8.4, only arrows and terms referring to the flow in the x direction are shown. Dupuit approximation is applied to the momentum conservation and is later introduced into the water mass balance calculation to develop the differential equation of flow through aquifers. In the schematic description of Figure 8.4, the water table rises by the amount dh during an elementary time interval dt. Therefore, in this figure, the water mass balance equation subject to Dupuit approximation is given by

$$\frac{\partial}{\partial x}\left(T\frac{\partial h}{\partial x}\right)+\frac{\partial}{\partial y}\left(T\frac{\partial h}{\partial y}\right)=\phi\frac{\partial h}{\partial t} \quad \text{where } T = Kh \tag{8.7}$$

where T is called the transmissivity of the aquifer. Whereas determining details of the hydraulic conductivity (and permeability) of the porous medium require complicated sampling of the geological formation, determining the transmissivity of the aquifer requires the analysis of a series of pumping tests (e.g., Fetter 1980). Such tests are based on applying Dupuit approximation in cylindrical coordinates to analyze the relationships between the pumping discharge and the water table drawdown under the unsteady state flow around the well, while ignoring the natural flow in the aquifer. Pumping tests are usually carried out prior to the final installation and operation of the water supply well. It should be noted that Dupuit approximation has led to the second-order differential equation (8.7) that refers to a domain described by the horizontal coordinates x and y. It means that for analyzing water balances in aquifer flows using 2D numerical models may provide proper results. Further, it is possible to consider and add terms to Equation 8.7 without changing its order, in order to refer to natural recharge (accretion), pumping, injecting, etc. Usually, regional balances water quantities are required for the evaluation of proper water supply to the regional water consumers. Such calculations should refer to all sources of regional water, including groundwater natural and possibly man-made recharge. Due to the storage of the aquifers some fluctuations of the difference between water recharge and consumptions are possible, but in the long run the groundwater consumption cannot be larger than the natural supply of water into the aquifers. Equation 8.7 is also used for calculating water balances in confined aquifers. However, for such aquifers, instead of the porosity, which appears at the right-hand-side (RHS) of Equation 8.7, appears another parameter that expresses the aquifer compressibility and is called storativity. This parameter is smaller than the porosity in several orders of magnitude. As stated earlier, using Dupuit approximation is very common for calculating water mass balances in aquifers. Equation 8.7 provides the basis for calculating aquifer management and optimizing pumping rates of groundwater via water supply wells. Such calculations are usually associated with reference to various types of constraints that should be considered during pumping groundwater by the regional consumers.

However, in cases of managing water quality and environmental issues of groundwater resources preservation, often the 2D differential equation given by Equation 8.7 cannot be applied and 3D modeling is needed. Then applying Darcy's law for calculating the flow through the porous medium by using Equation 8.5 is needed. Further, for environmental issues of multiphase flow, as shown later, Darcy's law should be subject to some changes originating from the presence of various fluid phases in the domain.

In the 1960s, the issue of dual-porosity porous media started to be of interest. Such a formation is usually characterized by porous blocks (of the rock matrix) embedding fractures. The presence of the fracture network leads to high bulk permeability of the formation, but most groundwater quantities are stored within the porous matrix blocks (Barenblatt et al. 1960). The analysis of pumping tests of wells drilled into dual-porosity aquifer is therefore usually different from the analysis of pumping tests applied to a homogeneous aquifer.

8.2.2 Transport of Dissolved Species in Groundwater

This section concerns issues associated with contamination by small quantities of contaminants released into the aquifer that are completely dissolved within the water phase. The extensive development of the human society, increasing number of population over the globe, and thereby releasing contaminants into the environment have led to increasing interest in phenomena associated with the fate of contaminants that are introduced into the vadose zone and aquifers. The equation of contaminant transport in aquifers is obtained by considering the contaminant

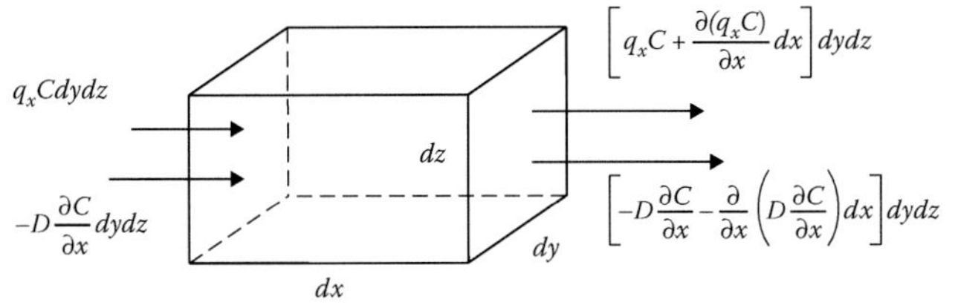

FIGURE 8.5 Schematic description of contaminant transport in an elementary control volume of the porous medium.

concentration is very low and does not affect the water density and flow in the aquifer. Such a contaminant is often called a neutrally buoyant tracer. We refer to an elementary volume of the porous medium, as shown in Figure 8.5. Again, in order to simplify the schematic description of Figure 8.5, only transport components in the x direction are shown in this figure. However, the contaminant is assumed to be advected in both horizontal directions and the vertical directions.

Figure 8.5 shows the advection flux q_xC, where q_x is the component of Darcy's velocity in the x direction and C is the contaminant concentration. It also shows the diffusive flux $-\phi D\partial C/\partial x$, where D is the coefficient of diffusion.

Considering temporal changes of the contaminant concentration in the elementary volume of Figure 8.5 and dividing the expression by the porosity, we obtain the following mass balance equation for the contaminant:

$$\frac{\partial C}{\partial t}+\frac{\partial(uC)}{\partial x}+\frac{\partial(vC)}{\partial y}+\frac{\partial(wC)}{\partial z}$$
$$=\frac{\partial}{\partial x}\left(D\frac{\partial C}{\partial x}\right)+\frac{\partial}{\partial y}\left(D\frac{\partial C}{\partial y}\right)+\frac{\partial}{\partial z}\left(D\frac{\partial C}{\partial z}\right) \tag{8.8}$$

where u, v, and w are components of the average interstitial flow velocity through the porous medium in the x, y, and z direction, respectively.

The first term of Equation 8.8 refers to temporal changes of the contaminant concentration. The following three terms represent the contaminant advection, and the RHS of Equation 8.8 represents the effect of contaminant diffusion as a Fickian process. Figure 8.5 and Equation 8.8 indicate that in many cases calculations of contaminant transport require 3D modeling, as contaminants usually penetrate into the aquifer through its top, and sometimes through the leaky bottom of the aquifer, like brines and saltwater intrusion. It is possible to consider the incompressible flow of the water in order to simplify the advection terms of Equation 8.8. Equation 8.8 can be applied provided Peclet number of the flow is very low, smaller than 10^{-1} (Bear 1972, p. 607). Peclet number is defined as the product of the interstitial flow velocity multiplied by the grain particle of the porous medium and divided by the molecular diffusion coefficient. As the molecular diffusion coefficient for most contaminants is of the order of magnitude of 10^{-9} or even less, Peclet number in most cases of groundwater flows is considerably larger than 10^{-1}. Under such conditions, the quantity D of Figure 8.5 and Equation 8.8 should be replaced by a tensor quantity, $\mathbf{D}_h$, called "hydrodynamic dispersion tensor." The hydrodynamic dispersion tensor expresses the effect of the growth of contaminant plumes due to molecular diffusion and tortuosity of the tracks of the fluid flowing through the porous medium. If Peclet number is larger than 10, then the effect of molecular diffusion on the magnitude of the hydrodynamic dispersion tensor components is very minor and can be neglected (Bear 1972, p. 607). If under such conditions, the porous medium is isotropic, then the principal directions of the hydrodynamic dispersion tensor are parallel and perpendicular to the velocity vector. In cases of high Peclet numbers, which are common in groundwater flows, the principal components of the hydrodynamic dispersion tensor for isotropic porous media are represented as

$$D_{h_L}=a_LV;\quad D_{h_T}=a_TV \tag{8.9}$$

where

D_{h_L} and D_{h_T} are the longitudinal and transverse principal components of the hydrodynamic dispersion tensor, respectively

a_L and a_T are the longitudinal and transverse dispersivity, respectively

The ratio of the longitudinal to transverse dispersivity is between 10 and 30. As Peclet number in groundwater flow is rarely smaller than 10, instead of using Equation 8.8 for calculating contaminant transport in porous media, like aquifers, we should apply the following differential equation:

$$\frac{\partial C}{\partial t}+\mathbf{V}\cdot\nabla C=\nabla\cdot(\mathbf{D}_h\cdot\nabla C) \tag{8.10}$$

This expression is often called "the diffusion–advection equation." In this equation decay of the contaminant, interphase mass transfer, and sorption–desorption are not considered, but such processes can easily be introduced into Equation 8.10 by additional proper terms. Further, a basic differential equation similar to Equation 8.10, in which the value of C expresses the biomass concentration instead of representing the contaminant concentration, can be used for simulating the transport of biomass in groundwater. However, in the differential equation of biomass conservation, terms referring to natural growth, decay of substrates, supply of nutrients, and sorption–desorption should be added to simulating biomass developing in porous media. Numerical simulations based on using the flow equation, contaminant concentration, and biomass developing in the porous media is of major importance for quantitative calculations of environmental aspects associated with groundwater quality. However, in such cases additional information about the biochemical processes should be taken into account by proper models (e.g., Fennell and Gossett 1998). The peculiar characteristics of organic compounds arriving at the aquifer as NAPLs which may comprise the substrate for the biomass development that should be taken into account in such simulations are as follows: (1) NAPLs are liquids of very low miscibility in water; (2) after their arrival at the groundwater, they may represent chronic sources of dissolved contaminant into the flowing groundwater; and (3) some of them are poisonous and also carcinogenic materials. Therefore, although the effect of NAPL penetration into groundwater has often minor effect on the reliability of using the basic equations of single-phase hydrogeology shown in this section, dealing with NAPL contaminants requires additional means of quantification relevant to the typical characteristics of these contaminants, as described in following sections of this chapter.

Contaminant transport in dual-porosity formations has unique properties. Birkhölzer et al. (1993) have suggested to classify dual-porosity formations according to the value of the mobility number N_M that represents the ratio of the aquifer flow rate carried out through the porous blocks to the aquifer discharge carried out through the fracture network. In cases of significant permeability of the porous blocks, like sand stone, they suggest the term "fractured permeable aquifer (FPA)," or "fractured permeable formation (FPF)," in which the range of the mobility number is between 0.1 and 25. If the mobility number is higher than 25, then the effect of the fractures on contaminant transport is very minor. In the fractured permeable formations, the mixing between the permeable block flow and the fracture flow, as well as mixing in fracture intersections, is a major mechanism leading to contaminant dispersion in the domain. Rubin and Buddemeier (1996) have shown that in permeable fractured aquifers the transverse dispersivity may be larger than the longitudinal dispersivity in cases of fracture segment orientation close to perpendicular to the aquifer longitudinal direction. Mixing in fracture intersections is always an important mechanism of contaminant dispersion in fractured formations. However, in cases of low values of the mobility number also the effect of contaminant diffusion from the fracture flow into the surrounding porous blocks should be taken into account as an important mechanism of the contaminant transport in the fractured porous formation (e.g., Tang et al. 1981). A major problem associated with simulating contaminant transport in fractured formation originates from difficulties in mapping the fracture network and the fracture aperture. Even statistical presentation of such parameters may incorporate errors of orders of magnitude. Tsang and Tsang (1987) have shown that in fractured rock matrix most quantities of groundwater flow through a small number of fractures of the largest aperture. This phenomenon has a significant effect on the transport of contaminants in groundwater flowing through fractured formations.

8.2.3 Multiphase Fluids in Porous Media

The presence of several fluid phases, like water, NAPLs, and air within the pores of the porous medium is quantified by applying the principles of mass and momentum conservation for each one of the fluid phases. Abriola (1989) and Miller et al. (1998) have provided comprehensive reviews of different approaches by which the multiphase flow in porous media can be quantified. The constituent mass balance equation in a phase α is

$$\frac{\partial}{\partial t}(\varepsilon_\alpha \rho^\alpha \omega_i^\alpha) + \nabla\cdot(\rho^\alpha \omega_i^\alpha \mathbf{q}^\alpha) - \nabla\cdot(\varepsilon_\alpha \rho^\alpha \mathbf{D}_{h_i}^\alpha \cdot \nabla\omega_i^\alpha) = \sum_{\beta\neq\alpha} E_i^{\alpha\beta} \tag{8.11}$$

where

ε_α is the volume fraction of the α phase
ω_i^α is the mass fraction of component i in phase α
$\mathbf{D}_{h_i}^\alpha$ is the hydrodynamic dispersion tensor
$\mathbf{q}^\alpha$ is the Darcy velocity (also called specific discharge vector) of the α phase, namely, the volumetric discharge per unit area of the porous medium
ρ^α is the density of the α phase
$E_i^{\alpha\beta}$ is the exchange of mass of component i between α and β phases

Considering the flow in the aquifer, Equation 8.11 refers to components $i = 1, \ldots, n_c$ where n_c is the total number of components, and α and β can represent the organic (NAPL), water, and solid phases. Miller et al. (1998) incorporate with Equation 8.11 also terms representing general reactions and a solute source. Basically, Equation 8.11 collapses to Equation 8.10 if a single fluid phase is present in the domain. It should be noted that Equation 8.11 may be relevant to all possible four phases present in the domain, namely, water, NAPL, gas, and solid. However, the major interest in multiphase flow in aquifers is concentrated with the water phase and the NAPL. Further, as shown later, in many (and probably most) cases of environmental interest the NAPL is almost stagnant, and groundwater is the only fluid phase subject to flow in the aquifer.

Summing up Equation 8.11 over all species within each phase yields

$$\frac{\partial}{\partial t}(\varepsilon_\alpha \rho^\alpha) + \nabla\cdot(\rho^\alpha \mathbf{q}^\alpha) = E^\alpha \tag{8.12}$$

where E^α is the sum of the RHS of Equation 8.11 over all species. Substituting Equation 8.12 into Equation 8.11 yields

$$\varepsilon_\alpha \rho^\alpha \frac{\partial \omega_i^\alpha}{\partial t} + \rho^\alpha \mathbf{q}^\alpha \nabla\cdot\omega_i^\alpha - \nabla\cdot\left(\varepsilon_\alpha \rho^\alpha \mathbf{D}_{h_i}^\alpha \cdot \nabla\omega_i^\alpha\right) = \sum_{\beta\neq\alpha} E_i^{\alpha\beta} - \omega_i^\alpha E^\alpha \tag{8.13}$$

This equation is subject to the following constraints originating from mass conservation:

$$\sum_i \omega_i^\alpha = 1; \quad \sum_\alpha \varepsilon_\alpha = 1 \tag{8.14}$$

A separate momentum balance equation is usually not included in the balance equations of formulating multiphase systems.

It is common in the literature dealing with environmental issues to substitute a multiphase form of Darcy's law into the mass balance equation given by Equation 8.13. This equation collapses to the diffusion–advection equation in cases of single fluid phase present within the aquifer. In most cases of multiphase flow in porous media, the energy balance is not considered. Considering the energy balance is justified only if temperature changes in the domain cannot be neglected. However, in general cases of interest in multiphase flow in aquifers, this issue is of very minor importance.

Equations 8.11 through 8.14 refer to the very general case where significant rates of mass transfer of all species take place among all phases. There are several types of common assumptions that simplify these general equations (Miller et al. 1998) to provide basic means for practical modeling of multiphase flow through porous media: (1) the solid phase is immobile, (2) the solid phase is inert, (3) the relevant species in the system is a smaller subset of lumped species, and (4) local chemical equilibrium exists among the phases. The assumption of immobile solid phase ignores changes of the porous medium porosity originating from the elasticity of the porous matrix and changes in the fluid phase pressure. This assumption is very often justified for calculating aquifer contamination and remediation, because the time period of aquifer remediation is much longer than the time period of unsteady state flow. Therefore, average flow conditions in the aquifer may very often provide good estimates for calculating the aquifer remediation. However, the general case of unsteady state flow of groundwater in aquifers (mainly in confined aquifers) is affected by the aquifer elasticity. The assumption of inert solid phase considerably simplifies the calculations. We should also consider that organic compounds are subject to sorption–desorption onto the solid phase. The number of relevant species in the domain depends on the type of the source of contaminants. Usually, an industrial source uses a single type of NAPL that may contaminate the aquifer. In Tel-Aviv area, for several years it was considered that a single factory contaminated the groundwater with various types of chlorinated hydrocarbons, and this idea has been accepted as the opinion of the Water Commission. However, that factory claimed it had been using only trichloroethylene (TCE) in its production line. It took several years for the Water Commission to identify many more sources of different types of chlorinated hydrocarbons in Tel-Aviv area. Therefore, in most cases, for a limited area of groundwater contamination there are a few important species whose spreading in the groundwater should be followed. About groundwater contamination by different types of fuels (like gasoline, kerosene, and diesel oil), the most risky original species are the BTEX (benzene, toluene, ethylbenzen, xylenes) compounds and possibly agents added to increase the octane of gasoline.

The information represented earlier indicates that Equations 8.11 through 8.14 can be simplified for cases of limited numbers of species and mass transfer only between the NAPL and the water phase in the domain of interest. In typical cases of NAPL released into the environment, it may arrive at the aquifer and subject to entrapment within the porous medium. Then the NAPL is an immobile liquid phase and the groundwater is the mobile phase. Under such conditions, applying Equation 8.12 to the NAPL and the water phase yields

$$\phi\frac{\partial}{\partial t}(s_n\rho^n) = E^{nw} \quad \text{where} \quad E^{nw} = \sum_i E_i^{nw} \tag{8.15a}$$

$$\phi\frac{\partial}{\partial t}(s_w\rho^w) + \nabla\cdot(\rho^w\mathbf{q}^w) = E^{wn} \tag{8.15b}$$

where

s_n and s_w are the saturations of NAPL and water, respectively
ϕ is the porosity of the matrix
ρ^n and ρ^w are the density of NAPL and water, respectively
$\mathbf{q}^w$ is the specific discharge of the water phase
$E^{wn} = -E^{nw}$ is the mass exchange between the NAPL and the water phase

Equation 8.15 refers to mass exchange between the two fluid phases present in the aquifer. As the wettability of NAPL is lower than that of the water, only mass exchange between the solid phase and the water phase is feasible (namely, sorption–desorption of dissolved contaminants). In other words, only organic material dissolved in the water phase can be subject to sorption by the solid matrix.

If NAPL is entrapped within the aquifer, then we should consider only the groundwater flow. Under such conditions, the specific discharge vector $\mathbf{q}^w = \mathbf{q}$ is expressed by applying the modified multiphase Darcy's law, which states

$$\mathbf{q} = -\frac{\mathbf{k}k_{rw}}{\mu_w}\cdot(\nabla P_w + g\rho^w\nabla z) \tag{8.16}$$

Again, this expression collapses to the single phase Darcy's law given by Equation 8.5 if only water phase is present in the aquifer. Introducing Equation 8.16 into Equation 8.15b yields

$$\phi\frac{\partial}{\partial t}(s_w\rho^w) - \nabla\cdot\left[\rho^w\frac{\mathbf{k}k_{rw}}{\mu_w}\cdot(\nabla P_w + g\rho^w\nabla z)\right] = E^{wn} \tag{8.17}$$

In Equations 8.15 through 8.17, the water phase is represented by subscripts and superscripts w, $\mathbf{k}$ is the permeability tensor, k_{rw} is the relative permeability of the medium to the water phase, μ_w is the water viscosity, P_w is the water pressure, and g is the gravitational acceleration. In Equation 8.16 the direct influence of interphase partitioning on the water phase flow has been neglected, as its effect is usually much smaller than the effect of the pressure gradient. Also because of the very low solubility of the organic material within the aqueous phase, usually ρ^w is not affected by the dissolved organic compounds. Various approaches have been applied to provide constitutive equations for the relative permeability for each fluid phase existing in multiphase systems (e.g., Parker et al. 1987).

Because only the water phase is in contact with the solid phase, the solid phase balance equation for the i-constituent is obtained from Equation 8.13 as

$$(1-\phi)\frac{\partial}{\partial t}(\rho^s\omega_i^s) = E_i^{sw} \quad (8.18)$$

Here superscript s refers to the solid phase. Equation 8.18 neglects the bulk movement of the solid phase and surface diffusion of the solute i. By incorporating Equation 8.18 with Equation 8.13, the following equation for the dissolved i-constituent transport is obtained:

$$\phi s_w\frac{\partial C_i}{\partial t} + \rho_b\frac{\partial S_i}{\partial t} + \mathbf{q}^w\cdot\nabla C_i - \nabla\cdot(\phi s_w\mathbf{D}_{h_i}\cdot\nabla C_i) = E_i^{wn} \quad (8.19)$$

where
- C_i is the concentration of the i-constituent in the water phase ($C_i = \rho^i\omega_i$)
- S_i is the sorbed phase concentration of the i-constituent ($S_i = \omega_i^s$)
- ρ_b is bulk density ($\rho_b = (1 - n)\rho^s$)
- E_i^{wn} is the mass exchange of the i-constituent between the NAPL and the water phase

This term is usually expressed by

$$E_i^{wn} = k_f(C_{eq_i} - C_i) \quad (8.20)$$

where
- k_f is the lumped mass transfer coefficient
- C_{eq_i} is the equilibrium concentration of the i-constituent

It is common to apply the Sherwood number to study the effect of the domain and the entrapped NAPL properties on the value of k_f. The definition of the Sherwood number is

$$\text{Sh} = \frac{k_f d^2}{D} \quad (8.21)$$

where
- Sh is the Sherwood number
- d is the characteristic grain diameter
- D is the free liquid diffusivity

Various experimental studies (e.g., Saba and Illangasekare 2000; Nambi and Powers 2003) have shown that the Sherwood number depends on the entrapped NAPL saturation and the Reynolds number. The Sherwood number also depends on some other quantities of the domain, which are not affected by the entrapped NAPL dissolution. According to those studies (see, e.g., Mayer and Miller 1996), it is possible to represent the value of Sh by

$$\text{Sh} = \beta_0\,\text{Re}^{\beta_1}\,s_n^{\beta_2} \quad (8.22)$$

Here, β_i (i = 0 − 2) are coefficients, and Re is the Reynolds number based on the interstitial flow velocity. The coefficient β_0 is a dimensionless quantity that depends on the porous medium grain size and size of the region contaminated by entrapped NAPL. According to the review made by Rubin et al. (2008b), the value of β_1 is close to zero and the value of β_2 is close to 1. Equations 8.20 through 8.22 are constitutive equations applied to calculating rates of interphase transfer in NAPL–water fluid system.

The dissolved contaminant sorption–desorption is usually tested by using Freundlich, Langmuir, or Frumkin isotherms. If surfactant agent is introduced into the aquifer, its sorption should also be taken into account, in order to assure proper remediation of the aquifer.

As indicated earlier, besides the basic equations originating from conservation principles, constitutive equations are needed. Equations 8.21 and 8.22 are constitutive equations that may provide proper values for the lumped mass transfer coefficient. Another set of constitutive equations refers to the effect of the fluid phases (water, NAPL, and/or gas) on the permeability of each fluid phase.

8.2.4 Characteristics of Multiphase Flow in Aquifers

As indicated earlier, multiphase fluid flows in aquifer may usually result from the release of NAPLs into the environment. We identify two types of NAPLs: (1) light NAPLs (LNAPLs), whose density is lower than that of water; and (2) dense NAPLs (DNAPLs), whose density is higher than that of water. LNAPLs are fossil fuels like gasoline, kerosene, and diesel oil; DNAPLs are various types of halogenated hydrocarbons. Generally, soil and aquifer contamination by NAPLs incorporates processes of percolation, entrapment, interphase mass transfer, advection, and dispersion of the dissolved organic compounds. Percolation of NAPL takes place in the vadose zone, where some quantities of NAPL are subject to entrapment and later evaporate and/or gradually dissolved by percolating rainwater. Due to their low density, LNAPLs cannot penetrate into aquifers, unless the water table is rising. However, as discussed later, seasonal and annual changes in the elevation of the water table are very common. Due to their high density, DNAPLs can penetrate into aquifers and subject to entrapment in the pores of the porous medium or accumulation as pools on top of impermeable lenses of silt and clay or at the impermeable bottom of the aquifer.

NAPL entrapped in the aquifer, LNAPL lenses floating on top of the water table, and DNAPL pools within the aquifer are usually chronic sources of dissolved organic compounds, whose plumes contaminate the groundwater for possibly large distances downstream of the site in which NAPL has been released into the environment. Ganglia of NAPL are entrapped when interfacial forces are sufficiently strong to overcome viscous and gravitational forces. It should be noted that some quantities of the volatile organic compounds (VOCs) incorporated with the dissolved organic plume, which migrates with the groundwater flow, may transfer into the gaseous phase of the vadose zone. Therefore, during the horizontal flow of the groundwater contaminated by the dissolved organic compounds of DNAPLs, it may release such compounds and contaminate the gaseous phase of the vadose zone in places that are located at long distances downstream of the sites of releasing DNAPL into the environment.

Due to the extensive use of LNAPLs (fuels like gasoline, kerosene, and diesel oil) and intensive activities of crude oil refineries, many reports concern field observations of LNAPL layers (or LNAPL lenses) floating on top of the water table (e.g., Rubin et al. 2004). However, in cases of deep water table and small quantities of released NAPL, NAPL plume is totally entrapped in the vadose zone. Then rainwater infiltrating through the vadose zone gradually dissolves the entrapped NAPL and leads to penetration of the dissolved organic compounds into the groundwater. In most cases, DNAPL quantities released into the environment are not large, and DNAPL plume is entrapped in the vadose zone. Later, the quantities of DNAPL entrapped in the vadose zone are subject to dissolution in infiltrating rainwater and gradual evaporation and release of high-risk gases (poisonous and sometimes carcinogenic) into the gaseous environment of the vadose zone. These gaseous compounds may move horizontally in the vadose zone and also diffuse upward toward the soil surface. As the quantities of DNAPL entrapped in the vadose zone are also subject to dissolution by rainwater percolating through the vadose zone, the risky organic compounds may contaminate the flowing groundwater. Entrapment of DNAPL in aquifers may result from uncommon accidental events or long time periods of improper handling of such liquids. Therefore, in many cases, phreatic aquifers located underneath DNAPL-releasing facilities include only the dissolved organic compounds. However, minute concentrations of chlorinated hydrocarbons make groundwater inappropriate for drinking. Contact of the human skin with such water can also be of great risk. The gases released from such water may affect natural habitats and cause significant risks to residential areas.

The elevation of the water table is subject to natural and man-made seasonal and annual fluctuations. If LNAPL lens exists on top of the water table, then such fluctuations lead to LNAPL entrapment in top layers of the aquifer. In the extreme case reported by Rubin et al. (2008b), an originally LNAPL layer 70 cm thick and horizontal extent of $300 \times 150\,\text{m}^2$ was subject to almost complete entrapment in part of the Coastal Plain Aquifer (CPA) of Israel aquifer. In this case, during a winter season rich with rain, the water table raised by 6 m has led to almost complete entrapment of LNAPL lens, which originally floated on top of the water table. Therefore, the top layers of that aquifer are presently contaminated with entrapped LNAPL. It should be noted that the characteristics of entrapment of LNAPL due to rising groundwater may be different from the characteristics of entrapment of DNAPL penetrating into the aquifer because of wettability differences of the three fluid phases, water, NAPL, and air, as discussed in the following.

Among the three phases, water, NAPL, and air, water has the highest wettability and air is of the lowest one. In the vadose zone, entrapment of LNAPL and DNAPL are similar. They are entrapped in pores that are not occupied by the residual water saturation. If LNAPL lens is created on top of the water table, then fluctuations of the water table elevation may bring to entrapment of LNAPL underneath the water table within pores of different sizes. The arrival of DNAPL at the water table is followed by its penetration into the water-saturated zone. However, as DNAPL wettability is lower than that of the water, it tends to be entrapped mainly in the largest pores, whereas in the vadose zone largest pores are occupied by air. Differences in wettability are combined with effects of the porous medium heterogeneity; both avoid uniform distribution of the entrapped NAPL ganglia in the porous medium. However, as stated earlier, DNAPL entrapment in aquifers is not a common event. Differences in wettability of the three phases may lead to important physical features of NAPL entrapment in heterogeneous porous media, like fractured rocks, as discussed later.

The organic compounds comprising NAPL may be adsorbed to the porous medium solid particles and mainly to clay minerals that are present within the matrix. The adsorption–desorption time scales are very long. Therefore, the soil and aquifer contamination is only partially represented by the quantities of NAPL subject to entrapment originating from surface tension effects. Remediation of the soil may often require elimination of the entrapped NAPL, sorbed contaminants, and treatment of the groundwater.

Heterogeneity of the porous medium has a major effect on the distribution of the entrapped NAPL in the porous medium. In the vadose zone, the downward flow of NAPL may be faster in the biggest pores; but later as NAPL wettability is larger than that of the air, NAPL migrates into the finer pores, allowing air to enter the bigger size pores. If the porous medium is a fractured rock, then the fractures may represent the bigger size pores of the heterogeneous porous medium. If the rock matrix is tight, then the organic compounds forming NAPL accumulated in the fracture network of the vadose zone may diffuse into the rock matrix, which often includes high saturation of water. Rainwater infiltrating through the NAPL plume entrapped within the vadose zone may flow through the medium large size pores, but later it migrates into the smallest size pores due to the highest wettability of the water phase. In cases of LNAPL lens floating on top of the water table and/or parts of the vadose zone contaminated by entrapped NAPL that are later flooded by rising water table, NAPL may be subject to entrapment within the fine pores of the top layers of the aquifer, leading to a chronic source of dissolved contaminants to the flowing groundwater (Rubin et al. 2008b). As the wettability of DNAPL is lower than that of water, following the event of DNAPL penetrating into a fractured aquifer, DNAPL may migrate downward only in the fractures in which it may be subject to molecular diffusion into the fine pores of the water-saturated matrix porous blocks (Parker et al. 1994). Therefore, during the downward migration of DNAPL plume through the fracture network of the fractured aquifer, it loses mass of organic compounds that are dissolved in the water phase which saturates the matrix blocks.

8.3 Methods of Analysis

8.3.1 Analytical Methods

Analytical methods used for calculating groundwater flow are based on using the potential flow theory. Polubarinova-Kochina (1962) has described many applications of the potential flow theory for calculating and analyzing flow phenomena in groundwater. The 2D potential flow model that is commonly applied to evaluate the groundwater flow

$$W = \Phi + i\Psi \tag{8.23}$$

where
- $i = \sqrt{-1}$
- W, Φ, and Ψ are the complex potential, potential, and stream functions, respectively

As shown by Equation 8.6

$$\Phi = -\frac{K}{\phi}h \tag{8.24}$$

where
- K is the hydraulic conductivity
- h is the piezometric head
- ϕ is the porosity

The flow velocity vector, **V**, comprises the gradient of the potential function, as

$$\mathbf{V} = \nabla\Phi \tag{8.25}$$

By applying proper expressions to the potential function Φ, many specific practical cases of groundwater flow can be calculated, quantified, and analyzed. Such analytical models can also be useful for optimizing pumping of groundwater.

In cases of significant change of the aquifer thickness or curvature of the streamlines of the groundwater flow, then sometimes Dupuit approximation can be useful; otherwise 3D potential flow methods can still be applied, provided the soil permeability is presumably constant.

According to Equation 8.13, the diffusion–advection equation that simulates the transport of dissolved contaminants in an aquifer contaminated by entrapped NAPL is given by

$$\frac{\partial(\phi s_w C)}{\partial t} + \nabla\cdot(\phi s_w \mathbf{V}C) = \nabla\cdot(\phi s_w \mathbf{D}\cdot\nabla C) + k_f(C_s - C) \tag{8.26}$$

where
- C is solute concentration
- C_s is the solute equilibrium concentration
- s_w is the water saturation
- t is the time
- **D** is the dispersion tensor
- k_f is the lumped mass transfer coefficient

In cases of small quantities of entrapped NAPL (i.e., $s_n \approx 0.1$), we may assume

$$s_w \approx 1; \quad \Rightarrow \phi s_w \approx \phi \tag{8.27}$$

Introducing this expression into Equation 8.26 and considering the conservation of the groundwater mass, the following expression is obtained:

$$\frac{\partial C}{\partial t} + \mathbf{V}\cdot\nabla C = \nabla\cdot(\mathbf{D}\cdot\nabla C) + \frac{k_f}{\phi}(C_s - C) \tag{8.28}$$

The analytical presentation of the groundwater flow by the potential flow solution can be used for solving Equation 8.27. In some cases, using curvilinear coordinates Φ and Ψ may lead to analytical solutions of Equations 8.25 and 8.27. However, in complicated cases of 3D flows, calculation of the groundwater flow and contaminant transport requires the employment of numerical codes.

8.3.2 Numerical Modeling

Since the last half of the twentieth century, the computer age has its impact on our capability to carry out complicated calculations and numerical simulations of flow and transport in aquifers. For such purposes, single-phase and multiphase flow and transport models have been developed. Within several decades, the capability of simulating transport phenomena in aquifers became only limited to the available proper information about the basic parameters needed for carrying out reliable numerical simulations. Unfortunately, gathering information about such parameters requires much effort and is not always economically justified. However, many studies concerning the development of finite difference, finite element, and other types of numerical models able to simulate single and multiphase flows in porous media have been presented in the scientific literature.

Single-phase groundwater flow modeling is usually carried out by applying the model MODFLOW, which has been developed and improved during many years by the United State Geohydrological Service (USGS). This numerical model and related codes are described in the website: http://water.usgs.gov/nrp/gwsoftware/modflow.html

Advanced versions of this code, which is of public domain, have been connected to GIS and graphical codes for impressive graphical presentations of the numerical simulation results. USGS is probably one of the largest producers of numerical codes applicable to flows in aquifers. The list of different codes of groundwater flow developed by USGS is given in the website: http://water.usgs.gov/software/lists/ground_water/

For contaminant transport the model MT3D is extensively used. This code is described in various websites, like: http://hydro.geo.ua.edu/mt3d/

However, for cases of multiphase flow the American Environmental Protection Agency (EPA), various universities and private firms have developed codes addressing particular issues of multiphase flow in porous media. Some websites that provide information and allow downloading of such codes are

http://www.scisoftware.com/environmental_software/
http://www.argusint.com
http://www.ehsfreeware.com/gwqclean.htm
http://www.waterloohydrogeologic.com/software/software_main.htm

Basically, nowadays the major limit of carrying out exact numerical simulation originates from lack of knowledge about boundary conditions and parameters that should be introduced into the numerical calculations. Therefore, every numerical simulation project should incorporate calibration and verification efforts. Such efforts should be based on comparing some results obtained by the numerical simulation with data gathered from the aquifer. If proper calibration and verification cannot be made, then possibly some approximate evaluation methods may provide information, whose reliability is sometimes not inferior to that of the complicated numerical simulation results. However, it should be noted that approximate methods should mainly be applied to get basic information about phenomena that may take part in the aquifer. Therefore, applying approximate methods is usually justified prior to carrying out advanced numerical simulations. Such approaches can also be useful whenever numerical simulations cannot provide proper and accurate information about phenomena taking place in the vadose zone and in the aquifer. In the following section, we provide some examples of using approximate methods to evaluate the performance of various actions associated with improving the state of aquifers and groundwater quality.

8.3.3 Approximate Evaluation Methods

Approximate evaluation methods are usually applied to get some overall approximate quantitative description of phenomena taking place in the vadose zone or the aquifer. Therefore, such methods often apply bulk parameters. Because of using bulk parameters in the approximate methods, it is convenient to apply characteristic properties of the domain in order to obtain dimensionless variables and coordinates involved in solving the relevant differential equations. Rubin et al. (2008a,b, 2009) have developed approximate methods aiming at analyzing and preliminary designing systems such as follows: (1) hydraulic barriers, (2) remediation of homogeneous aquifers, and (3) remediation of fractured permeable aquifers. All these topics consider using pump-and-treat methods. Implementing hydraulic barriers is of major importance for containing the contaminant plumes developed in deep aquifer parts. Operating the hydraulic barrier leads to gradual remediation of the aquifer. Higher rates of pumping may intensify the aquifer remediation and shorten the total remediation time. The reference to fractured permeable aquifers originates from the great impact of the aquifer heterogeneity on its remediation. A formation made of permeable blocks embedding fractures represents a dual-porosity matrix. Quantifying the remediation of such an aquifer may also provide an example of the effect of aquifer heterogeneity on its remediation.

8.4 Applications

8.4.1 Monitoring Groundwater Quality

Monitoring the water table in phreatic aquifers and monitoring the piezometric head distribution in confined aquifers usually provide important basic data for evaluating the availability of groundwater resources. Such data have been used for calculating water rights of different consumers and how groundwater quantities should be divided between the regional consumers. However, during the last three decades of the twentieth century, the importance of monitoring groundwater quality has been identified because intensive development of agricultural fertilization and use of pesticides as well as urbanization and industrial developments have led to continuous degradation of groundwater quality in many places. Monitoring of groundwater quality is needed for avoiding supply of water of low quality to the water consumers. Further, it should provide information leading to identifying point sources and non-point sources of pollutants. As examples, most cases of increasing chloride concentrations and nitrate concentrations are attributed to non-point sources of agricultural activities like irrigation and fertilization. On the other hand, groundwater contamination by specific species of organic compounds, like VOCs, is usually attributed to specific industries using DNAPLs for degreasing, dry cleaning, etc. Shlomi et al. (2010) developed an optimization method for sampling of a limited number of water supply wells for alarming risks of contaminated groundwater and identifying sources of contaminants. This method is based on applying Dupuit approximation for the groundwater flow and contaminant advection combined with a genetic algorithm for selecting the sampled wells.

Indigenous microflora may be very helpful in the remediation of soil and aquifer. Such biological processes are termed "natural attenuation." McCarty and Ellis (2002) emphasize the need for proper methodologies and methods aiming at monitoring the effectiveness of natural attenuation. Also with regard to this issue, adequate optimized management approaches should be applied to provide reliable results (maximum utility) with minimum investments.

8.4.2 Remediation of Soil and Aquifer

Contamination of the unsaturated (vadose) zone by NAPLs may incorporate the following physicochemical processes: percolation, evaporation of flowing NAPLs, entrapment of the NAPL phase, multiphase flow of NAPL-water-air, dissolution, solubilization, adsorption–desorption, chemical degradation, interphase mass transfer from the entrapped NAPL into the water phase, and interphase mass transfer from the entrapped NAPL into the air phase (evaporation). The soil surface comprises a boundary of zero organic gas concentration. Therefore, the gases of VOCs created in the vadose zone from the percolating and entrapped NAPL diffuse mainly upward with some horizontal spreading and diffusion (e.g., Abriola et al. 2004). It should be noted that the air hydraulic conductivity of soil is around 15 times smaller than that of water, as the kinematic viscosity of air is 15 times larger than that of water. Therefore, horizontal flow of VOCs in the vadose zone is quite limited. On the other hand, the boundary condition of soil surface with zero concentration of VOCs enhances diffusion of the gases into the atmosphere. Therefore, residential areas built on soils contaminated by VOCs are subject to high risks of air pollution by poisonous and often carcinogenic gases. Further, parts of the VOCs dissolved in the flowing groundwater may diffuse into the gaseous phase of the vadose zone. Therefore, increased levels of VOCs in the vadose zone may indicate the aquifer contamination by hydrocarbons. The physicochemical properties of NAPL that are relevant to remediation of the vadose zone are: density, viscosity, volatility, and surface tension. The soil properties relevant to remediation of the unsaturated zone are: heterogeneity (like presence of fractures), permeability, porosity, presence of adsorbing minerals. Percolation of NAPLs through the unsaturated zone of homogeneous soil, like sand, is basically similar to percolation of water (Kessler and Rubin 1987). However, the presence of clay minerals may cause significant differences between water percolation and NAPL percolation. Clay particles expand due to their contact with water, but they do not expand after contact with NAPLs. Therefore, fuel percolation through loam and clay soils may be faster than water percolation (Rubin and Narkis 2001). This phenomenon is extremely important for the possible remediation of soils rich with clay particles. The presence of fractures in the unsaturated zone of rock formation may considerably change the division of quantities of the three phases, water, NAPL, and air, between the fracture network and the porous blocks of the matrix (Berger and Braester 1998).

In the saturated zone, the following phenomena may take place: multiphase flow of NAPL and water, NAPL entrapment, interphase mass transfer from NAPL into the water phase (dissolution and solubilization), and adsorption–desorption, as well as degradation of the organic compounds. As stated earlier, at the contact film between the groundwater and the vadose zone, VOCs may evaporate leading to VOC gases diffusing from the water phase that includes dissolved organic compounds into the air phase of the vadose zone. Heterogeneity of the porous matrix may cause difficulties in aquifer remediation. In cases of fractured permeable aquifers, contaminated by entrapped LNAPL originating from changes in the elevation of the water, most quantities of NAPL may be entrapped in the permeable blocks (Rubin et al. 2008b). Under such conditions, applying pump-and-treat methods is associated with groundwater flow bypassing the contaminated permeable blocks via the fracture network. Laboratory studies of Nambi and Powers (2000, 2003) have shown that under high degrees of saturation of the entrapped NAPL, the reduced water permeability of the porous medium may cause significant groundwater flow bypassing the contaminated part of the aquifer. DNAPLs penetrating into a fractured aquifer tend to percolate through the fracture network from which the organic compounds diffuse into the porous blocks of the rock matrix (Parker et al. 1994). Therefore, the remediation of such contaminated rock formations may require extraction of the contaminant by diffusion. Such remediation may take very long time periods.

As NAPL entrapment originates from the surface tension existing between fluids of different wettability (water, NAPL, and air), various types of surfactants have been developed and can be used for enhancing the remediation of aquifers. Such agents are introduced into groundwater via injecting wells and later be collected with the solubilized NAPL by pumping wells. Methods for recycling of surfactants have been developed, and careful use of surfactants that should not contaminate the aquifer is required.

Experimental and field studies have indicated that heterogeneity of the porous medium is of major importance for considering the mechanisms of contamination of soils and aquifers by NAPLs and the implementation of remediation actions. Therefore, prior to designing the remediation actions, information about the aquifer should be collected, in order to introduce proper values into the different parameters of Equations 8.16 through 8.20. Such information is crucial for carrying out numerical simulations relevant to the hydrogeological environmental problem of interest.

8.4.3 Biological Aspects

Biological processes are involved in the degradation of NAPLs in the unsaturated and saturated zones. Natural attenuation (which is discussed later) is attributed to biological degradation of hydrocarbons by indigenous microorganisms, with no particular involvement of agents enhancing the remediation. However, various technologies of enhancing microbial activity have been used to convert the NAPL contaminants into less harmful species. Microbial activity may create natural surfactants and emulsifiers, which release the entrapped NAPL to biodegradation. Technologies for the bioremediation of excavated soil, unsaturated soil, or groundwater have been used and in many cases seemed to be feasible and economical. Bioremediation offers the most realistic and economic approach for taking care of general problems of wastewater treatment. Therefore, scientists and engineers have been attracted to practical uses of bioremediation methods with introduced acclimated microflora for the cleanup of soils and aquifers (e.g., Rubin and Narkis 2001). In situ bioremediation usually involves the enhancement of indigenous soil microflora (e.g., McCarty and Ellis 2002). Langwaldt and Puhakka (2000) have provided a comprehensive review of on-site biological remediation of groundwater while referring to economic issues and criteria for the necessity of microbial augmentation. Bioremediation of oil distillate hydrocarbons is efficiently carried out by aerobic microorganisms that require the availability of nutrients and oxygen. However, in most cases oxygen is the limiting component of the hydrocarbon degradation by aerobic microorganisms. Usually, the efficiency of the bioremediation requires proper supply of nitrogen, phosphorous, and oxygen to bacteria that are present in the contaminated soil and aquifer. If the site is contaminated by chlorinated aliphatic hydrocarbons, then very often a primary substrate, like toluene, supplied to the system enhances the degradation of that hydrocarbon by methanotrophs (e.g., Roberts et al. 1990) and phenol-utilizing microorganisms (Hopkins et al. 1993). McCarty and Ellis (2002) have provided a comprehensive review and recommendations about the parameters needed to achieve efficient remediation of soils and aquifers contaminated by DNAPLs. Such remediation is often carried out solely by anaerobic microorganisms.

Bioremediation systems should insure optimal metabolic activity of the biomass, which utilizes NAPL as a primary substrate in cases of crude oil distillates or as a secondary substrate in cases of halogenated hydrocarbons (McCarty and Semprini 1994). However, various studies reviewed by McCarty and Ellis (2002) have provided details about the anaerobic degradation of halogenated hydrocarbons. In such cases, the complex molecule of the halogenated hydrocarbon is subject to consecutive stages of degradation by different types of microorganisms. Each stage produces halogenated hydrocarbon whose molecule is less complex than that of the previous stage. Fennell and Gossett (1998) describe and provide models for the anaerobic biodegradation processes involved with the remediation of sites contaminated by halogenated hydrocarbons. Studies referring to anaerobic cleanup of halogenated hydrocarbons address the necessity of electron acceptors different from oxygen to allow the degradation processes. The objectives of bioremediation projects should be well planned and represented by quantitative criteria like target concentrations of the contaminants for the definition of "complete remediation" (e.g., Rubin and Narkis 2001). However, codes for achieving soil remediation vary from country to country, and in the United States even from state to state. These differences probably originate from different information about the effects of contaminants on ecosystems and on specific species. Such information is required to facilitate ecological risk assessments, which may lead to proper definition and validation of soil remediation criteria. Rice et al. (1995) have reviewed the effect of leaking fuel tanks on groundwater quality in California. Their review has been considered as a major source of information about possible assessing environmental issues associated with oil distillate contamination in groundwater, and whether natural degradation can be sufficient to avoid risks from the dissolved contaminant plume. DNAPLs (halogenated hydrocarbons) are often poisonous and carcinogenic materials. Their effects on human health and the environment may be significant even in concentrations of parts per billion (ppb). Therefore, developing of residential areas on soils contaminated with such compounds is not permitted prior to removing the contaminated soil and/or carrying out proper site remediation.

8.4.4 Technologies of Remediation

Economic calculations started approaches for developing technologies addressing handling of oil spills (LNAPLs). Such technologies were justified by the cost-benefit calculations of possible pumping of oil spills accumulated in significant quantities on top of the water table, mainly underneath refineries. In such places, pumping of the oil spill provides a commercial product that gives good return for the pumping expenses. Various types of equipment addressing pumping lenses of LNAPL are presently available, as described by various websites:

http://www.freepatentsonline.com/5013218.html
http://www.patentstorm.us/patents/5013218-description.html

Underneath many transportation terminals and airbases there are LNAPL lenses that extend to large horizontal distances. However, in many such cases cost-benefit calculations do not justify pumping of the oil spill as a commercial product. On the other hand, these lenses of LNAPL represent an environmental pollution issue and a serious problem of water resources quality that should be considered, evaluated, and handled very carefully. Probably, in most cases at least some effort should be given toward reducing environmental risks and avoiding the migration of contaminant plumes into areas of the aquifer that are free of contaminants. Also gradual remediation of the aquifer should be evaluated and taken into account by designing proper means.

Pumping DNAPLs after their penetration into the aquifer and accumulation on top of silt or clay lenses may be feasible. However, remaining small quantities of DNAPLs and dissolved VOCs within aquifers are still involved with high risks to the environment and the human health. Therefore, often remediation and/or containing of the organic solute sources and contaminant plumes are required.

Various books (e.g., Hyman and Dupont 2001) concern implementing different types of technologies for the remediation of soils and aquifers. Here only brief information is provided about most useful remediation technologies.

Groundwater remediation is often achieved by pump-and-treat methods. These methods incorporate treating the pumped water by using various methods, like air-stripping, according to codes and needs of the consumer and/or specific environmental requirements. Generally, pump-and-treat methods implement sets of pumping wells and often also injecting wells. The injecting wells may introduce some agents (surfactants) that enhance the aquifer remediation. Theoretical and practical case studies have shown that traditional pump-and-treat methods are often inefficient for removing the residual NAPL (e.g., Mackay and Cherry 1989). Nevertheless, water authorities continue using pump-and-treat methods for aquifer restoration purposes. Causes contributing to the inefficiencies of pump-and-treat cleanup may include no equilibrium interphase partitioning at high water velocities or with decreasing interfacial contact area between the entrapped NAPL and the flowing groundwater, and dilution and bypassing effects resulting from heterogeneous permeability and NAPL distribution (Anderson et al. 1992; Brusseau 1992a,b). Rubin et al. (2008b) analyze the effect of groundwater bypassing permeable blocks contaminated by entrapped LNAPL via the fracture network embedded within the matrix permeable blocks. They refer to an aquifer whose top layers are contaminated by entrapped LNAPL due to the rising elevation of the groundwater. In such a case, they consider that the LNAPL entrapped within the fracture network is quite quickly released and solubilized by the flowing groundwater. On the other hand, the dissolution and solubilization of LNAPL entrapped within the permeable blocks of the matrix are much slower. Therefore, their calculations have indicated high flow rates via the fracture network that bypass the major contaminated part of the domain, namely, the permeable blocks of the matrix (like sandstone). Hydraulic barriers aiming at controlling the spread of contaminants in aquifers apply pump-and-treat principles at low rates of pumping (Rubin et al. 2008a, 2009).

Soil vapor extraction (SVE) is an in situ unsaturated (vadose) zone soil remediation technology in which a vacuum is applied to the soil to induce the controlled flow of air and remove volatile organic compounds (VOCs) and some semivolatile contaminants from the soil. The gas leaving the soil may be treated to recover or destroy the contaminants, depending on local and state regulations. Vertical extraction vents are typically used at depths of 1.5 m or greater and have been successfully applied as deep as 90 m. Horizontal extraction vents (installed in trenches or horizontal borings) can be used as warranted by contaminant zone geometry, drill rig access, or other site-specific factors.

Air leading to evaporation of the VOCs can also be injected into an aquifer contaminated with such compounds. This method is called "air sparging," and it may be attractive in cases of LNAPL contamination of the aquifer due to seasonal and/or man-made water table fluctuations, and spills of DNAPLs that create contaminated porous media below the water table. According to this method, air is injected into the saturated zone via injecting wells below the areas of contamination. The contaminants dissolved in the groundwater and sorbed onto soil particles partition into the advective air phase. Also entrapped NAPL ganglia may gradually evaporate and get transferred into the upward moving air bubbles. The process of air sparging is basically simulating an in situ air-stripping system. The stripped contaminants are transported in the gas phase to the vadose zone, within the radius of influence of a vapor extraction and vapor treatment system. In situ air sparging is a complex multifluid phase process, which has been applied successfully since the mid-1980s. Major design parameters to be considered include contaminant type, gas injection pressures and flow rates, site geology, bubble size, injection interval (horizontal and vertical), and equipment specifications.

8.4.5 Natural Attenuation

Natural attenuation represents the natural characteristics of the hydrogeological systems to reduce and possibly eliminate the man-made contamination. The natural attenuation is based on the biological balances between nutrients and microflora consumers. The presence of increased quantities of hydrocarbons in the soil and aquifer enhances the growth of indigenous biomass that uses the hydrocarbon as a substrate. Unfortunately, we cannot always expect for the balancing Mother Nature to eliminate the contaminant at the required rate. Therefore, studies of natural attenuation concern and assess the potential long-term impacts of indigenous biomass to restore the natural soil and aquifer conditions existing prior to events of their contamination.

Various agencies (e.g., http://toxics.usgs.gov/topics/attenuation.html) are associated with evaluating the potential of remediation by natural attenuation and designing systems to monitor the performance of natural attenuation. Natural attenuation is often the only option of gradual site remediation. Substantial parts of the study and review carried out by McCarty and Ellis (2002) are devoted to natural attenuation of soils contaminated by chlorinated hydrocarbons and chains of processes involved in degradation of these compounds.

8.4.6 Physical–Chemical Barriers

NAPLs penetrating into aquifers gradually release dissolved organic compounds that migrate with the flowing groundwater and create contaminant plumes. Therefore, the area contaminated by the free phase NAPL represents a source of dangerous dissolved contaminants to aquifer areas that are free of NAPL. In order to reduce and possibly eliminate the migration of the dissolved contaminant plumes into areas free of the contaminants, some barriers should be employed. Such barriers avoid the expansion of contaminant plumes and contain the area that incorporates the entrapped NAPL, and possibly other sources of contaminants. Various types of physical barrier technologies have been developed, particularly vertical barriers such as soil-bentonite (SB) cutoff walls, slurry walls, etc., are useful. The design of such barriers emphasizes the need for low hydraulic conductivity and high absorption capacity of the physical barrier material, to reduce the contaminant transport by advection and diffusion from the polluted area into other parts of the aquifer. Chemical materials added to the physical barriers can intercept the contaminant plume and prevent the migration of the contaminant by transferring the contaminants from the groundwater to the immobile solids. With regard to hydrocarbon contaminants, the chemical material target is making a sorption barrier. This material can be emplaced in an aquifer cutoff wall or can be injected into a contaminant plume. Physical barriers that include steel sheets are also used to contain areas of the contaminant sources and avoid the development of the contaminant plume (http://www.oceta.on.ca/profiles/wbi/barrier.html).

8.4.7 Hydraulic Barriers

Hydraulic barriers are based on using sets of pumping and possibly also injecting wells to control the migration of contaminant plumes in aquifers.

Hydraulic barriers can be used in situations where excavating for physical barriers is too expensive or is not feasible. Hydraulic barriers contain the contaminant sources and avoid the spreading of contaminant plumes in the aquifer. The dissolved contaminant source might be an exogenous source that is above the ground (like landfills, transportation terminals), or entrapped NAPL and material adsorbed to the solid matrix of the aquifer. Pump-and-treat is usually considered as the proper definition for both aquifer remediation and containment of the contaminant plume. However, containment is expected to need smaller rates of extraction pumping, injection, and treatment than aquifer remediation. From the point of view of technologies, hydraulic barriers usually apply pump-and-treat technologies at minimum levels of pumping and injecting rates. Rubin et al. (2008a, 2009) have developed a simplified method of preliminary design of hydraulic barriers. Their method is based on applying the curvilinear coordinate system Φ and Ψ, where Φ and Ψ are the potential and stream functions, respectively, to solve Equations 8.25 and 8.28. For solving Equation 8.28, they also applied the boundary layer approximation to evaluate the effect of transverse dispersion on the contaminant plume. Further, by applying bulk coefficients characterizing the aquifer remediation properties, they predict the effect of operating the hydraulic barrier on the aquifer remediation. The studies of Rubin et al. (2008a, 2009) indicate that in many cases renewal of pumping groundwater from contaminated abandoned water supply wells may provide the required conditions of operating hydraulic barriers.

8.5 Major Challenges

Major challenges incorporated with soil and aquifer management are mainly connected with methods of analysis and applications.

8.5.1 Methods of Analysis

Currently, some top journals devoted to hydrology and contaminant fate in soil and aquifer require that articles incorporating innovative approaches for simulating field phenomena should include experimental results. This requirement originates from the large quantity of theoretical studies concerning the development of analytical and numerical methods that are presently available for evaluating, analyzing, and predicting phenomena associated with soil and aquifer management. On the other hand, a very small number of studies has provided experimental results and/or field measurements that may provide good basis for using the theoretical results. Further, very often results of theoretical studies seem to be very far from being of any practical use.

With regard to environmental fluid dynamics relevant to soil and aquifer management, in most cases boundary conditions of the simulated domain and coefficients of constitutive equations are not clear for all relevant concerned physical quantities. Therefore, in many cases applying simplified methods with bulk coefficients can be developed and used with reliability, which is not lower than that of very sophisticated modeling approach. Problems with domain boundary conditions and coefficients of constitutive equations may often justify developing site-specific modeling approaches.

In many cases of large gradient of dissolved species, numerical modeling approaches may lead to significant errors originating from numerical dispersion. In such cases, possibly approximate approaches using, e.g., boundary layer approximations, may be very useful.

Adopting contemporary approaches of optimal management for managing the quality of soil and aquifer is an important mission. Such approaches are very much needed for carrying out tasks of monitoring the quality of groundwater and following processes like natural attenuation.

8.5.2 Applications

The earlier sections concern challenges calling for theoretical studies whose results should be justified by experimental studies and field measurements. However, major challenges in topics of soil and aquifer management are incorporated with the extension and applications of theoretical results to engineering know-how. Preservation of the quality of the environment and groundwater depends to a great extent on our capability to apply results of the basic research to proper management of all topics comprising soil and aquifer management. Such topics are briefly reviewed in Section 8.4.

References

Abriola, LM. 1989. Modeling multiphase migration of organic chemicals in groundwater system—A review and assessment. *Environmental Health Perspectives* 83:117–143.

Abriola, LM, Bradford, SA, Lang, J, Gaither, CL. 2004. Volatilization of binary nonaqueous phase liquid mixtures in unsaturated porous media. *Vadose Zone Journal* 3:645–655.

Anderson, MR, Johnson, RL, Pankow, JF. 1992. Dissolution of dense chlorinated solvents into groundwater. 1. Dissolution from a well-defined source. *Ground Water* 30:250–256.

Barenblatt, GI, Zheltov, YP, Kochina, LN. 1960. Basic concepts in the theory of seepage of homogeneous liquids in fissured rocks. *PMM-Sov. Applied Mathematics and Mechanics* 24:852–864.

Bear, J. 1972. *Dynamics of Fluids in Porous Media*. American Elsevier, New York.

Berger, DG, Braester, C. 1998. Problems of flow through fractured rock formations related to contamination of aquifers. *Soil and Aquifer Pollution Non-Aqueous Phase Liquids Contamination and Reclamation* (eds. H. Rubin, N. Narkis, J. Carberry), pp. 274–288. Springer, Berlin, Germany.

Birkhölzer, J, Rubin, H, Daniels, H, Rouvé, G. 1993. Contaminant advection and spreading in a fractured permeable formation. 1. Parametric evaluation and analytical solution. *Journal of Hydrology* 144:1–33.

Brusseau, ML. 1992a. Rate-limited mass transfer and transport of organic solutes in porous media that contain immobile immiscible organic liquid. *Water Resources Research* 28:33–45.

Brusseau, ML. 1992b. Transport of rate-limited sorbing solutes in heterogeneous porous media: Applications of a one-dimensional multifactor nonideality model to field data. *Water Resources Research* 28:2485–2497.

Fennell, DE, Gossett, JM. 1998. Modeling the production of and competition for hydrogen in a dechlorinating culture. *Environmental Science and Technology* 32:2450–2460.

Fetter, CW. 1980. *Applied Hydrology*. Charles E. Merrill, Columbus, OH.

Hopkins, CD, Semprini, L, McCarty, PL. 1993. Microcosm and in-situ field studies of enhanced biotransformation of trichloroethylene by phenol utilizing microorganisms. *Applied and Environmental Microbiology* 59:2277–2285.

Hyman, M, Dupont, RR. 2001. *Groundwater and Soil Remediation Process Design and Cost Estimating of Proven Technologies*. ASCE Press, Reston, VA.

Kessler, A, Rubin, H. 1987. Relationships between water infiltration and oil spill migration in sandy soils. *Journal of Hydrology* 91:187–204.

Langwaldt, JH, Puhakka, JA. 2000. On-site biological remediation of contaminated groundwater a review. *Environmental Pollution* 107:187–197.

Mackay, DM, Cherry, JA. 1989. Groundwater contamination: Pump-and-treat remediation. *Environmental Science and Technology* 23:630–636.

Mayer, AS, Miller, CT. 1996. The influence of mass transfer characteristics and porous media heterogeneity on nonaqueous phase dissolution. *Water Resources Research* 32:1551–1567.

McCarty, PL, Ellis, DE. 2002. Natural attenuation. *Innovative Approaches to the On-Site Assessment and Remediation of Contaminated Sites* (eds. R. Danny, K. Demnerova), pp. 141–181. Kluwer, Amsterdam, the Netherlands.

McCarty, PL, Semprini, L. 1994. Groundwater treatment for chlorinated solvents. *Handbook of Bioremediation* (eds. R.D. Norris et al.), pp. 87–116. Lewis, Ann Arbor, MI.

Miller, CT, Christakos, G, Imhoff, PT, McBridge, JF, Pedit, JA. 1998. Multiphase flow and transport modeling in heterogeneous porous media: Challenges and approaches. *Advances in Water Resources* 21:77–120.

Nambi, IM, Powers, SE. 2000. NAPL dissolution in heterogeneous systems: An experimental study in simplified heterogeneous system. *Journal of Contaminant Hydrology* 44:161–184.

Nambi, IM, Powers, SE. 2003. Mass transfer correlations for nonaqueous phase liquid dissolution from regions with high initial saturations. *Water Resources Research* 39:1030–1040.

Parker, PL, Gilham, RW, Cherry, JA. 1994. Diffusive disappearance of immiscible-phase organic liquids in fractured geologic media. *Ground Water* 32:805–820.

Parker, JC, Lenhard, RJ, Kuppusamy, T. 1987. A parametric model for constitutive properties governing multiphase flow in porous media. *Water Resources Research* 23:618–624.

Polubarinova-Kochina, PY. 1962. *Theory of Ground Water Movement*. Princeton University Press, Princeton, NJ.

Rice, DW, Grose, RD, Michaelson, JC, Dooher, B, MacQueen, DH, Cullen, SJ, Kastenberg, WE, Everett, LG, Marino, MA. 1995. *California Leaking Underground Fuel Tanks (LUFT) Historical Case Analysis*. Lawrence Livermore National Laboratory, University of California, Livermore, CA.

Roberts, PV, Hopkins, GD, Mackay, DM, Semprini, L. 1990. A field evaluation of in-situ biodegradation of chlorinated ethenes part 1 methodology and field site characterization. *Ground Water* 28:591–604.

Rubin, H, Buddemeier, RW. 1996. Transverse dispersion of contaminants in fractured permeable formations. *Journal of Hydrology* 176:133–151.

Rubin, H, Narkis, N. 2001. Feasibility of on-site bioremediation of loam soil contaminated by diesel oil. *Journal of Environmental Science and Health A* 36:1549–1558.

Rubin, H, Narkis, N, Carberry, J (eds.). 1998. *Soil and Aquifer Pollution Non-Aqueous Phase Liquids Contamination and Reclamation*. Springer, Berlin, Germany.

Rubin, H, Rathfelder, K, Spiller, M, Köngeter, J. 2004. Continuum quantifying remediation of fractured permeable formations. *ASCE Journal of Environmental Engineering* 130:1345–1356.

Rubin, H, Shoemaker, CA, Köngeter, J. 2008a. Screening of one-well hydraulic barrier design alternatives. *Ground Water* 46:743–754.

Rubin, H, Shoemaker, CA, Köngeter, J. 2009. Implementing a method of screening hydraulic barrier design alternatives. *Ground Water* 47:306–309.

Rubin, H, Yaniv, S, Spiller, M, Köngeter, J. 2008b. Parameters that control the cleanup of fractured permeable aquifers. *Journal of Contaminant Hydrology* 96:128–149.

Saba, T, Illangasekare, T. 2000. Effect of groundwater flow dimensionality on mass transfer from entrapped nonaqueous phase liquid contaminants. *Water Resources Research* 36:971–979.

Shlomi, S, Ostfeld, A, Rubin, H, Shoemaker, CA. 2010. Optimal groundwater contamination monitoring using pumping wells. *Water Science and Technology* 62:556–569.

Tang, DH, Frind, EO, Sudicky, EA. 1981. Contaminant transport in fractured porous media; Analytical solution for a single fracture. *Water Resources Research* 17:555–564.

Tsang, YW, Tsang, CF. 1987. Channel model of flow through fractured media. *Water Resources Research* 23:467–479.

9

Security and Environmental Fluid Dynamics

Malcolm J. Andrews
Los Alamos National Laboratory

Fernando F. Ginstein
Los Alamos National Laboratory

Edward Kwicklis
Los Alamos National Laboratory

Rodman Linn
Los Alamos National Laboratory

9.1 Introduction

The role of "security" in environmental fluid dynamics (EFD) is as a cross-cutting topic that spans from flows in urban environments to groundwater flows, and then on to energy security and beyond. Indeed, the field is so wide that the best we can hope for is to give a flavor of EFD and related security research that is taking place at Los Alamos National Laboratory, as an indication of the range of research that is currently being performed at practically every research institution in the world. A complimentary review from Settles (2006) discusses the role of fluid dynamics in homeland security, and provides a useful extension beyond the EFD processes of interest here.

In the following sections you will find discussed ongoing research that considers the interplay between EFD and high energy density fluid dynamics of inertial confinement fusion (ICF) by M. Andrews; urban flow and contaminant transport, which considers the flow of hazardous materials due to intentional or accidental release in urban environments (complex urban geometries) by F. Grinstein; an overview of groundwater flow issues as they relate to national and international interests by E. Kwicklis; and wildfires (the Cerro Grande fire of 2000 is still fresh in the minds of Los Alamos residents) and urban fire conflagrations by R. Linn.

Aside from our efforts here to bring together various EFD topics, and their relationship to security, we have little doubt you will find abundant chapters within this book that will overlap with our descriptions, and extend our discussion, of EFD and security. In particular, we refer the reader to Chapter 24 of *Handbook of Environmental Fluid Dynamics, Volume Two*; Chapter 29 of *Handbook of Environmental Fluid Dynamics, Volume Two*; and Chapter 11. We feel strongly that this is an exciting time for EFD and security since we are applying our fundamental knowledge of EFD to new challenges and opportunities, that in turn feedback to fundamental progress and understanding of EFD.

9.2 Buoyant Interplay between EFD and ICF

9.2.1 Introduction

Aside from the immediate application of EFD to security, there is a strong relationship between fundamental fluid dynamics and other areas of security that include energy and, of particular interest at Los Alamos, the fluid physics associated with ICF. The compression of an ICF capsule causes material shells within the capsule to mix with the capsule fuel, and

Handbook of Environmental Fluid Dynamics, Volume One, edited by Harindra Joseph Shermal Fernando. © 2013 CRC Press/Taylor & Francis Group, LLC. ISBN: 978-1-4398-1669-1.

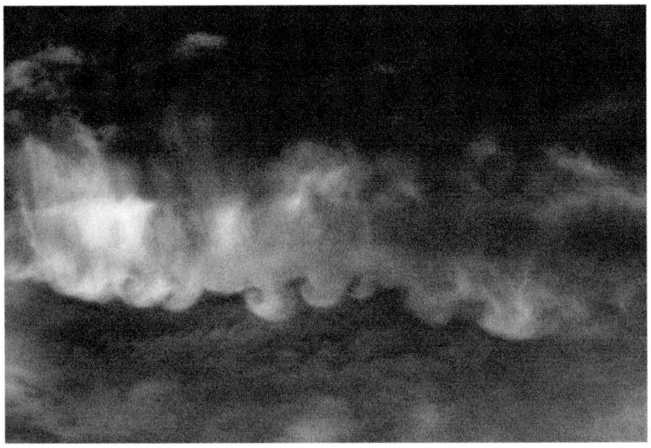

FIGURE 9.1 RT Cirrus clouds. (Photograph courtesy of David Jewitt, University of California, Los Angeles, CA.)

degrade the final fusion energy release. The mixing process is driven by pressure gradients that act on different density material shells that results in a buoyancy-driven Rayleigh–Taylor (RT) turbulent mix (Sharp, 1984). This same buoyancy mechanism occurs in many environmental flows that include cloud dynamics (Mellado et al., 2009, and see Figure 9.1, Jewitt, 2005), oil-trapping salt dome formation (Ribe, 1998), porous media (Zoia et al., 2009), sedimentation (Voltz et al., 2002), and even icicles (deBruyn, 1997).

9.2.2 Rayleigh–Taylor (Buoyancy) Driven Mixing Mechanisms

Indeed, the "strength" of this buoyant mixing mechanism is perhaps well demonstrated by the "hot/cold" water channel experiment of Ramaprabhu and Andrews (2004), in which cold water at 20°C is introduced over hot water at 25°C off the end of a splitter plate as shown in Figure 9.2.

In Figure 9.2, the photograph shows that within 6 s the two streams mix to a width of 20 cm, with a growth rate that goes as t^2. This water channel experiment illustrates the dominant role of buoyancy at small density differences, typical of EFD. More precisely, the RT mixing total width (W) has been found at late-time, or small density differences to evolve as

$$W = 2h = 2\alpha A g t^2 \tag{9.1}$$

where

- h is the height of the rising lighter "bubbles" (there are similar heavier falling "spikes")
- A is the Atwood number defined as $A = \rho_1 - \rho_2/\rho_1 + \rho_2$ (a nondimensional density contrast, with the density of fluid 1 greater than fluid 2)
- g the gravitational acceleration (for ICF an equivalent pressure gradient is used)
- t the time from the start of the mixing

FIGURE 9.2 Rayleigh–Taylor mixing around a fixed obstacle.

The coefficient "α" has measured values between 0.02 and 0.1, and is still an open area of research, but recent work suggests that initial conditions play a significant role in its determination. A recent review by Andrews and Dalziel (2010) of small Atwood RT experiments provides a more complete discussion, and references, of the fascinating fluid dynamics associated with RT. For Atwood numbers below 0.1, it is known that this RT mix is practically symmetrical about the initial density interface, but for Atwood number above 0.1, the mix becomes asymmetric and the formula (1.1) needs modification to account for different bubble and spike α's. However, an Atwood number of 0.1 corresponds to a density ratio of 1.22 well above most density ratios that occur in EFD. Perhaps another useful result from the study of RT instabilities is the nonlinear development of buoyancy-driven single-wavelength perturbations. Recent (Goncharov, 2002) progress has been made on this problem with some useful results at small EFD Atwood numbers. In particular, the following formula describes the terminal velocity (as a result of a balance between buoyancy and drag) of a 2-D wave, and a 3-D bubble as

$$V_{b,\infty}^{2D} = 1.025\sqrt{\frac{2A_t}{1+A_t}\frac{g}{3k}} \quad \text{and} \quad V_{b,\infty}^{3D} = 1.02\sqrt{\frac{2A_t}{1+A_t}\frac{g}{k}} \tag{9.2}$$

where

- $k = 2\pi/\lambda$ is the wave number
- subscript "b" refers to the "bubble" vertex velocity (which is the same for the spike in small Atwood EFD applications)

These formulas correspond to similar expressions for plume rise and fall, or the rise of a bubble in a tube. One way to look at these formulas is to recognize that they can be roughly obtained by considering a buoyancy/drag problem, in which the release of potential energy of a fluid as it falls is transferred into kinetic energy and drag (dissipation), the resulting terminal velocity results in Equation 9.2.

Moreover, the RT "plume" exhibits formation of Kelvin–Helmholtz (KH) roll-ups at its vertex, with a "mushroom" appearance, and can also appear as shearing interface instabilities when a large 2-D disturbance is superimposed on a RT mix layer, the so-called overturning problem (c.f. overturning in solar ponds). We leave the discussion of KH to other chapters.

Thus, the fundamental fluid dynamics that occur in EFD have broad application not only to immediate security applications but also indirectly to the study of fluid dynamic problems in technology and even high energy density research.

9.3 Urban Flow and Dispersal Predictions

9.3.1 Introduction

Hazardous chemical, biological, or radioactive (CBR) releases in urban environments may occur (intentionally or accidentally) during urban warfare, or as part of terrorist attacks on military bases or other facilities. The associated contaminant dispersion is complex and semi-chaotic. Urban predictive simulation capabilities can have direct impact in many threat-reduction areas of interest, including urban sensor placement and threat analysis, contaminant transport (CT) effects on surrounding civilian population (dosages, evacuation, shelter-in-place), education and training of rescue teams and services, onsite contaminant mitigation, assessing strategies, pyroclastic flows (visibility), and predicting CT from targets or missile intercepts. Detailed simulations for the various processes involved are in principle possible, but generally not fast.

Predicting urban airflow accompanied by CT presents extremely challenging modeling requirements (Britter and Hanna, 2003). Because of the flow configurations associated with very complex geometries, and unsteady buoyant flow physics involved, widely varying temporal and spatial scales quickly exhaust current modeling capabilities. Crucial technical issues include both turbulent fluid transport and boundary condition modeling, and post-processing of the simulation results for practical consequence management (Boris, 2002; Patnaik et al., 2007; Grinstein et al., 2009). Relevant fluid dynamic processes to be simulated include detailed energetic and contaminant sources, complex building vortex shedding, flows in recirculation zones, and modeling dynamic subgrid-scale (SGS) turbulent and stochastic backscatter. The simulation model must also incorporate a consistent stratified urban boundary layer with realistic wind fluctuations, solar heating including shadows from buildings and trees, aerodynamic drag and heat losses due to the presence of trees, surface heat variations, and turbulent heat transport.

Because of the short time spans and large air volumes involved, modeling a pollutant as globally well mixed is typically not appropriate. It is important to capture the effects of unsteady, buoyant flow, on evolving pollutant concentration distributions. In typical urban scenarios, both particulate and gaseous contaminants behave similarly insofar as transport and dispersion are concerned, so that the contaminant spread can usually be simulated effectively based on appropriate pollutant tracers with suitable sources and sinks. In some cases, the full details of multigroup particle distributions are required. In such cases, additional physics to be modeled include particulate fallout, as well as deposition, resuspension, and evaporation of contaminants. Despite the inherent physical uncertainties and model trade-offs, some degree of reliable prediction of CT within urban areas appears to be possible (Patnaik et al., 2007). Other crucial issues in the urban CT simulation process include modeling building damage effects due to eventual blasts, addressing appropriate regional and atmospheric data reduction, and feeding practical output of the complex combined simulation process into "urbanized" fast-response models—capable of translating flow and atmospheric variability into local (neighborhood) CT variability (Boris, 2002; Brown, 2004; Patnaik et al., 2007; Grinstein et al., 2009).

The industrial standard for plume prediction technology presently in use is based on idealized sources, rough urban canopy models, and Gaussian similarity solutions (puffs)—originally intended to model effects of large-scale flow dynamics over flat terrain. Diffusion is used in such models to emulate the effects of turbulent dispersion caused by complex building geometry and by wind gusts of comparable and larger size. Plume/puff models provide fast but unrealistic predictions for typical urban scenarios (e.g., Pullen et al., 2005) where *they cannot capture dispersal driven by inherently unsteady vortex dynamics*. A practical example is given in Figure 9.3 that shows

FIGURE 9.3 **(See color insert.)** CT from an instantaneous release in Times Square, New York City, as predicted by the FAST3D-CT MILES model. (From Patnaik, G. et al., *J. Fluids Eng.*, 129, 1524, 2007.) Concentrations shown at 3, 5, 7, and 15 min after release. The figure demonstrates the typical complex unsteady vertical mixing patterns caused by building vortex and recirculation patterns, and predicts endangered regions associated with the particular release scenario. Vortex dynamics driven CT cannot be captured by plume/pluff modeling.

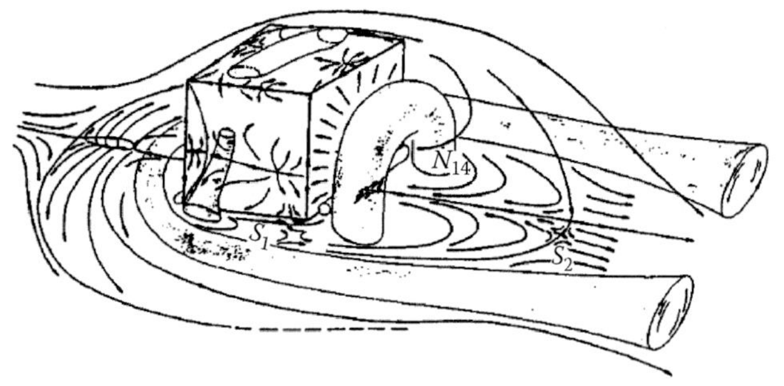

FIGURE 9.4 Arch vortices governing the flow dynamics past a surface-mounted cube. (From Hussein and Martinuzzi., *Phys. Fluids*, 8, 764–780, 1996.)

a simulated CT in Times Square, New York City (Patnaik et al., 2007). The figure depicts the so-called *fountain effect* that occurs behind tall buildings. The fountain effect is the systematic migration of contaminant from ground level up the downwind side of buildings, followed by continuous ejection into the air flowing over their tops. It has been observed in field experiments (Rappolt, 2002) and reported in wind tunnel studies (Brown et al., 2001), and is believed to be driven by arch vortices (Hussein and Martinuzzi, 1996) lying behind the buildings, as shown in Figure 9.4.

9.3.2 Urban Flow Simulation

The advantages of a computational fluid dynamics (CFD) representation to simulate CT transport and dispersion include the ability to quantify complex geometry effects, predict dynamic nonlinear processes faithfully, and treat turbulent problems reliably in regimes where experiments, and therefore model validations, are impossible or impractical. Solving for urban flow and dispersion is a problem for time-dependent, aerodynamic CFD methods. Computing urban aerodynamics accurately is a time-intensive, high-performance computing problem. Using this technology for the emergency assessment of blasts, industrial spills, transportation accidents, or terrorist CBR attacks requires very-tight time constraints that suggest the use of simple approximations, but which can unfortunately produce inaccurate results.

Unavoidable trade-offs demand choosing between fast (but inaccurate) and much slower (but accurate) models. Relevant time domains can be identified which require appropriate corresponding time-accurate ("full physics") simulation codes, involving physical processes that occur in microseconds-to-milliseconds and seconds-to-one-hour ranges. Linking codes between the various time domains allows the results of one to be used as the initial conditions for the next (Grinstein et al., 2009). The suite of *full-physics* simulations is used to develop source term, buoyant rise, and flow field parameterizations in urban environments for later use with fast-response high-fidelity analytical model tools (Brown, 2004; Grinstein et al., 2009).

9.3.3 LES Approach for Urban Contaminant Transport

Direct Numerical Simulation (DNS)—resolving all relevant space/time scales—is prohibitively expensive for most practical flows at moderate-to-high Reynolds number, and especially so for urban CT studies. On the other end of the CFD spectrum are the standard industrial methods such as the Reynolds-Averaged Navier-Stokes (RANS) approach, which simulate the mean flow and approximately model the effects of turbulent scales. These are generally unacceptable for urban CT modeling because they are not able to capture unsteady vortex-driven plume dynamics. Large-eddy-simulation (LES) (Sagaut, 2005) constitutes an effective intermediate approach between DNS and the RANS methods. In LES, the large energy containing structures are resolved whereas the smaller, presumably more isotropic, structures are filtered out and unresolved SGS effects are modeled. LES is capable of simulating the key unsteady flow features that cannot be handled with RANS (or Gaussian plume methods) and provides higher accuracy than RANS. Implicit LES (ILES, MILES) (Grinstein et al., 2007) effectively *addresses the seemingly insurmountable issues posed to LES by under-resolution*—truncation terms due to discretization comparable with SGS models in typical LES strategies—*by* relying on the use of SGS modeling and filtering provided implicitly by *physics capturing* numerics.

A state-of-the-art ILES urban CT model is the FAST3D-CT code (Patnaik et al., 2007, and references therein). It involves a scalable, low-dissipation, fourth-order phase-accurate flux-corrected transport convection algorithm, implementing direction-split convection, second-order predictor-corrector temporal integration, and time-step splitting techniques. The relevant system of equations for the problem under consideration involves the time-dependent buoyant flow equations for mass and momentum conservation; pollutant concentration tracers (convected separately) are used to model different contaminants and/or release scenarios. The flow equations must be supplemented with an equation of state, appropriate inflow, outflow, wall boundary-condition models, and the relationship between fluid and potential temperatures (Arya, 2001). The model uses rough-wall-boundary condition models for the surface stress and the heat transfer from the walls, and convective conditions at outflow boundaries. Other required physical models are discussed in Boris (2002).

9.3.4 Urban Geometry Specification

An efficient and readily accessible data stream is used to specify the building geometry database. High-resolution (1 m or smaller) vector geometry data are typically available for many major cities. From these data, building heights are determined on a regular mesh of horizontal locations with relatively high resolution (e.g., 1 m). Similar tables for terrain, vegetation, and other land use variables can be extracted. These tables are interrogated during the mesh generation to determine which cells in the

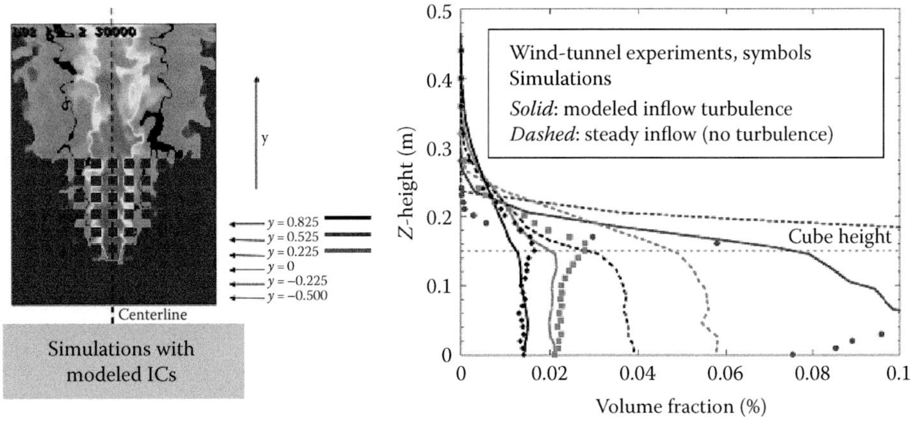

FIGURE 9.5 **(See color insert.)** ILES studies of flow and dispersion over an urban (cube arrangement) model. (From Patnaik, G. et al., *J. Fluids Eng.*, 129, 1524, 2007.) Predicted and measured dispersal of tracer volume fractions in the first few urban canyons indicated by color lines on the left.

computational domain are filled with building, vegetation, or terrain. This process is a very efficient way to convert a simple geometric representation of an urban area to a computational grid. A grid-masking approach is used to indicate which computational cells are excluded from the calculation as well as to determine where suitable wall boundary conditions are to be applied.

9.3.5 Atmospheric Boundary Layer Specification and Predictability

The upwind atmospheric boundary layer characterization directly affects the inflow condition prescription required in urban scenario simulations. Wind fluctuation specifics are major factors in determining urban CT. The important length scales (tens of meters to kilometers) and time scales (seconds to minutes) in wind gusts can in principle be resolved. However, the flow data from actual field trials or wind-tunnel experiments are typically inadequate and/or insufficient to fully characterize the boundary layer conditions required in the urban flow simulation model. In recent simulations of flow and dispersal experiments over an urban model (cube arrangement, see Patnaik et al. [2007]) in wind tunnels (Brown et al., 2001), the available datasets from the laboratory experiments consisted of high-quality, spatially dense (but not time-resolved) single-point statistical data. The inflow velocity in the simulations modeled mean profiles and superimposed fluctuations. An iterative process was used, in which a phenomenological unsteady-wind model was calibrated to provide a fit to the experimentally observed *rms* values at locations upstream of the urban model. Agreement was achieved by adjusting amplitude, spatial wavelength, and temporal frequency of the imposed wind fluctuations in an unsteady-wind model. This approach provided a practical approximation to the turbulent inflow boundary condition specification problem, consistent with the available laboratory data. Particularly relevant insights follow from this work, when comparing predicted and measured volume fractions of a tracer scalar in the first few urban model *canyons* using various different inflow condition models—shown in Figure 9.5.

Prescribing *some reasonable inflow turbulence—as opposed to prescribing steady inflow,* was found to be critical. On the other hand, the fluid dynamics within the cube arrangement, that is, beyond the first canyon, becomes somewhat insulated from flow events at the boundaries, that is, it is less dependent on the precise details of the modeled inflow turbulence and largely driven by the urban geometry specifics within the urban arrangement.

Establishing the credibility of the solutions is one of the stumbling blocks of urban CFD simulations. The goal of validating a numerical model is to build a basis of confidence in its use, and to establish likely bounds on the error that a user may expect in situations where correct answers are not known. A primary difficulty is the effective calibration and validation of the various physical models, since much of the input needed from experimental measurements of these processes is typically insufficient, or even nonexistent. Further, even though the individual models can all be validated separately, the larger problem of validating the overall simulation code has to be tackled as well. Validation studies with experiments require well-characterized datasets, with information content suitable to initiate and evaluate unsteady simulation models as well as the cruder steady-state models. In principle, such high-quality datasets can be made available to urban models in the context of carefully controlled and diagnosed wind-tunnel experiments.

9.4 National Security Implications of Groundwater in the United States and Worldwide

9.4.1 Introduction

The availability of sustainable water resources for drinking, agriculture, and energy production is fundamental to the well-being and economic development of societies around the world. In arid or semi-arid portions of the world, where surface water supplies are seasonal at best, and where societies are most vulnerable to climate fluctuations, groundwater can provide the only reliable supply of water for human consumption and agriculture. Groundwater provides an additional benefit in that it is generally less susceptible to pollution than surface-water supplies, and some forms of contamination are removed from the groundwater by filtration or sorption onto minerals in rocks and sediments, so that even when it becomes contaminated, there are natural attenuation mechanisms that remediate the contaminated groundwater. On the downside, there are countless instances where industrial or agricultural activities have rendered groundwater unusable for human consumption without costly remediation measures, and many instances where overexploitation of groundwater has brought into question the long-term sustainability of the societies that depend upon it. The long-term sustainability of societies that rely primarily on groundwater is especially uncertain in arid areas, where much of the groundwater currently in use may have been recharged during past glacial climates when temperatures and precipitation patterns were much different from those that exist today.

Superimposed on natural fluctuations in climate are climate changes related to the buildup of greenhouse gases in the atmosphere. Although the exact nature of climate change will be different in different parts of the world, and remains the subject of ongoing research, it is likely that societies that have evolved in response to long-term precipitation and temperature patterns will experience some disruption, and that the disruption will be most severe where societies are already vulnerable because of poverty, or because they already exist at the margins of what their environment can sustain. Whereas developed nations like the United States and those of western Europe may have the resources to mitigate the worst impacts of climate change, large parts of the developing world will be left exposed to the consequences of those changes.

Groundwater impacts the national security of the United States domestically because of the reliance of agriculture, energy production, and human consumption on groundwater. In addition, U.S. national security interests are at stake when groundwater shortages abroad threaten the stability of key allies or countries whose governments, and societies, could become destabilized and form the breeding grounds for radical ideologies. The following sections provide an overview of domestic and international water issues that are broadly intertwined with the national security concerns of the United States.

9.4.2 Domestic Water Issues

This section reviews the use of water in the United States and the availability and sustainability of groundwater throughout the nation. In doing so, it highlights the impact of potential shortages, and the vulnerability of groundwater supplies to contamination. Most of the information regarding water use in the United States is taken from Kenny et al. (2009), which is the latest in a series of U.S. Geological Survey reports that describe water use in the United States every 5 years since 1950. That report summarizes water use information provided by local, state, and national agencies on a county-by-county basis, and presents data summaries by state in tabular and graphical form. Temporal trends in water use since national data were first compiled in 1950 are also presented in the Kenny et al. (2009) report. Information on groundwater availability is taken largely from Reilly et al. (2008), which summaries major aquifer systems in the country and provides reconnaissance-level analyses of groundwater recharge, water-level declines since groundwater development began, and proposes strategies to study the sustainability of groundwater supplies in a nationally comprehensive fashion.

9.4.2.1 Water Use

As of 2005, approximately 410,000 million gallons per day (Mgal/day) of groundwater and surface water were used for a variety of purposes including thermoelectric-power generation, public and domestic water supply, irrigation and livestock, industry, mining, and aquaculture (Kenny et al., 2009). Of this total, approximately 85% or 349,000 Mgal/day were fresh surface water or fresh groundwater withdrawals, and the remainder was saline water, primarily ocean water used for cooling by thermoelectric power plants along the coast. Groundwater withdrawals totaled about 82,600 Mgal/day of which 96% were freshwater. Consumptive use—that is, the amount of water lost to evapotranspiration or incorporated into farm or manufactured products—was not estimated by Kenny et al. (2009), but some portion is known to be returned to either surface or groundwater by cooling tower discharge, excess irrigation, or treated wastewater.

Thermoelectric power, irrigation, and public water supply together consumed more than 90% of the total water used in the United States, with industry, mining, domestic water supply, and livestock consuming the remainder. Thermoelectric-power generation used 41% of all freshwater withdrawals (143,000 Mgal/day), nearly all of which was surface water used for cooling at power plants. Irrigation was the second largest category of water use, consuming 37% of all freshwater withdrawals and nearly two-thirds of fresh groundwater withdrawals. Irrigation was the largest consumer of groundwater in 25 states, and groundwater was the primary source of irrigation in Nebraska, Arkansas, Texas, Kansas, Mississippi, and Missouri.

About 86% of the U.S. population derive their water from the public water supply, with the remainder (primarily in rural areas) relying on domestic wells. Water use from the public supply tends to correlate directly with the size and population of the state. Fifteen states obtained more than half their public water supply from groundwater, and three states—Florida, Texas, and California—together accounted for about one-third of the total groundwater withdrawn for the public water supply. Florida relies on groundwater for 87% of its public water supply, making it the most reliant state on groundwater for this purpose. Although deriving far smaller fractions of their public water supply than Florida, California, and Texas each also derived more than 1000 Mgal/day of their public water supply from groundwater.

As in other categories, the amounts of surface water and groundwater withdrawn for industry, mining, livestock, and aquaculture needs tend to vary in importance by state. Nationally, however, each of these categories was responsible for just a few percent of the total water consumed. Whereas mining (which includes fossil fuel extraction) relies primarily on saline groundwater and surface water, other use categories rely primarily on freshwater.

Since 1950, population growth, demographic shifts from the east coast to other parts of the nation and from rural to urban areas, combined with changes to the U.S. economic profile, have changed the way the nation uses its water (Kenny et al., 2009, Table 14). Both fresh surface water and groundwater use essentially doubled between 1950 and 2005, but most of this growth in demand occurred between 1950 and 1980, after which demand stabilized or even declined slightly despite continued population growth and increasing electricity production. The flattening of water demand is primarily a reflection of more efficient irrigation practices and more efficient use of water in power plants, and to a lesser degree a reflection of the decline in the U.S. manufacturing base.

9.4.2.2 Groundwater Availability

Groundwater availability depends on many factors, including the volume of water present in the pore spaces and fractures of the sediments or rock; the ease with which this water moves toward a well or surface water body such as a river, lake, or stream; the quality of the water relative to the requirements of the intended purpose; and the economic cost of obtaining the water. Water in subsurface storage that cannot be recovered economically or is in a formation that cannot transmit the water readily to the recovery well, or whose quality is insufficient for its intended purpose, is effectively not available for the intended use (Reilly et al., 2008). An additional limitation to groundwater availability is sometimes imposed by the links between groundwater and surface water and the legal restrictions related to the surface water. Where groundwater discharges to streams and rivers, and sustains stream flow during dry seasons—a component of stream flow known as *base flow*—a small decrease in water levels adjacent to the stream or river can reverse flow directions and cause the stream to lose water to the rock or sediment rather than gain water. Legal requirements to maintain a minimum stream flow for ecosystem maintenance or for downstream water users can limit groundwater use to a small fraction of what is physically or economically recoverable.

The linkage between surface water and groundwater can blur the distinction between the amounts of surface water and groundwater used for irrigation, thermoelectric power generation, and other uses cited in an earlier section of this chapter. During dry seasons, when base flow supplied by groundwater is the dominant or sometimes only source of water in a stream or river, the user is effectively utilizing groundwater. Given that electricity production, manufacturing, or even agriculture continues during times when groundwater is the only source of water in the stream, the amount of fresh groundwater used by the nation tends to be underestimated by the water use summary cited earlier.

The report by Reilly et al. (2008, Figure 23) shows the nation's 30 principal aquifers which together accounted for about 94% of the nation's total groundwater withdrawals for public supply, irrigation, and self-supplied industrial uses combined in 2000 (Figure 9.6). The largest water-producing aquifer by far is the High Plains Aquifer extending from Texas to Nebraska which is dominantly used for irrigation, followed by the Central Valley aquifer system in California, the Mississippi River Valley alluvial aquifer which straddles parts of Arkansas, Louisiana, and Mississippi, the Basin and Range basin-fill and carbonate rock aquifer which span most of Nevada and parts of Utah, Arizona, and California, and the glacial sand and gravel aquifers, which cover most of the upper part of the country.

9.4.2.3 Implications of U.S. Groundwater Use to National Security

Groundwater is being overexploited in a nonsustainable fashion in many parts of the country, causing significant declines in water levels in many of the nation's most important aquifer systems (Reilly et al., 2008, Figure 12) (Figure 9.5). Nationally, the undesirable consequences of excessive pumping can locally include (1) increased pumping costs as water must be brought from greater depths, (2) land subsidence due to pressure declines in compressible fine-grained sediments, with the potential for fissures to develop at land surface, (3) degradation of water quality due to dewatering of adjacent fine-grained sediments into primary aquifers or upcoming of deeper, more saline water, (4) intrusion of sea water into coastal aquifers where fresh water pressure heads have been lowered by pumping, and (5) disruption of aquatic ecosystems along streams and wetlands that rely on groundwater to sustain them (Reilly et al., 2008). Overexploitation of the nation's aquifers has long-term social and economic implications because potentially costly alternatives to groundwater resources such as water diversion projects must be implemented to sustain the communities that have come to rely on these resources.

Degradation of water quality because of the discharge of industrial effluents, infiltration of fertilizers, herbicides, and pesticides in agricultural areas, salinity buildup in groundwater beneath irrigated croplands, the effects of urban and suburban sprawl on watersheds, and military activities have all taken their toll locally on the nation's groundwater quality. In some areas, pollution has rendered the groundwater unusable for some purposes, and in others, groundwater contamination has precipitated remedial measures whose cost will ultimately run into billions of dollars.

Public drinking water supplies are generally routinely tested for contaminants and pathogens and, in most areas, groundwater movement toward water-supply wells is generally slow and through sediments that will filter and sorb contaminants that are introduced either intentionally or unintentionally into an aquifer. Elsewhere, however, fractured aquifers or limestone aquifers with solution cavities can provide potential short circuits for flow to bypass most of the pore space in the aquifer and

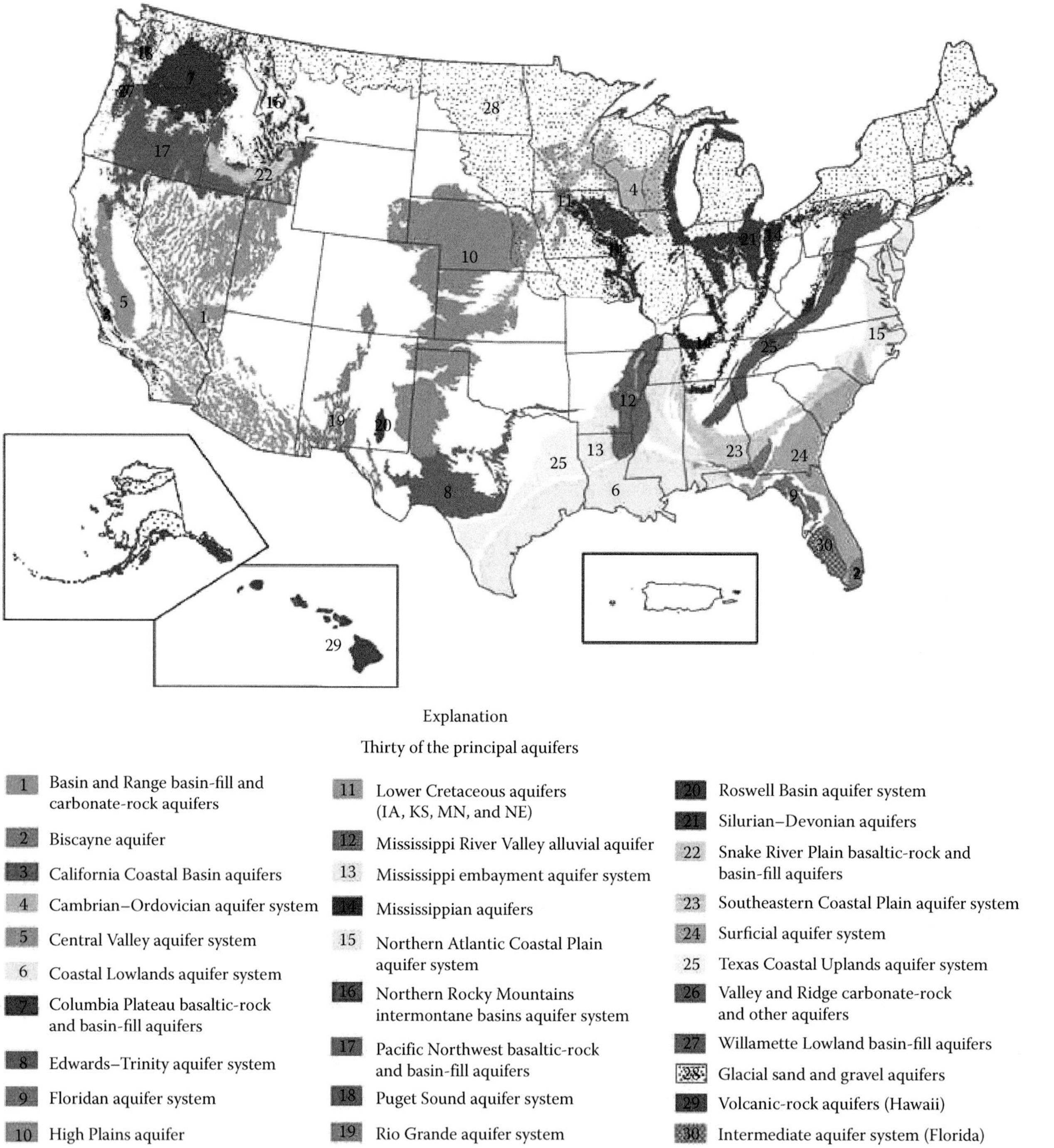

FIGURE 9.6 **(See color insert.)** Principal aquifers of the United States. (Reproduced from Reilly, T.E. et al., *U.S. Geol. Surv. Circ.*, 1323, 70, 2008, Figure 23.)

arrive quickly at a water supply well. Public supply wells that draw water from these aquifers are inherently more susceptible to contamination than aquifers composed of sand or gravel. A recent study by Landon et al. (2009) of deep water supply wells in different geologic settings found that pumping stresses could accentuate the potential for contaminants to bypass discontinuous confining layers that might otherwise protect deep aquifers from surficial contamination.

The effects of climate change on groundwater resources are the subject of ongoing study by federal and state agencies (e.g., Gurdak et al., 2009). It is generally anticipated, however, that some regions of the country will become dryer and others wetter as long-term weather patterns shift. Although groundwater responds much more slowly than surface water to changing weather patterns, particularly in the western United States where water tables are often deep below ground surface, eventually shifts in weather will affect groundwater because of changing rainfall and snowfall amounts, the timing of spring snowmelt and peak runoff, and the health and mortality of trees in the watersheds, which hold soil in place and promote infiltration.

In summary, responsible stewardship of the nation's groundwater resources is necessary to avoid the economic, social, and environmental costs associated with overexploitation and contamination of the nation's groundwater. The risk of unwise stewardship is the slow degradation of the quality of life, the need for the development of costly alternatives to groundwater or costly remediation, and the fraying of the social fabric in rural or agricultural communities whose existence is tied to its groundwater resource. Given the dispersed nature of the resource and the generally slow movement of groundwater, threats from terrorism to the nation's groundwater supply are remote.

9.4.3 International Water Issues

Globally, it is estimated that groundwater constitutes about 20% of the world's overall water use in agriculture, industry, energy production, and water supply (UNESCO, 2009), but with groundwater supplying close to 50% of the world's drinking water. Outside of North America and Europe, where most water withdrawals are surface water withdrawn for industry and energy production, most water use is surface water withdrawn for agricultural purposes. Worldwide, a nation's water use per person reflects both its affluence and its climate, with relatively arid countries with higher irrigation needs requiring more water than similarly developed countries in a tropical climate.

The availability of reliable supplies of water for agriculture and water supply is linked to the well-being and prosperity of societies around the world. Conversely, improvements in the living standards of some developing nations are placing increasing stresses on water supplies of those nations as a result of increased agricultural and energy production and population growth (UNESCO, 2009). In developing countries, effective water management is complicated by the relative sparseness of water-related scientific data, the lack of resources for developing or maintaining water-bearing infrastructure, the lack of legal frameworks for water sharing within and across borders, and national economic goals and priorities that may indirectly impact how water resources are used. In some countries, ineffective or even corrupt political leadership can affect economic development overall, with indirect impacts on water management goals and effectiveness. As emphasized in UNESCO (2009), water management takes place within the larger political and economic environment of a country, so that even well-informed water managers sometimes implement less than ideal water-management solutions either because of limited economic resources or due to overriding political or economic objectives.

In sub-Saharan Africa, oil-poor Arab countries, and large urban slums in Asia and South America, population growth will create an increasing demand for food and for economic opportunities that require energy development, thereby stressing existing water supplies and increasing the potential for political and social destabilization, conflict and population dislocation. As the recent decades have shown, America's security is linked to the well-being of societies halfway around the world in ways that until recently could not have been imagined, as failed states become breeding grounds for radical ideologies that are counter to America's national security interests or havens for warlords whose activities threaten the stability of neighboring states. In these fragile parts of the world, sustainable groundwater development could be an important part of a larger, long-term strategy for economic development that will ultimately make radical ideologies less appealing—a Marshall Plan for a new age.

9.4.4 Groundwater Management through Application of Numerical Groundwater Flow and Transport Models

To assess the consequences of groundwater development on groundwater levels or the impact of groundwater contamination on future groundwater quality, numerical groundwater flow and transport models are often employed as groundwater management tools. These models rely on the principle of mass conservation of water and solute both at the scale of single grid blocks or nodes of a computational grid and over the computational domain as a whole. A numerical groundwater flow or transport model is developed in such a way that it incorporates the essential features of the flow system, including (1) its initial state at the beginning of the period of interest, (2) inflows and outflows of water or solute at the boundaries of the model domain, (3) internal groundwater or solute sources and sinks, and (4) hydraulic and transport property distributions within the model domain. The USGS maintains a library of groundwater flow codes based on the MODFLOW (Harbaugh, 2005) code that have been adapted for specific applications involving solute transport, preferential flow through conduits such as karst features, variable density flow, surface-water/groundwater coupling, vadose-zone flow, and farm-specific conditions (http://water.usgs.gov/nrp/gwsoftware/modflow.html). These can be applied with calibration and optimization tools provided with MODFLOW, or with other parameter estimation software such as PEST (Doherty, 2007) or JUPITER (Banta et al., 2006). Enhancements to the basic (i.e., single-phase, single-component) groundwater flow and transport models can involve multicomponent liquid and gas flow, the transport of sensible and latent heat, and reactive transport (e.g., Lichtner, 2001; Zyvoloski, 2007). Because subsurface groundwater flow is generally assumed to be laminar, the physics represented in these codes are simpler than in other codes described in this chapter that consider turbulent fluid flows. Nonetheless, real-world applications of groundwater flow and transport models are often complicated by incomplete knowledge of the subsurface properties due to their inherent heterogeneity and our inability to directly observe or measure properties of the subsurface except through a limited number of often widely spaced wells. Although remote geophysical techniques can characterize some aspects of the subsurface relevant to hydrologic problems (e.g., Kirsch, 2006), many uncertainties remain that inhibit our ability to reliably predict flow and transport in the subsurface. Recognizing this limitation, many recent applications of subsurface flow and transport codes have attempted to characterize the uncertainty in possible outcomes via Monte Carlo techniques that consider uncertainty in the input parameters. The more advanced of these Monte Carlo techniques require that the model be constrained to the extent possible with calibration data, with only the model parameters that are uninformed by the calibration data free to vary according to the assumed parameter input distributions (Tonkin and Doherty, 2009).

Countless examples illustrating the application of groundwater flow and transport models to domestic groundwater management problems are available. Many of the groundwater issues confronting the nation described in the previous sections are already benefiting from the application of groundwater modeling tools at scales ranging from locally contaminated sites to regional aquifer systems spanning multiple states. The application of groundwater models to groundwater management problems in the developing world has historically been limited by lack of widespread access to computer resources, relevant experience, and the relative lack of subsurface data. However, with the widespread availability of inexpensive personal computers and increasing computer literacy, the lack of subsurface data may be the primary factor limiting the application of numerical groundwater models to groundwater management problems in the developing world.

9.4.5 Summary

Domestically, groundwater resources are critical to the nation's food production and water supply, and probably play an underestimated role in energy production because groundwater sustains stream flow during dry seasons when stream flow would otherwise cease. Additionally, groundwater sustains many aquatic habitats along rivers and wetlands that would be adversely affected by declining water levels or contamination. The economic vitality, social fabric of rural and agricultural communities, and ecosystem health are locally tied to the wise management of the nation's groundwater resources.

Because sustainable water development is critical to the economic growth and well-being of societies around the world, the United States has a broad national security interest in promoting sustainable water development projects as part of a larger initiative to promote economic development in impoverished states whose destabilization would promote humanitarian crises or regional conflicts.

9.5 Fire and Environmental Fluid Dynamics

9.5.1 Introduction

Another example of EFD with impacts on security occurs in the event of landscape-scale fire scenarios. These fires are those ranging in size from tens of meters to tens of kilometers. This class of fires includes both wildland fires and fires in urban setting and result from natural, accidental, and meditated events. Two-way coupled interactions at a variety of scales between the fire and its environment dictate the nature of the fire behavior including spread rates, fire intensity, plume dynamics, and emissions transport. EFD is one of the key vehicles through which fires are coupled with their environments through its influences on heating and cooling of unburned fuel, supply of oxygen to these often mixing-limited fires, buoyant plume dynamics, and the airborne transport of lofted burning solids that land in front of the fires. The length scales associated with the fluid mechanics processes that dictate fire behavior range between those of weather patterns, atmospheric boundary layer dynamics, flow through and around vegetation or structures, buoyancy-induced turbulence, and flame-sheet-scale mixing.

9.5.2 Wildland Fires

Wildland fire behavior is a direct result of the multiscaled coupled interaction between fire and winds. A brief examination of the ways that the fire-atmosphere coupling affects the basic spread mechanisms will illustrate the multiscale interaction. Various balances exist between influences of the ambient wind on the fire and the influences of the fire on the wind field that exist at a variety of length scales. Because these balances are also affected by ambient wind and fire intensity, there is a nonlinear feedback. At the scales of the entire fire there is a larger competition between the forces of the ambient wind and fire-induced buoyancy, which in turn leads to two major regimes of wildfire behavior, wind-driven and plume-dominated fires. In general, wind-driven fires exist when the ambient wind is strong enough to overcome the buoyant forces, yielding a predictable direction for fire spread and plume trajectories. Plume-dominated fires occur when the buoyant column overpowers the ambient winds and the local flow patterns become dominated by a largely vertical plume and the indrafts that occur at the base of these plumes.

The overall-fire-scale flow patterns directly influence the fire-line-width-scale flows and subsequently to the fuel-particle-scale flows, which together determine the mechanisms leading to fire spread: convective and radiative heating of unburned fuel (vegetation). In the case of wind-driven fires, where convective heat exchange is dominant, the air flow patterns typically include (1) ambient winds intermittently penetrating through the combustion zone, heating up as they intermingle with hot combustion gases, and then passing over the unburned fuels; (2) ambient winds being entrained around the curved flanking edges of wildfire perimeter as a result of the coupled effects of rising plume and strong ambient winds, entraining hot gases, that intermittently blow over unburned fuel. The resulting intermittent advection of hot gases over unburned fuel, which occur at fireline-width-scales but are driven by larger scale flows, becomes stronger with increasing ambient winds and thus heading-fire spread generally increases with wind speed.

For heading-fire spread in wind-driven wildfires, radiation heat transfer is far less important than it is in the combustion of structures because of the extremely high surface area to volume ratios of foliage, often on-the-order of 5000 m^{-1}. For these fine fuels, natural and forced convection at the scales of the fuel bed and fuel particle are able to negate significant amounts of radiative heat transfer as long as the fuel elements do not get engulfed in hot gases. When the fuel particles become surrounded by hot gases, the convective cooling ceases, and the strong heating effects of convective and radiative heating become additive. This should be contrasted to the radiative heating of large objects including structures, where the surface area per volume ratios are orders of magnitude smaller, conduction and thermal inertia become more important, wind speeds are reduced at the surface due to larger boundary layers, and the convective cooling is not as effective. This is most easily illustrated by the radiative heating of objects near a campfire. Blades of grass near the fire are often much cooler than one's skin, even when sitting farther away.

The convective and radiative balances are different in flanking-fire spread situations (fireline is nominally aligned with the ambient wind, and spreads perpendicular to wind). In this case, the average or mean wind moving through the unburned fuel bed is composed largely of ambient air because on average this air has not penetrated through the fireline to reach the unburned fuel. The combined ambient wind and fire-induced indraft that exist at fireline scales can cool the fuel adjacent to flanking portions of fires; thus, even though the fuels next to a flanking fireline are heated by radiation, convective cooling suppresses the lateral spread. This competition between radiative heating and convective cooling is similar to that which occurs on the grass blades at the edge of a campfire, where a fire-induced indraft continually cools the blades of grass and they normally do not ignite by radiation. This balance can be disrupted by turbulence in the atmosphere with scales ranging from meters to kilometers. When the winds waver from side to side or even lull momentarily, the hot gases can be momentarily ejected from the fire in irratic directions. When hot gases are ejected normal to a flanking fire, the convective cooling is temporarily eliminated. The net result is similar to a short-term head fire when the wind oscillates toward unburned fuel. The momentary combined effects of convective heating and the persistent radiative heating can be enough for the vegetation to begin to combust in a self-sustaining way, thus spreading the fire normal to the ambient wind.

In the case of a backing fire, a fire that is attempting to spread against the wind, the combined ambient wind and fire-induced indraft result in sufficiently large convective cooling that wildland fires are often unable to spread against the wind. In plume-dominated fire conditions, where the fluid dynamics has many similarities to oversized campfires, the spread is much less predictable than for wind-driven fires. The fire-spread mechanisms in plume-dominated fires are similar to those of the flanking or backing fires, since there is minimal ambient wind penetration through the plume, and the local flow fields near the fireline are dominated by indrafts which on average draw cool air toward the fireline. In the limit as the ambient-wind strength approaches zero compared with that of the buoyant plume on flat ground, the spread of the fire becomes dominated by fluctuations and instabilities in the flow field. Intermittent fluctuations in the wind field disturb this balance and allow fire to lunge forward in unpredictable directions, and occasional downwashes can lead to quick bursts and dangerous conditions.

9.5.3 Urban Fire Conflagrations

Urban conflagrations are another example of multiscale interaction between atmospheric fluid dynamics and fire, which has significant security implications. One possible origin of an urban fire conflagration could be the detonation of an explosive device in a large city. For this scenario there are many factors that would contribute to the nature of the ignition pattern, including energy released by the device, height of detonation, building construction, contents, and arrangement, etc. Once a large urban fire is ignited, the behavior of the fire is strongly affected by the coupled fire/atmosphere dynamics just as in the wildland fire scenario. Significant differences between the fires in urban and wildland settings include the effects of street canyons, much larger fuel loads, and very different fuel structures in urban settings. Two potential ignition patterns and fire evolutions would be (1) a central region of large fires and surrounding spot fires that begin to spread due to local winds and topography or (2) a set of fires that coalesce with enough buoyancy to create a firestorm. This is a direct analogy to the wind-driven and plume-dominated fire scenarios described for wildland fires. The impacts and ways to manage the effects of these fires can be naturally divided into two regimes: those associated with a firestorm and those that are not. A critical point is that if the firestorm occurs, the pattern of damage is likely to be significantly different from the patterns that would result from a spreading fire.

A firestorm causes a chimney effect where the quickly rising air gives rise to a low-pressure region underneath it and thus draws air in radially from its surroundings. This type of fire is illustrated in Figure 9.6 (although this test was in a wildland setting). This figure illustrates a firestorm resulting from a circular ignition of trees as a part of an experiment in Russia. Another significant aspect of the firestorm or plume-dominated fires in the urban setting is the channeling of the indrafts that is caused by the urban street canyons. Numerical simulations of urban firestorms have been conducted using HIGRAD/FIRETEC. The simulation, whose wind fields are illustrated in Figure 9.7, was conducted in low ambient wind conditions and therefore the fire-induced buoyancy is able to dominate the flow. The rising air draws a significant indraft which is channeled by street canyons and reaches magnitudes nearly 50 m/s at 3 m above the ground. Such strong horizontal inflow from the sides of the fire increases the oxygen supply to burning regions (aiding the complete consumption of everything inside the fire perimeter), but reduces the fire's ability to spread to new locations as discussed earlier for the plume-dominated wildland fires. The extent of

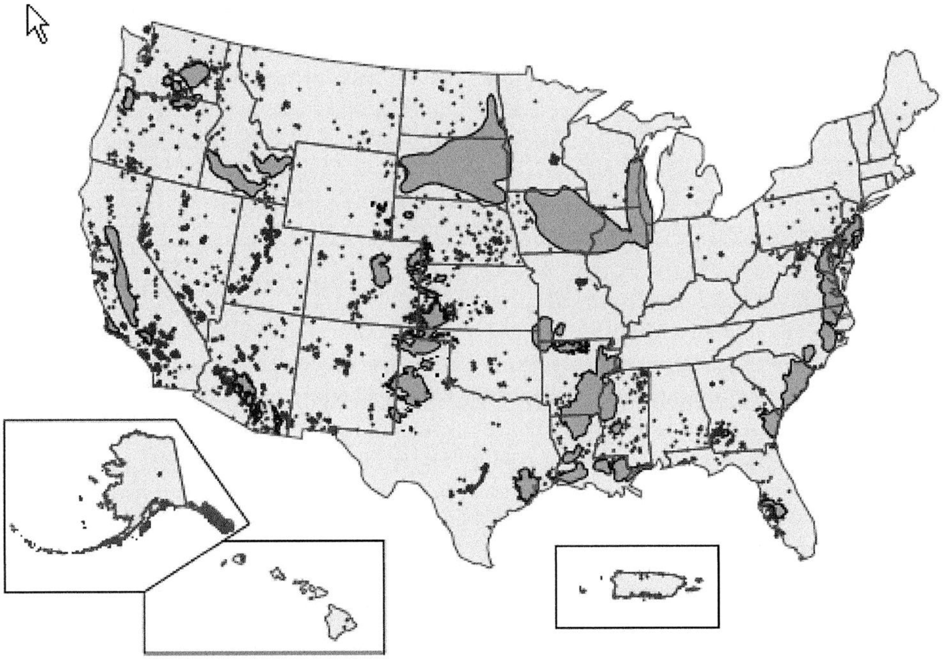

FIGURE 9.7 Water level declines in the United States relative to estimated predevelopment conditions. Red regions are areas in excess of 500 square miles, with a water-level decline of 40 ft in a confined aquifer or 25 ft in an unconfined aquifer. Blue dots are wells in the USGS National Water Information System with water-level declines of at least 40 ft. (Reproduced from Reilly, T.E. et al., *U.S. Geol. Surv. Circ.*, 1323, 70, Figure 12, 2008.)

noticeable indrafts is expected to be multiple miles within the street canyons. Moreover, the strength of these indrafts could affect evacuation efforts and potentially resuspend hazardous material that was dispersed during the blast event. The incineration of everything inside a firestorm will be much more complete than it would be for a fire that is propagating, because in a propagating fire convective cooling quenches smoldering material before it combusts completely. However, the total area that is severely burned could be much less in the case of a firestorm than a spreading wind-driven fire (Figures 9.8 and 9.9).

FIGURE 9.8 Firestorm in natural environments resulting from nominally circular ignition of a vegetation island. Flame heights near the epicenter of the fire were reported as being greater than 200 m in height, and the smoke column was more than 6 km high. (Image from Fire Research Campaign Asia-North (FIRESCAN)—Siberia, Kranoyarsk Region, 1993, coordinated by Global Fire Monitoring Center. Pictures courtesy of Global Fire Monitoring Center, Freiburg, Germany.)

FIGURE 9.9 Landscape view of firestorm-induced winds 3 m above street level in HIGRAD/FIRETEC simulation of urban firestorm. White vectors indicate horizontal flows, gold vectors indicate rising air, and purple vectors indicate falling air.

References

Andrews, M.J. and Dalziel, S.B., A review of Rayleigh-Taylor small Atwood number experiments, *Phil. Trans. R. Soc. A*, **368**, 1663–1679, March 2010.

Arya, S.P., *Introduction to Micrometeorology*, Academic Press, San Diego, CA, p. 65, 2001.

Banta, E.R., Poeter, E.P., Doherty, J.E., and Hill, M.C., *JUPITER: Joint Universal Parameter Identification and Evaluation of Reliability—An Application Programming Interface (API) for Model Analysis*, U.S. Geological Survey Techniques and Methods 6-E1, Denver, CO, 2006.

Boris, J.P., The threat of chemical and biological terrorism: Preparing a response, *Comp. Sci. Eng.*, **4**, 22–32, 2002.

Britter, R.E. and Hanna, S.R., Flow and dispersion in urban areas, *Ann. Rev. Fluid Mech.*, **35**, 469–496, 2003.

Brown, M.J., Reisner, J., Smith, S., and Langley, D., High fidelity urban scale modeling, LA-UR-01-1422, Los Alamos National Laboratory, Los Alamos, NM, 2001.

Brown, M., Urban dispersion—Challenges for fast response modeling, in *Fifth AMS Symposium on the Urban Environment*, LA-UR-04-5129, Los Alamos National Laboratory, Los Alamos, NM, 2004.

deBruyn, J.R., On the formation of periodic arrays of icicles, *Cold Reg. Sci. Technol.*, **25**(3), 225–229, 1997.

Doherty, J., *User's Manual for PEST Version 11*, Watermark Numerical Computing, Brisbane, Queensland, Australia, 339p., 2007.

Goncharov, V.N., Analytical model of nonlinear, single-mode, classical Rayleigh-Taylor instability at arbitrary Atwood numbers, *Phys. Rev. Lett.*, **88**(13), 134502, 2002.

Grinstein, F.F., Bos, R., and Dey, T., LES based urban dispersal predictions for consequence management, *ERCOFTAC Bull.*, **78**, 11–14, 2009.

Grinstein, F.F., Margolin, L.G., and Rider, W.J., *Implicit Large Eddy Simulation: Computing Turbulent Fluid Dynamics*, Cambridge University Press, Cambridge, U.K., 2007.

Gurdak, J.J., Hanson, R.T., and Green, T.R., Effects of climate variability and change on groundwater resources of the United States, U.S. Geological Survey Fact Sheet 2009-3074, Denver, CO, 2009.

Harbaugh, A.W., MODFLOW-2005: The U.S. Geological Survey modular ground-water model—The ground-water flow process, U.S. Geological Survey Techniques and Methods 6-A16, Reston, VA, 2005.

Hussein, H.J. and Martinuzzi, R.J., Energy balance for turbulent flow around a surface mounted cube placed in a channel, *Phys. Fluids*, **8**, 764–780, 1996.

Kenny, J.F., Barber, N.L., Hutson, S.S., Linsey, K.S., Lovelace, J.K., and Maupin, M.A., Estimated use of water in the United States in 2005, *U.S. Geol. Surv. Circ.*, **1344**, 52, 2009.

Kirsch, R., *Groundwater Geophysics: A Tool for Hydrogeology*, Springer Press, New York, 493p., 2006.

Landon, M.K., Jurgens, B.C., Katz, B.G., Eberts, S.M., Burow, K.R., and Crandall, C.A., Depth-dependent sampling to identify short-circuit pathways to public supply wells in multiple aquifer settings in the United States, *Hydrogeol. J.*, **17** (published online DOI 10.1007/s10040-009-0531-2), 2009.

Lichtner, P.C., FLOTRAN user's manual, Los Alamos National Laboratory, Report LA-UR-02-2349, Los Alamos, NM, 2001.

Mellado, J.P., Stevens, B., and Schmidt, H., Buoyancy-reversal incloud-top mixing layers, *Q. J. R. Meteorol. Soc.*, **135**(641, Pt. B), 963–978, 2009.

Patnaik, G., Boris, J.P., Young, T.R., and Grinstein, F.F., Large scale urban contaminant transport simulations with MILES, *J. Fluids Eng.*, **129**, 1524–1532, 2007.

Pullen, J., Boris, J.P., Young, T.R., Patnaik, G., and Iselin, J.P., A comparison of contaminant plume statistics from a Gaussian puff and urban CFD model for two large cities, *Atmos. Environ.*, **39**, 1049–1068, 2005.

Ramaprabhu, P. and Andrews, M.J., Experimental investigation of Rayleigh-Taylor mixing at small Atwood numbers, *J. Fluid Mech.*, **502**, 233–271, March 2004.

Rappolt, T.J., Measurements of atmospheric dispersion in the Los Angeles urban environment summer 2001, Project Report 1322, Tracer ES&T, San Marcos, CA, 2002.

Reilly, T.E., Dennehy, K.F., Alley, W.M., and Cunningham, W.L., Ground-water availability in the United States, *U.S. Geol. Surv. Circ.*, **1323**, 70, 2008.

Ribe, N.M., Spouting and planform selection in the Rayleigh-Taylor instability of miscible viscous fluids, *J. Fluid Mech.*, **377**, 27–45, 1998.

Sagaut, P., *Large Eddy Simulation for Incompressible Flows*, 3rd edn., Springer, Berlin, Germany, 2005.

Settles, G.S., Fluid mechanics of homeland security, *Ann. Rev. Fluid Mech.*, **38**, 87–110, January 2006.

Sharp, D.H., An overview of Rayleigh-Taylor instability, *Phys. D*, **12**, 3, 1984.

Tonkin, M. and Doherty, J., Calibration-constrained Monte Carlo analysis of highly parameterized models using subspace techniques, *Water Resour. Res.*, 45, doi:10.1029/2007WR006678, 2009.

United Nations Educational, Scientific and Cultural Organization (UNESCO), World water development report 3—Water in a changing world, UNESCO Publishing, Paris, France, 2009.

Voltz, C., Pesch, W., and Rehberg, I., Rayleigh-Taylor instability in a sedimenting suspension, *Phys. Rev. E*, **65**(1) 011404/1–011404/7, 2002.

Zoia, A., Latrille, C., Beccantini, A., and Cartadale, A., Spatial and temporal features of density-dependent contaminant transport: Experimental investigation and numerical modeling, *J. Contam. Hydrol.*, **109**(1–4), 14–26, 2009.

Zyvoloski, G.A., FEHM: A control volume finite element code for simulating subsurface multi-phase multi-fluid heat and mass transfer, Los Alamos National Laboratory, Report LA-UR-07-3359, Los Alamos, NM, 2007.

10

Large-Scale Disasters

Mohamed Gad-el-Hak
Virginia Commonwealth University

The subject of large-scale disasters is broadly introduced in this chapter. Many such disasters involve fluid transport in the oceans and atmosphere and hence the inclusion of the chapter in a book about environmental fluid mechanics. Both the art and science of predicting, preventing, and mitigating natural and human-caused disasters are discussed. A universal, quantitative metric that puts all natural and human-caused disasters on a common scale is proposed. The laws of nature govern the evolution of any disaster. In some cases, as for example, environmental or extreme-weather events, the first-principles laws of classical mechanics could be written in the form of field equations, but exact solutions of these often nonlinear differential equations are impossible to obtain, particularly for turbulent flows, and heuristic models together with intensive use of supercomputers are necessary to proceed to a reasonably accurate forecast. In other cases, as for example, earthquakes, the precise laws are not even known and prediction becomes more or less a black art. Management of any type of disaster is more art than science. Nevertheless, much can be done to alleviate the resulting pain and suffering. The present expansive presentation of the broad field of large-scale disasters precludes a detailed coverage of any one of the many topics touched upon. Three take-home messages are conveyed, however: a universal metric for all natural and human-caused disasters is presented, all facets of the genre are described, and a proposal is made to view all disasters as dynamical systems governed for the most part by the laws of classical mechanics.

10.1 Introduction

In this chapter, the subject of large-scale disasters is broadly introduced. Both the art and science of predicting, preventing, and mitigating natural and human-caused disasters are discussed. A universal, quantitative metric that puts all natural and human-caused disasters on a common scale is proposed. Issues of prediction, control, and mitigation of catastrophes are presented. The expansive presentation of the many facets of disaster research precludes a detailed coverage of any one of the many topics covered. We merely scratch the surface of a broad subject that may be of interest to all those who view the world mechanistically. The hope is that few readers of this book who are not already involved in disaster research would want to be engaged in this exciting endeavor whose practical importance cannot be overstated. The chapter is excerpted from Chapter 2 of the book edited by Gad-el-Hak [1].

10.1.1 Are Disasters a Modern Curse?

Although it appears that way when the past few years are considered, large-scale disasters have been with us since *Homo sapiens* set foot on this third planet from the Sun. Frequent disasters struck the Earth even before then, as far back as the time of its formation around 4.5 billion years ago. In fact, the geological Earth that we know today is believed to be the result of agglomeration of the so-called planetesimals and subsequent

Handbook of Environmental Fluid Dynamics, Volume One, edited by Harindra Joseph Shermal Fernando. © 2013 CRC Press/Taylor & Francis Group, LLC.
ISBN: 978-1-4398-1669-1.

impacts of bodies of similar mass [2]. The planet was left molten after each giant impact, and its outer crust was formed on radiative cooling to space. Those were the "good" disasters perhaps. On the bad side, there have been several mass extinctions throughout the Earth's history. The dinosaurs, along with about 70% of all species existing at the time, became extinct because a large meteorite struck the Earth 65 million years ago and the resulting airborne dust partially blocked the Sun, thus making it impossible for cold-blooded animals to survive. However, if we concern ourselves with our own warm-blooded species, then starting 200,000 years ago, ice ages, famines, infections, and attacks from rival groups and animals were constant reminders of human vulnerability. On average, there are about three large-scale disasters that strike the Earth every day, but only a few of these natural or human-caused calamities make it to the news. Humans have survived because we were programmed to do so.

10.1.2 Disaster Scope

There is no easy answer to the question of whether a particular disaster is large or small. The mild injury of one person may be perceived as catastrophic by that person or by his or her loved ones. What we consider herein, however, is the adverse effects of an event on a community or an ecosystem. What makes a disaster a large-scale one is the number of people affected by it and/or the extent of the geographic area involved. Such disaster taxes the resources of local communities and central governments. Under the weight of a large-scale disaster, a community diverges substantially from its normal social structure. Return to *normalcy* is typically a slow process that depends on the severity, but not the duration, of the antecedent calamity as well as the resources and efficiency of the recovery process.

The extreme event could be natural, human-caused, or a combination of the two in the sense of a natural disaster made worse by human's past actions. Examples of naturally occurring disasters include earthquakes, wildfires, pandemics, volcanic eruptions, mudslides, floods, droughts, and extreme weather phenomena such as ice ages, hurricanes, tornadoes, and sandstorms. Human foolishness, folly, meanness, mismanagement, gluttony, unchecked consumption of resources, or simply sheer misfortune may cause war, energy crisis, economic collapse of a nation or corporation, market crash, fire, global warming, famine, air/water pollution, urban sprawl, desertification, deforestation, bus/train/airplane/ship accident, oil slick, or terrorist act. Citizens suffering under the tyranny of a despot or a dictator can also be considered a disaster, and, of course, genocide, ethnic cleansing, and other types of mass murder are gargantuan disasters that often test the belief in our own humanity. Although technological advances exponentially increased human prosperity, they also provided humans with more destructive power. Human-caused disasters have led to the death of at least 200 million people during the twentieth century, a cruel age without equal in the history of man [3].

In addition to the degree or scope of a disaster, there is also the issue of the rapidity of the calamity. Earthquakes, for example, occur over extremely short time periods measured in seconds, whereas anthropogenic catastrophes such as global warming and air and water pollution are often slowly evolving disasters, their duration measured in years and even decades or centuries, although their devastation, over the long term, can be worse than that of a rapid, intense calamity [4]. The painful, slow death of a cancer patient who contracted the dreadful disease as a result of pollution is just as tragic as the split-second demise of a human at the hands of a crazed suicide bomber. The latter type of disaster makes the news, but the former does not. This is quite unsettling because the death of many spread over years goes unnoticed for the most part. The fact that 100 persons die in a week in a particular country as a result of starvation is not a typical news story. However, 100 humans perishing in an airplane crash will make CNN all day.

For the disaster's magnitude, how large is large? Much the same as is done to individually size hurricanes, tornadoes, earthquakes, and, very recently, winter storms, we propose herein a universal metric by which all types of disaster are sized in terms of the number of people affected and/or the extent of the geographic area involved. This quantitative scale applies to both natural and human-caused disasters. The suggested scale is nonlinear, logarithmic in fact, much the same as the Richter scale used to measure the severity of an earthquake. Thus, moving up the scale requires an order of magnitude increase in the severity of the disaster as it adversely affects people or an ecosystem. Note that a disaster may affect only a geographic area without any direct and immediate impact on humans. For example, a wildfire in an uninhabited forest may have long-term adverse effects on the local and global ecosystem, although no human is immediately killed, injured, or dislocated as a result of the event.

The *scope* of a disaster is determined if at least one of two criteria is met, relating to either the number of displaced/tormented/injured/killed people or the adversely affected area of the event. We classify disaster types as being of Scopes I–V, according to the scale pictorially illustrated in Figure 10.1.

For example, if 70 persons were injured as a result of a wildfire that covered $20\,km^2$, this would be considered Scope III, large disaster (the larger of the two categories II and III). However, if 70 persons were killed as a result of a wildfire that covered $2\,km^2$, this would be considered Scope II, medium disaster.

The quantitative metric introduced herein is contrasted to the conceptual scale devised by Fischer [5,6], which is based on the degree of social disruption resulting from an actual or potential disaster. Fischer's ten disaster categories are based on the scale, duration, and scope of disruption and adjustment of a normal social structure, but those categories are purely qualitative. For example, Disaster Category 3 (DC-3) is indicated if the event partially strikes a small town (major scale, major duration, and partial scope), whereas DC-8 is reserved for a calamity massively striking a large city (major scale, major duration, and major scope).

The primary advantage of having a universal classification scheme such as the one proposed herein is that it gives officials a quantitative measure of the magnitude of the disaster

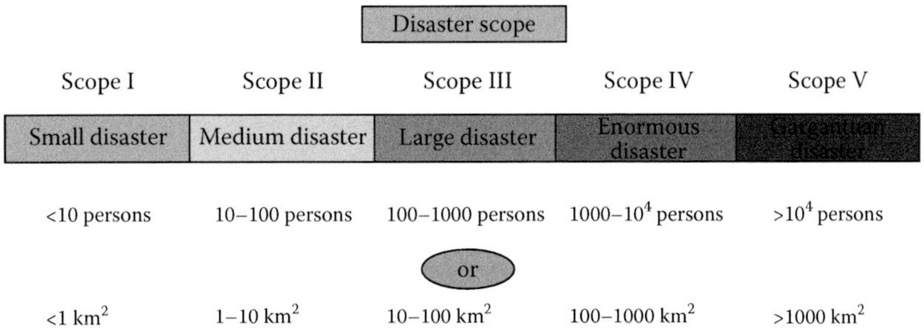

FIGURE 10.1 Classification of disaster severity.

so that proper response can be mobilized and adjusted as warranted. The metric suggested applies to *all* types of disaster. It puts them on a common scale, which is more informative than the variety of scales currently used for different disaster types: the Saffir–Simpson scale for hurricanes, the Fujita scale for tornadoes, the Richter scale for earthquakes, and the recently introduced Northeast Snowfall Impact Scale (notable, significant, major, crippling, extreme) for the winter storms that occasionally strike the northeastern region of the United States. Of course, the individual scales also have their utility; for example, knowing the range of wind speeds in a hurricane as provided by the Saffir–Simpson scale is a crucial piece of information to complement the number of casualties the proposed scale supplies. In fact, a prediction of wind speed allows estimation of potential damage to people and property. The proposed metric also applies to disasters, such as terrorist acts or droughts, where no quantitative scale is otherwise available to measure their severity.

In formulating all scales, including the proposed one, a certain degree of arbitrariness is unavoidable. In other words, none of the scales is totally objective. The range of 10–100 persons associated with a Scope II disaster, for example, could very well be 20–80, or some other range. What is important is the relative comparison among various disaster degrees; a Scope IV disaster causes an order of magnitude more damage than a Scope III disaster, and so on. One could arbitrarily continue beyond five categories, always increasing the influenced number of people and geographic area by an order of magnitude, but it seems that any calamity adversely affecting more than 10,000 persons or 1,000 km^2 is so catastrophic that a single Scope V is adequate to classify it as a gargantuan disaster. The book *Catastrophe* is devoted to analyzing the risk of and response to unimaginable but not impossible calamities that have the potential of wiping out the human race [7]. Curiously, its author, Richard A. Posner, is a judge in the U.S. Seventh Circuit Court of Appeals.

In the case of certain disasters, the scope can be predicted in advance to a certain degree of accuracy; otherwise, the scope can be estimated shortly after the calamity strikes with frequent updates as warranted. The magnitude of the disaster should determine the size of the first-responder contingency to be deployed; which hospitals to mobilize and to what extent; whether the military forces should be involved; what resources, such as food, water, medicine, and shelter, should be stockpiled and delivered to the stricken area; and so on. Predicting the scope should facilitate the subsequent recovery and accelerate the return to normalcy.

10.1.3 Facets of Large-Scale Disasters

A large-scale disaster is an event that adversely affects a large number of people, devastates a large geographic area, and taxes the resources of local communities and central governments. Although disasters can naturally occur, humans can cause their share of devastation. There is also the possibility of human actions causing a natural disaster to become more damaging than it would otherwise. An example of such an anthropogenic calamity is the intense coral reef mining off the Sri Lankan coast, which removed the sort of natural barrier that could mitigate the force of waves. As a result of such mining, the 2004 Pacific tsunami devastated Sri Lanka much more than it would have otherwise [1]. A second example is the soil erosion caused by overgrazing, farming, and deforestation. In April 2006, wind from the Gobi Desert dumped 300,000 tons of sand and dust on Beijing, China. Such gigantic dust tempests—exasperated by soil erosion—blew around the globe, making people sick, killing coral reefs, and melting mountain snow packs continents away [1]. Examples such as those incited the 1995 Nobel laureate and Dutch chemist Paul J. Crutzen to coin the present geological period as *anthropocene* to characterize humanity's adverse effects on global climate and ecology, http://www.mpch-mainz.mpg.de/~air/anthropocene/

What could make the best of a bad situation is to be able to predict the disaster's occurrence, location, and severity. This can help prepare for the calamity and evacuating, as needed, large segments of the population out of harm's way. For certain disaster types, their evolution equations can be formulated as a dynamical system, mostly from a mechanistic viewpoint. Predictions can then be made to different degrees of success using heuristic models, empirical observations, and giant computers. Once formed, the path and intensity of a hurricane, for example, can be predicted to a reasonable degree of accuracy up to 1 week in the future. This provides sufficient warning to evacuate several small, medium, and large cities in the path of the extreme event. However, smaller-scale severe weather such as tornadoes can only be predicted up to 15 min in the future, giving very little window for action. Earthquakes cannot be predicted beyond

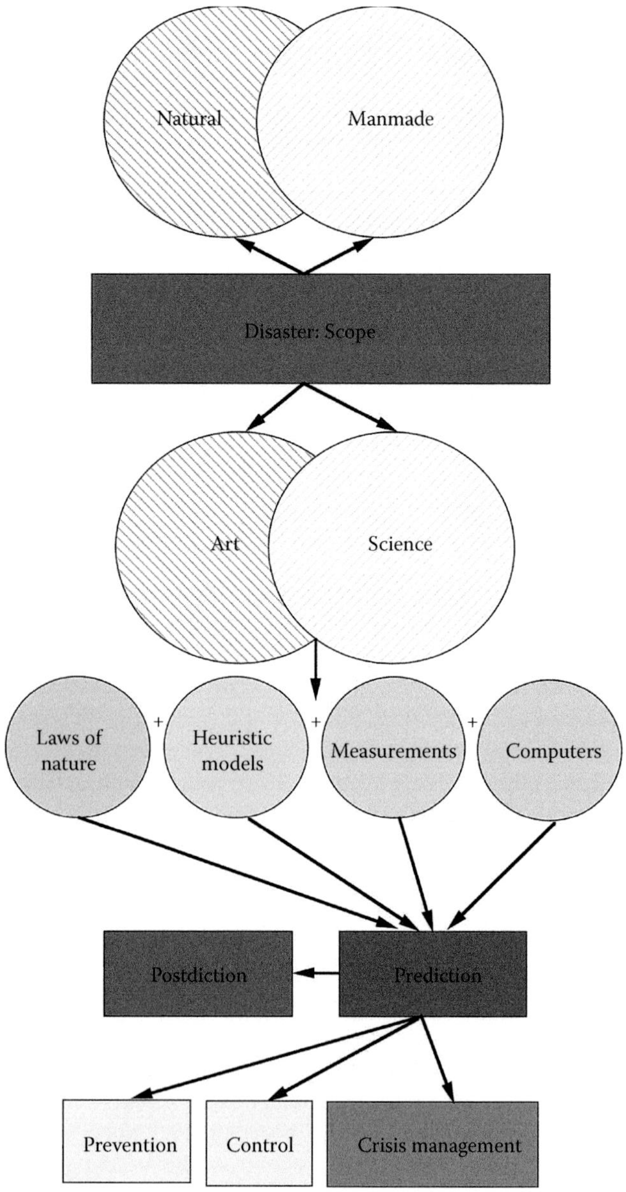

FIGURE 10.2 Schematic of the different facets of large-scale disasters.

stating that there is a certain probability of occurrence of a certain magnitude earthquake at a certain geographic location during the next 50 years. Such predictions are almost as useless as stating that the Sun will burn out in a few billion years.

Once disaster strikes, mitigating its adverse effects becomes the primary concern: how to save lives, take care of the survivors' needs, and protect properties from any further damage. Dislocated people need shelter, water, food, and medicine. Both the physical and the mental health of the survivors, as well as relatives of the deceased, could be severely jeopardized. Looting, price gouging, and other law-breaking activities need to be contained, minimized, or eliminated. Hospitals need to prioritize and even ration treatments, especially in the face of the practical fact that the less seriously injured tend to arrive at emergency rooms first, perhaps because they transported themselves there. Roads need to be operable and free of landslides, debris, and traffic jams for the unhindered flow of first responders and supplies to the stricken area, and evacuees and ambulances from the same. This is not always the case, especially if the antecedent disaster damages most if not all roads as occurred after the 2005 Kashmir Earthquake. Buildings, bridges, and roads need to be rebuilt or repaired, and power, potable water, and sewage need to be restored.

Figure 10.2 depicts the different facets of large-scale disasters. The important thing is to judiciously employ the finite resources available to improve the science of disaster prediction, and to artfully manage the resulting mess to minimize loss of life and property.

10.2 Scientific Principles

Science, particularly classical mechanics, could help predict the course of certain types of disaster. When, where, and how intense would a severe weather phenomena strike? Are the weather conditions favorable for extinguishing a particular wildfire? What is the probability of a particular volcano erupting? How about an earthquake striking a population center? How much air and water pollution is going to be caused by the addition of a factory cluster to a community? How would a toxic chemical or biological substance disperse in the atmosphere or in a body of water? Below a certain concentration, certain hazardous substances are harmless, and "safe" and "dangerous" zones could be established based on the dispersion forecast. The degree of success in answering these and similar questions varies dramatically. Once formed, the course and intensity of a hurricane (tropical cyclone), which typically lasts from inception to dissipation for a few weeks, could be predicted about 1 week in advance. The path of the much smaller and short-lived, albeit more deadly, tornado could be predicted only about 15 min in advance, although weather conditions favoring its formation could be predicted a few hours ahead. Earthquake prediction is far from satisfactory but is seriously attempted nevertheless. The accuracy of predicting volcanic eruptions is somewhere in between those of earthquakes and severe weather.

Patanè et al. [8] report on the ability of scientists to "see" inside Italy's Mount Etna and forecast its eruption using seismic tomography, a technique similar to that used in computed tomography scans in the medical field. The method yields time photographs of the 3D movement of rocks to detect their internal changes. The success of the technique is in no small part due to the fact that Europe's biggest volcano Mount Etna is equipped with a high-quality monitoring system and seismic network, tools that are not readily available for most other volcanoes.

Science and technology could also help control the severity of a disaster, but here the achievements to date are much less spectacular than those in the prediction arena. Cloud seeding to avert drought is still far from being a routine, practical tool. Nevertheless, it has been tried since 1946. In 2008, Los Angeles county officials used the technique as part of a drought-relief project that used silver iodide to seed clouds over the San Gabriel Mountains to ward off fires. China employed the same technology to bring some rain and clear the air before the 2008 Beijing Summer Olympics. Despite the difficulties, cloud seeding is still a notch more rational than the then Governor of Texas George W. Bush's 1999 call in the midst of a dry period to "pray for rain."

Slinging a nuclear device toward an asteroid or a meteor to avert its imminent collision with Earth remains solidly in the realm of science fiction (in the 1998 film *Armageddon*, a Texas-size asteroid was courageously nuked from its interior!). In contrast, employing scientific principles to combat a wildfire is doable, as is the development of scientifically based strategies to reduce air and water pollution, moderate urban sprawl, evacuate a large city, and minimize the probability of accident for air, land, and water vehicles. Structures could be designed to withstand an earthquake of a given magnitude, wind of a given speed, and so on. Dams could be constructed to moderate the flood–drought cycles of rivers, and levees/dikes could be erected to protect land below sea level from the vagaries of the weather. Storm drains; fire hydrants; fire-retardant materials; sprinkler systems; pollution control; simple hygiene; strict building codes; traffic rules and regulations in air, land, and sea; and many other examples are the measures a society should take to mitigate or even eliminate the adverse effects of certain natural and human-caused disasters. Of course, there are limits to what we can do. Although much better fire safety will be achieved if a firehouse is erected, equipped, and manned around every city block, and less earthquake casualties will occur if every structure is built to withstand the strongest possible tremor, the prohibitive cost of such efforts clearly cannot be justified or even afforded.

At the extreme scale, geoengineering is defined as options that would involve large-scale engineering of our environment in order to combat or counteract the effects of changes in atmospheric chemistry. Along those lines, Nobel laureate Paul J. Crutzen has proposed a method of artificially cooling the global climate by releasing particles of sulfur in the upper atmosphere, which would reflect sunlight and heat back into space. The controversial proposal is being taken seriously by scientists because Crutzen has a proven track record in atmospheric research. Sponsored by the U.S. National Science Foundation, a scientific meeting was held in 2008 to explore far-fetched strategies to combat hurricanes and tornadoes.

In contrast to natural disasters, human-caused ones are generally somewhat easier to control but more difficult to predict. The war on terrorism is a case in point. Who could predict the behavior of a crazed suicide bomber? A civilized society spends its valuable resources on intelligence gathering, internal security, border control, and selective/mandatory screening to prevent (control) such devious behavior, whose dynamics (i.e., time evolution) obviously cannot be distilled into a differential equation to be solved. However, even in certain disastrous situations that depend on human behavior, predictions could sometimes be made; crowd dynamics being a prime example where the behavior of a crowd in an emergency could to some degree be modeled and anticipated so that adequate escape or evacuation routes could be properly designed [9]. Helbing et al. [10] write on simulation of panic situations and other crowd disasters modeled as nonlinear dynamical systems. All such models are heuristic and do not stem from the first-principles laws of classical mechanics.

The tragedy of the numerous human-caused disasters is that they are all preventable, at least in principle. We cannot prevent a hurricane, at least not yet, but using less fossil fuel and seeking alternative energy sources could at least slow down global warming trends. Conflict resolution strategies could be employed between nations to avert wars. Speaking of wars, the Iraqi–American poet Dunya Mikhail lamenting on the many human-caused disasters, calls the present period "The Tsunamical Age." A bit more humanity, commonsense, selflessness, and moderation, as well as a bit less greed, meanness, selfishness, and zealotry, and the world will be a better place for having fewer human-caused disasters.

10.3 Methods of Analysis

For disasters that involve (fluid) transport phenomena, such as severe weather, fire, and release of toxic substance, the governing equations can be formulated subject to some assumptions, the less the better. Modeling is usually in the form of nonlinear partial differential equations with an appropriate number of initial and boundary conditions. Integrating those field equations leads to the time evolution, or the dynamics, of the disaster. In principle, marching from the present (initial conditions) to the future gives the potent predictability of classical mechanics and ultimately leads to the disaster's forecast.

However, the first principles equations are typically impossible to solve analytically, particularly if the fluid flow is turbulent, which unfortunately is the norm for the high Reynolds number flows encountered in the atmosphere and oceans. Furthermore, initial and boundary conditions are required for both analytical and numerical solutions, and massive amounts of data need to be collected to determine those conditions with sufficient resolution and accuracy. Computers are not big enough either, so numerical integration of the instantaneous equations (direct numerical simulations) for high Reynolds number natural flows is computationally prohibitively expensive if not outright impossible at least for now and the foreseeable future. Heuristic modeling then comes to the rescue but at a price. Large eddy simulations, spectral methods, probability density function models, and the more classical Reynolds stress models are examples of such closure schemes that are not as computationally intensive as direct numerical simulations, but are not as reliable either. This type of second-tier modeling is phenomenological in nature and does not stem from first principles. The more heuristic the modeling is, the less accurate the expected results are.

Together with massive ground, sea, and sky data to provide at least in part the initial and boundary conditions, the models are entered into supercomputers that come out with a forecast, whether it is a prediction of a severe thunderstorm that is yet to form, the future path and strength of an existing hurricane, or the impending concentration of a toxic gas that was released in a faraway location sometime in the past. The issue of nonintegrability of certain dynamical systems is an additional challenge and opportunity.

For other types of disasters such as earthquakes, the precise laws are not even known mostly because proper constitutive relations are lacking. Additionally, deep underground data are difficult to gather to say the least. Predictions in those cases become more or less a black art.

In the next three sections, we focus on the prediction of disasters involving fluid transport. This important subject has spectacular successes within the past few decades, for example, in being able to predict the weather a few days in advance. The accuracy of today's 5 day forecast is the same as the 3 and 1.5 day ones in 1976 and 1955, respectively. The 3 day forecast of a hurricane's strike position is accurate to within 100 km, about a 1 h drive on the highway [11].

The painstaking advances made in fluid mechanics in general and turbulence research in particular together with the exponential growth of computer memory and speed undoubtedly contributed immeasurably to those successes.

The British physicist Lewis Fry Richardson was perhaps the first to make a scientifically based weather forecast. Based on data taken at 7:00 am, May 20, 1910, he made a 6 h "forecast" that took him 6 weeks to compute using a slide rule. The belated results* were totally wrong as well! In his remarkable book, Richardson [12] wrote "Perhaps someday in the dim future it will be possible to advance the computations faster than the weather advances and at a cost less than the saving to mankind due to the information gained. But that is a dream" (p. vii). We are happy to report that Richardson's dream is one of the few that came true. A generation ago, the next day's weather was hard to predict. Today, the 10 day forecast is available 24/7 on www.weather.com for almost any city in the world. Not very accurate perhaps, but far better than the pioneering Richardson's 6 h forecast.

The important issue is to precisely state the assumptions needed to write the evolution equations, which are basically statements of the conservation of mass, momentum, and energy, in a certain form. The resulting differential equations and their eventual analytical or numerical solutions are only valid under those assumptions. This seemingly straightforward fact is often overlooked, and wrong answers readily result when the situation we are trying to model is different from that assumed. Much more details of the science of disaster's prediction are provided in a book edited by the present author [1].

10.3.1 Fundamental Transport Equations

Each fundamental law of fluid mechanics and heat transfer—conservation of mass, momentum, and energy—are listed first in their raw form, that is, assuming only that the speeds involved are nonrelativistic and that the fluid is a continuum. In nonrelativistic situations, mass and energy are conserved separately and are not interchangeable. This is the case for all normal fluid velocities that we deal with in everyday situations—far below the speed of light. The continuum assumption ignores the grainy (microscopic) structure of matter. It implies that the derivatives of all the dependent variables exist in some reasonable sense. In other words, local properties such as density and velocity are defined as averages over large elements compared with the microscopic structure of the fluid but small enough in comparison with the scale of the macroscopic phenomena to permit the use of differential calculus to describe them. The resulting equations therefore cover a broad range of situations, the exception being flows with spatial scales that are not much larger than the mean distance between the fluid molecules, as for example in the case of rarefied gas dynamics, shock waves that are thin relative to the mean free path, or flows in micro- and nanodevices. Thus, at every point in space–time in an inertial (i.e., nonaccelerating/nonrotating), Eulerian frame of reference, the three conservation laws of mass, momentum, and energy for nonchemically reacting fluids read in Cartesian tensor notations:

$$\frac{\partial \rho}{\partial t} + \frac{\partial}{\partial x_k}(\rho u_k) = 0 \tag{10.1}$$

$$\rho\left(\frac{\partial u_i}{\partial t} + u_k \frac{\partial u_i}{\partial x_k}\right) = \frac{\partial \Sigma_{ki}}{\partial x_k} + \rho g_i \tag{10.2}$$

$$\rho\left(\frac{\partial e}{\partial t} + u_k \frac{\partial e}{\partial x_k}\right) = -\frac{\partial q_k}{\partial x_k} + \Sigma_{ki}\frac{\partial u_i}{\partial x_k} \tag{10.3}$$

where

ρ is the fluid density
u_k is an instantaneous velocity component (u,v,w)
Σ_{ki} is the second-order stress tensor (surface force per unit area)
g_i is the body force per unit mass
e is the internal energy
q_k is the sum of heat flux vectors due to conduction and radiation

The independent variables are time t and the three spatial coordinates x_1, x_2, and x_3 or (x,y,z). Finally, the Einstein's summation convention applies to all repeated indices. Gad-el-Hak [13] provides a succinct derivation of the above conservation laws for a continuum, nonrelativistic fluid.

10.3.2 Closing the Equations

Equations (10.1) through (10.3) constitute 5 differential equations for the 17 unknowns ρ, u_i, Σ_{ki}, e, and q_k. Absent any body couples, the stress tensor is symmetric having only 6 independent components, which reduces the number of unknowns to 14. Obviously, the continuum flow equations do not form a determinate set. To close the conservation equations, relation between

* Actually delayed by a few years due to World War I and relocation to France. Richardson chose that particular time and date because upper air and other measurements were available to him some years before.

the stress tensor and the deformation rate, relation between the heat flux vector and the temperature field and appropriate equations of state relating the different thermodynamic properties are needed. The stress–rate of strain relation and the (conductive) heat flux–temperature gradient relation are approximately linear if the flow is not too far from thermodynamic equilibrium. This is a phenomenological result but can be rigorously derived from the Boltzmann equation for a dilute gas assuming the flow is near equilibrium. Thermodynamic equilibrium implies that the macroscopic quantities have sufficient time to adjust to their changing surroundings. In motion, exact thermodynamic equilibrium is impossible because each fluid particle is continuously having volume, momentum, or energy added or removed, and so in fluid dynamics and heat transfer we speak of quasi-equilibrium. The second law of thermodynamics imposes a tendency to revert to equilibrium state, and the defining issue here is whether the flow quantities are adjusting fast enough. The reversion rate will be very high if the molecular time and length scales are very small as compared to the corresponding macroscopic flow scales. This will guarantee that numerous molecular collisions will occur in sufficiently short time to equilibrate fluid particles whose properties vary little over distances comparable to the molecular length scales. Gas flows are considered in a state of quasi-equilibrium if the Knudsen number, which is the ratio of the mean free path to a characteristic length of the flow, is less than 0.1. In such flows, the stress is linearly related to the strain rate, and the (conductive) heat flux is linearly related to the temperature gradient. Empirically, common liquids such as water follow the same laws under most flow conditions. Gad-el-Hak [14,15] provides extensive discussion of situations in which the quasi-equilibrium assumption is violated. These may include gas flows at great altitudes, flows of complex liquids such as long-chain molecules, and even ordinary gas and liquid flows when confined in micro- and nanodevices, or when subjected to unusually high strain rates or temperature gradients.

For a Newtonian, isotropic, Fourier,* ideal gas, for example, those constitutive relations read

$$\Sigma_{ki} = -p\delta_{ki} + \mu\left(\frac{\partial u_i}{\partial x_k} + \frac{\partial u_k}{\partial x_i}\right) + \lambda\left(\frac{\partial u_j}{\partial x_j}\right)\delta_{ki} \tag{10.4}$$

$$q_i = -\kappa\frac{\partial T}{\partial x_i} + \text{Heat flux due to radiation} \tag{10.5}$$

$$de = c_v\, dT \quad \text{and} \quad p = \rho R T \tag{10.6}$$

where
- p is the thermodynamic pressure
- μ and λ are the first and second coefficients of viscosity, respectively
- δ_{ki} is the unit second-order tensor (Kronecker delta)
- κ is the thermal conductivity
- T is the temperature field
- c_v is the specific heat at constant volume
- R is the gas constant which is given by the Boltzmann constant divided by the mass of an individual molecule $k = mR$

The Stokes' hypothesis relates the first and second coefficients of viscosity, thus $\lambda + \frac{2}{3}\mu = 0$, although the validity of this assumption for other than dilute, monatomic gases has occasionally been questioned [16]. With the aforementioned constitutive relations and neglecting radiative heat transfer,† Equations (10.1) through (10.3), respectively, read

$$\frac{\partial \rho}{\partial t} + \frac{\partial}{\partial x_k}(\rho u_k) = 0 \tag{10.7}$$

$$\rho\left(\frac{\partial u_i}{\partial t} + u_k\frac{\partial u_i}{\partial x_k}\right) = -\frac{\partial p}{\partial x_i} + \rho g_i + \frac{\partial}{\partial x_k}\left[\mu\left(\frac{\partial u_i}{\partial x_k} + \frac{\partial u_k}{\partial x_i}\right) + \delta_{ki}\,\lambda\frac{\partial u_j}{\partial x_j}\right] \tag{10.8}$$

$$\rho\left(\frac{\partial e}{\partial t} + u_k\frac{\partial e}{\partial x_k}\right) = \frac{\partial}{\partial x_k}\left(\kappa\frac{\partial T}{\partial x_k}\right) - p\frac{\partial u_k}{\partial x_k} + \phi \tag{10.9}$$

The three components of the vector equation (10.8) are the Navier–Stokes equations expressing the conservation of momentum for a Newtonian fluid. In the thermal energy equation (10.9), ϕ is, according to the second law of thermodynamics, the always-positive dissipation function expressing the irreversible conversion of mechanical energy to internal energy as a result of the deformation of a fluid element. The second term on the right-hand side of Equation (10.9) is the reversible work done (per unit time) by the pressure as the volume of a fluid material element changes. For a Newtonian, isotropic fluid, the viscous dissipation rate is given by

$$\phi = \frac{1}{2}\mu\left(\frac{\partial u_i}{\partial x_k} + \frac{\partial u_k}{\partial x_i}\right)^2 + \lambda\left(\frac{\partial u_j}{\partial x_j}\right)^2 \tag{10.10}$$

There are now six unknowns, ρ, u_i, p, and T, and the five coupled Equations (10.7) through (10.9) plus the equation of state relating pressure, density, and temperature. These six equations together with sufficient number of initial and boundary conditions constitute a well-posed, albeit formidable, problem. The system of

* Newtonian implies a linear relation between the stress tensor and the symmetric part of the deformation tensor (rate of strain tensor). The isotropy assumption reduces the 81 constants of proportionality in that linear relation to two constants. Fourier fluid is that for which the conduction part of the heat flux vector is linearly related to the temperature gradient, and again isotropy implies that the constant of proportionality in this relation is a single scalar.

† An extremely important term in environmental fluid mechanics.

Equations (10.7) through (10.9) is an excellent model for the laminar or turbulent flow of most fluids such as air and water under many circumstances, including high-speed gas flows for which the shock waves are thick relative to the mean free path of the molecules.

Considerable simplification is achieved if the flow is assumed incompressible, usually a reasonable assumption provided that the characteristic flow speed is less than 0.3 of the speed of sound. The incompressibility assumption, discussed in great detail in Ref. [17], is readily satisfied for almost all liquid flows and many gas flows. In such cases, the density is assumed either a constant or a given function of temperature (or species concentration). The governing equations for such flow are

$$\frac{\partial u_k}{\partial x_k} = 0 \tag{10.11}$$

$$\rho\left(\frac{\partial u_i}{\partial t} + u_k \frac{\partial u_i}{\partial x_k}\right) = -\frac{\partial p}{\partial x_i} + \frac{\partial}{\partial x_k}\left[\mu\left(\frac{\partial u_i}{\partial x_k} + \frac{\partial u_k}{\partial x_i}\right)\right] + \rho g_i \tag{10.12}$$

$$\rho c_p\left(\frac{\partial T}{\partial t} + u_k \frac{\partial T}{\partial x_k}\right) = \frac{\partial}{\partial x_k}\left(\kappa \frac{\partial T}{\partial x_k}\right) + \phi_{\text{incomp}} \tag{10.13}$$

where ϕ_{incomp} is the incompressible limit of Equation 10.10. These are now five equations for the five dependent variables u_i, p, and T. Note that the left-hand side of Equation 10.13 has the specific heat at constant pressure c_p and not c_v. It is the convection of enthalpy—and not internal energy—that is balanced by heat conduction and viscous dissipation. This is the correct incompressible flow limit—of a compressible fluid—as discussed in detail in section 10.9 of Panton [17]; a subtle point perhaps but one that is frequently missed in textbooks. The system of Equations (10.11) through (10.13) is coupled if either the viscosity or density depends on temperature; otherwise, the energy equation is uncoupled from the continuity and momentum equations, and could therefore be solved *after* the velocity and pressure fields are determined from solving (10.11) and (10.12). For most geophysical flows, the density depends on temperature and/or species concentration, and the previous system of five equations is coupled.

For both the compressible and the incompressible equations of motion, the transport terms are neglected away from solid walls in the limit of infinite Reynolds number. The fluid is then approximated as inviscid and nonconducting, and the corresponding equations read (for the compressible, ideal gas case)

$$\frac{\partial \rho}{\partial t} + \frac{\partial}{\partial x_k}(\rho u_k) = 0 \tag{10.14}$$

$$\rho\left(\frac{\partial u_i}{\partial t} + u_k \frac{\partial u_i}{\partial x_k}\right) = -\frac{\partial p}{\partial x_i} + \rho g_i \tag{10.15}$$

$$\rho c_v\left(\frac{\partial T}{\partial t} + u_k \frac{\partial T}{\partial x_k}\right) = -p\frac{\partial u_k}{\partial x_k} \tag{10.16}$$

The Euler equation (10.15) could be integrated along a streamline, and the resulting Bernoulli's equation provides a direct relation between the velocity and the pressure.

10.3.3 Other Complexities

Despite their already complicated nature, the transport equations introduced earlier could be further entangled by other effects. We list herein a few examples. Geophysical flows occur at such large length scales as to invalidate the inertial frame assumption made previously. The Earth's rotation affects these flows, and such things as centrifugal and Coriolis forces enter into the equations rewritten in a non-inertial frame of reference fixed with the rotating Earth. Oceanic and atmospheric flows are more often than not turbulent flows that span the enormous range of length scales of nine decades, from a few millimeters to thousands of kilometers [18,19].

Density stratification is important for many atmospheric and oceanic phenomena. Buoyancy forces are produced by density variations in a gravitational field, and those forces drive significant convection in natural flows [20]. In the ocean, those forces are further complicated by the competing influences of temperature and salt [18]. The competition affects the large-scale global ocean circulation and, in turn, climate variability. For weak density variations, the Bousinessq approximation permits the use of the coupled incompressible flow equations, but more complexities are introduced in situations with strong density stratification, such as when strong heating and cooling are present. Complex topography further complicates convective flows in the ocean and atmosphere.

Air–sea interface governs many of the important transport phenomena in the ocean and atmosphere, and plays a crucial role in determining the climate. The location of that interface is itself not known a priori and thus is the source of further complexity in the problem. Even worse, the free boundary nature of the liquid–gas interface, in addition to the possibility of breaking that interface and forming bubbles and droplets, introduces new nonlinearities that augment or compete with the customary convective nonlinearity [21]. Chemical reactions are obviously important in fires and are even present in some atmospheric transport problems. When liquid water or ice is present in the atmasphere, two-phase treatment of the equations of motion may need to be considered, again complicating even the relevant numerical solutions.

However, even in the complex situations described in this section, simplifying assumptions could be made rationally to facilitate solving the problem. Any spatial symmetries in the problem must be exploited. If the mean quantities are time independent, then that too could be exploited.

10.3.4 Earthquakes

Thus far in this section, we discussed prediction of the type of disasters involving fluid transport phenomena, weather-related disasters being the most rampant. Predictions are possible on those cases, and improvements in forecast's accuracy and extent are continually being made as a result of enhanced understanding of flow physics, increased accuracy and resolution of global measurements, and exponentially expanded computer power. Other types of disaster do not fare as well, earthquakes being calamities that thus far cannot be predicted. Prediction of weather storms is possible in part because the atmosphere is optically transparent, which facilitates measurements that in turn provide not only the initial and boundary conditions necessary for integrating the governing equations but also a deeper understanding of the physics. The oceans are not as accessible, but measurements there are possible as well, and scientists learned a great deal in the past few decades about the dynamics of both the atmosphere and the ocean [18,19]. Our knowledge of *terra firma*, in contrast, does not fare as well mostly because of its inaccessibility to direct observation [2]. What we know about the Earth's solid inner core, liquid outer core, mantle, and lithosphere comes mainly from inferences drawn from observations at or near the planet's surface, which include the study of propagation, reflection, and scatter of seismic waves. Deep underground measurements are not very practical, and the exact constitutive equations of the different constituents of the "solid" Earth are not known. All that inhibits us from writing down and solving the precise equations, and their initial and boundary conditions, for the dynamics of the Earth's solid part. That portion of the planet contains 3 orders of magnitude more volume than all the oceans combined and 6 orders of magnitude more mass than the entire atmosphere, and it is a pity that we know relatively little about the solid Earth.

The science of earthquake basically began shortly after the infamous rupture of the San Andreas Fault that devastated San Francisco a little more than a century ago. Before then, geologists had examined seismic faults and even devised primitive seismometers to measure shaking. However, they had no idea what caused the ground to heave without warning. A few days after the Great Earthquake struck on April 18, 1906, Governor George C. Pardee of California charged the state's leading scientists with investigating how and why the Earth's crust had ruptured for hundreds of miles with such terrifying violence. The foundation for much of what is known today about earthquakes was laid 2 years later, and the resulting report [22] carried the name of the famed geologist Andrew C. Lawson.

Earthquakes are caused by stresses in the Earth's crust that build up deep inside a fault until it ruptures with a jolt. Prior to the Lawson Report, most scientists believed earthquakes created the faults instead of the other way around. The San Andreas Fault system marks the boundary between two huge moving slabs of the Earth's crust: the Pacific Plate and the North American Plate. As the plates grind constantly past each other, strain builds until it is released periodically in a full-scale earthquake. A few small sections of the San Andreas Fault had been mapped by scientists years before 1906, but Lawson and his team discovered that the entire zone stretched for more than 950 km along the length of California. By measuring land movements on either side of the fault, the team learned that the earthquake's motion had moved the ground horizontally, from side to side, rather than just vertically as scientists had previously believed.

A century after the Lawson Report, its conclusions remain valid, but it has stimulated modern earthquake science to move far beyond. Modern scientists have learned that major earthquakes are not random events—they apparently come in cycles. Although pinpoint prediction remains impossible, research on faults throughout the San Francisco Bay Area and other fault locations around the world enables scientists to estimate the probability that strong quakes will jolt a region within the coming decades. Sophisticated broadband seismometers can measure the magnitude of earthquakes within a minute or two of an event and determine where and how deeply on a fault the rupture started. Orbiting satellites now measure within fractions of a centimeter how the Earth's surface moves as strain builds up along fault lines, and again how the land is distorted after a quake has struck.

"Shakemaps," available on the Internet and by e-mail immediately after every earthquake, can swiftly tell disaster workers, utility companies, and residents where damage may be greatest. Supercomputers, simulating ground motion from past earthquakes, can show where shaking might be heaviest when new earthquakes strike. The information can then be relayed to the public and to emergency workers.

One of the latest and most important ventures in understanding earthquake behavior is the borehole drilling project at Parkfield in southern Monterey County, California, where the San Andreas Fault has been heavily instrumented for many years. The hole is about 3.2 km deep and crosses the San Andreas underground. For the first time, sensors can actually be inside the earthquake machine to catch and record the earthquakes right where and when they are occurring.

The seismic safety of any structure depends on the strength of its construction and the geology of the ground on which it stands—a conclusion reflected in all of today's building codes in the United States. Tragically, the codes in some earthquake-prone countries are just as strict as those in the United States, but are not enforceable for the most part. In other nations, building codes are not sufficiently strict or nonexistent altogether.

10.4 Applications

It is always useful to learn from past disasters and to prepare better for the next one. Losses of lives and property from recent years are staggering. Not counting the human-caused disasters that were tallied in Section 10.1.2, some frightening numbers from natural calamities alone are

- Seven hundred natural disasters in 2003, which caused 75,000 deaths (almost seven times the number in 2002), 213 million people adversely affected to some degree, and $65 billion in economic losses.
- In 2004, 244,577 persons were killed globally as a result of natural disasters.

TABLE 10.1 Scope of 13 Disasters

Disaster	Date	Scope	Descriptor	Basis
San Francisco earthquake	April 18, 1906	V	Gargantuan	3,000 death; 300,000 homeless
Hyatt Regency walkway collapse	July 17, 1981	III	Large	114 deaths; 200 injured
Loma Prieta earthquake	October 17, 1989	IV	Enormous	66 deaths; 3,757 injured
Izmit earthquake	August 17, 1999	V	Gargantuan	17,000–35,000 deaths
September 11	September 11, 2001	IV	Enormous	2,993 deaths; 6,291 injured
Pacific tsunami	December 26, 2004	V	Gargantuan	283,100 deaths
Hurricane Katrina	August 25, 2005	V	Gargantuan	1,836 deaths; 1.2 million evacuated
Kashmir earthquake	October 8, 2005	V	Gargantuan	73,276 deaths; 3.3 million homeless
Hurricane Wilma	October 18, 2005	V	Gargantuan	63 death (in the United States); 500,000 evacuated (in Cuba)
Hajj Stampede	January 12, 2006	III	Large	346 deaths; 289 injured
Al-Salam Boccaccio 98	February 3, 2006	IV	Enormous	1,094 deaths
Bird flu	2003–present	III	Large	Number of stricken in the hundreds
Energy crises/global warming	Since industrial revolution	V	Gargantuan	Covers entire Earth

- In 2005, $150 billion in economic losses, with hurricanes Katrina and Rita, which ravaged the Gulf Coast of the United States, responsible for 88% of that amount.
- Within the first half of 2006, natural disasters already caused 12,718 deaths and $2.3 billion in economic damages.

In Table 10.1, we evaluate the scope of 13 case studies used as examples of natural and human-caused disasters. More details about each of those cases can be found in Ref. [1]. The metric introduced in Section 10.1.2 is utilized to rank those disasters. Recall, the scope of a disaster is based on the number of people adversely affected by the extreme event (killed, injured, evacuated, etc.) or the extent of the stricken geographical area.

Of note is the scope of the "September 11" human-caused disaster, which is less than the scope of, say, hurricane Katrina. The number of people directly and adversely affected by September 11 is less than those in the case of Katrina (number of deaths is not the only measure). On the other hand, September 11 has a huge after effect in the United States and elsewhere, shifting the geopolitical realities and triggering the ensuing War on Terrorism that still rages many years later. The number of people adversely affected by that war is not considered in assigning a scope to September 11.

The wide range of casualties in the 1999 Izmit earthquake reflects the contradictory reports between the government and the news media. This is not a good indication of a responsible government.

The avian influenza is still in its infancy and fortunately has not yet materialized into a pandemic, hence the relatively low scope. The energy crises and its intimately related global warming problem have not yet resulted in widespread deaths or injuries, but both events are global in extent essentially affecting the entire world. Thus the descriptor gargantuan assigned to both is based on the size of the adversely affected geographical area.

10.5 Major Challenges

Disasters research—the prediction, control, and mitigation of both natural and human-caused disasters—is a vast field of research that no one chapter or even whole book can cover in any meaningful detail. In this chapter, we defined what constitutes a large-scale disaster, introduced a metric to evaluate its scope, and briefly described the different facets of disasters research. Basically, any natural or human-caused event that adversely affects many humans or an expanded ecosystem is a large-scale disaster. Such catastrophes tax the resources of local communities and central governments and disrupt social order. The number of people tormented, displaced, injured, or killed and the size of the area adversely affected determine the disaster's scope.

In this chapter, we showed how science could help predicting different types of disaster and reducing their resulting adverse effects. We listed a number of recent disasters to provide few examples of what can go right or wrong with managing the mess left behind every large-scale disaster.

The laws of nature, reflected in the science portion of any particular calamity, and even crisis management, reflected in the art portion, should be the same, or at least quite similar, no matter where or what type of disaster strikes. Humanity should benefit from the science and the art of predicting, controlling, and managing large-scale disasters, as discussed in this chapter and the book it is excerpted from Ref. [1].

The last *annus horribilis*, in particular, has shown the importance of being prepared for large-scale disasters, and how the world can get together to help alleviate the resulting pain and suffering. In its own small way, this chapter better prepare scientists, engineers, first responders, and, above all, politicians to deal with human-caused and natural disasters.

The most significant contribution of this chapter is perhaps the proposal to consider all natural and human-caused disasters as dynamical systems. Though not always easy, looking at the problem from that viewpoint and armed with the modern tools of dynamical systems theory may allow better prediction, control, and mitigation of future disasters. It is hoped that few readers of this book who are not already involved in disasters research would want to be engaged in this exciting endeavor whose practical importance cannot be overstated.

References

1. M. Gad-el-Hak, *Large-Scale Disasters: Prediction, Control, and Mitigation*, Cambridge University Press, London, U.K., 2008.
2. H. E. Huppert, Geological fluid mechanics, in *Perspectives in Fluid Dynamics: A Collective Introduction to Current Research*, eds. G. K. Batchelor, H. K. Moffatt, and M. G. Worster, pp. 447–506, Cambridge University Press, London, U.K., 2000.
3. J. de Boer and J. van Remmen, *Order in Chaos: Modelling Medical Disaster Management Using Emergo Metrics*, LiberChem Publication Solution, Culemborg, the Netherlands, 2003.
4. P. McFedries, Changing climate, changing language, *IEEE Spectr.* **43**, 60, August 2006.
5. H. W. Fischer, III, The sociology of disaster: Definition, research questions, measurements in a post-September 11, 2001 environment, presented at the *98th Annual Meeting of the American Sociological Association*, Atlanta, GA, August 16–19, 2003.
6. H. W. Fischer, III, The sociology of disaster: Definition, research questions, and measurements. Continuation of the discussion in a post-September 11 environment, *Int. J. Mass Emerg. Disasters* **21**, 91–107, 2003.
7. R. A. Posner, *Catastrophe: Risk and Response*, Oxford University Press, Oxford, U.K., 2004.
8. D. Patanè, P. de Gori, C. Chiarabba, and A. Bonaccorso, Magma ascent and the pressurization of Mount Etna's volcanic system, *Science* **299**, 2061–2063, 2006.
9. A. Adamatzky, *Dynamics of Crowd-Minds: Patterns of Irrationality in Emotions, Beliefs and Actions*, World Scientific, London, U.K., 2005.
10. D. Helbing, I. J. Farkas, and T. Vicsek, Crowd disasters and simulation of panic situations, in *The Science of Disasters: Climate Disruptions, Heart Attacks, and Market Crashes*, eds. A. Bunde, J. Kropp, and H. J. Schellnhuber, pp. 331–350, Springer, Berlin, Germany, 2002.
11. R. Gall and D. Parsons, It's hurricane season: Do you know where your storm is? *IEEE Spectr.* **43**, 27–32, 2006.
12. L. F. Richardson, *Weather Prediction by Numerical Process*, reissued in 1965 by Dover, New York, 1922.
13. M. Gad-el-Hak, *Flow Control: Passive, Active, and Reactive Flow Management*, Cambridge University Press, London, U.K., 2000.
14. M. Gad-el-Hak, The fluid mechanics of microdevices—The Freeman Scholar Lecture, *J. Fluids Eng.* **121**, 5–33, 1999.
15. M. Gad-el-Hak, *The MEMS Handbook*, 2nd edn., Vols. I–III, Taylor & Francis Group, Boca Raton, FL, 2006.
16. M. Gad-el-Hak, Questions in fluid mechanics: Stokes' hypothesis for a Newtonian, isotropic fluid, *J. Fluids Eng.* **117**, 3–5, 1995.
17. R. L. Panton, *Incompressible Flow*, 2nd edn., Wiley-Interscience, New York, 2006.
18. C. Garrett, The dynamic ocean, in *Perspectives in Fluid Dynamics: A Collective Introduction to Current Research*, eds. G. K. Batchelor, H. K. Moffatt, and M. G. Worster, pp. 507–556, Cambridge University Press, London, U.K., 2000.
19. M. E. McIntyre, On global-scale atmospheric circulations, in *Perspectives in Fluid Dynamics: A Collective Introduction to Current Research,* eds. G. K. Batchelor, H. K. Moffatt, and M. G. Worster, pp. 557–624, Cambridge University Press, London, U.K., 2000.
20. P. F. Linden, Convection in the environment, in *Perspectives in Fluid Dynamics: A Collective Introduction to Current Research*, eds. G. K. Batchelor, H. K. Moffatt, and M. G. Worster, pp. 289–345, Cambridge University Press, London, U.K., 2000.
21. S. H. Davis, Interfacial fluid dynamics, in *Perspectives in Fluid Dynamics: A Collective Introduction to Current Research*, eds. G. K. Batchelor, H. K. Moffatt, and M. G. Worster, pp. 1–51, Cambridge University Press, London, U.K., 2000.
22. A. C. Lawson, The California earthquake of April 18, 1906: Report of the State Earthquake Investigation Commission, Vols I & II, Carnegie Institution of Washington Publication 87, Washington, DC, 1908.

11

Risk Assessment

C.L. Winter
University of Arizona

D.M. Tartakovsky
University of California, La Jolla

11.1 Introduction

Policy makers, managers, and the public must frequently assess risks to society and the environment caused by catastrophic flows of environmental fluids as well as the mass and heat transported by them. Examples range from hurricane damage to groundwater contamination and prominently include climate change due to the accumulation of greenhouse gases in the atmosphere. Environmental risk has been succinctly defined as the probability that something bad will occur (Brillinger, 2003). More formally, risk is quantified as the expected loss due to the failure of a critical system (Bedford and Cooke, 2003):

$$R = \int_{\omega \subset \Omega} L(\omega)\, dP(\omega)$$

Hence the environmental risk assessment (ERA) problem consists of the following:

1. Evaluating the loss, $L(\omega)$, caused by system failure associated with a given realization, ω, of the system. This is primarily an economic problem.
2. Representing the structure of the system through a quantitative model, μ, of its essential elements and their logical dependencies.
3. Developing a probability model, $P_\mu(\omega|D)$, conditioned on μ and observed data, D. When the dependency on μ is obvious, we write $P(\omega|D) \overset{def}{=} P_\mu(\omega|D)$; and when the dependence on D is obvious, we drop the explicit conditioning and just write $P(\omega)$ or $P_\mu(\omega)$.

In this chapter we focus on the physical aspects (2) and (3) of ERA due to fluids. The probabilistic nature of ERA comes from uncertainty about the structure of the system, the degree of fidelity needed to model it, the values of model parameters, system nonlinearities (if any), and numerical approximations (Figure 11.1). Alternative approaches to quantifying uncertainty (e.g., fuzzy logic, interval analysis, Dempster–Shafer theory, and plausibility) are not discussed here because they do not have the straightforward relation to system dynamics that probability does (through Fokker–Planck and related stochastic PDEs). We further restrict our focus by concentrating on the influence of model and parametric uncertainties since they are frequently so large as to limit the need for exact dynamics based on advanced numerical techniques.

The system state description is nearly always composed of multiple elements, for instance, a contaminated drinking water aquifer may be described as resulting from the joint failure of engineered controls and geochemical, biological, and hydrologic processes that are ineffective at naturally attenuating contaminant concentration. The model μ can correspond to a set of stochastic PDEs for flow and transport, in which case $P_\mu(\omega)$ is usually approximated by its first few moments or through Monte Carlo simulation, or on models of reduced complexity that represent $P_\mu(\omega)$ as a joint distribution of abstracted system states corresponding to statements like "engineered remediation fails before time t_R." In this case the abstraction is still based on physical principles or expert judgment, but the dynamics are collapsed into transitions among coarsely defined states. In addition to μ, the distribution $P_\mu(\omega|D)$ depends on estimates of model parameters, $\hat{\Pi}_\mu(D)$, based on limited observations, D, of multiple random variables and fields. The parameter fields are usually highly heterogeneous in space, and often in time.

Handbook of Environmental Fluid Dynamics, Volume One, edited by Harindra Joseph Shermal Fernando. © 2013 CRC Press/Taylor & Francis Group, LLC. ISBN: 978-1-4398-1669-1.

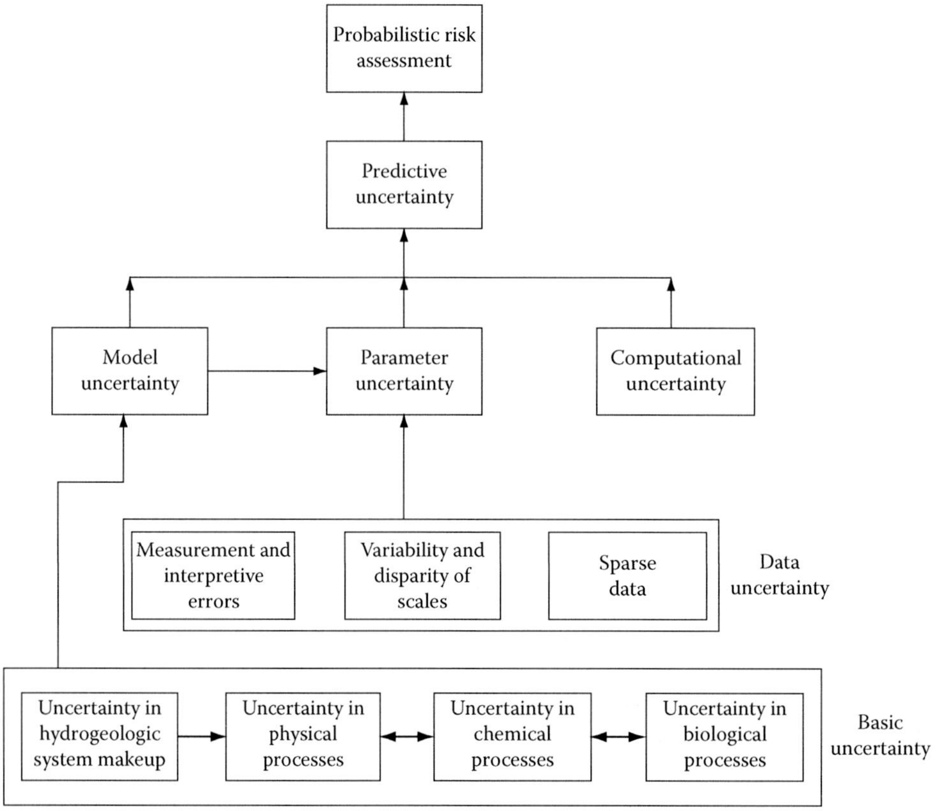

FIGURE 11.1 Different sources of uncertainty encountered in the modeling of subsurface flow and transport. Their quantification and propagation through a modeling process are essential ingredients of a quantitative probabilistic risk assessment.

The effects of natural catastrophes and failures of engineered systems impacting the environment have grown significantly as society makes increasing demands on environmental systems. Losses due to hurricanes (Pielke and Landsea, 1998), earthquakes (Klugel et al., 2006), floods (Pielke and Downton, 2000), climate change (IPCC, 2007), and drought (Wilhite et al., 2007) are enormous and are increasing at rates that demand a new generation of catastrophe modelers and risk managers schooled in the most current methods of ERA (Rollins, 2008; Elsner et al., 2009). Empirical statistical analyses have been extensively applied to estimates of risk from hurricanes (Webster et al., 2005), climate change (Katz, 2002), flooding (Kubal et al., 2009), waste storage (Merkhofer and Keeney, 1987; Otway and von Winterfeldt, 1992), and earthquake hazards (Klugel et al., 2006). More closely related to fluid dynamics are models of reduced complexity based on Markovian models of hurricane tracks (Emanuel et al., 2006; Hallegatte, 2007), climate change (Katz, 2002), earthquake hazard (Singhal and Kiremidjian, 1996), and groundwater contamination (Winter and Tartakovsky, 2008). Applications of Monte Carlo-based simulations of stochastic PDEs to ERA are almost too numerous to mention (Rugen and Callahan, 1996), including applications in groundwater (Zhang, 2002; Rubin, 2003), weather forecasting (Mullen and Baumhefner, 1994), earthquake hazard (Liang and Hao, 2009), and climate change (Cubasch et al., 1994).

We illustrate the application of environmental fluid dynamics to ERA with the problem of assessing the threats to a drinking water aquifer posed by a single chemical spill at a waste storage site. This is an important problem in its own right with an extensive literature to support it. While conceptually simple (linear flow and mass transport initiated by a localized event occurring at a specific time), the problem of aquifer contamination nonetheless illustrates the main features of ERA including uncertain model definition and parameterization, and a corresponding need for simplification.

11.2 Principles: Sources of Predictive Uncertainty in Subsurface Flow and Transport

Most fluids-based ERA problems have a strong physical basis that in principle would allow a high degree of predictability in the presence of complete information and unlimited computational resources. Even without that knowledge, physical models are an important basis for ERA because they help delineate the nature of the kinds of uncertainty that arise in a given domain and the effects of simplifying assumptions. In this section, we use the groundwater example to discuss the physical basis for different scales of predictive models first and then point out sources of uncertainty and methods for quantifying them.

Flow and transport in natural porous media are usually described by pore-scale (microscopic) models or continuum (macroscopic, or Darcy-scale) models. Pore-scale models represent a porous volume as a network of pores whose complement consists of solid grains. Approaches to dynamics range from lattice-Boltzmann modeling (Sukop and Thorne, 2007), smoothed particle hydrodynamics (Meakin and Tartakovsky, 2008), to direct numerical simulations of Navier–Stokes equations (Smolarkiewicz and Winter, 2010). Although pore-scale models have a solid physical foundation (e.g., Stokes equations for fluid flow and Fick's law of diffusion for solute transport), for predictive purposes they require the knowledge of pore geometry and properties that is seldom available at laboratory scales, much less field scales, so they are impractical at scales much larger than a few thousand pore segments.

Upscaling from the pore-scale to the continuum scale often enables one to establish the connection between the two modeling scales. Mathematical approaches to upscaling include the method of volume averaging (Whitaker, 1999), generalizations of Brenner's method of moments (Shapiro et al., 1996), homogenization via multiple-scale expansions (Adler, 1992), pore-network models (Acharya et al., 2005), and thermodynamically constrained averaging (Gray and Miller, 2005). Earlier approaches to upscaling are reviewed in Brenner (1987).

Nonlinearity of governing equations complicates the upscaling of the most reactive transport phenomena, requiring a linearization and/or other approximations, whose accuracy and validity cannot be ascertained a priori. This is especially so for a large class of transport processes, such as mixing-induced precipitation, which exhibit highly localized reacting fronts and consequently defy macroscopic descriptions that are completely decoupled from their microscopic counterparts (Ochoa-Tapia et al., 1991; Kechagia et al., 2002; Battiato et al., 2009).

The localized nature of reaction fronts suggests a role for hybrid simulations (van Leemput et al., 2007; Tartakovsky et al., 2008) that resolve a small reactive region with a pore-scale model coupled to a phenomenological counterpart in the rest of the porous medium which is represented as a continuum. Usually, the resolved pore space must be represented as a stochastic process since detailed information about the pore geometry of actual porous media is still available from just a few small sample volumes. Heterogeneity presents a number of obvious problems for these models. Hybrid simulations of flow and transport in porous media are in their infancy, and are not applicable to transport phenomena for which continuum models fail globally rather than locally either because "the connectivity of the pore space or a fluid phase plays a major role" or because of "long-range correlations in the system" (Sahimi, 1993, p. 1396).

Continuum-level models overcome the practical limitations of pore-scale models at the cost of relying on phenomenological descriptions. These models are defined on a "porous" continuum with parameters and state variables defined at every point in a medium, thus requiring some sort of upscaling (at least implicitly) from an actual porous medium. Examples of such models are Darcy's law and mass conservation for fluid flow in saturated porous media,

$$q = -K\nabla h, \quad S\frac{\partial h}{\partial t} = -\nabla \cdot q \tag{11.1}$$

and a corresponding transport equation, for example, an advection-dispersion equation (ADE) for transport of conservative (passive) solutes:

$$\frac{\partial c}{\partial t} = -\nabla \cdot (D\nabla c) - \nabla \cdot (vc) \tag{11.2}$$

In Equations 11.1 and 11.2, x is a point in $\mathfrak{R}^n$ and $n = 1, 2,$ or 3. The Darcy flux, $q(x,t) \in \mathfrak{R}^n$, and scalar hydraulic pressure head, $h(x,t)$, depend on boundary and initial conditions (IBCs), the hydraulic conductivity, $K(x)$, which in general is a second-order positive-definite tensor, and the specific storage, $S(x)$, a scalar accounting for mild compressibility of both a fluid and a porous medium. The scalar concentration of solute, $c(x,t)$, depends on its own IBCs, a dispersion coefficient, $D(x,v)$, another second-order positive-definite tensor, and the average microscopic velocity of water, $v(x,t) = q(x,t)/\varphi(x)$ where $\varphi(x) \leq 1$ is a medium's porosity.

11.3 Methods of Analysis: Reducing and Quantifying Uncertainty

Uncertainty in ERA arises from fundamental sources (e.g., nonlinearity) and lack of knowledge (Figure 11.1). The effects of lack of knowledge are seen in uncertainty about (1) the appropriate structure of dynamic risk models, μ, and (2) their parameters, $\hat{\Pi}_\mu(D)$, including IBCs.

11.3.1 Model Uncertainty

Although phenomenological physical models (e.g., Equations 11.1 and 11.2) are standard in groundwater and other fields, problems have increasingly come up with applying them to settings where parameters are heterogeneous. They fail to capture some important experimentally observed transport features of porous media, including a difference between fractal dimensions of the diffusion and dispersion fronts (isoconcentration contours) (Maloy et al., 1998), long tails in breakthrough curves (Neuman and Tartakovsky, 2009), and the onset of instability in variable density flows (Tartakovsky et al., 2008). ADE-based models of transport of (bio-)chemically reactive solutes can significantly overpredict the extent of reactions in mixing-induced chemical transformations (Knutson et al., 2007 and references therein). Even when continuum-level models provide an adequate description of processes at one local scale, support volumes (that might or might not represent Representative Elementary Volumes) of experiments used to infer various parameters in these models (e.g., porosity, hydraulic conductivity, reaction rate constants, and retardation coefficients) might be dramatically different.

Two conceptual frameworks have emerged for dealing with model uncertainty in groundwater modeling and other disciplines like climatology: model averaging (MA) and model selection (MS). Both require commensurability among model outputs in the sense that essentially the same states must be simulated for the same time intervals and volumes. Competing approaches to MA and MS are discussed in detail in Claeskens and Hjort (2008). MA, especially Bayesian MA (BMA), has been used in statistical analysis of data generated by multiple conceptual models of an underlying phenomenon (e.g., Hoeting et al., 1999). It provides a mechanism for making predictions of a system's behavior by averaging (deterministic or probabilistic) solutions of competing models, each of which is assigned a probabilistic weight (typically based on an information criterion, e.g., Akaike information criterion or Bayesian information criterion) consistent with the available data and other information (e.g., expert knowledge). Applications of various flavors of MA to subsurface flow and transport include Neuman (2003), Poeter and Anderson (2005), and Vrugt et al. (2008). Rather than averaging over competing models, MS attempts to identify the "best" model consistent with available measurements and general knowledge of a phenomenon under consideration. Examples of selecting the "best" model—rather than averaging over several models—in environmental sciences include Christakos (2003), Foglia et al. (2007), and McDonald and Urban (2010). MA and MS need not be mutually exclusive in that the former often relies on the latter to assign probabilistic weights to competing models.

11.3.2 Parameter Uncertainty

Subsurface heterogeneity and data scarcity preclude detailed knowledge of spatial distributions of the system parameters (including IBCs) in Equations 11.1 and 11.2. These parameters are typically sampled at only a few locations $x_i \in \Gamma$, ($i = 1, \ldots, N$ with $N < 100$) throughout a flow domain Γ, and their values at other locations are unknown. Hence Equations 11.1 and 11.2 are underdetermined. (An additional complication stems from possible measurement errors, see Figure 11.1.) Scaling is one of the chief complications of estimating parameters for phenomenological models, becoming especially acute when spatial heterogeneity and nonlinearities must be taken into account. Parameter estimates are often based on measurements made at scales smaller than the scale at which transport is modeled, and frequently parameters are measured at multiple scales using multiple techniques. For example, the support volume of a pumping test used to estimate the hydraulic conductivity of an aquifer is orders of magnitude larger than its counterpart used to measure geochemical parameters in the laboratory. For instance, a number of experimental and theoretical studies (e.g., Paces, 1983; Lichtner, 1993; Sverdrup and Warfvinge, 1995) have found a discrepancy of several orders of magnitude between field- and laboratory-estimated rate constants.

Parameter values at points where measurements are not available can be interpolated through geostatistics, which depends on the correlation structure of the interpolated parameters. The method regularizes corresponding boundary-value problems by treating available parameter data as realizations drawn from a sampling space characterized by a joint multivariate probability function or joint ensemble moments (Isaaks and Srivastava, 1990). This is the basis for both Monte Carlo simulation and direct extraction of statistical moments of system states by ensemble averaging the governing equations. In this representation, all hydraulic and transport parameters in Equations 11.1 and 11.2 depend not only on the space coordinate x but also on $\omega \in \Omega$, a realization drawn from the ensemble Ω of all possible $\Gamma(\omega)$ with fixed values at points of observation, x_i. For example, uncertain hydraulic conductivity is often treated as a random field $K(x, \omega) = \bar{K}(x) + K'(x, \omega)$ that is the sum of a deterministic mean field, $\bar{K}(x)$, and a zero-mean random field $K'(x,\omega)$. The overbar indicates ensemble averaging. Geostatistics based on the observations x_i are used to estimate $\bar{K}(x)$ and the correlation structure of $K'(x,\omega)$. This ordinarily requires assuming statistical regularity properties like stationarity and ergodicity.

Whereas spatial moments are obtained through sampling in real space, ensemble moments require sampling in probability space. Since only one realization of an actual reservoir or aquifer exists in reality, it is common to invoke an ergodicity hypothesis, which allows for interchanging ensemble and spatial moments. Ergodicity cannot be proved unless Γ falls under the purview of well-established ergodic theorems such as the law of large numbers. However, it has been demonstrated by Yaglom (1987) that "in any application, non-ergodicity usually just means that the random function concerned is, in fact, an artificial union of a number of distinct ergodic stationary functions." This is an approach that lies at the heart of the random domain decomposition (RDD) of Winter and Tartakovsky (2000, 2002), which takes advantage of the fact that a high degree of heterogeneity in environmental systems usually arises from the presence of different geologic units (understood here in a very broad sense to include hydrofacies, fractured regions, inclusions, layering, etc.) in the subsurface.

11.3.3 Stochastic PDEs

Statistical treatment of flow and transport parameters renders the corresponding flow and transport equations, for example, Equations 11.1 and 11.2, stochastic. Solving such equations is equivalent to propagating the parametric uncertainty (which is expressed in terms of probability density functions for the transport coefficients and their statistical moments) through a modeling process, thus quantifying the predictive uncertainty in terms of probability density functions for state variables (e.g., hydraulic head h and solute concentration c). Of particular interest from the practical point of view are effective flow and transport equations that describe the evolution of the mean (averaged) state variables. For instance, taking the ensemble mean of the Darcy equation yields

$$\bar{q} = -\overline{K\nabla h} = -(\bar{K}\nabla\bar{h} + \overline{K'\nabla h'}) = -(\bar{K}\nabla\bar{h} + r) \qquad (11.3)$$

The existence of the average residual flux, $r = \overline{K'\nabla h'}$, demands some sort of closure. Furthermore, it highlights the serious methodological restrictions that must be imposed to develop effective equations, for example, $q = -K_{eff}\nabla h$ (except when $r = 0$, which is generally not the case). Attempts to reduce stochastic flow and transport equations to their traditional counterparts through so-called effective coefficients, for example, K_{eff}, which are often dictated by the practical necessity of using legacy codes, require approximations, for example, localization. Effective coefficients are typically obtained through empirical optimization (or calibration) techniques that match estimates of state variables to a small number of observations. Since the potential surface used for optimization usually has multiple minima, it is difficult to avoid a local minimum so optimization techniques often produce results that cannot be verified outside the range of observations. The resulting effective parameters, such as K_{eff}, effective retardation coefficient, and effective kinetic rate constant (e.g., Lichtner and Tartakovsky, 2003), are typically space-time and/or scale dependent. Various forms of effective flow and transport equations can be found in Zhang (2002) and Rubin (2003). In general, such equations are fundamentally different from their local-scale deterministic counterparts: advection equations give rise to advection–dispersion equations, and traditional advection–dispersion equations become nonlocal (see a recent review by Neuman and Tartakovsky, 2009).

Much research in stochastic hydrogeology has focused on the first two statistical moments of state variables, for example, mean, $\bar{c}(x, t)$, and variance, $\sigma_C^2(x,t)$, of solute concentration. The first moment (ensemble mean) is used as an unbiased estimate of a system state, and the second moment (ensemble variance) is used as a measure of uncertainty. These statistical moments can be obtained either indirectly by solving stochastic flow and transport equations with (accelerated) Monte Carlo simulations and other numerical techniques (e.g., stochastic finite elements or polynomial chaos expansions and stochastic collocation methods) or directly by deriving deterministic equations for each statistical moment (Rubin, 2003, and references therein). While providing a measure of predictive uncertainty, such approaches are insufficient for probabilistic risk analyses that require the knowledge of a state variable's probability density function (PDF).

Monte Carlo simulations and other numerical approaches for solving stochastic transport equations can be used in principle to compute PDFs of system variables. However, 3D transient nonlinear problems render these approaches computationally expensive. Another possibility is to assume that the PDF of a system state is either Gaussian or a given transform thereof, that is, that it is fully described by its mean and variance (Amir and Neuman, 2004). Even for relatively simple nonlinear problems (unsaturated flow described by the Richards equation), such Gaussian closure approaches have proved to be limited to mildly heterogeneous formations (Tartakovsky and Guadagnini, 2001; Tartakovsky et al., 2003). Applications of the assumed-PDF method (Girimaji, 1991) to solute transport in heterogeneous subsurface environments can be found in, for example, Caroni and Fiorotto (2005) and Bellin and Tonina (2007). Assessment of the accuracy and robustness of such methods requires one to solve stochastic transport equations by Monte Carlo simulations or other means.

PDF methods (e.g., Pope, 2000) provide an attractive alternative to these approaches by deriving a deterministic equation for the PDF of a system variable. An additional benefit of PDF methods is that they enable one to avoid a linearization of reactive transport equations—advection–dispersion equations similar to Equation 11.2 with a reaction term $R(c)$ added. This nonlinearity arises through, for example, kinetic rate laws describing precipitation/dissolution of solids and kinetic sorption/desorption rates, as well as through incorporation of local equilibrium relations derived from the law of mass action for homogeneous and heterogeneous reactions. In subsurface hydrology, equations for the PDFs of the concentration of a conservative solute advected by a random velocity field were derived by Indelman and Shvidler (1985) and Shvidler and Karasaki (2003). Lichtner and Tartakovsky (2003) and Tartakovsky et al. (2009) obtained PDF equations for reactive transport in porous media with uncertainty chemical properties (effective reaction rates).

11.3.4 Models of Reduced Complexity

The difficulty of parameterizing stochastic PDEs on the basis of a few observations and the relatively weak conclusions that can be drawn from them (estimates of just a few moments without strict confidence intervals or, alternatively, small numbers of Monte Carlo simulations with correspondingly broad empirical confidence intervals) has led to interest in models of reduced complexity based (1) on measurements that are relatively easy to make and/or (2) on the subjective opinions of experts. Conclusions drawn from these models are not much stronger (or weaker) than those based more directly on physical models, but they are cheaper to compute (Hill, 2006).

A model of reduced complexity represents system failure as the result of interactions among coarsened basic states, or events, corresponding to statements like "engineered remediation fails by time t_R," and decomposes their joint probability into independent constituent probabilities based on physical properties, engineering principles, and expert judgment. As such, μ is usually expressed through Boolean logical operations on the basic events with independence/dependence among events expressed via probabilistic conditioning. It is most convenient if the logical analysis results in a disjunctive normal form (DNF) representing system failure as the union of mutually exclusive conjunctions of elementary events. A number of structuring methodologies have this property, and most others can produce DNFs as long as a little care is exercised.

Tartakovsky et al. (2007) and Bolster et al. (2009) represent the compound event C that "a drinking water aquifer will be contaminated by a chemical spill" as a logical combination of basic events including (R) success of systems engineered to intercept a contaminant plume, (N) the ability of hydrogeochemical reactions to naturally attenuate a plume, and (F) enhanced mechanical transport of the contaminant through high permeability volumes, or fast paths, in the aquifer. From Tartakovsky et al. (2007), for instance, it is possible to derive $C = R'N'F + R'N'F'$. Here "+" indicates logical or (union), juxtaposition indicates logical and (intersection), and the prime indicates negation so that R' indicates failure of efforts to engineer remediation, etc.

Independence (dependence) relations among system elements are analyzed to build a simple probability model for the event C. First, the existence of a fast path can be taken to imply engineering failure ($F \subset R'$), so $C = N'F + R'N'F'$. Since $N'F$ and $R'N'F'$ are mutually exclusive (disjoint) events, $P(C) = P[N'F] + P[R'N'F']$. Next the definition of conditional probability and reasonable independence assumptions based on physical properties of contaminant control systems yield

$$P(C) = P[N'|F]\, P[F] + P[R'|F']\, P[N'|F']\, P[F'] \qquad (11.4)$$

for the overall structure of the reduced probability model. The final step in model development is to abstract simple forms for constituent probabilities from either expert opinion, or the underlying physics, or both. Bolster et al. (2009), for instance, derive simple forms for probabilities like $P[N'|F']$ and $P[F']$ from underlying physical principles and obtain $P[R'|F']$ from experts. They give an insightful analysis of the sensitivity of $P(C)$ to system properties like the distance of the spill from the location of a spill to a remediation barrier, the relative size of the barrier, and so on.

Graphical representations of model structure can further clarify relations among model elements and indicate hierarchical relations among levels of system abstraction (e.g., Figure 11.2). For example, the logic of both Tartakovsky et al. (2007) and Bolster et al. (2009) is initially represented through fault-trees (Bedford and Cooke, 2003), and only later is conditioned on a probabilistic analysis of independence among system elements. Fault-tree decompositions are convenient because their natural outcome is a DNF. Alternative methods for structuring joint probability models include binary decision diagrams (BDDs), influence diagrams, and Bayesian networks. BDDs are equivalent to fault-trees, while influence diagrams and Bayesian networks incorporate conditioning relations directly in the graphical structure. None of these approaches represent time explicitly, which limits their utility somewhat as a basis for settings that demand estimates of failure before a given regulatory time.

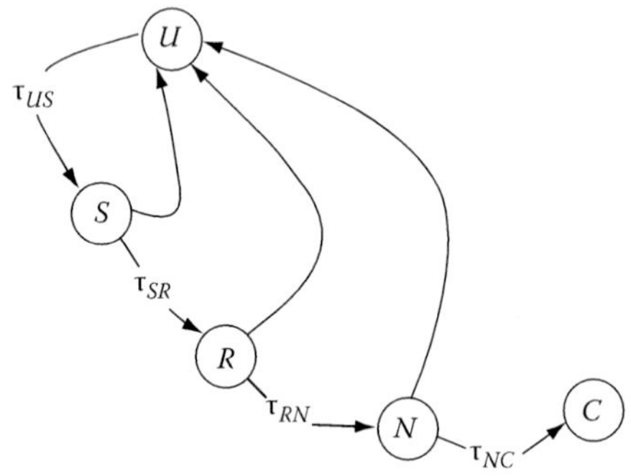

FIGURE 11.2 Transitions and transition times of finite-state model of aquifer contamination.

TABLE 11.1 Basic Events of the Containment System

U	The system is uncontaminated
S	A spill occurs
R'	At least one barrier well fails
N'	Natural attenuation fails
B_i'	The ith barrier well fails

Winter and Tartakovsky (2008) address this issue by representing aquifer contamination as a finite-state machine that includes rates of transition among system states directly in the model. They consider a waste storage site where a contaminant spill is known to have occurred at time t_S and a containment system consists of a zone of engineered remediation based on n barrier wells meant to intercept the contaminant plume and a zone further downstream where natural geochemical processes may attenuate the concentration of contaminants escaping the barrier wells. The containment system fails, again called C, if the contaminant reaches a drinking water aquifer at concentration $c(t) > c^*$ at $t < t^*$. The limits c^* and t^* are set by a regulatory body.

Basic events of the system are defined in Table 11.1. The logic of system failure is $C = SR'N' = SB_1'N' + \cdots + SB_n'N'$, that is, the system fails if a spill occurs, any barrier well fails, and natural attenuation is ineffective. In each case, failure while in a state means that $c(t) > c^*$ when the system is in that state. Because of the time element, failure also depends on the sequence of transitions, $U \rightarrow S \rightarrow R \rightarrow N \rightarrow C$, with each state, $\sigma \in \{U, S, R, N\}$ requiring a (random) time τ_σ to occur. For instance, $\tau_N = \tau_{US} + \tau_{SR} + \tau_{RN} + \tau_{NC}$ is the time required for a system, starting from an initially uncontaminated state to leave natural attenuation and become contaminated. A transition time (also random) is the time required to go from one state to another, for example, τ_{SR} is the time required to transit from the occurrence of a spill to the failure of engineered remediation. Since the transition $U \rightarrow S$ is based on the industrial properties of a waste site, calculations based on environmental flows can usually be assumed to start with a spill. If not, techniques like those in Bedford and Cooke (2003) must be applied. The sequence of events and times leading to system failure, for example, $S \rightarrow R \rightarrow N \rightarrow C$, or protection of the aquifer (any transition from an intermediate state to U) can be illustrated through a state transition diagram (Figure 11.2).

The probability that the system transitions from state $\sigma \rightarrow \sigma'$ at $\tau_\sigma < t$ is $P_{\sigma\sigma'}\,[\tau_\sigma < t]$. If (1) the state transitions are Markovian, that is, the transition from the current state to the next state is independent of past transitions, and (2) τ_σ does not depend on the next state, the system is a *jump process*, $P_{\sigma\sigma'}[\tau_\sigma < t] = F_\sigma(t; q_\sigma)\, Q_{\sigma\sigma'}$ (Doob, 1953). Here $Q_{\sigma\sigma'}$ is the probability that the system ever jumps from $\sigma \rightarrow \sigma'$. It is equivalent to the transition probabilities of Bolster et al. (2009), and therefore can be based on the same kinds of measurements. Transition times are exponentially distributed with parameter $q_\sigma = 1/\mathrm{E}[\tau_\sigma]$, which can be estimated from breakthrough curves.

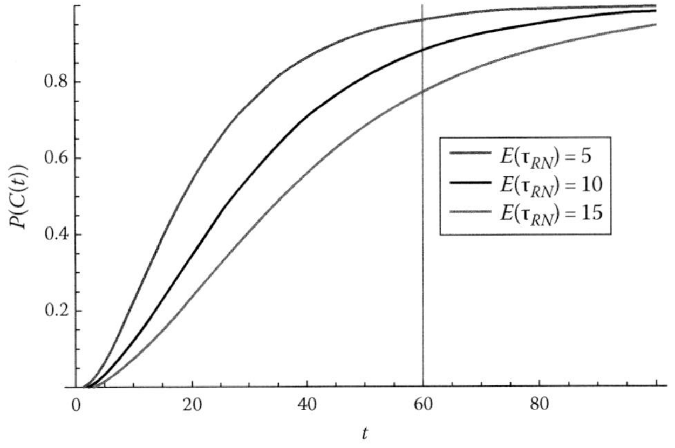

FIGURE 11.3 Probabilities of failure against time. The vertical line is at the regulatory limit $t^* = 60$ and the different curves correspond to different levels of remediation effectiveness.

TABLE 11.2 Probabilities of Failure before $t^* = 60$ for Different Levels of Remediation Effectiveness

1. $E[t_{RN}]$	2. $P_{SC}[60]$
5	0.96
10	0.88
15	0.77

The upshot is a simple model based on measurements for the probability of aquifer contamination starting with a spill,

$$P_{SC}[\tau_C < t^*] = Q_{SR}Q_{RN}Q_{NC}I(t^*) \tag{11.5}$$

that depends on the timing of transitions between a sequence of coarse states. The term $I(t^*)$ is the integral over all feasible sets of intermediate times of the product $F_S(\tau_S;q_S)F_R(\tau_R;q_R)F_R(\tau_N;q_N)$. When $t^* \to \infty$, $I(t^*) = 1$ and $P_{SC}[\tau_N < t^*] = Q_{SR}\, Q_{RN}\, Q_{NC}$, which is essentially the same expression as in Bolster et al. (2009). The jump from engineered remediation to natural attenuation, $R \to N$, is the only transition that can be controlled by human agency.

The other transitions, $S \to R$ and $N \to C$ depend on chemical, biological, and mechanical properties of the natural hydrogeological system that are fixed for a given setting. The effect of delaying the $R \to N$ jump is shown in Figure 11.3 which is based on relative times $t^* = 60$, $E[\tau_{SR}] = 1$, $E[\tau_{NC}] = 10$, and the values given in the figure for $E[\tau_{RN}]$. It is clear from Figure 11.3 and Table 11.2 that the probability of contamination by $t^* = 60$ is reduced as $E[\tau_{RN}]$ increases.

11.4 Major Challenges

ERA supports the decisions made to protect human and natural systems from actual or potential threats. Threats to the environment may be distributed or localized in space and/or time, and they may result from natural processes, human activities, or frequently a combination of both. Risk assessment methodologies were originally developed to analyze industrial sectors like nuclear energy, space exploration, and chemical processing (Bedford and Cooke, 2003) where extensive databases are often available to narrow the ranges of μ, $L(\omega)$, and $P(\omega)$. ERA, on the other hand, is almost always based on limited observations, D, of complex, highly heterogeneous systems with a corresponding level of uncertainty about μ, $L(\omega)$, and $P(\omega)$. ERA for systems influenced by environmental flows can be additionally complicated by feedbacks among multiple system elements (natural and human) operating at different scales of space and time. ERA is a predictive endeavor that must often project system states for ranges of parameters that have not been observed. Hence, while there is a need for simplified models, empirical models based on statistical correlations will not do if they are restricted to ranges of observed data. The challenge is to obtain model simplifications that are directly based on rigorous abstractions of the underlying physical, biological, and chemical processes.

References

Acharya, R. C., S. E. A. T. M. Van der Zee, and A. Leijnse. Transport modeling of nonlinearly adsorbing solutes in physically heterogeneous pore networks. *Water Resour. Res.*, 41: W02020, 2005, doi:10.1029/2004WR003500.

Adler, P. M. *Porous Media: Geometry and Transports*. Butterworth-Heinemann, Stoneham, MA, 1992.

Amir, O. and S. P. Neuman. Gaussian closure of transient unsaturated flow in random soils. *Transp. Porous Media*, 54(1): 55–77, 2004.

Battiato, I., D. M. Tartakovsky, A. M. Tartakovsky, and T. Scheibe. On breakdown of macroscopic models of mixing-controlled heterogeneous reactions in porous media. *Adv. Water Resour.*, 32: 1664–1673, 2009, doi:10.1016/j.advwatres.2009.08.008.

Bedford, T. and R. Cooke. *Probabilistic Risk Analysis: Foundations and Methods*. Cambridge University Press, New York, 2003.

Bellin, A. and D. Tonina. Probability density function of non-reactive solute concentration in heterogeneous porous formations. *J. Contam. Hydrol.*, 94: 109–125, 2007.

Bolster, D., M. Barahona, M. Dentz, D. Fernandez-Garcia, X. Sanchez-Vila, P. Trinchero, C. Valhondo, and D. M. Tartakovsky. Probabilistic risk analysis of groundwater remediation strategies. *Water Resour. Res.*, 45(10): W06413, 2009.

Brenner, H. *Transport Processes in Porous Media*. McGraw-Hill, New York, 1987.

Brillinger, D. R. Three environmental probabilistic risk problems. *Stat. Sci.*, 18(4): 412–421, 2003.

Caroni, E. and V. Fiorotto. Analysis of concentration as sampled in natural aquifers. *Transp. Porous Media*, 59(1): 19–45, 2005.

Christakos, G. Critical conceptualism in environmental modelling and prediction. *Environ. Sci. Technol.*, 37: 4685–4693, 2003.

Claeskens, G. and N. L. Hjort. *Model Selection and Model Averaging*. Cambridge University Press, Cambridge, U.K., 2008.

Cubasch, U., B. D. Santer, A. Hellbach, G. Hegerl, H. Hock, E. Maierreimer, U. Mikolajewicz, A. Stossel, and R. Voss. Monte-Carlo climate-change forecasts with a global coupled ocean-atmosphere model. *Clim. Dynam.*, 10: 1–19, July 1994.

Doob, J. L. *Stochastic Processes*. John Wiley, New York, 1953.

Elsner, J. B., R. K. Burch, and T. H. Jagger. Catastrophe finance: An emerging discipline. *Eos*, 90(33): 281–282, August 2009.

Emanuel, K. Increasing destructiveness of tropical cyclones over the past 30 years. *Nature*, 436(7051): 686–688, August 2005.

Emanuel, K., S. Ravela, C. Risi, and E. Vivant. A statistical deterministic approach to hurricane risk assessment. *BAMS*, 87(3): 299–314, March 2006.

Foglia, L., S. W. Mehl, M. C. Hill, P. Perona, and P. Burlando. Testing alternative ground water models using cross-validation and other methods. *Ground Water*, 45(5): 627–641, 2007.

Girimaji, S. S. Assumed β-pdf model for turbulent mixing: Validation and extension to multiple scalar mixing. *Combust. Sci. Technol.*, 78: 177–196, 1991.

Gray, W. G. and C. T. Miller. Thermodynamically constrained averaging theory approach for modeling flow and transport phenomena in porous medium systems: 1. Motivation and overview. *Adv. Water Resour.*, 28(2): 161–180, 2005.

Hallegatte, S. The use of synthetic hurricane tracks in risk analysis and climate change damage assessment. *J. Appl. Meteorol. Climatol.*, 46: 1956–1966, November 2007.

Hill, M. C. The practical use of simplicity in developing ground water models. *Ground Water*, 44(6): 775–781, November–December 2006.

Hoeting, J. A., D. Madigan, A. E. Raftery, and C. T. Volinsky. Bayesian model averaging: A tutorial. *Stat. Sci.*, 14(4): 382–417, 1999.

Indelman, P. V. and M. I. Shvidler. Averaging of stochastic evolution equations of transport in porous media. *Fluid Dynam.*, 20(5): 775–784, 1985.

Isaaks, E. H. and R. M. Srivastava. *An Introduction to Applied Geostatistics*. Oxford University Press, New York, January 1990.

Katz, R. W. Techniques for estimating uncertainty in climate change scenarios and impact studies. *Clim. Res.*, 20(2): 167–185, February 2002.

Kechagia, P. E., I. N. Tsimpanogiannis, Y. C. Yortsos, and P. C. Lichtner. On the upscaling of reaction-transport processes in porous media with fast or finite kinetics. *Chem. Eng. Sci.*, 57(13): 2565–2577, 2002.

Klugel, J. U., L. Mualchin, and G. F. Panza. A scenario-based procedure for seismic risk analysis. *Eng. Geol.*, 88: 1–22, November 2006.

Knutson, C., A. Valocchi, and C. Werth. Comparison of continuum and pore-scale models of nutrient biodegradation under transverse mixing conditions. *Adv. Water Resour.*, 30(6–7): 1421–1431, 2007.

Kubal, C., D. Haase, V. Meyer, and S. Scheuer. Integrated urban flood risk assessment—Adapting a multicriteria approach to a city. *Nat. Hazards Earth Syst. Sci.*, 9(6): 1881–1898, 2009.

van Leemput, P., C. Vandekerckhove, W. Vanroose, and D. Roose. Accuracy of hybrid lattice Boltzmann/finite difference schemes for reaction-diffusion systems. *Multiscale Model. Simul.*, 6(3): 838–857, 2007.

Liang, J. Z. and H. Hao. Influence of uncertain source parameters on strong ground motion simulation with the empirical Green's function method. *J. Earthquake Eng.*, 13(6): 791–813, 2009.

Lichtner, P. C. Scaling properties of time-space kinetic mass transport equations and the local equilibrium limit. *Am. J. Sci.*, 293: 257–296, 1993.

Lichtner, P. C. and D. M. Tartakovsky. Upscaled effective rate constant for heterogeneous reactions. *Stoch. Environ. Res. Risk Assess.*, 17(6): 419–429, 2003.

Maloy, K. J., J. Feder, F. Boger, and T. Jossang. Fractal structure of hydrodynamic dispersion in porous media. *Phys. Rev. Lett.*, 61(82): 2925, 1998.

McDonald, C. P. and N. R. Urban. Using a model selection criterion to identify appropriate complexity in aquatic biogeochemical models. *Ecol. Model.*, 221(3): 428–432, 2010.

Meakin, P. and A. M. Tartakovsky. Modeling and simulation of pore scale multiphase fluid flow and reactive transport in fractured and porous media. *Rev. Geophys.*, 47: RG3002, 2008, doi:10.1029/2008RG000263.

Merkhofer, M. W. and R. L. Keeney. A multiattribute utility analysis of alternative sites for the disposal of nuclear waste. *Risk Anal.*, 7(2): 173–194, 1987.

Mullen, S. L. and D. P. Baumhefner. Monte-Carlo simulations of explosive cyclogenesis. *Mon. Weather Rev.*, 122(7): 1548–1567, July 1994.

Neuman, S. P. Maximum likelihood Bayesian averaging of uncertain model predictions. *Stoch. Environ. Res. Risk Assess.*, 17(5): 291–305, 2003.

Neuman, S. P. and D. M. Tartakovsky. Perspective on theories of anomalous transport in heterogeneous media. *Adv. Water Resour.*, 32(5): 670–680, 2009, doi:10.1016/j.advwatres.2008.08.005.

Ochoa-Tapia, J. A., P. Stroeve, and S. Whitaker. Facilitated transport in porous media. *Chem. Eng. Sci.*, 46: 477–496, 1991.

Otway, H. and D. von Winterfeldt. Expert judgement in risk analysis and management: Process, context and pitfalls. *Risk Anal.*, 12(1): 83–93, 1992.

Paces, T. Rate constant of dissolution derived from the measurements of mass balances in catchments. *Geochim. Cosmochim. Acta*, 47: 1855–1863, 1983.

Pielke, R. A. and M. W. Downton. Precipitation and damaging floods: Trends in the United States, 1932–1997. *J. Clim.*, 13(20): 3625–3637, October 2000.

Pielke, R. A. and C. W. Landsea. Normalized hurricane damages in the United States: 1925–1995. *Weather Forecast.*, 13(3): 621–631, September 1998.

Poeter, E. and D. Anderson. Multimodel ranking and inference in ground water modeling. *Ground Water*, 43(4): 597–605, 2005.

Pope, S. B. *Turbulent Flow*. Cambridge University Press, Cambridge, U.K., 2000.

Rollins, J. The growing need for talent in catastrophe modeling and risk management. *AIR Currents*. AIR Worldwide Corp., Boston, MA, 2008.

Rubin, Y. *Applied Stochastic Hydrogeology*. Oxford University Press, New York, 2003.

Rugen, P. and B. Callahan. An overview of Monte Carlo, a fifty year perspective. *Hum. Ecol. Risk Assess.*, 2(4): 671–680, 1996.

Sahimi, M. Flow phenomena in rocks: From continuum models to fractals, percolation, cellular automata, and simulated annealing. *Rev. Mod. Phys.*, 5(4): 1393–1534, 1993.

Shapiro, M., R. Fedou, J.-F. Thovert, and P. M. Adler. Coupled transport and dispersion of multicomponent reactive solutes in rectilinear flows. *Chem. Eng. Sci.*, 51(22): 5017–5041, 1996, doi:10.1016/0009-2509(96)00269-2.

Shvidler, M. and K. Karasaki. Probability density functions for solute transport in random field. *Transp. Porous Media*, 50(3): 243–266, 2003.

Singhal, A. and A. S. Kiremidjian. Method for probabilistic evaluation of seismic structural damage. *J. Struct. Eng. ASCE*, 122(12): 1459–1467, December 1996.

Smolarkiewicz, P. and C. L. Winter. Pores resolving simulation of Darcy flows. *J. Comput. Phys.*, 229: 3121–3133, January 4, 2010.

Solomon, S., D. Qin, M. Manning, Z. Chen, M. Marquis, K. B. Averyt, M. Tignor, and H. L. Miller, ed., *Climate Change 2007: The Physical Science Basis*, Contribution of Working Group I to the Fourth Assessment Report of the Intergovernmental Panel on Climate Change, Cambridge University Press, ISBN 978-0-521-88009-1.

Sukop, M. C. and D. T. Thorne, Jr. *Lattice Boltzmann Modeling: An Introduction for Geoscientists and Engineers*. Springer, New York, 2007.

Sverdrup, H. and P. Warfvinge. Estimating field weathering rates using laboratory kinetics. *Rev. Mineral.*, 31: 485–541, 1995.

Tartakovsky, D. M., M. Dentz, and P. C. Lichtner. Probability density functions for advective-reactive transport in porous media with uncertain reaction rates. *Water Resour. Res.*, 45, 2009, doi:10.1029/2008WR007383.

Tartakovsky, D. M. and A. Guadagnini. Prior mapping for nonlinear flows in random environments. *Phys. Rev. E*, 64: 5302(R)–5305(R), 2001.

Tartakovsky, D. M., A. Guadagnini, and M. Riva. Stochastic averaging of nonlinear flows in heterogeneous porous media. *J. Fluid Mech.*, 492: 47–62, 2003.

Tartakovsky, A. M., D. M. Tartakovsky, and P. Meakin. Stochastic Langevin model for flow and transport in porous media. *Phys. Rev. Lett.*, 101(4): 044502, 2008, doi:10.1103/PhysRevLett.101.044502.

Tartakovsky, A. M., D. M. Tartakovsky, T. D. Scheibe, and P. Meakin. Hybrid simulations of reaction-diffusion systems in porous media. *SIAM J. Sci. Comp.*, 30(6): 2799–2816, 2007.

Vrugt, J. A., C. G. H. Diks, and M. P. Clark. Ensemble Bayesian model averaging using Markov Chain Monte Carlo sampling. *Environ. Fluid Mech.*, 8(5–6): 579–595, 2008.

Webster, P. J., G. J. Holland, and J. A. Curry. Changes in tropical cyclone number, duration, and intensity in a warming environment. *Science*, 309(5742): 1844–1846, September 2005.

Whitaker, S. *The Method of Volume Averaging*. Kluwer Academic Publishers, Dordrecht, the Netherlands, 1999.

Wilhite, D. A., M. D. Svoboda, and M. J. Hayes. Understanding the complex impacts of drought: A key to enhancing drought mitigation and preparedness. *J. Water Resour. Management*, 5: 763–774, May 2007.

Winter, C. L. and D. M. Tartakovsky. Mean flow in composite porous media. *Geophys. Res. Lett.*, 27(12): 1759–1762, 2000.

Winter, C. L. and D. M. Tartakovsky. Groundwater flow in heterogeneous composite aquifers. *Water Resour. Res.*, 38(8): W1148, 2002.

Winter, C. L. and D. M. Tartakovsky. A reduced complexity model for probabilistic risk assessment of groundwater contamination. *Water Resour. Res.*, 44: W06501, 2008.

Yaglom, A. M. *Correlation Theory of Stationary and Related Random Functions I, Basic Results*. Springer, Berlin, Germany, 1987.

Zhang, D. *Stochastic Methods for Flow in Porous Media: Coping with Uncertainties*. Academic Press, San Diego, CA, 2002.

12

Environmental Law

Patrick J. Paul
Snell & Wilmer L.L.P.

12.1 Introduction

Any understanding of how fluid dynamics impacts environmental law necessarily must start with a basic understanding of environmental law. This chapter will provide a general overview of some of the important environmental laws in the United States and throughout the world. In the United States, these laws predominantly come in the form of statutes (laws enacted by Congress and the states) and regulations (rules created by regulatory agencies).

12.2 Principles

12.2.1 Environmental Protection in the United States

12.2.1.1 Environmental Protection Agency

In 1970, Congress created the United States Environmental Protection Agency ("EPA" or the "Agency") to protect human health in, and the physical environment of, the United States. Currently, the Agency's five main goals are (1) clean air and global climate change, (2) clean and safe water, (3) land preservation and restoration, (4) healthy communities and ecosystems, and (5) compliance and environmental stewardship. EPA primarily accomplishes these goals by writing and enforcing environmental regulations. As of 2009, the Agency employed approximately 17,000 people, including a staff largely made up of scientists, engineers, lawyers, and policy analysts. EPA implements and ensures compliance with its programs through a variety of mechanisms, including civil and criminal penalties as well as cleanup programs.

12.2.1.2 National Environmental Policy Act

In 1970, the United States Congress enacted the National Environmental Policy Act (NEPA). NEPA establishes national environmental policies and goals for "the protection, maintenance, and enhancement of the environment, and provides a process for implementing these goals within the federal agencies." Under NEPA, the federal government is required to use all practicable means to foster conditions in which "man and nature can exist in productive harmony." NEPA also established the Council for Environmental Quality which serves the function of monitoring environmental policies to ensure that such policies are consistent with NEPA.

12.2.1.3 Environmental Impact Statement

Section 102 of NEPA imposes obligations on federal agencies to disclose all anticipated environmental effects of major federal actions or proposed legislation that significantly affects the environment. A major federal action is not limited to government-funded projects, but includes private projects which use government funding or require government approval. Any project which meets this criteria must prepare an environmental impact statement (EIS). An EIS is a detailed statement that outlines the impacts and alternatives of a federal-proposed action. NEPA does not require the government to choose the least environmentally intrusive alternative, but only requires that the government take a serious look at the environmental

Handbook of Environmental Fluid Dynamics, Volume One, edited by Harindra Joseph Shermal Fernando. © 2013 CRC Press/Taylor & Francis Group, LLC.
ISBN: 978-1-4398-1669-1.

impacts and alternatives to a proposed federal action and disclose these consequences to the public.

Proposed federal actions fall into three categories: (1) categorically excluded, (2) EIS, and (3) EA/FONSI. If EPA determines the federal action is of the type which will definitely not have a significant environmental impact, EPA will make a finding that the federal action is categorically excluded from the EIS requirement. If EPA initially believes the federal action will have a significant impact on the environment, EPA will require an EIS. If, however, it is unclear whether a federal action will have significant effects on the environment, the federal action will fall under the EA/FONSI category. In this category an environmental assessment (EA), which provides an initial evaluation of the environmental impact, will be required. If the EA demonstrates that there will be no significant environmental impact, then a Finding of No Significant Impact (FONSI) is issued and the proposed federal action may proceed without any additional NEPA requirements. However, if after conducting the EA it becomes apparent that the environmental impact will be significant, then an EIS is required. The basic structure of an EIS includes (1) a summary, (2) an explanation of the purpose and need for the proposed federal action, (3) a description and comparative assessment of the alternatives, (4) a description of the environment that will be affected by the federal action, and (5) an analysis of the environmental consequences of the proposal and its alternatives.

12.2.1.4 Clean Air Act

The Clean Air Act (CAA) was originally enacted by Congress in 1967 and since then has been substantially amended. The express purpose of the CAA is to promote health and welfare through enhancing the quality of the nation's air resources. Under the CAA, EPA is tasked with the regulation of air pollutants. EPA accomplishes this goal primarily through regulating the quantity of air pollutants that may lawfully be emitted in the United States. The CAA divides sources of air pollution into three categories: (1) stationary sources, (2) mobile sources, and (3) area-wide sources. Stationary sources include such things as large industrial sources and landfills. The CAA contains lengthy and specific requirements for stationary sources. Mobile sources include planes, trains, automobiles, and boats. Under the CAA, most of the mobile source restrictions are placed on manufactures and not on end users. Area-wide sources consist of multiple small sources such as gas stations and dry cleaners, and the CAA does not contain many restrictions for these sources.

12.2.1.5 Ambient Air Quality Standards

Under the CAA, EPA sets National Ambient Air Quality Standards (NAAQS), which are numerical criteria that are applied uniformly throughout the United States. EPA is required by the CAA to set NAAQS at a level in which no adverse human or environmental effects will be experienced. NAAQS have primary and secondary standards. The primary standard is designed to prevent any adverse impacts to human health within an adequate margin of safety. This standard takes into account sensitive populations such as asthmatics and the elderly. The secondary standard is designed to protect public welfare. Public welfare is broadly defined and includes everything from effects on visibility to impact on crops and wildlife.

EPA has established standards for six criteria air pollutants: ozone, particulate matter, carbon monoxide, sulfur dioxide, nitrogen oxides, and lead. In most instances, EPA has set both short-term and long-term NAAQS. Short-term standards are numerical concentrations averaged over short periods of time (e.g., 1, 8, and 24 h). Long-term standards are concentrations averaged over long periods of time (e.g., quarterly or annual averages).

12.2.1.6 State Implementation Plans

Although EPA is tasked with the job of setting the NAAQS for criteria pollutants, the CAA requires states to play an important role in helping to reduce air pollution. States develop and submit air quality plans (State Implementation Plans or SIPs) to EPA. A SIP must provide a detailed outline of how the state plans to control air pollution through regulations and policies, and therefore, states are given latitude to develop ways to limit air pollution. Control measures adopted in the SIPs do not need to be uniform amongst all states. EPA is obligated to approve a SIP if it meets the requirements set forth in CAA § 110(a)(2). Ultimately, if a SIP fails to meet the appropriate level of air quality control, EPA will deny the SIP and may sanction the state or even take over enforcement of the CAA in the area.

To achieve the NAAQS, the CAA requires each state to submit to EPA designations of those areas in the state that do not meet the NAAQS for one or more of the criteria pollutants. Areas are divided into either "attainment" or "nonattainment" areas. An area where the level of pollution is below the NAAQS is classified as an attainment area. If the pollution levels consistently are above the NAAQS, the area will be classified as a nonattainment area and will be subject to more stringent regulatory requirements. A nonattainment area remains a nonattainment area, regardless of actual air quality, until EPA takes action to redesignate the area. An area can be an attainment area for some pollutants while at the same time a nonattainment area for other pollutants. Areas with insufficient data are identified as areas that "cannot be classified" and are treated like attainment areas.

12.2.1.7 New Source Performance Standards

Section 111 of the CAA requires EPA to promulgate a list of categories of stationary sources that cause or contribute significantly to air pollution and may reasonably be anticipated to endanger public health or welfare. After these categories of sources have been identified, EPA must create standards for each category. A "standard of performance" must establish "allowable emission limitations that reflect the degree of emission limitation achievable through application of the best technological system of continuous emissions reduction" which EPA determines has been adequately demonstrated. These new source performance standards ("NSPS") apply to the construction, modification, and reconstruction of specified "affected facilities." Modifications do not include such things as routine maintenance, repair, replacement, or an increase in the hours of operation.

The NSPS program also imposes monitoring, recordkeeping, and reporting requirements on facilities. Specifically, a continuous monitoring system is mandated and periodic reports must be submitted that detail excess emissions. After the initial start up of an affected facility, performance tests must be completed. The CAA stipulates that the tests be conducted using specific test methods and procedures.

12.2.1.8 Hazardous Air Pollutant Program

The Hazardous Air Pollutant (HAP) Program addresses the emissions of HAPs from major and area sources. Section 112 of the CAA contains a list of chemicals which are presently regulated as HAPs. A "major source" is a stationary source or group of sources which have the potential to emit 10 tons per year (tpy) or more of a single HAP or 25 tpy or more of a combination of HAPs. An "area source" is any stationary source of HAPs that is not a major source. For major sources, EPA is required to set emissions standards at the maximum achievable control technology (MACT). For area sources, EPA may only require generally available control technology (GACT).

MACT is the maximum degree of reduction in emissions achievable, taking into account economic cost, non-air quality health and environmental impacts, and energy requirements. If achievable, MACT may include a total prohibition on emissions. For existing sources, an average reduction of 90% (95% for particulates) from uncontrolled levels is "an appropriate benchmark" for setting MACT standards. For most sources, MACT standards may require the use of measures to reduce the volume of pollutants through process changes and substitution of materials. GACT takes into account the methods and techniques which are commercially available by the sources in the category in light of the economic impacts and technical capabilities of the firms to control emissions.

12.2.1.9 Accidental Release Prevention Program

Under CAA Section 112(r), facilities that produce, process, handle, or store certain HAPs listed by EPA above certain threshold quantities will be required to conduct a hazardous assessment to anticipate the consequences of sudden accidental releases, and prepare a Risk Management Plan (RMP). The hazardous assessments must be updated every 5 years and within 6 months of any alteration that could substantially increase the likelihood of an accidental release. Under the RMP, sources that manage listed chemicals in excess of the threshold quantities are required to (1) register with EPA, (2) develop and implement a RMP, and (3) submit an RMP to the Chemical Safety and Hazard Investigation Board as well as to state and local agencies.

12.2.1.10 New Source Review

Under the new source review program, a preconstruction permit is required for new major stationary sources and significant modifications to existing major stationary sources in nonattainment and attainment areas. Under the Prevention of Significant Deterioration (PSD) program, each SIP must contain measures, such as emissions limitations, to prevent significant deterioration of air quality. The PSD review program only applies in attainment and unclassified areas. The Nonattainment Area (NA) new source review program requires new sources of nonattainment pollutants to comply with more stringent emission limitations to bring the air quality into compliance with the NAAQS.

1. *Stationary sources subject to the PSD program must*
   a. Determine and install the best available control technology (BACT) for the new source or major modification
   b. Prepare an air quality impact analysis which includes an assessment of existing air quality (using ambient monitoring data) and a prediction of future ambient concentrations (using dispersion modeling)
   c. Prepare an additional impact analysis to assess the impact of the source on air, ground, and water pollution of soils, vegetation, and visibility
2. *Stationary sources subject to the nonattainment program must*
   a. Obtain required emissions reductions or offsets from other sources that impact the same area
   b. Conduct an analysis of alternative sites, processes, and control techniques which demonstrate that the benefits of the proposed source outweigh the environmental and social costs imposed as a result of its location, construction, or modification
   c. Comply with the lowest achievable emission rate (LAER)
   d. Certify compliance (or adherence to a schedule for compliance) with all emissions limitations and standards by all other sources owned or operated by the same person in the state.

12.2.1.11 Enforcement

The CAA authorizes multiple methods of enforcement. If a source violates a provision of an applicable SIP or of a new federal operating permit program, EPA has the authority to issue a Notice of Violation (NOV). If the source does not cure the violation within 30 days after issuance of the NOV, EPA has the choice of issuing an administrative compliance order, bringing a civil enforcement action, or issuing an administrative penalty order (APO). A violation of the CAA may also result in criminal enforcement by EPA. Criminal penalties vary significantly, and for egregious violations, penalties can be serious. The knowing release of a HAP that results in an imminent danger of death or serious bodily injury to a person can be subject to up to 15 years in prison. The CAA also authorizes individuals to sue to enforce certain provisions of the CAA or SIPs. These citizen suits may be brought against any person including the United States and any governmental agency to the extent permitted by the Eleventh Amendment of the U.S. Constitution. Citizens may also sue to compel EPA to perform any act or duty that is not discretionary. However, under the CAA, citizen suits are forbidden if a governmental enforcement action is being diligently prosecuted in a state or federal court.

12.2.1.12 Clean Water Act

The Clean Water Act (CWA) is a compilation of several different regulatory programs. It is the primary federal regulation governing water pollution in the United States. The CWA is a media-based regulation which means that it regulates through placing restrictions on emissions. The CWA governs the discharge of pollutants from a point source into a navigable water. Navigable water has been interpreted to include all waters with a "significant nexus" to a navigable water. The term pollutant has been broadly interpreted to include almost anything from dredged soil to solid waste to sand. Under the CWA a point source is "any discernible, confined and discrete conveyance." This can include a ditch, construction equipment, or variety of other objects. Courts have even held that a discharge from an underground storage tank can be considered a point source. The CWA imposes strict liability for all discharges covered under the law; therefore, even if the polluter's actions were legal, reasonable, and consistent with standard industry practices, the polluter will still be liable.

12.2.1.13 National Pollutant Discharge Elimination System

The CWA provides some exceptions to the prohibition against discharging pollutants into navigable waters. One major exception is the National Pollutant Discharge Elimination System (NPDES) permit program. Under the CWA a party may lawfully discharge pollutants only after first obtaining a NPDES permit. The NPDES permit program mandates various standards including a requirement that the discharger meet effluent limitations. If EPA makes a determination that the facility requesting a permit is a "new source," then the applicant is required to prepare an environmental assessment under NEPA.

EPA has developed technology-based effluent limits for point source discharges. These standards are based on the performance of pollution control technology and not on the condition of the water body to be effected. Best Available Technology (BAT) is required. EPA has also promulgated water-based effluent limits. If after the technology-based effluent limits are implemented, the water quality is still not at the required level, EPA may then add water-based effluent limits which would provide stricter requirements for polluters.

12.2.1.14 Stormwater Discharges

Stormwater discharges are regulated under the CWA by the NPDES Stormwater Program. This program regulates stormwater discharges from municipal sewer systems, construction activities, and industrial activities. Stormwater is broadly defined to include "stormwater runoff, snowmelt runoff, and surface runoff and discharge." There is often a need to obtain a NPDES prior to discharging stormwater due to the fact that most stormwater are considered point sources. The permitting requirement is set up to prevent stormwater discharges from transferring dangerous pollutants into local streams and rivers.

12.2.1.15 Dredged and Fill Material Permits

Section 404 of the CWA establishes a permit program for the discharge of fill and dredged materials into United States' waters. The discharge of fill and dredged material is broadly defined and includes any fill or dredged material that affects the bottom elevation of a water body of the United States. The United States Army Corps of Engineers works with EPA to develop, administer, and enforce this policy.

12.2.1.16 Spill Prevention Control and Countermeasure

The Spill Prevention Control and Countermeasure (SPCC) program of the CWA requires the reporting of releases and the cleanup of discharges of hazardous substances and harmful quantities of oil to the "waters of the United States." Owners or operators of facilities that reasonably could be expected to discharge oil in harmful quantities, into waters of the United States, are required to prepare a written SPCC plan. This plan must include a prediction of the direction, rate of flow, and total quantity of oil that could be discharged, as well as containment and diversion structures to prevent discharges from reaching waters of the United States.

12.2.1.17 Comprehensive Environmental Response, Compensation, and Liability Act

The Comprehensive Environmental Response, Compensation, and Liability Act ("CERCLA" or "Superfund") was originally enacted in 1980 in an effort to clean up abandoned hazardous substance sites. CERCLA is triggered whenever there is a release or a substantial threat of a release of a "hazardous substance" into the environment. The release or threatened release must be from a vessel or facility to fall under CERCLA. Once CERCLA is triggered, EPA has the authority to take response actions, enforcement actions, or can even require responsible parties to take the necessary steps to remediate the contamination. Under CERCLA, EPA and private parties may recover cleanup costs from responsible parties.

CERCLA provides four categories of "responsible parties" ("RPs"): (1) current owners and operators, (2) any person who at the time of disposal of any hazardous substance owned or operated the facility, (3) any person who arranged for transportation or disposal of hazardous substances, and (4) any person who accepts hazardous substances for transportation or disposal.

CERCLA liability is somewhat unique and therefore important to understand. Under CERCLA, liability is strict, joint, and several. As discussed in the context of the Clean Water Act, strict liability means that a party is still liable even if their actions were legal, reasonable, and consistent with standard industry practices. Under a joint and several liability scheme, each party responsible for the violation could be liable for the total damage award, no matter how insignificant their culpability.

Although a large number of parties can be found liable under CERCLA, courts use the Gore factors when allocating judgments between joint and severally liable parties. The Gore factors include examining (1) contaminants distinguished as originating from one party, (2) the amount of the contaminants, (3) toxicity of the contaminants, (4) amount of involvement, (5) care exercised by the party, and (6) cooperation with the government.

Under CERCLA's framework, a private party may initiate a CERCLA claim to recover money for cleanup costs from another responsible party. In order for a private party to establish CERCLA liability, the plaintiff must be shown that (1) the site is a facility, (2) there has been a release of a hazardous substance, (3) that has caused the plaintiff to incur response costs, and (4) that the defendant falls within one of the four categories of responsible parties.

Although CERCLA liability is very broad, CERCLA does provide an Innocent Purchaser Defense. The defense may not be used by any party that had a contractual relationship in the transfer of the property, unless the party claiming the defense did not know or had no reason to know of the release of hazardous substances. Further, in order to claim the defense the party must have preformed proper due diligence and carried out all of the appropriate inquires into the previous ownership and uses. Among the requirements is a Phase I study, which includes a site visit, review of historical records, and interviews with facility employees to determine if there has ever been a release of hazardous materials. If the Phase I study returns questionable results, then a Phase II study is required which will take the necessary environmental samples to determine if there has been a release of hazardous matters.

12.2.1.18 Resource Conservation and Recovery Act

The Resource Conservation and Recovery Act (RCRA) was enacted by Congress in 1976 and is a classic command and control regulatory instrument that deals with hazardous waste disposal. Up until 1976, the disposal of such waste in the United States was virtually unregulated. Although RCRA was drafted primarily as a waste disposal law, the implementation of RCRA focuses on the management of hazardous waste from generation to disposal. Any person or entity who generates, transports, treats, stores, or disposes of a hazardous waste must comply with the applicable hazardous waste requirements. Violators of RCRA may be subject to an administrative order, a civil action, and/or a criminal enforcement.

Under RCRA, a material must first fit the definition of a "solid waste" before it can be considered a "hazardous waste." A solid waste is a solid, liquid, or contained gaseous material that is discarded. Once it is established that a material is a solid waste, it can be designated as a hazardous waste in two ways: (1) it exhibits a hazardous waste characteristic (ignitability, corrosivity, reactivity, or toxicity); or (2) it is listed as a hazardous waste or mixed with or derived from a hazardous waste.

12.2.1.19 Toxic Substances Control Act

Congress enacted the Toxic Substance Control Act (TSCA) in 1976 as a means of protecting human health and the environment from unreasonable risks presented by the use of untested chemicals. TSCA provides requirements for testing, manufacturing, reporting, and recordkeeping of information related to hazardous chemical substances. Under TSCA, the manufacturing or importation of any chemical that is not on the TSCA Inventory of Chemical Substances (Inventory), or subject to an exception, is prohibited. One notable exception to TSCA is that any substance subject to another U.S. statute is not regulated under TSCA. This includes such things as food additives, nuclear materials, and pesticides. The manufacturer or importer of a new chemical bears the responsibility for developing and reporting adequate data regarding the chemical's effect on heath and the environment. Applications for new chemicals must be submitted to EPA prior to manufacturing or importation of the chemical. If EPA determines that the chemical poses an "unreasonable risk to human health or the environment," EPA may then regulate the substance accordingly.

12.2.1.20 Emergency Planning and Community Right to Know Act

The Emergency Planning and Community Right to Know Act (EPCRA) was enacted in 1986 in large part as a response to the December 1984 release of methyl isocyanate from the Union Carbide plant in Bhopal, India. EPCRA attempts to avert a similar disaster by (1) requiring state emergency plans for environmental releases, (2) establishing the public dissemination of information, and (3) addressing the communities' right to know about potential hazards to the public posed by a facility. It is important to note that the EPCRA does not regulate the storage or disposal of chemicals but rather mandates reporting and planning.

Each state is required to establish a State Emergency Response Commission (SERC). The SERC establishes emergency planning districts and appoints Local Emergency Planning Committees (LEPCs). The LEPCs are responsible for developing an emergency plan which provides for, among other things, the proper notification of public and evacuations plans.

12.2.1.21 Occupational Safety and Health Act

The Occupational Safety and Health Act (OSH Act) was enacted in 1970 in an effort to reduce workplace injuries and illnesses. The OSH Act establishes standards that enhance safe and healthy working conditions in places of employment throughout the United States. The primary mechanism for accomplishing this goal is through the creation of the Occupational Safety and Health Administration (OSHA), which is a part of the U.S. Department of Labor. OSHA is the source of a wide range of regulations designed to promote workplace safety. Virtually, all private employers (including the self-employed) are covered by the federal and state OSHA statutes. Federal and state employees are specifically excluded from coverage, but the OSH Act permits the adoption of measures extending substantially similar requirements to them.

The OSH Act includes a "general duty clause" which imposes workplace safety and health obligations even when there is no specific OSHA standard on a matter. The general duty clause provides that "[e]ach employer shall furnish to each of his employees employment and a place of employment which are free from recognized hazards that are causing or are likely to cause death or serious physical harm to his employees."

Employers must also comply with various other rules and regulations found in the OSH Act.

OSHA requires a Hazard Communication Standard (HCS) which is similar to the Emergency Planning and Community Right to Know Act. The HCS is aimed primarily at ensuring the dissemination of information regarding potentially hazardous substances. The focus of the HCS is to alert workers to the existence of potentially dangerous substances in the workplace and provide them with the proper means and methods to protect themselves against such hazards. The Standard applies to any chemical "which is known to be present in the workplace in such a manner that employees may be exposed under normal condition of use or in a foreseeable emergency."

OSHA mandates permissible exposure limits (PELs), which are the maximum legal limits for exposure of a chemical substance or agent to an employee. Typically, determinations for employee exposure limits are representative of a 30 min short-term exposure and an 8 h time-weighted average (TWA).

12.2.2 Environmental Law throughout the World

Each country uses different methods to enact and enforce its environmental laws. When working in a world that is experiencing an intense amount of globalization, it is important to have a basic understanding of how some of these environmental schemes work. Therefore, this section will provide a brief overview of some different methods of environmental protection used outside of the United States.

12.2.2.1 China's Environmental Law

Similar to the United States a majority of China's environmental laws are media-based regulations and therefore focus on restricting emissions from sources. However, unlike the United States, rather than placing certain tight parameters and limits on emissions, Chinese laws set broad policy statements and goals which oftentimes leads to a great deal of ambiguity. Enforcement of China's environmental law is delegated to the local authorities. However, because the emphasis of the local authorities is typically on economic matters, environmental enforcement is often very sporadic.

Unlike the United States, in China there is no mechanism to allow for citizen suits. Therefore, when a company fails to properly follow the environmental laws, the only method of enforcement is the government. In the United States, Non-Governmental Organizations (NGOs) play a large role in influencing the government to enact and enforce comprehensive environmental laws. However, China does not have strong environmental NGOs to place similar pressure on the government; and therefore, without effective watchdog and lobbying organizations, many of China's environmental laws go unenforced.

12.2.2.2 China's Marine Pollution Regulation

One recent example of Chinese environmental regulation is the Marine Pollution Regulation. In 2009, China passed its first comprehensive system of marine pollution regulation. Similar to many of the environmental laws in the United States, the regulation establishes rules for the prevention, response, and cleanup of pollution within China's waters. The regulation also requires the reporting of any incident in which hazardous pollution is released. One interesting requirement is that prior to entering into China's ports, any ship which contains hazardous cargo must contract with pollution response companies and obtain compulsory insurance.

12.2.2.3 European Union's Environmental Laws

Since the European Union (EU) was established in 1993, the governance of the European Union has grown significantly. The European Union has developed a set of environmental laws which require compliance by member states. In the interest of brevity, this section will only touch on a few specific EU environmental laws. Both laws discussed are examples of information-gathering laws.

12.2.2.4 Registration, Evaluation, Authorisation, and Restriction of Chemicals

The Registration, Evaluation, Authorisation and Restriction of Chemicals (REACH) was enacted in 2006 and is being phased into the EU regulatory scheme over an 11 year period. Similar to the United States' TSCA, the goal of REACH is to regulate the production and use of chemical substances so as to minimize any harmful effects on human health and the environment. REACH requires all companies that manufacture or import chemicals into the EU, in a quantity of 1 ton or greater, to register the substance with a European Chemicals Agency (ECHA). After compiling the information, ECHA enters that information into a public database allowing the public access to in-depth information on chemical substances. REACH requires a phasing out of the most dangerous chemicals once a safer alternative is identified; if no safer chemical is currently available, the regulation mandates that the applicant work to find a safer chemical.

12.2.2.5 European Pollutant Release and Transfer Register

The European Pollutant Release and Transfer Register (E-PRTR) requires member states to produce an annual report detailing the release of pollutants to air, water, and land. The information is compiled by the E-PRTR and then made accessible to the public through an electronic database. As of 2009, the database contained data for over 24,000 industrial facilities. For each facility the database compiles information related to the release and transfer of 91 key pollutants. The goal of the E-PRTR is to increase transparency and public access to information on the environment and therefore assist in the long-term prevention and reduction of pollution.

12.3 Methods of Analysis

Legislators and regulators use a variety of methods to analyze potential laws in an effort to ascertain the most efficient use of environmental resources. This section will describe some of the methods of analysis that law makers use.

12.3.1 Cost–Benefit Analysis

One of the most basic and widely used tools is cost–benefit analysis (CBA). Under CBA, law makers can analyze a variety of environmental alternatives to determine which method maximizes benefits while minimizing costs. In this analysis a value is given to every cost and every benefit experienced by any stakeholder. The alternative with the highest total benefit relative to cost is the most efficient alternative.

12.3.2 Kaldor–Hicks

Another method of analysis law makers typically use is Kaldor–Hicks efficiency. An outcome is Kaldor–Hicks efficient when monetary wealth is maximized and there is no more preferable recollection. An environmental example of Kaldor–Hicks efficiency would occur when a polluter receives a benefit from an exchange that is positive enough such that even if the polluter is required to fully compensate the victims of the pollution, the polluter would still engage in the transaction.

12.3.3 Cap and Trade Program

A cap and trade program is an example of the government using a market system to attempt to more efficiently allocate limited environmental resources. Under this approach the government sets a limit on the total amount of pollutants that may lawfully be emitted. The government then provides pollution credits to companies through an auction or some other mechanism. Each credit is equivalent to a specific amount of pollution, and a polluter is only allowed to release the amount of pollution each to its credits. A market system develops as companies are able to sell and trade these credits back and forth. In theory, this results in the most efficient use of resources as the company that places the highest value on the credit will ultimately purchase the credit.

12.4 Applications

12.4.1 Effects of Environmental Fluid Mechanics on Environmental Law

As an understanding of fluid mechanics continues to increase, its application to environmental law will grow exponentially. A greater understanding of the science behind the environment will allow more accurate modeling for predicting, tracking, and remediating pollutants. Although the applications of fluid dynamics in environmental law are rapidly increasing, this section will provide some brief examples of current uses.

12.4.1.1 Industry

One way that industry uses fluid mechanics is to assess the environmental impact of proposed or current developments. Environmental Assessments (EAs) or Environmental Impact Studies (EISs), which are often required under NEPA, typically use a great deal of modeling to determine the impact of a proposed action on the environment.

An Environmental Assessment was required when a North American developer proposed plans to place two adjacent highway tunnels approximately 500 ft apart. Concerns were raised that airflow moving from one tunnel to the next would cause dangerously high levels of emissions to be circulated in the area. A Computational Fluid Dynamics study was performed to determine the impact of the tunnel design. Under NEPA, the study was required to look at the environmental impact of the tunnel design and the environmental impact of the alternatives. The most serious considerations were how and where the emissions would be released into the atmosphere as a result of the proposed tunnel layout. To accurately model the impact, assumptions needed to be made regarding ambient wind and airflow from the tunnels. Ultimately after examining the alternatives using fluid mechanic modeling, the determination was made that the total emissions generated were comparable between the alternative design; however, impacts to the surrounding areas were significantly different. The fluid dynamics analysis demonstrated that the adjacent tunnel structure could be used to strategically direct pollution away from sensitive areas such as homes and parks.

12.4.1.2 Regulators

When agencies create regulations, they do so with fluid mechanics in mind. Because the Earth's environmental systems are fluid, regulators are required to create and enforce rules with a macro-perspective as pollution in one state will likely impact the environment in other states. For instance, in regulating air quality under the CAA, EPA must take into account that wind flow results in pollutants traveling long distances and having an effect on areas outside of the immediate boundaries. Further, when in remediation efforts fluid mechanics is necessary to determine if a contaminant has migrated. In particular when dealing with underground storage tank discharges, parties must trace the pollutant in order to ensure that it is adequately removed.

12.4.1.3 Litigation

Fluid dynamics is often used in toxic tort litigation to prove or disprove the cause of an injury alleged to be the result of exposure to a toxin. In litigation a party claiming injury, as a result of exposure to some contaminant, must show that their particular injury would not have occurred "but for" the alleged exposure.

To prove causation a plaintiff must prove (1) the characteristic of the plaintiff's medical conditions; (2) identification of the specific substance at issue and the duration and levels of the exposure including the dose of the exposure received by the person, whether by inhalation, dermal contact, or ingestion; (3) general causation—that the alleged exposure (based on nature and amount) can cause personal injury of the type alleged by the plaintiffs; and (4) specific causation—that the weight of the evidence supports the conclusion that the toxin caused the specific injuries involved.

Fluid dynamics can play a large role in proving or disproving the causation of an injury. If a plaintiff is able to demonstrate exposure to a specific toxin through modeling, he is one significant set closer to proving causation. Alternatively, defendants will also attempt to prove a lack of causation through modeling that shows in adequate exposure.

12.5 Major Challenges

12.5.1 Admissibility of Evidence at Trial

In environmental law, experts are often required to testify in court on topics such as migration, exposure, and effects on human health. As discussed earlier, fluid dynamic modeling is used to determine, explain, and predict the effects of toxins and pollutants. However, not all scientific technology is admissible as evidence or testimony in court as courts are weary that the misrepresentation of scientific evidence will pervert justice. In order to present expert evidence, a party must meet the applicable federal or state standard of demonstrating that the scientific principles are appropriate for admission into court.

In all federal and in the majority of state court proceedings, the *Daubert* standard is used to determine if scientific expert opinion is admissible. The *Daubert* standard is given its name from a Supreme Court case that established the standard, *Daubert v. Merrell Dow Pharmaceuticals*. Under the *Daubert* test, if expert testimony will assist to understand the evidence, a witness qualified as an expert may testify if the theory or technology (1) is empirically tested, (2) has been subjected to peer review and publication, (3) has a known error rate, and (4) is acceptable in the relevant scientific community. The issue of reliable principles and methods can be a very contentious issue when parties are seeking to admit or exclude expert opinion based on cutting edge technology. When using the *Daubert* standard, the judge is placed in the role of "gatekeeper" and is tasked with the job of preventing testimony which fails to reach this standard from being heard by the jury. The goal is to preclude unreliable evidence from affecting the outcome of a trial.

Some states use the *Frye* standard, or some version of the *Frye* standard, to determine the admissibility of expert testimony. The *Frye* test is given its name from the test a court formulated in *Frye v. United States*. Under the *Frye* standard the scientific evidence must be "generally accepted by a meaningful segment of the scientific community." The outcome of this test is that an expert seeking to testify must provide a number of other experts who affirm the science behind his or her conclusion. Therefore, if a cutting edge technology has been accepted by some meaningful segment of the scientific community, it will be admissible in trial.

Proponents of the *Daubert* standard praise the higher threshold of admissibility in cases where expert testimony is needed. Specifically in toxic tort litigation, many argue that the *Daubert* test has been instrumental in reducing the so-called junk science in the courtroom, while others believe that the *Daubert* test excludes important and relevant testimony and therefore they find the *Frye* standard to be the more favorable of the two. Even though the *Daubert* standard is stricter than the *Frye* standard, under both standards, cutting edge technology may be excluded. Therefore, it is important to understand which test a court will administer and make sure to present scientific testimony accordingly.

References

Clean Air Act, 42 U.S.C. § 7401-7671 (1970).

Clean Water Act, 33 U.S.C. § 1251-1376 (1948).

Comprehensive Environmental Response, Compensation, and Liability Act, 42 U.S.C. § 9601-9675 (1980). *Daubert v. Merrell Dow Pharmaceuticals, Inc.*, 509 U.S. 579 (1993).

Emergency Planning and Community Right to Know Act, 42 U.S.C. § 11001-11050 (1986). *Frye v. United States*, 293 F. 1013 (D.C. Cir. 1923).

National Environmental Policy Act, 42 U.S.C. § 4321-4347 (1970).

Occupational Safety and Health Act, 29 U.S.C. § 651-677 (1970).

Resource Conservation and Recovery Act, 42 U.S.C. § 6901-6991 (1976).

Toxic Substances Control Act, 15 U.S.C. § 2601-2692 (1976).

II

Focus Areas for Study of Natural Systems

13

Physical Limnology

Alfred Wüest

Eawag—Swiss Federal Institute of Aquatic Science and Technology

Martin Schmid

Eawag—Swiss Federal Institute of Aquatic Science and Technology

13.1 Introduction

The interest in *physical limnology* as a subject of *environmental fluid dynamics* is threefold: (1) physical processes as an avenue for understanding lake ecosystems, (2) management of natural water resources, and (3) lakes as natural scale-up "laboratories" for stratified environmental flow studies.

First, studying aquatic ecosystems, and lakes in particular, calls for interdisciplinary approaches. Even very specific natural in situ processes can hardly ever be viewed independent of the hosting environment. As a practical example, the dynamics of an algae species can not be understood without considering the distribution of nutrients (and other biogeochemical constituents), the stratification and mixing of the water column, as well as baroclinic motions and subsequent modulations of the light regime. In this sense, physical limnology is a crucial discipline for supporting the interpretation of in situ observations of any property, which is always evolving along the fundamental balance

$$\frac{\partial}{\partial t}(\text{property}) = \text{Rates of transformations} - \text{div}(\text{property fluxes}) \tag{13.1}$$

In short, spatial and temporal changes of a property within the water column have always a transport component.

Second, on a global scale, natural water resources are intensely utilized and under increasing anthropogenic pressure. Many of the 110,000 lakes larger than 1 km^2—covering an area of 2.3 million km^2—are used for various purposes, including as recipients for polluted urban effluents. As habitations and infrastructure are often close to lakes, human impacts are strong and many lakes have been subject to enormous changes. To minimize detrimental effects, we have to strive for best environmental engineering practices for management of water resources (such as fisheries, water supply, irrigation, or electricity production, etc.). Therefore, we need to understand how geochemical and ecological processes are related to hydrodynamics and how anthropogenic influences, such as the currently much debated climate change (Section 13.3.2), affect lake ecosystem processes.

Third, lakes are often used as natural "laboratories" of stratified water bodies at intermediary scales for environmental fluid dynamics studies, for example, in studying small-scale turbulence, internal and surface waves, or density currents (Imberger 1998). Lakes as "laboratories" have many advantages: (1) enormous variety and variability of the physical, hydrogeochemical, and meteorological boundary conditions (Table 13.1), (2) intermediate spatial dimensions which are on geophysical scales but more easily accessible than those of ocean basins, (3) the possibility to derive budgets of tracers such as temperature, salinity, and density under natural conditions.

In this chapter, the most relevant lake-specific hydrodynamic phenomena are reviewed. Due to limited space, the review is largely incomplete and we refer to more comprehensive overviews, such as various sections in the *Encyclopedia of Inland Waters* (Likens 2009), textbooks (Imboden and Wüest 1995; Imberger 1998), and reviews (Imberger and Patterson 1990; Wüest and Lorke 2003).

Handbook of Environmental Fluid Dynamics, Volume One, edited by Harindra Joseph Shermal Fernando. © 2013 CRC Press/Taylor & Francis Group, LLC. ISBN: 978-1-4398-1669-1.

TABLE 13.1 Variability of the Physical Characteristics of Lakes

Property, Quantity	Approx. Extreme Values	Comments, Reference
Volume	23,000 km^3	Lake Baikal
Area	82,400 km^2	Lake Superior
Depth	1,637 m	Lake Baikal
Residence time-scale	~10,000 years	Deep-water of Powell Lake (BC)
Salinity	0.01–485 g kg^{-1}	Granite mountain lakes; Don Juan Pond (Antarctica)
Gas concentrations	CO_2: 0.36 mol L^{-1}	Lake Nyos
	CH_4: 0.022 mol L^{-1}	Lake Kivu
Suspended solids	0.1–200 > 1,000 mg L^{-1}	Ultra-oligotrophic lakes; alpine reservoirs below glaciers; near inflows during storms
Stratification, N^2	10^{-10}–10^{-2} s^{-2}	Bottom layer of Black Sea, thermoclines of ponds
Energy dissipation	10^{-12}–10^{-5} W kg^{-1}	Interior of quiescent hypolimnia; wave-affected SL during storms
Vertical diffusivities	~Molecular to 10^{-2} m^2 s^{-1}	Extreme (quiescent) stratification; convective surface layer
Horizontal diffusivities	~10^{-1} m^2 s^{-1}	In very large lakes
Kolmogorov scale	mm to few cm	Wave-affected SL; very weak turbulence and strongly-stratified
Ozmidov scale	cm to ~100 m	Strongly-stratified small lakes; Lake Baikal (Section 13.2.7)

13.2 Principles

In this section we provide a short overview of those hydrodynamic processes which we consider as specific and relevant for understanding the physical functioning of standing (stratified) inland waters.

13.2.1 Large-Scale Horizontal Flows

Natural waters are never at rest. Although often weak, various types of motions with spatial scales from basin dimensions (Table 13.1) to the smallest possible eddies at the Kolmogorov length (Chapter 1) are ubiquitous in lakes. Vertical motions are generally hindered by stratification and "inefficient" forcing (perpendicular to wind), and therefore restricted to slow and small baroclinic dislocations (Figure 13.9). Horizontal flows, however, are largely unbounded. Without much restoring forces, they follow the immediate driving forces of winds and pressure (density) gradients.

The question whether (latitude-dependent) circulation patterns are general and consistent between the northern and the southern hemisphere has long been challenging the environmental fluid scientists. Lake-wide motions mainly depend on the horizontal gradients of density (pressure) and wind stress, and in particular its curl (vorticity; Chapter 30). However, it appears intriguing that the circulation in large- and medium-sized lakes is, according to observations (Emery and Csanady 1973), mostly cyclonic (counterclockwise in the northern hemisphere). For the summer season—periods of strong stratification—this phenomenon is usually explained by the cyclonic geostrophic circulation developing around a "domed" density stratification with lighter surface water piling up at the shore, as is often observed in the North American (NA) Great Lakes (Schwab et al. 1995).

A comparative analysis of models and observations for Lake Michigan by Beletsky and Schwab (2008) showed indeed that, during stratified summer periods, baroclinic effects are primarily responsible for the predominantly cyclonic flow (Figure 13.1a). During nonstratified winter seasons, however, when density differences within the lake are small (Chapter 1), the circulation patterns (Figure 13.1b) are determined by the cyclonic wind stress curl. Interestingly, this curl over the Great Lakes is often a result of the lake–atmosphere interaction (e.g., in winter relatively warm lakes tend to generate their own meso-scale low-pressure system with cyclonic winds).

Wind stress curl can also be generated in a uniform wind field by the surrounding topography. In large lakes, this effect is usually much smaller than that from meso-scale atmospheric circulation. In small lakes, however, orographic effects by mountains or vegetated surroundings cause lateral wind gradients. In those lakes, the synoptic-scale wind stress curl over short distances is less important. Consequently, the overall

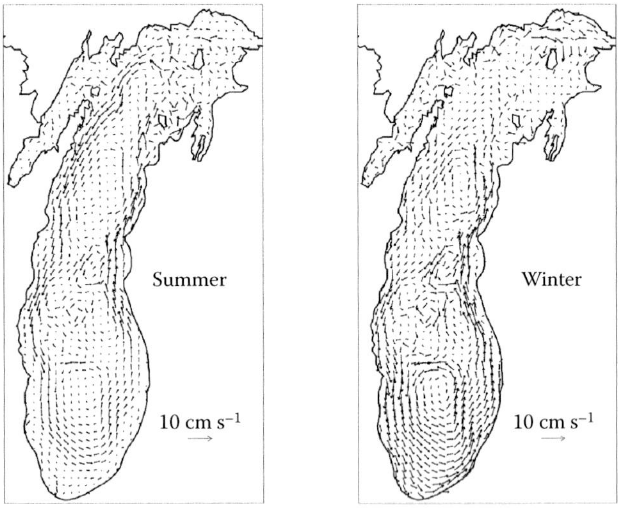

(a) (b)

FIGURE 13.1 Modeled mean 1998–2007 summer (a) and winter (b) depth-averaged circulation in Lake Michigan based on 10 years of continuous modeling using meteorological observations as forcing. The simulations reveal a remarkably stable large-scale cyclonic circulation pattern, confirmed by long-term current observations during both seasons. (From Beletsky, D. and Schwab, D., *Geophys. Res. Lett.*, 35, L21604, 2008.)

circulation can be either cyclonic or anticyclonic, as observed in Lakes Geneva and Biwa, respectively (Lemmin and D'Adamo 1996; Akitomo et al. 2009).

Unidirectional wind of limited duration moves surface water downwind. If the wind lasts for at least a quarter period of the basin-scale surface oscillation, then the water level tilts until the restoring pressure approaches the equilibrium with the wind stress, and the so-called set-up is reached (Stevens and Imberger 1996). If the winds blow longer than this timescale for set-up, the density layers become increasingly tilted, and downwelling (at the downwind end) and upwelling (at the upwind end) set in.

In Figure 13.2, an extraordinarily large up-/downwelling event, recorded in Lake Tahoe (United States), is documented as an instructive example. Driven by winds of ~10 m s^{-1} for about 5 days, this particular upwelling moved half of the lake's surface water downwind to the east, dislocated water masses as much as 500 m vertically—almost over the entire lake depth—and left an originally two-layered thermal structure as a diffuse (weak) gradient behind (Figure 13.2; Schladow et al. 2004). Such nonlocal and advective vertical exchange flows play an important role in the long-term stratification and nutrient balance of deep lakes such as Crater Lake (Crawford and Collier 2007) or Lake Baikal (Shimaraev et al. 1994).

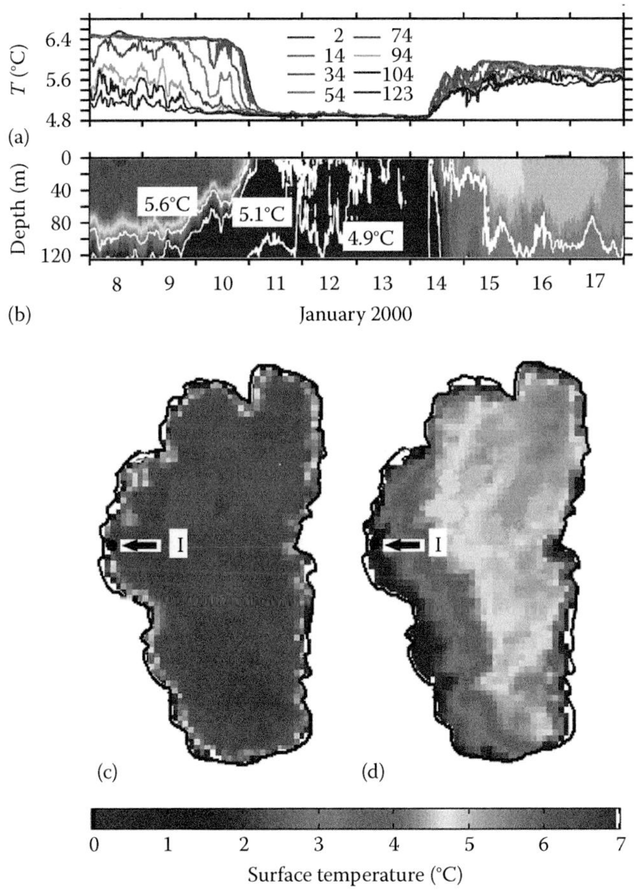

FIGURE 13.2 **(See color insert.)** Evolution of temperature stratification in Lake Tahoe as a result of an extraordinary wind-driven upwelling from January 9 to 14, 2000. Panels (a) and (b) show the temperature record in 2–123 m depths and a few isotherms in the top 120 m at the upwind location "I" (39°/6.7440′N, 120°/8.9320′W). Whereas before the event, temperatures in the top 120 m ranged from 4.9°C to 6.4°C, this layer was almost homogenous afterward (a and b). Panels (c) and (d) show space-borne ATSR images of surface temperatures on January 4 (c; before event) and on January 13 at 22:09 (d; during upwelling). The enormous cooling is due to vertical mixing and heat loss resulting from strong winds. (From Schladow, S.G. et al., *Geophys. Res. Lett.*, 31, L15504, 2004.)

13.2.2 Density Stratification and Water Column Stability

Density gradients oppose vertical mixing. A stronger density stratification results in higher resistance against mixing by wind (Section 13.2.3) or convection (Section 13.2.4). In combination with solar radiation, heat and gas exchange at the surface, as well as chemical and biological processes, the density stratification also leads to a vertical stratification (segregation) of physical and chemical properties, and consequently to different ecological niches for organisms ranging from bacteria to plankton and fish. A review on density stratification in lakes was recently published by Boehrer and Schultze (2008).

In the majority of lakes, density stratification is mainly determined by water temperature (T). At atmospheric pressure, the temperature of maximum density, T_{MD}, is ~4°C. Consequently, there are two different types of stable stratification: warm water overlying cold water at $T > 4°C$, and colder water overlying less cold water at $T < 4°C$. This is different to the ocean (and saline lakes with $S > 24.7‰$), where T_{MD} is always below freezing point and no inverse stratification exists. Due to the nonlinear temperature-dependence, a certain temperature difference ΔT has a much larger effect on density at higher temperatures. Therefore, smaller ΔT are required to produce convective mixing or density currents in tropical lakes than in cold lakes. Close to 4°C, the effect of ΔT variations on density vanishes.

The strength of the density stratification is usually expressed by the stability N^2 (s^{-2}), the square of the Brunt–Väisälä frequency N. It is defined as

$$N^2 = -\frac{g}{\rho} \cdot \frac{\partial \rho}{\partial z} \approx g \cdot \left[\alpha \cdot \left(\frac{\partial T}{\partial z} - \Gamma \right) - \beta_S \cdot \frac{\partial S}{\partial z} \right] \quad (13.2)$$

where

g (m s^{-2}) is the gravitational acceleration
ρ (kg m^{-3}) is the density of water (with effect of pressure removed)
z (m) is the vertical coordinate (positive upward)
α (K^{-1}) is the thermal expansivity
T (K) is the in situ temperature
S (g kg^{-1}) is the salinity
β_S (kg g^{-1}) is the coefficient of haline contraction
Γ (K m^{-1}) is the adiabatic lapse rate

Values for α, Γ, and ρ are calculated as functions of temperature, pressure, and salinity (Chen and Millero 1986). β_S depends on the composition of the dissolved species (Wüest et al. 1996) and typically ranges between 0.70 and 0.85×10^{-3} kg g^{-1} for almost all natural waters. N^2 varies over ~10 orders of magnitude in natural waters from ~10^{-11} s^{-2} (e.g., in perfectly well-mixed bottom layers) to ~0.1 s^{-2} in extreme haloclines, where freshwater overlies salty water (Table 13.1). In some cases, significant contributions of dissolved gases (such as in volcanic lakes; Figure 13.3) or suspended solids (such as in cold mountain reservoirs with glaciated watersheds) need to be considered for both density and stability.

Another concept for assessing the stability of a lake is the Schmidt stability. It is defined as the necessary work to transform (mix) the density stratification of a water body into a completely mixed layer of homogenous density (Schmidt 1928). This quantity is of practical relevance, as it provides a measure of the mechanical energy needed to destratify a water body and allows to quantitatively compare the potential energy stored in different stratifications.

The Wedderburn number and the Lake number (Imberger 1998), both dimensionless, are often-used quantities that relate the stabilizing force due to density stratification to the destabilizing forces supplied by wind. These numbers provide quantitative measures of whether external wind forces are sufficiently strong to overcome density stratification and cause substantial mixing. An example where these two numbers become small (strong external forcing) is provided in the upwelling event discussed in Figure 13.2.

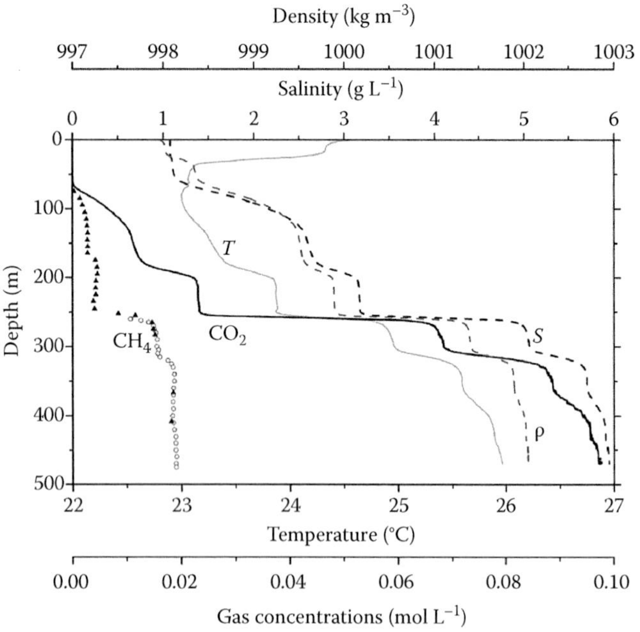

FIGURE 13.3 Vertical profiles of temperature (*T*), salinity (*S*), concentrations of CO_2 and CH_4, and density (ρ) in February 2004 in meromictic Lake Kivu. Whereas salinity and CO_2 contribute to the stability of the density stratification, temperature and CH_4 reduce the stability. The step-like structure of the profiles is caused by subaquatic springs entering the lake. (From Schmid, M. et al., *Geochem. Geophys. Geosyst.*, 6(7), Q07009, 2005.)

Stratification in lakes usually displays a strong seasonal variability, especially in temperate and boreal regions with a large seasonal variation of air temperature and irradiation. Cooling or warming of the surface water toward T_{MD}, in combination with wind forcing, leads to seasonal deep-mixing. Lakes are classified depending on their mixing regimes (Lewis 1983). *Dimictic* lakes mix completely twice a year with stratified periods in summer and winter, the latter usually under ice cover. *Monomictic* lakes become homogenized only once a year, either in winter if their surface temperature is $>T_{MD}$ during the stratified period (warm monomictic) or in summer if their surface temperature is $<T_{MD}$ during the stratified period (cold monomictic). In deep lakes with surface temperatures always $>T_{MD}$, complete mixing often occurs only every few years (called *oligomictic*), depending on the meteorological conditions during the cold season. *Meromictic* lakes, such as Lake Kivu (Figure 13.3), are permanently stratified by dissolved substances. Seasonal convective deep-mixing reaches only a certain depth. A special case of meromixis can exist in very deep cold lakes. In these lakes, the pressure-dependence of T_{MD}, decreasing by ~0.2°C per 100 m depth, becomes important. The deep-water of such lakes can be permanently stratified with temperatures <4°C but increasing upward. As a consequence, seasonal mixing reaches only to a certain depth, typically on the order of a few 100 m, depending on the energy introduced by the wind. Examples are Lake Baikal (Shimaraev et al. 1994) or Crater Lake (Crawford and Collier 2007). In these two lakes, also the phenomenon of *thermobaric instability/convection* can be observed, which causes cooler surface water to sink to the deepest reaches, after strong winds have pushed it far enough down, such that the cooler water can sink freely in the deep waters with positive α. Finally, shallow lakes often mix completely at each stronger wind event or even daily due to nocturnal cooling. Such lakes are called *polymictic*.

Observations from Soppensee are displayed in Figure 13.4 as a typical example of the seasonal pattern of stratification in a temperate monomictic lake. Here, irradiation and heat exchange with the atmosphere in summer heat up the epilimnion to temperatures >20°C while the deep-water temperature never exceeds 6°C. Between the surface mixed layer (Section 13.2.4) and the deep-water, a strong temperature gradient, the thermocline, builds up. At the same time, dissolved substances, especially nutrients and calcium carbonate, are consumed or precipitated in the surface layer (SL) and are redissolved in the deep-water. Consequently, conductivity decreases at the surface and increases in the deep-water, building up an additional weak salinity stratification (Figure 13.4). The transport of heat or dissolved substances through the thermocline is limited. Thus, the deep-water remains cool throughout the summer and anoxic conditions may develop in productive and shallow lakes.

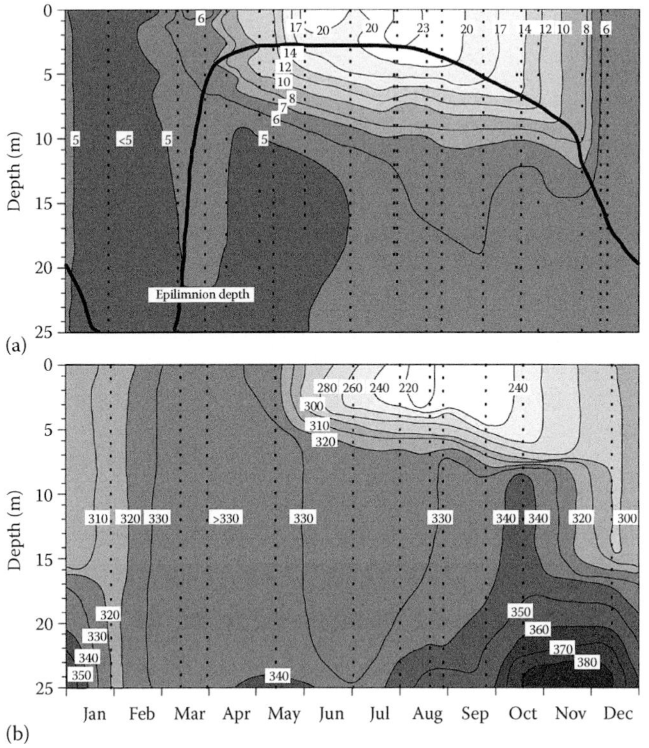

FIGURE 13.4 Seasonal evolution of the vertical profiles of (a) temperature and (b) conductivity κ_{20} (μS cm^{-1}) at T = 20°C (salinity [g kg^{-1}] ≈ (0.9/1000) κ_{20}) in the 26 m deep Soppensee. Epilimnion depth indicates the approximate depth to which cooling thermals reach at night (Section 13.2.4). The data (indicated by dots) are aggregated over 14 years. (From Gruber, N. et al., *J. Paleolimnol.*, 24(3), 277, 2000.)

13.2.3 Wind-Forced Surface Layer

The structure and dynamics of the SL of a "standing" water body evolve as the result of the interaction of wind stress, heating (especially radiative), and cooling. The physical features of the SL, such as currents, mixing, and stratification, can be understood as a competition of stabilizing and destabilizing processes. Wind stress and convective turbulence maintain mixing and thereby tend to remove stratification and to homogenize the SL. Absorption of solar radiation, however, creates turbulence-counteracting stratification and thus hinders mixing and tends to strengthen the stratification in the SL. Both of these antagonistic processes are self-fortifying. Predicting the build-up of the SL in spring can therefore be especially challenging. As mixing of the density-stratified water underneath the SL is strongly suppressed, we usually observe a sharp and distinct transition to the underlying quiescent and well-stratified hypolimnion. The depth of this SL base (epilimnion thickness in Figure 13.4) depends therefore on the "history" of heating, cooling, and winds.

Wind blowing over a lake exerts a stress τ_0, that is, a force per unit area (N m^{-2}), on the water surface. As for other drags in turbulent flows, τ_0 is formulated as proportional to the square of the wind speed U_{10} (measured 10 m above surface):

$$\tau_0 = \rho_{air} C_{10} U_{10}^2 \qquad (13.3)$$

The source of this stress is the downward turbulent flux of horizontal momentum $\tau_0 = \rho_{air}\overline{U'W'}$ from the atmosphere, of which a part of the stress $\tau_{SL} = \rho\overline{u'w'}$ entering the underlying water, where U, W (u, w) are the horizontal and vertical velocities of air (water) (m s^{-1}), ρ_{air} (ρ) is air (water) density (kg m^{-3}), and '(prime) denotes fluctuations. The momentum flux τ_{SL} into the SL, however, is smaller than the total stress τ_0, especially for stronger winds. This reduction is caused by surface waves consuming parts of τ_0 by accelerating and maintaining the (breaking) waves (wave stress: τ_{Wave}; Chapter 31). The momentum conservation ("constant stress") implies that the two fluxes on the water-side add to the total wind stress by $\tau_0 = \tau_{SL} + \tau_{Wave}$.

All other wind- and surface-related effects (or processes) are included in the nondimensional wind drag coefficient C_{10} (–), which becomes a very complex quantity as a function of wind speed U_{10}, the state of the waves (height, steepness, wave age, etc.) and the wind fetch (Csanady 2001). As summarized in Wüest and Lorke (2003), C_{10} increases almost proportionally to U_{10}^{-1} at weak winds below ~4 m s^{-1} (laminarization) and increases gradually for wave-generating higher winds, leaving a minimum of the surface roughness and therefore of the wind drag coefficient near $U_{10} \approx 4$ m s^{-1}.

As an effect of τ_{SL}, downwind currents develop in the SL (Section 13.2.1; Figure 13.5). Empirically, it has been found that the currents—if averaged over some minimum time-scale—decrease logarithmically (Law-Of-the-Wall, LOW) with depth from the surface velocity $u_0 \approx$ (0.02–0.03) U_{10} (Monismith and MacIntyre 2009). Even though LOW is a massive simplification, it offers many useful relations between the forcing wind and the turbulence level ($\varepsilon = u_*^3/(kz)$), turbulent velocity $\left(u_* = \sqrt{\tau_{SL}/\rho}\right)$, turbulent diffusivities (zku_*), and the mixed-layer depth in equilibrium with the momentary wind and SL heating (Monin–Obukhov scale: $L_{MO} = u_*^3/(kJ_{b,S})$, where k = 0.41 is the von Karman constant and $J_{b,S}$ is the buoyancy flux due to heating, see Section 13.2.4). Despite obvious limitations, all those approximations, which are expressed as a function of depth z and of the wind U_{10}, are of great practical value in estimating (atmosphere–water) fluxes and vertical distributions of water constituents.

Under strong wind forcing, the top ~2 m of the SL deviates from LOW structure because additional turbulent kinetic energy is injected into the very surface as a result of wave breaking. In this so-called wind-affected SL, turbulence production and mixing exceed LOW level, with the largest excess right at the surface. Langmuir circulation introduces another deviation from LOW: if a persistent wind faster than several m s^{-1} blows for several tens of minutes, then rows of vertical circulation, perpendicular to the wind direction, set in. These circulation cells, with the axis almost parallel to the wind, can reach approximately to the base of the mixed layer Z_{SL} (Section 13.2.4) and show a lateral spacing of ~2 Z_{SL} between the zones of divergence (upward) and convergence (downward). The convergence lines can be seen from above, as floating particles (such as debris, bubbles, leaves, etc.) accumulate there.

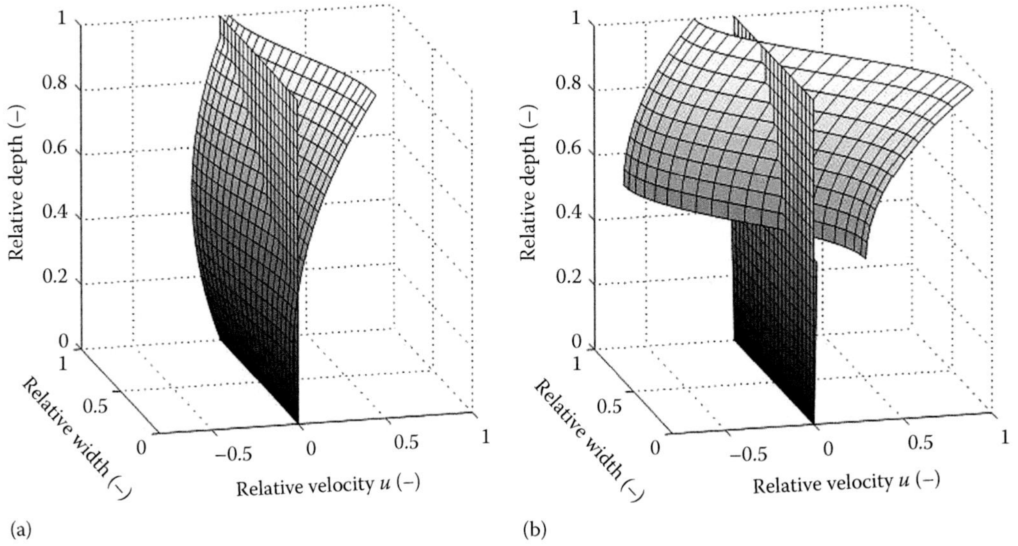

FIGURE 13.5 Analytically calculated reaction of a rectangular lake to a laterally inhomogenous wind stress for nonstratified conditions (a) and for strong stratification (b) of the SL. In a homogenous SL, the wind energy is distributed over a large depth range, whereas for stratified conditions, only the mixed SL is affected and the thermocline prevents further penetration of the wind energy. As the affected volume is smaller, this results in higher velocities. Baroclinic effects are neglected for short-term reactions. (From Toffolon, M. and Rizzi, G., *Adv. Water Res.*, 32, 1554, 2009.)

13.2.4 Convective Surface Layer Mixing

The SL in lakes can be found to be completely homogenous, especially early in the morning after a cold night. Different to daytime, when net heat input causes warming and subsequent stratification of the SL, the heat loss H_S (W m^{-2}) at night cools the uppermost millimeter of the water (index "S" for surface; fluxes positive upward). As a result, slightly cooler (heavier) water parcels are formed which sink as "thermals" until they reach heavier (stratified) layers, where the sinking comes to a halt. For reasons of continuity, equivalent volumes of slightly warmer (lighter) water are lifted up (Figure 13.6). The corresponding net downward (negative) density flux $F_{\rho,T}$ reduces the potential energy of the lake stratification. This energy is released at the very top of the water column at a rate of

$$J_{b,S} = \frac{\alpha g}{\rho c_p} H_S = -gF_{\rho,T} (\text{W kg}^{-1}) \qquad (13.4)$$

and is transformed to kinetic energy of the thermals, which sink/rise with vertical velocities of approximately $w_C \approx \pm(J_{b,S} Z_{SL})^{1/3}$ (m s^{-1}) within the well-mixed SL of depth Z_{SL}, where c_p is the heat capacity of water. The net effect of the thermals is to transport density downward, as expressed by the temperature-related density flux $F_{\rho,T}$. Two examples of convective SLs with different α (and therefore different $J_{b,S}$) as well as different thicknesses Z_{SL} and subsequently different thermal velocities w_C are illustrated in Figure 13.6.

Examining the essence of Equation 13.4 indicates that cooling of a warm SL, with $T > T_{MD}$, is just one among various processes producing a downward density flux F_ρ and causing convective mixing. Any process generating higher density at the top of the water column produces potential energy—expressed as surface buoyancy flux $J_{b,S}$ (W kg^{-1})—which can maintain convective plumes. An example is the warming of a cold SL of temperature $T < T_{MD}$ (where α is negative). This process is important in all dimictic and cold monomictic lakes (Section 13.2.2). Convective mixing usually starts already under ice cover, as soon as solar radiation penetrates the ice. Under-ice convection is weak due to the low value of α at low temperatures and because the buoyancy flux is not produced directly at the surface but, depending on the absorption characteristic of the water, within the convective layer itself.

In saline lakes where salinity S dominates density variations (Section 13.2.2), surface evaporation or freezing E (m s^{-1}) concentrates the SL water, increasing density and subsequently generating a surface buoyancy flux $J_{b,S} = g\beta SE$. Also the input of particles can locally (near inflows) increase the density in the SL and cause sinking of those particle-laden water parcels.

As evident from Equation 13.2, convective mixing is determined by the relevant fluxes (such as H_S and E) and by the density coefficients (such as α, and β). Therefore, convective mixing in temperate lakes is strongest, and the most homogenous SLs are found during fall when heat loss reaches its maximum and the SL temperature (and therefore α) is still high. The convective SL is expanding in thickness as surface temperature decreases and as impinging thermals entrain water from the stratified layer underneath (so-called penetrative convection). As SL temperature approaches T_{MD}, the convection ceases. If deeper layers are stratified by an (even slight) geochemical gradient, the seasonal deep-mixing may come to a halt before reaching maximum depth. Depending on the strength of episodic winds during this period of decaying convection, the

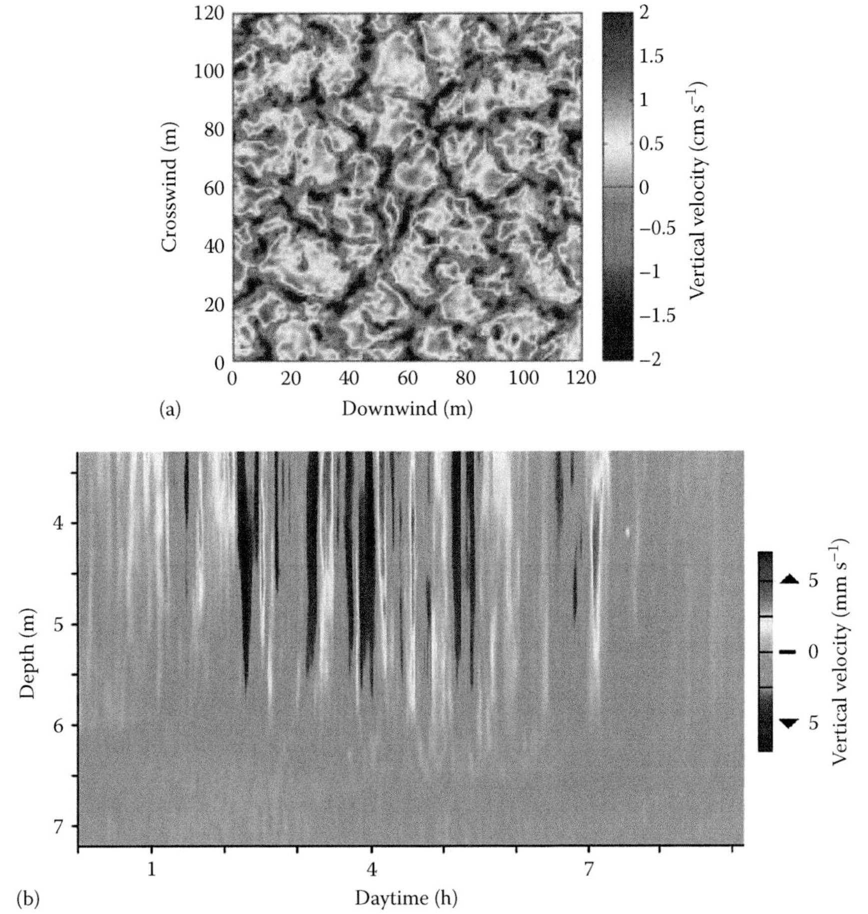

FIGURE 13.6 **(See color insert.)** (a) Horizontal cross section at 3 m depth showing vertical velocities of thermals as modeled by large-eddy simulations (e.g., see Chapter 21 of *Handbook of Environmental Fluid Dynamics, Volume Two*) for a 50 m thick convective SL loosing heat at 200 W m^{-2}. The model was run for oceanic water and no wind forcing with 1 m grid resolution and periodic horizontal boundary conditions. Upward (red) and downward (blue) thermals reach velocities of ~2 cm s^{-1}. (From Li, M. et al., *Deep Sea Res. I*, 52, 259, 2005.) (b) Time series of vertical velocity profiles observed in the top 6 m of the small wind-protected Lake Soppensee during a night of cooling in fall (September). The contour plot reveals rising (red) and sinking (blue) thermals from midnight to 8 a.m. (~sun rise). Due to lower α (and therefore weaker $J_{b,S}$) and smaller Z_{SL} (Equation 13.4) than in the upper panel, the water parcels reach velocities of only ±0.5 cm s^{-1}. (From Jonas, T. et al., *J. Geophys. Res.*, 108(C10), 3328, 2003.)

depth of seasonal deep-mixing can vary substantially from winter to winter. Details on the seasonal deep-mixing are provided in Section 13.2.2.

The relative importance of convection and wind stirring changes with depth z: the effect of the wind is inversely proportional to depth ($\varepsilon(z) = u_*^3/(kz)$), whereas the effect of convection, expressed by $J_b(z)$, decreases only gradually. The depth L_{MO} at which the energy dissipation of both processes are approximately equal is, analogous to the thermally stratified case (Section 13.2.3), defined as the Monin–Obukhov length $L_{MO} = u_*^3/(kJ_{b,S})$, although the physical interpretations are different (Wüest and Lorke 2003). If the mixed-layer depth is $>L_{MO}$, then SL deepening is dominated by convection.

Besides the SL, various other types of convective mixing occur in lakes. The bottom boundary experiences convective mixing caused by geothermal warming (such as by lacustrine springs; Figure 13.3) or forced by shear-induced convection over sloping sediments (Section 13.2.6, BBL mixing). Warm and saline deep inflows can also cause double-diffusive mixing, discussed in the next section. Finally, thermobaric instability (Section 13.2.2) is a convective process, which is, due to α, changing its sign at the depth where temperature crosses the line of $T_{MD}(z)$.

13.2.5 Double-Diffusion

Double-diffusive convection can occur in lakes in the case of opposing contributions of temperature and dissolved substances to the density gradient. The processes leading to this phenomenon are described in detail in Chapter 39. Two different regimes exist: the *diffusive regime* if temperature is the destabilizing component; and the *finger regime* if dissolved substances are the destabilizing component. In lakes, the finger regime has only been observed right at the top of evaporating saline lakes (Hanjalic and Musemic 1997).

Ideal conditions for the occurrence of the diffusive regime, first observed in Antarctic Lake Vanda, can exist in some deep meromictic lakes and especially mining lakes (von Rohden et al. 2010). The most spectacular case is Lake Kivu, where double-diffusive staircases of more than 300 convective layers have been observed over a vertical range of more than 300 m (Newman 1976; Schmid et al. 2010). An instructive example is Lake Nyos, where the formation and subsequent expansion of a double-diffusive staircase (Figure 13.7) was triggered by a strong seasonal cooling of the SL (Schmid et al. 2004). In both these lakes, dissolved gases contribute to the density stratification in addition to temperature and salts. Double-diffusion allows removing heat without a corresponding removal of dissolved substances, and therefore might have supported the accumulation of large amounts of gases in these lakes (Figure 13.3). In Lake Banyoles, suspended particles as another third component with different transport properties were involved in the formation of a double-diffusive staircase (Sánchez and Roget 2007).

Differential diffusion due to the different molecular diffusivities of heat and salt is also responsible for other phenomena observed in lakes: convective mixing of a hypolimnion by surface cooling due to heat flow through a strong chemocline without direct exchange of water masses between the two layers (Boehrer et al. 2009); preferential transport of heat compared to dissolved substances through diffusive boundary layers, for example, at the sediment–water interface; exchange of heat between water layers with different properties in the absence of significant turbulence. An instructive example of the latter is the long-term development of qualitatively different vertical profiles of temperature and salts in Powell Lake, which was formerly a fjord (Sanderson et al. 1986).

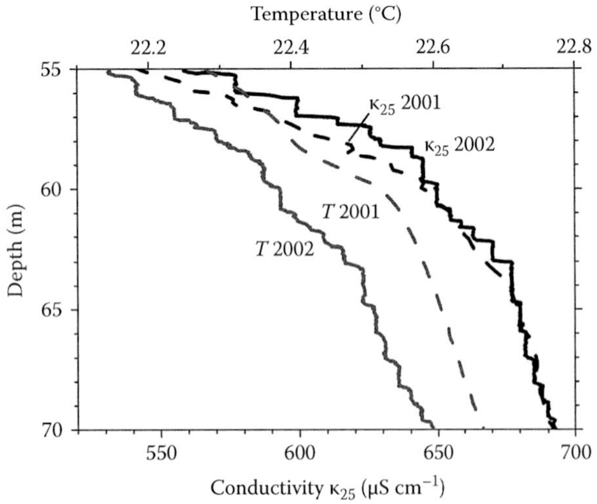

FIGURE 13.7 Vertical profiles of conductivity and temperature observed in Lake Nyos between 55 and 70 m depth on November 3, 2001 and December 8, 2002. The formation of the double-diffusive staircase is clearly visible as is the resulting heat loss due to the increased vertical heat flux (whereas conductivity was hardly affected). (From Schmid, M. et al., *Deep Sea Res. Part I*, 51, 1097, 2004.)

13.2.6 Bottom Boundary Mixing and Sediment Interaction

Winds over lakes drive currents of various frequencies and scales (Section 13.2.1). In combination with the lakes' stratification and restricting topography, basin-scale internal waves (such as seiches, Kelvin waves, etc.; Antenucci 2009) develop and maintain corresponding currents over the sediment (Lorke and MacIntyre 2009). Within the lowest few meters, friction-induced stresses $\tau_{BBL} = \rho\overline{u'w'}$ balance the kinetic energy input and limit the currents to velocities of a few cm s^{-1}, hardly ever >10 cm s^{-1} (Figure 13.8). Currents decrease toward the local bottom and come to a halt at the sediment surface. The near-sediment layer of typically a few meters thickness, where the (steep) current gradients $\partial u/\partial h$ are controlled by bottom friction, is considered as the bottom boundary layer (BBL).

As in the SL, the stress τ_{BBL} is nearly constant through the BBL and the currents and turbulence largely follow the LOW, if neither waves nor surface undulations disturb the BBL dynamics. The friction velocity $u_* = \sqrt{\tau_{BBL}/\rho}$ and the height h above the sediment are the two relevant parameters, which allow estimating the shear $\partial u/\partial h = u_*/(kh)$, the corresponding logarithmic velocity profile $u(h)$, the eddy viscosity khu_*, the turbulent energy production (equal to dissipation) $\varepsilon = u_*^3/(kh)$ and other related LOW properties. The friction-induced stress τ_{BBL} is equal to the shear $u_*/(kh)$ times the turbulent viscosity khu_* times ρ. When the bottom is approached, the shear $\partial u/\partial h$ rises drastically toward the sediment, and therefore, the Richardson number $Ri = N^2(\partial u/\partial h)^{-2}$ decreases and the energy dissipation $\varepsilon = u_*^3/(kh)$ increases sharply. As a result, vertical profiles of the water column often reveal well-mixed layers immediately above the sediment.

More recently, evidence has been provided that the so-called shear-induced convection (Lorke et al. 2005) also contributes to

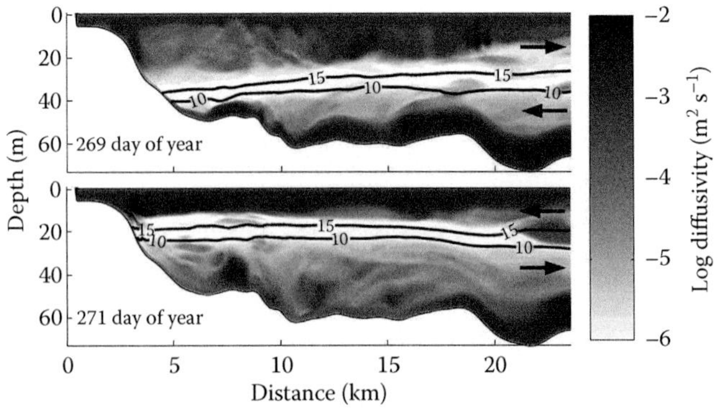

FIGURE 13.8 2D transect of 3D modeled vertical diffusivity along the bottom boundary of a ~5 km wide channel-like sub-basin of Lake Geneva for day 269 and 271 in 2002. In the upper panel (day 269), an upslope current causes shear-induced convection in the BBL and higher near-sediment diffusivities. In the lower panel (day 271), a down-slope current causes near-sediment stratification and lower diffusivities. For both flows, the mixing intensity is much higher in the BBL than in the stratified interior (especially low in the thermocline between 10°C and 15°C). (From Umlauf, L. and Lemmin, U., *Limnol. Oceanogr.*, 50(5), 1601, 2005.)

BBL mixing. Due to the LOW, a bottom-parallel current $u(h)$ has a larger velocity at a few meters height h than immediately above the sediment ($h \sim 0$). If the bottom is sloping, the water column stratified, and the flow unidirectional up-slope for some time, then—for purely geometrical reasons—denser (usually cooler) water is moved over slightly lighter (usually warmer) water above the sediment. The resulting unstable density profiles induce convective mixing and enhanced turbulence within the BBL (Figure 13.8, upper panel). Conversely, down-slope currents move lighter layers on top of denser water enhancing stratification above the sediment (Figure 13.8, lower panel). As most deep-water currents (such as seiching) have wave-type oscillatory components with basin-scale baroclinic periods of hours to days, the BBL is usually exposed to a periodic sequence of those two flows, which are overlaid onto the regular friction-induced turbulence. The combination of these two processes (friction and convection) explains that BBLs are usually well-mixed with an often highly variable thickness, as observed in Figure 13.8.

The density in the well-mixed BBL typically differs slightly from that in the interior stratified water at the same depth z. The resulting small horizontal pressure gradients cause intrusions of well-mixed BBL water parcels into the stratified interior layers (Wain and Rehmann 2010). This exchange implies that at a certain depth, vertical basin-scale mixing is a composition of weak turbulence in the quiescent interior (Section 13.2.7) and intense turbulence in the BBL. The buoyancy flux $J_b(x,y)$ (potential energy stored per time by mixing per mass of water) produced in the BBL contributes to the basin-wide turbulent diffusivity by

$$K_v(z) = N(z)^{-2} A(z)^{-1} \int_{area\,A(z)} J_b(x,y)dxdy \qquad (13.5)$$

where $A(z)$ is the lake cross-sectional area at depth z and the horizontal integration covers the BBL as well as the interior waters at depth z. As shown in Figure 13.8, most of the contributions in Equation 13.5 come from the x–y locations directly above the sediment (Lorke et al. 2008). $K_v(z)$ (m^2s^{-1}) varies strongly as a function of z, depending on the ratio of the BBL volume to the lake cross-sectional area within that particular layer at z. Usually, the BBL contribution increases progressively with z, and vertical diffusivity in the deep hypolimnion is typically about two orders of magnitude larger than in the upper hypolimnion (thermocline; Section 13.2.8).

As indicated earlier, eddy viscosity decreases toward the sediment. At a distance of ~1 cm, that is, smaller than the Kolmogorov scale (Chapter 1), turbulence is drastically suppressed as also indicated by small Reynolds numbers. In this viscous sublayer, momentum is transferred solely by molecular viscosity ν and the flow becomes laminar. Within the lowest ~10% of this laminar layer, that is, at ~1 mm, also all the concentration fluctuations cease, and the transition to the sediment is characterized by a linear concentration gradient $\partial C/\partial z$ and a purely molecular flux $-D_C \cdot \partial C/\partial z$ of dissolved compounds. The exchange of heat and dissolved solids and gases between the sediment and the open water is eventually controlled by this diffusive sublayer with a height of typically ~1 mm directly at the sediment–water interface. The thickness of and the gradients in this diffusive boundary layer are modified by the hydrodynamics (turbulence) above the sediment. For a detailed discussion of the diffusive boundary layer, we refer to Glud (2008) and Boudreau and Jørgensen (2001).

13.2.7 Interior Stratified Mixing

As SL mixing (Sections 13.2.3 and 13.2.4) is of limited vertical extent, a sharp transition to the stratified water underneath (hypolimnion) develops during the onset of seasonal stratification. The stability of the interior $N^2(z)$ (Equation 13.2) usually reaches a maximum at the top of the hypolimnion (typical: ~$10^{-3}\,s^{-2}$) and decreases with z to the lowest N^2 in the deepest reaches (typical: ~$10^{-6}\,s^{-2}$ or less; Table 13.1). The buoyancy forces related to vertical movements (proportional to N^2) suppress turbulent mixing. Density inversions are restratified gravitationally, and the interior appears calm with little structures from stirring. Typically, only ~1% to a few % of a probed hypolimnia show signs of density fluctuations related to turbulent mixing.

Due to episodic wind energy input to the oscillatory stratification, the interior is subject to continuous baroclinic motions. Internal waves (Chapter 28; Figure 13.9) can exist in the bandwidth between the inertial frequency f and the buoyancy frequency N with wavelengths extending from tens of meters to the

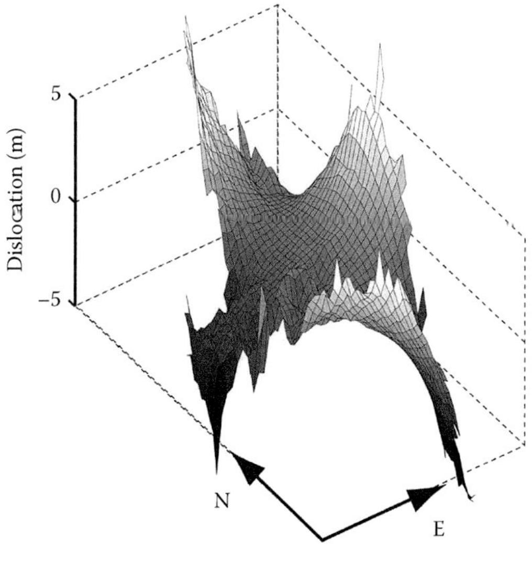

FIGURE 13.9 Spatial structure of the modeled 5.5°C isotherm displacement for the internal H3V1 (third horizontal, first vertical mode) Kelvin wave, as an example of various types of internal waves observed in Lake Tahoe (United States). The vertical dislocations of several meters are based on a combination of observations and 3D modeling, but for clarity only the modeled structure is shown. Such internal waves are observed during times of weak stratification in winter and the passage of storm fronts. Under these conditions, internal waves with periods greater than 5 days and displacements of 30 m are possible. (From Rueda, F.J. et al., *J. Geophys. Res.*, 108(C3), 3097, 2003.)

basin-scale (Antenucci and Imberger 2001). The internal waves carry oscillatory energy E_{IW} (J m^{-2}), defined by the internal dislocations $d(z, t)$ relative to the equilibrium levels

$$E_{IW}(t) \approx A_S^{-1}\rho \int_{\max z}^{surface} N^2(z,t)\overline{d^2(z,t)}A(z)dz \qquad (13.6)$$

which is typically in the range of $E_{IW} \approx 1$ J m^{-3} times the depth of the water body.

Shearing in the interior is very weak compared to the boundary layers (SL, BBL), where currents experience strong gradients near the interface. Combined with the strong stratification N^2, the interior Richardson numbers $Ri = N^2(\partial u/\partial z)^{-2}$ are much larger than 1 and typically in the range of ~10–1000. Therefore, the interior is quiescent and turbulence is weak, as indicated by low dissipation rates observed (typically ~10^{-10} to 10^{-9} W kg^{-1}). Consistently, the Ozmidov scale $L_O = (\varepsilon/N^3)^{1/2}$, which quantifies the maximum vertical extent of turbulent eddies, is in the range of a few centimeters only. Turbulent diffusion in the interior is slow, close to the molecular diffusivity of heat (temperature) for strongly stratified hypolimnia, as confirmed by tracer studies (Wüest and Lorke 2003). The weak mixing that is present is thought to originate from the sporadic superposition and subsequent breaking of internal waves.

According to Equation 13.6, E_{IW} varies with time. Given typical energy contents and the dissipation rates in the interior, E_{IW} would be expected to decay with a time-scale of weeks to months. In reality, the energy E_{IW} disappears much faster after a storm (~1 d per 40 m of lake depth; for example, ~40 d in Lake Baikal; Wüest and Lorke 2003), which is commonly explained by mixing (friction) in the BBL (Section 13.2.6). This is supported by the "apparent" vertical diffusivity in the hypolimnion as a function of depth z, which is usually about an order of magnitude larger than in the interior. Inspecting Equation 13.5 explains that the rate of vertical mixing of the entire basin is fueled by buoyancy flux $J_b(x, y, t)$ produced in the BBL (large contribution) and in the interior (small contribution). The mentioned tracer studies showed that, indeed, mixing in the BBL exceeds mixing in the interior by far, and the integral of Equation 13.5 is dominated by the BBL (Section 13.2.6).

As indicated, the variety of internal waves is enormous (Chapter 28) with complex dynamics due to variable forcing (storms) and phasing as well as amplification and annihilation by superposition. In terms of energetics, the large basin-scale internal waves are the most important. Figures 13.9 and 13.10 show two examples of the first vertical mode (two-layer model) and the second vertical mode (three-layer model).

13.2.8 Multi-Basin Exchange Processes

In multi-basin lakes, typical for complex (mountainous) topography, sills and narrows restrict lateral exchange between their sub-basins (Figure 13.11). As a result, sub-basins may develop individual physical and geochemical properties, especially in

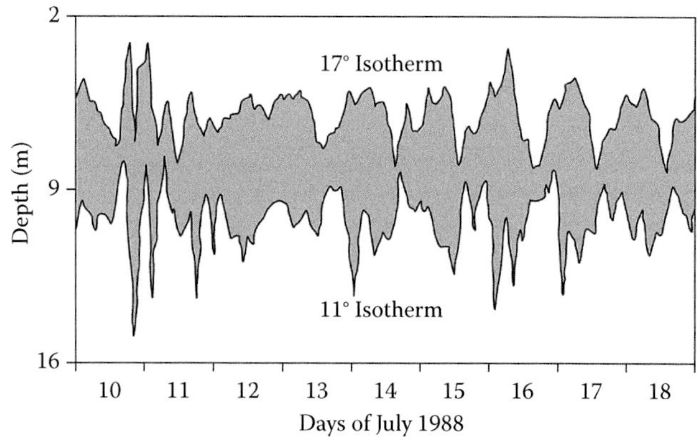

FIGURE 13.10 Record of 11°C and 17°C isotherms (low-pass filtered at 2.5 h) in the 33 m deep Lake Alpnach showing two types of basin-scale internal seiches: the first vertical mode (period of ~8 h) and second vertical mode (period ~24 h). The latter is strongly expressed for the mid-depth layer (metalimnion) between the two isotherms showing vertical variations from ~0 to 10 m thickness. (From Münnich, M. et al., *Limnol. Oceanogr.*, 37(8), 1705, 1992.)

the layers below sills (Rueda and MacIntyre 2009). The reasons for such inter-basin differences (gradients) are (1) heterogeneous inputs, (2) differential warming/cooling and other non-uniform transformations, and (3) differential mixing.

Heterogeneities of the catchments lead to spatially and temporally non-uniform inputs, such as river loading of heat, particles, and geochemical constituents. This affects the density and intrusion depths of inflows and creates lateral density gradients. Differential cooling is usually caused by varying water depth. Shallower layers cool off faster than deeper ones, and subsequently lateral temperature gradients develop. Differential warming is mainly due to varying exposure to and penetration of solar radiation caused by topographic shading, vegetation (e.g., reed stands) and variable algae contents. Similarly, non-uniform transformation rates within a water body cause biogeochemical gradients. Finally, differential mixing (Umlauf and Lemmin 2005) due to inhomogenous forcing (such as wind exposure) can create lateral density gradients.

Although these density differences drive large-scale currents (Section 13.2.1), which tend to remove those gradients, the exchange between sub-basins is restricted by topographic features (Rueda and MacIntyre 2009). The extent of the resulting heterogeneity depends on the relative values of the differential forcing, the hydraulic residence times in the sub-basins, and the time-scales for transforming the constituent of interest.

A typical example for multi-basin exchange flows is Lake Lucerne, which consists of four sub-basins (Aeschbach-Hertig et al. 1996). As exemplified in Figure 13.11, in late winter/early spring substantial lateral gradients develop, as Urnersee (U in Figure 13.11a), is experiencing the well-know Foehn (south wind), which causes efficient seasonal deep-mixing (Figure 13.11a). Also salinity varies by a factor of 4 among the tributaries (inhomogenous catchment), causing considerable inter-basin

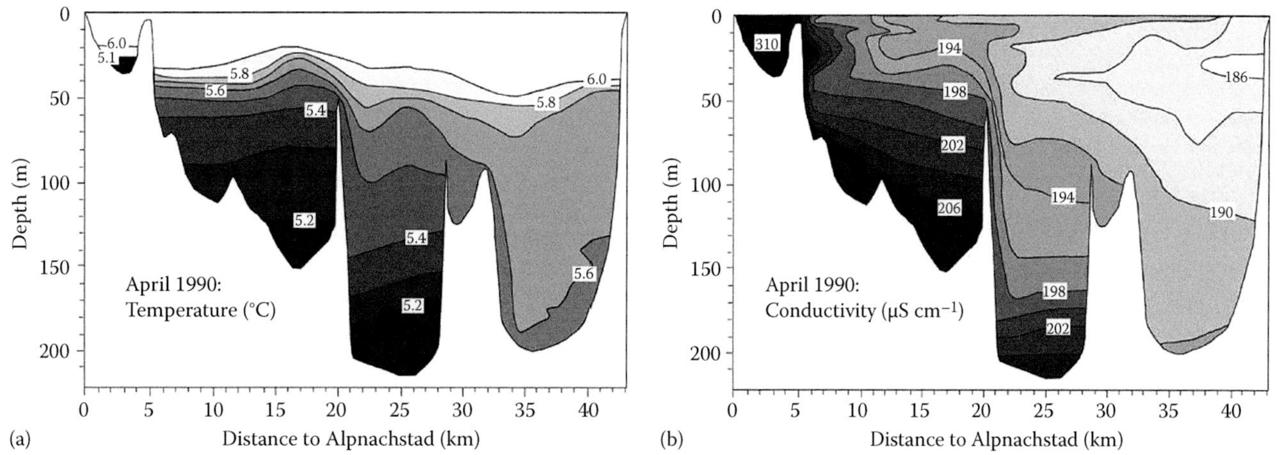

FIGURE 13.11 2D contours of temperature (a) and salinity (b) in spring 1990 in Lake Lucerne. The weaker vertical temperature gradients in Urnersee (rightmost) (a) are due to stronger seasonal winds over this sub-basin. The lateral salinity gradients are caused by inhomogenous river loadings, with salinities of inflows >300 mg L^{-1} into left-most basin (Lake Alpnach) and <100 mg L^{-1} into the right-most basin (Urnersee) leaving an ~50% salinity residual in Lake Lucerne. The slopes of the isolines indicate the down-flow direction over the sills. (From Aeschbach-Hertig, W. et al., *Limnol. Oceanogr.*, 41(4), 707, 1996.)

density gradients (Figure 13.11b). These gradients drive currents across the sills and contribute significantly to the deep-water exchange (especially in sub-basins V and U, Figure 13.11).

13.3 Examples

13.3.1 Methane Emissions from Lakes and Reservoirs

The release of methane from lakes and reservoirs has recently been shown to contribute significantly to the global methane budget (Bastviken et al. 2004). Of special interest are methane emissions from reservoirs, since they must be considered as anthropogenic and should be included in greenhouse gas emission budgets. Methane can be emitted by gas exchange at the lake surface, by ebullition of sediment-borne gas bubbles, or by plant-mediated fluxes in shallow areas (Bastviken et al. 2004). All these fluxes can vary in time by several orders of magnitude and are therefore not easily quantified by measurements. Analyzing physical mixing and transport processes in combination with mass balances can help constraining methane fluxes and in the design of measurement campaigns. The major processes that need to be considered are methane formation and consumption, gas exchange at the lake surface, formation, growth and dissolution of bubbles, as well as turbulent mixing and advective transport within the water body. This is illustrated in the following with a few examples.

Lake Tanganyika (Durisch-Kaiser et al. submitted) and Lake Baikal (Schmid et al. 2007) are two contrasting examples of meromictic lakes of comparable size and form, but located in completely different climate regions. Methane accumulates in the permanently stratified anoxic deep-water of Lake Tanganyika. It is slowly transported upward by turbulent diffusion and uplift, driven by deep-water formation. To a large extent it is consumed by both aerobic and anaerobic methane-oxidizing bacteria. Only a small fraction of the methane produced is emitted to the atmosphere. In Lake Baikal, the permanently stratified deep-water is completely oxic and therefore a sink of methane which is transported downward by intense turbulent diffusion and to a lesser extent by downwelling of surface water. Methane is mainly supplied from shallow sources, and methane emission occurs at the time when ice is thawing.

In Rotsee (Schubert et al. 2010), a seasonally stratified lake, methane accumulates under anoxic conditions during the stratified period and disappears during seasonal deep-mixing. During the stratified period, methane is largely oxidized in the metalimnion, similar to Lake Tanganyika. During deep-mixing events, larger amounts of methane may be transferred in short time periods from the hypolimnion to the epilimnion. However, it is not clear what fractions of this methane are emitted to the atmosphere or oxidized in the epilimnion. In order to quantify diffusive methane emissions from lakes, it is therefore important to measure emissions during storm events as well as during periods of convective mixing.

In reservoirs, water is often withdrawn from the hypolimnion where methane is seasonally accumulating. This methane is then partially emitted in the spillways, turbines, and downstream rivers. For three reservoirs in Brazil, emissions from downstream rivers have been estimated to be 9%–33% of emissions from the lake surface (Guérin et al. 2006). These can potentially be mitigated by withdrawing the water from the epilimnion rather than the hypolimnion.

Methane emissions by ebullition are much less dependent on lake-internal mixing and transport processes. Nevertheless, they can be highly variable and may show a diel cycle or depend on lake level variations. For Wohlensee, a small run-of-the-river reservoir, bubbling from the sediment has been shown to be the most important source of methane (DelSontro et al. 2010). However, the methane flux and the corresponding emissions to the atmosphere were more reliably quantified by measuring concentrations in the lake water and estimating fluxes based on assumptions for bubble dissolution and gas exchange, rather than by directly measuring the spatially and temporally variable bubble fluxes.

13.3.2 Climate Change

All major external driving forces of physical transport and mixing processes in lakes are to some extent effected by climate change. In order to predict the impacts of climate change on lake ecosystems, a thorough understanding of the relationship between these driving forces and the mixing and transport processes, as well as the interactions between physical processes and biogeochemical cycles, is required. The current knowledge on the effects of climate change on European Lakes has recently been reviewed (George 2010). An overview of some expected changes for lakes in Northern Europe is given in Figure 13.12 (Blenckner et al. 2010).

The most direct and best investigated effect is that of increasing air temperature. This typically leads to a prolongation of the stratified period in summer as well as a strengthening of the density gradient—resulting in a higher tendency to develop anoxic conditions in the deep-water. In meromictic lakes, stronger stratification may decrease the internal supply of nutrients by turbulent mixing and subsequently primary productivity, as exemplified by Verburg et al. (2003) for Lake Tanganyika. The interactions between the external forcing, the lake-internal physical and geochemical processes and the biotic community may be very complex and the resulting impacts on the ecosystem are difficult to predict. For example, it has been shown for Lake Tahoe that increased stratification led to a decrease in diatom cell size and respective changes in internal nutrient recycling due to lower settling velocities (Winder et al. 2007).

A type of lake that is potentially very sensitive to climate change is deep-monomictic (or oligomictic) lakes, because the seasonal deep-mixing in these lakes depends on the surface temperature cooling down to the hypolimnion temperature. Since global warming affects SL faster than the hypolimnion, such lakes might turn meromictic, at least for some time, until the deep-water temperature has also adapted to a potential new equilibrium in a future climate (Matzinger et al. 2007).

Increased air temperatures also result in shortening the ice cover duration, an effect which is more pronounced for warmer than for colder regions (Weyhenmeyer et al. 2004). However, because of the isolating ice cover, under-ice water temperatures

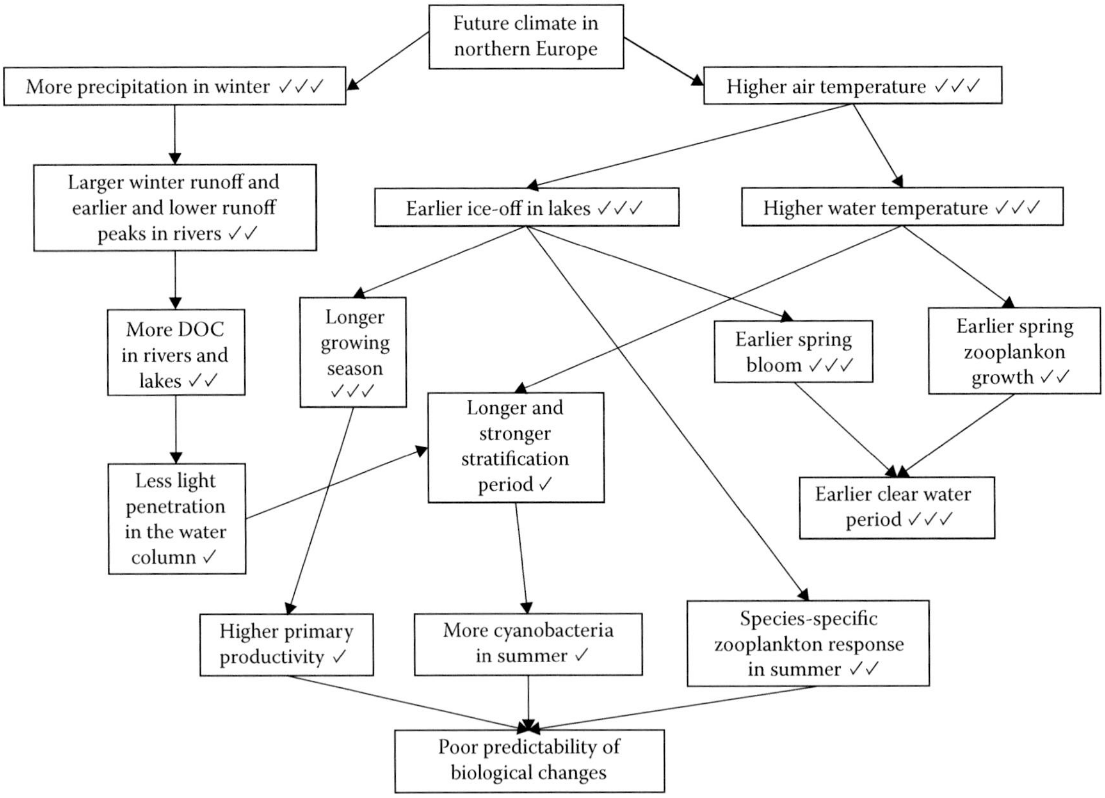

FIGURE 13.12 Schematic overview of the main climatic responses expected in northern European lakes. The number of ticks indicates the estimated level of confidence in a particular response. "Earlier" refers to the seasonal development. (From Blenckner, T. et al., The impact of climate change on lakes in northern Europe, In *The Impact of Climate Change on European Lakes*, G. George, ed., Springer, Dordrecht, the Netherlands, pp. 339–358, 2010.)

depend mostly on the meteorological conditions before freezing. They are therefore not directly related to warming of the atmosphere (Bengtsson 1996).

Changes in other external forcing parameters, such as wind speed and humidity, may have similar or even stronger effects on stratification and mixing than air temperature. These effects have up to now only insufficiently been investigated because of inconsistent predictions for these forcing parameters in future climate scenarios (IPCC 2007). The example of Clearwater Lake (Canada) showed that lake temperatures can decrease in a warming climate in case of reduced wind speed, in this case due to reforestation (Tanentzap et al. 2008). Other expected impacts are changes in frequency of flood events, impacts of changes in precipitation on lake level and salinity (especially for lakes with long water residence times), and land cover alterations that may, for example, reduce the penetration of visible and/or UV radiation by increasing the amount of dissolved organic matter flowing into lakes.

Thanks to their sensitivity to changes in climatic conditions, observations in lakes can also be used as indicators for climate change, despite the often confounding effects of other anthropogenic forcing such as land-use change, eutrophication or hydrological changes (Adrian et al. 2009). For large lakes, even a feedback between the lake surface and the local climate may exist that needs to be considered in climate studies (see Section 13.2.1). Wind speeds over Lake Superior have been shown to increase in response to changes in the air–water temperature gradients (Desai et al. 2009).

13.4 Methods of Analysis

Observations and modeling, as well as a combination of these two approaches, are the traditional methods for studying natural processes and systems. As new techniques and methods often lead to breakthroughs, we briefly list some promising new developments.

Profiling from ships is expensive, inflexible, and limited by weather and other uncontrollable conditions. Autonomous underwater vehicles (AUVs; Doble et al. 2009) provide measurements over a wide range of conditions and sampling strategies: AUVs operate in dangerous conditions such as thawing ice and severe weather, reach inaccessible regions such as the underside of thick ice sheets, and collect data over long time intervals and distances. Microstructure measurements are already being taken with autonomous gliders using a self-contained package of shear probes, fast thermistors, and accelerometers. They can operate for several dozen days and reach noise levels (5×10^{-11} W kg^{-1}) that match the best resolution obtained with traditional ship-based vertical profilers (Wolk et al. 2009).

Recent developments provide increasingly fast, robust, and of low-drift sensors of scalar water properties (such as temperature, dissolved oxygen, etc.). With simultaneous measurements of velocity fluctuations, these sensors can be used to resolve the complete eddy flux (co-variance) of scalar constituents (Berg et al. 2003). This eddy correlation technique, which opens a new field of in situ experimentation, has many advantages compared to traditional flux estimates; among them are the measurement of locally and temporally resolved fluxes over long intervals, and the simultaneous recording of currents and turbulence.

Also sensors for biogeochemical constituents are experiencing a drastic progress regarding temporal and spatial resolution, opening up the perspective of applying them in profilers and eddy-correlation devices. In situ automated wet-chemical sensor packages allow continuous analysis of nutrients (nitrate, nitrite, ammonium, phosphate, silicate) with low detection limits through small-scale gradients. Measurements of chlorophyll pigments by fluorescence techniques and properties of living particles by in situ flow cytometry with laser detection provide unprecedented high-resolution (temporal–spatial) biological information.

The ongoing development of new generations of air- and space-borne remote sensors widens their potential for aquatic applications, especially for monitoring purposes. With satellite sensors providing data of 300 m spatial resolution and biweekly coverage (clear weather conditions), surface water constituent measurement accuracies almost comparable to in situ measurements can be achieved even in optically challenging meso- to oligotrophic waters (Odermatt et al. 2010). Within the next few years, earth observation sensors at higher spectral resolution (more wavelengths) will be launched by the European, German, and Italian Space Agencies, enabling the identification of surface water properties at an even higher quality.

13.5 Major Challenges

As indicated in the introduction, the future challenge is to balance the human-induced pressure on aquatic systems with the importance of maintaining the ecological integrity of ecosystems. As living standards of the global population progress, the demands for water resources services will increase. Compensating the increasing water consumption by abstraction and diversions will change hydrological regimes and thereby the character of affected lakes, ponds, and reservoirs. In most cases the effects will not be as drastic as in the Dead Sea, where Jordan River water abstraction has lowered the water level and turned the former meromictic character into a monomictic regime. But in regions with critical hydrological balances, such as in the Mediterranean, we expect that not only water scarcity may be exacerbated by climate change but also various physical processes and interactions within the ecosystem will be altered (Section 13.3.2). As mentioned in Section 13.3.2, those alterations will be superimposed on eutrophication and the intensified use of water resources for irrigation, heating, and cooling, as well as hydropower production in many parts of the world (Africa, Asia). At the same time, access to safe drinking water needs to be ensured. One of the main challenges in *Physical Limnology* is therefore to support the development of management practices, minimizing detrimental effects on natural waters while providing economic and quality services.

References

Adrian, R., O'Reilly, C. M., Zagarese, H. et al. 2009. Lakes as sentinels of climate change. *Limnology and Oceanography* 54: 2283–2297.

Aeschbach-Hertig, W., Kipfer, R., Hofer, M., Imboden, D. M., and Baur, H. 1996. Density-driven exchange between the basins of Lake Lucerne (Switzerland) traced with the 3H-3He method. *Limnology and Oceanography* 41(4): 707–721.

Akitomo, K., Tanaka, K., Kumagai, M., and Jiao, C. 2009. Annual cycle of circulations in Lake Biwa, part 1: Model validation. *Limnology* 10(2): 105–118.

Antenucci, J. P. 2009. Currents in stratified water bodies 3: Effects of rotation. In *Encyclopedia of Inland Waters*, Vol. 1, ed. G. E. Likens, pp. 559–567, Oxford, U.K.: Elsevier.

Antenucci, J. P. and Imberger, J. 2001. Energetics of long internal gravity waves in large lakes. *Limnology and Oceanography* 46(7): 1760–1773.

Bastviken, D., Cole, J., Pace, M., and Tranvik, L. 2004. Methane emissions from lakes: Dependence of lake characteristics, two regional assessments, and a global estimate. *Global Biogeochemical Cycle*, 18: GB4009, doi:10.1029/2004GB002238.

Beletsky, D. and Schwab, D. 2008. Climatological circulation in Lake Michigan. *Geophysical Research Letters* 35: L21604, doi:10.1029/2008GL035773.

Bengtsson, L. 1996. Thermal regime of ice covered Swedish lakes. *Nordic Hydrology* 27:39–56, doi:10.2166/nh.1996.2003.

Berg, P., Roy, H., Janssen, F., Meyer, V., Jørgensen, B.B., Huettel, M., and de Beer, D. 2003. Oxygen uptake by aquatic sediments measured with a novel non-invasive eddy-correlation technique. *Marine Ecology Progress Series* 261: 75–83.

Blenckner, T., Adrian, R., Arvola, L., Järvinen, M., Nõges, P., Nõges, T., Pettersson, K., and Weyhenmeyer, G. A. 2010. The impact of climate change on lakes in Northern Europe. In *The Impact of Climate Change on European Lakes*, ed. G. George, pp. 339–358, Dordrecht, the Netherlands: Springer, doi: 310.1007/1978-1090-1481-2945-1004_1018.

Boehrer, B., Dietz, S., Von Rohden, C. et al. 2009. Double-diffusive deep water circulation in an iron-meromictic lake. *Geochemistry, Geophysics, Geosystems* 10: Q06006, doi:10.1029/2009GC002389.

Boehrer, B. and Schultze, M. 2008. Stratification of lakes. *Reviews of Geophysics* 46: RG2005, doi:10.1029/2006RG000210.

Boudreau, B. P. and Jørgensen, B. B. 2001. *The Benthic Boundary Layer: Transport Processes and Biogeochemistry*, Oxford, U.K.: Oxford University Press.

Chen, C. T. and Millero, F. J. 1986. Precise thermodynamic properties for natural waters covering only the limnological range. *Limnology and Oceanography* 31: 657–662.

Crawford, G. B. and Collier, R. W. 2007. Long-term observations of deepwater renewal in Crater Lake, Oregon. *Hydrobiologia* 574: 47–68, doi:10.1007/s10750-006-0345-3.

Csanady, G. T. 2001. *Sea-Air Interaction: Laws and Mechanisms*, Cambridge, U.K.: Cambridge University Press.

DelSontro, T., McGinnis, D. F., Sobek, S., Ostrovsky, I., and Wehrli, B. 2010. Extreme methane emissions from a Swiss hydropower reservoir: Contribution from bubbling sediments. *Environmental Science and Technology* 44: 2419–2425, doi:10.1021/es9031369.

Desai, A. R., Austin, J. A., Bennington, V., and McKinley, G. A. 2009. Stronger winds over a large lake in response to weakening air-to-lake temperature gradient. *Nature Geoscience* 2: 855–858.

Doble, M. J., Forrest, A., Wadhams, P., and Laval, B. 2009. Through-ice AUV deployment: Operational and technical experience from two seasons of Arctic fieldwork. *Cold Regions Science and Technology* 56: 90–97.

Durisch-Kaiser, E., Schmid, M., Peeters, F., Kipfer, R., Dinkel, C., Diem, T., Schubert, C. J., and Wehrli, B. 2011. What prevents out-gassing of methane to the atmosphere in Lake Tanganyika? *Journal of Geophysical Research-Biogeosciences* 116: G02022, doi:10.1029/2010JG001323.

Emery, K. O. and Csanady, G. T. 1973. Surface circulation of lakes and nearly land-locked seas. *Proceedings of the National Academy of Sciences U S A* 70: 93–97.

George, G. 2010. *The Impact of Climate Change on European Lakes*, Dordrecht, the Netherlands: Springer.

Glud, R. N. 2008. Oxygen dynamics of marine sediments. *Marine Biology Research* 4: 243–289.

Gruber, N., Wehrli, B., and Wüest, A. 2000. The role of biogeochemical cycling for the formation and preservation of varved sediments in Soppensee (Switzerland). *Journal of Paleolimnology* 24(3): 277–291, doi:10.1023/A:1008195604287.

Guérin, F., Abril, G., Richard, S., Burban, B., Reynouard, C., Seyler, P., and Delmas, R. 2006. Methane and carbon dioxide emissions from tropical reservoirs: Significance of downstream rivers. *Geophysical Research Letters* 33: L21407, doi:10.1029/2006GL027929.

Hanjalic, K. and Musemic, R. 1997. Modeling the dynamics of double-diffusive scalar fields at various stability conditions. *International Journal of Heat and Fluid Flow* 18(4): 360–367.

Imberger, J. 1998. Physical processes in lakes and oceans. *Coastal and Estuarine Studies*, Vol. 54, Washington, DC: AGU Publication.

Imberger, J. and Patterson, J. C. 1990. Advances in applied mechanics. *Physical Limnology* 27: 303–475.

Imboden, D. and Wüest, A. 1995. Mixing mechanisms in lakes. In *Physics and Chemistry of Lakes*, eds. A. Lerman, D. Imboden, and J. Gat, pp. 83–138, Berlin, Germany: Springer.

IPCC. 2007. Climate change 2007: The physical science basis. Contribution of Working Group I to the Fourth Assessment Report of the IPCC, Cambridge, U.K.: Cambridge University Press.

Jonas, T., Stips, A., Eugster, W., and Wüest, A. 2003. Observations of a quasi shear-free lacustrine convective boundary layer: Stratification and its implications on turbulence. *Journal of Geophysical Research* 108(C10): 3328, doi:10.1029/2002JC001440.

Lemmin, U. and D'Adamo, N. 1996. Summertime winds and direct cyclonic circulation: Observations from Lake Geneva. *Annales Geophysicae* 14: 1207–1220.

Lewis, W. M. 1983. A revised classification of lakes based on mixing. *Canadian Journal of Fisheries and Aquatic Sciences* 40: 1779–1787.

Li, M., Garrett, C., and Skyllingstad, E. 2005. A regime diagram for classifying turbulent large eddies in the upper ocean. *Deep-Sea Research I* 52: 259–278.

Likens, G. E. (ed.). 2009. *Encyclopedia of Inland Waters*, Oxford, U.K.: Elsevier.

Lorke, A. and MacIntyre, S. 2009. The benthic boundary layer (in rivers, lakes, and reservoirs). In *Encyclopedia of Inland Waters*, Vol. 1, ed. G. E. Likens, pp. 505–514, Oxford, U.K.: Elsevier.

Lorke, A., Umlauf, L., and Mohrholz, V. 2008. Stratification and mixing on sloping boundaries. *Geophysical Research Letters* 35: L14610, doi:10.1029/2008GL034607.

Lorke, A., Wüest, A., and Peeters, F. 2005. Shear-induced convective mixing in bottom boundary layers on slopes. *Limnology and Oceanography* 50: 1612–1619.

Matzinger, A., Schmid, M., Veljanoska-Sarafiloska, E., Patceva, S., Guseska, D., Wagner, B., Müller, B., Sturm, M., and Wüest, A. 2007. Eutrophication of ancient Lake Ohrid—Global warming amplifies detrimental effects of increased nutrient inputs. *Limnology and Oceanography* 52(1): 338–353.

Monismith, S. G. and MacIntyre, S. 2009. The surface mixed layer in lakes and reservoirs. In *Encyclopedia of Inland Waters*, Vol. 1, ed. G. E. Likens, pp. 636–650, Oxford, U.K.: Elsevier.

Münnich, M., Wüest, A., and Imboden, D. M. 1992. Observations of the second vertical mode of the internal seiche in an alpine lake. *Limnology and Oceanography* 37(8): 1705–1719.

Newman, F. C. 1976. Temperature steps in Lake Kivu: A bottom heated saline lake. *Journal of Physical Oceanography* 6: 157–163.

Odermatt, D., Giardino, C., and Heege, T. 2010. Chlorophyll retrieval with MERIS Case-2-Regional in perialpine lakes. *Remote Sensing of Environment* 114: 607–617.

von Rohden, C., Boehrer, B., and Ilmberger, J. 2010. Evidence for double diffusion in temperate meromictic lakes. *Hydrology and Earth System Sciences* 14: 667–674, doi:10.5194/hess-14-667-2010.

Rueda, F. J. and MacIntyre, S. 2009. Flow paths and spatial heterogeneity of stream inflows in a small multibasin lake. *Limnology and Oceanography* 54(6): 2041–2057.

Rueda, F. J., Schladow, S. G., and Palmarsson, S. Ó. 2003. Basin-scale internal wave dynamics during a winter cooling period in a large lake. *Journal of Geophysical Research* 108(C3): 3097, doi:10.1029/2001JC000942.

Sánchez, X. and Roget, E. 2007. Microstructure measurements and heat flux calculations of a triple-diffusive process in a lake within the diffusive layer convection regime. *Journal of Geophysical Research* 112: C02012, doi:10. 1029/2006JC003750.

Sanderson, B., Perry, K., and Pedersen, T. 1986. Vertical diffusion in meromictic Powell Lake, British Columbia. *Journal of Geophysical Research* 91: 7647–7656.

Schladow, S. G., Palmarsson, S. O., Steissberg, T. E., Hook, S. J., and Prata, F. E. 2004. An extraordinary upwelling event in a deep thermally stratified lake. *Geophysical Research Letters* 31: L15504, doi:10.1029/2004GL020392.

Schmid, M., Busbridge, M., and Wüest, A. 2010. Double-diffusive convection in Lake Kivu. *Limnology and Oceanography* 55: 225–238.

Schmid, M., De Batist, M., Granin, N. G., Kapitanov, V. A., McGinnis, D. F., Mizandrontsev, I. B., Obzhirov, A. I., and Wüest, A. 2007. Sources and sinks of methane in Lake Baikal: A synthesis of measurements and modeling. *Limnology and Oceanography* 52: 1824–1837.

Schmid, M., Halbwachs, M., Wehrli, B., and Wüest, A. 2005. Weak mixing in Lake Kivu: New insights indicate increasing risk of uncontrolled gas eruption. *Geochemistry, Geophysics, and Geosystems* 6(7): Q07009, doi:10.1029/2004GC000892.

Schmid, M., Lorke, A., Dinkel, C., Tanyileke, G., and Wüest, A. 2004. Double-diffusive convection in Lake Nyos, Cameroon. *Deep Sea Research Part I* 51: 1097–1111.

Schmidt, W. 1928. Über Temperatur- und Stabilitätsverhältnisse von Seen. *Geografiska Annaler* 10: 145–177.

Schubert, C. J., Lucas, F. S., Durisch-Kaiser, E., Stierli, R., Diem, T., Scheidegger, O., Vazquez, F., and Müller, B. 2010. Oxidation and emission of methane in a monomictic lake (Rotsee, Switzerland). *Aquatic Sciences* 72(4):455–466, doi:10.1007/s00027-010-0148-5.

Schwab, D. J., O'Connor, W. P., and Mellor, G. L. 1995. On the net cyclonic circulation in large stratified lakes. *Journal of Physical Oceanography* 25: 1516–1520.

Shimaraev, M. N., Verbolov, V. I., Granin, N. G., and Sherstyankin, P. P. 1994. *Physical Limnology of Lake Baikal: A Review*, Irkutsk, Okayama, Japan: BICER.

Stevens, C. L. and Imberger, J. 1996. The initial response of a stratified lake to surface shear stress. *Journal of Fluid Mechanics* 312: 39–66.

Tanentzap, A. J., Yan, N. D., Keller, B., Girard, R., Heneberry, J., Gunn, J. M., Hamilton, D. P., and Taylor, P. A. 2008. Cooling lakes while the world warms: Effects of forest regrowth and increased dissolved organic matter on the thermal regime of a temperate, urban lake. *Limnology and Oceanography* 53(1): 404–410.

Toffolon, M. and Rizzi, G. 2009. Effects of spatial wind inhomogeneity and turbulence anisotropy on circulation in an elongated basin: A simplified analytical solution. *Advances in Water Resources* 32: 1554–1566.

Umlauf, L. and Lemmin, U. 2005. Interbasin exchange and mixing in the hypolimnion of a large lake: The role of long internal waves. *Limnology and Oceanography* 50(5): 1601–1611.

Verburg, P., Hecky, R. E., and Kling, H. 2003. Ecological consequences of a century of warming in Lake Tanganyika. *Science* 301: 505–507.

Wain, D. J. and Rehmann, C. R. 2010. Transport by an intrusion generated by boundary mixing in a lake. *Water Resources Research* 46, W08517, doi:10.1029/2009WR008391.

Weyhenmeyer, G. A., Meili, M., and Livingstone, D. M. 2004. Nonlinear temperature response of lake ice breakup. *Geophysical Research Letters* 31: L07203, doi:10.1029/2004GL019530.

Winder, M., Reuter, J. E., and Schladow, S. G. 2007. Lake warming favours small-sized planktonic diatom species. *Proceedings of the Royal Society B* 276(1656): 427–435, doi:10.1098/rspb.2008.1200.

Wolk, F., Lueck, R. G., and St. Laurent, L. 2009. Turbulence measurements from a glider. *OCEANS-IEEE* 1–3: 470–475.

Wüest, A. and Lorke, A. 2003. Small-scale hydrodynamics in lakes. *Annual Review of Fluid Mechanics* 35: 373–412.

Wüest, A., Piepke, G., and Halfman, J. D. 1996. Combined effects of dissolved solids and temperature on the density stratification of Lake Malawi. In *The Limnology, Climatology and Paleoclimatology of the East African Lakes*, ed. T. C. Johnson and E. O. Odada, pp. 183–202, Toronto, Ontario, Canada: Gordon and Breach.

14

Microbial and Ecological Fluid Dynamics

Miki Hondzo
University of Minnesota

14.1 Introduction

Microscopic organisms including autotrophs and heterotrophs are integral constituents of aquatic ecosystems and are continuously exposed to a variety of climatic, physical, and chemical conditions. Physical and chemical processes control a fundamental property of aquatic ecosystems: the balance of organic material production (autotrophy) and decomposition (heterotrophy). The autotrophic–heterotrophic (A–H) balance can refer to either masses or processes: the ratio of autotrophic biomass to heterotrophic biomass or the ratio of primary production (mostly oxygenic photo-autotrophic) to respiration. Therefore, the A–H balance also is called the photosynthesis: respiration ratio (P:R), since these metabolic processes underlie autotrophy and heterotrophy, respectively. The planktic organisms that contribute to the A–H balance span a size range from bacteria (~10^{-6}m) to phytoplankton (~10^{-5}m) and zooplankton (~10^{-3}m). The net A–H balance in aquatic ecosystems is fundamentally important for our understanding of local and global carbon cycling. Autotrophy apparently has exceeded heterotrophic production of organic carbon through the history of life on Earth. This is inferred from the occurrence of an oxygenated atmosphere, oxidized crustal rocks, and buried organic carbon and pyrite. Most permanent carbon burial on Earth occurs in the aquatic realm; an understanding of processes that lead to net aqueous heterotrophy and/or autotrophy therefore has great relevance to understanding how the Earth system functions. More localized problems relevant to A–H balance are water quality, dissolved oxygen net balance, fisheries, and eutrophication in both marine and freshwater ecosystems.

Existing aquatic circulation models are designed for large, basin-size scales and therefore cannot resolve the planktic organism scales of A–H balance. The nonlinear kinetics of biological responses to physical and chemical variables are more appropriately described at micro-scales, at the level of molecular diffusion. It is striking to realize that typical computation grids in 3D numerical models for lakes and oceans are more than 10^6 times larger than the dimensions of planktic organisms. Therefore, there is a major difference between the typical computational grid scale with associated scale of resolved fluid motions and the majority of phytoplankton and bacteria. Further, both fluid motions and categories of plankton have significant temporal and spatial heterogeneities in aquatic ecosystems (Durham et al. 2009; Wüest and Lorke 2009). To what extent do physical and chemical processes guide formation, location, and duration of such heterogeneities? How do individual organisms "feel" the physical environment as they respond to it? Parameterization for smaller scales is therefore necessary, and there is an urgent need for modeling A–H processes at scales that are much smaller than those addressed by existing aquatic circulation models.

Fluid motions operate over a wide range of spatial and temporal scales over landscapes and associated aquatic ecosystems. A conceptual model of the hierarchy of organization between physical and biological processes in aquatic ecosystems was suggested by Nikora (1999) and Biggs et al. (2005). The effect of physical processes on biological responses could be depicted by interacting hydrological and hydraulic temporal and spatial scales. The hydrologic scales are mediated by climatic processes

Handbook of Environmental Fluid Dynamics, Volume One, edited by Harindra Joseph Shermal Fernando. © 2013 CRC Press/Taylor & Francis Group, LLC. ISBN: 978-1-4398-1669-1.

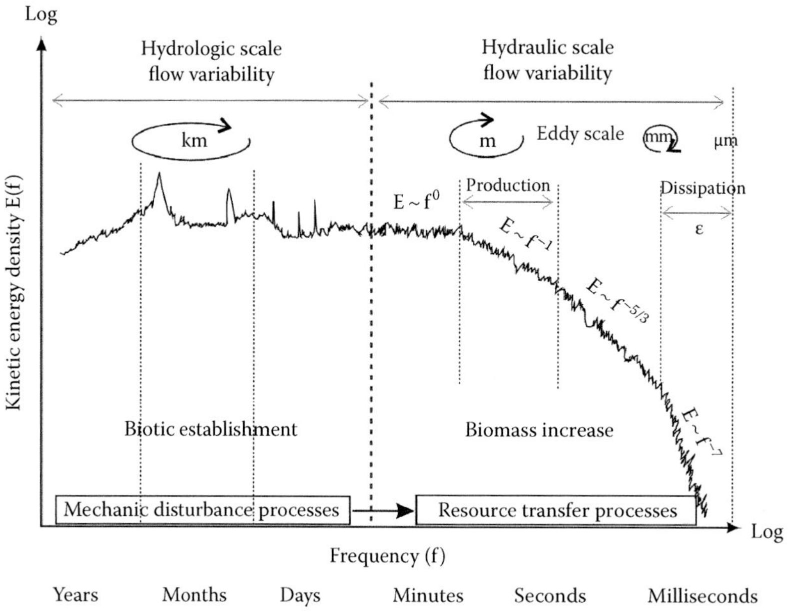

FIGURE 14.1 A kinetic energy spectral density of fluid-flow velocity over temporal scales from milliseconds to years. Biotic establishment is facilitated by hydrologic time-scale variabilities and biomass increase is mediated hydraulic time- and spatial-scale variabilities. Planktic nutrient related biomass increase occurs in and beyond the dissipation range.

and associated landscape characteristics and generally determine biotic establishment and composition in aquatic ecosystems (Figure 14.1). The hydraulic scales mediate the rate of resource transfer to biological organisms and determine the biomass increase over the range of scales. The hierarchy of temporal and spatial hydraulic scales in a turbulent flow can be effectively visualized through the kinetic energy spectral density with a characteristic "–1" slope, that is, production range, "–5/3" slope in the intertial subrange, and "–7" slope in the dissipation range (Figure 14.1). The majority of energy is generated at the production time-scale of eddies on the order of seconds and minutes and dissipated at dissipation time-scale on the order of milliseconds. Through the kinetic energy spectral density, energy is systematically transferred from production scales to dissipation scales. The rate of energy dissipation (ε) represents physically based fluid flow variable that, through the energy cascade, integrates large-scale fluid motion with metabolic responses of microorganisms. In the production and inertial subregions, the corresponding large-scale eddies can export planktic organisms through a variety of nutrient concentrations and therefore setup a potential background for biomass increase in the moving fluid. The fluid flow-related nutrient transfer and corresponding biomass increase at the cell scale is determined in and beyond the corresponding dissipation range. Planktic organisms, whether they live in a turbulent or nonturbulent fluid, absorb nutrients from the ambient fluid across the diffusive sublayer, where presumably molecular diffusion is the dominant transport process, surrounding the cell. The rate of energy dissipation controls the thickness of diffusive sublayer and corresponding nutrient uptake by planktic organisms (e.g., Hondzo and Wüest 2009).

14.2 Principles

Mass transport of nutrients in the proximity of planktic organism is governed by the advection–diffusion equation:

$$\frac{\partial C}{\partial t} + \mathbf{U} \cdot \nabla C = D\nabla^2 C \tag{14.1}$$

where

- C is the concentration of nutrient
- **U** is the velocity vector
- D is the molecular diffusion coefficient of nutrient

If the concentration of nutrient is uniform at the planktic organism surface, C_c, and far from the cell the concentration of nutrient is C_∞, we can rescale the dependent variable in a nondimensional form as

$$\tilde{C} = \frac{C - C_\infty}{C_c - C_\infty} \tag{14.2}$$

so that $\tilde{C} = 1$ at the cell surface, r = a, and $\tilde{C} = 0$ as $r \to \infty$ where r is the distance from the center of microorganism in spherical coordinates. We introduce the scaling $\tilde{U} = \mathbf{U}/U_\infty$ where U_∞ is the characteristic velocity and $\tilde{L} = L/L_c$ with L_c equals a characteristic length scale of planktic organism. In the steady-state, (14.1) and (14.2) with the scaled velocity and length scale in nondimensional form read

$$Pe(\tilde{\mathbf{U}} \cdot \tilde{\nabla}\tilde{C}) = \nabla^2\tilde{C} \tag{14.3}$$

where $Pe = U_\infty L_c/D$ is the Péclet number which is the ratio of advective nutrient transport to molecular diffusion nutrient transport. An alternative formulation is $Pe = t_{diff}/t_{ad} = (L_c^2/D)/(L_c/U_\infty)$, where t_{diff} is the diffusion time, and t_{ad} is the advection time. The traditional scaling argument in a turbulent flow (e.g., Lewis 2005) employs $U_\infty = c(\varepsilon/\nu)^{1/2}L_c$, where "c" is a constant, and ν is the kinematic viscosity of fluid. The Péclet number in an energetic turbulent flow is $Pe = (c(\varepsilon/\nu)^{1/2}L_c\, L_c)/D \sim 0.0001$ for a planktic organism of $L_c \sim 2\,\mu m$, $\varepsilon \sim 10^{-7}\,m^2/s^3$, $\nu_{20°C} \sim 10^{-6}\,m^2/s$, and $D \sim 10^{-9}\,m^2/s$. The relatively small Péclet number implies almost no effect of small-scale turbulent flow on planktic organisms. Recent laboratory measurements on the effect of turbulence on *Escherichia coli* cells (e.g., Al-Homoud et al. 2007), *Selenastrum capricornutum* cells (e.g., Warnaars and Hondzo 2006), and *Alexandrium minutum* (Berdalet et al. 2007) indicate significant effects of small-scale turbulence on microorganism growth and metabolic rates. Therefore, the experimental evidence questions the validity of traditional scaling arguments in the definition of the Péclet number.

Consider a planktic organism of a spherical geometry in a turbulent flow. Assume an asymptotic solution for the concentration of nutrient exists. In the form of a regular perturbation expansion in the inner region for $Pe < 1$, the solution reads

$$\tilde{C}^i = \tilde{C}_o^i + \tilde{C}_1^i Pe + \tilde{C}_2^i Pe^2 + \cdots \tag{14.4}$$

where $\tilde{C}^i$ is the nondimensional expression for nutrient concentration in the inner region. Substituting (14.4) into (14.3) with the boundary conditions, $\tilde{C} = 1$ at the cell surface, $\tilde{r} = r/a = 1$, and $\tilde{C} = 0$ as $\tilde{r} \to \infty$. Neglecting the left-hand side of (14.3) leads to Laplace's equation with solution to "zero" order:

$$\tilde{C}_o^i = \frac{1}{\tilde{r}} \tag{14.5}$$

According to (14.5), the concentration decreases from microorganism cell with distance from the sphere in all directions. The ratio of total nutrient flux to microorganism cell in a moving fluid (Q) versus the total nutrient flux to the cell by molecular diffusion (Q_D) can be expressed by the Sherwood number as

$$Sh = \frac{Q}{Q_D} = \frac{\int_{As} -D(\nabla C \cdot n)\big|_s \, dAs'}{(D/a)(C_c - C_\infty)As} \tag{14.6}$$

where

As is the surface area of sphere

"a" is the radius of sphere

The zero-order approximation of nutrient flux to planktonic organism implies diffusional mass transport in the numerator of (14.6) and therefore Sh = 1.

In the outer region diffusion and advection terms should be the same order of magnitude, $Pe(\mathbf{U} \cdot \nabla\tilde{C}^o) \sim \nabla^2\tilde{C}^o$, even as $Pe \to 0$. The dimensionless concentration in the outer region is denoted by $\tilde{C}^o$. The physical dimension of a planktonic organism, L_c, does not provide a characteristic length scale. An outer nondimensional variable is selected as

$$\tilde{\rho} = \left(\frac{U_\infty L_c}{D}\right)^m \frac{r}{a} = \left(\frac{E\eta_K L_c}{D}\right)^m \frac{r}{a} = Pe^m \tilde{r} \tag{14.7}$$

where

$\eta_K = (\nu^3/\varepsilon)^{1/4}$ is the Kolmogorov length scale or a length scale where turbulent velocity fluctuations are smeared by ν

"m" is an unknown exponent

A notable difference with respect to the conventional scaling is that the shear rate, E, is multiplied by the Kolmogorov scale. Traditionally, the enhanced shear rate is multiplied by L_c. Since advection terms should be balanced by the diffusion terms, flow field around a sphere in a simple shear flow should be specified to analyze the change of $\tilde{C}^o$ in the outer region. Normalizing the velocity components by fluid flow strain rate, $U_\infty = E\eta_K$, considering (14.7), and conducting a second-order asymptotic analysis in the outer region yields

$$Sh = 1 + \alpha\left(\frac{E\eta_K a}{D}\right)^{1/2} + \alpha^2\left(\frac{E\eta_K a}{D}\right) = 1 + \alpha Pe_K^{1/2} + \alpha^2 Pe_K \tag{14.8}$$

The rate of strain velocities, $E\eta_K$, in the direction of the local ambient vorticity vector is $E_{zz} = |c_1|(\varepsilon/\nu)^{1/2}$, where $c_1 = 0.46$ is the proportionality constant which is typically reported in the range 0.3–0.6 (Davidson 2004). Combing E_{zz} and (14.8) yields

$$Sh \approx 1 + \alpha\left(\frac{\tilde{u}_K L_c}{D}\right)^{1/2} + \alpha^2\left(\frac{\tilde{u}_K L_c}{D}\right) \tag{14.9}$$

where $\tilde{u}_K = 0.5\,(\varepsilon\nu)^{1/4}$ is the layer-averaged Kolmogorov velocity. This result suggests that nutrient transport to planktic organism in a turbulent flow is determined by cell geometry ($\alpha \sim O(1)$, L_c), the rate of energy dissipation (ε), and molecular diffusion coefficient in the cell proximity. The rate of energy dissipation propagates through the energy spectrum from large turbulence production scales to planktic organism scale and provides a physical linkage between large and microscopic scales in aquatic ecosystems.

14.3 Examples

14.3.1 Cell Metabolism: *Escherichia coli* and *Selenastrum capricornutum*

We demonstrate the effects of small-scale turbulence on *E. coli*, a single-celled motile heterotroph, and *S. capricornutum*, a single-celled nonmotile autotroph. Experiments were conducted in closed bioreactors specifically designed to study the effects of small-scale turbulence on algal and bacterial growth and nutrient uptake (Warnaars et al. 2006; Al-Homoud et al. 2007). The experimental setups provided desktop controllable fluid flow reactors where ε and turbulence characteristics were reproducible in separate experiments. The velocities and corresponding ε in the

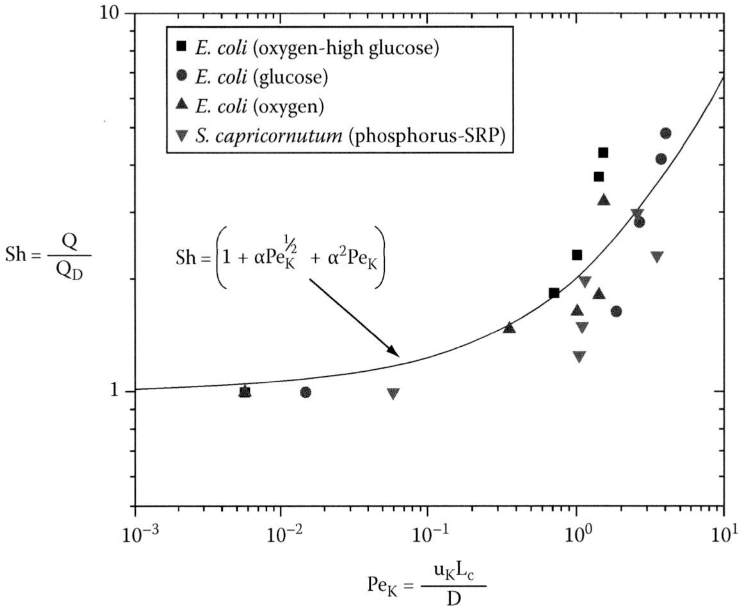

FIGURE 14.2 Total mass flux to *Selenastrum capricornutum* and *Escherichia coli* cells in a turbulent fluid, Q, normalized by the total mass flux to the *S. capricornutum* and *E. coli* cells in a stagnant fluid, Q_D, as a function of the Péclet number. The *x*-axis represents independent variable of the prediction function given by Equation 14.8, where $\alpha = 0.62$, $L_c = 2\,\mu m$ (*E. coli*), $L_c = 5\,\mu m$ (*S. capricornutum*), $D = 2.1 \times 10^{-9}\,m^2/s$ is the molecular diffusion coefficient for oxygen, $D = 6.7 \times 10^{-10}\,m^2/s$ (glucose), $D = 7.8 \times 10^{-10}$ (SRP), $\nu = 1 \times 10^{-6}\,m^2/s$, and $u_K = (\varepsilon\nu)^{1/4}$ is the Kolmogorov velocity.

reactors were resolved at the Kolmogorov scale. Turbulent flow conditions including spatial and temporal variabilities of velocity fluctuations and the existence of inertial subrange in velocity spectra were quantified in the reactors. Apart from fluid stirring, all other environmental variables were kept constant. The stirring magnitude with associated ε in the bioreactors was comparable to the reported field measurements of $10^{-4} > \varepsilon > 10^{-11}\,m^2/s^3$ in aquatic ecosystems (Wüest and Lorke 2009).

Concurrent experiments were conducted under turbulent and stagnant or still-water conditions. The still-water experiments provided controls to the experiments with stirred fluid. Ratio between nutrient uptake in a turbulent bulk fluid, Q, and the corresponding nutrient uptake in the still-water control, Q_D, is depicted in Figure 14.2. Turbulent flow facilitated the uptake of nutrients up to five times more in comparison to the still-water control. The solid line, intended to display overall trend rather than a regression function, is defined by the Kolmogorov velocity, plankton length-scale, and molecular diffusion coefficient. This implies a power law scaling relationship $Q/Q_D \sim (\varepsilon\nu)^{1/4}L_c/D \sim Pe_K$, where ε is the only fluid-flow property. Furthermore, the result depicted in Figure 14.2 supports the proposed scaling of fluid flow velocity in the proximity of cell $u_{cK} = u_K = (\varepsilon\nu)^{1/4}$, which determines the rate of uptake of nutrients over the physically meaningful range of Pe_K. Mechanistic models of plankton population dynamics (Denman and Gargett 1995; Lewis 2005) customarily define the advective velocity scale by $u_{cp} \approx \beta(\varepsilon/\nu)^{1/2}L_c$ where $\beta \approx 0.1$. For sample values of $\varepsilon = 10^{-6}\,m^2/s^3$, $\nu = 10^{-6}\,m^2/s$, $L_c = 2 \times 10^{-6}\,m$, and $D = 10^{-9}\,m^2/s$, the advective velocities ($u_{cK} = 1$ mm/s and $u_{cp} = 2 \times 10^{-4}$ mm/s) are significantly different. This implies that the traditional definition of advective velocity (u_{cp}) defines $Pe_{cp} = \beta(\varepsilon/\nu)^{1/2}L_c^2/D \approx 4 \times 10^{-10}$ that is much smaller than the proposed $Pe_K = (\varepsilon\nu)^{1/4}L_c/D \approx 2$ defined by the velocity scale u_{cK}. Since $Pe_{cp} \ll Pe_K$, the u_{cp} cannot explain the enhanced growth and nutrient uptake of *E. coli* and *S. capricornutum* in the $Pe_K > 1$ (Figure 14.2).

14.3.2 Ecosystem Metabolism: Autotrophy and Heterotrophy

The in situ measurement of dissolved oxygen concentration has been customarily used to quantify the A–H metabolism of flowing water communities (Odum 1956). A diurnal change in DO concentration is determined by photosynthetic primary production, respiration, and gas exchange with the atmosphere. Algae and other aquatic plants are responsible for gross primary production (GPP), while total respiration (R) measures the rates of respiration by aquatic plants, algae, fish, invertebrates, and microbes. Since abiotic aquatic variables including geomorphology and hydraulics are easier to estimate than biological variables, the scaling relationship could be instrumental for GPP and R prediction over a rage of scales. The ratio of GPP/R classifies aquatic environments according to their predominantly autotrophic (GPP/R > 1) or heterotrophic (GPP/R < 1) characteristics. The estimated GPP and R values, based on DO measurements, reported by studies across a range of geomorphic and climatic conditions are summarized in Figure 14.3. GPP to R ratio followed a power law scaling relationship $GPP/R \sim (B/H)^{4/5}(UB/u_*H)^{2/5}$ where B is the wetted stream width, H is the cross-sectional averaged depth, and u_* is the bed shear stress velocity. Since longitudinal and lateral dispersion coefficients in streams scale as $K \sim u_*H$, the proposed functional relationship can be rewritten as $GPP/R \sim (B/H)^{4/5}(Pe_{P/R})^{2/5}$ where

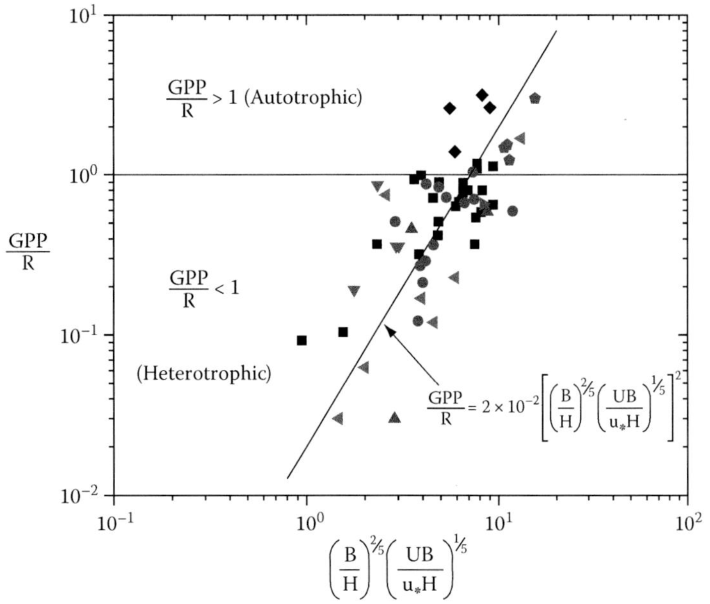

FIGURE 14.3 Running waters functional classification according to the total metabolism of autotrophic (GPP) and heterotrophic (R) communities versus stream geomorphology and hydraulic conditions.

$Pe_{P/R}$ = (UB/K) is the cross-sectional Péclet number, which can be interpreted as the ratio of advective transport along the stream to dispersive transport across the stream cross section. The proposed scaling relationship implies that stream geomorphology and hydraulics are major determinants of stream A–H metabolic balance. In forested watersheds, stream width can be interpreted as a surrogate for photosynthetically active radiation (PAR) exposure and stream depth represents the attenuation of PAR. The corresponding hydraulic conditions, quantified by $Pe_{P/R}$, contributed to the modulation of A–H balance. GPP to R ratio, which is an indicator of A–H balance and ecosystem metabolism, was empirically predictable from local stream geomorphic and hydraulic conditions. The proposed scaling relationship of GPP/R is a nonlinear function of local geomorphic and hydraulic driver variables, including B, H, U, and u_*, which inherently possess spatial and temporal variabilities in flowing waters. This implies a need for quantifying the driver variables at local scales and consequently estimating the GPP/R ratios over several locations along the stream reach that can provide data for estimating the average GPP/R ratio of the stream reach. Since geomorphic and hydraulic data are increasingly available at high spatial resolution from airborne laser altimetry, the proposed scaling relationship facilitates the prediction of A–H balance over a range of scales in flowing waters.

14.4 Methods of Analysis

While the responses of microorganisms to turbulence are likely species-specific, the fundamental and challenging questions to the majority of microorganisms are (1) How can we introduce fluid motion and turbulence in the laboratory protocols for the examination of physiological responses of microorganisms? and (2) How can we conceptually couple turbulence and microorganism growth in the models of population dynamics? Under laboratory conditions, microorganisms are customarily either shaken in flasks on tables or fluid motion is introduced by magnetic spinning bars (Eaton et al. 2005; Hurst et al. 2007). Mechanically generated turbulence characteristics and the rate of dissipation of kinetic energy per unit mass, ε, in these enclosures are unsteady and three-dimensional with high spatial heterogeneities (Sucosky et al. 2003). The most widely used functional relationship for the nutrient uptake and growth of microscopic organisms is the hyperbolic function of external growth-limiting nutrient concentration (e.g., Saito et al. 2008). The relationship is empirically described by equation

$$\mu = \mu_{max} \frac{C}{K_c + C} \tag{14.10}$$

where

- μ is the specific growth rate or rate of nutrient uptake
- μ_{max} is the maximum specific growth rate or maximum rate of nutrient uptake
- K_c is the half-saturation substrate concentration
- C is the substrate concentration

When applied to growth rates, (14.10) is referred to as the Monod equation, but when applied to nutrient uptake the term Michaelis–Menten is used. In either case, the functional relationship does not explicitly account for the effects of fluid motion. Following the approach of Edelstein-Keshet (2005) and extending the analysis by introducing a diffusive sublayer at the water side of microorganisms, the following expression could be derived

$$\mu = \mu_{max} \frac{C}{K_c/(1+\theta) + C} \tag{14.11}$$

where θ is the dimensionless constant that designates the ratio of nutrient mass transfer rate through diffusive sublayer thickness surrounding microorganism versus nutrient uptake rate through the receptors in the cell. If fluid surrounding the cell is stagnant, diffusive sublayer thickness is not determined by moving fluid, that is, it is determined by the swimming velocity of motile organism or for nonmotile organism $\theta = 0$, and (14.11) is reduced to the familiar format of (14.10). In order to facilitate a simple estimation procedure of constants in (14.11), the expression can be rearranged as

$$\frac{1}{\mu} = \frac{K_c}{\mu_{max}(1+\theta)}\left(\frac{1}{C}\right) + \frac{1}{\mu_{max}} \tag{14.12}$$

On a 2D plot, $1/\mu$ versus $1/C$, (14.12) depicts a straight line. The best-fit line through the experimental data provides an estimate of μ_{max} at the intercept. A mediating effect of fluid motion, θ, can be estimated by conducting laboratory experiments in a stagnant fluid (control) and in turbulent fluid at specified ε. For each C, a specific growth rate could be estimated during the exponential growth or nutrient uptake growth phase. The resulting growth rates will provide data for estimating the maximum specific growth rate or nutrient uptake in a turbulent fluid ($\mu_{max\,t}$), and the corresponding coefficient in the still-water control ($\mu_{max\,s}$). The estimated coefficients could be related through the following equality $\mu_{max\,t} = \mu_{max\,s}(1 + \theta)$ which implies $\theta = (\mu_{max\,t}/\mu_{max\,s}) - 1$. The magnitude and sign of θ provide an experimental quantification of the role of turbulence on microorganism growth and nutrient uptake, that is, $\theta > 0$ implies a facilitated role of turbulence and $\theta < 0$ designates inhibitory effects on microorganisms. For the given estimates of μ_{max} and θ, K_c can be calculated from a slope of the line depicted by (14.12).

14.5 Challenges

The classical theory of turbulence assumes the velocity and concentration fluctuations are dissipated at the Kolmogorov and Batchelor micro-scales, respectively. Recent studies have revealed the existence of sub-Kolmogorov-scale velocity fluctuations and regions of very strong and intermittent gradients concentrated in 3D sheets (Zeff et al. 2003; Schumacher 2007). The enhanced growth and nutrient uptake of *E. coli* ($L_c \approx 2\,\mu m$) and *S. capricornutum* ($L_c \approx 5\,\mu m$) demonstrate the mediating effects of turbulence on planktonic organisms at the sub-Kolmogorov micro-scales. Furthermore, planktic organisms depict turbulence-type structures that are induced by swing organisms by themselves in a stagnant fluid (Figure 14.4). *Dunaliella primolecta*, unicellular motile green alga, depicts significantly different swimming signatures in a stagnant fluid

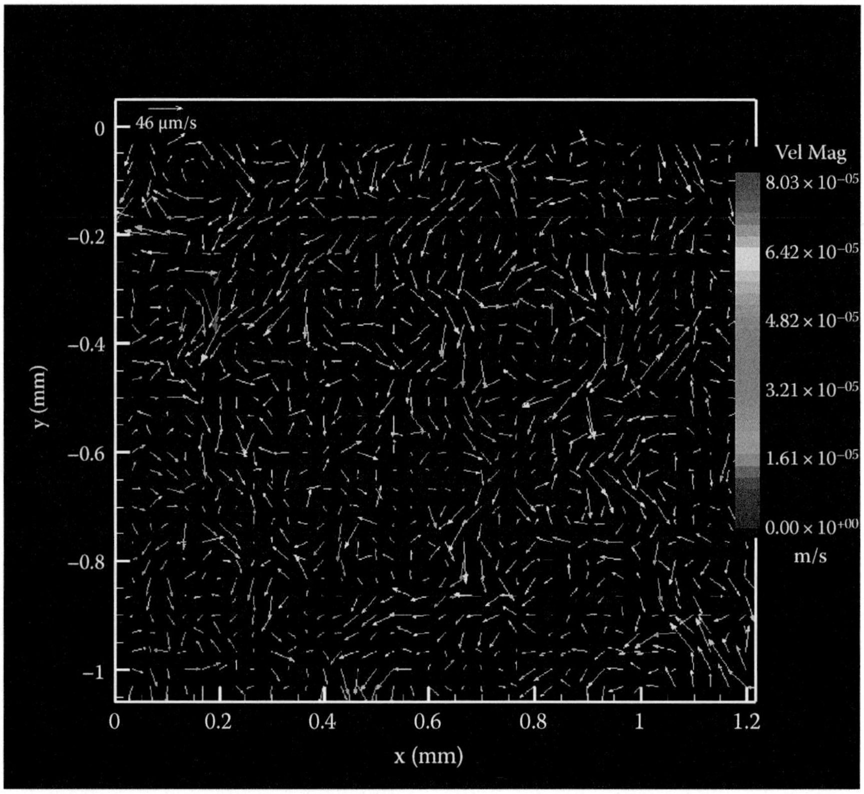

FIGURE 14.4 **(See color insert.)** Vector plot of *Dunaliella primolecta* (average cell size ~8 to 10 μm) velocity magnitudes in a stagnant fluid captured by a 4× objective of the microscope using the double-frame CCD camera with 1376 × 1024 pixel resolution. The field of view is 1.2 mm × 1.03 mm. A set of 250 double-frame image pairs were recorded with the timing between each frame (ΔT) adjusted to have a *Dunaliella primolecta* displacement of approximately 8–10 pixels.

versus moving fluid with $\varepsilon \sim 10^{-7}\, m^2/s^3$ which is commonly observed in aquatic ecosystems. The corresponding metabolic pathways including accumulation of lipid and polyunsaturated fatty acids changed in a moving fluid versus stagnant fluid control. Potential challenging research questions are (1) What is the resulting flow field at sub-Kolmogorov-scale in the proximity of motile planktic organisms? (2) How does the corresponding microscopic flow structure influence the physiology, metabolic pathways, and swimming signatures of planktic organisms? (3) What large-scale fluid flow variables mediate microscopic flow structure that is significant to the physiology of planktic organisms?

Laboratory studies examining the effects of turbulence on planktonic organisms should be conducted at the magnitude and range of ε observed in nature. A Monod-type kinetics of functional dependence is proposed that explicitly accounts for the effect of turbulence on planktic organism growth and nutrient uptake. Equation 14.11 can be used in conjunction with laboratory measurements and the prediction models of autotrophic and heterotrophic biomasses in aquatic ecosystems. We have demonstrated a functional dependence for nutrient uptake, Q/Q_D, and Péclet number determined by the Kolmogorov velocity and associated ε. We envision similar relationships between θ and Pe for a variety of physiological characteristics of microorganisms.

Reported data imply that physical processes, including turbulence at the planktic organism scale, mediate nutrient uptake, growth, and the balance of organic material production and decomposition in aquatic ecosystems. The auto–heterotrophic balance determines whether a lake or ocean is a source or sink for atmospheric CO_2. This balance is relevant to a more general question of whether marine productivity is generally subsidized by organic material produced on continents.

References

Al-Homoud, S.A., Hondzo, M., and LaPara, T. (2007). The impact of fluid dynamics on bacterial physiology: Implications for the measurements of biochemical oxygen demand. *Journal of Environmental Engineering*, 133: 226–236.

Berdalet, E., Peters, F., and Koumandou, V.L. (2007). Species-specific physiological response of dinoflagellates to quantified small-scale turbulence. *Journal of Phycology*, 43: 965–977.

Biggs, B.J.F., Nikora, I.V., and Snelder, H.T. (2005). Linking scales of flow variability to lotic ecosystem structure and function. *River Research and Applications*, 21: 283–298.

Davidson, P.A. (2004). *Turbulence: An Introduction for Scientists and Engineers*. Oxford University Press, New York.

Denman, K.L. and Gargett, A.E. (1995). Biological-physical interactions in the upper ocean: The role of vertical and small scale transport processes. *Annual Review of Fluid Mechanics*, 27: 225–256.

Durham, M.W., Kessler, O.J., and Stocker, R. (2009). Disruption of vertical motility by shear triggers formation of thin phytoplankton layers. *Science*, 323: 1067–1070.

Eaton, A.D., Clesceri, L.S., Rice, E.W., Greenberg, A.E., and Franson, M.A.H. (2005). *Standard Methods for the Examination of Water and Wastewater*, 21th edn. American Public Health Association (APHA), American Water Works Association (AWWA), Water Environment Federation (WEF), Washington, DC.

Edelstein-Keshet, L. (2005). *Mathematical Models in Biology*. SIAM, New York.

Hondzo, M. and Wüest, A. (2009). Do microscopic organisms feel turbulent flows? *Environmental Science and Technology*, 43: 764–768.

Hurst, J.C., Crawford, L.R., Garland, L.J., Lipson, A.D., Mills, L.A., and Stetzenbach, D.L. (eds.). (2007). *Manual of Environmental Microbiology*, 3rd edn. ASM Press, Washington, DC.

Lewis, D.M. (2005). A simple model of plankton population dynamics coupled with a LES of the surface mixed layer. *Journal of Theoretical Biology*, 234: 565–591.

Nikora, V.I. (1999). Origin of the "−1" spectral law in wall-bounded turbulence. *Physical Review Letters*, 83: 734–737.

Odum, H.T. (1956). Primary production in flowing waters. *Limnology and Oceanography*, 1: 102–117.

Saito, M.A., Goepfert, T.J., and Ritt, J.T. (2008). Some thoughts on the concept of colimitation: Three definitions and the importance of bioavailability. *Limnology and Oceanography*, 53: 276–290.

Schumacher, J. (2007). Sub-Kolmogorov-scale fluctuations in fluid turbulence. *Europhysics Letters*, 80: 54001–54006.

Sucosky, O., Osorio, D.F., Brown, J.B., and Neitzel, G.P. (2003). Fluid mechanics of spinner-flask bioreactor. *Biotechnol and Bioengineering*, 85: 34–46.

Warnaars, T., and Hondzo, M. (2006). Small-scale fluid motion mediates growth and nutrient uptake rate of *Selenastrum capricornutum*. *Freshwater Biology*, 51:999–1015.

Warnaars, T., Hondzo, M., and Carper, M.A. (2006). A desktop apparatus for studying interactions between microorganisms and small-scale fluid motion. *Hydrobiologia*, 563: 431–443.

Wüest, A. and Lorke, A. (2009). Small-scale turbulence and mixing: Energy fluxes in stratified lakes. In *Encyclopedia of Inland Waters*, ed. G.E. Likens, Vol. 1, pp. 628–635. Elsevier, Oxford, U.K.

Zeff, B.W., Lanterman, D.D., McAllister, R., Roy, R., Kostelich, J.E., and Lathrop, D.P. (2003). Measuring intense rotation and dissipation in turbulent flows. *Nature*, 421: 146–149.

15

Micro- and Nano-Scale Flows Relevant to the Environment

Howard A. Stone
Princeton University

15.1 Introduction

Fluid dynamics principles are relevant to flow and transport processes at all length scales in the environment, from the natural environment of the atmosphere, oceans and other bodies of water, to the subsurface flows of water and oil, and in the built environment of our homes, work places, and vehicles for transportation. This chapter focuses on topics where flow occurs on the nano- and micro-length scales, so from nanometers to hundreds of micrometers, which is often a regime not considered when presenting environmentally relevant themes in traditional courses and research articles. The presentation is in the form of sketches of themes and application areas, which serve to highlight the breadth of this range of environmental fluid mechanics problems.

Admittedly, such micro- and nano-scale flows are necessarily a subset of the major environmental, societally relevant fluid dynamics questions frequently mentioned. From the viewpoint of atmosphere and ocean dynamics, micron and nanometer dimensions are probably only mentioned when discussing sub-Kolmogorov length scales. Nevertheless, flows at small length scales arise in a surprisingly broad range of environmentally important situations simply because such flows are typically relevant wherever surfaces or interfaces are important to a transport process. As we shall see later, such interfaces can be air–water interfaces where aerosol droplets form and are then transported to the atmosphere, the three-phase systems that characterize oil and water trapped in a porous matrix, the drying of building materials or soils, or the surfaces of materials where biofilms form. Indeed, biological processes constitute many micro- and nano-scale science and engineering questions that have environmental consequences. Moreover, small-scale flows can have significant impact on understanding and treating various aspects of pollution, for example, thin films of oil on water following an oil spill, and contamination, for example, leaching of chemicals. As a consequence, one idea advanced in this chapter is the importance of including this set of small-scale fluid mechanics themes in the educational repertoire considered in environmental fluid mechanics. For example, only a few of the topics discussed here are even mentioned in standard modern textbooks covering environmental fluid mechanics (e.g., Rubin and Atkinson 2001, Shen et al. 2002).

15.1.1 Spectrum of Micro- and Nano-Scale Flows in the Environment

Where do micro- and nano-scale flows occur in the environment? A partial listing of such flows and applications is given

Handbook of Environmental Fluid Dynamics, Volume One, edited by Harindra Joseph Shermal Fernando. © 2013 CRC Press/Taylor & Francis Group, LLC.
ISBN: 978-1-4398-1669-1.

TABLE 15.1 Overview of the Environmental Systems and Problems Involving Micro- and Nano-Flows

Environmental Nano- and Micro-Flows	Applications
Single- and multiphase flows in porous media (including sand, soil, filters, materials packaging)	Filtration; clogging of porous materials Colloidal transport; swelling of clay Water contamination by heavy metals (e.g., arsenic) Oil recovery; CO_2 sequestration Remediation of soils Nonaqueous phase liquids (NAPL) Mining waste, disposal of tailings (rheology)
Microorganisms	Bacteria and phytoplankton in oceans; patchiness Bacteria in soils; bacterial adhesion to surfaces biofilms; formation of bacterial streamers Spreading of biofilms and bacterial swarming
Surface tension	Organisms that live at surfaces (e.g., water striders) Particles trapped at interfaces Bubble and droplet breakup; aerosol generation Wetting and drying of materials and associated problems involving transport of solutes
Small devices for environmental measurements	Monitoring and diagnostics (sensing and detection); chemical delivery Experimental testing at the pore scale
Electrical manipulation of small-scale flows	Desalination; soil remediation
Particulate materials	Toxicity of nanoparticles Generation of particulate materials (e.g., combustion, wear and erosion, storms)

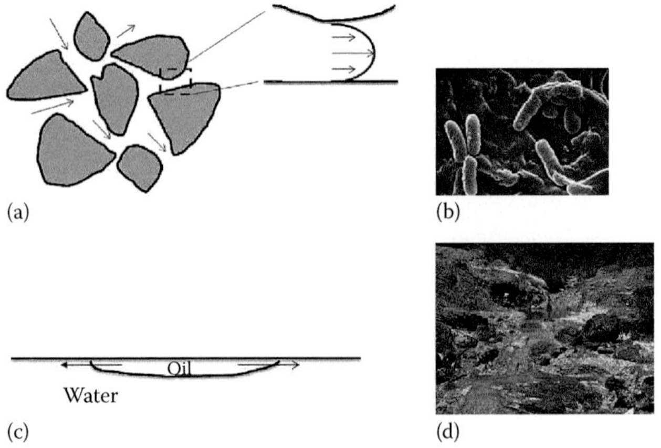

FIGURE 15.1 Examples of small-scale flows. (a) Sketch of flow in a porous material, where at the scale of a few grains the local pressure-driven velocity profile is expected to be approximately parabolic; (b) an image of the bacteria *Pseudomonas aeruginosa*, common in soils, since fluid dynamics at the scale of the microorganism (typically a micron or so) can be significant in biofilm formation, disease transmission, etc. (image available for public use); (c) a sketch of a film of oil spreading on water, as occurs in oil spills; and (d) an image of the waste from a mining operation in the Rio Tinto River in southwestern Spain (NASA image available for public use).

in Table 15.1 and we indicate in Figure 15.1 examples such as a sketch of the flow in the pores of a material, an image of bacteria since fluid dynamics at the scale of microorganisms can be significant in a wide range of flow problems (examples given in the following), a sketch of a film of oil spreading on water, as occurs in various environmental disasters, and an image of the waste from a mining operation. The spectrum of environmental themes impacted by small-scale flows is actually quite large. Traditional areas include the science and engineering of oil recovery (this theme has environmental consequences and leads to possible locations for carbon sequestration, even if the usual consideration is that it is an industrial problem), flow of ground water, the roles of bacteria and other small organisms in natural environmental processes and the ways the cells interact with the flow, and the many processes affected by surface tension such as formation of small droplets and the use of various methods for cleaning contaminated soils or retaining and recovering oil following oil spills. In addition, such problems as the waste from mining operations and the disposal of tailings from the mineral processing raise difficult questions about the rheology of the slurries, which complicate how to respond to the engineering and environmental challenges. Not surprisingly, a wide range of environmental remediation strategies raise questions of micro- and nano-scale fluid mechanics.

In recent years, a wide variety of new kinds of questions involving small-scale fluid mechanics have arisen: the themes include CO_2 sequestration, the toxicity and transport of nanoparticulate materials (Murr and Garza 2009), the rationalization of some aspects of bacterial spreading and swarming as a consequence of fluid motions driven by surface tension gradients (Daniels et al. 2006), and cleaning of water contaminated by heavy metals. These various problems have in common that they incorporate themes of fluid mechanics, physical chemistry, and particulate matter. Thus, the topical area known as physicochemical hydrodynamics (e.g., Probstein 2003) has much to say about studying environmentally significant small-scale flows.

To take the theme of water contamination by heavy metals as an example, arsenic contamination in natural waters is a worldwide problem with particular relevance to the developing world. The problem demands the attention of the science and engineering communities because of the significant health issues that are engendered. Other heavy metals that can occur in groundwater pose similar threats to human health. As a treatment strategy, for example, nanocrystalline magnetite is effective in removing arsenic from solution (Yavuz et al. 2006). Although it has been demonstrated that the effectiveness of the separation of nanometer-sized magnetite particles from solution is dependent on particle size, there is little understanding of the physicochemical transport processes that control the process. Hence, progress toward applying these ideas for effective and widespread water treatment, implementable in the developing world, have so far been limited.

15.1.2 Examples of Small-Scale Environmental Flows: From Oil Fields to Natural Waters

The movement of oil through micron dimension pores in subsurface rock to the surface of the earth, where oil (usually with water and gas) flows through large pipelines for transportation to refineries, is a common illustration of one micro-flow, that is, the pore scale, of environmental interest. The fact that a significant fraction, often estimates are 30%–40%, of the oil originally in place in a petroleum reservoir remains trapped after an oil field is retired provides one manner of motivation for understanding micron-scale fluid flows in an environmental context. It also motivates methods to enhance oil recovery. Oil reservoirs that are no longer active are considered one possible site for storing CO_2, and this topic of sequestration requires understanding pore-scale multiphase flows (e.g., Celia and Nordbottten 2009, Neufeld and Huppert 2009). Other porous materials such as the ocean sediments have also been considered for CO_2 storage (e.g., House et al. 2006). It is thus important to assess possible drainage of the stored CO_2 through these porous media and such research questions are being pursued actively.

The movement of microorganisms provides several examples of micro-flows relevant to environmental processes. There are many types of microorganisms including bacteria, algae, protists, and fungi, which play important roles in aquatic environments, including soil, streams, lakes, and oceans. Here we give one other example of how movements at micron dimensions can impact macroscopic flows by considering the swimming of algae (Kessler 1985) and how the swimming direction is established by interaction of the cells with the local flow (Figure 15.2).

Many algae cells are biflagellated and swim with a "breaststroke" movement of the flagella. Because some organelles in the cell are denser than the rest of the cellular components and are located in the half of the cell further from the flagella, the cells can be bottom heavy. In any shear-like flow the cells adopt a preferred orientation relative to the flow direction: the orientation is determined from a balance of a gravitational torque (the center of mass is below the center of buoyancy) and a viscous torque from the local shear flow about the particle (Figure 15.2b). Consequently, the cells swim toward the center of a down-flow, which focuses cells into columns (Figure 15.2a and c, left), whereas in any up-flow the cells are readily dispersed (Figure 15.2a and c, right). This "gyrotactic" phenomenon is relevant to explaining the movement of microorganisms in natural waters. Moreover, as Figure 15.2c demonstrates there are cases where the flow and the microorganisms interact to localize biomass to specific regions; such flow phenomena are also relevant in open waters where up- and down-welling flows occur, for example, see Section 15.4.

We next discuss some of the underlying fluid mechanics to better understand the physical features relevant to flow at the micron (and smaller) length scale.

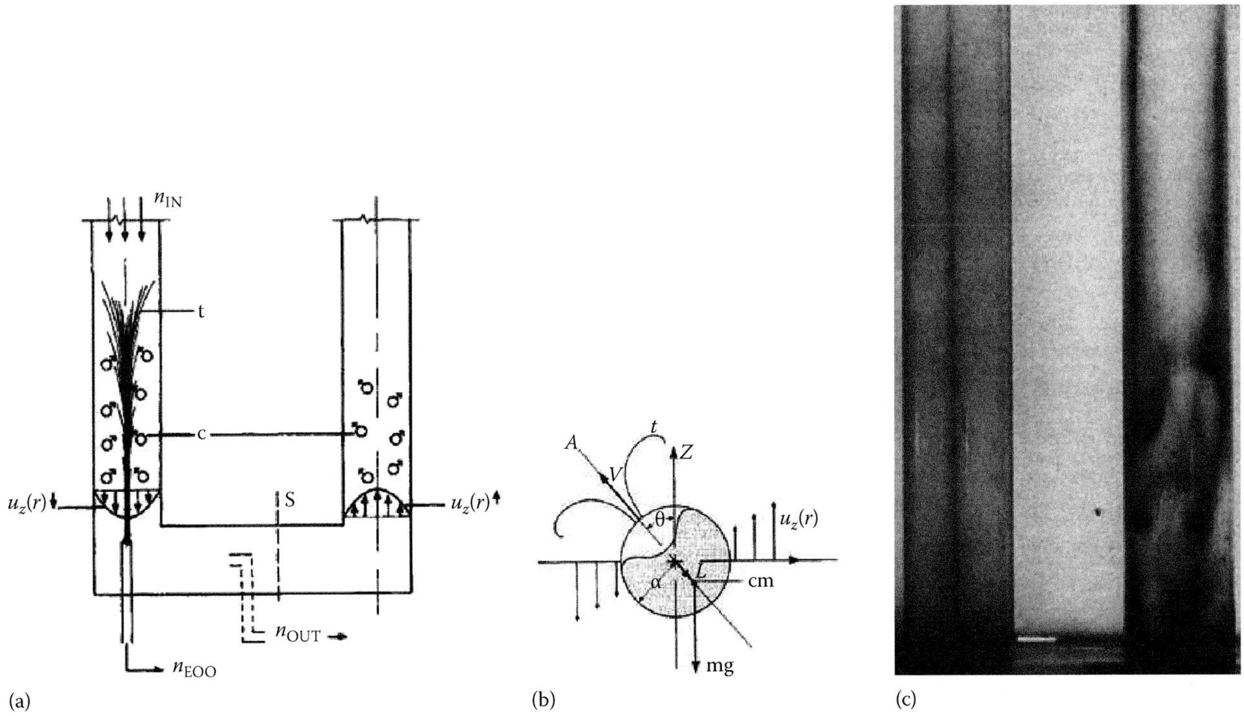

FIGURE 15.2 When bottom-heavy cells swim in a non-uniform flow, they adopt an orientation that balances gravitational torques and hydrodynamic torques. (a) A schematic of the orientation of cells in a parabolic flow in a circular pipe. (b) Torques acting on an orientable cell in a shear flow. (c) Experimental realization of gyrotaxis. In a down-flow (left, as in (a)) the cells (*Chlamydomonas nivalis*) are oriented so that they swim to the centerline, which focuses the cells into a column, while in an up-flow (right) the cells swim away from the centerline. (From Kessler, J., *Nature*, 313, 218, 1985.)

15.2 Principles

Analyses of fluid motions begin with the continuity and Navier-Stokes equations. Since micro- and nano-scale flows are almost always incompressible in the contexts relevant to the environment, the fluid velocity $\boldsymbol{u}$ and pressure p fields satisfy

$$\nabla\cdot\boldsymbol{u}=0$$

$$\rho\left(\frac{\partial \boldsymbol{u}}{\partial t}+\boldsymbol{u}\cdot\nabla\boldsymbol{u}\right)=-\nabla p+\rho\boldsymbol{g}+\mu\nabla^2\boldsymbol{u} \qquad (15.1)$$

where we have assumed that the fluid (usually water or oil in all of the applications indicated in Table 15.1) is Newtonian in its stress versus rate-of-strain response and the density (ρ) and viscosity (μ) are constants, and $\boldsymbol{g}$ is the acceleration due to gravity.

In environmental transport problems involving colloidal materials and situations where ions influence the flow, an electrical body force term, $\rho_{el}\boldsymbol{E}$, where ρ_{el} is the charge density (electrical charge per unit volume) and $\boldsymbol{E}$ is the local value of the electric field, must be added to the right-hand side of the Navier-Stokes equation. The subject of electrohydrodynamic phenomena raises a host of new and challenging questions beyond those encountered for single- and multiphase flows, and the reader is referred to specialized textbooks for an introduction to the important ideas (e.g., Probstein 2003). In addition, we note that osmotic stresses that occur due to the presence of ions, even in the absence of externally applied electrical fields, are known to have important in the swelling of clays, colloidal transport, etc.

15.2.1 Reynolds Number

To characterize the dynamics of a flow, we must identify characteristic length and velocity scales. We can either consider the characteristic length ℓ of a porous medium, which has a typical scale of tens of microns or smaller (e.g., the typical radius of a pore), or we can consider the particles that are transported to have this micron size. A characteristic Reynolds number for a flow is defined as $Re = \rho u_c \ell/\mu$, where u_c is a typical speed. Most often the representative Reynolds number for these micro-flows is order one or smaller, which implies that the dynamics are those for a laminar, viscously dominated flow. For flows involving water, we can expect $Re = O(1)$ on the length scale of 10 μm when the flow speeds are no larger than about 0.1 m/s. In the case of swimming microorganisms, where ℓ is say 10 μm and the swimming speed is 10 body lengths per second, then $Re = 10^{-3}$. For those cases where the Reynolds number is small, then the left-hand side of the linear momentum equation (15.1) can be neglected (we are assuming that there are no time-periodic forces being applied to the fluid). It is generally convenient to absorb the gravitational body force term into a dynamic pressure $p_d = p - \rho\boldsymbol{g}\cdot\boldsymbol{x} = p + \rho g z$, where z is directed vertically upward, and so consider Stokes equations:

$$\nabla\cdot\boldsymbol{u}=0 \quad \text{and} \quad 0=-\nabla p_d+\mu\nabla^2\boldsymbol{u} \qquad (15.2)$$

These equations have been studied for a wide range of particle transport problems, thin-film flows, colloidal flows, porous media flows, etc. The fluid density may reappear in boundary conditions where the actual pressure is required, for example, a fluid–fluid interface.

As a final remark, we note that there is a subclass of small-scale flows that occur sufficiently rapidly that the local Reynolds number is large (typically 100 or larger). Many of these kinds of problems of potential environmental interest involve surface tension (see the following, de Gennes et al. 2004). The bursting of a bubble at an air–water interface (e.g., Bird et al. 2010, see Figure 15.3a), which leads to the formation of small aerosol drops, is an example. In such cases, inertial effects dominate for which it is generally reasonable to model the flow as irrotational at the early times relevant to the dynamics.

15.2.2 Boundary Conditions

The subject of boundary conditions for micro- and nano-flows has received considerable attention in recent years with some reports of significant violations of the no-slip boundary condition. Consider the case of water, which has molecular dimensions about 3 Å, or 0.3 nm. If we consider a water-wet pore, which is 1 μm in radius, then the ratio of the "macroscopic" length scale to the molecular scale is >1000 so we expect a continuum treatment with a no-slip boundary condition to be very good indeed. Nevertheless, when the substrate is non-wetting, there is some evidence at smooth boundaries that microscopic slip lengths in the tens of nanometers are possible; note that this slip length is still typically 30–100 times smaller than a pore with a 1 μm radius, and so the no-slip condition should be an excellent approximation. However, when surfaces are textured on the nano- and micro-scales, larger slip lengths are possible (Bocquet and Charlaix 2010).

This short summary makes clear that for the majority of common cases involving small-scale flows, the no-slip condition remains an excellent approximation. In any event, in most environmental flows of interest, there are normally many other sources of significant uncertainty than the boundary conditions. Only in the case of electrically driven flows, for the limit of small Debye screening lengths, is it typically necessary to consider slip at a solid boundary at the length scale of a few microns and above.

15.2.3 Description of Darcy

Because of the geometric complexity of many small-scale flows, such as flow in sand, packed beds, fibrous filter media, etc., we often recognize that the details of the flow at the level of individual grains are much less important than the average flow. In terms of the average velocity $\langle\boldsymbol{u}\rangle$, the appropriate averaged form of the Stokes equations (15.2) is

$$-\nabla p_d=\frac{\mu}{k}\langle\boldsymbol{u}\rangle \qquad (15.3)$$

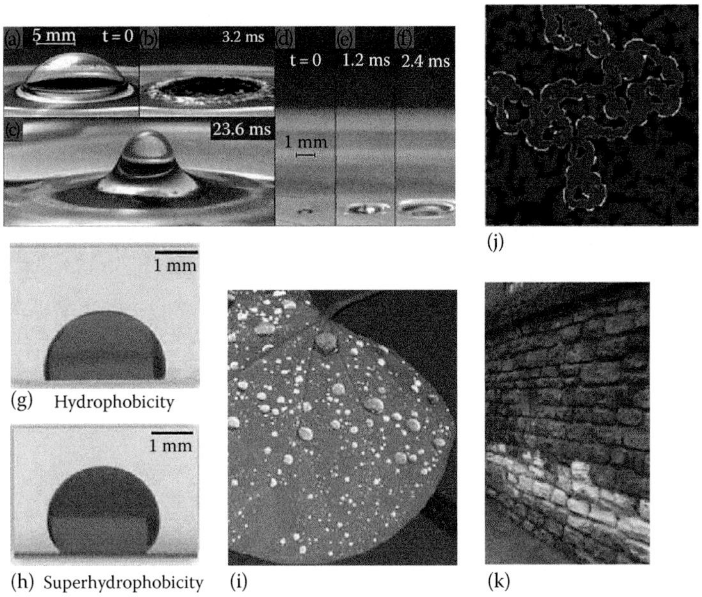

FIGURE 15.3 **(See color insert.)** Examples where surface tension affects the statics and dynamics of a fluid system. (a–f) A bubble bursts (a) at the air–water interface, forming an array of even smaller bubbles (b–c); when a smaller bubble bursts, a small aerosol droplet is released. (From Bird, J.C. et al., *Nature*, 465, 759, 2010.) (g–i) Wetting of surfaces with (g) a partial wetting liquid drop on a smooth substrate, (h) the same system when the surface is roughened displays an even higher contact angle, and (i) such roughness-induced superhydrophobicity is common in the leaves of plants as droplets are readily displaced from the surface. (j) Two-phase flow in a porous sample with an example shown here of decane (blue) and nitrogen gas (maroon) in a polymeric micromodel. (From Cheng, J.-T. et al., *Geophys. Res. Lett.*, 31, L08502, 2004.) during drainage of the fluids. (k) Salt weathering, or efflorescence, of building materials as evaporation of water occurs at the surface of the sample. (From Hall, C. et al., *Proc. R. Soc. A*, 2010.)

which is known as Darcy's law. This equation is then solved together with the continuity equation. The permeability k characterizes the geometric features of the resistance; k has dimensions length squared, so that we may expect that the order of magnitude of k is the square of the typical pore size through which flow occurs.

In such porous media the permeability also depends on the packing of the particles, which is characterized by the porosity, that is, the volume or pore space per unit volume of material. A common correlation for the relation of the permeability k to the porosity, ϕ, for a packing of spheres of diameter d is given by the Carman-Kozeny relation $k = (d^2/180)\varphi^3/(1 - \varphi)^2$ (note that there is some variability in the literature for the constant chosen in this equation). For local flow speeds high enough that inertial effects are significant, corrections to Darcy's law are known. The generalization of the Darcy description to multiphase flow is a topic of continual interest and models exist in the literature.

Remark about microfluidics: Microfluidic devices are small systems for handling fluids where controlled flow occurs in channels typically 100 μm and smaller (e.g., Stone et al. 2004). These devices are being applied to environmental monitoring (Marle and Greenway 2005) as discussed a little more in the following. The flows are almost always laminar in these devices and are often pressure driven, in which case we expect the flow profile transverse to the flow direction to be approximately parabolic. The devices typically have a structure consisting of a network of channels, and the analysis of such networks can be estimated using Darcy's law to connect the pressure drops and the average flows speeds (or flow rates) in the various channels.

15.2.4 Multiphase Flows, Surface Tension, and the Capillary Number

The subject of multiphase flow in such porous media raises many questions and challenges. The interface between air and water, or between any two immiscible liquids, can be important in dynamical situations, for example, the movement of bubbles or drops through a porous media or microfluidic network, or the formation of aerosol drops (Figure 15.3). We mention a few ideas relevant to thinking about such flows. The most important parameter is the surface, or interfacial tension, γ, which characterizes the energy ΔE needed to increase the interfacial area by ΔA, that is, $\Delta E = \gamma \Delta A$. It is often convenient to think of γ as the force per unit length acting perpendicular to a curve in the interface. A second important parameter is the contact angle, which characterizes the tendency of a liquid to want to wet (or coat) the surface. A third important idea is that across a curved interface, the pressure difference (for a static fluid) is $\Delta p = 2\gamma\kappa_m$, where κ_m is the mean curvature of the interface. This classical result is associated with the names of Young and Laplace.

In the context of dynamics, when the Reynolds number is small and the typical length scale is ℓ, we compare viscous stresses with stresses from surface tension. This ratio yields the capillary number, $Ca = \mu u_c/\gamma$. In many applications, $Ca < 1$, and is often $\ll 1$, in which case we expect the interfaces to have shapes that are close to the equilibrium shapes that minimize the surface area for a given volume. Nevertheless, many of these multiphase flow problems are characterized by thin films where the relevant dynamics is established by the balance of viscous stresses, pressure gradients, and capillary stresses (e.g., de Gennes et al. 2004).

15.2.5 Relative Importance of Convection and Diffusion

There are a wide variety of transport processes that occur at small scales, which are important for understanding the sequestration of chemicals, the bioavailability of nutrients, etc. For example, chemicals undergo adsorption/desorption processes, chemical reactions occur at interfaces and in bulk solution, and simultaneously the chemical species diffuse and are advected by the flow. The relative importance of convection to diffusion is measured by the Peclet number, $Pe = u_c\ell/D$, where u_c and ℓ are, respectively, velocity and length scales as defined as in the Reynolds number, and D is the diffusion coefficient. For small molecules in aqueous solutions, $D \approx 10^{-5}\,\text{m}^2/\text{s}$; and so if ℓ is smaller than the length scale D/u_c, then diffusion dominates on the scale of ℓ. For speeds 0.1 cm/s, diffusion dominates at length scales smaller than 1 cm, while for higher speeds of 10 cm/s, diffusion only dominates at length scales of 100 μm or smaller. These ideas can be refined through "boundary layer" analyses of the interplay of convection and diffusion near boundaries, as covered in standard textbooks on transport phenomena. It follows that sufficiently close to a surface where chemical reactions occur we can expect diffusion and reaction to have comparable orders of magnitude.

15.3 Methods of Analysis

The analysis of small-scale flow problems typically begins with the Stokes or Darcy equations, (15.2) or (15.3), respectively. We have introduced earlier the two common scenarios concerning dynamics within narrow pores or the movement of small particles. In this section we summarize briefly (1) ideas useful for thinking about dynamics of particles in fluids, including the response of individual particles as well as the effective properties of a suspension of particles, and (2) dynamics involving multiphase flows, where either or both of surface tension and viscosity contrasts are important. Finally, we close with a few remarks about microfluidic devices. The discussion is inevitably too terse, but it is hoped that the brief remarks will be useful in pointing a reader toward relevant ideas and literature that may be useful.

15.3.1 Transport Properties of Particles

There are several classic results of low Reynolds number hydrodynamics of particular systems that are worthwhile to record because of their use in basic modeling of particulate systems. Some of these ideas should be expected to be useful when considering particle-laden systems (colloids, slurries, etc.) in environmental applications. An excellent advanced reference for this topic of "microhydrodynamics" is Batchelor (1977). First, consider a spherical particle translating at velocity $\boldsymbol{V}$ through a fluid. When Re $\ll 1$, the hydrodynamic drag force $\boldsymbol{F}^H$ that resists the motion is $\boldsymbol{F}^H = -6\pi\mu a\boldsymbol{V}$ (Stokes drag). Similar results are available for other standard shapes relevant to particles found in the environment, such as disk-shaped clays and rod-shaped particles. For such non-isotropic particles, the drag force depends on the orientation of the particle, but in all cases the order of magnitude of the drag force is the product of a shear stress and the surface area of the particle. The former is estimated using Newton's law of viscosity as $\mu V/\ell_{small}$, where ℓ_{small} refers to the smallest geometric length-scale of the particle (this estimate only misses an order-one factor involving the logarithm of the particle's aspect ratio). For long slender particles, of length ℓ, with surface area $O(\ell\ell_{small})$, we observe that the force on a long slender particle will be proportional to the longest dimension of the particle, that is, the force is approximately $O((\mu V/\ell_{small})\,\ell\ell_{small}) \approx O(\mu V\ell)$. At higher Reynolds numbers, some analytical, numerical, and empirical results are available.

A second feature of suspended particles in flows is that the effective viscosity of the medium, treated as an equivalent single-phase fluid, is higher than the viscosity of the solvent alone. For spherical particles in the dilute limit, as measured by the volume fraction ϕ of spheres, the effective viscosity μ_{eff} of a suspension is (this result was first given by Einstein in 1906)

$$\frac{\mu_{eff}}{\mu} = 1 + \frac{5}{2}\phi \quad \text{for } \phi \ll 1 \tag{15.4}$$

For higher volume fractions, the fluid is in general viscoelastic, but a common approximation is to treat it as an equivalent Newtonian fluid with an effective viscosity given by (there are other similar correlations; Stickel and Powell 2005)

$$\frac{\mu_{eff}}{\mu} = \left(1 - \frac{\phi}{\phi_m}\right)^{-(5/2)\phi_m} \tag{15.5}$$

As the volume fraction approaches a maximum packing density ϕ_m the viscosity increases rapidly, while for small volume fractions the Einstein viscosity is recovered.

Another property of small particles in solution concerns their effective diffusivity. For small enough particles, that is, colloidal materials, thermal agitation, or Brownian motion, causes inevitable spreading of an initial collection of particles. On the other hand, in any suspension, where there is a shear flow, the particles cause disturbances to the flow and these disturbances produces fluctuations in the trajectories of nearby particles. Thus, there is a hydrodynamic contribution to the diffusivity, in addition to the fluctuations produced by Brownian motion. For motions at low Reynolds numbers, the hydrodynamic, or shear-induced, diffusivity of a suspension of non-Brownian spheres is proportional to the shear rate and the square of the

TABLE 15.2 Transport Properties of Suspended Particles

Forces on particles (effects of size and shape)	Spheres, radius a: $6\pi\mu a V$ for Re $\ll 1$ Other results available for slender objects and disk-shaped objects
Effective viscosity, μ_{eff} (spheres)	$\mu_{eff} = \mu\left(1+\frac{5}{2}\phi\right)$ for $\phi \ll 1$; see also Equation 15.5
Shear-enhanced diffusivity (spheres)	$D = \dot{\gamma}a^2 f(\phi)$ with $f(\phi)$ $= \begin{cases} \propto \phi^2 \text{ (hydrodynamic interactions only)} \\ \propto \phi \text{ (non-hydrodynamic interactions)} \end{cases}$

particle size, as basically demanded by dimensional arguments. Therefore, the shear-induced diffusivity is expected to have the form $D = \dot{\gamma}a^2 f(\phi)$. The volume fraction dependence for dilute systems is summarized in Table 15.2 and otherwise generally requires experimental measurements or simulations.

15.3.2 Surface Tension

We introduced earlier the surface or interfacial tension. The contribution to local stresses in a fluid is potentially important whenever a fluid–fluid interface is important and there are many applications (e.g., see Figure 15.3). On small length scales the interface tries to minimize its energy and so, roughly speaking, adopts shapes that are sections of spheres or circles. Thus, when a hydrophilic surface is placed vertically into a liquid, the liquid forms a meniscus whose rise height is a balance between the interfacial energy and the gravitational potential energy. The length scale that characterizes the rise height and interface curvature is the capillary length $\ell_c = (\gamma/g\Delta\rho)^{1/2}$, where $\Delta\rho$ is the density difference. With $\Delta\rho \approx 1000\,\text{kg/m}^3$, $g \approx 10\,\text{m/s}^2$ and $\gamma \approx 0.07\,\text{N/m}$, this length scale is a few millimeters for the air–water interface. The same physical balances set the typical size of rain drops. This typical length can be on the scale of centimeters for the oil–water interface. On the other hand, when an air–liquid interface is pulled up a vertical capillary of radius a, as described in introductory physics books, the equilibrium rise height is approximately $\ell_{rise} = \gamma/\rho g a = \ell_c^2/a$.

One familiar case where surface tension plays a role in dynamics of liquid occurs when a liquid is contacted with a dry porous hydrophilic material, for example, consider dipping a paper towel or your shirt into water. The liquid imbibes the porous media with a speed that decreases monotonically in time due to the increased viscous resistance as the distance imbibed increases. The classical result, often associated with the names of Washburn and Lucas (though the first identification of the basic result was given by Bell and Cameron in 1906), is that the distance imbibed increases as $\ell(t) \approx (\gamma a t/\mu)^{1/2}$, where a is a typical pore radius. Note that for the case that the capillary or porous media is vertical we can estimate the "rise time" t_{rise} to reach the equilibrium rise height ℓ_{rise} by solving for time according to $\ell_{rise} \approx (\gamma a t_{rise}/\mu)^{1/2}$, which yields the estimate $t_{rise} \approx \mu\gamma/(\rho g)^2 a^3$.

As an example of an environmental application, we note that imbibition dynamics are significant for understanding moisture dynamics in the walls of buildings (e.g., Hall et al. 2010). In this problem, ground water flow and the moisture profile in the soil are linked to evaporative transport processes to the local atmosphere. Characterizing such environmental transport problems are important for estimating and mitigating damage rates to structures, including common building materials, architecture landmarks, etc. (Figure 15.3k).

Finally, as mentioned earlier, although viscously controlled, surface-tension-driven dynamics are common at the micron length scales, there are also cases where the surface-tension-driven dynamics have a large Reynolds number so that inertial effects are significant. The latter typically occur in the rapid motions associated with the rearrangements of thin liquid films (de Gennes et al. 2004). In such cases it is a good estimate to treat the fluid domain as a mass on a spring. Then, the typical time scale $\tau_{inertia}$ of the response is $\tau_{inertia} \approx (m/\gamma)^{1/2}$, where m is the mass of fluid set in motion.

15.3.3 Multiphase Flows and Viscous Fingering Instability

In many environmental processes there are multiple fluid phases present. For example, water is often used to help push oil out of the pore spaces in oil fields. Alternatively, some strategies for carbon dioxide sequestration focus on forcing high pressure CO_2 back into the pore matrix of a partially drained oil reservoir, or the porous sediments at the bottom of the ocean, or other porous substrates. In such cases it is of interest to understand displacement processes of one fluid phase by another of different viscosity. At the pore scale it is necessary to consider motion of individual slugs of fluid which may or may not wet the solid substrate (e.g., Figure 15.3j). Because this topics is directly relevant to oil recovery there is a large literature, but owing to the geometric and hydrodynamic complexities (small capillary numbers, complicated geometries, viscosity ratio between the phases, wettability), it is difficult to give simple generalizations.

In the simpler case of a displacement process of one fluid by another in a homogeneous geometry (either a porous media or the experimental prototype of a Hele-Shaw cell), there is a well-known instability, the Saffman-Taylor or viscous fingering instability, which has been investigated widely since the 1950s. When a fluid of low viscosity is driven by a pressure gradient into a fluid of higher viscosity, the interface becomes unstable and fingers of low viscosity fluid propagate past the basically stationary higher viscosity fluid (Homsy 1987). On the other hand, a higher viscosity fluid can simply push the lower viscosity fluid out of the way by maintaining a uniform front during the displacement process.

In the case of propagation of the two phases in a porous material with variations in the permeability k, it is most common to work with the mobility parameter, k/μ. In this case, we expect the front to be unstable whenever the interface moves from a region of high mobility to a region of low mobility. Any small disturbance then locally places the high-mobility fluid ahead of the low-mobility fluid and the fingering instability grows.

15.3.4 Microfluidics for Environmental Monitoring

Microfluidic devices are tools for handling fluids in networks of channels that have cross-sectional dimensions that are typically tens to hundreds of microns. The field has grown rapidly in the past 10–15 years and there are many laboratory-scale studies possible (e.g., Stone et al. 2004). Portable devices have been made and offer great potential for environmental monitoring and assessment.

Microfluidic devices allow taking of small samples, rapid mixing of reagents, optical detection, electrochemical measurements, control of the thermal and chemical environment necessary for various chemical steps, etc. Thus, a wide range of analytical chemical and biological assays and tool can be readily adapted for in situ real-time and continual measurements. Marle and Greenway (2005) provide a review of ideas specifically focused on environmental monitoring. The methods can also be used for addressing basic scientific questions associated with flows at the pore scale in model porous materials, studying the movement of bacteria in confined spaces, and the spatial and temporal development of a biofilm (see the following), etc. There are many new opportunities to be explored in environmental science and engineering using microfluidic approaches.

15.4 Examples and Applications

As mentioned in the introduction, small-scale flows have significant environmental impact in the many different situations involving microorganisms. In this section we summarize two applications, with one relevant to patchiness (biomass accumulation) in the oceans and the other significant for understanding biofilm development in channel flows.

15.4.1 Oceans

In Figure 15.2 we illustrated a laboratory scale example of gyrotaxis where bottom-heavy algal cells were focused in the vertical direction because of horizontal gradient in the shear rate caused by an up- or down-flow. In natural waters, such as oceans and lakes, it is frequently observed that microorganisms accumulate in narrow horizontal regions. This "patchiness" is ecologically significant since it influences feeding and growth rates of organisms that rely on the microorganisms as a food source, though this mechanism also can lead to the formation of harmful algal blooms. Such localized cell populations also produce variations in acoustic and optical signals in the water.

To illustrate the idea we consider a bottom-heavy cell, which is representative of common phytoplankton in the ocean (Durham et al. 2009). When exposed to a horizontal shear rate, the cell orients itself to swim upward (Figure 15.4a). If the shear rate is

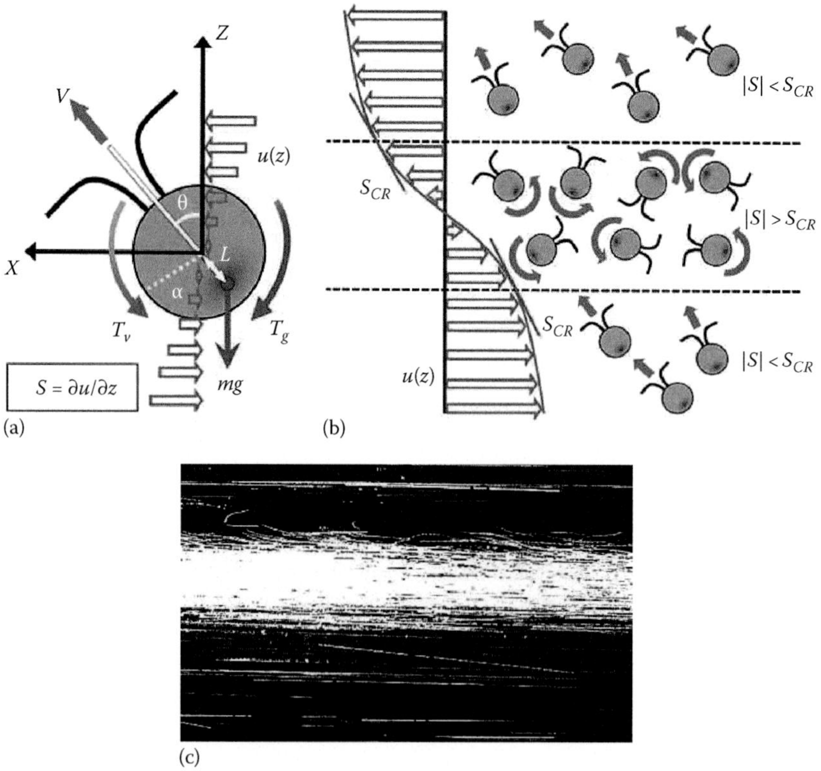

FIGURE 15.4 Formation of thin horizontal layers in a flow with a vertical gradient of shear rate. (a) Orientation of a bottom-heavy cell in a horizontal shear flow. (b) Where there is a vertical gradient of shear, cells tumble if the shear rate exceeds a critical value. Below the critical shear rate, the cells are oriented so that they swim upward. (c) Multiple exposure image of phytoplankton accumulating in a horizontal layer in a flow cell with a vertical gradient of shear rate. (From Durham, W.M. et al., *Science*, 323, 1067, 2009.)

too strong, there is no steady orientation, and the directed cell motion is disrupted. Therefore, we can anticipate that where there is a gradient of shear in the vertical direction, spanning shear rates below and above the critical shear rate to induce tumbling, the cells that are in a region below the critical shear rate continually swim upward; in this way a layer of higher cell density is produced, which is localized to shear rates below the critical value for tumbling (Figure 15.4b). Experiments in laboratory flow cells, where a vertical gradient of shear rate is generated, support this idea, as illustrated in Figure 15.4c by the multiple-exposure image of a solution of the algal cell *Chlamydomonas nivalis* (Durham et al. 2009).

15.4.2 Biofilms and Bacterial Streamers

Another place where microorganisms are important in the micro-environment concerns bacteria, which are present everywhere around us (and inside our bodies). Bacteria influence many aspects of life, including biodegradation in soils and natural waters, water purification in industrial operations, etc. A common structure in which we find bacteria is as biofilms, which frequently form on surfaces. Biofilms are bacterial communities that organize themselves in a self-secreted matrix of exopolysaccharides and appear to be capable of adhering to almost any kind of surface. In fact, the common view is that bacterial biofilms develop on surfaces, which while true is not the only places they can be localized.

Since many fluid systems involve flow, we recently studied how biofilms develop in a model channel flow (Rusconi et al. 2010). For our experiments we used a microfluidic approach, as was mentioned earlier, with the bacteria *Pseudomonas aeruginosa* (Figure 15.5a). At early times, the cells (labeled with green fluorescent protein) grew as a biofilm on the surface (the light background in Figure 15.4b). In fact, when the channel was straight, the surfaces were the only locations where we observed the biofilm. However, in experiments with channels that contained bends, or corners, at longer times we discovered a significant fraction of the biomass accumulating as a

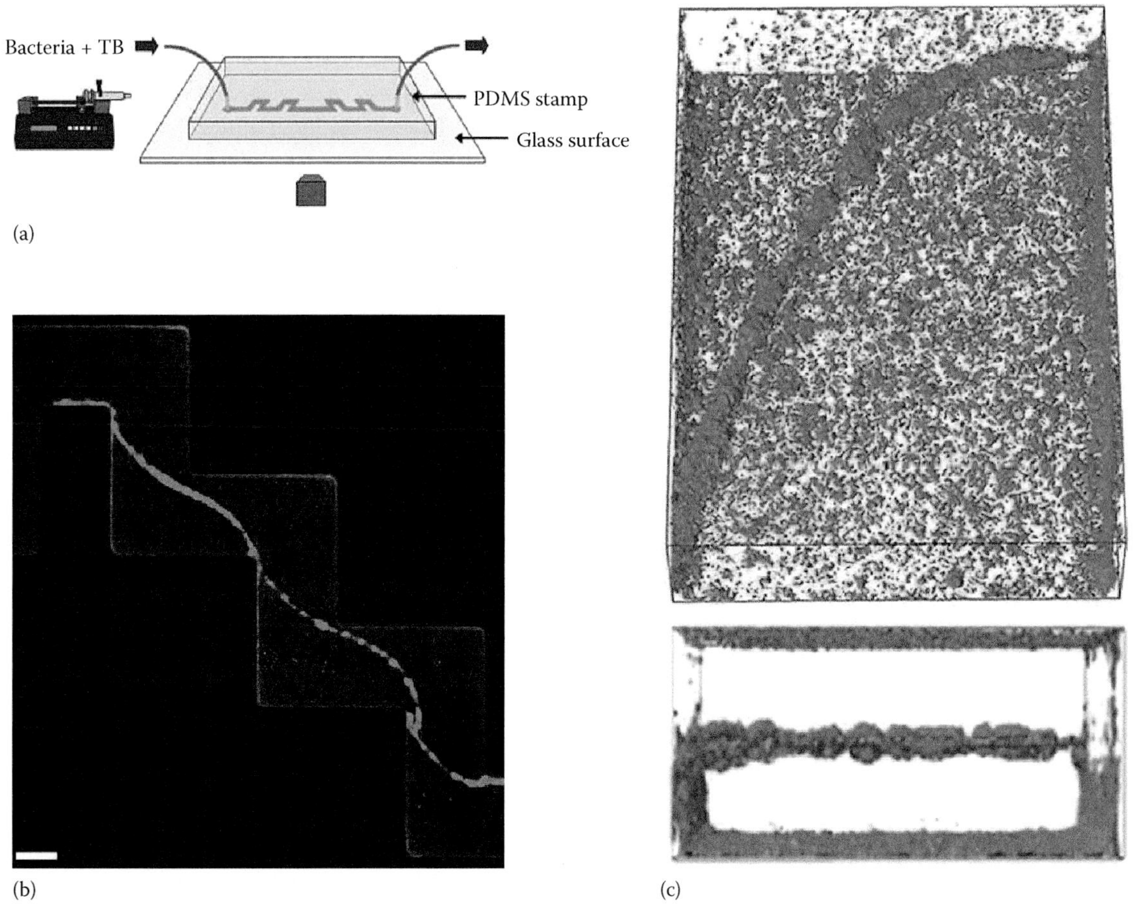

FIGURE 15.5 When a solution of bacteria flows through microchannels, biofilms not only grow on the surfaces but can also grow as filamentous structures, or streamers, in the flow. (a) Controlled experiments using microfluidic channels where the flow is visualized using different forms of microscopy. (b) Following many hours of flow of *Pseudomonas aeroginosa* PA14 containing green fluorescent protein, an image taken exactly in the middle of a channel shows a biofilm streamer, attached just downstream of many corners, and extending along the length of the channel. (c) A 3D confocal reconstruction shows both the (expected) biofilm on the channel boundaries and the (unexpected) biofilm streamer in the middle of the channel. Both perspective and horizontal views are shown. (From Rusconi, R. et al., *J. Roy. Soc. Interface,* 7, 1293, 2010.)

thread that was located exactly in the middle of the channel, where it was attached to the channel boundaries just downstream of almost every corner (Figure 15.5b and c). Such bacterial streamers will obviously have a significant impact on the flow. We have provided evidence that we interpret as strongly suggestive of an underlying hydrodynamic mechanism for the rearrangement of the biofilm to allow this development of the bacterial streamer (Rusconi et al. 2010). Hence, it is possible that bacterial streamers, that is, biomass that is not simply growing flat on surfaces but rather grows as thread-like structures in a flow, are much more common than previously imagined and may be a regular feature of many complicated flow systems.

15.4.3 Nano- and Micro-Particles That Affect Human Health

The rapid expansion of nanofabrication methods, including the many routes for synthesizing nanoparticles, have raised questions about the generic safety of these materials. In part, the concerns about toxicity and the impact on human health are associated with the reactivity of these materials, their ready dispersal since Brownian motion maintains the suspension of the material, and experimental observations of the uptake of various nanomaterials by cells. One area of nanoparticle science that has received significant attention for many years is themes associated with aerosols, in large part because of the adverse health effects caused by air pollution (e.g., see the report of the 2003 NSF Workshop on "Emerging Issues in Nanoparticle Aerosol Science and Technology").

In recent years, there have been systematic efforts to collect various types of nanoparticles and assess whether or not they are toxic to cells (e.g., Murr and Garza 2009). Much work remains in this area. The impact of fluid flows on these problems has not been addressed widely. Such flows may have some effect simply because we expect the flux of particles to a substrate to increase as the local shear rate increases.

15.5 Major Challenges

Perhaps one of the most significant challenges in the research community in this area of environmental transport is to link the different length scales relevant to many problems of interest. To take one example, transport of dangerous chemical or colloids in groundwater or sediments involves processes, possibly also involving bacteria and biofilms, at the scale of the granules in the soil or sediment to large-scale processes that spans tens to hundred of meters or more. These questions arise when considering pollution prevention and cleanup. Sometimes even knowing the relevant material and transport properties is difficult, which may introduce significant uncertainty into any solution strategy.

A second major theme concerns availability of clean water. The most common news in this area emphasizes the problem of the elimination of hazards to water sources and purification in the developing world. With respect to health threats posed by heavy metals in water, although basic kinetics and reactivity of surfaces is studied, systematic understanding appears to be lacking of the transport processes important in the natural environment or in treatment processes.

As the two examples involving microorganisms in Section 15.4 should make clear, there remains much to be learned about the manner in which microorganisms can interact with fluid flow. In one case, the interaction of the local shear rate and the swimming direction of the bottom-heavy algae produced localization of biomass, that is, it is a source of patchiness. In a second case, the interaction of the flow with a biofilm on a surface produced a bacterial streamer in the middle of a channel flow. Thus, surprises remain both in the natural environment as well as in engineered systems. Since such effects impact transport processes, these examples emphasize the need to recognize important roles that small-scale flows have in the natural and built environment around us.

After being invited to contribute a chapter on the subject of "micro- and nano-scale flows," I asked myself how such small-scale flows contribute to environmentally significant processes and I proceeded to structure the discussion around such topics. I am sure that I have missed significant themes, but perhaps the reader will find this attempt valuable in some ways for broadening either their view of environmental fluid mechanics, or their view of micro- and nano-scale flows.

Acknowledgments

I am grateful to my research group for their continual input and support. They have been a continual sounding board and source of inspiration. In particular, some of the ideas discussed in this chapter are based on research done with Laurent Courbin on wetting of surfaces; Jacy Bird, Laurent Courbin, and Rielle de Ruiter on bursting bubbles; and Laura Guglielmini, Sigolene Lecuyer, and Roberto Rusconi on biofilms. I would also like to thank Ian Griffiths for conversations about heavy metals in water; Anthony Aldykiewicz, Alexandre Gontcharov, Laura Guglielmini, and George Scherer for discussions about water transport in building materials; and Ian Griffiths and Janine Nunes for feedback on the draft of the chapter.

References

Batchelor, G.K. 1977. Developments in microhydrodynamics. In *Theoretical and Applied Mechanics: Proceedings of the 14th International Congress*, Delft, the Netherlands, pp. 33–55. North-Holland, Amsterdam, the Netherlands.

Bird, J.C., R. de Ruiter, L. Courbin, and H.A. Stone. 2010. Daughter bubble cascades produced by folding of ruptured thin films. *Nature* 465:759–762.

Bocquet, L. and E. Charlaix. 2010. Nanofluidics, from bulk to interfaces. *Chemical Society Reviews* 39:1073–1095.

Celia, M.A. and J.M. Nordbotten. 2009. Practical modeling approaches for geological storage of carbon dioxide. *Ground Water* 47:627–663.

Cheng, J.-T., L.J. Pyrak-Nolte, D.D. Nolte, and N.J. Giordano. 2004. Linking pressure and saturation through interfacial areas in porous media. *Geophysical Research Letters* 31:L08502–L08505.

Daniels, R., S. Reynaert et al. 2006. Quorum signal molecules as biosurfactants affecting swarming in *Rhizobium etli*. *Proceedings of the National Academy of Sciences USA* 103:14965–14970.

Durham, W.M., J.O. Kessler, and R. Stocker. 2009. Disruption of vertical motility by shear triggers formation of thin phytoplankton layers. *Science* 323:1067–1070.

de Gennes, P.G., F. Brochard-Wyart, and D. Quere. 2004. *Capillarity and Wetting Phenomena: Drop, Bubbles, Pearls, Waves*. Springer, New York.

Hall, C., A. Hamilton, W.D. Hoff et al. 2010. Moisture dynamics in walls: Response to micro-environment and climate change. *Proceedings of the Royal Society A*. Published online.

Homsy, G.M. 1987. Viscous fingering in porous media. *Annual Review of Fluid Mechanics* 19:271–311.

House, K.Z., D.P. Schrag, C.F. Harvey, and K.S. Lackner. 2006. Permanent carbon storage in deep-sea sediments. *Proceedings of the National Academy of Sciences USA* 103:12291–12295.

Kessler, J. 1985. Hydrodynamic focusing of motile algal cells. *Nature* 313:218–220.

Marle, L. and G.M. Greenway. 2005. Microfluidic devices for environmental monitoring. *TrAC Trends in Analytical Chemistry* 24:795–802.

Murr, L.E. and K.M. Garza. 2009. Natural and anthropogenic environmental nanoparticulates: Their microstructural characterization and respiratory health implications. *Atmospheric Environment* 43:2683–2692.

Neufeld, J. and H. Huppert. 2009. Modelling carbon dioxide sequestration in layered strata. *Journal of Fluid Mechanics* 625:353–370.

Probstein, R.F. 2003. *Physicochemical Hydrodynamics*. Wiley-Interscience, New York.

Rubin, H. and J. Atkinson. 2001. *Environmental Fluid Mechanics*. Marcel Dekker Inc., New York.

Rusconi, R., S. Lecuyer, L. Guglielmini, and H.A. Stone. 2010. Laminar flow around corners triggers the formation of biofilm streamers. *Journal of the Royal Society Interface* 7:1293–1299.

Shen, H.H., A.H.D. Cheng, K.-H. Wang et al. (eds.). 2002. *Environmental Fluid Mechanics: Theories and Applications*. American Society for Civil Engineers, Reston, VA.

Stickel, J.J. and R.L. Powell. 2005. Fluid mechanics and rheology of dense suspensions. *Annual Review of Fluid Mechanics* 37:129–149.

Stone, H.A., A.D. Stroock, and A. Ajdari. 2004. Engineering flows in small devices: Microfluidics toward a lab-on-a-chip. *Annual Review of Fluid Mechanics* 36:381–411.

Yavuz, C.T., J.T. Mayo et al. 2006. Low-field magnetic separation of monodisperse Fe_3O_4 nanocrystals. *Science* 314:964–967.

16

Volcanic Flows

J.H. Fink
Portland State University

A.B. Clarke
Arizona State University

16.1 Introduction

Volcanic eruptions are spectacular natural phenomena (Figure 16.1) that produce some of the most extreme physical and chemical conditions found on Earth. Many of the most critical data needed for predicting how volcanoes are going to behave are impossible to obtain because they are concealed underground, inside violent eruption clouds, within blistering hot lava flows, or at the bottom of the ocean. Some types of eruptions occur so infrequently that they have never been subject to direct scientific observation. Others take place hidden from view on the surfaces of other planetary bodies. Volcanologists and civil defense officials thus must rely on theoretical models and analyses to help infer these crucial states. As a result of this societal need to assess hazards, the fluid dynamics of volcanic activity has been subjected to intense scrutiny.

In this chapter, we use a fluids perspective to survey the life cycle of magma (molten rock) from its origin at depth to its ascent through volcanic conduits, its eruption into the atmosphere, and its flow across the Earth's surface. Each of these stages involves strong gradients in temperature, pressure, chemical composition, and rheology, leading to a dizzying number of variables and highly complex boundary conditions. Geologists must simplify these systems in their analytical and numerical models, in order to come up with useful relationships that can help determine the beginning, end, and intensity of eruptions. Here we describe a sampling of this fascinating set of behaviors, organized into three sections: (1) What happens at depth where magma forms, mobilizes, and rises to the surface, (2) what takes place when gas-rich magma erupts explosively, and (3) what occurs during relatively quiet effusive eruptions that produce lava flows and domes (Figure 16.2). In each section, we begin with a description of the geologic phenomena, then outline the relevant physical processes, and end with a few specific examples of how knowledge of fluid dynamics can help us interpret the geologic record and predict transitions in eruption style.

In contrast to many other applications of fluid dynamics, volcanology is directly tied to policy implications, because authorities responsible for assessing the dangers of erupting volcanoes must make decisions no matter how limited their scientific information. Being able to decipher physical and chemical clues about future eruptive behavior based on first principles, laboratory simulations, and numerical models is critical.

16.2 Principles

Eruption dynamics can be described by conservation equations, and each volcanic process or regime is best approximated by its own set of simplifying assumptions and constitutive relationships. For example, ascent of a bubbly, crystal-rich magma through the Earth's crust is often treated as a single isothermal fluid because all phases (gas bubbles, liquid magma, and crystals) have the same temperature, which remains approximately constant until it erupts at the surface (as is assumed in many studies such as Melnik and Sparks, 2002; de' Michieli Vitturi et al., 2008). Exceptions exist, such as near lateral boundaries of the flow where thermal exchange with surrounding rock is significant, or when temperatures evolve during ascent due to phase changes such as partial solidification via crystallization.

Handbook of Environmental Fluid Dynamics, Volume One, edited by Harindra Joseph Shermal Fernando. © 2013 CRC Press/Taylor & Francis Group, LLC. ISBN: 978-1-4398-1669-1.

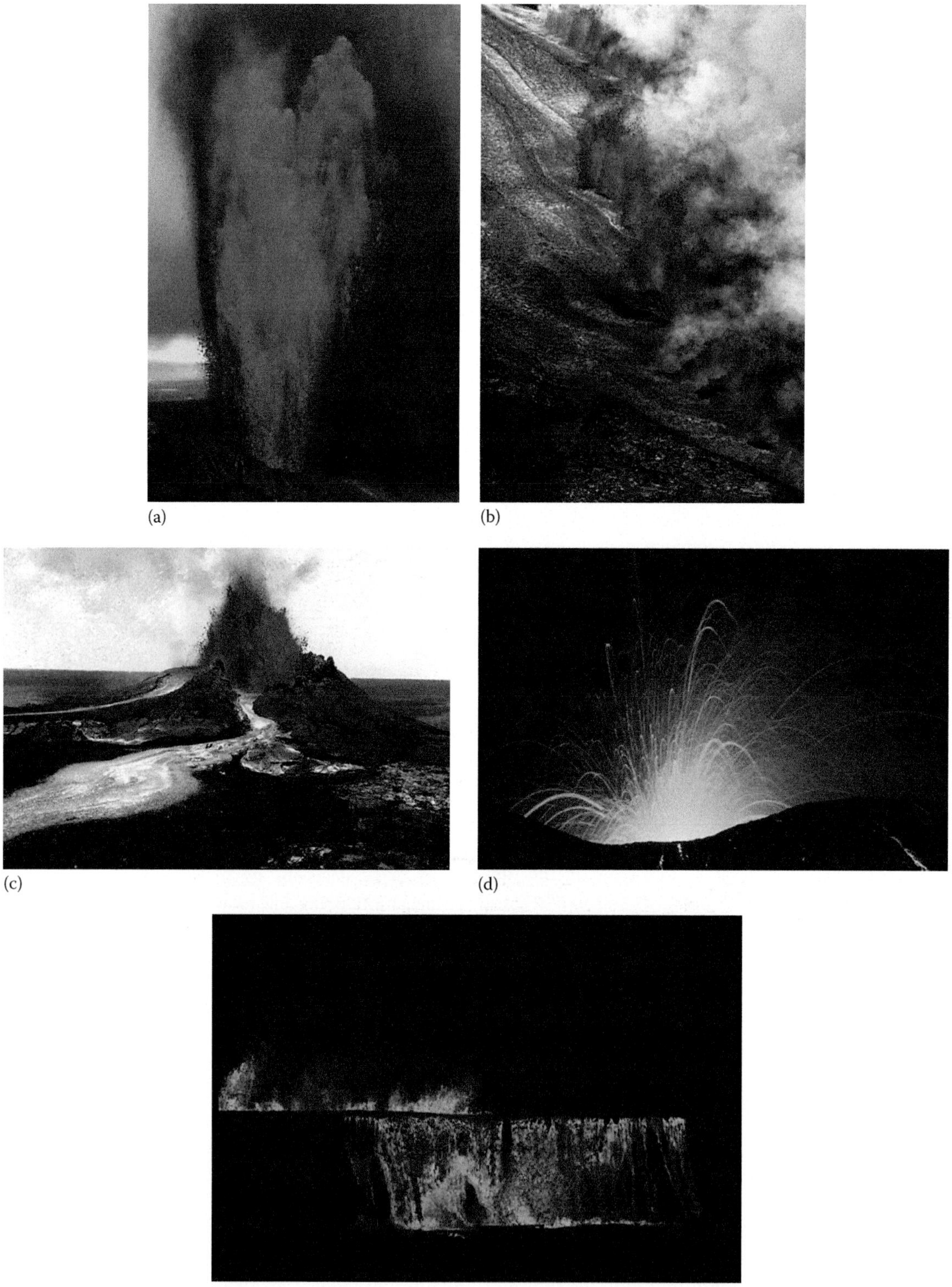

(a) (b) (c) (d) (e)

FIGURE 16.1 **(See color insert.)** (a) Spectacular fire-fountain at Kilauea Iki vent, Kilauea Volcano, 1959. (Photo courtesy of U.S. Geological Survey, Reston, VA.) (b) Aerial view of "curtain of fire" eruption, East Rift Zone, Kilauea Volcano, Hawaii. Gas-charged magma erupts along linear fissures, collects near the vent, and flows away as streams of solidifying lava, seen here as dark, subparallel channels moving to the upper left. Compare with Figure 16.3. (Photo courtesy of U.S. Geological Survey, Reston, VA.) (c) Vent of 1984 eruption of Mauna Loa volcano, Hawaii. Gas-rich, fluid magma rises many meters into the air, falls back, and builds up a rampart of solid clasts around the vent. The fluid lava flow emerges from a breach in the crater. (Photo courtesy of U.S. Geological Survey, Reston, VA.) (d) Classic Strombolian behavior at Stromboli Volcano. (e) Eruption of Kilauea Volcano showing high fountains in the background feeding extensive lava flows that are cascading over a cliff in the foreground. (Photo courtesy of U.S. Geological Survey, Reston, VA.)

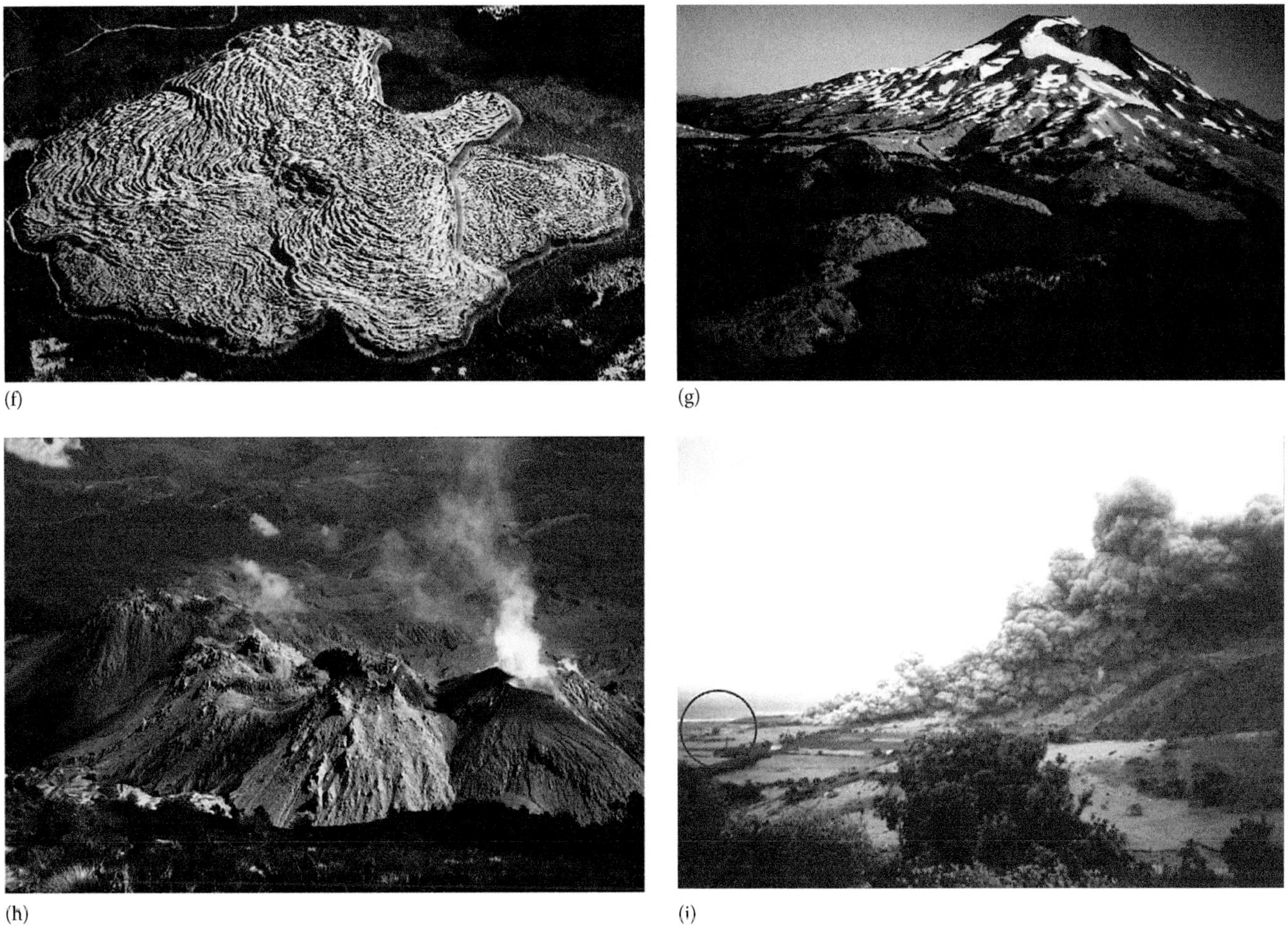

FIGURE 16.1 (continued) (f) Little Glass Mountain rhyolite lava flow, Medicine Lake Volcano, northern California. Approximately, 1000 year old, viscous lava flow showing prominent surface ridges formed by flow-parallel compression, amplified by buoyant rise of bubble-rich zones in the flow interior. Snow accentuates ridges. Diameter of flow is about 2 km. Ridges are about 10 m apart. (Photo by J.H. Fink.) (g) Devil's Hill rhyolite lava domes on the south flank of South Sister Volcano in the Cascade Range of central Oregon. Domes are assumed to have been fed by a planar, vertically oriented dike connected to a magma chamber at depth. Larger domes probably emerged from wider zones along the dike. (Photo by J.H. Fink.) (h) Santiaguito dacite lava dome complex, Santa Maria Volcano, western Guatemala. Coalesced chain of steep-sided domes, inferred to have been fed by an elongate dike that developed in the scar left by a huge eruption of Santa Maria in 1902. The Santiaguito domes have been erupting continuously since 1922. Steam marks site of active vent. Image is approximately 2.5 km across. Compare with (j). (Photo by J.H. Fink.) (i) Pyroclastic density current formed by a June 25, 1997, collapse or explosion of the Soufrière Hills volcano lava dome on the island of Montserrat, British West Indies. This pyroclastic flow generally followed topographically low drainages from the volcanic vent to this location approximately 7 km down slope. However, at various points along its path, the dilute and turbulent overriding surge surmounted topography and escaped the channel. Buoyant plumes are forming above the current and are particularly well developed approximately 1.5 km behind the flow front. Airport terminal building is circled at the left edge of the image for scale. (Photograph by A.B. Clarke and R.B. Watts.)

(continued)

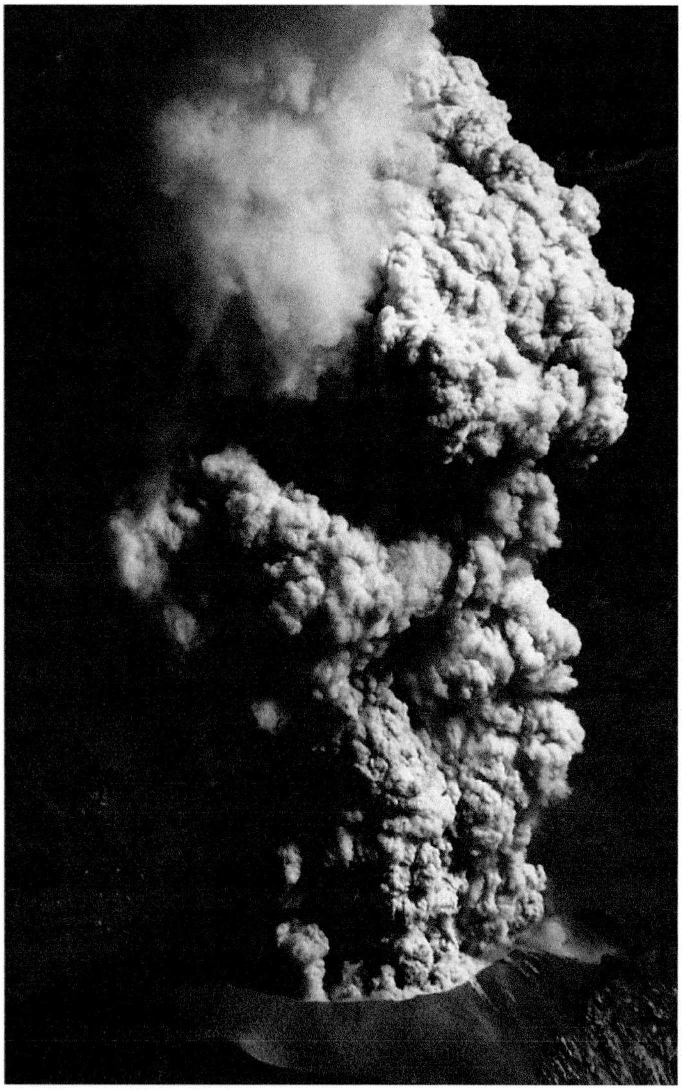

(j)

(k)

(l)

FIGURE 16.1 (continued) (j) Vulcanian eruption of ash from vent at the top of one of the Santiaguito lava domes shown in (h). Such explosions can occur as frequently as several times per hour and the resulting plume can rise several kilometers above the top of the dome. (Photo by J.H. Fink.) (k) Ascending eruption cloud from Redoubt Volcano as viewed from Kenai Peninsula (to the east). Gravitational collapse of Redoubt's lava dome produced pyroclastic density currents and a plume that reached nearly 10 km above sea level. The corresponding umbrella cloud at the top of the plume spread in a nearly circular pattern and distributed volcanic particles over a region many kilometers in diameter. (Photo courtesy of J. Warren, April 21, 1990 and the U.S. Geological Survey, Reston, VA.) (l) Vulcanian eruption with pyroclastic density currents (Soufrière Hills volcano). In this instance, the eruption simultaneously produced a buoyant plume that reached 15 km above sea level and pyroclastic density currents that initially spread radially in all directions and were then channeled into topographic lows to distances of up to 5 km. (Photograph by A.B. Clarke.)

Once eruption occurs, cooling at exposed surfaces can be rapid and plays a critical role in controlling the dynamics of explosive eruptions and lava flows and domes.

During many stages of magma or lava movement, it is often assumed that all phases move together with one velocity, requiring only a single momentum equation for the mixture as a whole. However, when melt viscosities are sufficiently low, crystals can sink and bubbles rise, which may have bearing on large-scale dynamics. Also, bubbles may become connected, forcing consideration of the movement of volatile phases through a permeable medium. Additionally, in explosive eruptions, small pieces of fragmented lava, called *pyroclasts*, are thought to be well-coupled to the gas phases, such that gas and fragments can be assumed to have the same velocity and temperature ($<$1 mm diameter; Woods, 1995). However, for larger grains, heat and momentum exchange between phases may affect large-scale

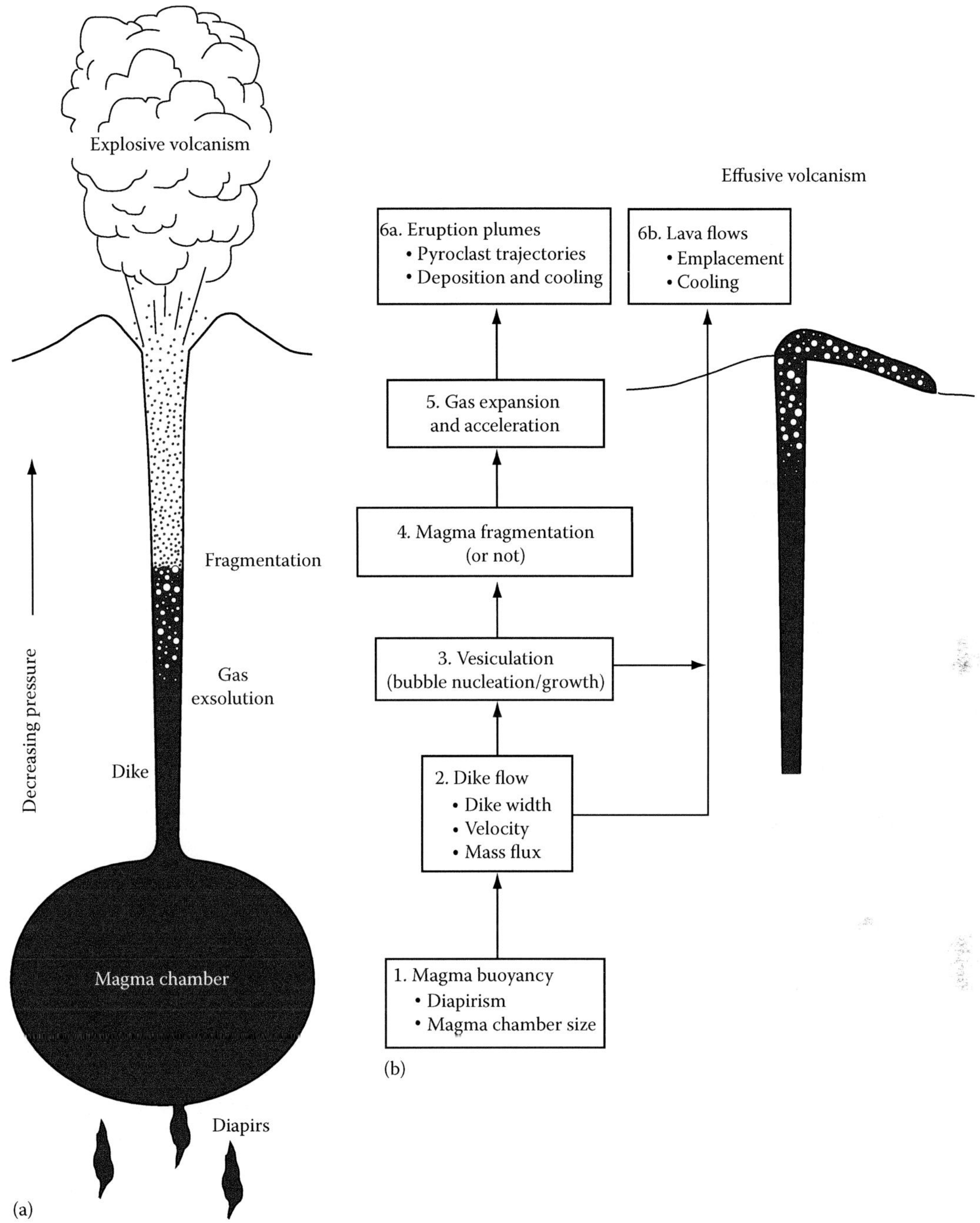

FIGURE 16.2 Schematic representation of the two main styles of volcanic eruptions. Explosive eruptions (a) are typically associated with volatile-rich, high-viscosity magmas, and are defined by run-away fragmentation which breaks bubbly magma into a gas–particle mixture. Gas-poor magma erupts effusively (b) forming either lava domes (high-viscosity magma) or lava flows (low-viscosity magma).

eruption evolution (Valentine and Wohletz, 1989; Wohletz and Valentine, 1990; Dobran et al., 1993; Neri and Macedonio, 1996). Empirically constrained constants and relevant flow variables are typically combined to construct inter-phase thermal and momentum exchange terms, but unfortunately relevant constants are difficult to define for complex volcanic systems.

Constitutive equations are needed for closure of any system of conservation equations. Magma viscosity, for example, is commonly assumed to be Newtonian, especially in the subsurface, and is a function of composition, temperature, dissolved water content, crystal fraction, and bubble size distribution and fraction (Murase, 1962; Murase and McBirney, 1973; Lejeune and Richet, 1995; Hess and Dingwell, 1996; Llewellin et al., 2002; Costa, 2005). Solubility of volatiles is a function of pressure and temperature (Moore et al., 1998) and dissolved volatiles vary accordingly, in both the vertical and horizontal directions depending on flow conditions. Crystals may grow as a result of magma degassing (Moore and Carmichael, 1998), and corresponding complex relationships affect large-scale dynamics (Sparks, 1997; Melnik and Sparks, 2002). Appropriate boundary and initial conditions are also required to obtain solutions and to define the system and scenario of interest. In most volcanic environments, fluid and particle characteristics, and flow regimes (particle concentration, Reynolds number, Richardson number, Mach number, Bingham number) change rapidly in space and time, making simplification an absolute must for efficient and practical solutions.

16.3 Methods of Analysis

Numerical models of eruption processes rarely consider 3D flow fields, and commonly ignore 3D topography, which is critical to realistic hazard assessment (exceptions include numerous lava flow studies and some explosive studies, e.g., Wadge et al., 1998; Esposti-Ongaro et al., 2008a,b). Therefore, at this stage, models are used largely to understand physical processes and to discover overarching relationships, with particular emphasis placed on transitions in eruption style and scale. As such, previous studies using mathematical (numerical and analytical) and analog modeling of volcanic processes have been invaluable in developing our understanding of highly time-dependent, eruptive behavior (e.g., Sparks, 1986), identifying parametric relationships (e.g., Wilson et al., 1978, 1980; Woods, 1988, 1995; Valentine and Wohletz, 1989), illuminating non-intuitive and unexpected behavior (e.g., Neri and Macedonio, 1996; Fink and Griffiths, 1998) and, in some cases, demonstrating an ability to reasonably replicate real eruption characteristics (e.g., Dobran et al., 1993; Griffiths and Fink, 1993; Wadge et al., 1998; Clarke et al., 2002; Esposti-Ongaro et al., 2002; Todesco et al., 2002; Widiwijayanti et al., 2009).

Fluid dynamics principles have greatly enhanced our understanding of volcanic eruptions, and may eventually play a critical role in hazard assessment by predicting inundation area and subsequent volcanic impact. However, demonstrating the ability of models to accurately capture time-dependent processes and serve in a predictive capacity to reduce risk and hazards remains a significant challenge to the community. Nonetheless, the physical relevance and practical utility of quantitative models of volcanic processes are continually improving via increased computational power, new solution algorithms, and a better understanding of the physics of complex fluid systems. But possibly the greatest progress may arise through formulation of constitutive relationships and development of new physical theory through laboratory experiments (e.g., Hulme, 1974; Huppert et al., 1982; Fink and Griffiths, 1990; Gladstone et al., 1998; Chojnicki et al., 2006), along with comparison and validation with real-time observations and deposits from natural events (e.g., Lipman and Banks, 1987; Clarke et al., 2002; Watts et al., 2002; Ripepe et al., 2005; Dufek and Bergantz, 2007a,b; Esposti-Ongaro et al. 2008b).

16.4 In the Beginning: Magma Movement at Depth

Much of our understanding of what happens beneath volcanoes must be inferred from observing the dissected interiors and dispersed remnants of ancient mountains, which have been eroded, transported, and exposed by rain, wind, and glaciers. Other insights come from the interpretation of geophysical measurements and geochemical analyses of volcanic products. The resulting picture is much more complicated than suggested by common, cartoon-like depictions showing spherical underground magma chambers connected to conical volcanoes through vertical soda straw-like conduits.

In reality, magma bodies form by an episodic combination of upward movement of hot material from the Earth's mantle and melting of cooler crustal rocks closer to the surface; coalescence and chemical evolution of the molten constituents into increasingly large subterranean pools and pockets; convective overturning and differential segregation within these bodies; buoyant migration of portions of this fluid by ductile displacement of overlying rocks, or brittle penetration along cracks; and eventual eruption of parts of these blob-like bodies at the surface. Upward movement can stop at any stage when magma encounters cool, brittle, and lighter crust; it may then accumulate and freeze at a level of neutral buoyancy.

Throughout this sequence, gases in the magma come out of solution (*exsolve*) in the form of bubbles that expand and deform as they or the surrounding magma rise; solid rinds form on the cooler solidifying margins of magma bodies; and crystals grow and settle to the bottom, float to the top, or remain in suspension. In different geographic regions defined largely by the slow movement of the Earth's tectonic plates and the ascent of hot plastic mantle plumes, some magma bodies may achieve tremendous size, with volumes of thousands of cubic kilometers of eruptible material, while others remain small because they evacuate more frequently or accumulate more slowly. At depth, the fluid dynamic factors of interest include the scale and intensity of magma convection and mixing, the causes and effects of crystallization, and the exsolution of gases. Intense temperature and rheological gradients develop near the margins. Hotter and less viscous magma from deeper in the Earth's crust or mantle can be injected into a chamber from below, triggering convection within one or both of the magma bodies, or mixing between the two different magmas. Eruption occurs when a magma develops sufficient pressure or buoyancy to rise through a new or existing path. The threshold for eruption depends not only on magma crystallization and volatile exsolution but also on the strength and elasticity of the rock surrounding the magma body (Phillips and Woods, 2002). In hotter sections where the magma remains most ductile, buoyant portions may extend upward as protrusions that rise toward the Earth's surface. In some cases, these intrusions may show a regular spacing that reflects Rayleigh–Taylor instabilities caused by density contrasts between the magma and the overlying zones.

For most models of magma formation, migration and eruption, all of this complexity needs to be greatly simplified; yet simple relationships can still be quite useful. For example, in interpreting magma body geometry, generalized shapes can be used to constrain the dynamic state of stress in the surrounding environment at the time of emplacement. In deeper portions of the crust and upper mantle, this stress tends to be relatively homogeneous, resulting in bodies that are more spherical or equant. Closer to the surface, crustal behavior becomes more brittle, and the shapes of magma bodies are more elongate to planar. In the shallowest zones beneath some volcanoes, magma travels through a scaffolding of subvertical and subhorizontal planar sheets referred to as *dikes* and *sills*, respectively.

Movement of magma through these channels tends to be incremental, with brittle fractures wedging open the surrounding *country rock*, and magma then filling the crack. In some cases, gases come out of solution in the void space ahead of the advancing magma fingers, applying additional pressure that forces the crack tip forward. This zone around the crack front tends to have very complicated, multiphase flow. The depth of the transition from ductile to brittle behavior depends on the ambient thermal gradient, heat flow from depth, and chemical composition of the host rock. Elastic processes may also accompany the brittle and ductile lateral advance of magma through sills. *Laccoliths* form when magma, slowly spreading horizontally along a crack, also pushes upward, causing the overlying layer to bend into a turtle-shaped intrusion; if unchecked by cooling, the bending of the overlying layers eventually exceeds their tensile strength, and vertically oriented fractures allow the piston-like rise of a circular plug of rock. The resulting intrusions (and subsequently uplifted landforms) can be hundreds of meters high, as in the Henry Mountains of southern Utah.

Away from the central vents of conical volcanic edifices, magma can emerge when dikes breach the surface, resulting in a segmented fissure, which may extend in a linear fashion as much as 10 km. Closer examination of this eruptive style and of eroded volcanic zones reveals that blade-like dikes can break down into smaller segments as they approach the surface. Individual sections tend to orient subparallel to each other, offset in a right- or left-stepping "en echelon" pattern, reflecting near-surface rotation of the stress field (Figure 16.3). This results in a "curtain of fire" eruptive style (Figure 16.1b), which can last for several hours before becoming concentrated in one or two more energetic, fountain-like sites along the dike

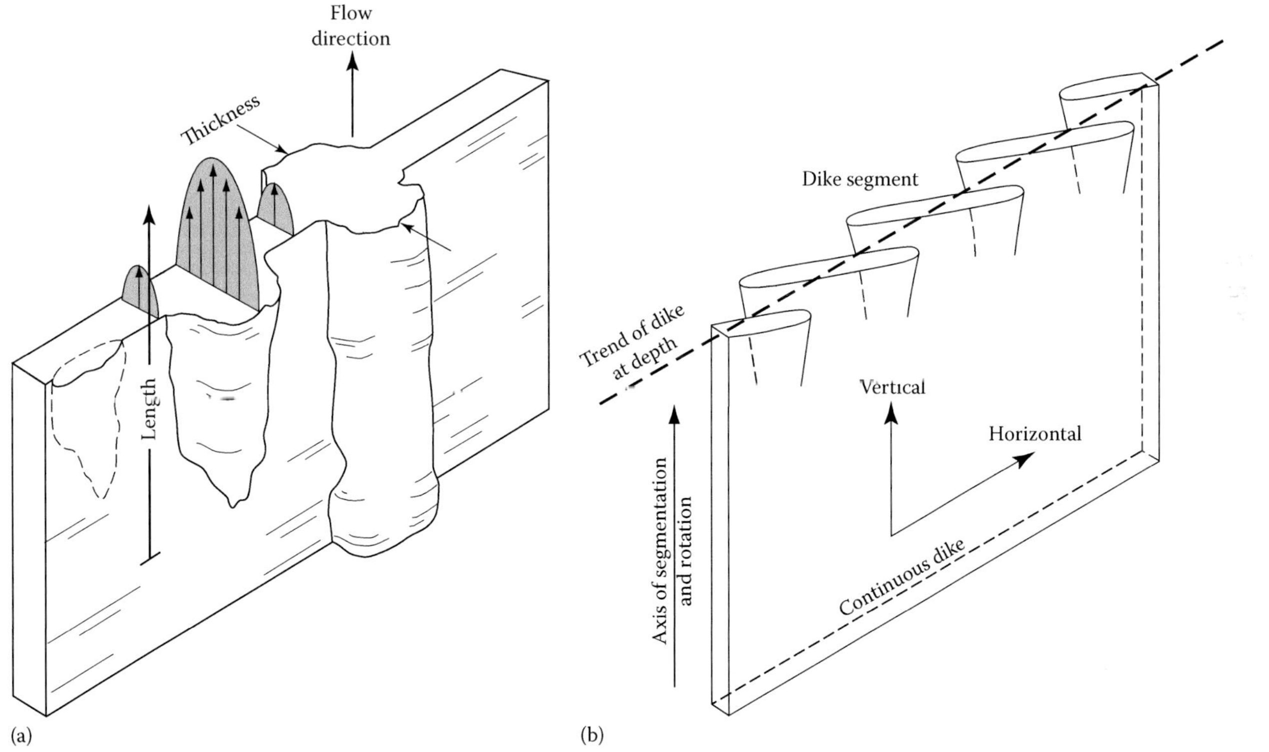

FIGURE 16.3 (a) Schematic diagram showing vertical flow of magma through a dike of variable thickness. Arrows indicate that higher flow velocity occurs in thicker portions of dike. Greater flux of magma keeps these thicker sections hotter, and thus flowing longer. (b) Schematic diagram showing how a continuous dike at depth can segment as it approaches the surface, with individual segments rotating around a vertical axis. Segment rotation results in subparallel orientation of the eruptive sections of the dike. Rotation of the segments is assumed to have been caused by vertical variations in the orientation of stresses. Curtain of fire eruptions commonly show this segmented structure. (Modified from Delaney, P.T. and Pollard, D.D., Deformation of host rocks and flow of magma during growth of minette dikes and breccia-bearing intrusions near Ship Rock, New Mexico, U.S. Geological Survey Professional Paper 1202, 61p., 1981.)

(Figure 16.1a, c, and d). The transition from elliptical to more circular eruptive vents is accompanied by erosion of brecciated or fragmented rock along the dike margins, and freezing of magma within narrower portions of the dike where cooling is enhanced.

These phenomena associated with ascent along a dike are most commonly associated with hotter and more fluid *basaltic* magma seen in places like Hawaii, but *silicic* compositions like at Mount St. Helens can also exhibit these kinds of intrusive relationships (other examples are shown in Figure 16.1g and h). Silica-rich *rhyolite* or *dacite* magmas have lower melting points and are more viscous than basalts, and thus have a greater tendency to congeal beneath the surface. This means that eruptions generally require thicker dikes in order to prevent freezing en route to the surface. Alternatively, exsolution of gases from the top of rising dikes may exert sufficient force to pry apart the overlying rocks, clearing a path to the surface (Blake and Fink, 1987). Silicic dikes exposed by erosion can also show echelon structure and segmentation, reflecting the state of stress in and around the volcanic edifice when it was active.

Intrusive magmatic processes thus involve a combination of ductile, brittle, and elastic behavior of multiphase materials with extreme variations in temperature and rheology, leading to highly complex geometries. Understanding what happens in the interiors of volcanoes requires simplifying assumptions about all of the key variables, coupled to fluid dynamic insights provided by numerical modeling and laboratory simulation.

16.5 Big Ones: Explosive Eruptive Processes

In a typical eruptive sequence, gases rise buoyantly toward the upper parts of a magmatic intrusion. This higher concentration of volatiles increases the likelihood for explosive eruptions, which continue until most of the gases driving the eruption have been expelled, at which point either the eruption ceases or its style shifts to quieter, more effusive behavior. Less commonly, lava-producing eruptions can switch back to explosive events due to local concentration of volatiles, or recharge of volatile-rich magma at depth. Determining when such transitions between explosive and effusive behavior are likely to take place is a major goal of volcanic hazard assessment. Depending on a volcano's vent geometry, volatile content, and the physical configuration of the magma body, explosive behavior can continue for days, weeks, months, or years.

When magma first breaches the Earth's surface, it is commonly charged with dissolved gases, increasing the tendency to erupt explosively. Low-viscosity basaltic magmas can yield spectacularly photogenic explosive eruptions (Figure 16.1a through e), but these generally do not pose great risks because they readily release their gas, tend to be short-lived, and are relatively localized. In contrast, silicic magmas hold their volatile phases more tightly, increasing the chances that ascent rate and internal pressure will reach dangerous levels.

For basalt lavas, which typically have viscosities on the order of 10^2–10^4 Pa s, a range of explosive eruption activity is feasible. *Strombolian* behavior occurs when bubbles coalesce, rise, grow, and finally burst near the surface, launching clots of partially molten bubble walls and finer-grained quenched lava fragments (pyroclasts) into the atmosphere. Bursting bubbles may be on the order of meters in diameter, and lava fragments typically fall tens to several hundreds of meters from the vent (Figure 16.1d). *Fire-fountaining* (Figure 16.1a), on the other hand, occurs when bubbles and magma rise at approximately the same rate, causing gas and lava to be ejected together at high speeds, creating fountain heights of up to several 100 m above the vent and distributing fine particles up to kilometers away. In general, the transition between Strombolian and fire-fountaining styles is controlled by the relative rise rates of magma and bubbles; for typical basalts, magma ascent slower than 0.5–1 m s^{-1} leads to Strombolian activity, whereas faster ascent causes fire-fountains (Wilson and Head, 1981). Deposits in the geologic record reflect eruption characteristics and thus provide information about unobserved volcanic events (Figure 16.4; Head and Wilson, 1989). For example, for basaltic eruptions, low pyroclast temperatures at the time of deposition result in stacks or cones of quenched loose cinders, whereas high pyroclast temperatures favor deposition of plastic and fluid grains. High accumulation rates and hot pyroclasts lead to welded collections of fluid lava clots, while still higher accumulation rates may result in secondary lava flows which form by remelting and remobilization of lava fragments (Figure 16.1e).

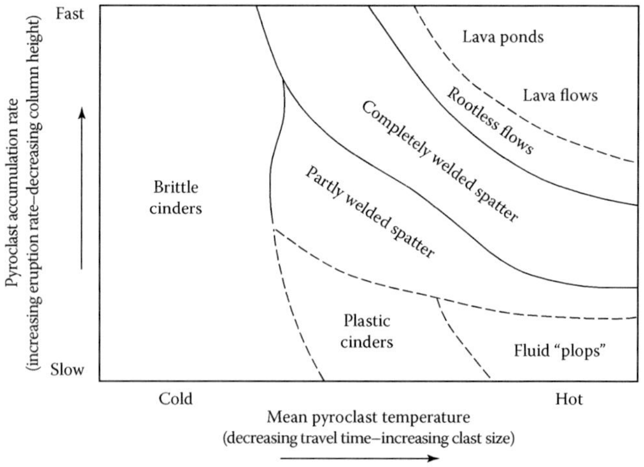

FIGURE 16.4 Relationship between eruption and deposit characteristics for explosive basaltic volcanism. Low pyroclast temperature at the time of deposition favors piles and cones of brittle, distinct cinders, whereas higher temperatures produce welded deposits. A combination of high pyroclast temperature and high accumulation rate may lead to secondary lava flows formed by wholesale remelting and remobilization of deposited pyroclasts. Brittle cinder deposits (left side of figure) are generally favored by the most violent basaltic eruptions, which tend to generate fine-grained pyroclasts and project them over long distances. (Modified from Head, J. and Wilson, L., *J. Volcanol. Geotherm. Res.*, 37, 261, 1989.)

Pyroclast temperature at deposition is proportional to particle size and inversely proportional to travel time; accumulation rate increases with increasing eruption rate and decreasing fountain height. Fine-grained pyroclasts and long travel times (high fountains) are indicative of highly explosive events, and accordingly, eruption violence generally decreases from left to right and from bottom to top in Figure 16.4.

In general, the most explosive volcanic eruptions (Figure 16.1i through l) are produced by a combination of high volatile contents and high viscosities (>10^3 Pa s) as occurring in rhyolite, dacite and crystal-rich basalt magmas. Under these conditions, magma and bubbles can be assumed to rise together with approximately the same velocity. Gas expansion and decreasing solubility of volatile phases increase the bubble content of magma as it ascends through the Earth's crust. These processes in turn accelerate magma ascent, which in some cases leads to high enough strain rates to break it apart. This defining feature of explosive volcanic activity, also called runaway *fragmentation* (Figure 16.2a), ruptures the bubbly magma into a mixture of solid quenched clots (pyroclasts) and expanding gas. Fragmentation occurs when high strain rates force magma through its *glass transition* (Dingwell, 1996, 2001), at which point it transforms from fluid-like behavior to brittle solid behavior. During ascent, magma may experience high strain rates due to large velocity gradients in both streamwise (Papale, 1999) and flow-perpendicular directions (Gonnerman and Manga, 2003). Fragmentation can also result from rapid bubble expansion due to large pressure differences between the gas phase and the surrounding magma (Alidibirov, 1994; Zhang, 1999; Spieler et al., 2004); high differential pressures can result from restricted bubble growth in a viscous medium (McBirney and Murase, 1971; Sparks, 1978) or due to rapid "uncorking" of a volcanic conduit (Alidibirov, 1994). The glass transition occurs when elongational or shear strain rate (typically expressed in terms of vertical and radial velocity gradients, respectively) exceeds the ratio of the magma's elastic or shear modulus to its viscosity (times an empirical constant). Moduli, viscosities, and constants depend on magma composition, temperature, and crystal and volatile contents, all of which can change rapidly in space and time. Fragmentation thresholds are thus difficult to constrain and quantify.

Once an explosive eruption initiates, the flow is divided into two zones: a lower bubbly region and a gas–particle mixture above a surface referred to as the *fragmentation front*. This front is commonly treated as a discontinuity across which the conservation of mass, momentum, and energy for the different phases of the mixture must be satisfied. In steady explosive eruptions (*Plinian* or *sub-Plinian* depending on degree of violence), the rising magma steadily (or nearly steadily) feeds the fragmentation front. In short-lived *vulcanian* explosions (commonly modeled as modified shock tubes), the fragmentation front is thought to be nonstationary, traveling down the conduit behind a decompression wave, disaggregating deeper and deeper portions of the magma column until it finally reaches magma that is too fluid or bubble-poor to break apart. Behind (above) the front, the gas–particle mixture exits the vent as a turbulent jet or plume, which may have Reynolds numbers ranging from 10^5 to 10^8 or even higher in some cases.

What happens next in the jet or plume depends largely on three conditions: (1) exit velocity, (2) mixture density at the vent, and (3) vent diameter. Velocity and mixture density in turn depend largely on the proportion and pressure of magmatic volatiles at the time of fragmentation, which represent the total compressed gas energy. The initial turbulent flow, commonly dominated by momentum (Figure 16.5, *gas thrust region*; Sparks, 1986), entrains the atmosphere at its edges, and this mixing leads to heating and expansion of the entrained air and to increasing buoyancy flux. Simultaneously, the flow decelerates due to gravity, drag forces, and waning expansion of compressed volcanic gases. At some point above the vent, the flow exhausts its momentum and gravity takes over. If buoyancy at this point is positive, the mixture enters the *convective region* (Figure 16.5) and forms a buoyant plume. Otherwise, the column collapses to form *pyroclastic density currents* (Figure 16.1i and l). Buoyant plumes are favored by high vent velocities, small vent diameters, and low initial mixture densities, all of which promote efficient entrainment and subsequent development of buoyancy (Figure 16.6a through c), (Wilson et al., 1980). The presence of large-sized particles (>1 mm), which are not thermally or physically well-coupled to the gas phase, pushes the system toward collapse (Wohletz and Valentine, 1990), as do overpressured vent conditions and supersonic flow, which suppress entrainment (Brown and Roshko, 1974; Valentine, 1998).

Buoyant volcanic plumes rise in the Earth's stratified atmosphere, and due to that stratification, eventually reach a height of neutral buoyancy where they spread radially in what is termed the *umbrella region* or cloud (Figures 16.1k and 16.5; Carey and Sparks, 1986; Sparks, 1986). By assuming that most of an eruption column is dominated by buoyancy effects, the system can be simplified as a thermal plume in a stratified environment (Morton et al., 1956). Accordingly, the final height is proportional to the fourth root of the energy flux at the vent, which can be expressed in terms of mass flux, thermal properties of the pyroclastic mixture, and the temperature difference between the atmosphere and the plume (Wilson et al., 1978). Although measurements of mass (or normalized volume) eruption rate are rare and contain significant uncertainty, this relationship is generally supported by field estimates (Wilson et al., 1978; Figure 16.6d) and therefore can be rearranged to constrain eruption rate for real eruptions using observable column heights. Upon reaching the height of neutral buoyancy, the umbrella region spreads radially and ash and pyroclasts fall out over long distances, which in some cases can span thousands of kilometers. In a still atmosphere, the resulting depositional footprint is circular, with particle size and deposit thickness decreasing in a bull's eye pattern with distance from source. In a windy atmosphere, the pattern can be highly elongate. The normalized extent of these dispersal patterns, along with pyroclast size, are used to qualitatively classify explosive eruptions (Walker, 1973), and can be used to reconstruct column height for unobserved volcanic events (Carey and

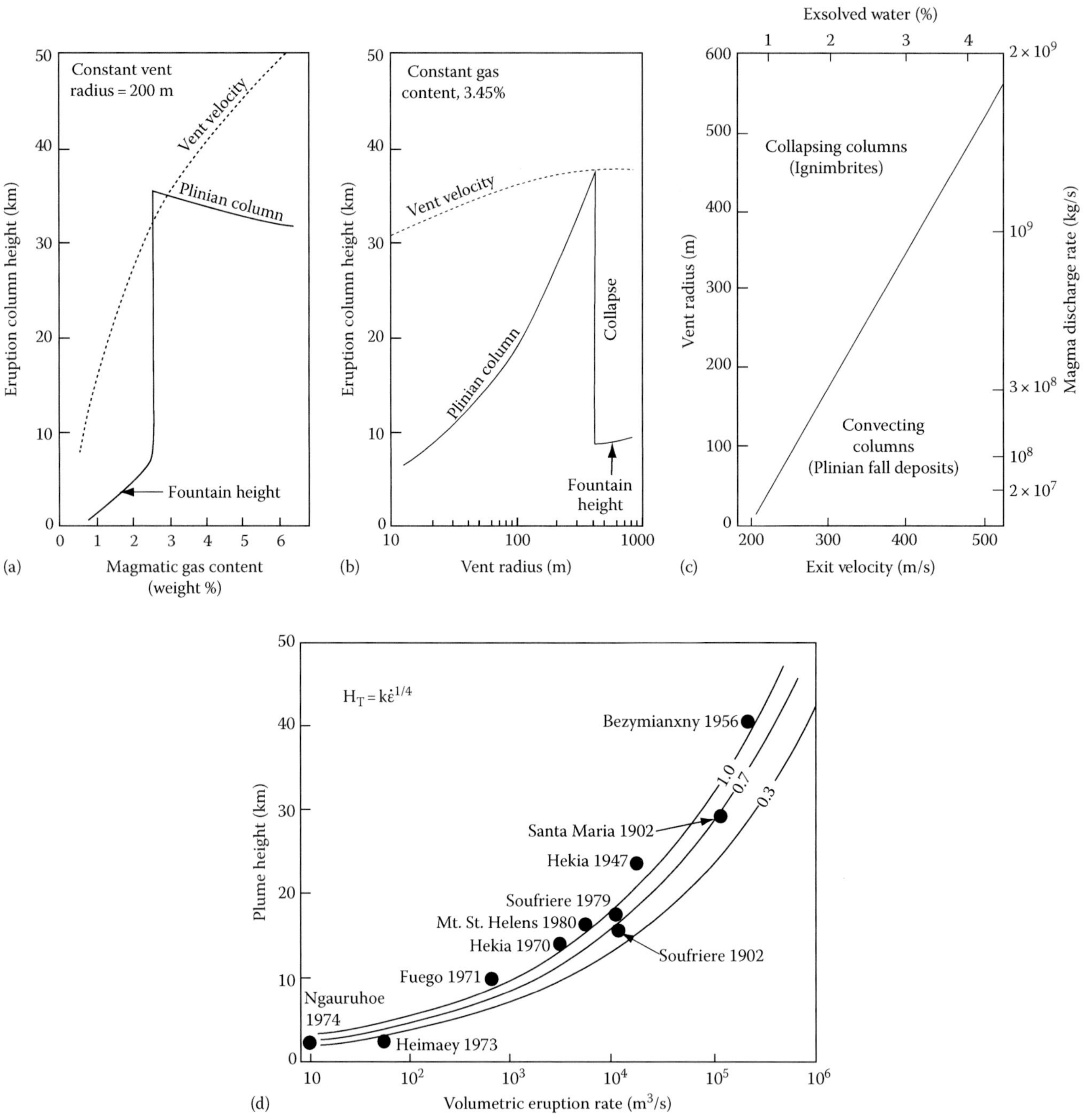

FIGURE 16.5 (a–c) General trends for large explosive eruption columns, derived from 1D, steady-state solutions of Modified from Wilson et al. (1980). (a) Eruption column height as a function of magma gas content for a constant vent radius of 200 m. Gas contents less than approximately 2.5 wt% result in column collapse and pyroclastic density currents. Note that vent velocity scales with gas content. (b) Eruption column height as a function of the log of vent radius, for a constant volatile content of 3.45 wt%. Column height increases with vent radius until nearly 400 m, beyond which the column collapses. (c) Threshold between two possible regimes, buoyant and collapsing columns, in terms of exit velocity (controlled by volatile fraction) and vent radius and corresponding mass discharge rate. Column collapse becomes more likely as radius and discharge rate increase, and as volatile content and vent velocity decrease. (d) Comparison between column height and eruption rate (normalized volume flux) for models of eruption plumes Modified from Wilson et al. (1978) and several real eruption data sets. Eruption data are generally captured by a simple semi-empirical relationship, $H = k\varepsilon^{1/4}$ where H is column height, k is a semi-empirical constant, and ε is the thermal energy flux. Slight variations in trends may be explained by differences in the proportion of pyroclasts in thermal equilibrium with the gas phase, as indicated by curve labels, and accounted for by an efficiency factor in the equation. (a–c: Modified from Wilson, C.J.N., *J. Volcanol. Geotherm. Res.*, 8(2–4), 231, 1980; d: Modified from Wilson, L. et al., *J. Geophys. Res.*, 83, 1829, 1978.)

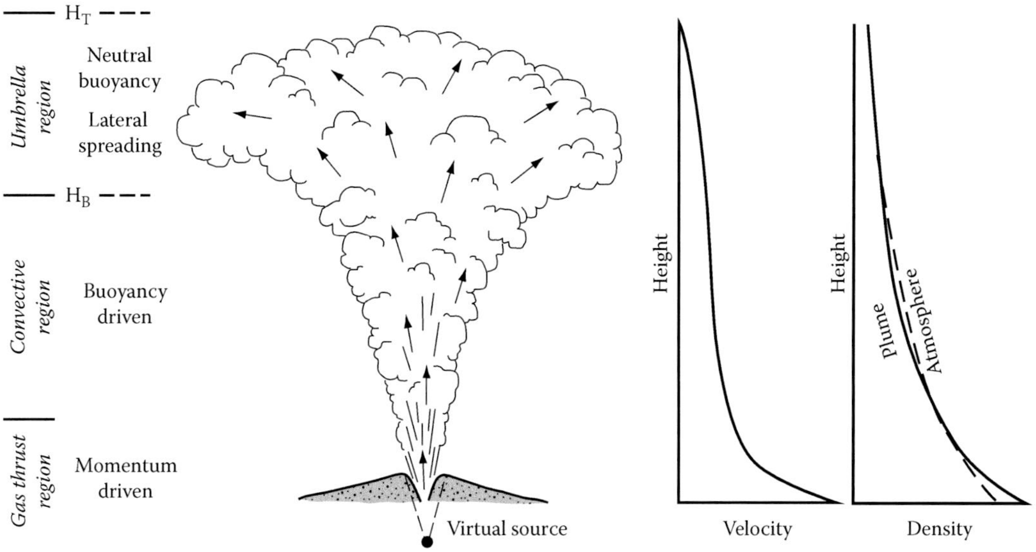

FIGURE 16.6 Conceptual model of eruption columns. Eruption columns are commonly modeled as buoyant or forced plumes and consist of three main regions. In the *gas thrust region*, nearest the vent, the gas–pyroclast mixture may be denser than the atmosphere and flow is driven primarily by momentum created by expanding volcanic gases. Due to waning expansion of volcanic volatiles and gravitational and drag forces, eruption columns decelerate in this zone. Due to entrainment and subsequent expansion of ambient air, they simultaneously gain buoyancy. When momentum is exhausted, the eruption column may be less dense than the ambient air (as shown in this figure) and enter the *convective region*, where flow is driven by positive buoyancy forces. Upon reaching a height of neutral buoyancy in a stratified atmosphere, an eruption column will spread radially in the *umbrella region*. (Modified from Sparks, R.S.J., *Bull. Volcanol.*, 48(1), 3, 1986; Carey, S. and Sparks, R.S.J., *Bull. Volcanol.*, 48(2–3), 109, 1986.)

Sparks, 1986). The far-reaching dispersal associated with the umbrella region also presents significant problems for air traffic, because the exact dispersal pattern is difficult to predict due to complex interaction with the atmosphere and because volcanic plumes cannot be detected by on-board instruments (Grindle and Burcham, 2003).

Pyroclastic density currents form a continuum between concentrated *pyroclastic flows* within which grain to grain collisions are frequent, and dilute turbulent *pyroclastic surges*. Observations of active flows and deposits suggest that under many circumstances both occur simultaneously (Figure 16.7), with the lower portion of the current (often called the *basal avalanche*) behaving as a granular flow, above which rides a dilute surge. At the interface between the current and the atmosphere, air is entrained and heated and buoyant plumes may develop (Figures 16.1i, 16.7, and 16.8d). Pyroclastic density currents can form by collapse or explosion of a lava dome or collapse of an eruption column as described earlier (Figure 16.8). Propagation of such currents ceases when they become less dense than the surrounding atmosphere; current density is reduced via particle sedimentation and entrainment of ambient air. Total distance traveled can range from a few to many tens of kilometers in the largest eruptions in the geologic record.

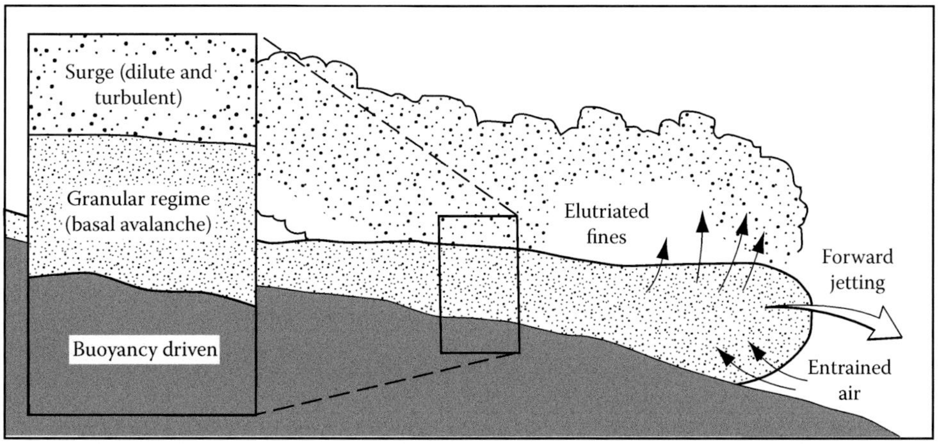

FIGURE 16.7 Pyroclastic density currents are thought to consist of a dense, granular lower zone (basal avalanche), over which a dilute turbulent surge rides. Entrainment of ambient air occurs at the margins, and the flow front and buoyancy may develop and *elutriate* fine particles into the upper regions of the current. (Modified from Francis, P. and Oppenheimer, C., *Volcanoes*, 2nd edn., 2004, Oxford University Press.)

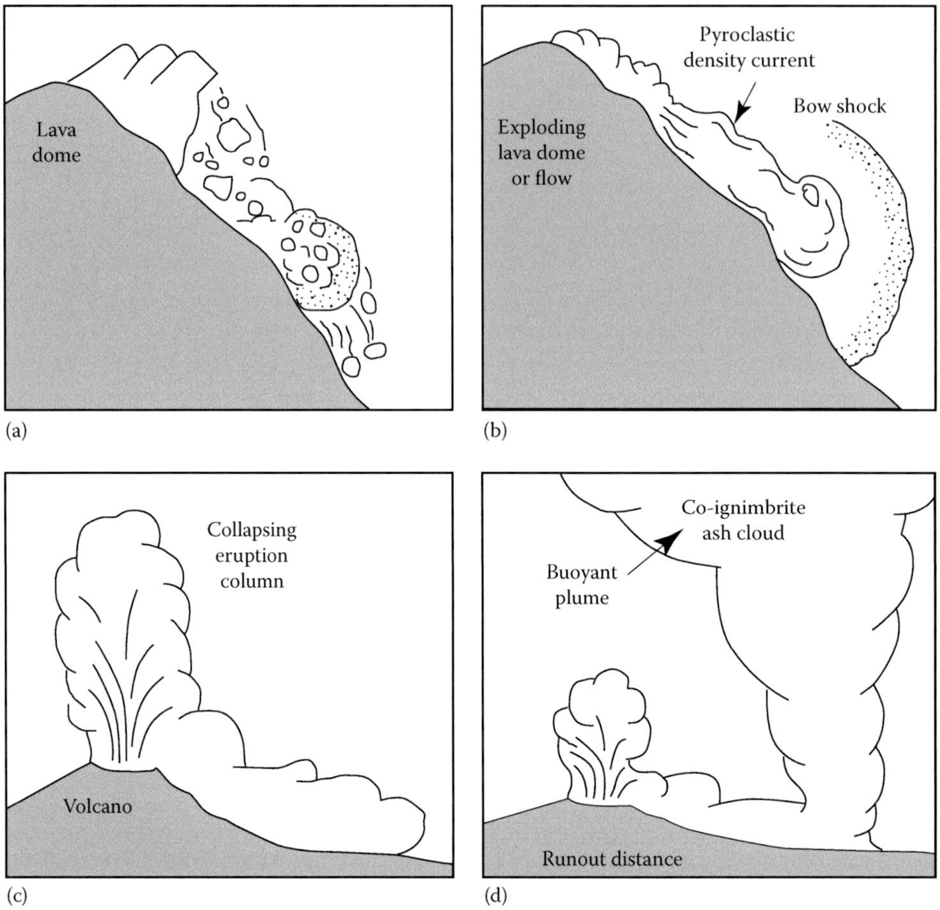

FIGURE 16.8 Four mechanisms for generating pyroclastic currents. (a) Passive (gravitational) collapse of a lava dome sends a cascade of rocks and ash down the slope of the volcano, with only minimal, localized explosive activity. (b) Explosive decompression of a pressurized lava dome interior releases gas, ash, and blocks in a violent, superheated, pyroclastic density current that can race down the side of the volcano at hundreds of kilometers per hour and at hundreds of degrees centigrade. (c) Gravitational collapse of an explosive eruption column can result in the formation of one or more pyroclastic currents that travel down the flanks of the volcano. (d) Pyroclastic flows emerging from a volcanic vent can segregate as they move down slope, with denser material hugging the ground, and hotter, finer-grained material rising as a buoyant plume. Propagation ceases when the current becomes less dense than the ambient air due to sedimentation and entrainment. (Modified from Francis, P. and Oppenheimer, C., *Volcanoes*, 2nd edn., 2004, Oxford University Press.)

Runout distance increases with increasing mass (or volume) and with decreasing particle size and entrainment rate. The basal avalanche portions of pyroclastic currents do not easily surmount topographic obstacles and are readily channeled by volcanic topography, which tends to significantly extend runout distance. Pyroclastic surges, on the other hand, are less easily contained by topography and have been observed to detach from the basal avalanche and scale topographic features, as occurred with fatal consequences at Mt. Lamington in Papua New Guinea in 1951, Mt. Unzen in Japan in 1991, and Merapi volcano in Indonesia in 1994.

Quantifying the complex relationships between explosive eruption phenomena and the deposits they leave behind is crucial because close observations of these violent processes are very difficult to make. Furthermore, several classes of eruptions (including the largest, which form giant depressions called *calderas*, as occurred in the area of Yellowstone National Park, Wyoming, USA), have only occurred in the prehistoric geologic past. Thus understanding the wide range of possible volcanic behavior requires interpretation of the geologic record in terms of eruption dynamics. For example, the degree to which a pyroclastic current is fluidized during transport and emplacement is reflected in the grain-size distribution of the resulting deposit (Figure 16.9) (Wilson, 1980, 1984). The highest-velocity flows tend to be the most fluidized and are often rich in fine particles because they were subject to highly energetic fragmentation. These flows tend to produce sorted deposits in which large, low-density *pumice* floats to the top whereas large, high-density grains sink to the bottom (Type 3, Figure 16.9). In contrast, slow-moving, non- or weakly fluidized flows show little sorting of coarse clasts (Type 1, Figure 16.9). Very large pyroclastic density current deposits associated with caldera formation, called *ignimbrites*, may be emplaced so quickly and at such high temperatures that the individual fragments within the deposit become fused or welded together. Large grains can become flattened into wavy flame-like shapes (*fiame*), and the deposits

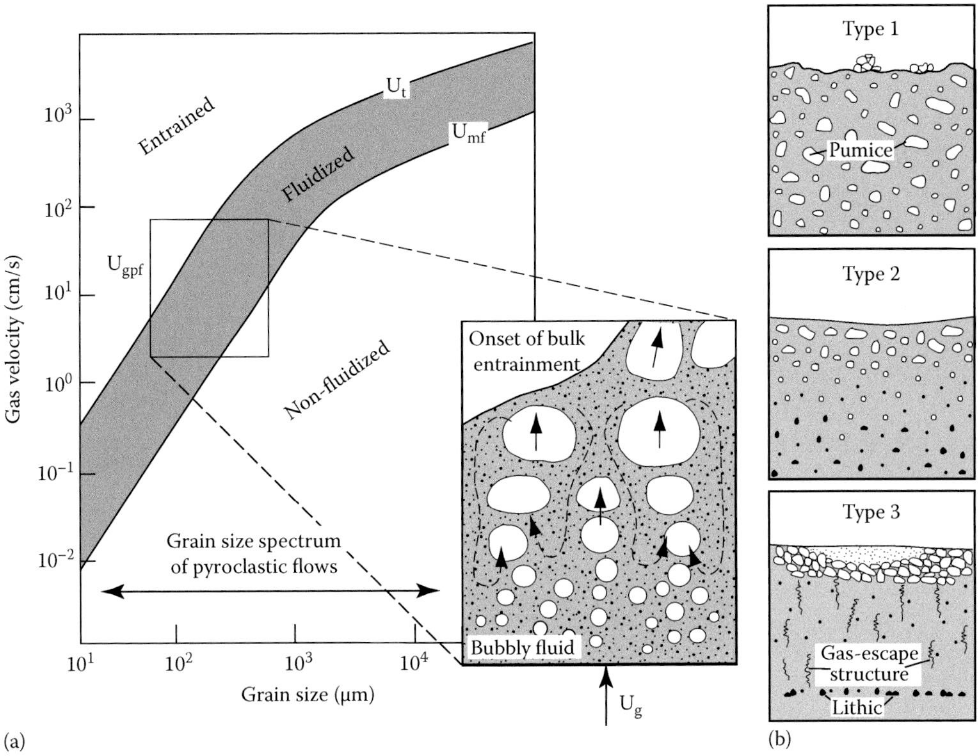

FIGURE 16.9 Relationship between pyroclastic density current deposit characteristics and degree of fluidization during emplacement, expressed in terms of vertical gas velocity and particle size (a). Fluidization occurs in the grey swath in the center of the plot, and the highest degrees of fluidization favor Type 3 deposits (b) (bottom right). Lower gas velocities and larger particles (bottom right sector of plot in part a) do not favor fluidization and result in Type 1 deposits (top, right). Type 2 deposits represent an intermediate case. A combination of high gas velocities and small particles (top left sector of plot in part a) lead to entrainment and lofting of particles into an overriding buoyant plume. (Modified from Wilson, C.J.N., *J. Volcanol. Geotherm. Res.*, 20(1–2), 55, 1984.)

can be difficult to discern from coherent quenched lava. As one might predict, welding tends to be most intense in the interior of a deposit, while the base and top of a *welded ignimbrite* may resemble an ordinary pyroclastic deposit.

Another important style of explosive volcanism, *phreatomagmatism*, occurs when magma and water interact. The resulting eruption behavior is controlled by the mass ratio of water to magma, which determines the amount of thermal energy available to vaporize liquid water to a violently expanding gas phase (Figure 16.10, Wohletz and McQueen, 1984). Very low ratios (<0.1) result in ordinary volcanic eruptions as described elsewhere in this chapter, whereas very high ratios (~100 or greater) suppress explosion due to cooling by the water; deep submarine eruptions represent this high-ratio end-member. Violence peaks for ratios between 0.1 and 1, as illustrated in highly energetic eruptions of the Icelandic volcano Surtsey in the mid-1960s.

16.6 Going with the Flow: Effusive Eruption of Lavas

Lava flows are the most closely observed form of volcanic activity because they are relatively safe and accessible in places like Hawaii and Italy. Lavas can be quite voluminous and can remain near their eruptive vents for thousands or even millions of years before being eroded away; thus they can provide unique preserved insights about magmatic processes that occurred hidden from view deep below the surface, or in prehistoric obscurity. Lava flows have distinctive surface structures, textures, and shapes, all of which provide useful evidence of eruptive behavior. Fluid dynamics plays a key role in lava flow interpretation, because such flows are relatively straightforward to simulate, both in the laboratory using analog materials and with theoretical and analytical models. Fluid properties also influence the growth of bubbles and crystals, which affect a flow's surface appearance and its transport behavior. Lavas of contrasting compositions can look quite different from each other, yet they may develop according to similar principles.

From a hazards standpoint, lavas of any composition pose relatively localized dangers by engulfing or overrunning populations and structures in their immediate paths. Basaltic lavas, with relatively low viscosities, advance rapidly along paths that are generally constrained by pre-existing topography. Silicic flows and domes, which generally move slowly and predictably, can place downstream communities and facilities at risk if they suddenly switch back to explosive behavior. Heavy rainfall on a fresh lava flow can generate destructive mudflows that can

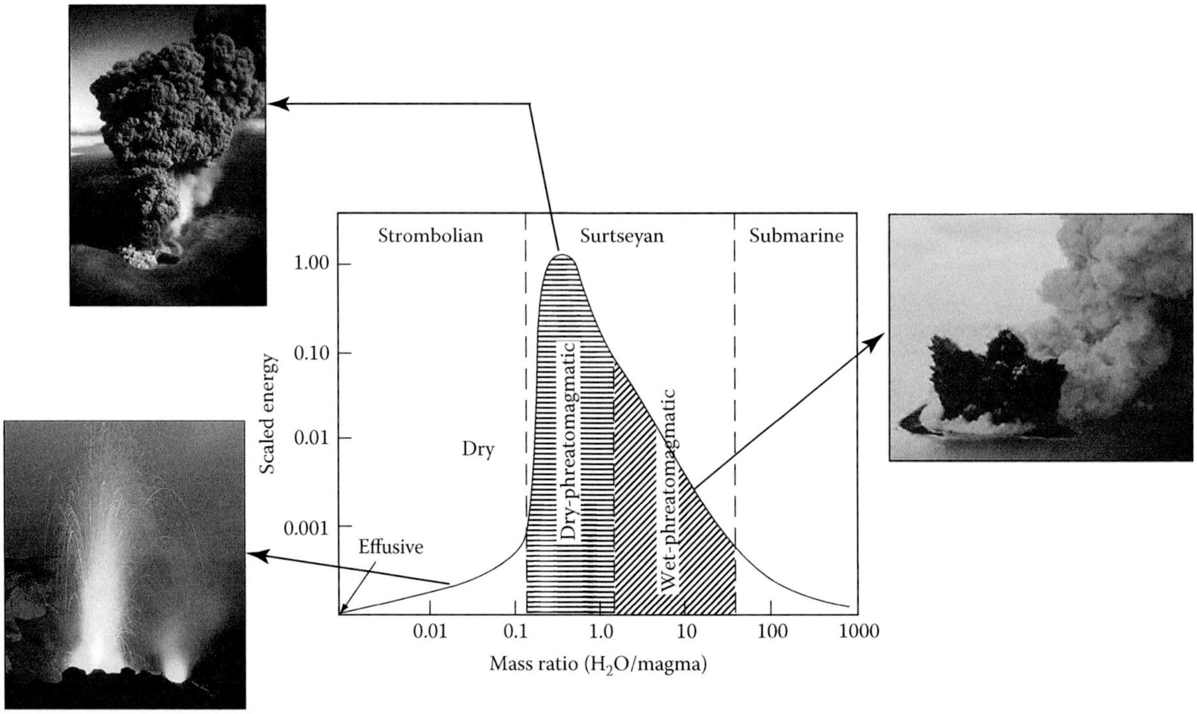

FIGURE 16.10 Proposed relationship between water–magma mass ratio and eruption energy and style. Low ratios (<0.1) produce typical magmatic eruptions, while large ratios (>100) result in rapid quenching of lava and suppression of explosions, as is characteristic of submarine eruptions. Intermediate ratios result in violent phreatomagmatic eruptions, with the lower ratios within this range producing the most violent phreatomagmatic eruptions, which are called Surtseyan after the famous Icelandic volcano and its typical eruptions of the 1960s. (Modified from Wohletz, K.H. and McQueen, R.G., Experimental studies of hydromagmatic volcanism, in *Explosive Volcanism: Inception, Evolution and Hazards*, National Research Council, National Academic Press, Washington, DC, 1984.)

rapidly travel long distances, knocking down bridges and other structures in their paths. Similarly, lava flows or pyroclastic flows erupted onto snow or ice can cause rapid melting and catastrophic flooding at great distances from the volcanic vent. The explosive eruption of Nevado del Ruiz Volcano in Colombia in 1985 partially melted a glacier, and the ensuing mudflow killed 29,000 people in towns and villages tens of kilometers away. Fluid dynamics can explain the mixing phenomena that control the extent of melting that occurs when hot magma encounters snow or ice.

One of the most widespread uses of fluid dynamics in volcanology is in the interpretation of the physical properties of lavas, which in turn constrain their hazards potential. For instance, the lengths and speeds of flows are controlled by the volume of eruptible magma, the eruption rate, underlying topography, and composition-dependent viscosity and yield strength. Most of these variables can be very difficult to estimate in field situations. Effusion rates can only be determined if one knows the cross-sectional geometry of the vent or of the advancing flow, both of which become obscured once an eruption begins. Rheological properties can be inferred through interpretations of surface structures and textures. For example, many flows of all compositions exhibit regularly spaced, transverse *surface ridges* that form in response to a fluid instability caused by flow-parallel compressive stresses. Ridge spacing can be used to back out the temperature- and depth-dependent gradients of viscosity, because wavelengths tend to be proportional to the thickness of the cooling surface crust. These gradients can in turn be related to composition (Fink, 1980a; Gregg et al., 1998). Similarly, many lava flows are bordered by elevated stationary *levee* structures, whose dimensions relate to the lava's temperature-dependent yield strength (Hulme, 1974; Figure 16.11). Levee dimensions have been used to infer the rheological and chemical properties of lavas on the Moon and Mars (Moore and Schaber, 1975; Moore et al., 1978) as well as on Earth (e.g., Fink and Zimbelman, 1990).

Although less destructive than primary explosive eruptions that emerge directly from a volcanic vent, secondary explosions that come out suddenly from the front of a silicic lava flow can be locally deadly. One explanation for why silicic flows revert to explosive behavior is that gases in the advancing flow migrate internally and become concentrated beneath the stiff surface crust, eventually building sufficient pressure to explode violently out of the flow front or through the surface forming, respectively, pyroclastic flows that race downhill, or explosion craters that can pockmark the flow surface. Although it is impossible to measure the internal volatile content of an active

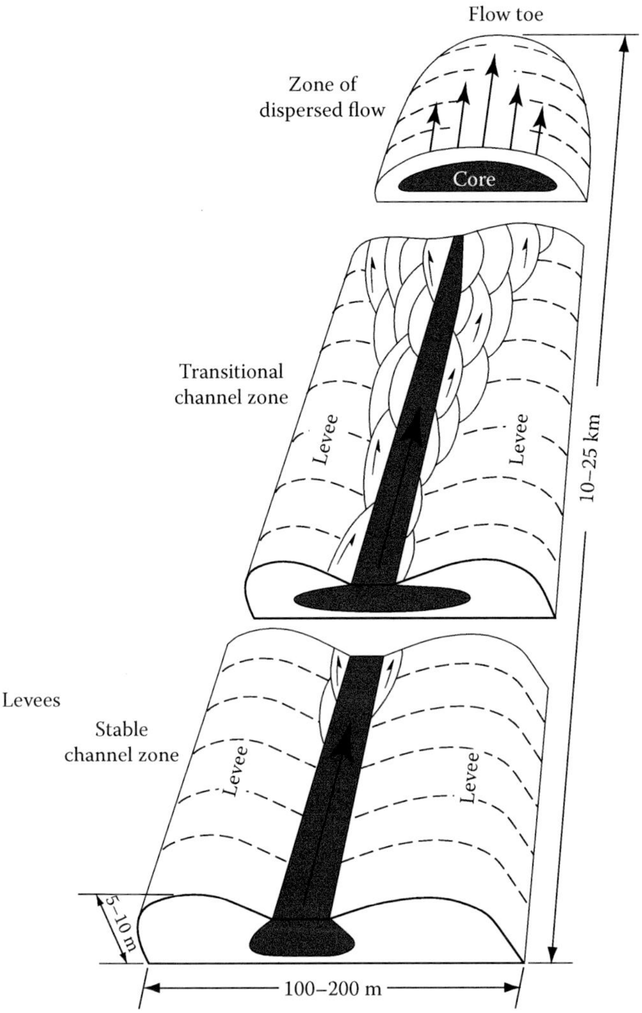

FIGURE 16.11 Schematic view of a typical, channelized basaltic lava flow (modeled after the 1984 Mauna Loa eruption; Lipman, P.W. and Banks, N.G., A flow dynamics, Mauna Loa 1984, in: Decker, R.W., Wright, T.L., and Stauffer, P.H. (eds.), *Volcanism in Hawaii*, U.S. Geological Survey Professional Paper 1350, pp. 1527–1567, 1987.). Cooler, slower-moving lava along the sides accretes to form marginal levees that confine the hotter flow interior. As the lava travels away from the vent, it cools, solidifies, and fractures, resulting in a transitional channel zone in which flow-parallel shear stresses break up the levee into elongate lenses of cooled lava. These solid wedges and blocks accumulate on the flow surface, eventually encasing the incandescent interior in a cooled crust. Near the flow front or "toe," the surface of the lava completely solidifies and the channel transitions to a zone of slow dispersed flow.

silicic flow or dome, the concentrating process just described may be reflected in the flow's surface textures, which can also result from the same redistribution process. For instance, volatile-rich zones in the flow interior may become buoyant and rise to the surface through a Rayleigh–Taylor instability, resulting in a regular spacing of bubble-rich (*pumiceous*) zones that punctuate a lava flow surface. The spacing of the pumice-rich intrusions reflects the thickness of the buoyant zone and its viscosity relative to that of the overlying material (Fink, 1980b; Manley and Fink, 1987; Baum et al., 1989).

Although the presence of bubble- or crystal-rich textures and the geometry of individual surface structures may be related to eruption conditions and magmatic properties, the most quantitatively useful way to infer this kind of information is to look at the overall morphology of a flow. Depending on how fast they erupt and cool, lavas develop surface crusts that strongly influence their final appearance. This is true for lavas of all chemical compositions, but the types of features that form differ qualitatively for basaltic and silicic lavas. For the basaltic case, under the coldest and slowest eruption conditions, surface crusts form rapidly enough to greatly inhibit flow advance, resulting in thick piles of short, tube- or bulb-like protrusions called *pillows*. These features most commonly develop in submarine or subglacial settings where seawater or ice quickly chills the flow exterior. Under conditions with slightly slower cooling rates and/or higher eruption rates, basalt flow forms are dominated by *rift-like fractures* where the solidifying crust gets stretched and cracked into smooth-sided plates. For more slowly cooled or rapidly erupted lavas, surface crusts remain continuous and ductile for longer periods and commonly exhibit transverse, regularly spaced *surface folds* or ridges, as described earlier. Still slower crust formation is associated with the development of marginal, flow-parallel *levees*, also described earlier. Finally, basalt flows with the least crust growth advance into topographic depressions and accumulate to form *lava ponds*, with smooth, flat, largely featureless surfaces. There is thus a continuum among basaltic lava morphologies, based on eruption and cooling rates, from pillows, to rift-like fractures, surface folds, levees, and smooth, featureless surfaces.

More silicic flows and domes show a comparable sequence of morphologies, although their characteristics and scales differ due to the lower eruption temperatures and effusion rates, thicker dimensions, and blockier flow surface. The progression of exhibited silicic flow and dome types includes *spiny* under most rapid cooling, *platy* under slightly slower cooling or higher eruption rates, *lobate* under still warmer conditions, and *axisymmetric* for the slowest rate of crust formation. Figure 16.12 shows examples of these four types.

Laboratory simulations using polyethylene glycol (PEG) wax to represent basalts and PEG mixed with kaolin powder to represent silicic lavas can generate all of the flow and dome types listed earlier. Such experiments (Fink and Griffiths, 1990, 1998) combined with careful field observations of active lavas and dimensional analyses allow quantification of the transitions from one flow type to another. Figure 16.13 plots data for the silicic case, with transitions among morphologies depending upon eruption rates and composition-dependent rheology. Comparable relationships can also be derived for basaltic lavas. Such correlations can be useful for inferring the conditions that occurred during unobserved eruptions. This approach has been used to estimate effusion rates for submarine basalt flows, the most common yet most difficult to

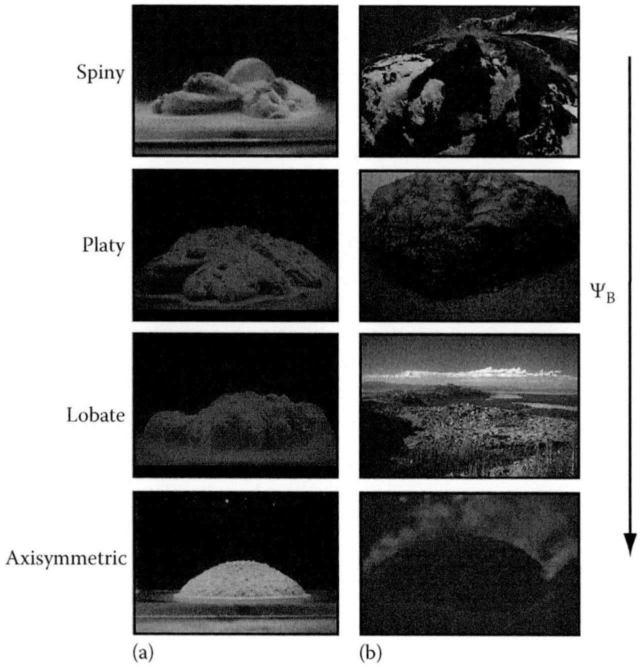

FIGURE 16.12 Morphologic progression of simulated (a) and natural (b) lava domes, displayed as a function of a dimensionless number (Ψ_B), which is proportional to the rate of growth of the solidified crust on the surface of a cooling flow. Simulated and natural domes toward the bottom of the figure have higher eruption rates, higher eruption temperatures, lower viscosities, and lower yield strengths. The natural domes are (from top to bottom) Mount St. Augustine Alaska, 1986; Mount St. Helens, Washington, June 1980; Inyo Dome, eastern California; Mount St. Helens, Washington, June 1980. Laboratory domes are approximately 4–10 cm high, and were formed by extrusion of slurries of kaolin powder mixed with polyethylene glycol into tanks of cold water. Each of the four types is associated with a specific range of values of Ψ_B. (Modified from Fink, J.H. and Griffiths, R.W., *J. Geophys. Res.*, 103(B1), 527, 1998.)

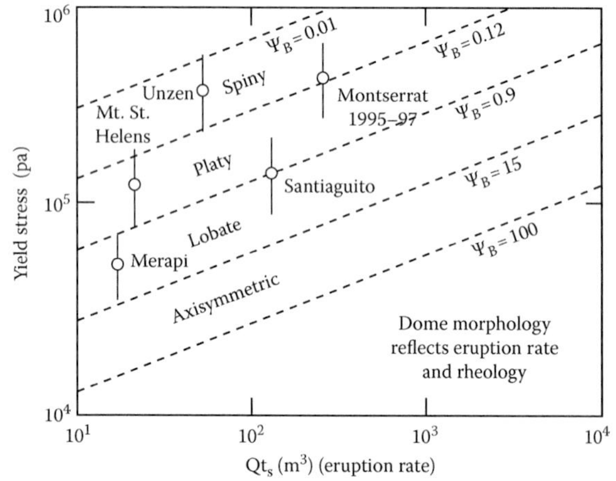

FIGURE 16.13 Plot showing how five scientifically observed, natural lava domes can be categorized in terms of the four morphologic types (spiny, platy, lobate, axisymmetric) defined in laboratory experiments. A graph like this allows either rheological or eruption rate information to be deduced if the other is known. (Modified from Fink, J.H. and Griffiths, R.W., *J. Geophys. Res.*, 103(B1), 527, 1998.)

observe on Earth (Griffiths and Fink, 1992b; Gregg and Fink, 1995). It can also be used to interpret flows on other planets (Griffiths and Fink, 1992a; Gregg and Fink, 1996).

16.7 Concluding Remarks

The examples provided in this chapter show how an appreciation of fluid dynamic principles can help geoscientists fill in critical gaps in knowledge about infrequent yet potentially catastrophic eruptive events. Volcanology is a relatively young science in which the most basic questions still remain to be solved. We do not know with any precision why eruptions start or end when they do, or why some are more violent than others. Once an eruption starts, we generally cannot say with assurance which areas are likely to be affected. This ignorance is especially frustrating for policy makers who need to make expensive decisions about evacuating populations, re-routing aircraft, and relocating or otherwise protecting key facilities. Volcanologists appreciate, perhaps more than any other natural scientists, how knowledge of fluid dynamics can provide unique and complementary insights to the other tools at their disposal for informing those responsible for making life and death interpretations of Nature's complexity.

Acknowledgment

We thank Susan Selkirk for assistance in drafting and constructing the figures for this chapter.

References

Alidibirov, M.A. (1994). A model for viscous magma fragmentation during volcanic blasts. *Bull. Volcanol.* 56(6–7), 459–465.

Baum, B.A., Krantz, W.B., Fink, J.H., and Dickinson, R.E. (1989). Taylor instability in rhyolite lava flows. *J. Geophys. Res.* 94(B5), 5815–5828.

Blake, S. and Fink, J.H. (1987). The dynamics of magma withdrawal from a density stratified dyke. *Earth Planet. Sci. Lett.* 85(4), 516–524.

Brown, G.L. and Roshko, A. (1974). Density effects and large structure in turbulent mixing layers. *J. Fluid Mech.* 64, 775–816.

Carey, S. and Sparks, R.S.J. (1986). Quantitative models of the fallout and dispersal of tephra from volcanic eruption columns. *Bull. Volcanol.* 48(2–3), 109–125.

Chojnicki, K., Clarke, A.B., and Phillips, J.C. (2006). A shocktube investigation of the dynamics of gas particle mixtures: Implications for explosive volcanic eruptions. *Geophys. Res. Lett.* 33, L15309, doi:10.1029/2006FL026414.

Clarke, A.B., Neri, A., Macedonio, G., Voight, B., and Druitt, T.H. (2002). Computational modeling of the transient dynamics of the August 1997 Vulcanian explosions at Soufrière Hills volcano, Montserrat: Influence of initial conduit conditions on near-vent pyroclastic dispersal. In: Druitt, T.H. and Kokelaar, B.P. (eds.), *The Eruption of Soufrière Hills Volcano, Montserrat, from 1995 to 1999.* Geological Society, London, U.K., Memoir, 21.

Costa, A. (2005). Viscosity of high crystal content melts: Dependence on solid fraction. *Geophys. Res. Lett.* 32, L22308, doi:10.1029/2005GL024303.

Delaney, P.T. and Pollard, D.D. (1981). Deformation of host rocks and flow of magma during growth of minette dikes and breccia-bearing intrusions near Ship Rock, New Mexico. U.S. Geological Survey Professional Paper 1202, 61p.

Dingwell, D.B. (1996). Volcanic dilemma: Flow or blow? *Science* 273, 1054–1055.

Dingwell, D.B. (2001). Magma degassing and fragmentation: Recent experimental advances. In: Freundt, A. and Rosi, M. (eds.), *From Magma to Tephra*. Elsevier, Amsterdam, the Netherlands, pp. 1–23.

Dobran, F., Neri, A., and Macedonio, G. (1993). Numerical simulation of collapsing volcanic columns. *J. Geophys. Res.* 98, 4231–4259.

Dufek, J. and Bergantz, G.W. (2007a). Dynamics and deposits generated by the Kos Plateau Tuff eruption: Controls of basal particle loss on pyroclastic flow transport. *Geochem. Geophys. Syst.* 21(2), 119–145.

Dufek, J. and Bergantz, G.W. (2007b). Suspended load and bed-load transport of particle-laden gravity currents: The role of particle-bed interaction. *Theor. Comput. Fluid Dynam.* 21(2), 119–145.

Esposti-Ongaro, T., Clarke, A., Neri, A., Voight, B., and Widiwijayanti, C. (2008a). Fluid dynamics of the 1997 Boxing Day volcanic blast on Montserrat, W.I. *J. Geophys. Res.* 113, B03211, doi:10.1029/2006JB004898.

Esposti-Ongaro, T., Neri, A., Menconi, G., de'Michieli Vitturi, M., Marianelli, P., Cavazzoni, C., Erbacci, G., and Baxter, P.J. (2008b). Transient 3D numerical simulations of column collapse and pyroclastic density current scenarios at Vesuvius. *J. Volcanol. Geotherm. Res.* 178(3), 378–396.

Esposti-Ongaro, T., Neri, A., Todesco, M., and Macedonio, G. (2002). Pyroclastic flow hazard assessment at Vesuvius (Italy) by using numerical modeling; II, Analysis of flow variables. *Bull. Volcanol.* 64(3–4), 178–191.

Fink, J.H. (1980a). Gravity instability in the Holocene Big and Little Glass Mountain rhyolitic obsidian flows, Northern California. *Tectonophysics* 66(1–3), 147–166.

Fink, J.H. (1980b). Surface folding and viscosity of rhyolite flows. *Geology* 8(5), 250–254.

Fink, J.H. and Griffiths, R.W. (1990). Radial spreading of viscous gravity currents with solidifying crust. *J. Fluid Mech.* 221, 485–509.

Fink, J.H. and Griffiths, R.W. (1998). Morphology, eruption rates, and rheology of lava domes: Insights from laboratory models. *J. Geophys. Res.* 103(B1), 527–545.

Fink, J.H. and Zimbelman, J. (1990). Longitudinal variations in rheological properties of lavas: Puu Oo Basalt Flows, Kilauea Volcano, Hawaii. In: Fink, J.H. (ed.), *Lava Flows and Domes: Emplacement Mechanisms and Hazard Implications. IAVCEI Proceedings in Volcanology*, Vol. 2, Springer Verlag, Heidelberg, Germany, pp. 157–173.

Francis, P. and Oppenheimer, C. (2004). *Volcanoes*, 2nd edn. Oxford University Press, New York, 392pp.

Gladstone, C., Phillips, J.C., and Sparks, R.S.J. (1998). Experiments on bidisperse, constant-volume gravity currents: Propagation and sediment deposition. *Sedimentology* 45(5), 833–843.

Gonnermann, H.M. and Manga, M. (2003). Explosive volcanism may not be an inevitable consequence of magma fragmentation. *Nature* 426, 432–435.

Gregg, T.K.P. and Fink, J.H. (1995). Quantification of submarine lava-flow morphology through analog experiments. *Geology* 23, 73–76.

Gregg, T.K.P. and Fink, J.H. (1996). Quantification of extraterrestrial lava flow effusion rates through laboratory simulations. *J. Geophys. Res.* 101(E7), 16891–16900.

Gregg, T.K.P., Fink, J.H., and Griffiths, R.W. (1998). Formation of multiple fold generations on lava flow surfaces: Influence of strain rate, cooling rate, and lava composition. *J. Volcanol. Geotherm. Res.* 80(3–4), 281–292.

Griffiths, R.W. and Fink, J.H. (1992a). The morphology of lava flows in planetary environments—Predictions from analog experiments. *J. Geophys. Res.* 97(B13), 19739–19748.

Griffiths, R.W. and Fink, J.H. (1992b). Solidification and morphology of submarine lavas—A dependence on extrusion rate. *J. Geophys. Res.* 97(B13), 19729–19737.

Griffiths, R.W. and Fink, J.H. (1993). The dynamics of lava flows with solidifying crusts. *J. Fluid Mech.* 252, 667–702.

Grindle, T.J. and Burcham, F.W. (2003). Engine damage to a NASA DC-8-72 airplane from a high-altitude encounter with a diffuse volcanic ash cloud. NASA TM-2003-212030, North Edwards, CA, 22p.

Head, J. and Wilson, L. (1989). Basaltic pyroclastic eruptions: Influence of gas-release patterns and on fountain structure, and the formation of cinder cones, spatter cones, rootless flows, lava ponds and lava flows. *J. Volcanol. Geotherm. Res.* 37, 261–271.

Hess, K.U. and Dingwell, D.B. (1996). Viscosities of hydrous leucogranitic melts: A non-Arrhenian model. *Am. Mineral.* 81, 1297–1300.

Hulme, A. (1974). Interpretation of lava flow morphology. *Geophys. J. R. Astron. Soc.* 39(2), 361–383.

Huppert, H.E., Shepherd, J.B., Sigurdsson, H., and Sparks, R.J.S. (1982). On lava dome growth, with application to the 1979 lava extrusion of the Soufrière of St. Vincent. *J. Volcanol. Geotherm. Res.* 14, 199–222.

Lejeune, A.-M. and Richet, P. (1995). Rheology of crystal-bearing silicate melts: An experimental study at high viscosities. *J. Geophys. Res.* 100(3), 4215–4229.

Lipman, P.W. and Banks, N.G. (1987). A flow dynamics, Mauna Loa 1984, In: Decker, R.W., Wright, T.L., and Stauffer, P.H. (eds.), *Volcanism in Hawaii.* U.S. Geological Survey Professional Paper 1350, pp. 1527–1567.

Llewellin, E.W., Mader, H.M., and Wilson, S.D.R. (2002). The rheology of a bubbly liquid. *Proc. R. Soc. Lond. A.* 458, 987–1016.

Manley, C.R. and Fink, J.H. (1987). Internal textures of rhyolite flows as revealed by research drilling. *Geology* 15(6), 549–552.

McBirney, A.R. and Murase, T. (1971). Factors governing the formation of pyroclastic rocks. *Bull. Volcanol.* 34(2), 372–384.

Melnik, O. and Sparks, R.S.J. (2002). Dynamics of magma ascent and lava extrusion at Soufrière Hills Volcano, Montserrat. *Geol. Soc. London Mem.* 21, 153–171.

de' Michieli Vitturi, M., Clarke, A.B., Neri, A., and Voight, B. (2008). Effects of conduit geometry on magma ascent dynamics in dome-forming eruptions. *Earth Planet. Sci. Lett.* 272, 567–578.

Moore, H.J., Arthur, D.W.G., and Schaber, G.G. (1978). Yield strengths of flows on the earth, mars, and moon. *Proceedings of the 9th Lunar and Planetary Science Conference*, The Woodlands, Texas, pp. 3351–3378.

Moore, G. and Carmichael, I.S.E. (1998). The hydrous phase equilibria (to 3 kbar) of an andesite and basaltic andesite from western Mexico: Constraints on water content and conditions of phenocryst growth. *Contribut. Mineral. Petrol.* 130, 304–319.

Moore, H.J. and Schaber, G.G. (1975). An estimate of the yield strength of the Imbrium flows. *Proceedings of the 6th Lunar and Planetary Science Conference*, Houston, TX, March 17–21, pp. 101–118.

Moore, G., Vennemann, T., and Carmichael, I.S.E. (1998). An empirical model for the solubility of H_2O in magmas to 3 kilobars. *Am. Min.* 83, 36–42.

Morton, B.R., Taylor, G.I., and Turner, J.S. (1956). Turbulent gravitational convection from maintained and instantaneous sources. *Phil. Trans. R. Soc. Lond. A* 234, 1–23.

Murase, T. (1962). Viscosity and related properties of volcanic rocks at 800° to 1400°. *Hokaido Univ. Fac. Sci. J.* 7, 487–584.

Murase, T. and McBirney, A.R. (1973). Properties of some common igneous rocks and their melts at high temperatures. *Geol. Soc. Am. Bull.* 84, 3563–3592.

Neri, A. and Macedonio, G. (1996). Numerical simulation of collapsing volcanic columns with particles of two sizes. *J. Geophys. Res.* 101(B4), 8153–8174.

Papale, P. (1999). Strain-induced magma fragmentation in explosive eruptions. *Nature* 397, 425–428.

Phillips, J.C. and Woods, A.W. (2002). Suppression of large-scale magma mixing by melt-volatile separation. *Earth Planet. Sci. Lett.* 204(1–2), 47–60.

Ripepe, M., Marchetti, E., Ulivieri, G., Harris, A.J.L., Dehn, J., Burton, M., Caltabiano, T., and Salerno, G. (2005). Coupled thermal oscillations in explosive activity at different craters of Stromboli volcano. *Geophys. Res. Lett.* 32(17), L17302.

Sparks, R.S.J. (1978). The dynamics of bubble formation and growth in magmas: A review and analysis. *J. Volcanol. Geotherm. Res.* 3, 1–37.

Sparks, R.S.J. (1986). The dimensions and dynamics of volcanic eruption columns. *Bull. Volcanol.* 48(1), 3–15.

Sparks, R.S.J. (1997). Causes and consequences of pressurization in lava dome eruptions. *Earth Planet. Sci. Lett.* 150, 177–189.

Spieler, O., Kennedy, B., Kueppers, U., Dingwell, D. B., Scheu, B., and Taddeucci, J. (2004). The fragmentation threshold of pyroclastic rocks. *Earth Planet. Sci. Lett.* 226, 139–148.

Todesco, M., Neri, A., Ongaro, T.E., Papale, P., Macedonio, G., Santacroce, R., and Longo, A. (2002). Pyroclastic flow hazard assessment at Vesuvius (Italy) by using numerical modeling. I. Large-scale dynamics. *Bull. Volcanol.* 64(3–4), 155–177.

Valentine, G.A. (1998). Eruption column physics. In: Freundt, A. and Rosi, M. (eds.), *From Magma to Tephra: Modeling Physical Processes of Explosive Volcanic Eruptions*. Elsevier, Amsterdam, the Netherlands, pp. 91–138.

Valentine, G.A. and Wohletz, K.H. (1989). Numerical models of Plinian eruption columns and pyroclastic flows. *J. Geophys. Res.* 94(B2), 1867–1887.

Wadge, G., Jackson, P., Bower, S.M., Woods, A.W., and Calder, E.S. (1998). Computer simulations of pyroclastic flows from dome collapse. *Geophys. Res. Lett.* 25(19), 3677–3680.

Walker, G.P.L. (1973). Explosive volcanic eruptions—A new classification scheme. *Geol. Rundsch.* 62(2), 431–446.

Watts, R.B., Herd, R.A., Sparks, R.S.J., and Young, S.R. (2002). Growth patterns and emplacement of the andesitic lava dome at Soufriere Hills Volcano, Montserrat. In: Druitt, T.H. and Kokelaar, B.P. (eds.), *Volcanic Processes, Products, and Hazards*. Geological Society, London, U.K., Memoirs 21, pp. 115–152.

Widiwijayanti, C., Voight, B., Hidayat, D., and Schilling, S.P. (2009). Objective rapid delineation of areas at risk from block-and-ash pyroclastic flows and surges. *Bull. Volcanol.* 71(6), 687–703.

Wilson, C.J.N. (1980). The role of fluidization on pyroclastic flow emplacement: An experimental approach. *J. Volcanol. Geotherm. Res.* 8(2–4), 231–249.

Wilson, C.J.N. (1984). The role of fluidization on pyroclastic flow emplacement 2: Experimental results and their interpretation. *J. Volcanol. Geotherm. Res.* 20(1–2), 55–84.

Wilson, L. and Head, J. (1981). Ascent and eruption of magma on the Earth and moon. *J. Geophys. Res.* 86, 2971–3001.

Wilson, L., Sparks, R.S.J., Huang, T.C., and Watkins, N.D. (1978). The control of volcanic eruption column heights by eruption energetics and dynamics. *J. Geophys. Res.* 83, 1829–1836.

Wilson, L., Sparks, R.S.J., and Walker, G.P.L. (1980). Explosive volcanic eruptions IV. The control of magma properties and conduit geometry on eruption column behavior. *Geophys. J. R. Astron. Soc.* 63, 117–148.

Wohletz, K.H. and McQueen, R.G. (1984). Experimental studies of hydromagmatic volcanism. In: *Explosive Volcanism: Inception, Evolution, and Hazards: Studies in Geophysics*. National Academic Press, Washington, DC, pp. 158–169.

Wohletz, K.H. and Valentine, G.A. (1990). Computer simulations of explosive volcanic eruptions. In: Ryan, M.P. (ed.), *Magma Transport and Storage*. Wiley, London, U.K., Chapter 8, pp. 113–135.

Woods, A.W. (1988). The fluid dynamics and thermodynamics of eruption columns. *Bull. Volcanol.* 50(3), 169–193.

Woods, A.W. (1995). A model of vulcanian explosions. *Nucl. Eng. Des.* 155, 345–357.

Zhang, Y. (1999). A criterion for the fragmentation of bubbly magma based on brittle failure theory. *Nature* 402, 648–650.

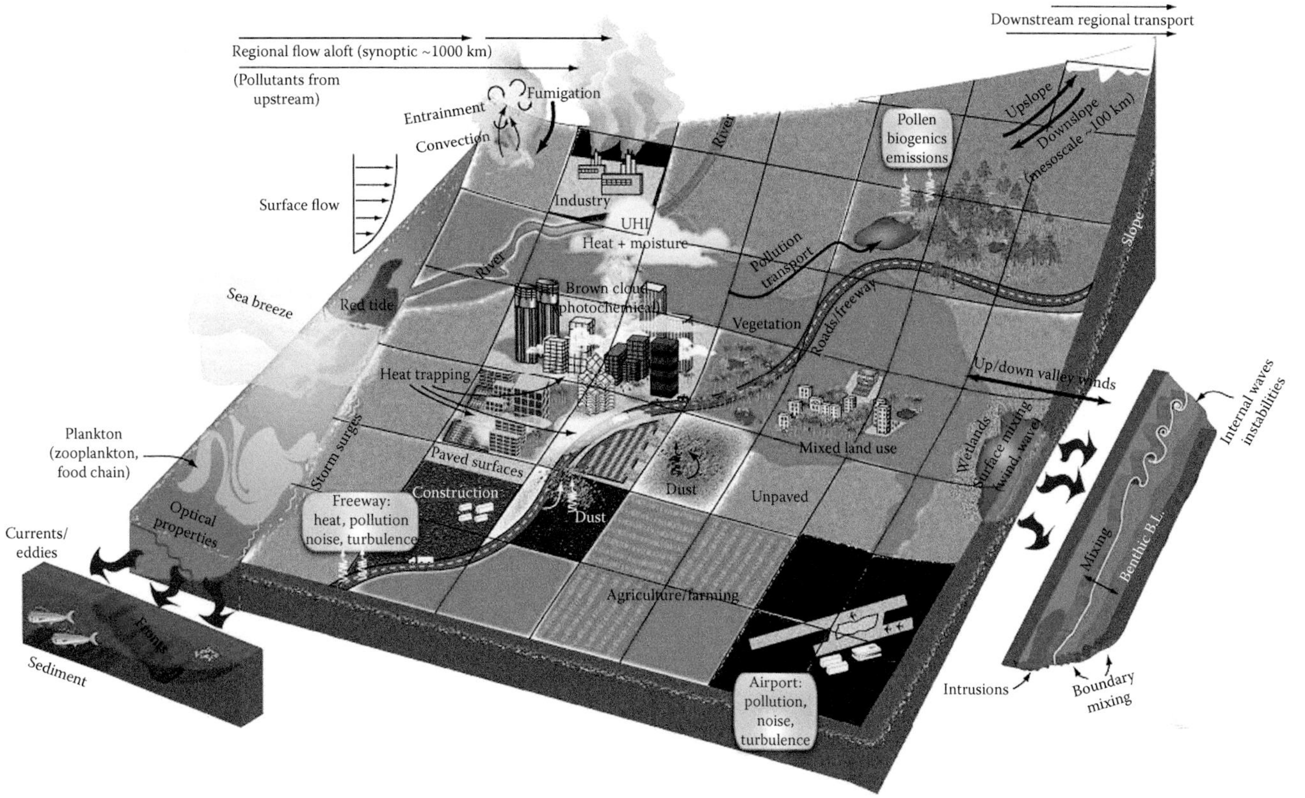

FIGURE 1.3 A schematic of anthropogenically modified environs, and pertinent fluid motions. Coasts, water bodies, cities, and terrain add Sisyphean complexity to modeling and prediction. (Adapted from Fernando, H.J.S. et al., *Phys. Fluids*, 22, 051301, 2010.)

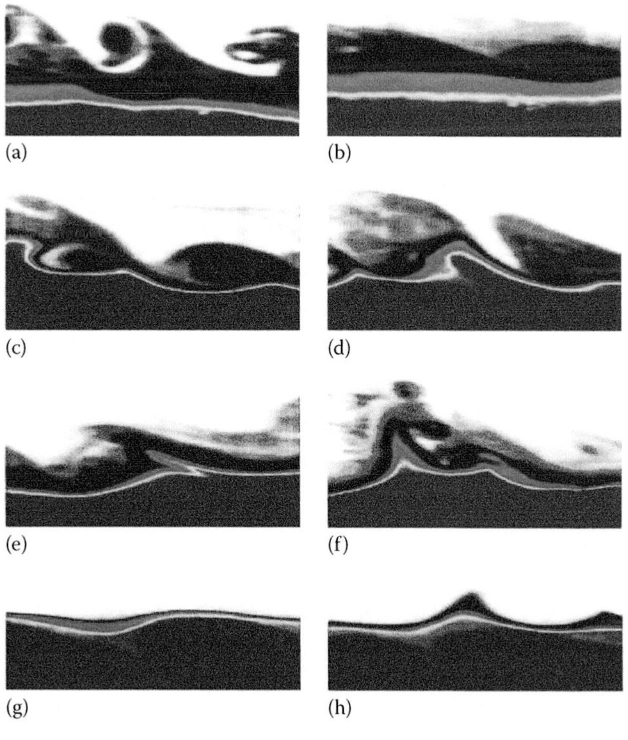

FIGURE 1.7 Laboratory modeling of turbulent mixing in a two-layer stably stratified fluid, with fluid layers separated by a sheared density interface. A generic flow configuration was used here to obtain results of general validity. To (h) represent cases with different Bulk Richardson number ⟦Ri⟧_B (Section 1.3.2) with L_V replaced by the depth of the upper turbulent layer. (a) ⟦Ri⟧_B = 1.8, (b) 3.2, (c) 4.5, (d) 5.5, (e) shortly after (d), (f) 5.8, (g) 9.2 and (h) shortly after (d). (From Strang, E.J. and Fernando, H.J.S., *J. Fluid Mech.*, 428, 349, 2001. With permission.)

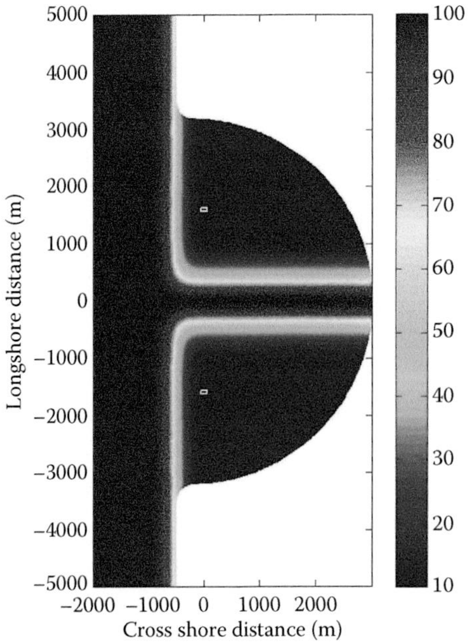

FIGURE 5.2 Idealized depth contours of the model embayment depicting a shallow shelf incised by a deep channel. Locations of model offshore fish pens are depicted by the white boxes. The depth is indicated by the color bar in meters.

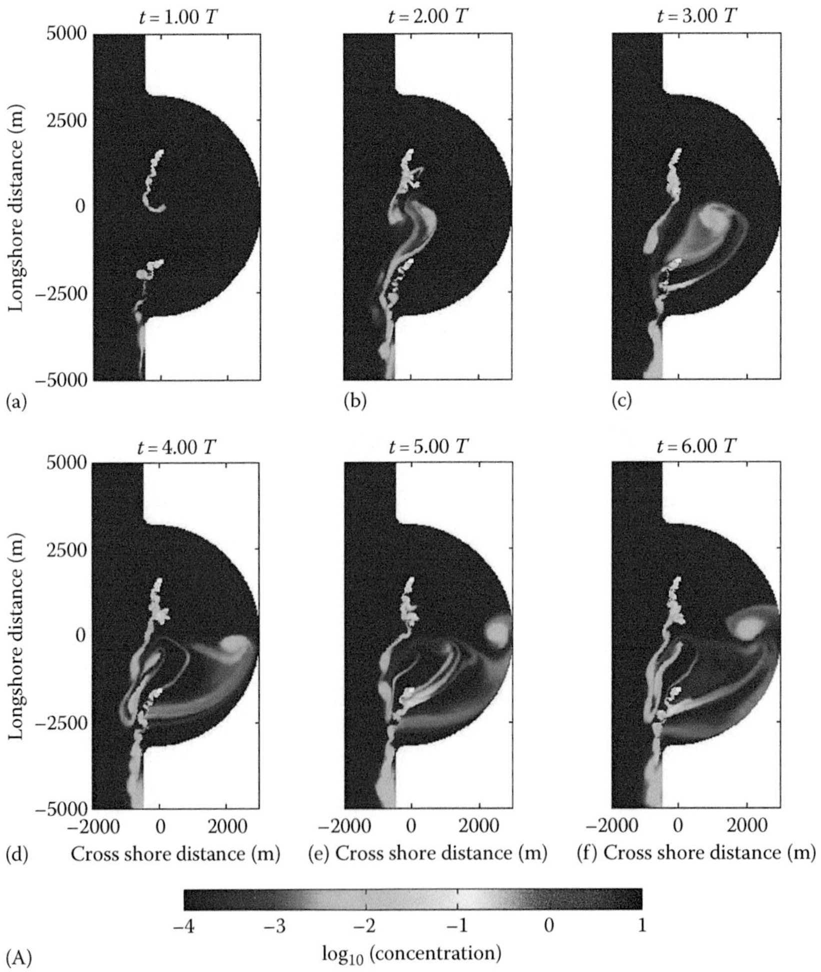

FIGURE 5.3 "Birds-eye-view" of the time history of the normalized concentration in the vicinity of two sets of six 20 m diameter pens (depicted as white boxes) releasing a passive scalar at the coastal embayment for (A) the "offshore base case" (upper panel) and

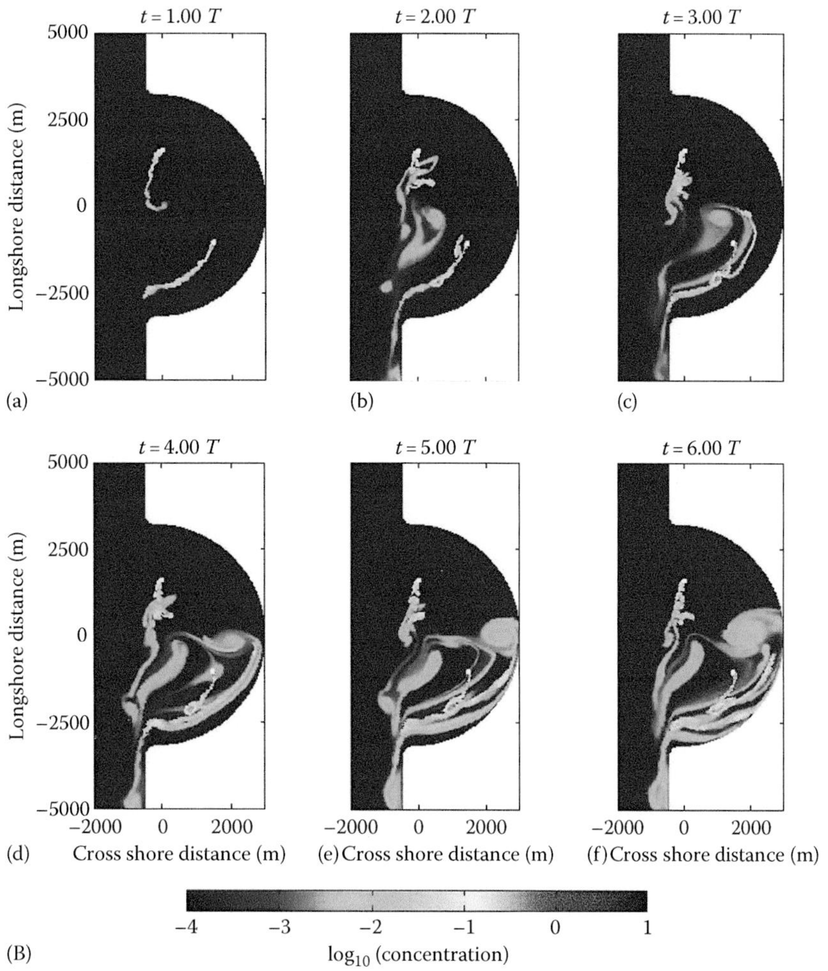

FIGURE 5.3 (continued) (B) "nearshore base case" (lower panel) shown in Table 5.1.

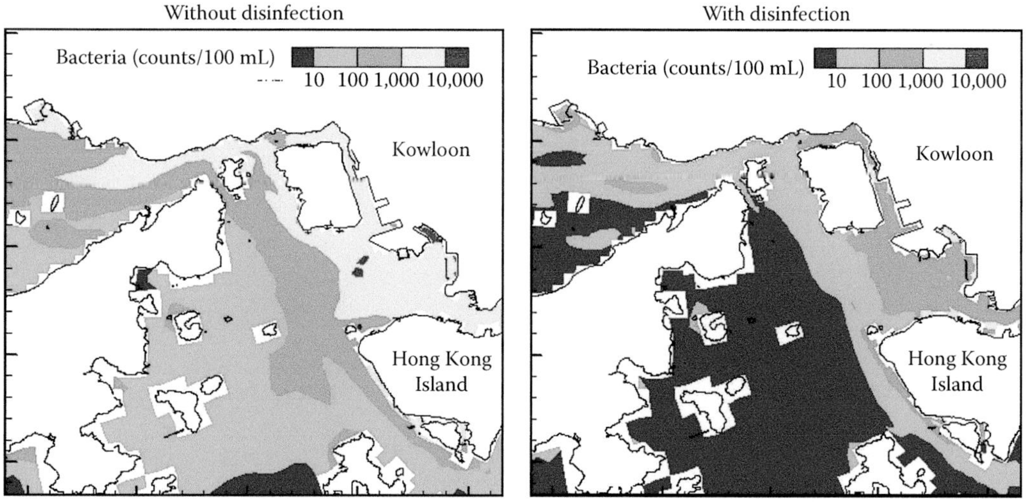

FIGURE 7.7 Predicted *Escherichia coli* concentrations in the vicinity of the HATS outfall.

FIGURE 9.3 CT from an instantaneous release in Times Square, New York City, as predicted by the FAST3D-CT MILES model. (From Patnaik, G. et al., *J. Fluids Eng.*, 129, 1524, 2007.) Concentrations shown at 3, 5, 7, and 15 min after release. The figure demonstrates the typical complex unsteady vertical mixing patterns caused by building vortex and recirculation patterns, and predicts endangered regions associated with the particular release scenario. Vortex dynamics driven CT cannot be captured by plume/pluff modeling.

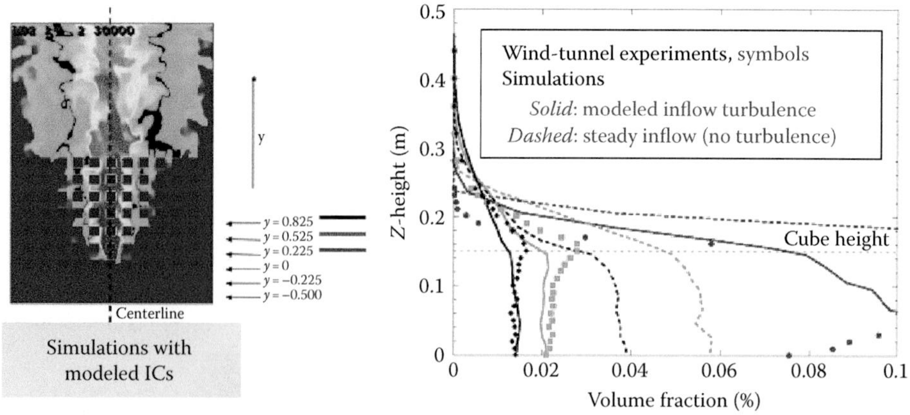

FIGURE 9.5 ILES studies of flow and dispersion over an urban (cube arrangement) model. (From Patnaik, G. et al., *J. Fluids Eng.*, 129, 1524, 2007.) Predicted and measured dispersal of tracer volume fractions in the first few urban canyons indicated by color lines on the left.

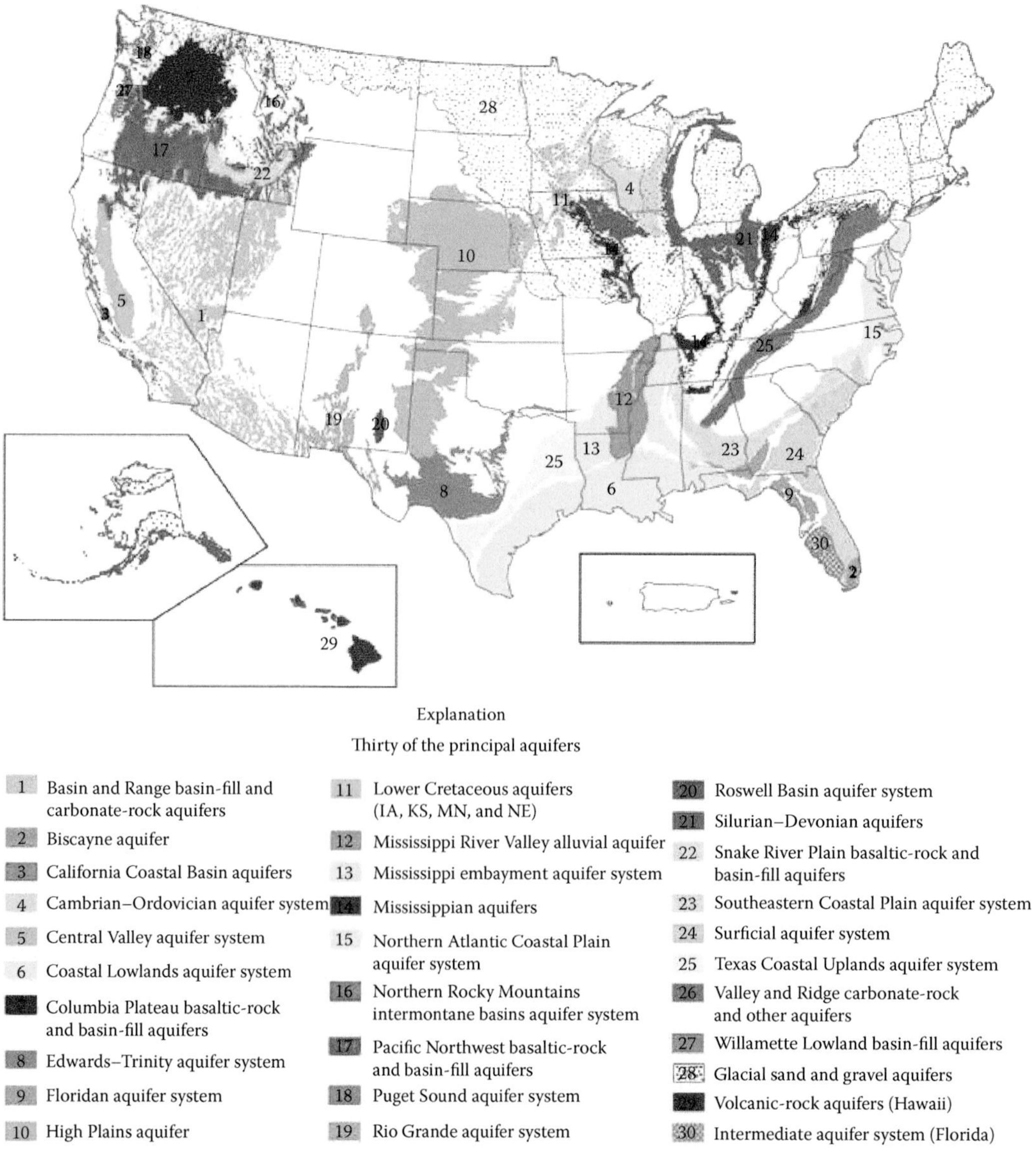

FIGURE 9.6 Principal aquifers of the United States. (Reproduced from Reilly, T.E. et al., *U.S. Geol. Surv. Circ.*, 1323, 70, 2008, Figure 23.)

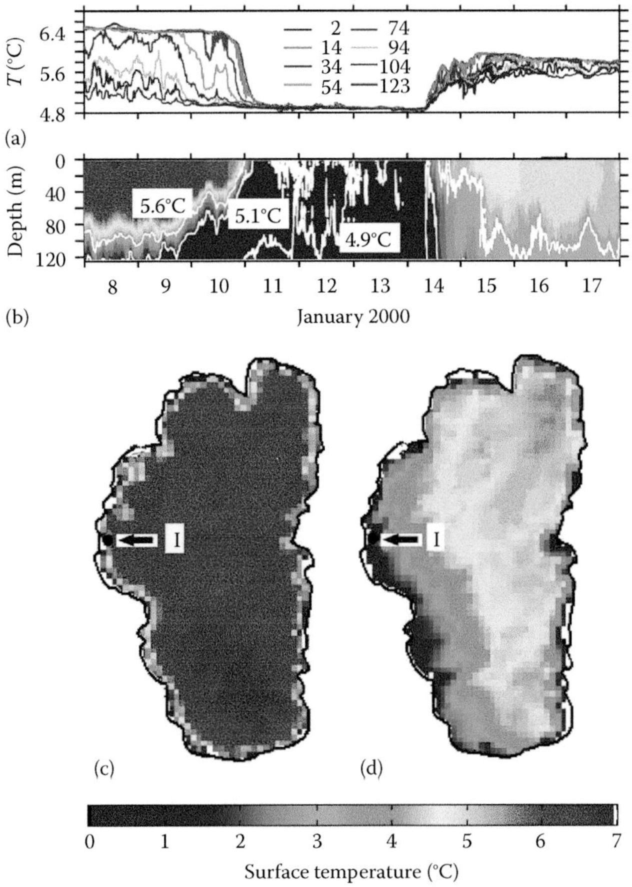

FIGURE 13.2 Evolution of temperature stratification in Lake Tahoe as a result of an extraordinary wind-driven upwelling from January 9 to 14, 2000. Panels (a) and (b) show the temperature record in 2–123 m depths and a few isotherms in the top 120 m at the upwind location "I" (39°/6.7440′N, 120°/8.9320′W). Whereas before the event, temperatures in the top 120 m ranged from 4.9°C to 6.4°C, this layer was almost homogenous afterward (a and b). Panels (c) and (d) show space-borne ATSR images of surface temperatures on January 4 (c; before event) and on January 13 at 22:09 (d; during upwelling). The enormous cooling is due to vertical mixing and heat loss resulting from strong winds. (From Schladow, S.G. et al., *Geophys. Res. Lett.*, 31, L15504, 2004.)

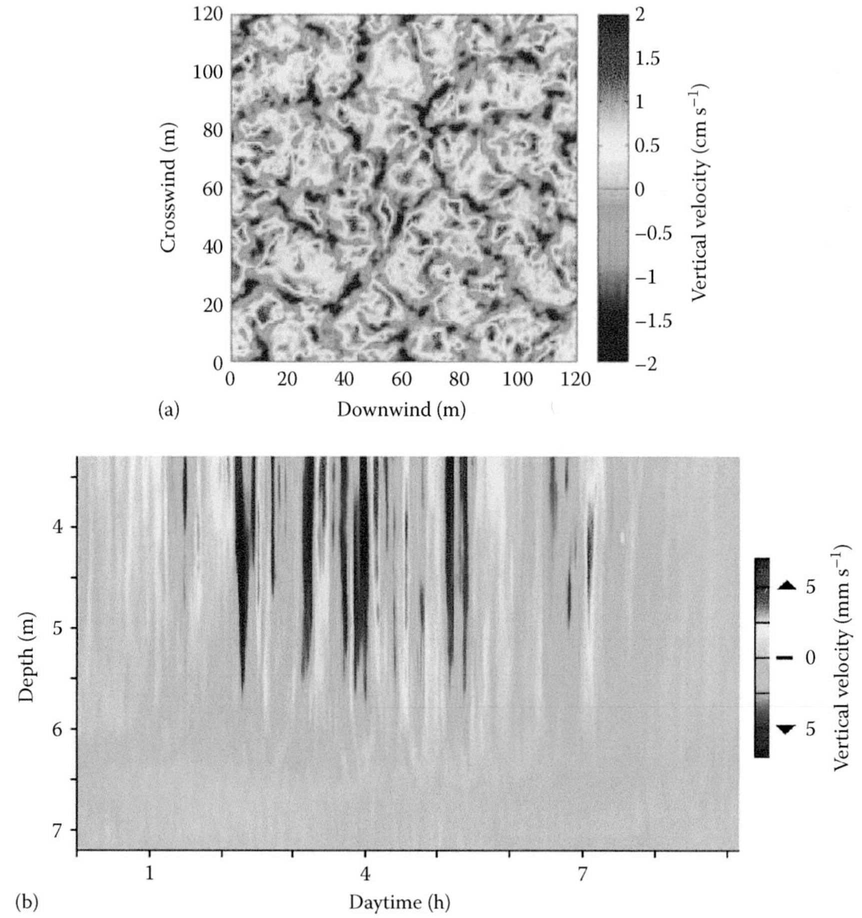

FIGURE 13.6 (a) Horizontal cross section at 3 m depth showing vertical velocities of thermals as modeled by large-eddy simulations (e.g., see Chapter 21 of *Handbook of Environmental Fluid Dynamics, Volume Two*) for a 50 m thick convective SL loosing heat at 200 W m^{-2}. The model was run for oceanic water and no wind forcing with 1 m grid resolution and periodic horizontal boundary conditions. Upward (red) and downward (blue) thermals reach velocities of ~2 cm s^{-1}. (From Li, M. et al., *Deep Sea Res. I*, 52, 259, 2005.) (b) Time-series of vertical velocity profiles observed in the top 6 m of the small wind-protected Lake Soppensee during a night of cooling in fall (September). The contour plot reveals rising (red) and sinking (blue) thermals from midnight to 8 a.m. (~sun rise). Due to lower α (and therefore weaker $J_{b,S}$) and smaller Z_{SL} (Equation 13.4) than in the left panel, the water parcels reach velocities of only ±0.5 cm s^{-1}. (From Jonas, T. et al., *J. Geophys. Res.*, 108(C10), 3328, 2003.)

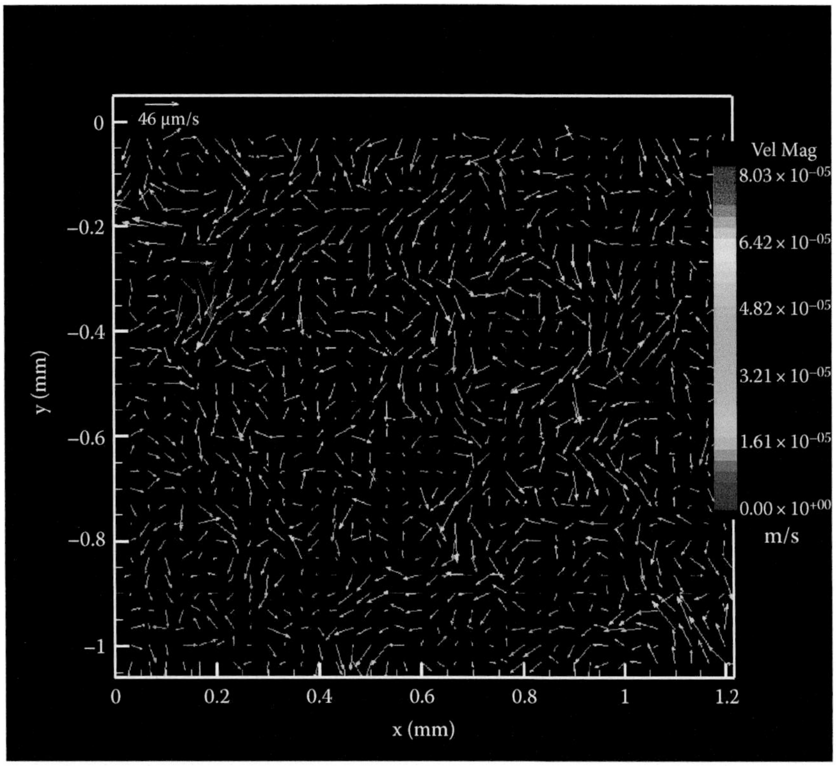

FIGURE 14.4 Vector plot of *Dunaliella primolecta* (average cell size ~8 to 10 μm) velocity magnitudes in a stagnant fluid captured by a 4× objective of the microscope using the double-frame CCD camera with 1376 × 1024 pixel resolution. The field of view is 1.2 mm × 1.03 mm. A set of 250 double-frame image pairs were recorded with the timing between each frame (ΔT) adjusted to have a *Dunaliella primolecta* displacement of approximately 8–10 pixels.

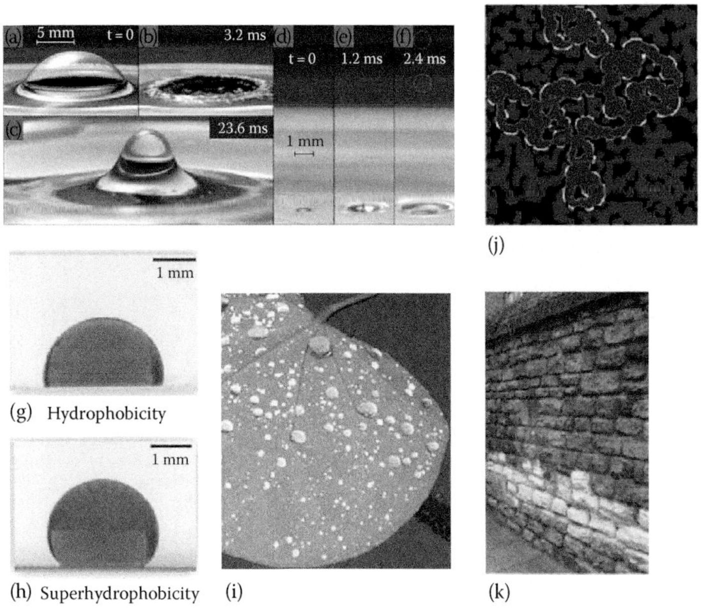

FIGURE 15.3 Examples where surface tension affects the statics and dynamics of a fluid system. (a–f) A bubble bursts (a) at the air–water interface, forming an array of even smaller bubbles (b–c); when a smaller bubble bursts, a small aerosol droplet is released. (From Bird, J.C. et al., *Nature*, 465, 759, 2010.) (g–i) Wetting of surfaces with (g) a partial wetting liquid drop on a smooth substrate, (h) the same system when the surface is roughened displays an even high contact angle, and (i) such roughness-induced superhydrophobicity is common in the leaves of plants as droplets are readily displaced from the surface. (j) Two-phase flow in a porous sample with an example shown here of decane (blue) and nitrogen gas (maroon) in a polymeric micromodel. (From Cheng, J.-T. et al., *Geophys. Res. Lett.*, 31, L08502, 2004.) during drainage of the fluids. (k) Salt weathering, or efflorescence, or building materials as evaporation of water occurs at the surface of the sample. (From Hall, C. et al., *Proc. R. Soc. A*, 2010.)

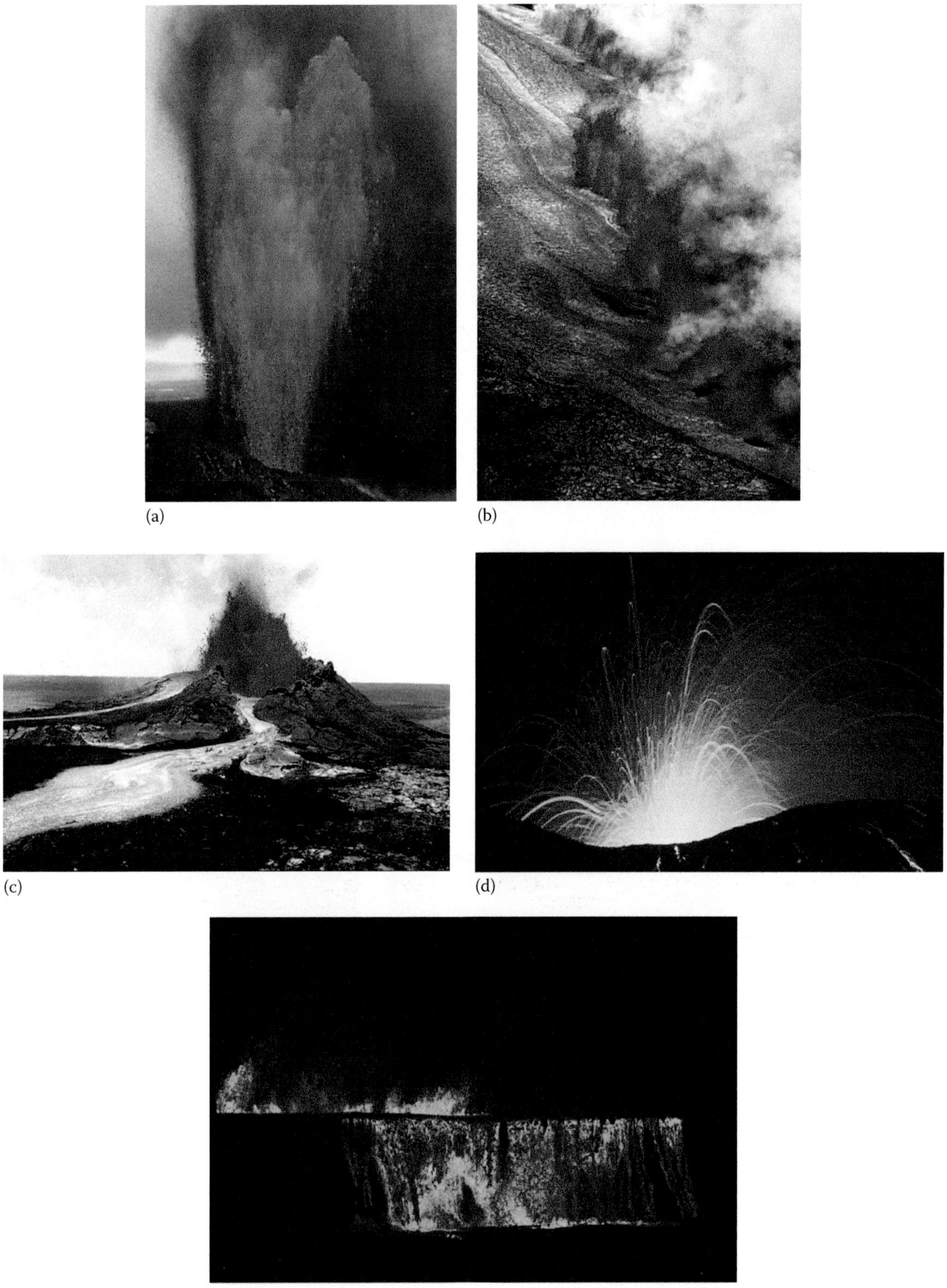

FIGURE 16.1 (a) Spectacular fire-fountain at Kilauea Iki vent, Kilauea Volcano, 1959. (Photo courtesy of U.S. Geological Survey, Reston, VA.) (b) Aerial view of "curtain of fire" eruption, East Rift Zone, Kilauea Volcano, Hawaii. Gas-charged magma erupts along linear fissures, collects near the vent, and flows away as streams of solidifying lava, seen here as dark, subparallel channels moving to the upper left. Compare with Figure 16.3. (Photo courtesy of U.S. Geological Survey, Reston, VA.) (c) Vent of 1984 eruption of Mauna Loa volcano, Hawaii. Gas-rich, fluid magma rises many meters into the air, falls back, and builds up a rampart of solid clasts around the vent. The fluid lava flow emerges from a breach in the crater. (Photo courtesy of U.S. Geological Survey, Reston, VA.) (d) Classic Strombolian behavior at Stromboli Volcano. (e) Eruption of Kilauea Volcano showing high fountains in the background feeding extensive lava flows that are cascading over a cliff in the foreground. (Photo courtesy of U.S. Geological Survey, Reston, VA.)

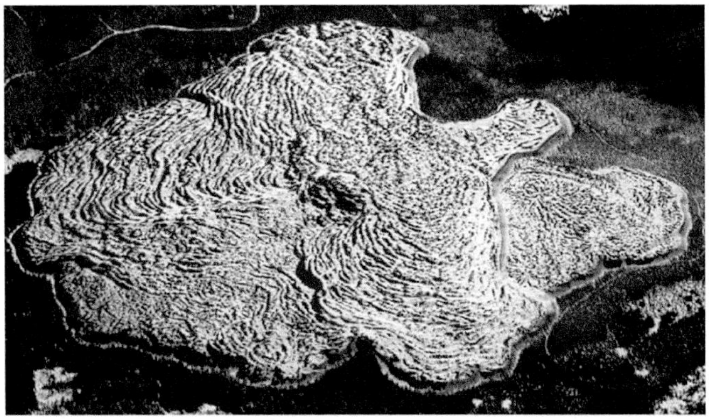

(f)

(g)

(h)

(i)

FIGURE 16.1 (continued) (f) Little Glass Mountain rhyolite lava flow, Medicine Lake Volcano, northern California. Approximately, 1000 year old, viscous lava flow showing prominent surface ridges formed by flow-parallel compression, amplified by buoyant rise of bubble-rich zones in the flow interior. Snow accentuates ridges. Diameter of flow is about 2 km. Ridges are about 10 m apart. (Photo by J.H. Fink.) (g) Devil's Hill rhyolite lava domes on the south flank of South Sister Volcano in the Cascade Range of central Oregon. Domes are assumed to have been fed by a planar, vertically oriented dike connected to a magma chamber at depth. Larger domes probably emerged from wider zones along the dike. (Photo by J.H. Fink.) (h) Santiaguito dacite lava dome complex, Santa Maria Volcano, western Guatemala. Coalesced chain of steep-sided domes, inferred to have been fed by an elongate dike that developed in the scar left by a huge eruption of Santa Maria in 1902. The Santiaguito domes have been erupting continuously since 1922. Steam marks site of active vent. Image is approximately 2.5 km across. Compare with (j). (Photo by J.H. Fink.) (i) Pyroclastic density current formed by a June 25, 1997, collapse or explosion of the Soufrière Hills volcano lava dome on the island of Montserrat, British West Indies. This pyroclastic flow generally followed topographically low drainages from the volcanic vent to this location approximately 7 km down slope. However, at various points along its path, the dilute and turbulent overriding surge surmounted topography and escaped the channel. Buoyant plumes are forming above the current and are particularly well developed approximately 1.5 km behind the flow front. Airport terminal building is circled at the left edge of the image for scale. (Photograph by A.B. Clarke and R.B. Watts.)

(continued)

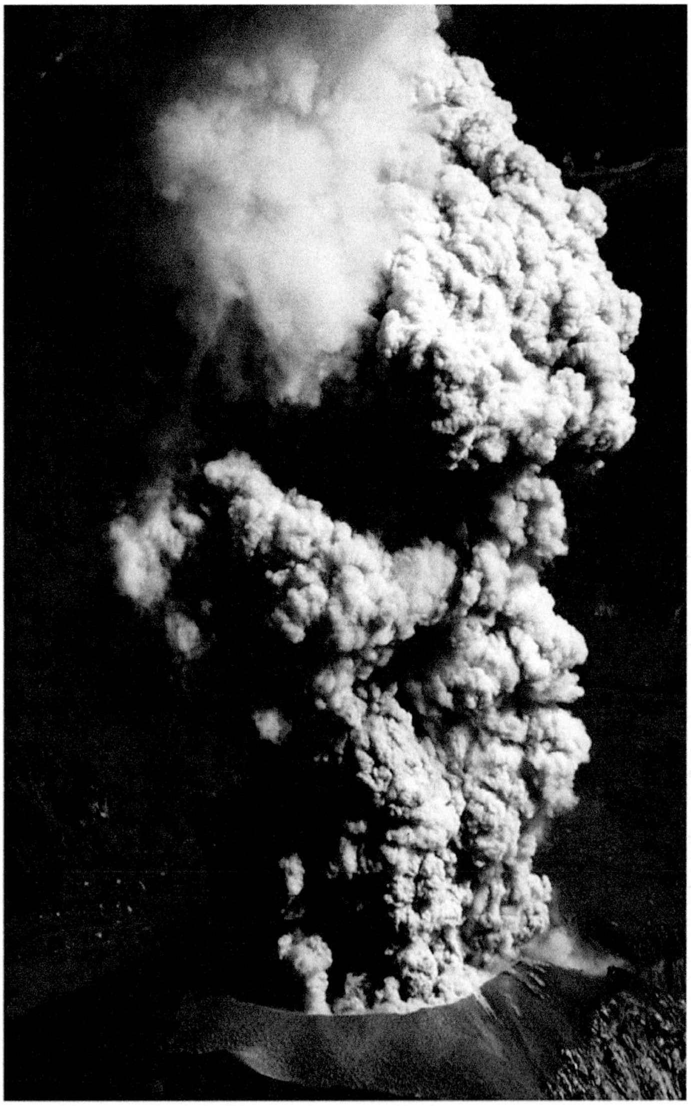

(j)

(k)

(l)

FIGURE 16.1 (continued) (j) Vulcanian eruption of ash from vent at the top of one of the Santiaguito lava domes shown in (h). Such explosions can occur as frequently as several times per hour and the resulting plume can rise several kilometers above the top of the dome. (Photo by J.H. Fink.) (k) Ascending eruption cloud from Redoubt Volcano as viewed from Kenai Peninsula (to the east). Gravitational collapse of Redoubt's lava dome produced pyroclastic density currents and a plume that reached nearly 10 km above sea level. The corresponding umbrella cloud at the top of the plume spread in a nearly circular pattern and distributed volcanic particles over a region many kilometers in diameter. (Photo courtesy of J. Warren, April 21, 1990 and the U.S. Geological Survey, Reston, VA.) (l) Vulcanian eruption with pyroclastic density currents (Soufrière Hills volcano). In this instance, the eruption simultaneously produced a buoyant plume that reached 15 km above sea level and pyroclastic density currents that initially spread radially in all directions and were then channeled into topographic lows to distances of up to 5 km. (Photograph by A.B. Clarke.)

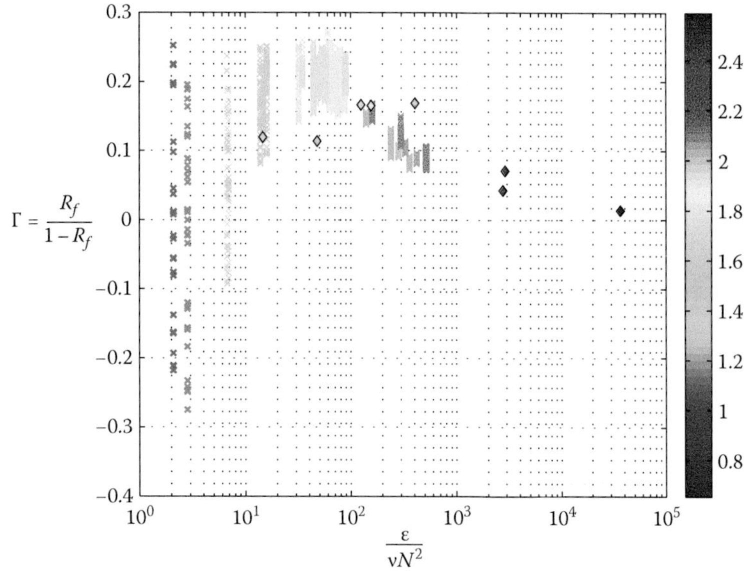

FIGURE 20.3 Mixing parameter $\Gamma = R_f/(1 - R_f)$ as a function of $\varepsilon/\nu N^2$. The color scale on the right represents the turbulent Reynolds number Re_T as it ranges from 5 to 4×10^3 (see also Figure 20.2). (Data are from (◊): Barry, M.E. et al., *J. Fluid Mech.*, 442, 267, 2001; (×): Shih, L.H. et al., *J. Fluid Mech.*, 525, 193, 2005.)

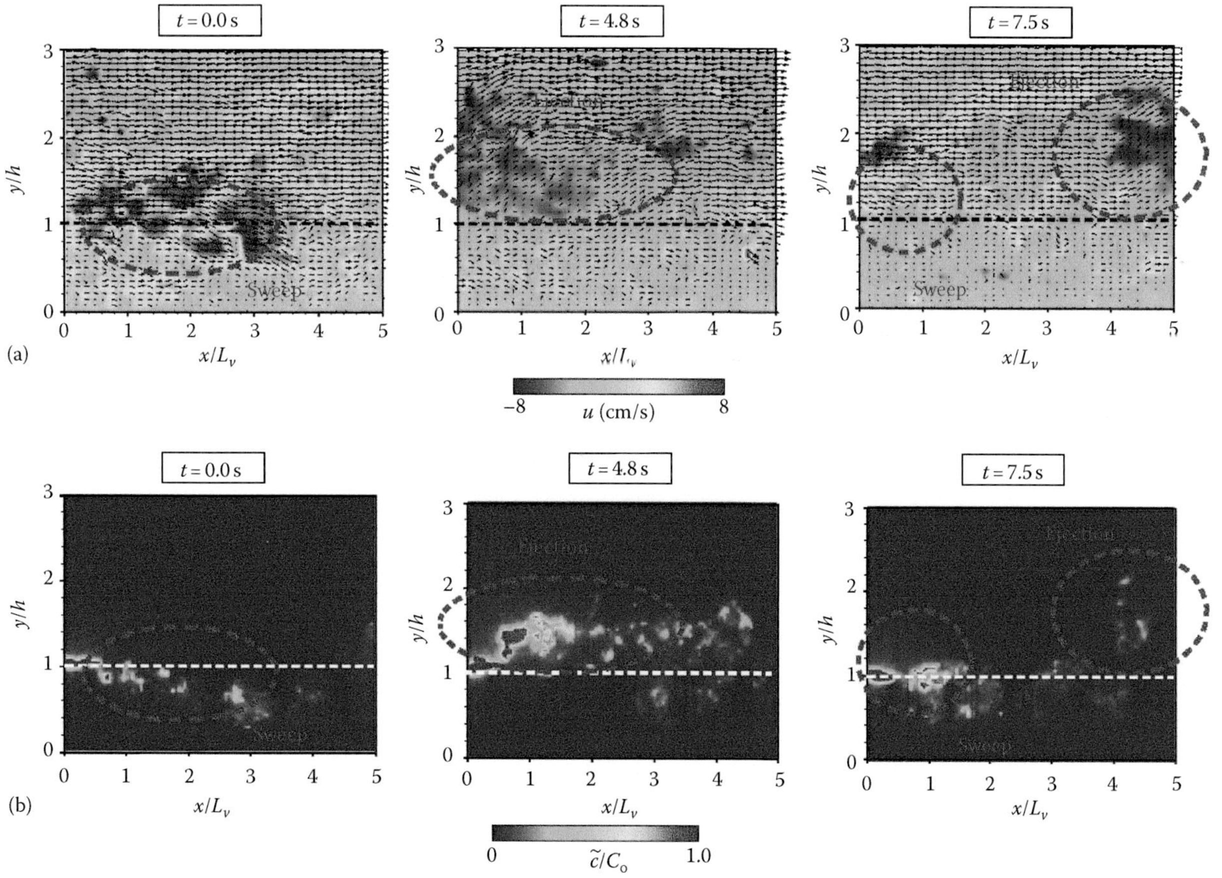

FIGURE 23.24 Simultaneous measurements of (a) instantaneous velocity vectors by PIV and (b) instantaneous concentration by LIF.

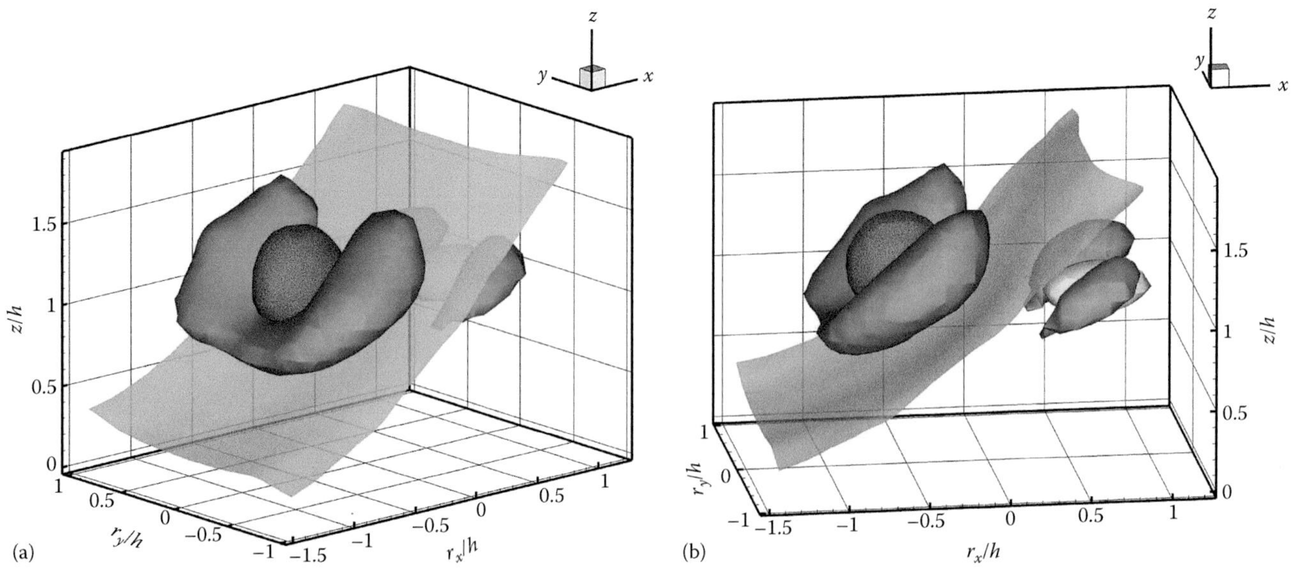

FIGURE 24.9 3D images depicting the composite eddy structure from canopy-resolving large-eddy simulation from two vantage points (Finnigan et al., 2009). (a) from above and behind right compared to the mean flow which is in the positive x-direction, and (b) from the side, looking in the positive y-direction. The isosurfaces shown in blue are of λ_2 at a value of −0.77 (See Jeong and Hussain (1995) for details of λ_2 analysis). The translucent sheet in green is an isosurface of zero scalar concentration perturbation from the horizontal average. The isosurface shown in orange is of $\overline{u'w'}$ at a value of $-0.6\,m^2\,s^2$ in the sweep region, while that in yellow is the same quantity at a value of $-0.15\,m^2\,s^2$ associated with the ejection. Note that the $\overline{u'w'}$ sweep isosurface is drawn at a value that is four times greater than that of the ejection.

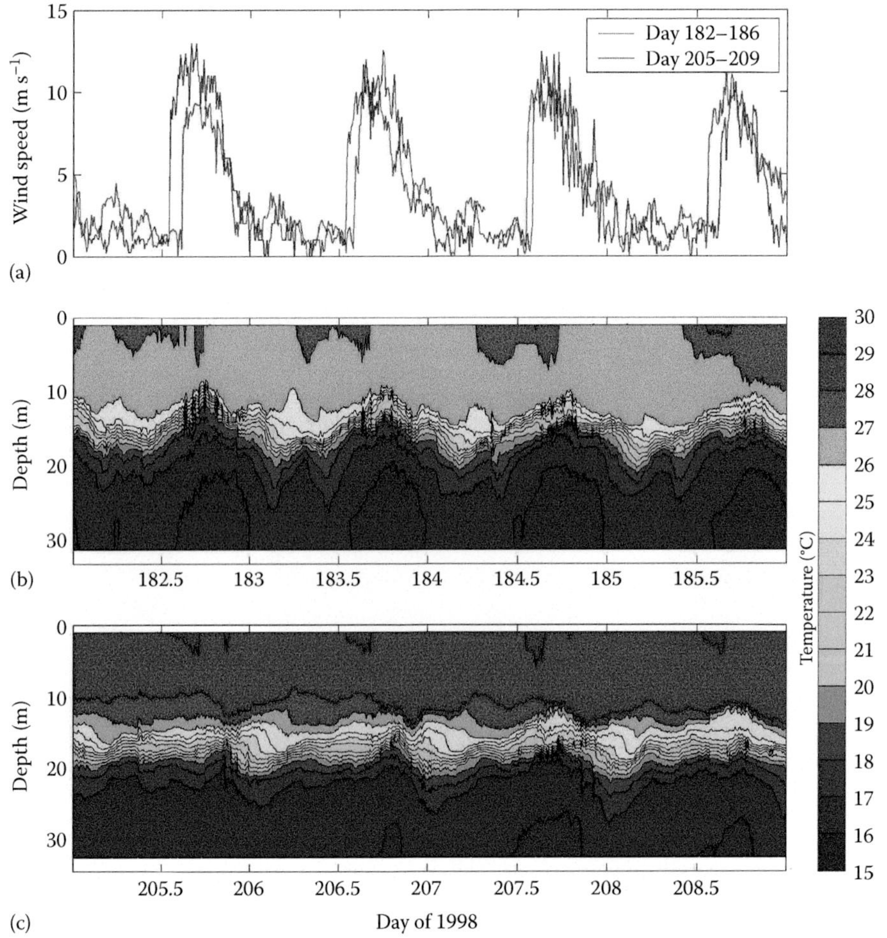

FIGURE 29.5 Internal Kelvin and Poincaré waves forced by periodic surface winds (a) as revealed by temperature measurements in Lake Kinneret, for resonant (b) and non-resonant (c) conditions. (Courtesy of J. Antenucci.)

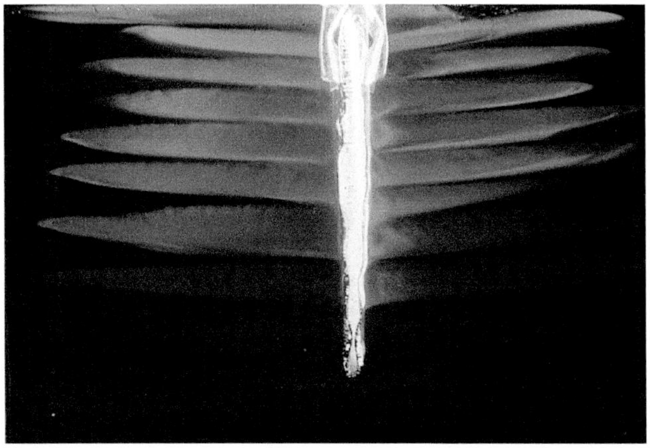

FIGURE 35.6 A vertical block of ice melting into a warm, stratified salt solution. Double-diffusive convection results in the formation of nearly horizontal, nearly uniform layers of water that are colder and fresher than the environment.

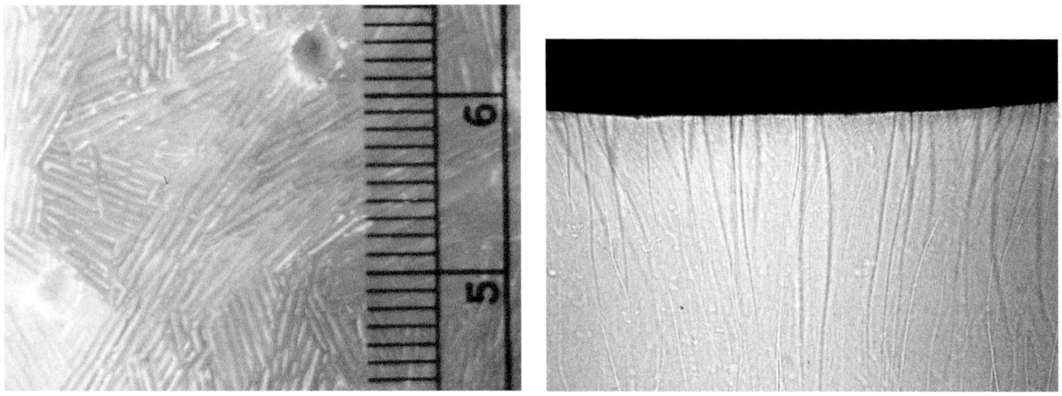

(a) (b)

FIGURE 35.7 (a) The underside of columnar sea ice, showing ice platelets spaced slightly less than a millimeter apart and two brine channels of approximately 3 mm diameter. (b) A shadowgraph image of the side view of growing sea ice (the black region at the top), showing the plumes of dense brine that emanate from brine channels and delver salt to the underlying ocean. The image is of a region approximately 20 cm wide

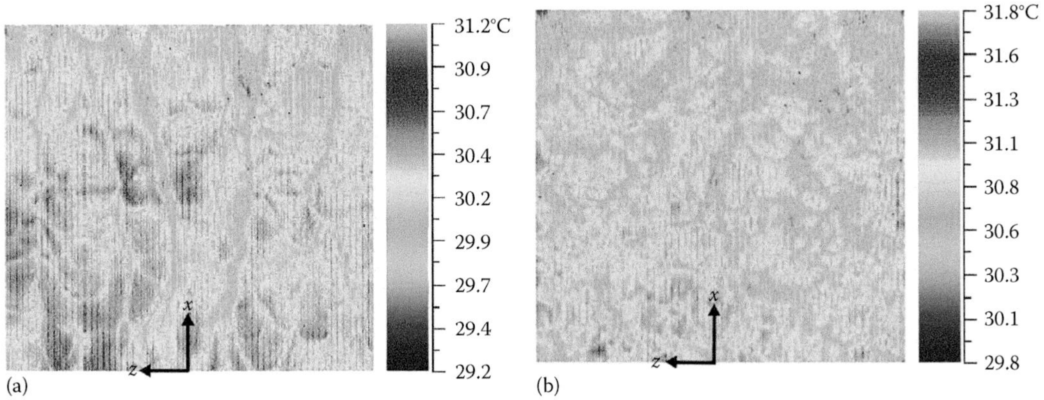

(a) (b)

FIGURE 36.6 Instantaneous surface temperature at (a) $U_\infty = 3.1$ m/s and (b) $U_\infty = 13.2$ m/s measured by a high-speed scanning infrared thermometer in a wind-wave tank. (From Komori, S. et al., Heat and mass transfer across the wavy sheared interface in wind-driven turbulence, In *Proceedings of 6th International Symposium on Turbulent Shear Flow and Phenomena*, Eds. Kasagi, N., Eaton, J.K., Friedrich, R., Humphrey, J.A.C., and Sung, H.J., Vol. II, pp. 399–407, Seoul National University, Seoul, Korea, 2009. With permission.)

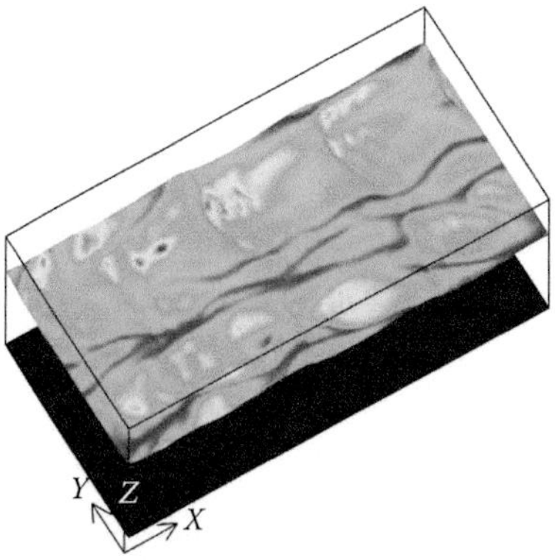

FIGURE 36.7 Instantaneous surface flux of virtual gas at $U_\infty = 5.2\,\text{m/s}$ predicted by the DNS. (From Komori, S. et al., Heat and mass transfer across the wavy sheared interface in wind-driven turbulence, In *Proceedings of 6th International Symposium on Turbulent Shear Flow and Phenomena*, Eds. Kasagi, N., Eaton, J.K., Friedrich, R., Humphrey, J.A.C., and Sung, H.J., Vol. II, pp. 399–407, Seoul National University, Seoul, Korea, 2009. With permission.)

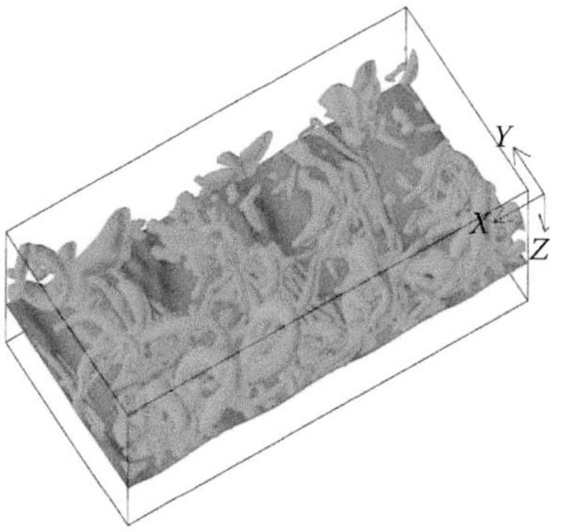

FIGURE 36.8 Instantaneous second invariant on the water side at $U_\infty = 5.2\,\text{m/s}$ predicted by the DNS. (From Komori, S. et al., Heat and mass transfer across the wavy sheared interface in wind-driven turbulence, In *Proceedings of 6th International Symposium on Turbulent Shear Flow and Phenomena*, Eds. Kasagi, N., Eaton, J.K., Friedrich, R., Humphrey, J.A.C., and Sung, H.J., Vol. II, pp. 399–407, Seoul National University, Seoul, Korea, 2009. With permission.)

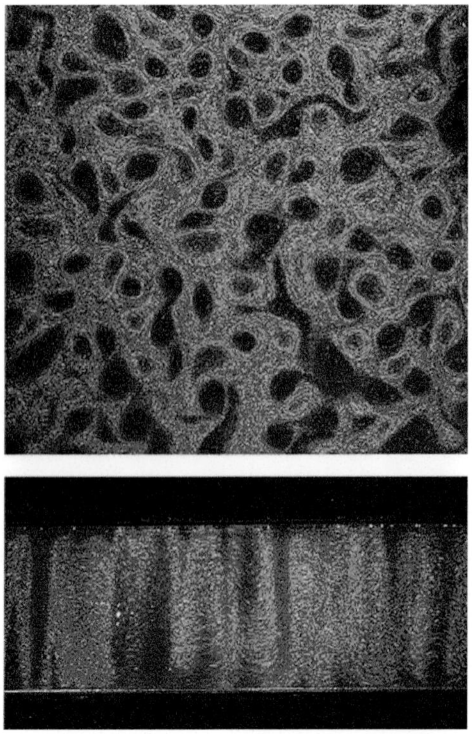

FIGURE 38.4 Thermochromic liquid crystal flow visualization of rotating convection in water with $Pr = 6$, $Ra = 8 \times 10^6$, $Ta = 10^8$, $Ro = 0.1$, $Ra/Ra_c = 8$. (Reproduced with permission from Sakai, S., *J. Fluid Mech.*, 333, 85, 1997. Copyright 2006 Cambridge Journals.)

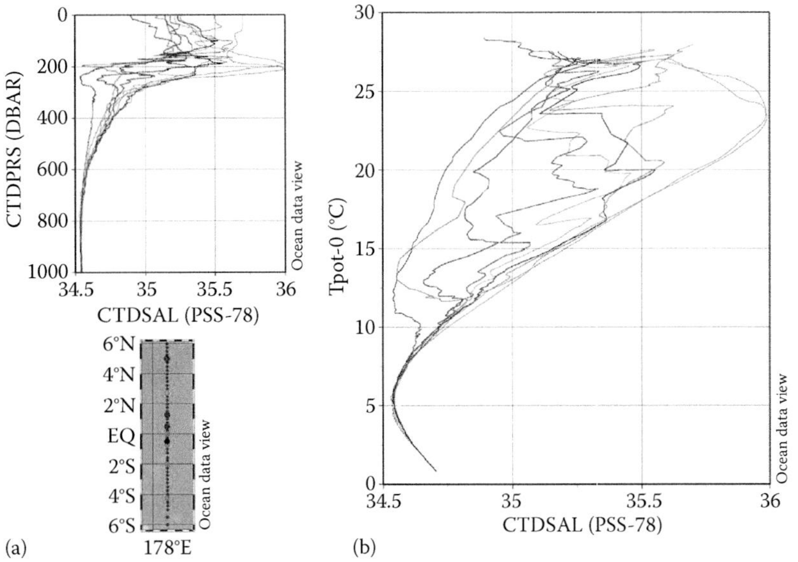

FIGURE 39.1 (a) Salinity (CTDSAL) versus depth (CTDPRS). (b) Potential temperature versus salinity, showing "sawtooth" characteristic of intrusions.

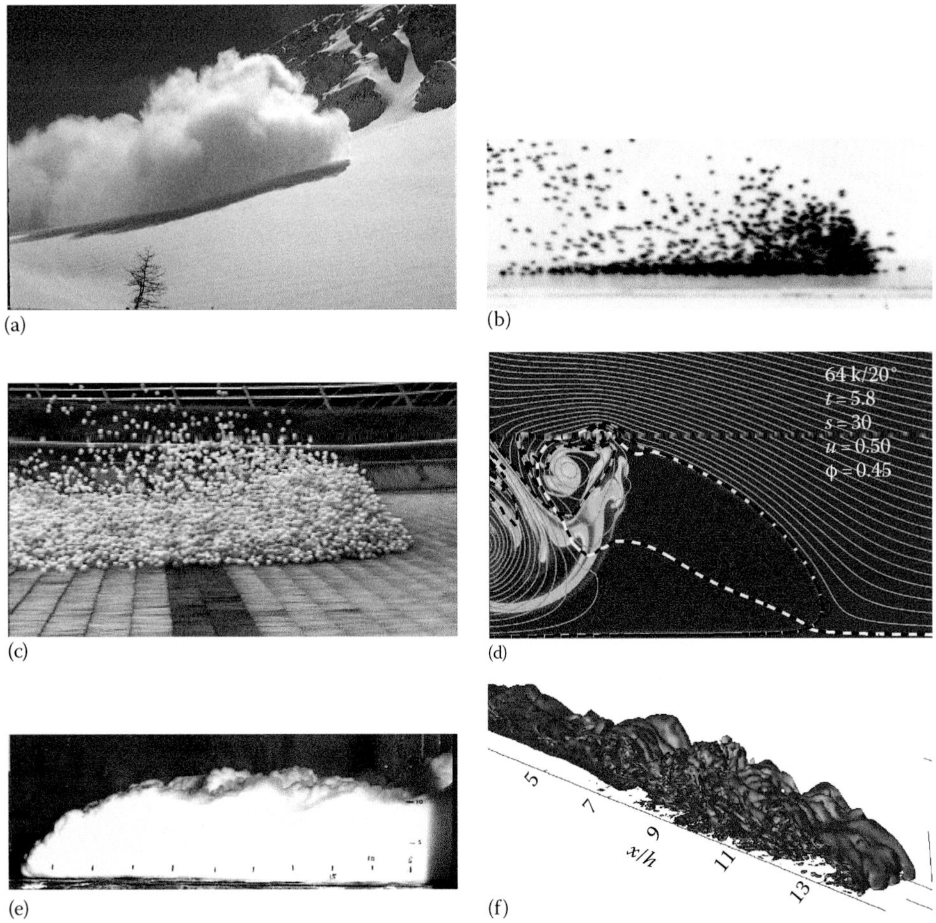

FIGURE 42.4 (a) Head of a powder snow avalanche (T. Castelle, Cemagref). (b) Head of a 1 L polystyrene ball avalanche 0.05 m high on 70° slope. (c) Head of a 550,000 ball ping pong ball avalanche approximately 0.5 m high. (d) Head of a 2D DNS on 20° slope at Re = 64,000. The color represents flow density and the streamlines in the frame of reference of the current are plotted. (e) Head of a turbidity current experiment. (From Baas, J.H. et al., *J. Geophys. Res.*, 110, C11015, 2005.) (f) Head of a 3D DNS showing iso-surfaces of concentration. (From Necker, F., et al., *Int. J. Multiphase Flow*, 28, 279, 2002.)

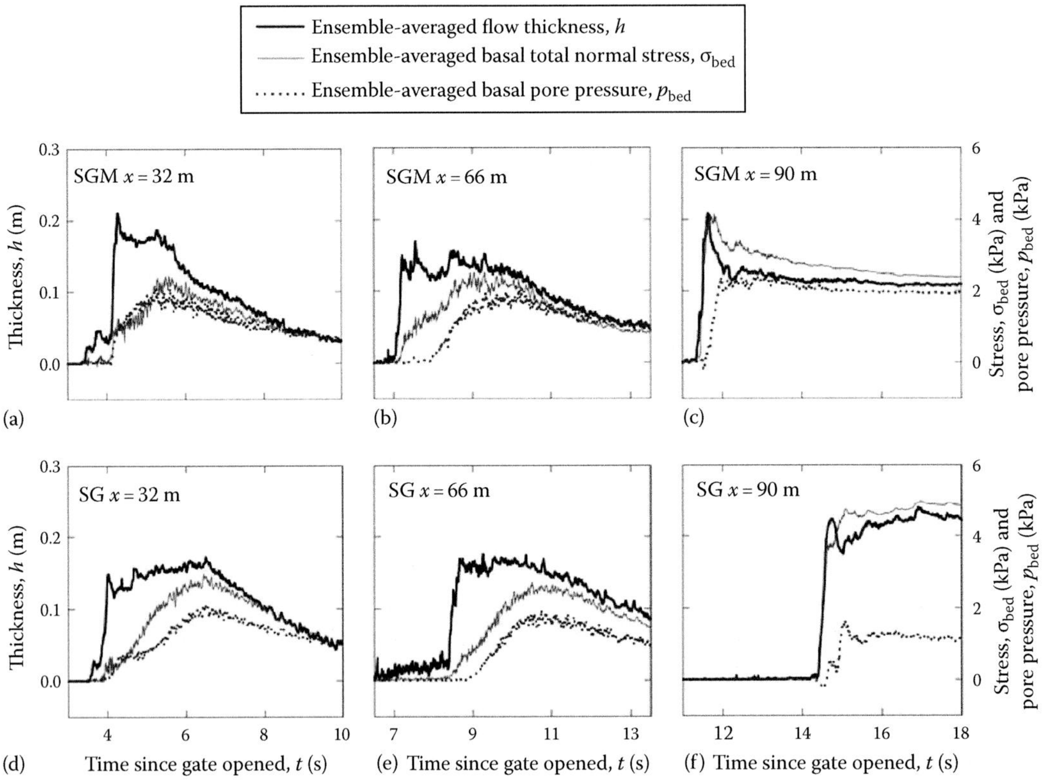

FIGURE 43.3 Ensemble averages of flow thicknesses, basal total normal stresses, and basal pore-fluid pressures measured at three distances from the headgate (x = 32, 66, and 90 m) in two sets of experimental debris flows. (a–c) Data from eight SGM flows containing 7% mud. (d–f) Data from nine SG flows containing <1% mud. (Reproduced by Iverson, R.M. et al., *J. Geophys. Res.*, 115, 2010, doi:10.1029/2009JF001514. With permission of the American Geophysical Union.)

17

Buoyant Outflows to the Coastal Ocean

Pablo Huq
University of Delaware

17.1 Introduction

Buoyant freshwater outflows from rivers, estuaries, or straits that results in extended offshore areas of low salinity waters. The influence of the low salinity strongly impacts the ecology of coastal waters, thereby affecting fisheries and the economies dependent on them. These buoyant outflows can form large recirculating bulges and downshelf propagating coastal currents. Here we review the progress in understanding the dynamics of bulges and coastal currents arising from buoyant outflows to the coastal ocean.

Typically, the waters of such freshwater outflows contain chemicals such as nitrates and phosphorous from agricultural runoff. Such chemicals adversely impact the health of the coastal ecosystem. Excessive amounts of nitrates and phosphorous can lead to the depletion of oxygen in coastal waters coupled to such outflows. The resulting anoxia and hypoxia of coastal waters is a worldwide problem. Thus, knowledge of transport rates and details of the structure of the freshwater flow through the coastal ocean is an important component of the dynamics as well as in the evaluation of fluxes of chemicals, biological materials, and sediments from land to the deep ocean.

Buoyant outflow to the coastal ocean is a multifaceted problem. Flows are driven principally by wind stress, density differences between the buoyant outflow and the ambient coastal ocean, and heat fluxes at the sea-surface. The Earth's rotation (i.e., Coriolis force), tides, and turbulent mixing further complicate the dynamics. These forcings were discussed in "The Sea Volume 10—Global Coastal Ocean," edited by Brink and Robinson (1998). The focus of this review of buoyant outflows to the coastal ocean is to discuss the role of the geometry of the outflow—in particular, the effects of the bottom slope, α, of the continental shelf; the angle, θ, of the longitudinal axis of the buoyant outflow to the downshelf coast; and the radius of curvature, r_c, of the coastline at the exit of the outflow. The interplay between these geometric variables influences the transport and path of the buoyant outflow.

The transport of suspended biota and sediments from the estuary to the coastal ocean depends on the "flow pathways" in the vicinity of the outflow exit. For example, observations have established that one flow pathway is the buoyancy-driven coastal current that can often form downshelf of the exit. (Downshelf is defined as the direction of Kelvin wave propagation. Downshelf is to the right in the northern hemisphere and to the left in the southern hemisphere.) Mean hydrographic properties of these coastal currents have been established from observations of the Alaskan coastal current, the Chesapeake Bay outflow, the Columbia River plume, the Connecticut River outflow, the Delaware coastal current, the Gaspe coastal current, the Hudson River outflow, the Rhine outflow, the Rio de la Plata outflow, the Scottish coastal current, the South Atlantic Bight river outflows, and the Vancouver Island coastal current. Figure 17.1 shows the coastal current that forms as the buoyant warm water flows through the Tsugaru Strait between Hokkaido and Honshu islands in Japan. It exits at Cape Shiriya and turns and flows southwards down the Sanriku coast in north-east Honshu.

Figure 17.1 also shows a second feature, a flow pathway that sometimes exists at the exit of buoyant outflows in the form

Handbook of Environmental Fluid Dynamics, Volume One, edited by Harindra Joseph Shermal Fernando. © 2013 CRC Press/Taylor & Francis Group, LLC. ISBN: 978-1-4398-1669-1.

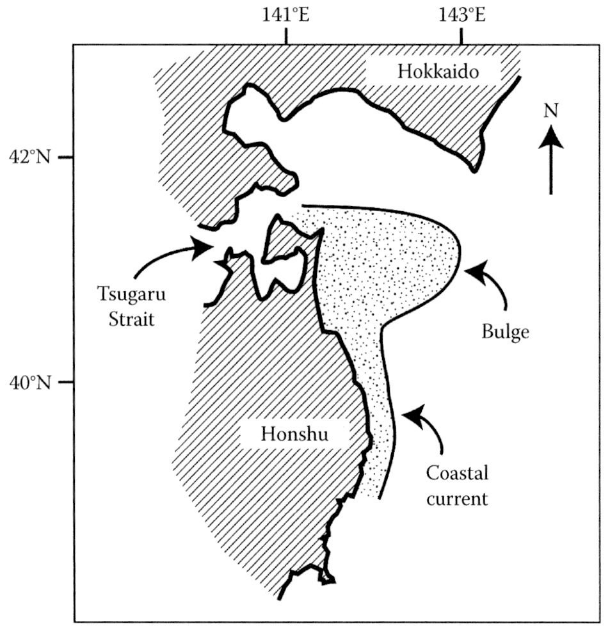

FIGURE 17.1 This shows the buoyant outflow exiting from the Tsugaru Strait in Japan in summer and autumn. Evident are a bulge and a coastal current further downshelf. (Adapted from Sugimoto, T., A review of recent physical investigation of straits around the Japanese Islands, in *The Physical Oceanography of Sea Straits*, ed., Pratt, L.J., Kluwer Academic, Dordrecht, the Netherlands, 1990.)

of a recirculating gyre or bulge. Details of this bulge feature are shown in Figure 17.2 of the contours of isotherms (°C) at a depth of 200 m of the bulge that forms during summer and autumn offshore of Cape Shiriya at the exit of Tsugaru Strait. The distribution of isotherms of the bulge is circular,

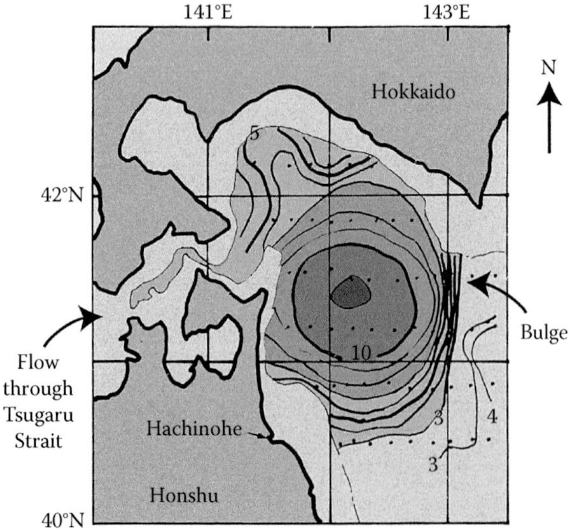

FIGURE 17.2 Detailed hydrography of the Tsugaru Bulge located offshore of Cape Shiriya. Shown are contours of temperature (°C) at 200 m. (Adapted from *Ocean Hydrodynamics of the Japan and East China Seas*, Kawasaki, Y. and Sugimoto, T., Experimental studies on the formation and degeneration process of the Tsugaru Warm Gyre, pp. 225–238, Copyright 1984, Elsevier.)

and of large scale—150 km in diameter and at least 200 m in depth. A consequence is that a large fraction of the buoyant water flowing through Tsugaru Strait is stored in the bulge. In the results discussed later in this review, it is demonstrated that such storage attenuates the scales of the coastal current. Bulges are characterized by anticyclonic vorticity, and their dynamics involve a cyclostrophic balance (Yankovsky and Chapman 1997).

It has been established that the geometry of the coastline, in particular the radius of curvature of the coastline and the outflow angle to the coastline, influences whether or not bulges form (Avicola and Huq 2003a,b; Horner-Devine et al. 2006). A schematic of a buoyant outflow to the coastal ocean is shown in Figure 17.3 to introduce the notation of this review. This shows a buoyant outflow of typical density ($\rho_o - \Delta\rho$) that exits from a river, estuary, or strait to a coastal ocean of typical density ρ_o. Thus the density anomaly is $\Delta\rho$. The offshore direction is y, and the downshelf direction is x, and the Coriolis parameter is f. The angle θ is the angle between the longitudinal axis of the buoyant outflow and the downshelf coastline. The local radius of curvature at the exit on the downshelf side is r_c. A region of extended offshore extent of buoyant water is designated as the bulge. Downshelf of the bulge is a propagating coastal current of offshore width γL and alongshore length L (the value of the slenderness ratio γ is less than one).

Hydrodynamic equations governing buoyant outflows are introduced in Section 17.2. A brief description of the rotating tank apparatus used for the laboratory experiments to simulate buoyant outflows follows in Section 17.3. Two sets of results are presented in Section 17.4. Discussed first are criterion for when bulges form and bulge characteristics. The scaling analysis of downshelf propagating coastal currents, and oceanic observations are considered in the second half of Section 17.4. Outstanding issues and challenges for the future are summarized in the concluding Section 17.5.

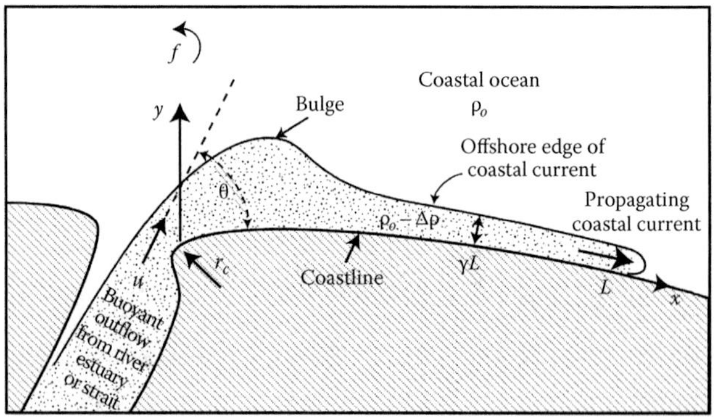

FIGURE 17.3 Schematic of bulge and coastal current arising from buoyant outflow from a river, estuary, or ocean. Buoyant outflow of density ($\rho_o - \Delta\rho$) exits at a velocity u to a coastal ocean of density ρ_o. x and y are the alongshore and offshore coordinates. The axis of the buoyant outflow is at an angle θ to the downshelf coastline. At the exit, the radius of curvature to the local coastline is r_c. The width γL of the coastal current is smaller than its length L. The Coriolis parameter is f.

17.2 Principles

Buoyant outflows to the coastal ocean are deflected by the influence of Earth's rotation and flow downshelf along a coastline in the direction of Kelvin wave propagation. The dynamics involves rotation, pressure gradients, and friction. Forcing can also include winds and tides. The conventional approach is to consider two layers of differing densities—($\rho_o - \Delta\rho$) for the buoyant layer of depth h, and ρ_o for the coastal ocean of depth H. For simplicity, a steady-state analysis with a constant outflow Q is undertaken. Tides and variability of winds are neglected. Garvine (1995) utilized a framework of continuity and vertically integrated continuity equations to describe the flow.

$$\frac{\partial}{\partial x}(hu)+\frac{\partial}{\partial y}(hv)=0 \tag{17.1}$$

This neglects entrainment or mixing into the buoyant layer h. Oceanic observations of buoyant outflows show a slender flow whose offshore width of the buoyant layer is smaller than its downshelf length (i.e., transverse velocity v is smaller than longitudinal velocity u). So

$$v \sim \gamma u \tag{17.2}$$

Vertically averaged equations in the longitudinal (x) and offshore (y) directions are

$$\frac{\partial}{\partial x}(u^2)+\frac{\partial}{\partial y}(uv)-fv=-\frac{1}{\rho_o}\frac{\partial p}{\partial x}+\frac{\tau_w^{(x)}}{\rho_o h}-\frac{\tau_b^{(x)}}{\rho_o h} \tag{17.3a}$$

$$\frac{\partial}{\partial x}(uv)+\frac{\partial}{\partial y}(v^2)+fu=-\frac{1}{\rho_o}\frac{\partial p}{\partial y}+\frac{\tau_w^{(y)}}{\rho_o h}-\frac{\tau_b^{(y)}}{\rho_o h} \tag{17.3b}$$

where

- f is the Coriolis parameter
- p is the pressure
- τ_w and τ_b are wind and bottom stress decomposed into the x and y direction

Scaling yields dimensionless numbers characterizing the flow. $\gamma \sim v/u$ is the slenderness ratio of the offshore to downshelf velocities. The Kelvin number $K = (\gamma L/R)$ is the ratio of the across-shelf length scale to the internal Rossby radius R. (Here $R = c/f$ with internal wave speed $c = (g'h)^{1/2}$ where h is the depth of the buoyant layer and g' is the reduced gravitational acceleration.) The internal Rossby radius R is the horizontal scale at which the effects of rotation become as important as restoring buoyancy. For outflows with $K > 1$ the dynamics are rotational, whereas for $K < 1$ the influence of the Coriolis force is small. The Rossby number $R_o = u/f\gamma L$ reflects the relative balance of nonlinear advection to Coriolis force. For $R_o \ll 1$, the dynamics of the flow field are linear. The Froude number $F = u/c$ is a measure of the importance of inertial and buoyancy forces. The value of the ratio (h/H) of the buoyant layer depth to the total oceanic depth establishes the buoyant outflows as either surface-advected ($h/H \sim 0.1$) or bottom-advected buoyant outflows ($h/H \sim 1.0$).

To summarize, the parameters describing the dynamics of rotating buoyant outflows include h/H, K, R_o, F. Wind and bottom stress may form Ekman layers at the top and bottom of the ocean. Thus friction contributes near boundaries; this is parameterized by the Ekman number $E_k \sim A_v/(fH^2)$, where A_v is a turbulent eddy viscosity. The different forcings and the large number of the parameters allow for the likelihood of future discovery of new flow phenomena. Presently known phenomena include flow separation, formation of bulges, and/or coastal currents.

17.3 Methods of Analysis

Numerical models such as ROMS (Regional Ocean Model System) are a popular analysis tool. They are based on the governing hydrodynamic equations. With regard to buoyant outflows exiting to the coastal ocean, there are little observational data on lateral velocity and density profiles across the exit to guide the specification of model boundary conditions. Thus a laboratory rotating tank model is useful to examine the role of geometry and bathymetry on the dynamics of buoyant outflows. Changes in exit angle θ, and radius of curvature r_c and fractional depth h/H are readily implemented in such studies. To isolate the effects of coastline geometry, we focus on buoyant outflows under conditions of weak winds and tides. Such conditions are set up by placing a Plexiglas sheet over the top of the tank to prevent wind-stress-induced motions, and balancing the entire rotating tank facility to minimize tidal motions.

The apparatus for two sets of laboratory experiments are described briefly. The first set examines flow separation, bulge formation, and their characteristics; the second set describes the scaling of coastal currents. For the first set, laboratory experiments are conducted on a large, flat-bottomed, rotating cylindrical tank. To simulate a river, estuary, or strait to the coastal ocean, a box is welded to the perimeter of the tank (see Figure 17.4b). The walls of the box make a 90° angle with the wall of the tank. False walls are available for placement within the box, which allows experiments to be performed with varying exit angles, θ, outflow widths, W, and radius of curvature, r_c, at the exit of the outflow to the coastal ocean. Buoyant outflow of density ($\rho_o - \Delta\rho$) is introduced through a pipe at the most upstream end of the box at a known outflow rate Q from a freshwater reservoir that co-rotates with the rotating tank. The tank rotates at a specified rotation rate, f, which can be carefully controlled. The tank (i.e., the coastal ocean) is filled with salt water of density ρ_o to a depth H, and allowed to spin up to solid body rotation before experiments are conducted. The value of the fractional depth h/H of the buoyant layer to the total depth was 0.1, so that the buoyant outflows of the first set of experiments are surface-advected; the exit Kelvin number $K_e = W/R = 1$.

The second set of experiments was similar to the first, except the box is closed off—so that the tank's plan geometry is circular. The source, a pipe, discharged parallel to the wall (i.e., coast).

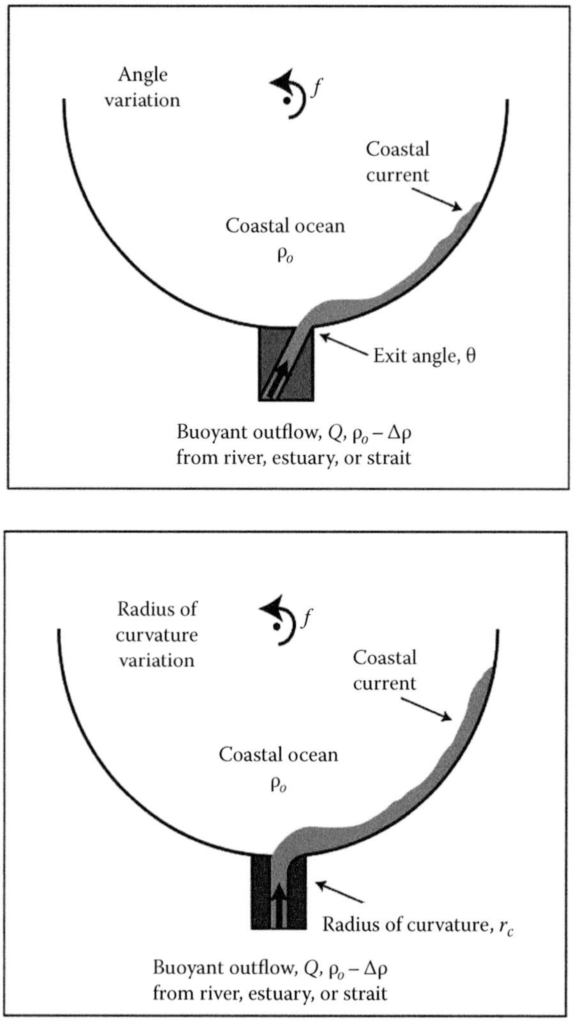

FIGURE 17.4 Schematics of rotating turntable experiment of buoyant outflows to the coastal ocean. The dark area is a box that is welded to the perimeter of a cylindrical tank. The panels (a) and (b) show buoyant outflow configurations with variation of exit angle θ or variation of radius of curvature. (Adapted from Avicola, G. and Huq, P., *J. Marine Res.*, 61, 435, 2003a.)

Experiments were conducted with a flat-bottomed tank and also a tank with a sloping-bottom for the second set of experiments so as to discriminate the effects of the slope (i.e., vortex stretching). The fractional depth h/H varied between about 0.1–1 spanning the range between surface- and bottom-advected outflows. (See Figure 17.11 for a schematic of the flow field of the second set of experiments.)

The rotating tank facility is equipped to interrogate buoyant outflows. Whole field velocity measurements are obtained through the Particle Image Velocimetry (PIV) technique utilizing surface drifters. A digital camera in conjunction with a PC is utilized to record visualizations and analyze the experiments. In addition, 3D density fields are determined from density transects using traversing micro-probes.

Vertical, horizontal, and velocity scales are nondimensionalized by the scale depth of the buoyant layer h, the internal Rossby radius R, and the internal wave speed c. Here,

$$h = \sqrt{\frac{2Qf}{g'}} \quad R = \frac{\sqrt{g'h}}{f} \quad c = \sqrt{g'h} \tag{17.4}$$

finally, time t is nondimensionalized by the rotation period of the rotating turntable $T = 4\pi/f$ so that $t/T = 1$ is equivalent to a day.

17.4 Applications/Results

A general configuration for buoyant outflow exiting to the coastal sea is shown in Figure 17.5. Results to follow can be used to make predictions for coastal buoyant outflows if scales such as the internal Rossby radius R and flow rate Q, etc., are known.

17.4.1 Flow Separation

The radius of curvature at the exit is r_c. The trajectory of the outflow is deflected by the Coriolis force: the exiting buoyant outflow follows an inertial circle of radius u/f. The ratio of the radius of curvature to the radius of the inertial circle is important in determining whether the buoyant outflow will separate from the coastline. If r_c is smaller than u/f, then the flow will separate from the right coastline. (In the southern hemisphere it will separate from the left coastline.) The ratio of radii forms the exit Rossby number $R_o = u/fr_c$; for values of R_o greater than one, flow separation occurs. The buoyant outflow impacts or reattaches to the coastline further downshelf from the exit at an angle φ.

The exit angle, θ, of the outflow to the coastline also influences flow separation. Buoyant outflows emerging perpendicular or at obtuse angles to the coastline are more likely to separate than flows exiting at acute angles to the coastline. Also the

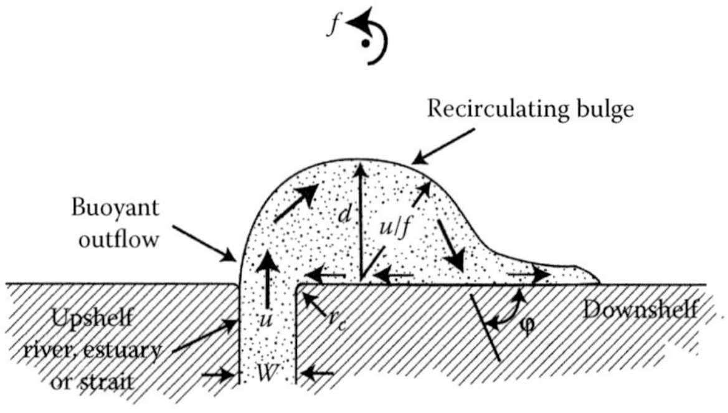

FIGURE 17.5 Schematic of a buoyant outflow from a river, estuary, or strait. The width of the exit is W; the velocity at the exit is u; the radius of curvature between the exit and the coastline is r_c. The buoyant outflow forms a recirculating, semicircular bulge of radius d. Also shown is an inertial circle of radius u/f. As drawn $d = u/f$. The buoyant outflow recirculates and impacts the downshelf coastline at an angle φ. A fraction of the buoyant outflow recirculates toward the source, with the remainder forming a coastal current which propagates downshelf.

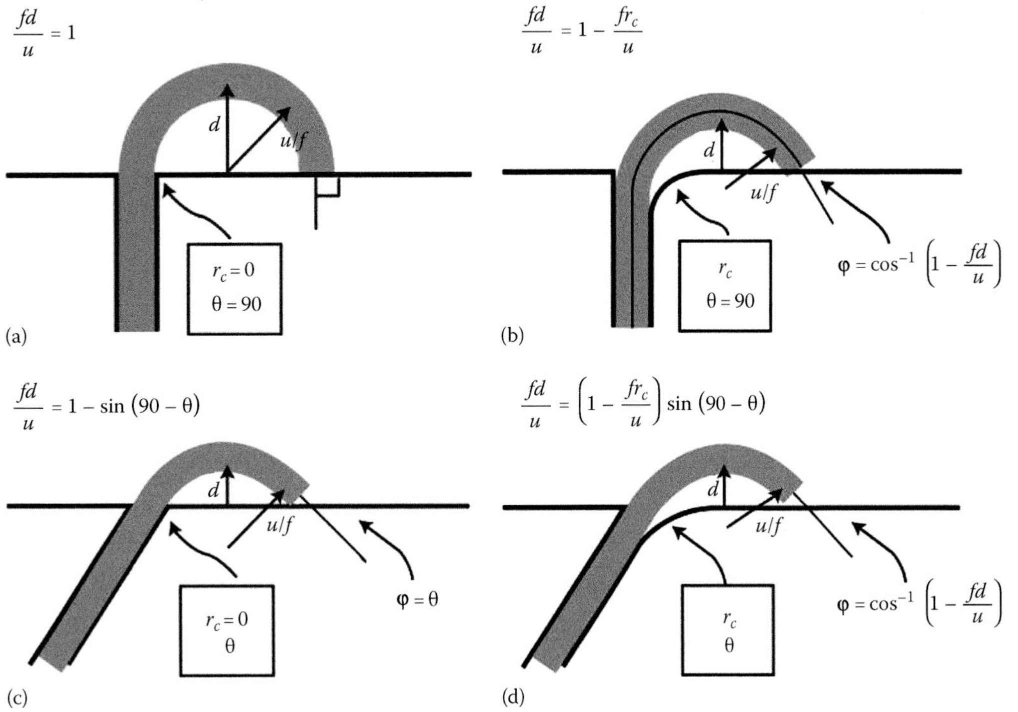

FIGURE 17.6 Schematics of the trajectory of the buoyant outflow (thick shaded line) are shown for various permutations of exit angle θ, radius of curvature r_c, and inertial circle *u/f*; these variables determine the downshelf impact angle φ. The upper panels depict an outflow exiting perpendicularly to the coastline (θ = 90°); the lower panels are for outflows exiting at angle (90 − θ) to the coastline. Panels (a) and (c) describe sharp-cornered exits with zero radius of curvature r_c; panels (b) and (d) have non-zero r_c. General expressions for the separation ratio *fd/u* and impact angle φ are also given. (Adapted from Avicola, G. and Huq, P., *J. Marine Res.*, 61, 435, 2003a.)

offshore extent, *d*, of outflows perpendicular to the coastline is greater than outflows exiting parallel to the coastline. Different scenarios or configurations arising from combinations of *u/f*, r_c, and θ are shown in Figure 17.6.

Figure 17.6a and b show configurations with an exit angle θ = 90°, and cases with negligible and finite values of radius of curvature r_c. The lower panels show the situation for cases when θ ≠ 90° with negligible and finite values of r_c. The combined effects of the exit angle and radius of curvature can be examined in terms of the ratio Γ = *df/u*. The value of Γ varies between zero and one; buoyant outflows do not separate for Γ = 0. Maximum values of Γ occur when the separation distance *d* is equal to the inertial turn radius r_c. If buoyant outflows do not separate from the coastline, then only coastal currents are likely. Conversely, if flow separation occurs, then a flow with both a bulge and a coastal current are likely. Experimental data suggest a critical value of Γ = 0.5 (see Figure 17.7).

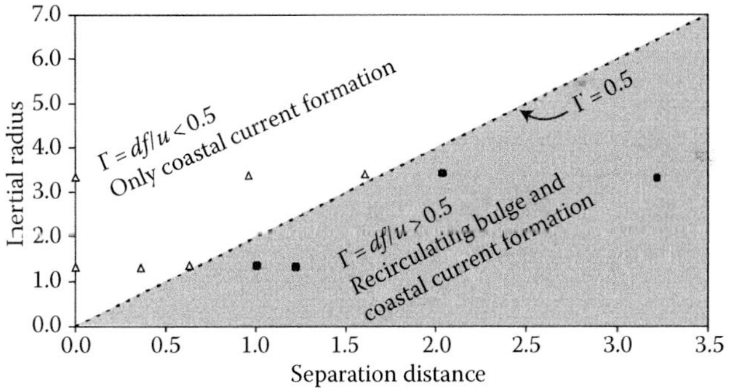

FIGURE 17.7 Comparison of predicted inertial radius *f/u* and separation distance *d*. The line Γ = 0.5 delineates flows where only coastal currents form from flows where both bulges and coastal currents form. Coastal currents only formed for Γ < 0.5 (triangles); both recirculating bulge and coastal currents form for Γ > 0.5 (circles in shaded region). (Adapted from Avicola, G. and Huq, P., *J. Marine Res.*, 61, 435, 2003a.)

The shaded trajectory of the buoyant outflow in Figure 17.6 suggests that the flow approach to the downshelf coastline is perpendicular when the separation distance *d* is equal to the inertial radius *f/u* (Figure 17.6a). In contrast, the approach is less than perpendicular and more parallel to the coastline for separation distances less than the inertial radius. Thus, the separation ratio Γ can also be interpreted in terms of the impact angle φ = cos^{-1}(1 − *fd/u*) at which the buoyant outflow impacts the coastline. This is useful as the magnitude of φ determines whether or not bulges form and so determines flow pathways. The dynamics of the point of impact has been analyzed as the deflection of a baroclinic jet by a wall in a rotating frame of reference (Whitehead 1985). Analysis showed that volume and momentum fluxes are increasingly recirculated back toward the outflow exit with increasing impact angle φ (see Figure 17.8). For φ > 60° more than 50% of the volume flux is recirculated: this promotes bulge formation. The balance or remaining volume

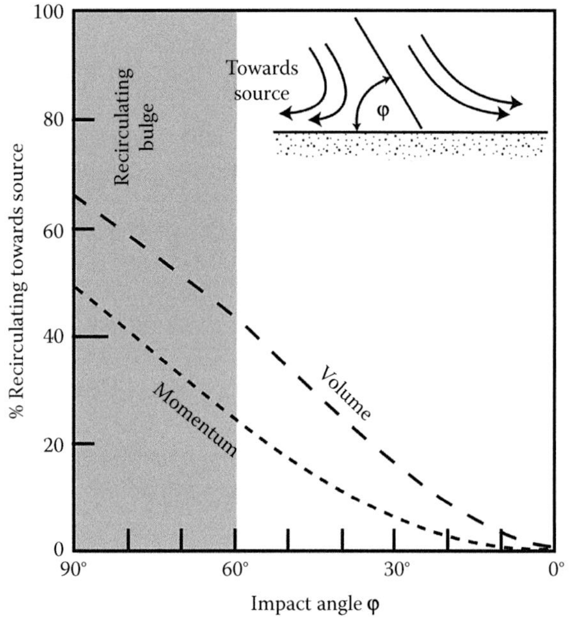

FIGURE 17.8 Schematic of fluxes for a buoyant flow impacting a wall. Shown is a plot of percentage of momentum and volume flux returning to the source region as a function of impact angle φ. Recirculating bulges form for impact angles greater than 60° as indicated by the shaded region.

flux propagates downshelf as a coastal current. The recirculation of the flow has obvious consequences. For example, the residence times of fluid in the bulge will be much longer than estimates based solely on advection.

17.4.2 Characteristics of Bulges

When bulges form, they are among the largest coherent structures in the coastal ocean. Figure 17.2 shows that the bulge forming at the exit of the Tsugaru Strait is approximately 150 km. They are crucially important to the dynamics and biology of the coastal ocean. Experimental data of the evolution of the bulge length, width, depth, and volume are presented in nondimensional form in Figure 17.9 so as to allow comparison with field data. Nondimensional length and width of the bulge evolved as $t^{0.40}$, and depth evolved as $t^{0.16}$. It is insightful to compare the dynamics of a recirculating bulge with that of a stable buoyant lens (or vortex) formed from a constant volume flux source. This yields growth rates of radius evolving as $t^{1/4}$ and depth evolving as $t^{1/2}$ (Nof and Pichevin 2001; Avicola and Huq 2003b). Comparison shows that the measured growth rates of horizontal scales are larger ($t^{0.40}$ compared to $t^{1/4}$) and that vertical scales are smaller ($t^{0.16}$ compared to $t^{1/2}$) than those of a stable, buoyant lens. The difference arises because bulges are unstable. Stability is evaluated by comparing the ratio, J, of rotational kinetic energy

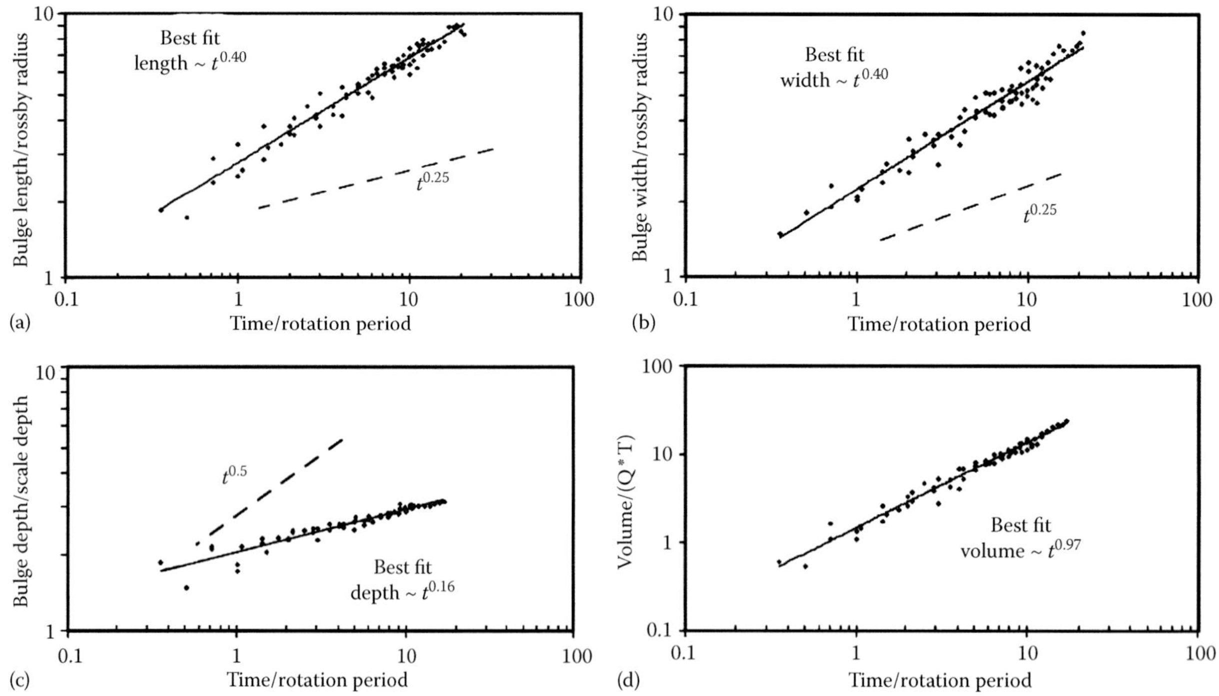

FIGURE 17.9 Evolution of the nondimensional length, width, depth, and volume of bulges. The abscissa is nondimensionalized by the rotation period (numbers correspond to days). In panels (a) and (b) the ordinate is nondimensionalized by the internal Rossby radius R; in panels (c) and (d) the ordinate is nondimensionalized by the scale depth h of the buoyant layer and the product of time t and flow rate Q. Solid lines show the best-fit power laws to the experimental data. Dashed lines are power laws for the evolution of a stable, buoyant lens. The different growth rates for the solid and dashed lines suggest that bulges are unstable. Exit Kelvin number $K_E = W/R = 1$. (Adapted from Avicola, G. and Huq, P., *J. Marine Res.*, 61, 411, 2003b.)

to the available potential energy in the lens ($J = r_b^2 f^2 / g' H_b$, where r_b and H_b are the measured bulge radius and depth at the bulge centerline). Values of the instability parameter J greater than one indicate instability. Measurements indicate that values of $J > 1$ arise as early as time $t < 1T$ (i.e., less than a day, see Figure 17.10 of Avicola and Huq 2003b). Instability augments growth rates in the horizontal direction, but attenuates in the vertical. Observations of the Tsugaru bulge show that it is unstable (Sugimoto 1990).

An important consequence of bulge formation that affects coastal dynamics is the fact that recirculation and growth of the bulge results in storage of outflowing buoyant waters. This storage delays and attenuates the volume of buoyant water propagating downshelf. Without bulge formation, the entire buoyant outflow forms a coastal current that flows downshelf along the coastline. With bulge formation, only about a half of the buoyant outflow is available to form a coastal current. Bulge formation thus impacts the scales of the downshelf propagating coastal current. Data showing this are presented in Figure 17.10. For identical experimental parameters, data are presented for downshelf propagating coastal currents with variations in either the exit angle θ (Figure 17.10a and b) or variations of radius of curvature r_c (Figure 17.10c and d). Comparison show similar evolutions. This provides further support that it is the combined effects of θ and r_c, expressed by the impact angle φ, that is dynamically important. Nondimensional distances (Figure 17.10a and c) propagated by the coastal current for bulge-less outflows are typically 30% larger. Similarly, Figure 17.10b and d show that nondimensional widths of coastal currents for bulge-less outflows are 25% greater than outflows with bulges.

These results can be used to assess aspects of the dynamics of bulges and coastal currents in the ocean. For example, regarding the buoyant outflow at the Tsugaru Strait between Hokkaido and Honshu, Figure 17.2 shows a bulge with circular temperature contours at 200 m with a diameter of about 150 km. Estimates of the internal Rossby radius R and scale depth h for the Tsugaru strait outflow are 16 km and 80 m, respectively, from Table 1 of Avicola and Huq 2002. The data of Figure 17.2 yield a bulge diameter of about 150 km; this corresponds to nondimensional length and width of about 10 R. Figure 17.9a and b indicate that nondimensional length and width of 10 R occur at nondimensional time of 20 days. At 20 days, Figure 17.9c indicates a nondimensional depth of 3 h corresponding to a bulge depth of about 240 m. As local ocean depths are about 300 m, this suggests that the bulge at the Tsugaru Strait is influenced by the proximity of the ocean bottom or may even extend to the ocean bottom.

17.4.3 Scaling of Downshelf Propagating Coastal Currents

The schematics of a downshelf propagating coastal current over a coastal ocean of bottom slope α are presented in Figure 17.11. The depth of the buoyant layer is h, and the depth of the coastal ocean is H. The upper panel shows a relatively thin buoyant layer $h/H < 1$—this configuration is termed surface-advected.

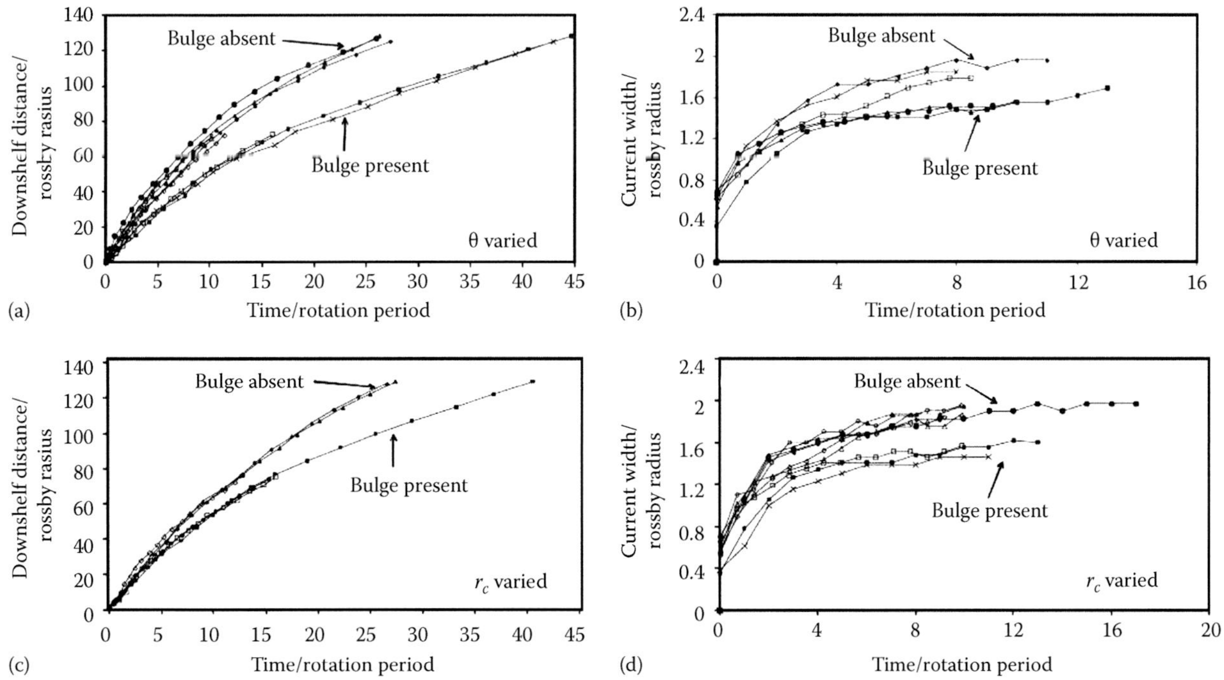

FIGURE 17.10 Evolution of the nondimensional length and width of coastal currents. Ordinate is nondimensionalized by the internal Rossby radius R, and abscissa is nondimensionalized by the rotation period. The exit angle θ was varied in the experiments of panels (a) and (b); the radius of curvature r_c was varied in the experiments of panel (c) and (d). Scales of coastal currents are smaller when bulges are present for all cases due to storage in bulge. Exit Kelvin number $K_E = W/R = 1$. (Adapted from Avicola, G. and Huq, P., *J. Marine Res.*, 61, 411, 2003b.)

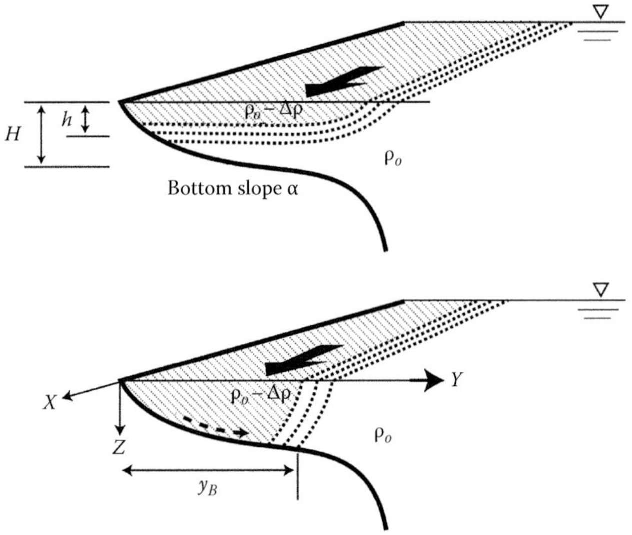

FIGURE 17.11 Schematic of a buoyant coastal current of density ($\rho_o - \Delta\rho$) propagating over a sloping-bottom continental shelf. The scale depth of the buoyant layer is h (see Equation 17.4); the depth of the coastal ocean of density ρ_o is H (defined at an offshore distance approximately equal to the internal Rossby radius R); the slope of the bottom is α. The dashed arrow indicates offshore Ekman transport. The upper panel shows a surface-advected coastal current ($h/H \ll 1$). The lower panel shows a bottom-advected coastal current ($h/H \sim 1$). The dashed arrow denotes offshore Ekman transport; also shown is y_B—the offshore extent at the bottom, or bottom-trapped width, of the buoyant coastal current. (Adapted from Chapman, D.C. and Lentz, S.J., *J. Phys. Oceanog.*, 24, 1464, 1994.)

The lower panel shows the case where the buoyant layer extends to the bottom so that $h/H = 1$—a bottom-advected case. The dashed arrow at the bottom of the lower panel indicates offshore transport of buoyant water by a thin Ekman layer. The buoyant layer above the Ekman layer is in geostrophic balance. The Ekman layer extends to a distance y_B offshore—the location at which the downshelf velocity of the geostrophic layer vanishes at the level of the Ekman layer.

The flow configuration of Figure 17.11 was analyzed by Lentz and Helfrich (2002) and Avicola and Huq (2002). They recognized the important role of the bottom Ekman layer as an additional flow pathway that transports buoyant water in the offshore direction. The dynamics of the flow were characterized by a two-variable nondimensional parameter space: the ambient depth parameter, h/H, and the bottom-slope parameter R/y_B. The ratio h/H is the fraction of the available depth H occupied by the buoyant layer. The scale depth, h, derived from geostrophic dynamics, is a measure of the depth of the buoyant layer. The bottom-slope parameter is the ratio of two horizontal scales, the internal Rossby radius, R, to the bottom-trapped width y_B. An estimate for y_B is

$$y_B \sim \frac{f\rho}{\alpha g}\frac{u_{max}}{\partial\rho/\partial y} \tag{17.5}$$

here u_{max} is the downshelf propagation speed of the leading edge or nose of the coastal current (see Figure 17.3). Scaling by Equation 17.4 reduces Equation 17.5 to the simple form $y_B \sim h/\alpha$. Thus in nondimensional form, the bottom-slope parameter is

$$\frac{R}{y_B} = \frac{\alpha R}{h} \tag{17.6}$$

Noting that the average isopycnal slope in the buoyant layer is h/R, we find that the bottom-slope parameter R/y_B can also be interpreted as the ratio between the bottom slope α and the average isopycnal slope h/R. (See Avicola and Huq [2002] for discussion and derivation of Equation 17.5.)

The two parameter space (h/H, R/y_B) together with observations of coastal currents values of the parameters are plotted on Figure 17.12. For an idealized coastal ocean with a single bottom slope α, R/y_B and h/H are related through the definition of H ($=\alpha R$). The only possible values in the parameter space falls on the line $h/H = (R/y_B)^{-1}$, which is shown as a dashed line on Figure 17.12. Values of coastal current parameters are constrained to lie below the line $h/H = (R/y_B)^{-1}$ for realistic coastal bottom bathymetry. Increasing values of the depth parameter h/H reflects increasing interaction between the coastal current and the bottom bathymetry. Coastal currents with $h/H \sim 0.1$ are isolated from the bottom (surface-advected); on the other hand, coastal currents with $h/H \sim 1$ are bottom-trapped with pronounced offshore transport of buoyant waters via the bottom Ekman layer.

A consequence of offshore transport of buoyant water is attenuation of downshelf propagation speed of coastal currents. Thus in the limit of bottom-trapped coastal currents, propagation speed u_{max} is set (slope-controlled) by the slope as ($\alpha g'/f$). Isopycnal slopes are flattened and isopycnal gradients $\partial\rho/\partial y$ are weakened if y_B is smaller than R. That is coastal currents undergo lateral expansion for $R/y_B < 1$. For $R/y_B > 1$ coastal currents are compressed closer to the shore.

The framework of the two parameter space (h/H, R/y_B) for the scaling of coastal currents has been tested and verified via both

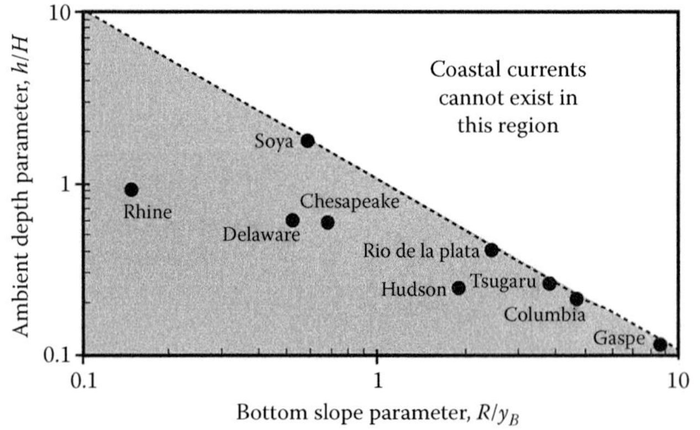

FIGURE 17.12 Log–log plot of field data of coastal currents plotted in the parameter space (h/H, R/y_B). Field data lie below the line $h/H = (R/y_B)^{-1}$ in agreement with scaling arguments. (Adapted from Avicola, G. and Huq, P., *J. Phys. Oceanog.*, 32(11), 3233, 2002.)

laboratory experiments and coastal ocean data. First, characteristics of surface-advected coastal currents ($h/H \sim 0.1$) are independent of the value of the bottom-slope parameter R/y_B. In contrast, the evolution of downshelf propagation distance and offshore extent varied with whether the value of R/y_B was greater or less than one for bottom-trapped coastal currents ($h/H \sim 1$). And second that coastal current propagation speeds varied with the slope as $(\alpha g'/f)$. Available data of coastal currents fall in the permissible half-space below the line $h/H = (R/y_B)^{-1}$ on Figure 17.12. Finally, density transects through both laboratory and oceanic coastal currents confirm that isopycnal slopes are of order (R/h).

17.5 Major Challenges

The fluid dynamics of buoyant outflows to the coastal ocean is a complex interplay between buoyancy, rotation, friction, pressure gradients, and coastline geometry. The approach of analysis of the transport of momentum and buoyancy based on steady, depth-averaged hydrodynamic equations combined with scaling analysis and exploration of the parameter space via laboratory experiments has yielded insights and useful results. However, a weakness of such approaches, as noted by Garvine (1995), is that it relies on the knowledge of coastal current scales such as the depth of the buoyant layer, h, the internal Rossby radius, R, and internal wave phase speed, c. A goal for the future is to develop approaches for predicting the characteristics of coastal currents based on the knowledge of far upstream hydrography. Some outstanding issues that seem within reach are discussed in the following to conclude this review.

An obvious shortcoming of the approach taken here is the assumption of a steady flow. Oceanographic observations of buoyant outflows show that flow rates vary seasonally through the year. Thus a time-averaged approach is an oversimplification. The bulge can also be unstable. More work on the mechanics of the instability is needed as instability influences transport and mixing between buoyant outflow and ambient coastal waters. Tidal time scales are also likely to be important. While it is known that tides stabilize bulge formation, the effect of tides on other aspects of the outflow is unknown.

Meteorological variability and its impact on the dynamics of bulges and coastal currents is another unresolved issue. The trajectory of the Columbia River buoyant outflow varies with seasonal wind direction—it flows to the north in the winter and to the south in the summer. Thus the effects of the variability of wind direction (and strength) need to be assessed. The structure of the buoyant outflows from Delaware Bay and Hudson River can be disrupted by upshelf (upwelling) winds, with the result that buoyant waters can move upshelf and offshore. Conversely, propagation speeds of coastal currents are augmented by winds blowing downshelf.

The depth of the wind-driven surface Ekman layer may differ from the depth of the buoyant layer. Thus it is likely that the effect of winds on very large-scale buoyant outflows such as the Tsugaru (flow rates $Q \sim 10^6\,m^3/s$) will differ from smaller-scale outflows such as the Rhine, Chesapeake, and Delaware outflows where $Q \sim 10^3\,m^3/s$. The effects of persistence or duration of winds is also unknown. A third issue of meteorological variability on the dynamics of buoyant outflows is the question of baroclinic forcing due to inputs of buoyancy at the sea surface. This has not been addressed; however, it is likely to alter the potential energy anomaly due to the density contrast between the buoyant waters and the ambient coastal ocean. Clearly, meteorological variability is a critical issue.

Another question that needs to be addressed is the interaction between buoyant outflows and the larger-scale coastal circulation. For example, how do ambient shelf currents impact bulge formation and propagation of buoyant coastal currents? The focus of recent work has concerned outflows equal in exit width, W, to the internal Rossby radius (i.e., $K_E = W/R = 1$). Typically, in these studies the velocity profile of the buoyant outflow is laterally uniform. Recent numerical and laboratory experiments of wide outflows $K_E > 1$ by Valle-Levinson (2008) and Huq (2009) show that lateral velocity shear can arise if frictional effects are significant. They also found that rotation produces separation of the buoyant outflow from the left-hand side (looking seaward) of the outflow. This results in dynamical isolation of the outflow and inflow—buoyant outflow occurs on the right-hand side (looking seaward) and inflow on the left. (Reverse left and right in the southern hemisphere.) Surprisingly, little attention has been paid to the dynamics associated with inflow. Inflow and attendant mixing will alter the density field upstream of the outflow exit. The density of the buoyant outflow will obviously depend on the inflow and mixing within the estuary. Wide estuaries have room for expansion: thus differential advection, tidal straining, and secondary circulation may play even more surprising roles than they do for a $K_E = 1$ buoyant outflow. To reiterate, flow pathways for wide estuaries ($K_E > 1$) differ from that of narrow estuaries ($K_E \sim 1$). It will be useful to determine the characteristics of and magnitude of transport through these flow pathways, and their dependence upon width and friction, parameterized by K_E and h/H.

The picture emerging from this review of buoyant outflows to the coastal ocean is of deterministic coherent structures such as bulges and coastal currents constituting flow pathways in the coastal ocean. Such flow pathways determine the transport and dispersion of pollution, and distribution and deposition of embedded sediments on the continental shelf. As they impact the biology and productivity of coastal waters, bulges and coastal currents are vital components of the health of the fragile coastal environment.

References

Avicola, G. and P. Huq, 2002, Scaling analysis for the interaction between a buoyant coastal current and the continental shelf: Experiments and observations, *Journal of Physical Oceanography*, 32(11), 3233–3248.

Avicola, G. and P. Huq, 2003a, The characteristics of the recirculating bulge region in coastal buoyant outflows, *Journal of Marine Research*, 61, 435–463.

Avicola, G. and P. Huq, 2003b, The role of outflow geometry in the formation of the recirculating bulge region in coastal buoyant outflows, *Journal of Marine Research*, 61, 411–434.

Brink, K.H. and A.R. Robinson, eds., 1998, *The Sea Volume 10: The Global Coastal Ocean*, John Wiley & Sons, New York, p. 604.

Chapman, D.C. and S.J. Lentz, 1994, Trapping of a coastal density front by the bottom boundary layer, *Journal of Physical Oceanography*, 24, 1464–1479.

Garvine, R.W., 1995, A dynamical system for classifying buoyant coastal discharges, *Continental Shelf Research,* 15, 1585–1596.

Horner-Devine, A., D.A. Fong, S.G. Monismith, and T. Maxworthy, 2006, Laboratory experiments simulating a coastal river inflow, *Journal of Fluid Mechanics*, 555, 203–232.

Huq, P., 2009, The role of Kelvin number on bulge formation from estuarine buoyant outflows, *Estuaries and Coasts,* 32, 709–719.

Kawasaki, Y. and T. Sugimoto, 1984, Experimental studies on the formation and degeneration process of the Tsugaru Warm Gyre, in *Ocean Hydrodynamics of the Japan and East China Seas*, ed. T. Ichiye. Oceanography Series. No. 39, Elsevier, Amsterdam, the Netherlands, pp. 225–238.

Lentz, S.J. and K.R. Helfrich, 2002, Buoyant gravity currents along a sloping bottom in a rotating fluid, *Journal of Fluid Mechanics*, 464, 251–278.

Nof, D. and T. Pichevin, 2001, The ballooning of outflows, *Journal of Physical Oceanography*, 31, 3045–3058.

Sugimoto, T., 1990, A review of recent physical investigation of straits around the Japanese Islands, in *The Physical Oceanography of Sea Straits*, ed. L.J. Pratt, Kluwer Academic, Dordrecht, the Netherlands.

Valle-Levinson, A., 2008, Density-driven exchange flow in terms of the Kelvin and Ekman numbers, *Journal of Geophysical Research*, 113(C5), doi:1029/2008JC004853.

Whitehead, J.A., 1985, The deflection of a baroclinic jet by a wall in a rotating fluid, *Journal of Fluid Mechanics*, 157, 79–93.

Yankovsky, A.E. and D.C. Chapman, 1997, A simple theory for the fate of buoyant coastal discharges, *Journal of Physical Oceanography*, 27, 1386–1401.

18

Hydrodynamics of Coastal Circulation

D. Haidvogel
Rutgers—The State University of New Jersey

R. Chant
Rutgers—The State University of New Jersey

18.1 Introduction

The coastal oceans are hydrodynamically very complex. They are forced in many contrasting ways and at multiple space–time scales—for example, by the *global tides*, by *atmospheric winds and pressure fields*, by *river outflows*, by *extra-coastal influences* (e.g., offshore current systems), as well as by the more subtle, longer-timescale influences of *global climate signals* (e.g., El Nino events). In addition, the coastal ocean is geometrically variable, with *irregular coastlines*, *order one variations in water depth* (i.e., from vanishing depth to thousands of meters), *submarine banks and canyons*, etc. The resulting coastal circulation patterns are therefore highly variable, and may feature both persistent and time-variable fronts, intense currents with strong spatial (offshore and/or vertical) dependence, coastal-trapped waves, internally generated mesoscale features, strong vertical stratification, and regions of intense turbulent mixing in both surface and bottom boundary layers. Fortunately, despite this complexity, substantial progress has been made in understanding observed coastal circulation patterns via theoretical and numerical modeling. We review some of this understanding in the following. Supplemental discussions may be found in the prior reviews prepared by Winant (1980), Huyer (1990), and Brink and Robinson (1998).

The chapter is organized as follows. The equations of motion used to study the coastal ocean circulation, as well as several simplifying assumptions, are briefly reviewed in Section 18.2. The theory of freely propagating coastal-trapped waves, and its extension to wind-forced motions on continental shelves, is addressed in Section 18.3. The consequences of forcing by the global tides are considered in Section 18.4. The subsequent two sections investigate processes associated with the evolution of the 3D coastal temperature and salt fields: wind-driven coastal upwelling (Section 18.5) and buoyancy-driven plumes (Section 18.6). The final section offers concluding remarks.

18.2 Physical Principles, Equations, and Approximations

The physical principles underlying our theoretical understanding of the coastal ocean and its circulation are the conservation of momentum, mass, and dynamic tracers (temperature and salinity). A relationship between the tracer fields and the density of the fluid (an "equation of state") must also be provided. Importantly, the conservation statements for momentum must properly account for the effects of the Earth's rotation. The resulting terms—the Coriolis forces—are essential contributors to circulation on the scale of the coastal oceans. A more complete treatment of the following concepts may be found in Holton (1992).

In a Cartesian coordinate system fixed to the rotating Earth, and with x (eastward), y (northward) and z (positive upward)

Handbook of Environmental Fluid Dynamics, Volume One, edited by Harindra Joseph Shermal Fernando. © 2013 CRC Press/Taylor & Francis Group, LLC.
ISBN: 978-1-4398-1669-1.

axes, the un-approximated, so-called primitive, equations take the following form:

$$u_t + uu_x + vu_y + wu_z = \left(\frac{-1}{\rho}\right)p_x + fv + F^x \tag{18.1a}$$

$$v_t + uv_x + vv_y + wv_z = \left(\frac{-1}{\rho}\right)p_y - fu + F^y \tag{18.1b}$$

$$w_t + uw_x + vw_y + ww_z = \left(\frac{-1}{\rho}\right)p_z - g + F^z \tag{18.1c}$$

$$u_x + v_y + w_z = 0 \tag{18.1d}$$

$$T_t + uT_x + vT_y + wT_z = F^T \tag{18.1e}$$

$$S_t + uS_x + vS_y + wS_z = F^S \tag{18.1f}$$

$$\rho = \rho(T, S, p) \tag{18.1g}$$

where subscript x, y, z, or t indicates a partial derivative of the preceding quantity. The seven dependent state variables in 18.1a through 18.1g are (u, v), the components of horizontal velocity; w, the vertical velocity component; p, pressure; ρ, fluid density; and T and S, temperature and salinity. The Coriolis parameter arising from the Earth's rotation, $f = 2\,\Omega \sin(\theta)$, is a function of the Earth's angular rotation rate (Ω) and the latitude (θ). Finally, g is the gravitational acceleration; and F^x, F^y, F^z, F^T, and F^S represent the influences of any sources and/or sinks. Important examples of these nonconservative influences include wind stress and friction in the momentum equations, and heating/cooling and freshwater input in the temperature and salinity equations. As an example, the forces acting in the u and v momentum equations can be written as

$$F^x = \left(\frac{1}{\rho}\right)(\tau^x)_z \quad \text{and} \quad F^y = \left(\frac{1}{\rho}\right)(\tau^y)_z \tag{18.2}$$

where τ^x and τ^y are any shear stresses arising from, for example, either an applied wind stress or a frictional process.

Upon specification of any forcing or friction terms, and with an appropriate set of initial and boundary conditions, these equations can in principle be solved (usually numerically) for the subsequent evolution of the coastal circulation. Nonetheless, much progress in understanding the resulting processes can be made with simplified versions of these equations. Five approximations are of particular value, and will be variously adopted in the following.

The momentum equations may be simplified in several ways. The most common is the *Boussinesq approximation*, which assumes that the density ρ can be replaced by a constant mean value in the horizontal pressure gradient terms in (18.1a) and (18.1b). Consider representing the density field as the sum of three terms: a constant mean value, a z-dependent vertical stratification, and finally a time- and space-variable perturbation:

$$\rho(x, y, z, t) = \rho_o + \bar{\rho}(z) + \rho'(x,y,z,t) \tag{18.3}$$

The Boussinesq approximation is then valid so long as the background stratification $\bar{\rho}(z)$ varies little about the mean value ρ_o, and the fluctuations ρ' in turn vary little about the basic state profile. These conditions are approximately satisfied in the coastal ocean.

In many circumstances (e.g., if sound waves are not a concern), the vertical momentum equation may be greatly simplified. The *hydrostatic approximation* assumes that the pressure at a point in the ocean is equal to the weight of the fluid in the vertical column above, that is, that (18.1c) may be replaced by

$$p_z = -\rho g \tag{18.4}$$

Under this assumption, given the density field, pressure can be obtained from a simple vertical integration downward from the sea surface:

$$p(x, y, z, t) = p^{atm} - \int_{-z}^{\eta} \rho g dz \tag{18.5}$$

where $\eta(x, y, z, t)$ is the sea-level height (vertical position above or below the resting sea level, the latter coincident with a geopotential surface) and p^{atm} is the atmospheric pressure at the sea surface.

An important consequence of the hydrostatic approximation is that the prognostic equation for the vertical velocity is replaced by a diagnostic equation for pressure. We are therefore left to determine the vertical velocity field in another way, typically by vertical integration of the continuity equation (18.1d), which relates w diagnostically to the horizontal convergence or divergence of fluid motion. Nonetheless, vertical acceleration has been entirely neglected. The nonlinear *Boussinesq and hydrostatic primitive equations* are not easily studied theoretically; however, they are the building blocks of many numerical ocean circulation models (Haidvogel and Beckmann 1999).

Two other customary simplifications of the horizontal momentum equations are noteworthy. In the first, it is assumed that the horizontal momentum budget is to a good approximation a balance between the horizontal pressure gradient and the Coriolis forces (the *geostrophic approximation*). The Boussinesq, geostrophic forms of Equations 18.1a and 18.1b, read

$$fv = \left(\frac{1}{\rho_o}\right)p_x \quad \text{and} \quad fu = \left(\frac{-1}{\rho_o}\right)p_y \tag{18.6}$$

An approximate state of geostrophic balance is often observed to hold away from the ocean surface and bottom boundaries where viscous processes become important.

Near the surface and bottom, where viscous processes mediate the transfer of momentum from the wind and/or to the sea bed, a useful approximation is that the x and y momentum balance is between the Coriolis force terms and the vertical turbulent exchange of momentum, that is

$$fv = \left(\frac{-1}{\rho_o}\right)(\tau^x)_z = -\upsilon u_{zz} \tag{18.7a}$$

$$fu = \left(\frac{1}{\rho_o}\right)\left(\tau^y\right)_z = -\upsilon v_{zz} \tag{18.7b}$$

Here, υ is the vertical eddy viscosity. This is the *Ekman layer approximation*. When solved, (18.7a) and (18.7b) yield one prescription for the vertical structure of surface currents driven by winds or of currents in the bottom boundary layer (e.g., Holton 1992, pp. 129–131). The resulting Ekman boundary layers are confined to the surface or bottom with a trapping scale given by $\delta = (2\upsilon/f)^{1/2}$. Note that at these scales ν is a property of the flow rather than a property of the fluid, as would be the case for molecular viscosity.

Lastly, it is often possible to replace Equation 18.1g—in reality an empirically based high-order polynomial relation between density and T, S, and p—with a *linearized equation of state*

$$\rho(x, y, z, t) = \rho_o(1 - \alpha T + \beta S) \tag{18.8}$$

where α and β are positive constants. Note that if we assume the linearized form in (18.8), and if in addition the temperature and salinity forcing terms F^T and F^S can be neglected, then the evolution equations for temperature and salinity imply a similar equation for density

$$\rho_t + u\rho_x + v\rho_y + w\rho_z = 0 \tag{18.9}$$

which states simply that density is conserved following the fluid motion.

Whether or not any of the foregoing approximations are justified will of course depend on the circumstances. A formal scaling of the primitive equations can be carried out (e.g., Holton 1992, pp. 38–47), from which various nondimensional parameters emerge which measure the relative importance of various terms in the complete equations. We will meet some of these parameters in more detail in the following. These include the *Rossby number*, which measures the importance of horizontal advection of momentum relative to the Coriolis force; the *Richardson number*, the ratio of the buoyancy frequency, N, $[N^2 = (-g/\rho_0)\overline{\rho_z}]$ to the vertical shear squared; the *Burger number*, N/f, the ratio of the buoyancy frequency to the Coriolis frequency; the *Froude number*, fluid velocity relative to wave speed; and the *Ekman number*, viscous relative to Coriolis forces.

18.3 Fluctuating Currents over the Continental Shelf and Slope

Fluctuating currents on continental shelves are produced in many ways. As an example, Figure 18.1 shows the frequency spectra of kinetic energy at two sites on the New England and Oregon shelves. Despite regional differences in shelf properties (e.g., a narrow shelf off Oregon, and a wider shelf off the U.S. East Coast), the spectra share several properties in common. Among these are that the spectra are red (that is, containing more variance at the lower frequencies) and that the spectra are peaked in narrow frequency bands. The most important of these are the bands associated with weather events (wind forcing; periods of roughly 2–20 days), the diurnal and semidiurnal periods arising from the global tides, and motions at the inertial frequency (f).

18.3.1 Theoretical Development for Low-Frequency Motions ($\omega \ll f$)

The combination of a coastal wall and the Earth's rotation is all that is necessary to support low-frequency, time-dependent coastal currents. In the absence of any complicating factors—such as wind forcing, variable bottom depth, stratification, and friction—and with a straight coastline, propagating wave solutions are easily determined. These *Kelvin waves* are geostrophic in the cross-shore momentum balance, travel preferentially with the coastline on the right in the northern hemisphere, propagate at the gravity wave phase speed $c = (gH)^{1/2}$ where H is the fluid depth, and are trapped to the coastal wall with a trapping scale of (c/f).

Unfortunately, unsteady motions over the continental shelf and slope are strongly linked to the winds, and are influenced by stratification, variable bathymetry, and bottom friction. Consequently, all these effects must be retained in any inclusive theoretical treatment. The theory of wind-driven time-dependent currents in the coastal ocean has a rich and lengthy history. Here, we can only give a brief introduction to the subject.

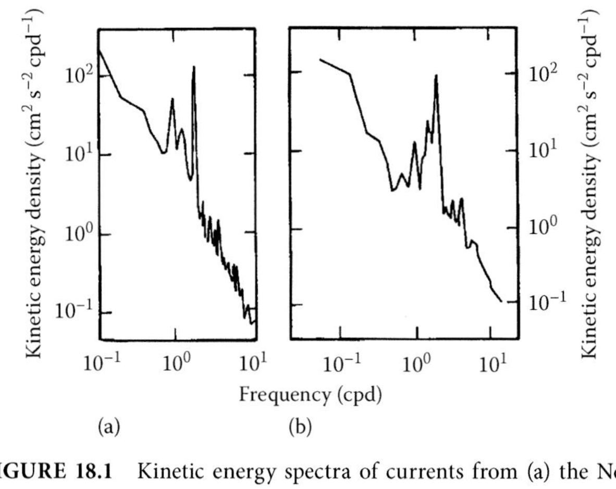

FIGURE 18.1 Kinetic energy spectra of currents from (a) the New England shelf and (b) the Oregon shelf. (Reprinted from Winant, C., *Ann. Rev. Fluid Mech.*, 12, 271, 1980. With permission.)

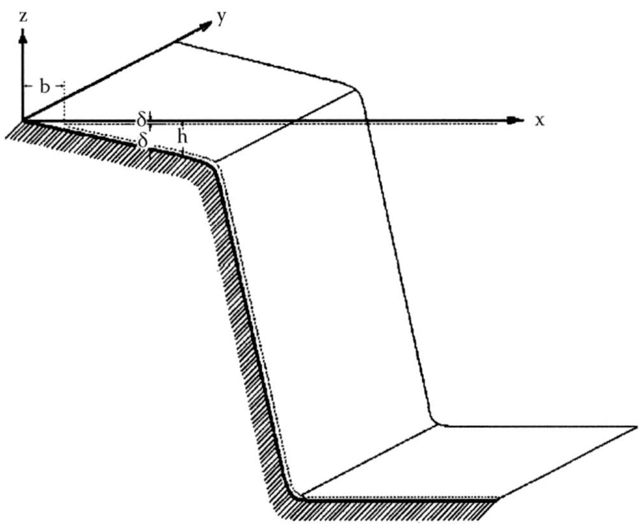

FIGURE 18.2 The simplified coastal geometry considered in the text.

The reader is referred to Clarke and Brink (1985) and Clarke and Van Gorder (1986) for further theoretical development.

Consider a coastal ocean at rest in the geometry depicted in Figure 18.2, where the coordinate axes are *x* (offshore), *y* (alongshore), and *z* (positive upward from the resting sea surface). The inner shelf is assumed to begin at x = b where the fluid depth is equal to three times the Ekman e-folding scale (δ). The density field is represented in the form (18.3), by the sum of a mean value, a background stratification, and a small perturbation.

Equations are developed for the "interior" flow, that is, the flow outside of the surface and bottom Ekman layers. Adopting the Boussinesq approximation, perturbations to Equations 18.1a, 18.1b, 18.1c, 18.4, and 18.9 lead, at lowest order, to

$$u_t - fv = -p_x + F^x \tag{18.10a}$$

$$v_t + fu = -p_y + F^y \tag{18.10b}$$

$$p_z = \left(\frac{-g}{\rho_o}\right)\rho' \tag{18.10c}$$

$$u_x + v_y + w_z = 0 \tag{18.10d}$$

$$(\rho')_t = \left(\frac{\rho_o N^2}{g}\right) w \tag{18.10e}$$

where for consistency with Clarke and Brink (1985), the symbol p will, for the remainder of this section, denote perturbation pressure divided by ρ_o.

Finally, Equations 18.10a through 18.10e are simplified by assuming that the frequencies (ω) of the time-dependent motions are less than the inertial frequency f, and that the scales of motion in the along-shelf direction exceed those in the cross-shelf direction (the *long-wave approximation*). The resulting equations can then be manipulated to form a single vorticity equation:

$$p_{xx} + f^2\left(\frac{p_z}{N^2}\right)_z = 0 \tag{18.11}$$

where the *buoyancy frequency*, N(z), is determined from

$$N^2(z) = \left(\frac{-g}{\rho_o}\right)\bar{\rho}_z \tag{18.12}$$

Solution of Equation 18.11 requires specification of boundary conditions at the surface (z = 0), bottom (z = −H), the inner shelf (x = b), and in the deep ocean (x → ∞). Among other things, these boundary conditions are needed to incorporate the effects of the surface wind stress and bottom frictional forces. The former will be assumed to act in the along-shelf direction only, that is, $\boldsymbol{\tau}^{wind} = (0, \tau^y)$. The latter is prescribed via a linear drag law; accordingly, the kinematic bottom stress in the y direction is assumed to be given by rv, where r is a turbulent-resistance coefficient.

The resulting boundary conditions are as follows (see Clarke and Brink [1985] for a discussion). At the sea surface it is permissible to apply the rigid-lid approximation, which gives

$$w = \frac{-p_{zt}}{N^2} = 0 \ (z = 0) \tag{18.13a}$$

The bottom boundary condition may be obtained by integration over the bottom boundary layer and subsequent application of the long-wave approximation, resulting in

$$\left(\frac{f^2}{N^2}\right) p_{zt} + h_x(p_{xt} + fp_y) + (rp_x)x - rh_x p_{xz} = 0 \ (z = -H) \tag{18.13b}$$

At the inner coastal edge, it is assumed that the depth-integrated velocity perpendicular to the shore must vanish, giving

$$p_{xt} + \left(\frac{r}{h}\right) p_x + fp_y = \left(\frac{f}{h}\right) \tau^y (x = b) \tag{18.13c}$$

Finally, the pressure disturbance is trapped to the coast, so that

$$p \to 0 \ (x \to \infty) \tag{18.13d}$$

Solutions must therefore be found to (18.11) subject to the conditions in (18.13a) through (18.13d).

18.3.2 Unforced Shelf Wave Modes and Some of Their Properties

Solutions to (18.11) in the absence of forcing or friction represent the free long-wave modes associated with the specified bathymetry H(x) and stratification $N^2(z)$. The free-mode structures

$M_n(x,z)$ and their associated eigenvalues c_n, the latter corresponding to the free-mode phase speeds, may be obtained from the associated eigenvalue problem

$$(M_n)_{xx} + f^2\left(\frac{(M_n)_z}{N^2}\right)_z = 0 \tag{18.14}$$

with the four applied conditions

$$(M_n)_z = 0 \ (z = 0) \tag{18.15a}$$

$$(M_n)_z + N^2 H_x f^{-2}(M_n)_x + f c_n^{-1} M_n = 0 \ (z = -H(x)) \tag{18.15b}$$

$$(M_n)_x + f c_n^{-1} M_n = 0 \ (x = b) \tag{18.15c}$$

$$M_n \to 0 \ (x \to \infty) \tag{18.15d}$$

The eigenvalue problem (18.14) and (18.15a) through (18.15d) may be solved in several alternative ways, for example, by the method of resonance iteration (Wang and Mooers 1976).

The resulting eigenfunctions comprise a complete set. As a consequence, a general solution to (18.11) may be written as follows:

$$p(x, y, z, t) = \sum_{i=1}^{\infty} \varphi_i(y,t) M_i(x,z) \tag{18.16}$$

where the functions φ_i describe the amplitude variation of the corresponding wave modes at a given alongshore position and time. Utilizing the orthogonality properties of the free modes, the amplitude functions of the unforced problem can be shown to obey the first-order wave equation (Clarke and Van Gorder 1986)

$$(\varphi_i)_t - c_i(\varphi_i)_y = 0 \tag{18.17}$$

In the unforced case, a general initial state $p(x, y, z, t = 0)$ will evolve in time as an uncoupled set of wave equations, with the contribution from each wave mode M_i advancing at its own phase speed c_i. Once the amplitudes φ_i have been determined, the pressure field can be reassembled from (18.16). Similar series representations can also be obtained for the other dependent variables: along- and cross-shelf velocity, and density perturbation [Chapman 1987, Equations (5) through (7)].

The coastal-trapped long-wave modes have several systematic properties (Wang and Mooers 1976; Clarke and Brink 1985). For example, if H is monotonic (increasing offshore), then it can be shown that all waves travel with the coast on their right in the northern hemisphere (e.g., $c_i > 0$ on the U.S. west coast), and with the coast on the left in the southern (e.g., $c_i < 0$ on the west coast of Chile). In addition, increases in the buoyancy frequency produce increases in phase speed for all modes.

18.3.3 Applications and Limitations of the Long-Wave Theory

With the inclusion of external forcing by both wind stress and bottom friction, as would be expected in any realistic application, the expansion in (18.16) is still complete. With the forcing and frictional terms, the associated wave equation for the amplitudes $\varphi_j(y,t)$ becomes

$$\left(\frac{-1}{c_j}\right)(\varphi_j)_t + (\varphi_j)_y + \sum_{i=1}^{\infty} a_{ji}\varphi_i = b_j \tau^y(y, t) \tag{18.18}$$

where

$$b_j = \frac{\left\{\int_{-h(b)}^{0} M_j(b, z)\, dz\right\}}{H(b)} \tag{18.19a}$$

$$a_{ji} = \left(\frac{1}{f}\right)\left\{\int_b^x M_j(x,-H)[r\, M_{ix}(x,-H)]_x\, dx + \left[\frac{r\, M_j M_{ix}}{H}\right]_{x=b} dz\right\} \tag{18.19b}$$

Note in particular the third term in (18.18). Besides representing frictional dissipation of the modal amplitude φ_i (the a_{ii} term), the influence of bottom friction results in a coupling of the modal amplitude equations (the a_{ji}, $j \neq i$).

Many authors have carried out careful comparisons of predictions obtained from this wind-driven, long-wave theory to observations of time-evolving coastal ocean pressure and velocity fields. The procedure is in principle straightforward. First, the designated coastal area is divided into multiple segments. For each segment, an offshore bathymetric profile H(x), Coriolis parameter, and stratification profile N(z) are defined; these parameters are assumed to have no alongshore variation within a given segment.

Second, having made these parametric decisions, the free-mode properties [$M_i(x, z)$ and c_i] are determined for each segment. From the free modes, and a specification of r(x) and τ^y (both assumed to be invariant alongshore within a segment), the coupling coefficients (18.19a and 18.19b) may be obtained. Finally, the coupled wave equations (18.18) are integrated in the alongshore direction and time, and the summation (18.16) is used to reconstruct the predicted pressure field. Note that integration of the first-order wave equation requires the specification of a boundary condition for pressure at some upcoast (in the sense of wave propagation) location and time.

As an example, this procedure was applied by Battisti and Hickey (1984) to the Oregon and Washington coasts, as shown

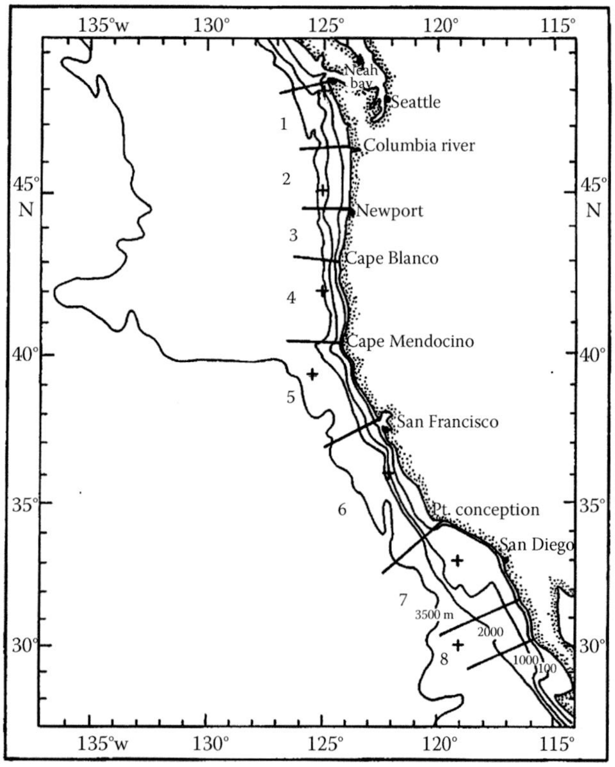

FIGURE 18.3 A map of the west coast of the United States showing the regions used in the forced wave model. (Reprinted with permission from Battisti, D. and Hickey, B., *J. Phys. Oceanogr.*, 14, 887, 1984, Copyright American Meteorological Society.)

in Figure 18.3. The authors divided the coastal region of interest into eight segments, and represented each by a buoyancy frequency N(z), assumed to have no cross-shelf dependence, as well as an average bottom topographic profile (their Figures 2 and 3). Further, they assumed the forced response to be dominated by the first shelf mode, thereby truncating the infinite sum in (18.16) after the first term.

The properties of the first shelf mode within each of the eight coastal segments, as determined by Battisti and Hickey, are shown in Table 18.1. The first mode is determined to travel northward at speeds ranging from 2.8 to 4.8 m s^{-1}, increasing to the north. Frictional decay timescales $[T_F = (c_1 a_{11})^{-1}]$ are approximately 10 days.

TABLE 18.1 Properties of the Lowest Wave Mode in Each of the Eight Coastal Segments: Phase Speed, c_1; Modal Coupling Coefficient, b_1; Frictional Decay Time, T_F; Length of Coastal Segment, Δy

Property	1	2	3	4	5	6	7	8
c_1 (m s^{-1})	4.8	4.8	3.8	3.2	3.0	2.8	2.8	2.8
$b_1 \times 10^2$ ($cm^{-1/2}$)	0.8	0.8	0.72	0.65	0.68	0.68	0.68	0.68
T_F (days)	9.0	9.0	9.0	10.0	10.0	11.0	12.0	13.0
Δy (km)	220	220	150	300	380	450	510	150

Source: Battisti, D. and Hickey, B., *J. Phys. Oceanogr.*, 14, 887, 1984.

To integrate (18.18) northward, two further pieces of information must be provided: the values of along-shelf wind stress within each coastal segment and the initial condition for the amplitude of the first mode at the southern extreme of the coastal domain (i.e., Section 8). In the absence of observed wind data for the entire coastal region, the authors used winds inferred from surface atmospheric pressure maps, in the manner of Bakun, by rotating the surface geostrophic wind by 15° to the left and reducing the resulting wind amplitude by 30% to account for frictional deceleration in the atmospheric boundary layer. Finally, the authors assumed vanishing pressure signal at the southern boundary.

Comparison of observed and predicted subsurface pressure (tidal height plus atmospheric pressure) at South Beach (south of Newport in Section 3; see Figure 18.3) and Neah Bay for a 60 day period in the summer of 1978 is shown in Figure 18.4. The visual similarity between observed and modeled pressure signals is striking. Coherence between the time series at frequencies lower than 0.2 cycles/day is significant at the 95% confidence level. The authors note that the most significant contributions to these pressure fluctuations are remotely produced in Section 5 (spanning Cape Mendocino to San Francisco), presumably due to the enhanced wind stress amplitudes in this subregion.

Despite successful applications such as these from Battisti and Hickey, and other authors not shown here, application of the wind-driven, long-wave theory has its limitations. Among these are: first, ambiguities in the specification of $N^2(z)$, H(x), the bottom resistance coefficient r, and the wind stress τ^y (all of which are assumed, to a greater or lesser degree incorrectly, to be uniform in the along-coast direction); and second, uncertainties in the application of the solution procedure itself. In the latter category, we note the uncertainties associated with the exact number of shelf modes to retain in expansion in (18.16) and in the boundary conditions to apply in the along-coast integration of (18.18). Finally, the quality of the long-wave predictions may (and indeed does) depend on where in the water

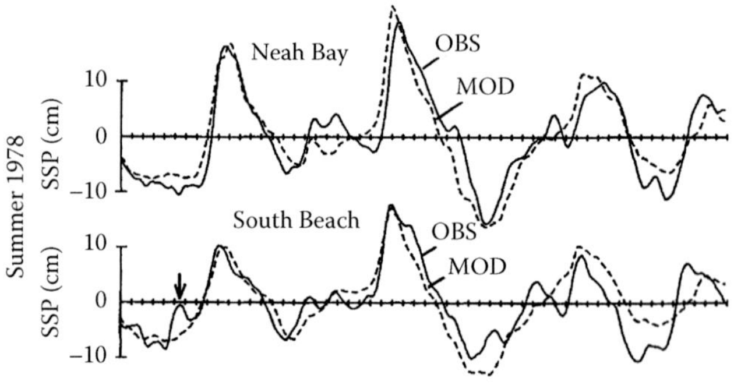

FIGURE 18.4 Observed and predicted subsurface pressure at Neah Bay and South Beach for summer 1978. (Reprinted with permission from Battisti, D. and Hickey, B., *J. Phys. Oceanogr.*, 14, 887, 1984, Copyright American Meteorological Society.)

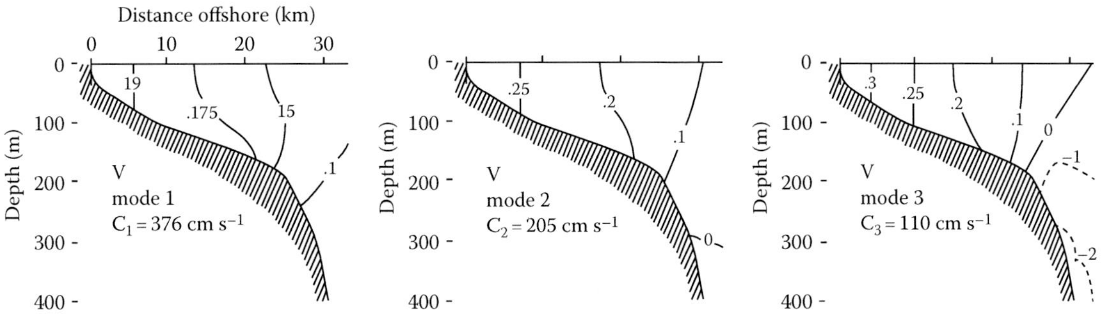

FIGURE 18.5 Modal structures and phase speeds for the first three wave modes at the CODE site. (Reprinted with permission from Chapman, D, *J. Geophys. Res.*, 92, 1798, 1987, Copyright American Geophysical Union.)

column the model–data comparisons are performed and what variables are compared.

Chapman (1987) performed an extensive examination of these complicating factors in an application of wind-driven, long-wave theory off the coast of California. For this purpose he was fortunate to have a relatively ample dataset obtained during the Coastal Ocean Dynamics Experiment (CODE 2) in the summer of 1982. Observations made during CODE 2 included measurements of pressure, velocity, and density (temperature) fluctuations at several cross-shelf locations, allowing a detailed comparison with the long-wave theory.

Unfortunately, we do not have the space for a detailed review of Chapman's detailed results! In brief, given the many "subjective decisions" involved in applying the long-wave theory, Chapman found many sources of uncertainty in his results. A brief synopsis of the findings is as follows.

Examples of the first three free-wave modal structures for velocity at the central CODE 2 site are shown in Figure 18.5. Note that modal phase speeds decrease roughly as (1/n), and that the cross-shelf trapping scale likewise decreases with mode number. All three modes are nearly depth-independent, suggesting that they will not easily account for any observed vertical shear in the velocity. Finally, in contrast to the experience of Battisti and Hickey, Chapman finds considerable along-shelf variation in the properties of his first three modes (c_i, b_i, a_{ii}; his Figure 7) that are in turn highly dependent on decisions made with respect to (e.g.) bottom topography.

As Chapman remarks, it is not obvious a priori how many free-wave modes should be retained to represent a given variable accurately. Indeed, the modal structures for pressure, velocity, and temperature each have different asymptotic properties, suggesting that a fixed number of modes may do relatively less well for velocity fluctuations than for pressure, and in turn less well for temperature signals. And indeed, Chapman (e.g., his Table 5) finds the predicted pressure field to be consistently better correlated with in situ observations than predictions of along-shelf velocity.

Finally, Figure 18.6 shows cross-shelf sections of the correlations obtained for along- and across-shelf velocity, and for temperature, obtained from Chapman's standard case. The along-shelf velocity component is well correlated with observations, especially at mid-shelf. In contrast, temperature fluctuations are only weakly correlated with concurrent observations, and the predicted cross-shelf velocity has essentially no skill at all.

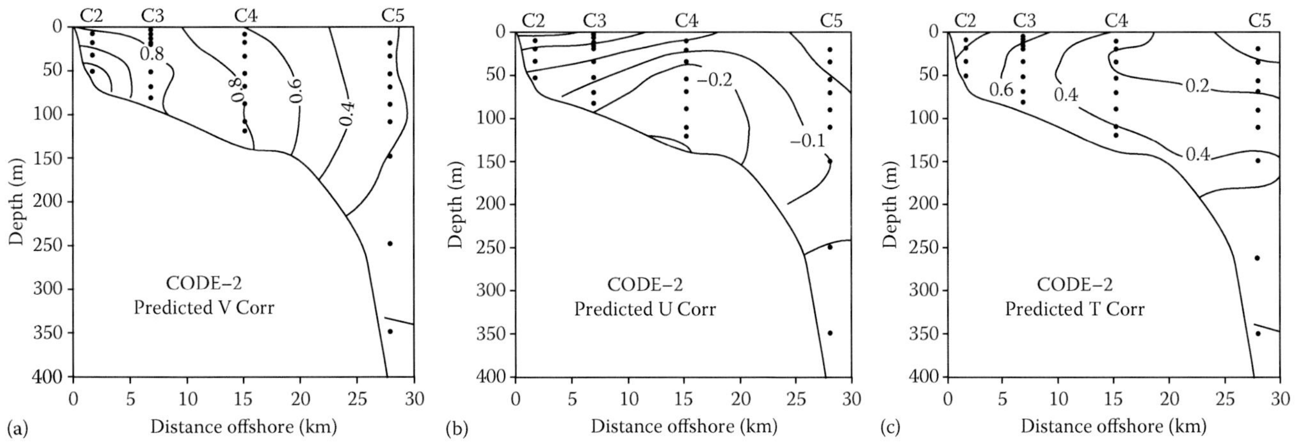

FIGURE 18.6 Cross-shelf sections of correlations between predicted and observed along-shelf velocity (a), cross-shelf velocity (b), and temperature fluctuations (c). (Reprinted with permission from Chapman, D., *J. Geophys. Res.*, 92, 1798, 1987, Copyright American Geophysical Union.)

18.4 Tides on Continental Shelves

18.4.1 Dynamics and Structure of Tides

Tidal motions on continental shelves are forced by global tides generated by astronomical forces associated with the Earth/Moon/Sun system. The dynamics of global tides are beyond the scope of this section other than to note that a significant fraction of global tidal dissipation occurs on continental shelves. Currents on many shelves are dominated by tidal motions which provide energy for mixing and may produce a mean rectified flow driven by nonlinear terms in the equations of motion. Tidal motion comprises numerous "tidal constituents," each with a specific frequency, and on most shelves tidal motion is dominated by semidiurnal (twice daily) motion which itself is dominated by the principle lunar tidal constituent, M2, with a period of 12.42 h. Other semidiurnal constituents such as the principle solar (S2) and lunar–solar (N2) have periods of 12.00 and 12.66 h and amplitudes of 10%–20% of the M2. The beating of the M2, S2, and N2 tidal frequencies gives rise to spring/neap variability at 14.8 and 27.3 days, respectively.

Tidal period motion in the diurnal bands are dominated by the K1 and O1 constituents which have periods of 23.96 and 25.83 h and can give rise to significant diurnal inequality to tidal period variability. Some shelves, such as the Alaskan Shelf, are dominated by diurnal tides while others, such as the south coast of Australia, have mixed tides whereby tidal period motion modulates between semidiurnal and diurnal over the tidal month. Tidal currents are characterized by their elliptical properties by defining a major axis, minor axis, and a sense of rotation. In the following analysis, tidal elevation is represented as $\eta = \eta_o e^{i(\omega t+kx+ly)}$ where ω, k, and l are tidal frequency, the cross-shore wave number and the alongshore wave number, respectively.

Since tidal motion on most shelves is dominated by the M2 tide, this discussion focuses on the tidal response of shelves to M2 tidal forcing. A fundamental dynamical difference between the semidiurnal and diurnal tides is that on most shelves the semidiurnal motion is super-inertial (ω > f) while the diurnal tide is sub-inertia (ω < f). Tidal period motion on most shelves can be characterized via a linearized version of the momentum equation (Battisti and Clarke 1982) and a depth-averaged version of the continuity equation, that is,

$$u_t - fv = -g\eta_x - \frac{\tau^x{}_B}{\rho H}$$

$$v_t + fu = -g\eta_y - \frac{\tau^y{}_B}{\rho H} \qquad (18.20)$$

$$\eta_t + (Hu)_x + (Hv)_y = 0$$

where τ_B^x and τ_B^y represent bottom stress terms. Assuming that the shelf is smooth, and that the along-shelf wave number (l) is small compared to the shelf width (W), that is, $l * W = \xi \ll 1$, an integration of the continuity equation across the shelf is accurate to O(ξ). In addition, since the cross-shelf scale of the tidal wave is generally large compared to the shelf width, there tends to be little cross-shelf variation in tidal height (μx), where $\mu \ll 1/W$. Thus, from (18.20) the cross-shelf tidal current amplitude (u_T) to accuracy $O(\xi + \mu x)$ is

$$u_T = -\frac{i\omega\eta(0)x}{H} \qquad (18.21)$$

where η(0) is the tidal amplitude at the coast. Next by linearizing bottom friction as $\tau_B = ru$, where $r(x) = C_d u_T$, and recognizing that $il\eta = \eta_y$, the alongshore tidal velocities can be expressed as

$$v_T = \frac{\eta(0)}{[1 - ir/H\omega]}\left(\frac{xf}{H} - \frac{gl}{\omega}\right) \qquad (18.22)$$

Equations 18.21 and 18.22 are operationally useful; they provide a simple way to estimate the magnitude and structure of tidal currents on continental shelves based on readily available coastal sea-level observations (Figure 18.7).

The structure of the observed tidal ellipse provides insights into the underlying dynamics. For example, if the alongshore wave number is zero (l = 0; no along-shelf variability) and friction is unimportant (r = 0) then the ratio of the major to minor ellipse of the tidal currents is simply f/ω and the major axis is directed onshore. This ellipse structure is consistent with observations on the wide mid-Atlantic Bight shelf because xf/H > gl/ω, although tidal flows here are also modified by friction. In contrast, on narrow shelves, such as the U.S. west coast, tidal motion is driven by an alongshore gradient; and because xf/H is not small relative to gl/ω, tidal ellipses are more rectilinear with the major axis directed along the coast. Finally, friction also plays an important role by making (18.21) complex, causing the ellipse to rotate clockwise on the wide and shallow MAB where frictional forces are important. Note, however, that coastal shape also modifies ellipse orientation, and small-scale coastal bathymetry along the Oregon coast is likely responsible for the failure of the model there.

18.4.2 Tidal Mixing Fronts

Tides on continental shelves drive vertical mixing. Simpson et al. (1978) describe tidal mixing on shelves in terms of the potential energy anomaly (Φ), defined as the difference between the potential energy of a stratified water column and that after it is mixed, that is,

$$\Phi = \frac{1}{h}\int_{-h}^{0} g(\rho - <\rho>)z\,dz \quad <\rho> = \frac{1}{h}\int_{-h}^{0} \rho\,dz \qquad (18.23)$$

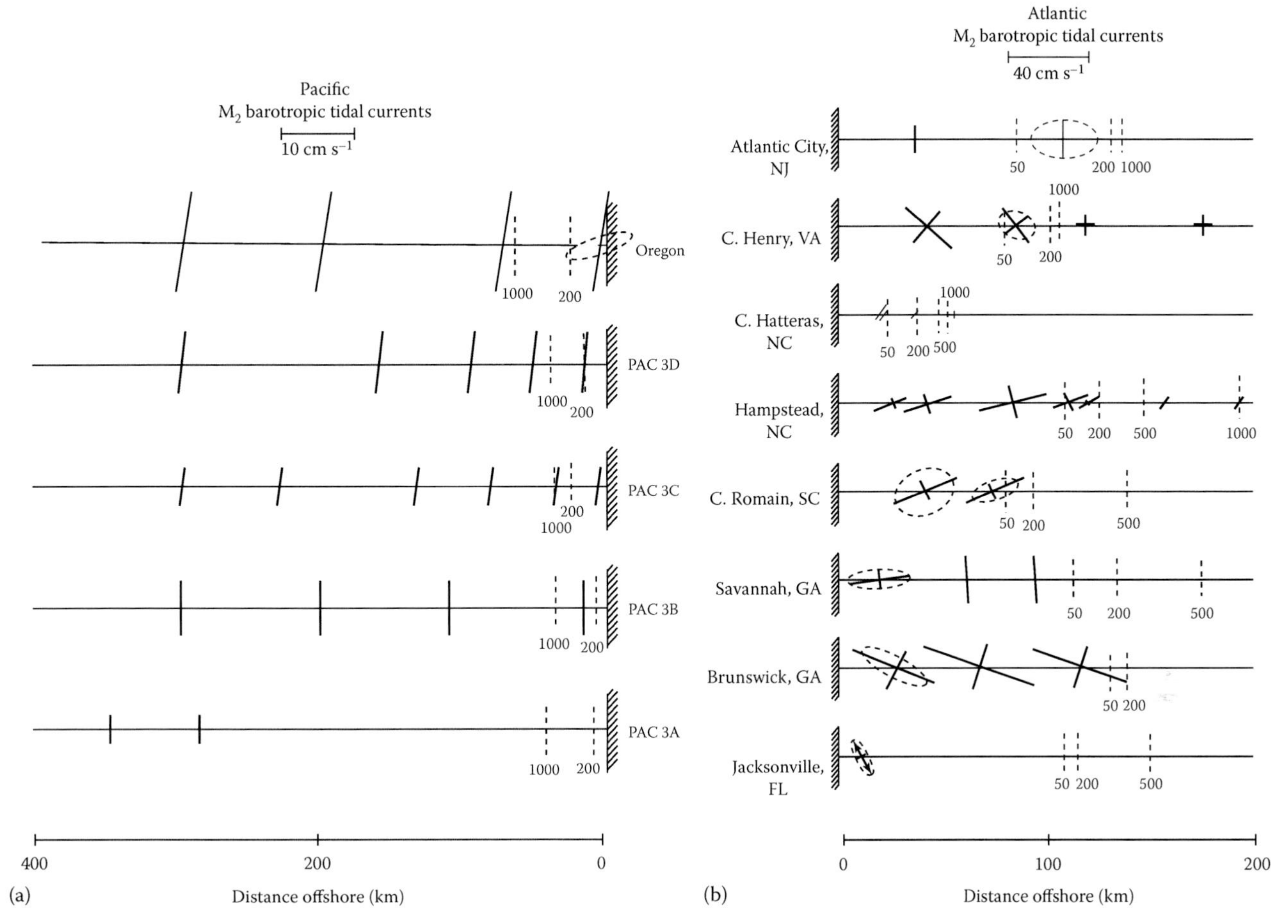

FIGURE 18.7 Tidal ellipses calculated using the Battisti and Clarke shelf tidal model (solid line) compared to observations (dashed ellipse). Dashed lines depict water depth. Panel (a) shows west coast while panel (b) shows east coast. Note model–data misfit on the inner shelf off Oregon. (Reprinted with permission from Battisti, D. and Clarke, A., *J. Phys. Oceanogr.*, 12, 8, 1982, Copyright American Meteorological Society.)

Away from buoyant discharges, stratification is driven by surface heat flux (Q) which decreases Φ at a rate $-\partial\Phi/\partial t = -(\alpha g Q/2c_p)$ where α ($1.6\ 10^{-4}$ °C^{-1}) and c_p (4200 J kg^{-1} °C^{-1}) are the volume expansion coefficient and the specific heat, respectively. In contrast, turbulence driven by the tidal motion increases Φ at a rate $\varepsilon k_b \rho \left(U_T^2/H\right)$, where u_T, k_b, and ε are the tidal current amplitude, the coefficient for quadratic bottom stress (0.002–0.003), and the mixing efficiency, respectively. The mixing efficiency represents the fraction of turbulent kinetic energy that goes into mixing stratification (buoyancy flux), with the remaining fraction dissipated as heat. Estimates of ε from field data are extremely low (0.003) indicating that the vast majority of TKE generated by tidal motion on shelves is dissipated as heat. Despite the inefficiency of tidal mixing, it is sufficient in many locations to overcome the surface heat flux and produce a well-mixed water column. This occurs where $U_T^3/H > \alpha g Q/\varepsilon k_b \rho 2c_p$.

Since Q is relatively constant on a given shelf, it is the variability of the quantity $\left(U_T^3/H\right)$ that determines where this condition is met. Where surface heat flux dominates tidal mixing, the water column remains stratified, while in regions where tidal mixing dominates, the water column is mixed. Between these two regions lies a "tidal mixing front" which is clearly evident in satellite imagery on continental shelves with strong tides (Figure 18.8). For typical values of Q (150–200 Wm^{-2}) on temperate shelves, the tidal mixing front occurs at approximately $\left(U_T^3/H\right) = .005$. Tidal mixing fronts are important because the pressure gradients associated with the frontal region drive strong geostrophic flows along the front, and the variable mixing across the front can drive strong vertical fluxes of nutrients that can fuel primary productivity.

18.4.3 Tidal Rectification

The inclusion of nonlinearity in the momentum equation introduces a mean circulation that to first order is proportional to the bottom slope times the tidal current amplitude squared

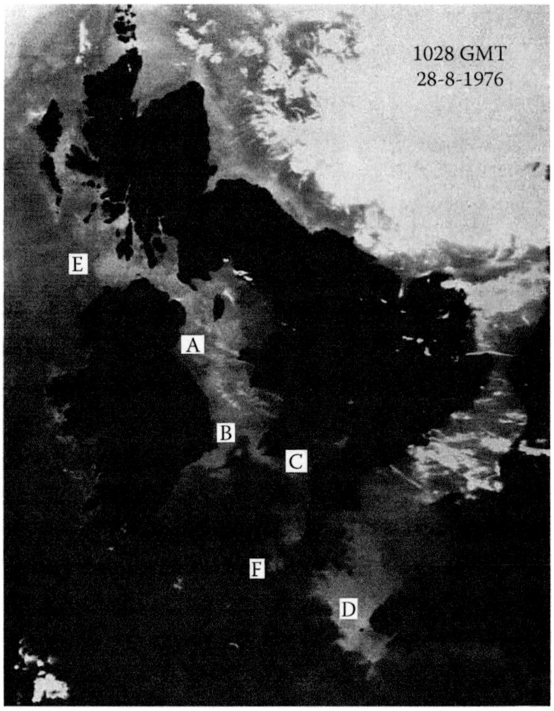

FIGURE 18.8 Infrared image from NOAA 5 at 1028 UT on August 20, 1976. Image darkness increases with temperature. Tidal mixing front areas are apparent where gradients are large. Letters represent the following frontal systems: A, Western Irish Sea; B and C, Celtic Sea; D, Ushant; E, Islay; and F, Scilly Isles. (Reprinted with permission from Simpson, J.H. et al., *J. Geophys. Res.*, 83, 4609, 1978. Copyright American Geophysical Society.)

(Loder 1980). Tidal rectification driven by tidal motion on a slope can be explained by considering a volume of fluid oscillating up and down a slope over a tidal cycle. During the flood tide, when the current is onto the slope, the leading edge of the volume is faster than the trailing edge and Coriolis acceleration torques this volume to produce a clockwise vorticity. The opposite happens during the ebb tide, whereby a counterclockwise vorticity is produced. In the absence of friction, however, the production of vorticity and its subsequent advection are 90° out of phase; and thus when averaged over a tidal cycle, there is no mean advection of vorticity. Frictional forces shift this phase difference out of quadrature, and the subsequent tidally averaged advection of vorticity across the bank leads to a mean Eulerian flow that is parallel to the isobaths with the shallow water to the right (in the northern hemisphere). Tidal rectification is further complicated by different trajectories of the Eulerian, Lagrangian, and the Stokes drift. The Stokes drift is in the opposite direction of the Eulerian mean flow, and thus the speed of a particle (i.e., the Lagrangian) is less than would be suggested by an Eulerian measurement. Tidal rectification contributes significantly to the mean circulation around Georges Bank (Figure 18.9), and more generally is found on many shelves where strong tidal currents flow over variable topography.

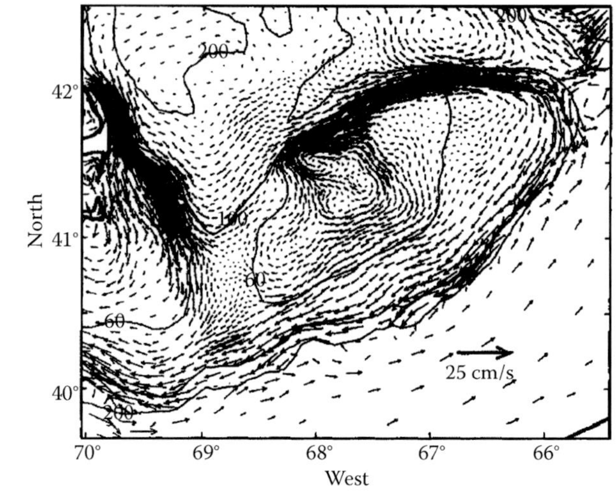

FIGURE 18.9 Tidally averaged, depth-averaged residual currents from a numerical model of Georges Bank. (Reprinted with permission from Werner, F. et al., *Deep Sea Res.*, 43, 1793, 1993, Copyright Deep-Sea Research.)

18.5 Coastal Upwelling and Associated Processes

Coastal upwelling brings cold nutrient-rich waters to the surface and fuels the base of the marine food web. Many of the world's most productive fisheries occur in upwelling regions and thus upwelling has received significant attention for over a century. Major upwelling regions around the world include the North and South American west coasts and the west coast of Africa. However, upwelling occurs on many coastlines and the dynamics discussed next are relevant to coastal systems worldwide. While coastal upwelling has been discussed in terms of an Ekman response to surface wind stress for over 100 years, only in the past few decades have details of its structure been characterized due to technological advances in both measurement and modeling technologies.

Upwelling is driven by winds blowing in the opposite direction of shelf wave propagation, which for the U.S. west coast would be a wind from the north. Such an "upwelling-favorable" wind drives an offshore Ekman transport in the surface boundary layer of $U_E = \tau^{sy}/\rho f$, where τ^{sy} is the alongshore wind stress. For example, based on standard parameterization of wind stress, an upwelling-favorable wind of 15 knots (7.5 m s^{-1}) would produce a wind stress of ~0.1 Pa and drive an offshore transport at 40N ($f = 9.7 \times 10^{-5}\,s^{-1}$) of approximately 1.0 $m^2\,s^{-1}$.

The offshore transport depresses coastal sea level, producing a cross-shore sea-level slope that drives a compensating onshore flow at depth that converges at the coast. Together with a divergent offshore flow in the upper layer, a mean upwelling flow is established to satisfy continuity (conservation of volume). Assuming that the upwelling is confined to an internal Rossby radius, the ratio of internal wave speed to f, ~10 km in the coastal ocean, a 7.5 m s^{-1} wind drives an upwelling velocity of 1.0×10^{-4} m s^{-1}, corresponding to about 8 m/day. This vertical motion in turn modifies the density field and feeds back on the coastal circulation.

The onshore flow in the lower layer is accelerated in the alongshore direction by Coriolis forces, eventually establishing a vertically sheared "thermal-wind" geostrophic along-shelf flow

$$u(z) = u^s + \frac{g}{f\rho_o}\rho_y z \qquad (18.24)$$

where u^s is the surface current. This simple geostrophic cross-shelf momentum balance is consistent with numerous observational and numerical studies of upwelling systems (e.g., Huyer 1980; Lentz and Chapman 2004).

Details of the along-shelf momentum balance are more complex and determine the depth from which the upwelled waters originate. This has implications for the cross-shelf transport of heat, nutrients, and biota. For example, long-term current meter mooring deployments off of Peru, Oregon, northern California, and North West Africa indicate that while the offshore transport in the surface layer is consistent with Ekman dynamics, the details of the return flow vary from site to site (Figure 18.10). Only off the coast of North West Africa is the return flow confined to a bottom Ekman layer. In contrast, at all other sites the onshore flow occurs in the interior of the water column. The dynamics that determine the vertical structure of the cross-shelf flow can be understood with a steady-state, 2D model (Lentz and Chapman 2004). The model is based on a depth-integrated, alongshore momentum equation:

$$\frac{\partial}{\partial y}\int_{-h}^{0} uv dz = -gh\eta_x + \frac{\tau^{sx}}{\rho_o} - \frac{\tau^{bx}}{\rho_o} \qquad (18.25)$$

where the terms on the right-hand side are the alongshore barotropic (depth-integrated) pressure gradient, the alongshore wind stress, and the alongshore bottom stress, respectively. The term on the left-hand side represents the depth-integrated cross-shore divergence of alongshore momentum flux. Bottom stress and surface stress would balance in the absence of both an alongshore pressure gradient and a cross-shore momentum flux, and in such a case the onshore transport in the bottom Ekman layer would equal that in the surface Ekman layer. However, if either of the first two terms tend to balance the wind stress, then bottom stress is reduced and an interior flow, $v_i = (\tau^{sx} - \tau^{bx})/\rho fh$, is required.

Note that upwelling circulation acting on the thermal wind shear in the along-shelf flow provides a mechanism to drive a cross-shelf momentum flux that tends to balance the wind stress. Lentz and Chapman estimated the size of the cross-shelf momentum flux using the following three assumptions: (1) the cross-shore flow is composed of a surface Ekman layer ($v^s = \tau^{sx}/\rho_o f\delta^s$), an interior flow ($v_i$ as above), and a bottom Ekman layer $v^b = \tau^{bx}/\rho_o f\delta^b$, where δ^s and δ^b are the surface and bottom boundary layer thicknesses; (2) there are no cross-shelf variations in δ^s and δ^b, τ^{sx}, τ^{bx}, and p_y; and (3) that the region of sloping isopycnals during upwelling is scaled by the baroclinic radius of deformation (Nh/f). With these assumptions they find

$$\frac{\partial}{\partial y}\int_{-h}^{0} uv dz \approx -b\frac{\tau^{sx}}{\rho_o}\frac{S_l}{2}\left(1+\frac{\tau^{bx}}{\tau^{sx}}\right) \qquad (18.26)$$

where S_l is the slope Burger number (bottom-slope times the Burger number) and b is a "constant" that contains both the scaling for the isopycnal slopes and deviations from the assumed structure of the alongshore flow. Neglecting an alongshore pressure gradient, the effect of this cross-shore momentum flux on the bottom stress is

$$\tau^{bx} \approx \tau^{sx}\left(\frac{1-bS_l/2}{1+bS_l/2}\right) \qquad (18.27)$$

from which the cross-shore divergence of momentum flux can be expressed as

$$\frac{\partial}{\partial y}\int_{-h}^{0} uv dz \approx \frac{\tau^{sx}}{\rho_o}\frac{bS_l}{1+bS_l/2} \qquad (18.28)$$

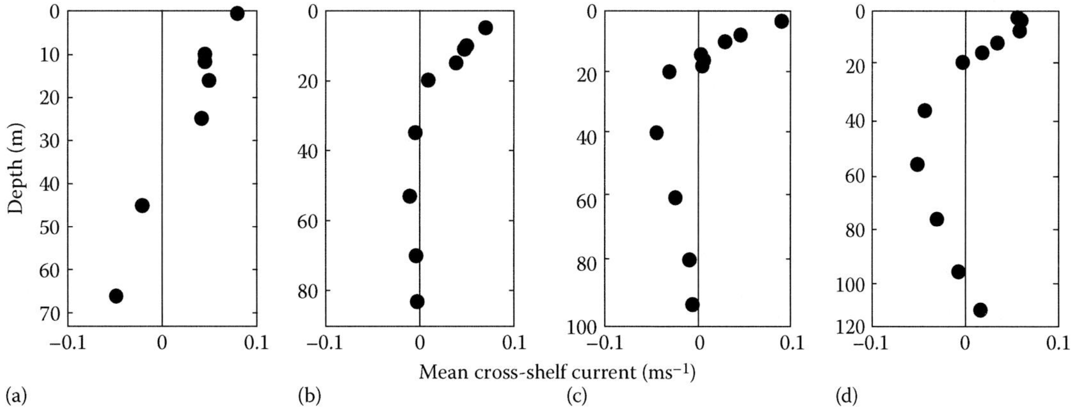

FIGURE 18.10 Long-term mean cross-shore flow from four major upwelling regions. (a) North West Africa, (b) North California, (c) Oregon, and (d) Peru. (Reprinted with permission from Lentz, S. and Chapman, D., *J. Phys. Oceanogr.*, 34, 2444, 2004, Copyright American Meteorological Society.)

Thus in the absence of an along-shelf pressure gradient, the bottom stress (and thus the onshore Ekman transport in the bottom boundary layer) and the cross-shore advection of alongshore momentum are determined by the wind stress and the slope Burger number. For small S_l, onshore transport is confined to the bottom Ekman layer and the interior cross-shore flow is weak. As (bS_l) increases, the onshore flow becomes increasingly concentrated in the interior. At $S_l = -2b^{-1}$ the bottom stress vanishes and all the onshore flow is the interior.

For $(bS_l) > 2$, however, the theory breaks down because it requires an offshore flow in the bottom Ekman layer which in turn requires isopycnals to tilt downward near the bottom while those in the interior are tilted upward, thus violating the assumption of a constant cross-shore density gradient. A second shortcoming of this model is that it does not include the evolution of the density field in response to the circulation but rather relies on an f/N scaling for the isopycnal slope. Nevertheless, the simple model provides a dynamical explanation for the vertical structure of the onshore flow across upwelling systems spanning a range of Burger numbers from 0.2 for the northwest African shelf, where onshore flow is confined to the bottom Ekman layer, to 1.5 for the Peruvian shelf where the onshore flow is in the interior.

In reality, many upwelling systems exhibit a fully 3D structure that is not captured by the 2D model (Barth and Wheeler 2005). For example, modeling and observational studies of upwelling circulation around a submarine bank off the Oregon coast depict a fully 3D upwelling circulation that includes different upwelling pathways upstream and downstream of the bank, significant recirculation, and enhanced cross-shelf transport downstream of the bank (e.g., Figure 18.11), none of which can be captured by a 2D model.

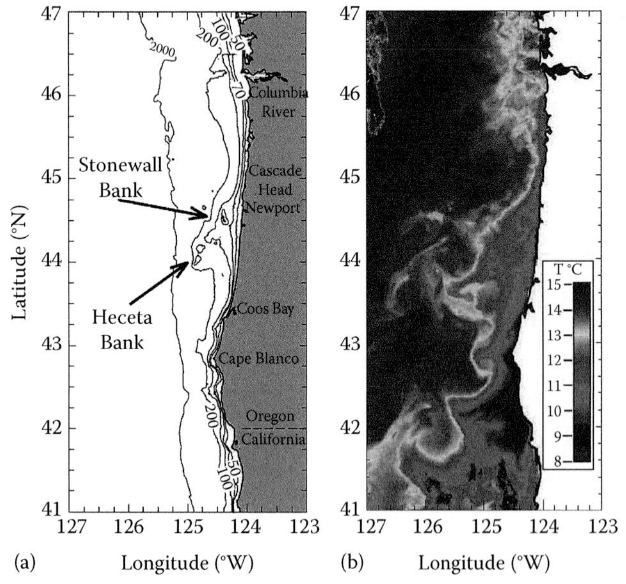

FIGURE 18.11 (a) Oregon coastal region with bottom topography in meters showing the locations of Heceta and Stonewall Banks; (b) Satellite sea surface temperature image from August 13, 1995. (Reprinted with permission from Barth, J. and Wheeler, P., *J. Geophy. Res.*, 110, C10S01, 2005, Copyright American Geophysical Union.)

Understanding the 3D structure of coastal upwelling with models and observational programs and elucidating the importance of this 3D structure on cross-shelf transport processes, coastal ecosystems and even on large-scale ocean-atmospheric processes, remain topics of contemporary research in oceanography.

18.6 Buoyancy-Driven Plumes

Freshwater debouching from rivers and estuaries onto the continental shelf is subject to a variety of forces that define its shape, structure, and the resultant flow field. Since these discharges represent a major link between terrestrial and marine ecosystems, it is critical to characterize their structure and the rate that these buoyant outflows mix into the coastal ocean. Buoyant outflows are among the most complex oceanographic physical processes. They can be fully nonlinear, time dependent, and are influenced by both estuarine and coastal processes. Their structure is strongly influenced by turbulent mixing, coastal bathymetry, ambient shelf circulation, and even nonlinear internal waves.

Garvine (1995) delineated "small-scale" outflows that tend to spread laterally from "larger-scale" outflows that form a coastally trapped current, which flows in the direction of Kelvin wave propagation. Based on dimensional analysis of the momentum equation, Garvine suggests that five independent variables characterize the dynamics of buoyant discharges. The first is the plume's "slenderness," γ, the ratio of the plume's width (W) to its length (L). The second is the Kelvin number, $K = \gamma L/(c/f)$, where c is the internal wave speed. The third is the Froude number. The fourth term (V_e/UH) represents the ratio of the cross-shore surface Ekman transport (V_e) to the down-shelf volume transport in the plume (UH). Finally, the fifth (r/fH) is the relative importance of bottom drag (based on a linearized drag coefficient r) and represents the ratio of the spin-down time due to bottom friction to the local inertial period.

Garvine's analysis is guided by scaling of the momentum equation. For example, the five terms in the steady, cross-shore momentum equation—momentum advection, Coriolis, pressure gradient, and surface and bottom stress—are scaled as $[U^2/L \quad f\gamma Y \quad c^2/L \quad \tau_w/\rho H \quad rU/H]$. The alongshore momentum balance is scaled similarly. Since these flows are forced by the pressure gradient, it must be a dominant term in the momentum equation. Thus, the horizontal momentum equations are rescaled by c^2/L to yield

$$F^2 \quad KF \quad 1 \quad \frac{KF}{\gamma}\left(\frac{V_E}{HU}\right) \quad \frac{KF}{\gamma}\left(\frac{V_E}{HU}\right) \tag{18.29a}$$

$$\gamma^2 F^2 \quad KF \quad 1 \quad KF\left(\frac{V_E}{HU}\right) \quad \gamma KF\left(\frac{V_E}{HU}\right) \tag{18.29b}$$

where (18.29a) comprises the representative magnitudes of the five scaled terms in the cross-shore momentum equation and (18.29b), the scaled terms in the alongshore momentum equation. This scaling permits a deduction of the dominant terms in the momentum equation for various plume regimes in Kelvin number/Froude number space.

Garvine (1995) considers two limiting cases based on the Kelvin number. The first case considers small Kelvin numbers which he termed "small-scale" discharges. From the definition of K, these plumes have a small cross-shore scale relative to the internal Rossby radius. The scaling indicates that the pressure gradient [O (1)] is balanced by advection and represents a Bernoulli-type flow that is characterized by strong boundary fronts and internal hydraulic jumps. In the limiting case of very small K, the along-plume momentum balance suggests that the Froude number is near unity, while the cross-shore momentum balance indicates that $\gamma \sim O(1)$, suggesting that cross-shore and alongshore length scales are similar, which implies a radially spreading plume.

On the other extreme, for large Kelvin numbers, the flow is considered large scale. The choice $K \gg 1$ suggests that $KF = O(1)$ and thus $F = K^{-1}$, indicating that the advective term can be neglected and the coastal current (cross-shore momentum balance) is in geostrophic balance. Observations indicate that in such flows the plume's length far exceeds its width and, thus, $\gamma < 1$. Based on observations of the buoyant outflow from Delaware Bay (Figure 18.12), Garvine (1995) estimates that

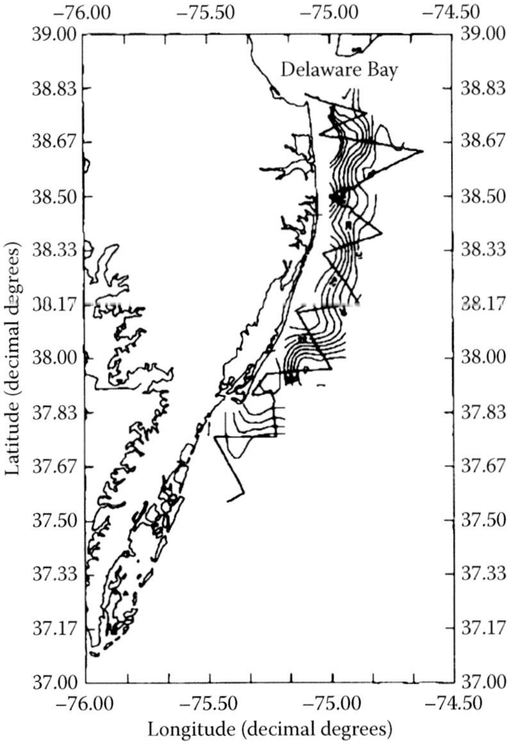

FIGURE 18.12 The Delaware coastal current May 1–2, 1993. Contour lines show near-surface salinity at intervals of 0.5 psu. The zigzag line marks the ship track. The buoyant layer at the top center is in the mouth of Delaware Bay. (Reprinted with permission from *Continental Shelf Research*, 15, Garvine, R., A dynamical system for classifying buoyant coastal discharges, 1585–1596, Copyright 1995, Elsevier.)

$K = 4$, $F = 0.1$, and $\gamma = 0.2$, thus consistent with a large Kelvin number plume. Interestingly, the outflow from the world's largest river, the Amazon, falls into the "small-scale" plume parameter space because its discharge occurs near the equator where $f \rightarrow 0$, while the Delaware with a discharge orders of magnitude smaller is classified as a "large-scale" discharge. More generally, the small-scale ($K \ll 1$) limit is approached for engineering structures, narrow river mouths, and equatorial outflows. Large-scale flows appear to occur for K as low as 2 and will occur for mid- and high-latitude outflows from wide river mouths. Intermediate cases ($K \sim 1$) have many dynamical features of the limiting cases and tend to be the most dynamically complicated.

The above scaling also provides insights for the relative importance of wind forcing. In particular, for buoyancy to be the dominant driver of the flow $V_E/(HU) \leq O(\gamma/KF)$. If the wind-driven Ekman transport exceeds this value, then the flow is dominated by wind forcing rather than buoyancy forcing. Similar reasoning can be applied to bottom friction.

Yankovski and Chapman (1997) characterized plumes by their vertical structure and distinguish "surface-advected plumes" from "bottom-attached plumes." Assuming that the plume's alongshore flow is in thermal wind balance and equating the transport in the plume to that being supplied by the estuary, it is easily shown that a river plume will extend to depth $h_b = \sqrt{2T_e f/g'}$, where T_e is the transport of fluid out of the estuary. Assuming a cross-shore bathymetry of $h(x) = h_o + \alpha x$ where h_o is the depth at the coastal wall, the plume will be surface advected when $h_b < h_o$ and is bottom attached when $h_b > h_o$. An important dynamical consequence of a bottom-attached plume is the existence of a bottom Ekman layer that drives a buoyancy flux offshore in the bottom Ekman layer which feeds diapycnal mixing at the foot of the offshore front at depth h_b. The Yankovski and Chapman theory can also be used to estimate the plume's cross-shelf length scale. For a bottom-attached plume, this is determined by bottom slope and the estuarine outflow, while surface-advected plumes have a width on the order of the internal Rossby radius.

While details of coastal geometry and discharge rates clearly influence the plume's structure, wind forcing can be dominant in defining the plume's configuration for a given geometry and discharge rate (i.e., a single point in Yankovski and Chapman parameter space); a plume may be bottom attached, surface advected, or even detached from the coast, depending on the alongshore wind stress forcing. Upwelling winds reverse the cross-shore (secondary) flows in the plume and the plume is advected offshore in a thinned surface layer. Secondary flows at the surface supply buoyant fluid to the seaward side of the plume where enhanced diapycnal mixing occurs. In contrast, downwelling favorable winds augment secondary flows in the plume, causing it to vertically thicken as it is compressed horizontally against the coast and diapycnal mixing is reduced. Interestingly, there has been little work on the effect of cross-shelf winds on river plumes.

A final complication of plume dynamics is that they are often unsteady. The unsteadiness can be due to the variable forcing associated with river discharge and winds. In addition, unsteadiness even occurs under steady forcing. This time dependency takes the form of a growing bulge in the outflow regions because, in contrast to Yankovski and Chapman's assumption, the volume transport from the estuary exceeds that in the coastal current. Numerical observations by Fong and Geyer (2002) suggest that the fraction of the estuarine outflow fed to the coastal current is dependent on the estuarine outflow Rossby Number (U/fB) where U is the velocity of fluid exiting the estuary and B is the width of the estuarine mouth. For $Ro \ll 1$ most of the discharge goes into the coastal current. However, for $Ro = 0.5$ only half of the outflow feeds the coastal current with the remaining fraction feeding the growing bulge. When $Ro > 1$, coastal current formation can cease and the entire outflow become incorporated in the growing bulge.

Note that the high-Rossby number growing bulge in the Fong and Geyer (2002) model simulations is similar to Garvine's low Kelvin number radially spreading small-scale plume. Similarly, the low-Rossby number coastal current is consistent with Garvine's high-Kelvin number large-scale plume. This emphasizes the relationship between the Kelvin, Rossby, and Froude numbers, which, given the above definitions, is $RoK = \frac{W}{B}F$. For the large-scale plume, Garvine suggests that $K = F^{-1}$ and thus $RoK^2 = W/B$. In large-scale plumes $B> = W$ and $K > 1$ implying $Ro < = 1$. Similarly, for small-scale plumes where $K < 1$, Garvine suggests that $F \sim 1$ and thus $RoK = W/B$, implying $Ro> = 1$. Thus, due to these interdependencies it appears nontrivial to unravel the specific role of the Rossby and Froude number on many buoyant discharges.

Many of the aforementioned features are apparent in observations of the Hudson River outflow (Chant et al. 2008). The outflow is surface advected as predicted by the Yankovksi and Chapman scaling. The plume is approximately an internal Rossby Radius in width, currents in the coastal current are in geostrophic balance, and the nose of the gravity current propagates down-shelf at the internal gravity current speed. In addition, the transport of freshwater in the plume is less than ½ of the discharge from the Hudson, consistent with the Fong and Geyer simulations given the outflow Rossby number of 0.5. This caused a bulge region near the outflow to grow over time and was observed to be highly responsive to wind forcing and ambient shelf circulation and resulted in rapid cross-shelf mixing of the Hudson River outflow.

18.7 Concluding Remarks

A knowledge of the coastal oceans, and their behavior, is of fundamental importance to many aspects of our daily society. While many examples could be given, we complete this chapter in the summer of 2010 while a disaster (an oil spill) of unprecedented dimensions unfolds in the Gulf of Mexico, promising unforeseeable economic and environmental impacts. The continued development and deployment of novel observing and modeling systems are essential to our ability in future to monitor and to predict the state of the coastal oceans and the fate of the geological, chemical, and ecological resources therein.

A review of coastal ocean prediction systems is beyond our intent here. We close, however, by noting the rapid progress being made on coupled modeling systems for the coastal ocean. Community models for the coastal oceans have now reached a state of some sophistication. In particular, current systems have gone well beyond the "simple" problem of the coastal circulation itself, and are now able to address issues related to (e.g.) sediment transport and ecosystem dynamics. Together with new data—assimilative technologies for the combination of numerical models with observational data streams—the requisite elements for reliable inter-disciplinary coastal ocean prediction systems are now in place.

References

Barth, J. and P. Wheeler. 2005. Introduction to special section: Advances in coastal transport. *Journal of Geophysical Research* **110**: C10S01, doi:10.1029/2005JC003124.

Battisti, D. and A. Clarke. 1982. A simple model for estimating tidal currents on continental margins with specific applications to the M2 tide off the Atlantic and Pacific coasts of the United States. *Journal of Physical Oceanography* **12**: 8–16.

Battisti, D. and B. Hickey. 1984. Application of remote wind-forced coastal trapped wave theory to the Washington and Oregon coasts. *Journal of Physical Oceanography* **14**: 887–903.

Brink, K. H. and A. R. Robinson. 1998. *The Sea, Volume 10: The Global Coastal Ocean: Processes and Methods.* John Wiley & Sons, Inc., New York.

Chant, R., J. Wilkin, W. Zhang, B. Choi, E. Hunter, R. Castelao, S. Glenn, J. Jurisa, O. Schofield, R. Houghton, and J. Kohut. 2008. Synthesis of observational and numerical studies during LaTTE. *Oceanography* **21**: 70–89.

Chapman, D. 1987. Application of wind-forced, long, coastal-trapped wave theory along the California coast. *Journal of Geophysical Research* **92**: 1798–1816.

Clarke, A. and K. Brink. 1985. The response of stratified, frictional flow of shelf and slope waters to fluctuating, large-scale, low-frequency wind forcing. *Journal of Physical Oceanography* **15**: 439–453.

Clarke, A. and S. Van Gorder. 1986. A method for estimating wind-driven frictional, time-dependent, stratified shelf and slope water flow. *Journal of Physical Oceanography* **16**: 1013–1028.

Fong, D. and W. Geyer. 2002. The along shore transport of fresh water in surface-advected river plumes. *Journal of Physical Oceanography* **32**: 957–972.

Garvine, R. 1995. A dynamical system for classifying buoyant coastal discharges. *Continental Shelf Research* **15**: 1585–1596.

Haidvogel, D. B. and A. Beckmann. 1999. *Numerical Ocean Circulation Modeling.* Imperial College Press, World Scientific Publishing Co, Ltd., London, U.K.

Holton, J. R. 1992. *An Introduction to Dynamic Meteorology* (3rd edn.). Academic Press, San Diego, CA.

Huyer, A. 1980. The offshore structure and subsurface expression of sea level variations off Peru, 1976–1977. *Journal of Physical Oceanography* **10**: 1755–1768.

Huyer, A. 1990. Shelf circulation. In: *The Sea*, B. LeMehaute and D. M. Hanes, eds., Vol. 9. John Wiley & Sons, Inc., New York.

Lentz, S. and D. Chapman. 2004. The importance of non-linear momentum flux during wind driven coastal upwelling. *Journal of Physical Oceanography* **34**: 2444–2457.

Loder, J. 1980. Topographic rectification of tidal currents on the sides of Georges Bank. *Journal of Physical Oceanography* **10**: 1399–1416.

Simpson, J. H., C. M. Allen, and N. G. C. Morris. 1978. Fronts on the continental shelf. *Journal of Geophysical Research* **83**: 4609–4614.

Wang, D.-P. and C. Mooers. 1976. Coastal-trapped waves in a continuously stratified ocean. *Journal of Physical Oceanography* **16**: 853–863.

Werner, F., R. Perry, R. Lough, and C. Namie. 1993. Trophodynamic and advective influences on Georges Bank larval cod and haddock. *Deep Sea Research* **43**: 1793–1922.

Winant, C. 1980. Coastal circulation and wind-induced currents. *Annual Review of Fluid Mechanics* **12**: 271–301.

Yankovski, A. and D. Chapman. 1997. A simple theory of the fate of buoyant coastal discharges. *Journal of Physical Oceanography* **27**: 1386–1401.

19

Ice Dynamics and Transport

Steven F. Daly
United States Army Corps of Engineers

Robert Ettema
University of Wyoming

19.1 Introduction

The phase change of water from liquid to solid ice and back to liquid is a seasonal feature of water bodies in the Earth's cold regions. Ice occurs in rivers, streams, ponds, lakes, reservoirs, and oceans. This chapter focuses primarily on ice in rivers and streams, and briefly considers ice in lakes and ponds. It outlines how ice formation, dynamics, and transport introduce considerable complexity because of the profound and immediate interactions between ice and water flow. This complexity is evident in the seasonal evolution of ice which starts with its first appearance and continues through growth, transport, metamorphosis, accumulation, ice cover formation, deterioration, and breakup. At each step, the physical and mechanical properties of ice and water, the flow hydrodynamics, heat transfer between ice, water, and the atmosphere, and the energetic efforts of engineers, all interact to determine the path of this evolution.

The chapter begins with a description of ice formation in river and stream flow and quiescent water. Formation starts with the first appearance of ice and then continues through the winter season to eventual ice breakup during spring. It entails the initial interaction of supercooled water, seed crystals, the anisotropic crystalline kinetics of ice, and fluid turbulence to produce the ubiquitous floating stationary ice covers of the cold regions of the Earth. The ways in which the ice covers modify channel flow characteristics then are described, along with the significant flow transients caused by ice cover formation and breakup. The formation of breakup ice jams is discussed subsequently (Figure 19.1). The chapter also briefly covers methods for analysis of topics in ice formation and transport, and ends with indicating several notable challenges associated with ice. The challenges include improved insight into ice formation, ice effects on other natural processes, and engineering activities.

The field of ice dynamics and transport began with the seminal papers of Pariset and Hausser (1961) and Pariset et al. (1966), which presented systematic and rational approaches for analyzing river ice jams. It has been the subject of considerable research since. Notable books include those by Michel (1978), Ashton (1986), and Beltaos (1996, 2008). Two recent review articles of interest are those by Hicks (2009) and Shen (2010). Steady progress has been reported in the biannual conferences organized by the International Association for Hydro-Environment Engineering and Research,* and the series of workshops by the Committee on River Ice Processes and the Environment (CRIPE), a section of the Canadian Geophysical Union. A substantial source of information is the extensive technical report series is the U.S. Army

* Formerly known as the International Association of Hydraulic Engineering and Research.

Handbook of Environmental Fluid Dynamics, Volume One, edited by Harindra Joseph Shermal Fernando. © 2013 CRC Press/Taylor & Francis Group, LLC. ISBN: 978-1-4398-1669-1.

FIGURE 19.1 Ice jam on the Elkhorn River at West Point, Nebraska. (Photo courtesy of U.S. Army Corps of Engineers by James Laney.)

Cold Regions Research and Engineering Laboratory. Several journals encompass this chapter's topic, for example, *Journal of Glaciology*, *Cold Regions Science & Technology*, and *Journal of Cold Regions Engineering*.

19.2 Principles

19.2.1 Ice Formation

19.2.1.1 First Appearance of Ice

19.2.1.1.1 Distribution of Supercooled Water

The generation of supercooled water through heat transfer from the water surface to the frigid atmosphere above creates the potential for ice to form in lakes, rivers, and other water bodies. The spatial distribution of supercooled water controls the initial spatial distribution of ice. The nucleation characteristics of ice crystals and the seeding rate of ice crystals from the atmosphere into the supercooled water determine the time delay from the onset of supercooling to the appearance of ice crystals.

The initial distribution of supercooled water in rivers, streams, lakes, and other water bodies is controlled by the vertical stability of the water column. The vertical stability reflects the balance between stratification due to heat loss and turbulent mixing (Turner 1979). Due to the density maximum of freshwater at 3.98°C, a surface heat loss at water temperatures less than 3.98°C is a stabilizing buoyancy flux into the water. Once the surface water temperature reaches 0°C, further cooling produces supercooled water and the vertical stability of the water column controls the vertical distribution of supercooled water. While it is not common to think of liquid water at temperatures below 0°C, it must be remembered that the only mechanism limiting the magnitude and duration of supercooling is the latent heat released by growing ice. Accordingly, the presence of supercooled water implies a lack of sufficient growing ice to counterbalance the heat loss to the atmosphere.

While supercooled water creates the conditions for ice formation, there is a time lag before enough latent heat released by growing ice can at first balance the heat loss to the atmosphere and then warm the water back to near 0°C. The time lag provides time for the turbulent mixing provided by the flow velocity and/or wind stress, if present and sufficiently strong, to overcome the density stratification due to the net heat loss at the surface and vertically mix the supercooled water. There are two limiting cases: at one extreme are lakes and ponds with no flow velocity and no wind mixing; the other limiting case is flowing rivers and streams with rapid vertical mixing sufficient to overcome any stratification. In the first case, heat transfer from the water leads to a stratified water column with the coldest and least dense water at the surface. Stratification limits the initial ice growth to a thin surface layer of supercooled water. In the second case, sufficiently energetic turbulent mixing leads to well-mixed water column with supercooled water advected from the surface to the bed. Vertical mixing enables the initial ice growth to proceed throughout the depth of flow, which results in the ubiquitous ice forms known as frazil and anchor ice (Daly 1984, 1994).

19.2.1.1.2 Ice Nucleation

The spontaneous formation of ice in natural water bodies, by homogeneous or heterogeneous nucleation, has never been shown to happen, or even to be possible. This uncertainty arises because no nucleating agent for water has been demonstrated to be effective at the very small supercooling levels found in natural water bodies (<0.1°C) (Osterkamp and Gilfilian 1975). Nor has spontaneous nucleation been observed in any carefully controlled laboratory experiment at these temperatures (Heneghan et al. 2002, Wilson et al. 2002). As a result, the hypothesis of spontaneous nucleation has been set aside and replaced by the concept of seed crystals.

Seed crystals are ice crystals that are introduced from outside the natural water body to begin the process of ice formation. Seed crystals can come from a number of different sources: vapor evaporating from the water surface on encountering cold air can sublimate into ice crystals, which fall back onto the water surface and are entrained by the turbulent motion of the flow; small water droplets generated by breaking waves, bubbles bursting at the water surface, and splashing; snow and sleet (Osterkamp 1977). The number density of seed crystals introduced into the thin supercooled layer at the surface of lakes and ponds determines the crystal fabric of the surface ice cover (Gow 1986, Muller-Stoffels et al. 2008). Seed crystals introduced into turbulent supercooled water begin the formation of frazil. In turbulent water, however, it is observed that once a very few seed crystals are introduced into turbulent supercooled water, very quickly many, many new crystals are created. The process by which these new crystals are created is called secondary nucleation (Botsaris 1976). Strictly speaking, secondary nucleation refers to the formation of new crystals because of the presence of ice crystals and does not refer to the particular mechanism that creates new crystals. The most likely mechanism of creation is thought to be the removal of the new crystals from the surface of the parent crystals through collisions of the crystals with hard surfaces (including other crystals). The collisions of crystals suspended in turbulence have long been described and provide estimates of the resulting collision energy (Evans et al. 1974a,b). The number of ice particles created per unit of collision energy is an aspect that has never been measured directly, although various values have been proposed (Mercier 1984, Hammar and Shen 1995, Clark and Doering 2009).

19.2.1.1.3 Ice Crystal Growth Habits

Ice crystal form or morphology is strongly influenced by its crystallization kinetics, defined as the rate at which the ice crystal can incorporate new water molecules into its structure. At the crystalline level, ice has two principal growth directions: *a*-axis and *c*-axis. The crystallization kinetics of each growth direction is different. The interface kinetics of *a*-axis growth appears to be very fast at the levels of supercooling found in water bodies; for all practical purposes the rate at which latent heat is transferred away from the ice/water interface controls *a*-axis growth (Shibkov et al. 2003). The *c*-axis growth is apparently controlled by the crystallization kinetics as it is much slower than growth along the *a*-axis for all sizes of crystals (Forest 1986). The anisotropic crystalline kinetics, combined with the distribution of supercooled water and the heat transfer rate, ultimately determine crystal morphology. Especially crucial is the formation of temperature gradients in the water at the ice/water interface when the surrounding water is supercooled. This condition can lead to the formation of dendrites that grow rapidly (Mullins and Serkerka 1964). Given the very small supercooling levels found in water bodies (normally much less than 0.1°C) and the ability of turbulence to suppress gradients through mixing, dendritic growth is typically a special case. It is commonly seen only in thin supercooled layers at the surface of quiescent ponds when the number density of seed crystals is low. In contrast, frazil ice crystals, suspended in turbulent flow at very small supercoolings, grow as perfect, thin, circular disks with a major diameter that is 5–15 times greater than their thickness (Daly and Colbeck 1986). The disk shape results from the anisotropic crystalline kinetics combined with the turbulent suppression of temperature gradients surrounding the crystals.

19.2.1.2 Ice Growth

19.2.1.2.1 Quiescent Water

In lakes and ponds where the surface buoyancy flux due to the heat transfer from the surface can counterbalance applied wind stress, the water column is stratified with the coldest and least dense water at the surface. Stratification limits the initial ice growth to a thin surface layer of supercooled water. The crystals grow downwards into non-supercooled water (their own melt) with latent heat removed through the ice crystals to the ice surface and ultimately to the atmosphere above. The origins of modeling this thermal ice growth are generally credited to Stefan (1891), who analyzed ice growth as a moving boundary problem with the growth rate primarily controlled by thermal conduction through the ice. Stefan (1889) and most subsequent studies (e.g., Ashton 1986) used a physical approximation to obtain the solution for the limiting case of very large latent heat compared to the heat capacity of the solid. This approximation, particularly relevant to ice, assumes that, behind the slowly moving solidification boundary, the temperature distribution is equivalent to the steady-state distribution that would occur if the boundary were to be fixed in position at that instant (Crank 1984). This approach is widely used in the field of ice engineering. It can include any number of layers, each with its own thermal properties, to represent the thermal influence of the ice cover, snow cover, and the lower boundary layer of the atmosphere (Ashton 1989).

The quasi-steady approach has been coupled with a variety of boundary conditions. When the temperature of the upper surface of the top layer is assumed equal to the air temperature, the quasi-steady model leads to the classic result that the ice thickness, during growth, is proportional to the square root of the accumulated freezing degree days (AFDD) (Ashton 1986, U.S. Army 2006). The fact that this approach is well suited to applications where information on the meteorology is limited to estimates of daily or longer period air temperature averages contributed greatly to its widespread use. One limitation of this approach is that its quantitative application requires the use of calibrated parameters to describe the effective values of the thermal properties of each model layer. Calibration based on field measurements of ice thickness is generally required as estimating these parameters from first principles is difficult. In general, the parameters can vary from location to location and year to year (Daly 1998).

19.2.1.2.2 Turbulent Water

Given the material properties of water, the change in water density with temperature is relatively small at 0°C, and, as a result, channels with any appreciable flow generally have enough turbulent mixing to overcome any stabilizing buoyancy flux. This makes the vertical distribution of supercooled water and the formation of frazil ice common in rivers located in cold regions. Arden and Wigle (1972), for example, describe a study in the Niagara River where frazil ice production began in water at −0.05°C during cold, clear nights. However, in channels with very low velocities turbulent mixing may not be able to completely overcome positive buoyancy flux. In these cases, the supercooled layer is confined to the surface and skim ice and other ice forms may result. The ice types that result from each particular balance of turbulent mixing and buoyancy flux has not been completely worked out (however, see Matousek 1984, Hammar et al. 2002). In midlatitude lakes and ponds at night with clear skies, winds can often mix supercooled water deep enough to reach the bed. This wind mixing of the surface supercooled water into the depths leads to the formation of frazil just as in rivers. In the case of both flowing rivers and wind-stressed lakes, the vertical distribution of supercooled water leads to the development of ice attached to the bottom substrate called anchor ice and to the blockage of water intakes by frazil (Daly and Ettema 2006, Kempema et al. 2008a,b).

FIGURE 19.2 Frazil ice accumulated as fine disks on a screen.

Frazil crystals suspended in turbulent supercooled water actively increase in size. The crystals are in nonequilibrium with the supercooled water—the ice/water phase diagram fixes the interface temperature at 0°C under ordinary pressures. The difference between the interface temperature and the bulk water temperature results in heat transfer from the crystal surfaces to the supercooled water. This heat transfer carries away latent heat, which causes the crystal to grow in size along the fast growth a-axis that forms the major diameter of the disk-shaped frazil ice crystals (Figure 19.2). The rate at which latent heat is transferred from the growing crystal controls the growth rate of the major diameter. This rate depends on the crystal size relative to the dissipation (Kolmogorov) length scale of the turbulence, the supercooling level, and the material properties of ice and water (Holland et al. 2007).

The size distribution of ice crystals suspended in turbulent supercooled water can be modeled using a crystal number continuity equation that includes descriptions of the crystal growth rate and secondary nucleation rate (Daly 1984), along with additional equations describing the fluid flow and turbulence (Mercier 1984, Svensson and Omstedt 1994, Hammar and Shen 1995, Ye and Doering 2004). It is generally assumed that frazil ice in suspension acts as a "passive tracer" and has no influence on the fluid turbulence. These models have shown that there are essentially four independent parameters that determine the size distribution of suspended frazil crystals: the heat transfer rate from the volume of interest; the introduction rate of seed crystals; the turbulence intensity; and the number of new crystals produced per unit of collision energy during secondary nucleation.

When the major diameter of the frazil ice crystals is small, on the order of 0.1 mm, the crystals are easily transported throughout the depth of flow. As the crystals increase in size, their buoyancy increases and at some point they will begin to rise and collect at the water surface. (Morse and Richard [2009] provide a thorough review of frazil rise velocities.) The balance of the upward crystal flux due to buoyancy and the downward flux due to the turbulence of the flow suggests an immediate, although spatially inverted, analogy with suspended sediment, and leads to a Rouse equation description of the frazil ice concentration. New acoustic sensors have been shown to detect suspended frazil ice crystals in situ in rivers and allowed direct estimation of the Rouse number of the suspended frazil crystals (Marko and Jasek 2010a,b, Richard et al. 2011). Further advances in acoustic technology promise the ability to measure the size distribution directly.

19.2.1.3 Anchor Ice

Anchor ice, always associated with supercooled water, forms on the bottom substrate material of rivers and lakes. Anchor ice is most commonly associated with gravel, cobble, and rock substrates (Hirayama et al. 1997, Kerr et al. 2002, Stickler and Alfredsen 2005, Qu and Doering 2007, Kempema et al. 2008, Bisaillon and Bergeron 2009) although it is occasionally associated with sandy bottoms (Kempema et al. 2008). Anchor ice typically has a milky white appearance when viewed through the water body, and sediment deposited in the anchor ice mass can darken its color.

The mass of anchor ice can increase through deposition of frazil ice crystals or through growth of deposited crystals by transport of latent heat to the supercooled flow by convection. In flowing rivers, it has been unclear whether anchor ice initiated and grew exclusively through the deposition and adhesion of frazil ice crystals on the bed (Hirayama et al. 1997, Doering et al. 2001, Kerr et al. 2002, Qu and Doering 2007). For example, Kerr et al. (2002) reported that "No evidence of in-situ thermal growth of frazil particles in the anchor ice accumulation … was found." However, other studies do report in situ ice growth. Kempema et al. (2008) described anchor ice observed in the Laramie River, Wyoming, as containing very large individual ice crystals that "grew through in-situ thermal growth; i.e., the anchor ice mass increased through growth of attached ice crystals rather than by the accumulation of frazil crystals"; Figure 19.3 depicts anchor ice reported by Kempema et al. (2008). Other reported instances of anchor ice that formed through thermal growth include those by Michel (1971), Osterkamp and Gosink (1983), Schaefer (1950), Wigle (1970).

FIGURE 19.3 Anchor ice crystals formed on the bed of the Laramie River, Wyoming.

The actual attachment of anchor ice to the substrate material can vary between tenuous to tenacious. While anchor ice has been observed to "not penetrate deeper than approximately one-half of the gravel diameter from the top of the gravel" (Kerr et al. 2002), the actual mechanisms of attachment are not known. Anchor ice often forms at night, and then lifts up and floats away at first light or soon after. Apparently, the anchor ice attachment responds to subtle changes in the water temperature resulting from the influx of solar radiation. Anchor ice can sometimes release abruptly and somewhat randomly as has been observed in the laboratory (Kerr et al. 2002) and in the field (Kempema 2008). In some cases, anchor ice can last for many days and even through an entire winter season, and attain thickness large enough to raise upstream water levels and cause flooding (Daly 2005) or impact upstream hydroelectric production (Girling and Groeneveld 1999). The buoyancy of anchor ice can sometimes lift significant amounts of sediment off of the bottom and "raft" it under the influence of currents and wind (Kempema et al. 2002, Kempema 2008). Rafted sediment, especially finer sediment, can become included in an ice cover downstream. Coarser sediment, gravel, eventually drops from the slush when it passed through regions of elevated turbulence.

19.2.1.4 Floe Formation

Supercooling is a temporary condition which reflects the balance between the surface heat loss and the latent heat released by the growing crystals. This balance is dynamic, and as the ice crystals increase in size and number, their release of latent heat inexorably drives the water temperature back to 0°C, ending the period of supercooling. In addition, changes in the meteorological conditions, the insulating presence of a floating ice cover, and other factors can also end the period of supercooling. In rivers and streams when the period of supercooling has ended, ice crystals are advected by the river flow velocity and are continuously in movement either in open water or under stationary ice covers. Most often the frazil ice appears as *slush* at the water's surface. Without supercooled water, crystal growth ceases and new crystals are no longer created, but changes in ice morphology continue on, both, the microscale (crystal level) and the macroscale (the formation of floes). Little is known about how the crystals evolve on the microscale. Their growth under nonequilibrium supercooled conditions combined with the anisotropic crystalline kinetics produces a crystal shape far from the spherical shape of an ice crystal at equilibrium—that is, with minimum surface energy for the volume enclosed. The process that pushes the crystal toward equilibrium shape, and leads to "equilibrium metamorphosis," is apparently overwhelmed during the period of growth, and exerts little influence on the crystal form during this stage (Colbeck 1992). The metamorphosis of disk-shaped crystals into their equilibrium form in rivers has not been studied and little is understood about the controlling mechanisms.

Slush ice is formed from the collection of frazil crystals, flocs, and anchor ice that have risen from the channel bottom. The existence of frazil flocs, collections of crystals joined together as larger masses, has long been recognized and observed in rivers and streams. The reduction in overall surface energy is thought to drive the crystals to join together, or sinter, into one larger floc (Martin 1981, Mercier 1984, Clark and Doering 2009). Observations and theory of ice sintering have a long history, going back to the observations by Faraday (1859, 1860). It is interesting to note that sintering of ice crystals in water is not thought possible by Colbeck (1997) and others, who view the bonds between ice crystals brought together in the presence of water as inherently unstable. This inherently unstable interface has been used to explain observations that snow crystals saturated with water, that is snow slush, have "virtually no mechanical strength" (Blackford 2007).

Ice floes are formed from frazil slush and at times anchor ice. The formation of floes of all sizes is governed by two separate but intertwined processes: the increase in floe area and floe strength. The increase in floe area reflects the accumulation of frazil ice into a single, moving unit. It is strongly influenced by the level of turbulence in the river reach. In highly energetic streams, the frazil slush at the surface never consolidates and the frazil remains in the form of slush. In less energetic reaches, in large lakes, and the ocean, frazil slush at the water surface shows a marked tendency to clump together into porous floes (Andreasson et al. 1998, Hammar et al. 2002). The initial clumps, if they remain on the surface long enough, gain strength through freezing of their interstitial water and form pans, or small floes. With time, these pans grind against one another, become roughly circular and gain upturned edges. They are then known as pancake ice. In slow-moving streams, very large floes can form with effective diameters on the order of the channel width. Osterkamp and Gosink (1983) described floe formation in interior Alaskan streams in detail. They identified five mechanisms that can produce floes from smaller frazil pans:

1. Contact, penetration, and bonding of drifting frazil pans
2. Compaction and flow-induced drag cutoff of pans
3. Compaction by convergent flow, with occasional cutoff by impact with incoming pans
4. Extrusive flow and drag cutoff
5. Agglomeration with flow cutoff controlled primarily by river curvature

Extrusive flow and drag cutoff produced very large floes, with maximum dimensions exceeding 1 km on the Yukon River.

19.2.2 Hydraulics of Ice-Covered Channels

The presence of an ice cover makes a portion of the channel unavailable for flow and changes the flow characteristics of the channel by presenting an additional stationary, rough boundary, which modifies the channel wetted perimeter and hydraulic radius. These effects impact steady flow. In addition, the inertia and stiffness of the ice cover can potentially interact with unsteady flow. However, if unsteady flow analysis is limited to "long" waves (where "long" is defined relative to the characteristic length of the ice cover), the inertia and stiffness of the cover can be neglected (Daly 1993). The characteristic length of an ice cover, l, reflects ice cover "stiffness," and is defined as

$$l = \left[\frac{\eta^3 E}{12\rho g(1 - \nu^2)} \right]^{1/4}$$

where

η is the ice thickness
E is the elastic modulus of the ice cover
ν is the Poisson's *ratio*
ρ is the water *density*
g is the gravity *acceleration*

Values estimated or measured in the field for the elastic modulus of intact freshwater ice range from about 0.4 to 9.8 GPa (Frankenstein 1966, Gow et al. 1978, Beltaos 1990). Values for extensively cracked or deteriorated ice may be much lower. The value of ν is estimated to be 0.35 (Ashton 1986). Observations show that the characteristic length ranges roughly from 5 to 20 times the ice thickness. Analysis of unsteady flows with wavelengths greater than 50–100 times the characteristic length need only account for the ice effects on steady flow and can treat the ice cover as an inertia-less, infinitely flexible, floating membrane on the water's surface. In practice, almost all river flood waves can be considered "long" by this measure, but waves resulting from ice cover breakup or jam release, as described in the following, may not be.

A floating ice cover exerts three effects on steady flow. These effects can significantly impact the discharge conveyance of a channel. The first is that a portion of the channel flow area is blocked by the area of the ice cover beneath the water surface. The second results from the increase in the channel wetted perimeter because of the presence of the stationary ice cover. This can significantly reduce the channel hydraulic radius, often by as much as a factor of 2 for channels that are much wider than deep, the usual case for many natural channels. The third is the modification of the effective channel hydraulic roughness. Some information is required about the flow conditions beneath the ice cover to determine the composite roughness of the ice cover and bed. It is known that beneath the cover the vertical water velocity profile approximates a parabola with zero velocity at the bottom of the ice cover and at the channel bed, and a velocity maximum located somewhere in-between. Since the fluid shear stress and other fluid properties are generally assumed proportional to the gradient of the velocity, the presence of the velocity maximum (where the gradient is zero) can be thought of as effectively dividing the flow under the ice cover into two layers: an ice cover layer controlled by the roughness of the ice cover and a bed layer controlled by the roughness of the bed. Proceeding along this line, formulations for the composite Manning's roughness coefficient have been developed that are widely applied (see Ashton [1986] for discussion). More recently, Beltaos (2001) noted that field measurements show that "the point of maximum velocity, which approximately delineates the ice and bed-controlled layers, is near the mid-depth, regardless of how rough the ice cover might be relative to the conventional bed roughness." This finding suggests that there is more communication between the layers beneath the ice cover than contemplated by simple application of the two-layer model. Finally, the roughness of the ice cover has been found to vary throughout the winter season, as the relative size of the roughness elements on the underside of the ice cover change in response to heat transfer and ice deposition.

Channel conveyance for an ice-covered channel, k, as described earlier can be written as

$$k = \frac{1}{n_c}(A_o - A_i)R_i^{2/3}$$

where

n_c is the composite Manning's roughness coefficient
A_o is the open water flow area
A_i is the flow area blocked by ice
R_i is the hydraulic radius of the ice-covered section

As can be seen by the proceeding discussion, floating ice covers significantly reduce the effective channel conveyance at a specific stage compared to open water. One consequence of this reduction in conveyance is that rapid ice cover formation or breakup can create transient flows that have significant impacts in the immediate river reaches and propagate large distances up- and downstream. The magnitude of the transients depends on the hydraulic conditions of the channel. Rapid ice cover changes in the backwater of a dam, for example, where the water level is much greater than needed to convey the river flow, result in little or no transients. However, in most river reaches, especially those without strong downstream control, rapid ice cover changes can create significant transients.

19.2.3 Creation of Stationary Ice Covers in Rivers and Streams

Starting from its initial inception point, river ice covers advance upstream by incorporating ice carried by the flow to its upstream (or *leading*) edge. Almost all river ice covers are formed in this way. Many different and separate processes may occur at the leading edge, depending on the hydraulic flow conditions and the form of the arriving ice. The various processes at the leading edge are described in the following.

Bridging: The downstream advection of floating ice can be arrested by ice control structures such as floating booms, weirs, or piers, obstacles such as bridge piers, intact ice covers, and hydraulic control structures, and other causes. Ice motion can also be spontaneously arrested through bridging. Ice moving at or near the flow velocity has little impact on the channel flow conditions. The possibility for ice impact on the flow begins when the surface concentration and strength of floating ice increases to the point where significant shear stresses can be transmitted through the surface ice to the banks. The shear stress causes the velocity of the floating ice to slow relative to the water velocity; this slowing in turn exerts stress on the flowing water, impacting the discharge rate and water level. When the surface ice stops moving altogether bridging occurs. The location and conditions under which bridging will occur are difficult to predict and are usually determined for specific reaches through observation.

Floe Stability: Juxtaposition and Under-Turning: At relatively low flow velocities, ice floes arriving at the leading edge may simply come to a stop, remain horizontal, and not under turn. In this case the ice cover will progress upstream by juxtaposition of floes. At higher flow velocities, the arriving floes may not be stable but may instead under-turn. When the motion of the floe is arrested at the ice cover leading edge, the floe is subjected to an under-turning moment caused by acceleration of the water velocity under the floe due to continuity and flow separation beneath the ice floe. This under-turning moment is resisted by the hydrostatic righting moment of the floe. As the floe rotates about its lower downstream edge, the maximum righting moment is quickly reached. If the under-turning moment exceeds this maximum, the floe is likely to under-turn.

Ice Cover Shoving: During shoving, the cover collapses in the downstream direction and becomes thicker. This happens when the downstream forces acting on the cover exceed its ability to withstand those forces. The downstream forces include the water drag on the underside of the cover and weight of the cover in the downstream direction. The ability to withstand the applied forces arises from the shear stress the ice cover can apply to the channel banks. The strength of an ice cover, that is, its ability to transmit forces to the banks as shear stress, is directly proportional to its thickness. When shoving takes place, the ice cover thickens and the strength of the ice cover is increased. An ice cover may repeatedly shove and thicken as it progresses upstream (Beltaos 1996).

Under-Ice Transport: Under-ice transport occurs when the ice arriving at the leading edge of the ice cover is submerged and carried beneath the ice cover. The transported ice may move downstream under the cover for considerable distances, especially if the ice is formed of relatively small frazil ice crystals and flocs. Under-ice transport suggests another spatially inverted analogy with sediment transport, in this case bed-load transport. Laboratory and field observations have been used to develop an under-ice frazil granule transport equation that is entirely analogous with bed-load transport of sediment (Shen and Wang 1995). The transport equation balances the buoyancy of the frazil particle against the stress of the moving flow to estimate the transport capacity of the flow for each particle size. Given sufficient supply of arriving ice, under-ice transport can result in channel reaches almost completely blocked by ice for long distances.

No Ice Cover Progression: The ice cover will not progress upstream if the flow at the leading edge causes 100% of the arriving ice to submerge and be transported downstream beneath the cover. Further upstream progression of the ice cover halts until the flow conditions at the leading edge change. The deposition of ice beneath the cover downstream of the leading edge or shoving and thickening of the cover may reduce the channel conveyance sufficiently to cause the upstream water levels to rise and the flow velocities at the leading edge to be reduced. Open water will remain upstream of the leading edge if the flow velocities cannot be sufficiently reduced. Frazil ice will be produced in the open water reach all winter, creating freeze-up jams, or other problems downstream.

Flow Transients Resulting from Ice Cover Progression: One important consequence of the reduction in channel conveyance caused by upstream progression of ice covers is that water levels tend to increase compared to the open water level for the same discharge. While the cover is progressing upstream, a portion of the flow arriving at the leading edge of the ice cover will go into increasing the channel water levels, reducing the downstream discharge and water levels. The maximum decrease in the downstream stages is determined by the upstream progression rate of the cover, its thickness, and roughness. The magnitude of the decrease in discharge can be significant and lead to exposure of water intakes and other impacts (Wuebben et al. 1995).

Ice Cover Thickening: Figure 19.4 summarizes the genesis of eventual ice cover formation, beginning with frazil and anchor ice (when it forms) growth. The slush and floes drift with water flow until accumulating at a location that builds back upstream. Wind-driven ice floes and slush on a lake or sea may cluster and form an ice cover.

An ice cover on a water body thickens by thermal growth with the original crystals growing downward in column form, as mentioned in Section 19.2.1. Thickening occurs at the cover bottom (discounting snow deposit on the top). It is customary (e.g., Ashton 1986, Michel 1978) to relate ice cover growth to the difference between the heat fluxes at the top and bottom of the cover. This approach simplifies the relationship by relating cover thickness to cumulative degree-days freezing using the Stefan equation, a simplification of a heat flux balance through the ice cover;

$$T = h\sqrt{ADDF}$$

where

T is the ice thickness
$ADDF$ is the accumulated degree-days of freezing (product of days times degrees below freezing temperature of water)
h is the coefficient depending on ice cover exposure, notably wind effects

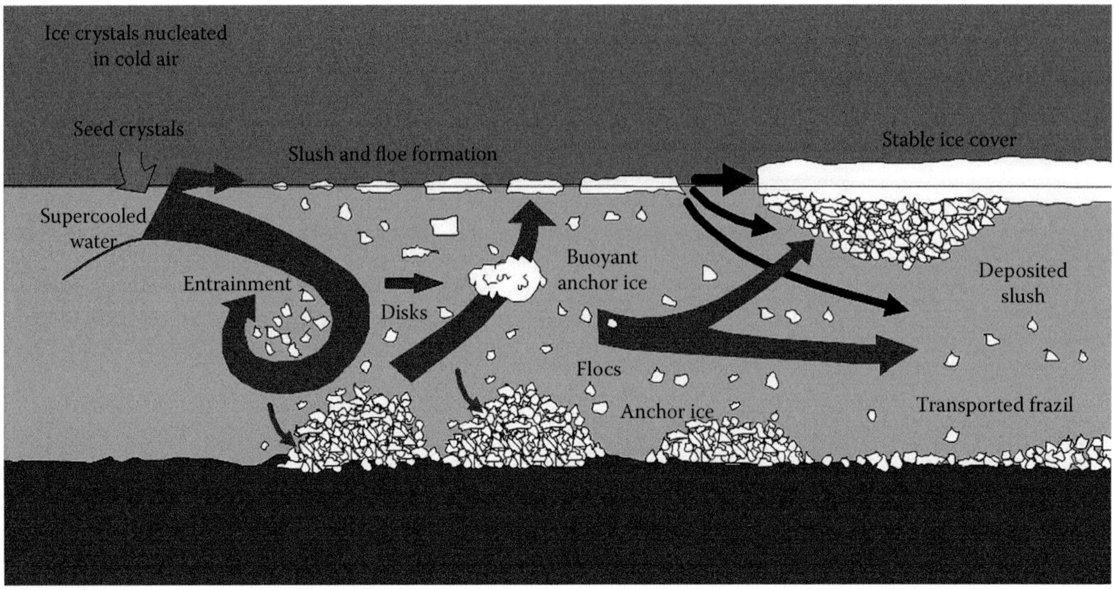

FIGURE 19.4 Schematic of ice formation, transport, and accumulation in a river or stream.

Thermal ice growth purges impurities (e.g., salts) from water, concentrating them at crystal boundaries. Impurity presence at boundaries within an ice cover may play an important role in the strength properties and later weakening of the ice cover.

19.2.4 Ice Cover Breakup

Breakup transforms an ice-covered river into an open river. Two ideal forms of breakup bracket the types of breakup commonly found throughout most of the globe. At one extreme is thermal breakup. During an ideal thermal breakup, the river ice cover deteriorates through warming and the absorption of solar radiation and melts in place, with no increase in flow and little or no ice movement. At the other extreme is the more complex and less understood mechanical breakup (also referred to as a dynamic breakup). Mechanical breakup requires no deterioration of the ice cover, but rather results from an increase in flow entering the river. The increase in flow induces stresses in the cover, and the stresses in turn cause cracks and the ultimate fragmentation of the ice cover into pieces that are carried by the channel flow. Ice jams take place at locations where the ice fragments stop; severe and sudden hydraulic transients can result when these ice jams form or when they release. As breakup happens most often during warming periods, when the ice cover strength deteriorates to some degree and the flow entering the river increases because of snow melt or precipitation, most river ice breakups actually fall somewhere in between these extremes of thermal and mechanical breakup. As a general rule, the closer that a breakup is to being a mechanical breakup, the more dramatic and dangerous it is because of the increase in flow and the large volume of fragmented ice produced.

Thermal Meltout: Thermal meltout does not take place at all points on a river simultaneously, but rather occurs at different locations at different rates, depending on the latitude, local climate, and exposure. It proceeds as the ice cover absorbs heat from the atmosphere and the water flowing beneath. Often the ice cover is snow covered by the end of winter and its albedo is relatively large, preventing penetration of shortwave radiation (sunlight). The transfer of heat from the water to the underside of the ice cover can be very substantial, especially if there is open water upstream, which with its relatively low albedo has much greater potential for heat transfer from the atmosphere. The creation of melt-water on the surface drastically lowers the albedo which promotes the absorption of sunlight. Penetration of solar radiation into the ice cover causes internal deterioration without much of a loss of thickness. The absorbed solar radiation causes melting in the interior of the ice, which results in a loss of structural integrity and strength of the cover. This is most likely to happen in ice covers composed of columnar crystals, because impurities accumulated between the crystals absorb solar radiation and hasten melting at the crystal boundaries. Fine-grained ice covers, composed of snow ice or frazil ice, are much less susceptible to internal deterioration through absorption of solar radiation.

Mechanical Breakup: As with thermal meltout, mechanical breakup does not happen simultaneously everywhere along a river network. Often breakup occurs first on smaller tributaries, and then proceeds haphazardly to the main stem rivers. Breakup can progress upstream or downstream. Generally, if it progresses upstream, less ice jams result.

The first indication of breakup is the formation of shore cracks. Shore cracks are longitudinal cracks running parallel to the banks of the rivers. Shore cracks form when the magnitude of the water level change in the river channel exceeds a limit determined by the material properties of the ice, the ice thickness, channel width, and the type of attachment of the ice cover to the channel bank (hinged or fixed). Only a small increase or decrease in discharge is necessary to cause shore cracks, and they are usually common soon after runoff into a river has begun to increase. The presence of shore cracks does not necessarily indicate the immediate onset of breakup as they may be present throughout the winter season. In the next step of ice cover breakup, the ice cover fractures into individual floes. Transverse cracks (across the channel) will appear soon after the river water level has begun to increase. The first cracks will generally create relatively large ice floes, a river-width wide, and many river-widths long (Beltaos 1985, 1990), but sometimes the ice covers are immediately broken into much smaller floes. These fractures are apparently caused by the propagation of waves. The length of the propagating wave influences the size of the floe produced (Daly 1995, Xia 1998). The final step of a mechanical breakup is the movement of floes in response to the increase in water levels. If the floes are relatively large, they may be kept from movement by geometric constraints, until a substantial increase in water level is reached. If the floes are relatively small, they may begin to move immediately after a small water level increase. Once the floes begin moving, they are reduced in size, eventually arriving at a size roughly four to six times the ice thickness.

Flow Transients Resulting from Ice Cover Breakup: In the transition from an ice-covered channel to an open water channel, the hydraulic radius and overall channel roughness are changing abruptly from their ice-covered values to those associated with open water. These abrupt changes produce flow transients that can propagate downstream and interact with the downstream ice cover. It has been proposed that this process can be reenforcing with the flow transients causing further breakup which produces more transients, but substantial supporting field observations remain to be taken.

19.2.5 Ice Jam Formation

Ice jams form where high concentrations of ice floes created by mechanical breakup exceed the river's transport capacity. This is most likely to occur at reaches where an intact ice cover remains in place, the slope of the river significantly decreases, or geometric constraints exist, such as sharp river bends or constrictions and bridge piers. At these locations, ice stops moving and a breakup ice jam forms. Breakup ice jams substantially reduce the channel flow conveyance, leading to sudden and substantial upstream flooding. Large, sharp flood waves can be created when the jams release.

The geometry of breakup ice jams is strongly influenced by river depth, width, and slope at the jam location; the flow conditions under which the jam formed; the volume of ice that arrived from upstream; and, to a lesser extent, the strength and thickness of the arriving floes. Jam features include a downstream limit or toe, a transition section upstream of the toe leading to an equilibrium section (where the longitudinal forces are in balance or equilibrium), and a transition section upstream of the equilibrium section leading to the upstream limit or head of the jam (Beltaos 1996). The toe of the jam may or may not be grounded on the stream bed. The river discharge will be conducted as porous flow through the jam at locations where it is grounded and completely blocks the flow area of the channel. In other areas, the discharge will flow under the floating ice cover created by the jam, and all the insight regarding the impact of river ice covers on the hydraulics of channel flow described earlier can be applied. Wherever the ice cover is floating at hydrostatic equilibrium, the vertical stress state of the floating ice can be estimated. The longitudinal and lateral stresses the jam can support can be estimated based on this vertical stress state by assuming that the ice jam can be treated as a granular material, the primary insight of Pariset and Hausser (1961) and Pariset et al. (1966). These stresses determine the strength of the jam, that is, its ability to transmit forces to the channel banks as shear stress. The required jam strength at each location can be estimated by determining the force balance over the length of the jam. The jam thickness can then be estimated at each river section as the ice cover strength is directly proportional to its thickness.

19.3 Methods of Analysis

As with many other areas of fluid mechanics, practitioners working on river or lake ice topics have seen an extensive growth in the use of numerical modeling, especially over the past two decades. Today, quite a few ice processes can be addressed using numerical models. Additionally, numerical models have benefited from advances in the collection of field data using remote sensing tools such as LIDAR, though extensive instrumentation development work remains to be done. Laboratory and hydraulic model investigation remain as useful approaches, though their chief difficulty is in the replication of thermal processes and, though to a lesser extent, ice-material properties.

19.3.1 Laboratory Investigation

Laboratory experiments and hydraulic modeling are used fairly extensively to investigate aspects of ice transport and accumulation, ice loads and movement at man-made structures and vessels, and the performance of vehicles and vessels in ice-covered water bodies. Most ice modeling is a straightforward extension of modeling free-surface flows, though made somewhat more difficult by the need to consider additional flow boundaries imposed by ice covers and the modeling requirements for the additional processes of ice-piece drift and accumulation. Complications arise when the thermal and strength properties of ice must be taken into account. Reproducing the effects of wind, which may propel drifting ice over broad widths of water, may further complicate modeling.

Common models are conventional free-surface flow models for which ice is a floating solid boundary retarding water flow, or ice occurs as solid pieces conveyed and accumulated by water flow. Such models primarily concern patterns and profiles of water flow, and possibly how they interactively affect patterns and profiles of ice movement and accumulation. The simplest process is flow in a channel with a level ice cover. The drift and accumulation of discrete ice pieces require simulation of additional processes, thereby adding similitude constraints to be met.

When the strength and deformation properties of ice, either consolidating accumulations of ice pieces or breakable ice sheets, have to be taken into account, the selection of an effective model ice material imposes quite tight constraints on model length scale and design. The same may be said for laboratory investigation of processes involving ice growth, phase change, and heat transfer. Those processes, and stress propagation, occur at rates that may differ from rates prescribed from hydrodynamic similitude criteria. The similitude constraints and conflicts are elaborated later.

The strength and deformation properties of monolithic ice sheets are of primary interest for modeling ice-sheet loading. Typical modeling situations concern ice loads against bridge piers, walls, and riprap embankments, and against ship hulls. Also of interest, and difficult to model, is ice sheet failure due to hydraulic effects produced by flood and other waves. Modeling requires a model ice that not only satisfies buoyancy and frictional requirements but also deforms and fails in the manner dominating ice behavior at full scale. By way of a cautionary note: considerable judgment and experience are needed when modeling many ice-load situations, because the full-scale conditions of ice loading and material behavior of ice are complex, still ill-defined, and subject to scientific discussion.

The important ice failure modes are flexure, shear, and crushing. All three modes may occur simultaneously during structure or ship interaction with ice, though one mode usually dominates. The waterline shape of a structure or ship, contact conditions, together with the strength and thickness properties of an ice sheet, determine which mode dominates. The most common dominant mode for hydraulic failure of ice sheets is flexure caused by change in water-surface profile of a flow or shoving of ice under or above the sheet. To ensure model ice deforms in the same manner as ice at full scale, it has been customary (e.g., Michel 1978, Schwarz 1978, Ashton 1986) to prescribe that the ratio of ice strength, σ, and elastic modulus, E, for a particular load mode be held constant at model and full scales; that is,

$$\left(\frac{E}{\sigma}\right)_r = 1$$

At both scales, E/σ exceeds a minimum value associated with brittle elastic failure. Many modeling guides (e.g., Michel 1978, Schwarz 1978, Ashton 1986) stipulate that the minimum be about 2000.

The Cauchy number, Ch, is often used as a similitude parameter for prescribing the load and deformation behavior of level sheets of ice. It is a convenient ratio of forces attributable to inertia and elastic deformation;

$$Ch = \frac{\rho U^2}{E}$$

where

U is a velocity of interest
ρ is ice density

Ideally, Ch should have the same value in the model and prototype; whence

$$(Ch)_r = \frac{\rho_r U_r^2}{E_r} = 1$$

As water normally is used to replicate water in model studies, and as Froude number equivalence prescribes velocity ratio U_r, the modulus of elasticity is equal to the length scale for undistorted models; that is,

$$E_r = X_r$$

In accordance with the earlier equations, the strength scale equals the geometric scale; that is,

$$\sigma_r = X_r$$

Vertical distortion is undesirable for models intended to simulate loading situations for which vertical forces are important, because vertical and horizontal forces and stresses would differ in scale and thereby unacceptably distort ice-failure patterns. Note that the Froude similitude criterion leads to the same result for scaling stress or pressure (force divided by contact area) when the density scale is unity.

The limitations on scales needed for adequate similitude constrain the utility of hydraulic modeling and laboratory experiments to the investigation of local processes, notably aspects of local-scale processes of ice formation or failure. Increased reliance is placed on numerical modeling for investigating ice formation and transport processes occurring at larger scales.

19.4 Numerical Modeling

As with many other areas of fluid mechanics, practitioners working on river or lake ice topics have seen an extensive growth in the use of numerical modeling, especially over the past two decades. In general, numerical models have developed as the understanding of ice processes has grown and been expressed in mathematical form, generally as differential equations. Most often, numerical models are used to address specific areas such as ice transport and ice jam impacts on flood levels. These *component river ice models* have also been combined into *comprehensive river ice models* (Shen 2010) that can simulate all the important ice processes from initial freeze-up through to ice jam formation and release. Important component models include models for addressing the impact of ice covers and ice jams on river water surface elevations (Daly et al. 1998) and ice passage at structures (Liu et al. 2002) such as locks and dams. These models assume the presence of ice as relatively small floes and then investigate the interaction of the ice and the flow to answer specific engineering questions. Comprehensive river ice models generally address questions related to the impact of ice cover formation and breakup on relatively long river reaches or large river systems. They are used to determine the operation of hydraulic structures such as dams to minimize ice impacts on flow, flood levels, hydropower production, etc.

Component models that address the interaction of ice and flow are probably the most widely used class of component model. This type of component model and comprehensive models combine flow simulation with ice simulation. A number of flow models have been used including steady-flow standard step backwater models (Daly et al. 1998), 1D finite difference and finite element unsteady flow models (Hicks et al. 1992; Chen et al. 2006), and 2D finite difference and finite element unsteady flow models (Liu et al. 2002). A number of ice models have been used including static (no time-dependent terms) stress representation (see, for example, Pariset and Hausser 1961, Pariset et al. 1966, Daly et al. 1998), dynamic stress representation with viscous–plastic (Shen et al. 1997) or viscoelastic–plastic law (Ji et al. 2005), or a discrete element description (Hopkins et al. 1996). The frequency at which information is passed between the hydraulic and ice model, and the use of the information passed, determines the degree of the coupling between the models. While the degree of coupling can be adjusted to meet the requirements of the problem addressed, most approaches are closely coupled, that is the flow and the ice models are solved simultaneously at each time step.

The input data requirements vary depending on the model type, but certain general classes of data are required by all models. The hydraulic models all require descriptions of the channel geometry including the distribution of hydraulic roughness. They require descriptions of the upstream and downstream flow boundary conditions such as flow and stage. Dynamic (time varying) models require time series descriptions of these boundary conditions. In addition, comprehensive models require time series descriptions of the upstream water temperature boundary condition, ice transport boundary condition, and the relevant meteorological conditions over all the river reaches. The data requirements of the ice models include the relevant material properties of the ice such as the angle of internal friction, porosity, bulk and shear viscosities, etc. Some parameters such as the material properties of the ice mass and the hydraulic roughness of the underside of the ice cover are difficult to estimate a priori and require adjustment based on field observations. This presents immediate difficulties because virtually none of these parameters can be measured in situ. In practice, almost all of models are adjusted by matching the most common observed field data: the observed water surface elevations. Assimilation of observed water surface elevations using the Kalman Filter has been shown to improve real-time operational use of ice models for forecasting (Daly 2002).

19.5 Applications

Two brief illustrations are offered here regarding the application of laboratory investigation and numerical simulation of ice processes.

19.5.1 Frazil and Anchor Ice Formation on Submerged Intakes

This application illustration shows how some ice growth and dynamics processes can be replicated in a refrigerated laboratory, in this case a large ice tank. It also indicates how numerical simulation can elaborate and extend laboratory experiments.

Submerged, water intakes in lakes, reservoirs, and coastal regions are used commonly to meet the water needs of urban communities, industry, and thermal-power stations. However, there presently are few guidelines (and little actually known) about designing such intakes to operate in frigid-water conditions subject to ice formation. Frazil accretion and anchor ice growth can cause intake blockage. A series of ice tank experiments were conducted (Chen et al. 2004) to elucidate frazil accumulation and ice growth in conically shaped, submerged water intakes. The insights and trends obtained from unique laboratory experiments were augmented by information from a numerical model of intake flow. The study's findings usefully showed how frazil can accumulate and eventually choke an intake, and how a modification to the intake's design can reduce, or at least delay, the amount of frazil ice and supercooled water entering an intake. The experiments worked well, and revealed that ice blockage, at least for some submerged intakes, occurs differently than was previously understood. Figure 19.5a and b illustrate the intake modeled and an example ice blockage formed in the model intake.

Subsequent field observations reported by Daly and Ettema (2006), also Kempema et al. (2008), show how ice growth within water intakes is the more likely major cause of blockage. Although frazil accumulation occurs at intakes, for an intake on a lake bed, anchor ice growth caused the eventual blockage. As explained in Section 19.2.1.3, anchor ice growth entails crystal growth fed by a supply of supercooled water convected to depth within a water body.

Useful engineering design information was obtained for situation modeled. The experiments in the large, refrigerated ice tank could replicate the processes of frazil formation and transport, anchor ice growth, and the overall of ice blockage of a conical water intake on a lake bed. However, the experiments could not actually scale crystal sizes relative to intake size; full-size crystals formed. The innate difficulties of simulating spatial and temporal scales associated with ice-related thermal and material processes in the laboratory lead to reliance on numerical simulation of ice dynamics and transport.

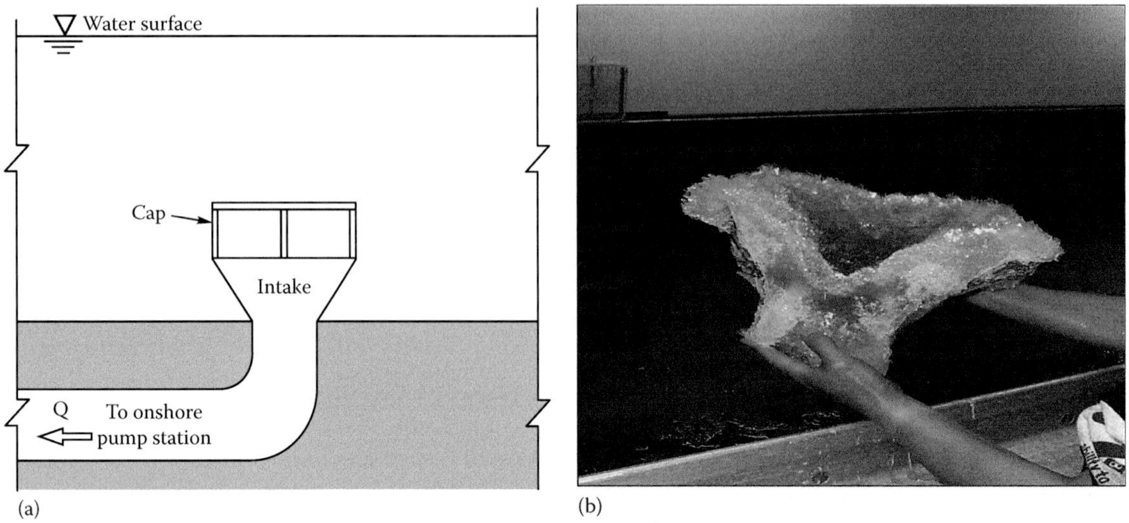

FIGURE 19.5 Submerged intakes in frigid water commonly face the problem of blockage by frazil and anchor ice; (a) a typical form of intake, (b) frazil and anchor ice blockage removed from model intake in an ice tank (the large crystals indicate anchor ice growth).

19.5.2 Discrete Element Models of River Ice

Discrete element models simulate an assemblage of ice floes as discrete elements. By resolving the contact and body forces acting on thousands of individual discrete elements at each time step, the motion of ice can be numerically simulated and studied. The total force acting on each particle is a combination of fluid drag, buoyancy, gravity, and interparticle contact forces. The resultant force acting on each element and the element's material properties of size, density, and stiffness determine its motion. The discrete element model (DEM) has been successfully coupled with a 1D unsteady flow model to simulate ice jam formation (Daly and Hopkins 1998, 1999, 2001) and a 2D flow model (Stockstill et al. 2009). The simulated hydrographs were directly compared with results of physical experiments using ice. Hopkins and Tuhkuri (1999) studied the compression of floating ice fields using DEM simulations and directly compared simulation results with results of parallel physical experiments. Hopkins and Tuthill (2002) document the accuracy of DEM simulation of ice boom forces by directly comparing the results of physical experiments and computer simulations.

A DEM is a computer program that explicitly models the dynamics of a system of discrete particles. The position, orientation, velocity, shape, and mass of each particle are stored in arrays. At each time step, the contact and body forces on each particle are calculated and the particles are moved to new locations with new velocities that depend on the resultant of the forces. A detailed description of the mechanics of the simulation used in this work is given in Hopkins et al. (1996) and Hopkins and Tuhkuri (1999). The discrete objects that make up the ice floes interact with each other and with the water. Forces between neighboring objects are created by relative motion. When two objects are pushed together, their shapes overlap. Wherever there is contact between two objects, a force is placed at the point of contact that is normal to the plane of contact and proportional to the amount of overlap between the objects. The force resists further overlap. The computer code checks for contact between neighboring particles at each time step.

The force between two colliding objects is determined by first finding the location where the objects touch. The overlap at the point of contact is interpreted as elastic deformation of the objects. A contact plane is defined tangent to each object. The incremental change in the frictional force is proportional to the relative tangential velocity. If the magnitude of the frictional force exceeds the Coulomb limit, then the frictional force is scaled such that it equals the Coulomb limit. A tangential damping term is also computed based on an effective displacement rate. A fluid drag force is applied to each particle that is exposed to the flow. The fluid velocity parallel to the water surface is estimated by the flow model; the vertical component of fluid velocity is estimated from the change in water surface elevation. The buoyancy force is assumed to act perpendicular to the local water surface.

Model startup includes the initial placement of the floating discrete elements within the domain. The simplest approach, introducing the particles at the upstream boundary of the domain and allowing them to drift into the area of interest with the current, often requires too many computational cycles before the particles have reached the area of interest. It is much more efficient to introduce the particles immediately upstream of the point of interest. The following procedure is used in placing the particles densely at the water surface. This procedure allows them to interact with each other and respond to the flow conditions. Initial placement of the ice elements in the simulation domain begins by defining the portion of the domain to be filled. The concentration of ice, the shape of the elements (usually flat disks in the case of ice floes), and the material properties of the elements including friction, density, and stiffness are specified. An initial calculation from the hydrodynamic model is used to define the water surface and velocity vectors throughout the solution domain. Ice floes are initially placed on the water surface as infinitesimal point elements with the desired spatial distribution. With the fluid drag set to zero, the elements are inflated to their final sizes with a specified aspect ratio. As they grow, they interact with their neighbors and the water surface and adjust their positions as necessary. Once the final element sizes are reached, the fluid drag is turned on and the elements are released with an initial velocity equal to the local flow velocity. After a short transient, the initial placement of ice moves steadily with the flow.

19.6 Major Challenges

Several major challenges attend ice formation, dynamics, and transport. Each challenge requires significant further investigation. Some related to understanding fundamental aspects of ice formation, and others concern engineering design issues. A selection of challenges is outlined herein:

1. Forecasting ice-related flooding
2. Climate-change effects on ice
3. River ice effects on alluvial channel morphology and stability
4. Ice blockage of water intakes

19.6.1 Forecasting Ice-Related Flooding

Ice jams, initiated when discrete ice pieces concentrate with a cumulative volume sufficient to hinder flow through a river reach, are widespread across the northern tier of the United States and Alaska. They can potentially occur in all ice-forming rivers. Ice jams may cause sudden and devastating rises in water level that can lead to flooding with all its attendant woes and potential for disaster. Forecasting ice jams is an important component of wintertime hydrologic forecasting.

Ice jams result from the complex interplay of weather, river flow, river geometry, and the mechanical properties of ice. Consequently, they are notoriously difficult to forecast. There is no standard procedure for forecasting ice jams at this time. A variety of techniques have been tried, including threshold models (e.g., Shulyakovskii 1963, Wuebben et al. 1995, Tuthill et al. 1996, White and Kay 1996, White and Daly 2002, Vuyovich et al. 2005), multiple regression models (e.g., Beltaos 1984), discriminant analysis (e.g., Zachrisson 1990, White and Daly 2002), artificial neural networks (ANN) (e.g., Massie et al. 2002), and fuzzy logic (e.g., Mahabir et al. 2002). This variety of technique reflects the fact that, at this time, there is no clear consensus, or reason to expect that any clear consensus will develop, on the proper technique to be used for forecasting ice jams. In fact, it is not yet clear that any one technique could be applied to all situations. Certainly, one major bar to the widespread application of any or all of the techniques developed in the previous studies is their near complete lack of universality. The previous studies concentrated on jam formation at specific locations where flooding or other serious impacts occurred. The forecasting procedures they developed while often successful, are difficult, if not impossible, to apply beyond the specific location where they were developed. A review of the previous studies will show that the threshold parameters and their threshold values; the discriminant functions, regressions, and the "training" data for the ANN; and fuzzy logic procedures all reflect very site-specific conditions and their utility over a broad range of river locations is unclear.

19.6.2 Climate-Change Effects

Ice formation and dynamics of rivers play important roles in the natural environment of watersheds and communities in mid- to high-latitude regions (regions above latitude 35° in the northern hemisphere). In various ways, ice regimes are a major physical facet of the annual biological and hydrological cycles of such watersheds and communities. The influences of climate changes on the ice regimes of rivers over vast areas like Siberia and North America have major consequences for the stability of environments evolved in response to the annual cycle of ice formation, presence, and breakup. The changes also will have major significance for communities of people residing along rivers subject to ice.

During the past several years, research has begun examining possible consequences of climate change (e.g., Magnuson et al. 2000, Beltaos and Prowse 2001, Andrishak and Hicks 2008, Prowse et al. 2007, Prowse et al. 2009), and fragments of information have begun emerging that global climate change is altering the ice regimes of many rivers in higher latitude regions. Some of the research is hypothetical, exploring possible consequences of climate change, notably warming (Andrishak and Hicks 2008). The emerging information indicates a shorter ice-formation season under some climate-change scenarios, and reduced magnitude of spring breakup flows (Beltaos and Prowse 2001). Some indications may seem counter-intuitive, and are still anecdotal lacking adequately documented. For example, some information suggests that reduced snowfall in some watersheds has led to increased river ice thicknesses on some northern rivers in Alaska, Canada, and Russia and that ice cover breakup events have been more problematic, leading to more severe ice jams than observed before. Moreover, for several rivers ice-cover breakup has coincided with rainfall rather than from runoff from snow melt (e.g., Canada's Athabasca River). The writers' informal discussions with Russian, and North American researchers indicate that changes in the ice regime of northern rivers are occurring, but there has not yet been a systematic study of sufficient duration to delineate the changes, and their likely consequences for northern watersheds and people living in them.

Although caution is needed when generalizing about the exact changes in river-ice regimes (such as a possible increased incidence and severity of ice jams), the changes will affect regional environments, and will pose concerns for societal infrastructure. Ice jams and runs, for instance, are more common with ascending latitude, but the risks to life, property, and commerce incurred with ice jams and runs increase with descending latitude, owing to increased human habitation along, and commercial reliance on, rivers in the lower latitude areas (below approximately 35°–45° latitude). Considerable scientific challenges arise when evaluating climate change effects on freshwater bodies subject to ice. Moreover, further challenges attend ecological and societal adaptation to the possible effects.

19.6.3 River Ice Effects on Fluvial Morphology and Stability

In various ways, some not well understood, most not quantified, river ice affects fluvial channels subject to seasonally frigid weather. The effects' significance depends on the temporal and spatial scales associated with the thermal and fluvial processes prevailing in fluvial channels subject to frigid weather. The thermal processes associated with ice in rivers generally act over scales that relate strongly to weather conditions and usually much less to factors governing water flow and channel boundary material. However, in the limiting condition when potential ice thickness exceeds flow depth, this statement does not entirely hold.

In river flow, the time scales for mixing are much faster than those for cooling of water, especially cooling by heat flux through ice (Fischer et al. 1979, Ashton 1986). Normally, the entire flow depth becomes more or less uniformly cooled once an ice cover forms over the flow. When the cover is formed, and then thickens, the rate of heat loss from water to air decreases markedly. Consequently, the maximum thickness of thermally grown ice covers normally is of the order of 1 m, sometimes more. Thicker covers can form by means of ice jamming.

For a given ice thickness (or cumulative heat loss to the air above a river), the importance of ice on channel morphology and bed-material transport diminishes as channel depth or hydraulic radius increases. The quantitative implications of this trend have to be investigated. For the moment, it can be said that a given ice thickness (say, 1 m) imposes less influence on flow as the channel depth increases. A 1 m-thick ice layer on a 1.5 m-deep river has a significant effect on the flow. By contrast, the same 1 m-thick ice cover has lesser effect on a 15 m-deep channel; the effect of the ice cover on the bulk flow is relatively small for a deep channel. Although a large, deep channel may produce more ice, and possibly produce thicker ice jams. For a shallow flow, an ice cover may dramatically increase flow resistance, even divert flow from the channel. The limiting condition in this regard is when aufeis (river icing) completely inundates a channel, choking its flow. These considerations suggest that a range of channel depths and ice cover thicknesses exists for which river ice actively affects channel dynamic equilibrium. As Figure 19.6 illustrates, an example reach where ice presence

FIGURE 19.6 Ice formation causes flow redistribution, thalweg shifting, and the bank destabilization along this reach of the Missouri River, Montana.

promotes channel instability by causing a thalweg shift and channel-bank erosion. Little research has been done to delineate quantitative impacts of ice on fluvial channels, and the ecosystems in and along them.

Some river ice effects add erosive mechanisms to a fluvial channel, whereas others dampen erosion. Therefore, the net impact of ice remains an open question. Erosive mechanisms include direct entrainment and transport of bed sediment by ice, adjustments in the distribution and transport capacity of ice-covered flow, localized increase in bed-material transport when ice accumulations increase local flow velocities, and ice gouging of channel bed and banks by moving broken ice. Additionally, thermal effects, especially freezing and thawing, weaken channel banks. Of importance for channel behavior, and yet to be examined thoroughly, are river ice effects on bank and floodplain vegetation; the limited extant literature indicates that ice abrasion inhibits growth of bank vegetation, thereby weakening banks (e.g., Hamelin 1979, Unila 1997, Prowse 2001). Erosion-damping effects include increased flow resistance due to ice-cover presence, ice-cover suppression of secondary currents in channel bends (Urroz and Ettema 1992), and ice-cover armoring of small ephemeral channels and channels filled with aufeis (McNamara and Kane 2009).

Detailed information addressing the overall effects of river ice on alluvial channels is quite sparse, partly because accessibility to channels during winter often is difficult, and because of the need for further developments in field instrumentation. Until fairly recently, as interests in wintertime fluvial ecology have grown, and as the extent of engineering works has increased in cold regions (such as the design and construction of water intakes and bridges), little attention has been devoted to ice effects on rivers.

19.6.4 Ice Blockage of Water Intakes

The sound operation of cylindrical, wedge-wire screens for submerged water intakes poses substantial engineering challenge. Their use has been mandated recently by the Environmental Protection Agency (in Section 316(b) of the Clean Water Act; EPA 2006) for the purpose of preventing fish from being drawn into water intakes. Cylindrical wedge-wire screens (Figure 19.7) comprise an efficient technology for decreasing fish impingement and entrainment losses at water intakes. Screens draw water through a fine mesh at low flow velocities. Though mitigating fish impingement and entrainment, wedge-wire screens in frigid water run a major risk of blockage by frazil and anchor ice. Fairly numerous case-study examples of blockage exist (Daly and Ettema 2006).

As traditional trashracks are replaced with cylindrical wedge-wire screens, intake operators will have to be prepared for a potential increase in frazil ice problems. The same characteristics that make wedge-wire screens successful protectors of fish make them efficient frazil collectors and prone to anchor ice growth within the intake cylinder. The small slot openings facilitate clogging of the openings by frazil. The large surface area of the screen elements provides a large area for frazil adherence.

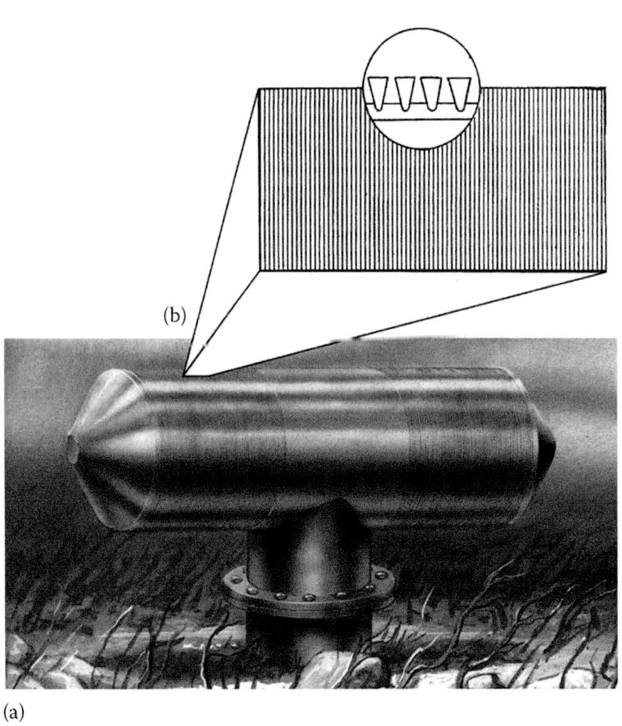

FIGURE 19.7 A wedge-wire screen installation, with detailed illustration of the wedge-wire screen elements attached to a reinforcing bar. (From Amaral, S., Laboratory evaluation of wedgewire screens for protecting fish at cooling water intakes, *Proceedings of Symposium on Cooling Water Intake Technologies to Protect Aquatic Organisms*, May 6–7, 2003, Arlington, VA, pp. 279–302, 2003. With permission.)

Limited observations of operational wedge-wire screens suggest that existing methods for mitigating ice blockage are ineffective. For example, airburst cleaning systems may not be capable of clearing frazil adhering to the intake screens, and anchor ice within the cylinder. Also, heating elements to warm intakes are usually not practicable, especially for intakes on lake beds. New methods are needed for frazil-blockage control. Kempema et al. (2008) outline potential options to ice blockage, and indicate the need for further development of intake design and operation.

References

Amaral, S. (2003) Laboratory evaluation of wedgewire screens for protecting fish at cooling water intakes. *Proceedings of Symposium on Cooling Water Intake Technologies to Protect Aquatic Organisms*, May 6–7, 2003, Arlington, VA, pp. 279–302.

Andreasson, P., Hammar, L., and Shen, H.T. (1998) Influence of surface turbulence on the formation of ice pans. In *Ice in Surface Waters*, H.T. Shen, Ed., *Proceedings of the 14th International Symposium on Ice*, Potsdam, New York, July 27–31, Vol. 1, pp. 69–76.

Andrishak, R. and Hicks, F. (2008) Simulating the effects of climate change on the ice regime of the Peace River. *Canadian Journal of Civil Engineering*, 35(5), 461–472.

Arden, R.S. and Wigle, T.E. (1972) Dynamics of ice formation in the upper Niagara River. *Proceedings of Banff Symposia on the Role of Snow and Ice in Hydrology*, IAHS-UNESCO-WMO, Banff, Alberta, Canada, pp. 1296–1313.

Ashton, G.D. (1986) *River and Lake Ice Engineering*. Water Resources Publications, Littleton, CO.

Ashton, G.D. (1989) Thin ice growth. *Water Resources Research*, 25(3), 564–566.

Beltaos, S. (1984) Study of river ice breakup using hydrometric station records. *Proceedings of Workshop on Hydraulics of River Ice*, Fredericton, New Brunswick, Canada, pp. 41–59.

Beltaos, S. (1985) Initial fracture patterns of river ice cover. Contribution No. 85-139, Natural Water Research Institute, Burlington, Ontario, Canada.

Beltaos, S. (1990) Fracture and breakup of river ice cover. *Canadian Journal of Civil Engineering*, 17(2), 173–183.

Beltaos, S. (Ed.) (1996) *River Ice Jams*. Water Resources Publications, Littleton, CO.

Beltaos, S. (2001) Hydraulic roughness of breakup ice jams. *Journal of Hydraulic Engineering*, 127(8), 650–656.

Beltaos, S. (Ed.) (2008) *River Ice Breakup*. Water Resources Publications, Littleton, CO.

Beltaos, S. and Prowse, T.D. (2001) Climate impacts on extreme ice-jam events in Canadian rivers. *Journal of Hydrology Science*, 46(1), 157–181.

Bisaillon, J.-F. and Bergeron, N.E. (2009) Modeling anchor ice presence–absence in gravel bed rivers. *Cold Regions Science and Technology*, 55, 195–201.

Blackford, J. (2007) Sintering and microstructure of ice: A review. *Journal Physics D: Applied Physics*, 40, R355–R385.

Botsaris, G.D. (1976) Secondary nucleation—A review. In *Industrial Crystallization*, J.W. Mullin, Ed., Plenum Press, New York.

Chen, Z., Ettema, R., and Lai, Y. (2004) Ice-tank and numerical study of frazil ingestion by submerged intakes. *Journal of Hydraulic Engineering*, 130(2), 101–111.

Chen, F., Shen, H.T., and Jayasundara, N.C. (2006) A one-dimensional comprehensive river ice model. *Proceedings of the 18th IAHR International Symposium on Ice*, Sapporo, Japan, Vol. 1, pp. 61–68.

Clark, S.P. and Doering, J.C. (2009) Frazil flocculation and secondary nucleation in a counter-rotating flume. *Cold Regions Science and Technology*, 55, 221–229.

Colbeck, S. (1992) The slow growth of ice crystals in water. USA Cold Regions Research and Engineering Laboratory, CRREL Report 92–3, Hanover, NH.

Colbeck, S. (1997) A review of sintering in seasonal snow. USA Cold Regions Research and Engineering Laboratory, CRREL Report 97–10, Hanover, NH.

Crank, J. (1984) *Free and Moving Boundary Problems*. Clarendon Press, Oxford, U.K.

Daly, S.F. (1984) Frazil ice dynamics. USA Cold Regions Research and Engineering Laboratory, CRREL Monograph 84–1, Hanover, NH.

Daly, S.F. (1993) Wave propagation in ice-covered channels. *Journal of Hydraulic Engineering*, 1199(8), 895–910.

Daly, S.F. (Ed.) (1994) Report on frazil ice. USA Cold Regions Research and Engineering Laboratory, CRREL Special Report 94–23, Hanover, NH.

Daly, S.F. (1995) Fracture of river ice covers by river waves. *Journal of Cold Regions Engineering*, 9(1), 41–52.

Daly, S.F. (1998) Thermal ice growth: Real-time estimation. *Journal of Cold Regions Engineering*, 12(1), 11–28.

Daly, S.F. (2002) Data assimilation in river ice forecasting. *Proceedings of the 16th International Symposium on Ice*, Dunedin, New Zealand, Vol. 1, pp. 204–210.

Daly, S.F. (2005) Anchor ice flooding. In *World Water Congress 2005 Impacts of Global Climate Change World Water and Environmental Resources Congress 2005*, W.Y. Jackson, Ed., Anchorage, AK. American Society of Civil Engineers, Reston, VA, Paper No. 241. http://www.ascelibrary.org/2005conferences/ASCECP000173040792

Daly, S.F., Brunner, G., Piper, S., Jensen, M., and Tuthill, A. (1998) Modeling Ice Covered Rivers Using HEC-RAS. In *Cold Regions Impact on Civil Works*, D.E. Newcomb, Ed., *Proceedings of the American Society of Civil Engineers 9th International Conference on Cold Regions Engineering*, Duluth, MN, Sep. 27–30, 1998, pp. 373–380.

Daly, S.F. and Colbeck, S.C. (1986) Frazil ice measurements in CRREL's flume facility. *Proceedings of the 8th IAHR Symposium on Ice*, Iowa City, IA, Vol. 1, pp. 427–438.

Daly, S.F. and Ettema, R. (2006) Frazil ice blockage of water intakes in the Great Lakes. *Journal of Hydraulic Engineering*, 132(8), 814–824.

Daly, S.F. and Hopkins, M.A. (1998) Simulation of river ice jam formation. In *Ice in Surface Waters. Proceedings of the IAHR, 14th International Symposium on Ice*, July 27–31, 1998, Clarkson University, Potsdam, New York 13699-5710, USA, Vol. 1, pp. 101–108.

Daly, S.F. and Hopkins, M.A. (1999) Modeling river ice using discrete particle simulation. *Proceedings of the 10th International Conference on Cold Regions Engineering*, ASCE, Reston, VA, pp. 612–622.

Daly, S.F. and Hopkins, M.A. (2001) Estimating forces on an ice control structure using DEM. *Proceedings of the 11th Workshop on the Hydraulics of Ice Covered Rivers CD-ROM*, Committee on River Ice Processes and the Environment, Canadian Geophysical Union-Hydrology Section, Calgary, Alberta, Canada.

Doering, J.C., Bekeris, L.E., Morris, M.P., Dow, K.E., and Girling, W.C. (2001) Laboratory study of anchor ice growth. *Journal of Cold Regions Engineering*, 15(1), 60–66.

Durand, G., Gillet-Chaulet, F., Svensson, A., Gagliardini, O., Kipfstuhl, S., Meyssonnier, J., Parrenin, F., Duval, P., and Dahl-Jenson, D. (2007) Change in ice rheology during climate variations—Implications for ice flow modelling and dating of the EPICA Dome C core. *Climate of the Past*, Vol. 3, pp. 155–167.

EPA. (2006) National Pollutant Discharge Elimination System; Establishing Requirements for Cooling Water Intake Structures at Phase III Facilities; Final Rule, Friday June 16, 2006, 40 CFR Parts 9, 122, 123, etc., Friday, July 16, 2006., Federal Register, Environmental Protection Agency, pp. 35006–35046, http://www.epa.gov/fedrgstr/EPA-WATER/2006/June/Day-16/w5218.pdf.

Evans, T.W., Margolis, G., and Sarofim, A.F. (1974a) Mechanism of secondary nucleation in agitated crystallizers. *AICHE Journal*, 20, 950–958.

Evans, T.W., Sarofim, A.F., and Margolis, G. (1974b) Models of secondary nucleation attributable to crystal-crystallizer and crystal-crystal collisions. *AICHE Journal*, 20, 959–966.

Faraday, M. (1859) On regelation, and on the conservation of force. *Philosophical Magazine*, 17, 162–169.

Faraday, M. (1860) Note on regelation. *Proceedings of Royal Society of London*, 10, 440.

Fischer, H.B., List, E.G., Koh, R.C.Y., Imberger, J., and Brooks, N.H. (1979) *Mixing in Inland and Coastal Waters*, Academic Press, New York, NY.

Forest, T.W. (1986) Thermodynamic stability of frazil ice crystals. *Proceedings of the 5th International Offshore Mechanics and Arctic Engineering Symposium*, Tokyo, Japan, April 13–18, Vol. 4, pp. 266–270.

Frankenstein, G. (1966) Strength of ice sheets. *Proceedings of the Conference on Ice Pressure against Structures*, Laval University, Quebec, Canada, pp. 79–87.

Girling W.C. and Groeneveld, J. (1999) Anchor ice formation below Limestone Generating Station. *Committee on River Ice Processes and the Environment, 10th Workshop on the Hydraulics of Ice Covered Rivers Winnipeg*, Manitoba, Canada, pp. 160–173.

Gow, A.J. (1986) Orientation textures in ice sheets of quietly frozen lakes. *Journal of Crystal Growth*, 74(2), 247–258.

Gow, A.J., Ueda, H.T., and Ricard, J. (1978) Flexural strength of ice grown on temperate lakes. CRREL Rep. 78-9, U.S. Army Cold Regions Research and Engineering Laboratory, Hanover, NH.

Hamelin, L.E. (1979) The Bechevnik: A river bank feature from Siberia. *The Musk Ox*, 25, 70–72.

Hammar, L. and Shen, H.T. (1995) Frazil evolution in channels. *Journal of Hydraulic Research*, 33(3), 291–306.

Hammar, L., Shen, H.T., Evers, K.U., Kolerski, T., Yuan, Y., and Sobczak, L. (2002) A laboratory study on freezeup ice runs in river channels. *Proceedings of the 16th International Symposium on Ice*, Dunedin, New Zealand, Vol. 3, pp. 22–29.

Heneghan, A.F., Wilson, P.W., and Haymet, A.D.J. (2002) Heterogeneous nucleation of supercooled water, and the effect of an added catalyst. *Proceedings of the National Academy of Sciences*, 99(15), 9631–9634.

Hicks, F. (2009) An overview of river ice problems. *Cold Regions Science and Technology*, 55(2), 175–185.

Hicks, F., Steffler, M., and Gerard, D.R. (1992) Finite element modeling of surge propagation and an application to the Hay River, N.W.T. *Canadian Journal of Civil Engineering*, 19, 454–462.

Hirayama, K., Terada, K., Sato, M., Hirayama, K., Sasamoto, M., and Yamazaki, M. (1997) Field measurements of anchor ice and frazil ice. *Proceedings of the 9th Workshop on the Hydraulics of Ice Covered Rivers*, Fredericton, NB, pp. 141–151.

Holland, P.R., Feltham, D.R., and Daly, S.F. (2007) On the Nusselt number for frazil ice growth. *Journal of Hydraulic Research*, 45(3), 421–424.

Hopkins, M.A., Daly, S.F., and Lever, J.H. (1996) Three-dimensional simulation of river ice jams. *Proceedings of the American Society of Civil Engineers 8th International Conference on Cold Regions Engineering*, August 13–16, 1996, Fairbanks, AK, pp. 582–593.

Hopkins, M.A. and Tuhkuri, J. (1999) Compression of floating ice fields. *Journal of Geophysical Research*, 104(C7), 15815–15825.

Hopkins, M.A. and Tuthill, A.M. (2002) Ice boom simulations and experiments. *Journal of Cold Regions Engineering*, 16(3), 138–155.

Ji, S., Shen, H.T., Wang, Z., Shen, H.H., and Yue, Q. (2005) A viscoelastic-plastic constitutive model with Mohr–Coulomb yielding criterion for sea ice dynamics. *Acta Oceanologica Sinica*, 24(4), 54–65.

Kempema, E., Daly, S., and Ettema, R. (2008a) Fish protection, wedge-wire intake screens, and frazil ice. *Proceedings of the 19th IAHR Symposium on Ice*, Vancouver, British Columbia, Canada.

Kempema, E.W., Ettema, R., and McGee, B. (2008b) Insights from anchor ice formation in the Laramie River, Wyoming. *Proceedings of the 19th IAHR Symposium on Ice*, Vancouver, British Columbia, Canada, pp. 63–76.

Kempema, E.W., Shaha, J.M., and Eisinger, A.J. (2002) Anchor-ice rafting of coarse sediment: Observations from the Laramie River, Wyoming, USA. *Proceedings of the 16th International Symposium on Ice*, Dunedin, New Zealand, pp. 27–33.

Kerr, D.J., Shen, H.T., and Daly, S.F. (2002) Evolution and hydraulic resistance of anchor ice on gravel bed. *Cold Regions Science and Technology*, 35(2), 101–114.

Liu, L., Tuthill, A.M., and Shen, H.T. (2002) Simulating ice passage at locks using the DynaRICE model. *Proceedings of the 16th IAHR International Symposium on Ice*, Dunedin, New Zealand, Vol. 1, pp. 8–13.

Mahabir, C., Hicks, F., and Robinson Fayek, J.M. (2002) Forecasting ice jam risk at Fort McMurray, AB using fuzzy logic. *Proceedings of the 16th IAHR International Symposium on Ice*, Dunedin, New Zealand, Vol. 1, pp. 112–118.

Magnuson, J.J., Robertson, D.M., Benson, B.J., Wynne, R.H., Livingstone, D.M., Arai, T., Assel et al. (2000) Historical trends in lake and river ice cover in the Northern Hemisphere. *Science*, 289, 1743–1746.

Marko, J.R. and Jasek, M. (2010a) Sonar detection and measurements of ice in a freezing river I: Methods and data characteristics. *Cold Regions Science and Technology*, 63(3), 121–134.

Marko, J.R. and Jasek, M. (2010b) Sonar detection and measurement of ice in a freezing river II: Observations and results on frazil ice. *Cold Regions Science and Technology*, 63(3), 135–153.

Martin, S. (1981) Frazil ice in rivers and oceans. *Annual Review of Fluid Mechanics*, 13, 379–397.

Massie, D.D., White, K.D., and Daly, S.F. (2002) Application of neural networks to predict ice jam occurrence. *Cold Regions Science and Technology*, 35(2), 115–122.

Matousek, V. (1984) Types of ice run and conditions for their formation. *Proceedings of the 7th IAHR International Symposium on Ice*, Hamburg, Germany, Vol. 1, pp. 315–327.

McNamara, J.P. and Kane D.L. (2009) The impact of a shrinking cryosphere on the form of arctic alluvial channels. *Hydrological Processes*, 23, 159–186.

Mercier, S. (1984) The reactive transport of suspended particles: Mechanics and modelling. PhD dissertation, Joint Program in Ocean Engineering, Massachusetts Institute of Technology, Cambridge, MA.

Michel, B. (1971) Winter regime of rivers and lakes. CRREL Report No: M III-B1a, U.S. Army Cold Regions Research and Engineering Laboratory, Hanover, NH.

Michel, B. (1978) *Ice Mechanics*. University of Laval, Quebec, Canada.

Morse, B. and Richard, M. (2009) A field study of suspended frazil ice particles. *Cold Regions Science and Technology*, 55, 86–102.

Muller-Stoffels, M., Langhorne, P.J., Petrich, C., and Kempema, E.W. (2008) Preferred crystal orientation in fresh water ice. *Cold Regions Science and Technology*, 56(1), 1–9.

Mullins, W.W. and Serkerka, R.F. (1964) Stability of a planar interface during solidification of a dilute binary alloy. *Journal of Applied Physics*, 35(7), 444–451.

Osterkamp, T.E. (1977) Frazil-ice nucleation by mass-exchange processes at the air-water interface. *Journal of Glaciology*, 19(81), 619–625.

Osterkamp, T.E. and Gilfilian, R.E. (1975) Nucleation characteristics of stream water and frazil ice nucleation. *Water Resources Research*, 11(6), 926–928.

Osterkamp, T.E. and Gosink, J.P. (1983) Frazil ice formation and ice cover development in interior Alaska streams. *Cold Regions Science and Technology*, 8(1), 43–56.

Pariset, E. and Hausser, R. (1961) Formation and evolution of ice covers on rivers. *Engineering Institute of Canada. Transactions*, 5(1), 41–49.

Pariset, E., Hausser, R., and Gagnon, A. (1966) Formation of ice covers and ice jams in rivers. *Journal of the Hydraulics Division*, 92(HY6), 1–24.

Prowse, T.D. (2001) River-ice ecology: I: Hydrologic, geomorphic, and water-quality effects. *Journal of Cold Regions Engineering*, 15(1), 1–16.

Prowse, T.D., Bonsal, B.R., Duguay, C.R., and Lacroix, M.P. (2007) River-ice break-up/freeze-up: A review of climate drivers, historic trends and future predictions. *Annals of Glaciology*, 46, 443–451.

Prowse, T.D., Brooks, R., Callaghan, T., de Rham, L., Dibike, Y., Harder, S., Saloranta, T., von de Wall, S., and Wrona, F.J. (2009) River and lake ice responses to climate variability and change. *ArcticNet 6th Annual Scientific Meeting*, Victoria, British Columbia, Canada, p. 21.

Qu, Y.X. and Doering, J. (2007) Laboratory study of anchor ice evolution around rocks and on gravel beds. *Canadian Journal of Civil Engineering*, 34, 46–55.

Richard, M., Morse, B., Daly, S.F., and Emond, J. (2011) Quantifying suspended frazil ice using multi-frequency underwater acoustic devices. *River Research and Applications*, 27(9), 1106–1117.

Schaefer, V.J. (1950) Formation of frazil and anchor ice in cold water. *Transactions on American Geophysical Union*, 31, 885–893.

Schwarz, J. (1977) New developments in modelling ice problems. *Proceedings of the 4th International Conference on Port and Ocean Engineering under Arctic Conditions*, September, St. Johns, Newfoundland, pp. 45–61.

Shen, H.T. (2010) Mathematical modeling of river ice processes. *Cold Regions Science and Technology*, 62, 3–13.

Shen, H.T., Lu, S.-A., and Crissman, R.D. (1997) Numerical simulation of ice transport over the Lake Erie-Niagara River ice boom. *Cold Regions Science and Technology*, 26, 17–33.

Shen, H.T. and Wang, D. (1995) Under cover transport and accumulation of frazil granules. *Journal of Hydraulic Engineering*, 121(2), 184–195.

Shibkov, A.A. et al. (2003) Morphology diagram of nonequilbrium patterns of ice crystals growing in supercooled water. *Physica A*, 319, 65–79.

Shulyakovskii, L.G. (Ed.) (1963) Manual of Ice-Formation Forecasting for Rivers and Inland Lakes, Israel Program for Scientific Translations TT 66-51016, Jerusalem, Israel.

Stefan, J. (1889) Uber einige Probleme der Theorie der Warmeleitung, Sitzungsberichte der Osterreichischen Akademie der Wissenschaften Mathematisch Naturwissenschaftliche Klasse. *Abteilung 2, Mathematik, Astronomie, Physik, Meteorologic und technic*, 98, 473–484.

Stefan, J. (1891) Über die Theorie der Eisbildung, insbesondere Über die Eisbildung im Polarmeere. *Annalen der Physik*, 3rd Ser., 42, 269–286.

Stickler, M. and Alfredsen, K. (2005) Factors controlling anchor ice formation in two Norwegian rivers. *CGU HS Committee on River Ice Processes and the Environment, 13th Workshop on the Hydraulics of Ice Covered Rivers*, Hanover, NH.

Stockstill, R.L., Daly, S.F., and Hopkins, M.A. (2009) Modeling floating objects at river structures. *ASCE Journal of Hydraulic Engineering*, 135(5), 403–414.

Svensson, U. and Omstedt, A. (1994) Simulation of supercooling and size distribution in frazil ice dynamics. *Cold Regions Science and Technology*, 22, 221–233.

Tesaker, E. (1994) Ice formation in steep rivers. *Proceedings of the 12th AHR Symposium on Ice*, Trondheim, Norway.

Turner, J.S. (1979) *Buoyancy Effects in Fluids*. Cambridge University Press, New York.

Tuthill, A.M., Wuebben, J.L., Daly, S.F., and White, K.D. (1996) Probability distributions for peak stage on rivers affected by ice jams. *American Society of Civil Engineers, Journal of Cold Regions Engineering*, 10(1), 36–57.

Urroz, G.E. and Ettema, R. (1992) Bend ice jams: Laboratory observations. *Canadian Journal of Civil Engineering*, 19(5), 855–864.

U.S. Army (2006) Ice Engineering, EM 1110-2-1612. http://www.usace.army.mil/publications/eng-manuals/em1110-2-1612/toc.htm

Uunila, L.S. (1997) Effects of river ice on bank morphology and riparian vegetation along the Peace River, Clayhurst to Fort Vermilion. *Proceedings of the 9th Workshop on River Ice*, Fredericton, New Brunswick, Canada, pp. 315–334.

Vuyovich, C.M., Tuthill, A.M., Daly, S.F., and White, K.D. (2005) Ice impact evaluation for the Lower Connecticut River. *Proceedings of the 13th Workshop on Ice Covered Rivers. Committee of River Ice Processes and the Environment, Canadian Geophysical Union, Hydrology, Section* 15–16, Ascutney, VT, Paper No. 5.

White, K.D. and Daly, S. (2002) Predicting ice jams with discriminate function analysis. *Proceedings of the 21st International Conference on Offshore Mechanics and Arctic Engineering*, Oslo, Norway, pp. 683–690.

White, K.D. and Kay, R.L. (1996) Ice jam flooding and mitigation, Lower Platte River Basin, Nebraska. USACRREL Special Report 96-1, Hanover, NH.

Wigle, T.E. (1970) Investigations into frazil, bottom ice and surface ice formation in the Niagara River. *Proceedings of Symposium on Ice and its Action on Hydraulic Structures*, Reykjavik, Iceland, International Association for Hydraulic Research.

Wilson, P.W., Heneghan, A.F., and Haymet, A.D.J. (2002) Ice nucleation in nature: Supercooling point (SCP) measurements and the role of heterogeneous nucleation. *Cryobiology*, 46(2003), 88–98.

Wuebben, J.L., Daly, S.F., White, K.D., Gagnon, J.J., Tatinclaux, J.-C., and Zufelt, J.E. (1995a) Ice impacts on flow along the Missouri River. U.S. Army Cold Regions Research and Engineering Laboratory. Special report, Hanover, NH. SR 95-13.

Wuebben, J.L., Gagnon, J.J., and Deck, D.S. (1995b) Ice jamming near the confluence of the Missouri and Yellowstone Rivers: Characterization and mitigation. U.S. Army Cold Regions Research and Engineering Laboratory. Special report, SR 95-19. Hanover, NH.

Xia, X. (1998) Interaction of shallow water waves with ice cover. PhD thesis. Clarkson U., Potsdam, New York.

Ye, S.Q. and Doering, J. (2004) Simulation of the supercooling process and frazil evolution in turbulent flows. *Canadian Journal Civil Engineering*, 31, 915–926.

Zachrisson, G. (1990) Severe break-ups in the River Tornealven: Measures to mitigate damages from ice jamming. *Proceedings of the International Association of Hydraulic Research Ice Symposium*, Espoo, Finland, Vol. 2, pp. 845–857.

III

Fundamental Flow Phenomena and Turbulence

20

Turbulence in the Environment

G.N. Ivey
University of Western Australia

20.1 Introduction

Turbulence and the enhanced mixing it causes, above and beyond molecular mixing rates, is of great importance in controlling the transport of heat, salt, momentum, chemical tracers, suspended particles, and biological matter in environmental flows. While some environmental flows can be steady, such as riverine flows or industrial jet and plume discharges into the atmosphere or ocean, most observations of turbulence in the environment show it to be characterized by its variability in both space and time. It is this variability, along with the inherently statistical nature of turbulence, which makes it challenging to quantify the effects of turbulent mixing in the environment.

Turbulent fluxes of quantities such as heat, salt, or tracers, result from the correlation between fluctuations of the fluid velocity and the quantity itself. The greater the intensity of the turbulence, the greater is the flux. As turbulence is an inherently dissipative process, it requires a continual supply of energy to sustain it, and in environmental flows this is typically provided by either a drain of kinetic energy from the mean flow or by potential energy losses associated with an unstable mean density stratification, as in the case of convective turbulence driven by surface cooling. If the rate of supply of energy from the mean flow or background density stratification is unsteady, then the small-scale turbulent mixing will also be unsteady. In the vertical, ambient density stratification is almost always present; buoyancy effects thus inhibit vertical motion, and, as a consequence, the characteristic vertical length scales of the motion are often small compared with the total depth. For this reason, as well as the generally small vertical depth compared to the lateral dimension of most water bodies, environmental flows usually have large horizontal scales of motion relative to the vertical scales. Mean currents are strong in the horizontal and weak or negligible in the vertical. As a consequence, the mean flow tends to be the dominant transport process for fluxes of tracers in the horizontal, whereas fluxes in the vertical are much more likely to be dominated by turbulent processes.

Turbulence in the environment is generated by a wide variety of mechanisms and Burchard et al. (2008) provide a recent review, with particular focus on the oceans. The surface boundary layer is directly in contact with the atmosphere and typically highly turbulent as it is stirred by the direct action of the wind, surface waves, and thermal fluxes across the air–water interface. As a consequence, a surface well-mixed layer is often present (e.g., Krauss 1977); and when nutrients are abundant, then due to the high light levels near surface, biological productivity is typically high. With the exception of very shallow or highly energetic systems, there is a region of stable density stratification below this surface layer due to decreasing temperature with depth and often saline effects, as in the ocean. While stable in density, this region experiences mean flows in the horizontal and is also inevitably perturbed by internal waves. The vertical shears associated with these flows can drive turbulent mixing in the region, and much effort is devoted to understanding the connection between the mean flow, internal waves, and turbulence (e.g., Burchard et al. 2008). Finally, adjacent to the sediment–water interface, a bottom boundary layer of well-mixed or weakly density stratified fluid is present where high levels of turbulent mixing can also be observed, forced by tidal flows, internal waves, and local mean flows. If the system is shallow, then the surface and bottom boundary layers overlap and this is typical of estuarine systems, for example (e.g., Jones and Monismith 2008).

The quantification of the vertical flux of active tracers, passive tracers, and momentum due to small-scale turbulent processes in these environments represents an ongoing challenge. In water

Handbook of Environmental Fluid Dynamics, Volume One, edited by Harindra Joseph Shermal Fernando. © 2013 CRC Press/Taylor & Francis Group, LLC.
ISBN: 978-1-4398-1669-1.

bodies such as lakes and oceans, in the density stratified interior direct estimates of vertical fluxes are generally an order of magnitude smaller than those required to close bulk estimates, suggesting that the majority of the mixing may occur elsewhere, either at boundaries or in under-sampled "hot spots" in the interior where mixing rates are considerably higher (e.g., Wunsch and Ferrari 2004). Questions regarding overall energetics and mixing efficiency (e.g., Ivey et al. 2008) are especially important in efforts to calculate the state of a particular environment under climate-change scenarios, for example. The issue arises both in the context of simple models of the natural environment as well as the need for quantitative descriptions of mixing in numerical models of circulation and mixing.

It is thus crucial to be able to describe small-scale turbulent mixing with high accuracy in the environment. Due to computational constraints, numerical models of the ocean circulation, for example, have typical model resolution in the range of 1–10 m in the vertical and often 1 km or coarser in the horizontal. Using information from resolved grid-scale motions, models must therefore employ parameterizations to describe the action of subgrid-scale turbulent processes.

Many subgrid-scale closure schemes have been proposed, and there is a particular focus on the performance of these models in boundary and coastal regions (e.g., Burchard et al. 2008). These regions of the ocean are particularly challenging to model due to the topographical complexity, strong density stratifications, the often intense local mean flow fields, and the rapid changes (compared to the interior) in these quantities that can occur in time as well as in space (both vertically and horizontally) in the density and velocity fields. Comparison of different closure schemes (e.g., Warner et al. 2005) showed they can produce significantly different solutions, and in the coastal region and in confined water bodies like estuaries and lakes these issues can become severe.

As the gravitationally influenced vertical component of the turbulent mixing is highly unsteady and inhomogeneous, this leads to fundamental measurement and interpretation problems. In the environment we have a mix of essentially quiescent and patchy turbulent regions (see the example in Figure 20.1),

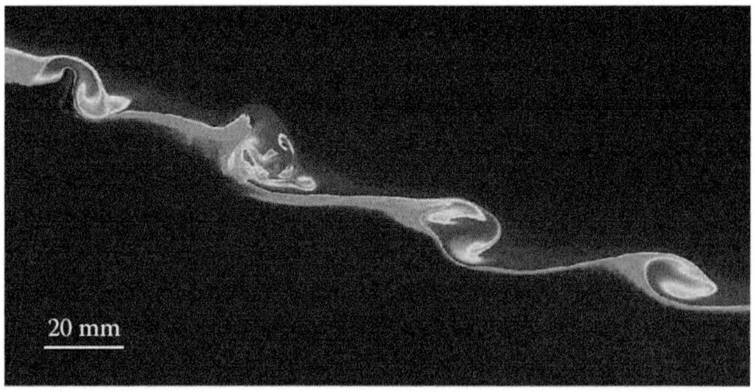

FIGURE 20.1 Planar laser–induced fluorescence image from a laboratory experiment showing developing turbulent mixing due to a breaking internal wave at a density interface in a two-layer fluid. (From Troy, C.D. and Koseff, J.R., *J. Fluid Mech.*, 53, 107, 2005. With permission.)

and the turbulence intensity is highly variable even within the turbulent patches. Historically, our understanding of turbulent mixing has come from the study of flows characterized by high Reynolds number, with no spatial or temporal variability. Away from the energetic surface and bottom boundary layers, environmental flows are often at the weaker end of the turbulent state, defined here to mean that the effective diffusivities are typically only 1 order of magnitude greater than molecular rates. This appears to be the state in the interior of many natural geophysical flows and is the range where the role of internal waves and turbulence coexist. In the time domain at a fixed location, the flow field can vary from laminar to turbulent and back again. There is, thus, an ongoing challenge to both describe these flows as well as to develop models that can capture the complex mixing processes in the environment. The field is vast, and the focus here is on the mixing of tracers and pollutants, of central importance to environmental flows.

20.2 Principles

It is useful to examine mixing in environmental flows in the context of some fundamental variables and simple dimensionless numbers. Consider the case where the largest turbulent eddy in the flow is characterized by a turbulent velocity scale q and length-scale l and the ambient fluid has a kinematic viscosity ν. Environmental flows are typically density stratified, and, hence, an additional necessary parameter is the strength of the vertical density stratification as quantified by the buoyancy frequency $N = ((-g/\rho_0)(d\rho/dz))^{1/2}$ where ρ_0 is a reference density. This parameter list suggests two independent dimensionless numbers are needed to characterize the flow and, with meaning analogous to their mean flow counterparts, these are the turbulent Reynolds number Re_T and the Froude number Fr_T

$$\mathrm{Re}_T = \frac{ql}{\nu} \tag{20.1}$$

$$\mathrm{Fr}_T = \frac{q}{Nl} \tag{20.2}$$

Introducing the turbulent kinetic energy dissipation rate ε, then an additional dimensionless parameter commonly referred to as the buoyancy Reynolds number Re_b may be defined as

$$\mathrm{Re}_b = \frac{\varepsilon}{\nu N^2} \tag{20.3}$$

The parameter Re_b has many interpretations in the literature, but here we may think of it as simply proportional to the ratio of the Ozmidov length-scale $l_O = (\varepsilon/N^3)^{1/2}$ to the Kolmogorov length-scale $l_K = (\nu^3/\varepsilon)^{1/4}$.

Providing one can write the turbulent dissipation rate as $\varepsilon = q^3/l$, then (20.1), (20.2), and (20.3) are related (Ivey and Imberger 1991) by

$$\mathrm{Re}_b = \mathrm{Re}_T\,\mathrm{Fr}_T^2 \quad (20.4)$$

While three dimensionless numbers can thus be defined, only two are necessary and sufficient to characterize these flows.

Unlike many engineering flows, owing to variability in the mean flow or the presence of internal waves in the environment, the forcing may be variable in time T. Furthermore, the turbulent fields are typically spatially inhomogeneous and we characterize this by the scale L, and as we are interested in the vertical fluxes we think of L as a vertical length-scale. Two further dimensionless numbers are thus potentially required, and one choice would be

$$\frac{l}{L} \quad \text{and} \quad \frac{(l/q)}{T} \quad (20.5)$$

Thus five dimensionless numbers (four of which are independent) can be important for characterizing the turbulent mixing. Given this complexity, one of the challenges is to try and simplify the dynamics by identifying the dominant parameters needed to quantify environmental mixing.

20.3 Methods of Analysis

Approaches to measuring turbulent mixing can conveniently be divided into two: indirect and direct measurements of the vertical flux. Indirect measurements do not measure the vertical turbulent density flux $-\overline{w'\rho'}$ (where primes denote turbulent quantities), rather they typically measure either ε or χ, the turbulent dissipation rates for momentum and temperature variance, respectively, and then use models to infer the vertical eddy diffusivities.

The first commonly used indirect model, due to Osborn and Cox (1972), is based on the temperature variance equation:

$$\frac{\partial\overline{\theta'^2}}{\partial t} + \overline{u}_j\frac{\partial\overline{\theta'^2}}{\partial x_j} + 2\overline{u_j'\theta}\frac{\partial\overline{\theta}}{\partial x_j} - \frac{\partial}{\partial x_j}\left(\overline{u_j'\theta'^2} + \kappa\frac{\partial\overline{\theta'^2}}{\partial x_j}\right) = -2\kappa\overline{\left(\frac{\partial\theta'}{\partial x_j}\right)^2} \quad (20.6)$$

where the subscript j denotes the coordinate direction and κ is the thermal diffusivity. If $l/L = 0$ and $(q/l)/T = 0$, then one can neglect unsteadiness, advection by the mean flow, and the divergence terms; hence, (20.6) simplifies to

$$2\overline{u_j'\theta}\frac{\partial\overline{\theta}}{\partial x_j} = -2\kappa\overline{\left(\frac{\partial\theta'}{\partial x_j}\right)^2} = \chi \quad (20.7)$$

Assuming further that the turbulence is isotropic at the smallest scales, that all mean and turbulent density gradients are taken in the direction n normal to the isothermal surfaces, and that the mean gradient is defined with respect to the sorted state of minimum potential energy, then from (20.7) the eddy diffusivity for heat K_θ is

$$K_\theta = \frac{-\overline{w'\theta'}}{\partial\overline{\theta}/\partial n} = 3\kappa\frac{\overline{(\partial\theta'/\partial n)^2}}{(\partial\overline{\theta}/\partial n)^2} \quad (20.8)$$

In the field n is always assumed to be in the z direction.

The second commonly used indirect model, due to Osborn (1980), is based on the turbulent kinetic energy (TKE) equation:

$$-\frac{1}{2}\frac{\partial\overline{u_i'^2}}{\partial t} - \frac{1}{2}\overline{u}_j\frac{\partial\overline{u_i'^2}}{\partial x_j} - \overline{u_i'u_j'}\frac{\partial\overline{u}_i}{\partial x_j}$$

$$-\frac{\partial}{\partial x_j}\left(\frac{1}{2}\overline{u_j'u_i'u_i'} + \frac{1}{\rho}\overline{u_j'p'} + \nu\overline{u_i'\frac{\partial u_i'}{\partial x_j}}\right) = -b + \varepsilon \quad (20.9)$$

where all of the terms on the left-hand side of (20.9) can produce TKE, the buoyancy flux $b = (g/\rho_0)\overline{w'\rho'}$ represents the loss of TKE to irreversible mixing of the background density stratification, and ε the irreversible loss of TKE by viscous dissipation. Again if $l/L = 0$ and $(q/l)/T = 0$, then one can neglect unsteadiness, transport of TKE by the mean flow, and all the divergence terms, and (20.9) reduces to a simple balance between local shear production, buoyancy flux, and dissipation:

$$-\overline{u_i'u_j'}\frac{\partial\overline{u}_i}{\partial x_j} = -b + \varepsilon \quad (20.10)$$

Osborn (1980) showed that if the flux Richardson number is defined as $R_f = -b/-\overline{u_i'u_j'}(\partial\overline{u}_i/\partial x_j)$, then the vertical eddy diffusivity for the density stratifying species K_ρ may be obtained from (20.10) as

$$K_\rho = -\frac{b}{N^2} = \left(\frac{R_f}{1-R_f}\right)\frac{\varepsilon}{N^2} \quad (20.11)$$

The mixing parameter is defined from (20.11) as $\Gamma = R_f/(1 - R_f)$. The earlier definition of the flux Richardson number R_f is overly restrictive as it assumes that only the shear production term $-\overline{u_i'u_j'}\frac{\partial\overline{u}_i}{\partial x_j}$ can produce TKE. As all of the terms on the left-hand side of (20.9) can produce TKE, a more fundamental definition of R_f (Ivey and Imberger 1991) may be defined from (20.9) as

$$R_f = \frac{-b}{-b+\varepsilon} \quad (20.12)$$

and this should be used in (20.11), which is still valid in this generalized case.

Entirely analogous to the derivation in (20.11), Crawford (1982) showed that an expression for the vertical eddy diffusivity for momentum K_v can also be developed from (20.9), and this is

$$K_v = -\frac{\overline{u'w'}}{S} = \left(\frac{1}{1-R_f}\right) Ri \frac{\varepsilon}{N^2} \quad (20.13)$$

where the gradient Richardson number $Ri = N^2/S^2$ and $S = dU/dz$ is the vertical shear in the mean horizontal velocity U. Instrumentation has been developed to measure ε directly in the field (e.g., Stips 2005), thus (20.11) and (20.13) are attractively simply to use. However, the flux Richardson number R_f must be independently determined to utilize either expression.

On the basis of laboratory experiments at the time, Osborn (1980) suggested $R_f \leq 0.17$ (and, hence, $\Gamma \leq 0.2$). If temperature is the sole contributor to density, it is tempting to equate K_θ in (20.8) with K_ρ in (20.11) and infer the unknown mixing efficiency R_f (Oakey 1982). However, apart from needing simultaneous and independent measurements of both the turbulent velocity and temperature fluctuations, this also requires that all the assumptions leading to (20.8) and (20.11) are satisfied at the same point in the environment, and this is not possible to evaluate in the field. While the mixing efficiency remains to be determined, note also that the formulations in both (20.11) and (20.13) do not answer the fundamental question: as the turbulence intensity decreases, when do the turbulent diffusivities K_ρ and K_v revert to their respective molecular values of κ and ν?

As suggested earlier, when the flow is steady and homogeneous, two independent dimensionless parameters are both necessary and sufficient to characterize the flow, and one choice is $\varepsilon/\nu N^2$ and Re_T. Barry et al. (2001) conducted laboratory experiments using a mechanical grid to steadily and uniformly mix a stratified fluid. Shih et al. (2005) conducted direct numerical simulations (DNS) experiments in a linearly sheared and stratified flow. Bluteau et al. (2011) conducted field experiments in the turbulent bottom boundary layer in 420 m of water on the Australian North West Shelf where turbulence was generated by an oscillatory tidal flow with a dominant period of 12.4 h. In the latter case, the data have been analyzed in segments of 17 min lengths—very small compared to the timescale of the background flow—and we thus assume that the flow is effectively steady and the background stratification does not change on such short timescales. In each of the three cases, turbulent dissipation rates and overturning length-scales were estimated, and in Figure 20.2 we plot the results from the three very diverse sets of experiments in $\varepsilon/\nu N^2$ vs Re_T parameter space.

There are several features of note in this plot. First, the very large range of parameter space covered with both $\varepsilon/\nu N^2$ and Re_T ranging from 10^0 to 10^5. The DNS experiments from Shih et al. (2005) have the most modest range of Re_T, with a maximum of about 10^2. Not surprisingly, the field data have the highest values of Re_T, with some values approaching 10^5. The laboratory experiments lie between these two extremes, although it is interesting to note that the most energetic cases have $\varepsilon/\nu N^2 > 10^4$ and $Re_T = 10^3$.

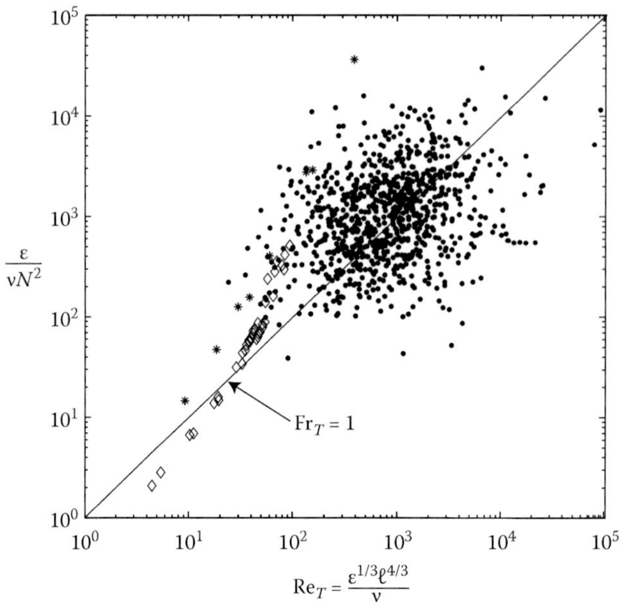

FIGURE 20.2 Results from three sets of observations in $\varepsilon/\nu N^2$ vs. Re_T parameter space. (The data are from (◇): Barry, M.E. et al., *J. Fluid Mech.*, 442, 267, 2001; (✱): Shih, L.H. et al., *J. Fluid Mech.*, 525, 193, 2005; (•): Bluteau, C. et al., *J. Geophys. Res.*, 116, 2011. With permission.)

While there are some trends evident in individual data sets, the most interesting aspect of the plot is the way the data points tend to cluster around the line shown where $Fr_T = 1$. Despite the diverse nature of the turbulent fields in the three datasets, there appears, on average, a preference for a turbulent state where $Fr_T \approx 1$, or equally $l \approx l_0 = (\varepsilon/N^3)^{1/2}$.

This suggests that the two optimal dimensionless numbers which display the most dynamic range are in fact $\varepsilon/\nu N^2$ and Re_T. Accordingly, in Figure 20.3 we examine the dependence of the mixing parameter Γ on both $\varepsilon/\nu N^2$ and Re_T with data from Shih et al. (2005) and Barry et al. (2001) (Bluteau et al. [2011] did not measure buoyancy flux). Dissipation is computed differently in the two cases, directly from the full dissipation tensor for Shih et al. (2005) and indirectly from an external energy balance by Barry et al. (2001). When $\varepsilon/\nu N^2 < 10$ and $Re_T < 10$, there are negative values of Γ in the DNS results, and these are a result of re-stratification effects at small levels of turbulent activity (see Shih et al. 2005). Above these very low levels of turbulence intensity, both experiments show there is an increase of Γ to a maximum of approximately 0.20, and then a subsequent decrease with a further increase in the turbulence intensity. Note that when $\varepsilon/\nu N^2 = 10^4$, the parameter Γ is only 0.02—an order of magnitude smaller than the mean value traditionally used. The peak in Γ is at a slightly higher value of $\varepsilon/\nu N^2$ for the laboratory experiments compared to the DNS experiments but, most importantly, above the threshold value of $Re_T \approx 10$, there appears to be no systematic dependence of mixing efficiency on Re_T. In the extreme when $\varepsilon/\nu N^2 \to \infty$, this limit corresponds to the case where there is no background density stratification; without a background stratification there can be no buoyancy flux, and, hence, $\Gamma \to 0$ (e.g., Linden 1979).

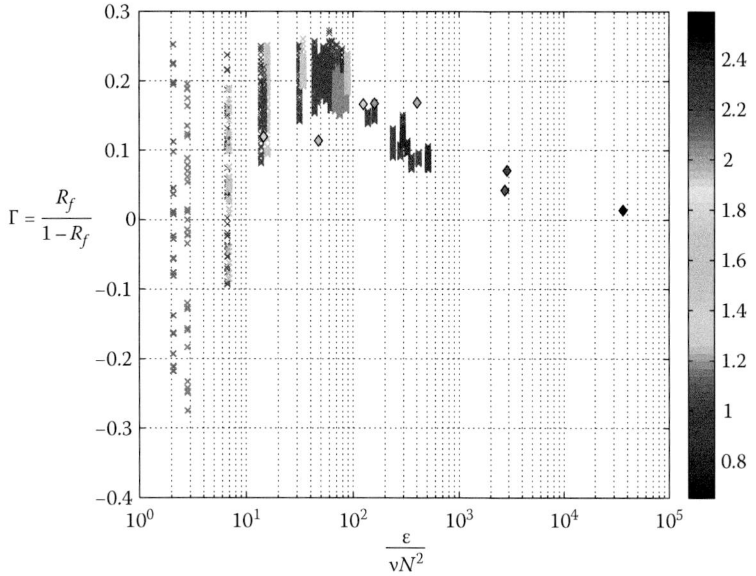

FIGURE 20.3 **(See color insert.)** Mixing parameter $\Gamma = R_f/(1 - R_f)$ as a function of $\varepsilon/\nu N^2$. The color scale on the right represents the turbulent Reynolds number Re_T as it ranges from 5 to 4×10^3 (see also Figure 20.2). (Data are from (◇): Barry, M.E. et al., *J. Fluid Mech.*, 442, 267, 2001; (×): Shih, L.H. et al., *J. Fluid Mech.*, 525, 193, 2005.)

This last point is significant as for natural environmental flows one would naturally expect large values of Re_T. Indeed, most field scale values will likely be in the region where both $Re_T > 100$ and $\varepsilon/\nu N^2 > 100$ (e.g., Bluteau et al. 2011). Recent field measurements of buoyancy flux by Davis and Monismith (2011), where the mixing efficiency was estimated out to values of $\varepsilon/\nu N^2 \approx 10^4$, appear consistent with the laboratory and DNS results. In general, however, there are very limited field measurements of the buoyancy flux, and hence the mixing parameter Γ, at large values of $\varepsilon/\nu N^2$ and Re_T. Available computing power restricts the use of DNS methods to relatively low values of $\varepsilon/\nu N^2$ and Re_T. However, the clear implication from Figure 20.3 is that the mixing parameter Γ, or equally the mixing efficiency R_f, has little dependence on turbulent Reynolds number Re_T but a strong dependence on $\varepsilon/\nu N^2$.

In summary, this result lends supports to the contention (e.g., Shih et al. 2005, Ivey et al. 2008) that the mixing efficiency can be described simply as a function of $\varepsilon/\nu N^2$. In Table 20.1, we summarize the predicted values for, both, vertical diffusivity of a tracer and, in the absence of any internal waves in the flow, the vertical diffusivity of momentum, from very weak to highly energetic stratified turbulent flows.

TABLE 20.1 Predicted Total Vertical Mixing Rates in Continuously Stratified Fluids Where $K_\rho^{tot} = \kappa + K_\rho$ and $K_\nu^{tot} = \kappa + K_\nu$

Regime	$\frac{\varepsilon}{\nu N^2}$ Range	K_ρ^{tot}	K_ν^{tot}
Molecular	$\frac{\varepsilon}{\nu N^2} < 7$	κ	ν
Transitional	$7 < \left(\frac{\varepsilon}{\nu N^2}\right) < 100$	$0.2\nu\left(\frac{\varepsilon}{\nu N^2}\right)^1$	$0.2\nu\left(\frac{\varepsilon}{\nu N^2}\right)^1$
Energetic	$\frac{\varepsilon}{\nu N^2} > 100$	$2\nu\left(\frac{\varepsilon}{\nu N^2}\right)^{1/2}$	$1.5\nu\left(\frac{\varepsilon}{\nu N^2}\right)^{1/2}$

Source: Shih, L.H. et al., *J. Fluid Mech.*, 525, 193, 2005. With permission.

20.4 Applications

The analysis described earlier yields direct insight into the nature of turbulent mixing at the process level. Providing one knows the parameters ε, *l*, ν, and *N* (and, hence, Re_T and $\varepsilon/\nu N^2$), an outcome of this analysis is that the vertical eddy diffusivity for a tracer can be computed to high accuracy. There are considerable challenges associated with the measurement of dissipation and length-scale in the field (e.g., Stips 2005), and there are also statistical averaging issues in order to get representative values over space and time. However, as our measurements improve, it is an ongoing challenge to apply this knowledge to model turbulence within the context of numerical circulation models of environmental flows.

There are a number of approaches to turbulence modeling and they can conveniently be divided into four categories: integral models, DNS, large eddy simulation (LES), and Reynolds-averaged Navier Stokes (RANS) models. Integral models (e.g., Krauss 1977) rely on good observations and are typically applied to describe the near-surface or near-bottom boundary layers. By integrating over the boundary layer thickness, they are able to provide estimates of layer-averaged or slab properties and their evolution in time. In direct contrast, DNS makes no assumptions about structure but rather models all scales of the flow from the largest to the smallest by solving the discretized versions of the Navier Stokes equations. Given the current levels of computing power, these are limited by the Reynolds numbers they can access and, while they cannot model field scale environmental flows, are finding increasing use at laboratory-scale modeling and to improve our process understanding (e.g., Shih et al. 2005). LES modeling does not attempt to model the full scale of motions, but rather resolves the larger energy containing scales of motion, spatially filters out the smaller scales of motion, and parametizes these unresolved scales (Ferziger and Peric 2001).

RANS models compute the mean flow by averaging, almost always in time, the turbulence components and thereby introducing into the equations of motion the Reynolds stress terms which must, in turn, be specified or parameterized. This is by far the most common approach. In recent decades, the development of 3D, free surface, hydrostatic, numerical circulation models has occurred. These models are useful for a wide variety of environmental flows and use RANS averaging to describe the turbulence. The particular RANS model used, along with the choice of numerical scheme, has important implications for the model results and can affect their ability to simulate environmental flows.

While many schemes have been proposed for RANS models, Umlauf and Burchard (2003) proposed a generic length-scale (GLS) scheme that included the most popular RANS schemes within a single numerical scheme. The GLS scheme solves two equations describing the turbulent flow and has been implemented in the Regional Ocean Model System (ROMS) (Warner et al. 2005). The GLS scheme describes turbulent mixing by solving a two-equation turbulence model: the TKE equation for q, and a second equation for a generic turbulent length-scale l, and then this is used to specify the eddy diffusivity from the classical Prandtl concept of eddy diffusivity as $K \propto q^{1/2}l$.

The GLS scheme of Umlauf and Burchard (2003) takes advantage of the similarities in four previous formulations to develop a single model with user selectability of scheme. Two things are worthy of note here. While there is no debate over the form of the TKE equation, there is debate over the second equation for the length-scale. Furthermore, the length-scale l is assumed to have an upper bound given by the Ozmidov scale $l_0 = (\varepsilon/N^3)^{1/2}$ (e.g., Warner et al. 2005). This assumes that $Fr_T = q/Nl = (l_0/l)^{2/3} \leq 1$, an assumption not consistent with the data shown in Figure 20.2, for example. There is also ongoing debate over the choice of which mixing schemes to use and this translates into some uncertainty in the coefficients that must be specified (e.g., Warner et al. 2005).

Applying these models to the particular case where there is a steady-state balance between shear production, buoyancy flux, and dissipation, they effectively apply a fixed mixing efficiency of approximately 10% (Warner et al. 2005). Again, that is not consistent with the data in Figure 20.3, and not consistent with the constant 17% (i.e., $R_f = 0.17$) often assumed in processing microstructure observations of turbulence in the field. When applying RANS closure schemes in models such as ROMS, there is of course the assumption that the diffusivities derived from the closure schemes are large compared to the numerical diffusivities inherent in the models; and this poses a particular challenge in flows with low levels of environmental turbulence. Given these challenges, it is no surprise to find that all models do not yield the same results (e.g., Warner et al. 2005). The choice of model, the choice of closure scheme to apply, and the comparison and evaluation with field data must be taken thoughtfully.

There have been attempts to simplify these turbulence closure schemes. For example, in a study of breaking internal waves, Klymak and Legg (2010) simply proposed that $l = l_0$ and the mixing parameter was fixed at $\Gamma = 0.2$. As Figures 20.2 and 20.3 show, this is an oversimplification, but reflects the ongoing debate about how best to incorporate the descriptions of the turbulent mixing into practical and simple models of environmental flows. Really none of the numerical models that are in use captures the dependence of the mixing on $\varepsilon/\nu N^2$ discussed earlier and summarized in Table 20.1.

The earlier analysis suggests that providing one is not close to a solid boundary or a very strong density interface, where the turbulent length-scale is controlled by distance from the boundary or density interface, then the earlier analysis is applicable. Second, if the turbulent Reynolds number is above a threshold, then as the results in Table 20.1 suggest the mixing rate is dependent only on dissipation and ambient properties and does not appear to depend on an independent length-scale. This concept is, of course, implicit in (20.11), above which may be written as $K_\rho = \Gamma\varepsilon N^{-2} = \Gamma\varepsilon^{1/3} l_0^{4/3}$. This suggests that perhaps solving only a single equation for dissipation is necessary in the interior of stratified water bodies. At the very least, the assumption that mixing efficiency is fixed is an over simplification.

20.5 Major Challenges

We have limited experimental measurement of buoyancy flux, and hence mixing efficiency, at high Re_T. This is clearly needed in order to examine the hypothesis that mixing efficiency can be determined as a function of $\varepsilon/\nu N^2$ only. This likely requires time series field measurements from a fixed array in either the ocean or the strongly stratified atmosphere, and these measurements are difficult to make and interpret. One difficulty in the measurements is the separation of the internal wave and turbulence components. This separation is not only an ongoing challenge for measurements but also a challenge in numerical models of turbulent mixing (e.g., Burchard et al. 2008). On the other hand, effecting a careful marriage between observations and models, and models that are as simple as possible but no simpler, appears to be the key to understanding turbulence in the environment.

References

Barry, M.E., Ivey, G.N., Winters, K.B., and Imberger, J. 2001. Measurements of diapycnal diffusivity in stratified fluids. *J. Fluid Mech.* 442:267–291.

Bluteau, C., Jones, N.L., and Ivey, G.N. 2011. Dynamics of a tidally-forced stratified shear flow on the continental slope. *J. Geophys. Res.* 116. DOI:10.1029/2011JC007214.

Burchard, H., Craig, P.D., Gemmrich, J.R. et al. 2008. Observational and numerical modelling methods for quantifying coastal ocean turbulence and mixing. *Prog. Oceanogr.* 76:399–442.

Crawford, W.R. 1982. Pacific equatorial turbulence. *J. Phys. Oceanogr.* 12:1137–1149.

Davis, K.A. and Monismith, S.G. 2011 The modification of bottom boundary layer turbulence and mixing by internal waves shoaling on a barrier reef. *J. Phys. Oceanogr.* 41:2223–2241.

Ferziger, J.H. and Peric, M. 2001. *Computational Methods for Fluid Mechanics*. Springer, New York.

Ivey, G.N. and Imberger, J. 1991. On the nature of turbulence in a stratified fluid, Part 1: The energetics of mixing. *J. Phys. Oceanogr.* 21:650–658.

Ivey, G.N., Winters, K.B., and Koseff, J.R. 2008 Density stratification, turbulence, but how much mixing? *Ann. Rev. Fluid Mech.* 40:169–184.

Jones, N.L. and Monismith, S.G. 2008. Modelling the influence of wave enhanced turbulence in a shallow tide- and wind-driven water column. *J. Geophys. Res.* 113:C03009. DOI: 10.1029/2007JC004246.

Klymak, J.M. and Legg, S. 2010. A simple mixing scheme for models that resolve breaking internal waves. *Ocean Model.* 33:224–234.

Krauss, E.B. Ed. 1977. *Modelling and Prediction of the Upper Layers of the Ocean*. Pergamon Press, Oxford, U.K.

Linden, P.F. 1979. Mixing in stratified fluids. *Geophys. Astro. Fluid Dyn.* 13:3–23.

Oakey, N.S. 1982. Determination of the rate of dissipation of turbulent energy from simultaneous temperature and velocity shear microstructure measurements. *J. Phys. Oceanogr.* 12:256–271.

Osborn, T.R. 1980. Estimates of the local rate of vertical diffusion from dissipation measurements. *J. Phys. Oceanogr.* 10:83–89.

Osborn, T.R. and Cox, C.S. 1972. Oceanic fine structure. *Fluid Dyn.* 3:321–345.

Shih, L.H., Koseff, J.R., Ivey, G.N., and Ferziger, J.H. 2005. Parameterization of turbulent fluxes and scales using homogenous sheared stably stratified turbulence simulations. *J. Fluid Mech.* 525:193–214.

Stips H. 2005. Dissipation measurement: Theory. In *Marine Turbulence: Theories, Observations and Models*. Eds. H.Z. Baumert, J. Simpson, and J. Sundermann, pp. 115–126. Cambridge University Press, Cambridge, U.K.

Troy, C.D. and Koseff, J.R. 2005 The instability and breaking of long internal waves. *J. Fluid Mech.* 543:107–136.

Umlauf, L. and Burchard, H. 2003. A generic length scale equation for geophysical turbulence models. *J. Mar. Res.* 61:235–265.

Warner, J.C., Sherwood, C.R., Arango, H.G., and Signell, R.P. 2005. Performance of four turbulence closure modes implemented using a generic length scale method. *Ocean Model.* 8:81–113.

Wunsch, C. and Ferrari, R. 2004. Vertical mixing, energy, and the general circulation of the oceans. *Ann. Rev. Fluid Mech.* 36:281–314.

21

Turbulent Dispersion

Benoit Cushman-Roisin
Dartmouth College

After recalling a few salient features of turbulent dispersion, this chapter reviews classical and relatively new theories. Following the observation that turbulent dispersion typically proceeds not only much faster but also in a qualitatively different manner than molecular diffusion, it is argued that the conventional approach of eddy diffusivity is inadequate and an alternative model ought to be developed to better represent observed behavior.

The new term in the diffusion equation, which is nonlocal, may be interpreted in terms of the probability density distribution of the turbulent velocity. Different assumptions about this distribution lead to a family of models, one of which is the model proposed here and another, the classical Fickian model of diffusion. A connection is also made with models using fractional calculus.

21.1 Turbulent Dispersion According to Observations

A primary consideration in environmental turbulent dispersion is the spread of a patch of pollution caused by ambient turbulence, and a good place to start is with observations of smokestack plumes as their white puffy signatures are ubiquitous and familiar to most of us. Due to their direct effect on air quality, smokestack plumes have been extensively studied (e.g., Boubel et al. 1994, and references therein), and the rate at which their size grows with distance from the tip of the smokestack has been summarized in the so-called Pasquill–Gifford curves (Gifford 1961, Pasquill 1961). Figure 21.1 shows the rate of widening of the plume as expressed by the standard deviation σ_y of the concentration distribution in the horizontal direction transverse to the wind as an observed function of downwind distance.*

We note that the plume's width increases nearly proportionally to the downwind distance, which at constant wind speed is a measure of time.

Turbulent dispersion in water is somewhat more difficult to observe, but data do exist. Worth mentioning is the tracer release experiment conducted in the Hudson River by Clark et al. (1996), which documented the spread of a passive tracer over almost 2 weeks along a distance of a 60 km reach of the river. Their data are re-plotted as Figure 21.2, which shows the standard deviation of the longitudinal (along-river axis) tracer concentration distribution. The straight line is a least-square fit to the data.

It is reasonable to infer from this graph that, in first approximation, the patch size (proportional to the standard deviation σ) grows proportionally to time.

As a final example among countless other observations, it is worth mentioning the compilation by Okubo (1974) of float scattering in the ocean, a case of 2D turbulent dispersion (Figure 21.3). These data clearly reveal that the vigor of dispersion, as measured by the estimated diffusion coefficient (see next section), increases with patch size. On Figure 21.3, a line of unit slope has

* There is a companion chart giving the vertical standard deviation, σ_z, of the plume as a function of distance, but this other chart is not shown here as it implicates buoyancy effects, which are not considered in the present chapter.

Handbook of Environmental Fluid Dynamics, Volume One, edited by Harindra Joseph Shermal Fernando. © 2013 CRC Press/Taylor & Francis Group, LLC. ISBN: 978-1-4398-1669-1.

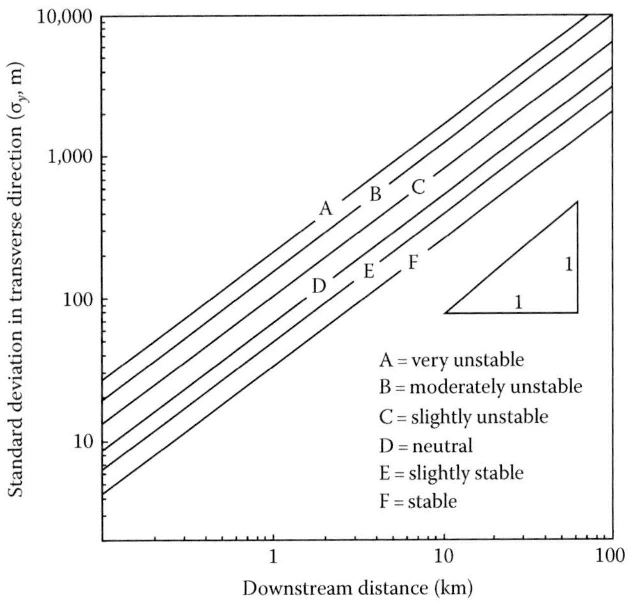

FIGURE 21.1 Pasquill–Gifford curves for the width of a smokestack plume as a function of downwind distance. The letter labels refer to various levels of atmospheric turbulence, and the triangular inset shows the slope of proportionality for comparison. Regardless of the turbulence level, the plume widens almost proportionally with distance.

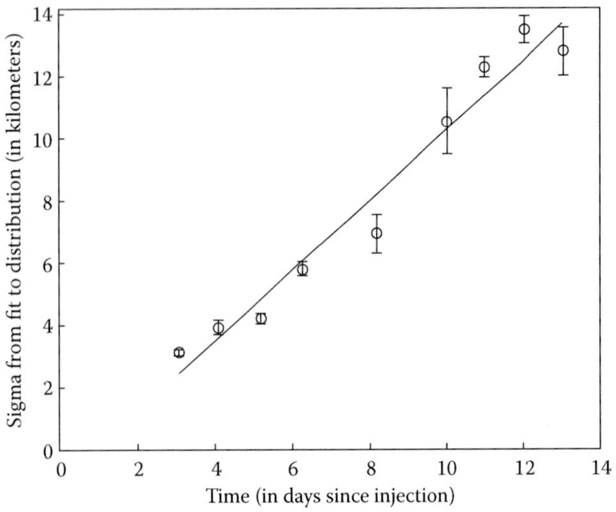

FIGURE 21.2 Standard deviation of tracer distribution along the Hudson River following the release of SF_6, as determined after fitting Gaussian distributions to each of the measured concentration profiles. (From Clark, J. F. et al., *Environ. Sci. Technol.*, 30, 1527, 1996.). The open circles with vertical bars indicate the data with associated errors, and the straight line is the least-square fit. The goodness of fit indicates that patch size grows proportionally to time (σ growing like 1.12*t*, with σ in kilometers and *t* in days).

been added for reference, although a line of slope equal to 4/3 is occasionally preferred (e.g., Fischer et al. 1979).

Numerous other examples (e.g., Min et al. 2002) have been reported in the literature, but the preceding ones are deemed sufficiently indicative of the spread of a patch of a passive tracer or set of particles in a natural turbulent environment. We now turn to theoretical modeling aimed at making sense of these observations.

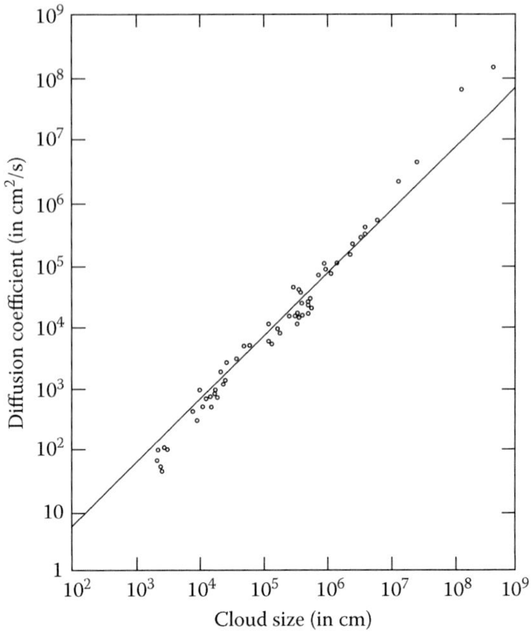

FIGURE 21.3 Estimated diffusion coefficient plotted against cloud size for several sets of float dispersion experiments in the upper ocean. Note the broad range of scales, spanning 6 orders of magnitude. Clearly, the vigor of diffusion increases with patch size; the wider the patch, the faster it disperses. Data compiled by Okubo (1974) with line of unit slope added by the present author. For a slightly augmented set of data, see Figure 3.5 of Fischer et al. (1979).

21.2 Eddy Diffusivity Model

Adolf Fick (1855) was first in proposing a formula for molecular diffusion, whereby the flux of substance is taken proportional to its concentration gradient, with the coefficient of proportionality being called the (molecular) diffusivity. Needless to say, the situation is different in turbulent flow, but the idea that chaotic swirls of turbulence are not all that unlike random motions of colliding molecules has led to the concept of enlarging the molecular diffusivity to become an eddy diffusivity. This early approach, which still receives much currency today (Roberts and Webster 2002), can be traced to Joseph Boussinesq (1877) as an expedient way to overcome the turbulence closure problem exposed a few years earlier by Osborne Reynolds. The mathematical formulation that ensues is that the flux q of concentration of a passive tracer in the x-direction is taken proportional to the gradient of the concentration $c(x, t)$ of the substance:

$$q = -D\frac{\partial c}{\partial x}, \qquad (21.1)$$

in which D is the eddy diffusivity and $\partial c/\partial x$ the concentration gradient.

The 1D mass budget for the passive tracer is

$$\frac{\partial c}{\partial t} = -\frac{\partial q}{\partial x}, \tag{21.2}$$

which simply states that any local convergence or divergence of the flux caused accumulation or depletion of the substance at that point. Note that budget (21.2) is based on sound physical principles (although it can be augmented to include additional processes such as advection by a mean flow, diffusion in more than one direction of space, and sources and sinks caused by chemical reactions). Using parameterization (21.1) for the flux in mass budget (21.2) yields the well-known diffusion equation in one dimension:

$$\frac{\partial c}{\partial t} = \frac{\partial}{\partial x}\left(D\frac{\partial c}{\partial x}\right). \tag{21.3}$$

This equation needs no particular discussion; the purpose of showing it here is only to contrast it later with an alternative formulation. Suffices it to mention that the solution of Equation 21.3 for the evolution of the concentration distribution following an instantaneous (at $t = 0$) and localized (at $x = 0$) release is

$$c(x,t) = \frac{M}{\sqrt{4\pi Dt}}\exp\left(-\frac{x^2}{4Dt}\right), \tag{21.4}$$

if the diffusivity D may be assumed constant over time. The spatial variation of this solution follows the Gaussian distribution, also called the "bell curve."

A salient feature is that the width of the resulting distribution grows in space like the standard deviation of the Gaussian distribution, namely

$$\sigma = \sqrt{2Dt}, \tag{21.5}$$

that is, the width of the patch is predicted to grow like the square root of time. In case of a continuous release with advection by a mean flow, such as the wind blowing a smokestack plume, travelled distance supersedes time, and the prediction is a plume width growing like the square root of downstream distance from the source.

21.3 Failure of Eddy Diffusivity Model

Clearly, this preceding result is in flagrant disagreement with the observed behavior. Figures 21.1 and 21.2 show σ growing more like t than $t^{1/2}$, if not a little faster, and Figure 21.3 shows that D does not remain constant but grows during the evolution of the patch. The inability of the eddy diffusivity model to capture key elements of turbulent dispersion has been recognized for a long time. Richardson (1926) circumvented the issue by proposing that the eddy diffusivity ought to grow like the 4/3 power of the size of the patch. However, a few years later, Richardson and Gaunt (1930) revised the exponent downward by suggesting that the eddy diffusivity ought instead to grow as the first power of the patch size. Richardson was still struggling with the issue when two decades later, with Stommel (Richardson and Stommel 1948), he remarked in the context of marine dispersion that "The variation of K depends on a geometrical quantity σ, and Fick's equation is also geometrical in so far as it contains $\partial^2/\partial x^2$. For this reason it is difficult to regard the variation of K as an outer circumstance detached from Fick's equation. There appears to be a fault in the equation itself." (*Note*: In this quote the quantity K stands for the diffusivity, noted D here.) Notwithstanding their own statement, Richardson and Stommel retained the eddy diffusivity concept and Fick's equation, preferring to suggest that the diffusivity ought to grow like a power of patch size and proposing (in the present notation) $D \sim \sigma^\alpha$ with α varying from 1.0 to 1.4. In a follow-up study of turbulent dispersion in the sea, Stommel (1949) reiterated the conclusion that the Fickian model fails to describe horizontal diffusion in the sea.

Similar comments continued to be voiced over the years casting doubt on the basic validity of the Fickian model, such as those words by Batchelor and Townsend (1956): "Turbulent diffusion is a non local effect ..., and a description of the diffusion with some kind of integral equation is more to be expected." Okubo (1962, 1971) made an effort to find a new law of diffusion but fell short of doing so, returning instead to the idea of a variable eddy diffusivity.

An eddy diffusivity that varies explicitly with patch size may be an adequate descriptor of collected observations, but is wholly inadequate in modeling. Indeed, how could a model be so constructed when the system includes multiple overlapping patches at various stages of development? What should the D value be at a given point in space when an older and larger patch overtakes a more recent and smaller patch? To be effective, a model of turbulent dispersion ought to include a manner by which the vigor of dispersion adapts to the evolution of the spatial distribution of the tracer's concentration.

21.4 Basic Phenomenology

To establish such a model, it is helpful to consider the essence of dispersion. The cause of the process are eddies of various sizes acting on moving, distorting, and stirring the patch (Figure 21.4).

The largest eddies sweep the patch around their orbits and merely displace the patch from place to place with little distortion, while the smallest eddies internally stir the patch without significantly enlarging its size. In contrast, eddies of size comparable to the patch greatly distort the patch, enlarge its overall extent, and cause the most dispersion. While the eddy population may remain statistically unchanged over time, the mere fact that the patch grows in size within this unchanging eddy population implies that eddies take their turn in effecting dispersion, with the smaller eddies acting first when the patch is small and increasingly larger eddies

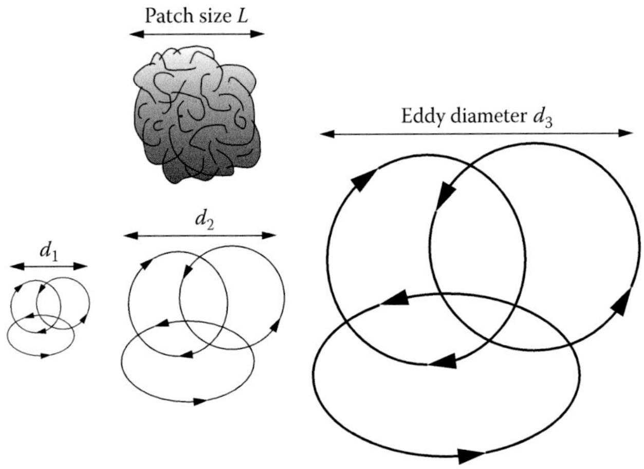

FIGURE 21.4 Contaminant patch of size d in a turbulent flow with eddies of various sizes L. Eddies much larger than the patch ($d_3 \gg L$) merely translate the patch around their orbits, causing displacement but little dispersion. Eddies of size comparable to that of the patch ($d_2 \sim L$) greatly distort the shape of the patch and bring in close proximity previously separated points with differing concentrations, thus causing effective mixing. Finally, eddies of diameter much shorter than the patch ($d_1 \ll L$) stir the inside of the patch while only marginally increasing the patch size along its edges. The conclusion is that the eddies that most contribute to the mixing are those of length scale comparable to the patch size.

acting in sequence as the patch size widens. Thus, the vigor of the dispersion changes with patch size.

The question to contemplate next is the structure of the turbulent field. How are the various eddy sizes characterized? More precisely, for a given eddy diameter d, what is the typical orbital velocity $u_\star$? Here, two primary avenues are offered to us. For homogeneous turbulence, we may invoke the Kolmogorov theory and write (Pope 2000, p. 187)

$$u_\star = (\varepsilon d)^{1/3}, \tag{21.6}$$

in which ε is the energy per mass and per time that is cascading through the turbulent flow field from the largest scale, where it is injected into the fluid, to the shortest scale, at which dissipation occurs. Alternatively, should the situation dictate, we may consider shear turbulence in the proximity of a wall, for which the eddy velocity $u_\star$ is the so-called friction velocity u_τ (Pope 2000, p. 269):

$$u_\star = u_\tau = \sqrt{\frac{\tau}{\rho}} = \text{a constant}, \tag{21.7}$$

where

τ is the stress exerted by the turbulent flow along the wall

ρ is the fluid density

Next, we consider the approximate time duration Δt during which the patch about doubles in size, from L to $2L$. The additional length L is caused by the distorting eddies of diameter d similar to L, which have an orbital velocity on the order of $u_\star(d = L)$. Thus, we write

$$L \sim u_\star(d = L)\Delta t. \tag{21.8}$$

Imagining for a moment that a diffusivity D existed, it would be such that the incremental growth of the patch would scale according to Equation 21.5:

$$L \sim u_\star(d = L)\Delta t \sim \sqrt{D\Delta t}. \tag{21.9}$$

The effective diffusivity that seems to operate on the patch is then

$$D \sim u_\star^2(d = L)\Delta t \sim u_\star(d = L)L \tag{21.10}$$

for the two types of turbulence, that is,

- Homogeneous turbulence

$$D \sim \varepsilon^{1/3}L^{4/3} \tag{21.11a}$$

- Wall turbulence

$$D \sim u_\tau L. \tag{21.11b}$$

Either of these behaviors agrees reasonably well with the data shown in Figure 21.3. ([21.11a] is a better fit when all data sets are considered together, while [21.11b] yields a better fit when each data set is fitted separately.) We note in passing that (21.11a) is none other than the so-called 4/3-law of Richardson mentioned earlier. The variation of the standard deviation σ and variance σ^2 over time, based on Equation 12.5, is again,

- Homogeneous turbulence

$$\sigma \sim \varepsilon^{1/2}\Delta t^{3/2}, \quad \sigma^2 \sim \varepsilon\Delta t^3 \tag{21.12a}$$

- Wall turbulence

$$\sigma \sim u_\tau\Delta t, \quad \sigma^2 \sim u_\tau^2\Delta t^2. \tag{21.12b}$$

With traveled distance playing the role of time, smokestack data (Figure 21.1) agree with (21.12a), and so do Hudson River data (Figure 21.2).

21.5 Heuristic Model

The next step is to seek an equation governing the spatiotemporal distribution of the concentration distribution that obeys either form of turbulence. For this, we again imagine for a moment that a diffusivity D actually exists, in which case the governing equation is (21.3). Passing to its Fourier transform, we write

$$\frac{d\hat{c}}{dt} = -k^2 D\,\hat{c} \tag{21.13}$$

in which $\hat{c}(k, t)$ is the Fourier transform of $c(x, t)$ and k is the wavenumber. If the patch size at time t is on the order of L, the wavenumber of the eddies effecting the dispersion is $k \sim 1/L$, and by virtue of (21.11a) and (21.11b), the operating diffusivity is

- Homogeneous turbulence

$$D \sim \varepsilon^{1/3} k^{-4/3} \tag{21.14a}$$

- Wall turbulence

$$D \sim u_\tau k^{-1} \tag{21.14b}$$

and substitution in budget equation (21.13) yields

- Homogeneous turbulence

$$\frac{d\hat{c}}{dt} = -\varepsilon^{1/3} k^{2/3}\,\hat{c} \tag{21.15a}$$

- Wall turbulence

$$\frac{d\hat{c}}{dt} = -u_\tau k^1\,\hat{c}. \tag{21.15b}$$

The governing equation in the original (x, t) variables is obtained by taking the inverse Fourier transform:

- Homogeneous turbulence

$$\frac{\partial c}{\partial t} = \varepsilon^{1/3}\,\Phi(c) \tag{21.16a}$$

- Wall turbulence

$$\frac{\partial c}{\partial t} = \frac{u_\tau}{\pi} \int_{-\infty}^{+\infty} \frac{c(x-\xi,t) - c(x,t)}{\xi^2}\, d\xi \tag{21.16b}$$

where $\Phi(c)$ is a linear integral operator that is difficult to express in closed form. The integral nature of these equations seems to correspond to what Batchelor and Townsend (1956) wrote, namely that turbulent diffusion is a nonlocal effect, and that "some kind of integral equation is to be expected." An integral equation is not new, however. Einstein (1905) already obtained an integral equation for the particle density in a random-walk process. The interested reader is referred to the extensive review paper on turbulent diffusion scalings by Bakunin (2004, especially Section 10).

21.6 A Somewhat More Formal Model

The preceding derivation of governing equations standing as alternatives to Fickian diffusion was rather ad hoc in the sense that the behavior gleaned from observations was force fed in the formalism to ensure that it would produce the expected outcome. A somewhat more rigorous model follows.

Considering that dispersion at its root is none other than movement by turbulent eddies, we may begin by stating that the concentration distribution $c(x, t)$ is governed by the most simple advection equation:

$$\frac{\partial c}{\partial t} + u\frac{\partial c}{\partial x} = 0, \tag{21.17}$$

in which the velocity u is a fluctuating variable. Let us consider it as a random variable. The solution to (21.17) is obviously

$$c(x, t+\Delta t) = c(x - u\Delta t, t),$$

stating that the distribution after interval of time Δt is the earlier distribution shifted over the distance $u\Delta t$ traveled during that same time interval. The ensemble average of this equation over all realizable values of the random velocity u is

$$c(x, t+\Delta t) = \int_{-\infty}^{+\infty} c(x - u\Delta t, t)\, f(u)\,du, \tag{21.18}$$

in which $f(u)du$ is the probability that the turbulent velocity has an instantaneous value falling in the interval $[u, u + du]$ at location x. Note how this operation is the reverse of what is traditionally done in the wake of Osborne Reynolds' approach to turbulence: while Reynolds first averaged the equation over the turbulent fluctuations and left us to solve it afterward, here the equation is first solved and then averaged over the turbulent fluctuations. No closure problem arises, but a probability distribution function ought to be formulated.

Since all realizable values of u add to a 100% probability of occurrence, it follows that the integral of $f(u)du$ over all possible values of u is unity, and we may write after subtracting $c(x, t)$ from both sides of the equation and dividing by Δt

$$\frac{c(x,t+\Delta t) - c(x,t)}{\Delta t} = \int_{-\infty}^{+\infty} \frac{c(x-u\Delta t,t) - c(x,t)}{\Delta t} f(u)\,du. \tag{21.19}$$

Passing from the velocity as the random variable to the displacement $\xi = u\Delta t$, we write

$$\frac{c(x, t+\Delta t) - c(x, t)}{\Delta t} = \int_{-\infty}^{+\infty} \frac{c(x-\xi, t) - c(x, t)}{\Delta t} g(\xi, \Delta t)\, d\xi, \qquad (21.20)$$

in which $g(\xi, \Delta t)\Delta\xi$ is the probability of a displacement of length within interval $[\xi, \xi + \Delta\xi]$ after interval of time Δt, which we shall call the jump probability distribution function.

Because time is a continuous variable, we could conceivably imagine shorter or longer time intervals and, at the limit, continuous-time random walk (Klafter et al. 1987). It is more expedient here, however, to consider that the jump probability distribution must possess a self-replicating property, which in the language of statistics is called the divisibility property. Mathematically, we require that the probability of a jump of length ξ after a double time interval $2\Delta t$ is equal to the probability of making a first partial step over distance ξ' during a first time interval Δt and then the complementary step $\xi - \xi'$ over a second time interval Δt, with the *same* probability distribution, that is,

$$g(\xi, 2\Delta t) = \int_{-\infty}^{+\infty} g(\xi', \Delta t) g(\xi - \xi', \Delta t)\, d\xi'. \qquad (21.21)$$

Dimensional analysis requires that the function $g(\xi, \Delta t)$ have the dimension of $1/\xi$, the inverse of a length. Therefore, the parameter Δt, which has the dimension of time, must be combined, somehow, with a physical variable that connects length and time. Returning to our two types of turbulence, we invoke the rate of energy cascade ε (with dimensions L^2T^{-3}) or the friction velocity u_τ (with dimension LT^{-1}). Adding the diffusivity D (of dimension L^2T^{-1}) for comparison, we define the following assortment of dimensionless quantities:

- Homogeneous turbulence

$$a = \frac{\xi}{\varepsilon^{1/2}\Delta t^{3/2}} \rightarrow g(\xi, \Delta t) = \frac{1}{\varepsilon^{1/2}\Delta t^{3/2}} G(a) \qquad (21.22a)$$

- Wall turbulence

$$a = \frac{\xi}{u_\tau \Delta t} \rightarrow g(\xi, \Delta t) = \frac{1}{u_\tau \Delta t} G(a) \qquad (21.22b)$$

- Diffusive model

$$a = \frac{\xi}{\sqrt{2D\Delta t}} \rightarrow g(\xi, \Delta t) = \frac{1}{\sqrt{2D\Delta t}} G(a). \qquad (21.22c)$$

We note, interestingly enough, that the power of Δt in the denominator is decreasing from 3/2 to 2/2 to 1/2 as we move from one choice to the next. Using the respective dimensionless variable, we recast the divisibility requirement in the following generic form:

$$\frac{1}{\lambda} G\left(\frac{a}{\lambda}\right) = \int_{-\infty}^{+\infty} G(a') G(a - a')\, da', \qquad (21.23)$$

in which the denominator λ takes the values $2\sqrt{2}$, 2, and $\sqrt{2}$ for the cases of homogeneous turbulence, wall turbulence, and diffusive model, respectively. For each value of λ, the solution is universal since there is no parameter remaining. Calculations are rather tedious, but the solutions in the last two cases can be expressed in closed form:

- Wall turbulence

$$G(a) = \frac{1}{\pi} \frac{a_0}{a^2 + a_0^2} \quad \text{with} \quad a = \frac{\xi}{u_\tau \Delta t} \qquad (21.24a)$$

- Diffusive model

$$G(a) = \frac{1}{\sqrt{2\pi} a_0} \exp\left(-\frac{a^2}{2a_0^2}\right) \quad \text{with} \quad a = \frac{\xi}{\sqrt{2D\Delta t}} \qquad (21.24b)$$

in each of which a_0 is an arbitrary dimensionless constant. These two functions are, respectively, the Cauchy and Gaussian probability distributions, each known to be divisible. The solution in the case of homogeneous turbulence has not been found to be expressible in closed form (Monin and Yaglom 1975), and we shall simply denote it $G_h(a)$, in which a stands for $a = \xi/\varepsilon^{1/2}\Delta t^{3/2}$.

Now, implementing these forms for the jump probability distribution function into Equation 21.20, we obtain

- Wall turbulence

$$\frac{\Delta c}{\Delta t} = \int_{-\infty}^{+\infty} \frac{c(x-\xi, t) - c(x, t)}{\Delta t} \frac{1}{\pi} \frac{a_0 u_\tau \Delta t}{\xi^2 + (a_0 u_\tau \Delta t)^2}\, d\xi \qquad (21.25a)$$

- Diffusive model

$$\frac{\Delta c}{\Delta t} = \int_{-\infty}^{+\infty} \frac{c(x-\xi, t) - c(x, t)}{\Delta t} \frac{1}{a_0\sqrt{4\pi D\Delta t}} \exp\left(-\frac{\xi^2}{4a_0^2 D\Delta t}\right) d\xi. \qquad (21.25b)$$

The limit of vanishing Δt is not singular by virtue of the divisibility property, and we find after some algebra (trivial in the first case but lengthy in the second case)

- Wall turbulence

$$\frac{\partial c}{\partial t} = \frac{a_0 u_\tau}{\pi} \int_{-\infty}^{+\infty} \frac{c(x-\xi, t) - c(x, t)}{\xi^2}\, d\xi \qquad (21.26a)$$

- Diffusive model

$$\frac{\partial c}{\partial t} = a_0^2 D \frac{\partial^2 c}{\partial x^2}. \qquad (21.26b)$$

The dangling constant a_0 can be absorbed in a trivial rescaling of the respective dimensional constant (u_τ or D), and we recover Equations (21.16b) and (21.3), respectively. That the second equation is none other than the Fickian model amounts to the conclusion that molecular diffusion is equivalent to random walk with jumps following a Gaussian (normal) distribution, a result echoing that of Einstein (1905). Needless to say, the treatment of the case of homogeneous turbulence has not been found amenable to algebra, and we shall here leave it at the symbolic notation:

- Homogeneous turbulence

$$\frac{\partial c}{\partial t} = \varepsilon^{1/3}\,\Phi(c), \qquad (21.27a)$$

in which the operator $\Phi(c)$ is to be obtained from the limit

$$\Phi(c) = \lim_{\Delta t \to 0} \int_{-\infty}^{+\infty} \frac{c(x-\xi,t) - c(x,t)}{(\varepsilon \Delta t^3)^{5/6}} G\left(\frac{\xi}{\sqrt{\varepsilon \Delta t^3}}\right) d\xi \qquad (21.27b)$$

in which the function G is the solution to the integral equation (21.23) with $\lambda = 2\sqrt{2}$.

21.7 Wall Turbulence

Since the Fickian model is inappropriate in turbulent dispersion studies and since the formalism in the case of homogeneous turbulence is not tractable analytically, let us here pursue only the case of wall turbulence. The governing equation which we obtained, namely (21.16b), has been found to correspond to dispersion accomplished by a random-walk process in which the steps follow the Cauchy probability distribution, in dimensional form:

$$g(\xi, \Delta t) = \frac{1}{\pi} \frac{u_\tau \Delta t}{\xi^2 + u_\tau^2 \Delta t^2}, \qquad (21.28)$$

in which the interval of time Δt does not matter because this distribution is infinitely divisible.* The corresponding probability distribution function for the random velocity fluctuations is

$$f(u) = \frac{1}{\pi} \frac{u_\tau}{u^2 + u_\tau^2}. \qquad (21.29)$$

* Note that although the factor $u_\tau \Delta t$ is a measure of the width of the distribution, the variance of this distribution is infinite because of its "heavy tails."

The spread of a patch following an instantaneous and localized release of mass M is

$$c(x,t) = \frac{M}{\pi} \frac{u_\tau t}{x^2 + (u_\tau t)^2}. \qquad (21.30)$$

The flux q that corresponds to the budget equation (21.2) is

$$q(x,t) = \frac{u_\tau}{\pi} \int_{-\infty}^{x} \int_{x}^{+\infty} \frac{c(x',t) - c(x'',t)}{(x'-x'')^2} dx' dx''. \qquad (21.31)$$

Note how this flux, which is nonlocal, includes all possible jumps across the position x where it is evaluated, from any position x' to the left of x to any position x'' to the right of x (Figure 21.5).

It is most interesting to apply the preceding formalism to the quintessential example of shear turbulence, namely to the turbulent velocity profile along a wall. To do this, we adapt the model as follows: the flow is considered to be statistically steady (vanishing time derivative $\partial/\partial t$), the spatial variable x is replaced by distance z from the wall and is limited to the semi-infinite domain $[0, +\infty]$, and the concentration variable $c(x, t)$ is now taken as the fluid momentum $u(z)$. The equation governing the spatial structure of $u(z)$ is obtained from (21.16b), which after the present adaptations becomes

$$0 = \frac{u_\tau}{\pi} \int_{0}^{+\infty} \frac{u(z-\xi) - u(z)}{\xi^2} d\xi. \qquad (21.32)$$

The solution to this equation is

$$u(z) = \frac{u_\tau}{\kappa} \ln \frac{z}{z_0}, \qquad (21.33)$$

in which, at this stage, κ and z_0 are two arbitrary constants. We readily recognize the logarithmic law of the wall for the velocity profile (Pope 2000, pp. 274, 296), with κ being the von Karman constant and z_0 the roughness height of the wall surface.

It ought to be noted that, when momentum plays the role of concentration, the "contaminant" is no longer a passive tracer. The velocity distribution defines the turbulent flow in which it mixes itself. To account for this realization, the model can be generalized to make the friction velocity u_τ depend on the flow field u (see Cushman-Roisin and Jenkins 2006). The logarithmic velocity profile survives this generalization.

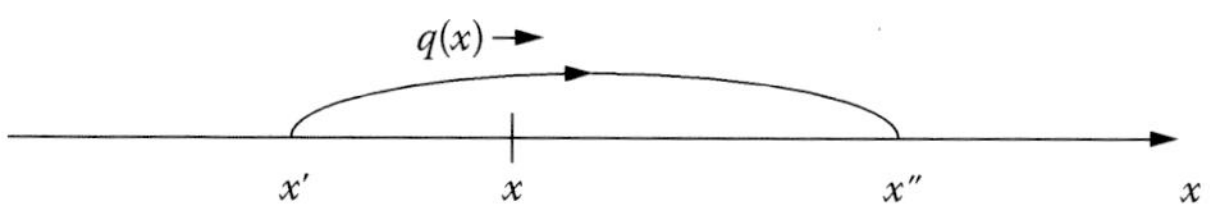

FIGURE 21.5 Depiction of the nonlocal flux associated with wall turbulence.

Generalization to higher dimensions is relatively evident, and we only cite the results here:

1D

$$\frac{\partial c}{\partial t} = \frac{u_\tau}{\pi} \int \frac{c(x-\xi, t) - c(x, t)}{\xi^2} d\xi. \quad (21.34)$$

2D

$$\frac{\partial c}{\partial t} = \frac{u_\tau}{2\pi} \iint \frac{c(x-\xi, y-\eta, t) - c(x, y, t)}{(\xi^2 + \eta^2)^{3/2}} d\xi\, d\eta. \quad (21.35)$$

3D

$$\frac{\partial c}{\partial t} = \frac{u_\tau}{\pi^2} \iiint \frac{c(x-\xi, y-\eta, z-\varsigma, t) - c(x, y, z, t)}{(\xi^2 + \eta^2 + \varsigma^2)^2} d\xi\, d\eta\, d\varsigma. \quad (21.36)$$

As one can readily see from these expressions, one needs to integrate over all possible jumps in 1D, 2D, or 3D space and with a probability distribution function that tapers with distance in a manner that is dimensionally consistent. The numerical factors in front of the integrals are chosen such that, if there is uniformity of c in one direction, performing the integration over that direction yields the coefficient used in the expression for the next lower dimension.

Addition of other terms, such as those representing source, decay, and advection by a mean flow, is straightforward.

21.8 Connection with Fractional Dispersion Equation

The preceding formalism with its three options (diffusive model, shear turbulence, and homogeneous turbulence) may be generalized into a single approach with a sliding factor λ and, by so doing, can be connected with the fractional dispersion equation studied by Schumer et al. (2001) and Meerschaert et al. (2006). That generalized equation may be cast in terms of fractional derivatives, with the term on the right-hand side of the budget equation (21.26a,b) being expressed as the αth derivative of c, with α being a fraction. These fractional derivatives, however, are actually nonlocal expressions consisting, as mentioned earlier, of an integral over space of concentration differences times a certain weighing coefficient vanishing at large distance more or less rapidly depending on the value of the exponent.

The α values of the three models pursued in Section 21.6 are related to the sliding factor λ by

$$\lambda = 2^{1/\alpha}, \quad (21.37)$$

and, thus, correspond to the various cases as follows: $\alpha = 2$ for the Fickian diffusive model with Gaussian probability distribution function, $\alpha = 1$ for the wall turbulence model with Cauchy probability distribution function, and $\alpha = 2/3$ for the Kolmogorov homogeneous turbulence model.

Fractional-derivative models have found applications in dispersion through porous media (e.g., Cushman et al. 2005), where dispersion is not effected by turbulence but rather by the fractal nature of the geometry in which many physical scales coexist. This multiplicity of scales may justify the use of empirically deduced fractional exponents, but in fluid turbulence as encountered in the atmosphere and natural water bodies, one would be hard pressed to justify the use of a physical quantity with fractional dimensions that are inexplicable powers of space and time.

21.9 Epilogue

As we continue to lack a solid theory for fluid turbulence (pun intended), it is clear that the last word on turbulent dispersion has not yet been said. The topic will remain the object of investigation and subject of lively debate for the undetermined future. It is hoped, however, that the preceding considerations will prove useful to the computer modeler whose task is not an inquiry into turbulent dispersion per se but the simulation or prediction of some more general system, in which turbulent dispersion plays some role and for which it is desirable to overcome the fundamental inadequacy of the eddy diffusivity model. Such applications abound in weather forecasting, coastal ocean dynamics, and contaminant transport and fate.

References

Bakunin, O. G. 2004. Correlation effects and turbulent diffusion scalings. *Rep. Prog. Phys.* 67: 965–1032.

Batchelor, G. K. and A. A. Townsend. 1956. Turbulent diffusion. In *Surveys in Mechanics*, eds. G. K. Batchelor and R. M. Davies, Cambridge University Press, Cambridge, U.K., pp. 352–399.

Boubel, R. W., D. L. Fox, D. B. Turner, and A. C. Stern. 1994. *Fundamentals of Air Pollution*. Academic Press, New York, 574pp.

Boussinesq, J. 1877. Essai sur la théorie des eaux courantes [Essay on the theory of flowing waters]. *Mém. l'Acad. Sci. Inst. France, Paris*, 23: 252–260.

Clark, J. F., P. Schlosser, M. Stute, and H. J. Simpson. 1996. SF^6–^3He tracer release experiment: A new method of determining longitudinal dispersion coefficients in large rivers. *Environ. Sci. Technol.* 30: 1527–1532.

Cushman, J. H., M. Park, and N. Kleinfelter. 2005. Super-diffusion via Lévy lagrangian velocity processes. *Geophys. Res. Lett.* 32. DOI: 10.1029/2005GL023645, pp. 1–4.

Cushman-Roisin, B. and A. D. Jenkins. 2006. On a non-local parameterization for shear turbulence and the uniqueness of its solutions. *Boundary Layer Meteor.* 118: 69–82.

Einstein, A. 1905. Über die von der molekularkinetischen Theorie der Wärme geforderte Bewegung von in ruhenden Flüssigkeiten suspendierten Teilchen (On the motion required by the molecular kinetic theory of heat of small particles suspended in stationary liquid). *Ann. Phys.*, 322.4 Folge, 17: 549–560.

Fick, A. 1855. On liquid diffusion. *Philos. Mag. J. Sci.* 10: 31–19.

Fischer, H. B., E. J. List, R. C. Y. Koh, J. Imberger, and N. H. Brooks. 1979. *Mixing in Inland and Coastal Waters*. Academic Press, New York, 483pp.

Gifford, F. A. 1961. Use of routine meteorological observations for estimating atmospheric dispersion. *Nucl. Safety* 2: 47–51.

Klafter, J., A. Blumen, and M. F. Schlesinger. 1987. Stochastic pathway to anomalous diffusion. *Phys. Rev. A* 35: 3081–3085.

Meerschaert, M. M., J. Mortensen, and S. W. Wheatcraft. 2006. Fractional vector calculus for fractional advection-dispersion. *Physica A* 367: 181–190.

Min, I. A., R. N. Abernathy, and H. L. Lundblad. 2002. Measurement and analysis of puff dispersion above the atmospheric boundary layer using quantitative imagery. *J. Appl. Meteorol.* 41: 1027–1041.

Monin, A. S. and A. M. Yaglom. 1975. *Statistical Fluid Mechanics: Mechanics of Turbulence*, Vol. 2. Mass MIT Press, Cambridge, MA, 896pp.

Okubo, A. 1962. A review of theoretical models for turbulent diffusion in the sea. *J. Oceanogr. Soc. Japan*, 20th Anniversary Volume: 286–320.

Okubo, A. 1971. Horizontal and vertical mixing in the sea. In *Impingement of Man on the Oceans*, ed. D. W. Wood, John Wiley & Sons, New York, pp. 89–168.

Okubo, A. 1974. Some speculations on oceanic diffusion diagrams. *Rapp. Proc. Verb. Cons. Int. Explor. Mer.* 167: 77–85.

Pasquill, F. 1961. The estimation of the dispersion of windborne material. *Meteorol. Mag.* 90: 33–49.

Pope, S. B. 2000. *Turbulent Flows*. Cambridge University Press, Cambridge, U.K., 771pp.

Richardson, L. F. 1926. Atmospheric diffusion shown on a distance-neighbour graph. *Proc. R. Soc. Lond. A* 110: 709–737.

Richardson, L. F. and J. A. Gaunt. 1930. Diffusion regarded as a compensation for smoothing. *Mem. R. Meteorol. Soc.* 3(30): 171–175.

Richardson, L. F. and H. Stommel. 1948. A note on eddy diffusion in the sea. *J. Meteorol.* 5: 238–240.

Roberts, J. W. and D. R. Webster. 2002. Turbulent diffusion. In *Environmental Fluid Mechanics—Theories and Applications*, eds. H. H. Shen, A. H. D. Cheng, K.-H. Wang, M. H. Teng, and C. C. K. Liu, American Social of Civil Engineers, Reston, VA, pp. 7–45.

Schumer, R., D. A. Benson, M. M. Meerschaert, and S. W. Wheatcraft. 2001. Eulerian derivation of the fractional advection-dispersion equation. *J. Contam. Hydrol.* 48: 69–88.

Stommel, H. 1949. Horizontal diffusion due to oceanic turbulence. *J. Mar. Res.* 8: 199–225.

22

Stratified Hydraulics

Peter G. Baines
University of Melbourne

22.1 Introduction

The science of hydraulics of density-stratified flows is concerned with the flow of density-stratified fluids through regions in which the conditions vary in the direction of flow. The subject is conceptually based on single-layer hydraulics of a homogeneous fluid that describes flows in channels and rivers. It is essentially concerned with 1D or channel flows which may have varying width and depth, in which the fluid has variable speed and density in the vertical direction, but with vertical and lateral scales that are much smaller than along-channel scales. Under such circumstances, the pressure within the moving fluid is very nearly hydrostatic, namely that it is given by the weight of the fluid lying above it.

Stratified flows of this type occur in a variety of situations in the environment. They may occur in rivers where lower level fluid is more dense (colder or saltier) than upper levels, in estuaries where river flow encounters the ocean, in long stratified lakes where wind-driven seiches occur, in shallow straits and channels in the ocean such as at Gibraltar and the Bosphorus, and in deep ocean channels that control the flow of bottom waters. In some cases such as at Gibraltar and the Bosphorus, the flow may be approximated by two homogeneous layers that flow in opposite directions. These are known as *exchange flows*. This approximation may also give a useful description of the exchange flow through open doorways and windows that separate two air masses with different temperatures.

Compared with the single-layer case, density stratification introduces a number of new concepts and processes: internal wave propagation and reflection, critical flow conditions for different wave modes, critical layers, rarefactions and supercritical leaps, and flow blocking. Two-layer flows contain some of these phenomena, and these are described first. Since density variations within the fluid are much smaller than those at the free surface, the dominant wave speeds are much smaller than long surface waves, and flow states take longer to evolve than the corresponding ones in single-layer hydraulics. The effects on horizontal velocities may, however, be just as large.

If the effect of rotation (specifically, the Earth's rotation) is significant in these flows, an extra dimension of complexity is added. This has been omitted from this review because it involves flows on larger length scales than is the focus of this volume, and the reader is referred to the volume by Pratt and Whitehead (2008).

22.2 Principles

Stratified hydraulics is concerned with quasi-horizontal flows that pass through a region in which the flow conditions vary (usually in a gradual manner) from one horizontal configuration to another. These conditions on each side may be the same, as for flow over an isolated topographic feature or through an isolated contraction in a channel, or they may differ. Further, the flow may be down a gentle slope, where the buoyancy force is balanced in some way by friction.

The internal waves that propagate within these flows are essential features of these dynamics. These waves consist of several modes with different vertical structure, but the ones with the most significance are those that have small speeds relative to the stationary features of the region of interest. Long internal waves have the effect of altering the structure and velocity of the flows through which they propagate, and this embodies the difficulty and interest of the subject. Under appropriate circumstances,

Handbook of Environmental Fluid Dynamics, Volume One, edited by Harindra Joseph Shermal Fernando. © 2013 CRC Press/Taylor & Francis Group, LLC.
ISBN: 978-1-4398-1669-1.

the flow may be controlled by a critical condition of an internal wave mode at, say, an obstacle crest or contraction minimum, as happens for single-layer hydraulics, but critical conditions may occur at other places also, and for different reasons. There is a wide variety of different possibilities, and these are described in the next section. The means of analysis are based on the equations for long disturbances of finite amplitude in stratified shear flows, with the equations for small-amplitude (linear) disturbances being invoked where appropriate.

A stratified shear flow may be represented by a finite number of homogeneous layers, each with uniform horizontal velocity. In most environmental flows this amounts to an approximation to a continuously stratified fluid, because immiscible fluids that maintain sharp density differences are rare in nature. Continuously stratified fluids are characterized by the buoyancy frequency N, and Richardson number R_i, where

$$N^2 = -\frac{g}{\rho}\frac{d\rho}{dz}, \quad R_i = \frac{N^2}{(dU/dz)^2}, \tag{22.1}$$

where
- g is gravity
- z the vertical coordinate
- ρ and U the fluid density and horizontal velocity

N and R_i are both functions of position. N is the natural frequency of vertical oscillations in the fluid, and the Richardson number compares it wth the horizontal shear, which tends to oppose internal wave propagation. If the shear is more than twice the value of N so that $R_i < ¼$, internal waves are not able to propagate through that region of the fluid (see Baines 1995). Under such circumstances the fluid may be unstable to shear flow instabilities, which introduces further complications.

The hydraulics of a continuously stratified miscible fluid is governed by the properties of long internal waves that propagate through these flows. In general, it may be shown that there are two denumerably infinite sets of normal modes of such internal waves, with speeds $c_{\pm 1}, c_{\pm 2}, c_{\pm 3}, \ldots$ that satisfy

$$0 < U_{max} < \cdots < c_3 < c_2 < c_1 < U_{max} + \frac{a_1 N_{max} D}{\pi},$$

$$U_{min} - \frac{a_2 N_{min} D}{\pi} < c_{-1} < c_{-2} < c_{-3} < \cdots < U_{min}, \tag{22.2}$$

where
- a_1, a_2 are constants
- $N_{max}, U_{max}, N_{min}, U_{min}$ denote the maximum and minimum values of buoyancy frequency N and fluid velocity U, respectively, in the range $0 < z < D$

Clearly, this means that the first set of normal wave modes have speeds that exceed the maximum fluid speed, with the latter as a limit point, and the velocities of the second set are all less than U_{min} with the latter as a limit point. No modes have speeds that lie within the range of speeds of the fluid. Internal waves that do have such speeds encounter critical layers, or levels, which occur at levels where the phase speed of the wave equals the fluid speed. If the wave amplitude is small, the wave is mostly absorbed at the critical layer, with its associated momentum being transferred to the fluid. Alternatively, if it is large enough mixing occurs, and this tends to result in a homogeneous layer at this level. Such difficulties may be accommodated by using layered models. Models with n layers have $2n$ modes (or $2(n - 1)$ with a rigid upper boundary), and all wave speeds are permitted. Wave modes with speeds equal to that of a given layer are confined to one side of that layer. An important parameter in many hydraulics problems is the Froude Number F, which is usually defined to be a fluid speed divided by a wave speed. In stratified hydraulics, these may be the mean fluid speed and the most relevant wave speed from the above set, but relative to the mean flow. The latter is likely to be the wave that is opposed to the mean flow with net speed (as given in [22.2]) closest to zero.

A notable feature of single-layer hydraulics is the presence of hydraulic jumps. These may also occur in stratified or layered flows, but only under certain circumstances that depend on the properties of linear disturbances. They have only been identified and analyzed in detail in simple flows, and mostly two-layer flows. Large amplitude stationary hydraulic jumps have been observed in the atmosphere downstream of mountain ranges (Grubišić et al. 2008), but these may be equivalent to two-layer flows with a deep upper layer. The inflows to these jumps may be highly sheared (with $R_i < ¼$), and these conditions do not permit internal wave propagation and associated baroclinic structure. Hydraulic jumps in stratified fluids dissipate energy and mix fluid but may be analyzed in the same manner as for single layers—by invoking the conservation of mass and momentum flux through the jump. Further, in single-layer hydraulics the introduction of some feature such as a sill or contraction may alter the flow on the upstream side, as well as downstream. The same happens in the stratified case, where an upstream density and velocity profile normally cannot be arbitrarily chosen, as this flow is affected by conditions further downstream. Given this, an appropriate way to obtain valid flow solutions is to solve the appropriate initial value problem, in which the net fluid transport is initially uniform and specified. The baroclinic features of the flow may then develop with time. The problem may be simplified by using the single-layer concepts of critical flow and hydraulic jumps, or rarefactions, as described in the following. We proceed to examine a variety of flow situations that illustrate the phenomena mentioned, beginning with two-layer flows.

22.3 Methods of Analysis

22.3.1 Two-Layer Hydraulics

The flow of two homogeneous layers of different densities provides a useful model for many situations of stratified hydraulics, and we consider the case where the fluid is assumed to have a rigid horizontal upper boundary. This gives similar results to those for situations where the upper boundary is a free surface, and removes the complexity (and additional algebra) associated with the presence of the external or barotropic mode. In a context where the motion of the free surface is important, one may proceed by treating the whole fluid as a single layer, and first establish the resulting barotropic flow. Once this is done, the internal hydraulics may be examined by assuming that the free surface and barotropic flow (i.e., the net mass transport) are determined.

We consider the equations for two-layer hydrostatic flow in a horizontal channel that has a rigid upper surface but varying depth and width with horizontal position x. Scale analysis shows that such flow may be assumed to be (mostly) hydrostatic if the horizontal length scale L of the variations and mean channel depth D and width W are such that $(D/L)^2 \ll 1$, $(W/L)^2 \ll 1$. The equations of motion are then

$$\frac{\partial u_1}{\partial t}+\frac{\partial}{\partial x}\left[\frac{1}{2}u_1^2+g'\eta+\frac{p_s}{\rho_1}\right]=0,\quad g'=\frac{g\Delta\rho}{\rho_1},\quad \Delta\rho=\rho_1-\rho_2, \tag{22.3}$$

$$\frac{\partial u_2}{\partial t}+\frac{\partial}{\partial x}\left[\frac{1}{2}u_2^2+\frac{p_s}{\rho_2}\right]=0, \tag{22.4}$$

$$\frac{\partial A_i}{\partial t}+\frac{\partial}{\partial x}(u_iA_i)=0,\quad i=1,2, \tag{22.5}$$

where

- $i = 1, 2$ denote the lower and upper layers, respectively
- g is gravity
- u_i, ρ_i, and A_i are the velocity density and cross-sectional area of the ith layer
- η is the vertical displacement of the interface from an initial reference level
- p_s denotes the pressure at the upper boundary

Both channel depth and width may vary along and across the channel, but the density in each layer is constant, and the layer velocities and interface level are assumed to be uniform across the channel. Since the barotropic flow is assumed to be known or determined, the total mass flux through all sections of the channel is

$$Q=u_1A_1+u_2A_2=\text{constant}, \tag{22.6}$$

and

$$A_1+A_2=A(x)=\int b(x,z)\mathrm{d}z, \tag{22.7}$$

where the total cross-sectional area $A(x)$ is specified by the varying breadth $b(x, z)$ of the channel, with z the vertical coordinate. Hence, these equations may be used to treat two-layer flows in channels with arbitrarily slowly varying cross sections. Specific situations are described in the following to illustrate various phenomena governed by these equations. For these the equations are not solved here, but the method of solution is indicated.

The behavior of hydraulic systems is dominated by the properties of long gravity waves in the system. From Equations 22.3 through 22.5 one may obtain the velocities of long linear waves in a two-layer system of total depth D. Relative to the mean velocity $\bar{U} = Q/D$, these wave speeds depend on the density difference, the position of the interface, and the velocities of the layers. With the Boussinesq approximation (in which the density difference between the layers is assumed to be small, and is only important in the buoyancy term), the two long wave speeds may be written as

$$c_{\pm}=\bar{U}+(g'D)^{1/2}[V(1-2r)\pm((1-V^2)r(1-r))^{1/2}], \tag{22.8}$$

where r denotes the interface position and V the shear, defined respectively by $r = d_1/D$, $V = (u_1 - u_2)/(g'D)^{1/2}$. When the interface is near mid-depth, the wave speed is dominated by propagation and is slowed by increasing shear; but when the interface is near a boundary, it is dominated by the advective speed of the thin layer near the (upper or lower) surface.

22.3.2 Two-Layer Hydraulic Jumps

Hydraulic jumps may occur in stratified flows through a convergence of internal wave energy to form a coherent finite-amplitude disturbance that maintains its overall structure. Such jumps may be stationary relative to some topographic feature, or propagate at their own speed. In general, hydraulic jumps dissipate energy in some manner, but relative to axes moving with the jump the fluxes of mass and momentum are conserved. Energy dissipation normally occurs through the process of mixing at the interface between the two fluids. This has the tendency to create a third, mixed layer, which tends to invalidate the two-layer model if the effect is large enough. Further, observed jumps tend to be spread out over a horizontal distance, and may involve an undular structure, with waves and mixing. However, for practical purposes we consider an idealized structure with notation as defined in Figure 22.1. Upper and lower boundaries are horizontal and frictionless. With flow in both layers from left to right, subscripts "u" and "d" denote upstream and downstream, respectively, and with total depth D from conservation of mass we have

$$d_{1d}+d_{2d}=D=d_{1u}+d_{2u}, \tag{22.9}$$

$$d_{id}u_{id}=q_i=d_{iu}u_{iu},\quad i=1,2, \tag{22.10}$$

where q_i denotes the volume flux in the ith layer relative to the jump. The conservation of momentum flux S across the jump, where S is defined by

$$S=\int_0^D p+\rho u^2 dz, \tag{22.11}$$

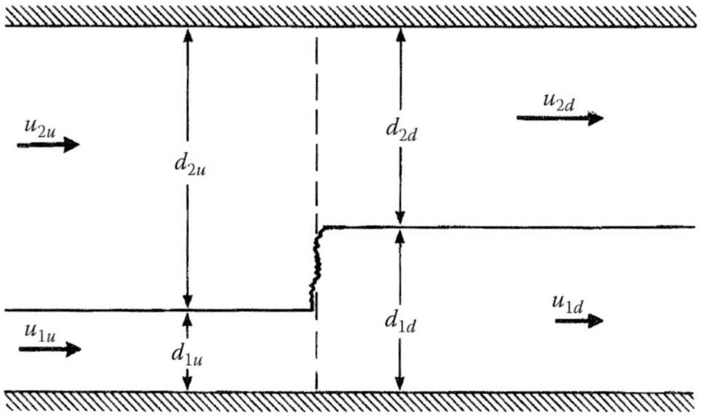

FIGURE 22.1 Schematic diagram of the flow showing notation in coordinates moving with a hydraulic jump.

yields

$$S = p_{su}D + \tfrac{1}{2}\Delta\rho g d_{1u}^2 + \rho_1 u_{1u}^2 d_{1u} + \rho_2 u_{2u}^2 d_{2u}$$
$$= p_{sd}D + \tfrac{1}{2}\Delta\rho g d_{2d}^2 + \rho_1 u_{1d}^2 d_{1d} + \rho_2 u_{2d}^2 d_{2d}, \quad (22.12)$$

where p_s denotes the pressure at the upper boundary. This introduces the new variable $p_{su} - p_{sd}$, and we need another equation in order to be able to determine the downstream variables in terms of the values upstream. Three different ways have been proposed for doing this, and the most appropriate depends on the circumstances. The first method (denoted YG, due to Yih and Guha [1955]) involves conserving mass and momentum flux in each layer and taking averages across the jump, and although the results are not radically different from those of the other methods, it is now the least preferred. The second method (denoted CBWS, due to Chu, Baddour, Wood and Simpson) assumes that all the energy dissipation in the jump occurs in the expanding layer, and the third method (denoted KRS, due to Klemp et al. [1997]) assumes that it all occurs in the contracting layer. In practice, one expects energy dissipation to occur in both layers, but it is reasonable that one layer may dominate; and when this assumption is made, the Bernoulli equation may be applied at the appropriate boundary in the layer where the dissipation does not occur. If we assume that the expanding layer is the lower one (the more common situation), the appropriate relations are

$$p_{su} - p_{sd} = \frac{1}{2}\rho_2\left(u_{2d}^2 - u_{2u}^2\right), \text{ CBWS, lower layer dissipation} \quad (22.13)$$

and

$$p_{su} - p_{sd} = \frac{1}{2}\rho_1\left(u_{2d}^2 - u_{2u}^2\right) + \Delta\rho g(d_{1d} - d_{1u}),$$
$$\text{KRS, upper layer dissipation.} \quad (22.14)$$

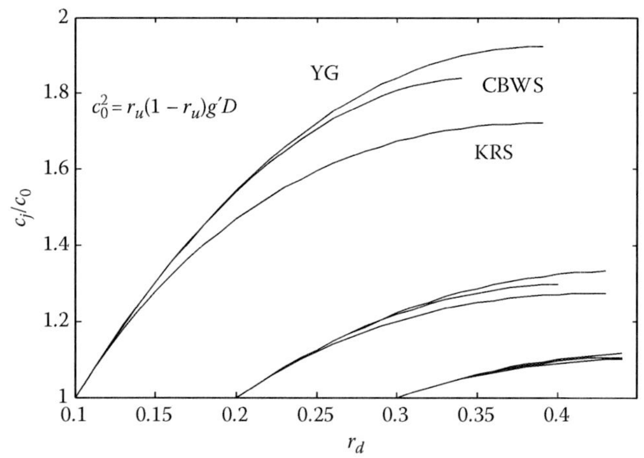

FIGURE 22.2 The speeds c_j of hydraulic jumps propagating in a two-layer Boussinesq fluid at rest, relative to the linear internal wave speed c_0, for the three formulations: KRS, CBWS, and YG. The lower layer is the more shallow, with $r_u = d_{1u}/D$, $r_d = d_{2d}/D$, and the curves are for $r_u = 0.1, 0.2, 0.3$.

In situations in which the hydraulic jump takes the form of a bore that is propagating into stationary or weakly sheared environments, the KRS form gives the better agreement with experimental observations, and is also consistent with expressions for long-wave and gravity current (where the upstream lower layer thickness is zero) speeds. But in situations where the upstream flow is highly sheared, such as where stationary hydraulic jumps occur on the lee side of a topographic feature, the CBWS form seems more appropriate. From Equations 22.9, 22.10, 22.12, and 22.13 or 22.14, the downstream flow states may be obtained in terms of the upstream ones, and some examples of the speeds for bores propagating into fluid at rest are given in Figure 22.2.

22.3.3 Flow over an Isolated Obstacle

We next consider flow over an obstacle in a channel of uniform breadth. The relevant equations are (22.3) through (22.5), with breadth $b(x, z)$ specified as

$$b(x,z) = B, \quad 0 \le h(x) < z < D,$$

so that

$$A_1 = B(d_{10} - h(x) + \eta), \quad A_2 = B(d_{20} - \eta), \quad (22.15)$$

in an overall channel of height D and breadth B, lower boundary specified by $z = h(x)$, and interface displacement η. d_{10} and d_{20} are the undisturbed thicknesses of layers 1 and 2. For steady flows governed by these equations, one may deduce the "hydraulic alternative," namely that at a point where $dh/dx = 0$, one must have

$$\frac{d}{dx}(u_1, u_2, \eta) = 0, \quad \text{or} \quad c_- = 0, \quad (22.16)$$

where c_- is here defined by (22.8) but using *local* values of D, r, U, and V above the point where $dh/dx = 0$. Hence, as for single-layer hydraulics, the flow must either be symmetric about such a point or else critical, meaning that the upstream-propagating (i.e., against the mean flow) wave is stationary there. This criterion is useful in describing the properties of these flows.

We consider the specific case of two-layer flow that is abruptly set into constant motion with speed U from a state of rest, which is dynamically equivalent to an obstacle being moved at speed U through a stationary two-layer fluid. This initial flow state may be characterized by three dimensionless parameters:

$$r = \frac{d_{10}}{D}, \quad H_m = \frac{h_m}{d_{10}}, \quad F_0^2 = \frac{U^2}{r(1-r)g'D}, \tag{22.17}$$

where

h_m is the maximum height of $h(x)$

F_0 is the Froude number of the undisturbed flow

As for single-layer hydraulics, this Froude number is defined to be the mean flow speed divided by the long-wave speed, here given by (22.8) with $V = 0$. A variety of different flow states may occur depending on the values of these parameters. For obstacles of sufficiently small height, a steady-state flow is established after initial transient disturbances propagate away, and the flow is symmetric about the obstacle crest. Regions where this occurs are depicted as clear regions in Figure 22.3, as functions of H_m and F_0 for a range of values of r. In the regions marked "subcritical," the transients propagate both upstream and downstream, whereas in the regions marked "supercritical," they are all on the downstream side. In the shaded regions the flow is more complex. On the solid lines marking the boundary with the sub- and supercritical regions, the propagation speed c_- from (22.8) is zero, so that the flow is "critical" in conventional terminology. Inside the shaded regions, the commencement of flow causes disturbances to propagate upstream, which permanently alter the upstream flow seen by the obstacle. This process is controlled by the critical condition $c_- = 0$ at the obstacle crest. If the lower layer thickness is small ($r << 1$) and the obstacle height is modest, these upstream disturbances accumulate to form a hydraulic

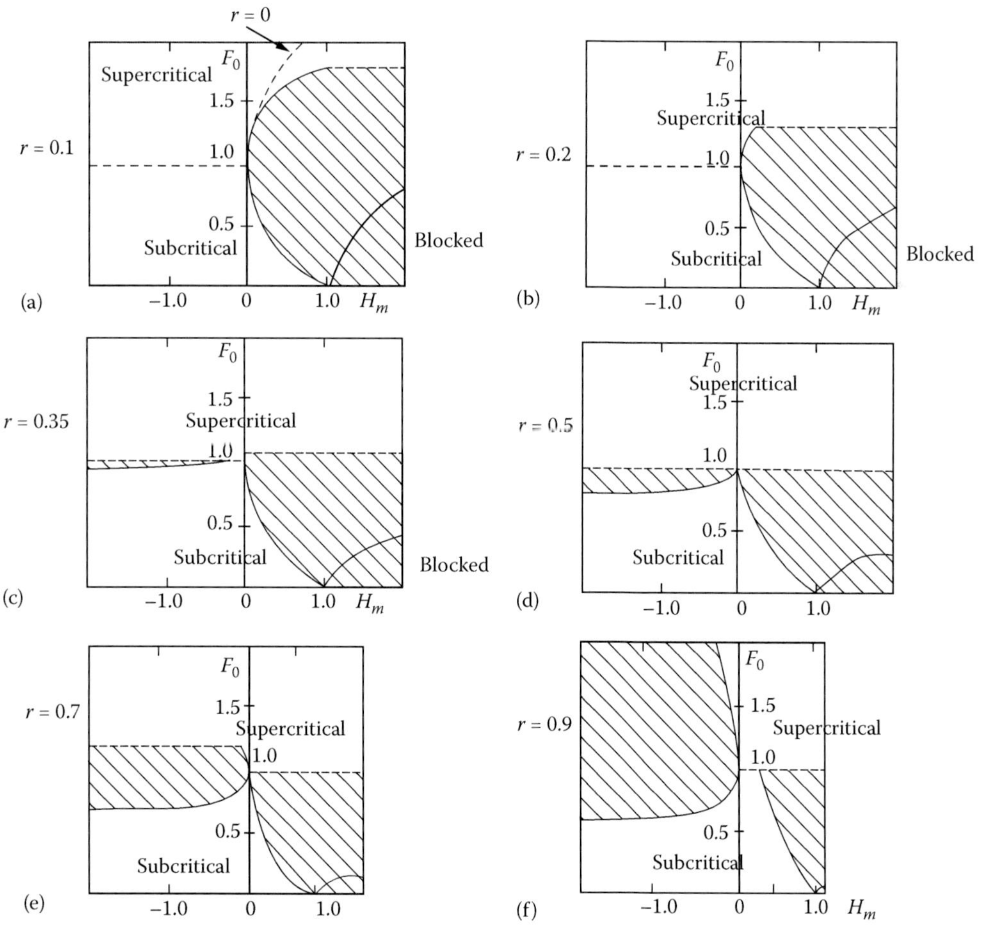

FIGURE 22.3 Boundary curves for wholly subcritical and supercritical flow as functions of initial Froude number F_0 and obstacle height H_m for two-layer Boussinesq flow over an obstacle at uniform undisturbed speed, for a variety of initial layer thicknesses specified by r. In the shaded regions, the presence of the obstacle alters the upstream flow. In the region below the solid curve in the bottom right-hand corner, the lower layer is blocked. The curves have been calculated using the YG formulation. The parameter $r = d_{10}/D$ has the values: (a) 0.1, (b) 0.2, (c) 0.35, (d) 0.5, (e) 0.7, (f) 0.9.

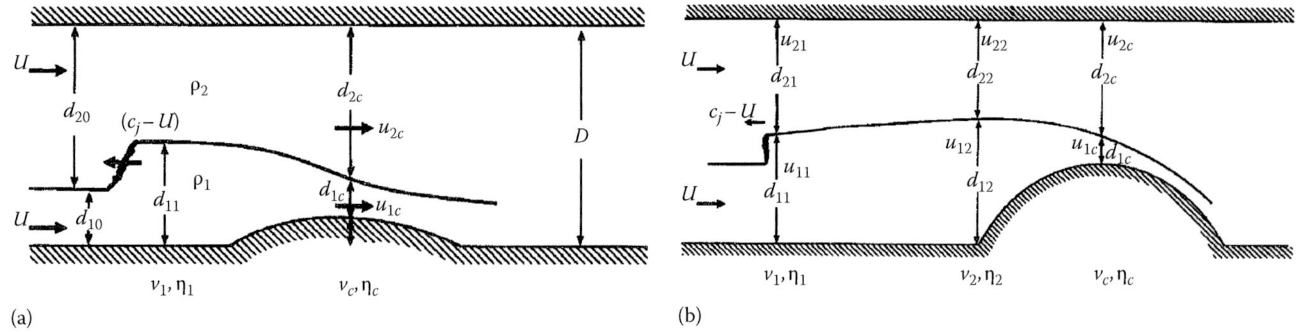

FIGURE 22.4 (a) Schematic of two-layer flow over an obstacle with upstream jump, with relevant variables for calculation; (b) corresponding schematic of two-layer flow over an obstacle with upstream jump and rarefaction.

jump (which may take the form of an undular bore) that propagates to upstream infinity (in an inviscid world) with a fixed structure and speed. In principle, the conditions across the bore (described earlier) and the critical condition at the obstacle crest may be used to solve for the bore amplitude and speed and the flow over the upstream side of the obstacle. Such a flow is shown schematically in Figure 22.4a (Baines 1984).

If the obstacle height is increased further, a point may be reached where the upstream lower layer is completely blocked, so that none of it passes over the obstacle. Regions where blocked fluid occurs are indicated in Figure 22.3. However, if blocking does not occur, with increasing obstacle height the amplitude of the upstream hydraulic jump increases if the additional wave disturbances are able to travel faster and catch it up. With further increases in obstacle height, however, the upstream wave speeds decrease, the jump reaches maximum amplitude, and further upstream disturbances trail behind it, becoming more elongated in the process, and this situation is depicted in Figure 22.4b. This time-dependent feature of gradual change being elongated further with time is called a *rarefaction*. This term is borrowed from gas dynamics, and is used to imply that the *disturbance* is being rarefied, rather than the fluid, as it is progressively stretched out with time. With further increases in obstacle height, the wave speed immediately upstream of the obstacle decreases further as h_m approaches the upper boundary, to the point where the flow is again critical just upstream of the obstacle, with $c_- = 0$ there (from (22.8)). This point marks the maximum amplitude of the upstream rarefaction. If the obstacle height is increased beyond this point, no more upstream disturbances can result, and instead the fluid reverts from the critical state to the symmetric state over the obstacle crest. This generally means that the fluid is supercritical over the lee side of the obstacle, with an elevated interface. This is incompatible with the downstream conditions and is followed by a novel feature termed a *supercritical leap*, in which the flow makes a smooth transition from one supercritical state with a thin upper layer to another with a thin lower layer (Baines 1995).

In the flows that generate upstream disturbances (i.e., in shaded regions in Figure 22.3), the flow is supercritical on the downstream side. This flow then makes the transition to the uniform flow further downstream (primarily) through a downstream hydraulic jump, which may be attached to the lee side of the obstacle (and the supercritical leap, if present), or swept away downstream at a uniform speed.

22.3.4 Flow through a Contraction

We next consider flow through a simple contraction in a channel of uniform depth, so that

$$b = b(x), \quad 0 < z < D, \quad A_1 = b(x)(d_{10} + \eta), \quad A_2 = b(x)(d_{20} - \eta), \tag{22.18}$$

with Equations 22.2 through 22.4. The steady-state form of these equations may then be expressed as

$$G^2 \frac{d}{dx}\begin{bmatrix} \eta \\ v \end{bmatrix} = -\begin{bmatrix} \dfrac{u_1 + u_2}{g'} \\ G^2 - \dfrac{(u_1 d_2 + u_2 d_1)(u_1 + u_2)}{g' d_1 d_2} \end{bmatrix} \frac{v}{b}\frac{db}{dx},$$
$$G^2 = F_1^2 + F_2^2 - 1, \tag{22.19}$$

where $F_i^2 = u_i^2/g'd_i$, $i = 1, 2$, and $v = u_1 - u_2$. Here $G^2 = 0$ denotes the critical condition that a wave speed c_+ or c_- (from [22.8]) vanishes. From [22.19] one sees that the condition $db/dx = 0$ implies the usual hydraulic alternative, namely that the flow conditions are symmetric about this point, or the flow is critical there. However, it also shows that the condition $v = 0$ of zero shear has the same implications. This can lead to features known as *virtual controls*. Since we are here concerned with baroclinic flows, we assume as earlier that the total barotropic transport Q is specified, where $Q = q_1 + q_2$, and the respective volume flux in each layer is $q_i = u_i d_i b$, $i = 1, 2$.

A contraction may be a region separating two large reservoirs of fluid in each of which the conditions are horizontally homogeneous and the fluid is at rest. The direction of the fluid velocity in the two layers between the reservoirs may be the same, or it may be the opposite, a situation known as *exchange flow*.

The steady-state form of Equations 22.3 through 22.5 and 22.18 may be integrated to give the relations

$$v = \frac{((q_1/d_1) - (q_2/d_2))}{b}, \quad \frac{vQ}{bD} + \frac{v^2(d_2 - d_1)}{2D} + g'\eta = 0. \qquad (22.20)$$

Since v and η vanish in the reservoirs where the fluid is notionally at rest, these relations apply to flow adjacent to the reservoirs that does not involve dissipative features such as hydraulic jumps.

There is a wide variety of possible flow situations, and only a brief discussion is possible here; a more detailed description is provided in Chapter 3 of Baines (1995). We first consider the case where the layer thicknesses in the two reservoirs are the same, with specified transports (in either direction) in each layer through a contraction that is sufficiently wide so that the flow velocities are small enough to be subcritical everywhere. One may then proceed by examining the steady flow states that result by successively narrowing the width of the contraction, in a similar manner to the method used earlier for flow over an obstacle. The earlier equations show that the transports in each layer are unchanged, until the point is reached where the flow becomes critical at the minimum width of the contraction. With further contraction the transports in each layer are altered, and the flow is governed by the critical condition at minimum width; this is accompanied by upstream disturbances that decrease with distance as the contraction widens, and a downstream hydraulic jump which becomes stationary at some point, to adjust the flow to the downstream reservoir (see Figure 22.5). In the unidirectional case the flow is still subcritically connected to the conditions upstream, so that the flow is described by Equation 22.20. As the contraction width is further decreased, a point is reached where the fluid velocities become identical ($v = 0$) at the contraction minimum (the "virtual control"), and the transports are unchanged with further contraction. In the exchange flow case with oppositely directed flows, the situation is a little more complicated, but the same method produces similar results. The flow is subcritically connected to one side or the other so that (22.20) applies there, and in most cases the flow eventually reaches a state where $u_1 = -u_2$ at the point of minimum contraction.

For the situation where the interface levels in the two reservoirs are different, with $r = r_l$ (left) and $r = r_r$ (right) say, the earlier procedure of progressively narrowing a contraction cannot be applied, because the contraction is necessary to separate the two reservoirs. Instead, one begins by considering the possible conditions at the point of minimum width, and we confine this discussion to the situation of exchange flows where $Q = 0$, and $q_1 < 0$. This implies that the ratio of transports, $q_r = -q_2/q_1 = 1$. From (22.19), one sees that one must have either $G = 0$ there, or $d/dx(\eta, v) = 0$. Since the flow is driven by the difference between the reservoirs, the latter condition seems most unlikely, and does not occur, so that the exchange of fluids between the reservoirs is

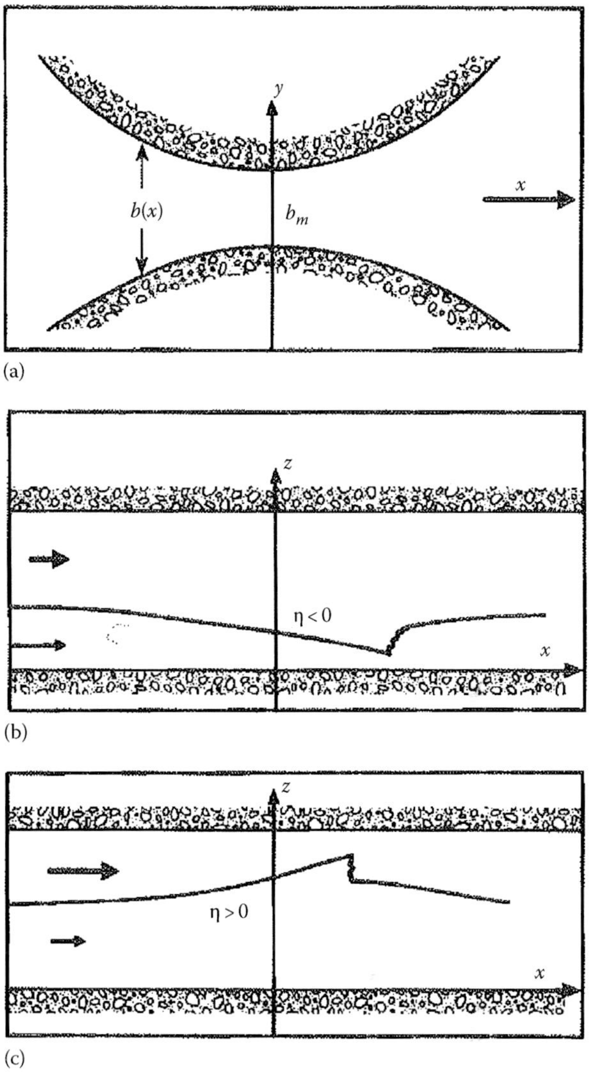

FIGURE 22.5 Schematics of unidirectional flow through a contraction. (a) Plan view of a contraction; (b) side view of flow with critical flow at the minimum width, with $\eta < 0$; (c) as for (b) but with $\eta > 0$.

controlled by the critical condition in the contraction. A useful tool for analyzing these flows is the plane in which the axes are F_1^2 and F_2^2. This is termed the "Froude number plane," because F_1 and F_2 would be the Froude numbers of the respective layers if the other layer were passive. Critical flow (where $G = 0$) is here represented by the straight line $F_1^2 + F_2^2 = 1$, with subcritical flow below the line, supercritical flow above. Solutions are given by the integral (22.20), which may be expressed in terms of F_1, F_2 to give (Armi 1986)

$$\frac{F_1^{-2/3}\left(1 + F_1^2/2\right) - q_r^{2/3} F_2^{4/3}/2}{F_1^{-2/3} + q_r^{2/3} F_2^{-2/3}} = r, \qquad (22.21)$$

where here $q_r^{2/3} = 1$, provided that the flow is subcritically connected to a reservoir with this value of r. An additional equation

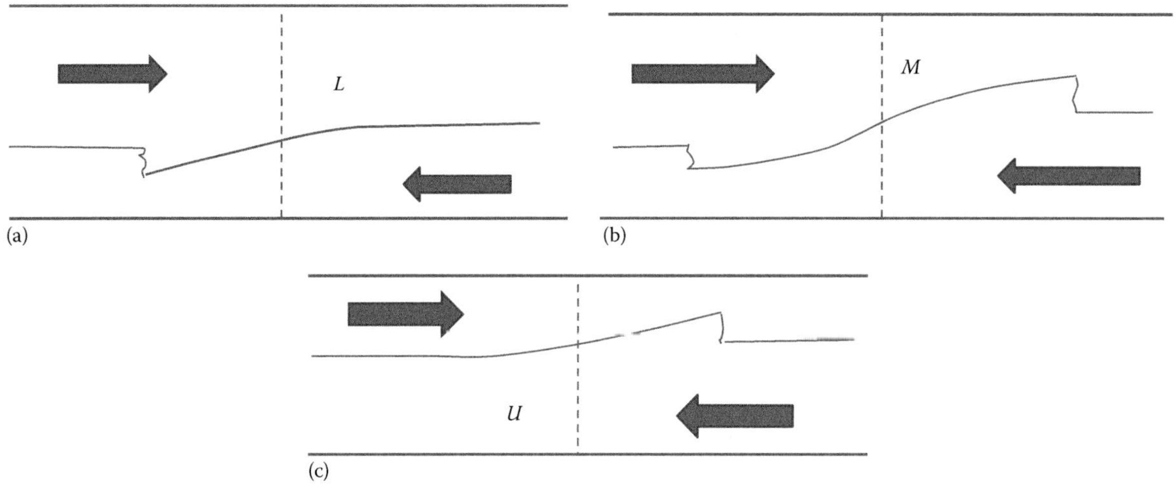

FIGURE 22.6 Schematic diagrams of two-layer flow types with zero net transport ($Q = 0$) through a contraction as shown in Figure 22.5. (a) Type L—lower layer subcritical at the contraction; (b) type M—maximal exchange; (c) type U—upper layer subcritical at the contraction.

is obtained from the uniformity of volume flux in each layer, which gives F_1, F_2 in terms of $b(x)$ in the form

$$F_1^{-2/3} + F_2^{-2/3} = \left(\frac{b(x)^2 g'D^3}{q_1^2}\right)^{1/3} = \hat{q}_1^{-2/3}. \tag{22.22}$$

Three different types of flow are found, termed types L for "lower subcritical," U for "upper subcritical," and M for "maximal exchange," respectively. In type L flows, the narrowest section (where $b = b_m$) is connected subcritically to the (right-hand) reservoir with the inflowing lower layer, in type U it is subcritically connected to the (left-hand) reservoir with the inflowing upper layer, and in type M it is subcritically connected to neither. The three types of flow are depicted in Figure 22.6. In type L flows, a stationary hydraulic jump exists to the left of the contraction minimum, and to the right of the jump equation 22.21 holds with $r = r_r$. Conditions across the jump are given by (22.22), to fluid that is subcritically connected with the left-hand reservoir, governed by (22.21) with $r = r_l$. These equations enable the determination of the strength and position of the jump, giving a flow as shown in Figure 22.6a. Type U flows are obtained in a similar manner, and depicted in Figure 22.6c. Curves in the Froude-number plane corresponding to (22.21) are shown as solid curves in Figure 22.7a, where the labels (0.4, 0.48. 0.5, 0.52, 0.6) denote the appropriate values of r. Note that the flow state emanates from the origin (the reservoir), crosses the critical line, changes solution curve at the hydraulic jump along a line given by Equation 22.22, and then returns to the origin (the other reservoir). The ranges of the parameters r_l, r_r in which types L and U flows are obtained are shown in Figure 22.7b. In both these flow types, the transport in each layer depends on the conditions in the reservoir that is subcritically connected to the control section (i.e., on r_l for type U, r_r for type L), and the minimum width, b_m.

Type M flows are also critical at $b = b_m$, and are supercritical in different ways on each side of it, as shown in Figure 22.6b. In most cases, each of these supercritical regions is terminated by a hydraulic jump, across which (22.22) applies, and the flow is connected to the appropriate reservoir on each side by (22.21). An example of a maximal exchange solution from $r_r = 0.52$ to $r_l = 0.48$ is given by the path in the Froude number plane in Figure 22.7a from the origin to point A on (22.21) with $r_r = 0.52$, from point A to point B on (22.22) with $\hat{q}_1 = 0.215$, from B to P to C on $r = 0.5$, and similarly from C to D and back to the origin on $r_l = 0.48$. The relevant regions for these flows are shown in Figure 22.7b. For all of these type M flows the transport is given by

$$q_2 = -q_1 = 0.25 b_m D (g'D)^{1/2}, \tag{22.23}$$

which is independent of conditions in the reservoirs, and depends only on the minimum width in the contraction. It is also greater than the transports in types L and U. For this reason, type M flows are termed "maximal exchange" flows, and types U and L are termed "submaximal." The situations where the reservoirs are filled with homogeneous fluid ($r_l = 0$, $r_r = 1$, and $r_l = 1$, $r_r = 0$) are special cases known as "lock exchange flow." These are indicated by the two corner points in Figure 22.7b. These flows do not contain any hydraulic jumps, and the $r = 0$ curve through points B-P-C in Figure 22.7a describes the whole solution, supercritical all the way.

If a mean flow (i.e., $Q \neq 0$) is added to the aforementioned exchange flows through contractions, types L, U, and M are still present but their regions in (r_l, r_r)-space vary. The flow is more complex but may be treated by the same methods. One general result is that, as for the unidirectional flows mentioned earlier, if the contraction width is decreased sufficiently, the baroclinic motion is squeezed out of the system at finite b_m, leaving only the uniform barotropic flow (Baines 1995, Section 3.11.2). These simple examples illustrate the methods and resulting phenomena in two-layer hydraulics. We next consider flows with more complex vertical structure.

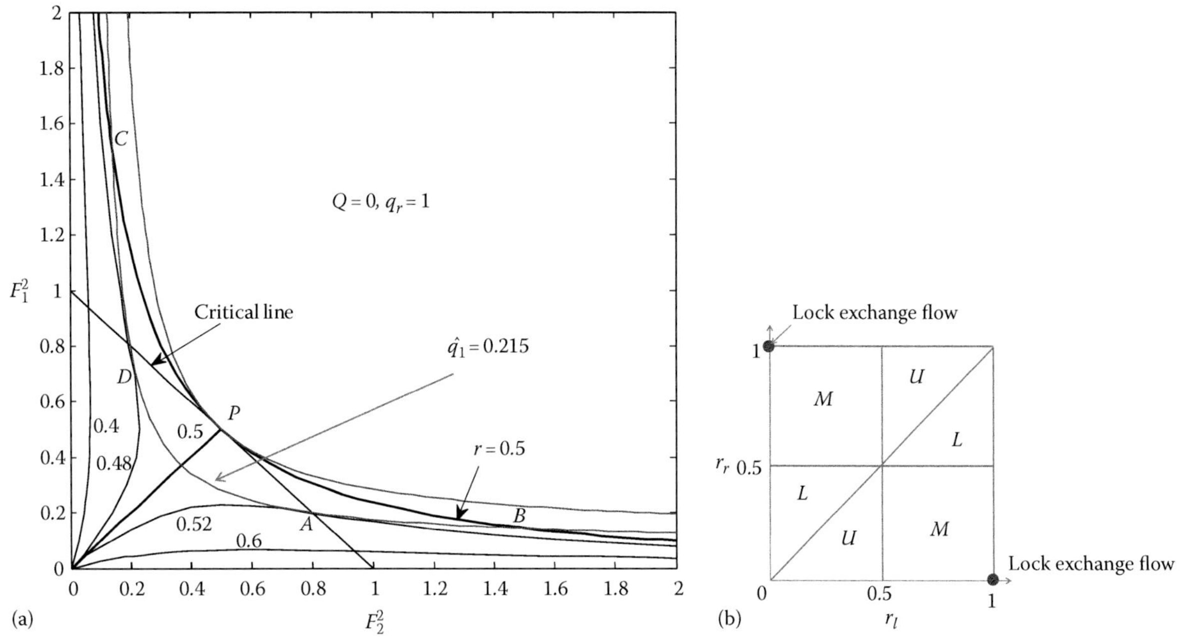

FIGURE 22.7 (a) Froude number plane with coordinates F_2^2, F_1^2, for Boussinesq two-layer exchange flow with zero net transport. Solid lines show solution curves for Equation 22.21, for given values of reservoir layer depth ratio r. The condition for critical flow is shown dashed. Two dot-dashed curves denote curves with constant $\hat{q}_1$ from (22.22) for $\hat{q}_1 = 0.25$ (upper), 0.215 (lower). Curves for smaller values lie below, higher values lie above. All flows with maximal exchange pass through point P. A maximal exchange solution with $r_l = 0.48$, $r_r = 0.52$ is given by the path from the origin along the $r = 0.52$ curve to point A, along the constant $\hat{q}_1$ curve from A to B, the curve $r = 0.5$ from A to P to C, and then a similar path from C to D and back to the origin. (b) Flow types L (lower subcritical), U (upper subcritical), and M (maximal exchange) in r_l, r_r space for the same flow conditions as for (a). Lock exchange flow (with no hydraulic jumps) occurs at the two corners indicated (Armi and Farmer 1986, Baines 1995).

22.3.5 Hydraulics of Multilayered Flows and Continuously Stratified Fluids

The earlier two-layer analysis may be extended to three or more layers to enable a better approximation to a more complex and continuously stratified flow. To illustrate the fundamentals we consider multilayered flow in a horizontal channel of uniform width with variations in bottom topography. For a flow with n layers, with thickness d_i and velocity u_i for the ith layer, with hydrostatic pressure the equations may be expressed as

$$\frac{\partial u_i}{\partial t} + \frac{\partial}{\partial x}\left(\frac{1}{2}u_i^2 + g\sum_{j=0}^{i} d_j + g\sum_{j=i+1}^{n} \frac{\rho_j}{\rho_i} d_j + \frac{p_s}{\rho_i}\right) = 0,$$

$$\frac{\partial d_i}{\partial t} + \frac{\partial}{\partial x}(d_i u_i) = 0, \quad i = 1,2,\ldots,n, \tag{22.24}$$

where $p_s(x, t)$ is the pressure the top of the nth layer (numbered from below), and $d_0 = h(x)$ denotes the height of bottom topography above the level $z = 0$. The upper boundary may be rigid as above, or a *pliant* boundary, where the nth layer is surmounted by a deep layer of density ρ_{n+1} (if $\rho_{n+1} = 0$ it is a *free* boundary). Here we assume that the upper boundary is pliant, so that

$$p_s = P_s - g\rho_{n+1}\left(\sum_{j=1}^{n} d_j - D\right), \quad D = \sum_{j=1}^{n} D_j, \tag{22.25}$$

where P_s, D_j denote the values of p_s, d_j in the undisturbed state, or sufficiently far upstream.

The steady-state form of (22.24) may be expressed as

$$\begin{bmatrix} \Delta_1\rho - t_1 & t_2 & 0 & . & 0 \\ \Delta_2\rho & \Delta_2\rho - t_2 & t_3 & & 0 \\ \Delta_3\rho & \Delta_3\rho & \Delta_3\rho - t_3 & . & 0 \\ . & . & . & . & . \\ \Delta_n\rho & \Delta_n\rho & \Delta_n\rho & . & \Delta_n\rho - t_n \end{bmatrix} \frac{d}{dx}\begin{bmatrix} d_1 \\ d_2 \\ d_3 \\ . \\ d_n \end{bmatrix} = -\frac{dh}{dx}\begin{bmatrix} \Delta_1\rho \\ \Delta_2\rho \\ \Delta_3\rho \\ . \\ \Delta_n\rho \end{bmatrix}, \tag{22.26}$$

where

$$\Delta_i\rho = \rho_i - \rho_{i+1}, \quad t_i = \frac{u_i^2}{gd_i}. \tag{22.27}$$

This shows that at a point where $dh/dx = 0$, we must have either $dd_i/dx = 0$, for all i, or the determinant of the matrix of coefficients must vanish. It may be shown that the latter implies that *one* of the long wave speeds must vanish, so that (22.26) is a generalization of the familiar "hydraulic alternative."

The treatment of hydraulic jumps in multilayered models follows the two-layer case. One assumes that the jump has a fixed structure that propagates at a constant speed and separates two flow states with layer velocity and thickness defined by U_i, D_i on the upstream side, and u_i, d_i on the downstream side, relative to the jump. Here it may not be obvious which layers are contracting and which are expanding, so that CBWS and KRS formulations may not be applicable, leaving the YG formulation as the only choice. In this formulation, the mass and momentum equations (22.24) are integrated across the jump, with the approximation that the layer thicknesses within the jump are equal to the mean of the values on each side of it. This leads to the relation (Su 1976):

$$\frac{U_i^2}{gD_i}\frac{2\xi_i}{(1+\xi_i)(2+\xi_i)} = \sum_{j=1}^{i}\frac{D_j}{D_i}\xi_j + \sum_{j=i+1}^{n}\frac{\rho_j D_j}{\rho_i D_i}\xi_j + \frac{p_s - P_s}{g\rho_i D_i},$$

$$\xi_i = \frac{d_i}{D_i} - 1, \quad i = 1,2,3,\ldots,n, \tag{22.28}$$

where P_s and p_s are the upstream and downstream pressures at the top of the nth layer. For a jump propagating into a given (upstream) flow, this defines the jump speed and the downstream structure. It may also be used as a first approximation, to identify which layers are expanding and contracting, and CBWS and KRS versions may be calculated from this.

We may now describe an outline for a general procedure for determining hydraulic channel flows that arise by perturbing a known state (details are given in Baines 1988, 1995). This closely follows the two-layer procedure described earlier. We first define an overall Froude number F_0 by

$$F_0 = \frac{\bar{U}}{\bar{U} - c_1}, \tag{22.29}$$

where $\bar{U}$ is the mean undisturbed velocity in the channel, the initial velocities U_i are positive, and c_1 is the velocity of the fastest linear wave mode propagating against the stream, in the frame of the topography ($c_1 < 0/c_1 > 0$ implies sub/supercritical flow for mode 1). One then proceeds by examining the changes that result as the bottom topography height is increased or the channel width is contracted, and we take the former for definiteness. As the obstacle height is increased, the quasi-static response of the flow over the topography is given by the steady-state solutions of (22.24). This continues until the flow becomes critical at $\mathrm{d}h/\mathrm{d}x = 0$, with respect to one of the internal wave modes: $c_j = 0$ there, say. If the obstacle height is increased by a further increment, the flow adjusts by sending a small disturbance upstream with the structure of the jth internal wave mode, maintaining the critical condition at $\mathrm{d}h/\mathrm{d}x = 0$. If the height is further increased, the same occurs except that the nature of the upstream disturbance depends on whether the upstream wavespeed of mode j increases (i.e., becomes more negative) or decreases with amplitude. If it increases so that subsequent disturbances catch up with earlier ones, a hydraulic jump results; and if it decreases, it is a rarefaction. This may be summarized by

$$\frac{\mathrm{d}c_j}{\mathrm{d}a} < 0 \Rightarrow \text{hydraulic jump}; \quad \frac{\mathrm{d}c_j}{\mathrm{d}a} > 0 \Rightarrow \text{rarefaction}. \tag{22.30}$$

In general, rarefactions are much more common than hydraulic jumps, and are easier to compute (Baines 1988). This process continues with further obstacle height increases, until one of a number of other factors occurs, as follows:

1. *Critical flow upstream.* With upstream rarefactions, c_j progressively increases to zero, so that further upstream disturbances cannot occur with this mode. The flow then becomes supercritical over higher obstacles, with respect to this mode.
2. *Another mode becomes critical at* $\mathrm{d}h/\mathrm{d}x = 0$. The earlier procedure is then repeated for this mode.
3. *Blocking.* Here the upstream disturbances cause the lowest layer to come to rest upstream, and its thickness and transport vanish at the crest of the obstacle. Further increase in obstacle height will not change this, and the upper level of the blocked upstream layer is thereafter treated as a pliant interface. Further increases in h_{max} may cause the second and higher layers to come to rest, and the process is repeated. Sometimes a layer other than the lowest comes to rest, but observations indicate that layers never reverse direction. Further increase in h_{max} then implies that two modes are generated upstream, one by the critical condition and the second as a "passenger mode" to prevent the stationary layer having negative velocity and transport.

Flow on the downstream side may then be computed as for two-layer flows, once the conditions at the obstacle crest are determined. The complexities that result from blocked fluid layers limit the usefulness of models when the number of layers is large, and they are most appropriate to models where $n \leq 4$. An example of three-layer flow over an obstacle is shown in Figure 22.8. Here the initial flow is uniform; the upper boundary is rigid; the two interfaces have the same density difference; the layer thicknesses are $0.4D$, $0.4D$, $0.2D$; and the Boussinesq approximation applies. There are two wave modes that give rise to the patchwork nature of the flow regimes, which is typical of three-layer flows. More layers give a $F_0 - H_m$ diagram with similar structure, but with more patches.

In the real world, flows of this nature are increasingly tackled by numerical grid-point models that aim to include all the relevant physics, to sufficient resolution. Such models may provide answers, but their interpretation and understanding of the dynamics involved should benefit from an accompanying analysis of the type described here.

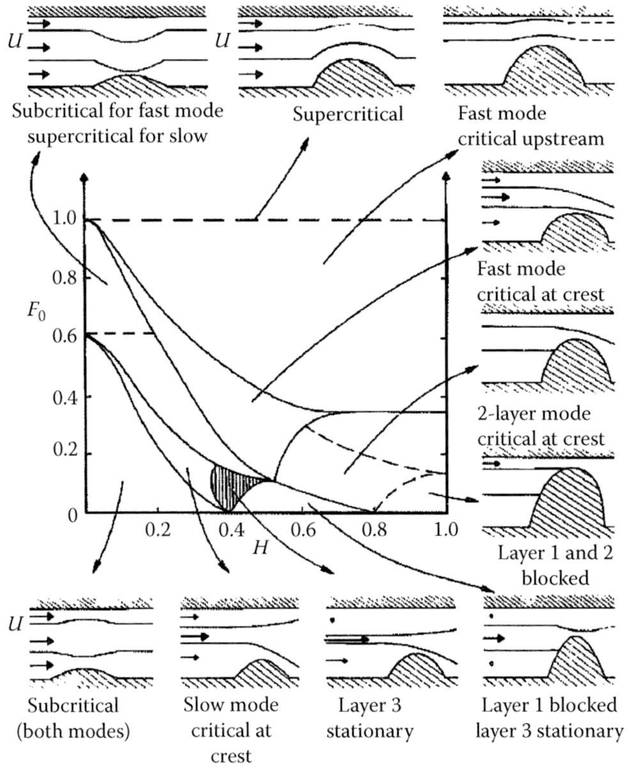

FIGURE 22.8 Regime diagram for the flow of three Boussinesq layers with uniform initial speed U and initial thicknesses ($0.4D$, $0.4D$, $0.2D$) numbered upwards over an obstacle. The upstream disturbances are mostly rarefactions.

22.4 Applications

To date, the main applications of this type of analysis have been to environmental flows in the mesoscale atmosphere and ocean, and in the urban environment. In the oceans, there are approximately 100 or more straits through which stratified water flows (Whitehead 1998), either unidirectionally or as exchange flow. Many of these are in the deep ocean, and are as yet unstudied. Two shallow straits that have received considerable attention are Gibraltar (Armi and Farmer 1988) and the Bosphorus (Gregg and Özsoy 2002). Both of these are exchange flows, and their geometry is considerably more complex than the simple sills and contractions described earlier. The strait of Gibraltar has a prominent sill and a prominent contraction, but at different locations. It is also complicated by a significant tidal component to the flow. The Bosphorus is more complex, with multiple features. In the atmosphere, the hydraulic model has been used to model the character of air flow over mountain ranges (e.g., Smith 1989).

In urban environments, situations often arise in which homogeneous air masses with different densities (through temperatures) exist at either ends of hallways or on opposite sides of doorways and windows (Linden et al. 1990, Linden 1999). Although the latter are not exactly slowly varying topography, the mean streamlines adopted by the flow are consistent with such topography, and the hydraulic approximation can reproduce the physics involved and provide useful answers.

22.5 Major Challenges

The main objective here is to obtain an adequate quantitative understanding of the flow of stratified fluids through regions of varying environment. The earlier hydraulic framework provides a means for doing this if suitable approximations may be made. There are many aspects of the flow of stratified fluids that are underexplored, but from the viewpoint of hydraulics we confine attention to the following areas.

Our understanding of hydraulic jumps in stratified fluids is limited to some situations in two layers. Large lee-side hydraulic jumps and rotors are observed in the atmosphere (e.g., Grubišić et al. 2008), but their detailed dynamics are poorly understood. To what extent do jumps have undular character, or involve turbulent mixing? Experimental studies are needed in order to improve parameterization.

The phenomenon of upstream blocking complicates the use of these models, particularly if many layers are involved, and the development of simplifying treatments would be useful. Blocking is reasonably well observed and understood for uniform and two-layer flows, but results for shear flows in general are lacking. Also, the dynamics of situations where upstream levels other than the lowest become stationary are unconfirmed.

Apart from two-layer exchange flows through oceanic straits, the models have yet to be seriously tested with application to complex flow situations such as multiple layers with highly sheared flows. Their application to flows in urban environments and buildings also appears to be under-explored.

References

Armi, L. 1986. The hydraulics of two flowing layers of different densities. *J. Fluid Mech.* 163: 27–58.

Armi, L. and D.M. Farmer. 1986. Maximal two-layer exchange through a contraction with barotropic net flow. *J. Fluid Mech.* 164: 27–51.

Armi, L. and D.M. Farmer. 1988. The flow of Mediterranean water through the Strait of Gibraltar. *Prog. Oceanogr.* 21: 1–105.

Baines, P.G. 1984. A unified description of two-layer flow over topography. *J. Fluid Mech.* 146: 127–167.

Baines, P.G. 1988. A general method for determining upstream effects in stratified flow of finite depth over long two-dimensional obstacles. *J. Fluid Mech.* 18: 1–22.

Baines, P.G. 1995, 1997. *Topographic Effects in Stratified Flows*. Cambridge University Press, New York, 482pp. (1997 is the updated paperback edition).

Gregg, M.C. and E. Özsoy. 2002. Flow, water mass exchanges and hydraulics in the Bosphorus. *J. Geophys. Res.* 107: C3, doi:10.1029/2000JC000485, 2002.

Grubišić, V., J.D. Doyle, J. Kuettner et al. 2008. The terrain-induced rotor experiment. *Bull. Am. Meteorol. Soc.* 89: 1513–1533.

Klemp, J.B., R. Rotunno, and W.C. Skamarock. 1997. On the propagation of internal bores. *J. Fluid Mech.* 331: 81–106.

Linden, P.F. 1999. The fluid mechanics of natural ventilation. *Ann. Rev. Fluid Mech.* 31: 201–238.

Linden, P.F., G.F. Lane-Serff, and D.A. Smeed. 1990. Emptying filling boxes: The fluid mechanics of natural ventilation. *J. Fluid Mech.* 212: 309–336.

Pratt, L.J. and J. Whitehead. 2008. *Rotating Hydraulics*. Springer, New York, 589pp.

Smith, R.B. 1989. Hydrostatic air-flow over mountains. *Adv. Geophys.* 31: 1–41.

Su, C.H. 1976. Hydraulic jumps in an incompressible fluid. *J. Fluid Mech.* 73: 33–47.

Whitehead, J. 1998. Topographic control of ocean flows in deep passages and straits. *Rev. Geophys.* 36: 423–440.

Yih, C.S. and C.R. Guha. 1955. Hydraulic jump in a fluid system of two layers. *Tellus* 7: 358–366.

23

Hydraulics of Vegetated Canopies

Iehisa Nezu
Kyoto University

Taka-aki Okamoto
Kyoto University

23.1 Introduction

Aquatic vegetation occurring in natural rivers has significant effects on hydrodynamic characteristics such as the velocity distributions, turbulence and coherent structures, as well as mass and momentum exchanges between the vegetated and non-vegetated zones. Aquatic vegetation canopies are responsible for water quality, nutrient and particulate removal, and can reduce turbidity and sediment transport significantly. Vegetation canopies in rivers and wetlands have recently been recognized to be of essential importance not only for water management but also for the river environment, and it is therefore necessary to investigate turbulence structures and their essential effects on scalar transport and the aquatic ecosystem, For example, see Poggi et al. (2004), Ghisalberti and Nepf (2006) and Nezu and Sanjou (2008).

Fortunately, open-channel flows with submerged vegetation may be similar to air flows over the forest and crop fields, although the former is confined by the free surface, whereas the latter is composed of unconfined terrestrial flows. For example, Finnigan (2000) has written a splendid review article about the aerodynamic characteristics and turbulence in plant canopies on the basis of recently measured and calculated data, and he pointed out the interesting analogy between the canopy flow and the plane mixing layer, which was first discussed by Raupach et al. (1996).

In contrast, the measurement of turbulence structures in vegetated rivers and open-channel flows has become possible only relatively recently with the advent of non-intrusive measurement devices such as laser Doppler anemometer (LDA), acoustic Doppler velocimetry (ADV), and particle-image velocimetry (PIV). Figure 23.1 shows the schematized patterns of aquatic vegetation flows. One of the simplest patterns is the homogeneous vegetation canopy shown in Figure 23.1a which has been investigated more intensively by many researchers than (Figure 23.1b) partly vegetation and (Figure 23.1c) heterogeneous vegetation patches, although the latter may be seen more often in natural rivers.

Nezu and Onitsuka (2001) conducted turbulence measurements in partly vegetated open-channel flows by using both LDA and PIV, and found (1) that the horizontal vortices near the free surface are generated by the inflection-point instability and (2) that turbulence near the free surface is transported laterally from the non-vegetation zone toward the vegetation zone by secondary currents. Nepf and Vivoni (2000) conducted some turbulence measurements in homogenous vegetation canopy open-channel flows using ADV. They highlighted the effect of water depth on turbulence structure, and suggested

Handbook of Environmental Fluid Dynamics, Volume One, edited by Harindra Joseph Shermal Fernando. © 2013 CRC Press/Taylor & Francis Group, LLC. ISBN: 978-1-4398-1669-1.

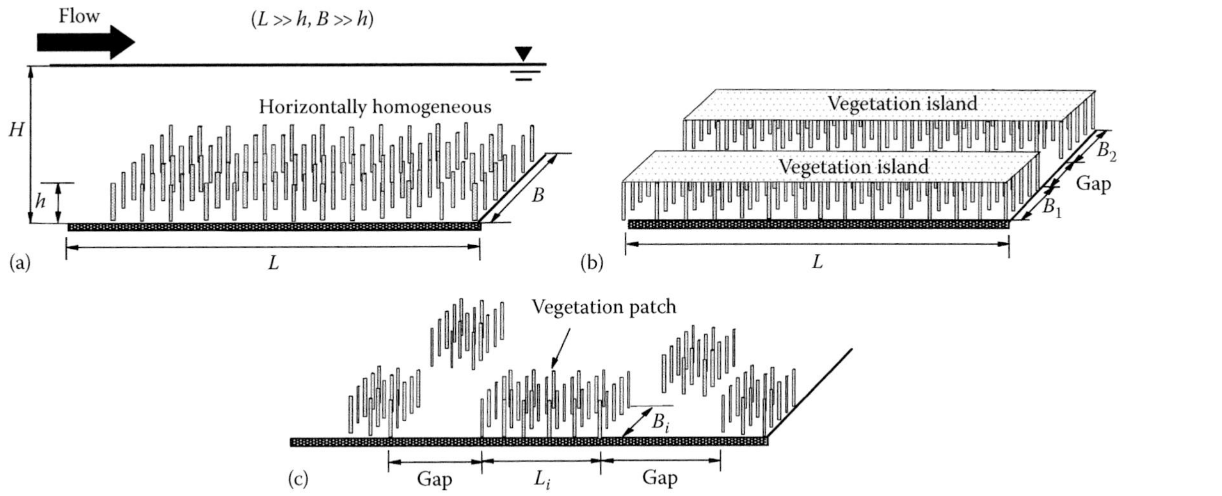

FIGURE 23.1 Schematized patterns of aquatic vegetation flows. (a) Homogeneous vegetation ($L \gg h$, $B \gg h$). (b) Partly vegetation $L \gg B$. (c) Heterogeneous vegetation patches ($L \sim B \sim O(h)$).

that turbulence structures in vegetated open-channel flows may be different from those in terrestrial canopy flows. Wilson et al. (2003) also examined open-channel turbulence through different forms of submerged flexible vegetation using ADV. Poggi et al. (2004) conducted some turbulence measurements in canopy open-channel flows with rigid vegetation elements by using LDA. They proposed a phenomenological model in which the whole flow-depth region was divided into three layers, i.e., the first is the lower layer in which a Karman vortex is generated due to vegetation stems, the second is the middle layer in which Kelvin-Helmholtz (K-H) waves are generated by the inflection-point instability, and the third is the upper layer which is similar to a boundary layer.

Recently, Ghisalberti and Nepf (2006) and Nepf and Ghisalberti (2008) conducted flume experiments with rigid and flexible vegetation models, and found that the waving vegetation, the so-called *Monami*, has significant effects on large-scale coherent turbulence, i.e., sweeps and ejections. They also pointed out that momentum transfer toward the within-canopy layer is dominated by sweeps. Nezu and Sanjou (2008) conducted turbulence measurements in submerged canopy open-channel flows with rigid vegetation models by using LDA and PIV, and examined the mean velocity, turbulence characteristics, and their dispersive properties due to space-averaging. They also revealed coherent structures such as sweeps and ejections, which were well predicted numerically by large-eddy simulation (LES). Furthermore, Okamoto and Nezu (2009) conducted simultaneous measurements of water velocity and vegetation motion by using a combination of PIV and PTV in flexible vegetation open-channel flows, and revealed the relation between coherent vortices and Monami motion in aquatic flows.

In the present study, we highlight homogenous vegetation canopies and examine fundamental hydrodynamic characteristics and coherent structure in open-channel flows with rigid and flexible vegetation on the basis of recently available experimental and numerical results.

23.2 Principles

23.2.1 Momentum Equation

We deal with fully developed and homogenous vegetation flow (Figure 23.1a). As shown in Figure 23.2, U, V, and W denote the components of mean velocity; u, v, and w denote the velocity fluctuations, and u', v', and w' denote the RMS values, i.e., the turbulence intensities, in the x-direction (streamwise with the origin at the vegetation-zone entrance), the y-direction (vertical with the origin at the channel bed), and the z-direction (spanwise with the origin at the channel center). Using the tensor notation, the continuity equation and the RANS equation (time-averaged N-S equation) for vegetated open-channel flows are as follows:

$$\frac{\partial U_i}{\partial x_i} = 0 \tag{23.1}$$

$$\frac{\partial U_i}{\partial t} + U_j \frac{\partial U_i}{\partial x_j} = g_i - \frac{1}{\rho}\frac{\partial P}{\partial x_i} + \frac{\partial \tau_{ij}}{\partial x_j} \tag{23.2}$$

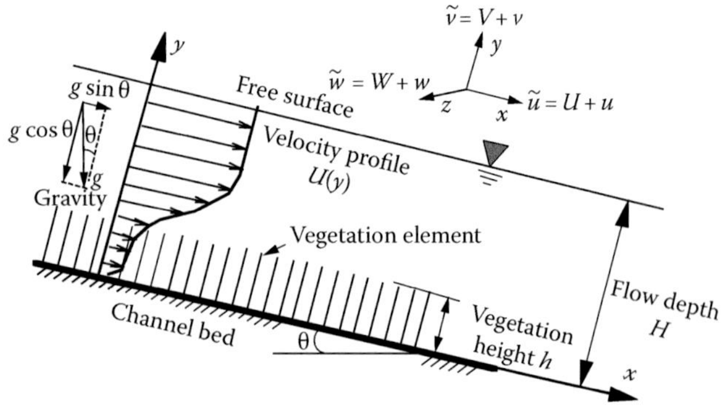

FIGURE 23.2 2-D vegetated canopy open-channel flows and coordinate system.

in which

$$\tau_{ij} \equiv -\overline{u_i u_j} + \nu \frac{\partial U_i}{\partial x_j} \quad (23.3)$$

τ_{ij} is the total shear stress, P is the mean pressure, and g_i is the gravity force vector.

In submerged canopy open-channel flows, the drag force due to vegetation elements decreases the velocity significantly within the canopy layer and the flow field becomes highly heterogeneous in space. The double-averaging method (DAM) was first developed by atmospheric scientists to provide a tool for assessing flow variables within the canopy in terrestrial canopy flows. Recently, the same attention has been addressed to a significant application of DAM to the obstructed flows such as gravel bed, irregular rough surface, and vegetation, as discussed by Nikora et al. (2007). The space-averaging of any time-averaged value Φ for obstructed flows is defined as

$$\langle \Phi \rangle = \frac{1}{V} \iiint_V \Phi \, dV \quad (23.4)$$

The averaging volume V represents water volume only and excludes all vegetation elements, and is large enough to eliminate variations in canopy structure on the horizontal plane. The deviation from the space- and time-averaged value (double-averaged) is denoted by a double prime, as follows:

$$\Phi'' = \Phi - \langle \Phi \rangle \quad (23.5)$$

When vegetation elements are stationary in flow such as rigid vegetation, the space-averaging of (23.2) is simplified as follows, For example, see Finnigan (2000) and Nikora et al. (2007).

$$\frac{\partial \langle U_i \rangle}{\partial t} + \langle U_j \rangle \frac{\partial \langle U_i \rangle}{\partial x_j} = g_i - \frac{1}{\rho}\frac{\partial \langle P \rangle}{\partial x_i} + \frac{\partial \langle \tau_{ij} \rangle}{\partial x_j} + f_{Fi} + f_{Vi} \quad (23.6)$$

$$\langle \tau_{ij} \rangle = \langle -\overline{u_i u_j} \rangle + \langle -U_i'' U_j'' \rangle + \nu \frac{\partial \langle U_i \rangle}{\partial x_j} \quad (23.7)$$

$$f_{Fi} = \frac{1}{V} \iint_S P \cdot n_i dS \quad \text{and} \quad f_{Vi} = -\frac{1}{V} \iint_S \nu \frac{\partial U_i}{\partial n_j} \cdot n_j dS \quad (23.8)$$

in which f_{Fi} and f_{Vi} are the form and viscous drag force vectors exerted on a unit mass of water within the averaging volume V, and n_i is the unit normal vector pointing away from the vegetation surface S into V. Equation 23.6 indicates that the space-averaged RANS equation has the form and viscous drag forces due to vegetation elements. Equation 23.7 also shows that the space-averaged shear stress $\langle \tau_{ij} \rangle$ has further a dispersive stress $\langle -U_i'' U_j'' \rangle$ due to the nonlinearity of advection term; this is the same reason as an appearance of Reynolds stress $-\overline{u_i v_j}$.

When Equation 23.6 is applied to steady and uniform 2-D submerged canopy open-channel flow with homogenous vegetations shown in Figure 23.2, the left-hand-side terms of (23.6) are reduced to zero, and as the result, the streamwise momentum equation is simplified:

$$\frac{\partial}{\partial y}\left(\langle -\overline{uv} \rangle + \langle -U''V'' \rangle + \nu \frac{\partial \langle U \rangle}{\partial y} \right) = -gI_e - f_{Fx} - f_{Vx} \quad (23.9)$$

in which I_e is the energy gradient, i.e., $I_e = \sin\theta - dH/dx \cos\theta$. H is the flow depth and θ is the channel slope (see Figure 23.2). In this case, the space-averaging means the horizontal averaging and the values of $\langle \Phi \rangle$ depend on only the vertical coordinate y.

Over the canopy ($y \geq h$), the drag force due to vegetation elements is zero and the dispersive stress is negligibly smaller than the Reynolds stress, as shown experimentally by Nezu and Sanjou (2008). Neglecting the dispersive stress and viscous stress, and integrating Equation 23.9 from y up to the free surface H, the following results are obtained.

$$\therefore \quad \frac{\langle -\overline{uv} \rangle}{U_*^2} = 1 - \xi, \quad \xi \equiv \frac{y-h}{H-h} \quad \text{(over the canopy, i.e., } y \geq h) \quad (23.10)$$

$$U_* \equiv \sqrt{gI_e(H-h)} \quad (23.11)$$

in which h is the vegetation height. The friction velocity U_* is defined as the value of Reynolds stress at the vegetation edge, i.e., $\langle -\overline{uv} \rangle = U_*^2$ at $y = h$. In the case of unobstructed flow such as over a flat bed, i.e., $h = 0$, Equations 23.10 and 23.11 are reduced to the familiar equations in 2-D open-channel flows, For example, see Nezu and Nakagawa (1993). As the depth of the over-canopy flow is approached to zero, i.e., $H \to h$, the friction velocity U_* becomes negligibly small. This infers that the turbulent structures are quite different between the submerged ($H > h$) and emergent ($H < h$) vegetation flows, which were examined experimentally by Nepf and Vivoni (2000).

In contrast, within the canopy($y < h$), the form drag force f_{Fx} is of essential importance for vegetated canopy flow. Neglecting the viscous terms, Equation 23.9 reads

$$\frac{\partial}{\partial y}\left(\frac{\langle -\overline{uv} \rangle}{U_*^2} + \frac{\langle -U''V'' \rangle}{U_*^2} \right) = -\frac{1}{H-h} - \frac{f_{Fx}}{U_*^2} \quad (23.12)$$

The form drag force is often given by

$$f_{Fx} = -\frac{1}{2} C_D \cdot a \cdot \langle U \rangle^2 \quad (23.13)$$

where

C_D is an element drag coefficient

a is the frontal area per vegetation volume and means the vegetation density

Recently, Nezu and Sanjou (2008) concluded from LDA measurements of aquatic canopy flows that the dispersive stress $\langle -U''V'' \rangle$ is negligibly small, within 5% of U_*^2 even within the canopy layer.

Neglecting the dispersive stress even within the canopy, Equation 23.12 is reduced to

$$\frac{\partial}{\partial y}\left(\frac{-\langle \overline{uv} \rangle}{U_*^2}\right) = \frac{1}{2} C_D \cdot a \cdot \left(\frac{\langle U \rangle}{U_*}\right)^2 - \frac{1}{H-h} \quad (23.14)$$

As the flow depth H becomes larger ($H \to \infty$ for terrestrial canopy flow), the following approximation is obtained:

$$\frac{\partial}{\partial y}\left(\frac{-\langle \overline{uv} \rangle}{U_*^2}\right) \approx \frac{1}{2} C_D \cdot a \cdot \left(\frac{\langle U \rangle}{U_*}\right)^2 \quad (23.15)$$

It is difficult to predict the accurate value of C_D for vegetation elements a priori, because it may depend on various factors such as sheltering effects, allocation patterns, and Reynolds numbers of vegetation elements. Therefore, it is necessary to evaluate the in situ value of C_D experimentally.

23.2.2 Turbulent Kinetic Energy Equation

Space-averaging of the turbulent kinetic energy (TKE) equation for 2-D vegetated flow simplifies to

$$\frac{\partial \langle k \rangle}{\partial t} = \underbrace{\langle -\overline{uv} \rangle \frac{\partial \langle U \rangle}{\partial y}}_{\equiv G_s} + \underbrace{\left\langle -\overline{u_i u_j}'' \frac{\partial U_i''}{\partial x_j} \right\rangle}_{\equiv G_w} + \underbrace{\frac{\partial}{\partial y}\langle -\overline{k \cdot v} \rangle}_{\equiv T_d}$$
$$+ \underbrace{\frac{\partial}{\partial y}\left\langle -\overline{k}'' \cdot V'' \right\rangle}_{\equiv T_{disp}} + \underbrace{\frac{\partial}{\partial y}\left\langle -\overline{\frac{p}{\rho} \cdot v} \right\rangle}_{\equiv P_d} + \underbrace{\nu \frac{\partial^2}{\partial y^2}\langle \overline{k} \rangle}_{\equiv V_d} - \langle \varepsilon \rangle \quad (23.16)$$

where

$k \equiv u_i u_i/2 = (u^2 + v^2 + w^2)/2$ is the instantaneous TKE
G_s is the shear generation of turbulence
G_w the wake generation of turbulence due to vegetation wakes
T_d is the TKE diffusion
P_d is the pressure diffusion
T_{disp} is the dispersive transport and is negligibly small
V_d is the viscous diffusion
ε is the dissipation of turbulence

It is difficult to evaluate the wake generation term G_w according to its definition. Raupach et al. (1996) and others assumed that the value of G_w should be equal to the conversion of mean-flow energy to turbulent energy in vegetation wakes, which is given by working of mean flow against the drag force in the followings.

$$G_w \equiv \left\langle -\overline{u_i u_j}'' \frac{\partial U_i''}{\partial x_j} \right\rangle \approx \langle U \rangle (f_{Fx} + f_{Vx}) \quad (23.17)$$

Neglecting the viscous drag force and using (23.14), Equation 23.17 simplifies to

$$G_w = \langle U \rangle \left(\frac{\partial \langle -\overline{uv} \rangle}{\partial y} + \frac{U_*^2}{H-h} \right) \quad \text{(within the canopy, } y \le h) \quad (23.18)$$

Also, neglecting the viscous diffusion, the final space-averaged equation of TKE is reduced to

$$G_s + G_w + (T_d + P_d) = \langle \varepsilon \rangle \quad (23.19)$$

The sum of turbulence generation due to both velocity shear and vegetation wakes balances with the space-averaged turbulence dissipation through the turbulent energy diffusion and pressure diffusion. Of course, in the case of unobstructed flow such as flat bed, i.e., $h = 0$, Equation 23.19 is reduced to

$$G_s + (T_d + P_d) = \langle \varepsilon \rangle \quad (23.20)$$

Nezu and Nakagawa (1993) discussed such a TKE budget of (23.20) in smooth and rough open-channel flows.

23.2.3 Phenomenological Model for Aquatic Canopy Flow

23.2.3.1 Two-Subzone Model

The double-averaged momentum Equation 23.9 cannot be solved without relevant turbulence closure models. For example, Lopes and Garcia (2001) have calculated turbulence structures in submerged vegetated open-channel flows by using two-equation turbulence model, and reproduced the mean-flow and turbulence characteristics, as well as the TKE budget.

In contrast, to solve Equation 23.9 analytically, it is useful to divide the flow field into subzones on the basis of phenomenological consideration. In earlier studies, submerged canopy flow is often divided into two subzones; one is the over-canopy layer ($y/h > 1.0$) and the other is the within-canopy layer ($y/h < 1.0$). For the terrestrial canopy ($H \to \infty$), Equation 23.10 suggests an existence of constant shear-stress layer, i.e., $\langle \overline{uv} \rangle \approx U_*^2$, over the canopy. Using a mixing-length closure model, the following well-known log-law distribution can be obtained.

$$\frac{\langle U \rangle}{U_*} = \frac{1}{\kappa} \ln\left(\frac{y-d}{y_0}\right) \quad (23.21)$$

in which, d and y_0 are the zero-plane displacement height and the roughness length, respectively. The von Karman constant κ is the universal value, i.e., $\kappa = 0.412$, in open-channel flows, as pointed out by Nezu and Nakagawa (1993). Equation 23.21 is the same as used in rough boundary layers. However, Raupach et al. (1991) reviewed that it may be difficult to determine the values of d and y_0 accurately from the experiments for canopy flows although y_0 can be evaluated a priori for sand roughness k_s, i.e., $y_0 = k_s/30$ (Nikuradse value). Jackson (1981), Brunet et al. (1994), and others defined the value of d as the mean level of momentum absorption due to vegetation, as follows:

$$d = \frac{\int_0^h \left(y\partial\langle\overline{uv}\rangle/\partial y\right)dy}{\int_0^h \left(\partial\langle\overline{uv}\rangle/\partial y\right)dy} \quad (23.22)$$

Evaluating the zero-plane displacement height d from Equation 23.22, the roughness length y_0 is determined so that the experimental values of $\langle U\rangle/U_*$ are best-fitted to the log-law of (23.21). Raupach et al. (1991) reviewed the available data of d/h and y_0/h in wind tunnel as well as field tests of forest canopies and found that these values are almost constant, i.e., $d/h \approx 0.8$ and $y_0/h \approx 0.1$ for terrestrial canopy flows in a wide range of canopy density. Similarly, in the case of vegetated open-channel flows, the mean velocity $\langle U\rangle$ obeys the log-law distribution of Equation 23.21, which was examined earlier using Pitot tube by Kouwen and Unny (1973) and recently using LDA by Nezu and Sanjou (2008).

On the other hand, within the canopy layer ($y/h < 1.0$), the drag force due to vegetation appears and governs the flow structure significantly. Assuming a constant mixing-length ℓ within the canopy, the following exponential velocity profile is obtained.

$$\frac{\langle U\rangle}{U_h} = \exp\left(\alpha\left(\frac{y}{h} - 1\right)\right) \quad (23.23)$$

in which $\alpha \equiv \left(C_D a h^3/(4\ell^2)\right)^{1/3}$ becomes a constant value and U_h is the velocity at the canopy edge, i.e., $U_h \equiv \langle U\rangle(y = h)$. If the log-law of (23.21) is extendedly applied up to the edge of the canopy, the velocity gradient of (23.21), i.e., the velocity shear, at $y = h$ must be equal to that of (23.23). If this assumption is applied, α is determined as follows:

$$\alpha = \frac{U_*/U_h}{\kappa(1 - d/h)} \quad (23.24)$$

23.2.3.2 Three-Subzone Model

In submerged canopy flows, the vertical discontinuity of drag force results in strong velocity shear at the top of the canopy and the inflection point appears in velocity profiles. Raupach et al. (1996) found that coherent vortices such as sweeps and ejections are generated due to the inflection-point instability of velocity, i.e., Kelvin–Helmholtz (K-H) instability, and govern mass and momentum transfer between the within- and over-canopies. Of particular significance is that this type of generation mechanism of coherent vortices for terrestrial canopy flows is more analogous to pure plane mixing layers than to boundary layers.

Ghisalberti and Nepf (2006) and Nezu and Sanjou (2008) found that coherent vortices are also generated in aquatic canopy flows in the same manner as terrestrial ones. Turbulence characteristics near the vegetation resemble those of mixing layers rather than flatbed open-channel flows. Therefore, the aquatic canopy flow can be divided into the following three subzones, as pointed out by Nezu and Sanjou (2008):

$$\left.\begin{array}{l}\text{Wake zone } (0 \le y \le h_p) \\ \text{Mixing-layer zone } (h_p \le y \le h_{\log}) \\ \text{Log-law zone } (h_{\log} \le y \le H)\end{array}\right\} \quad (23.25)$$

A phenomenological model for submerged canopy flows is shown in Figure 23.3. In the wake zone ($0 \le y < h_p$), the vertical turbulent transport of momentum, i.e., the Reynolds shear stress $-\langle\overline{uv}\rangle$, is negligibly small due to the strong wake effects behind vegetation stems. That is to say, the Karman vortex appears predominately in the wake zone and governs the longitudinal transport of momentum more significantly than the vertical one. Nepf and Vivoni (2000) defined the penetration depth h_p at which the Reynolds stress has decayed to 10% of its maximum value. Nezu and Sanjou (2008) also adopted the penetration depth h_p as the border of the wake zone. Assuming $\langle\overline{uv}\rangle/U_*^2 \cong 0$ in the wake zone, Equation 23.14 is reduced to

$$\langle U\rangle = \sqrt{\frac{2gI_e}{C_D a}} = \text{const.} \quad (23.26)$$

Equation 23.26 means that the double-averaged mean velocity $\langle U\rangle$ becomes constant in the wake zone, and that the drag force balances with the gravity force in the momentum Equation 23.14.

The mixing-layer zone ($h_p \le y \le h_{\log}$) is defined between the penetration depth h_p and the lower limit position $h_{\log}$ of the log-law zone, as shown in Figure 23.3. A large-scale coherent vortex is generated near the vegetation edge due to the inflection-point instability. Consequently, the vertical turbulent exchange contributes largely to the momentum transfer between the over- and

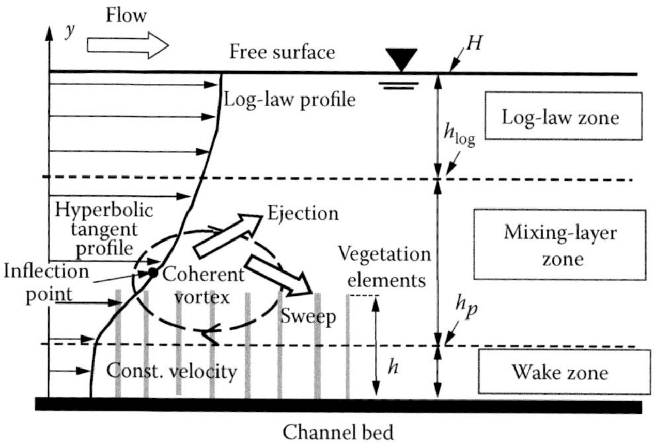

FIGURE 23.3 Three-subzone model for aquatic canopy flow.

within-canopies, as discussed by Raupach et al. (1996) for terrestrial canopy flows and by Nezu and Sanjou (2008) for aquatic ones. In terrestrial canopy flows ($H/h \to \infty$), the coherent vortices may be highly 3-D due to an interaction with larger boundary-layer turbulence, which may enhance secondary instabilities in flows. In contrast, for aquatic canopy flows, the submergence depth H/h is not so large and thus large-scale boundary-layer turbulence may not be significant. Consequently, the K-H vortices dominate the entire mixing-layer zone and the limited depth affects the mean-flow and turbulence characteristics, as pointed out by Nepf and Vivoni (2000) and Nezu and Sanjou (2008).

The momentum thickness θ of a pure mixing layer is defined by

$$\theta = \int_{-\infty}^{\infty} \left[\frac{1}{4} - \left(\frac{\langle U \rangle - \overline{U}}{\Delta U} \right)^2 \right] dy \tag{23.27}$$

The velocity difference of the mixing layer, ΔU, is defined as $\Delta U = U_2 - U_1$. U_1 and U_2 are the margin velocities at the lower and upper margins y_1 and y_2 of the mixing layer, respectively. $\overline{U} = (U_1 + U_2)/2$ and $\overline{y} = (y_1 + y_2)/2$. Nezu and Sanjou (2008) assumed that y_1 and y_2 are equal to h_p and $h_{\log}$, respectively. The following hyperbolic tangent-type profile is often used to describe the mean velocity in plane mixing layers, For example, see Raupach et al. (1996) and Ghisalberti and Nepf (2002).

$$\frac{\langle U \rangle - \overline{U}}{\Delta U} = \frac{1}{2} \tanh \left(\frac{y - \overline{y}}{2\theta} \right) \tag{23.28}$$

Nezu and Sanjou (2008) found that the experimental values of $\langle U \rangle$ in the mixing-layer zone are approached to the mixing-layer profile of (23.28) as the submergence depth H/h increases.

In the log-law zone ($h_{\log} \le y \le H$), turbulence structures are analogous to ones in boundary layers and open-channel flows with rough beds, and the log-law of (23.21) for rough beds is reasonably applied even to vegetated flows. Therefore, the log-law zone is most familiar to researchers, and thus, it is possible to deal with smooth and rough boundary layers including the vegetation canopies in the same point of view, as really reviewed by Raupach et al. (1991).

Similar three-subzone models were also proposed by Nepf and Vivoni (2000) and Poggi et al. (2004), although these definitions are slightly different from each other. Nezu and Sanjou (2008) pointed out that the log-law zone disappeared at smaller submergence depth than the critical depth H_c/h and found $H_c/h = 1.5 - 2.0$ in rigid vegetation canopy. In other words, when the flow depth is smaller than H_c, the canopy flow is divided into the wake zone and the mixing-layer zone.

23.2.4 Flexible Vegetation Flow

Most aquatic plants in natural rivers and wetlands have great flexibility, and move in response to the turbulent flow. In particular, the downstream advection of the coherent vortices is possible to cause a large-scale organized waving of aquatic plants, which is termed the "*Monami*" phenomena and corresponds well to the "*Honami*" phenomena in terrestrial canopy flows, as pointed out by Raupach et al. (1996), Ghisalberti and Nepf (2006), Okamoto and Nezu (2009), and others. However, it is very difficult to analyze the flexible canopy flows theoretically because the momentum equation 23.6 and associated equations are influenced significantly by interactions between the flow and the moving vegetation elements. Therefore, turbulent structures in flexible vegetation flows should be compared with those in rigid ones that can be considered even theoretically, as discussed previously.

For practical purposes, the resistance law in flexible vegetation flows has been investigated intensively since Kouwen and Unny (1973). Using a dimensional analysis, they proposed the following functional expression of the resistance law.

$$\sqrt{\frac{8}{f}} \equiv \frac{U_m}{U_*} = fun.\left(\frac{Bio}{h}, \frac{\overline{h_d}}{h}, \frac{\overline{h_d}}{H} \right) \tag{23.29}$$

$$Bio \equiv \left(\frac{mEI}{\rho U_*^2} \right)^{1/4} \tag{23.30}$$

in which f is the friction factor, U_m (= Q/A) is the bulk mean velocity, Q is the discharge, U_* is the friction velocity, h is the undeflected (or erect) vegetation height, $\overline{h_d}$ is the time-averaged deflected vegetation height, m is the number of vegetation stems per unit area of channel bed, which means the vegetation density, E is the stiff modulus, and I is the inertial moment of stem cross-sectional area. The product $J \equiv EI$ means the flexural rigidity of vegetation stems.

Of particular significance is that the biomechanical property of vegetation may be described simply by Equation 23.30, which means the ratio of aggregate stiffness of vegetation to the shear stress of underlying flow. An increase of vegetation density m has the same effects on flow resistance as increasing the aggregate flexural rigidity J of vegetation. Later, Kouwen and Li (1980) re-analyzed the flume experiments of Kouwen and Unny (1973) including the available field data in natural vegetated rivers, and found that the value of mEI is the most suitable parameter to describe the flexibility of vegetation; mEI is sometimes called the "*aggregate stiffness*."

Flexible vegetated open-channel flows are classified into the following four types on the basis of the biomechanical property, as schematized in Figure 23.4.

a. *Erect* or *Rigid* type: vegetation elements are erect and do not change their tip position in time.
b. *Swaying* type: vegetation elements are independently waving each other (*Gently Swaying*) without organized plant motion.
c. *Monami* type: vegetation elements wave in an organized fashion in response to a coherent vortex.
d. *Prone* type: vegetation elements are prone due to large drag force.

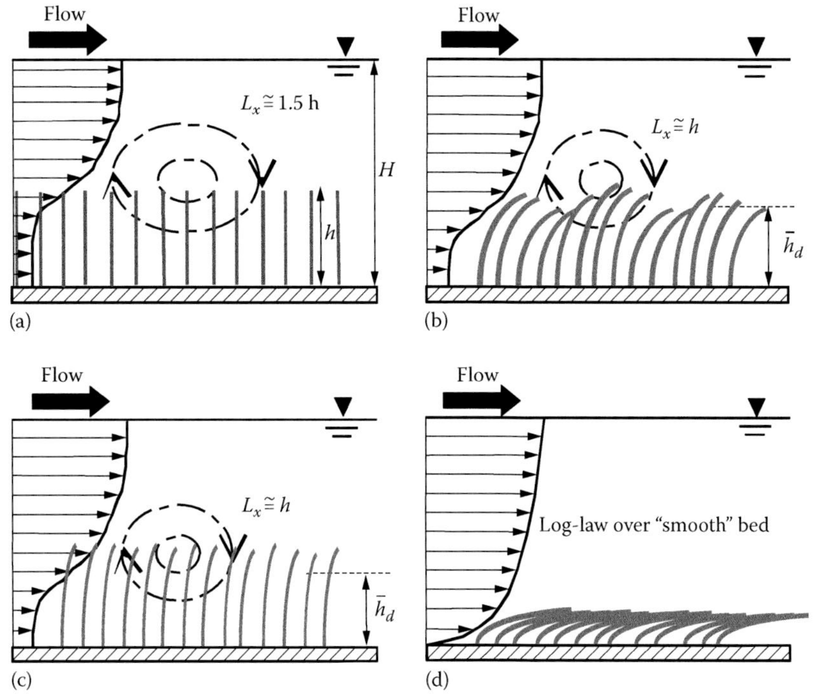

FIGURE 23.4 Flow patterns in dense flexible vegetation. (a) Erect or rigid. (b) Swaying (no organized). (c) Monami (organized). (d) Prone.

In contrast, Kouwen and Unny (1973) and Kouwen and Li (1980) classified the flexible vegetation into three types, i.e., erect, waving, and prone. For larger values of *Bio*, i.e., low flow velocity and high aggregate stiffness, the flexible vegetation is erect and analogous to *rigid* vegetation, which is discussed in previous sections. For smaller values of *Bio*, the flexible vegetation will be prone due to larger drag force and low flexural rigidity. Kouwen and Unny (1973) found experimentally that the *prone* type occurred when the friction velocity U_* was larger than a critical value U_{*c}, and proposed the following empirical criterion for artificial vegetation elements,

$$U_{*c} = 0.028 + 6.33(mEI)^2 \quad \text{for} \quad 0.01 < mEI < 0.2 \ (\text{m} - \text{s unit}). \tag{23.31}$$

However, Kouwen and Li (1980) found that Equation 23.31 may not be applied to long and stiff natural vegetation for $mEI > 0.2$.

Of particular significance is the waving types, in which the organized plant motion, i.e., *Monami*, should be distinguished from non-organized swaying motion, as pointed out by Okamoto and Nezu (2009). As the vegetation density becomes larger, an inflection-point instability, i.e., K-H instability, occurs near the vegetation edge and a large-scale coherent vortex is generated in the mixing-layer zone (Figure 23.3), which governs mass and momentum transfer between the within- and over-canopies. For aquatic flexible vegetation, such a coherent vortex will trigger an organized plant motion, i.e., *Monami*, in the same mechanism as *Honami* in terrestrial canopy flows, which was reviewed by Raupach et al. (1996) and Finnigan (2000).

For practical purposes, the erect and waving types may be classified by flow over a rough bed, whereas the prone type may be analogous to flow over a smooth bed. It is, therefore, expected that all of the flows satisfy the log-law distribution of (23.21) over the canopies, although the values of d and y_0 may be influenced by biomechanical properties of vegetation.

23.3 Methods of Analysis

23.3.1 Experimental Techniques

There are a lot of flume studies on aquatic vegetated open-channel flows. Typical experiments are indicated in Table 23.1. Unlike studies on terrestrial canopy flows, little field measurements in vegetated rivers and wetlands are available because such velocity measurements of water flows are much more difficult than those of air flows in terrestrial canopies. Until the 1990s, a Pitot tube and hot-film anemometers were often used for velocity measurements in aquatic vegetated flows. For example, Kouwen and Unny (1973) conducted velocity measurements in flexible vegetated open-channel flows by Pitot tube and examined that the log-law distribution of (23.21) or its equivalent equation is satisfied over the vegetation canopies. Since the 1990s, particle-tracking velocimetry (PTV) and PIV have been used for water flow measurements. For example, Ikeda and Kanazawa (1996) first examined the Monami characteristics in flexible vegetated open-channel flows using PIV. Since the 2000s, ADV and LDA have become popular instruments for turbulence measurements in vegetated open-channel flows.

ADV and LDA are very powerful for point measurements in vegetated canopy flows because they are non-intrusive instruments and thus feasible to measure velocities even within the canopy layer accurately, irrespective of rigid and flexible vegetation. For example, Nepf and Vivoni (2000) successfully conducted

TABLE 23.1 Typical Experiments of Aquatic Vegetation

Authors	References	Devices	Vegetation	Bulk Velocity U_m (cm/s)	Density a (1/m)	Vegetation Length h (cm)	Submergence Depth H/h
Kouwen and Unny (1973)	*JHD, ASCE*	Pitot tube	F	<60.0	0.25–1.69	10, 15	1.0–4.0
Ikeda and Kanazawa (1996)	*JHE, ASCE*	PIV	F	34.0–60.0	dense	5.0	3.0–3.6
Nepf and Vivoni (2000)	*J. Geophys. Res.*	ADV	F	2.0–8.0	5.5	16.0	1.0–2.75
Ghisalberti and Nepf (2002)	*J. Geophys. Res.*	ADV, LDA	F	2.0–8.0	5.5	12.7	1.0–2.75
Stephan and Guktnecht (2002)	*J. Hydrology*	ADV	F	4.0–10.0	dense	20–200	1.0–2.6
Wilson et al. (2003)	*JHE, ASCE*	ADV	F	10.0	1.7	15.5	1.5–3.4
Poggi et al. (2004)	*B.L. Meteor.*	LDA	R	30.0	0.26–4.26	12.0	5.0
Carollo et al. (2005)	*JHE, ASCE*	ADV	F	2.0–10.0	0.36–0.57	13.0	1.0–2.1
Ghisalberti and Nepf (2006)	*E. Fluid Mech.*	ADV	R, F	1.5–8.0	2.5–8.0	13.8, 20.3	3.4
Nezu and Sanjou (2008)	*JHER, LAHR*	LDA, PIV	R	12.0	7.6–30.4	5.0	3.0
Okamoto and Nezu (2009)	*JHER, LAHR*	PIV, PTV	F	8.0–20.0	7.6	7.0	1.25–4.0

F = flexible, R = rigid.

turbulence measurements in flexible vegetated open-channel flows by ADV, although they had to remove some vegetation for the down-looking ADV sensors. In contrast, LDA is the most suitable and accurate instrument if laser beams are not blocked by vegetation elements; for the much denser vegetation, LDA may not be able to measure velocities within the canopies. For example, Poggi et al. (2004) successfully conducted turbulence measurements even within the canopies by using a forward-scattering two-component LDA. Recently, Nezu and Sanjou (2008) conducted turbulence measurements more intensively in rigid vegetated open-channel flows by using an innovative back-scattering two-component fiber-optic LDA, the measured data of which are discussed later in comparison with the theoretical considerations of Section 23.2.

On the other hand, PIV and PTV are very useful to measure the coherent structure of water flow and the associated instantaneous plant motion. Nezu and Sanjou (2008) conducted PIV measurements in rigid vegetated open-channel flows. Further, Okamoto and Nezu (2009) analyzed the instantaneous tip motion of flexible vegetation elements by a discriminator PTV, in which turbulence measurements were conducted simultaneously by a discriminator PIV. A small fluorescence marker (Nylon-12 particle of 1.0 mm diameter and 1.02 specific density) was attached on the corner edge of the vegetation element and the laser light sheet was projected on the particle markers from the free surface. Consequently, PTV could analyze the tip motion of the vegetation for markers. These significant results of the interactions between turbulence and vegetation motion will be reviewed later in Section 23.4.5.

A laser-based PIV was further developed to simultaneously measure mass concentration by using a combination with a laser-induced fluorescence (LIF) method. Such a challenging combination of PIV and LIF is feasible to examine mass and momentum transfer in vegetated canopy flows, which will also be discussed later in Section 23.5.1.

23.3.2 Vegetation Model

The vegetation density a was defined as the total frontal area A per vegetation volume V in Equation 23.13. When the vegetation elements are allocated homogenously, a is given simply by

$$a \equiv \frac{A}{V} = \frac{m \cdot h_d b}{h_d S} = m(b/S) \quad \text{and} \quad \lambda \equiv ah \tag{23.32}$$

in which m is the number of vegetation elements allocated in the unit area $S = 1\text{ m} \times 1\text{ m}$ on the channel bed and b is the element width. Nezu and Sanjou (2008) have used $\lambda = ah$ as a vegetation density for aquatic rigid vegetation because λ is the dimensionless factor and thus can be compared with results of terrestrial canopy flow in which λ was often used by, for example, Raupach et al. (1996) and Finnigan (2000).

However, for flexible vegetation the frontal area A is a function of flow speed, and the vegetation height is deflected, i.e., $h_d < h$ (see Figure 23.4). Wilson et al. (2003) and Ghisalberti and Nepf (2006) have directly used a (with dimension) as a flexible vegetation density. According to Nepf and Ghisalberti (2008), real aquatic canopies exhibit a wide range of geometry and the vegetation density a changes from $a = 1$ (marsh grasses) to $20\,\mathrm{m}^{-1}$ (mangroves). As indicated in Table 23.1, typical flume experiments have $a = 0.25$ to $30\,\mathrm{m}^{-1}$ and cover well the real aquatic vegetation. In an earlier study, Kouwen and Unny (1973) used the number of stems m simply as vegetation density and they obtained Equation 23.30.

23.3.3 Flexural Rigidity

For flexible vegetation, it is important to evaluate the flexural rigidity EI or the aggregate stiffness mEI for predicting any type of vegetation motion, i.e., erect, waving, and prone. From a cantilever beam theory, EI of a strip plate can be calculated by

$$EI = \frac{F}{\delta}\frac{h^3}{3} \tag{23.33}$$

in which h is the height or length of a strip plate. When the force F loads on the plate, the deflection δ is measured easily and then the value of EI can be evaluated from (23.33).

Figure 23.5 shows the biomechanical characteristics Bio/h against the time-averaged deflected height $\bar{h}_d/h$ in flexible vegetation experiments, which were obtained for plastic strips by Kouwen and Unny (1973) and recently by Okamoto and Nezu (2009). Kouwen and Li (1980) proposed the following empirical relation from data of Kouwen and Unny (1973).

$$mEI = \left(3.4h\left(\frac{\bar{h}_d}{h}\right)^{0.63}\right)^4 \left(\rho U_*^2\right) \tag{23.34}$$

The calculated curve of (23.34) is included in Figure 23.5. The data of Kouwen and Unny (1973) and Okamoto and Nezu (2009) are in good agreement with Equation 23.34 in a wide range of $EI = 2.0 \times 10^{-6}$ to 1.3×10^{-4}(N m^2) although there are some scatters in the experimental data; these scatters may be due to the different attachment of flexible elements to the bed. Kouwen and Li (1980) suggested that Equation 23.34 also describes natural vegetation such as grasses.

23.3.4 Drag Coefficient

The drag coefficient C_D may be a measure of drag force due to individual vegetation elements such as stems, fronds, and branches, which was defined by Equation 23.13. Thus, the value of C_D depends on vegetation geometry. The simplest vegetation models are circular, cylinder (simulated stem), and strip plate (simulated grass). Nezu and Sanjou (2008) evaluated the value of C_D for strip vegetation from Equation 23.14 using the measured data of Reynolds stress $\langle -\overline{uv} \rangle$ and mean velocity $\langle U \rangle$. As the results showed, C_D is almost constant in the wake zone and decreases toward the vegetation edge. These features are similar to the data of Nepf and Vivoni (2000), although the latter vegetation elements were quite different from the former. Nepf and Vivoni (2000) explained that the decrease of C_D toward the vegetation edge is due to a relaxation of form drag as the flow bleeds around the free end.

Nevertheless, the value of C_D may be regarded simply as a constant. Nezu and Sanjou (2008) obtained that the averaged value of C_D is about 2.0 for strip vegetation, which is in good agreement with that of single isolated strip plate.

23.4 Applications

23.4.1 Mean-Flow Structure

Nezu and Sanjou (2008) have recently conducted turbulence measurements in rigid vegetated open-channel flows by using two-component LDA in order to compare the theoretical considerations mentioned in Section 23.2 and also the available data in terrestrial

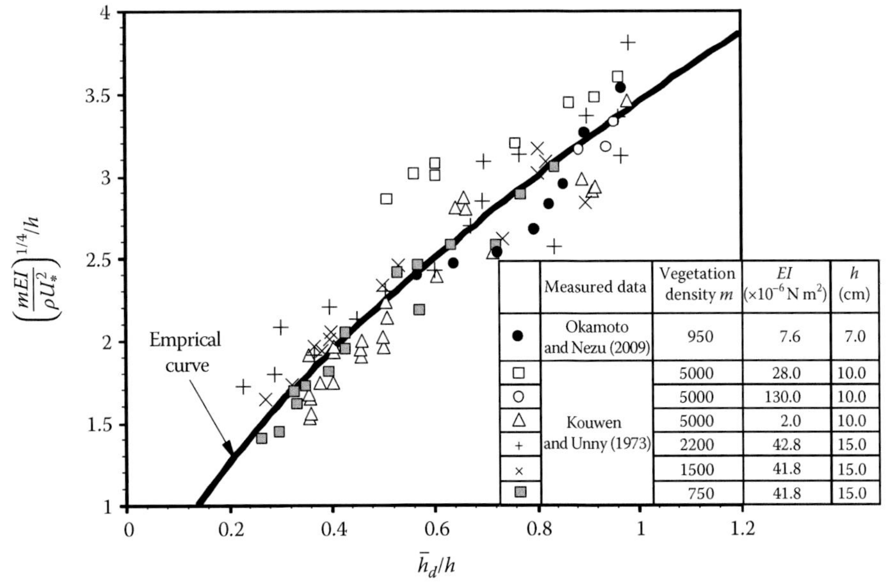

FIGURE 23.5 Relation between aggregate stiffness mEI and the deflected height $\bar{h}_d$.

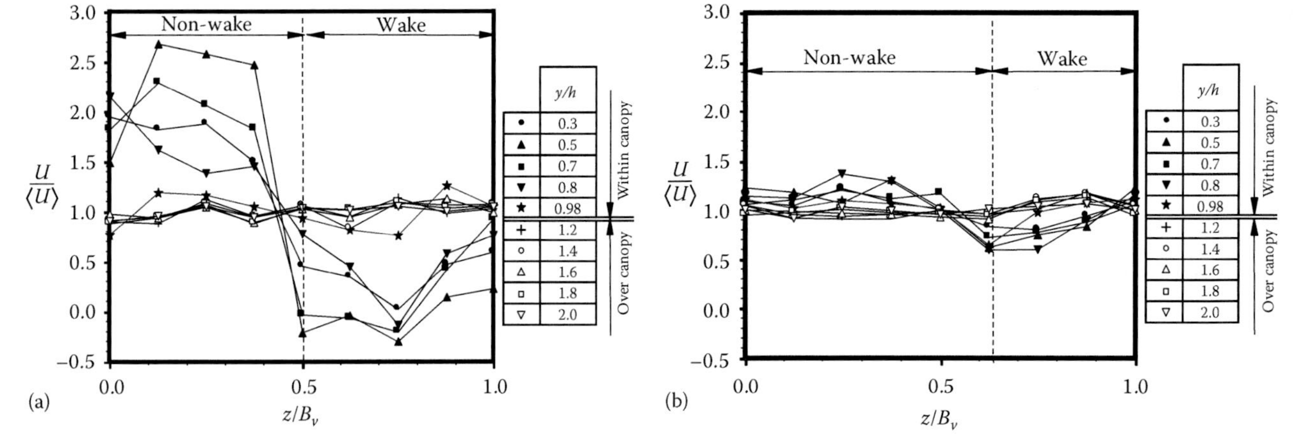

FIGURE 23.6   Spanwise distribution of mean velocity within and over canopy. (a) $\lambda = 1.55$, $a = 31\,\mathrm{m^{-1}}$ (dense). (b) $\lambda = 0.39$, $a = 7.8\,\mathrm{m^{-1}}$ (sparse).

canopy flows. The vegetation density $\lambda \equiv ah$ was changed as 0.39, 0.78, and 1.55 at the constant submergence depth $H/h = 3.0$. The following are the main results of Nezu and Sanjou (2008).

23.4.1.1 Spatial Heterogeneity and Dispersive Stress

Figure 23.6a and b show the lateral distributions of the streamwise component of time-averaged mean velocity $U(z)$ at several elevations y/h within and above the canopy for dense ($\lambda = 1.55$) and sparse ($\lambda = 0.39$) vegetation densities, respectively. The mean velocity U could be measured accurately even within the canopy by LDA, and is normalized by its space-averaged (i.e., horizontally averaged) velocity $\langle U \rangle$. It is clearly found that the streamwise velocity $U(z)$ is varied significantly in the spanwise direction within the canopy ($y/h \leq 1$), due to the local deceleration behind the vegetation elements; it is called here the "wake region." In the non-wake region within the canopy, the value of $U/\langle U \rangle$ is greater than 1.0. These lateral variations of $U(z)$ are less significant in the sparser vegetation density. In particular, it is observed that for dense vegetation, the velocity near the channel bed in the wake region is almost zero and attains even a negative value; that is to say, the reverse flow is generated behind the vegetation elements for dense vegetation. These results conform that the flow structures within the canopy have the spatial heterogeneity and the space-averaging of Equation 23.4 should be conducted reasonably within the canopy. In contrast, over the canopy ($y/h > 1$), these lateral variations become much smaller with an increase of y/h, because the local effects of vegetation on flow are negligibly smaller. It is strongly suggested that turbulent structure over the canopy becomes 2-D and thus turbulence theory of 2-D open-channel flows can be applied to the present flow over the canopy, as pointed out by Nezu and Nakagawa (1993).

The vertical component $V(z)$ of mean velocity was also obtained by LDA measurements and it varied laterally in the same manner as $U(z)$. Nezu and Sanjou (2008) calculated the space-averaged momentum deviation, i.e., the dispersive stress $\langle -U''V'' \rangle$ and found that the values of $\langle -U''V'' \rangle$ over the canopy were negligibly small as compared with the Reynolds stress $\langle -\overline{uv} \rangle$, which confirmed the validity of Equation 23.10. Within

the aquatic canopy, the values of $\langle -U''V'' \rangle$ were within 5% of U_*^2. This is in good agreement with the estimation of Raupach et al. (1996) in a wind tunnel with plant canopy. Therefore, the approximation of (23.14) is reasonably applied to aquatic canopy flows.

23.4.1.2 Mean Velocity Distribution

Figure 23.7 shows the typical examples of double-averaged streamwise mean velocity $\langle U(y) \rangle$, which is normalized by its velocity at the vegetation edge, $U_h \equiv \langle U(h) \rangle$. The drag effects of vegetation decelerate the streamwise velocity more significantly in the canopy layer of $y/h \leq 1$ as the vegetation density λ becomes larger. On the contrary, the velocity over the canopy is larger in the denser vegetation because of the continuity equation with constant discharge. Consequently, a significant inflection point (indicated in Figure 23.7) appears just near the canopy edge $y/h = 1$, which is in good agreement with Nepf and Vivoni

Case	λ	h_p/h	$h_{\log}/h$
● A-10	1.55	0.78	1.80
□ B-10	0.78	0.52	1.65
▲ C-10	0.39	0.50	1.70

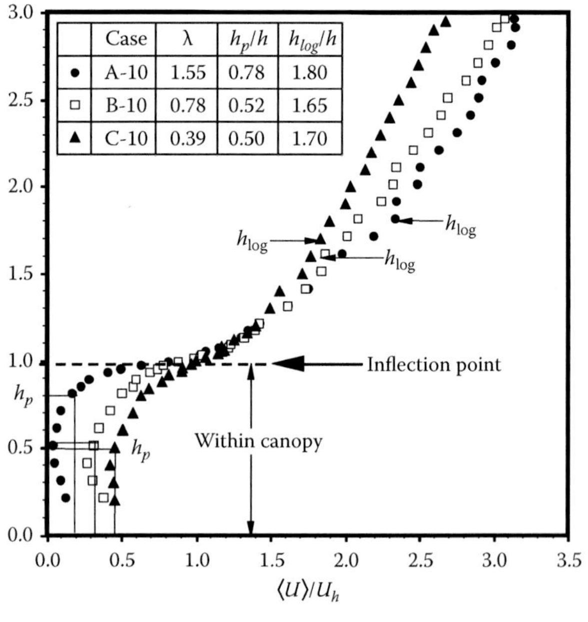

FIGURE 23.7   Space-averaged mean velocity profiles in the whole region.

(2000) for aquatic canopies. These features are also consistent with those of terrestrial canopy flows, as reviewed by Finnigan (2000). Therefore, it is considered that an appearance of the inflection point in velocity profiles is essentially important for canopy flows, irrespective of aquatic and terrestrial canopies.

The penetration depth h_p and the lower limit position $h_{\log}$ of the log-law zone introduced in Equation 23.25 are indicated for each vegetation density in Figure 23.7. Nezu and Sanjou (2008) and Okamoto and Nezu (2009) found that the mean velocity $\langle U \rangle$ satisfies well the log-law distribution of (23.21) in the log-law zone, $y \geq h_{\log}$ (see Figure 23.3), irrespective of rigid and flexible vegetation, and also that $h_{\log}/H$ are nearly equal to a constant of 1.5–2.0 in a wide range of vegetation density. Therefore, when the submergence depth H/h is smaller than H_c/h = 1.5–2.0, the log-law zone disappears and the resistance law of (23.29) is different from the log-law-based formula. As the submergence depth increases from this critical value H_c/h, the log-law of Equation 23.21 is satisfied in the same manner as boundary layers.

Nezu and Sanjou (2008) found that the values of d/h and y_0/h in the log-law of (23.21) are almost constant, irrespective of the vegetation density λ. As the averaged values, $d/h = 0.76$ and $y_0/h = 0.16$ were obtained in a range of λ = 0.4–1.5. These values are consistent with those for aquatic canopy flows by Nepf and Vivoni (2000), who obtained $d/h = 0.7$ and $y_0/h = 0.11$ in the case of $H/h = 2.75$ and $\lambda = 0.9$. Raupach et al. (1991) reviewed the available data of d/h and y_0/h in wind tunnel as well as field tests of forest canopies and found that these values are almost constant, i.e., $d/h \approx 0.8$ and $y_0/h \approx 0.1$ for terrestrial canopy flows in a wide range of λ = 0.1–10.0. Therefore, the log-law for aquatic canopy flows is almost same as that for terrestrial ones although the log-law zone of the latter is much wider than that of the former because H/h is extremely large in terrestrial canopies.

In contrast, the vegetation density significantly influences the penetration depth h_p. The momentum and the Reynolds stress is penetrated more significantly toward the canopy bed as the vegetation density is sparser, as discussed later. In the wake zone ($0 \leq y < h_p$), the mean velocity becomes roughly constant and satisfies Equation 23.26. However, an appearance of the counter-gradient velocity profiles is recognized in the wake zone, which was also suggested by Nepf and Vivoni (2000).

In the mixing-layer zone ($h_p \leq y \leq h_{\log}$) of Figure 23.7, the mean velocity resembles the mixing layer type. The data of Nezu and Sanjou (2008) for $H/h = 3.0$ are compared with the hyperbolic tangent profile of Equation 23.28 in Figure 23.8. In a wide range of vegetation density, the mean velocity of the mixing-layer zone is described well by Equation 23.28. Nezu and Sanjou (2008) further found that the velocity of the mixing-layer zone resembles well the hyperbolic tangent profile of Equation 23.28 as the submergence depth increases, and thus, the deeper aquatic canopy flows at $H/h \geq$ 3–4 have the same structures as the terrestrial canopy flows which correspond to $H/h \to \infty$. The mixing-layer zone of the terrestrial canopy flows resembles pure mixing layers much more reasonably than boundary layers, as pointed out by Raupach et al. (1996) and Finnigan (2000). Nezu and Sanjou (2008) also examined

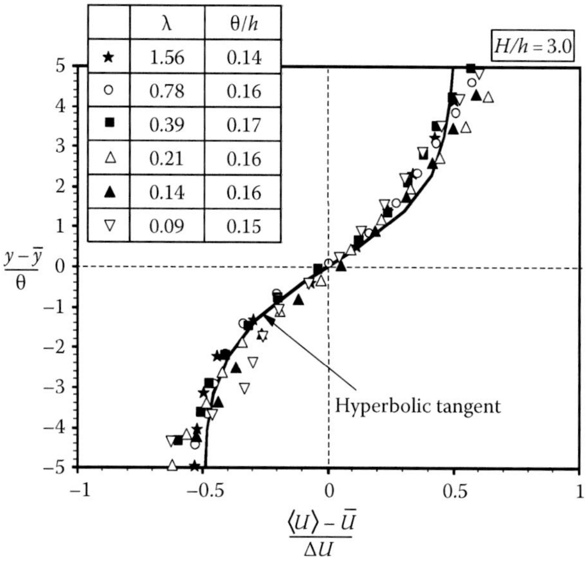

FIGURE 23.8 Mean velocity in the mixing-layer zone.

the validity of the exponential type of Equation 23.23 within the canopy ($y/h < 1.0$) and found that Equation 23.23 is satisfied well only in the mixing-layer zone ($h_p \leq y \leq 1.0$).

To sum up, the mean velocity in the mixing-layer zone resembles one in pure plane mixing layers and thus the inflection-point instability theory in mixing layers is applied to canopy flows, as pointed out by Raupach et al. (1996). However, coherent vortices generated due to such an inflection-point instability may be influenced significantly by the submergence depth H/h, especially in shallower flows of $H < H_c$, which will be discussed later.

23.4.2 Turbulence Structure

23.4.2.1 Reynolds Shear Stress

Figure 23.9 shows the space-averaged Reynolds shear stress $\langle -\overline{uv} \rangle$, which is normalized by the friction velocity $U_*^2 = \langle -\overline{uv} \rangle_{y=h}$. The values of $\langle -\overline{uv} \rangle$ satisfy the theoretical line of Equation 23.10 well although there is some scattering among data. The Reynolds stress attains a maximum very near the vegetation edge, i.e., $y/h = 1$, which corresponds well to the elevation of the inflection point as shown in Figure 23.7. Therefore, it is recognized that the flow over the canopy forms almost 2-D and fully developed flow, although slightly negative values are seen very near the free surface due to secondary currents because of the comparatively small aspect ratio, i.e., $B/(H-h) = 4$, as pointed out by Nezu and Nakagawa (1993).

Of particular significance is the sharp decrease in the Reynolds stress within the canopy. This is caused by the drag force due to vegetation elements. In other words, the momentum transfer from the over-canopy layer toward the bed is obstructed by vegetation elements. Nepf and Vivoni (2000) also observed such properties of $\langle -\overline{uv} \rangle$ within the aquatic canopy by using ADV, and they defined the penetration depth h_p at which the Reynolds stress decays to 10% of its maximum value.

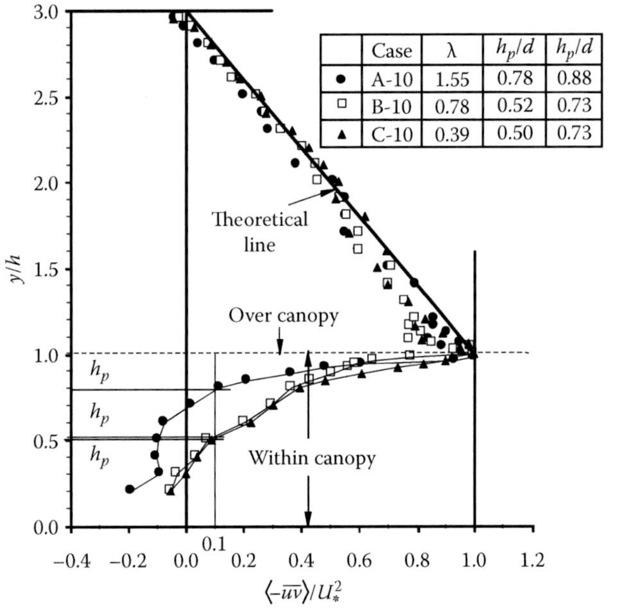

FIGURE 23.9 Vertical distribution of space-averaged Reynolds stress.

The value of h_p is evaluated from Figure 23.9 and indicated in the legend of this figure. As the vegetation density λ becomes smaller, the penetration of the Reynolds stress into the canopy becomes larger. The concept of the penetration depth h_p may be the same as that of momentum transfer (mean level of momentum absorption), which is described by Equation 23.22. The values of h_p/d are indicated in the legend of Figure 23.9, and seem to be roughly constant. In the sparse case of λ = 0.39–0.78, the value of h_p/d is 0.73, which is in good agreement with the data of Nepf and Vivoni (2000) data; $h_p/d = 0.7$ for λ = 0.9.

23.4.2.2 Turbulent Kinetic Energy Budget

In order to reveal the turbulence structure within and above the canopies, it is necessary to examine the turbulent kinetic energy (TKE) budget even approximately, the importance of which is discussed in 2-D open-channel flows by Nezu and Nakagawa (1993). According to the TKE equation of (23.16), the shear generation of turbulence G_s is evaluated accurately from the measured LDA data, whereas the wake generation G_w is estimated from (23.18). The turbulent diffusion T_d of TKE is evaluated experimentally by using an approximation of $\overline{k \cdot v} = \overline{u^2 v}/2 + \overline{v^3}$. On the other hand, the pressure diffusion P_d is sometimes estimated as the residuals of TKE equation, i.e., Equation 23.19, because it cannot be obtained experimentally even at present.

Finally, the dissipation rate ε of turbulence is evaluated reasonably from the following Kolmogoroff's −5/3 power law of the frequency power spectrum $F(f)$, For example, see Nezu and Nakagawa (1993).

$$f \cdot F(f) = C \cdot \varepsilon^{2/3} \cdot (2\pi f / U_c)^{-2/3} \tag{23.35}$$

in which f is the frequency and C is the Kolmogoroff's universal constant, i.e., C = 0.5. U_c is the convection velocity of Taylor's frozen turbulence. Equation 23.35 is obtained from the wave-number spectrum using the Taylor's frozen turbulence hypothesis, i.e., the wave-number $k_w = 2\pi f / U_c$. As a first approximation, the mean velocity U is often used as the convection velocity U_c.

Nezu and Sanjou (2008) evaluated the values of ε from (23.35) assuming $U_c = U$, because the spectral distributions obeyed the −5/3 power law well in the inertial subrange for aquatic canopy flows. Figure 23.10 shows the TKE budget for dense (λ = 1.55)

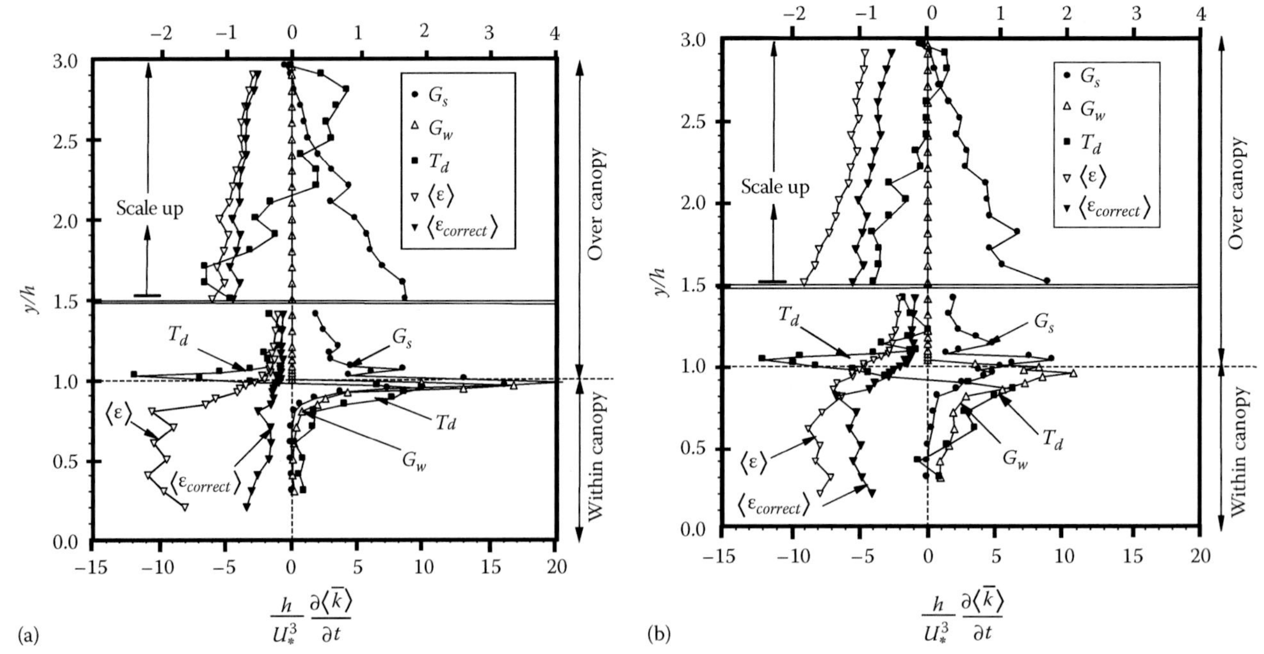

FIGURE 23.10 Turbulent kinetic energy (TKE) budget. (a) Case A-10, λ = 1.55 (dense vegetation). (b) Case C-10, λ = 0.39 (sparse).

and sparse ($\lambda = 0.39$) vegetation. The unit scale in the longitudinal axis $h/U_*^3 \partial\langle \bar{k} \rangle / \partial t$ is enlarged by five times (this scale is indicated at the upper margin line) in the region of $y/h \geq 1.5$, which corresponds to the log-law zone of $y \geq y_{\log}$ (Figure 23.7).

Near above and within the canopy, the TKE budget is fairly complicated. Of particular significance is that the wake generation G_w becomes the same order of magnitude or larger rather than the shear generation G_s within the canopy. It should be noted that the space-averaged dissipation rate $\langle \varepsilon \rangle$, in which ε was evaluated from (23.35) using $U_c = U$, is abnormally too large to balance with the turbulence generation $(G_s + G_w)$ even if the turbulent diffusion is taken into account. These discrepancies have suggested strongly that the assumption of $U_c = U$ was not valid even approximately, and motivated Nezu and Sanjou (2008) to measure the convection velocity U_c directly from the two-point space-time correlations. As the result, the values of U_c were really significantly larger than the mean velocity U. Using these directly evaluated values of U_c, the dissipation rate was recalculated from (23.35) and denoted as $\varepsilon_{correct}$. The space-averaged values of $\langle \varepsilon_{correct} \rangle$ are also plotted in Figure 23.10. It is judged from the TKE balance that the corrected values $\langle \varepsilon_{correct} \rangle$ are much more reasonable than the non-corrected values $\langle \varepsilon \rangle$.

Both of the shear generation G_s and the wake generation G_w attain the maximum near the vegetation edge, i.e., $y/h = 1.0$, and decrease toward the channel bed within the canopy. In contrast, the dissipation $\langle \varepsilon_{correct} \rangle$ becomes constant deeply within the canopy and rather decreases toward the vegetation edge. Consequently, the TKE diffusion T_d plays an essentially important role in the energy budget, as seen in Figure 23.10. Evidently, the values of T_d become significantly negative near the vegetation edge ($y/h = 1.0$), whereas they become significantly positive within the canopy. This implies that the turbulent energy near the vegetation edge is diffused toward the within-canopy as well as toward the free surface. In the case of canopy flow, such an energy diffusion toward the within-canopy layer is essentially important, and compensates the turbulent dissipation deeply within the canopy. As discussed later, the sweep motion of large-scale coherent vortices plays an essentially important role on the downward transport of turbulent energy and momentum into the canopy, which has the same mechanism as found in terrestrial canopy flows by Raupach et al. (1996) and Finnigan (2000).

Above the canopy, the turbulent energy diffusion is also an important key factor to understand turbulence structure. T_d is still negative just above the canopy, whereas it becomes positive near the free surface. This means that the turbulent energy is transported from the canopy boundary toward the free surface, which is the same mechanism as observed in open-channel flows with smooth and rough flat beds, For example, see Nezu and Nakagawa (1993). It is well known that there exists an equilibrium region of $G_s \cong \varepsilon$ in the intermediate range from the near-wall region up to $\xi \cong 0.6$ in smooth and rough open-channel flows. $\xi \cong 0.6$ corresponds to $y/h = 2.2$ from Equation 23.10 for the present vegetation flows. In contrast, it should be recognized from Figure 23.10 that T_d becomes the same order of magnitude as ε in the region of $y/h \leq 2.2$ and thus the equilibrium state of $G_s \cong \varepsilon$ is not satisfied for canopy flows, which is quite different from open-channel flows with smooth and rough flat beds. Such a strong transport of turbulent energy from the canopy boundary toward the free surface may be caused by ejection motions. Each distribution of the TKE budget in Figure 23.10 is not so significantly affected by the vegetation density in spite that the vegetation density of $\lambda = 1.55$ is four times larger than $\lambda = 0.39$. The present distributions of the TKE budget are also in good agreement with those of Nepf and Vivoni (2000) who measured a flexible vegetation canopy flow for $H/h = 2.75$ and $\lambda = 0.9$ by using ADV.

To sum up, the TKE budget in the mixing-layer zone of aquatic canopy flows has the same mechanism as that of terrestrial canopy flows, and the turbulent diffusion plays an essentially important role on the mechanism of TKE budget. This suggests strongly that the ejections and sweeps govern turbulence structure and coherent motion in aquatic canopy flows as well as terrestrial canopy flows, which will be discussed in next section.

23.4.2.3 Quadrant Conditional Analysis

It was inferred from an appearance of inflection point in velocity profiles (Figures 23.7 and 23.8) as well as the TKE budget (Figure 23.10) that coherent vortices are generated by the inflection-point instability, i.e., K-H instability near the vegetation edge. Consequently, they cause significant momentum exchanges between the within- and over-canopies, as pointed out by Raupach et al. (1996) for terrestrial canopy flows. In order to evaluate the contributions of coherent vortices such as bursting events quantitatively, Nakagawa and Nezu (1977) proposed a quadrant conditional analysis for the instantaneous Reynolds stress $-u(t)v(t)$. The quadrant Reynolds stress RS_i is defined as follows:

$$RS_i = \lim_{T \to \infty} \frac{1}{T} \int_0^T u(t) \cdot v(t) I_i(t) dt / (\overline{uv}) \quad (i = 1, 2, 3, \text{ and } 4) \tag{23.36}$$

If the (u, v) exists in a quadrant i, then $I_i(t) = 1$, and otherwise $I_i(t) = 0$. Each quadrant event of (u, v) corresponds to the following coherent one:

$$i = 1 \quad (u > 0, v > 0): \text{Outward interaction}$$

$$i = 2 \quad (u < 0, v > 0): \text{Ejection}$$

$$i = 3 \quad (u < 0, v < 0): \text{Inward interaction}$$

$$i = 4 \quad (u > 0, v < 0): \text{Sweep}$$

RS_i means the contribution rate of each event to the Reynolds stress, and of course,

$$RS_1 + RS_2 + RS_3 + RS_4 = 1 \tag{23.37}$$

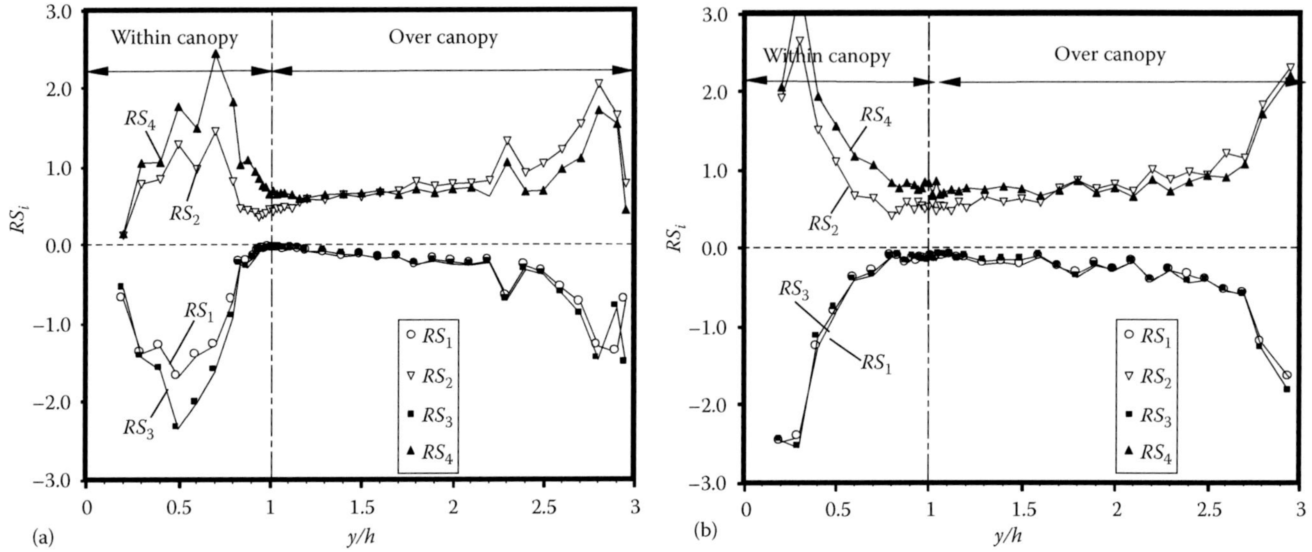

FIGURE 23.11 Quadrant Reynolds stress RS_i for aquatic canopies. (a) Case A-10, λ = 1.55 (dense). (b) Case C-10, λ = 0.39 (sparse).

Figure 23.11 shows the vertical distributions of the quadrant Reynolds stress RS_i in the center of the non-wake region. It is found that RS_2 is larger than RS_4 in the log-law zone ($y/h \geq h_{\log}/h \approx 2$), and thus the ejection event is pronounced in the same manner as observed in open-channel flows and boundary layers, For example, see Nakagawa and Nezu (1977) and Raupach et al. (1991). The rapid increase of each RS_i near the free surface is not caused by coherent vortices, but rather caused by the relatively smaller Reynolds stress $-\overline{uv}$ in the denominator of (23.36) near the free surface.

In contrast, RS_4 is much larger than RS_2 in the mixing-layer zone ($h_p \leq y \leq h_{\log}$), and evidently the sweeps are much more dominant than the ejections. It should be noticed that the coherent sweep motion transports the high-speed fluid from the over-canopy layer toward the within-canopy layer. The values of RS_4 within the canopy for sparser vegetation (C-10) become larger than those for denser one (A-10), and this suggests strongly that the high-speed sweep motion is more in-rushed toward the channel bed in sparser vegetation. These properties of quadrant Reynolds stress are consistent with those of the TKE diffusion T_d discussed in 23.4.2.2.

Furthermore, it should be recognized that the outward and inward interactions, RS_1 and RS_3, become much smaller than RS_2 and RS_4 near the vegetation edge ($y/h = 1$). This suggests strongly that the sweeps and ejections near the vegetation edge are much more organized motions which do not quite contain the less-organized motion such as the outward and inward interactions. In order to exclude the less-organized motion from the raw signals of velocities, a threshold level, the so-called hole value, is often used in the discriminator function $I_i(t)$. For example, the bursting phenomena in open-channel flows and boundary layers are analyzed using the hole value, for example, see Nezu and Nakagawa (1993). In contrast, such a threshold level may not be needed to analyze coherent motions in the mixing-layer zone of aquatic canopy flows because the contribution of less-organized motion is significantly small.

23.4.2.4 Third-Order Moments

Nakagawa and Nezu (1977) derived theoretically that the ejections and sweeps are correlated with the third-order moments (skewness factor) and the TKE diffusion term T_d, using Gram-Charlier probability density function $p(u, v)$. Subsequently, this theory is often applied to canopy flows as well as to open-channel flows and boundary layers, For example, see a review article of Raupach et al. (1991). The skewness factors of $u(t)$ and $v(t)$ are defined as follows:

$$S_u \equiv \overline{u^3}/u'^3, \quad \text{and} \quad S_v \equiv \overline{v^3}/v'^3 \tag{23.38}$$

Figure 23.12 shows the variations of S_u and S_v for dense (λ = 1.55) and sparse (λ = 0.39) vegetation. It is found that the distributions of S_u are almost symmetrical to those of S_v with regard to the zero-value line of $S_u = S_v = 0$ (Gaussian value). Of particular significance is that $S_u > 0$ and $S_v < 0$ in the mixing-layer zone, whereas $S_u < 0$ and $S_v > 0$ in the log-law zone ($y \geq h_{\log}$). The latter characteristics are consistent with those of bursting phenomena in open-channel flows, which were revealed by Nakagawa and Nezu (1977).

In contrast, the former characteristics are peculiar to the canopy flows. The values of S_u and S_v attain maximum just below the vegetation edge and become larger with an increase of vegetation density λ. It should be noticed that these maximum values of $S_u \approx 1.0$ and $S_v \approx -1.0$ are significantly too large in comparison with bursting phenomena in open-channel flows. These features of aquatic canopy flows are in good agreement with those of terrestrial canopy flows measured by Brunet et al. (1994). This shows existence of strong coherent vortices near the canopy vegetation, which are quite different from those arising during the bursting phenomena over a solid wall in boundary layers and open-channel flows, for example, see Nezu and Nakagawa (1993). These differences motivated Raupach et al. (1996) and Finnigan (2000) to compare the canopy flow with pure mixing layer flow, and they concluded that coherent vortices near the canopy vegetation are much more analogous to those of mixing layers rather

(a)   (b)

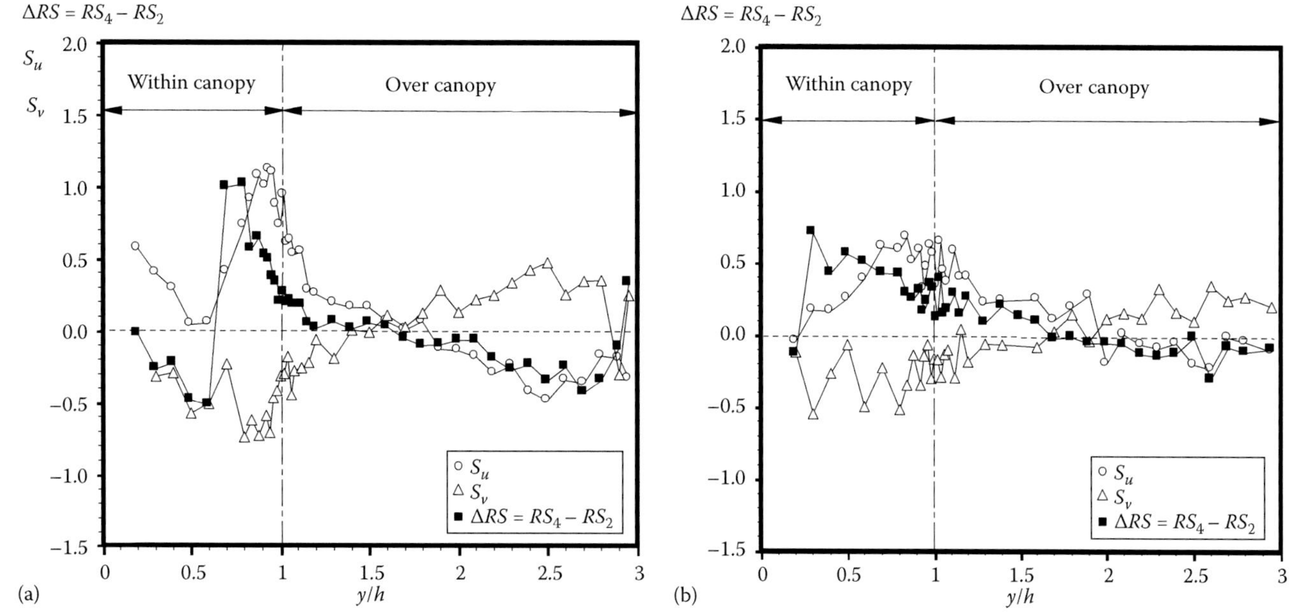

FIGURE 23.12 Skewness factors S_u, S_v and the difference ΔRS between sweeps and ejections. (a) Case A-10, $\lambda = 1.55$ (dense). (b) Case C-10, $\lambda = 0.39$ (sparse).

than boundary layers. Ghisalberti and Nepf (2002), Poggi et al. (2004), and Nezu and Sanjou (2008) recognized that this mixing-layer analogy is also applied to aquatic canopy flows.

Furthermore, it is interesting to reveal the difference between sweeps and ejections, which is defined as $\Delta RS \equiv (RS_4 - RS_2)$. Raupach et al. (1991) obtained the following relation from Nakagawa and Nezu's (1977) quadrant theory:

$$\Delta RS \equiv (RS_4 - RS_2) = C \cdot S_u \qquad (23.39)$$

in which C is constant. Raupach et al. (1991) obtained $C = 0.37$ experimentally in smooth and rough boundary layers. Figure 23.12 also includes the values of ΔRS. It is seen that Equation 23.39 is almost satisfied in aquatic canopy flows. However, the value of C is significantly larger than 0.37; that is to say,

$$\Delta RS \cong S_u \quad \text{(aquatic canopy flows)} \qquad (23.40)$$

The value of ΔRS shows clearly that the sweeps are more predominant than the ejections in the mixing-layer zone ($h_p \leq y \leq h_{\log}$).

23.4.3 Flexible Canopies

23.4.3.1 Classification of Vegetation Motion

As mentioned in Section 23.2.4, Kouwen and Unny (1973) conducted earlier experiments on resistance law in flexible vegetated open-channel flows (see Table 23.1). They classified flexible vegetation motions into three types, i.e., (1) *erect*, (2) *waving*, and (3) *prone*, and distinguished the *erect* and *waving* types as rough boundaries from the *prone* type as smooth boundaries by the critical friction velocity U_{*c}, which was given experimentally by Equation 23.31. The present study divides the *waving* type

further into the *gently swaying* (non-organized motion) and the *Monami* (organized motion) (see Figure 23.4), on taking into account an essential importance of coherent motions.

We calculated the value of U_{*c} from Equation 23.31 using Okamoto and Nezu's (2009) data (Figure 23.5) and obtained $U_{*c} = 2.8\,\text{cm/s}$, which was almost independent of aggregate stiffness mEI because of too small value of $mEI = 0.0072\,\text{Nm}^2$. Further, we judged that the value of $U_{*c} = 2.8\,\text{cm/s}$ evaluated from Equation 23.31 was too small to generate the prone-type vegetation. Therefore, we re-defined the critical friction velocity U_{*c} as the maximum of U_*, in which U_* varied with the vegetation Reynolds number $U_m h/\nu$. Figure 23.13 shows the values of

	h(cm)	U_{*c}(cm/s)	M	S
Present data ($H/h = 3.0$)	5.0	5.8	■	□
	7.0	5.0	●	○
	9.0	3.7	▲	△
	10.5	3.3	▼	▽

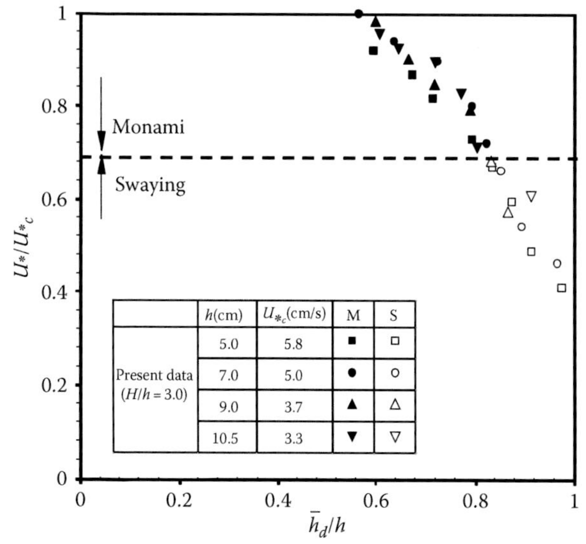

FIGURE 23.13 Classification of swaying type and Monami type in flexible vegetation.

U_*/U_{*_c} against the mean deflected height $\bar{h}_d/h$. We distinguished the Monami from the swaying visually using video pictures and judged tentatively that the Monami occurred when the mean deflected height $\bar{h}_d$ was larger than about 80% of vegetation height h. This visual conclusion was confirmed later by spectral analysis of vegetation motion (Figure 23.20). Of particular significance is that the Monami occurs at $0.7 \leq U_*/U_{*_c} \leq 1.0$, whereas the swaying occurs at the lower friction velocity, i.e., the lower flow-resistance force.

23.4.3.2 Reynolds Stress and Penetration Thickness

Plant flexibility and the presence of Monami may affect significantly the turbulent exchanges between the over- and within-canopies. Figure 23.14 shows the vertical distribution of the Reynolds stress $\langle -\overline{uv} \rangle$ normalized by bulk mean velocity U_m, in which the vegetation height h is varied systematically. The Reynolds stress distributions for rigid (R) and swaying (S) types have a sharp peak near the vegetation edge ($y/h = 1.0$), which is in good agreement with previous data of rigid vegetation shown in Figure 23.9.

In contrast, the value of $\langle -\overline{uv} \rangle$ for Monami (M) type has a constant-stress region near the deflected vegetation edge ($y = \bar{h}_d$). Thus, the sharpness of $\langle -\overline{uv} \rangle$ distributions decreases and these peak values become smaller for Monami than rigid and swaying canopies. This suggests that the oscillations of Monami increase the momentum absorption near the canopy much higher compared to the rigid and swaying canopies. Consequently, the friction factor of Equation 23.29 is reduced for Monami canopies and these characteristics are consistent with Ghisalberti and Nepf (2006).

Of particular significance is that the higher Reynolds stress near the vegetation edge penetrates more deeply into the canopy layer, and results in the vertical transport of momentum

	Type	h(cm)	U_m(cm/s)
□	R	5.0	20.0
■	S	5.0	20.0
○	S	7.0	10.0
□	M	7.0	20.0
•	M	9.0	20.0
▲	M	10.5	20.0

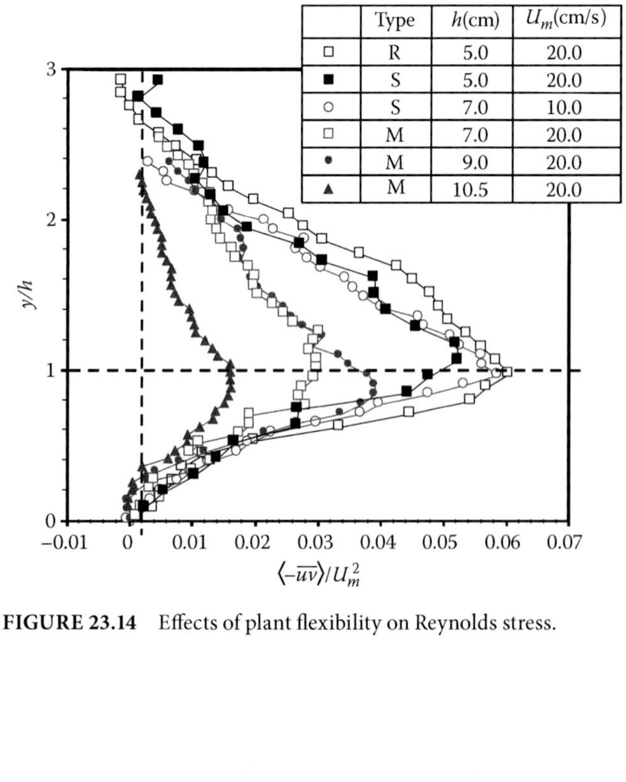

FIGURE 23.14 Effects of plant flexibility on Reynolds stress.

toward the wake zone, which is consistent with diffusion properties of turbulent energy as discussed in Figure 23.10. In the same manner as rigid vegetation in Figure 23.9, Okamoto and Nezu (2009) evaluated the penetration depth h_p for flexible vegetation, and found that the value of h_p becomes smaller (i.e., larger momentum transfer toward the within-canopy) for rigid canopy than flexible one in a wide range of the submergence depth H/h when the vegetation density a is constant. These properties are consistent with Nepf and Vivoni (2000) and Wilson et al. (2003).

However, the definition of h_p, i.e., the height at 10% of peak Reynolds stress defined originally by Nepf and Vivoni (2000), may be arbitrary more or less. Instead, Nepf and Ghisalberti (2008) have recently used the following shear length scale L_s as a penetration thickness, which is defined by

$$\frac{1}{L_s} = \frac{\partial}{\partial y}\left(\frac{\langle U \rangle}{U_h}\right)_{y=h} \tag{23.41}$$

Raupach et al. (1996) have already recognized the importance of L_s in the generation mechanism of coherent vortices near the canopies using the inflection-point instability theory and found that the mean streamwise spacing Λ_x of coherent vortices is correlated well with L_s, as follows:

$$\Lambda_x = n \cdot L_s \tag{23.42}$$

Raupach et al. (1996) obtained from a lot of field plant tests and laboratory experiments that $n = 8.1$ is reasonable for terrestrial canopy flows. Nezu and Sanjou (2008) also examined the validity of (23.42) and suggested that the instability theory of pure mixing layers is slightly less accurate when applied to aquatic canopy flows rather than to terrestrial ones and has to be slightly modified because the former mixing-layer zone is much confined by the free surface and thus much more complicated than the unconfined terrestrial flows.

We can also use the shear length scale L_s as a penetration thickness. Figure 23.15 shows the values of L_s against the vegetation density $\lambda = ah$. Nepf and Ghisalberti (2008) proposed a function of L_s using turbulent energy balance of large vortices, as follows:

$$L_s/h = \frac{K}{C_D ah} \tag{23.43}$$

Nepf and Ghisalberti (2008) evaluated from previous experimental data that the value of K was almost constant, i.e., $K = 0.23 \pm 0.06$ in a range of $C_D ah = 0.2$ to 1.0 for cylinder vegetation elements ($C_D \approx 1.0$), in which their data included rigid and flexible vegetation. The calculated curves of (23.43) with $K/C_D = 0.17$ and 0.29 for cylinder vegetation are compared in Figure 23.15. The values of L_s for rigid strip vegetation of Nezu and Sanjou (2008) are coincident with the cylinder data of Nepf and Ghisalberti (2008) and Equation 23.43 although the former drag coefficient is $C_D \approx 2.0$ (see Section 23.3.4).

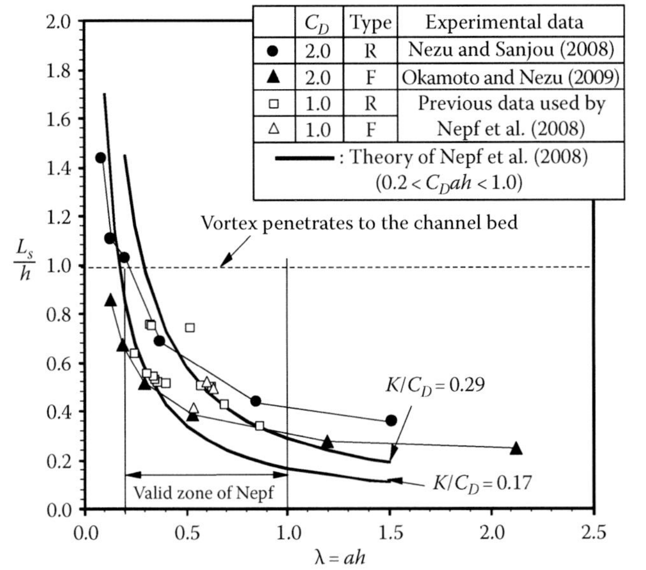

FIGURE 23.15 Shear length scale (penetration thickness) against vegetation density.

Of particular significance is that the momentum transfer toward the channel bed becomes smaller for flexible vegetation than for rigid. When the vegetation density $\lambda = ah$ is smaller than 0.2, the coherent vortex penetrates to the channel bed and the inflection point disappears near the vegetation edge. Nepf and Ghisalberti (2008) and Okamoto and Nezu (2009) also confirmed that the penetration thickness L_s is coincident well with the value of $(h - h_p)$, in which the penetration depth h_p is evaluated from the Reynolds stress.

23.4.4 Coherent Vortex

23.4.4.1 Visualization of Coherent Motion

PIV techniques are much more powerful for capturing coherent motions and examining trajectories of such coherent vortices than point measurements such as LDA and ADV. Nezu and Sanjou (2008) have conducted PIV measurements in rigid aquatic vegetation and analyzed the space-time contours of the instantaneous Reynolds stress $-u(x, y, t)v(x, y, t)$ in the same way as the quadrant conditional analysis of (23.36). They found that the sweep motion and ejection motion are generated periodically in the same manner as observed in terrestrial canopy flows by Raupach et al. (1996). The sweep period T_S and the ejection period T_E for rigid vegetation were in good agreement with the period T_{ML} of coherent vortices in mixing layers, which is obtained from the K-H instability theory in the followings.

$$T_S \cong T_E \cong T_{ML} = \frac{\theta}{0.032\bar{U}} \tag{23.44}$$

The momentum thickness θ and the averaged velocity $\bar{U}$ are defined in Equation 23.27 for the mixing-layer zone. Ikeda and Kanazawa (1996) and Ghisalberti and Nepf (2006) also found the validity of (23.44) in flexible aquatic vegetation. These analogous characteristics between submerged canopy flows and pure mixing layers should be noted about coherent vortex period as well as the mean velocity profile of (23.28) shown in Figure 23.8.

When the vegetation is flexible, the coherent structure may be much more complicated than for rigid and erect vegetation because any interaction between coherent vortices and fluctuations of vegetation elements is essentially significant in the mixing-layer zone. Very recently, Okamoto and Nezu (2009) conducted such a challenging investigation using simultaneous measurements of water velocity by PIV and the motion of vegetation elements by PTV, and they examined the interaction between the coherent vortices and the vegetation motions.

Figure 23.16 shows some examples of the instantaneous velocity vectors $(\tilde{u}, \tilde{v})$ at time $t = 0.0$, 1.43, and 2.16 s for the Monami canopy ($H/h = 2.0$), which were obtained from PIV. The contours indicate the magnitude of turbulent fluctuations $u = (\tilde{u} - U)$ for the streamwise velocity component. The positions (Δx, Δy) of tips of vegetation elements are also indicated by a triangular symbol. The downward vectors, i.e., the *sweep motions* ($u > 0$ and $v < 0$), appear near the vegetation edge and high-speed fluid parcels are transported into the canopy layer (indicated by dashed circles A, C, D, F, and H in Figure 23.16). These sweep motions are followed by *ejection motions*, which are the upward vectors of low-speed fluid

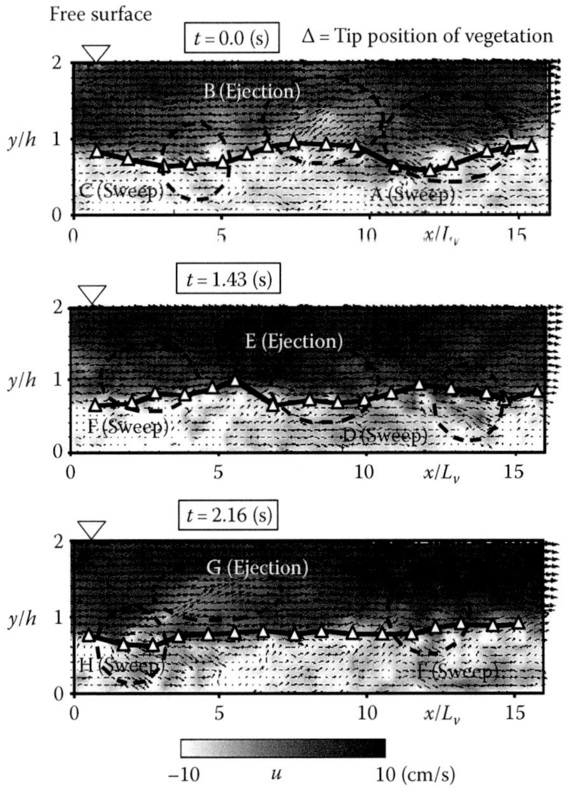

FIGURE 23.16 Instantaneous velocity vectors and vegetation motions for Monami of a = 7.6 [1/m]. L_v is the streamwise spacing between vegetations.

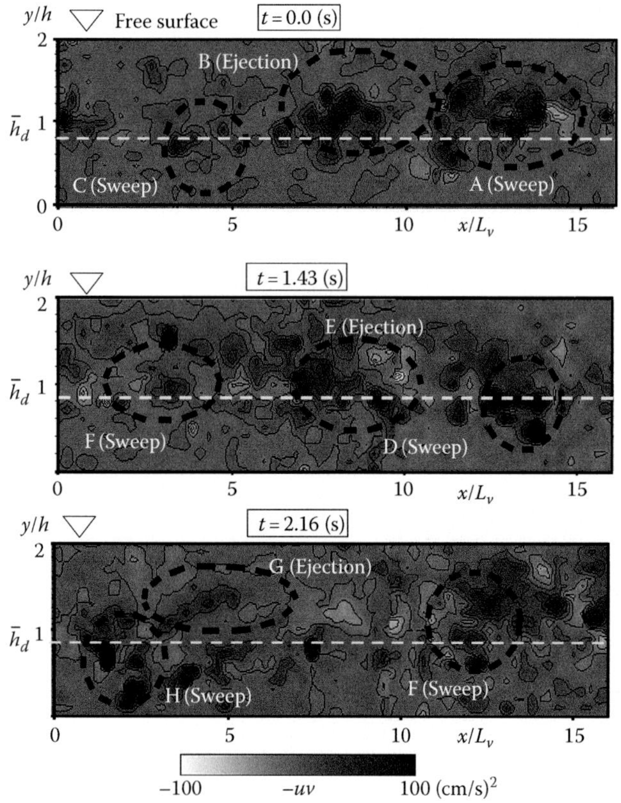

FIGURE 23.17 Contours of instantaneous Reynolds stress (correspond to Figure 23.16).

parcels (indicated by dashed circles B, E, and G in Figure 23.16). Figure 23.16 shows a periodical generation of sweeps and ejections for Monami canopy, and Okamoto and Nezu (2009) confirmed the validity of (23.44) for swaying and Monami types (Figure 23.4). Of particular significance is that the instantaneous flexible vegetation height Δy decreases in the sweep motions. This indicates that the flexible vegetation elements are deflected most significantly by the sweeps. In contrast, the value of Δy increases in the ejection motions. These results suggest that the downstream convection of the coherent vortex causes the coherent waving of flexible vegetation, i.e., Monami phenomena, which is in good agreement with the flow visualization of Ikeda and Kanazawa (1996).

Figure 23.17 shows the contours of the instantaneous Reynolds stress $-u(t)v(t)$, which corresponds to Figure 23.16. It is recognized that large values of the instantaneous Reynolds stress are located near the vegetation edge ($y = \bar{h}_d$) and these locally high distributions of Reynolds stress correspond well to the ejection and sweep motions of Figure 23.16. This implies that the sweep and ejection motions govern the momentum transfer near the canopies significantly for Monami. These significant features for flexible canopies are consistent with those of rigid canopies, the latter of which was examined in detail by Nezu and Sanjou (2008).

23.4.4.2 Convection Velocity and Spatial Scale of Vortex

The length scales of mean eddies are often defined as integral scales, in the followings:

$$L_x = \int_0^\infty \frac{\overline{u(x_0, y_0) \times u(x_0 + x, y_0)}}{u'(x_0, y_0) \times u'(x_0 + x, y_0)} dx \quad \text{and}$$

$$L_y = \int_0^\infty \frac{\overline{u(x_0, y_0) \times u(x_0, y_0 + y)}}{u'(x_0, y_0) \times u'(x_0, y_0 + y)} dy \tag{23.45}$$

in which L_x and L_y are the streamwise and vertical length-scales, respectively. Subscript 0 denotes a reference point, i.e., (x_0, y_0). $u' \equiv \sqrt{\overline{u^2}}$ is the turbulence intensity and the overbar means the time-averaged operator. The upper infinite (∞) of the integrals in Equation 23.45 was replaced by the cross point for the present calculation. Figure 23.18 shows the distributions of L_x and L_y at the vegetation edge for rigid and flexible vegetation canopies, which were evaluated from Equation 23.45 using PIV data. Both L_x and L_y increase monotonically with the submergence depth H/h, and seem to approach constant values at $H/h > 3.0$. It is considered that a large-scale vortex develops fully in such a large submergence depth ($H/h > 3.0$) and becomes identical as observed in terrestrial canopies, as pointed out by Nepf and Vivoni (2000) and Nezu and Sanjou (2008). Such a fully developed vortex structure becomes smaller for flexible than for rigid canopies, that is to say, $L_x \cong 1.5h$ for a rigid canopy and $L_x \cong h$ for a flexible one (see Figure 23.4). Figure 23.18 also shows that the submergence effect on the mean-eddy scales of coherent structures is more significant for a rigid canopy than for a flexible one. This suggests again that an interaction between flow and vegetation elements becomes much stronger for flexible than for rigid canopies and consequently induces vegetation waving.

The convection velocity U_c of mean eddies is evaluated directly by conventional two-point space-time correlation analysis as

$$U_c = \frac{r}{\tau_{max}} \tag{23.46}$$

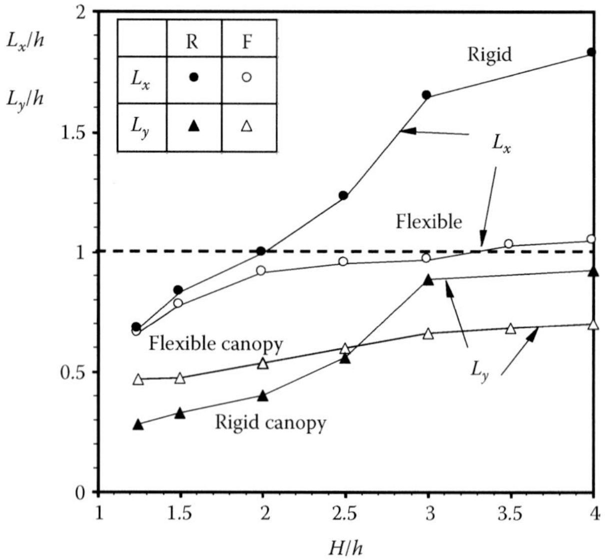

FIGURE 23.18 Mean-eddy scales for vegetation density $a = 7.6$ [1/m].

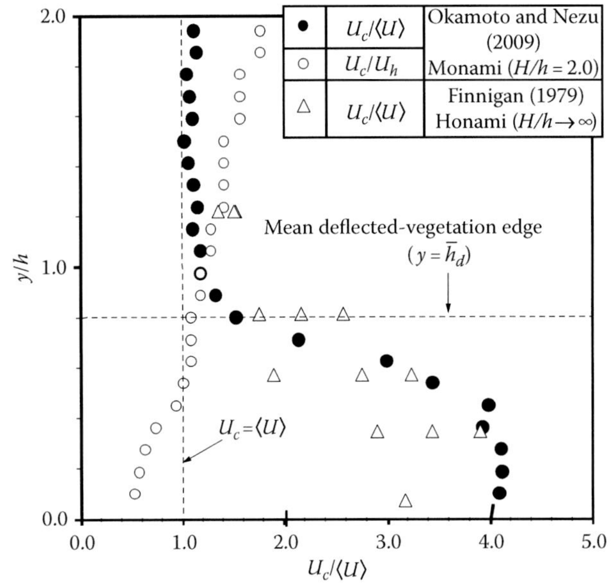

FIGURE 23.19 Distribution of convection velocity.

where r is the streamwise distance between the reference and the movable points, at which the conventional velocity–velocity correlation becomes maximum as the time lag τ_{max} passes. Figure 23.19 shows the vertical distribution of $U_c(y)$ for Monami canopy, which is normalized by the double-averaged mean velocity $\langle U\rangle$. For a comparison, Figure 23.19 also includes the flexible vegetation data of Finnigan (1979), who estimated the value of U_c from single-point hot-wire measurements, also see Brunet et al. (1994).

Of particular significance is that the convection velocity U_c becomes larger than the local mean velocity $\langle U\rangle$ in the vegetation layer ($y/h < 1$) and the value of $U_c/\langle U\rangle$ attains 1.5 at the flexible vegetation edge ($y = \bar{h}_d$), which is in good agreement with Finnigan (1979) and Raupach et al. (1996) for terrestrial canopy flows. This implies that a large-scale coherent vortex representing mean eddies penetrates into the vegetation layer for Monami canopy. It should not be surprising that the value of $U_c/\langle U\rangle$ increases rapidly within the canopy and attains large values of about 4.0 in the wake zone of $y \le h_p$, because the mean velocity $\langle U\rangle$ decreases significantly within the canopy (Figure 23.7). For a comparison, the value of U_c/U_h is also plotted in Figure 23.19. It is found that the convection velocity U_c of mean eddies is fairly constant over the canopy and becomes larger than the mean velocity U_h at the vegetation edge. These results strongly suggest that coherent eddies are convected faster than the mean velocity near the vegetation edge and thus induce vegetation waving. Consequently, Monami and Honami phenomena probably occur as suggested by Raupach et al. (1996).

23.4.5 Interaction between Flow Field and Vegetation Motion

23.4.5.1 Spectral Analysis of Velocity and Vegetation Motion

Coherent vortices should induce periodical fluctuations of turbulent velocity components in the mixing-layer zone of aquatic vegetation. Figure 23.20a shows an example of the power spectrum of the streamwise velocity fluctuations $u(t)$ at $y/h = 1.3$ (above the canopy) for swaying and Monami types. For a swaying canopy, no significant spectral peak property is observed. In contrast, for a Monami canopy, a predominant spectral peak f_p occurs near 0.2 Hz. This predominant peak f_p was in good agreement with the theoretical frequency $f_{ML} \equiv 1/T_{ML}$ of Equation 23.44 proposed in mixing layers. These findings confirm that the generation of coherent structure near the canopies is due to the K-H instability for flexible vegetation in the same manner as for rigid vegetation, the latter of which was discussed by Nezu and Sanjou (2008).

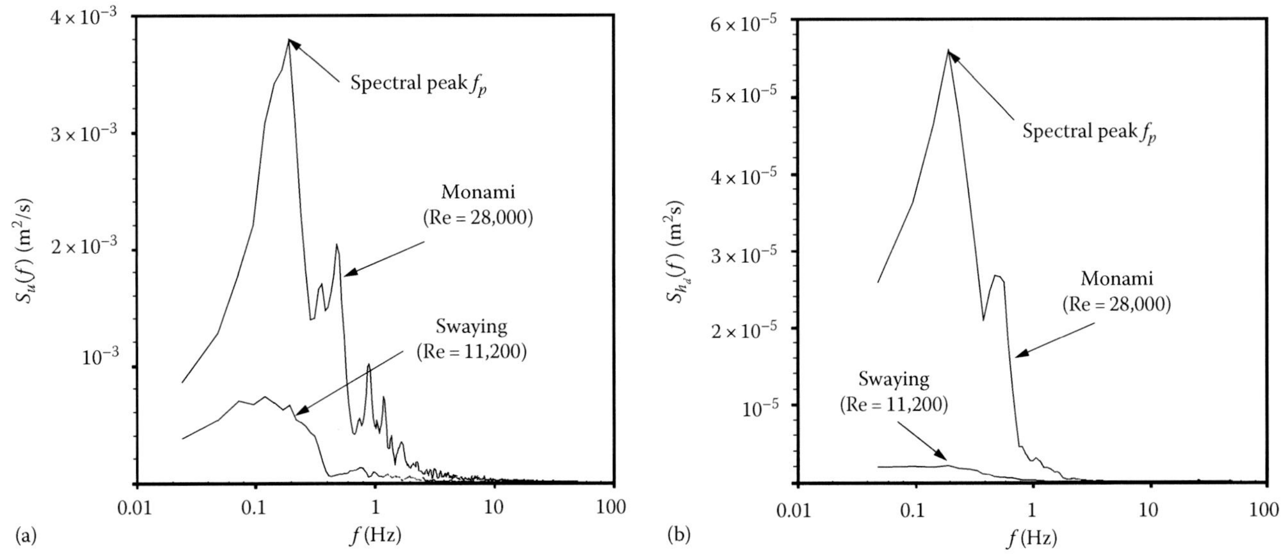

FIGURE 23.20 (a) Spectrum of streamwise velocity component $u(t)$ at $y/h = 1.3$, and (b) spectrum of deflected vegetation height $h_d(t)$ for flexible vegetation.

Figure 23.20b shows the power spectrum of the fluctuations $h_d(t)$ of the flexible vegetation height for swaying and Monami canopies, which were calculated from the PTV data by Okamoto and Nezu (2009). The Monami peak frequency of vegetation stems is in good agreement with that of the velocity spectral peak f_p and the K-H frequency f_{ML} for the Monami canopy. These results imply that the coherent waving motion of vegetation stems is associated with the K-H instability, i.e., the large-scale coherent vortices in the mixing-layer zone. Consequently, it is considered that coherent vortices should induce the Monami motion of flexible vegetation. Figure 23.20 also confirms that the present visual classification of the Monami and swaying (Figure 23.13) was reasonable in comparison between these two spectra of vegetation fluctuations.

23.4.5.2 Space-Time Correlation between Vegetation Motions

No previous investigations are available to evaluate the Monami phenomena of vegetation quantitatively. To examine the coherent motion of flexible vegetation, the space-time correlation $C_{h_d h_d}(x, \tau)$ between the fluctuations of the deflected stem heights, $h_d(x_0, t)$ and $h_d(x_0+x, t+\tau)$, was analyzed as

$$C_{h_d h_d}(x,\tau) = \frac{\overline{h_d(x_0,t) \times h_d(x_0+x,t+\tau)}}{h'_d(x_0) \times h'_d(x_0+x)} \tag{23.47}$$

in which x_0 is the reference point of vegetation stem, $(x_0 + x)$ is the movable point, and τ is the time lag. Figure 23.21 shows the values of $C_{h_d h_d}$ for swaying and Monami canopies. The integer of x/L_v indicates the position of flexible vegetation and thus the distance lag between the two vegetation stems. The time lag is selected to be $\tau = 0.0$, 1.6, and 3.2 s. It is found from Figure 23.21a that the large correlation is observed significantly near the reference point ($x/L_v = 0.0$) for Monami canopy. This correlation peak is convected with an increase of τ, which means that the coherent waving motion of flexible vegetation is convected in the downstream direction for Monami canopy.

In contrast, the values of $C_{h_d h_d}$ for swaying are much smaller than those for Monami. The convection of the correlation peak is not observed significantly for swaying canopy, which is coincident with the spectral properties (Figure 23.20) and suggests that the vegetation stems are waving independently of each other, i.e., without organized motion for swaying canopy (see Figure 23.4).

23.4.5.3 Interaction between Coherent Vortex and Vegetation Motion

Figure 23.22 demonstrates an interaction between the velocity field and the associated vegetation-stem motion $h_d(t)$ in the mixing-layer zone of Monami canopy. The instantaneous velocity vectors $(\tilde{u}, \tilde{v})$ and the contours of the instantaneous Reynolds-stress $-u(t)v(t)$ are shown in Figure 23.22, together with the time series of the deflected vegetation height $h_d(t)$, which is indicated by a solid line in Figure 23.22.

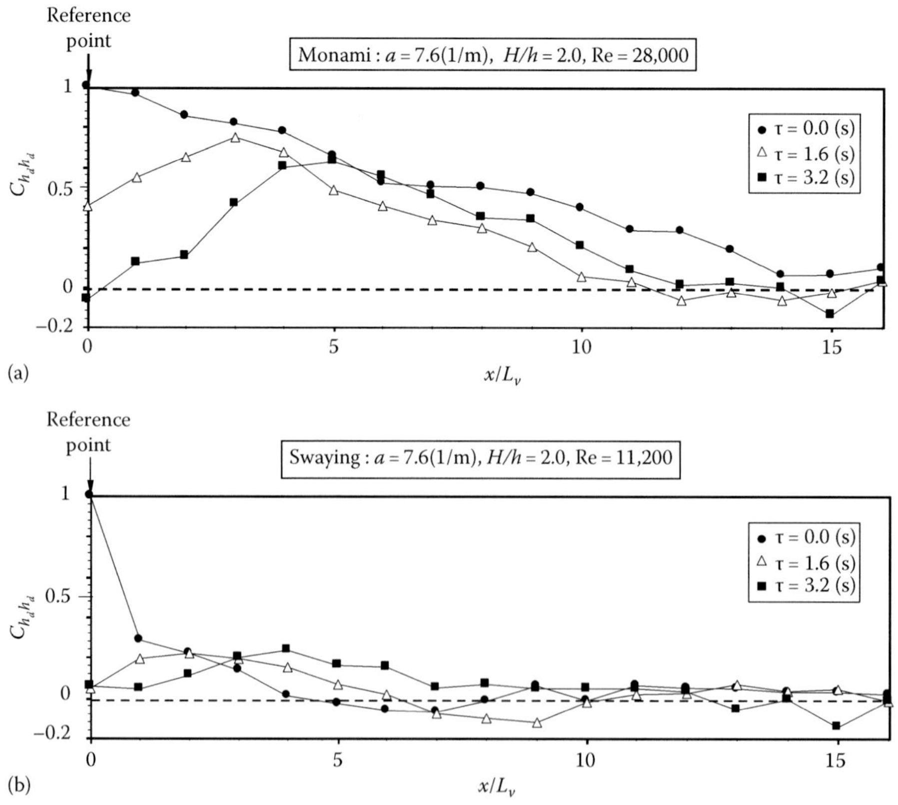

FIGURE 23.21 Space-time correlations of deflected vegetation height for (a) Monami and (b) Swaying canopies.

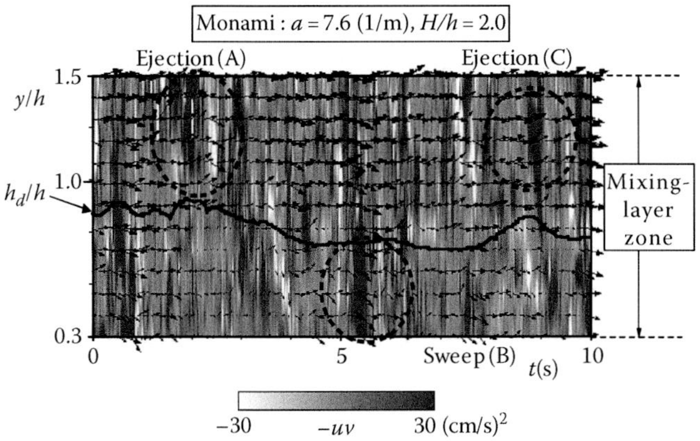

FIGURE 23.22 Time series of instantaneous velocity vectors, Reynolds-stress contours, and the associated deflected-vegetation height, $h_d(t)$, in mixing-layer zone.

At t = 2.0 s, the ejection motion appears above the canopy top, which is indicated by a dashed circle (A). A high-value distribution of the Reynolds stress is observed locally in the circle (A), which is consistent with Figure 23.17. The value of $h_d(t)$ decreases with $t > 2.0$ s and the vegetation stem is deflected most significantly at about t = 5.0–6.0 s. The particular significance is that the sweep motion indicated by a dashed circle (B) appears within the canopy when the vegetation is deflected the greatest. At t = 9.0 s, $h_d(t)$ increases and the ejection parcel (indicated by a dashed circle (C)) appears again above the vegetation top. These features imply a close interaction between the coherent vortex and the vegetation motion in the mixing-layer zone of Monami vegetation.

23.4.5.4 Phase-Average Analysis

To evaluate such an interaction between the flow field and the vegetation motion, Ghisalberti and Nepf (2006) conducted a synchronization between the video footage of vegetation motion and velocity signals of ADV in flexible vegetated open-channel flows, and then examined a phase-averaged turbulence structure on the basis of vegetation motion, which was divided into eight phases. They found that the vegetation height $h_d(t)$ was inversely correlated with the fluctuating velocity values; that is to say, the sweep appeared at the minimum vegetation height (Phase 4–5) whereas the ejection appeared at the maximum vegetation height (Phase 8 = cyclic Phase 1).

In contrast, Okamoto and Nezu (2009) divided the values of $h_d(t)$ into ten phases in the same manner as Ghisalberti and Nepf (2006). The phase sequence represents a phase-averaged Monami cycle, which begins from the time of the maximum height of vegetation, as shown in Figure 23.23a. The mean Monami period was 10.0 s. Any variable Φ is phase-averaged on the basis of these different phases, as follows:

$$\langle \Phi(\tau) \rangle = \frac{1}{N} \sum_{i=1}^{N} \frac{1}{\Delta t} \int_{t_i}^{t_i+\Delta t} \Phi(t)\,dt, \quad \tau \equiv \frac{t - t_i}{T_i} \tag{23.48}$$

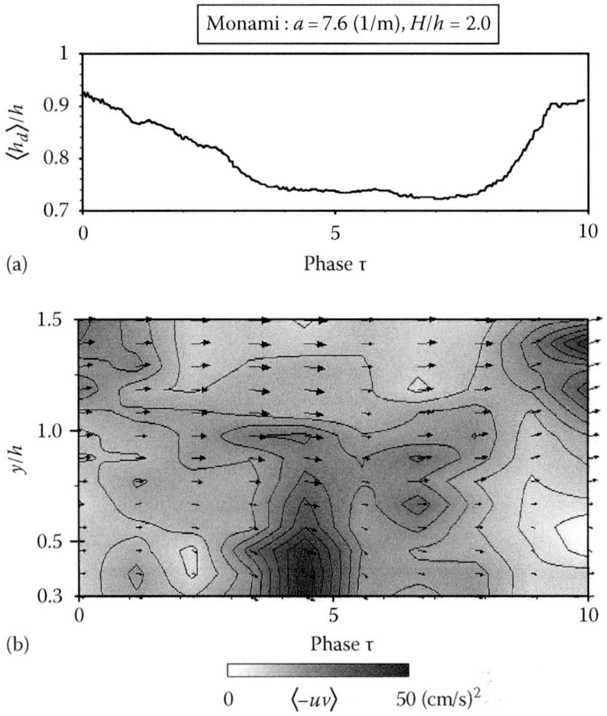

FIGURE 23.23 (a) Phase-averaged deflected height $\langle h_d \rangle$ and (b) phase-averaged Reynolds stress $\langle -uv \rangle$.

in which N is the number of Monami cycles, T_i is the duration time of the i-th Monami cycle, τ is the phase time, and Δt is the bin-averaged time. In this study, N = 40 and $\Delta t = T_i/10$ were chosen. Figure 23.23b shows the phase-averaged velocity vectors $(\langle \tilde{u} \rangle, \langle \tilde{v} \rangle)$ and the corresponding contours of phase-averaged Reynolds stress $\langle -uv \rangle$. When the vegetation height $\langle h_d(\tau) \rangle$ becomes maximum (Phase 0 and Phase 10), the ejection motion occurs and the phase-averaged Reynolds stress $\langle -uv \rangle$ attains its maximum above the vegetation canopy. When the vegetation is deflected most significantly during Phase 4.0–5.0, the largest positive region of $\langle -uv \rangle$ appears below the vegetation top. This indicates that the large-scale sweep motion deflects the flexible vegetation stems most significantly. The present results are in good agreement with those of Ghisalberti and Nepf (2006), and provide a definite evidence that large positive regions of the instantaneous Reynolds stress are consistent with the coherent sweep and ejection motions and that the vertical transport of momentum occurs most significantly when the vegetation height attains minimum and maximum.

23.4.6 Comparison of Flow Properties between Rigid and Flexible Vegetation

As mentioned earlier, the flexibility of vegetation and the presence of Monami affect the mean flow and turbulence structure in aquatic submerged vegetation flows. Table 23.2 compares typical flow properties between the rigid and the flexible vegetation

TABLE 23.2 Comparison of Flow Properties between Rigid and Flexible Vegetation Flows

Items	Rigid	Flexible
Flow resistance	Large	Small
Infection point in velocity profile	Appear	Appear
Mixing layer-type velocity profile	Appear near vegetation	Appear near vegetation
Reynolds stress	Large peak vegetation edge	Mild small peak
Penetration of momentum within canopy	Large	Small
Mean-eddy scale L_x	$L_x = 1.5h$	$L_x = 10h$
Convection velocity U_c	$U_c = (1.5–2.0)U$ at h	$U_c = (1.5–2.0)U$ at h
Coherent vortices	Sweeps > Ejections within canopy	Sweeps > Ejections within canopy
Interaction between flow and vegetation	Weak	Strong
Mass transport	Large?	Small?

flows. The mean velocity profile has an inflection point at the vegetation edge ($y = h$ for rigid vegetation and $y = \overline{h}_d$ for flexible vegetation), and approaches the hyperbolic tangent profile of Equation 23.28 in the mixing-layer zone for both rigid and flexible vegetations. Consequently, a large-scale coherent vortex is generated due to the inflection-point instability in the same manner as pure mixing layers.

The values of Reynolds stress $-\overline{uv}$ within the canopy become smaller for flexible vegetation than for rigid, and the Monami canopy decreases the turbulent momentum transport toward the bed. The diminished momentum exchange suggests that the coherent vortices may become weaker and smaller in the presence of Monami. This is confirmed by the observed mean-eddy length scale L_x. The value of L_x is larger for rigid vegetation than for flexible (Figure 23.18).

Quadrant analysis revealed that the momentum transfer is more dominated by sweeps within the canopy, and the contributions of sweep events are larger for rigid vegetation than for flexible. The Monami occurs for flexible vegetation because the convection velocity U_c of the coherent vortex is much faster than the mean velocity at the vegetation edge. The value of U_c for a Monami canopy attains $(1.5 - 2.0)U_h$ at the vegetation edge and is on the same order of magnitude as that for a rigid canopy. These results suggest that when the vegetation is waving, the instantaneous drag distribution may not contribute much to the vortex formation. Consequently, the vortex structure would lose coherence in the presence of Monami as compared with rigid vegetation.

Lastly, the effects of flexible canopy on scalar transports, such as various water quality, dissolved gases (O_2 and CO_2), and sediment transport, are not fully investigated as yet. Some researchers suggest that submerged flexible canopies protect the coastal bed by reducing the friction forces near the water–sediment interface, and thus may protect the channel bed from erosive processes. It is further necessary to investigate such effects within the canopy by simultaneous measurements of velocity and scalar concentration including water temperature.

23.5 Major Challenges

23.5.1 Mass Transport in Submerged Vegetation Flow

In aquatic vegetated flows, it is suggested strongly that large-scale coherent vortices govern the vertical exchanges of mass and momentum between the over- and within-canopies. Therefore, it is very important for the river environment to reveal such a mass transport mechanism in vegetated open-channel flows. However, it has been difficult to measure the distributions of scalar concentration such as mass and temperature within canopies in aquatic flows.

Fortunately, the recently developed LIF technique provides a non-intrusive method of high-resolution measurements of mass concentration in water channels. Measuring the concentration field with the LIF technique involves exciting a fluorescent dye tracer with light at a wavelength within its absorption range. The dye re-emits a longer-wavelength light with a measurable intensity. After careful calibration, the digital images of the instantaneous emitted intensity are converted to the concentration field $\tilde{c}(x,y,t)$. If the LIF is successfully combined with PIV, the instantaneous velocity vectors $(\tilde{u}(x,y,t), \tilde{v}(x,y,t))$ and dye concentration $\tilde{c}(x,y,t)$ can be measured simultaneously in space and time, which are very challenging topics to investigate the relation between the coherent vortices and the associated mass transport in aquatic flows, especially in vegetated open-channel flows. These non-intrusive techniques are superior to probe (point) measurements even if the latter may be more accurate than the former.

We tried to conduct such challenging simultaneous measurements with a combination of PIV and LIF in aquatic rigid vegetation flow. Figure 23.24a shows some examples of the instantaneous velocity vectors $(\tilde{u}, \tilde{v})$, which were obtained by PIV. The contours indicate the magnitude of streamwise velocity fluctuations $u(x, y, t)$. Figure 23.24b shows the corresponding simultaneous concentration field $\tilde{c}(x,y,t)$ by LIF, and reveals a high spatial variability of the instantaneous concentration.

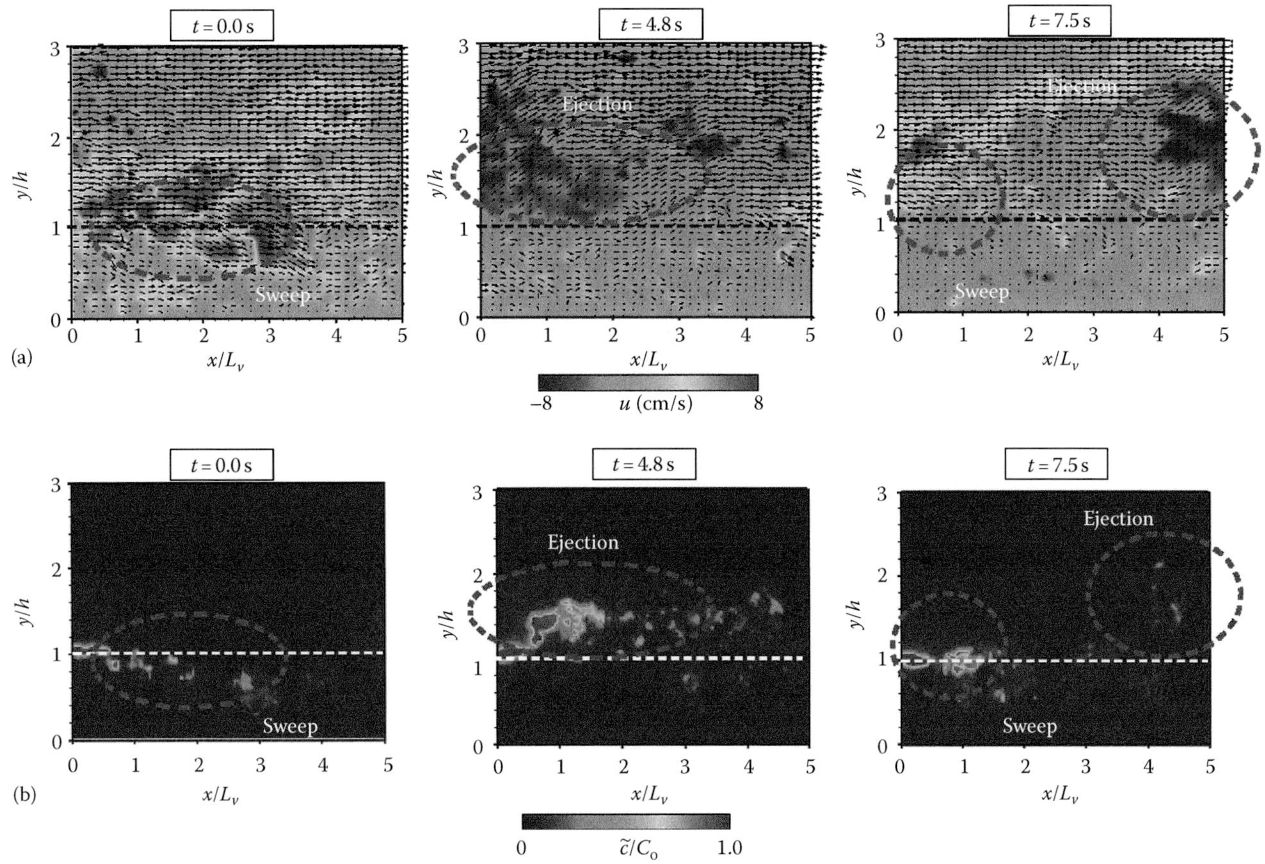

FIGURE 23.24 **(See color insert.)** Simultaneous measurements of (a) instantaneous velocity vectors by PIV and (b) instantaneous concentration by LIF.

At t = 0.0(s), a sweep motion appears near the vegetation edge. It is observed that the large distribution of the instantaneous concentration is transported toward the within-canopy by this sweep motion. At t = 4.8(s), an ejection motion appears and causes the local increase of the concentration above the canopy. At t = 7.5(s), the ejection motion is convected downstream and the other sweep motion appears at the upstream side. The individual filaments of the dye concentration are observed clearly and the highly intermittent nature of the dye plume is found in Figure 23.24b. These images also show that the locally high distributions of dye concentration correspond well to the coherent structure zones. This confirms that large-scale organized motions govern mass transport in aquatic vegetated open-channel flows.

Of particular significance is the evaluation of turbulent scalar flux of a passive contaminant. The PIV–LIF measurements will afford us to evaluate the local covariance between the concentration and the velocity fluctuations, $\overline{uc}$ and $\overline{vc}$. Our preliminary evaluation indicated that the high distribution of concentration ($c > 0$) is transported upward by the ejection motion ($u < 0$). In contrast, $\overline{uc}$ becomes positive within the canopy, which implies that the sweep motion ($u > 0$) transports the dye concentration into the canopy layer ($c > 0$).

23.5.2 Computational Simulation

In this chapter, we treated the fully developed and homogenous aquatic canopy flows by changing the vegetation density and plant flexibility. We investigated turbulence structures and coherent motion in vegetated canopy open-channel flows, on the basis of non-intrusive LDA and PIV measurements as well as recent LIF.

Theoretical consideration revealed that the double-averaging techniques of both time-average and space-average are important for canopy flows and consequently that the wake generation of turbulence due to vegetation is produced within the canopy. The present experimental results of aquatic vegetated flows are in good agreement with theoretical ones and also compared with those of terrestrial canopy flows which have been investigated intensively in meteorological and agriculture sciences.

Of particular significance is that large-scale coherent vortices due to K-H instability govern the turbulence dynamics and promote the mass and momentum exchanges between the vegetated and the non-vegetated zones. One of the most important characteristics of a coherent vortex is its 3-D flow patterns, which show organized structures in the horizontal plane (x–z plane) as well as in the vertical plane (x–y plane). However, it is very difficult at present to measure 3-D flow

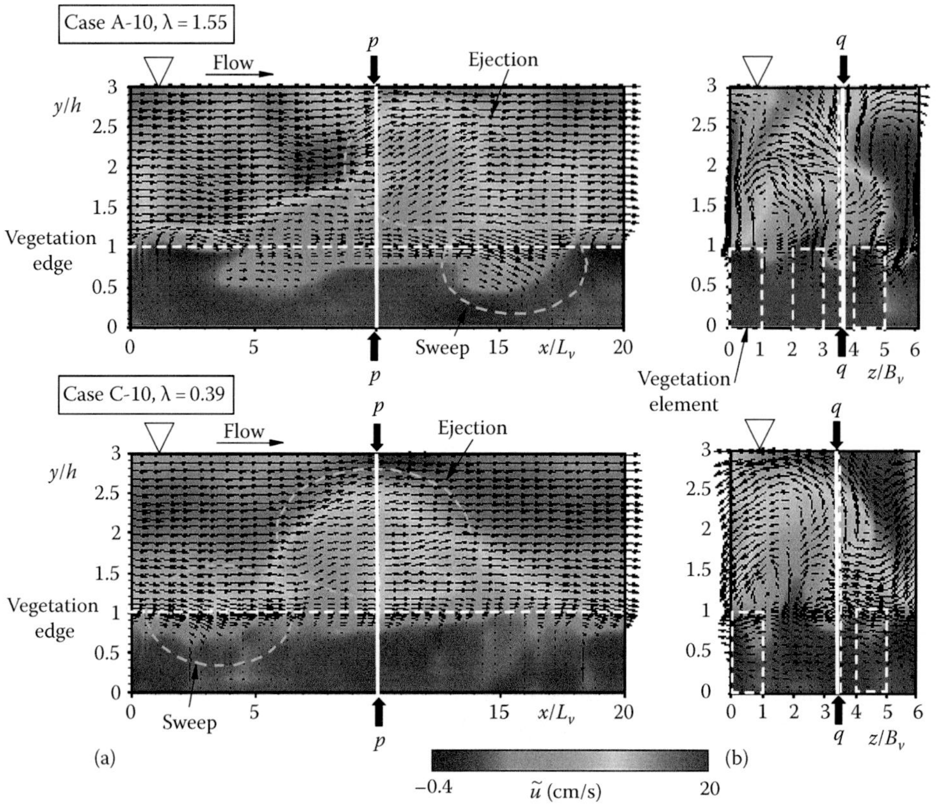

FIGURE 23.25 3-D coherent structures in vegetated canopy flow predicted by LES. (a) x–y plane at the q–q section and (b) y–z cross plane at the p–p section.

patterns quantitatively and simultaneously even by using an innovative stereoscopic PIV as well as multipoints measurements of 3-D LDA and 3-D ADV.

Such essential difficulties of experimental techniques may be overcome by introducing computational simulation. Large-eddy simulation (LES) may be much more powerful to simulate coherent vortices in aquatic vegetated flows even if the accuracy of LES is lower than that of the direct numerical simulation (DNS). Nezu and Sanjou (2008) tried to calculate the characteristics of 3-D coherent vortices generated in rigid vegetated canopy open-channel flow by using an LES, in which a modified Smagorinsky model was used. The calculated values of mean velocity and the second-order moments such as Reynolds stress were in good agreement with the measured ones. Of particular significance is that typical sweep and ejection motions were reproduced reasonably by LES.

Figure 23.25 shows some examples of all three components ($\tilde{u}$, $\tilde{v}$, $\tilde{w}$) of the instantaneous velocity vectors for Case A-10 (dense vegetation $\lambda = 1.55$) and C-10 (sparse vegetation $\lambda = 0.39$). Figure 23.25a indicates the description of velocity vectors ($\tilde{u}$, $\tilde{v}$) in the x–y vertical plane at the non-wake region ($z/B_v = 3.5$), indicated by the q–q section. In Case A-10, the downward vectors, i.e., the sweep motion, appear near the vegetation edge and the high-speed fluid parcel is transported into the within-canopy. The ejection motion, which is upward vectors of low-speed fluid parcel, appears at its upstream side. It is recognized in computational simulation that there is a periodical generation mechanism of sweeps and ejections in vegetated open-channel flows.

On the other hand, Figure 23.25b indicates the description of velocity vectors ($\tilde{v}$, $\tilde{w}$) in the y–z cross sectional plane at $x/L_v = 10$, indicated by the p–p section. These results show that the coherent motions are associated with large-scale secondary flows. It is inferred from Figure 23.25 that a complicated 3-D large-scale organized motion dominates the turbulence structure and thus significant turbulent transports of mass, momentum, and energy in vegetated open-channel flows.

The present LES can calculate turbulence structures and coherent vortices reasonably in aquatic rigid vegetation flows because the boundary conditions of vegetation are not waving. The LES also predicted that large-scale coherent vortices behave in a 3-D organized fashion involving secondary currents, which needs to be confirmed experimentally. However, actual vegetation flows are much more complicated, and non-homogeneous vegetation patterns are often observed. One of the most challenging topics is the LES simulation of flexible vegetation, in which the vegetation elements are waving depending on local fluid velocities. In particular, it is further necessary to predict Monami phenomena reasonably by LES and DNS, in which a relevant interaction mechanism between the flow and the vegetation motion should be introduced.

23.5.3 Further Challenges

In this chapter, we have treated the hydrodynamics of aquatic rigid and flexible vegetation canopy flows in comparison with terrestrial ones. However, actual vegetated flows are much more complicated, and various non-homogeneous vegetation patterns are often observed (Figure 23.1). The hydraulic resistance in vegetated flows depends on many factors, including vegetation density, plant morphology, patchiness, and mechanical properties of plants.

These canopy flows are significantly responsible for water quality, nutrient, and particulate removal. They can also reduce turbidity and sediment transport significantly in rivers and wetlands. Therefore, it is further necessary in hydro-environmental sciences to investigate the following fundamental topics on the basis of the present vegetation researches and furthermore to apply to vegetated rivers and wetlands.

1. Effects of non-homogeneous vegetation patterns: If the vegetation zone and non-vegetation zone exist laterally or partially as well as in patch patterns, the time-averaged secondary currents should be generated due to anisotropy of turbulence, For example, see Nezu and Onitsuka (2001).
2. Effects of transition from upstream non-vegetated flat bed to downstream fully developed vegetation canopy. "Boundary layer theory" is applied to open-channel flows over non-vegetated flat beds, whereas "mixing layer theory" is applied to fully developed canopy flows. It is very challenging to investigate any interrelation between bursting phenomena in boundary layers and Monami phenomena in canopy flows.
3. Effects of reduced Reynolds stress on channel bed and application to scour protection. In the wake zone, the Reynolds stress and bed shear stress are negligibly small and thus suspended sediment will settle down on the channel bed. A Karman vortex street governs such streamwise exchanges in momentum and mass transfer.
4. Finally, effects of water waves induced by wind. When there exists an interfacial shear between air and water, typical coherent vortices such as downbursts are generated by the interfacial shear instability. These coherent vortices of the air/water interfacial shear layer may resemble those of the vegetated canopy shear layer treated in this study. It is one of the most challenging and important topics in environmental hydraulics to investigate gas transfer and dynamic interrelation of coherent vortices between canopy layer and air/water interfacial layer in shallow vegetated open-channel flows with induced water waves.

References

Brunet, Y., Finnigan, J. J., and Raupach, M. R. 1994. A wind tunnel study of air flow in waving wheat: Single-point velocity statistics, *Boundary-Layer Meteorology*, 70, 95–132.

Finnigan, J. J. 1979. Turbulence in waving wheat. I. Mean statistics and Honami, *Boundary-Layer Meteorol.* Vol. 16, 181–211.

Finnigan, J. 2000. Turbulence in plant canopies, *Annu. Rev. Fluid Mech.*, 32, 519–572.

Ghisalberti, M. and Nepf, H. 2002. Mixing layers and coherent structures in vegetated aquatic flows, *J. Geophysical Research - Oceans*, Vol. 107 (C2), doi:10.1029/2001JC000871.

Ghisalberti, M. and Nepf, H. 2006. The structure of the shear layer in flows over rigid and flexible canopies, *Environ. Fluid Mech.*, 6, 277–301.

Ikeda, S. and Kanazawa, M. 1996. Three-dimensional organized vortices above flexible water plants, *J. Hydraul. Eng.*, 122, 634–640.

Jackson, P.S. 1981. On the displacement height in the logarithmic velocity profile, *J. Fluid Mech.*, 111, 15–25.

Kouwen, N. and Li, R. 1980. Biomechanics of vegetative channel linings, *J. Hydraul. Div.*, ASCE, 106, 1085–1103.

Kouwen, N. and Unny, T. E. 1973. Flexible roughness in open channels, *J. Hydraul. Div.*, ASCE, 99, 713–728.

Lopes, F. and Garcia, M. 2001. Mean flow and turbulence structure of open-channel flow through non-emergent vegetation, *J. Hydraul. Eng.*, 127, 392–402.

Nakagawa, H. and Nezu, I. 1977. Prediction of the contributions to the Reynolds stress from the bursting events in open-channel flows, *J. Fluid Mech.*, 80, 99–128.

Nepf, H. M. and Ghisalberti, M. 2008. Flow and transport in channels with submerged vegetation, *Acta Geophysica*, 80, 99–128.

Nepf, H. M. and Vivoni, E. R. 2000. Flow structure in depth-limited, vegetated flow, *J. Geophys. Res.*, 105, 28547–28557.

Nezu, I. and Nakagawa, H. 1993. *Turbulence in Open-Channel Flows*, IAHR-Monograph, Balkema, Rotterdam, the Netherlands.

Nezu, I. and Onitsuka, K. 2001. Turbulent structure in partly vegetated open-channel flows with LDA and PIV measurements, *J. Hydraul. Res.*, 39, 629–642.

Nezu, I. and Sanjou, M. 2008. Turbulence structure and coherent motion in vegetated canopy open channel flows, *J. Hydro-environ. Res.*, 2, 62–90.

Nikora, V., McEwan, I., McLean, S., Coleman, S., Pokrajac, D., and Walters, R. 2007. Double-averaging concept for rough-bed open-channel and overland flows: Theoretical background, *J. Hydraul. Eng.*, 133, 873–883.

Okamoto, T. and Nezu, I. 2009. Turbulence structure and "Monami" phenomena in flexible vegetated open-channel flows, *J. Hydraul. Res.*, 47, 798–810.

Poggi, D., Porpotato, A., and Ridolfi, L. 2004. The effect of vegetation density on canopy sub-layer turbulence, *Boundary-Layer Meteorol.*, 111, 565–587.

Raupach, M. R., Antonia, R. A., and Rajagopalan, S. 1991. Rough-wall turbulent boundary layers, *Appl. Mech. Rev.*, ASME, 44, 1–25.

Raupach, M. R., Finnigan, J. J., and Brunet, Y. 1996. Coherent eddies and turbulence in vegetation canopies: The mixing-layer analogy, *Boundary-Layer Meteorol.*, 78, 351–382.

Wilson, C. A. M. E., Stoesser, T., Bates, P. D., and Pinzen, A. B. 2003. Open channel flow through different forms of submerged flexible vegetation, *J. Hydraul. Eng.*, 129, 847–853.

24

Canopy Turbulence

E.G. Patton
National Center for Atmospheric Research

J.J. Finnigan
Commonwealth Scientific and Industrial Research Organisation

24.1 Introduction

Vegetation covers nearly 30% of the Earth's land surface and influences climate through the exchanges of energy, water, carbon dioxide, and other chemical species with the atmosphere (Bonan, 2008). Through the process of photosynthesis (using the Sun's energy to convert carbon dioxide from the atmosphere into sugars), vegetation generally grows wherever there is available water and light (Monteith and Unsworth, 2008). The complement of photosynthesis is transpiration, the movement of water from the soil through the plants and into the air, a process which plants use both to access nutrients and to regulate their temperature. Compared to evaporation from bare soils, transpiration is extremely efficient at moving water between layers deep in the soil and the atmosphere.

The Earth's vegetation plays a critical role in the hydrological, nitrogen, and phosphorus cycles, as well as providing habitat and shelter for biota that deliver essential ecosystem services such as pollination. Living foliage also produces a variety of chemical compounds that can significantly influence the oxidation capacity (cleansing ability) of the atmosphere (e.g., Fuentes et al., 2000; Guenther et al., 2006). In addition, plant–atmosphere interactions can have negative effects with billions of dollars each year lost from wind damage to forests and crops. At the most fundamental level, therefore, understanding the processes controlling vegetation–atmosphere exchange is of critical importance for weather, climate, and environmental forecasting as well as for agricultural and natural resource management.

This chapter aims to provide the reader with a basis for embarking upon further studies into the complex phenomena associated with canopy turbulence by detailing the various ways in which it is different from turbulence elsewhere in the atmospheric boundary layer (ABL) and by explaining the reasons for the differences. The chapter assumes that the reader has a basic knowledge of the dynamics of atmospheric turbulence, particularly surface layer turbulence such as is set out in texts like Garratt (1992) or Kaimal and Finnigan (1994), although some of the basic principles will be recapitulated. No special knowledge of canopy flows is required.

First, some essential features of turbulence in the ABL near the ground but well above the canopy (Section 24.2) are introduced, and then Section 24.3 goes on to set out in broad terms the ways that turbulence in the canopy layer is different. The discussion becomes more formal in Section 24.4, where the canopy flow equations are derived and the extra terms affecting motions in the canopy are discussed. In Sections 24.5 through 24.8, increasingly detailed evidence for the unique features of canopy turbulence is presented, drawing on data from field measurements, wind tunnels, and numerical simulations. Finally, in Section 24.9 these various lines of evidence are brought together by setting out the current picture of canopy turbulence dynamics, which successfully explains most of the unique features.

Handbook of Environmental Fluid Dynamics, Volume One, edited by Harindra Joseph Shermal Fernando. © 2013 CRC Press/Taylor & Francis Group, LLC. ISBN: 978-1-4398-1669-1.

Space dictates that the discussion will focus on momentum and energy exchange in horizontally homogeneous canopies under neutrally stratified conditions, but pointers to more complex situations are provided in Section 24.10. Emphasis throughout is placed on the basic physics of canopy flows with a clear focus on explaining the dynamics of "organized" turbulent motions, as it is this feature more than any other that sets canopy turbulence apart from that in the boundary layer above.

24.2 Atmospheric Turbulence near Rough Walls

The ABL is the atmospheric layer which responds directly to changes in the surface fluxes of momentum, energy, and other scalars. Its depth varies from a few hundred meters in mid-latitudes at night to several kilometers in the daytime tropics. Under most conditions, the ABL is fully turbulent. "Turbulent fluxes" are the average correlations between random fluctuations in velocity and scalars, which transport momentum, energy, and other scalars efficiently from the Earth's surface into the overlying troposphere. Vertical mixing in the troposphere is generally nonturbulent and is therefore several orders of magnitude slower than in the ABL. For a thorough discussion of the ABL and its dynamics see Garratt (1992).

In the lowest 100–200 m (i.e., the lowest 10% or so) of the ABL, a series of simplifying assumptions can be made regarding the way the turbulent wind behaves; this region of the ABL is called the "atmospheric surface layer" (ASL). The incompressible Boussinesq equations are the starting point for describing fluid motion in the ASL (Businger, 1982; Wyngaard, 2010). Careful consideration of the scales of various terms in these equations (Tennekes, 1982) can be used to obtain the equations for time-averaged velocity, continuity, and virtual potential temperature:

$$\bar{u}_j \frac{\partial \bar{u}_i}{\partial x_j} = -\frac{1}{\rho_\circ}\frac{\partial p}{\partial x_i} - \frac{\partial \overline{u_i' u_j'}}{\partial x_j} + \frac{g}{\theta_\circ}\bar{\theta}\delta_{i3} - \nu\frac{\partial^2 \bar{u}_i}{\partial x_j \partial x_j}, \tag{24.1}$$

$$\frac{\partial \bar{u}_i}{\partial x_i} = 0, \tag{24.2}$$

$$\bar{u}_j \frac{\partial \bar{\theta}}{\partial x_j} = -\frac{\partial \overline{\theta' u_j'}}{\partial x_j} - \alpha\frac{\partial^2 \bar{\theta}}{\partial x_j \partial x_j}, \tag{24.3}$$

where these equations are written in tensor notation for compactness and Einsteins's summation convention assumed. x_i is the position vector, and u_i the corresponding velocity vector in a right-handed Cartesian coordinate system, where x_1 is the streamwise, x_2 the lateral, and x_3 the vertical direction. In Equations 24.1 through 24.3, $\rho_\circ$ is a reference density, $\theta_\circ$ is a reference temperature, p is pressure, g is the gravitational acceleration, θ is the virtual potential temperature, ν is the kinematic fluid viscosity, α is the thermal molecular diffusivity, δ_{i3} is the Dirac delta function. As necessary, the equivalent notation $x_i \equiv (x_1, x_2, x_3) \equiv (x, y, z)$, and $u_i \equiv (u_1, u_2, u_3) \equiv (u, v, w)$ will be used, especially when referring to published data.

Flow in the ASL is usually turbulent so that time-averaged quantities are distinguished with an overbar and turbulent fluctuations around these averages with a prime, hence $u_i(x_i,t) = \bar{u}(z) + u_i'(x_i,t)$. The average products of fluctuations which appear on the right-hand side (RHS) of Equations 24.1 and 24.3 are the turbulent or "Reynolds" fluxes, which are the result of splitting quadratic terms in the original Boussinesq equations into means and fluctuations and then time averaging them. For example, the expression $\overline{u_1' u_3'} = \overline{u'w'}$ should be interpreted as the vertical kinematic turbulent flux density of x-direction (streamwise) mean momentum ($\rho\bar{u}$). Because the surface is a momentum sink, the flux of mean streamwise momentum is toward the surface (i.e., in the negative z-direction) so that $\overline{u'w'}$ is negative.

Away from the immediate vicinity of the individual elements making up the rough ground surface, turbulent mixing ensures that only the broad features of the surface are reflected in gradients of mean quantities. Hence in the canonical case of horizontally homogeneous steady flow, all horizontal gradients of mean quantities can be set to zero relative to vertical gradients. With this assumption, the continuity equation (Equation 24.2) demands that $\bar{w}/\bar{u} \ll 1$ and the streamwise momentum equation (24.1) becomes

$$0 = -\frac{\overline{u'w'}}{\partial z} + \nu\frac{\partial^2 \bar{u}}{\partial z^2}. \tag{24.4}$$

The second term on the RHS of Equation 24.4 represents the viscous stress, that is, the vertical flux of mean streamwise momentum carried by molecular motion. Except very close to the surfaces of leaves and other canopy elements (as will be discussed in the next section), the viscous stress is negligible compared to the turbulent shear stress. So, when the ASL is turbulent, $\overline{u'w'}(z) \simeq$ constant. Except during stably stratified, low-wind conditions at night, the ASL is invariably turbulent so that the turbulent shear stress, $\overline{u'w'}$ and turbulent fluxes of other quantities like heat and moisture are also approximately constant with height in the ASL. This deduction allows the use of these constant fluxes as parameters in a set of relationships known as Monin–Obukhov Similarity Theory (MOST).

In neutrally stratified conditions, MOST takes the well known "logarithmic-law" form, which relates the height dependence of the mean horizontal velocity $\bar{u}(z)$ to the constant shear stress $\overline{u'w'}$ and to parameters characterizing the surface. The "log-law" for velocity is conventionally written:

$$\bar{u}(z) = \frac{u_*}{k}\ln\left(\frac{z-d}{z_\circ}\right), \tag{24.5}$$

where

$u_* = \sqrt{-\overline{u'w'}/\rho}$ is known as the "friction velocity"

z_o is known as the roughness length, and characterizes the momentum absorbing capacity of the rough surface

d is called the displacement height, because over tall roughness (canopies are a typical example) the origin of the log-law coordinate system is not the ground surface but at a height d which occurs somewhere between the ground and the top of the roughness elements (Jackson, 1981)

k is a universal constant called von Kármán's constant which is usually taken as equal to 0.4

Differentiating Equation 24.5 with respect to height z yields

$$\frac{\partial \bar{u}}{\partial z} = \frac{u_*}{k(z-d)} \quad \text{or} \quad \tau = \rho u_*^2 = [ku_*(z-d)]\frac{\partial \bar{u}}{\partial z}. \tag{24.6}$$

In other words, the log-law is equivalent to a "flux–gradient" relationship linking the turbulent momentum flux $\overline{u'w'}$ to the gradient of mean velocity via an "eddy diffusivity," $K_M = [ku_*(z-d)]$.

Analogous log-law or flux–gradient relationships can be found for the various scalars, and these (with suitable assumptions) allow turbulent fluxes to be estimated from the gradients of their mean values; this is important for quantities whose turbulent fluctuations are difficult to measure directly. These log-laws can be readily extended to diabatic flows by incorporating another length scale into the similarity analysis, that is, the Obukhov length L (Obukhov, 1946, 1971; Monin and Obukhov, 1954). The Obukhov length is related to the turbulent buoyancy flux $\overline{w'\theta'}$. Detailed discussions regarding the derivation and application of MOST in the ASL, and how profiles of velocity and scalars and their eddy diffusivities vary with diabatic stability can be found in Tennekes and Lumley (1972), Garratt (1992), and Kaimal and Finnigan (1994). So pervasive are these logarithmic relationships that the ASL is often referred to simply as the *log layer* and is why MOST is often regarded as the basis of conventional surface-layer micrometeorology.

24.3 Atmospheric Turbulence within a Canopy

As discussed in Section 24.2, vertical turbulent fluxes through the ASL are not strictly constant but approximate this condition closely because sources or sinks of momentum such as synoptic pressure gradients and Coriolis forces play a negligible role in ASL dynamics. Within a plant canopy, however, this situation changes. The downward turbulent momentum flux $\overline{u'w'}$ continually decreases with descent into the canopy because mean streamwise momentum $\rho\bar{u}$ is absorbed through aerodynamic drag on the plants. Equivalent to Equation 24.4, this relationship can be expressed as

$$0 = -\frac{\partial \left\langle \overline{u'w'} \right\rangle}{\partial z} - F(z), \tag{24.7}$$

where F is the aerodynamic drag force on the foliage. Note that Equation 24.7 introduces new notation. ⟨⟩ denotes that the shear stress has been spatially averaged as well as time-averaged, recognizing that the complex quasi-random nature of the canopy airspace requires spatial averaging to produce statistics reflecting the average properties of the vegetation. Therefore, within the canopy, the similarity scaling leading to the log-law and MOST no longer applies and the direct connection between turbulent fluxes and local gradients (embodied in flux–gradient relationships like Equation 24.6) is lost, which leads to surprising phenomena such as "counter-gradient" diffusion of scalars and momentum (Denmead and Bradley, 1985; Finnigan, 1985). To appreciate how this comes about requires a basic understanding of the nature of canopy turbulence.

When the study of plant canopy turbulence was in its infancy, researchers assumed that it could be treated like boundary layer turbulence with the addition of the fine-scale eddies generated in the wakes of the leaves, twigs, and branches. By the 1970s, it had become clear that the dominant eddies in plant canopy turbulent flows are much larger than plant element size. However, it took a further two decades of research to show that these eddies are generated by a hydrodynamic instability associated with an inflection point in the mean velocity profile; where, this inflection results from the absorption of the fluid's momentum throughout the entire canopy depth as drag on the plant elements rather than as friction on the ground, as is the case on a plane rough surface (Finnigan, 2000).

The distribution of mean velocity in the canopy airspace is the result of the interaction between the downward turbulent transport of momentum and canopy drag. Canopy drag varies with height and depends on both the foliage distribution and the velocity field itself. Similarly, the within-canopy distribution of scalar concentrations results from a balance between turbulent transfer and the distribution of scalar sources and sinks, where these scalar sources/sinks are determined by (1) solar radiation as it filters through the foliage, (2) the biological state of the plants such as their access to soil water, and (3) the ambient concentrations of temperature, humidity, CO_2, and other scalars in the canopy airspace.

Because canopy turbulence is dominated by relatively large eddies which are generated near the canopy top and affect the flow in the lower part of the log layer above, MOST flux–gradient relationships actually start to become inapplicable at heights well above the canopy (Figure 24.1). This region of the ASL, where standard MOST relationships are modified by the canopy, is called the *roughness sublayer* and extends up to $z/h = 2$ or 3, where h is the canopy height. Within the upper reaches of the roughness sublayer, MOST requires adjustment (e.g., Harman and Finnigan, 2007); within the canopy itself, MOST must be completely abandoned.

24.4 Canopy Equations

Much of our understanding of canopy turbulence arises through observations. Measurements taken at a particular location within the canopy airspace necessarily reflect the interaction of the wind with nearby canopy elements. This adds enormous

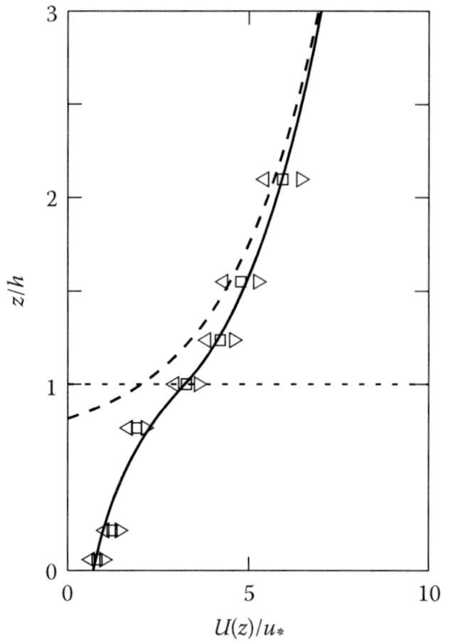

FIGURE 24.1 Vertical profiles of mean wind speed in neutral conditions. Comparison with observations from Moga forest (+, where the triangles denote one standard deviation). Dashed line: surface-layer MOST profile, solid line: profile predicted by Harman and Finnigan's (2007) theory, where MOST is modified to include the influence of a vegetation canopy. Dotted line indicates the canopy top. Note the departure of the standard MOST profile from the measurements and the measured profile through the canopy but also in the RSL above.

spatial complexity to the temporal fluctuations characteristic of all turbulent flows. Therefore, to produce equations representative of the canopy flow on average, Raupach and Shaw (1982), Finnigan (1985), Raupach et al. (1996), and Finnigan and Shaw (2008) introduced a "double-averaging strategy." In this approach, the flow equations at a single point are averaged first in time and then over a representative plane or thin slab parallel to the ground. This second operation captures the fundamental vertical heterogeneity of the canopy but smooths out random plant-to-plant variations in the horizontal.

As before, let an overbar ($\bar{\ }$) represent the time-averaging operation and a single prime (′) denote deviations from that average. Let angle brackets ($\langle\rangle$) denote a volume-averaging operation over a thin horizontal slab spanning numerous canopy elements and a double prime (″) denote deviations from that average. In the presence of canopy elements, the volume-averaging operator ($\langle\rangle$) does not commute with spatial-differentiation operators ($\partial/\partial x_i$), implying that

$$\left\langle \frac{\partial \bar{\phi}}{\partial x_i} \right\rangle = \frac{\partial \langle \bar{\phi} \rangle}{\partial x_i} - \frac{1}{V} \iint_{S_I} \phi n_i \mathrm{d}S. \tag{24.8}$$

In Equation 24.8,

ϕ is any random variable

S_I is the part of the bounding surface S of the averaging volume V that coincides with the plant surfaces

n_i is the unit normal vector pointing away from S into V.

Time-averaging Equations 24.1 through 24.3, substituting the equivalent of $\bar{\phi} = \langle \bar{\phi} \rangle + \bar{\phi}''$ for all time-averaged variables, and then volume averaging results in

$$\langle \bar{u}_j \rangle \frac{\partial \langle \bar{u}_i \rangle}{\partial x_j} + \frac{\partial \langle \overline{u_i' u_j'} \rangle}{\partial x_j} + F_{\tau_i} = -\frac{1}{\rho_\circ} \frac{\partial \langle \bar{p} \rangle}{\partial x_i} + \frac{g}{\theta_\circ} \langle \bar{\theta} \rangle \delta_{i3} - \nu \frac{\partial^2 \langle \bar{u}_i \rangle}{\partial x_i \partial x_j} + F_{p_i} + F_{v_i} \tag{24.9}$$

$$\frac{\partial \langle \bar{u}_i \rangle}{\partial x_i} = 0, \tag{24.10}$$

$$\langle \bar{u}_j \rangle \frac{\partial \langle \bar{\theta} \rangle}{\partial x_j} + \frac{\partial \langle \overline{u_j' \theta'} \rangle}{\partial x_j} + F_{\tau\theta} = -\alpha \frac{\partial^2 \langle \bar{\theta} \rangle}{\partial x_j \partial x_j} + S_\theta, \tag{24.11}$$

where the terms which would arise through plant motion have been ignored because they are usually very small and can be incorporated into modifications of the canopy drag terms (F_{pi} and F_{vi}, to be discussed next). However, for a more complete discussion of flexible plant canopies and their interplay with turbulence, see Finnigan (1985), Ghisalberti and Nepf (2006) and Dupont et al. (2010), and references therein. Equations 24.9 and 24.11 reveal that spatially averaging the equations in the multiply connected canopy airspace generates terms (e.g., F_{p_i}, F_{v_i}, F_{τ_i}, $F_{\tau\theta}$, and S_θ) which do not appear in the well-known Reynolds equations (Equations 24.1 through 24.3) that apply in the free-air above the canopy. These terms will now be discussed in more detail.

24.4.1 Canopy Drag Forces

Viscous drag is generated in the thin viscous boundary layers which develop as air flows over and is brought to rest at each solid surface in the canopy by the "no-slip" condition of viscous fluid flow. Pressure or "form" drag results from the front–back pressure differences that develop around each canopy element. The viscous (F_{vi}) and pressure drag (F_{pi}) terms represent the two mechanisms by which momentum is transferred to the foliage and arise through the application of the noncommutation relationship (Equation 24.8) to the pressure gradient and viscous flux terms:

$$F_{p_i} = \frac{1}{V} \iint_{S_I} \bar{p} n_i \mathrm{d}S, \tag{24.12}$$

$$F_{v_i} = -\frac{\nu}{V} \iint_{S_I} \frac{\partial \bar{u}_i}{\partial n} \mathrm{d}S. \tag{24.13}$$

The tangential viscous stresses and normal pressure forces on all the plant elements in the averaging volume V sum to the total aerodynamic force so that

$$F_i = F_{p_i} + F_{v_i}. \tag{24.14}$$

As mentioned in Wilson and Shaw (1977) and reiterated in Finnigan and Shaw (2008), the drag terms can only appear explicitly in equations which have been spatially averaged over a volume containing the object exerting the drag.

When the Reynolds number of an individual leaf, twig, or branch is high enough, the drag force is dominated by pressure drag, which is usually the case in moderate to strong winds through the canopy. However, in low winds or deep in the canopy, viscous drag can dominate (Finnigan and Raupach, 1987). Using direct measurements on individual branches and leaves, Thom (1968) determined that when canopy winds are moderate or strong, pressure drag is at least three times larger than viscous drag. Therefore, canopy drag is typically parameterized using a quadratic velocity dependence, namely,

$$F_i = -C_d a V \bar{u}_i, \tag{24.15}$$

where C_d is a drag coefficient representing the elements' momentum absorption efficiency, a is a frontal plant area density (one-sided plant area per unit canopy volume) which varies spatially but under horizontally homogeneous conditions describes the average canopy element density profile, and the square of the velocity is written as the product of the scalar wind speed ($V = \sqrt{\bar{u}_i \bar{u}_i}$) and the local mean velocity vector $\left(\langle \bar{u}_i \rangle\right)$ so that the canopy drag force (F_i) is a vector which is always directed against the local wind direction. Note that F in Equation 24.7 is equivalent to F_1.

Consistent with the changing contributions of pressure and viscous drag as element Reynolds numbers change, the magnitude of the canopy-level drag coefficient C_d varies slightly with wind speed. Attempts to relate the whole canopy C_d to the drag coefficients of individual canopy elements measured in isolation have generally been unsuccessful. An apparent *shelter effect* ensures that whole-canopy drag is generally lower than the sum of the drag of individual canopy elements in isolation (Finnigan and Raupach, 1987). In practice, the combination $L_c = [C_d\, a]^{-1}$ always appears in the canopy flow equations rather than C_d or a alone, and has a direct interpretation: L_c represents the *e-folding* distance or adjustment length required for a steady airflow encountering a canopy to come into equilibrium with canopy drag (Belcher et al., 2008). The key message from this discussion is that compared to turbulent shear layers over plane rough walls, porous vegetation canopies extract momentum from the flow throughout the canopy's depth at rates dependent upon the geometry of the canopy elements (captured in the parameter L_c) and the local wind speed.

24.4.2 Scalar Sources and Sinks in the Canopy

Vegetation also absorbs or scatters incoming solar radiation across a broad range of the incident spectrum, which prevents a portion of that radiation from reaching the ground surface. Canopy turbulence is indirectly but strongly impacted by the spatial distribution of this radiation absorption. Depending on the spatial distribution of the elements, some regions of the canopy absorb or emit more radiation than others. The air contacting these elements is subsequently warmed or cooled through sensible heat exchange. Photosynthesis and transpiration ensure that not all radiant energy absorbed by the elements becomes immediately available to warm air around them. A small amount of that energy (~1%) is used in photosynthesis, while a much larger fraction may be used to move water from the soil, through the plant roots, and then to transpire water vapor through the leaf stomata. Through transpiration, biology therefore directly controls radiant energy partitioning into sensible and latent heat exchange at the plant's surfaces, thereby regulating the plant's temperature, the buoyancy of the air in the canopy, and ultimately affects the state of the canopy turbulence.

Double-averaging the conservation equations for heat (e.g., Equation 24.3), moisture, and CO_2 (and any other scalar produced/consumed by vegetation) generates additional terms representing the net result of the exchange of that scalar across all the vegetation surfaces in the control volume V (e.g., S_θ in Equation 24.11):

$$S_\theta = \frac{\alpha}{V} \iint_{S_I} \frac{\partial \bar{\theta}}{\partial n}\, dS \tag{24.16}$$

and represents the scalar sources and sinks in the canopy. In analogy to the drag terms, S_θ arises through the application of the noncommutation relationship (Equation 24.8) to the molecular diffusion term. It should also be noted that in deriving Equation 24.11, the radiative flux divergence within the airspace was ignored as this is usually negligible.

24.4.3 Dispersive Fluxes

The double-averaging process also produces another additional term in each of Equations 24.9 and 24.11, that is, $F_{\tau i}$ and $F_{\tau\theta}$, respectively. These terms are of the form

$$F_{\tau_i} = \frac{\partial \langle \bar{u}_i'' \bar{u}_j'' \rangle}{\partial x_j} \tag{24.17}$$

$$F_{\tau_\theta} = \frac{\partial \langle \bar{u}_j'' \bar{\theta}'' \rangle}{\partial x_j} \tag{24.18}$$

and are called dispersive fluxes. They represent the spatial correlation of quantities averaged in time but varying with position (Raupach and Shaw, 1982; Finnigan and Shaw, 2008) so that they bear the same relationship to spatial averaging that the more familiar Reynolds fluxes do to time averaging. Therefore, within a plant canopy, the total spatially averaged flux is the sum of the Reynolds and dispersive fluxes.

In practice, dispersive fluxes are difficult to observe and have by default been assumed to be small. However, Böhm et al. (2000) and Poggi et al. (2004) presented evidence from wind tunnel and water flume measurements within canopies of sparse bluff bodies showing that (1) dispersive fluxes can be as much as 10% of the total momentum flux, and (2) dispersive scalar fluxes are significantly more important than those for momentum. This finding is particularly relevant for urban canopies, but it is not yet clear if it carries over to natural vegetation.

24.5 Turbulence Moments

In deriving Equation 24.9, the continuous Boussinesq equations were first time-averaged at a point, and were then spatially averaged. However, in practice, most canopy measurements are typically limited to time-averages from vertical arrays of sensors at a single location. Exceptions to this have become more common in recent years, though, with explicit spatial averaging in wind tunnel model canopies (e.g., Böhm et al., 2000), multiple tower arrays in the field (e.g., Feigenwinter et al., 2008), and dense 3D sensor arrays dedicated to educing turbulent spatial structure in the field (Patton et al., 2011). In most cases, however, there is an implicit assumption that time-averaged statistics are representative of the spatially averaged moments at a given height in the foliage. In contrast, high-resolution turbulence-resolving 3D numerical models typically take advantage of spatially available information by combining explicit horizontal and time averages (e.g., Patton et al., 2003).

With these caveats, a *family portrait* of turbulence statistics from within and above a wide variety of vegetation types reveals systematic differences between turbulence moments in the ASL and those within and just above a canopy, as well as emphasizing distinct similarities across a range of canopy types (Figure 24.2, adapted from Raupach et al., 1996). Many of the remaining differences between canopies can be attributed to variations in the vertical profile of foliage area density, a (Figure 24.2f).

In Figure 24.2, all lengths are normalized by canopy height (h), and all velocity moments by either the canopy-top mean wind ($\langle \underline{u}(h) \rangle$) or by the canopy-top friction velocity $u_* = [-\langle \overline{u'w'} \rangle(h)]^{1/2}$. The canopies presented in Figure 24.2 span a wide range of foliage density (a, defined as the total frontal area of canopy elements per unit ground area, Figure 24.2f) and a 500-fold height range from wind tunnel model canopies to forests.

As foreshadowed in Section 24.3, momentum absorption through the canopy depth ensures that canopy mean velocity profiles (Figure 24.2a) are distinctly different from profiles in the ASL. Mean streamwise velocity profiles follow a standard logarithmic surface layer form well above the canopy, while within the canopy space the profile can be roughly described as exponential (e.g., Cionco, 1965). The two types of profiles merge within the "roughness sublayer" and a clear inflection point (where the streamwise velocity shear is at a maximum) appears at the top of all but the sparsest canopies. Some of the mean velocity profiles also exhibit a secondary streamwise velocity maximum at $z/h \simeq 0.25$. The sub-canopy secondary wind maximum is limited to canopies where the understory is less dense than the overstory and can arise from a variety of processes such as vertical transport of momentum from aloft or through horizontal pressure gradients across forests of limited fetch (e.g., Shaw, 1977; Holland, 1989).

Above the canopy, the momentum fluxes are nearly constant with height (typical of the ASL) but rapidly decay with descent into the canopy as streamwise momentum is absorbed through canopy drag (Figure 24.2b); note that $-\langle \overline{u'w'} \rangle / u_*^2$ reduces nearly to zero by ground level, indicating that nearly all horizontal momentum has been absorbed by the canopy elements even though the ensemble spans a wide range of canopy density.

Normalized standard deviations of streamwise and vertical velocity (Figure 24.2c and d) do not collapse as well as the momentum fluxes, but like the momentum fluxes they decrease rapidly with height within the canopy. Just above the ground σ_u/u_* has a non-zero value across the range of canopy types, which suggests the presence of horizontal motion at this level; since there is little momentum flux here, the motions are considered "inactive" (e.g., Katul et al., 1998). σ_w/u_* tends toward zero with descent toward the surface as the vertical velocity fluctuations are blocked by the wall below.

Above the canopy, the correlation coefficient ($r_{uw} = \langle \overline{u'w'} \rangle / (\sigma_u \sigma_w)$, Figure 24.2e) tends toward its standard surface layer value of about −0.35 (Garratt, 1992). Near the canopy top, however, r_{uw} increases to about −0.5, indicating a much higher efficiency of momentum transport compared to the overlying ASL.

Skewness is a measure of the asymmetry of a random variable's probability distribution.

Above the canopy, streamwise and vertical velocity skewness profiles ($Sk_u = \langle \overline{u'^3} \rangle / \sigma_u^3$ and $Sk_w = \langle \overline{w'^3} \rangle / \sigma_w^3$, respectively) tend toward the near-zero values expected in the ASL (Figure 24.2f and g). However, as the canopy is approached from above, Sk_u becomes positive (and Sk_w negative) reaching canopy-top values of about 1 and −1, respectively.

Thirty-minute time traces of the end point of the u', w' fluctuating velocity vector at two heights above and within a deciduous mixed-hardwood forest (Shaw and Patton, 2003) clearly illustrate the contrast between the pattern of turbulence well above the canopy and that just within the stand (Figure 24.3). The density of the lines traced by this vector is effectively the joint probability distribution of u' and w'. The u' and w' velocity signals near treetop level reveal the increased correlation and increased skewness not present in the surface layer above. It is clear from these figures that turbulent momentum transport near the canopy top is effected by frequent upward motions, bringing low momentum fluid upward (ejections) with infrequent but stronger downward penetrations of high momentum fluid from aloft (sweeps). In contrast to surface layer flows, where the contributions from sweeps (Q4) and ejections (Q2)

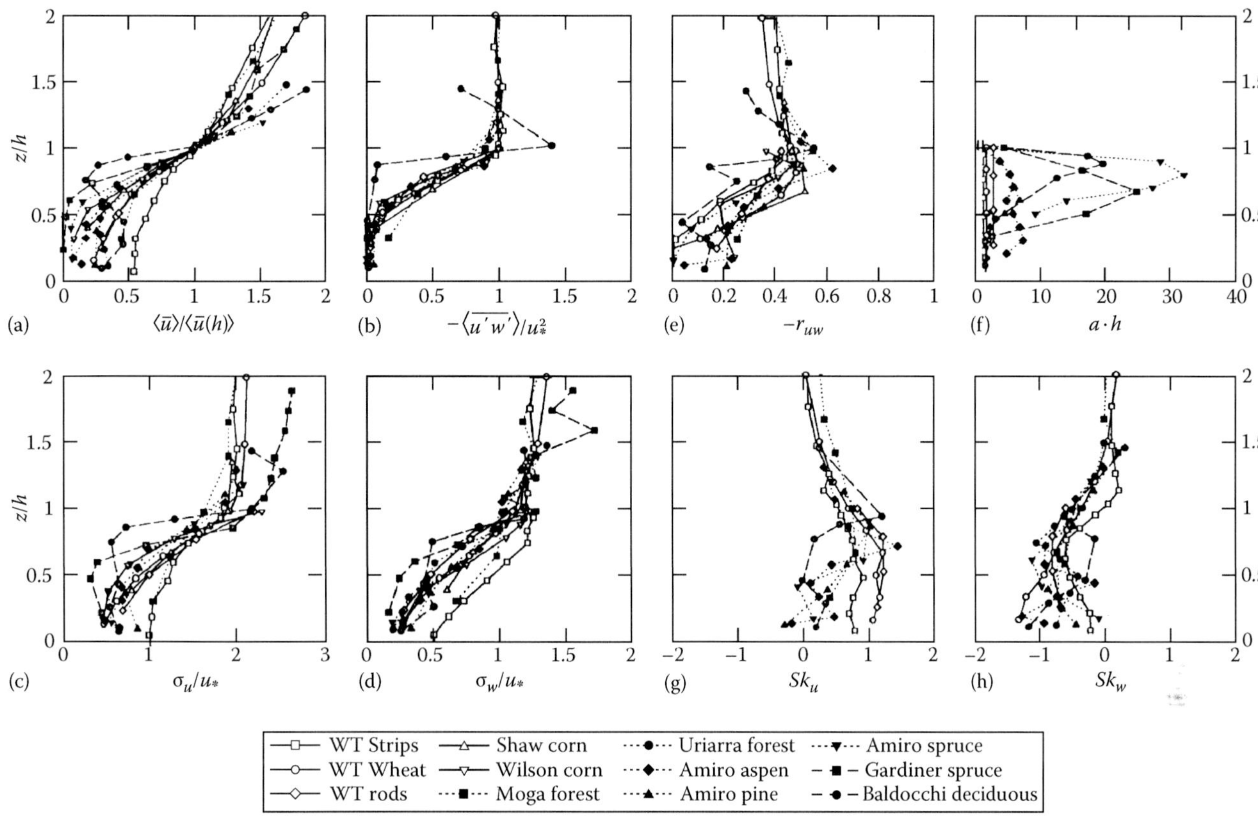

FIGURE 24.2 A family portrait of canopy turbulence statistics in near neutral stability conditions across a wide range of canopy types, showing profiles with normalized height (z/h) of (a) normalized mean streamwise velocity, $\langle\bar{u}\rangle/\langle\bar{u}(h)\rangle$; (b) normalized vertical momentum flux, $-\langle\overline{u'w'}\rangle/u_*^2$; (c) normalized streamwise velocity standard deviation, $\sigma_u/u_* = \langle\overline{u'^2}\rangle^{1/2}/u_*$; (d) normalized vertical velocity standard deviation, $\sigma_w/u_* = \langle\overline{w'^2}\rangle^{1/2}/u_*$; (e) streamwise and vertical velocity correlation coefficient, $-r_{uw} = -\langle\overline{u'w'}\rangle/(\sigma_u\sigma_w)$; (f) normalized vegetation element density distribution, $a(z)\,h$; (g) streamwise velocity skewness, $Sk_u = \langle\overline{u'^3}\rangle/\sigma_u^3$; (h) vertical velocity skewness, $Sk_w = \langle\overline{w'^3}\rangle/\sigma_w^3$. (Adapted from Raupach, M.R. et al., *Bound. Lay. Meteorol.*, 78(3), 351, 1996.)

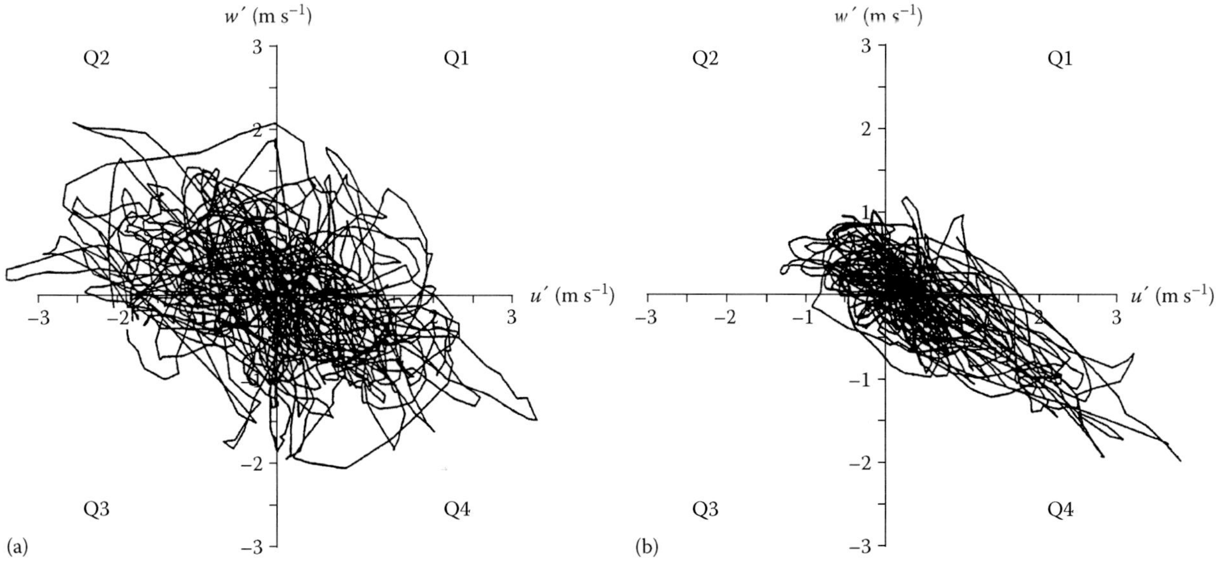

FIGURE 24.3 A 30 min time trace of the perturbation u- and w-velocity signals near a deciduous mixed-hardwood forest in Camp Borden, Ontario, Canada, at two heights: (a) z/h = 2.4, and (b) z/h = 0.86. The quadrants Q2 and Q4 depict the ejection and sweep phases, respectively. (Adapted from Shaw, R.H. and Patton, E.G., *Agric. For. Meteorol.*, 115, 5, 2003.)

are of comparable magnitude or in which ejections dominate (Raupach and Thom, 1981), quadrant analysis of canopy turbulence reveals that in the roughness sublayer, sweeps provide a greater contribution to the total exchange of momentum and scalars than do the ejections (Finnigan, 1979; Shaw et al., 1983; Finnigan et al., 2009).

The essential features of these second moments and joint probability distributions depart significantly from those in the ASL. However, they reveal characteristics which are similar in many ways to those of turbulence in plane mixing layers (i.e., shear flows that develop at the interface between two flows of different speeds which are allowed to mix). One such characteristic is the canopy-top inflection in the mean velocity profile, which led Raupach et al. (1996) to propose the "mixing layer hypothesis," stating that canopy flows are much closer to mixing layer flows than to standard rough wall boundary layers. This idea will be elaborated upon more thoroughly in Section 24.9.

24.6 Budgets of Turbulent Second Moments

Budgets of turbulent second moments illustrate the relative importance of the processes contributing to the production and destruction of turbulent kinetic energy (TKE) and of turbulent momentum transport. Within a plant canopy, extra processes must be considered in addition to those that contribute to the turbulence budgets in the ASL.

Deriving the full second-moment equations (i.e., the conservation equation for $\langle \overline{u_i'u_j'} \rangle$) in the presence of vegetation requires a number of steps (see, e.g., Brunet et al., 1994). One-half the trace of this full second-moment equation ($i = j$) is TKE budget (Raupach and Shaw, 1982):

$$\left(\frac{\partial}{\partial t}+\langle \bar{u}_j \rangle \frac{\partial}{\partial x_j}\right)\frac{\langle \overline{u_i'u_i'} \rangle}{2}$$

$$= \underbrace{-\langle \overline{u_i'u_j'} \rangle \frac{\partial \langle \bar{u}_i \rangle}{\partial x_j}}_{P_s} + \underbrace{\frac{g}{\theta_\circ}\langle \overline{u_i'\theta'} \rangle \delta_{i3}}_{P_b} \underbrace{- \frac{\partial}{\partial x_j}\frac{\langle \overline{u_i'u_i'u_j'} \rangle}{2}}_{T_t}$$

$$\underbrace{-\frac{\partial}{\partial x_j}\frac{\langle \overline{u_i'u_i'}'' \bar{u}_j'' \rangle}{2}}_{T_d} \underbrace{-\frac{1}{\rho_\circ}\frac{\partial \langle \overline{u_j'p'} \rangle}{\partial x_j}}_{T_p}$$

$$\underbrace{+\nu\left\langle \overline{u_i'\frac{\partial u_i'}{\partial x_j \partial x_j}} \right\rangle}_{D_\nu} + \underbrace{\left\langle \overline{u_i'u_i'}'' \frac{\partial \bar{u}_i''}{\partial x_j} \right\rangle}_{W_p}. \tag{24.19}$$

The left-hand side (LHS) of Equation 24.19 is the total derivative and represents the change in TKE $\left(\langle \overline{u_i'u_i'} \rangle/2\right)$ following the mean flow. It contains both the time rate of change of TKE and any "advection" of TKE into the control volume by the mean flow. In steady, horizontally homogeneous conditions, the LHS is identically zero. The terms on RHS of Equation 24.19 can be interpreted in the following way. TKE is produced as the mean shear does work against the mean Reynolds stresses (called shear production, P_s) and as the turbulent buoyancy flux does work against the mean buoyancy gradient (called buoyant production, P_b). TKE is transported vertically primarily by turbulent transport (T_t), dispersive transport (T_d), and pressure transport (T_p). The fundamental difference between the production and transport terms is that the latter can be written as divergences, which means that they must integrate to zero over a control volume whose boundaries are outside the turbulent region.

The molecular term (D_ν) can be split into two terms:

$$\nu\left\langle \overline{u_i'\frac{\partial u_i'}{\partial x_j \partial x_j}} \right\rangle = \underbrace{\nu\frac{1}{2}\frac{\partial^2 \langle \overline{u_i'u_i'} \rangle}{\partial x_j \partial x_j}}_{T_\nu} \underbrace{-\nu\left\langle \overline{\frac{\partial u_i' \partial u_i'}{\partial x_j \partial x_j}} \right\rangle}_{\varepsilon}, \tag{24.20}$$

where T_ν represents transport of TKE by molecular diffusion and ε the destruction of TKE as local fluctuating velocity gradients do work against fluctuating viscous stresses, thereby dissipating TKE as heat. The last term in Equation 24.19 (W_p) is the local analogue of P_s and represents the rate at which local turbulent stresses do work against local time-mean rates of strain. Because these local variations are associated with the flow around individual canopy elements, this term is known as "wake production" and can be interpreted (approximately) as the work done by the mean flow against the pressure drag of the canopy.

Well above the canopy in the ASL, the TKE budget (Equation 24.19) is generally in local equilibrium, that is, with shear production balances dissipation at any height. However, within and just above a canopy, TKE produced in the high streamwise velocity shear region at the canopy-top is exported by pressure and turbulent transport to regions away from its source (Figure 24.4). Hence, the local rates of production and dissipation are not in balance and local equilibrium is lost. All canopy second-moment budgets exhibit this departure from a local balance between production and dissipation, which occurs because the dominant turbulent eddies in the canopy span most of the canopy depth and can efficiently transport TKE through the entire canopy. Many of the important differences between turbulence in the canopy and the ASL are a result of this fact that canopy turbulence is dominated by eddies that are larger-scale than the scale over which gradients of mean quantities vary in the canopy. In the next section, we will look more closely at the evidence for these large eddies.

24.7 Two-Point Correlations: Integral Length and Velocity Scales

Integral length scales are an important statistical description of a turbulent flow, and represent the physical separation in space over which signals remain correlated (Shaw et al., 1995). These

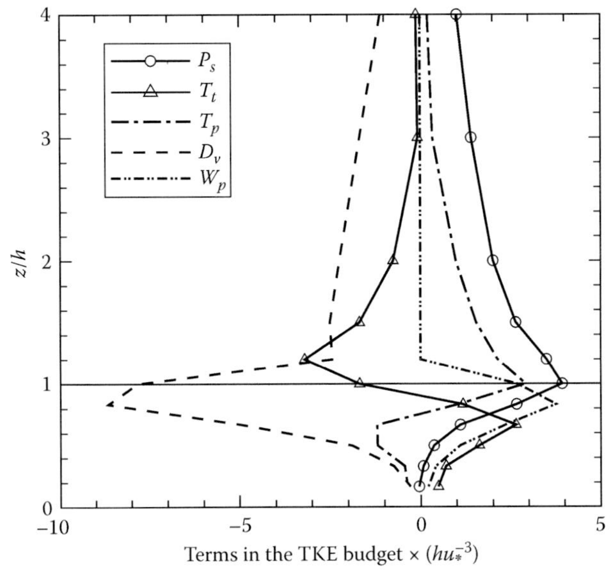

FIGURE 24.4 Vertical profiles of the terms in the TKE budget within and above a model wheat canopy in a neutrally stratified wind tunnel simulation. (From Brunet, Y. et al., *Bound. Lay. Meteorol.*, **70**(1), 95, 1994.) The labels refer to the terms presented in Equation 24.19, and all terms have been multiplied by hu_*^{-3}. Due to the difficulties of measuring pressure fluctuations and dispersive transport, T_p is calculated as the residual of all the other terms. The thin solid line is the canopy top. (Adapted from Finnigan, J.J., *Ann. Rev. Fluid Mech.*, 32(1), 519, 2000.)

length scales are commonly interpreted as representative of the typical eddy size. Most observations in turbulent flows are made at single points in space, and streamwise length scales can only be estimated from these by utilizing Taylor's (1938) "frozen eddy" hypothesis. Taylor's (1938) hypothesis assumes that the patterns of velocity comprising turbulent eddies do not change much in the time they take to be convected past a sensor by the mean wind, which allows integral space scales to be deduced from integral time scales. Unfortunately, this is not always a good assumption in canopies, so Shaw et al. (1995) performed two-point wind tunnel observations to allow spatial length scales to be measured without invoking Taylor's hypothesis, and Su et al. (2000) followed that work using 3D turbulence-resolving numerical simulations to evaluate spatial length scales directly.

Following Shaw et al. (1995), the two-point space–time correlation coefficients of two random velocity variables u_i and u_j in a horizontally homogeneous flow field can be expressed as

$$R_{ij}(r_x,r_y,z,z',\tau)=\frac{\overline{u_i(x,y,z,t)u_j(x+r_x,y+r_y,z',t+\tau)}}{\left[\overline{u_i^2}(x,y,z,t)\overline{u_j^2}(x+r_x,y+r_y,z',t)\right]^{1/2}},\quad(24.21)$$

where r_x and r_y are separations in the streamwise and lateral directions, respectively; z is the height of the reference observation; z' is the height of spatially varying observation, t is time; and, τ is a time lag imposed on the spatially varying observation. When $i = j$, R_{ii} is the autocorrelation function. When the two variables u_i and u_j are collocated (i.e., at $r_x = 0$, $r_y = 0$, and $z = z'$), Equation 24.21 represents a single-point time-lagged correlation.

The contours of two-point streamwise velocity autocorrelations describe an elliptical pattern which is elongated in the streamwise direction and displays bilateral symmetry in the y-direction. On the x–z plane of symmetry, the two-point streamwise velocity autocorrelation pattern exhibits a downwind tilt (Figure 24.5), suggesting that u-perturbations (u') at lower heights lag those higher in the flow. While two-point vertical velocity autocorrelation contours in the same x–z plane are roughly circular suggesting that w-perturbations (w') are vertically in phase with each other, scalar autocorrelation contours tilt downstream with a similar pattern to streamwise velocity (Su et al., 2000).

Integral length scales from the two-point correlations are calculated using

$$L_x(u_i,u_j,z)=\frac{1}{R_{ij}(0,0,z,z)}\int_0^{\infty}R_{ij}(r_x,0,z,z)dr_x.\quad(24.22)$$

Well above the canopy ($z/h > 2$), integral length scales computed from two-point measurements (L_x) match single-point length scales calculated using Taylor's hypothesis (Figure 24.6). However, two-point integral length scales diminish much more slowly with descent into the canopy than do single-point estimates (Shaw et al., 1995) and do not appear to be

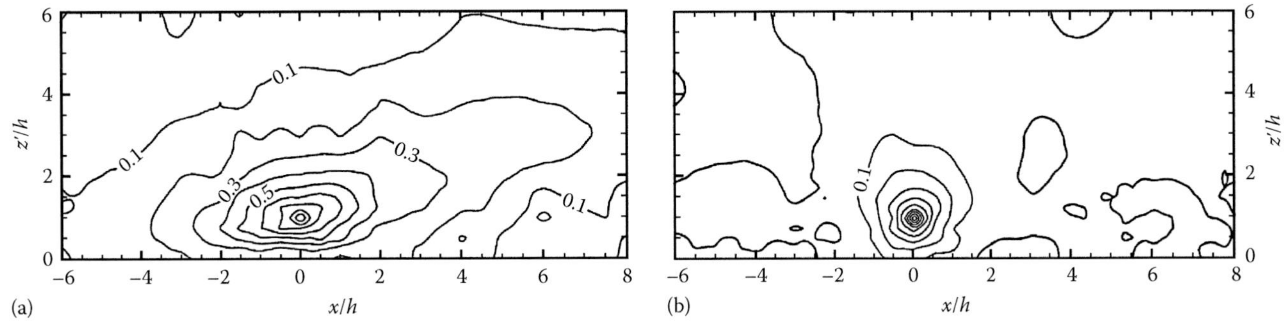

FIGURE 24.5 Autocorrelation contours of streamwise (a) and vertical (b) velocity for an x–z slice with no lateral displacement with the reference probe fixed at $z/h = 1$. Contour interval is 0.1. (Adapted from Shaw, R.H. et al., *Bound. Lay. Meteorol.*, 76(4), 349, 1995.)

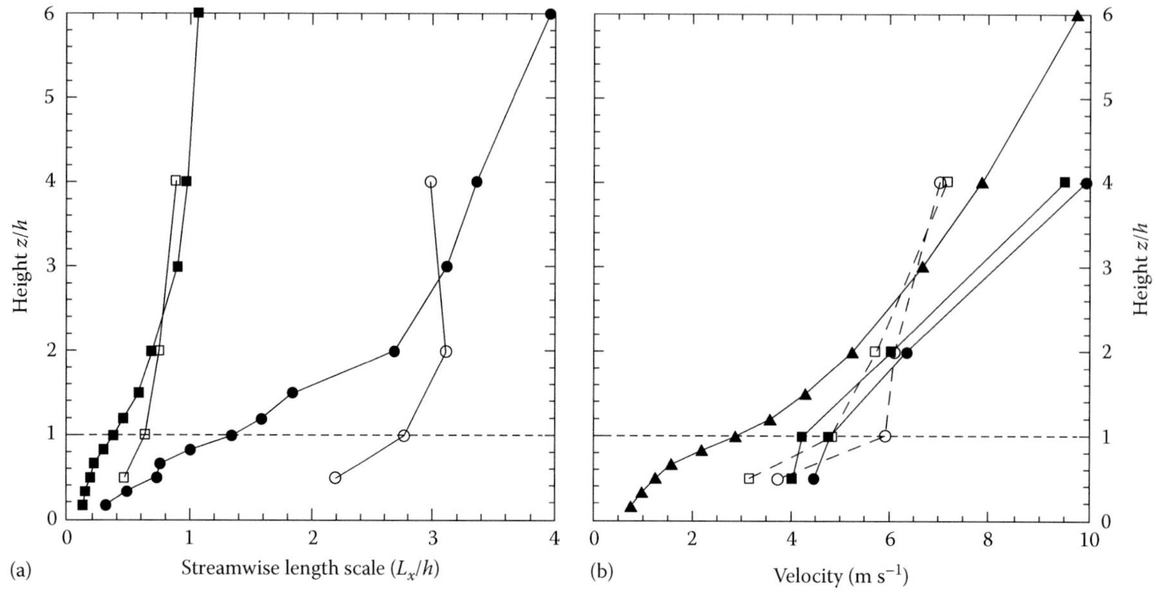

FIGURE 24.6 (a) Streamwise integral length scales computed from spatial correlations for streamwise velocity (open circles) and for vertical velocity (open squares). For reference, the lines with solid symbols depict length scales estimated using single-point time-series and the assumption Taylor's hypothesis. (b) Convective velocities computed from the time lag required to optimally restore correlations lost by increasing streamwise separation. Solid line and closed circles are for streamwise velocity, and solid line with closed squares is for the vertical velocity signal. Closed triangles depict the mean velocity profile, and the dashed lines with open symbols represent convective velocities calculated by dividing two-point integral length scales by single-point integral time scales. (Adapted from Shaw, R.H. et al., *Bound. Lay. Meteorol.*, 76(4), 349, 1995.)

related to the distance from the surface. It is very clear from Figures 24.5 and 24.6 that the coherent motions in the roughness sublayer dominating these length scales span the entire canopy height.

Streamwise convection velocities calculated from the correlation patterns ($U_c(z)$) suggest that the turbulent structures which produce these patterns travel much faster in and just above the canopy than the local mean velocity [$(U_c(h)) \simeq 1.8\langle\overline{u}\rangle(h)$, Shaw et al., 1995; Su et al., 2000; Figure 24.7], suggesting that their structure is related in some way to processes higher in the boundary layer where mean velocities are higher.

24.8 Spectra and Dissipation

Fourier spectra provide a window into the distribution of turbulence across space and time scales in a way which is complementary to the information contained in the two-point correlation fields. The power spectral densities of u_i' (labeled as S_{u_i}) are the Fourier transforms of the instantaneous spatial autocovariance functions $R_{ii}(r_x, r_y, z, z, 0)$, such that the scalar energy spectrum $E(k)$ is defined as

$$\frac{1}{2}\langle\overline{u_i'u_i'}\rangle = \int_0^\infty \left(S_{u_1}(k) + S_{u_2}(k) + S_{u_3}(k)\right)dk = \int_0^\infty E(k)dk, \quad (24.23)$$

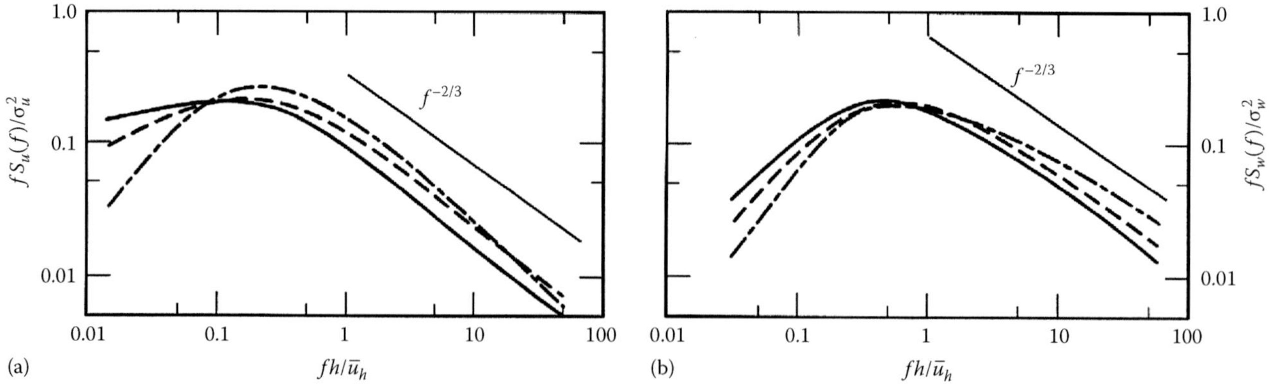

FIGURE 24.7 Normalized velocity spectra from measurements above the canopy (solid), near the top of the canopy (dashed), and within the canopy (dash-dot), for the streamwise (a) and vertical (b) velocities. (Adapted from Kaimal, J.C. and Finnigan, J.J., *Atmospheric Boundary Layer Flows: Their Structure and Measurement*, 1994. Oxford University Press.)

where k is the scalar wavenumber. For further information regarding the use of spectral analysis in turbulence, see Tennekes and Lumley (1972), Kaimal and Finnigan (1994), or Pope (2000). Note also that most spectral theory is derived in terms of wavenumbers. However, measured canopy spectra are not typically wavenumber spectra. Rather, they are actually frequency spectra obtained from time series of velocity and scalars which are then converted to wavenumber spectra by applying Taylor's hypothesis and setting $k = \overline{u}/(\text{frequency})$. It has already been noted that this is not a good assumption in canopies so canopy frequency spectra must be interpreted with care. See, for example, Wyngaard and Clifford (1977).

Power spectra in canopies differ from those in the inertial sublayer above in three important ways: (1) the scaling of the spectral peak, (2) the slope of the inertial subrange, and (3) the rate of viscous dissipation. When plotted in the form $kS_{u_i}(k)$ vs. $ln(k)$, the peak of the spectrum indicates the wavenumbers at which TKE is being created by shear or buoyant production processes (P_s and P_b in Equation 24.19, respectively). In the ASL well above the canopy, this peak is inversely proportional to the height above the displacement level (z–d, see Section 24.2). In other words, the size of the most energetic eddies increases with height (Kaimal and Finnigan, 1994). However, with descent through the roughness sublayer, this height dependence vanishes and the spectral peak approaches a constant wavenumber, suggesting that the size of the dominant canopy eddies is height independent.

Kolmogorov's (1941, 1962) spectral theory is the basic reference to turn to for interpreting atmospheric turbulence spectra. A fundamental assumption of this theory is that there exists an "inertial subrange" (ISR) of wavenumbers between the low wavenumber spectral peak and the high wavenumber "dissipation scales" where TKE is turned into heat by work against viscous stresses. In the ISR there is negligible creation or dissipation of TKE; rather, TKE is simply passed from lower to higher wavenumbers and the energy density decreases at a rate proportional to $k^{-5/3}$.

In canopy flows, the assumptions of Kolmogorov's (1941, 1962) theory are violated in three ways. First, the wake production process (W_p, Equation 24.19) turns mean kinetic energy into TKE directly at the scale of individual canopy elements—leaves, twigs, and branches—and these scales are usually at the high wavenumber end of the ISR. Second, the larger scale turbulent eddies created by P_s and P_b (Equation 24.19) also interact with canopy elements just as the mean flow does, and their large-scale TKE is converted into element-scale TKE. This "spectral-shortcut" process tends to remove TKE from the low wavenumber end of the ISR and to inject that TKE at the high wavenumber end. Third, the work eddies do against the viscous drag of the canopy removes energy at all scales, including the scales comprising the ISR. This third process changes the overall magnitude of canopy dissipation as explained next. The net result is that within canopies the "ISR" becomes a region where $E(k)$ decreases faster than $k^{-5/3}$ at low wavenumbers and more slowly at high wavenumbers.

In the free air above the canopy layer, the rate at which TKE is dissipated is controlled by the rate at which fine-scale shear layers (whose transverse scale is on the order of the Kolmogorov microscale $\eta = \nu^3/\epsilon$, where η is a few millimeters in the atmosphere) can be produced by the mutual stretching of turbulent vortices. Within canopies, the thin viscous boundary layers on the surfaces of all the canopy elements provide an abundant source of shear layers at the Kolmogorov microscale so that the dissipation rate in canopies can be several times higher than seen in free air flows with shear of comparable magnitude.

Finnigan (2000) presented a thorough review of the dynamics of canopy spectra contrasting them with spectra in the boundary layer above, which obey Kolmogorov theory (e.g., Amiro, 1990). Finnigan (2000) shows that when inferring dissipation from within-canopy spectra, it is essential to use formulae modified for canopy effects because applying standard Kolmogorov theory gives dissipation values which are far too low.

24.9 Turbulence Structure

Evidence for large-scale organized motion (or coherent eddies) in plant canopies comes from many sources (Finnigan, 2000). We have already seen direct evidence from two-point correlations and indirect evidence from the transport terms in the TKE budget. However, before focusing on the dynamic process which generates these eddies, a further set of observations will be presented which played an important role in developing our current understanding.

Gao et al. (1989) presented a time series of high-frequency temperature and velocity fluctuations at multiple heights within and above a forest canopy (Figure 24.8a). They noted ramp-like patterns in the time series which were coherent across the range of heights observed ($0.3 < z/h < 2.4$). Using the sharp transition between warm and cold temperatures as their trigger, Gao et al. (1989) composited many such events into a time-height cross-section of an average structure (Figure 24.8b). A coherent sloping temperature microfront is revealed, which occurs at earlier times above and at later times within the canopy. Prior to the passage of this sharp temperature microfront, the entire canopy depth experiences weak *ejection* motion ($u' < 0, w' > 0$) bringing low-momentum and relatively warm air upward, which is rapidly terminated by a *sweep* motion ($u' < 0, w' > 0$) bringing relatively cool high-momentum air down from aloft. The sweep is notably stronger than the ejection. Gao et al. (1989) found that these organized structures occur both day and night and that they account for between 60% and 80% of the scalar exchange between the canopy layers and aloft. Other researchers (e.g., Collineau and Brunet, 1993; Lu and Fitzjarrald, 1994, among others) used more objective wavelet-based methods to identify, scale, and composite these ramp structures and found their percentage contribution to vertical transport to range between 40% and 80%. The scalar microfronts have been found to convect downwind faster than $\langle\overline{u}\rangle$ (h) (Zhang et al., 1992; Shaw et al., 1995; Su et al., 2000), suggesting a link with the velocity patterns revealed by two-point correlations (Section 24.7, Shaw et al., 1995).

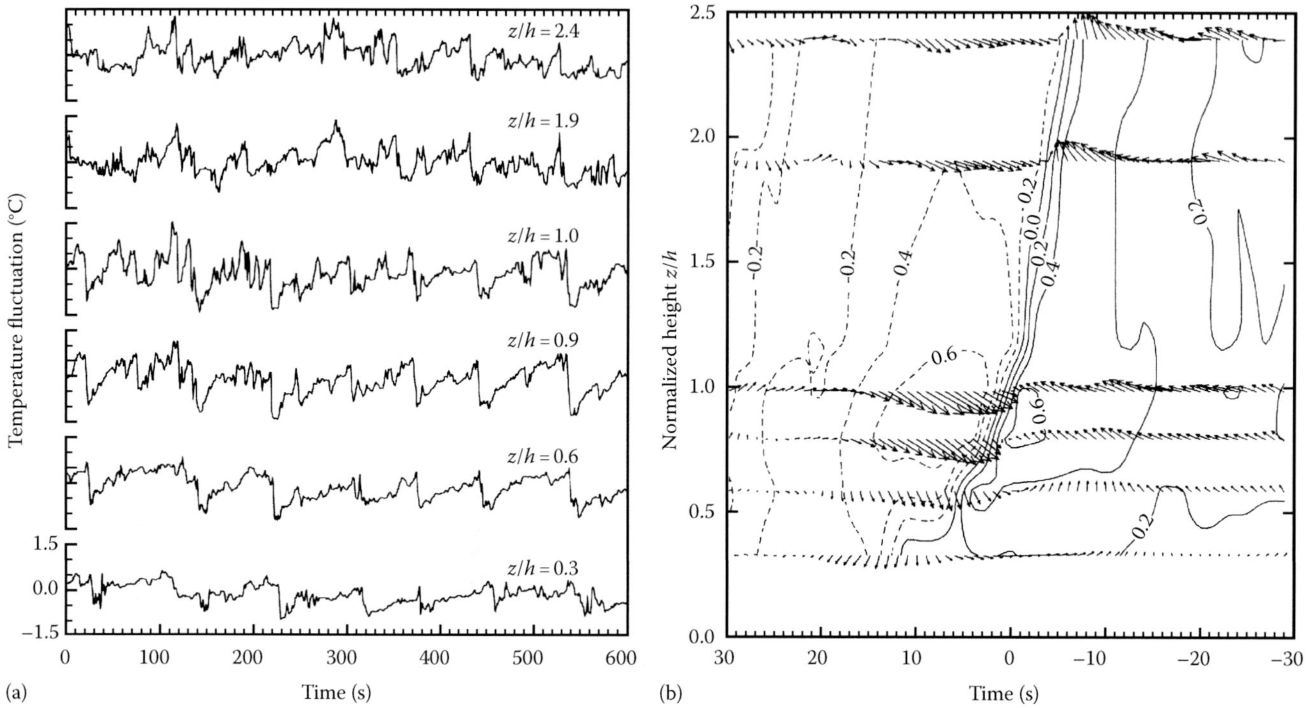

FIGURE 24.8 (a) Ten minute time series of temperature fluctuations at six heights within and above a deciduous forest canopy. The canopy height (h) is about 18 m. The ordinate scale is the same for all time series, ranging from −1.5 → 1.5°C. (b) A time-height cross section from tower-based observations of composite averaged 2D wind vectors (maximum arrow length represents 1.9 m s^{-1}) and temperature fluctuations (contours,°C) centered about the occurrence of a sharp temperature microfront at canopy-top. (Adapted from Gao, W. et al., *Bound. Lay. Meteorol.*, 47(1), 349, 1989.)

Numerous observationally based efforts have attempted to deduce the 3D spatial flow structures underlying the observed scalar ramp patterns (e.g., Zhang et al., 1992; Shaw et al., 1995; Finnigan and Shaw, 2000); however, the difficulty in making the necessary observations prompted researchers to turn to numerical methods. Shaw and Schumann (1992) had the insight to apply the large-eddy simulation technique (Deardorff, 1970) to canopy flows.* Following Shaw and Schumann (1992), Fitzmaurice et al. (2004), and Watanabe (2004) concatenated numerous canopy-scale organized structures from canopy-resolving LES results. Fitzmaurice et al. (2004) used a pulse in static pressure at canopy-top to identify the structure location, while Watanabe (2004) used the wavelet approach suggested by Collineau and Brunet (1993) and Lu and Fitzjarrald (1994) to detect scalar ramp patterns. Both Fitzmaurice et al. (2004) and Watanabe (2004) found bilateral symmetry in their educed composite structure.

Using a similar strategy to Fitzmaurice et al. (2004), Finnigan et al. (2009) deduced the 3D structure by triggering on canopy-top pressure maxima to ensemble average the flow field. They found that in neutrally stratified flow the organized structure consists of a combination of "head-up" and "head-down" hairpin vortices, which are both inclined in the downstream direction (Figure 24.9). The ejections and sweeps are generated as induced flows between the counter-rotating legs of the hairpins. The scalar microfront lies between them in the convergence region between the downstream ejection and upstream sweep. This convergence is also responsible for the static pressure pulse.

This dual-hairpin eddy structure is quite different from the eddy structure inferred by some researchers who have applied a range of conditional-sampling techniques to smooth-wall boundary-layer and channel flows, where they produced a model consisting of packets of head-up ejection-generating hairpins which straddle and synergistically generate coherent low-speed streaks (e.g., Adrian et al., 2000). However, the dual-hairpin structure corresponds with the eddy structure deduced by other conditional sampling techniques in boundary layers and shear layers (e.g., Kim and Moin, 1986; Rogers and Moser, 1992).

The model deduced by Finnigan et al. (2009) extends, and in some senses completes, the *mixing-layer hypothesis* of Raupach et al. (1996) who argued that the inflected mean-velocity profile associated with the canopy's distributed momentum absorption

* In large-eddy simulation (LES), a spatially filtered set of three-dimensional time-dependent equations for fluid motion are solved numerically, where the large energy-containing scales of turbulent motion are explicitly captured by the numerical resolution of the simulation and the fine-scale unresolved motions are modeled. Deardorff (1970) argued that if the filter-scale resides well within the inertial subrange, then the unresolved (subfilter scale) motions should act mainly to dissipate energy as it cascades down in scale. Over the years since Deardorff (1970) first presented the ideas surrounding LES, significant refinement of the methods used, assumptions made, and subfilter scale modeling has occurred. After much scrutiny and verification against observations LES has become something of a direct counterpart to observations.

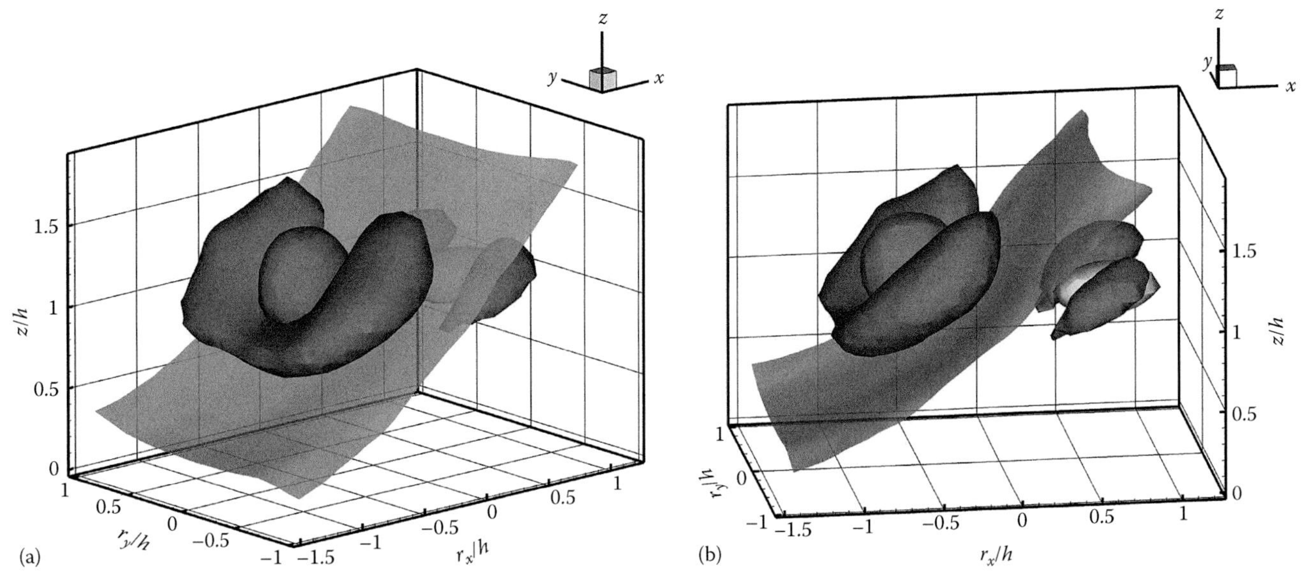

FIGURE 24.9 **(See color insert.)** 3D images depicting the composite eddy structure from canopy-resolving large-eddy simulation from two vantage points (Finnigan et al., 2009). (a) From above and behind right compared to the mean flow which is in the positive x-direction, and (b) from the side, looking in the positive y-direction. The isosurfaces shown in blue are of λ_2 at a value of −0.77 (See Jeong and Hussain (1995) for details of λ_2 analysis). The translucent sheet in green is an isosurface of zero scalar concentration perturbation from the horizontal average. The isosurface shown in orange is of $\overline{u'w'}$ at a value of $-0.6\,\text{m}^2\,\text{s}^2$ in the sweep region, while that in yellow is the same quantity at a value of $-0.15\,\text{m}^2\,\text{s}^2$ associated with the ejection. Note that the $\overline{u'w'}$ sweep isosurface is drawn at a value that is four times greater than that of the ejection.

plays a critical role in generating the coherent eddies which dominate canopy turbulence structure. The inflection point is inviscidly unstable with the initial instability taking the form of a linear Kelvin–Helmholtz (K–H) wave. This K–H wave emerges preferentially when larger-scale motions from higher in the ASL bring high momentum fluid down to the canopy top and augment the canopy-top velocity shear (Figure 24.10a). This is consistent with Shaw et al.'s (1995) finding that the canopy-scale eddies convect at speeds faster than the local mean wind. The K–H wave can be interpreted as alternating regions of positive and negative spanwise vorticity perturbations with wavelength λ, which is proportional to the vorticity thickness δ_ω. For sufficiently dense canopies, the vorticity thickness can be assumed to be $\delta_\omega = 2\langle\overline{u}\rangle/[\partial\langle\overline{u}\rangle/\partial z]$ evaluated at canopy height *h*.

The mean background flow with this wave-like perturbation superimposed upon it is itself unstable; nonlinear analysis shows that the K–H wave can evolve into finite-amplitude transverse "Stuart" vortices, which are spaced in the streamwise direction with the original wavelength λ (see Figure 24.10b and Stuart, 1967). Finnigan et al. (2009) postulated that in response to random turbulent perturbations two "Stuart" vortices could approach and rotate around each other (e.g., Figure 24.10c), with the head-up hairpins deflected upward and forward by their downwind neighbors and the head-down hairpins deflected downward and backward by their upwind neighbors. This process matches the *helical-pairing* instability described by Pierrehumbert and Widnall (1982). Finally, stretching of the hairpin legs by the mean shear and self-induction by the two hairpins completes the production of trains of hairpin vortex pairs comprising a leading head-up and trailing head-down hairpin. Flow convergence between the hairpins produces a positive static pressure pulse (Figure 24.10d).

Far from any walls or vegetation, Rogers and Moin (1987) and Gerz et al. (1994) find that plane-mixing layers produce head-up and head-down hairpin vortices with equal likelihood, which suggests an equal likelihood and importance of ejections and sweeps. However, as discussed in Section 24.5, canopy turbulence is characterized by relatively weak ejections and strong intermittent sweeps. Finnigan et al. (2009) proposed that there are two symmetry-breaking mechanisms distinguishing canopy-mixing layers from Gerz et al.'s (1994) plane-mixing layers. First, downward motions with horizontal scales comparable to the distance to the wall are effectively blocked (Hunt and Morrison, 2000) such that at any height, the probability of large head-up deflections of hairpin vortices away from the wall exceeds that of deflections toward the wall. Second, the presence of a porous canopy allows downward deflections by eddy motions with scales smaller than the distance to the wall. At the canopy-top, a roughly logarithmic velocity profile above is joined to a roughly exponential within-canopy profile (see Figure 24.1), resulting in downwardly deflected hairpins experiencing stronger strain and vortex amplification than hairpins deflected upward. In the ASL well above the canopy, the first symmetry-breaking mechanism applies resulting in the dominance of ejections, while the second symmetry-breaking mechanism dominates near the canopy-top resulting in the dominance of sweeps.

Although some details of Finnigan et al.'s (2009) theory remain somewhat speculative, it successfully explains most observations in neutrally stratified flows. For example, rescaling

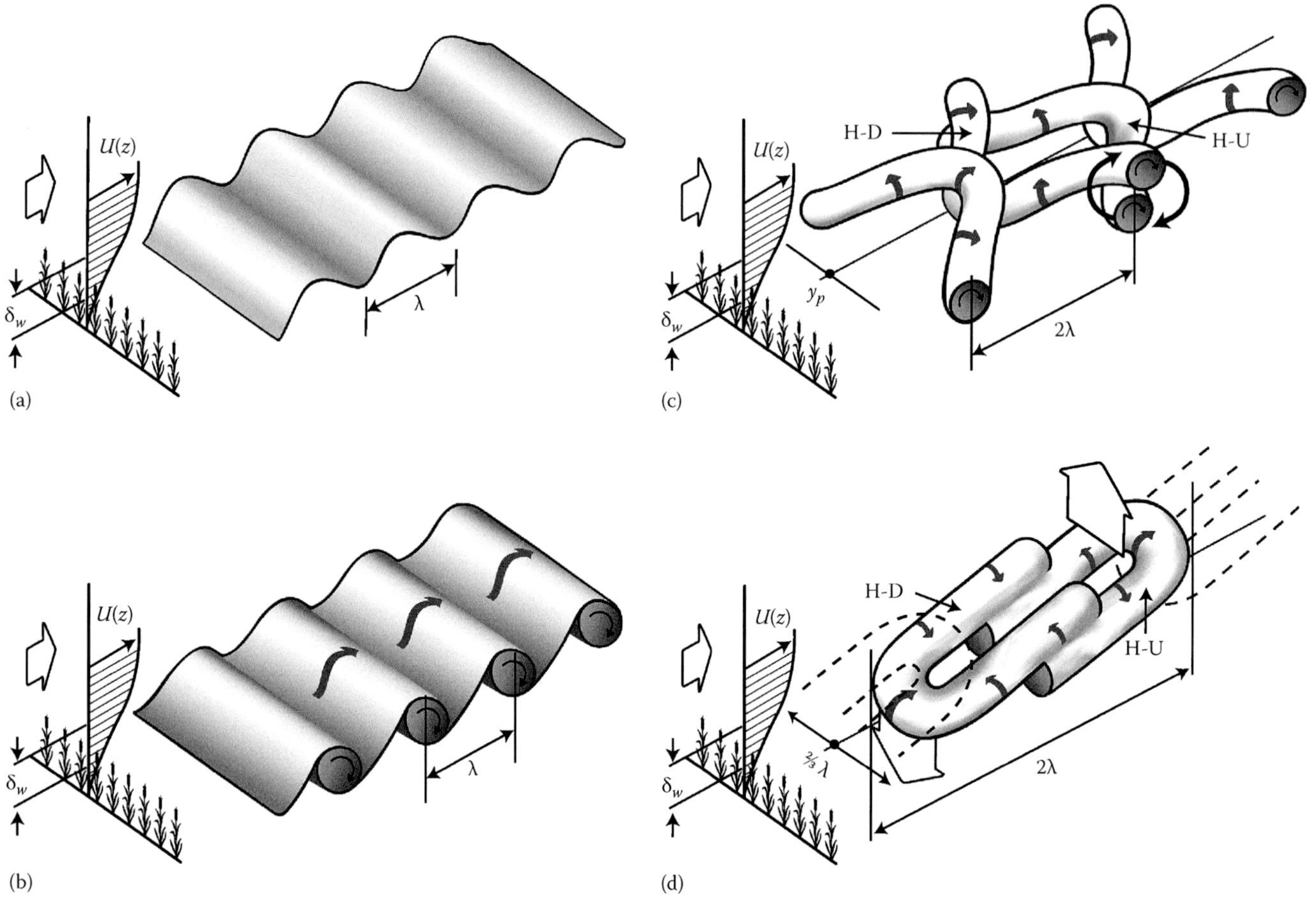

FIGURE 24.10 Schematic diagram of the formation of the dual-hairpin eddy: (a) The initial instability is a Kelvin–Helmholtz wave of wavelength λ, which develops on the inflected mean velocity profile at canopy top. (b) The resulting velocity field is nonlinearly unstable and successive regions of alternating spanwise vorticity clump into coherent "Stuart" vortices, which retain the wavelength λ. (c) Two successive Stuart vortices are moved closer together at some y-location by the ambient turbulence. The mutual induction of their vorticity fields causes them to approach more closely and rotate around each other. Vortex pairing doubles the wavelength of the disturbance to 2λ. Note that this disturbance of the streamwise symmetry of the induced velocity fields of successive vortices will propagate upwind and downwind at the same y location. (d) As the initial hairpins are strained by the mean shear, most of the vorticity accumulates in the legs and self-induction by the vortex legs dominates the motion of the hairpins. As a result, the head-down hairpin, which was initially below, moves up while the head-up hairpin moves down. The broad arrows in panel (d) indicate the direction of hairpin movement under this self-induction. (Adapted from Finnigan, J.J. et al., *J. Fluid Mech.*, 637, 387, 2009.)

Raupach et al.'s (1996) data reveals an eddy spacing consistent with the theory (i.e., $\lambda \simeq 15\delta_\omega$) and spanwise/streamwise eddy spacing educed from LES that is comparable to Pierrehumbert and Widnall's (1982) most probable helical-pairing mode. Furthermore, this cascade of instabilities automatically produces a scalar microfront between the hairpins as required to explain the field measurements and numerical simulations reported earlier (see Figure 24.9).

Indirect evidence for the fundamental role played by this canopy-top instability comes from Harman and Finnigan's (2007, 2008) extension of MOST to the roughness sublayer. In their MOST extension, Harman and Finnigan (2007) used the vorticity thickness (δ_ω) as an additional scaling length and were able to match observed mean-velocity and scalar profiles through the roughness sublayer across a wide range of canopy types and diabatic stabilities.

24.10 Final Remarks

This chapter has attempted to provide an overview of the characteristics and dynamics of canopy turbulence by detailing the various ways in which it is different from turbulence elsewhere in the ABL and by explaining the reasons for the differences. "Organized" turbulent motions in canopies are shown to set canopy turbulence apart from that in the boundary layer above.

Although diabatic influences on canopy-turbulence statistics are well established, this chapter focuses mainly on neutral stratification because the community's understanding of diabatic influences on canopy turbulence structure remain in its infancy. This chapter's discussion also assumes that the canopy consists of horizontally homogeneously distributed elements which are located on flat terrain. However, in reality these conditions are rarely met. Rather, heterogeneously distributed vegetation is the norm, whether as a result of human agency through shelter-belts, clear-cuts, forest edges, transitions from one crop to another, etc., or through natural transitions in surface cover. Currently, there is no universal theory describing how canopy flow reacts to stratification, heterogeneity, or topography although knowledge is accumulating in incremental steps. Kaimal and Finnigan (1994) give a general introduction to flow over heterogeneous surfaces, while Belcher et al. (2008) introduce the effects of stable stratification. A range of particular situations are covered in more detail by Patton et al. (1998), Flesch and Wilson (1999), Lee (2000), Mahrt et al. (2000), Wang et al. (2001), Morse et al. (2002), Belcher et al. (2003), Yang et al. (2006), Belcher et al. (2008), Dupont and Brunet (2008), Huang et al. (2009), Dupont and Brunet (2009), and references therein.

Another interesting twist comes from the community studying aquatic canopies (e.g., Ghisalberti and Nepf, 2006; Stoesser et al., 2009). Although much of the fluid mechanics of aquatic canopy turbulence mimics that which occurs in terrestrial ecosystems, due to the rapid attenuation of solar radiation by liquid water ensures that most aquatic canopies typically grow in water of limited depth, a depth which can be similar to the depth of the entire roughness sublayer. This leads to interesting and illuminating departures from the terrestrial situation described here, where the canopy invariably sits at the bottom of deep turbulent boundary layers (Finnigan, 2010).

There is much left to learn about canopy turbulence.

References

Adrian, R. J., Meinhart, C. D., and Tomkins, C. D., 2000, Vortex organization in the outer region of the turbulent boundary layer, *J. Fluid Mech.* **422**, 1–54.

Amiro, B. D., 1990, Drag coefficients and turbulence spectra within three boreal forest canopies, *Bound. Lay. Meteorol.* **52**(3), 227–246.

Belcher, S. E., Finnigan, J. J., and Harman, I. N., 2008, Flows through forest canopies in complex terrain, *Ecol. Appl.* **18**, 1436–1453.

Belcher, S. E., Jerram, N., and Hunt, J. C. R., 2003, Adjustment of a turbulent boundary layer to a canopy of roughness elements, *J. Fluid Mech.* **488**, 369–398.

Böhm, M., Finnigan, J. J., and Raupach, M. R., 2000, Dispersive fluxes and canopy flows: Just how important are they? *Twenty Fourth Conference on Agricultural and Forest Meteorology*, American Meteorological Society, Davis, CA, pp. 106–107.

Bonan, G. B., 2008, Forests and climate change: Forcings, feedbacks, and the climate benefits of forests, *Science* **320**(5882), 1444–1449.

Brunet, Y., Finnigan, J. J., and Raupach, M. R., 1994, A wind tunnel study of air flow in waving wheat: Single-point velocity statistics, *Bound. Lay. Meteorol.* **70**(1), 95–132.

Businger, J. A., 1982, Equations and concepts, in F. T. M. Nieuwstadt and H. van Dop (eds.), *Atmospheric Turbulence and Air Pollution Modelling*, D. Reidel Publishing Company, Dordrecht, the Netherlands, pp. 1–36.

Cionco, R., 1965, A mathematical model for air flow in a vegetative canopy, *J. Appl. Meteorol.* **4**(4), 517–522.

Collineau, S. and Brunet, Y., 1993, Detection of turbulent coherent motions in a forest canopy. Part II: Time-scales and conditional averages, *Bound. Lay. Meteorol.* **66**, 49–73.

Deardorff, J. W., 1970, A numerical study of three-dimensional turbulent channel flow at large Reynold's numbers, *J. Fluid Mech.* **41**, 453–480.

Denmead, O. T. and Bradley, E. F., 1985, Flux-gradient relationships in a forest canopy, in B. A. Hutchison and B. B. Hicks (eds.), *The Forest-Atmosphere Interaction*, D. Reidel Publishing Company, Dordrecht, the Netherlands, pp. 421–442.

Dupont, S. and Brunet, Y., 2008, Influence of foliar density profile on canopy flow: A large-eddy simulation study, *Agric. For. Meteorol.* **148**(6–7), 976–990.

Dupont, S. and Brunet, Y., 2009, Coherent structures in canopy edge flow: A large-eddy simulation study, *J. Fluid Mech.* **630**, 93–128.

Dupont, S., Gosselin, F., Py, C., de Langre, E., Hemon, P., and Brunet, Y., 2010, Modelling waving crops using large-eddy simulation: Comparison with experiments and a linear stability analysis, *J. Fluid Mech.* **652**, 5–44.

Feigenwinter, C., Bernhofer, C., Eichelmann, U., Heinesch, B., Hertel, M., Janous, D., Kolle, O., Lagergren, F., Lindroth, A., and Minerbi, S., 2008, Comparison of horizontal and vertical advective CO_2 fluxes at three forest sites, *Agric. For. Meteorol.* **148**(1), 12–24.

Finnigan, J. J., 1979, Turbulence in waving wheat. I. Mean statistics and honami, *Bound. Lay. Meteorol.* **16**, 181–211.

Finnigan, J. J., 1985, Turbulent transport in flexible plant canopies, in B. A. Hutchison and B. B. Hicks (eds.), *The Forest-Atmosphere Interaction*, D. Reidel Publishing Company, Dordrecht, the Netherlands, pp. 443–480.

Finnigan, J. J., 2000, Turbulence in plant canopies, *Ann. Rev. Fluid Mech.* **32**(1), 519–571.

Finnigan, J. J., 2010, Waving plants and turbulent eddies, *J. Fluid Mech.* **652**, 1–4.

Finnigan, J. J. and Raupach, M. R., 1987, Transfer processes in plant canopies in relation to stomatal characteristics, in E. Zeiger, G. D. Farquhar, and I. R. Cowan (eds.), *Stomatal Function*, Stanford University Press, Stanford, CA, pp. 385–429.

Finnigan, J. J. and Shaw, R. H., 2000, A wind-tunnel study of airflow in waving wheat: An EOF analysis of the structure of the large-eddy motion, *Bound. Lay. Meteorol.* **96**(1), 211–255.

Finnigan, J. J. and Shaw, R. H., 2008, Double-averaging methodology and its application to turbulent flow in and above vegetation canopies, *Acta Geophys.* **56**(3), 534–561.

Finnigan, J. J., Shaw, R. H., and Patton, E. G., 2009, Turbulence structure above a vegetation canopy, *J. Fluid Mech.* **637**, 387–424.

Fitzmaurice, L., Shaw, R. H., Paw U. K. T., and Patton, E. G., 2004, Three-dimensional scalar microfront systems in a large-eddy simulation of vegetation canopy flow, *Bound. Lay. Meteorol.* **112**, 107–127.

Flesch, T. K. and Wilson, J. D., 1999, Wind and remnant tree sway in forest cutblocks. I. Measured winds in experimental cutblocks, *Agric. For. Meteorol.* **93**(4), 229–242.

Fuentes, J. D., Lerdau, M., Atkinson, R., Baldocchi, D., Bottenheim, J. W., Ciccioli, P., Lamb, B. et al., 2000, Biogenic hydrocarbons in the atmospheric boundary layer: A review, *Bull. Am. Meteorol. Soc.* **81**(7), 1537–1575.

Gao, W., Shaw, R. H., and Paw U, K. T., 1989, Observation of organized structure in turbulent flow within and above a forest canopy, *Bound. Lay. Meteorol.* **47**(1), 349–377.

Garratt, J. R., 1992, *The Atmospheric Boundary Layer*, Cambridge University Press, Cambridge, U.K.

Gerz, T., Howell, J., and Mahrt, L., 1994, Vortex structures and microfronts, *Phys. Fluids* **6**(3), 1242–1251.

Ghisalberti, M. and Nepf, H., 2006, The structure of the shear layer in flows over rigid and flexible canopies, *Environ. Fluid Mech.* **6**(3), 277–301.

Guenther, A., Karl, T., Harley, P., Wiedinmyer, C., Palmer, P. I., and Geron, C., 2006, Estimates of global terrestrial isoprene emissions using MEGAN (model of emissions of gases and aerosols from nature), *Atmos. Chem. Phys.* **6**, 3181–3210.

Harman, I. N. and Finnigan, J. J., 2007, A simple unified theory for flow in the canopy and roughness sublayer, *Bound. Lay. Meteorol.* **123**, 339–363.

Harman, I. N. and Finnigan, J. J., 2008, Scalar concentration profiles in the canopy and roughness sublayer, *Bound. Lay. Meteorol.* **129**(3), 323–351.

Holland, J. Z., 1989, On pressure-driven wind in deep forests, *J. Appl. Meteorol.* **28**, 1349–1355.

Huang, J., Cassiani, M., and Albertson, J. D., 2009, The effects of vegetation density on coherent turbulent structures within the canopy sublayer: A large-eddy simulation study, *Bound. Lay. Meteorol.* **133**(2), 253–275.

Hunt, J. C. R. and Morrison, J. F., 2000, Eddy structure in turbulent boundary layers, *Eur. J. Mech.: B* **19**(5), 673–694.

Jackson, P. S., 1981, On the displacement height in the logarithmic velocity profile, *J. Fluid Mech.* **111**, 15–25.

Jeong, J. and Hussain, F., 1995, On the identification of a vortex, *J. Fluid Mech.* **285**, 69–94.

Kaimal, J. C. and Finnigan, J. J., 1994, *Atmospheric Boundary Layer Flows: Their Structure and Measurement*, Oxford University Press, New York.

Katul, G. G., Geron, C. D., Hsieh, C.-I., Vidakovic, B., and Guenther, A. B., 1998, Active turbulence and scalar transport near the forest–atmosphere interface, *J. Appl. Meteorol.* **37**(12), 1533–1546.

Kim, J. and Moin, P., 1986, The structure of the vorticity field in turbulent channel flow. Part 2. Study of ensemble-averaged fields, *J. Fluid Mech.* **162**, 339–363.

Kolmogorov, A. N., 1941, The local structure of turbulence in incompressible viscous fluid for very large Reynolds number, *C. R. Acad. Sci., U.R.S.S.* **30**, 301–305.

Kolmogorov, A. N., 1962, A refinement of previous hypotheses concerning the local structure of turbulence in a viscous incompressible fluid at high Reynolds number, *J. Fluid Mech.* **13**(01), 82–85.

Lee, X., 2000, Air motion within and above forest vegetation in non-ideal conditions, *For. Ecol. Manage.* **135**(1–3), 3–18.

Lu, C. H. and Fitzjarrald, D. R., 1994, Seasonal and diurnal variations of coherent structures over a deciduous forest, *Bound. Lay. Meteorol.* **69**(1), 43–69.

Mahrt, L., Lee, X., Black, A., Neumann, H., and Staebler, R. M., 2000, Nocturnal mixing in a forest subcanopy, *Agric. For. Meteorol.* **101**, 67–78.

Monin, A. S. and Obukhov, A. M., 1954, Osnovnye zakonomernosti turbulentnogo pere-meshivanija v prizemnom sloe atmosfery (Basic turbulent mixing laws in the atmospheric surface layer), *Trudy Geofiz. Inst. AN SSSR* **24**(151), 163–187.

Monteith, J. L. and Unsworth, M. H., 2008, *Principles of Environmental Physics*, 3rd edn., Elsevier, London, U.K.

Morse, A. P., Gardiner, B. A., and Marshall, B. J., 2002, Mechanisms controlling turbulence development across a forest edge, *Bound. Lay. Meteorol.* **103**(2), 227–251.

Obukhov, A. M., 1946, Turbulentnost v temperaturnoj neodnorodnoj atmosfere, *Trudy Inst. Theor. Geofiz. AN SSSR* **1**, 95–115.

Obukhov, A. M., 1971, Turbulence in an atmosphere with a nonuniform temperature, *Bound. Lay. Meteorol.* **2**, 7–29.

Patton, E. G., Horst, T. W., Lenschow, D. H., Sullivan, P. P., Oncley, S., Burns, S., Guenther, A. et al., 2011, The canopy horizontal array turbulence study, *Bull. Am. Meteorol. Soc.* **92**(5), 593–611.

Patton, E. G., Shaw, R. H., Judd, M. J., and Raupach, M. R., 1998, Large-eddy simulation of windbreak flow, *Bound. Lay. Meteorol.* **87**, 275–306.

Patton, E. G., Sullivan, P. P., and Davis, K. J., 2003, The influence of a forest canopy on top-down and bottom-up diffusion in the planetary boundary layer, *Quart. J. R. Meteorol. Soc.* **129**, 1415–1434.

Pierrehumbert, R. T. and Widnall, S. E., 1982, The two- and three-dimensional instabilities of a spatially periodic shear layer, *J. Fluid Mech.* **114**, 59–82.

Poggi, D., Katul, G. G., and Albertson, J. D., 2004, A note on the contribution of dispersive fluxes to momentum transfer within canopies, *Bound. Lay. Meteorol.* **111**, 615–621.

Pope, S. B., 2000, *Turbulent Flows*, Cambridge University Press, Cambridge, U.K.

Raupach, M. R., Finnigan, J. J., and Brunet, Y., 1996, Coherent eddies and turbulence in vegetation canopies: The mixing-layer analogy, *Bound. Lay. Meteorol.* **78**(3), 351–382.

Raupach, M. R. and Shaw, R. H., 1982, Averaging procedures for flow within vegetation canopies, *Bound. Lay. Meteorol.* **22**(1), 79–90.

Raupach, M. R. and Thom, A. S., 1981, Turbulence in and above plant canopies, *Ann. Rev. Fluid Mech.* **13**(1), 97–129.

Rogers, M. M. and Moin, P., 1987, The structure of the vorticity field in homogeneous turbulent flows, *J. Fluid Mech.* **176**, 33–66.

Rogers, M. M. and Moser, R. D., 1992, The three-dimensional evolution of a plane mixing layer: The Kelvin–Helmholtz rollup, *J. Fluid Mech.* **243**, 183–226.

Shaw, R. H., 1977, Secondary wind speed maxima inside plant canopies, *J. Appl. Meteorol.* **16**(5), 514–521.

Shaw, R. H., Brunet, Y., Finnigan, J. J., and Raupach, M. R., 1995, A wind tunnel study of air flow in waving wheat: Two-point velocity statistics, *Bound. Lay. Meteorol.* **76**(4), 349–376.

Shaw, R. H. and Patton, E. G., 2003, Canopy element influences on resolved- and subgridscale energy within a large-eddy simulation, *Agric. For. Meteorol.* **115**, 5–17.

Shaw, R. H. and Schumann, U., 1992, Large-eddy simulation of turbulent flow above and within a forest, *Bound. Lay. Meteorol.* **61**(1–2), 47–64.

Shaw, R. H., Tavangar, J., and Ward, D. P., 1983, Structure of the Reynolds stress in a canopy layer, *J. Clim. Appl. Meteorol.* **22**, 1922–1931.

Stoesser, T., Salvador, G., Rodi, W., and Diplas, P., 2009, Large eddy simulation of turbulent flow through submerged vegetation, *Transp. Porous Media* **78**(3), 347–365.

Stuart, J. T., 1967, On finite amplitude oscillations in laminar mixing layers, *J. Fluid Mech.* **29**(3), 417–440.

Su, H. B., Shaw, R. H., and Paw U. K. T., 2000, Two-point correlation analysis of neutrally stratified flow within and above a forest from large-eddy simulation, *Bound. Lay. Meteorol.* **94**(3), 423–460.

Taylor, G. I., 1938, The spectrum of turbulence, *Proc. R. Soc. Lond. A* **164**, 476–490.

Tennekes, H., 1982, Similarity relation, scaling laws and spectral dynamics, in F. T. M. Nieuwstadt and H. van Dop (eds.), *Atmospheric Turbulence and Air Pollution Modelling*, D. Reidel Publishing Company, Dordrecht, the Netherlands, pp. 37–68.

Tennekes, H. and Lumley, J. L., 1972, *A First Course in Turbulence*, Number 0, MIT Press, Cambridge, MA.

Thom, A. S., 1968, The exchange of momentum, mass, and heat between an artificial leaf and the airflow in a wind-tunnel, *Q. J. R. Meteorol. Soc.* **94**(399), 44–55.

Wang, H., Takle, E. S., and Shen, J., 2001, Shelterbelts and windbreaks: Mathematical modeling and computer simulations of turbulent flows, *Ann. Rev. Fluid Mech.* **33**(1), 549–586.

Watanabe, T., 2004, Large-eddy simulation of coherent turbulence structures associated with scalar ramps over plant canopies, *Bound. Lay. Meteorol.* **112**, 307–341.

Wilson, N. R. and Shaw, R. H., 1977, A higher order closure model for canopy flow, *J. Appl. Meteorol. Clim.* **16**(11), 1197–1205.

Wyngaard, J. C., 2010, *Turbulence in the Atmosphere*, Cambridge University Press, Cambridge, U.K.

Wyngaard, J. C. and Clifford, S. F., 1977, Taylors hypothesis and high-frequency turbulence spectra, *J. Atmos. Sci.* **34**(6), 922–929.

Yang, B., Raupach, M. R., Shaw, R. H., Paw U, K. T., and Morse, A. P., 2006, Large-eddy simulation of turbulent flow across a forest edge. Part I: Flow statistics, *Bound. Lay. Meteorol.* **120**(3), 377–412.

Zhang, C., Shaw, R. H., and Paw U, K. T., 1992, Spatial characteristics of turbulent coherent structures within and above an orchard canopy, in S. E. Schwartz and W. G. N. Slinn (eds.), *Precipitation Scavenging and Atmosphere-Surface Exchange*, Richland, Washington, DC, pp. 603–1172.

25

Jets and Plumes

Scott A. Socolofsky
Texas A&M University

Tobias Bleninger
Federal University of Paraná

Robert L. Doneker
MixZon, Inc.

25.1 Introduction

Turbulent buoyant jets are a fundamental flow class in the natural and engineered environment and span the full asymptotic range of jet and plume behavior: round jets, line jets, momentum puffs, negative jets, round plumes, line plumes, and thermals. In summary, buoyant jets occur whenever fluid is discharged with an excess of or deficit in momentum and/or buoyancy through a constriction into a receiving fluid body. They occur in a whole host of applications, and we would like to quote Gerhard Jirka from his (2004) paper in the journal *Environmental Fluid Mechanics* where he outlined his view of the topic:

> Buoyant jet motions (sometimes called forced plumes) are prevalent in the natural environment and in engineering applications. They are most spectacular in volcanic gas eruptions, they occur as hydrothermal vent flows in the deep ocean or as fresh groundwater plumes in the coastal zone. They are a key feature in society's fluid waste disposal methods, be it in the form of gaseous emissions into the atmosphere from industrial and domestic smokestacks, from mobile exhausts and from cooling towers, or of liquid releases into water bodies from industrial, municipal and agricultural sources or mining and oil extraction operations. They are an integral part of building ventilation and air conditioning systems. And they play a central role as mixing and injection devices in chemical reactors, waste and sewage treatment plants, desalination plants, combustion chambers, jet engines, or heat exchangers as well as stratification control and oxygenation devices in lakes or reservoirs.

In this chapter, we will focus our attention on the most common tools used to analyze buoyant jet behavior in the environment and highlight areas where research is ongoing.

Though the study and observation of buoyant jets began hundreds of years ago, the quantitative analysis of their behavior began in earnest with the introduction of the boundary layer theory by Prandtl and colleagues (Görler 1942; Tollmien 1926). Jirka (2004) presents a detailed history of turbulent buoyant jet analysis; here, we highlight some of the critical stages of development from his survey. Initial work employed similarity solutions based on different formulations of the Prandtl mixing length theory (see also Schlichting [1960] for a summary). The more generalized integral model approach began its development through the work of Reichardt (1941), who showed that the Gaussian profile was an acceptable approximation to the cross-sectional shape of the jet integral properties. Using this approximation, early models relied on a turbulent diffusion model for jet expansion until the seminal paper by Morton et al. (1956) who introduced Taylor's concept of jet entrainment. The entrainment model hypothesizes that jets grow by incorporating ambient fluid into the jet by turbulent motion and further that the inward velocity of the entraining fluid at the jet edge is proportional to a characteristic velocity scale in the jet, taken as the time-average

Handbook of Environmental Fluid Dynamics, Volume One, edited by Harindra Joseph Shermal Fernando. © 2013 CRC Press/Taylor & Francis Group, LLC.
ISBN: 978-1-4398-1669-1.

centerline velocity. The advantage of the entrainment approach is in the robustness of its applicability, making it capable of modeling quite complicated flows, including ambient stratification and crossflow.

General buoyant jet models based on the entrainment hypothesis and capable of simulating several different source and ambient conditions later developed, initiated by the work of Fan (1967). This model development was also supplemented by the excellent dimensional analysis of Wright (1977), and a detailed summary of these early activities is presented in Fischer et al. (1979). A more recent historical perspective on the entrainment hypothesis is also presented in Turner (1986). Since then, models have been generalized (see, e.g., Doneker and Jirka 1991; Jirka 2004; Jirka and Doneker 1991) and have benefitted from a wide range of new validation data, particularly from laboratory methods utilizing particle image velocimetry (PIV) and laser-induced fluorescence (LIF) (see, e.g., Davidson and Pun 1999; Tian and Roberts 2003; Yu et al. 2006, and other chapters in this book).

While our understanding of turbulent buoyant jet behavior is becoming quite detailed, their analysis remains challenging due to the wide range of time and space scales involved in their dynamic evolution. Figure 25.1 presents an example laboratory experiment of a buoyant jet along with a summary of the scales involved in coastal wastewater discharges. In the near-field region, close to the source, the buoyant jet exhibits the canonical self-similarity behavior and jet entrainment. As the jet is either arrested by the stratification (as in Figure 25.1) or encounters a boundary (side walls, free surface, or reservoir bottom), it enters an intermediate regime of rapid spreading in which the boundary layer assumption critical to the turbulent jet analysis breaks down, and the flow is no longer classified as a buoyant jet. At the end of this spreading region, the discharge enters a far-field

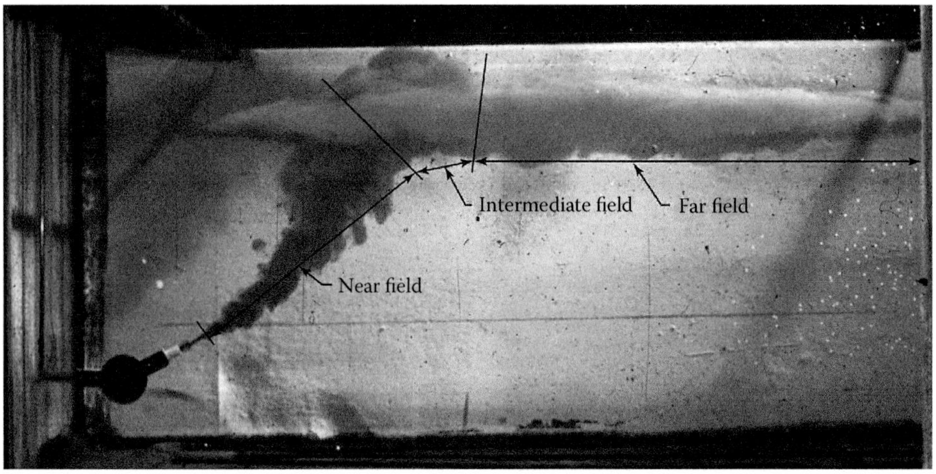

(a)

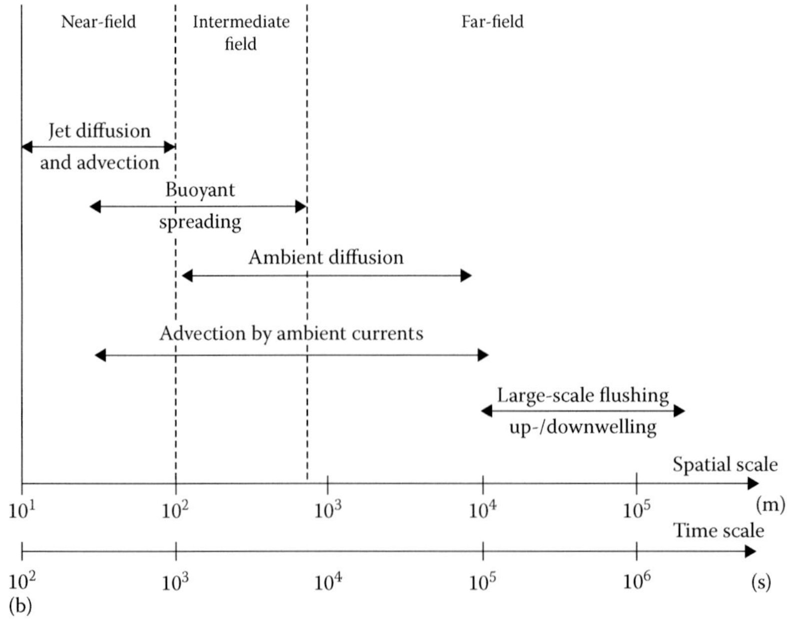

(b)

FIGURE 25.1 (a) Laboratory experiment of an inclined, turbulent buoyant jet in a linearly stratified quiescent reservoir and (b) a sketch of the characteristic length and time scales affecting coastal wastewater discharges.

behavior, dominated by ambient currents, turbulent diffusion, and large, basin-scale motion. Because the height of rise of the plume and the initial dilution entering the far field is strongly affected by processes in the near-field, models of buoyant jet behavior are critical to the fate and transport modeling of environmental discharges. Hence, the near-field jet model is of fundamental importance.

In the remainder of this chapter, we present an overview of the methods used to analyze buoyant jets in the environment. The Principles section outlines the quantitative parameters used to describe buoyant jets and introduces the boundary layer approximation and the details of the entrainment hypothesis as it applies to turbulent buoyant jets. In the Methods of Analysis section, we begin with a discussion of dimensional analysis and the scaling relationships that have been developed for many asymptotic and transitional regimes of jet flow and conclude with a presentation of the governing equations for the integral model of a general turbulent buoyant jet in a flowing and density-stratified ambient reservoir. The Applications section highlights how such models are used in regulatory mixing zone analysis, particularly through the U.S. Environmental Protection Agency (EPA) model CORMIX, and introduces example extensions of the integral model to multiport diffusers and surface and bottom jets. The final section, Challenges, highlights several areas of ongoing research to improve and extend buoyant jet models for an ever increasing range of applications.

25.2 Principles

25.2.1 Governing Variables

Jets and plumes are typically described mathematically by a set of flux variables expressed in kinematic form. Figure 25.2 is a sketch of the general case of a buoyant jet discharged at an arbitrary angle (σ relative to the x-axis and θ relative to the horizontal plane) into a flowing and density-stratified ambient reservoir. For this case, the independent flux variables are the volume flux Q, momentum flux M, buoyancy flux J, and the mass flux of passive tracers Q_{ci}. Although J is the important forcing parameter for buoyancy, it is often more convenient to track the fluxes of the state variables X_i (e.g., heat, salinity, etc.) affecting the density and to compute the density from an equation of state of the form $\rho = f(X_i)$—once the density is known, it is substituted to compute the buoyancy flux J separately.

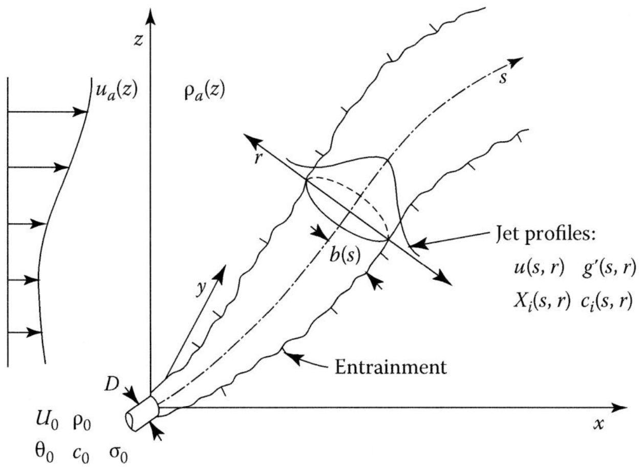

FIGURE 25.2 Schematic of a general round buoyant jet into a flowing and density-stratified ambient reservoir.

At the jet nozzle, these governing flux quantities are given by their initial values

$$Q_0 = U_0 a_0$$

$$M_0 = U_0 Q_0$$

$$J_0 = g_0' Q_0$$

$$Q_{Xi0} = X_{i0} Q_0$$

$$Q_{c_{i0}} = c_{i0} Q_0 \quad (25.1)$$

where

U_0 is the jet exit velocity

a_0 is the nozzle cross-sectional area (for round jets with pipe diameter D, $a_0 = (\pi/4)D^2$)

X_{i0} is the concentration of state variables in the effluent

c_{i0} is the concentration of passive tracers in the effluent

g_0' is the reduced gravity of the effluent, given by $g_0' = g(\rho_a - \rho_0)/\rho_r$, g is the acceleration of gravity, ρ_a is the ambient fluid density at the elevation of the jet exit, ρ_0 is the density of the effluent, ρ_r is a reference density, generally taken as a constant equal to the average density in the receiving fluid

The subscript "i" is used to keep track of multiple tracers

The values of these flux variables also evolve with distance from the nozzle along the jet trajectory s. This is true for both laminar and turbulent jets, though for turbulent jets, we average over an appropriate integral time scale to obtain statistically stationary results. The local flux variables can be computed by integrating profiles of velocity $u(s, r)$, reduced gravity $g'(s, r)$, and tracer concentrations $X_i(s, r)$ and $c_i(s, r)$ over the jet cross section (perpendicular to the jet axis). Thus, for a round jet, we obtain

$$Q(s) = 2\pi \int_0^\infty u(s,r) r dr$$

$$M(s) = 2\pi \int_0^\infty u^2(s,r) r dr$$

$$J(s) = 2\pi \int_0^\infty u(s,r) g'(s,r) r dr$$

$$Q_{Xi}(s) = 2\pi\int_0^{\infty} u(s,r)X_i(s,r)rdr$$

$$Q_{ci}(s) = 2\pi\int_0^{\infty} u(s,r)c_i(s,r)rdr \tag{25.2}$$

These variables form the state space of unknown quantities that are solved for by jet integral models. For line jets, the equations are similar, with the integrals taken over a Cartesian coordinate system and the results expressed as per unit length along the jet slot.

When the receiving fluid has non-zero values of some of the variables (as due to an ambient current or background tracer concentration), care must be taken to distinguish between the relative value in the jet compared to the background value. We have already seen this fact in the reduced gravity, which expresses the difference between ρ_a, the absolute density in the ambient fluid outside the jet at s, and $\rho(s, r)$, the absolute density in the jet: $g'(s,r) = g(\rho_a - \rho(s,r))/\rho_r$. Similarly, our interest is to solve for the excess velocity and concentration in the jet above the ambient value. Hence, we may express the local variables as

$$u(s,r) = u_j(s,r) + u_a$$

$$X_i(s,r) = X_{ij}(s,r) + X_{ia}$$

$$c_i(s,r) = c_{ij}(s,r) + c_{ia} \tag{25.3}$$

Here, the subscript "j" indicates the excess value in the jet and the subscript "a" indicates the value for the ambient reservoir outside the jet at s; for the vector velocity, we take u_a as the component along the jet axis.

Away from the nozzle, the buoyant jet is also described by characteristic values of its dynamic jet properties. These variables include the jet half-width $b(s)$ and centerline values of the velocity $u_c(s)$, reduced gravity $g'_c(s)$, and passive tracer concentrations $X_{ic}(s)$ and $c_{ic}(s)$. Because the cross-sectional profiles of the jet properties asymptotically approach zero, the half-width is usually defined as the lateral distance from the centerline to a point where the mean velocity is a fixed fraction of the average centerline velocity: $b = r(\delta u_c)$. Common values for δ are 1/2 or $1/e$. Profiles of concentration are wider than that of velocity by a constant spreading factor λ, such that dissolved constituents have a width λb.

From these parameters describing the jet motion, an important property for environmental applications of jets and plumes known as the dilution S may also be derived, defined as the total volume of a sample divided by the volume of effluent in the sample. Several processes affect the effluent volume in the plume, including physical mixing, chemical transformation, and boundary interaction processes. In this chapter, we will focus on the pure hydrodynamic dilution, which for a conservative tracer (nonreacting) can be defined as

$$S_i = \frac{c_{i0}}{c_{ic}} \tag{25.4}$$

Because c_{ic} is a jet excess concentration, the dilution computed by this formula will give the true dilution of the effluent regardless of concentrations in the ambient reservoir.

25.2.2 Boundary Layer Approximation

Jets and plumes are among the canonical free shear flows that satisfy the boundary layer approximation, where the width of the jet is much less than the longitudinal length scale along the jet centerline. For an axisymmetric jet, the Reynolds averaged boundary layer equations are (refer to Figure 25.2)

$$\frac{\partial u}{\partial s} + \frac{1}{r}\frac{\partial rv}{\partial r} = 0 \quad \text{(Continuity equation)}$$

$$u\frac{\partial u}{\partial s} + v\frac{\partial u}{\partial r} = -\frac{1}{r}\frac{\partial}{\partial r}\left(r\overline{u'v'}\right)$$

$$\text{(Momentum equation along the jet axis)}$$

$$0 = \frac{\partial p}{\partial r} \quad \text{(Momentum equation perpendicular to jet axis)}$$

where u' and v' are the fluctuating turbulent velocities in the s and r directions, respectively. For a pure jet in a quiescent and uniform unbounded domain, the boundary conditions at $r = 0$ are $v = 0$ and $\partial u/\partial r = 0$ and at $r \to \infty$ are $u \to 0$ and $\overline{u'v'} \to 0$.

For certain simplified turbulence closure models, such as the Prandtl mixing length hypothesis with

$$\overline{u'v'} = \varepsilon\frac{\partial u}{\partial r} \tag{25.5}$$

where ε is a constant, similarity solutions can be found (see e.g., Kundu and Cohen 2008; Schlichting 1960). For more complicated situations, numerical solutions may be obtained using various advanced turbulence models. A simple, yet flexible third alternative is to convert the aforementioned system of partial differential equations to ordinary differential equations by integrating over the plume cross section using an assumed, self-similar shape of the local variables, yielding the integral model equations for the flux variables defined in the previous section. The general form of the integration for a round jet is $2\pi\int_0^{\infty}(\cdot)rdr$, and applying this operation to the continuity equation and using the boundary conditions gives

$$\frac{d}{ds}\left(2\pi\int_0^{\infty} u(z,r)rdr\right) = -2\pi v(s,r)r\Big|_{r=\infty} \tag{25.6}$$

The integral on the left-hand side of the equation is just $Q(s)$. The term on the right-hand side of the equation is not zero and introduces a turbulent closure problem through the fact that the solution for v depends on the turbulence model used in the momentum equation. This term is usually evaluated at the edge of the plume at $r = b$ since its contribution is essentially constant beyond this radius, and the velocity obtained there is called the entrainment velocity v_e directed toward the jet centerline ($-r$ direction), yielding

$$\frac{dQ}{ds} = 2\pi b v_e \tag{25.7}$$

as the continuity equation. To obtain an equation for the momentum flux, we combine the continuity and momentum equations and then integrate over the cross section to obtain (Kundu and Cohen 2008)

$$\frac{dM}{ds} = 0 \tag{25.8}$$

hence, the momentum flux is preserved for a pure jet experiencing no ambient forcing or boundary interaction.

25.2.3 Turbulence Closure: Entrainment Hypothesis

As we see in the previous section, the solution for the flow in a jet requires a turbulence closure model, either for the kinematic Reynolds stress $\overline{u'v'}$ or for an entrainment velocity v_e. Two popular closure models are the spreading and entrainment hypotheses. The spreading hypothesis is based on the experimental observation that the jet width grows linearly with distance from the source, so that

$$\frac{db}{ds} = \text{const} \tag{25.9}$$

This closure is particularly useful for analytical solutions based on Prandtl's mixing length theory. Schlichting (1960) explains that the mixing length l for a jet is proportional to the local width l/b = const and that this constant leads to the linear growth in the spreading hypothesis. With this model, the solution to the boundary layer equations takes the same form as the laminar solution, but with the molecular kinematic viscosity replaced by the turbulent viscosity ε = const, which may be obtained experimentally.

The second closure model, the entrainment hypothesis, leads to the same solution for a simple jet as the spreading hypothesis, but from a more mechanistic perspective, explaining the reason for the observed spreading. In the entrainment hypothesis, the entrainment velocity is assumed to depend on the turbulence intensity, which scales with the centerline velocity, resulting in the relationship

$$\frac{v_e}{u_c} = \alpha \tag{25.10}$$

where α is a constant, called the entrainment coefficient and which is obtained by experiment. Substituting this relationship in the continuity Equation 25.7 for the 1D integral model and using the definitions in (25.2) leads to a closed system of equations for Q and M.

The entrainment hypothesis has been shown to be a very robust relationship, valid from laboratory scales of a few centimeters up to geophysical scales of several kilometers, as created by volcanic eruptions (Turner 1986). Because of its physical interpretation as the inflow of ambient fluid along the edge of the jet, the entrainment hypothesis is also a much more flexible tool than the spreading hypothesis for developing models of more complicated jets involving forcing from ambient currents and stratification. It is also important to note that the entrainment hypothesis is a turbulence closure model; hence, it is only appropriate for turbulent jets and plumes, at a Reynolds number based on the jet width above about 500.

25.3 Methods of Analysis

The three classes of methods for buoyant jet flows used to solve for the governing parameters described earlier are *empirical*, *integral*, and *numerical* methods. The empirical methods are introduced in the following in the Dimensional Analysis section, followed by the integral methods in the section on Buoyant Jet Integral Models. These methods are generally adequate for each of the asymptotic flow regimes defined in the introduction (e.g., pure jet, pure plume, momentum puff, etc.) and in gradual transitions among these regimes.

For highly complex situations, such as nontrivial jet merging processes, unsteady flow analysis, complex discharge geometries, and boundary interactions, numerical solutions are required using advanced (at least two equation) turbulence closures. As distinct from the empirical or integral approach, numerical solutions calculate all flow characteristics at every point of the flow domain, thus they need to resolve the entire jet-induced velocity field and its interaction with the surrounding fluid. For the buoyant jet, the hydrodynamic and constituent transport equations are dynamically coupled. In addition, the sharp gradients and small scales in the jet near field require nondispersive numerical schemes for the convective terms and specialized turbulence closure models. For the same reasons, grid generation and computations are time-consuming, limiting the opportunity to apply numerical solutions to many designs. Furthermore, uncertainties exist in prescribing appropriate boundary conditions for the ambient flow, and calibration is needed for the parameters of the turbulence model, limiting the predictive capabilities of these models (Xiao et al. 2005); thus, a more detailed discussion of computational fluid dynamics models applied to jets and plumes is beyond the scope of this chapter. We will show, however, that the empirical and integral solutions are adequate for the design of most single and multiport discharges in the environment.

25.3.1 Dimensional Analysis

Because of the self-similarity behavior of the asymptotic types of the buoyant jet, dimensional analysis is a powerful tool to both predict relationships among the governing variables and to classify buoyant jet flows into different asymptotic regimes.

As an example of empirical solutions to the jet flow, consider a pure jet discharging horizontally in the x-direction in a stagnant, uniform ambient reservoir. The local unknown quantities (dependent variables) are the velocities u and v, and the concentration of passive tracer c. Characteristic scales of interest that evolve along the jet trajectory are $b(x)$, $u_c(x)$, $Q(x)$, and $S_c(x)$, each a different function of several independent variables, (x, M_0, a_0, etc.). Dimensional analysis for a round jet then results in the parameters

$$\frac{b}{x} = k_1, \quad \frac{u_c x}{\sqrt{M_0}} = k_2, \quad \frac{Q}{\sqrt{M_0 x}} = k_3, \quad \frac{S_c Q_0}{\sqrt{M_0 x}} = k_4 \qquad (25.11)$$

The width b evolves following a linear dependence on x with the proportionality factor $k_1 = 0.11$, determined by experiments (Jirka 2004), and the centerline velocity decays inversely with the distance with the proportionality factor $k_2 = 7.25$ (Jirka 2004). The flow rate and dilution increase linearly with distance (from experiments $k_3 = 0.27$, and $k_4 = 0.1623$; Jirka 2004).

Because of the self-similarity of the flow, these four constants can also be expressed in terms of two other experimental constants, an entrainment coefficient $\alpha_{jet} = 0.055$, and a tracer spreading rate $\lambda_{jet} = 1.20$, yielding

$$k_1 = 2\alpha_{jet}, \quad k_2 = \frac{1}{\sqrt{2\pi}\alpha_{jet}}, \quad k_3 = 2\sqrt{2\pi}\alpha_{jet},$$

$$k_4 = \frac{\lambda_{jet}^2}{1+\lambda_{jet}^2}\frac{2\sqrt{2\pi}\alpha_{jet}}{1+\lambda_{jet}^2} \qquad (25.12)$$

Similar analysis is possible for other jet properties and each of the asymptotic buoyant jet regimes. These are summarized in Table 25.1. However, with increasing complexity, the number of independent parameters increases, and difficulties arise in defining consistent relations on the one hand and elaborating necessary laboratory studies on the other. Thus, correlation equations resulting from dimensional analysis are restricted to steady asymptotic cases with some slight extensions.

As a second example of dimensional analysis, we consider more complicated cases and attempt to classify regimes of the flow as, for instance, jet, plume, crossflow, or stratification dominated, among others. Consider the jet of the previous example with the added effect of an initial buoyancy flux. Early attempts to predict the trajectory of a buoyant jet used the exit geometry as the important scales to nondimensionalize the data. However, as shown in the left plot of Figure 25.3, this scaling is unable to collapse the data, and different lines are obtained for different initial Froude numbers

$$F_0 = \frac{U_0}{\sqrt{g_0' D_0}} \qquad (25.13)$$

Instead of the nozzle exit, another characteristic length scale for the jet can be defined that accounts for differences in the initial Froude number, by taking a ratio of the initial momentum and buoyancy fluxes yielding

$$L_M = \frac{M_0^{3/4}}{J_0^{1/2}} \qquad (25.14)$$

When applied to predict the buoyant jet trajectory, this new scale efficiently collapses the data to a single line (right plot of Figure 25.3). Because L_M is a measure of the length scale of the jet-dominated region of a buoyant discharge, it also allows distinguishing the jet-dominated region of more complicated flows, as illustrated in Figure 25.4 and expanded in Section 25.4.1.

A consistent length-scale-based categorization of the different buoyant jet regimes in the presence of crossflow and/or stratification is summarized in Fischer et al. (1979) and modified by Jirka and Akar (1991), giving the following length scales:

- $L_M = M_0^{3/4}/J_0^{1/2}$: The jet to plume transition length scale, which denotes a scaling for the transition from jet to plume behavior in a stagnant ambient (note that the product D_0F_0 is proportional to L_M).
- $L_M = M_0^{1/2}/u_a$: The jet to crossflow length scale, which denotes a scaling for the distance of transverse jet penetration beyond which strong deflection by the crossflow occurs.
- $L_b = J_0/u_a^3$: The plume to crossflow length scale, which denotes a scaling for the distance of plume penetration beyond which strong deflection by the crossflow occurs.
- $L_b' = J_0^{1/4}/\varepsilon^{3/4}$: The plume to stratification length scale, which denotes a scaling for the distance at which the jet becomes strongly affected by the stratification (defined by $\varepsilon = -(g/\rho_r)(d\rho_a/dz)$, the ambient buoyancy gradient), leading to terminal layer formation and horizontally spreading in a stagnant linearly stratified ambient.

For a comprehensive discussion of these and other scales for buoyant jet classification, see Jirka and Doneker (1991). Similar analyses for the dilution of a buoyant jet in quiescent, flowing, and stratified ambient reservoirs is presented in Fischer et al. (1979) and Jirka and Lee (1994).

TABLE 25.1 Equations for the Five Self-Similar Asymptotic Regimes of the Round Buoyant Jet

Regime (All $\lambda = 1.20$)	Width b	Centerline Velocity u_c or Trajectory Elevation z	Volume Flux Q	Centerline Dilution $S_c = c_o/c_c$	Remarks
Jet	$b = 2\alpha_{jet} x$	$u_c = \frac{1}{\sqrt{2\pi}\,\alpha_{jet}} \frac{M_o^{1/2}}{x}$	$Q = \sqrt{2\pi}\, 2\alpha_{jet} M_o^{1/2} x$	$S_c = \frac{\lambda^2}{1+\lambda^2} \frac{\sqrt{2\pi}\, 2\alpha_{jet}}{1+\lambda^2} \frac{M_o^{1/2} x}{Q_o}$	Assumes discharge in x-direction
$\alpha_{jet} = 0.055$	$b = 0.11x$	$u_c = 7.25 \frac{M_o^{1/2}}{x}$	$Q = 0.27 M_o^{1/2} x$	$S_c = 0.162 \frac{M_o^{1/2} x}{Q_o}$	
Plume	$b = \frac{6}{5}\alpha_{plume} z$	$u_c = \frac{1}{2}\left(\frac{25}{3\pi} \frac{1+\lambda^2}{\alpha_{plume}^2}\right)^{1/3} \frac{J_o^{1/3}}{z^{1/3}}$	$Q = \frac{6}{5}\left(\frac{9}{5}(1+\lambda^2)\pi^2\alpha_{plume}^4\right)^{1/3} J_o^{1/3} z^{5/3}$	$S_c = \frac{\lambda^2}{1+\lambda^2} \frac{6}{5}\left(\frac{9}{5}\frac{\lambda^6}{(1+\lambda^2)^2}\pi^2\alpha_{plume}^4\right)^{1/3} \frac{J_o^{1/3} z^{5/3}}{Q_o}$	
$\alpha_{plume} = 0.083$	$b = 0.10z$	$u_c = 4.9 \frac{J_o^{1/3}}{z^{1/3}}$	$Q = 0.152 J_o^{1/3} z^{5/3}$	$S_c = 0.090 \frac{J_o^{1/3} z^{5/3}}{Q_o}$	
Wake	$b = \left(\frac{3\alpha_{wake}}{2\pi}\right)^{1/3} \frac{M_{oe}^{1/3} x^{1/3}}{u_a^{2/3}}$	$u_c = \frac{1}{\pi}\left(\frac{2\pi}{3\alpha_{wake}}\right)^{2/3} \frac{M_{oe}^{1/3} u_a^{1/3}}{x^{2/3}}$	$Q = 2\pi\left(\frac{3\alpha_{wake}}{2\pi}\right)^{2/3} \frac{M_{oe}^{2/3} x^{2/3}}{u_a^{1/3}}$	$S_c = \lambda^2\pi\left(\frac{3\alpha_{wake}}{2\pi}\right)^{2/3} \frac{M_{oe}^{2/3} x^{2/3}}{u_a^{1/3} Q_o}$	
$\alpha_{wake} = 0.11$	$b = 0.37 \frac{M_{oe}^{1/3} x^{1/3}}{u_a^{2/3}}$	$u_c = 2.27 \frac{M_{oe}^{1/3} u_a^{1/3}}{x^{2/3}}$	$Q = 0.88 \frac{M_{oe}^{2/3} x^{2/3}}{u_a^{1/3}}$	$S_c = 0.63 \frac{M_{oe}^{2/3} x^{2/3}}{u_a^{1/3} Q_o}$	
Advected line puff	$b = \left(\frac{3\alpha_{puff}}{4\pi}\right)^{1/3} \frac{M_{ot}^{1/3} x^{1/3}}{u_a^{2/3}}$	$z = \left(\frac{6}{\pi\alpha_{puff}^2}\right)^{1/3} \frac{M_{ot}^{1/3} x^{1/3}}{u_a^{2/3}}$	$Q = \left(\sqrt{\frac{\pi}{2}}\, 3\alpha_{puff}\right)^{2/3} \frac{M_{ot}^{2/3} x^{2/3}}{u_a^{1/3}}$	$S_c = \frac{\lambda^2}{2}\left(\sqrt{\frac{\pi}{2}}\, 3\alpha_{puff}\right)^{2/3} \frac{M_{ot}^{2/3} x^{2/3}}{u_a^{1/3} Q_o}$	Assumes transverse momentum flux M_{ot} in z-direction
$\alpha_{puff} = 0.5$	$b = 0.49 \frac{M_{ot}^{1/3} x^{1/3}}{u_a^{2/3}}$	$z = 1.96 \frac{M_{ot}^{1/3} x^{1/3}}{u_a^{2/3}}$	$Q = 1.52 \frac{M_{ot}^{2/3} x^{2/3}}{u_a^{1/3}}$	$S_c = 1.10 \frac{M_{ot}^{2/3} x^{2/3}}{u_a^{1/3} Q_o}$	
Advected line thermal	$b = \left(\frac{3\alpha_{thermal}}{8\pi\lambda}\right)^{1/3} \frac{J_o^{1/3} x^{2/3}}{u_a}$	$z = \left(\frac{3}{\pi\lambda\alpha_{thermal}^2}\right)^{1/3} \frac{J_o^{1/3} x^{2/3}}{u_a}$	$Q = \left(\sqrt{\frac{\pi}{2}} \frac{3\alpha_{thermal}}{2\lambda}\right)^{2/3} \frac{J_o^{2/3} x^{4/3}}{u_a}$	$S_c = \frac{\lambda^2}{2}\left(\sqrt{\frac{\pi}{2}} \frac{3\alpha_{thermal}}{2\lambda}\right)^{2/3} \frac{J_o^{2/3} x^{4/3}}{u_a Q_o}$	
$\alpha_{thermal} = 0.5$	$b = 0.37 \frac{J_o^{1/3} x^{2/3}}{u_a}$	$z = 1.47 \frac{J_o^{1/3} x^{2/3}}{u_a}$	$Q = 0.85 \frac{J_o^{2/3} x^{4/3}}{u_a}$	$S_c = 0.61 \frac{J_o^{2/3} x^{4/3}}{u_a Q_o}$	

Source: Reproduced from Jirka, G.H., *Environ. Fluid Mech.*, 4(1), 1–56, 2004.

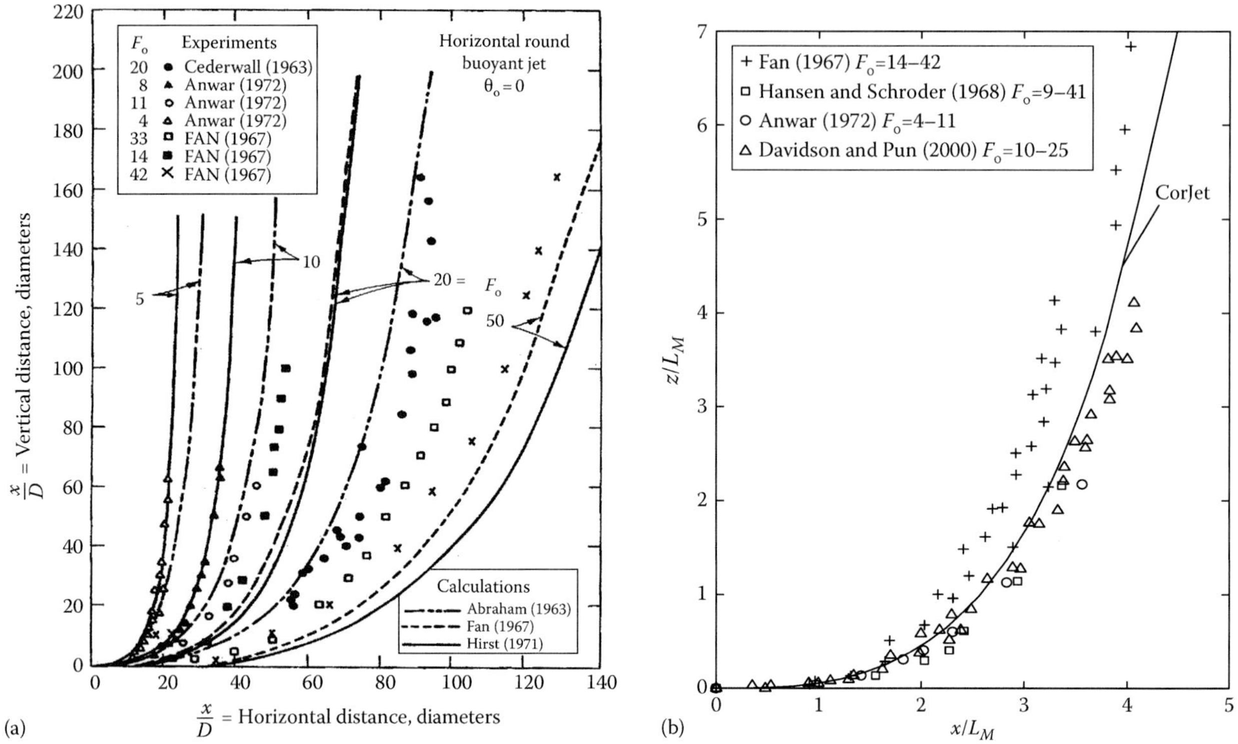

FIGURE 25.3 Three-dimensional horizontal buoyant jet trajectories for a single-port discharge into a stagnant reservoir. Comparison between predictions and experimental data. (a) normalized with port diameter. (b) Normalized with the momentum length scale L_M. (Reproduced from Jirka, G.H., *Environ. Fluid Mech.*, 4(1), 1–56, 2004. With permission.)

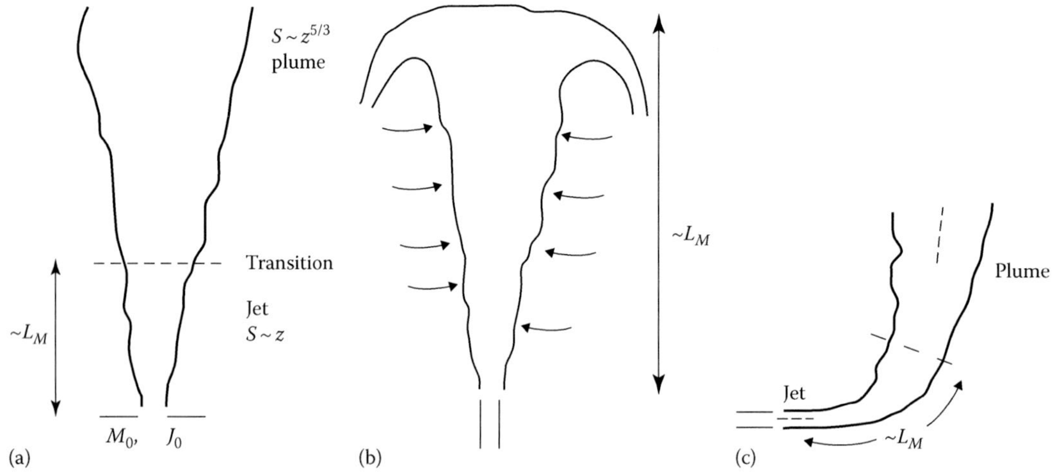

FIGURE 25.4 Schematic representation of the jet length scale L_M for (a) a vertical buoyant jet in an unstratified ambient reservoir, (b) a vertical buoyant jet in stratification, and (c) a horizontal buoyant jet.

25.3.2 Buoyant Jet Integral Models

For the design and more detailed analysis of real discharges in the environment, integral jet models are the most common tools for predicting the evolution in the near field. Strictly speaking, these tools are based on the similarity assumption: profiles of velocity, concentration, and other jet properties have the same shapes at all heights; only their magnitudes change. Self-similarity is valid in unbounded domains where there is no characteristics length scale for the flow. Stratification, crossflows, and the existence of lateral (side-walls) and horizontal (bottom and free-surface) boundaries all act to introduce external length scales to the problem and break down the self-similarity of the jet. Nonetheless, because of the robustness of the entrainment hypothesis, acceptable results may be obtained in most of these situations provided experimental data are used to validate their accuracy and range of applicability.

In addition to self-similarity, the buoyant jet integral equations derived here are constrained by a few other assumptions. Their derivation starts with the Reynolds averaged Navier-Stokes equations and applies the boundary layer approximation; hence, the fluids must be Newtonian, the flow must be thin in lateral extent compared to distance along the flow path, and the results should be interpreted as time averages over the turbulent fluctuations. Jirka (2004) summarized the implications of these limitations in six principles that must be met by any jet integral model to obtain a reasonably accurate solution. Briefly, for a round jet, these are as follows:

1. Solutions are valid only for the five asymptotic conditions exhibiting strict self-similarity, a constant internal force balance, and invariant turbulence properties. These are the pure jet, pure plume, pure wake, advected line puff, and advected line thermal. As a corollary, all jet integral models must be validated to show that they accurately reproduce these five asymptotic regimes.
2. Models of the transitions among these five asymptotic solutions are arbitrary (they cannot be derived from first principles) and should be guided by good model agreement with available data.
3. Model solutions cannot be trusted when the boundary layer assumption is violated. This occurs whenever the jet undergoes strong spreading (as in the arrest due to ambient stratification) or strong curvature (as in jets directed into an opposing ambient current).
4. The initial zone of flow establishment (ZOFE) lacks self-similarity (the velocity profile shape transitions from top-hat pipe profiles to the Gaussian jet profile); hence, the ZOFE should not be modeled by the jet integral equations. Instead, empirical relations in this region (extended 5–10 times the exit diameter) should be used to establish an equilibrium initial condition for the jet integral model at the end of the ZOFE.
5. The jet integral equations should be formulated in terms of the flux quantities (Q, M, J, etc.) because these quantities are conservative and generally do not exhibit strong changes or singularities as some of the local variables do (e.g., b, u, g', etc.). This principle acts to preserve solution accuracy.
6. The model formulation must be accompanied by limits of applicability. In particular, the model must terminate when boundary interactions occur or the flow transitions to an intermediate- or far-field behavior dominated by turbulent diffusion over entrainment-related spreading.

One integral model that meets each of these principles is the CorJet module as implemented in the CORMIX system for the analysis of regulatory mixing zones (Doneker and Jirka 1991; Jirka 2004; Jirka and Doneker 1991; see also the Application section of this chapter). In the following, we derive the fundamental equations solved by this model and presented in Jirka (2004).

25.3.2.1 Coordinate System

Here, we derive the jet integral equations in an Eulerian sense along the jet centerline. Figure 25.5 shows a sketch of the jet trajectory and overlaying coordinate systems. The local, cylindrical coordinate system (s, r) is fixed to the jet trajectory, with s tangent to the local jet centerline and r the radial coordinate from s. The (s, r) coordinate system is also mapped to the fixed, Cartesian coordinate system (x, y, z), in which x points in the direction of the ambient current $\vec{u}_a = (u_a(z),0,0)$, z points opposite to the gravity vector $\vec{f}_g = (0,0,-g)$, and the origin is at the jet exit (CORMIX solves for the more general case of a 2D, planar velocity field; however, we limit this chapter to the simplified

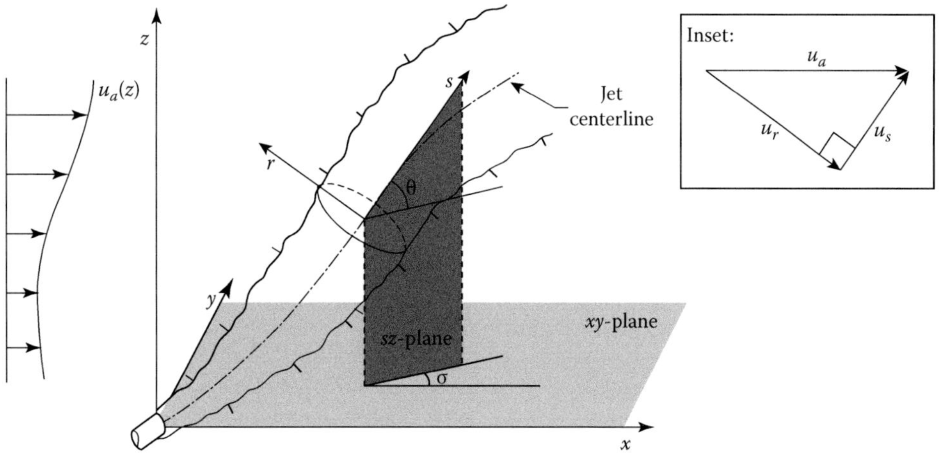

FIGURE 25.5 Schematic of the local (s, r) and fixed Cartesian (x, y, z) coordinate system for the buoyant jet integral model.

case defined here). A transform between the two coordinate systems is also depicted in the figure, where σ is the angle between the s- and x-axes, and θ is the angle between the s- and z-axes. Thus, a unit vector in the s-direction at the origin of the (x, y, z) coordinate system has Cartesian coordinates

$$e_s = (\cos\sigma\cos\theta, \sin\sigma\cos\theta, \sin\theta) \quad (25.15)$$

Because entrainment fluxes into the jet are perpendicular to the s-axis in the model, an important quantity is the component of the ambient velocity vector transverse to s. Because $\vec{u}_a$ depends on z only, the projection of $\vec{u}_a$ on s is obtained by $u_s = \vec{u}_a \cdot e_s = u_a(z)\cos\sigma\cos\theta$. Then, the ambient velocity component transverse to s is found using the Pythagorean theorem (see inset in Figure 25.5), yielding

$$u_t = u_a\sqrt{1-\cos^2\sigma\cos^2\theta} \quad (25.16)$$

These and similar geometric transforms are used throughout the equations given in the following.

25.3.2.2 Lateral Profiles of the Local Variables

The choice of the shape of the lateral profiles for velocity, concentration, etc. is essentially arbitrary—square, so-called top-hat profiles perform equally well to more realistic profile shapes—because the model tracks only the flux resulting from an integral over the profile shape. Here, we use Gaussian profiles, both for their ease of use and because they most closely match the experimental data. We will develop the model for the general case of a stratified, flowing ambient condition, as sketched in Figure 25.5. For this model, Gaussian profiles give (Jirka 2004)

$$u(s,r) = u_c(s)\exp\left(-\frac{r^2}{b^2}\right) + u_a(z)\cos\sigma\cos\theta,$$

$$g'(s,r) = g_c'(s)\exp\left(-\frac{r^2}{(\lambda b)^2}\right)$$

$$X_i(s,r) = X_{ic}(s)\exp\left(-\frac{r^2}{(\lambda b)^2}\right) + X_{ia}(z),$$

$$c_i(s,r) = c_{ic}(s)\exp\left(-\frac{r^2}{(\lambda b)^2}\right) + c_{ia}(z)$$

where the centerline quantities are jet excess quantities above their ambient values.

25.3.2.3 Flux Equations

The state space of model fluxes are obtained by substituting these profiles into Equation 25.2 to obtain

$$Q(s) = 2\pi\int_0^\infty u(s,r)r\,dr = \pi b(s)^2(u_c(s) + 2u_a(z)\cos\sigma\cos\theta) \quad (25.17)$$

$$M(s) = 2\pi\int_0^\infty u(s,r)^2 r\,dr = \frac{1}{2}\pi b(s)^2(u_c(s) + 2u_a(z)\cos\sigma\cos\theta)^2 \quad (25.18)$$

$$Q_{Xi}(s) = 2\pi\int_0^\infty u(s,r)(X_i(s,r) - X_{ia}(z))r\,dr = \pi b(s)^2\left(u_c(s)\frac{\lambda^2}{1+\lambda^2} + \lambda^2 u_a(z)\cos\sigma\cos\theta\right)X_{ic}(s) \quad (25.19)$$

$$Q_{ci}(s) = 2\pi\int_0^\infty u(s,r)(c_i(s,r) - c_{ia}(z))r\,dr = \pi b(s)^2\left(u_c(s)\frac{\lambda^2}{1+\lambda^2} + \lambda^2 u_a(z)\cos\sigma\cos\theta\right)c_{ic}(s) \quad (25.20)$$

So far we have shown explicitly the dependence of each variable on s and r. After integrating as done earlier, the model unknowns become 1D along s; hence, we will drop this notation going forward and all flux and local variables may be assumed to depend on s only.

25.3.2.4 Model Conservation Equations

Figure 25.6 shows a sketch of a differential control volume along the jet trajectory for which we will derive the model equations. Flux quantities F enter the element at face 1 and exit at face 2. Ambient fluid is entrained along the sides of the jet with effective entrainment flow rate E per unit height. And the forces acting on the control volume are the drag force f_d due to the crossflow

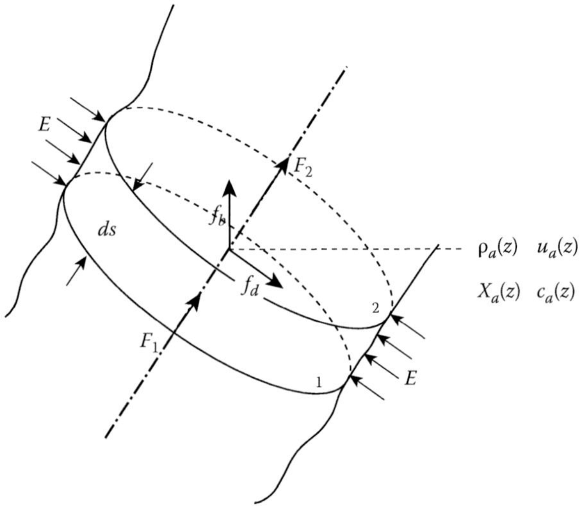

FIGURE 25.6 Schematic of the control volume of the buoyant jet used to derive the governing conservation equations of the integral jet model.

and the net buoyancy f_b of the control volume. The model equations are obtained by applying conservation laws to this control volume.

The entrainment rate E accounts for several shear mechanisms that result in entrainment of ambient fluid into the body of the turbulent buoyant jet. Similar mechanisms are at work in each of the asymptotic flow regimes (principle 1), and a general model is sought that accounts for an appropriate transition among these asymptotic solutions (as required by principle 2). Jirka (2004) suggests a general entrainment model of the form

$$E = 2\pi b(\alpha_e u_c + \alpha_4 |\cos\sigma\cos\theta| u_t) \tag{25.21}$$

The term $2\pi b$ is the local circumference of the jet. The first term in (25.21) accounts for the entrainment from jets, plumes and wakes, and an acceptable empirical model to achieve smooth transition among these asymptotic regimes is suggested by Jirka (2004) as

$$\alpha_e = \alpha_1 + \alpha_2 \frac{\sin\theta}{F_l^2} + \alpha_3 \frac{u_a |\cos\sigma\cos\theta|}{u_c + u_a} \tag{25.22}$$

The first term in (25.22) gives the entrainment coefficient for a pure jet, the second term adds the effect of the pure plume, and the third term incorporates the effect of the pure wake. The plume contribution is shown by experiment to depend on the local densimetric Froude number $F_l = u_c / \sqrt{g_c' b}$ and the vertical angle θ since buoyancy acts in the vertical direction. The wake effect is proportional to the wake parameter $u_a/(u_c + u_a)$ and the projection of the jet velocity on the x axis since the wake propagates in the direction of the ambient current (x-direction). The final term in the entrainment model (25.21) accounts for line puffs and thermals: the entrainment velocity is proportional to the ambient velocity transverse to the jet and the entrainment contribution depends on the component of the jet velocity in the ambient current direction: $|\cos\sigma\cos\theta|$. Hence, vertically oriented buoyant jets are dominated by jet and plume entrainment while the wake and crossflow entrainment mechanisms dominate as the jet bends over and transitions to a buoyant line thermal. Although crossflow entrainment is zero for the vertical jet ($\cos\theta = 0$ for $\theta = \pi/2$), the plume still bends over as regular jet entrainment engulfs ambient fluid having x-direction momentum and as the jet experiences a drag force in the x-direction.

The conservation of mass is obtained by a volume balance on the flows into and out of the control volume, giving

$$Q_2 - Q_1 = dQ = E ds \tag{25.23}$$

or, rearranging:

$$\frac{dQ}{ds} = E \tag{25.24}$$

Likewise, the conservation of mass for the state parameters and passive tracers are a simple flux balance over the control volume. However, because the flux variables are calculated in terms of the excess concentration, there are a few more steps. Consider the conservation of mass for the state variables over the control volume

$$\frac{d}{ds}\int X_i u dA = E X_{ia} \tag{25.25}$$

To arrange this equation in terms of concentration excess, substitute the conservation of volume equation 25.24 to replace E in (25.25), giving

$$\frac{d}{ds}\int X_i u dA = X_{ia} \frac{d}{ds}\int u dA \tag{25.26}$$

We use the product rule on the right-hand side to bring X_{ia} into the derivative

$$\frac{d}{ds}\int X_i u dA = \frac{d}{ds}\left[X_{ia}\int u dA\right] - \frac{dX_{ia}}{ds}\int u dA \tag{25.27}$$

and bringing the first two terms of (25.27) together on the left-hand side of the equation to obtain

$$\frac{d}{ds}\left[\int X_i u dA - X_{ia}\int u dA\right] = -\frac{dX_{ia}}{ds}\int u dA \tag{25.28}$$

It is now possible to bring X_{ia} inside the integral in the second term of (25.28) since the integration is over r and not z or s. On the right-hand side we also apply the chain rule to dX_{ia}/ds and substitute $\int u dA = Q$, yielding

$$\frac{d}{ds}\int (X_i - X_{ia}) dA = -Q \frac{dX_{ia}}{dz}\sin\theta \tag{25.29}$$

We now recognize the quantity inside the integral as the mass flux of the excess concentration; hence,

$$\frac{dQ_{Xi}}{ds} = -Q \frac{dX_{ia}}{dz}\sin\theta \tag{25.30}$$

By similar algebra, the conservation equation for passive tracers gives

$$\frac{dQ_{ci}}{ds} = -Q \frac{dc_{ia}}{dz}\sin\theta \tag{25.31}$$

The advantage of the excess concentration formulation is that a simple equation for conservation of buoyancy is obtained when the equation of state can be linearized (see, e.g., Jirka 2004). When the ambient concentration of passive tracers is zero, the concentration of tracer mass flux simply becomes

$$\frac{dQ_{ci}}{ds} = 0 \quad (25.32)$$

To derive the momentum equation, we require expressions for the forces acting on the control volume. The buoyancy contribution is the net reduced gravity across the centroid of the control volume, giving

$$f_b = 2\pi \int_0^\infty g'(s,r) r dr = \pi(\lambda b)^2 g'_c \quad (25.33)$$

in kinematic force per unit length dimensions, acting exclusively in the z-direction. The drag force is formulated in analogy to the drag around a cylinder of diameter $2\sqrt{2}b$ using the velocity transverse to s as the characteristic scale; hence,

$$\left|\vec{F}_d\right| = \frac{1}{2} c_D 2\sqrt{2} b u_a^2 (1 - \cos^2\sigma\cos^2\theta) \quad (25.34)$$

and the F_d vector is assumed to act in the same direction as the transverse velocity u_t. Decomposing the drag force into (x, y, z) components yields

$$\vec{f}_d = \left|\vec{F}_d\right| \left(\sqrt{1-\cos^2\sigma\cos^2\theta},\ -\frac{\sin\sigma\cos\sigma\cos^2\theta}{\sqrt{1-\cos^2\sigma\cos^2\theta}},\ -\frac{\sin\sigma\cos\sigma\cos\theta}{\sqrt{1-\cos^2\sigma\cos^2\theta}} \right) \quad (25.35)$$

To test this formulation, consider three simple cases:

1. For a vertical buoyant jet, $\theta = \pi/2$ and $\sigma = 0$ so that the drag force acts in the x-direction, causing the plume to bend over.
2. For a horizontal jet with centerline along the x-axis, $\theta = 0$ and $\sigma = 0$. Here, the transverse velocity is zero; thus, F_d is zero. If the jet deviates slightly in the y-direction so that σ is small, the drag force is small, but almost all of it is directed in the negative y-direction; hence, causing the jet to correct its path and move back to the x-axis.
3. For a horizontal jet with centerline along the y-axis, $\theta = 0$ and $\sigma = \pi/2$, the drag force is large and is directed solely in the x-direction, similar to case 1.

With these forces, the conservation of momentum flux in the x-, y-, and z-directions follows from Newton's first law as (Jirka 2004)

$$\frac{d(M\cos\sigma\cos\theta)}{ds} = Eu_a + \left|\vec{F}_D\right|\sqrt{1-\cos^2\sigma\cos^2\theta} \quad (25.36)$$

$$\frac{d(M\sin\sigma\cos\theta)}{ds} = -\left|\vec{F}_D\right| \frac{\sin\sigma\cos\sigma\cos^2\theta}{\sqrt{1-\cos^2\sigma\cos^2\theta}} \quad (25.37)$$

$$\frac{d(M\sin\theta)}{ds} = \pi(\lambda b)^2 g'_c - \left|\vec{F}_D\right| \frac{\cos\sigma\cos\theta\sin\theta}{\sqrt{1-\cos^2\sigma\cos^2\theta}} \quad (25.38)$$

where the term Eu_a accounts for the change in momentum due to entrainment of ambient fluid with velocity u_a in the x-direction and the other terms on the right-hand side of the equations are buoyancy and drag contributions.

To close the system of equations, the trajectory is obtained from

$$\frac{dx}{ds} = \cos\sigma\cos\theta \quad (25.39)$$

$$\frac{dy}{ds} = \sin\sigma\cos\theta \quad (25.40)$$

$$\frac{dz}{ds} = \sin\theta \quad (25.41)$$

and the density necessary to compute the reduced gravity is obtained from the equation of state.

25.3.2.5 Values of the Empirical Coefficients

To solve the model equations presented earlier, values for the five model parameters are required. These are obtained by validation to available data, and because of the robustness of the similarity solution, are constant over a very wide range of physical scales and also in many transition states between asymptotic flow solutions. Jirka (2004) reports the values used in CorJet as

$$\alpha_1 = 0.055,\quad \alpha_2 = 0.6,\quad \alpha_3 = 0.055,\quad \alpha_4 = 0.5 \quad (25.42)$$

$$\lambda = 1.20,\quad c_D = 1.3 \quad (25.43)$$

It is important to note here that these values depend on the shape of the lateral profiles (these values are for Gaussian profiles) and the chosen definition of the plume half-width b (taken as $1/e$ for the entrainment coefficients). Models using other profiles or definitions may result in the same solution as this model, but using different values for these five model parameters.

25.3.2.6 Validation

Integral models based on the equations presented earlier have been validated to all of the asymptotic regimes as well as many transition cases in which different parts of the jet trajectory are dominated by different asymptotic dynamics, as in a buoyant jet transitioning to a plume. The purpose of model validation is both to check the accuracy of the numerical routines and computer code to ensure that the right equations are solved well and to check that the governing equations are indeed an adequate representation of the jet physics. In this section we highlight a few of the validation cases that have been applied to the CorJet model. For a more comprehensive suite of validation data, please see Jirka (2004).

1. *Pure jet*: The pure jet has no ambient density stratification or crossflow, and the effluent is of neutral density with the environment. Since validation data span laboratory and field experiments of a wide range of scales, comparisons are made for nondimensional variables. Figure 25.7 shows the decay of the centerline velocity u_c/U_0 and the bulk dilution S versus the distance from the source x/D (as for a jet discharge with $\sigma = 0$, $\theta = 0$). The model comparison is limited to the self-similar region beyond the ZOFE. Over this region, the model predictions track through the mean of the experimental data, and the scatter in measured data is typically ±5%. Because the model assumptions are wholly valid within all of the asymptotic regimes, model comparisons for pure plumes, wakes, line puffs, and line thermals show similar levels of agreement.
2. *Buoyant jet in stratification*: As a first example of a transitional regime, consider a buoyant jet in a quiescent reservoir with linear density stratification. The comparison in Figure 25.8 shows the predicted centerline trajectory and half-width overlain on line tracings from images of the laboratory dye studies for an inclined discharge in the x-direction ($\sigma = 0$ and $\theta = \pi/4$). $T = g_0'/(\varepsilon D)$ is the stratification parameter. Note that the simulation terminates at the start of the intermediate field as the jet undergoes rapid spreading and begins to violate the boundary layer approximation. The model accurately predicts the evolution in the jet-like region of the flow and the height of maximum rise.
3. *Buoyant jet in crossflow*: Another important transition region considers the buoyant jet in an unstratified crossflow. The dominant length scale for this case is L_b; Figure 25.9 presents the model validation for a vertically discharging jet ($\sigma = 0$ and $\theta = \pi/2$). From this log–log plot, it is apparent that the model captures the correct scale-law behavior in the weakly bent ($z \sim x^{3/4}$ power law) and strongly bent ($z \sim x^{2/3}$) regions. The model also tracks the experimental data throughout the near field, but slightly underpredicts the rise height (overpredicts the degree of bending) for $x/L_b > 100$. This is due to an overestimate of the drag force in the latest stages of the jet, indicating that the drag force term becomes less important as the jet transitions from a coherent column to a diffusion-dominated puff or thermal.
4. *Buoyant jet in stratified crossflow*: Among the more complicated cases that can be analyzed using the CorJet model equations is a buoyant jet into a flowing and density-stratified ambient reservoir. Figure 25.10 shows the case of a vertical buoyant jet ($\sigma = 0$ and $\theta = \pi/2$) into a uniform current with linear density stratification. In these experiments, the initial momentum was small so that the flow was dominated by the discharge buoyancy. The model predicts well the intrusion level of the plume and is particularly capable of matching the measured dilution, as indicated by the variation of ρ_c/ρ_a versus x/D. Again, the model simulation is terminated when the jet enters to collapsing region of the intrusion formation.

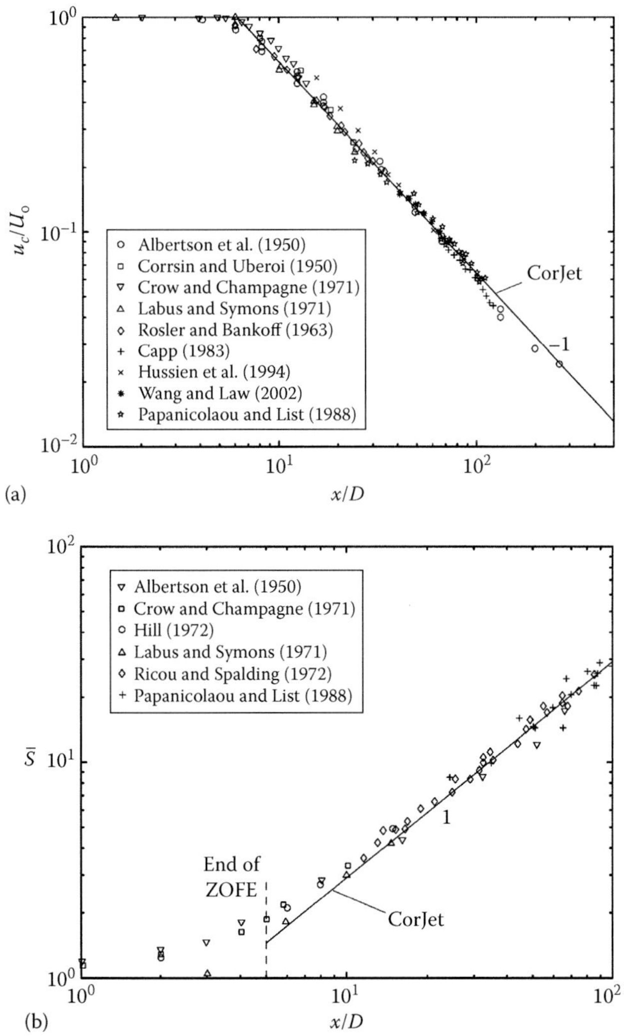

FIGURE 25.7 Validation of the pure jet showing the decay of the centerline velocity and the evolution of the bulk dilution as a function of nondimensional distance from the jet nozzle. (Reproduced from Jirka, G.H., *Environ. Fluid Mech.*, 4(1), 1–56, 2004. With permission.)

25.4 Applications

25.4.1 Discharge Analysis Using CORMIX

Environmental impact assessment is often required for a wide range of point source discharges—including municipal wastewaters, cooling waters, industrial wastes, oil, and gas produced waters, dredging operations, and desalination brines—into

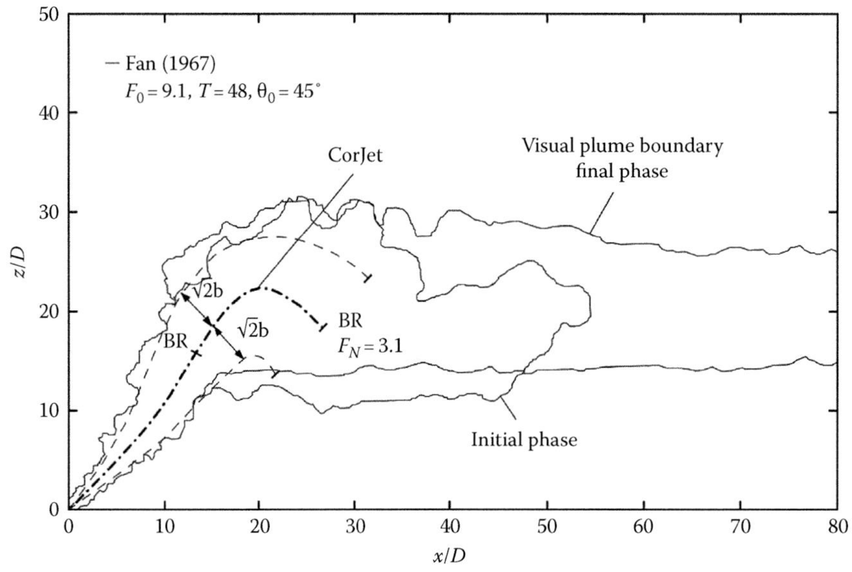

FIGURE 25.8 Validation of the trajectory and half-width for an inclined buoyant jet into a linearly stratified ambient reservoir. BR is buoyancy reversal. (Reproduced from Jirka, G.H., *Environ. Fluid Mech.*, 4(1), 1–56, 2004. With permission.)

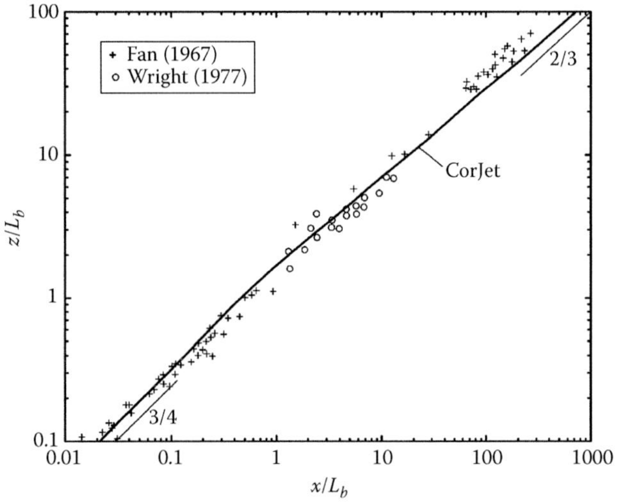

FIGURE 25.9 Validation of the trajectory for a buoyant jet in crossflow using the buoyancy length scale as the normalization parameter. (Reproduced from Jirka, G.H., *Environ. Fluid Mech.*, 4(1), 1–56, 2004. With permission.)

rivers, lakes, estuaries, and the coastal ocean. Design optimization of wastewater disposal systems using multiport diffusers can often mitigate negative environmental impacts. Because of this, predictive simulation models are used to design, manage, and regulate wastewater disposal systems.

The previous sections outlined the physical concepts and theoretical basis for buoyant jet modeling into an unbounded (infinite) ambient. However, water bodies always have limitations; be it the surface, bottom, shorelines, or boundaries caused by internal ambient density stratification. This section addresses the practical application of buoyant jet modeling to water quality management within the regulatory mixing zone, where boundary interaction is often important. The regulatory mixing zone is an administrative concept which includes the region near the discharge where the initial mixing occurs and generally spans an area from the discharge through the intermediate field and to the start of the far field (refer to Figure 25.1). Although the potential for water quality impacts are greatest within the mixing zone, the opportunity also exists to mitigate adverse conditions through outfall design optimization.

The CORMIX modeling system (Version 6.0 or higher) is a software system for the analysis, prediction, and design of aqueous toxic or conventional pollutant discharges into diverse water bodies. It contains a buoyant jet integral model as previously described along with enhancements to account for boundary interaction within the mixing zone. The system emphasizes steady-state ($\partial Q/\partial t = 0$), near-field mixing and contains methods to predict conditions within the regulatory mixing zone which typically occurs after boundary interaction in the far-field region (where ambient conditions dominate mixing). While CORMIX was originally developed under the assumption of steady ambient conditions, Version 6.0 includes application to highly unsteady environments, such as tidal reversal conditions, in which transient recirculation and pollutant build-up effects can occur. Figure 25.11 gives the problem domain covered by the CORMIX methodology in relationship to ambient and discharge conditions.

Among available water quality simulation models, CORMIX has a unique data-driven approach to simulation model selection based primarily on the characteristic scales defined in the Dimensional Analysis section and on other similar scales for more complicated cases. To do this, CORMIX employs a rule-based expert system to screen input data and select the appropriate core hydrodynamic simulation model to simulate the physical mixing processes contained within a given

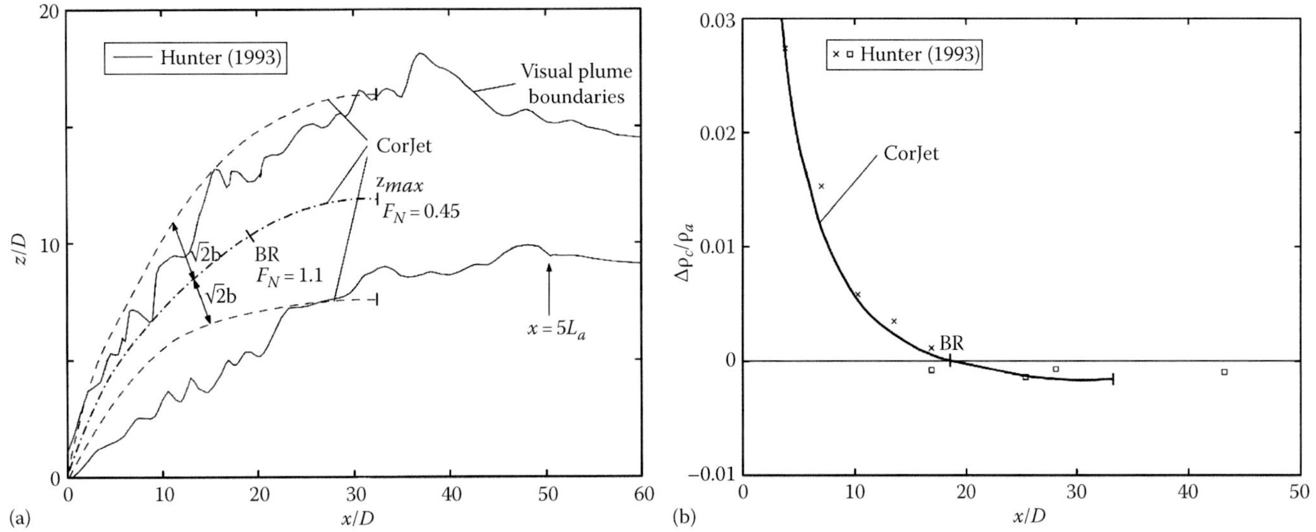

FIGURE 25.10 Validation of the buoyant jet in stratified crossflow showing the jet trajectory, half-width, and centerline decay of the density difference. BR is buoyancy reversal, F_N is local Froude number. (Reproduced from Jirka, G.H., *Environ. Fluid Mech.*, 4(1), 1–56, 2004. With permission.)

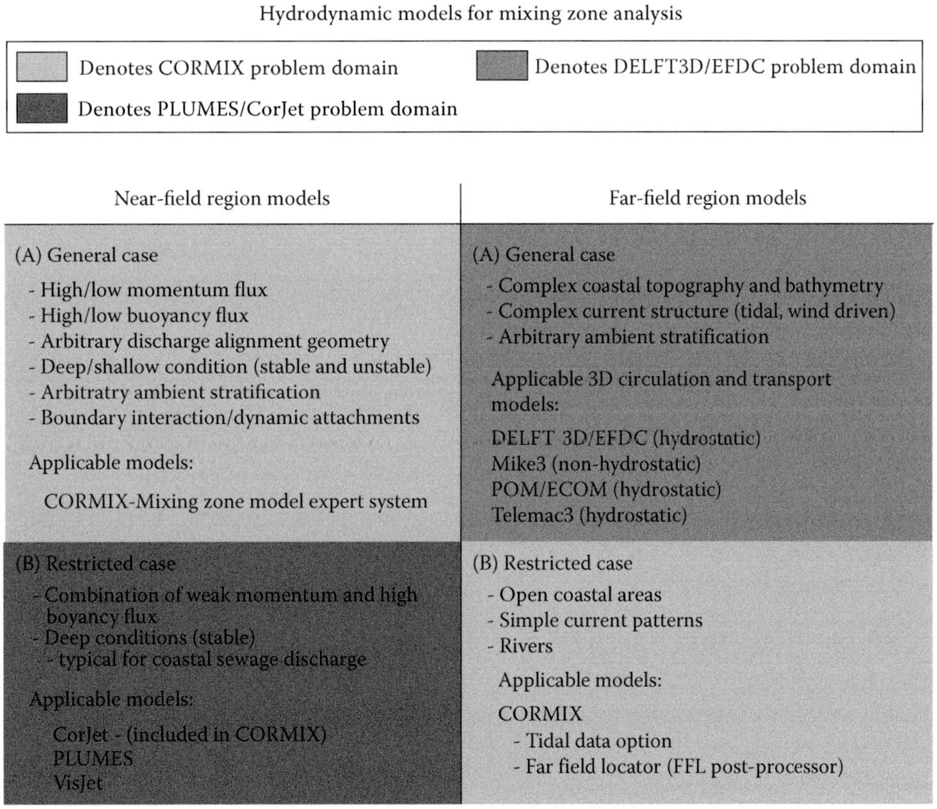

FIGURE 25.11 CORMIX mixing zone model problem domain.

discharge–environment interaction. The methodology also comprehensively documents the model selection process and interprets the physical mixing processes present in relation to applicable regulatory criteria. Since the model selection process is data-driven and explicitly contained within the methodology, the approach facilitates a discussion between regulators and the regulated on the physical processes contained within the mixing zone, limits the potential for model misapplication, and helps to develop consensus on appropriate model selection. Figure 25.12 shows an example of the CORMIX flow classification system based upon length scale analysis of the boundary interaction process. CORMIX determines discharge stability and

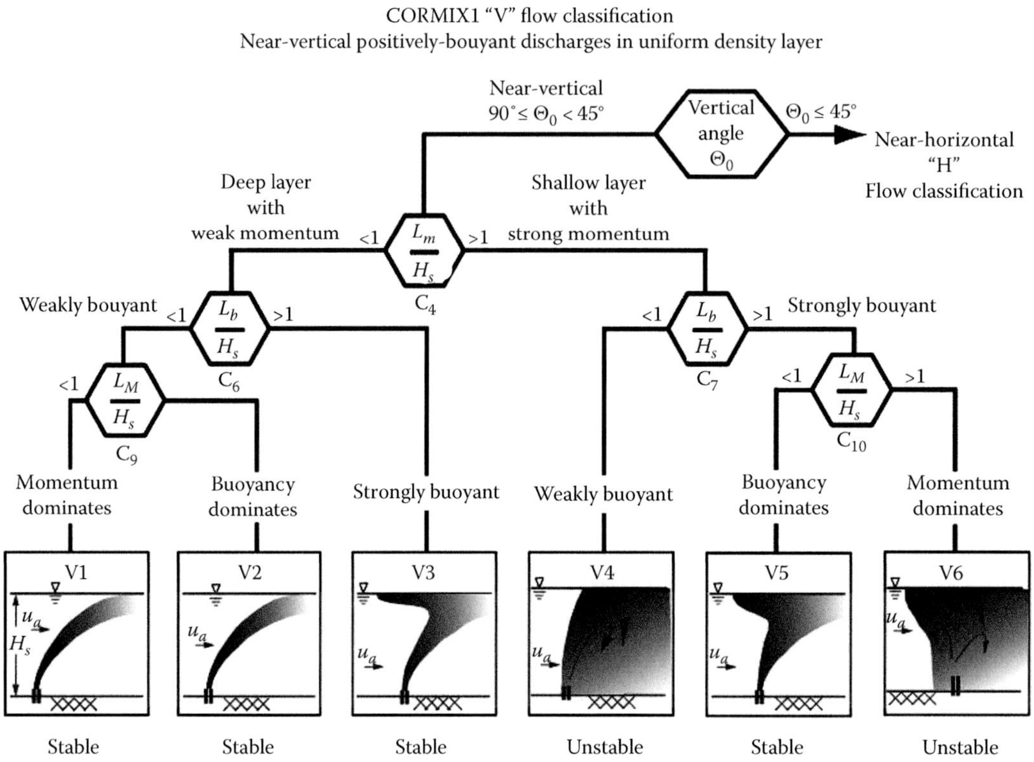

FIGURE 25.12 The CORMIX flow classification.

the boundary interaction process before executing a series of regional flow models to complete the near-field and far-field simulation. CORMIX also determines if dynamic wake or Coanda attachments occur in the near-field. The entire CORMIX system contains about 60 flow classifications similar to V1 to V6 in the figure.

A User Manual gives a comprehensive description of the CORMIX system (Doneker and Jirka 1990). Multiple publications describe the scientific basis for the CORMIX system and demonstrate comparison and validation with field and laboratory data (Akar and Jirka 1991; Bleninger et al. 2002; Doneker 2003, 2006; Doneker and Jirka 1990; Jones et al. 1996).

25.5 Extensions for Multiport Diffusers

For larger flow rates, single jet discharges may not be sufficient to achieve dilution requirements, for example, if applied for municipal treated wastewater effluent or cooling water discharges where large volume flow rates must be accommodated. Multiple jets, called multiport diffusers, are used in such cases to distribute the effluent through many single ports discharges along a diffuser pipe. Such diffusers can generally be modeled as a continuous line source, aligned at an oblique angle to the ambient current, discharging from a theoretical slot of finite length resulting in a plane jet (see Figure 25.13). Provided that the slot length is considerably larger than the local jet width, the resulting jet motion is then characterized by a locally 2D plane geometry. The overall behavior of the resulting jet motion (including its complete trajectory) may, however, still be 3D. Jirka (2006) described the extension from single jets to plane jets and the details of a plane jet integral model. The definitions conform to those given previously for the round jet geometry. In addition, there are the diffuser length L_D, port spacing l, port and diffuser orientations β, γ, θ, and the equivalent slot width B, calculated to achieve diffuser flux quantities equal to those caused by the sum of the individual multiport jets. The assumption of the equivalent slot width concept is generally valid for most of the existing closely spaced multiport diffuser installations, where merging takes place after short distances and in a rather uniform manner. Exceptions are for tunneled outfalls with only a few largely spaced long risers and rosette-like port arrangements. Individual jet quantities are then no longer the right measure for calculating the equivalent slot width and more complicated jet interactions must be considered.

Turbulent fluctuations caused by turbulence shearing mechanisms lead to a gradual growth of the characteristic jet thickness $2b$ and characteristic width $L_D + 2B$. The relatively large diffuser lengths compared to plume thickness $L_D/(2B) \gg 1$ generally allow neglecting the entrainment at the lateral plume ends. Plane plume growth, thus, is dominated by 2D processes, and quantities per unit jet length can be described for the initial fluxes as (Jirka 2006)

$$q_0 = \frac{Q_0}{L_D} = U_0 B \tag{25.44}$$

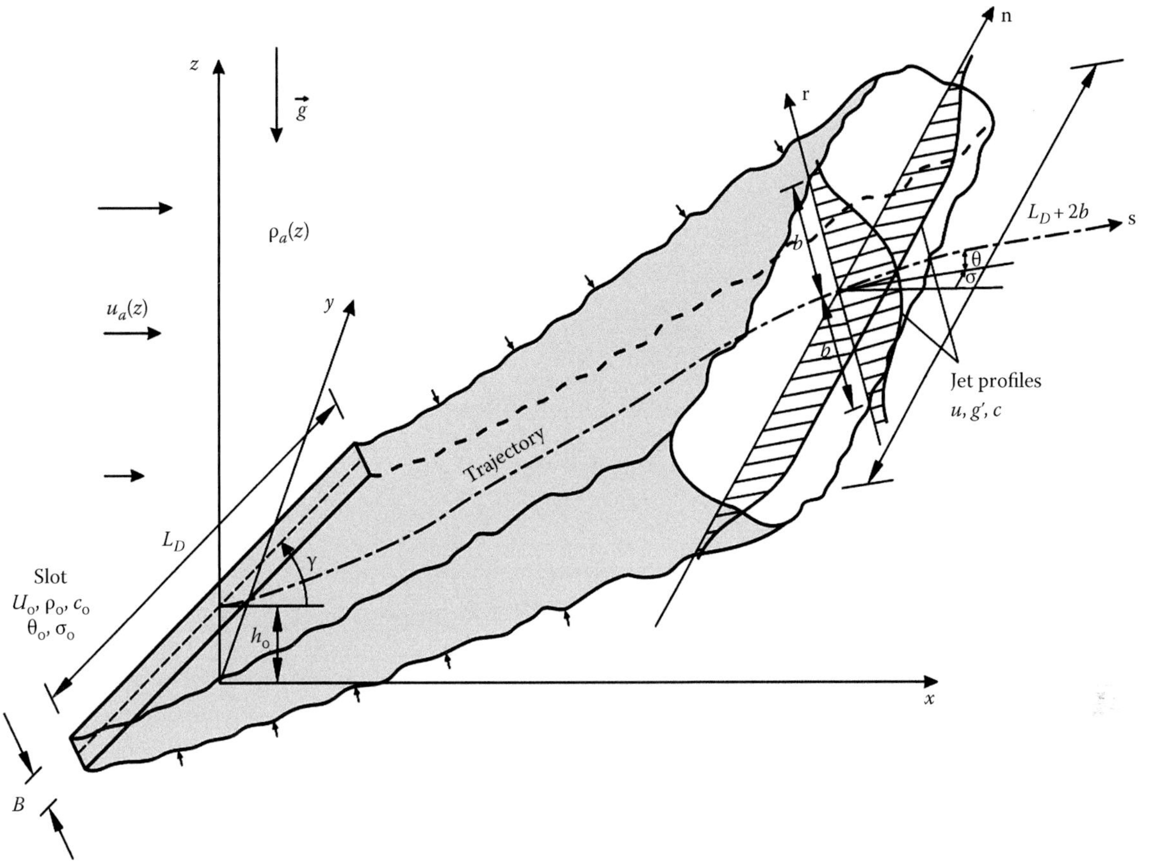

FIGURE 25.13 Schematization for merging jets discharged unidirectionally by multiport diffusers. (Reproduced from Jirka, G.H., *Environ. Fluid Mech.*, 6(1), 43–100, 2006. With permission.)

$$m_0 = \frac{M_0}{L_D} = U_0^2 B \tag{25.45}$$

$$j_0 = \frac{J_0}{L_D} = U_0 g_0' B \tag{25.46}$$

$$q_{c0} = \frac{Q_{c0}}{L_D} = U_0 c_0 B \tag{25.47}$$

The development of the plane jet fluxes q, m, and j along the centerline can be computed in integral values as described in Jirka (2006).

It has been demonstrated both theoretically and experimentally that the maximum mixing can be achieved with closely spaced ports that allow some interference of adjacent jets (Fischer et al. 1979). In relatively shallow coastal waters of typical depth 5–15 m, however, it is often the case that, given practical considerations (e.g., in order to maintain a minimum jet velocity and minimum diameter), multiport diffusers are designed to minimize interference of adjacent plumes. In such cases, the required spacing is about $H/3$.

25.5.1 Extensions for Surface Jets

Buoyant surface discharges occur in many environmental and water treatment processes, including inflows to treatment facilities, discharges to receiving bodies, and riverine inflows to larger rivers, lakes, of the coastal ocean. When lighter than the receiving water, the discharge forms a buoyant surface jet, as sketched in Figure 25.14, where a river discharges at an angle into a flowing coastal region.

Buoyant jet integral models have recently been adapted to accurately model the jet-like near field, the transition to plume-like buoyant spreading, and the buoyancy-dominated intermediate field. Jones et al. (2007) present scale relationship to classify the dominant inflow dynamics, and Jirka (2007) presents the integral model equations. A 3D numerical model with an imbedded surface jet are also presented by Hetland (2010) and Chen et al. (2009) and compared to field data of the Merrimack river plume near Boston, Massachusetts, by Hetland and McDonald (2008). Laboratory data are also available from Nash et al. (1995).

The general physics of a buoyant jet is analogous to the round jet detailed previously. An initial ZOFE takes place at the start of the jet as the nearly uniform velocity profile of the channel rapidly adapts to a nearly round jet with typical Gaussian profiles (though

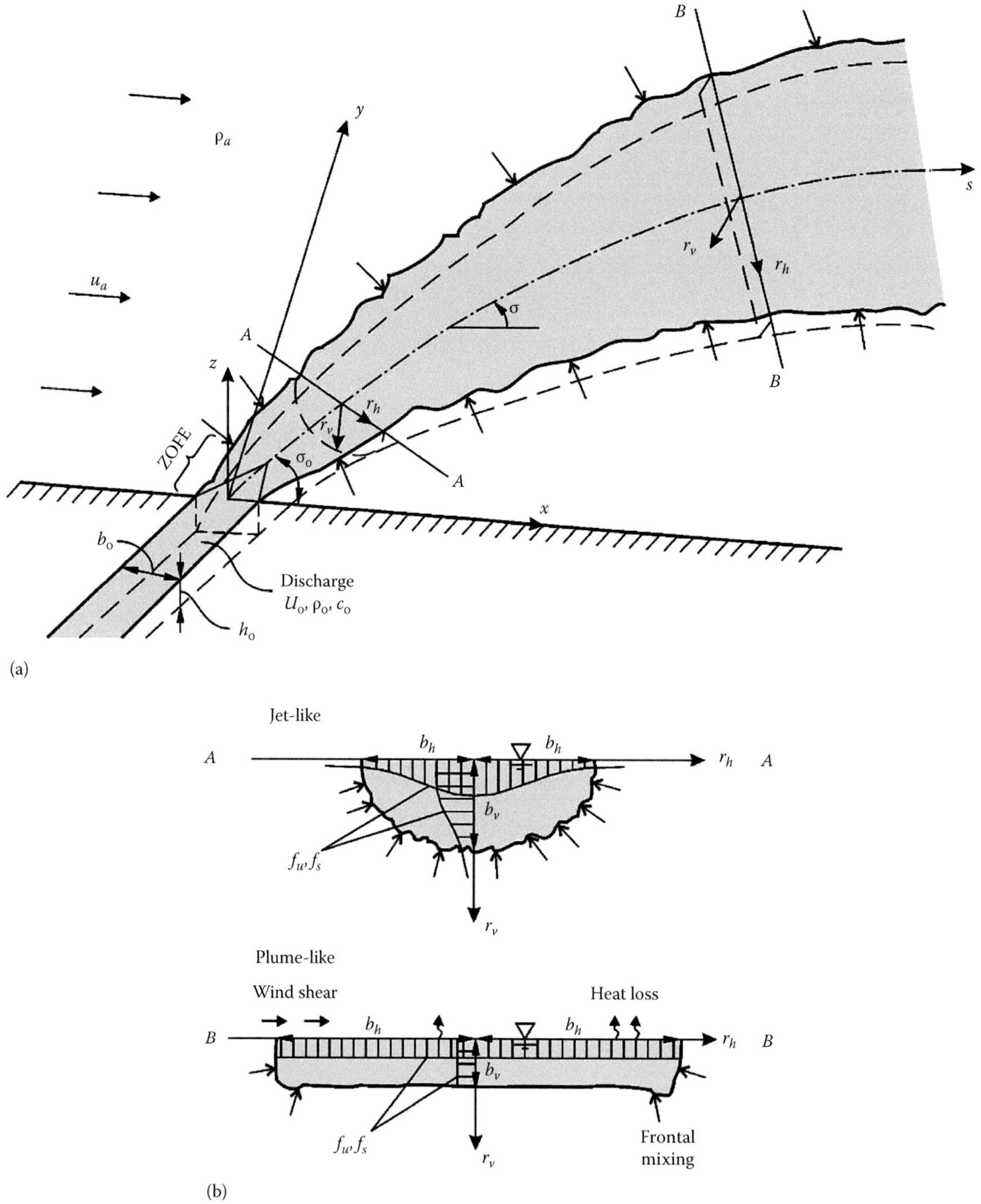

FIGURE 25.14 Schematic of a surface discharge: (a) perspective view and (b) cross sections. (Reproduced from Jirka, G.H., *J. Hydraul. Eng-Asce*, 133(9), 1021–1036, 2007. With permission.)

influenced by the surface as shown in inset A of Figure 25.14). As the jet flow slows down, buoyancy begins to dominate and the flow collapses to form a plume-like lateral intrusion. In this latter region, vertical diffusion is damped by density stratification on the bottom of the plume, and the velocity profiles steepen, approaching a more top-hat profile (see inset B of Figure 25.14). Wind shear can enhance mixing, and coastal currents can deflect the plume; eventually, the flow enters the far-field, where it is dominated by turbulent diffusion and passive advection.

Throughout the buoyant jet near and intermediate field, several different entrainment fluxes engulf ambient fluid. These rates are additive, and are summarized by

$$E = E_h + E_v + E_p + E_f + E_i$$

where the subscripts stand for horizontal, vertical, advected puff, frontal, and interfacial entrainment, respectively (Jirka 2007). E_h enters at the jet edges, depends on the centerline velocity, and is comprised of jet and wake entrainment. The vertical entrainment E_v on the jet bottom has a similar form to E_h, but with an added density term to account for damping at the sharp interface in the collapsed jet. As in the round jet, the advected puff entrainment E_p depends on the ambient velocity and the angle of the surface jet. In the plume region, an additional frontal entrainment E_f is proportional to the lateral spreading velocity of the surface plume sides. The final term E_i acts on the plume bottom, modeling the effect of wind mixing, and is proportional to the wind shear velocity, the ambient current velocity, and the lateral spreading. The drag force is also formulated similarly to that in a round jet, following a quadratic law drag model that depends on the ambient current velocity. No additional drag is modeled due to the wind since it is assumed that this effect is already included in the ambient current.

25.6 Challenges

Although an extensive and diverse range of buoyant jets types in the environment have been studied in great detail, as described briefly in this chapter and elsewhere, there are numerous challenges remaining to further our understanding and description of jets and plumes. A few of these challenges are highlighted as follows.

A rapidly expanding field of buoyant jet discharges result from membrane-based seawater desalination plants, and generate plumes with positive momentum and negative buoyancy, similar to a fountain. As long as the discharge angle is not vertical (which would violate principle 3 similarly to a jet in a counterflow), a buoyant jet following the boundary layer approximation results, and the equations presented earlier accurately track its trajectory from the source until it plunges to the sea floor, where boundary interaction dominates. There, the dense discharge interacts with the nonhorizontal sea bed and creates further gravity currents down slope, which need more detailed consideration than pure buoyant spreading processes at the water surface for positively buoyant effluents. A related desalination byproduct are thermal saline discharges resulting from distillation plants, where double diffusion effects may occur. Because these plumes interact with the flora and fauna of the sea floor, their environmental impact can be significant, and their design requires new models, experiments, and methods.

Another challenge area in jets and plumes are those resulting from multiphase discharges, such as direct ocean CO_2 sequestration plumes, particle-laden jets from dredging operations or untreated wastewater discharges, lake aeration and mixing plumes, among others (see also Chapter 5 of *Handbook of Environmental Fluid Dynamics, Volume Two*). Because of the immiscibility of the phases, there exists the possibility that the entrained fluid separates from the dispersed phase (e.g., air, oil, liquid CO_2, sediment, etc.). And because the dispersed phase often contains much of the buoyancy in a multiphase jet, the result is a nonuniform loss of buoyancy to the jet as the dispersed phase settles or rises out of the jet, thus depleting the driving force of the jet motion. When separation of phases is weak, the integral approach described here can be effective; however, when separation is important (notably in the presence of crossflow or ambient stratification), new models are required. Because of the difficulty to make measurements in multiphase flows, data are often lacking for more complicated cases, and much work is needed to understand the fundamental flow dynamics.

Other ambient forcing conditions beyond those discussed here may also play a major role in some buoyant jet applications. These may include the influence of waves, ambient turbulence, counter flows, and others. These processes can affect both the jet trajectories and dilution. In many of these cases and also in cases of the other traditional flow classes as they interact with boundaries, no experimental data are available for model validation; hence, there remain many buoyant jet flow types that require fundamental observation in the laboratory. These may also be explored in the near future as numerical solutions can be increasingly applied to jets and plumes. Certainly, this field will be vibrant for years to come.

Dedication

This chapter is dedicated to the memory of Professor Gerhard H. Jirka (1944–2010). He was intended to be the author of this chapter, a topic that he loved, and he had very much looked forward to writing it. The text that follows attempts to follow his original outline for the chapter, making use of his lecture notes on the subject, and is inspired by his insights on the topic as recorded in this papers and imprinted on the minds of the current authors through their collaboration with him on this and many other topics of environmental fluid mechanics.

Gerhard Jirka was truly a giant in the field of environmental hydraulics. He received his Dipl-Ing with honors from the Hochschule für Bodenkultur, Vienna, Austria, in 1969 and his MS and PhD from Massachusetts Institute of Technology (MIT) in 1971 and 1973, respectively, working with Professor Donald Harleman, among others. After completing his studies, he continued to work at MIT as a research engineer and lecturer and then accepted a faculty position at Cornell University in 1977. At Cornell, he rose to the rank of full professor and was the founding director of the DeFrees Hydraulics Laboratory (1984–1995). In 1995, he accepted the position as professor and director of the Institute for Hydromechanics at the University of Karlsruhe, where he continued to work following his retirement in 2009 until his death on February 14, 2010. He also served as vice president of the International Association of Hydro-Environment Engineering and Research (IAHR) from 2004 to 2009. Although he will be greatly missed, his legacy lives on through his many students (more than 30 PhD students) and publications, including over 230 journal and conference publications as well as 17 books and invited chapters.

Professor Jirka took great interest in the topic of this book. In fact, during a recent discussion on this topic, he pointed out that the term "environmental fluid mechanics" was coined by him in the 1970s while at MIT. His love for the topic was also very contagious, as he was always friendly and open for discussion. He inspired others, was full of creativity, and remained ever willing to learn. Whenever we read one of his many publications, we found new ideas and hidden secrets. Each of us had the good fortune and privilege of working closely with him for many years as student, colleague, and friend. He was an exceptional and patient mentor, and we always looked forward to working with him on new topics and continuing projects. His great personal charisma was matched by his exceptional intellectual gifts. We will deeply miss his advice and counsel and hope to promote his remarkable contributions to environmental fluid mechanics through this brief chapter and through the work we do in the areas he pioneered.

May the insights and knowledge of Professor Jirka on this topic shine through this brief summary of his own work and his research on jets and plumes.

References

Akar, P. J. and Jirka, G. H. (1991). CORMIX2: An expert system for hydrodynamic mixing zone analysis of conventional and toxic submerged multiport diffuser discharges. *EPA-600/3-19/073*, USEPA, Athens, GA.

Bleninger, T., Lipari, G., and Jirka, G. H. (2002). Design and optimization program for internal diffuser hydraulics. *Proceedings of International Conference on Marine Waste Water Discharges 2002*, Istanbul, Turkey, pp. 16–20.

Chen, F., MacDonald, D. G., and Hetland, R. D. (2009). Lateral spreading of a near-field river plume: Observations and numerical simulations. *J. Geophys. Res. Oceans*, 114, 12.

Davidson, M. J. and Pun, K. L. (1999). Weakly advected jets in cross-flow. *J. Hydraul. Eng. ASCE*, 125(1), 47–58.

Doneker, R. L. (2003). Systems development for concentrate disposal: CorVue and corSpy interactive visualization tools for CORMIX mixing zone analysis. *Final Report No. 98*, U.S. Bureau of Reclamation, DSRP, Denver, CO.

Doneker, R. L. (2006). Systems development for environmental impact assessment of concentrate disposal: Development of density current simulation models, rule base, and graphic user interface. Department of the Interior Bureau of Reclamation.

Doneker, R. L. and Jirka, G. H. (1990). CORMIX1: An expert system for mixing zone analysis of conventional and toxic single port aquatic discharges. *EPA-600/3-90/012*, USEPA, Athens, GA.

Doneker, R. L. and Jirka, G. H. (1991). Expert systems for mixing-zone analysis and design of pollutant discharges. *J. Water Res. Plann. Manage.*, 117(6), 679–697.

Fan, L. N. (1967). Turbulent buoyant jets into stratified or flowing ambient fluids. Report No. KH-R-15, W.M. Keck Laboratory of Hydrology and Water Resources, California Institute of Technology, Pasadena, CA.

Fischer, H. B., List, E. J., Koh, R. C. Y., Imberger, J., and Brooks, N. H. (1979). *Mixing in Inland and Coastal Waters.* Academic Press, New York.

Görler, H. (1942). Berechnung von Aufgaben der freien Turbuleny aus Grund eines neuen Näherungsansatzes. *ZAMM*, 22, 244–254.

Hetland, R. D. (2010). The effects of mixing and spreading on density in near-field river plumes. *Dyn. Atmos. Oceans*, 49(1), 37–53.

Hetland, R. D. and MacDonald, D. G. (2008). Spreading in the near-field Merrimack River plume. *Ocean Model*, 21(1–2), 12–21.

Jirka, G. H. (2004). Integral model for turbulent buoyant jets in unbounded stratified flows. Part I: Single round jet. *Environ. Fluid Mech.*, 4(1), 1–56.

Jirka, G. H. (2006). Integral model for turbulent buoyant jets in unbounded stratified flows, Part 2: Plane jet dynamics resulting from multiport diffuser jets. *Environ. Fluid Mech.*, 6(1), 43–100.

Jirka, G. H. (2007). Buoyant surface discharges into water bodies. II: Jet integral model. *J. Hydraul. Eng. ASCE*, 133(9), 1021–1036.

Jirka, G. H. and Akar, P. J. (1991). Hydrodynamic classification of submerged single-port discharges. *J. Hydraul. Eng. ASCE*, 117(HY9), 1095–1111.

Jirka, G. H. and Doneker, R. L. (1991). Hydrodynamic classification of submerged single-port discharges. *J. Hydraul. Eng. ASCE*, 117(9), 1095–1112.

Jirka, G. H. and Lee, J. H. W. (1994). Waste disposal in the ocean. In: *Water Quality and Its Control*, Vol. 5, M. Hino, ed. A.A. Balkema Publishers, Rotterdam, the Netherlands.

Jones, G. R., Nash, J. D., Doneker, R. L., and Jirka, G. H. (2007). Buoyant surface discharges into water bodies. I: Flow classification and prediction methodology. *J. Hydraul. Eng. ASCE*, 133(9), 1010–1020.

Jones, G. R., Nash, J. D., and Jirka, G. H. (1996). CORMIX3: An expert system for mixing zone analysis and prediction of buoyant surface discharges. DeFrees Hydraulics Laboratory, Cornell University, Ithaca, NY.

Kundu, P. K. and Cohen, I. M. (2008). *Fluid Mechanics*, 4th Edn. Academic Press, New York.

Morton, B. R., Taylor, G., and Turner, J. S. (1956). Turbulent gravitational convection from maintained and instantaneous sources. *Proc. R. Soc. Lond. A Mat. Phys. Sci.*, 234(1196), 1–23.

Nash, J. D., Jirka, G. H., and Chen, D. (1995). Large-scale planar laser-induced fluorescence in turbulent density-stratified flows. *Exp. Fluids*, 19(5), 297–304.

Reichardt, H. (1941). Gestzmäßigkeiten der freien Turbulenzen. *VDI-Forschungsheft*, 414.

Schlichting, H. (1960). *Boundary Layer Theory.* McGraw-Hill Book Company, Inc., New York.

Tian, X. D. and Roberts, P. J. W. (2003). A 3D LIF system for turbulent buoyant jet flows. *Exp. Fluids*, 35(6), 636–647.

Tollmien, W. (1926). Berechnung turbulenter Ausbreitungsvorgänge. *ZAMM*, 6, 468–478.

Turner, J. S. (1986). Turbulent entrainment: The development of the entrainment assumption, and its application to geophysical flows. *J. Fluid Mech.*, 173, 431–471.

Wright, S. J. (1977). Mean behavior of buoyant jets in a crossflow. *J. Hydraul. Div. ASCE*, 103(HY5), 499–513.

Xiao, Y., Lee, J. H. W., Tang, H., and Yu, D. Y. (2005). Numerical study of multiple tandem jets in crossflow. *XXXI IAHR Congress*, Seoul, South Korea. Paper E11-7.

Yu, D. Y., Ali, M. S., and Lee, J. H. W. (2006). Multiple tandem jets in cross-flow. *J. Hydraul. Eng. ASCE*, 132(9), 971–982.

26

Stratified Wakes and Their Signatures

S.I. Voropayev
University of Notre Dame

26.1 Introduction

In this chapter, we review the latest research activities concerning wakes generated by bodies moving in stratified fluids and resulting wake "signatures" which are of importance in designing, detecting, and stealth of large underwater dirigibles. Special attention will be given to surface signatures of wakes, their longevity and survival in background conditions (shear, turbulence) typical of the upper ocean. The emphasis is mostly on late or far wakes of self-propelled bodies in steady and unsteady motion. In the latter case, large (much larger than the body size) pancake eddies become important which dynamics are also discussed. To make the material self-sufficient, the general problem of flows generated by localized forcing is also formulated. Different aspects of such flows were studied in the past by using mostly laboratory experiments and theoretical analysis. Besides scientific interest, these basic flows have applications related to stratified wakes, oceanic mushroom-like currents, atmospheric/oceanic jets and quasi-2D, geophysical, turbulence. Because of strict space limitations, results of numerical simulations were not included into this brief review.

26.2 Theoretical Preliminaries

When a solid body moves in a fluid, it generates vorticity, because of no-slip conditions on the body surface, and leaves behind a narrow vortical wake surrounded by fluid were the motion is mostly potential. At large times the body is far from the observer, and for mathematical modeling it may be considered as a localized or "point" forcing (Swain 1929; Tennekes and Lumley 1972; Schlichting 1979). Without specifying the details of force distribution, the forcing can be modeled by a system of basic singularities at the origin, with their intensities determined by the integral moments of applied forces. The fundamental singularity is a momentum source $\mathbf{f}(\mathbf{x},t) = \mathbf{f}(t)\delta(\mathbf{x})$ [$\delta(\mathbf{x})$—delta function, $\mathbf{x}$—position vector, t—time], which exerts on the fluid a kinematic momentum flux $\mathbf{J}(t) = \int \mathbf{f}(\mathbf{x}', t)d\mathbf{x}' = \mathbf{f}(t)$. To construct higher order singularities, for example, a force doublet $Q_{ij}\nabla_j\delta(\mathbf{x})$ or quadruplet $M_{ijk}\nabla_j\nabla_k\delta(\mathbf{x})$, standard limiting procedure can be used.

From the equations of motion of viscous (kinematic viscosity ν) homogeneous (density ρ = const) incompressible fluid with proper boundary conditions (e.g., initially and at infinity the fluid is at rest)

$$\frac{\partial u_i}{\partial t} + (u_j\nabla_j)u_i = -\frac{1}{\rho}\nabla_i p + \nu\nabla^2 u_i + J_i\delta(\mathbf{x}) + Q_{ij}\nabla_j\delta(\mathbf{x}) + M_{ijk}\nabla_j\nabla_k\delta(\mathbf{x}) + \cdots \tag{26.1}$$

$$\nabla\mathbf{u} = 0,\quad |\mathbf{u}|,\quad p = 0 \quad \text{at } (t = 0) \quad \text{and} \quad \text{at } (|\mathbf{x}| \to \infty) \tag{26.2}$$

Handbook of Environmental Fluid Dynamics, Volume One, edited by Harindra Joseph Shermal Fernando. © 2013 CRC Press/Taylor & Francis Group, LLC.
ISBN: 978-1-4398-1669-1.

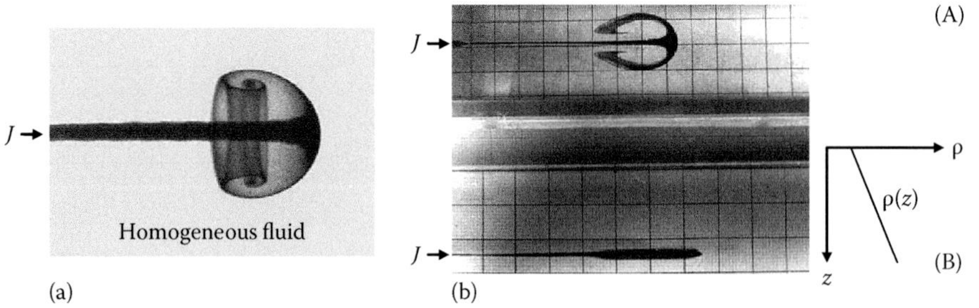

FIGURE 26.1 (a) Momentum source J acts in a homogeneous fluid generating round laminar jet with starting spherical vortex. (b) The same source acts in a density-stratified fluid where the fluid density ρ increases linearly with depth z: (A) top, (B) side view, Re ~ 200. (Reproduced from Voropayev, S.I. and Afanasyev, Y.D., *Vortex Structures in a Stratified Fluid: Order from Chaos*, Chapman & Hall, London, U.K., 1994. With permission of Taylor & Francis Group LLC.)

follows that the equation for the vorticity $\boldsymbol{\omega} = \nabla \times \mathbf{u}$ ($\boldsymbol{u}$ is the fluid velocity) has characteristic "source" terms $\mathbf{J} \times \nabla\delta(\mathbf{x}) + \cdots$

$$\frac{\partial \boldsymbol{\omega}}{\partial t} + (\mathbf{u}\nabla)\boldsymbol{\omega} - (\boldsymbol{\omega}\nabla)\mathbf{u} = \nu\nabla^2\boldsymbol{\omega} - \mathbf{J} \times \nabla\delta(\mathbf{x}) + \cdots \quad (26.3)$$

Thus, the resulting flows are essentially vortical. Starting and steady flows induced by such singularities have been studied in the past and useful analytical solutions were derived (e.g., Cantwell 1986; Voropayev and Afanasyev 1994).

In practice, a "point" momentum source of intensity J can be realized simply by using a jet from small nozzle of diameter d. In this case J can be estimated as (e.g., Schlichting 1979)

$$J = \frac{q^2}{s} = U_0^2 s \quad (26.4)$$

where $q = U_0 s$ is the volume flux and U_0—velocity at the nozzle exit, $s = \pi d^2/4$. By increasing U_0 and decreasing d in such a way that J remains constant, one can make $q \approx J^{1/2} d$ small and neglect it, thus producing exclusively the momentum flux (force) J of intensity (26.4). From (26.4), it follows that the jet Reynolds number (which represents the ration of inertial and viscous forces) is equal to $Re = U_0 d/\nu \approx J^{1/2}/\nu$.

In applications, the momentum source $J(t)$ can act continuously or impulsively, during short time interval Δt, thus generating a net momentum of $I = J\Delta t$. It can be at rest or moving in the direction of force action or in opposite direction, thus generating variety of vortical flows. When fluid is homogeneous in density and forcing possesses an axial symmetry, the resulting flows are also symmetric (Figure 26.1). Classical examples are vortex rings (impulsive forcing), steady round jets (continuous forcing), and axisymmetric wakes behind moving bodies (or moving momentum forcing).

26.3 Effect of Stratification and Characteristic Examples

In many liquid systems (shallow flows, thin soap films, conducting fluids in a magnetic field, etc.) the motion in one direction is suppressed. Typical example is a density-stratified fluid where $\rho = \rho(z)$ and the vertical motion is suppressed by the buoyancy force. The intensity of stratification is characterized by the buoyancy frequency $N = \sqrt{g\rho_0^{-1} d\rho/dz}$, where g is the gravity acceleration, z positive down vertical coordinate, ρ_0 reference density.

As can be seen in Figure 26.1, in a stratified fluid momentum flows loose the axial symmetry and in many cases are effectively horizontal. The characteristic feature of such flows in a stratified fluid is the formation of compact pancake eddies with dipolar vertical vorticity distribution—two-dimensional analog of vortex rings. In the examples shown in Figure 26.1, the flow Re number is small and the motion is laminar. At large Re the initial flow is turbulent, but nevertheless with time the buoyancy forces

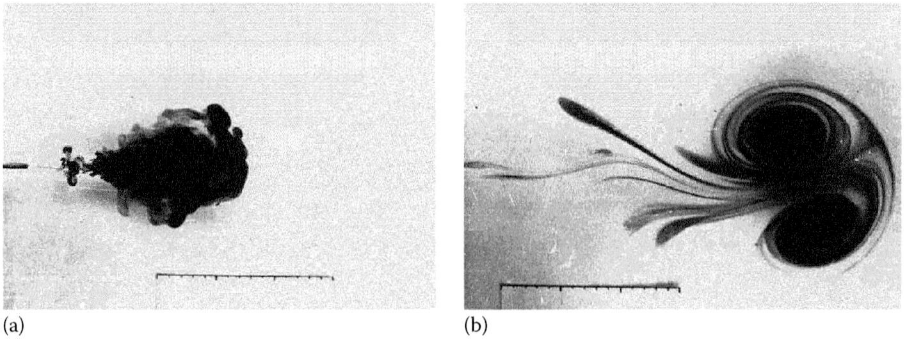

FIGURE 26.2 Formation with time (a and b) of pancake dipolar eddy from initial turbulent jet puff in a stratified fluid. Top view, $Re \sim 1000$. (Reproduced from Voropayev, S.I. and Afanasyev, Y.D., *Vortex Structures in a Stratified Fluid: Order from Chaos*, Chapman & Hall, London, U.K., 1994. With permission of Taylor & Francis Group LLC.)

prevail and similar pancake dipolar eddies are formed in a stratified fluid under the action of impulsive forcing (Figure 26.2).

Large (from tens to hundreds kilometers in horizontal size) dipolar eddies are frequently observed in the ocean (e.g., Mied et al. 1991; DeFelice et al. 2000). They transport momentum, heat, salt, and biological products, and at some conditions their signatures can be traced at the water surface by remote methods, thus giving the information on processes which occur in the water depth.

Dipolar eddies have been extensively studied and general mechanisms of their formation and evolution explained in detail (e.g., Voropayev and Afanasyev 1994). In particular, it was shown that during a prolonged intermediate asymptotic stage such eddies develop in a self-similar regime and their typical horizontal size D^* increases with time t as

$$D^* \approx I^{*1/4} N^{1/12} t^{1/3} \tag{26.5}$$

($I^* = J^* \Delta t$—flow momentum), while the propagating velocity decreases as $U^* \propto t^{-2/3}$. Thus, the Reynolds and Froude numbers for these flows decrease with time as $Re_F = U^* D^* / \nu \propto 1/t^{1/3}$ and $Fr_F = U^*/ND^* \propto 1/t$, and this leads to the flow laminarization and increasing stability with time (see Riley and Lelong 2000 for a comprehensive review).

26.4 Stratified Wakes Classification

Although the literature on wakes of towed bodies is voluminous (e.g., Pao and Kao 1977; Lin and Pao 1979; Lofquist and Purtell 1984; Spedding et al. 1996) and the problem itself lends much information on fundamental aspects of wakes, information so obtained has limited utility in self-propelled body applications, given that there are fundamental differences between the ways the momentum is imparted to the fluid. This is illustrated in Figure 26.3, where pictures of different types of wakes are shown based on Voropayev et al. (1999).

Note the wide differences between these wakes, which could be explained by the nature of momentum forcing resulting from the body–fluid interaction. This interaction leads to momentum exchange between the two, which, as far as the late wake is concerned, can be considered as occurring in a compact area compared to the size of the late wake, that is, the effect of the body on the fluid can be considered as occurring due to "point" forcing. A straightforward case is the *momentum* wake of a towed body (Figure 26.3A) with characteristic vortex street (e.g., Pao and Kao 1977), where the momentum flux J imparted into the wake becomes the same as the drag J on the body, the effect of which can be represented by a point momentum source.

If the body is self-propelled and cruising *steadily* (Figure 26.3B) and internal wave drag is negligible, the viscous and form drags ($J/2 + J/2$) balance the engine thrust J, and there is no net momentum imparted to the wake (*zero momentum* wake). Such wakes decay much faster than the momentum wakes shown in Figure 26.3A, and large eddies are not formed in this case (Figure 26.3B). Forcing in such wakes can be considered as equivalent to a moving force doublet that produces zero net momentum. When the self-propelled body is in *unsteady* motion, the thrust $2J$ and drag ($J/2 + J/2$) are not balanced (Figure 26.3C) and forcing therein can be considered as a combination of a momentum source and a force doublet, the latter decaying faster and leaving behind an intense vortex street due to the momentum wake.

In practice, self-propelled bodies frequently move unsteadily, for example, accelerate during a relatively short time interval Δt. Estimates also show (Meunier and Spedding 2006) that after acceleration the body reaches its new terminal velocity

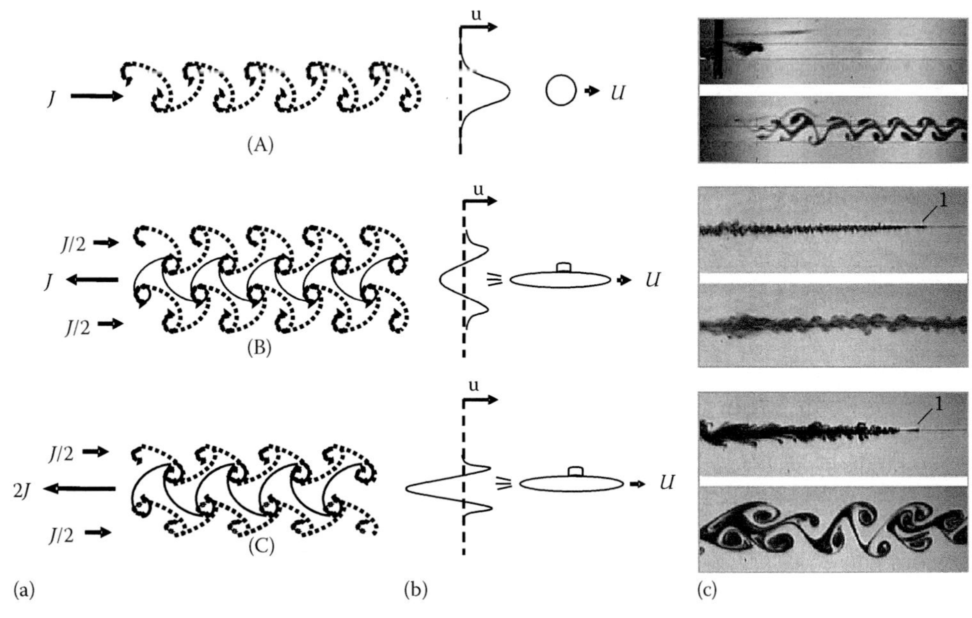

FIGURE 26.3 In columns (a–c) we show (a) simplified top view schematics, (b) velocity profiles, and (c) wake patterns for three basic cases (A–C) of far wakes in a stratified fluid: (A) momentum wake behind a towed body or moving concentrated momentum source (jet), (B) zero-momentum, and (C) momentum (overthrusted) wake behind an accelerating self-propelled body (1).

only asymptotically at large time and most of the time it moves unsteadily. During the period of unsteady motion, the body acquires a net momentum of $I = J\Delta t$ (here J is the difference between the thrust reaction and drag forces on the body) and, by definition, the same momentum I with the opposite sign is transported to the fluid. The balance of momentum for submerged self-propelled body of volume V moving with velocity U is

$$(1 + k)V\frac{dU}{dt} = J_T - J_D = J \quad (26.6)$$

where k is the virtual mass coefficient, J_T the thrust (engine) reaction *on the body*, $J_D = J_D^{\star} + J_D^W$ the net drag *on the body* that includes the viscous and form drag $J_D^{\star}$ and the wave drag J_D^W as a result of momentum flux radiated as internal waves (which is not associated with the narrow wake itself; see schematic in Figure 26.4). For steady motion $J = J_T - J_D = 0$, and the net momentum flux applied to the fluid is zero. When $J = J_T - J_D > 0$, the body accelerates, for example, from U_- to U_+, during the time interval Δt and acquires a net (kinematic) momentum

$$I = \int^{\Delta t} J dt = (1+k)V(U_+ - U_-) = (1+k)V\Delta U \quad (26.7)$$

This gives the estimate $J \approx I/\Delta t$ for the net momentum flux transported *to the fluid* in the direction opposite to the direction of body acceleration. Note that *the history of body acceleration* is not important and the net *flow momentum I* can easily be estimated from (26.7) if ΔU is known.

By definition the wake behind a moving body is a narrow conical region (Figure 26.4c) where the fluid motion is vortical and

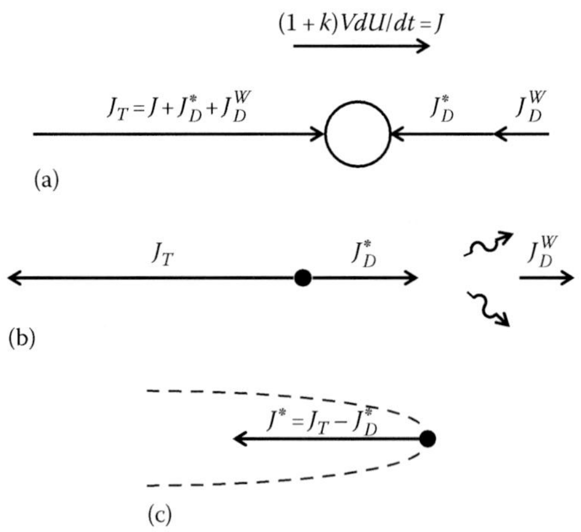

FIGURE 26.4 Schematic showing forces acting on (a) accelerating self-propelled body (circle), (b) surrounding fluid, and (c) wake, shown by a dashed line.

the wake intensity is characterized by the *wake momentum flux* $J^{\star}$ in this narrow region

$$J^{\star} = J_T - J_D^{\star} = J + J_D^W \quad (26.8)$$

The engine thrust J_T that needs to be supplied to overcome $J_D^{\star}$, however, has to be imparted into the wake. For steadily moving self-propelled body from (26.8) follows that although the net momentum flux J transported to the fluid is equal to zero, the *wake momentum flux* $J^{\star}$ is equal to the wave drag on the body

$$J^{\star} = J_D^W \quad (26.9)$$

From this analysis, it follows that the action of an accelerating self-propelled body on a fluid is equivalent to the action of a momentum source of intensity (26.8) that acts impulsively during a time interval Δt. Thus, one may predict that the action of an accelerating self-propelled body on a stratified fluid will result in the formation of large pancake dipolar eddy, of size given by (26.5) and in general similar to that generated by a "point" force in Figure 26.2.

26.5 Large Wake Vortices of Maneuvering Bodies

This idea was verified by Voropayev and Fernando (2010) in experiments with real self-propelled body. To reach high *Re* numbers (>10^4), experiments were conducted in large tank with stratified fluid (Figure 26.5). To model oceanic pycnocline, approximately linear salinity stratification was used, and dye visualization and Particle Image Velocimetry (PIV) flow diagnostics were conducted at the level of body pass, at the mid-depth of the fluid (Figure 26.5c). Different basic cases of unsteady body motion were studied and explained.

For example, in the experiment shown in Figure 26.6, the model starts from the rest, accelerates and then moves with constant velocity (Figure 26.6f). During this maneuver the body deposits on the fluid a momentum of *I*. Initially, the propeller generates an intense jet-like flow with a sharp vorticity front that propagates away from the body. Soon, however, the buoyancy effects become important and the front collapses in the vertical direction and expands horizontally forming a pancake structure (Figure 26.6b). The self-propagation velocity of the vorticity front is less than that of the fluid velocity of the jet-like flow behind the front. As a result, the (vertical) vorticity in this flow is advected to the front region, forming patches of concentrated vorticity of opposite signs in the form of conjunct two to three dipoles (Figure 26.6c), which move in the background potential *dipolar* flow, induced by the pressure forces arising during the body acceleration, and merge together (Figure 26.6d) forming a large dipole of the size $D^{\star}$ (Figure 26.6e and g) that is much larger than the model diameter D. Similar results were obtained in cases when the body moves steadily and then strongly/weakly accelerates.

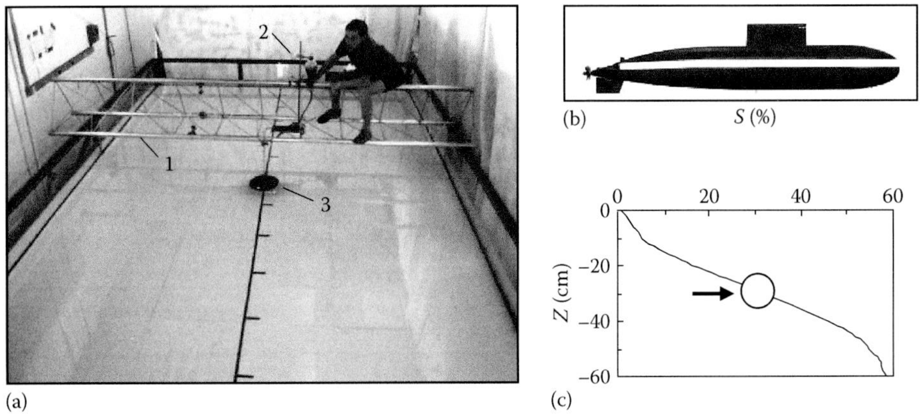

FIGURE 26.5 (a) Tank (8 × 4 × 0.8 m) with removable bridge (1) and device (2) to produce a dyed spot (3). (b) Radio-controlled submarine model (length L = 75 cm, diameter D = 11 cm). (c) Typical vertical density (salinity, S) distribution, circle—the body position; arrow—level where the dye spot or tracer particles (for PIV) were seeded.

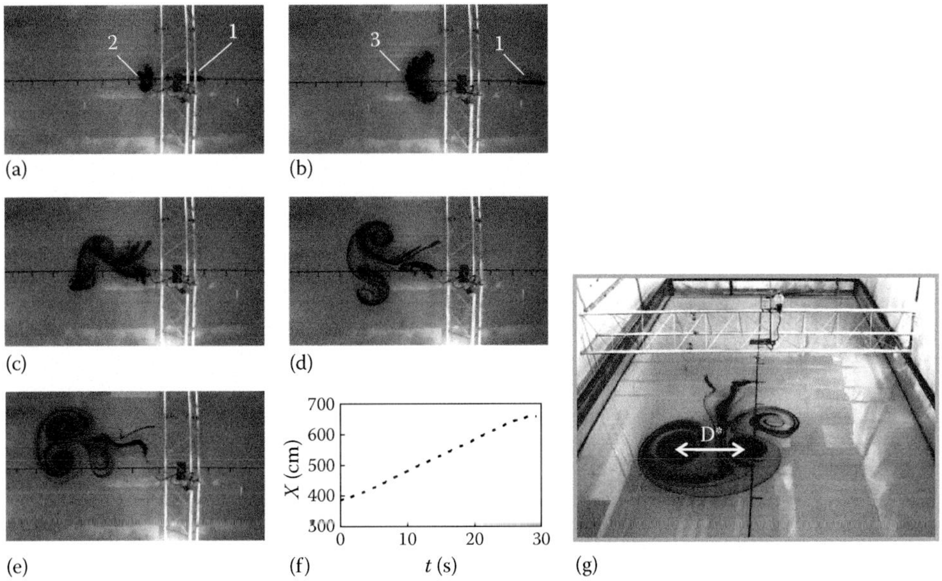

FIGURE 26.6 (a) The body (1) is at rest and positioned behind a dyed spot (2). (b) The body (1) starts moving to the right, generating the vorticity front (3); after a number of bifurcations (c), (d), a large dipolar eddy is forming in the late wake (e). (f) The position X of the body as a function of time. (g) An enlarged image of the resulting eddy at Nt = 630. Experimental parameters: Re = 10,500, Nt = 0 (a), 16.5 (b), 140 (c), 230 (d), 630 (e).

Thus, the acceleration of a self-propelled body, moving with high Re number in a stratified fluid, leads to impulsive momentum wake with a system of eddies, which merge and asymptotically form large and long-lived dipolar vortex. Comparison of measurements with the model predictions show that in all cases of the body acceleration, the size D^* of the resulting eddy at late times can be correctly estimated using (26.5), whence the net wake momentum $I^* = J^*\Delta t$ is estimated by (26.7) and (26.8). Using these results, useful predictions can be made for an oceanic submarine, which, for example, changes its speed only by 10%. Unusually large and long-living eddies with typical size of 1–2 km and decay time of several days may be expected as a result of such maneuver.

26.6 Surface Signatures of Underwater Vortices

The upper ocean is frequently homogeneous in density and can be considered as a shallow layer (typical depth h = 50–100 m) that rests aloft denser (more salty/cold) water. In a shallow fluid, when the flow typical length $D^* \gg h$, the motion is restricted in the vertical direction and may be considered as quasi-horizontal. In such fluids one may expect that under certain conditions large momentum disturbances, for example, generated in the upper ocean by a maneuvering submerged self-propelled body, may penetrate the shallow upper layer and produce detectable signatures at the water surface. This idea was verified by Voropayev et al. (2007) in experiments with an accelerating self-propelled body, which generates a controllable momentum disturbance of intensity $J \approx I/\Delta t$ given by (26.8) in the upper fluid layer of depth h. Of interest were the conditions under which a momentum disturbance generated at depth penetrates to the water surface and leaves a detectable surface signature.

Typical example of the flow at the *water surface* is shown in Figure 26.7. In this experiment the model was initially at rest and located near the tank center in the mid-depth of the upper layer. At $t = 0$ the model accelerates and then moves to the right with a constant velocity. At early times (Figure 26.7a), there is no detectable signature near the water surface. A compact dipolar eddy becomes clearly visible in the vertical vorticity field after $t \approx 10$ s (Figure 26.7b). It propagates in the direction opposite to the accelerating body (Figure 26.7c and d), and for clarity the eddy is highlighted by white contour in Figure 26.7c and d.

To explain the results of experiments, a semi-empirical model was advanced, which includes two main parameters: *confinement* number (dimensionless forcing) $C = J^{1/2}\Delta t/h^2$ and *contrast* number $Cn = \omega^*/\langle\omega\rangle$ (ratio of the maximum vorticity amplitude ω^* of the eddy to the *root-mean-square* background vorticity $\langle\omega\rangle$). Following a large number of experiments, empirical scales were introduced to rank the intensity (contrast) of surface signatures as "high," $Cn > 50$, "significant," $5 \le Cn \le 50$, and "insignificant," $Cn < 5$.

The resulting experimental dependence $Cn(C)$ clearly showed a direct relationship between the two, and this dependence was parameterized as

$$Cn = \frac{\omega^*}{\langle\omega\rangle} = Cn^*\left[1 - \exp2(C_0 - C)\right], \quad Cn^* \approx 65, \quad C_0 \approx 0.65 \tag{26.10}$$

Using this relation, useful estimates were derived for the conditions under which a momentum disturbance generated at depth penetrates to the water surface and leaves a detectable signature. These conditions were formulated in terms of the critical values for the depth of the upper layer

$$h^* \approx 0.84J^{1/4}\Delta t^{1/2}, \quad h^{**} \approx 1.2J^{1/4}\Delta t^{1/2} \tag{26.11}$$

where, for a given disturbance intensity, there will be either a high surface signature contrast ($h < h^*$), a significant contrast ($h^* \le h \le h^{**}$), or an insignificant contrast ($h > h^{**}$). Estimates show that for a typical oceanic self-propelled body making an underwater maneuver in a two-layer ocean (e.g., at low latitudes), the critical depth h^{**} at which the eddy is still detectable with a significant surface contrast is $h^{**} \approx 110$ m.

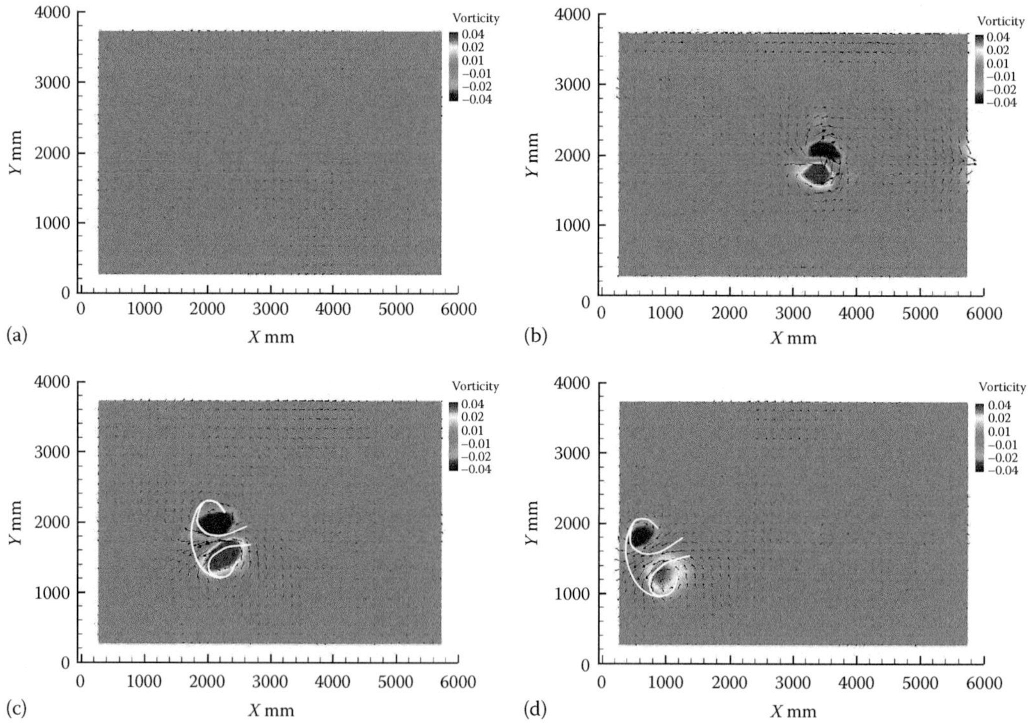

FIGURE 26.7 Typical top view PIV data (vertical vorticity—different colors, velocity—small black arrows) showing significant penetration of wake signature of submerged self-propelled body to the water surface. Initially, the model is at rest near the tank center in the mid-depth of the upper layer, and at $t = 0$ it starts moving from left to right: $t = 4$ (a), 10 (b), 150 (c), 540 s (d). (Reprinted with permission from Voropayev, S.I., Fernando, H.J.S., Smirnov, S.A., and Morrison, R., On surface signatures generated by submerged momentum sources, *Phys. Fluids*, 19(7), 076603. Copyright 2007, American Institute of Physics.)

26.7 Survival of Wake Signatures in Shear Flow and Turbulent Environment

The models and estimates discussed earlier were obtained based on the results of experiments in which the main body of stratified fluid was at rest. The real stratified ocean is never at rest, and background shear and turbulence are the most important factors that may efface eddy signatures in the upper ocean and make these models/predictions not very useful for practical purposes. Hence, an attempt was made to reproduce both these effects in laboratory experiments.

Background shear: In the oceans, the horizontal shear is typically two to three orders of magnitude smaller than the vertical one and may be neglected. Therefore, it is of practical interest to investigate the effect of the background vertical shear $s = du/dz$ (u—background horizontal velocity) on the dynamics of pancake eddies. To produce controllable vertical shear in a stratified fluid, Voropayev et al. (2001) used a circular water channel similar to "Odell–Kovasznay" water tank. Linearly stratified (by salt) water was driven around the channel by two vertical stacks of thin horizontal Plexiglas disks. The upper and lower parts of each stack rotate in opposite directions, and water in the upper and bottom layers moves horizontally in opposite directions with controllable vertical shear. Momentum eddies with controllable intensity were generated using standard impulsive jet technique.

Considering the flow at $Re \gg 1$ and using the idea of flow similarity at intermediate times, the results of experiments, conducted in a broad range of external parameters, were explained. A theoretical flow-regime diagram and estimates for the eddy lifetimes were derived and compared with the results of measurements. Extrapolating the model predictions to the oceanic conditions, it was concluded that for typical oceanic values, the background shear does not affect significantly the dynamics of eddies, and the lifetime of large eddies (e.g., mushroom-like currents, momentum wake eddies) will remain significant (of the order 2–5 days) at typical oceanic shear.

Background turbulence: The nature of turbulence in the oceans varies substantially according to the location. Hence, only the decaying turbulence in a linearly stratified layer was considered recently by Voropayev et al. (2008). The methodology in this experimental study was to produce a controlled eddy whilst decaying horizontal eddying motions, which mimic remnants of a decaying field of 3-D turbulence in the oceanic thermocline, was generated by a towed grid in a stratified fluid. The interaction between two flows were studied, paying particular attention to the survival of momentum eddies.

Three series experiments were conducted (Figure 26.8). First, dipolar eddies were generated using a momentum source. As this

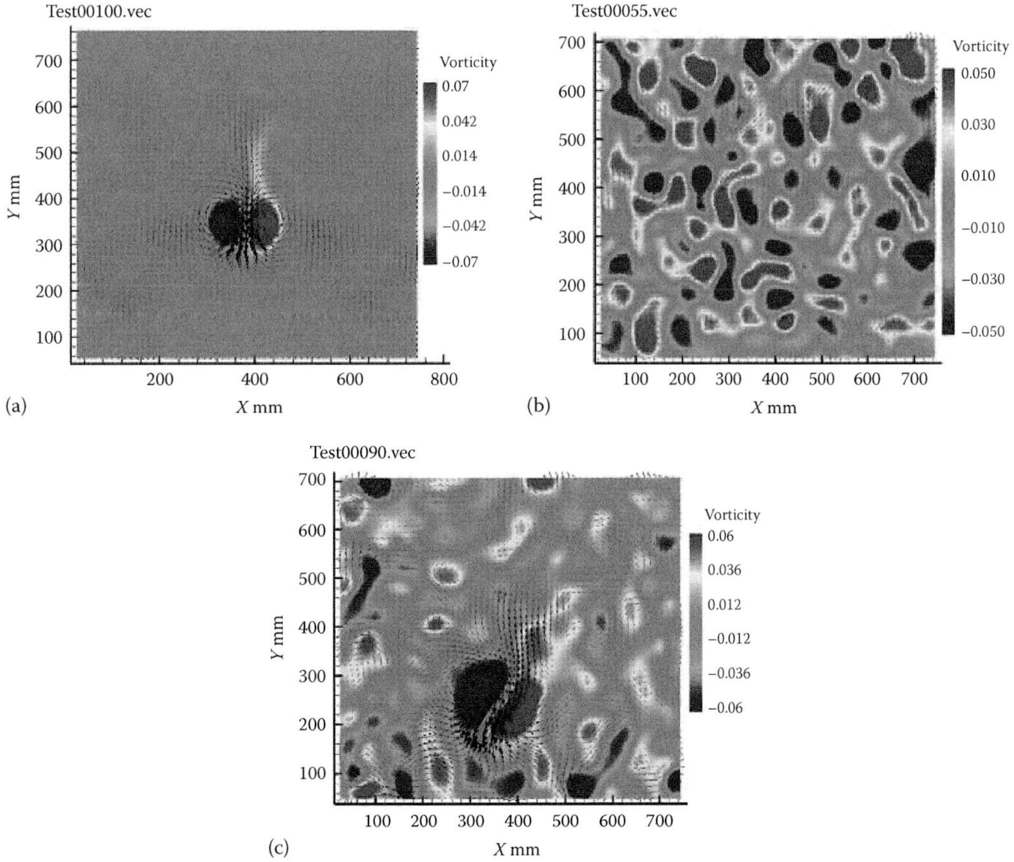

FIGURE 26.8 Typical top view PIV vorticity/velocity data showing: (a) a dipole that propagates in a linearly stratified fluid, (b) decaying grid generated turbulence, and (c) dipole that survives in the field of background turbulence. (Reprinted with permission from Voropayev, S.I., Fernando, H.J.S., and Morrison, R., Dipolar eddies in a decaying stratified turbulent flow, *Phys. Fluids*, 20(2), 026602. Copyright 2008, American Institute of Physics.)

eddy propagated along the tank, its characteristics were measured and compared with theoretical estimates. Second, experiments with grid turbulence were conducted, wherein the flow characteristics were documented and explained theoretically. Finally, experiments with the momentum eddy source and the background grid turbulence were conducted. Data on vorticity/velocity fields of wake dipoles and background flow were collected in the presence or absence of the other. Theoretical explanations, based on similarity arguments, were advanced to explain the flow behavior. In particular, it was predicted that the maximum vorticity in a dipole

$$\omega^* = \frac{A}{t} \tag{26.12}$$

where $A = const$ irrespective to forcing intensity. The measurements confirmed all predictions and yielded $A \approx 17$. The self-similar evolution of dipoles was directly verified by plotting vorticity in self-similar time-dependent coordinates. Applying similar similarity arguments to grid-generated turbulence (Figure 26.8b), the dependence of basic flow characteristics (the *rms* horizontal kinetic energy, integral length scale, and vertical vorticity $<\omega>$) on main governing parameters (grid speed, mesh size, width of grid rods) was derived and verified experimentally. In particular, it was predicted that

$$<\omega> \approx \frac{A^*}{t} \tag{26.13}$$

where $A^* = const$ irrespective to forcing intensity. The measurements confirmed all predictions and yielded $A^* \approx 1.7$. When both types of flows are superimposed (Figure 26.8c), they tend to act rather independently over a significant period of time, indicating negligible nonlinear interactions between them (i.e., the decay laws remain the same for a period of time as if one or the other flow is absent). The "observable" signatures of dipoles in the field of background turbulence were characterized by the contrast number

$$Cn = \frac{\omega^*}{<\omega>} \approx \frac{A}{A^*} \tag{26.14}$$

which was shown to be sufficiently large, so as to retain detectable imprints of dipoles for a long time period. Given that the oceanic thermocline can be characterized by intermittent generation and decay of turbulence, the results are expected to have implications for survival of dipolar wake eddies in ocean thermocline, and give assurance that momentum eddies generated due to maneuvering self-propelled bodies may survive turbulence of oceanic thermocline and retain significant vorticity signatures for long time periods.

26.8 Wave Drag of Self-Propelled Body

When a self-propelled body moves in a stratified fluid and the body Froude number $Fr = U/DN$ (which represents the ration of inertial and buoyancy forces) is large, $Fr \gg 1$, the internal wave drag in (26.8) is negligible. This case is shown in Figure 26.3b where the wake has zero momentum and there are no large eddies in the late wake. At small Fr, however, the buoyancy force becomes significant, and from studies with towed sphere it follows that at $Fr \approx 1$ the wave drag may be comparable with the form and viscous drags and cannot be neglected (Gorodtsov and Teodorovich 1982; Lofquist and Purtell 1984; Greenslade 2000). To our knowledge, existing literature on the wave drag has dealt with only the case of a steadily towed body and the case of self-propelled bodies was not considered.

Experiment shown in Figure 26.9 illustrates this case. In this example, a self-propelled model moves steadily at small Fr in

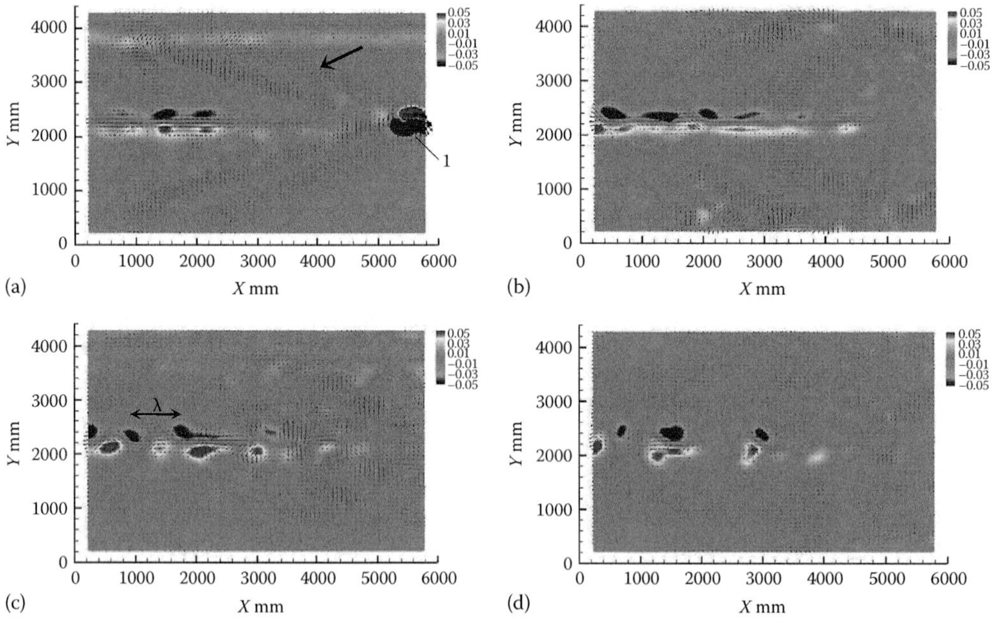

FIGURE 26.9 PIV data showing the formation of a vortex street in the wake of a steady moving (from left to right) submerged self-propelled body (1). Internal waves (large arrow) are also visible in (a). Experimental parameters: $Re = 15{,}500$, $Fr = 0.8$, $Nt = 27$ (a), 81 (b), 162 (c), 270 (d).

a linearly stratified fluid and $J = 0$ in (26.6), (26.8). As can be seen, the late-wake flow demonstrated a different, compared to Figure 26.3b, behavior, in that the PIV measurements show a vortex street that slowly drifts to the left in the wake of a steadily moving from left to right self-propelled body. This means that the wake possesses momentum, and the wave drag in (26.9) is important. To maintain steady self-propelled motion, therefore, the propulsion system should generate a thrust that is equal to the sum of the form, viscous, and wave drags. Since internal gravity waves radiate the momentum (related to the wave drag) away from the source region, the momentum flux associated with form and viscous drag remains unbalanced in the vicinity of the body. Thus, the resultant wake in accordance with (26.9) has a momentum flux $J^* = J_D^W$ (overthrusted wake); and a vortex street, similar to that shown in Figure 26.3c, is generated in the flow with typical wave length λ (Figure 26.9c).

To quantify this effect, indirect method was used. The difficulty here is as follows. For a towed body the system body-fluid is *not closed* and the wave drag may be directly estimated from the force measurements. For a self-propelled body the system body-fluid is *closed* and such direct measurements are hardly possible without disturbing the body motion. Thus, indirect method was used and the main idea is as follows.

In Voropayev and Smirnov (2003), the equivalence of vortex streets generated by either *towed bodies* or moving "point" *momentum sources* was demonstrated. In particular, it was shown that for a momentum source of intensity J^* which moves with velocity U the characteristic length scale l is

$$l = \frac{2.3\sqrt{J^*}}{U} \tag{26.15}$$

instead of the body diameter D, as in the case of a towed body. Using these two length scales, the Strouhal and Froude numbers for body (St, Fr) and momentum source (St^*, Fr^*) take the form

$$St = \frac{D}{\lambda}, \quad Fr = \frac{U}{DN} \tag{26.16}$$

$$St^* = \frac{l}{\lambda}, \quad Fr^* = \frac{2.3U}{lN} \tag{26.17}$$

where λ is the wavelength of the primary vortex street. The dependence of St^* (Fr^*) for the moving momentum source was parameterized in the mentioned paper as

$$St^* = \frac{C^*}{Fr^{*1/3}}, \quad C^* = 0.28 \tag{26.18}$$

When a self-propelled body moves steadily in a stratified fluid with the wave drag J^*, the momentum wake behind the body may be considered as equal to the wake generated by a moving *momentum source* of the same intensity J^*. The use of (26.15), (26.17), (26.18) gives the estimate $J^* = (C^*/2.3)^3 NU\lambda^3$. Normalizing this by the viscous and form drag $J_D^* = (\pi/8)C_D U^2 D^2$ of a self-propelled body (C_D is the body drag coefficient), the dimensionless wave drag A may be estimated as

$$A = \frac{J^*}{J_D^*} = \frac{8C^{*3}}{2.3^3 \pi C_D}\left(\frac{\lambda}{D}\right)^3 \frac{ND}{U} = 0.026\frac{1}{FrSt^3} \tag{26.19}$$

To find A, the function $FrSt^3$ for a *self-propelled body* is needed. Note that for a towed body this function is equal to constant in a broad range of parameters (Voropayev and Smirnov 2003), but for a self-propelled body it needs to be determined from experiments with self-propelled body. Using the results of experiments, similar to that shown in Figure 26.9, the empirical function $FrSt^3$ was derived with the following best fit

$$A = \frac{J^*}{J_D^*} = \frac{1.84}{Fr^{3/2}}, \quad 0.2 < Fr < 3.5 \tag{26.20}$$

Thus, the wave drag of self-propelled body increases when the body Fr number decreases, and may be larger than the viscous and form drag. Relatively large, compare to a towed sphere, values of A at small Fr can be explained by much smaller value of C_D for a long elliptical self-propelled body compared to that for a sphere. Note also that for a self-propelled body the income into the net wave drag comes from the body itself as well as from propulsion system (propeller), and at present the role of the latter remains unclear.

26.9 Thermal Wakes and Surface Signatures

Surface ocean layer (depth 1–10 m) is frequently warmer than the water below (e.g., diurnal thermocline). When submerged or surface self-propelled body moves in such temperature-stratified fluid, it mixes the water; and when observed from above, a temperature "signature" (1°C–2°C) may be detected by infrared (IR) camera at the water surface (Peltzer et al. 1987.). Such thermal signatures are of great value in identifying and characterizing submerged wakes. To quantify this idea, scaled experiments were conducted recently by Voropayev et al. (2012) using researcher series cooled IR camera (FLIR Systems) that measures the water temperature in the upper 10–15 μm (skin layer) with high sensitivity <0.02°K. Both surface and submerged (see Figure 26.10) self-propelled bodies were used with a variety of temperature stratifications. For example, for the case of moderate winds, the surface layer is characterized by its depth h_0 and temperature jump $\Delta_0 = T_S - T_0$ (Figure 26.11a). Because of its penetration to a depth h, the temperature in the wake T_W will be lower than the background surface temperature T_S, and this difference, $\Delta_C = T_S - T_W$, could be detected by the remotely located IR camera (Figure 26.10).

To characterize the intensity of thermal wake signature relative to the background, the wake contrast number,

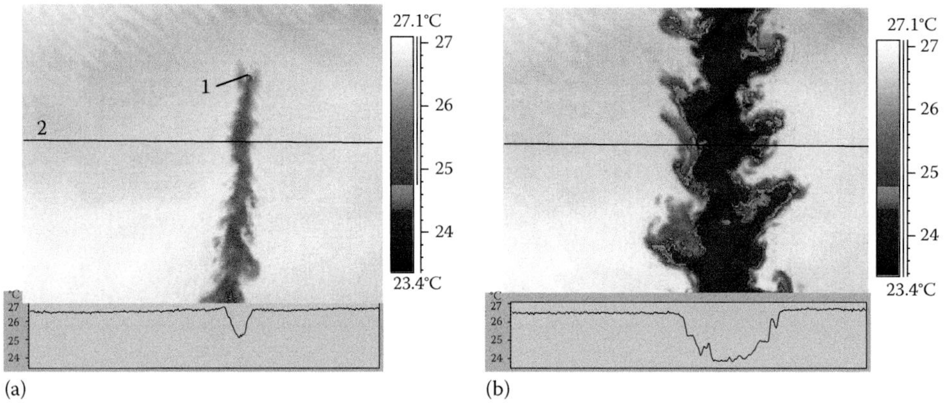

FIGURE 26.10 Surface thermal signature generated by submerged self-propelled body (shown in Figure 26.5b) with its periscope (1) piercing the surface: (a) near wake and (b) far wake. Temperature traces along horizontal lines (2) are shown below each image.

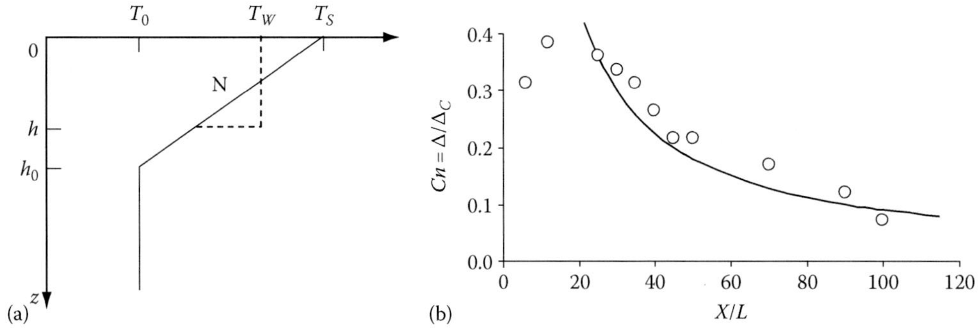

FIGURE 26.11 (a) Schematic showing undisturbed temperature profile in the ocean surface layer under moderate winds (solid line) and profile in the wake (dashed line). (b) Typical IR wake contrast number, *Cn*, as a function of dimensionless distance, *X*/*L*, from body: symbols—measurements; solid line—asymptotic theory.

$Cn = \Delta_C/\Delta_0$, was used in Voropayev et al. (2012). In a series of experiments with stratification similar to that shown in Figure 26.11a, the temperature measurements were made and downstream evolution of thermal wake characteristics was mapped, including mean temperature profile, wake width, wake signature contrast. To explain the results of experiments, a physical model was proposed and the main idea is as follows. Conservation of heat content in the upper layer (see Figure 26.11a) gives the relation between the wake contrast *Cn* and wake mixing depth *h* as

$$Cn = \frac{(T_S - T_W)}{(T_S - T_0)} = \frac{\Delta_C}{\Delta_0} = \begin{cases} h/2h_0, & h \le h_0 \\ 1 - h_0/2h, & h \ge h_0 \end{cases} \tag{26.21}$$

The mixing depth in its turn depends on the body forcing and background stratification. Using the conservation of energy and proper parameterization for body forcing, the wake mixing depth *h*, and hence the contrast number *Cn*, can be estimated via the most important external parameters. For example, for the case of momentum wake generated by a self-propelled body which moves with velocity *U* at the water surface with the (surface) wave drag C_W, the mixing depth can be estimated as

$$\frac{h}{h_0} \approx \frac{2^{4/3} C_W^{1/3} \Gamma^{1/3} U}{h_0 N (X/L)^{2/3}} = \frac{2^{4/3} C_W^{1/3} \Gamma^{1/3} Fr_W}{(X/L)^{2/3}} \tag{26.22}$$

Using this approach, the system of dimensional external parameters was reduced to identify four salient dimensionless parameters, which could be matched between laboratory and oceanic conditions. These include *modified* Froude number for warm surface layer, $Fr_W = U/h_0N$; *modified* Froude number for body, $Fr_B = U/NL$; body geometric parameter, Γ; and dimensionless distance from body, *X*//*L* (*L* is the body length). Using these parameters such wake characteristics as contrast, *Cn*, critical distance, X^*/L, below which *Cn* is significant, and critical distance, X_0/L, above which *Cn* is insignificant, may be estimated. The model proposed by Voropayev et al. (2012) predicts the dependencies of *Cn*, X^*/L, and X_0/L on the parameters Fr_B, Fr_W, Γ, and *X*/*L*, and these dependencies were verified experimentally (see, e.g., Figures 26.11b and 26.12). Estimates for typical ocean body-wake scenarios were also made, which

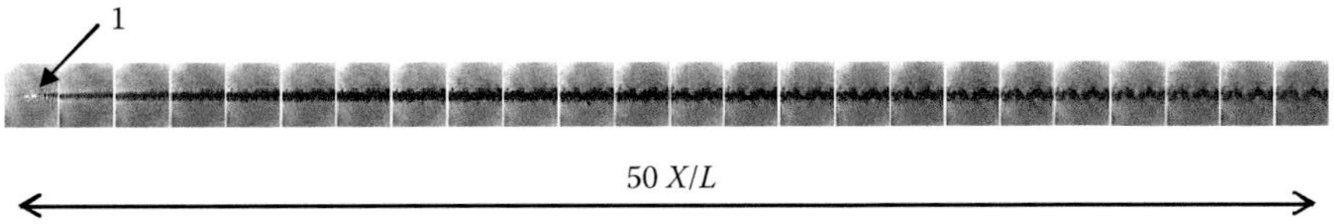

FIGURE 26.12 Thermal IR surface signature behind a ship (1) of length L, which moves from right to left in laboratory. This image closely resembles oceanic signatures reported by Peltzer et al. 1987.

are supported by available field observations. In particular, estimates show that for a body moving steadily in the warm surface layer of ocean (depth 10 m, temperature drop 1°C), the wake contrast will be significant for large ($X/L \approx 200-250$) distances downstream of the body.

26.10 Conclusions

While the progress in submerged stratified wake research has been significant over the past years a new crop of research issues has arisen due to increasing role of satellite-enabled platforms. But no single sensor/platform has all the answers, and only some of them with computer-based data mining can provide "true undersea picture." Such developments call for not only a better understanding of submerged wakes but also their surface manifestation, research on which leaves much to be desired.

References

Cantwell B.J. 1986. Viscous starting jets. *J. Fluid Mech.* **173**, 159–189.

DeFelice T.P., Meyer D.J., Xian G., Christopherson J., and Cahalan R. 2000. Landsat-7 reveals more than just surface features in remote areas of the globe. *Bull. Am. Meteorol. Soc.* **81**, 1047–1049.

Gorodtsov V.A. and Teodorovich E.V. 1982. Study of internal waves in the case of rapid horizontal motion of cylinders and spheres. *Fluid Dyn.* **17**(6), 893–898.

Greenslade M.D. 2000. Drag on a sphere moving horizontally in a stratified fluid. *J. Fluid Mech.* **418**, 339–354.

Lin J.-T. and Pao Y.-H. 1979. Wakes in stratified fluids. *Ann. Rev. Fluid Mech.* **11**, 317–338.

Lofquist K.E.B. and Purtell L.P. 1984. Drag on a sphere moving horizontally through a stratified liquid. *J. Fluid Mech.* **148**, 271–284.

Meunier P. and Spedding G.R. 2006. Stratified propelled wakes. *J. Fluid Mech.* **552**, 229–256.

Mied R.P., McWilliams J.C., and Lindemann G.J. 1991. The generation and evolution of mushroom-like vortices. *J. Phys. Oceanogr.* **21**(4), 489–510.

Pao H.-P. and Kao T.W. 1977. Vortex structures in the wake of a sphere. *Phys. Fluids* **20**(2), 187–191.

Peltzer R.D., Garret W.D., and Smith P.M. 1987. A remote sensing study of a surface ship wake. *Int. J. Remote Sens.* **8**(5), 689–704.

Riley J.J. and Lelong M.P. 2000. Fluid motions in the presence of strong stable stratification. *Ann. Rev. Fluid Mech.* **32**, 613–657.

Schlichting H. 1979. *Boundary Layer Theory*, 7th edn. McGraw-Hill, New York.

Spedding G.R., Browand F.K., and Fincham A.M. 1996. Turbulence, similarity scaling, and vortex geometry in the wake of a sphere in a stable-stratified fluid. *J. Fluid Mech.* **314**, 53–103.

Swain L.M. 1929. On the turbulent wake behind a body of revolution. *Proc. R. Soc. Lond. A* **125**, 647–659.

Tennekes H. and Lumley J.L. 1972. *A First Course of Turbulence*. MIT Press, Cambridge, MA.

Voropayev S.I. and Afanasyev Y.D. 1994. *Vortex Structures in a Stratified Fluid: Order from Chaos*. Chapman & Hall, London, U.K.

Voropayev S.I. and Fernando H.J.S. 2010. Wakes of maneuvering body in stratified fluids. *Progress in Industrial Mathematics at ECMI 2008* (Eds.: A.D. Fitt, J. Norbury, H. Ockendon, and E. Wilson). Springer-Verlag, Berlin, Germany, pp. 259–264.

Voropayev S.I., Fernando H.J.S., and Morrison R. 2008. Dipolar eddies in a decaying stratified turbulent flow. *Phys. Fluids* **20**(2), 026602.

Voropayev S.I., Fernando H.J.S., Smirnov S.A., and Morrison R. 2007. On surface signatures generated by submerged momentum sources. *Phys. Fluids* **19**(7), 076603.

Voropayev S.I., McEachern G.B., Fernando H.J.S., and Boyer D.L. 1999. Large vortex structures behind a maneuvering body in stratified fluids. *Phys. Fluids* **11**(6), 1682–1684.

Voropayev S.I., Nath C., and Fernando H.J.S. 2012. Thermal surface signatures of ship propeller wakes in stratified waters. *Phys. Fluids*, resubmitted after revision.

Voropayev S.I. and Smirnov S.A. 2003. Vortex streets generated by a moving momentum source in a stratified fluid. *Phys. Fluids* **15**(3), 618–624.

Voropayev S.I., Smirnov S.A., and Brandt A. 2001. Dipolar eddies in a stratified shear flow. *Phys. Fluids* **13**(12), 3820–3823.

27

Gravity Currents and Intrusions

Marius Ungarish
Technion—Israel Institute of Technology

27.1 Introduction

A gravity current (GC) appears when fluid of one density, ρ_c, propagates into another fluid of a different density, ρ_a, and the motion is mainly in the horizontal direction, say x. A schematic description of four typical GCs in two-fluid systems is given in Figure 27.1. In our discussion x_N is the length of the GC; h_N and u_N are height of the nose and speed of propagation; g is the gravity acceleration.

A GC is formed when we open the door of a heated house and cold air from outside flows over the floor into the less dense warm air inside. A GC is formed when we pour honey on a pancake and we let it spread out on its own. GCs originate in many natural and industrial circumstances and are present in the atmosphere, lakes, and oceans as winds, cold or warm streams or currents, volcanic clouds, various intentional or accidental discharges (oil slicks, poisonous, and toxic gases), etc. The GC is sometimes visible on its own (e.g., an oil slick on water), or outlined by some additional material carried with the flow (e.g., sand particles or airborne pollution; or dye added in laboratory experiments). In other cases, the rather invisible GC is manifest by some associated effect, like change of temperature, odor, erosion of surface, etc. The presence, shape, and motion of the GC can be detected and recorded by photographs and various sensors (of pressure, temperature, salinity, etc.). Radars and satellites provide useful images of various environmental GC manifestations.

The study of GCs has generated a quite large literature; the presentation here is based mostly on the books of Simpson (1997) and Ungarish (2009) and references therein.* The start of the theoretical-analytical study (or modeling) of the GC is usually attributed to von Karman in 1940. However, about a century earlier Saint-Venant solved the strongly related flow of a reservoir of water of height h_0 above the ground (in air, or vacuum) released by dam break.

The first important *dimensionless parameter* of our problem is *the density ratio of current to ambient*, ρ_c/ρ_a. We are mostly interested in systems with $\rho_c/\rho_a \approx 1$ (in other words, the relative density difference, $\Delta\rho/\rho$, is small, less than 20% say). Systems of this type are called *Boussinesq*. The typical GC is of Boussinesq type, because (1) many geophysical and environmental flows are associated with small density differences. In the atmosphere a typical temperature difference ΔT = 10°K about T = 300°K produces $\Delta\rho/\rho \approx \Delta T/T$ = 3%. Oil sleaks in water are driven by $\Delta\rho/\rho \approx$ 10%–20%. The presence of salt (NaCl) in water, which is relevant to oceans, can produce a density increase of only about 20% before reaching saturation (at room temperature). (2) The aforementioned Boussinesq systems of saltwater in freshwater are a widely used, convenient and cheap, setup for laboratory GC tests (some dye is added for visualization); non-Boussinesq setups may require more expensive fluids and equipment. (3) The mathematical analysis and physical interpretation of Boussinesq systems is simpler.

* Due to space limitation, only a very short list of references is given at the end of this chapter. Additional relevant references can be found in these books.

Handbook of Environmental Fluid Dynamics, Volume One, edited by Harindra Joseph Shermal Fernando. © 2013 CRC Press/Taylor & Francis Group, LLC. ISBN: 978-1-4398-1669-1.

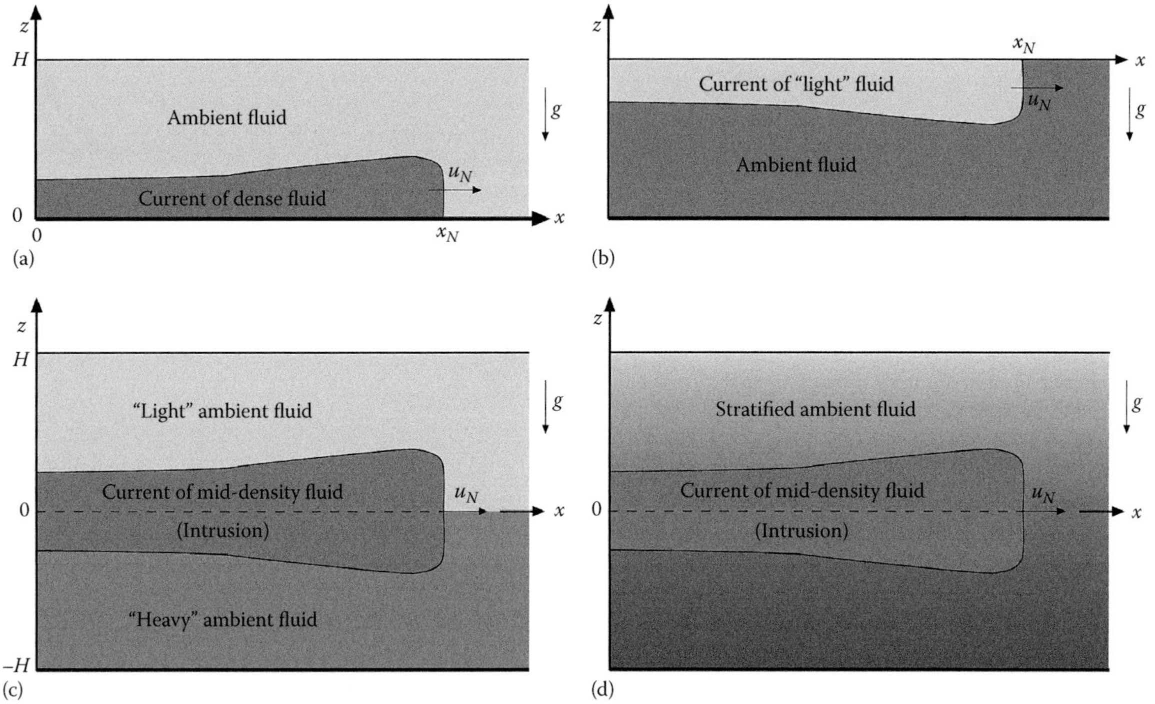

FIGURE 27.1 Schematic description of typical GC configurations: (a) bottom current of more dense (heavy) fluid, $\rho_c > \rho_a$; (b) top (surface) current of less dense (light) fluid, $\rho_c < \rho_a$; (c) intrusion of "mixed" fluid in a sharply stratified ambient; (d) intrusion of "mixed" fluid in a linearly stratified ambient, $\rho_c = \rho_a(z = 0)$. A solid "backwall" is located at $x = 0$, and the system is unbounded in the positive x direction. The local thickness of the current is $h(x, t)$, and the length of the current is $x_N(t)$. The symmetric intrusion is, geometrically, a combination of two currents about the neutral-buoyancy plane $z = 0$.

We see that the Saint-Venant classical water-in-air dam-break problem is an extreme $\rho_c/\rho_a \rightarrow \infty$ non-Boussinesq case; von Kármán's more recent study was for a two-gas system, which is relevant to the more typical Boussinesq GC.

27.2 Principles

27.2.1 Fundamental Cases

The basic rectangular 2D configuration for the understanding of these flows is sketched in Figure 27.1a. For definiteness, we consider a bottom GC with $\rho_c > \rho_a$, on the boundary $z = 0$; in the Boussinesq case, the top GC, see Figure 27.1b, is a mirror image about the surface of propagation. A fairly sharp interface, $z = h(x, t)$, separates between the GC and the ambient fluid. The length of the GC is $x_N(t)$. The motion starts at $t = 0$ with $x_N(0) = x_0$.

The typical GC is thin and long. By this we mean that the ratio of the typical thickness of the current to the horizontal length is small, say (average $h)/x_N \ll 1$. This geometric property has significant dynamical implications. In particular, the small h/x_N implies that the typical motion in the x direction produces, relatively, very small dynamic accelerations in the z direction. Consequently, as good approximations: (1) Both the current and the ambient are in *hydrostatic quasi-equilibrium in the z direction*. This can be expressed as $\partial p_i/\partial z = -\rho_i g$, for $i = a, c$; p denotes the pressure, and the a, c subscripts the ambient and current. (2) The *pressure is continuous at the interface* $z = h(x, t)$ between the fluids (except for points of discontinuity, or jumps, of h).

These features provide justification to useful mathematical simplifications. One may wonder why such facilitation or reduction is needed in the age of supercomputers. Indeed, formally, the flow field can be obtained from the full set of the governing Navier–Stokes equations of motion (given later). This formidable formulation is not amenable to analytical methods, but the appropriate solution can be attempted by a numerical discretization method which is then solved on a computer. Such numerical computer solutions, or "simulations," work fairly well for 2D cases which require 10^4–10^5 grid points. However, the more reliable 3D solutions use millions of grid points and still require weeks of runtime on powerful computers. The real flow field is irregular, often turbulent; the interface contains Kelvin–Helmholtz vortices and instabilities. However, many of these small-scale details provided by the simulations (and experiments) are rarely needed.

For the qualitative understanding and systematic mathematical analysis, and for casting the numerical and experimental data into manageable patterns, it is necessary to employ a simplified description of the GC. In the statistical sense, the flow is 2D (or axisymmetric). The description of the GC flow in terms of a layer of length $x_N(t)$, thickness (height) $h(x, t)$, and depth-averaged speed $u(x, t)$ (where t is time and x the horizontal coordinate), is of interest, and often sufficient, in numerous applications.

The governing equations for these variables, approximated from the Navier–Stokes ones, provide two powerful formulations (models, theories, approaches): the shallow-water (SW) approximation for the "inviscid" GC and the lubrication approximation for the "viscous" GC, respectively. These theories are the backbone of the analytical investigation. A bolder simplification, called box-model, confines the GC to a predetermined simple shape (box), which spreads due to an approximate balance of the global buoyancy-inertial (or buoyancy-viscous) forces. The box-model $x_N(t)$ result is a useful quick estimate, but should be used with great care when not supported by the more rigorous models.

The GC is driven by the gravity body force in a rather subtle way. The gravitational acceleration acts in the vertical ($-z$) direction, while the most significant motion of the GC is the propagation in the horizontal direction x. This is a very common occurrence: the cold outside air which enters a warm room when we open the door, or syrup on the pancake, spreads out in the horizontal direction because they are more dense (heavier) than the embedding air. The governing mechanism is as follows. The pressure p in the fluids satisfies (approximately) the hydrostatic $\partial p/\partial z = -\rho_i g$ ($i = a, c$). The difference in ρ between the fluids thus produces a pressure excess, $\propto(\rho_c - \rho_a)gh$, in the more dense fluid. The fact that h is not constant in x-direction (e.g., the layer of cold air outside is thicker) gives rise to pressure difference in the x direction. In other words, $\partial p/\partial x \propto(\rho_c - \rho_a)g(\partial h/\partial x)$ appears, and the fluid particles are pushed into the x-direction by what can be called a (horizontal) buoyancy force. A solid dam (or lock, or gate) which contains the dense fluid can oppose the pressure (buoyancy) force and block this tendency for motion; otherwise, the $\partial p/\partial x$ in the fluid is balanced by the dynamic reaction of the fluid (i.e., a non-trivial velocity field with a major horizontal component which we denote by u). The gravity effect associated with the density difference is referred to as the *reduced gravity*

$$g' = \left(\frac{|\rho_c - \rho_a|}{\rho_a}\right)g. \tag{27.1}$$

For Boussinesq systems $g'/g \ll 1$; a typical value is 50 cms^{-2}. (For more complex and non-Boussinesq cases, some care in the interpretation of g' and even redefinition may be needed; see Ungarish [2009], Section 10, e.g., note the difficulty of the $\rho_c/\rho_a \to \infty$ system).

In some circumstances g' drives a vertical motion, for example, when warm smoke is released into a calm atmosphere; this vertical current is referred to as a *plume*. Buoyancy-driven vertical plumes and horizontal GCs are often connected components in large-scale environmental flows, in particular volcanic eruptions. However, the plume and GC obey different scalings and interactions with the ambient, and must be analyzed separately. Here we focus our attention on the GC which is driven by g' in the horizontal direction (allowing also for some small-angle inclination).

The typical velocity of propagation of the GC can be estimated from energy conversion arguments. Suppose that the current is released from rest in a lock of height (thickness) h_0. The mean potential energy of a particle in the lock is $(1/2)g'h_0$ per unit mass. If this is converted to the kinetic energy $(1/2)U^2$, we obtain the *typical speed*

$$U = (g'h_0)^{1/2}. \tag{27.2}$$

The obvious restriction on the applicability of (27.2) is that viscous friction is negligible.

Indeed, the $\partial p/\partial x$ buoyancy forcing can produce either an inertial (inviscid) or viscous flow, depending on the circumstances and the fluid. The indicative dimensionless parameter is the *Reynolds number* given by

$$Re = \frac{Uh_0}{\nu} = (g'h_0)^{1/2}\frac{h_0}{\nu}, \tag{27.3}$$

where ν is the typical kinematic viscosity coefficient. Roughly, the GC flow is *inviscid, or inertial, when* $Re \gg 1$, *and viscous when* Re *is small.* This distinction is essential.

A GC released from rest from a reservoir of moderately small (say 0.1–0.5) height-to-length ratio is typically in the inviscid flow regime, unless thicknesses of less than 1 cm and high-viscosity fluids are concerned. Consider, for example, a GC created in the laboratory in a rectangular tank filled with freshwater. We mix some salt in a sub-volume of 10 cm height and length to obtain a density difference of about 1%. Using $g' = 10$ cms^{-2}, $h_0 = 10$ cm, and $\nu = 10^{-2}$ cm^2 s^{-1} we estimate $Re = 10^4$. A larger height, or more salt, will increase Re. On the other hand, consider a pool of honey or syrup released on a plate and surrounded by air, with $h_0 = 1$ cm, $\nu = 30$ cm^2 s^{-1}, and $g' \approx g = 980$ cms^{-2}. Now $Re = 1$, and the subsequent flow is in the viscous regime. In this case the inviscid velocity estimate (27.2) cannot be trusted; a more relevant estimate is given later (after [27.31]).

We emphasize that the GC is a *time-dependent phenomenon*, and hence, the same GC may pertain to different types or regimes at different times. Typically, as the current spreads out, both its speed and thickness decay. An initially inviscid GC may become viscous, with a remarkable drop of the subsequent rate of spread u_N. For reliable interpretation of a GC which propagates over a significant distance the value of the initial Re, given by (27.3), is not sufficient; it is necessary to monitor the ratio of inertia to viscous effects in the flow as a function of t [see (27.38)], or x_N. Surprisingly perhaps, there are circumstances in which the propagation is initially in the viscous regime, and after some time the flow becomes inertial. This unexpected behavior concerns GCs produced by strong inflow, not by the prototype lock-release mechanism. Indeed, it is necessary to distinguish between *fixed-volume* and *variable-volume (inflow)* GCs. Unless stated otherwise, we consider the former.

The inviscid GC is little affected by no-slip conditions, while the viscous GC is dominated by them (in this case the averaged u represents a rather parabolic z/h profile). The GC is governed by partial differential equations (PDE) which contain h_t, u_t and spatial derivatives of these variables. The thickness of the viscous GC is governed by a parabolic PDE, and hence h decreases continuously with x to $h_N = 0$ (although a steep h_x may develop at the nose). On the other hand, the inviscid GC is governed by a hyperbolic system of PDEs, and the nose is well modeled by a "jump" (shock): a vertical portion of the interface of height h_N, which supports a discontinuity of pressure. The mathematical solution of hyperbolic systems may also indicate internal discontinuities (which can be interpreted as hydraulic jumps and bores); this renders these problems more fascinating, but also more difficult and less conclusive for analysis.

The inviscid GC obeys the *front condition* $u_N = Fr(g'h_N)^{1/2}$. Here Fr is the *Froude number* whose typical value is 1. (Actually Fr is a function of the height ratio of current to ambient, h_N/H^*, see [27.13], [27.15], and Figure 27.5.) The height of the nose, h_N, is not a clear-cut defined quantity in real flows, and some average over distance and time is usually taken.

As the GC passes on $z = 0$, a return flow in the ambient above the current, of relative magnitude h/H^*, appears (H^* is the height of the ambient, or channel). The viscous GC is usually so thin that the return flow is insignificant. On the other hand, the return flow may be of importance for the inviscid GC, in particular in lock-release problems. We introduce the dimensionless parameter

$$H = \frac{H^*}{h_0}, \quad (27.4)$$

which turns out to be very influential in the analysis and interpretation of high-Re GC.

27.2.1.1 Typical Behavior of the GC

We focus on a 2D rectangular case. The GC is released from a lock of length x_0 and height h_0. The *reference, or scaling, quantities* are x_0 for horizontal lengths, h_0 for vertical lengths, speed $U=(g'h_0)^{1/2}$, and time $T=x_0/U$. The *dimensionless variables*, which we use in this section, are scaled with these values.

The typical GC has $Re \gg 1$, and hence it is first in the inviscid regime. Here the propagation displays three stages. The first is called "slumping," from the release to $t = 3$ at least. The opening of the gate (similar to a dam-break occurrence) produces, instantaneously, a forward-moving nose of height $h_N \approx 0.5$ and speed $u_N \approx 0.5$. This forward propagation draws fluid from the reservoir and, eventually, sets in motion the whole bulk of dense fluid. Evidently, in the slumping stage the shape of the interface and the velocity field change dramatically. However, *the nose maintains a constant height, h_N, and a constant speed, u_N.* For shallow ambient, $H < 2$, a hydraulic jump propagates from the lock into the reservoir, and is then reflected from the backwall $x = 0$ to the nose. For a deep ambient, the internal shocks are replaced by smooth rarefaction-compression waves. When the reflected bore (or wave) reaches the nose, the slumping ends. Due to the slower propagation of the jumps, the typical x_N at the end of slumping is 10, 5, and 3 for $H = 1$, 2 and $H > 3$; the u_N value during slumping also increases with H, from 0.5 to 0.7. The slumping stage is strongly influenced by the initial conditions; a significant change of the shape of the reservoir (e.g., the cylinder $h = (1 - x^2)^{1/2}, 0 \le x \le 1$) will produce a different behavior, with no constant speed.

The second stage (sometimes ignored) is a relatively short transition to the next stage, without special features. The influence of the initial conditions and of the upper boundary (via the return flow) diminish.

The next stage is of *self-similar motion*. This pattern occurs often in the propagation of GCs: x_N behaves like Ct^β, where the constants C and β are determined by global properties (inertial or viscous regime, total volume, 2D or axisymmetric geometry, etc.); the initial conditions are forgotten. For the present inviscid, 2D, fixed volume case, $\beta = 2/3$. The thickness and speed maintain their x-dependent shape up to a t-depending scaling function, that is, $h = \varphi(t)\ \mathcal{H}(y)$ and $u = \psi(t)\ \mathcal{U}(y)$, where $y = x/x_N(t)$. See Figure 27.2a. Note that $\mathcal{H}$, and hence h, has a positive inclination, with maximum at $y = 1$; this is typical of self-similar propagation in the inviscid regime. Here u_N decreases like $t^{-1/3}$ or $x_N^{-1/2}$.

With time, the self-similar inviscid GC becomes longer, thinner, and slower. The importance of the viscous effect (triggered

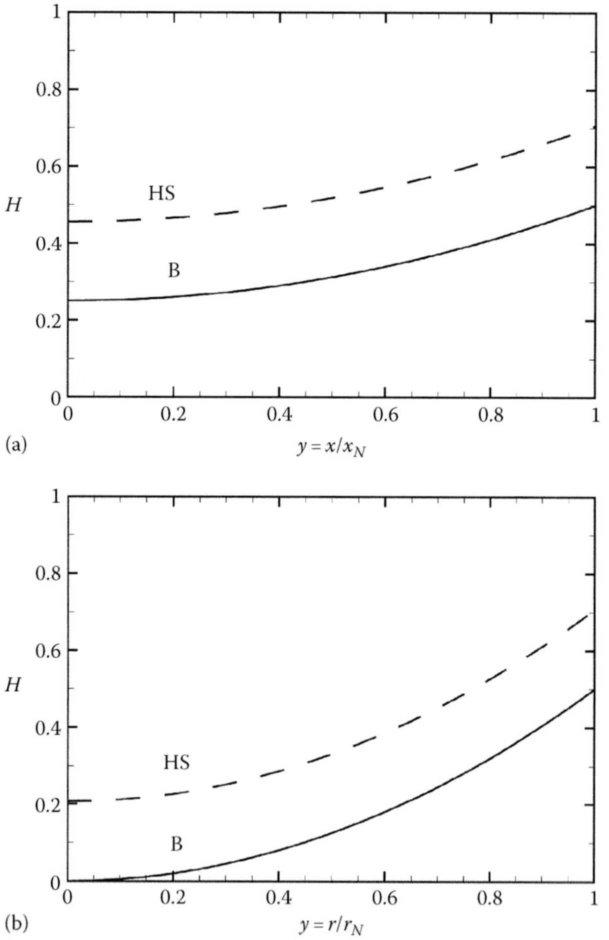

FIGURE 27.2 Similarity thickness profiles for inviscid GC of fixed volume. $H(y)$ is given for Benjamin's $Fr = \sqrt{2}$ (solid line, label B), and HS $Fr = 1.19$ (dashed line). (a) 2D geometry; (b) axisymmetric geometry. The speed has the similarity $U = y$ in all shown cases.

by the no-slip condition on the wall $z = 0$) increases continuously, and becomes dominant after propagation to the position

$$x_v \approx \left(\frac{Re\,h_0}{x_0}\right)^{2/7}. \tag{27.5}$$

For example, for $Re = 10^4$ and $h_0/x_0 = 0.5$ we obtain $x_V \approx 11$ (measured in lock-lengths x_0). Subsequently, for $x_N(t) > x_V$, the GC is of viscous type. In this regime a negative $\partial h/\partial x$ is necessary to create a pressure push against the shear $-\rho_c \nu(\partial u/\partial z)$. Again, a self-similar motion appears, $x_N(t) = Ct^{1}/^{5}$. The viscous self-similar motion is very different from the inviscid self-similar flow; in particular, in the viscous case h has a negative inclination, with $h_N = 0$ at the edge $x = x_N(t)$, see Figure 27.3a (in contrast to the positive $\partial h/\partial x$, with a finite h_N, in the inviscid case). The viscous u_N is of the order of $g'(h_0x_0)^3/(\nu x_N^4)$ (dimensional); note the fast decrease with distance at power 4 (in contrast to the power 1/2 in the inertial case).

The switch from inertial to viscous propagation behavior, that is, from $x_N \sim t^2/^3$ to $x_N \sim t^1/^5$, is readily detected in practice. For example, a log–log record of x_N vs. t shows a quite dramatic change of slope from 2/3 to 1/5 at the position x_V.

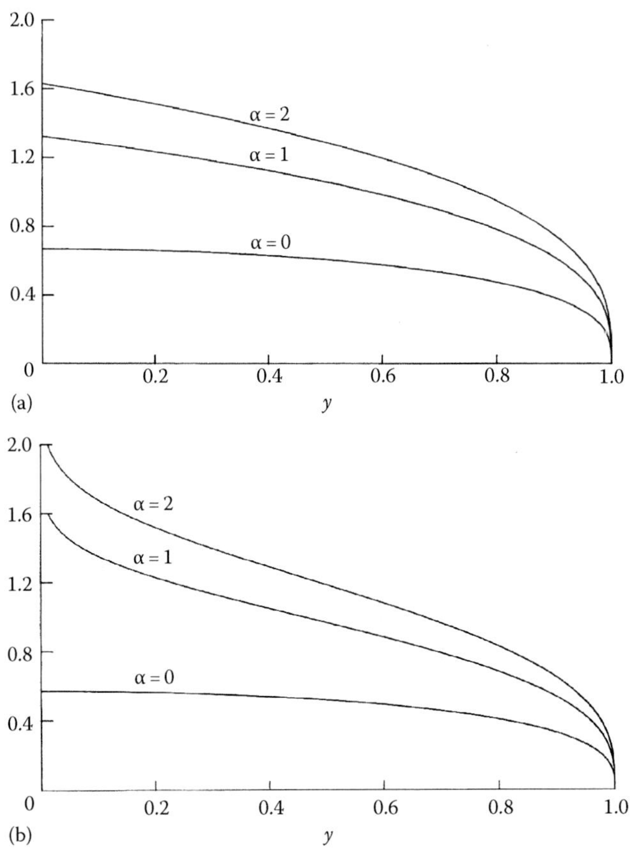

FIGURE 27.3 Similarity thickness profiles for viscous GCs of volume qt^{α}, for various values of α. (a) 2D case, $y = x/x_N(t)$; and (b) axisymmetric case, $y = r/r_N(t)$. (From Huppert, H.E., *J. Fluid Mech.*, 121, 43, 1982. With permission.)

We must keep in mind that this description of the GC is a big simplification and idealization of a rather complex flow phenomenon. Various discrepancies may appear in real (especially natural) circumstances. Perhaps the most prominent effect is that the interface between the inertial GC and ambient is not sharp, but rather a domain of shear, small billows, and mixing; the nose is rather rounded than a sharp jump. The front of a high-Re GC displays a plethora of tiny lobe-and-cleft instabilities (see Simpson [1997], Figure 11.12). Broadly speaking, there is *entrainment* of ambient fluid into the GC. The discussion of these small-scale details is beyond our scope (see Ungarish [2009], Section 9.6).

The "end" of the GC process depends on the circumstances. As explained later, the propagation of rotating axisymmetric GCs is stopped by Coriolis effects; in particle-driven cases, or flow over a porous boundary, the GC simply runs out of particles, or fluid, at some x_{max}. In many other cases the "end" of the GC is vague. The viscous GC becomes, eventually, insignificantly thin and slow; surface tension and small-scale fluctuation in the ambient may further slow down and dissipate the current. On the other hand, GCs created by strong inflow may evolve into rather thick bulks which obey different governing balances. The presence of obstacles or other flows in the domain of propagation is also relevant, and in extreme cases the propagation of the GC changes direction due to reflection from a vertical wall, or a strong counter-current. For example, forced ventilation of coal mines and long vehicular tunnels can push back "natural" GCs of methane or smoke which appear on the roof. On the other hand, a barrier of less than $2h_0$ is usually unable to fully stop and reflect an inviscid GC; some flow continues beyond the barrier in the original direction (see Simpson [1997], Section 11.6).

27.2.2 Cylindrical-Axisymmetric GC

In this counterpart of the 2D rectangular case, z is the axis of symmetry and the GC propagates outwardly in the horizontal-radial direction, r, over a full circle, or in a wedge. The release is from a cylinder of radius r_0 and height h_0. The *scaling* is as before, but with r_0 instead of x_0.

The axisymmetric high-*Re* GC undergoes the same phases of propagation as the 2D counterpart, but the details are different. When the current is released (dam-break instant), the initial speed and nose height are like in the 2D case. However, in the following slumping stage, u_N and h_N decrease due to the diverging geometry. The self-similar propagation appears at about $t = 3$, with $r_N(t) = Ct^{1/2}$. The self-similar shape is shown in Figure 27.2b. In this geometry the self-similar u_N decays like t^{-1} or r_N^{-1}.

The axisymmetric GC is significantly more prone to viscous influence than the 2D case because the thickness decrease like r_N^{-2}, and the area of frictional contact with the bottom increases like r_N^2. Viscous dominance appears after spread to radius

$$r_V \approx \left(\frac{Re\,h_0}{r_0}\right)^{1/6}. \tag{27.6}$$

For example, for $Re = 10^4$, $h_0/r_0 = 0.5$ we obtain the quite modest $r_V \approx 4$ (measured in lock-lengths r_0). Subsequently, for $r_N(t) > r_V$, the GC is of viscous type. Again, a self-similar motion appears, $r_N(t) = Ct^{1/8}$. In the viscous self-similar motion, h has a negative inclination, with $h_N = 0$ at the rim $r = r_N(t)$, see Figure 27.3b (in contrast to the positive $\partial h/\partial r$, with a finite h_N, in the inviscid case). Now u_N is of the order of $g'(h_0 r_0^2)^3/(\nu r_N^7)$ (dimensional). The speed decrease with r_N^{-7} is remarkable: when r_N doubles, u_N decays to less than 1% of previous value.

The typical cylindrical GC, as considered earlier, is divergent. However, in certain circumstances an axisymmetric GC may propagate from the periphery of the container to the center. In this convergent GC situation u_N and h_N first decrease, then increase (see Hallworth et al. 2003). This peculiar behavior has no counterpart in the 2D rectangular case.

27.2.3 More Complex Cases

The GC appears in numerous more complex forms than sketched earlier, due to possible natural or designed variations of the boundary and initial conditions. Here we mention some of these effects and additional governing dimensionless parameters. In this section we use *dimensional* variables.

27.2.3.1 Stratified Ambient

In natural occurrences the density ρ_a of the ambient fluid may be a function of the vertical coordinate z. This stratification, assumed stable (i.e., less dense layers lay above more dense layers), may be either sharp, that is, a jump between two relatively thick layers of homogeneous fluid, or a continuous (typically linear) smooth transition.

Consider a GC of density ρ_c which propagates at the bottom ($z = 0$) of a container filled with an ambient fluid whose density ρ_a decreases linearly from ρ_b at the bottom to ρ_0 at the top (often an open) surface, that is, $\rho_a(z) = \rho_b - (\rho_b - \rho_0)(z/H)$. The dimensionless *stratification parameter* is defined as $S = (\rho_b - \rho_o)/(\rho_c - \rho_o)$. Now $S = 0$ reproduces the homogeneous ambient, and $S = 1$ the maximal stratification ($\rho_b = \rho_c$). The density contrast between current and ambient, at the typical mid-position $z = h/2$, decreases with S, and hence the driving force and the speed of propagation, u_N, are reduced as compared with the $S = 0$ case. This is the major initial effect of stratification. The release and motion of the GC generate in the stratified ambient internal gravity waves of frequency $\mathcal{N} = [(\rho_b - \rho_o)g/(\rho_o H)]^{1/2}$ and speed $u_W = \mathcal{N}H/\pi$. If $u_N > u_W$ (supercritical GC), the effect of the waves is small; but in the subcritical case ($u_N < u_W$) the waves interact with the head of the GC, reduce u_N, and may even stop the propagation (for a while, or even completely). This remarkable effect has been observed in laboratory and numerical tests, but still defies analytical understanding and modeling. There are indications that cylindrical-axisymmetric GCs are less influenced by this interaction than the 2D rectangular counterpart.

Intrusions are associated with a stratified ambient. A GC propagates on a well-defined lower or upper geometric boundary of the system; see Figure 27.1a and b. An *intrusion* propagates horizontally inside a stratified fluid, typically like an isolated wedge which does not touch the horizontal boundaries; see Figure 27.1c and d. The guiding surface, $z = 0$ say, is the plane of neutral buoyancy for the intruding fluid, $\rho_c \approx 1/2[\rho_a(z=0^+)+\rho_a(z=0^-)]$. Many results developed for the GC can be applied to intrusions. Actually, for an inviscid Boussinesq system subject to some simple symmetries about the plane of neutral buoyancy, the intrusion can be reduced to a superposition of a bottom GC above the neutral buoyancy plane, and a top GC below this plane; the currents are mirror-images and propagate with the same speed. A symmetric intrusion into a linearly stratified ambient corresponds to GCs with $S = 1$. Since at the center of the nose there is no density difference, the effective reduced gravity can be expressed as $\mathcal{N}^2 h_0$, and the typical speed $U = \mathcal{N}h_0$ replaces the estimate in (27.2). (In the atmosphere the typical value is $\mathcal{N} = 2\cdot 10^{-2}\ \mathrm{s}^{-1}$.)

27.2.3.2 Rotating Frame

Suppose that the system in which the GC propagates is rotating with Ω about the vertical axis z. The centrifugal acceleration of the system $\Omega^2 x_N$ (say) is typically small as compared to the gravitational acceleration g, and hence, the usual g' is the dominant driving effect. However, the driving gradient of the pressure excess $g'h$ is now balanced not only by the inertial terms but also (or even mostly) by the Coriolis acceleration $-2\Omega v\hat{x} + 2\Omega u\hat{y}$. (We consider an inviscid GC; $u\hat{x} + v\hat{y}$ is the horizontal speed of the current, x is the major direction of propagation.) The consequences are remarkable. It is convenient to distinguish between two typical cases:

1. The axisymmetric (fully cylindrical) GC, while spreading in the radial r direction, develops an azimuthal counter-rotation (in the rotating frame) with negative speed v. The Coriolis acceleration $-2\Omega v$ in the r direction competes with the $u_t + uu_r$ term in the balance of $-\partial p/\partial r$. This first reduces the speed u_N (as compared with the nonrotating case), and eventually the r-propagation stops. The "current" becomes a steady-state *lens* with finite r_N, and $h_N = u_N = 0$; this dense fluid is counter-rotating ($v < 0$). (These structures are relevant to geophysical eddies, vortices, rings, etc.) The underlying effect of this peculiar shape is $\partial h/\partial r \propto \partial p/\partial r \propto \Omega v < 0$. The relevant dimensionless parameter is $\mathcal{C} = 2\Omega r_0/(g'h_0)^{1/2}$, which expresses the ratio of Coriolis to inertial accelerations. When $\mathcal{C}$ is small, the radius of the lens is $r_N = 2r_0/\mathcal{C}^{1/2}$, and the thickness is $\mathcal{C}h_0[1-(r/r_N)^2]/2$ (approximately). The relationship $u_N = Fr\cdot(g'h_N)^{1/2}$ holds also for the rotating GC with small $\mathcal{C}$; the propagation starts with the same (u_N, h_N) as in the non-rotating case and ends with (0, 0) at the rim of the lens. When $\mathcal{C}$ is large the propagation is small, $r_N - r_0 \approx r_0/\mathcal{C} = (g'h_0)^{1/2}/(2\Omega)$ (this length is called the Rossby radius), and the flow that occurs over this small distance is not a true GC. The lens is formed during the first revolution of the rotating frame (for any C). The process involves overshoot-contraction oscillations,

and triggers viscous and instability mechanism which dissipate the lens over many revolutions of the system.

2. A GC in an x-long channel bounded by vertical side walls, $y = 0, D$. These walls block lateral motion and hence $v \approx 0$. The main propagation is in the x direction, with little direct Coriolis contribution. However, the x propagation requires $2\Omega u$ Coriolis acceleration in the direction y. To achieve this component, the current agglomerates near the side wall, $y=0$ say, that can provide the necessary pressure "push," $g'(\partial h/\partial y) \propto 2\Omega u$. This renders a boundary current, that is, one that is thicker near the "pushing" vertical wall and thinner (or even detached) at the opposite wall of the channel. (An observer attached to the channel may say that the GC is deflected toward a sidewall by a Coriolis force, like in a centrifugal pump.) The importance of this deflection is represented by the dimensionless parameter $\mathcal{W} = 2\Omega D/(g'h_0)^{1/2}$. When $\mathcal{W} \ll 1$, a lock-released GC behaves like in the classical 2D case; when $\mathcal{W}$ is large the thickness h depends strongly on the lateral y; the $\partial h/\partial y$ may be so strong that the current detaches from the $y = D$ wall. Roughly, the lateral extent of the Coriolis-affected GC is $(g'h_0)^{1/2}/(2\Omega)$ (again, called the Rossby radius). The relationship $u_N = Fr\cdot(g'h_N)^{1/2}$ holds, essentially, for the main propagation; h_N is taken at the "pushing" wall, and Fr is well approximated by Benjamin's formula (27.13), plus a slight increase with $\mathcal{W}$.

27.2.3.3 Particle-Driven GC

The density difference of ρ_c to ρ_a is typically a result of different fluids (e.g., oil and water), or of different concentrations of a dissolved material like salt, or of different temperatures. This is referred to as a *compositional driving* effect.

Suppose that the current is made up of a mixture of small particles of density ρ_p which are suspended, with small volume fraction α, in the fluid of density ρ_a. Let ρ_p be significantly larger than ρ_a (e.g., sand and water). Now $\rho_c = \alpha\rho_p + (1 - \alpha)\rho_a$. The density difference between the layer of suspension (where $\alpha > 0$) and the clear ambient fluid ($\alpha = 0$) generates a *particle-driving* effect. In geological and environmental applications, the particle-driven GCs are called turbidity currents (when the interstitial fluid is a liquid, usually water) and pyroclastic currents (when the interstitial fluid is a gas, usually air). Now the reduced gravity is $g' = \alpha(\rho_p/\rho_a - 1)g$. In general, α is a function of space and t, and particles settle out from the suspension with speed W_s (which is approximated by the Stokesian formula $(1/18)(\rho_p/\rho_a - 1)\, d_p^2 g/\nu$, where d_p is the diameter of the particle). The settling of the particles causes dilution (i.e., decay of α) and/or descent of the interface. This reduces the driving effect, and eventually, at some finite t_{ro}, the GC disappears (runs out of particles, hence the subscript "ro"). This means that either $\alpha \to 0$ due to dilution, or $h \to 0$ due to settling of the interface. Suppose that the GC is released from a 2D lock with initial particle volume fraction $\alpha = \alpha(0)$, and that $Re = \nu h_0/(g_0'h_0)^{1/2} \gg 1$ (g_0' is calculated with the initial $\alpha(0)$). The relationship $u_N = Fr\cdot(g'h_N)^{1/2}$ holds during the propagation; note that the decay of α reduces g', and that $u_N = 0$ at t_{ro}. The dimensionless parameter $\beta = W_s x_0/[(g_0'h_0)^{1/2}h_0]$, typically a small number, is now relevant. Theory and experiments show that the inviscid GC runs out of particles after propagation to $x_{\max} \approx 2x_0/\beta^{2/5}$. (The axisymmetric counterpart is $r_{\max} \approx 2r_0/\beta^{1/4}$.) We keep in mind that the "disappearing" GC usually leaves behind a thin, but long and persistent sediment (deposit). Consequently, it may be possible to re-trace particle-driven GCs long time (sometimes many years) after the occurrence of the flow process; this effect has important environmental and geological applications.

Combinations of the compositional- and particle-driving effects may also occur. The "heavy" particles may be suspended in an interstitial fluid of density ρ_I, different from ρ_a of the ambient. The effective density difference which drives the GC is $\Delta\rho = [\alpha\rho_p + (1 - \alpha)\rho_I] - \rho_a$. Now the particles contribute to the driving effect, but the flow continues after the particle runout, $\alpha = 0$, occurrence. A fascinating *reversing buoyancy* phenomenon appears when $\rho_I < \rho_a$, because $\Delta\rho$ may change sign during propagation. Suppose that a sufficiently large initial particle volume fraction, $\alpha(0)$, generates a "heavy" suspension so that $\Delta\rho > 0$, and a bottom GC. Settling and dilution reduce continuously α; when $\alpha < (\rho_a - \rho_I)/(\rho_p - \rho_I)$ the suspension becomes "light," $\Delta\rho < 0$. This light fluid (suspension) lifts off from the bottom (as a plume) to the top boundary, and then propagates as a top GC in the same direction as before; see, for example, Sparks et al. (1993).

A GC *over a porous boundary* bears some similarity with the particle-driven case: the dense fluid seeps slowly into the pores, and the GC disappears (runs out of fluid) at some $x_{\max}$. On the other hand, a GC *inside a porous medium* behaves like a slow viscous layer: the u_t and uu_x terms are negligibly small, while the negative $\partial h/\partial x$ creates a pressure-drive against the Darcy hindering force.

27.2.3.4 Non-Boussinesq

Upon a careful scaling, the Boussinesq results, derived for $\rho_c/\rho_a \approx 1$, can provide some useful guiding lines for non-Boussinesq systems. Roughly, departure from the Boussinesq case to the "heavy" side $\rho_c/\rho_a > 1$ makes the GC thinner and faster, and to the "light" side $\rho_c/\rho_a < 1$ makes it thicker and slower. However, some special feature may appear for truly large and small values of ρ_c/ρ_a; see Ungarish (2009), Section 27.10 and Ungarish (2011). Some numerical simulation codes may be unsuitable for non-Boussinesq cases because of the built-in formulation, or loss of numerical accuracy.

27.3 Methods of Analysis

The analysis of the GC is performed by experiments, numerical solutions of the Navier–Stokes equations of motion, and approximate models. In many investigations a combination of these means is employed.

The experiments are mostly performed in the laboratory. Typically, a transparent tank of 300 × 50 × 20 cm^3 (length × height × width) is used; the ambient is made of freshwater and the GC of salt-water; tiny amounts of dye are added for visualization. Suitable background illumination means and reference grids are used to sharpen the contrasts and mark the position of the GC. Laser sheets, particle-image-velocity (PIV), and other sensors are used to increase the resolution of the measurements. The motion is recorded by video cameras. The images and sensor records are analyzed by simple eye inspection and, for fine details, with special computer software. By changing the dimensions (and the shape) of the lock, and the concentration of salt in the current and in the ambient, various configurations, values of g', Re and H can be produced; also, by filling the ambient domain from two reservoirs of different densities, it is possible to produce desired continuous stratifications and parameters $\mathcal{N}$ and S. For the investigation of axisymmetric GCs, wedge-like and fully cylindrical containers can be used. The tank can be mounted on a rotating table to study the effects of the parameters $\mathcal{C}$ and $\mathcal{W}$. Field experiments on larger scales (tens or hundreds of meters) are also used (see Simpson [1997], Section 6.2 and Ungarish [2009], Section 18.2.4), but the control on the parameters and accuracy of measurements are problematic. The experiments are very important for the verification of the theoretical tools, for the quantitative estimation and calibration of effects which defy simple calculations (e.g., entrainment and turbulence), and for detecting novel features in un-investigated ranges of parameters or scales. Experiments also have a strong psychological impact: direct observation is exciting, convincing, and indisputable (these attributes should not be carried over to the straightforward interpretations). However, the experimental method encounters many practical limitation (e.g., the setup and operation of non-Boussinesq and rotating systems is expensive and difficult), and the interpretations and generalizations of the recorded data may be a challenge. Like in other branches of fluid mechanics, the laboratory and field experiments are more and more supplemented, or even replaced, by computer simulations.

The simulations are numerical solutions of the Navier–Stokes (usually incompressible) equations, which express the volume, momentum, and mass balances. In an inertial frame of reference, this reads

$$\nabla \cdot \boldsymbol{v} = 0; \tag{27.7}$$

$$\rho \frac{D\boldsymbol{v}}{Dt} = -\nabla \mathcal{P} - (\rho - \bar{\rho}) g \hat{z} + \mu \nabla^2 \boldsymbol{v}; \tag{27.8}$$

$$\frac{\partial \rho}{\partial t} + \boldsymbol{v} \cdot \nabla \rho = \kappa \nabla^2 \rho. \tag{27.9}$$

$\boldsymbol{v}(x, y, z, t)$ and $\rho(x, y, z, t)$ are the velocity and density fields, $\bar{\rho}$ is the mean density of the unperturbed ambient ($=\rho_a$ in non-stratified cases), $\mathcal{P}(x, y, z, t)$ is the reduced pressure ($p + \bar{\rho} g z$), $\mu(= \rho \nu)$ and κ are the viscosity and diffusion coefficients. Rotation, stratification, particle-settling, and boundary inclination can be efficiently incorporated in this formulation. These equations must be solved in a fixed domain, with appropriate initial and boundary conditions. The simulation codes compute the velocity components, density, and pressure numbers which satisfy these equations (within some small numerical error) at certain spatial grid points and time intervals. In this formulation there is no clear-cut separation between the domains of the "current" and "ambient"; the initial conditions and solution for $\rho(x, y, z, t)$ tell us which is which.

The numerical simulation code is a more accessible and flexible tool than the laboratory experiment, and can be used for both exploratory and prediction purposes. The numerical codes allow us to "experiment" with fluids of densities and viscosities which are very difficult, or even impossible, to attain in practice; to set the container into strong rotation without the need to dynamically balance the loads and the danger of being strangled by an entangled tie; also, we can turn on and off effects like boundary friction and particle settling. This provides great help for gaining insights and testing the accuracy of simplified models. The codes allow us to attempt calculations for realistic scales, geometries, and conditions which are beyond the capabilities of laboratory and models. In other words, the code is expected to be the most versatile and reliable tool of analysis. However, the potential of these simulations is only partly realized in practice. To reproduce accurately the observed features of real GC systems, and in particular the sharp moving interface, it is necessary to use small grid intervals, small time steps, and high-resolution 3D codes. The appropriate computations are presently feasible only on expensive powerful number-crunching computer arrays, and the clock run time for one solution is several weeks. For GCs in straight geometry, 2D simulations seem to provide a good approximation to the 3D results, with just a fraction of computational resources and run time. For GCs in cylindrical geometry, however, the situation is less encouraging. 2D-axisymmetric codes reproduce well the real flow for a limited distance of propagation (about $3r_0$) only.

Both laboratory and computer simulations indicate the importance and indispensability of the solutions based on approximate equations, called models. This method of analysis is used before and after any experiment and number-crunching simulation, for guidance and processing of the acquired data. In many cases of interest, this is the only practical means for obtaining quantitative information on the propagation of the GC. We shall therefore focus in some detail on this method of analysis. The main objective is to obtain simple (preferably analytical) solutions for the position of the nose $x_N(t)$, thickness $h(x, t)$, and local mean speed $u(x, t)$. In the axisymmetric case r replaces x; in more complex situations additional variables (like particle volume fraction, lateral speed v, angle of inclination of the boundary, and dependency on the lateral coordinate) are incorporated. The equations of motion of the models are derived from the Navier–Stokes balances.

A widely used robust simplification is to replace the z-component of (27.8) by the hydrostatic $\partial p/\partial z = \rho_i g$ ($i = a, c$), subject to continuity of pressure at the interface $z = h(x,t)$. The justification is that in long (almost) horizontal streams of fluid the z acceleration of the moving particles is negligible compared to g. Another powerful simplification, called the one-layer model, is to discard the dynamic effect of the return flow in the ambient; this reaction is indeed weak when the ambient is thick (or when $\rho_a \ll \rho_c$).

The models illustrated here assume a thin-layer bottom-current of dense (heavy) fluid in non-stratified (homogeneous) ambient. In the corresponding one-layer model, the aforementioned hydrostatic pressure considerations can be integrated with respect to z, and then we obtain for the current

$$\frac{\partial p}{\partial x} = \Delta\rho g \frac{\partial h(x,t)}{\partial x}, \tag{27.10}$$

$\Delta\rho = \rho_c - \rho_a$. This fundamental relationship allows us to eliminate the pressure from the x-component of the momentum equations. It models in a very convenient analytical way the (horizontal) buoyancy-driving effect: a decrease of the height h with x, either moderate or by jump, provides a forward driving effect to the GC. This useful formula applies to both inviscid and viscous GCs, in both Boussinesq and non-Boussinesq systems.

27.3.1 Shallow-Water (SW) Models for Inviscid ($Re \gg 1$) GC

In this framework the viscous effects are discarded inside the fluid and at the boundaries. The speed does not satisfy no-slip conditions.

27.3.1.1 Steady-State GC and Nose Jump Conditions

The pertinent discussion and results are in general referred to as Benjamin's analysis, following the classical paper Benjamin (1968). We use *dimensional variables* unless stated otherwise.

Consider a steady-state half-infinite GC in an infinite horizontal channel of height H; see Figure 27.4a. The GC is assumed to move like a slug (except for a small zone behind the head which may be influenced by interfacial wave-breaking), and the tail is assumed of constant thickness h. The origin of the coordinate system is attached to the foremost point of the current, O. In this system the ambient fluid appears to be moving over an obstacle (the GC) from right to left say; the foremost point of this obstacle is at the bottom and is also a point of stagnation for the ambient fluid. In the far upstream (right, denoted by subscript r) and the far downstream (left, subscript l) domains, the flow is uniform and parallel–horizontal. The uniform and parallel flow in the r and l domains is an exact solution of the steady-state Navier-Stokes inviscid (or Euler) equations. The corresponding pressure is exactly z-hydrostatic, $\partial p/\partial z = -\rho_i g$ ($i = a, c$) in these domains.

This steady-state situation can prevail only when the flow fields in the two fluids are compatible. In other words, U must depend on other variables of the system, and a physically acceptable solution may even be impossible in some cases. We attempt to determine these conditions. We use a rectangular control volume which is bounded by the bottom and top walls of the channel, $z = 0, H$, and by two vertical planes in the far right and left positions. Let $a = h/H$. Here we use a two-layer model; we take into account the return flow above the current, and hence it is not necessary to restrict the analysis to small values of a.

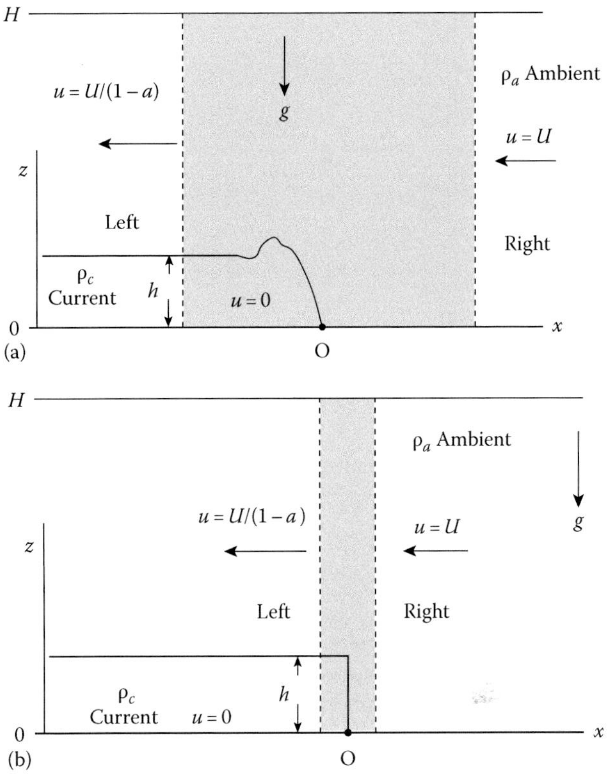

FIGURE 27.4 The system attached to O for the derivation of Fr and dissipation. (a) Control volume for the steady-state current; (b) control volume about the front shock for SW model.

First, we balance the volume fluxes. On the right side the uniform speed is $u_r = U$; on the left $u_l = U/(1 - a)$; see Figure 27.4a.

Next we use the momentum balances. We calculate the z-hydrostatic pressures p_l and p_r in the fluids, subject to the condition that O is a stagnation point. Since no viscous and turbulent stresses are present, the x-momentum balance is between the flow forces on the left and right, expressed as

$$\int_0^H (\rho u^2 + p)_l dz = \int_0^H (\rho u^2 + p)_r dz. \tag{27.11}$$

We substitute the previously calculated u and p into (27.11). Solving for U we obtain the classical Benjamin result

$$U = Fr(a)(g'_a h)^{1/2}, \tag{27.12}$$

where

$$Fr(a) = \left[\frac{(2-a)(1-a)}{(1+a)}\right]^{1/2}; \tag{27.13}$$

$g'_a = (\Delta\rho/\rho_a)g$ is the reduced gravity with respect to the ambient, and for the Boussinesq case $g'_a = g'$.

The so-called Froude number, Fr, is a function of a, see Figure 27.5. Benjamin's Fr increases from $1/\sqrt{2}$ to $\sqrt{2}$ when the relative thickness a decreases from 1/2 to 0. (The $a > 1/2$ cases are

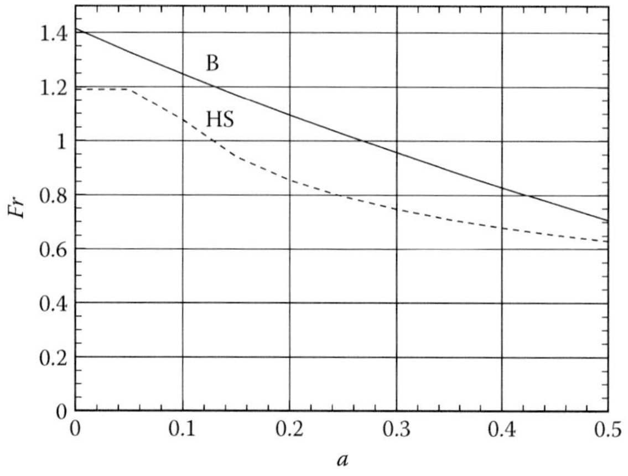

FIGURE 27.5 Froude functions of Benjamin (B) and Huppert–Simpson (HS), given by Equations 27.13 and 27.15, respectively.

irrelevant as shown in the following analysis; $a = 0$ means a GC of given h in a very deep ambient.)

This steady-state situation does not conserve mechanical energy, in general. The stringent requirement is that no external supply of energy should be necessary to sustain the flow which satisfies (27.12). On the other hand, an energy loss (dissipation) can be vindicated by the non-uniform velocity field about the nose inside the control volume. Consider the rate of energy flow, $u(p+\frac{1}{2}\rho u^2+\rho gz)$ integrated from $z = 0$ to H. The difference between the leaving (to the left) and arriving (from the right) terms is the rate of energy dissipation, $\dot{D}$, inside the control volume (actually, in the ambient fluid; the overdot denotes time derivative). Using the foregoing results for pressure and speed, we obtain

$$\dot{\mathcal{D}} = -\frac{1}{2}\rho_a g_a' UHha\frac{1-2a}{1-a^2} = -\frac{1}{2}\rho_a (g_a')^{3/2} h^{5/2} Fr(a)\frac{1-2a}{1-a^2}. \quad (27.14)$$

The "head loss" on a streamline of the ambient fluid is defined as $-\dot{\mathcal{D}}/(\rho_a g_a' UH)$. The only energy-conserving, that is, non-dissipative ($\dot{\mathcal{D}} = 0$), flow is obtained for $a = 1/2$. For $a > 1/2$ we find $\dot{\mathcal{D}} > 0$, or negative head loss, which is unacceptable in normal circumstances. The conclusion is that the GC is restricted to $a \le 1/2$, that is, the GC can occupy at most half of the depth of the channel.

The same analysis can be performed for a thin control volume attached to a simplified front-shock in a general GC, as depicted in Figure 27.4b. Moreover, this analysis is valid for both Boussinesq and non-Boussinesq cases. Consequently, the results (27.12) through (27.14), $a \le 1/2$, and the associated insights are general fundamental components, *as boundary conditions at* x_N, in the analysis of the (almost) inviscid GC. These nose, or front, or Froude (jump) conditions, are applicable (as a good approximation) in a wide range of circumstances, including time-dependent and non-Cartesian flows. (This is irrelevant to the viscous GC.)

Experiments and simulations indicate that in realistic lock-release flows (with large but finite *Re*) the value of $Fr(a)$ is below the analytical prediction (Equation 27.13). A difficulty in these tests is how and where to define the height h of a real GC; this must be taken further behind the top of the head, where the flow is rather horizontal. The presence of some friction and mixing vindicates a decrease of *Fr*. A practical remedy in the modeling of the GC is to use an adjusted value of *Fr*, such as the empirical formula of Huppert and Simpson (HS),

$$Fr(a) = \begin{cases} 1.19 & (0 \le a < 0.075) \\ \frac{1}{2}a^{-1/3} & (0.075 \le a \le 1). \end{cases} \quad (27.15)$$

Various SW solutions of lock-release GCs (see next section) which use the HS *Fr* (27.15) show better agreements of propagation vs. t with experiments than solutions obtained with Benjamin's *Fr* (27.13) (which usually overpredicts the speed).

We note that (27.12) is one equation for two unknowns, U and h. To obtain reliable results for a certain GC, it is necessary to also consider the initial and boundary conditions, as explained in the following. We also remark that (1) The dissipation (27.14) should not be confused with the real viscous irreversible dissipation of a Newtonian fluid. (2) The derivation of (27.13) assumes that there is no net volume flow in the channel at any x; a larger *Fr* may develop in a source-sink case which prevents the return flow in the ambient above the GC.

27.3.1.2 SW One-Layer Equations and Solutions

Here we use *dimensionless variables*. The reference length is x_0, height h_0, speed $U = (g'h_0)^{1/2}$, and time $T = x_0/U$. The dimensionless variables are scaled with these quantities.

The simplified equations of continuity and x-momentum are

$$h_t + uh_x + hu_x = 0; \quad (27.16a)$$

$$u_t + uu_x = -h_x. \quad (27.16b)$$

Here $u(x,t)$ is the z-averaged speed, and $h(x,t)$ the thickness. The LHS of (27.16b) is simply the inertial acceleration; the RHS is the pressure gradient $-(\partial p/\partial x)$ replaced by (27.10). The system is hyperbolic. The characteristic balances read

$$h^{-1/2}dh \pm du = 0 \quad \text{on} \quad \frac{dx}{dt} = c_{\pm} = u \pm h^{1/2}. \quad (27.17)$$

A nose (front) condition is needed at $x = x_N(t)$. Here we apply (27.12), which in scaled form reads:

$$\dot{x}_N = u_N = Fr\left(\frac{h_N}{H}\right)\cdot h_N^{1/2}, \quad (27.18)$$

where the overdot denotes time derivative, and *Fr* is given by (27.13), (27.15), or similar.

The other initial and boundary conditions are provided by the particular problem we wish to solve. Here we consider the classical release from rest from a rectangular lock of unit length and height (in scaled form). This is expressed as

$$h(x,t=0)=1;\quad u(x,t=0)=0;\quad (0\le x\le 1);\qquad (27.19)$$

$$u(x=0,t)=0,\quad (t\ge 0);\quad x_N(t=0)=1.\qquad (27.20)$$

The x domain of solution is $[0, x_N(t)]$ which increases with time. It is convenient to keep track of it in a standard domain, $[0, 1]$. This is achieved by introducing a stretched horizontal coordinate

$$y=y(x,t)=\frac{x}{x_N(t)}\quad (0\le y\le 1).\qquad (27.21)$$

(not to be confused with the lateral Cartesian y coordinate, which is not used in the 2D geometry).

A numerical solution of (27.16) through (27.20) is easily obtained. A standard finite-difference scheme (like Lax-Wendroff) with 500 intervals in the domain $0\le y\le 1$ calculates the propagation to $x_N = 20$ (say) in several CPU seconds on simple computers (see Ungarish [2009], Section 2.4).

27.3.1.3 Dam Break (Initial Motion)

The flow for $0 < t \le 1$, see Figure 27.6, can be calculated analytically by the method of characteristics, because (27.17) is amenable to simple integration and manipulations.

In particular, along the c_+ characteristic with the reservoir initial condition $u=0$, $h=1$, we obtain $u=2(1-\sqrt{h})$. This is intersected with (27.18) to obtain at the nose, formally

$$h_N=\left(\frac{1}{1+Fr/2}\right)^2;\quad u_N=\frac{Fr}{1+Fr/2}.\qquad (27.22)$$

For a deep ambient ($H > 10$ say) the value of Fr approaches a known constant (see Figure 27.5) and the result (27.22) is explicit: for $Fr = 1.19$, $h_N = 0.393$ and $u_N = 0.746$; for $Fr = \sqrt{2}$ the values are $h_N = 0.343$ and $u_N = 0.828$. Otherwise, since Fr decreases when $a = h_N/H$ increases, the (numerical) results of (27.22) depend on H, and this is displayed in Figure 27.7.

The values of h_N and u_N are time-independent in this stage. Further manipulation of the characteristics yields the

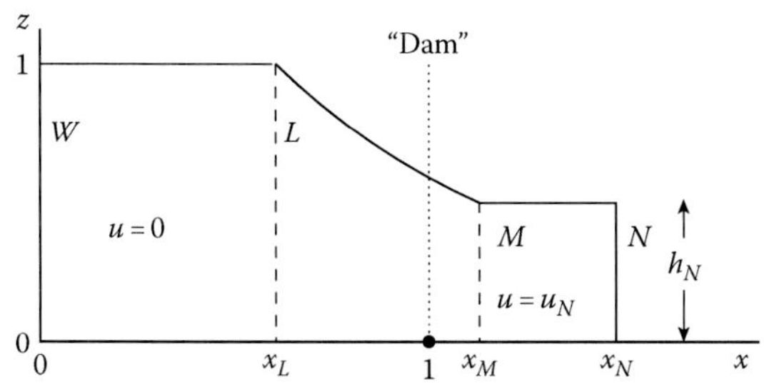

FIGURE 27.6 The SW GC during the initial dam-break stage. Height and length scaled with h_0 and x_0 of the lock.

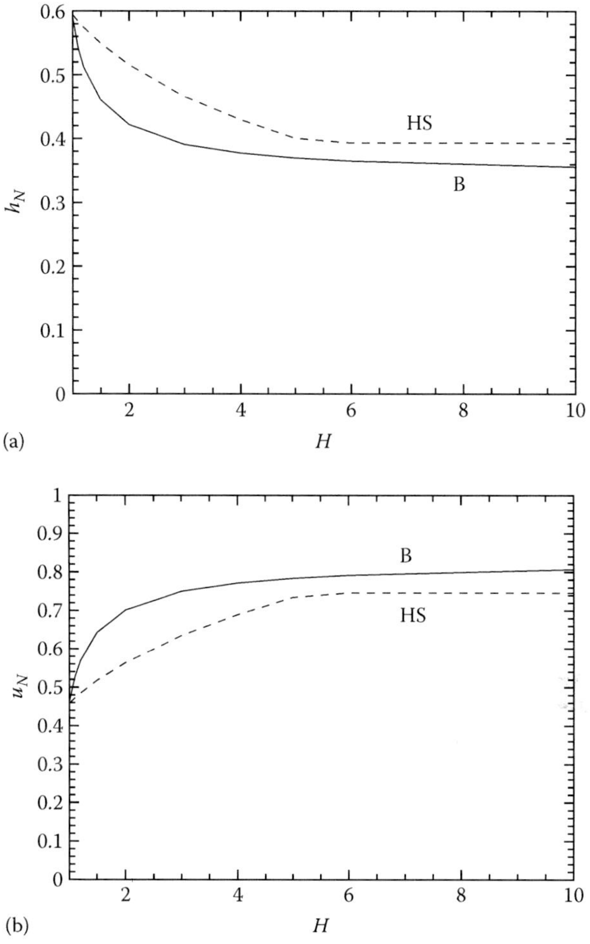

FIGURE 27.7 Nose height (scaled with h_0, upper frame) and speed (scaled with $(g'h_0)^{1/2}$, lower frame) as functions of ambient H (scaled with h_0), for Fr functions of Benjamin (B) and Huppert–Simpson (HS). Predicted by dam-break solution of the one-layer SW model for Boussinesq case.

following: the fluid in the reservoir is unperturbed for $x<x_L=1-t$; the parabolic-shaped interface LM domain satisfies

$$h=\frac{1}{9}\left(2-\frac{x-1}{t}\right)^2;\quad u=\frac{2}{3}\left(1+\frac{x-1}{t}\right);\quad (x_L\le x\le x_M)\qquad (27.23)$$

where $x_M = 1 + (2-3\sqrt{h_N})t$; then there is a rectangular domain of $h = h_N$ between x_M and $x_N = 1 + u_N t$; see Figure 27.6.

This solution becomes invalid after $t = 1$ when point L (which marks the propagation of the rarefaction wave into the reservoir) hits the backwall. However, the propagation with the aforementioned constant h_N, u_N prevails until the reflected wave reaches the nose at about $t = 3$. This marks the end of the slumping stage.

The solution considered earlier is for Boussinesq systems. The closely related solution of the classical dam break water-in-air (vacuum) solution can be recovered as follows. First, we replace g' by g in our scaling. The governing equations (27.16) and (27.17) are valid. However, for this $\rho_c/\rho_a\to\infty$ case the Boussinesq nose condition (27.18) must be reconsidered; the outcome is that the solution $u = 2(1-\sqrt{h})$ on the c_+ characteristic is correct until h diminished, and hence $h_N = 0$ and $u_N = 2$ appear [instead of the Boussinesq (27.22)]. With these changes, (27.23) recovers Saint-Venant dam break result. In particular, now points M and N coincide, and $x_N = 1 + 2t$.

27.3.1.4 Similarity Solution (Developed Motion)

Similarity (or self-similar) solutions are the result of some transformations which reduce the system of the PDEs dependent on x and t to a system of ODEs of a single variable, which we denote y (not to be confused with the Cartesian coordinate). Such a solution of the GC equations means that h and u attain and maintain a generic fixed shape, modified by some time-dependent scale factors, and the propagation is with t at some power. This occurs after some significant propagation, when the GC "forgets" the initial conditions of h and u. Here we present a short derivation of the essential results.

We seek an exact solution of (27.16), subject to $y = x/x_N(t)$, the boundary condition

$$\dot{x}_N = u(y=1,t) = Fr \cdot [h(y=1,t)]^{1/2}, \tag{27.24}$$

and prescribed total volume, $\mathcal{V} = \int_0^1 h(y,t)(x_N dy)$. The solution is relevant for large times after release, when the GC is thin and therefore Fr is a constant. We assume

$$x_N(t) = Kt^{\beta}; \quad h(y,t) = \varphi(t)\mathcal{H}(y); \quad u(y,t) = \dot{x}_N \mathcal{U}(y); \tag{27.25}$$

where K and β are some positive constants. The similarity coordinate is $y = x/(Kt^{\beta})$ in the range [0, 1]. The objective is to determine $\varphi(t)$, β, $\mathcal{U}(y)$, $\mathcal{H}(y)$, and K.

First, we use the boundary condition (27.24) to deduce $\varphi(t) = \beta^2K^2t^{2\beta-2} = \dot{x}_N^2$, and further use the total volume condition to find that $\beta = 2/3$. Next we substitute the postulated variables into the equations of motion (27.16). We obtain the continuity and momentum equations as

$$(\mathcal{U}-y)\mathcal{H}' / \mathcal{H} + \mathcal{U}' = 2(\beta^{-1}-1); \tag{27.26a}$$

$$\mathcal{H}' + (\mathcal{U}-y)\mathcal{U}' + (1-\beta^{-1})\mathcal{U} = 0; \tag{27.26b}$$

where the prime denotes y derivative. In general, the simplified system (27.26) must be solved numerically subject to appropriate boundary conditions (in the illustrated case, $\mathcal{U}(1) = 1$, $\mathcal{H}(1) = Fr^{-2}$). However, for our value of $\beta = 2/3$ the equations can be decoupled and simple solutions follow: $\mathcal{U}(y) = y$ and $\mathcal{H}(y) = Fr^{-2} + (y^2 - 1)/4$.

Finally, we use again the volume conservation. Letting $K = K_1\mathcal{V}^{1/3}$, we obtain

$$K_1 = \left[\frac{27Fr^2}{12-2Fr^2}\right]^{1/3}. \tag{27.27}$$

To be specific $K_1 = 1.61$ for $Fr = 1.19$ and 1.89 for $Fr = \sqrt{2}$. We summarize

$$x_N = K_1\mathcal{V}^{1/3}t^{2/3}; \quad h = \dot{x}_N^2\left[Fr^{-2} + \frac{(y^2-1)}{4}\right]; \quad u = \dot{x}_N y. \tag{27.28}$$

See Figure 27.2a. The numerical solution of (27.16) confirms that (27.28) develops from realistic initial conditions. The advantage is that (27.28) is an exact solution of the SW one-layer model; also, it applies to a given volume, $\mathcal{V}$ (scaled with h_0x_0), irrespective of the shape of the lock. The disadvantage is the "virtual origin" difficulty: the solution (27.28) admits the addition of any constant γ to t. For a practical use of the self-similar solutions with constant $\mathcal{V}$, it is necessary to determine this γ by applying an additional condition, such as $x_N(t_1) = K(t_1 + \gamma)^{2/3}$, where t_1 is a time when the self-similar behavior is relevant. (The simplification $\gamma = 0$ is often used, but this can be justified only for $t \gg 1$ after release.)

27.3.1.5 Axisymmetric Case

The axisymmetric GC is analyzed by an analogous one-layer model, with x replaced by r. The continuity equation (27.16a) contains the additional $-uh/r$ on the RHS, due to the diverging geometry (or curvature), and the momentum equation is unchanged. The system (27.16) remains hyperbolic. The characteristic balances on $c_\pm$, see (27.17), contain the additional curvature term $-uhdt/r$ on the RHS, which defies the simple integration used for the 2D case. The boundary conditions (27.18) through (27.21) remain relevant. The numerical solution can be obtained by the same method as for the 2D case (see Ungarish [2009], Section 6.1).

A simple analytical dam-break solution is not available. At release ($t = 0+$), the values h_N,u_N are like in (27.22), but the subsequent slumping motion is with decaying h_N and u_N.

A self-similar analytical solution of the form (27.25) (now $y = r/r_N(t)$) subject to (27.24) and a prescribed total volume, $\mathcal{V} = \int_0^1 h(y,t)(r_N^2 y dy)$, exists. Here, $\mathcal{V}$ is the volume per radian scaled with $r_0^2h_0$. In this case, due to the diverging geometry (or curvature terms) $\beta = 1/2$, and the reduced continuity equation (27.26a) contains the additional $-\mathcal{U}/y$ on the RHS; the momentum balance (27.26b) is unchanged. The calculations yield

$$r_N = K_1\mathcal{V}^{1/4}t^{1/2}; \quad h = \dot{r}_N^2\left[Fr^{-2} + \frac{(y^2-1)}{2}\right]; \quad u = \dot{r}_N y; \tag{27.29}$$

where $K_1 = [32Fr^2/(4 - Fr^2)]^{1/4}$ (=2.05 for $Fr = 1.19$ and 2.38 for $Fr = \sqrt{2}$) (see Figure 27.2b). Again, the "virtual origin" difficulty appears.

27.3.2 Lubrication-Type Models for Viscous (Re < 1) GC

The typical viscous GC propagates as a thin layer over a solid wall in a relatively thick ambient. The fluid sticks to the wall at $z = 0$, but the further-away fluid particles are set in motion by the pressure gradient like in a Couette flow. The shear $\rho_c \nu(\partial u/\partial z)$ diminishes to practically zero at the top of the current $z = h$, because the adjacent layer of ambient is thick (and/or less viscous) and cannot support shear. Simplifications of the type used in the lubrication theory (see Batchelor 1981) are relevant. The results also apply to Boussinesq and non-Boussinesq systems with non-large values of ρ_a/ρ_c.

Here we use *dimensional variables*. The pressure is hydrostatic in the z direction, and (27.10) is applicable. This x-driving effect is balanced by the shear, and hence we write

$$\frac{1}{\rho_c}\frac{\partial p}{\partial x} = g'\frac{\partial h(x,t)}{\partial x} = \nu\frac{\partial^2 u}{\partial z^2}, \tag{27.30}$$

where $g' = (\Delta\rho/\rho_c)g$. We integrate (27.30) twice to obtain u. We apply the conditions of no-slip on $z = 0$, and no-shear at $z = h(x, t)$. We find

$$u(x,z,t) = \frac{g'}{\nu}\frac{\partial h}{\partial x}\left(\frac{1}{2}z^2 - hz\right); \quad \bar{u}(x,t) = \frac{1}{h}\int_0^h u dz = -\frac{1}{3}\frac{g'}{\nu}h^2\frac{\partial h}{\partial x}. \tag{27.31}$$

The typical speed of the viscous GC is therefore $g'h_0^3/(\nu x_0)$.

The continuity equation (27.16a) (in which u is the z-averaged speed in the layer, i.e., $\bar{u}$) is a kinematic relationship, and hence remains valid for the viscous GC. Substitution of u from (27.31) into (27.16a) yields

$$\frac{\partial h}{\partial t} - \frac{1}{3}\frac{g'}{\nu}\frac{\partial}{\partial x}h^3\frac{\partial h}{\partial x} = 0. \tag{27.32}$$

This is a single PDE of parabolic type for $h(x, t)$, and therefore the solution is simpler than that of the counterpart SW hyperbolic system (27.16) which represents the inviscid GC. The boundary condition at $x = x_N(t)$ is also more straightforward: just $h_N = 0$. This follows from (27.31): a negative inclination $\partial h/\partial x$ is needed along the GC to push against the shear and maintain a positive u. However, h must have a $(x - x_N)^{1/3}$ singularity, to allow a finite u_N in spite of the vanishing h_N. Another physical condition which determines the behavior of the viscous GC is the prescribed total volume $\mathcal{V}$.

An appropriate self-similar solution of (27.32) can be obtained for the quite general case $\mathcal{V} = qt^\alpha$, with constant q and α (the source is at $x = 0$). We seek a solution of the form

$$x_N = Kt^\beta; \quad h = \frac{\mathcal{V}}{x_N}\mathcal{H}(y) = \frac{q}{K}t^{\alpha-\beta}\mathcal{H}(y); \tag{27.33}$$

where K and β are positive constants, and $y = x/x_N(t)$ is the similarity variable, in the range [0, 1]. Substitution into (27.32) gives $\beta = (3\alpha + 1)/5$, and renders an ODE of second order for $\mathcal{H}(y)$, with the $\mathcal{H} = C(1-y)^{1/3}$ for $y \to 0$ boundary condition. From this we calculate $\mathcal{H}(y)$, then use the given $\mathcal{V}$ to determine K.

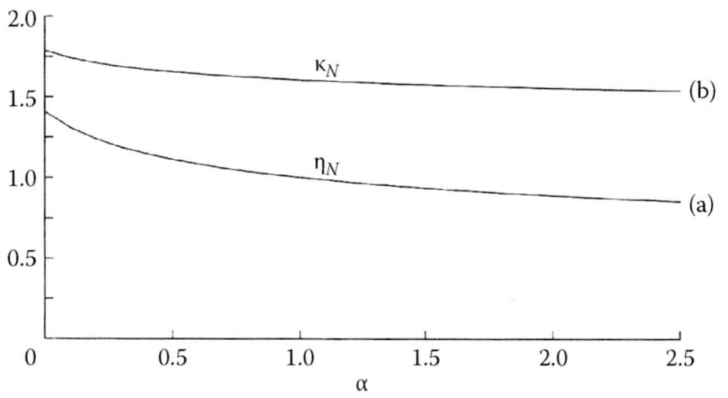

FIGURE 27.8 Coefficients for the self-similar viscous GC as functions of α. (a) η_N for the 2D case and (b) κ_N for the axisymmetric case. (From Huppert, H.E., *J. Fluid Mech.*, 121, 43, 1982. With permission.)

This closes the exact self-similar solution of the viscous GC of volume $\mathcal{V} = qt^\alpha$. (The speed can be calculated from (27.31).)

The main result is

$$x_N(t) = \eta_N(\alpha)\left[\frac{g'q^3}{(3\nu)}\right]^{1/5} t^{(3\alpha+1)/5}, \tag{27.34}$$

where $\eta_N(\alpha)$ is a coefficient of order 1 given in Figure 27.8. For $\alpha = 0$, the GC of fixed volume ($\mathcal{V} = q$) is recovered. This, again, poses the "virtual origin" difficulty: we can add a constant γ to t in (27.34). For practical use, we must determine this γ with an additional conditions, for example, $x_N(t_1) = K(t_1 + \gamma)^{1/5}$.

The interface profiles are smooth decreasing functions of y; see Figure 27.3. For $\alpha = 0$ an analytical result is available, and shows that, interestingly, u is linear with y.

The axisymmetric counterpart viscous GC can be solved by the same approach. Equations 27.30 and 27.31, with x replaced by r, are valid; then the continuity equation can be expressed as a parabolic PDE for $h(r,t)$. The curvature terms introduce some differences with (27.32), but the negative inclination of h, with decay to $h_N = 0$ and $(r - r_N)^{1/3}$ singularity remains. An exact similarity solution for a GC of volume $\mathcal{V} = qt^\alpha$ (per radian) can be obtained. The thickness profiles are displayed in Figure 27.3b; for $\alpha > 0$ the source domain near the center is, as expected, not well reproduced. The main result is

$$r_N(t) = \kappa_N(\alpha)[g'q^3/(3\nu)]^{1/8} t^{(3\alpha+1)/8}; \tag{27.35}$$

where $\kappa_N(\alpha)$ is a coefficient of order 1 given in Figure 27.8. For $\alpha = 0$, the GC of fixed volume ($\mathcal{V} = q$ per radian) is recovered, and, again, the "virtual origin" difficulty must be relaxed. Note the very slow spread with $t^{1/8}$: during a 10fold increase of t, r_N expands only 33%.

27.3.3 Box Models

A quick estimate of the behavior of the GC (in both inviscid and viscous regimes) can be obtained as follows. We postulate that the GC has a simple fixed shape, a well defined "box" of height

$h_N(t)$, length $x_N(t)$, and prescribed volume. The aim is to obtain simple equations, and solutions, for these variables. The first balance equation is $x_N(t)h_N(t) = \mathcal{V}(t) = qt^\alpha$. The second equation usually expresses the dynamic effect which governs the propagation. This box expands due to the action of global force balances which are either (1) buoyancy-inertial (i.e., inviscid) or (2) buoyancy-viscous. With some additional simplifications we can express the driving effect as an equation for $u_N = \dot{x}_N$ in terms of x_N and h_N $(=\mathcal{V}/x_N)$. By analytical or numerical integration of $\dot{x}_N$, we obtain $x_N(t)$. Remarkably, for many cases of interest an explicit solution of the form $x_N = Kt^\beta$ (with constant β and K) emerges. This is illustrated in the following (for Boussinesq systems). In this section we use *dimensional variables*.

The forces which act on the box GC are estimated as follows. The x-buoyancy force of the box, F_B, is given by the integral over h_N of the hydrostatic pressure excess, $\rho_c g'(h_N - z)$; we obtain $F_B = (1/2)\rho_c g' h_N^2$. Assume the simple $u = u_N x/x_N$. Now the integrals of $\rho_c u u_x h_N$ and $\rho_c \nu u/h_N$ over the GC domain $[0, x_N]$ provide the inertial and viscous forces as $F_I = (1/2)\rho_c u_N^2 h_N$ and $F_V = (1/2)\rho_c \nu u_N x_N/h_N$, respectively.

For the *inviscid*, or buoyancy-inertial, box-model solution, we use directly the nose-speed formula (27.12)

$$u_N = \dot{x}_N = Fr \cdot [g' h_N(t)]^{1/2}. \tag{27.36}$$

Assume a constant Fr (i.e., a deep GC). We recall the postulated $h_N = qt^\alpha/x_N(t)$ and assume $x_N = Kt^\beta$. Substitution of these simplifications into (27.36) yields the values of β and K. The main result for the deep 2D inviscid GC in the box model description is

$$x_N(t) = \left[\frac{3Fr}{(\alpha+2)}\right]^{2/3} [g'q]^{1/3} t^{(\alpha+2)/3}. \tag{27.37}$$

The value of Fr for this approximation is about 1. Note that for $\alpha = 0$ (with $q = \mathcal{V}$), (27.37) is in good agreement with the self-similar $x_N(t)$ given by (27.28); in this case, the box solution also has the virtual origin difficulty. $h_N(t)$ and $u_N(t)$ are easily calculated with (27.37).

The previous results also render the inertial to viscous force ratio behavior with t as

$$\frac{F_I}{F_v} = \frac{1}{\nu}\left[\left(\frac{\alpha+2}{3}\right)^7 \left(\frac{q}{Fr}\right)^4 \frac{1}{g'^2}\right]^{1/3} t^{(4\alpha-7)/3}. \tag{27.38}$$

The GC is expected to be in the viscous regime when $F_I/F_v < 1$ (roughly), and inertial otherwise. We notice an interesting behavior: the ratio F_I/F_V decreases with t for $\alpha < \alpha_{crit} = 7/4$, and increases with t for $\alpha > \alpha_{crit}$. This "critical α" prediction has been confirmed by experiments.

For the *viscous* GC box model description, let $F_B = F_V$. This yields $u_N = g'h_N^3/(\nu x_N)$. Combining with the volume balance and $x_N = Kt^\beta$, we obtain again explicit results for β and K. The $x_N(t)$ result is like the self-similar outcome (27.34), with $\eta_N = [15/(3\alpha+1)]^{1/5}$.

For the GC in *axisymmetric* geometry, a cylinder-box of volume $\mathcal{V} = (1/2)r_N^2(t)h_N(t) = qt^\alpha$ (per radian) is used. For the deep *inviscid* GC, (27.36) is employed, and manipulations like in the 2D case yield

$$r_N(t) = 2\left[\frac{Fr}{\alpha+2}\right]^{1/2} \left[2g'q\right]^{1/4} t^{(\alpha+2)/4}, \tag{27.39}$$

and

$$\frac{F_I}{F_V} = \frac{1}{\nu}\frac{(\alpha+2)^3}{32Fr^2}\frac{q}{g'}t^{\alpha-3}. \tag{27.40}$$

Now $\alpha_{crit} = 3$.

The *viscous* case box model r_N is like (27.35), with $\kappa_N = [288/(3\alpha+1)]^{1/8}$.

To summarize, the box model approach provides some simple and quick approximations to the spread, (mean) thickness, and force regime of the GC as functions of time. Additional effects (like porous bottom and particle settling) can be incorporated to obtain estimates of the appropriate influence (like runout time). We must keep in mind that this is a rough momentum-integral type of solution, based on a postulated (and usually unrealistic) behavior of the $h(x, t)$ and $u(x, t)$ variables, not a *bona fide* asymptotic model based on systematically simplified equations of motion. The fact that the box GC is governed by ODEs is not only computationally advantageous but also an indication that some essential physical features are lost. Indeed, the simple box description misses the constant-u_N slumping stage of the 2D inviscid GC, and the negative $\partial h/\partial x$ of the viscous GC. The solution of the one-layer PDEs formulation (SW or lubrication simplifications), if available, is more reliable and accurate than the box-model result. The latter may still be convenient for scaling arguments or for initial estimates of experimental and simulation data.

27.4 Applications

Applications may be concerned with predictions, analysis, assessment of risks, and prevention of environmental–geophysical flows and hazards.* The GC tools can be used to predict the arrival time of the front of an oil slick (see Hoult 1986), or the maximum spreading distance of a bulk of fluid which forms a lens in a rotating system. Another relevant application is the accidental release of widely stored and transported chemicals like propane, butane, and chlorine that generate bottom GCs. The prediction of $x_N(t)$ (or $r_N(t)$) requires a fair knowledge of the initial conditions and properties of the fluids which will form the GC and ambient. This information is not clear-cut in many practical problems. Consequently, various scenarios must be considered and solved for a given (real or potential) event; here the "insights" and the computational efficiencies of the GC tools are essential.

* Avalanches are a special case, discussed in another part of this book.

The GC tools are important in the backward analysis, or reconstruction, of past phenomena. For example, from the analysis of presently available geological gravity-surge deposits, which are believed to be the results of particle-driven (turbidity) GCs (either 2D or axisymmetric), it is possible to estimate the dimension and duration of the parent flows. Although this motion occurred a long time ago, we can now discuss with some confidence the necessary trigger events; see, for example, Dade and Huppert (1995).

The volcanic umbrella clouds can be viewed as axisymmetric GCs (intrusions) which spread radially (as from a point source) at the level of neutral buoyancy. The GC tools are essential for relating reliable field observations (such as height and speed of propagation from satellite images) to the source conditions at the vent (such as mass discharge); see, for example, Suzuki and Koyaguchi (2009). In the same spirit, the GC interpretations can be applied to giant volcanic ash clouds from "super-eruption" (the mass of the magma $>10^{15}$ kg). In this case the Coriolis effect of the Earth's rotation limits the spread to a lens-like structure, which helps to maintain the integrity again stratospheric winds; this, in turn, explains how these clouds can cover continent-sized areas, in agreement with geological evidence of deposited ash from known eruptions (Baines and Sparks 2005).

The theory of GC over a porous boundary can be applied to the safety assessment of chemical and fuel reservoirs embedded in gravel beds. The injection of liquid or gas into the ground is an important mechanism in large-scale processes like recovery of oil in natural reservoirs, and carbon capture and storage (CCS). The knowledge on GC inside a porous medium is applied for the analysis and optimization of the propagation effects of the injected fluid (Ungarish [2009], Section 11.3, Lyle et al. 2005; Neufeld and Huppert 2009).

27.5 Major Challenges

The major challenge is to enhance the applicability of the theoretical tools to practical environmental problems, and to increase the benefits gained by the practical user. This requires efforts in several directions. First, it is necessary to increase the exposure of students, at both undergraduate and graduate levels, to the topic of GCs and intrusions. This can be achieved by either a dedicated course, or as a major sub-topic in a course on environmental fluid dynamics; the availability of books facilitates the dissemination task.

Second, it is necessary to improve and assess the analytical and numerical tools. This is a topic for further research and development work. Some pressing issues are (1) for the GC in a stratified ambient (a) how to improve the SW model to account for the wave-head interaction, and for the return flow in the ambient (i.e., how to progress from the one-layer to a two-layer model); and (b) determine the meaning and practical implications of the multiple solutions of the front *Fr* (see Ungarish [2009], Section 14). (2) For the current in axially symmetric (cylindrical) propagation, develop 2D axisymmetric Navier–Stokes simulations which maintain stability for long distances. (3) The understanding of non-Boussinesq systems; here the analytical tools seem to be more advanced than the numerical and experimental body of knowledge, and a more balanced situation is needed for progress (see Ungarish 2010, 2011). (4) In general, clarify the "entrainment" effect. This is associated with the important topic of flow over an inclined bottom or topography, which, due to space limitation, has not been considered here (see Ungarish [2009], Section 9.6). In real environmental problems, the GC is mixed with other effects: not-calm ambient, heat transfer, evaporation and condensation, compressibility, under- and overflows, complex topography and obstacles, etc. There are numerous gaps of knowledge about these interactions. The modeling and prediction of these effects, as well as the extraction of the GC backbone from a composed phenomenon, are also of interest and significance.

References

Baines, P. G. and R. S. J. Sparks. (2005). Dynamics of volcanic ash clouds from supervolcanic eruptions. *Geophys. Res. Lett. 32,* L24808.

Batchelor, G. K. (1981). *An Introduction to Fluid Dynamics.* Cambridge University Press, London, U.K.

Benjamin, T. B. (1968). Gravity currents and related phenomena. *J. Fluid Mech. 31,* 209–248.

Dade, W. and H. Huppert. (1995). Runout and fine sediment deposits of axisymmetric turbidity currents. *J. Geophys. Res. 100,* 18, 597–18, 609.

Hallworth, M. A., H. E. Huppert, and M. Ungarish. (2003). On inwardly propagating axisymmetric gravity currents. *J. Fluid Mech. 494,* 255–274.

Hoult, D. P. (1986). Oil spreading on the sea. *Ann. Rev. Fluid Mech. 4,* 341–368.

Huppert, H. E. (1982). The propagation of two-dimensional and axisymmetric viscous gravity currents over a rigid horizontal surface. *J. Fluid Mech. 121,* 43–58.

Lyle, S., H. E. Huppert, M. Hallworth, M. Bickle, and A. Chadwick. (2005). Axisymmetric gravity currents in a porous medium. *J. Fluid Mech. 543,* 293–302.

Neufeld, J. A. and H. E. Huppert. (2009). Modelling carbon dioxide sequestration in layered strata. *J. Fluid Mech. 625,* 353–370.

Simpson, J. E. (1997). *Gravity Currents in the Environment and the Laboratory.* Cambridge University Press, Cambridge, U.K..

Sparks, R. S. J., R. T. Bonnecaze, H. E. Huppert, J. R. Lister, M. A. Hallworth, H. Mader, and J. C. Phillips. (1993). Sediment-laden gravity currents with reversing buoyancy. *Earth Planet. Sci. Lett. 114,* 243–257.

Suzuki, Y. J. and T. Koyaguchi. (2009). A three-dimensional numerical simulation of spreading umbrella clouds. *J. Geophys. Res. 114,* 1–18.

Ungarish, M. (2009). *An Introduction to Gravity Currents and Intrusions.* Chapman & Hall/CRC Press, New York.

Ungarish, M. (2010). The propagation of high-Reynolds-number non-Boussinesq gravity currents in axisymmetric geometry. *J. Fluid Mech. 643,* 267–277.

Ungarish, M. (2011). Two-layer shallow-water dam-break solutions for non-Boussinesq gravity currents in a wide range of fractional depth. *J. Fluid Mech. 675,* 27–59.

28

Internal Gravity Waves

Bruce R. Sutherland
University of Alberta

28.1 Introduction

Ocean waves move due to gravity: the water at crests is heavier than the air surrounding it and so it falls, overshoots its equilibrium position, and then at a trough feels an upward restoring force. The same phenomena occur below the ocean surface. Cold water lifted upward into warmer surroundings will feel a downward buoyancy force and downward-displaced warm water will feel an upward buoyancy force. If this motion is periodic in both space and time, it is referred to as an internal gravity wave. Internal gravity waves are manifest in two qualitatively different forms, which we will refer to here as "interfacial waves" and "internal waves." (The dynamics of internal waves are broadly discussed in the textbook "Internal Gravity Waves" (Sutherland 2010).)

Interfacial waves exist at the interface between dense and less dense fluid such as the thermocline, which refers to the interface between warm and cold water, or an atmospheric inversion, which refers to the interface between warm and cold air. In the ocean and in laboratory experiments, they can also exist at a halocline, which is the interface between fresh and salty water. An example of such a wave is shown in Figure 28.1a, in which a subsurface disturbance has launched a wave beneath a nearly flat surface. Looking down on the surface, the presence of these waves is evident from fluctuating horizontal flows and, on large scales, changing surface roughness.

Large-amplitude interfacial waves at the thermocline in the ocean and launched onto the continental shelf by tides provide enhanced transport of fluid and biology. For this reason, they are of interest to marine biologists, sediment geologists, as well as theoretical modellers.

The second form of internal gravity wave exists in continuously stratified fluid, which means that the effective density of the fluid gradually decreases with height. These waves again move up and down due to buoyancy, but they are not confined to an interface; they can move vertically as well as horizontally through the fluid. These are shown in Figure 28.1b in which model hills create a perturbation that launches waves moving upward through the fluid whose density decreases linearly with height as the salinity decreases.

Though not of primary importance, internal waves have a non-negligible influence upon weather and climate through the vertical transport of energy and momentum. In the atmosphere, for example, the waves launched by flow over mountains exert drag on the air far above where they break. In the ocean, wave breaking is a source of deep-ocean mixing, which is an important means of vertically redistributing heat. On the mesoscale, breaking large-amplitude internal waves are a source of clear-air-turbulence, which is a threat to air traffic.

The properties of periodic small-amplitude interfacial and internal waves are well-established, but less is known about the dynamics of large-amplitude waves. In this chapter, we will review the theory of these two types of waves having small and moderately large amplitude. The waves that influence the environment occur on small spatial and fast temporal scales, so that the effects of the Earth's rotation can be ignored.

Handbook of Environmental Fluid Dynamics, Volume One, edited by Harindra Joseph Shermal Fernando. © 2013 CRC Press/Taylor & Francis Group, LLC. ISBN: 978-1-4398-1669-1.

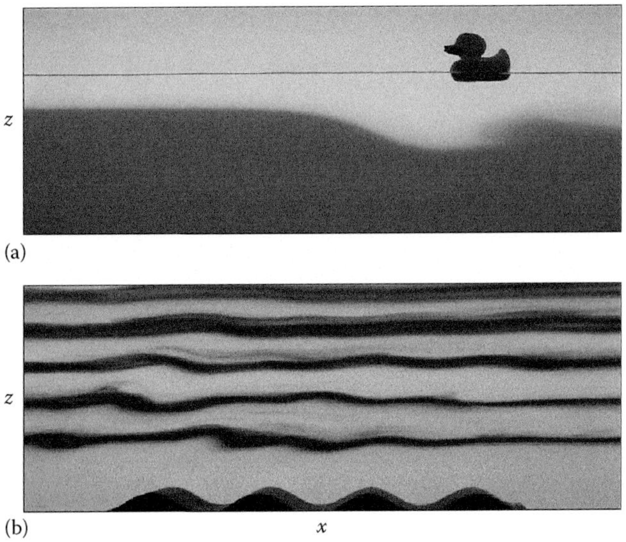

FIGURE 28.1 (a) Laboratory experiment in which the interface between (dark-dyed) salty fluid and (clear) freshwater is displaced by an interfacial internal wave that passes beneath a nearly flat surface. (b) Laboratory experiment in which towed model-hills launch vertically propagating internal waves in continuously stratified fluid. The waves are visualized by the displacement of dye-lines equally spaced by 5 cm.

28.2 Equations Describing Interfacial Waves

We will consider the propagation of interfacial waves in a two-layer fluid of upper depth H_1 and lower depth H_2, as shown in Figure 28.2. The density, ρ_1, of the upper layer is assumed to be moderately smaller than the lower-layer density, ρ_2. This allows us to make the Boussinesq approximation, in which the density affects buoyancy forces but not the momentum of the fluid.

An unsheared, uniform density, Boussinesq fluid is irrotational and incompressible. This means that the velocity everywhere in the fluid can be prescribed in terms of gradients of a velocity potential, $\phi(x, z, t)$, so that

$$(u,w)=\left(\frac{\partial\phi}{\partial x},\frac{\partial\phi}{\partial y}\right), \qquad (28.1)$$

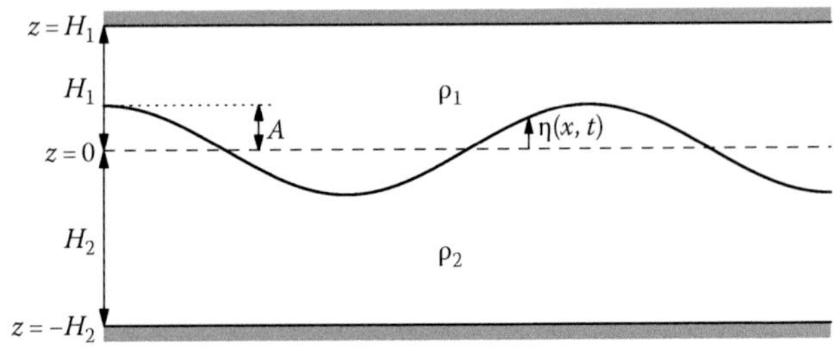

FIGURE 28.2 Schematic showing the definition of variables and scales used to model interfacial waves.

and ϕ itself satisfies

$$\frac{\partial^2\phi}{\partial x^2}+\frac{\partial^2\phi}{\partial z^2}=0. \qquad (28.2)$$

Insisting that the vertical velocity field is zero at the upper and lower boundaries, we also have the conditions

$$\left.\frac{\partial\phi}{\partial z}\right|_{z=-H_2}=\left.\frac{\partial\phi}{\partial z}\right|_{z=H_1}=0. \qquad (28.3)$$

The physics of buoyancy forces that drives the waves are included in interface conditions. Before invoking these, we can make some progress in solving the boundary value problem (28.2) with (28.3). Seeking solutions that are periodic in horizontal space and in time, we assume $\phi(x,z,t)=\hat{\phi}(z)\sin(kx-\omega t)$, in which $k = 2\pi/\lambda$ is the horizontal wavenumber for waves with wavelength λ and $\omega = 2\pi/T$ is the frequency for waves with period T. Substituting this expression into (28.2) is equivalent to a Fourier Sine transform. Thus we get an ordinary differential equation for $\hat{\phi}$:

$$-k^2\hat{\phi}+\frac{d^2\hat{\phi}}{\partial z^2}=0\,. \qquad (28.4)$$

General solutions are a superposition of the exponential functions e^{kz} and e^{-kz}. The particular combination of these functions that satisfies the boundary conditions (28.3) is written succinctly in terms of the hyperbolic cosine ($\cosh(\theta) = (e^{\theta} + e^{-\theta})/2$):

$$\phi=\begin{cases}A_1\cosh[k(z-H_1)]\sin(kx-\omega t) & z>0\\ A_2\cosh[k(z+H_2)]\sin(kx-\omega t) & z<0\end{cases}. \qquad (28.5)$$

Here we have assumed the amplitude of the waves is so small that it is sufficient to define ϕ above and below $z = 0$ rather than above and below the displaced interface at $z = \eta$. There is virtually no difference between $\phi(x,\eta,t)$ and $\phi(x,0,t)$ as η goes to zero.

Our problem reduces to finding the connection between the amplitudes A_1 and A_2 of the velocity potential in the upper and lower fluids as well as connecting these to the amplitude (and structure) of the interfacial displacement η.

The effects of buoyancy forces driving the wave motion are accounted for by Bernouilli's condition that $\rho(\partial_t\phi+(1/2)|\vec{u}|^2+g\eta)$ should not change across the interface. That is, there is no pressure jump across the interface. For small-amplitude waves, we can neglect the small kinetic energy term. Therefore, using (28.5) and evaluating ϕ at the height of the interface where $z \simeq 0$, we have

$$\eta(x,t)=-\frac{\omega}{g'}(A_1\cosh(kH_1)-A_2\cosh(kH_2))\cos(kx-\omega t), \qquad (28.6)$$

in which we have defined the reduced gravity to be $g' \equiv g(\rho_2 - \rho_1)/\rho_2$.

Another condition at the interface requires that there should be no jump in the vertical velocity, $w = \partial_z\phi$, across it. From (28.5), this relates the values A_1 and A_2 for small amplitude waves by the condition

$$-A_1 \sinh(kH_1) = A_2 \sinh(kH_2), \tag{28.7}$$

in which $\sinh\theta = (e^{\theta} - e^{-\theta})/2$ is the hyperbolic sine function. The vertical velocity field is also given in terms of the interface displacement for small-amplitude waves by $\partial_t\eta$. And so we have a second equation relating A_1 and A_2. Using (28.6) to equate $\partial_t\eta$ to $\partial_z\phi$ just above the interface gives

$$\frac{\omega^2}{g'}(A_1\cosh(kH_1) - A_2\cosh(kH_2)) = kA_1\sinh(kH_1). \tag{28.8}$$

Eliminating A_2 from (28.7) and (28.8) gives a single equation for A_1:

$$A_1\sinh(kH_1)\left[\frac{\omega^2}{g'}\left(\frac{\cosh(kH_1)}{\sinh(kH_1)} + \frac{\cosh(kH_2)}{\sinh(kH_2)}\right) - k\right] = 0.$$

Because we are assuming A_1, k, and H_1 are non-zero, we must have that the expression in square brackets is zero.

Thus we have determined the dispersion relation for an interfacial wave:

$$\omega^2 = g'k\left[\coth(kH_1) + \coth(kH_2)\right]^{-1}, \tag{28.9}$$

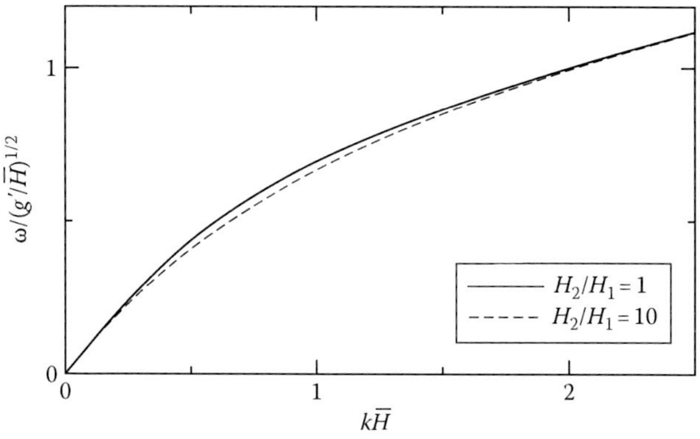

FIGURE 28.3 Plot of dispersion relation (28.9) for small-amplitude interfacial waves with $H_2/H_1 = 1$ and $H_2/H_1 = 10$ (or, equivalently $H_1/H_2 = 10$). Frequency is normalized by its characteristic value $\sqrt{g'/\bar{H}}$ and wavenumber by its characteristic value $1/\bar{H}$ in which $\bar{H} = H_1H_2/(H_1 + H_2)$ is the harmonic mean of the depths of the upper and lower layer fluids.

in which we have introduced the hyperbolic cotangent. This is plotted in Figure 28.3. Effectively we have just solved an eigenvalue problem. In its solution we have found that the frequency of the wave depends upon the wavenumber k. For a given value of the amplitude A_1, we can then find the value of A_2 through (28.7) and so find interfacial displacement (28.6). These and other polarization relations are listed in Table 28.1 in which the results have been recast to represent the relationship between fields in terms of the interfacial displacement amplitude A such that $\eta = A\cos(kx - \omega t)$.

TABLE 28.1 Relationships between the Interface Displacement (η), Velocity Potential (ϕ), the Horizontal (u) and Vertical (w) Components of Velocity, and the Vertical Displacement Field for Interfacial Waves in General and in the Long Wave Limit

	General Formula	Long Wave Limit ($k\bar{H} \ll 1$)
$\eta =$	$A\cos(kx - \omega t)$	$\rightarrow A\cos(kx - \omega t)$
$\phi =$	$\begin{cases} -A\dfrac{\omega}{k}\dfrac{\cosh[k(z-H_1)]}{\sinh(kH_1)}\sin(kx-\omega t) \\ A\dfrac{\omega}{k}\dfrac{\cosh[k(z+H_2)]}{\sinh(kH_2)}\sin(kx-\omega t) \end{cases}$	$\rightarrow \begin{cases} -A\dfrac{\omega}{k^2H_1}\sin(kx-\omega t) & z>0 \\ A\dfrac{\omega}{k^2H_2}\sin(kx-\omega t) & z<0 \end{cases}$
$u =$	$\begin{cases} -A\,\omega\dfrac{\cosh[k(z-H_1)]}{\sinh(kH_1)}\cos(kx-\omega t) \\ A\,\omega\dfrac{\cosh[k(z+H_2)]}{\sinh(kH_2)}\cos(kx-\omega t) \end{cases}$	$\rightarrow \begin{cases} -A\dfrac{\omega}{kH_1}\cos(kx-\omega t) & z>0 \\ A\dfrac{\omega}{kH_2}\cos(kx-\omega t) & z<0 \end{cases}$
$w =$	$\begin{cases} -A\,\omega\dfrac{\sinh[k(z-H_1)]}{\sinh(kH_1)}\sin(kx-\omega t) \\ A\,\omega\dfrac{\sinh[k(z+H_2)]}{\sinh(kH_2)}\sin(kx-\omega t) \end{cases}$	$\rightarrow \begin{cases} -A\,\omega\left(\dfrac{z}{H_1}-1\right)\sin(kx-\omega t) & z>0 \\ A\,\omega\left(\dfrac{z}{H_2}+1\right)\sin(kx-\omega t) & z<0 \end{cases}$
$\xi =$	$\begin{cases} -A\dfrac{\sinh[k(z-H_1)]}{\sinh(kH_1)}\cos(kx-\omega t) \\ A\dfrac{\sinh[k(z+H_2)]}{\sinh(kH_2)}\cos(kx-\omega t) \end{cases}$	$\rightarrow \begin{cases} -A\left(\dfrac{z}{H_1}-1\right)\cos(kx-\omega t) & z>0 \\ A\left(\dfrac{z}{H_2}+1\right)\cos(kx-\omega t) & z<0 \end{cases}$

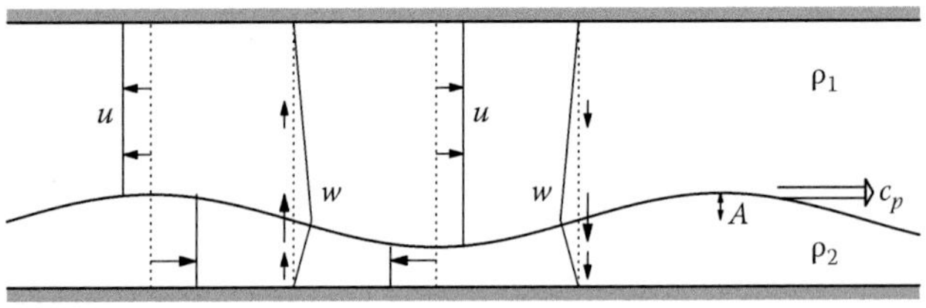

FIGURE 28.4 Schematic showing the velocity fields associated with a rightward-propagating shallow interfacial wave in a two-layer fluid. Here $H_1 = 0.75H$ and $H_2 = 0.25H$. For clarity of the illustration, the displacement amplitude has been exaggerated, though in reality such a value of A would constitute a large-amplitude interfacial wave.

Of particular interest are the properties of waves that are long compared to the fluid depth, so that both kH_1 and kH_2 are small. These are called "shallow water waves." For small θ, coth $\theta \simeq 1/\theta$. So the general dispersion relation (28.9) becomes

$$\omega^2 = c^2k^2 \tag{28.10}$$

in which $c = \sqrt{g'\bar{H}}$ and $\bar{H} = H_1H_2/(H_1 + H_2)$ is the harmonic mean of the two depths. So long waves are non-dispersive: all wavelengths travel at the same phase and group speed, c.

In the shallow-water limit the horizontal velocity is uniform within each layer but moves in opposite directions above and below the interface, as shown in Figure 28.4. The presence of shear at the interface suggests the possibility of breakdown into turbulence for sufficiently large amplitude waves (Troy and Koseff 2005). The vertical velocity changes linearly within each layer having maximum upward velocity at the inflection point leading the crest.

28.3 Equations Describing Internal Waves

Although it is often mathematically convenient to represent the thermocline as a sharp interface, for many wave phenomena the details of the ambient vertical density profile non-negligibly affect the structure and evolution of internal waves. In the extreme case of a uniformly stratified fluid, in which the density decreases linearly with height, the ambient fluid is no longer irrotational because the density gradients introduce vorticity within the ambient through so-called baroclinic torques.

The evolution of Boussinesq internal waves in the x–z plane is governed by the conservation of momentum and internal energy for an incompressible fluid:

$$\rho_0 \frac{Du}{Dt} = -\frac{\partial p}{\partial x}, \tag{28.11}$$

$$\rho_0 \frac{Dw}{Dt} = -\frac{\partial p}{\partial z} - g\rho, \tag{28.12}$$

$$\frac{D\rho}{Dt} = -w\frac{d\bar{\rho}}{dz}, \tag{28.13}$$

$$\frac{\partial u}{\partial x} + \frac{\partial w}{\partial z} = 0, \tag{28.14}$$

in which ρ_0 is the characteristic density, $\bar{\rho}\,(z)$ is the ambient density, and ρ is the fluctuation density resulting from the passage of waves. We have assumed there is no background flow so that the horizontal motion u is entirely due to waves.

If the waves are small amplitude, we may neglect the advective terms in the material derivative and so replace D/Dt with $\partial/\partial t$. The equations can then be combined into a single equation in one variable. Because the fluid is incompressible, we may define the streamfunction ψ so that $u = -\partial_z\psi$ and $w = \partial_x\psi$. Putting this into the momentum equations, taking the curl to eliminate pressure and finally eliminating ρ, we have the following:

$$\frac{\partial^2}{\partial t^2}\left(\frac{\partial^2}{\partial x^2} + \frac{\partial^2}{\partial z^2}\right)\psi + N^2\frac{\partial^2}{\partial x^2}\psi = 0, \tag{28.15}$$

in which $N^2(z) = -(g/\rho_0)d\bar{\rho}/dz$ is the squared buoyancy frequency.

We seek solutions that are periodic in horizontal space and time and so write $\psi(x,z,t) = \hat{\psi}(z)\cos(kx - \omega t)$. Substituting this into (28.15) gives the equation describing the vertical structure of the waves:

$$\frac{d^2\hat{\psi}}{dz^2} + k^2\left(\frac{N^2}{\omega^2} - 1\right)\hat{\psi} = 0. \tag{28.16}$$

The structure is oscillatory in z where $\omega < N$ and changes exponentially in z where $\omega > N$. For given ω and k the solution depends upon the upper and lower boundary conditions as well as the prescribed value of $N(z)$.

In a uniformly stratified fluid ($N = N_0$, constant), explicit analytic solutions can be found: $\hat{\psi}$ is the superposition of the functions $\sin(mz)$ and $\cos(mz)$ in which

$$m = k(N_0^2/\omega^2 - 1)^{1/2}. \tag{28.17}$$

In an unbounded fluid it is usual to form the superposition so that the streamfunction describes waves either with upward moving crests, $\psi(x,z,t) = A_{\psi+}\cos(kx + mz - \omega t)$, or with downward moving crests, $\psi(x,z,t) = A_{\psi-}\cos(kx - mz - \omega t)$. If the fluid is bounded above and below by horizontal boundaries, the superposition must be taken to ensure no vertical motion at top and bottom:

$$\psi(x,z,t) = A_{\psi i}\sin(m_i z)\cos(kx - \omega t), \quad i = 1,2,3,\ldots \tag{28.18}$$

in which the domain is assumed to extend from 0 to H, and $m_i = i\pi/H$ for positive integers i ensures that ψ, and hence, w is zero at $z = 0$ and H.

In either case, the dispersion relation for internal waves in uniformly stratified fluid is given by rearranging (28.17) to isolate ω:

$$\omega^2 = N_0^2 \frac{k^2}{k^2 + m^2} \tag{28.19}$$

in which m varies continuously for waves in an unbounded domain but is discrete-valued in a vertically finite domain. Whereas there is a single dispersion relation for interfacial waves with given k, internal waves have an infinite set of dispersion relations depending upon the value of m.

For a given mode, we may go on to find the velocity fields and other fields of interest. In particular, the vertical ($\hat{w}$) and horizontal ($\hat{u}$) velocities are proportional to $\hat{\psi}$. These polarization relations are listed in Table 28.2. For comparison with the polarization for interfacial waves, listed in Table 28.1, these have been cast in terms of the amplitude, A, of the vertical displacement field. This is the maximum displacement of isopycnals in the domain.

The dispersion relation for internal waves in a fluid of finite depth is shown in Figure 28.5a. The corresponding structure of the lowest two vertical modes is illustrated in Figures 28.5b and c through lines indicating the isopycnal displacements at snapshot in time.

For all vertical modes, the waves become nondispersive in the limit of long horizontal waves for which $k \ll m_i$. In this case the horizontal phase and group speed associated with the waves is $c = N_0/m_i = N_0H/(i\pi)$ for positive integers, i. The fastest speed is associated with the lowest ($i = 1$) mode.

The velocity fields associated with the mode-1 internal wave is illustrated in Figure 28.6. Here the thick black lines represent four initially evenly spaced isopycnals (lines of constant density) which have been lifted upward and downward from their equilibrium positions due to the passage of the waves. In an unbounded stratified fluid, the vertical and horizontal velocity fields associated with a plane wave are in phase. However, as with interfacial waves, the velocity fields of bounded internal wave modes are 90° out of phase. Comparing this figure with Figure 28.4, we see that the most significant qualitative difference is that u varies continuously with z and w varies smoothly with z.

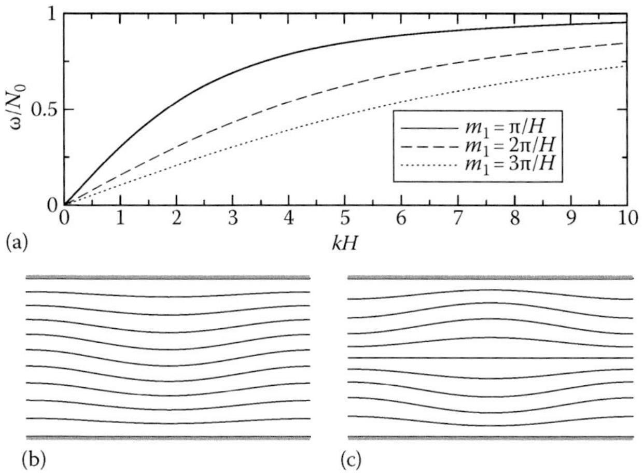

FIGURE 28.5 (a) Dispersion relation of lowest three vertical modes of internal waves in uniformly stratified fluid with buoyancy frequency N_0 in a domain of depth H. Isopycnal displacements are shown for (b) mode-1 and (c) mode-2 waves.

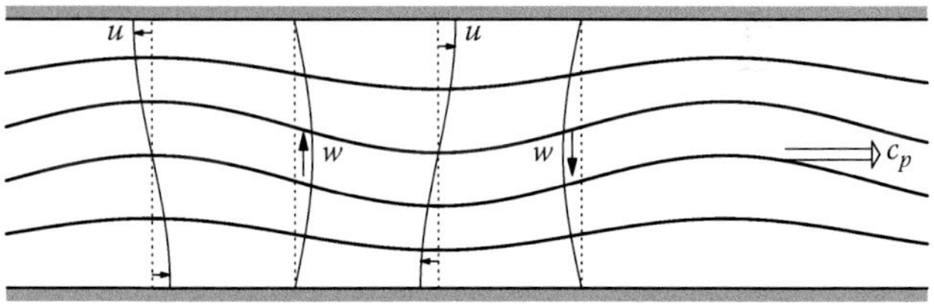

FIGURE 28.6 Schematic showing the velocity fields associated with a rightward-propagating mode-1 internal wave in a uniformly stratified fluid of finite depth.

TABLE 28.2 Relationships between the Isopycnal Displacement (ξ), Streamfunction (ψ), and the Horizontal (u) and Vertical (w) Components of Velocity for Internal Wave Modes in a Uniformly Stratified Fluid in a Channel with $0 \le z \le H$ and for Vertically Propagating Internal Waves in an Unbounded Domain

	Modes	Vertically Propagating Waves
$\xi =$	$A\sin(m_i z)\cos(kx - \omega t)$	$A\cos(kx + mz - \omega t)$
$\psi =$	$-A\dfrac{\omega}{k}\sin(m_i z)\cos(kx - \omega t)$	$-A\dfrac{\omega}{k}\cos(kx + mz - \omega t)$
$u =$	$-A\dfrac{\omega m_i}{k}\cos(m_i z)\sin(kx - \omega t)$	$-A\dfrac{\omega m}{k}\sin(kx + mz - \omega t)$
$w =$	$A\,\omega\sin(m_i z)\cos(kx - \omega t)$	$A\,\omega\sin(kx + mz - \omega t)$

The ith vertical mode has vertical wavenumber $m_i = i\pi/H$ for positive integers i.

We can draw the connection between interfacial waves and vertically bounded internal waves by considering modes in a cavity with ambient density given by

$$\bar{\rho}(z) = \begin{cases} \rho_1 & \sigma H_1 < z \le H_1 \\ \dfrac{\rho_1 H_2 + \rho_2 H_1}{H} - \dfrac{\rho_2 - \rho_1}{\sigma H} z & -\sigma H_2 \le z \le \sigma H_1 \\ \rho_2 & -\sigma H_2 > z \ge -H_2 \end{cases} \tag{28.20}$$

in which $H = H_1 + H_2$ is the total fluid depth and σ is a measure of the interface thickness. We recover the case of interfacial waves in the limit $\sigma \to 0$ and we recover the case of internal waves in uniformly stratified fluid as $\sigma \to 1$. The corresponding N^2 profile is zero in the upper and lower layers and holds the constant value

$$N_0 = \sqrt{\frac{g}{\rho_0}\frac{\rho_2 - \rho_1}{\sigma H}} = \sqrt{\frac{g'}{\sigma H}}, \tag{28.21}$$

for $-\sigma H_2 < z < \sigma H_1$, in which ρ_0 is the characteristic density.

Solving (28.16) in each layer and ensuring $w = 0$ (hence $\psi = 0$) at the upper and lower boundaries, we find

$$\hat{\psi}(z) = \begin{cases} A_1 \sinh[k(H_1 - z)] & \sigma H_1 < z \le H_1 \\ B_1 \sin(mz) + B_2 \cos(mz) & -\sigma H_2 \le z \le \sigma H_1 \\ C_1 \sinh[k(z + H_2)] & -H_2 \le z < -\sigma H_2 \end{cases}, \quad (28.22)$$

in which m is given by (28.17). The constants A_1, B_1, B_2, and C_1 are interrelated through conditions at $z = \sigma H_1$ and $z = -\sigma H_2$ that require continuity of pressure and vertical velocity. In the absence of background shear, this amounts to continuity of $\hat{\psi}(z)$ and its derivative. Thus we have an eigenvalue problem formed from the four equations in the four unknown constants.

After some algebra, we derive the following implicit formula for m:

$$\begin{aligned} \tan(mH\sigma) &= -\frac{m}{k}\left(\tanh[kH_1(1-\sigma)] + \tanh[kH_2(1-\sigma)]\right) \\ &\quad \times \left(1 - \frac{m^2}{k^2}\tanh[kH_1(1-\sigma)]\tanh[kH_2(1-\sigma)]\right)^{-1} \\ &\simeq -mH(1-\sigma)\left(1 - m^2 H_1 H_2 (1-\sigma)^2\right)^{-1}, \end{aligned} \quad (28.23)$$

where in the second expression we have made the long wave approximation ($kH \ll 1$). As $\sigma \to 1$, corresponding to the limit of uniform stratification, we have $\tan mH \simeq 0$. Therefore, m holds the values $m_i \equiv i\pi/H$, for integers i, as we found earlier. The corresponding dispersion relation is given by (28.19). As $\sigma \to 0$, corresponding to the limit of a two-layer fluid, we have $mH\sigma \simeq -mH/(1 - m^2 H_1 H_2) \simeq H/(mH_1H_2)$. That is, $m^2 H\sigma = 1/\bar{H}$, in which $\bar{H} = H_1H_2/(H_1 + H_2)$ is the harmonic mean of the upper- and lower-layer depths. Using this result in (28.19) with N_0 given by (28.21) and assuming $k \ll m$, we find $\omega^2 = g'\bar{H}k^2$, which is the dispersion relation for long waves given by (28.10).

In general, having found m through the empirical solution of (28.23), we can go on to find the interrelationship between the coefficients A_1, B_1, B_2, and C_1. Thus to within an arbitrary, though necessarily small, amplitude we have found the corresponding eigenfunction $\hat{\psi}(z)$. The polarization relations are then used to find the other fields of interest. In particular, from $w = \partial\psi/\partial x$ and the fact that the vertical displacement field ξ satisfies $w = \partial\xi/\partial t$, we have

$$\xi(x,z,t) = \hat{\xi} A \cos(kx - \omega t) \quad \text{with} \quad \hat{\xi}(z) = -\frac{k}{\omega}\hat{\psi}(z), \quad (28.24)$$

in which $\hat{\psi}$ is given by (28.22). Here we have defined A to be the maximum vertical displacement and it is implicitly assumed that the constants in $\hat{\psi}$ have been normalized so that $\hat{\xi}$ has a maximum value of unity.

Figure 28.7 shows the dispersion relation and vertical velocity amplitudes of the lowest vertical-mode internal waves for ambient profiles given by (28.20), with four different values of the interface thickness σH. So we see that in all three cases the vertical velocity is greatest in the interior of the domain with its value peaking closer to the mid-point of the interface at $z = 0$ as the interface becomes thinner ($\sigma \to 0$).

28.4 Dispersion of Small-Amplitude Waves

So far we have focused upon the structure and evolution of plane waves, with a single wavenumber and corresponding frequency. These are sometimes referred to as "monochromatic" waves. In reality, waves cannot be periodic out to infinity but have finite spatial extent. In many circumstances, however, the waves in a wavepacket have an approximately constant wavelength which is much smaller than the size of the wavepacket itself. This is

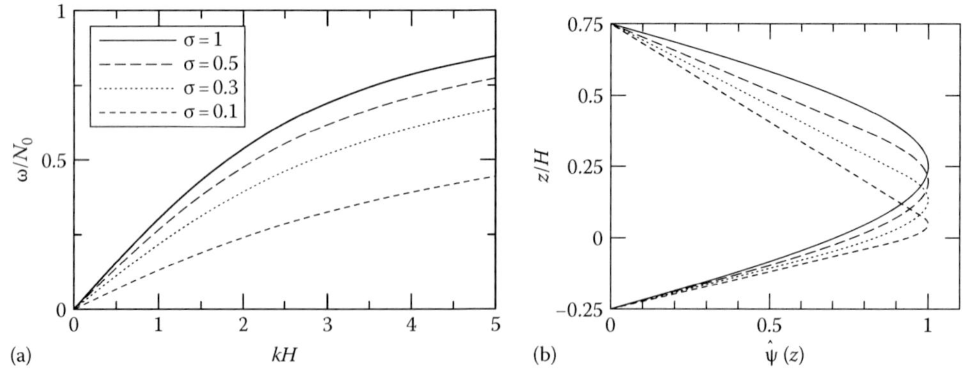

FIGURE 28.7 (a) Dispersion relation of the lowest mode vertically bounded internal waves in an ambient with density profile given by (28.20). The plots correspond to different interface thicknesses, σH with $H = H_1 + H_2$, as indicated. The upper and lower layers have relative depth $H_1/H = 0.75$ and $H_2/H = 0.25$, respectively. Frequency is normalized by the buoyancy frequency of the interface N_0 given by (28.21). (b) Streamfunction amplitudes for the lowest mode with $kH = 0.1$ normalized so that its maximum value is unity. Each plot corresponds to different values of σ as indicated in (a).

called a "quasi-monochromatic" wavepacket because it behaves similarly to monochromatic waves.

For simplicity, we will consider the evolution of a horizontally propagating packet of interfacial waves, though we will see in Section 28.6 how these ideas extend to describe vertically propagating internal wavepackets. The initial interface displacement field of a quasi-monochromatic wavepacket is represented by

$$\eta(x,0) = A(x,0)\cos(k_0 x),$$

in which the amplitude is not constant but varies in the horizontal as $A(x, t = 0)$. This describes the initial amplitude envelope of the wavepacket. For example, a Gaussian wavepacket centered at the origin has $A(x,0) = A_0\exp(-x^2/2L^2)$ in which L is the width of the wavepacket. In the limit of infinitely large L, $A(x,0) \to A_0$ and we recover the representation of plane waves having wavenumber k_0.

In what follows, it is more convenient to write η in terms of complex exponentials through

$$\eta(x,z,0) = \frac{1}{2}A(x,0)e^{\imath k_0 x} + c.c. = \Re\{Ae^{\imath k_0 x}\}, \qquad (28.25)$$

in which $c.c.$ denotes the complex conjugate of the first expression on the right-hand side of (28.25) and the symbol $\Re$ denotes the real-part of the expression it contains. Adding the complex conjugate has the effect of eliminating the imaginary part and doubling the real part. Because $e^{\imath\theta} = \cos\theta + \imath\sin\theta$ we recover $\eta = A\cos(k_0x)$ if A is real-valued. If $A = A_r + \imath A_i$ is complex-valued, then its argument $\phi(x) = \tan^{-1}(A_i/A_r)$ describes the variation in space of the phase of the waves.

The horizontal structure of the wavepacket can be seen as the superposition of plane waves through the Fourier transform

$$\eta(x,0) = A(x,0)e^{\imath k_0 x} = \int_{-\infty}^{\infty} \hat{\eta}(k)e^{\imath kx}\,dk, \qquad (28.26)$$

in which the amplitude of the plane wave with wavenumber k is

$$\hat{\eta}(k) = \frac{1}{2\pi}\int_{-\infty}^{\infty}\left(A(x,0)e^{\imath k_0 x}\right)e^{-\imath kx}\,dx. \qquad (28.27)$$

In (28.26) we have adopted the convention that η is the real part of the expression on the left-hand side without explicitly writing the symbol $\Re$.

For example, in the case of the Gaussian wavepacket we have

$$\hat{\eta}(k) = \left(\sqrt{2\pi}\,L\right)A_0\exp\left[-L^2\frac{(k_0-k)^2}{2}\right].$$

The amplitude is largest if $k = k_0$, the wavenumber of waves contained within the amplitude envelope. If $|k_0 - k| \gg 1/L$ the waves have negligible amplitude. So, although the wavepacket is a superposition of the whole spectrum of waves, only those with wavenumber near k_0 have any significant contribution to the structure of the waves.

For small-amplitude waves, each plane wave component of the wavepacket evolves in time through the change in the phase $-\omega t$. That is

$$\eta(x,t) = \int_{-\infty}^{\infty} \hat{\eta}(k)e^{\imath(kx-\omega t)}\,dk, \qquad (28.28)$$

in which $\hat{\eta}$ is given by the initial condition (28.27) and $\omega = \omega(k)$ is itself a function of k through the dispersion relation.

For nondispersive waves the dispersion relation is $\omega = ck$ with c constant. Substituting this into (28.28) and using (28.26), we immediately find $\eta(x, t) = \eta(x - ct, 0)$; the waves translate at speed c without change in structure of the initial wavepacket.

For dispersive waves the phase speed ω/k is not independent of k. So different wavelengths propagate at different speeds, and this results in a change of the shape of the wavepacket through a process referred to as "dispersion."

The effect of dispersion is described by the evolution of the amplitude envelope in time. We implicitly define the amplitude envelope, $A(x, t)$, of the evolving wavepacket by

$$\eta(x,t) = A(x,t)e^{\imath(k_0x-\omega_0 t)}, \qquad (28.29)$$

in which $\omega_0 = \omega(k_0)$ is the frequency of plane waves contained within the wavepacket, and it is understood η is the real part of the right-hand side of (28.29). Substituting this into (28.28) and isolating A gives an integral expression for the amplitude envelope:

$$A(x,t) = \int_{-\infty}^{\infty} \hat{\eta}(k)e^{\imath[(k-k_0)x-(\omega(k)-\omega_0)t]}\,dk, \qquad (28.30)$$

Given $\hat{\eta}(k)$ and the dispersion relation $\omega(k)$, the integral can be evaluated numerically and the interface displacement at any time is then given by $\Re\{A(x,t)\exp[\imath(k_0x - \omega_0 t)]\}$.

Insight into the evolution of A is found by noting that for a quasi-monochromatic wavepacket $\hat{\eta}$ is negligibly small except where $k \simeq k_0$. This inspires us to perform a Taylor series expansion of ω about $k = k_0$. Truncating at third-order in $k - k_0$ gives

$$\omega(k) \simeq \omega_0 + \omega'(k_0)(k-k_0) + \frac{1}{2}\omega''(k_0)(k-k_0)^2 + \frac{1}{6}\omega'''(k_0)(k-k_0)^3.$$

Substituting this into (28.30) gives a seemingly ugly expression. But nice things happen upon taking derivatives on both sides of the expression. Taking a time derivative brings a factor $-\imath\omega'(k-k_0) - (\imath/2)\omega''(k-k_0)^2 - (\imath/6)\omega''(k-k_0)^3$ into the integrand and taking successive x-derivatives brings factors $\imath(k-k_0)$, $-(k - k_0)^2$, and $-\imath(k-k_0)^3$ into the integrand. Thus we may combine these results to eliminate the integrals on the right-hand side, leaving an equation that relates time and space derivatives:

$$\frac{\partial A}{\partial t}+\omega'(k_0)\frac{\partial A}{\partial x}-\frac{\imath}{2}\omega''(k_0)\frac{\partial^2 A}{\partial x^2}-\frac{1}{6}\omega'''(k_0)\frac{\partial^3 A}{\partial x^3}=0. \quad (28.31)$$

The first two terms on the left-hand side describe the translation of the amplitude envelope at speed $\omega'(k_0)$. So we have shown that the wavepacket translates at the group speed $c_g = \omega'(k_0)$. The third term on the left-hand side of (28.31) shows that the amplitude-envelope changes in time where it has greater curvature. This describes the dispersion of the wavepacket. In some cases the coefficient $\omega''(k_0)\simeq 0$ in which case the fourth term on the left-hand side of (28.31) describes the dominant influence of dispersion.

Equation 28.31 is an extension of the well-known Schrödinger equation. The usual form of this equation neglects that last term on the left-hand side and is written in a frame of reference $X = x - c_g t$ moving at the group velocity. Hence

$$\frac{\partial A}{\partial t}=\frac{\imath}{2}\omega''(k_0)\frac{\partial^2 A}{\partial X^2}. \quad (28.32)$$

For a nondispersive wave with $\omega = ck$, $c_g = c$, and $\omega''(k) = 0$. So a wavepacket translates at the same speed as the phase speed and does not disperse. For interfacial waves that are long, but not too long, the dispersion relation (28.9) for rightward-propagating waves is given approximately for small k by

$$\omega\simeq ck-\frac{1}{6}cH_1H_2k^3, \quad (28.33)$$

in which c is the long wave speed, as given in the following (28.10). Substituting this into (28.31), thereby gives an equation describing the evolution of the amplitude envelope of moderately long interfacial waves:

$$\frac{\partial A}{\partial t}+c\frac{\partial A}{\partial x}+\frac{1}{6}cH_1H_2\frac{\partial^3 A}{\partial x^3}=0. \quad (28.34)$$

In this case, the dispersion is given by third-order spatial derivatives of A. For shorter waves, $\omega''(k_0)$ is non-zero so the second-order spatial derivatives in (28.31) dominate dispersion.

Likewise, for vertically propagating internal waves in uniformly stratified fluid, the dispersion in z of a horizontally periodic, vertically localized wavepacket with peak vertical wavenumber m_0 is given by the equation for the amplitude envelope $A(z, t)$:

$$\frac{\partial A}{\partial t}+\omega_m(m_0)\frac{\partial A}{\partial z}-\frac{\imath}{2}\omega_{mm}(m_0)\frac{\partial^2 A}{\partial z^2}=0. \quad (28.35)$$

Here the m subscripts denote derivatives of ω with respect to the vertical wavenumber, and ω given by the dispersion relation (28.19). The second term in (28.35) describes the vertical translation of the wavepacket at the group velocity $c_g = \partial\omega/\partial m$. Dispersion is given at leading order by the A_{zz} term which is valid provided $\omega_{mm} \neq 0$; the waves do not move near the fastest vertical group speed. Because the coefficient of the third term on the left-hand side of (28.35) is complex, we see that the curvature of the amplitude envelope first acts to change the relative phase of the waves and this then changes the magnitude of the amplitude.

28.5 Internal Solitary Waves

28.5.1 Solitary Waves in a Two-Layer Fluid

The term solitary wave originally referred to a single hump-shaped, moderately large amplitude wave that travels faster than the long wave speed and maintains its shape due to a balance between linear dispersion, which tends to spread out the wave, and nonlinear steeping, which tends to sharpen the wave crests. Its evolution in terms of the surface displacement, η, was originally formulated for surface waves through the Korteweg-de Vries (KdV) equation

$$\eta_t+c_0\eta_x+\frac{3c_0}{2H}\eta\,\eta_x+\frac{1}{6}c_0H^2\eta_{xxx}=0, \quad (28.36)$$

in which $c_0=\sqrt{gH}$ is the shallow water wave speed based upon gravity, g, and the water depth H. Consistent with (28.34), the second and fourth terms on the left-hand side of (28.36) describes the advection and linear dispersion of the wavepacket. The third (nonlinear) term of (28.36) describes the steepening of the wave, an effect that is larger for waves of larger amplitude.

The solution of (28.36) for waves that have no upstream or downstream disturbance is

$$\eta=A\,\mathrm{sech}^2\left(\frac{x-Ut}{\lambda}\right), \quad (28.37)$$

in which the width of the solitary wave is

$$\lambda=\sqrt{\frac{4H^3}{3A}} \quad (28.38)$$

and its speed is

$$U=c_0\left(1+\frac{1}{2}\frac{A}{H}\right), \quad (28.39)$$

as illustrated in Figure 28.8. This last formula confirms that solitary waves propagate faster than the fastest speed c_0 associated with long, small-amplitude waves.

In (28.37), A is the amplitude of the wave which measures the maximum deflection of the surface from its far-upstream depth. Unlike small-amplitude waves, the speed of a solitary wave

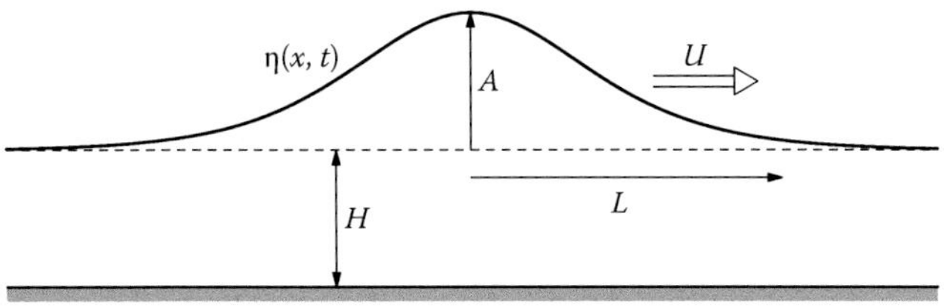

FIGURE 28.8 Schematic showing the structure of a solitary wave in a one-layer fluid of upstream-depth H. The surface displacement is represented by η with the maximum surface displacement being A.

increases as the amplitude increases. Also, the width of the wave changes as $H\sqrt{H/A}$; it has narrower extent as the amplitude increases relative to the fluid depth H.

The KdV equation has been adapted to describe internal solitary waves in stratified fluid (Benney 1966; Grimshaw et al. 2002). In this case the vertical displacement field η, which describes the displacement of isopycnals, is a function of z as well as x and t. Seeking the description of waves that propagate at constant speed, we assume η is separable so that we can write

$$\xi(x,z,t) = \hat{\xi}(z)\eta(x,t). \quad (28.40)$$

For example, in the small-amplitude limit for mode-1 waves in uniformly stratified fluid, we would have $\hat{\xi}(z) = \sin(m_1 z)$ and $\eta(x,t) = A\cos(kx - \omega t)$ (see Table 28.2). So, just as η describes the displacement of the interface for waves in a two-layer fluid, here η describes the maximum isopycnal deflection over the vertical extent of the domain. For the case of small-amplitude internal waves at a thick interface given by (28.20), the vertical displacement field is given by (28.24). That is $\hat{\xi}(z) = -(k/\omega)\hat{\psi}$ with $\hat{\psi}$ given by (28.22), normalized so that $\max(\hat{\xi}) = 1$.

For moderately large amplitude waves, we assume the vertical mode structure is unchanged from the linear theory values and so derive a formula that describes modification in $\eta(x,t)$ from sinusoidal behavior due to weakly nonlinear effects.

In general, the evolution equation is

$$\eta_t + c_0\eta_x + \alpha\eta\,\eta_x + \beta\eta_{xxx} = 0, \quad (28.41)$$

in which c_0 is the long wave speed from linear theory and the constants α and β are given in terms of vertical integrals over the domain of $\hat{\xi}$:

$$\alpha = \frac{3}{2}c_0\frac{\int \bar{\rho}(\hat{\xi}')^3\,dz}{\int \bar{\rho}(\hat{\xi}')^2\,dz} \quad \text{and} \quad \beta = \frac{1}{2}c_0\frac{\int \bar{\rho}\hat{\xi}^2\,dz}{\int \bar{\rho}(\hat{\xi}')^2\,dz}, \quad (28.42)$$

in which the primes denote z-derivatives. A generalization of these coefficient formulae that include background vertical shear was first determined by (Benney 1966) and later extended by Grimshaw et al. (2002) to include higher-order nonlinear forcing terms and allowing for a free surface. In the Boussinesq approximation, the ambient density profile, $\bar{\rho}(z)$, can be taken as constant so eliminating the density from the integrals altogether.

As for a one-layer fluid, the solution of (28.41) having no upstream or downstream displacement is given by (28.37) except now the width and speed of the solitary wave is given by

$$\lambda = \sqrt{\frac{12\beta}{A\alpha}} \quad (28.43)$$

and

$$U = c_0\left(1 + \frac{1}{3}A\alpha\right), \quad (28.44)$$

respectively. This can be confirmed through substitution of (28.37) into (28.41).

In particular, for a two-layer fluid, Table 28.1 gives $\hat{\xi} = 1 - z/H_1$, for $0 < z \le H_1$, and $\hat{\xi} = 1 + z/H_2$, for $-H_2 \le z \le 0$. So (28.42) gives

$$\alpha = \frac{3}{2}c_0\frac{-\frac{\rho_1}{H_1^2} + \frac{\rho_2}{H_2^2}}{\frac{\rho_1}{H_1} + \frac{\rho_2}{H_2}} \quad \text{and} \quad \beta = \frac{1}{6}c_0\frac{\rho_1 H_1 + \rho_2 H_2}{\frac{\rho_1}{H_1} + \frac{\rho_2}{H_2}} \quad (28.45)$$

For a one-layer fluid ($\rho_1 \to 0$), these reduce to $\alpha = 3c_0/2H_2$ and $\beta = c_0 H_2^2/6$, consistent with the coefficients in (28.36). Substituting these into (28.43) and (28.44) gives the width and speed and the solitary wave consistent with (28.38) and (28.39).

In the Boussinesq approximation, where the density difference between the upper and lower layers is small, (28.45) simplifies to

$$\alpha = \frac{3}{2}c_0\frac{H_1 - H_2}{H_1 H_2} \quad \text{and} \quad \beta = \frac{1}{3}c_0 H_1 H_2. \quad (28.46)$$

In particular, we see that $\alpha = 0$ if $H_1 = H_2$. That is, the effect of nonlinear steepening as it is captured by the KdV equation vanishes if the upper and lower layer depths are equal. This is true even if the interface between the two fluids has finite thickness, as shown in Figure 28.9.

Although the speed and spatial-scale of the wave changes with amplitude, the sech-squared shape does not. However, observations show that the crests of internal solitary waves tend to flatten as they grow to large amplitude. This effect has been captured in the so-called extended KdV equation, which includes higher-order linear and nonlinear dispersion effects (Lee and Beardsley 1974; Lamb and Yan 1996; Grimshaw et al. 2002).

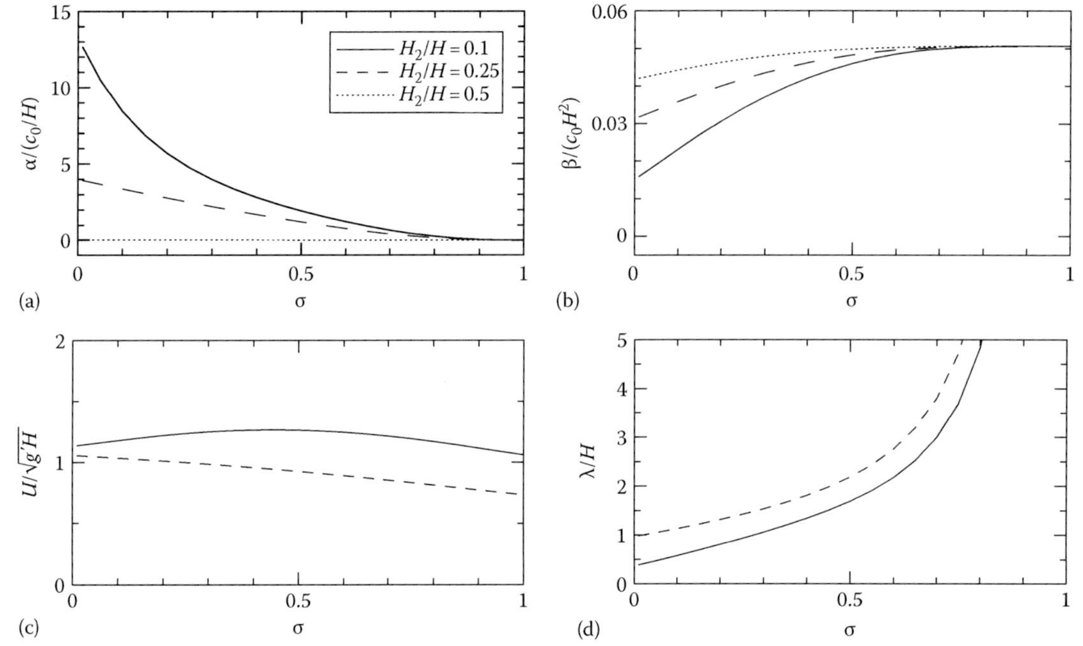

FIGURE 28.9 Numerically calculated coefficients of (a) α and (b) β in the KdV equation computed for the lowest mode of a shallow wave with $kH = 0.1$ in a fluid with ambient stratification given by (28.20). The corresponding (c) speed U computed for $A = 0.1H$ and (d) width λ of the wave. Values are given as a function of the relative thickness of the interface, σ, for different relative lower layer thicknesses H_2/H as indicted in (a). Note that the speed and width of the wave are shown only for $H_2/H = 0.1$ and 0.25, for which solitary wave solutions exist.

28.6 Weakly Nonlinear Internal Waves

In an unbounded stratified fluid, internal waves can propagate vertically as well as horizontally, as discussed for small-amplitude waves in Section 28.3. Here we will examine how the evolution of the waves changes when they have moderately large amplitude. We will focus upon 2D horizontally periodic internal waves in a uniformly stratified Boussinesq fluid that have limited vertical extent. Unlike horizontally propagating solitary waves which exist in steady state through a balance between nonlinear steepening and linear dispersion, we will show that weakly nonlinear effects cause the shape of the amplitude envelope of internal waves to change, either spreading out faster than linear dispersion predicts or narrowing and growing in amplitude.

As discussed in Section 28.4, the evolution of a horizontally periodic, vertically localized wavepacket can be described in terms of the change in the amplitude envelope $A(z, t)$ through (28.48). For a wavepacket with central wavenumber $\vec{k}_0 = (k_0, m_0)$ and initial vertical displacement field $\Re\{A(z,0)\exp[\imath(k_0x + m_0z)]\}$, the displacement at a later time is

$$\xi(x,z,t) = \Re\left\{A(z,t)e^{\imath(k_0x+m_0z-\omega_0t)}\right\} \quad (28.47)$$

in which $\omega_0 = \omega(k_0, m_0)$ is the dispersion relation given by (28.19) and the amplitude envelope changes in time according to (28.35).

In a frame of reference moving at the vertical group speed this equation is

$$\frac{\partial A}{\partial t} = \frac{1}{2}\gamma\frac{\partial^2 A}{\partial Z^2}, \quad (28.48)$$

in which $Z = z - c_g t$, $c_g \equiv \omega_m(m_0) = -N_0 m_0 k_0/|\vec{k}_0|^{3/2}$, and $\gamma \equiv \omega_{mm}(m_0) = N_0 k_0(2m_0^2 - k_0^2)/|\vec{k}_0|^{5/2}$.

Horizontally periodic internal waves induce a mean flow, akin to the Stokes drift of surface waves. It turns out that the weakly nonlinear evolution of vertically propagating internal waves is determined primarily through interactions between the waves and this mean flow, which we will denote by $\langle u \rangle_L$. The mean flow acts to advect the waves horizontally just as a background ambient flow would. Including the effects of this advection is done simply by replacing x in (28.47) with $x + \langle u \rangle_L t$. And so at leading-order, the time-derivative of ξ introduces a new term $\imath k_0 \langle u \rangle_L \xi$. This adjusts the evolution Equation 28.48 for the amplitude envelope so that

$$\frac{\partial A}{\partial t} = \frac{\imath}{2}\gamma\frac{\partial^2 A}{\partial Z^2} - \imath k_0\langle u \rangle_L A. \quad (28.49)$$

It remains to find an explicit expression for $\langle u \rangle_L$ in terms of the amplitude A. From the horizontal momentum Equation 28.11, if we expand out the material derivative and average horizontally over one wavelength we get $\partial_t\langle u \rangle_L = -\partial_z\langle uw \rangle$; the flow accelerates due to the divergence in the vertical flux of horizontal momentum per unit mass. In this case the flux changes with height because the amplitude envelope changes in the vertical. Far above and below the wavepacket there is no flux because there are no waves whereas the flux is large at the centre of the wavepacket. Knowing that the wavepacket moves upward at the group speed c_g, we can write $\partial_t \langle u \rangle_L + c_g \partial_z \langle u \rangle_L \simeq 0$. Thus we have $\langle u \rangle_L = \langle uw \rangle / c_g$. Finally, if we use the polarization relations in Table 28.2 that relate the horizontal and vertical velocity amplitudes for vertically propagating waves to the vertical displacement, we have

$$\langle u \rangle_L = -\frac{1}{2c_g} |A|^2 \, \omega_0^2 \frac{m_0}{k_0} = \frac{1}{2} N_0 \, | \vec{k}_0 \, | |A|^2 . \tag{28.50}$$

Putting this result into (28.49), we have the weakly nonlinear evolution equation for vertically propagating internal wavepackets

$$\frac{\partial A}{\partial t} = \frac{\imath}{2} \gamma \frac{\partial^2 A}{\partial Z^2} - \imath \omega_2 \, |A|^2 \, A, \tag{28.51}$$

in which $\omega_2 = N_0 \, | \vec{k}_0 \, | \, k_0/2$ is the finite-amplitude correction to the dispersion relation: $\omega = \omega_0 + \omega_2 |A|^2$. Equation 28.51, generally known as a "nonlinear Schrödinger equation," describes the evolution of finite-amplitude dispersive wavepackets.

Given an initial amplitude envelope $A(z,0)$, (28.51) describes how the amplitude and phase of the wavepacket changes in time through changes to the real and imaginary part of $A(z, t)$. Extracting the real part of $A(z,t)\exp\left[\imath(k_0 x + m_0 z - \omega t)\right]$ gives the vertical displacement field ξ. This is shown in Figure 28.10 for small and large amplitude wavepackets having relative vertical wavenumbers of $m_0 = -0.5k_0$ and $m_0 = -k_0$. With $k_0 > 0$, the negative sign assures that the wavepacket propagates upward.

The figure shows that wavepackets having initial maximum amplitude as small as 1% of the horizontal wavelength ($A_0 = 0.01\lambda_x$) undergo linear dispersion as would be predicted by (28.35). The influence of the nonlinear term in (28.51) is evident for waves having amplitude $A_0 = 0.05\lambda_x$. Depending upon the vertical wavenumber, the wavepacket either narrows and grows in amplitude ($m_0 = -0.5k_0$) or it broadens more quickly than linear theory predicts ($m_0 = -k_0$). In the former case the waves are said to be "modulationally unstable" and in the latter case the waves are "modulationally stable."

The transition from instability to stability occurs for wavepackets with m_0 satisfying $\omega_{mm}(m_0) = 0$. Explicitly, this occurs when $|m_0| = |k_0|/\sqrt{2}$, corresponding to waves of fixed k_0 moving at the fastest vertical group velocity (Sutherland 2006b).

The consequent dynamics of large amplitude internal waves is not as well captured by (28.51). The growth in amplitude of modulationally unstable wavepackets means that higher-order terms in the nonlinear Schrödinger equation, which we have neglected, have a non-negligible contribution to the wavepacket evolution. At later times still, wave–wave interaction give rise to superharmonic waves through what is known as parametric subharmonic instability (Sutherland 2006a). A detailed investigation of these weakly and fully nonlinear dynamics upon the evolution of large-amplitude internal wavepackets remains under investigation.

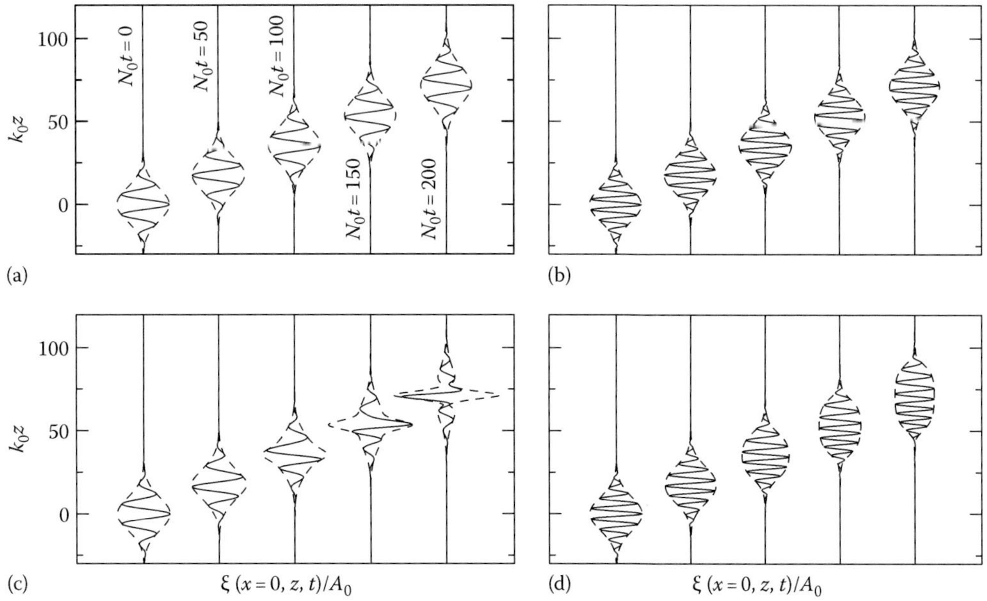

FIGURE 28.10 Numerically computed solution of (28.51) giving the evolution of a vertically propagating internal wavepacket with relatively small initial vertical displacement amplitude, $A_0 = 0.01/\lambda_x$, and central vertical wavenumber (a) $m_0 = -0.5k_0$ and (b) $m_0 = -k_0$, and with moderately large initial amplitude, $A_0 = 0.05/\lambda_x$ and wavenumbers (c) $m_0 = -0.5k_0$ and (d) $m_0 = -k_0$. The initial amplitude envelope is a Gaussian centered at the origin with width $L = 10k_0^{-1}$. The waves propagate upward in uniformly stratified fluid with buoyancy frequency N_0. In each plot, horizontally offset profiles of the normalized vertical displacement are shown at the five times indicated in (a).

28.7 Conclusions

In this chapter we have attempted to draw the connection and distinctions between finite-amplitude interfacial waves that propagate horizontally in a vertically confined domain and internal waves that propagate vertically in unbounded uniformly stratified fluid. The evolution of the former is given by the KdV equation 28.41 whereas that of the latter is prescribed by the nonlinear Schrödinger equation 28.51.

Generally, the KdV equation describes moderately long amplitude waves that have permanent form while translating at a speed moderately larger than the speed of long interfacial waves. Examination of this equation provides a useful starting point in understanding the dynamics of internal solitary waves such as those generated by tidal flows over sills and the continental shelf. However, the equation makes approximations that assume the amplitude is not too large. To describe internal solitary waves of large amplitude, higher-order terms can be included as in the Extended KdV equation, which predicts the wave crests flatten (Helfrich and Melville 2006). At larger amplitude still, the solitary waves can develop closed cores in which fluid is transported along with the wave. Such dynamics have been described by a different approach through the solution of the Dubreil-Jacotin-Long equation (Dubreil-Jacotin 1937; Long 1953; Brown and Christie 1998). These considerations lie beyond the scope of the material presented here.

Generally, the nonlinear Schrödinger equation describes the influence of large-amplitude effects upon dispersive waves. Rather than providing steady-state solutions for wavepackets, the equations predict that the amplitude envelope of the waves spread or narrow from their initial state. Spreading, modulationally stable, waves occur if $\omega_2\omega'' > 0$, in which ω_2 is the order amplitude-squared correction to the frequency ω predicted for small-amplitude waves. For internal waves, this occurs if the magnitude of the vertical wavenumber $|m_0|$ exceeds $|k_0|/\sqrt{2}$ in which $|k_0|$ is the magnitude of the horizontal wavenumber. The implications of this result remain to be explored. At first glance it would seem that large-amplitude internal waves should break at higher levels because the amplitude decreases as the wavepacket spreads nonlinearly. However, nonlinear effects neglected by the nonlinear Schrödinger equation may introduce other modes of instability that could cause breaking at lower levels. Whatever the case these results show that wave breaking in the atmosphere assessed by extrapolation of linear theory, likely incorrectly predicts where momentum is actually deposited by the waves.

References

Benney, D. J. 1966. Long nonlinear waves in fluid flows. *J. Math. Phys.*, 45:52–63.

Brown, D. J. and Christie, D. R. 1998. Fully nonlinear solitary waves in continuously stratified incompressible Boussinesq fluids. *Phys. Fluids*, 10:2569–2586.

Dubreil-Jacotin, M. L. 1937. Sur les théoremes d'existence relatifs aux ondes permanentes pérodiques à deux dimensions dans les liquides hétérogènes. *J. Math. Pures Appl.*, 16:43–67.

Grimshaw, R. H. J., Pelinovsky, E., and Poloukhina, O. 2002. Higher-order Korteweg-de Vries models for internal solitary waves in a stratified shear flow with a free surface. *Nonlinear Proc. Geophys.*, 9:221–235.

Helfrich, K. R. and Melville, W. K. 2006. Long nonlinear internal waves. *Annu. Rev. Fluid Mech.*, 38:395–425.

Lamb, K. G. and Yan, L. 1996. The evolution of internal wave undular bores: Comparison of a fully-nonlinear numerical model with weakly nonlinear theories. *J. Phys. Oceanogr.*, 26:2712–2734.

Lee, C.-Y. and Beardsley, R. 1974. The generation of long nonlinear internal waves in a weakly stratified shear flow. *J. Geophys. Res.*, 79:453–462.

Long, R. R. 1953. Some aspects of the flow of stratified fluids: A theoretical investigation. *Tellus*, 5:42–58.

Sutherland, B. R. 2006a. Internal wave instability: Wave-wave vs wave-induced mean flow interactions. *Phys. Fluids.*, 18:Art. No. 074107. doi:10.1063/1.2219102.

Sutherland, B. R. 2006b. Weakly nonlinear internal wavepackets. *J. Fluid Mech.*, 569:249–258.

Sutherland, B. R. 2010. *Internal Gravity Waves*. Cambridge University Press, Cambridge, U.K.

Troy, C. D. and Koseff, J. R. 2005. The instability and breaking of long internal waves. *J. Fluid Mech.*, 543:107–136.

29

Rotation Effects in Environmental Flows

Peter A. Davies
University of Dundee

Yakun Guo
University of Aberdeen

29.1 Context

For the present purposes, *Environmental Fluid Mechanics* (and the role of rotation in environmental flows) may be considered conveniently in terms of fluid motions and associated mass and heat transport processes that occur in the Earth's atmosphere and hydrosphere on local or regional scales. In particular, and to distinguish such flows from those within the larger (global) domain covered by geophysical fluid dynamics (see, e.g., Cushman-Roisin 2004; McWilliams 2006; Vallis 2006), emphasis is placed on the influence of the flows upon (and their interaction with) man-made facilities and structures and their responses to anthropogenic releases of mass and heat. Though the effects of the Earth's rotation are known to be important for the global circulations of the atmosphere and oceans (see, e.g., Gill 1982), smaller-scale environmental flows may also be affected by the presence of the Coriolis acceleration associated with the background rotation of the Earth. Examples (Bowden 1983; Mann and Lazier 1991; Rubin and Atkinson 2001) are (a) tidal flows in coastal waters, (b) estuarine and river outflow plumes, (c) waves and currents in large lakes and reservoirs, (d) boundary layer flows and sediment transport in coastal regions, and (e) coastal upwelling, though there are many others. Many important global-scale processes (e.g., the oceanic transport of pollutants (Dahlgaard 1995; Kershaw and Baxter 1995)) have environmental consequences on a local scale (and *vice versa*), but they are not included herein. For hydrospheric flows, attention is limited to rotationally-influenced processes operating inshore of the continental shelf break in the marine/estuarine environment or within the interiors/boundary layers of large lakes and reservoirs in the freshwater environment. Rotational effects associated with atmospheric flows (and the atmospheric boundary layer, in particular) are treated elsewhere in this volume and will not be considered explicitly here.

It is useful to note at this stage that confusion can arise over definitions of length scales for environmental flows (Boyer and Davies 2000), particularly with regard to the use of the terms *global-scale* and *meso-scale* motions (Røed 1996). The latter refer conventionally to horizontal length scales between 10 and 100 km in extent; for the atmosphere, such scales are typically much less than the internal Rossby deformation radius (see the following), whereas for estuarine, coastal, and near-shore marine phenomena, these length scales are comparable with the internal Rossby radius.

29.2 Fundamental Aspects

The equations of continuity, momentum (Navier–Stokes equation), and mass conservation for an incompressible fluid in a rotating frame are conveniently written (e.g., Gill 1982; Pedlosky 1987) in Boussinesq form as

$$\nabla \cdot \boldsymbol{u} = 0 \tag{29.1}$$

$$\rho_0\left[\frac{\partial \boldsymbol{u}}{\partial t} + (\boldsymbol{u}\cdot\nabla)\boldsymbol{u} + 2\boldsymbol{\Omega}\times\boldsymbol{u}\right] = -\nabla(P) + \rho\boldsymbol{g} + \rho_0\nu\nabla^2\boldsymbol{u} \tag{29.2}$$

$$\frac{\partial \rho}{\partial t} + (\boldsymbol{u}\cdot\nabla)\rho = \kappa_D\nabla^2\rho \tag{29.3}$$

Handbook of Environmental Fluid Dynamics, Volume One, edited by Harindra Joseph Shermal Fernando. © 2013 CRC Press/Taylor & Francis Group, LLC. ISBN: 978-1-4398-1669-1.

where the fluid has density $\rho(x,y,z,t)$, kinematic viscosity ν, and mass diffusion coefficient κ_D and motions are referred to a Cartesian frame of reference (x,y,z) rotating uniformly about a vertical (z) axis with angular velocity $\boldsymbol{\Omega}$ $(=0,0,\Omega)$. Here, $\boldsymbol{u}$ $(=u, v, w)$ is the fluid velocity, t is time, $\boldsymbol{g}$ $(=0,0,-g)$ is the gravitational acceleration, and P is the reduced pressure defined by $P = p - (1/2)\rho_0\Omega^2r^2$, with r being the radial distance from any point in the fluid to the axis of rotation and p being the pressure at that point. The term $(1/2)\rho_0\Omega^2r^2$ is the so-called centrifugal acceleration and for most environmental flows plays no dynamical role, so long as it does not cause the gravitational acceleration vector to be significantly nonvertical. The important additional term associated with the transformation from a non-rotating to a rotating frame of reference is the Coriolis acceleration term $2\boldsymbol{\Omega} \times \boldsymbol{u}$.

For most environmental flows of interest here (and even for large-scale atmospheric and oceanic flows), the dominant terms in (29.2) are the gravitational force per unit volume term $\rho\boldsymbol{g}$ and the vertical component of the pressure gradient term $-\nabla P$. Accordingly, it is convenient to decompose the pressure and density terms into equilibrium and perturbation components (assuming that centrifugal effects can be neglected) as

$$P = p_0(z) + p' \tag{29.4}$$

$$\rho = \rho_0(z) + \rho' \tag{29.5}$$

with the equilibrium values p_0 and ρ_0 being related by the hydrostatic equation

$$\frac{dp_0}{dz} = -\rho_0 g \tag{29.6}$$

an equation that is satisfied here, since $\partial w/\partial t \ll g$ for the environmental flows of interest.

Following the aforementioned decomposition, Equations 29.1 through 29.3 become

$$\nabla \cdot \boldsymbol{u} = 0 \tag{29.7}$$

$$\rho_0\left[\frac{\partial \boldsymbol{u}}{\partial t} + (\boldsymbol{u}\cdot\nabla)\boldsymbol{u} + 2\boldsymbol{\Omega}\times\boldsymbol{u}\right] = -\nabla P' + \rho'\boldsymbol{g} + \rho_0\nu\nabla^2\boldsymbol{u} \tag{29.8}$$

$$\frac{\partial \rho'}{\partial t} + (\boldsymbol{u}\cdot\nabla)\rho' = \kappa_D\nabla^2\rho' \tag{29.9}$$

29.2.1 Coriolis Term

The magnitude of the Coriolis acceleration term $2\boldsymbol{\Omega} \times \boldsymbol{u}$ does not depend upon position r, but only upon the magnitude of both Ω and $\boldsymbol{u}$. It acts in the horizontal plane in a direction at right angles to the direction of $\boldsymbol{u}$. In the northern hemisphere it causes any horizontal flow to deflect to the right—the larger the velocity the larger the Coriolis acceleration. Note that it is a so-called fictional acceleration; the associated Coriolis force does no work.

29.2.2 Spherical Earth

The preceding equations are derived for a Cartesian frame in which the gravity acceleration vector $\boldsymbol{g}$ and the angular velocity vector $\boldsymbol{\Omega}$ are coincident and antiparallel. On a rotating spherical Earth, no such alignment occurs (except at the poles) and approximations are introduced to accommodate this effect in the dynamics of the flow. If a local Cartesian coordinate system is adopted for any latitude φ, with z the local vertical and x and y the local northward and eastward directions respectively, then

$$\boldsymbol{\Omega} = (\Omega_x, \Omega_y, \Omega_z) = (0, \Omega\cos\varphi, \Omega\sin\varphi) \tag{29.10}$$

and the Coriolis acceleration $2\boldsymbol{\Omega} \times \boldsymbol{u}$ may be written in component form as

$$2\boldsymbol{\Omega}\times\boldsymbol{u} = 2\boldsymbol{\Omega}(w\cos\varphi - v\sin\varphi, u\sin\varphi, -u\cos\varphi) \tag{29.11}$$

Since hydrospheric and atmospheric flows occur in relatively thin layers ($H/L \ll 1$, where H and L represent, respectively, typical vertical and horizontal scales of the flow in question), application of this condition to the continuity equation (29.7) allows the assumption $w \ll (u, v)$ to be made, such that (away from the equator $\varphi = 0$)

$$w\cos\varphi \ll v\sin\varphi \tag{29.12}$$

and the Coriolis acceleration term in (29.8) becomes

$$2\boldsymbol{\Omega}\times\boldsymbol{u} = (-fv, +fu, -fu\cot\varphi) \tag{29.13}$$

where

$$f = 2\Omega\sin\varphi \tag{29.14}$$

is the so-called Coriolis parameter, incorporating the latitudinal variation in the Coriolis acceleration. (Note that the z component (namely, $-fu \cot \varphi$) of the Coriolis acceleration is many orders of magnitude smaller than the gravitational acceleration $-g$, so may be neglected).

For environmental flows that are of limited latitudinal extent, the Coriolis parameter f can be regarded as a constant—the so-called f-plane approximation. For cases in which the environmental flow occupies a significant latitudinal range (e.g., poleward-flowing coastal currents, coastal upwelling flows), the variation of f within the latitudinal range has to be taken into account. This is achieved by expanding f about a reference latitude φ_0 to give (Gill 1982; van Heijst 1994)

$$f(\varphi) = f_0 + \beta y - \gamma y^2 + O(y^3) \tag{29.15}$$

where

$$f_0 = 2\Omega\sin\varphi_0,\quad \beta = \left(\frac{1}{R}\right)(2\Omega\cos\varphi_0)\quad \text{and}\quad \gamma = \left(\frac{1}{R^2}\right)\Omega\sin\varphi_0 \tag{29.16}$$

Usually, it is sufficient only to include the β term for such flows and to ignore higher-order terms in (29.15) (the beta plane approximation). The approximations in effecting the aforementioned transformation of the equations have been formalized by Veronis (1963).

In the laboratory, flows for which the *f*-plane approximation is valid can be modeled with a container of constant depth. Pedlosky and Greenspan (1967) showed that the beta (β) effect, namely the creation of relative vorticity within the beta plane by poleward changes in the horizontal Coriolis acceleration, can be modeled in laboratory flows by incorporating a sloping bottom on the container (van Heijst 1994) to induce vortex stretching by motion across bottom contours. The conditions under which the sloping bottom configuration mimics the beta effect have been analyzed by Beardsley (1969), Baker and Robinson (1967), and Robinson (1970).

29.2.3 Diffusion

The Navier–Stokes and density equations, as written in Equations 29.8 and 29.9, respectively, include terms that represent the diffusion of momentum and mass by viscous action and molecular diffusion, respectively. Such formulations are applicable to laminar flows. The environmental flows of interest here are turbulent and the relevant coefficients ν and κ_D are then conveniently replaced by eddy viscosity $A_{V,H}$ and eddy diffusivity $D_{V,H}$ coefficients, respectively, that have values significantly greater than ν and κ_D. These coefficients are representations of the action of turbulence and are not physical properties of the fluid. The values of the eddy viscosity and eddy diffusivity coefficients to apply to environmental flows are difficult to assign (and measure).

29.2.4 Dimensional Parameters

In order to gauge the circumstances under which rotational effects will become significant in any given flow, it is convenient to estimate the order of magnitude of the terms in the momentum equation (29.8) by introducing typical velocity and horizontal length scales *U* and *L*, respectively. Then

$$\text{Unsteady term:}\quad \left|\rho_0\left[\frac{\partial \boldsymbol{u}}{\partial t}\right]\right| \sim \frac{\rho_0 U}{\tau} \tag{29.17a}$$

$$\text{Advective (inertial) term:}\quad |\rho_0[(\boldsymbol{u}\cdot\nabla)\boldsymbol{u}]| \sim \frac{\rho_0 U^2}{L} \tag{29.17b}$$

$$\text{Coriolis term:}\quad |\rho_0[2\boldsymbol{\Omega}\times\boldsymbol{u}]| \sim 2\rho_0\Omega U \tag{29.17c}$$

$$\text{Pressure gradient term:}\quad |[\nabla p']| \sim \frac{(\Delta P)}{L} \tag{29.17d}$$

$$\text{Gravitational (buoyancy) term:}\quad |[\rho'\boldsymbol{g}]| \sim (\Delta\rho)g \tag{29.17e}$$

$$\text{Viscous term:}\quad |[\rho_0\nu\nabla^2\boldsymbol{u}]| \sim \frac{\rho_0\nu U}{L^2} \tag{29.17f}$$

Here, Δρ and (Δρ) are typical pressure difference and density difference scales, respectively. From these equations, the individual terms can be compared in magnitude with the Coriolis term to provide the dimensionless parameters $T = 1/(2\Omega\tau)$; $Ro = U/2\Omega L$ (the Rossby number); $\Pi = (\Delta P)/2\Omega LU$; $B = (\Delta\rho)g/2\rho_0\Omega U$; $Ek = \nu/2\Omega L^2$ (the Ekman number).

For rotational effects to be dynamically dominant, it is reasonable to suppose that each of the preceding terms must be $O(10^{-1})$ or less, except for the pressure gradient term that typically adjusts to balance the largest of the other terms (in this case, the Coriolis term). Rotation effects will be important when the time scale of a given flow is larger than the rotation period ($\sim\Omega^{-1}$) such that $T \ll 1$ and the flow may be regarded as steady. For a typical river, estuarine flow or coastal current, with a flow velocity of, say, $U = 10^{-1}$ m s^{-1}, at a mid-latitude location with the Earth's angular velocity $\Omega_{Earth} = 7.27 \times 10^{-5}$ rad s^{-1}, a typical value of *Ro* would be $O(10^{-1})$ or less only if $L > 9$ km. (We show later that estimates of length scales based upon the Rossby deformation radius provide a better estimate of the conditions for which rotation plays an important dynamical role in the environmental flow dynamics context).

Estimates of the relevant Ekman number *Ek* for environmental flows of the aforementioned type depend critically upon the value of the eddy viscosity used in the numerator. Typical values vary from about 10^{-3} m^2 s^{-1} (Maas and van Haren 1987) to 3.0×10^{-2} m^2 s^{-1} (Nihoul et al. 1978) to 1×10^2 m^2 s^{-1} (Bowden 1983) for shallow seas, estuary, and coastal zones, so that length scales of greater than 50 m and 3 km, respectively, ensure that frictional effects may be neglected in comparison with Coriolis effects. (In the laboratory, the conditions $Ro \ll 1$ and $Ek \ll 1$ are relatively easy to achieve.)

The parameter *B* expresses the ratio of gravitational to Coriolis accelerations in the flow. For cases of homogeneous flow ($(\Delta\rho) = 0$), the parameter ceases to be meaningful since the gravitational acceleration can then be absorbed within the reduced pressure term. For estuarine, river or coastal current flows that contain salinity- and thermally-generated density differences, the parameter *B* is typically much greater than unity. For example, with the values of *U*, φ, and Ω_{Earth} adopted earlier and $(\Delta\rho)/\rho_0 \sim 10^{-2}$, a typical value of *B* is 10^4, confirming that buoyancy dominates completely over Coriolis effects for such cases (see earlier). However, the dominance enters through the *vertical* component of Equation 29.8 and does not affect significantly the control exerted by rotation on the horizontal flows. Note that the parameter *B* can be expressed as $B = g(\Delta\rho)/2\rho_0\Omega U = (Ri)(Ro)(L/H)$, where $Ri = (\Delta\rho)gH/\rho_0 U^2 = Fr^{-2}$ is a bulk Richardson number and *Fr* is a densimetric Froude number defined by $Fr^2 = \rho_0 U^2/[g(\Delta\rho)H]$. Maxworthy and Browand (1975) show that the combination $(Ri)(Ro)$ may be regarded conveniently as a dividing parameter for many rotating, stratified flows of environmental interest, with rotationally-dominated and stratification-dominated cases being characterized by ranges $[(Ri)(Ro)] < 1$ and $[(Ri)(Ro)] > 1$, respectively.

29.2.5 Geostrophic Flow

For cases in which $T \ll 1$, $Ro \ll 1$ and $Ek \ll 1$, a geostrophic balance exists between the horizontal pressure gradients and the horizontal components of the Coriolis acceleration associated with the steady flow. That is, from (29.8)

$$2\rho_0 \boldsymbol{\Omega} \times \boldsymbol{u} = -\nabla p' \tag{29.18}$$

With the hydrostatic balance being expressed here by

$$0 = -\nabla p' + \rho' \boldsymbol{g} \tag{29.19}$$

In component form, Equations 29.18 and 29.19 become

$$-2\Omega\rho_0 v = \frac{-\partial p'}{\partial x} \tag{29.20a}$$

$$+2\Omega\rho_0 u = \frac{-\partial p'}{\partial y} \tag{29.20b}$$

$$0 = \frac{-\partial p'}{\partial z} - \rho g \tag{29.20c}$$

By cross-differentiation and use of the continuity equation

$$\frac{\partial(u, v, w)}{\partial z} = 0 \tag{29.21}$$

such that a geostrophic flow $\boldsymbol{u}$ is independent of the direction (z) of the angular velocity vector—the Proudman Taylor constraint (e.g., Tritton 1988). (If such a flow is confined by horizontal impermeable boundaries at $z = 0$ and $z = H$, the application of the kinematic boundary condition

$$w = 0 \text{ at } z = 0, H \tag{29.22}$$

assures that $w = 0$ throughout the fluid depth H and the flow is forced to be horizontal everywhere.) This control implies that fluid is unable to flow over any isolated topographic obstruction and must flow around it in a pattern that is invariant with height. The "Taylor column" (Boyer and Davies 2000; Velasco Fuentes 2008), representing the volume of relatively stagnant fluid trapped above the obstacle (and around which all incident flow is diverted as if deflected by a solid, full-depth cylinder circumscribing the real obstacle), is an extreme case of topographic steering of rotating flows.

A consequence of Equation 29.20 is that a transverse pressure gradient is required to maintain a mean geostrophic flow. Thus, for example, a free-surface, fully mixed, geostrophic flow of, say, $U = 0.3\,\text{m s}^{-1}$ in a straight channel of width 10 km at a latitude φ of 45°N, will have a value of Ro of 0.3 and a difference of 3 cm in elevation of the water surface across the width of the channel as a consequence of the Earth's rotation.

29.2.6 Ekman Layers

Steady boundary layers in rotating systems differ fundamentally from their non-rotating counterparts, because the essential balance within the former is between Coriolis, pressure gradient, and viscous terms in the momentum equation, whereas in the non-rotating systems, the Coriolis contribution is absent. For the present purposes, it is useful to focus attention upon the boundary layers formed on horizontal (or, more generally, non-vertical) solid boundaries with homogeneous fluids. For simplicity, the analysis is posed in terms of laminar (i.e., viscous) boundary layers, on the assumption (Pedlosky 1987) that the results can be carried over to turbulent boundary layers after suitable replacement of ν by A_V or A_H (Caldwell et al. 1972).

Assuming *a priori* that the boundary layer (henceforth denoted the Ekman layer) is thin compared with the overall length scale dimension L of the fluid domain, it may be seen that within the boundary layer ("E")

$$\left(\frac{\partial^2}{\partial z^2}\right)_E \gg \left(\frac{\partial^2}{\partial x^2}\right), \left(\frac{\partial^2}{\partial y^2}\right)_E \tag{29.23}$$

Subject to this condition, and assuming that (a) viscous effects are confined to the boundary layer, (b) the interior ("I") flow (u_I, v_I) outside the boundary layer is horizontal and geostrophic, Equations 29.17 and 29.18 can be written in component form as follows

$$\frac{\partial u}{\partial x} + \frac{\partial v}{\partial y} + \frac{\partial w}{\partial z} = 0 \tag{29.24}$$

$$-2\Omega\rho_0 v_I = \frac{-\partial p'}{\partial x} \tag{29.25a}$$

$$-2\Omega\rho_0 u_I = \frac{-\partial p'}{\partial y} \tag{29.25b}$$

$$-2\Omega\rho_0 v_E = -\frac{\partial p'}{\partial x} + \rho_0 \nu \left(\frac{\partial^2 u_E}{\partial z^2}\right) \tag{29.26a}$$

$$-2\Omega\rho_0 v_E = -\frac{\partial p'}{\partial y} + \rho_0 \nu \left(\frac{\partial^2 u_E}{\partial z^2}\right) \tag{29.26b}$$

with boundary conditions

$$u_E = v_E = 0 \text{ (no slip)}$$

$$w_E = 0 \text{ (impermeability)}$$

$$u_E \rightarrow u_I : z \rightarrow \infty$$

$$v_E \rightarrow v_I : z \rightarrow \infty$$

The solution to the preceding equations, subject to the aforementioned boundary conditions, is (Pedlosky 1987)

$$u_E = u_I - \exp\left(\frac{-z}{\delta}\right)\left[\left(v_I \sin\left(\frac{z}{\delta}\right) + u_I \cos\left(\frac{z}{\delta}\right)\right)\right] \quad (29.27)$$

$$v_E = v_I - \exp\left(\frac{-z}{\delta}\right)\left[\left(v_I \cos\left(\frac{z}{\delta}\right) + u_I \sin\left(\frac{z}{\delta}\right)\right)\right] \quad (29.28)$$

where

$$\delta = \left(\frac{\nu}{2\Omega}\right)^{1/2} \quad (29.29)$$

is the Ekman layer thickness. The properties of (29.27) and (29.28) are most easily appreciated by considering an interior geostrophic flow solely in the x direction (i.e., u_I).

The structure of the boundary layer for this case is seen in Figure 29.1a. Far from the boundary all of the flow is in the x direction, but closer to the boundary there is a flow component to the left (i.e., in the y direction). When the horizontal velocities are plotted against each other for different values of z/δ, the familiar Ekman spiral is shown (Figure 29.1b). The total volume flux per unit width in the y direction can be easily calculated as

$$V_E = \int v_E dz = u_I\left(\frac{\delta}{2}\right) \quad (29.30)$$

so a geostrophic current above a solid horizontal surface produces a net volume flux to the left of the current, within the Ekman boundary layer.

Such a flow generates a vertical velocity w_E at the edge of the boundary layer. Its value can be calculated from the continuity equation, since

$$\int\left[\frac{\partial u_E}{\partial x} + \frac{\partial v_E}{\partial y}\right]dz = \int\left[\frac{-\partial w_E}{\partial z}\right]dz = -w_E \quad (29.31)$$

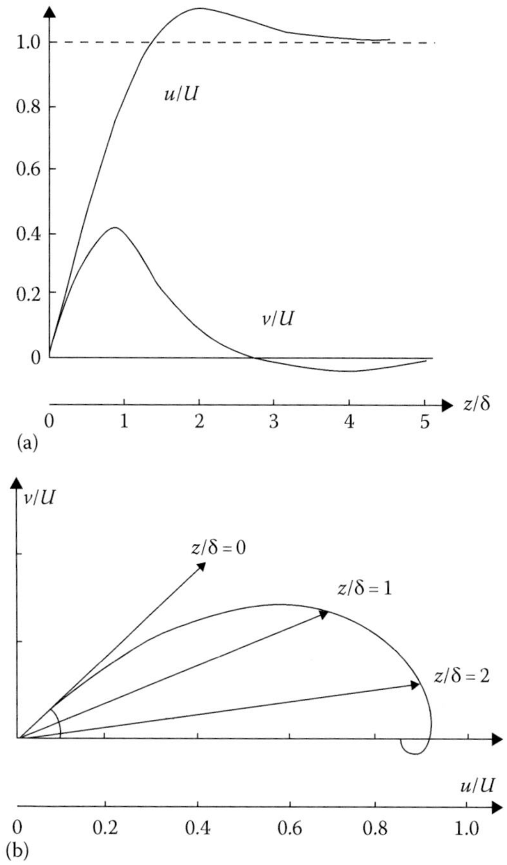

FIGURE 29.1 (a) Ekman layer solution for normalized velocity components u/U and v/U versus z/δ for geostrophic flow $(U,0,0)$ over a solid boundary at $z = 0$ and (b) Ekman spiral shown as hodograph of u/U versus v/U for various z/δ.

Substituting for u_E and v_E from (29.28) into (29.30), w_E is calculated (Pedlosky 1987) as

$$w_E = -\left(\frac{\delta}{2}\right)(\xi_I) \quad (29.32)$$

where ξ_I is the interior vorticity, given by

$$\xi_I = \frac{\partial v_I}{\partial x} - \frac{\partial u_I}{\partial y} \quad (29.33)$$

This analysis applies to the case of a geostrophic flow over a solid boundary (e.g., a wind blowing over the ground, a current flowing over the sea floor). A similar analysis can be carried out for a wind stress applied to the sea surface. As with the earlier analysis, an Ekman layer solution is found (see Figure 29.2), with the horizontal velocities within the Ekman layer lying on a spiral hodograph. For this wind stress configuration, the volume transport is to the right of the applied stress direction. If the fluid is bounded by a vertical wall, the Ekman transport either accumulates surface water at the wall or causes it to flow away

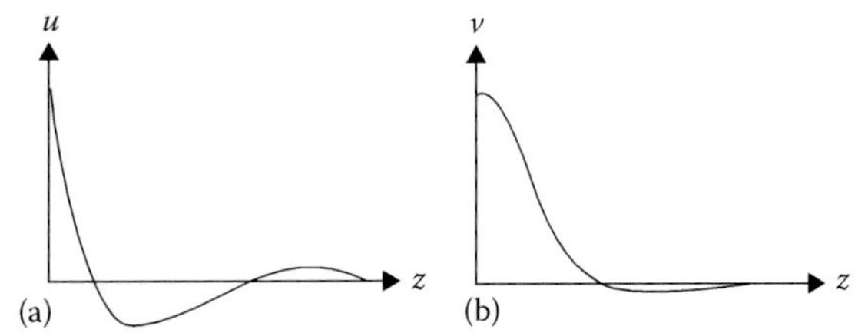

FIGURE 29.2 Sketched Ekman layer solutions for velocity components (a) u and (b) v versus depth z for the case of a surface stress (acting in x direction) over a stationary rotating fluid.

from the wall (depending on the direction of the wind stress), with the transport setting up an offshore/onshore pressure gradient that generates a geostrophic current below the Ekman layer. An offshore Ekman flux causes cold water upwelling to occur at the coast and frontal zones to form offshore. Such frontal regions inhibit offshore transport of pollutants and nutrients (Gocke et al. 1983) and promote primary production and mussel growth close to shore (Figueiras et al. 2002), as well as occasionally creating conditions for the formation of toxic red tides in the affected regions (Tilstone et al. 1994).

Steady and unsteady Ekman layers on solid boundaries are intrinsically unstable (Faller 1963, 1991; Faller and Kaylor 1966; Tatro and Mollo-Christensen 1967; Caldwell and van Atta 1970) with important consequences for mixing and the distribution of resuspended bottom sediments in coastal regions (Aelbrecht et al. 1999b).

29.2.7 Spin-Up/Spin-Down Adjustment

The existence of Ekman layers in rotating fluids has consequences for the manner in which a rotating fluid responds to either changes in external forcing or interior flow disturbances (Greenspan and Howard 1963). A good example of this behavior is provided by the response of a rotating fluid system to changes in background rotation rate. For purely viscous adjustment, the relevant time scale is L^2/ν, with successive layers of fluid undergoing viscous drag until solid body rotation is achieved at the new rotation rate. Such a time scale is much less than that observed and correct estimates for the spin up/down time can only be achieved if the assumptions are made that (a) viscous (frictional) effects are confined to the Ekman layers and (b) the fluid interior behaves as if inviscid and angular momentum is conserved.

Consider a cylindrical container of fluid of radius R and depth H rotating as a solid body with angular velocity Ω_0, before the rotation rate is increased by $\Delta\Omega$. Following the change in rotation rate, there is a radial outflow of fluid in the Ekman layers on the top and bottom bounding surfaces of the container and a pumping of fluid into the boundary layers from the interior. This, in turn, causes a radial inflow in the interior with rings of interior fluid spinning faster as they preserve angular momentum. Conservation of angular momentum for a ring of fluid initially at radius r with angular velocity Ω and moving inward to position $(r - \Delta r)$ to acquire angular velocity $(\Omega_0 + \Delta\Omega)$ gives

$$r^2(\Omega_0) = (r - \Delta r)^2(\Omega_0 + \Delta\Omega) \tag{29.34}$$

$$\downarrow$$

$$\Delta r = \left(\frac{r}{2\Omega_0}\right)(\Delta\Omega) \tag{29.35}$$

If the ring moves with inward velocity u_r, the time t_r taken for the ring to acquire the new rotation rate is

$$t_r = \frac{\Delta r}{u_r} = \frac{r(\Delta\Omega)}{2\Omega_0 u_r} \tag{29.36}$$

If the spin up time T_s is the time for the outermost ring to achieve the new solid body rotation, then

$$T_s = \frac{R(\Delta\Omega)}{2\Omega_0 u_r} \tag{29.37}$$

But, by continuity,

$$V_E \sim w_E R \sim u_r H \tag{29.38}$$

Using the expressions for V_E from (29.30) and putting $u_I \sim R(\Delta\Omega)$, Equations 29.37 and 29.38 can be combined to give

$$T_s \sim \frac{H}{\Omega_0 \delta} \sim (Ek_H)^{-1/2}(\Omega_0)^{-1} \tag{29.39}$$

where Ek_H ($=\nu/2\Omega H^2$) is an Ekman number based upon the length scale H. Measured spin up times for homogenous fluid systems have verified that the time scale T_s ($\ll L^2/\nu$) is the appropriate one to use to characterize the response of well-mixed environmental flow systems influenced by the Earth's rotation. The additional presence of density stratification modifies considerably the mechanism described earlier (Walin 1969; Spence et al. 1992; Duck and Foster 2001), not least because of the role of density stratification in suppressing vertical motion. For such systems, a so-called penetration distance $S^{-1/2}H$ exists, beyond which spin-up/down dynamics do not operate (Walin 1969; Spence et al. 1992), where $S = (H/L)^2(N_0/2\Omega)^2 = B(Ro)(H/L)$ and N_0 is the buoyancy frequency of the stratified water column. The adjustment takes place in several stages until quasi-steady conditions are achieved, with the attainment of a quasi-steady state of vertical shear in the interior flow in a time less than the homogenous spin up time $(Ek_H)^{-1/2}(\Omega_0)^{-1}$, but with solid body rotation being reached only after the diffusive time scale $(Ek_H)^{-1}(\Omega_0)^{-1}$ has elapsed.

29.2.7.1 Rectified Tidal Flow along a Vertical Coastline

The analysis in Section 29.2.6 has considered a *steady* current flowing over the sea floor. The consequences of Ekman-layer-induced transport on the behavior of unsteady (i.e., tidal), along-shore currents has been studied by Aelbrecht et al. (1993, 1999a) and Zhang et al. (1993, 1994), to show that such oscillating tidal currents can be rectified to give a mean flow, solely as a consequence of the role played by the background rotation of the Earth (through the formation of Ekman boundary layers) and the presence of bottom and lateral (coastline) boundaries. The rectification occurs because of the asymmetry in Ekman transport conditions between the ebb and flow phases of the tidal cycle (Figure 29.3). During one half cycle, when the Ekman transport is toward the coast, low momentum fluid is advected into the interior by the Ekman pumping process, while during the other half cycle, the Ekman transport is away from the coast and high offshore momentum is drawn into the interior flow. The rectified flow is in the along-shore direction defined by the case in which the coastline is to the right, with a lateral extent that is limited to a dimensionless distance $\ell_y/H \sim (Ek_H)^{1/2}(Ro_H)(Ro_t)^{-1}$, where $Ro_H = U/2\Omega H$, U is the maximum tidal velocity and Ro_t is a so-called temporal Rossby number defined for a tidal frequency σ by $Ro_t = \sigma/2\Omega$. The normalized residual rectified current strength u_r is predicted to have the dependence $u_r/U \sim (Ek_H)/Ro_t)^{1/2}(Ro_L/Ek_H)$, where $Ro_L = U/2\Omega L$. For the French coast bordering the Eastern Channel, the rectified current has an estimated width and speed of 4 km and 15 cm s^{-1}, respectively, for a maximum tidal current of 1 m s^{-1}—values that are associated with a volume flow rate that is a significant contribution (about 10%) to the estimated total wind- and tidal-induced residual flux through the Channel (Prandle et al. 1996).

This is one example of the rectification of tidal currents in the presence of background rotation, but there are others for which the background rotation forces a residual current; for example, shallow tidal flow over irregular bottom topography (Komen and Riepma 1981a,b).

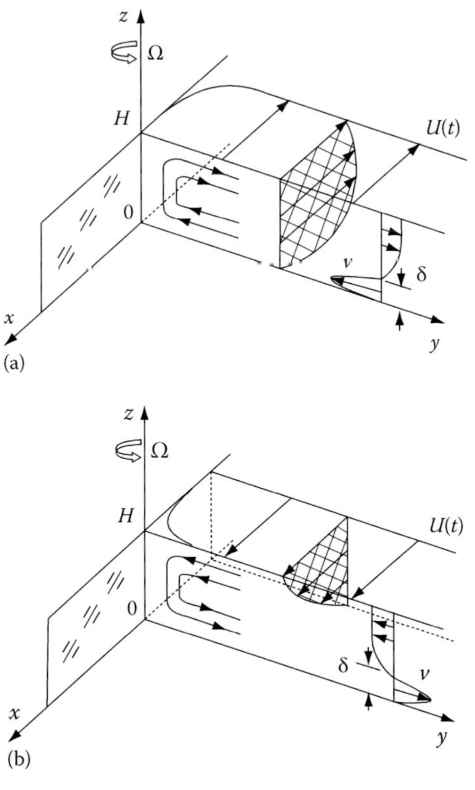

FIGURE 29.3 Ekman transport asymmetry between (a) ebb and (b) flood phases for tidal flow $U(t)$.

29.2.8 Wave Motions in Rotating Systems

The Proudman Taylor constraints (Equation 29.21) imposed on a rotating fluid basin such as a lake, coastal region, estuary, or shallow, semi-enclosed sea confer elastic properties, since any perturbation of a fluid particle that causes it to move horizontally away from its equilibrium position will result in a restoring force that will initiate interior wave motions or oscillations with natural frequency 2Ω (Tritton 1988; Pedlosky 2003). The existence of inertial waves can be demonstrated by including the unsteady term in the geostrophic equation (29.18) and seeking 2D ($\partial/\partial y = 0$, say) linear, *inertial wave* motions with forcing frequency σ and wave numbers k and n in the x and z directions. The analysis generates a dispersion relationship

$$\sigma^2 = \frac{(2\Omega)^2 n^2}{(n^2 + k^2)} = (2\Omega)^2 \,|\cos\theta|^2 \qquad (29.40)$$

where θ is the angle between the wave number vector and the rotation axis. From (29.40), waves can exist in a rotating fluid only for values of σ between 0 and 2Ω. [A similar result

$$\sigma^2 = \frac{N^2 n^2}{(n^2 + k^2)} \qquad (29.41)$$

is obtained for *internal waves* in a stratified, non-rotating fluid having constant buoyancy frequency N, illustrating one manifestation of a well-known and useful analogy (Veronis 1970) between the effects of background rotation and density stratification on environmental hydrospheric systems. Note that here the buoyancy (i.e., Brunt-Väisälä or stability) frequency $N(z)$ is defined as $N^2(z) = (g/\rho_0)(\partial\rho/\partial z)$.]

Waves on the free surface of a well-mixed, homogeneous sea, large lake, or coastal zone differ fundamentally from those propagating in basins such as small lakes and rivers having length scales too small for the Earth's rotation to be relevant to the flow dynamics. To illustrate this property, consider linear plane waves in an infinitely long (in the x direction) horizontal channel of width W and constant undisturbed depth H, formed between two vertical walls at $y = 0$ and $y = W$ and a horizontal bottom at $z = -H$, rotating uniformly about the vertical (z) axis with angular velocity Ω. Starting with (29.24) and (29.8), but neglecting the viscous, advective acceleration and buoyancy terms, the equations to be solved (Howard, 1963; Gill, 1982) are

$$\frac{\partial u}{\partial x}+\frac{\partial v}{\partial y}+\frac{\partial w}{\partial z}=0 \tag{29.42}$$

$$\frac{\rho_0 \partial u}{\partial t}+(-2\Omega\rho_0 v)=\frac{-\partial P'}{\partial x} \tag{29.43a}$$

$$\rho_0 \partial v/\partial t+(+2\Omega\rho_0 u)=\frac{-\partial P'}{\partial y} \tag{29.43b}$$

transformed to

$$\frac{\partial \zeta}{\partial t}+H\left(\frac{\partial u}{\partial x}+\frac{\partial v}{\partial y}\right)=0 \tag{29.44a}$$

$$\frac{\partial u}{\partial t}-2\Omega v=-g\frac{\partial \zeta}{\partial x} \tag{29.44b}$$

$$\frac{\partial v}{\partial t}+2\Omega u=-g\frac{\partial \zeta}{\partial y} \tag{29.44c}$$

in terms of the surface displacement $\zeta(x, y, t)$, after (a) using the hydrostatic equation to relate the pressure and displacement terms and (b) integrating the continuity equation in the vertical from the free surface to the bottom ($z = -H$) where the boundary condition $w_{z=-H} = 0$ is applied. Wave solutions to the preceding set of equations are conveniently divided into those in which $v = 0$ everywhere (Kelvin waves) (Figure 29.4) and those for which $v = 0$ only at the walls $y = 0,W$ (Poincaré waves). Both types of barotropic waves are observed in large lakes, reservoirs, and well-mixed coastal regions.

Kelvin waves have the solution

$$\zeta=\zeta_0\exp\left(\frac{-y}{a}\right)\cos(kx-\sigma t) \tag{29.45}$$

$$u=\left(\frac{g}{H}\right)^{1/2}\zeta_0\exp\left(\frac{-y}{a}\right)\cos(kx-\sigma t) \tag{29.46}$$

$$v=0 \tag{29.47}$$

where

$$a=\frac{(gH)^{1/2}}{2\Omega}=\frac{c}{f}=R_d \tag{29.48}$$

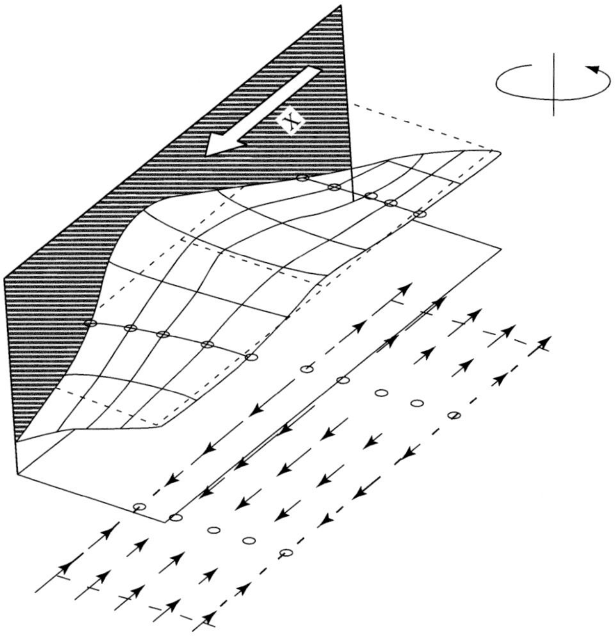

FIGURE 29.4 Kelvin waves in a homogeneous rotating fluid (free surface).

is the *Rossby radius of deformation* and $f(=2\Omega)$ is the Coriolis parameter. For length scales L small compared with a, effects of rotation on the flow are negligible but for $L > a$ effects of rotation on the flow are significant. For the North Sea, for example, the value of the Rossby radius of deformation for a fully mixed water column of depth H, say 50 m, is approximately 125 km (typical values for the North Sea lie between 100 and 300 km). The Kelvin wave has maximum amplitude ζ_0 at the lateral boundary (wall) of the flow domain and it travels in the northern hemisphere with the boundary to the right as it propagates cyclonically around the large lake, reservoir, or shelf sea in which it is formed. Estimates indicate that more than two-thirds of the semi-diurnal and half the diurnal tidal components along the California coast can be attributed to travelling barotropic Kelvin waves (Wang 2002).

Poincaré waves have the solution

$$\zeta=\zeta_0\cos(kx-\sigma t) \tag{29.49}$$

$$u=\left(\frac{\sigma\zeta_0}{kH}\right)\cos(kx-\sigma t) \tag{29.50}$$

$$v=\left(\frac{f\zeta_0}{kH}\right)\sin(kx-\sigma t) \tag{29.51}$$

showing that the effects of rotation enter here through the transverse velocity component v.

29.2.9 Effects of Density Stratification

Environmental water bodies of the type being discussed here may be stably stratified at certain locations and/or under particular seasonal conditions. For example, fresh or brackish river discharged into saline coastal regions spreads radially from the discharge source as a surface layer until reaching a horizontal scale sufficiently large (see following) for the effects of the Earth's rotation to be felt. At this stage, the river plume is deflected to the right (left) in the northern (southern) hemisphere and it flows along (and stays attached to) the shoreline boundary as a surface gravity current (Griffiths 1986), with a frontal boundary along which shore contaminants may collect and across which offshore transport is inhibited. In the regions of the coastal zone unaffected by river outflows or estuarine exchanges, the water column may experience seasonal stratification through surface heating, though well-mixed and homogeneous conditions may persist according to the vigor of any coastal mixing processes.

Seasonal thermal stratification also affects large lakes and reservoirs (Imberger and Hamblin 1982). In all of the aforementioned cases, the rotation effects described earlier are modified significantly by the buoyancy effects of the density stratification. In particular, the Rossby radius of deformation R_d loses its direct significance and relevance, to be replaced by the internal Rossby radius of deformation R_{id} given by $R_{id} = c_i/f$, where, for linear conditions, it is convenient to separate the adjustment of the rotating, stratified body of water into a set of normal modes, each of which has its own wave speed c_n (Gill 1982) and its own Rossby radius of deformation $(R_d)_n = c_n/f$. (For two layer fluids of total depth H and upper and lower thicknesses H_u and H_l, the wave speed is given by $c_i^2 = g'(H_uH_l)/(H_u + H_l)$, though for the realistic condition $H_u \ll H_l$ that is valid for coastal regions and large lakes and reservoirs (Imberger and Hamblin 1982), the internal Rossby radius of deformation becomes $(g'H_u)^{1/2}/f$). Because $g' \ll g$ and $H_u \ll H$, it is clear that the internal Rossby radius of deformation R_{id} is always much less than its barotropic equivalent R_d for a particular flow.

Subject to the earlier modifications, internal Kelvin and Poincaré waves having the properties outlined in the previous sections are able to travel on the interface between the two constituent layers, with R_d replaced by R_{id}. Physically, the internal Rossby radius of deformation represents the horizontal scale on which the effects of background rotation become of the same order as those due to buoyancy. Specifically, it represents the horizontal length scale for which the gravitational pressure gradient acceleration is of the same order as the Coriolis acceleration. For a continuously stratified fluid of constant buoyancy frequency N_0 and depth H, the internal Rossby radius of deformation may be conveniently estimated as N_0H/f. It is thus the natural length scale to expect to observe for coastal currents, river plumes, and large stratified lakes and reservoirs (Power and Grimshaw 1990). For these reasons, the internal Rossby radius of deformation R_{id} has been adopted as the fundamental length scale for rotating stratified flows of relevance to environmental flow problems. In dimensionless terms, the value of the Kelvin number $K = L/R_{id}$, the ratio of the horizontal length scale of the flow in question (e.g., the width of an estuary) to the Rossby radius, determines the degree to which the background rotation of the Earth plays a significant dynamical role in the flow development (Garvine 1995; Huq 2009). Kasai et al. (2000) have demonstrated, however, that even for estuaries having widths much greater than R_{id}, it is necessary to include Coriolis effects through an Ekman number dependence in order to simulate successfully the observed transverse structure of many freshwater outflows. Winant (2004) also invokes Ekman number dependence as the principal indicator of the significance of background rotation for wind-driven flows in elongated basins. Valle-Levinson (2008) has developed the approach of Kasai et al. (2000) in which Coriolis effects are incorporated in the model via the Ekman number dependence, in order *inter alia* to classify the 3D structure (transverse horizontal shear, vertical shear) of estuarine exchange flows in channels of different sectional shapes, in terms of both Kelvin and Ekman numbers. The classification resulting from the model into horizontally-sheared, vertically-sheared, and combined horizontally and vertically sheared states confirms a relatively weak dependence on the Kelvin number (but a strong dependence on the appropriate Ekman number $Ek_V = A_V/2\Omega H^2$), for a wide range of channel shapes.

Internal Kelvin and Poincaré wave motions have been measured in many environmental flows. For example, Walin (1972) reported temperature measurements in the Baltic, showing the response of the semi-enclosed sea to sustained wind stress (followed by relaxation of the forcing) to be the cyclonic propagation of internal Kelvin waves around the basin. In a series of very intensive and comprehensive modeling and field observation programmes (e.g., Antenucci et al. 2000; Antenucci and Imberger 2001, 2003; Boegman et al. 2003) in Lake Kinneret, the detailed characteristics of internal Kelvin and Poincaré wave fields (and, e.g., the occurrence of resonances between these wave with the periodicity of the driving wind) were determined. The wind and temperature data shown in Figure 29.5 illustrate the occurrence of internal Kelvin and Poincaré waves for resonant and non-resonant conditions. In these data, the 24 h heaving is the internal Kelvin wave, the smaller heaving seen in the troughs of the 24 h signal (most obvious about 0.25 into each day, that is, around 06.00) is the internal Poincaré wave signal.

Measurements by Umlauf and Lemmin (2005) in Lake Geneva demonstrate the role played by internal Kelvin waves generated by episodes of strong surface winds in causing significant bottom turbulence and irreversible exchange of passive tracers between the main and side basins of the lake. In cases where surface wind forcing on lakes is sustained for several days, significant distortion of the pycnocline occurs and steady geostrophic boundary jets traveling around the lake are generated (Csanady 1971; Rizk 2010). For an up-to-date review of internal wave motions in inland waters affected by the Earth's rotation see Antenucci (2009).

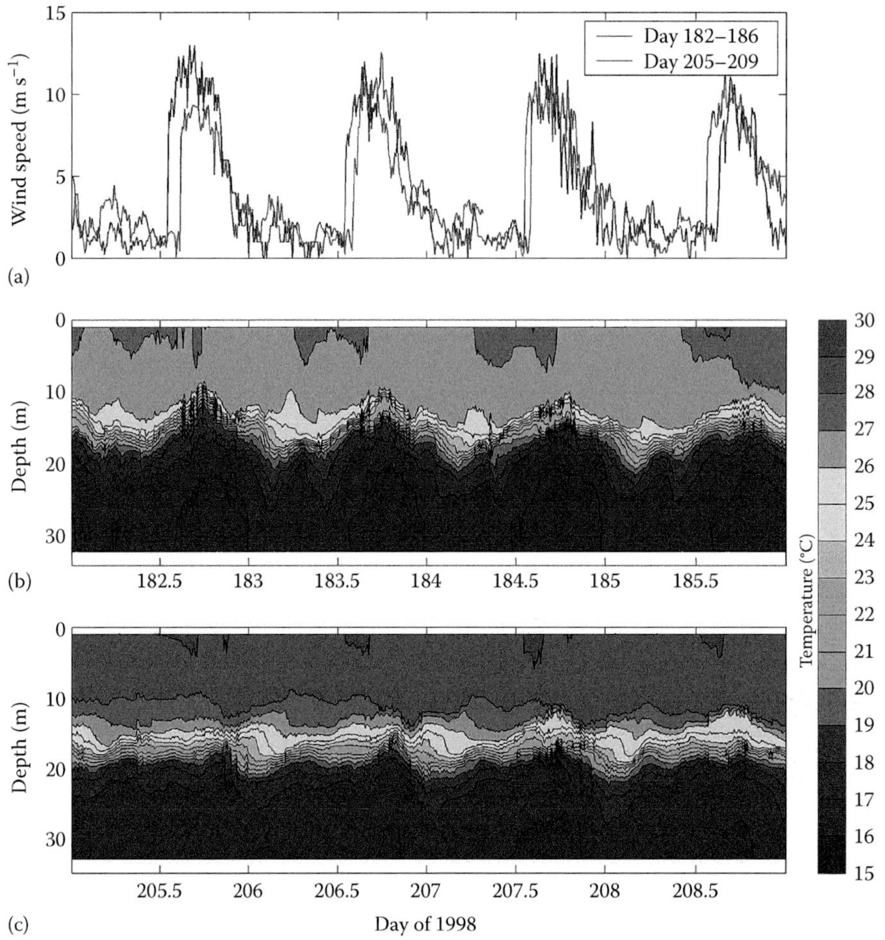

FIGURE 29.5 **(See color insert.)** Internal Kelvin and Poincaré waves forced by periodic surface winds (a) as revealed by temperature measurements in Lake Kinneret, for resonant (b) and non-resonant (c) conditions. (Courtesy of J. Antenucci.)

29.3 Boundary Currents and Fronts

As shown in Section 29.2.9, buoyancy-induced spreading in rotating systems has a natural horizontal length scale Ro_{id} in an unbounded domain. However, for many buoyancy-driven environmental flow problems affected by background rotation (e.g., coastal currents, river, and estuary plumes), coastal wall boundaries play an important role in the flow dynamics. The boundary condition is that there can be no flow normal to these boundaries, so, by definition, there can be no Coriolis acceleration *parallel* to the boundary. In consequence (Griffiths 1986), since the Coriolis acceleration vanishes at the wall the requirement that angular momentum is conserved is removed and the constraints imposed by the background rotation no longer apply. A buoyancy-driven gravity current, therefore, flows along the boundary wall. The width of the boundary current (Figure 29.6) adjusts to typically one internal Rossby radius, as it attains geostrophic equilibrium.

River and estuary plumes (see Section 29.2.9) provide good examples of buoyancy-driven boundary currents of direct environmental importance. In the European context, rivers such as the Seine, Rhine, Humber are intimately tied to anthropogenic activity; as such they discharge into the coastal zone water that has been contaminated from (a) agricultural run-off (providing nutrients for the growth of phytoplankton), (b) industrial pollution (toxic chemicals, heavy metals, organic micropollutants), and (c) urban wastewater disposal (bacterial, viral concentrations). In addition, such rivers deposit sediment loads (some of which may also be contaminated) within the coastal zone, with

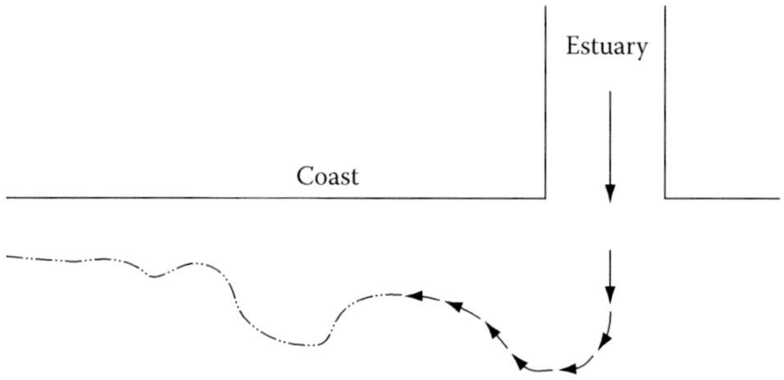

FIGURE 29.6 Schematic representation of river plume discharge into coastal waters (N. hemisphere), showing characteristic inertial bulging, the formation of a surface gravity current, and the generation of whirls and meanders by baroclinic instability.

distributions that are influenced by the form of the river plume flow. The river discharge constitutes a source of momentum and buoyancy as it debouches into the coastal zone. The associated pressure gradients, coupled with Coriolis effects of the Earth's rotation, establish the along-shore, buoyancy-driven surface flow bounded laterally by an offshore, convergent frontal boundary (Mann and Lazier 1991) that prevents offshore transfer and thereby controls the fate of anthropogenically derived contaminants discharged into the coastal zone. Surface boundary currents of this type are known to maintain their integrity for hundreds of kilometers (De Boer et al. 2006). Models (see, e.g., Griffiths 1986; Chabert d'Hières et al. 1991; De Kok 1997) show that, where frictional effects are negligible, some of the river plumes of the aforementioned type are susceptible to baroclinic instability, with the frontal boundary between the river-origin boundary current and undisturbed offshore receiving waters evolving into whirls and meanders on the scale of the internal Rossby radius of deformation. In all cases, the boundary between the fresh/brackish surface water of the river plume and the saline offshore waters constitutes a frontal interface across which the salinity changes significantly in a very short distance (Mann and Lazier 1991). In many locations, the location of the interface may have delineated the localized appearance of surface foam/debris lines and/or suspended sediment concentration gradients. Laboratory simulations of river plume discharges, boundary current generation, and frontal instability have reproduced well the main characteristics of the flow structure and scaling arguments (Davies et al. 1993; Lentz and Helfrich 2002) and geostrophic flow modeling analysis (Thomas and Linden 2007) have been applied successfully to predict the speed of the propagating current, its offshore extent, and its depth profile, for different external conditions.

The presence of along-shore, oscillatory tidal currents and a tidal residual flow (typically 0.2–0.6 m s^{-1} in peak magnitude and 0.03 m s^{-1}, respectively, for the Rhine plume region (Ruddick 1995; De Kok, 1996; van der Giessen et al. 2003) modify significantly the behavior of the buoyancy-driven, surface boundary current flow. In an attempt to generalize river plume dynamics in tidally affected coastal zones, Simpson et al. (1993) introduced the concept of a region of freshwater influence (ROFI) to incorporate effects of tidal straining (differential advection resulting from vertical shear in the tidal flow), vertical mixing processes and seasonal stratification in the receiving waters (see also Simpson et al. 1990, 1993, 2005; Simpson 1997; Souza and Simpson 1997). In the near-field, close to the region of discharge, these mechanisms dominate effects of background rotation in many cases. Instead, the formation of fronts in regions of freshwater influence is characterized primarily by interactions between the stratifying effects of the buoyancy input from river outflow(s) and combined destratifying effects of winds, waves, and tidal stirring (Simpson 1997). Such competing, contributory processes are temporally and spatially varying, with the consequence that frontal interfaces are formed episodically and at different locations along a typical coastline, demarcating regions in which the water column is either stratified or well-mixed. Some of the temporal variation may be ascribed to variations in the strength of the tidal stirring between neap and spring tide conditions, but shorter time scale variations (within a single tidal cycle) are associated with switching between stratified and mixed water column conditions as a result of variations in the degree of tidal straining between ebb and flow states (Sharples and Simpson 1993; Simpson 1997; Rippeth et al. 2001). Effects of rotation enter into the determination of the frontal structure when the expansion of the ROFI proceeds until the horizontal scale is sufficiently large for geostrophic conditions to be attained. Such a state is associated with periods of weak vertical mixing by winds and tides, leading to an expanding stratified region bounded by an offshore front. For such conditions, the space and time scales of the frontogenesis are, as anticipated, the internal Rossby radius and the inertial period, respectively (Czitrom and Simpson 1998), though model results (Kasai et al. 2000) indicate a determining role for the relevant Ekman number of the flow (see the text given earlier).

Tidal incursions and subsequent flushings respectively into and out of the river estuaries generate intermittent (pulsed) patch-like features of relatively fresh water that are embedded in the main plume. In order for this process to occur, the net discharge in the tidally affected river outflow must be halted at some stage in the tidal cycle and the pulse when formed must be sufficiently far from the river mouth that it does not interfere with the formation of successive pulses. Necessary conditions (De Ruijter et al. 1997) for the generation of such pulsed river plumes are that (a) the maximal tidal velocity must be greater than the river discharge velocity u and (b) the inertial radius u/f must be greater than the width of the river mouth. Measurements suggest that these patches are found up to about 10 km from the river mouth, for typical large rivers like the Rhine (Ruddick et al. 1994) or Columbia rivers (Cudaback and Jay 1996; Hickey et al. 1998) and that, under extreme conditions, a train of freshwater lenses can be transported along the affected coast (De Ruijter et al. 1997). Studies (see, e.g., van Alphen et al. 1988; Ruddick 1995; Simpson and Souza 1995) of the Rhine outflow suggest that the value of R_{id} is approximately 2 km for this site, a value that compares with the lateral scale of the river mouth, the inertial length scale u/f (see Equation 29.42b) of about 10 km, and the tidal excursion and tidal drift length scales of typically 5 km. For the Cap Gris Nez region, where the Seine discharges into the English Channel, salinity fronts are observed (Brylinski and Lagadeuc 1990) typically 2–10 km from the French coast.

29.4 Summary and Discussion

This chapter has demonstrated that the fluid dynamics of many important natural water bodies that are either utilized directly by (or provide access to) human activities can only be understood fully if the effects of the background rotation of the Earth are incorporated into the flow descriptions. Examples have been presented from water bodies in which there is high anthropogenic activity (e.g., estuaries, coastal zones) or where the water body provides an important source of supply of water or leisure opportunities (lakes, reservoirs) for urban populations. It has been estimated that 60% of the globe's population live within 100 km of the sea and almost two-thirds of the world's largest cities are coastal. For these reasons, at least, the environmental and economic demands of coastal zone management, clean water supply and security, ecological sustainability, and high standards of water quality assume a high priority and require the development of environmental flow models that can predict reliably and precisely the fate of contaminants and the hydrodynamic and density structure of the affected hydrospheric zones. In order to achieve these requirements, all contributory physical mechanisms and processes need to be understood and incorporated in the models. Though the importance of the background rotation of the Earth for global scale geophysical flows in the atmosphere and hydrosphere is well recognized, its role in the contexts outlined earlier may be less clear, *ab initio*.

The chapter has provided a comprehensive description of the fundamental effects of background rotation on the dynamics of stratified, rotating, and shallow flows and has explored the consequences of these effects for idealized limiting conditions. The application to the environmental (hydrospheric) flows in question has demonstrated the role played by the rotational constraint in the important processes determining the fate and transport of contaminants, sediments, chemicals, nutrients, both onshore and offshore. In many cases of coastal and freshwater flows and/or point discharges in the open sea, the primary and determining length scale indicating the importance of background rotation is the internal Rossby radius of deformation, Ro_{id}, with flows having horizontal scales comparable with (or greater than) Ro_{id} being affected significantly by the rotation. It is noted that for most cases of interest here, the value of Ro_{id} is a few tens of kilometers (sometimes only several kilometers) and this length scale is typical of, say, the width of an estuary, the dimensions of a large lake, or the length of a long submarine outfall (Roberts et al. 2010) discharging wastewater into the coastal zone. For these reasons, the role of the Coriolis acceleration is significant for many environmental flows relevant to anthropogenic activities.

This contribution is dedicated to the memory of Gerhard Jirka (1944–2010), a good friend and inspirational scientist, who was the driving force behind the establishment of the Summer Schools on *Environmental Fluid Mechanics—Theory, Experiments, Applications* that have brought the subject so effectively to many young researchers over the past 12 years.

References

Aelbrecht, D., Chabert d'Hières, G., and Renouard, D., 1999a. Experimental study of the Ekman layer instability in steady or oscillating flows. *Cont. Shelf Res.*, 19: 1851–1867.

Aelbrecht, D., Chabert d'Hières, G., and Renouard, D., 1999b. A 3-D coastal tidal rectification process: Observations, theory and experiments. *Cont. Shelf Res.*, 19: 1869–1903.

Aelbrecht, D., Chabert d'Hières, G., and Zhang, H., 1993. Generation of a residual current by interaction between the coastal boundary layer and the Ekman layer in a tidal motion. *Oceanol. Acta*, 16: 479–488.

Alphen van, J.S.L.J., de Ruijter, P.M., and Borst, J.C., 1988. Outflow and three-dimensional spreading of Rhine River water in the Netherlands coastal zone. In *Physical Processes in Estuaries* (eds. J. Dronkers and W. van Leussen), pp. 70–92. Springer Verlag, Berlin, Germany.

Antenucci, J.P., 2009. Currents in stratified water bodies—Effects of rotation. In *Encyclopaedia of Inland Waters* (ed. G.E. Likens), Vol. 1, pp. 559–567. Elsevier Science, Oxford, U.K.

Antenucci, J.P. and Imberger, J., 2001. Energetics of long internal gravity waves in large lakes. *Limnol. Oceanogr.*, 46: 1760–1773.

Antenucci, J.P. and Imberger, J., 2003. The seasonal evolution of wind/internal wave resonance in Lake Kinneret. *Limnol. Oceanogr.*, 48: 2055–2061.

Antenucci, J.P., Imberger, J., and Saggio, A., 2000. Seasonal evolution of the basin-scale internal wave field in a large stratified lake. *Limnol. Oceanogr.*, 45: 1621–1638.

Baker, D.J. and Robinson, A.R., 1969. A laboratory model for the general ocean circulation. *Philos. Trans. R. Soc. Lond. A*, 265(1168): 533–566.

Beardsley, R.C., 1969. A laboratory model of the wind-driven ocean circulation. *J. Fluid Mech.*, 38(2): 255–271.

Boegman, L., Imberger, J., Ivey, G.N., and Antenucci, J.P., 2003. High frequency internal waves in large stratified lakes. *Limnol. Oceanogr.*, 48: 895–919.

Bowden, K.F., 1983. *Physical Oceanography of Coastal Waters*. Ellis Harwood, Chichester, U.K.

Boyer, D.L. and Davies, P.A., 2000. Laboratory studies of orographic effects in rotating and stratified fluids. *Annu. Rev. Fluid Mech.*, 32: 165–202.

Brylinski, J.-M. and Lagaduec, Y., 1990. L'interface eaux côtières/eau du large dans le Pas de Calais (côte française): Une zone frontale. *C. R. Acad. Sci. Paris*, t.311(Série II): 535–540.

Caldwell, D.R. and van Atta, C.W., 1970. Characteristics of Ekman boundary layer instabilities. *J. Fluid Mech.*, 44: 79–96.

Caldwell, D.R., van Atta, C.W., and Helland, K.N., 1972. A laboratory study of the turbulent Ekman layer. *Geophys. Fluid Dyn.*, 3: 125–160.

Chabert d'Hières, G., Didelle, H., and Obaton, D., 1991. A laboratory study of surface boundary currents: Application to the Algerian current. *J. Geophys. Res.*, 96: 12539–12548.

Csanady, G.T., 1971. Baroclinic boundary currents and long edge waves in basins with sloping shores. *J. Phys. Oceanogr.*, 1: 92–104.

Cudaback, C.N. and Jay, D.A., 1996. Formation of the Columbia River plume—Hydraulic control in action. In *Buoyancy Effects on Coastal and Estuarine Dynamics* (eds. D.G. Aubrey and C.T. Friedrichs), Coastal and estuarine studies, Vol. 53, pp. 139–154. American Geophysical Union, Washington, DC.

Cushman-Roisin, B., 1994. *Introduction to Geophysical Fluid Dynamics*. Prentice-Hall, New York.

Czitrom, S.P.R. and Simpson, J.H., 1998. Intermittent stability and frontogenesis in an area influenced by land runoff. *J. Geophys. Res. (Oceans)*, 103: 10369–10376.

Dahlgaard, H.H., 1995. Transfer of European coastal pollution to the Arctic: Radioactive tracers. *Mar. Pollut. Bull.*, 31(1–3): 3–7.

Davies, P.A., Jacobs, P.T.G.A., and Mofor, L.A., 1993. A laboratory study of buoyant fresh water boundary currents in tidal crossflows. *Oceanol. Acta*, 16: 489–503.

De Boer, G.J., Pietrzak, J.D., and Winterwerp, J.C., 2006. On the vertical structure of the Rhine region of freshwater influence. *Ocean Dyn.*, 56: 1616–7341.

De Kok, J.M., 1996. A two-layer model of the Rhine plume. *J. Mar. Syst.*, 8: 269–284.

De Kok, J.M., 1997. Baroclinic eddy formation in a Rhine plume model. *J. Mar. Syst.*, 12: 35–52.

De Ruijter, W.P.M., Visser, A.W., and Bos, W.G., 1997. The Rhine outflow: A prototypical pulsed plume in a high energy shallow sea. *J. Mar. Syst.*, 12: 263–276.

Duck, P.W. and Foster, M.R., 2001. Spin up of homogeneous and stratified fluids. *Annu. Rev. Fluid Mech.*, 33: 231–263.

Faller, A.J., 1963. An experimental study of the instability of the laminar Ekman boundary layer. *J. Fluid Mech.*, 230: 245–269.

Faller, A.J., 1991. Instability and transition of the disturbed flow over a rotating disk. *J. Fluid Mech.*, 230: 245–269.

Faller, A.J. and Kaylor, R.E., 1966. A numerical study of the instability of laminar Ekman boundary layer. *J. Atmos. Sci.*, 23: 466–480.

Figueiras, F.G., Labarta, U., and Fernández Reeiriz, M.J., 2002. Coastal upwelling, primary production and mussel growth in the Rías Baixas of Galicia. *Hydrobiologia*, 484: 121–131.

Garvine, R.W., 1995. A dynamical system for classifying buoyant coastal discharges. *Cont. Shelf Res.*, 15: 1585–1596.

Gill, A.E., 1982. *Atmosphere-Ocean Dynamics*. Academic Press, London, U.K.

Gocke, K., Hoppe, H.G., and Bauerfeind, S., 1983. Investigations on the influence of coastal upwelling and polluted rivers on the microflora of the Northeastern Atlantic off Portugal. II. Activity and biomass production of the bacterial population. *Bot. Mar.*, 26: 189–199.

Greenspan, H.P. and Howard, L.N., 1963. On a time-dependent motion of a rotating fluid. *J. Fluid Mech.*, 17: 385–404.

Griffiths, R.W., 1986. Gravity currents in rotating systems. *Ann. Rev. Fluid Mech.*, 18: 59–89.

Hickey, B.M., Pietrafesa, L.J., Jay, D.A., and Boicourt, W.C., 1998. *J. Geophys. Res. (Oceans)*, 103: 10339–10368.

Howard, L.N., 1963. Fundamentals of the theory of rotating fluids. *Trans. ASME. J. Appl. Mech.*, 30: 481–485.

Huq, P., 2009. The role of Kelvin number on bulge formation from estuarine buoyant outflows. *Estuar. Coast*, 32: 709–719.

Imberger, J. and Hamblin, P.F., 1982. Dynamics of lakes, reservoirs and cooling ponds. *Annu. Rev. Fluid Mech.*, 15: 153–187.

Kasai, A., Hill, A.E., Fujiwara, T., and Simpson, J.H., 2000. Effect of the Earth's rotation on the circulation in regions of freshwater influence. *J. Geophys. Res. (Oceans)*, 105: 16961–16966.

Kershaw, P.J. and Baxter, A.J., 1995. The transfer of reprocessing wastes from north-west Europe to the Arctic. *Deep Sea Res.*, 42: 1413–1448.

Komen, G.J. and Riepma, H.W., 1981a. Residual vorticity induced by the action of tidal currents in combination with bottom topography in the Southern Bight of the North Sea. *Geophys. Astrophys. Fluid Dyn.*, 18: 93–110.

Komen, G.J. and Riepma, H.W., 1981b. The generation of residual vorticity by the combined action of wind and bottom topography in a shallow sea. *Oceanolog. Acta*, 3: 267–277.

Lentz, S.J. and Helfrich, K.R., 2002. Buoyant gravity currents along a sloping bottom in a rotating fluid. *J. Fluid Mech.*, 464: 251–278.

Maas, L.R.M. and van Haren, J.J.M., 1987. Observations on the vertical structure of tidal and inertial currents in the central North Sea. *J. Mar. Res.*, 45: 293–318.

Mann, K.H. and Lazier, J.R.N., 1991. *Dynamics of Marine Ecosystems*. Blackwell Scientific Publications, Oxford, U.K.

Maxworthy, T. and Browand, F.K., 1975. Experiments on rotating and stratified fluids with oceanographic application. *Annu. Rev. Fluid Mech.*, 7: 273–305.

McWilliams, J.C., 2006. *Fundamentals of Geophysical Fluid Dynamics*. Cambridge University Press, Cambridge, U.K.

Nihoul, J.C.J., Ronday, F.C., Peters, J.J., and Sterling, A., 1978. Hydrodynamics of the Scheldt estuary. In *Hydrodynamics of Estuaries and Fjords* (ed. J.C.J. Nihoul), pp. 27–54. Elsevier Scientific Publishing Co., Amsterdam, the Netherlands.

Pedlosky, J., 1987. *Geophysical Fluid Dynamics*, 2nd edn. Springer-Verlag, New York.

Pedlosky, J., 2003. *Waves in the Ocean and Atmosphere*. Springer-Verlag, New York.

Pedlosky, J. and Greenspan, H.P., 1967. A simple laboratory model for the oceanic circulation. *J. Fluid Mech.*, 27: 291–304.

Power, S.B. and Grimshaw, R.H.J., 1990. Free waves in stratified lakes. *Geophys. Astrophys. Fluid Dyn.*, 55: 47–69.

Prandle, D., Ballard, G., Flatt, D. et al., 1996. Combining modelling and monitoring to determine fluxes of water, dissolved and particulate metals through the Dover Strait. *Cont. Shelf Res.*, 16: 237–257.

Rippeth, T.P., Fisher, N.R., and Simpson, J.H., 2001. The cycle of turbulent dissipation in the presence of tidal straining. *J. Phys. Oceanogr.*, 31: 2458–2471.

Rizk, W.F., 2010. The response of a contained rotating two-layer fluid to an applied surface stress. PhD dissertation, University of Dundee, Dundee, U.K.

Roberts, P.J.W., Salas, H.J., Reiff, F.M. et al., 2010. *Marine Wastewater Outfalls and Treatment Systems*. IWA Publishing, London, U.K.

Robinson, A.R., 1970. Boundary layers in ocean circulation models. *Annu. Rev. Fluid Mech.*, 2: 293–312.

Røed, L.P., 1996. Modelling mesoscale features in the ocean. In *Waves and Nonlinear Processes in Hydrodynamics* (eds. J. Grue, B. Gjevik, and J.E. Weber), pp. 383–396. Kluwer Academic, Dordrecht, the Netherlands.

Rubin, H. and Atkinson, J., 2001. *Environmental Fluid Mechanics*. Marcel Dekker Inc., New York, USA.

Ruddick, K.G., 1995. Modelling of coastal processes influenced by the freshwater discharge of the Rhine. PhD dissertation, University of Liège, Liège, Belgium.

Ruddick, K.G., Deleersnijder, E., De Mulder, T., and Luyten, P.J., 1994. A model study of the Rhine discharge front and downwelling circulation. *Tellus*, 46A: 149–159.

Sharples, J. and Simpson, J.H., 1993. Periodic frontogenesis in a region of freshwater influence. *Estuaries*, 16: 74–82.

Simpson, J.H., 1997. Physical processes in the ROFI regime. *J. Mar. Syst.*, 12: 3–15.

Simpson, J.H., Bos, W.G., Schirmer, F. et al., 1993. Periodic stratification in the Rhine ROFI in the North Sea. *Oceanol. Acta*, 16: 23–32.

Simpson, J.H., Brown, J., Matthews, J., and Allen, G., 1990. Tidal straining, density currents, and stirring in the control of estuarine stratification. *Estuaries*, 13: 125–132.

Simpson, J.H. and Souza, A.J., 1995. Semidiurnal switching of stratification in the region of freshwater influence of the Rhine. *J. Geophys. Res. (Oceans)*, 100: 7037–7044.

Simpson, J.H., Williams, E., Brasseur, L.H., and Brubaker, J.M., 2005. The impact of tidal straining on the cycle of turbulence in a partially stratified estuary. *Cont. Shelf Res.*, 25: 51–64.

Souza, A.J. and Simpson, J.H., 1997. Controls on stratification in the Rhine ROFI system. *J. Mar. Syst.*, 12: 311–323.

Spence, G.R.M., Foster, M.R., and Davies, P.A., 1992. The transient response of a contained rotating stratified fluid to impulsively-started surface forcing. *J. Fluid Mech.*, 243: 33–50.

Tatro, P.R. and Mollo-Christensen, E.L., 1967. Experiments on Ekman layer instability. *J. Fluid Mech.*, 28: 531–543.

Thomas, P.J. and Linden, P.F., 2007. Rotating gravity currents: Small-scale and large-scale laboratory experiments and a geostrophic model. *J. Fluid Mech.*, 578: 35–65.

Tilstone, G.H., Figueiras, F.G., and Fraga, F., 1994. Upwelling-downwelling sequences in the generation of red tides in a coastal upwelling system. *Mar. Ecol. Prog. Ser.*, 112: 241–253.

Tritton, D.J., 1988. *Physical Fluid Dynamics*. Clarendon Press, Oxford, U.K.

Umlauf, L. and Lemmin, U., 2005. Inter-basin exchange and mixing in the hypolimnion of a large lake: The role of long internal waves. *Limnol. Oceanogr.*, 50: 1601–1611.

Valle-Levinson, A., 2008. Density-driven exchange flow in terms of the Kelvin and Ekman numbers. *J. Geophys. Res.*, 113: C04001.

Vallis, G.K., 2006. *Atmospheric and Oceanic Fluid Dynamics: Fundamentals and Large-Scale Circulation*. Cambridge University Press, Cambridge, U.K.

van der Giessen, A., De Ruijter, W.P.M., and Borst, J.C., 1990. Three-dimensional current structure in the Dutch coastal zone. *Neth. J. Sea Res.*, 25: 45–55.

van Heijst, G.J.F., 1994. Topography effects on vortices in a rotating fluid. *Meccanica*, 29: 431–451.

Velasco Fuentes, U.O., 2008. Kelvin's discovery of Taylor columns. *Eur. J. Mech.: B, Fluid*, 28: 469–472.

Veronis, G., 1963. On the approximations involved in transforming the equations of motion from a spherical surface to the β plane. *J. Mar. Res.*, 21: 110–124.

Veronis, G., 1970. The analogy between rotating and stratified fluids. *Ann. Rev. Fluid Mech.*, 2: 37–66.

Walin, G., 1969. Some aspects of time-dependent motion of a stratified rotating fluid. *J. Fluid Mech.*, 36: 289–307.

Walin, G., 1972. Some observations of temperature fluctuations in the coastal region of the Baltic. *Tellus*, 24: 187–198.

Wang, B., 2002. Kelvin waves. In *Enyclopedia of the Atmospheric Sciences* (eds. J.R. Holton, J. Pyle, and J. Curry), 2nd edn., pp. 1062–1068. Elsevier Science, Amsterdam, the Netherlands.

Winant, C.D., 2004. Three-dimensional wind-driven flow in an elongated, rotating basin. *J. Phys. Oceanogr.*, 34: 462–476.

Zhang, X., Boyer, D.L., Chabert d'Hières, G., and Aelbrecht, D., 1994. Rectified flow of a rotating fluid along a vertical coastline. *Phys. Fluid*, 6: 1440–1453.

Zhang, X., Boyer, D.L., Chabert d'Hières, G., Aelbrecht, D., and Didelle, H., 1993. Rectified flow along a vertical coastline. *Dyn. Atmos. Oceans*, 19: 115–146.

30

Vortex Dynamics

G.J.F. van Heijst
Eindhoven University of Technology

30.1 Introduction

Vortices are observed in nature in many different situations and in various types of appearance. For example, in the atmosphere organized vortex motion is observed in large-scale depressions, cyclones, tornadoes, and on smaller scales in dust devils and water spouts. Less-organized vortex structures are also abundant, for example, in the form of turbulent eddies in developing cumulus clouds, in the turbulent flow of smoke emanating from active volcanoes or industrial chimneys, and also in the vertical motion around large buildings on a windy day.

Coherent vortex structures are also found in the oceans, ranging from the large-scale Gulf Stream rings formed by pinching-off from the meandering principal current to vortex motion in the wake of islands and the vortices produced during the tidal exchange between an estuary and the open sea through one or more narrow openings. As in the atmosphere, less structured vortex motion in the form of turbulent eddies is also observed in seas, rivers, and lakes, in many different forms of appearance. In either case—as coherent vortex structures or as eddies in a turbulent flow—vortices play an essential role in the transport of properties such as heat, chemical and biological tracers, and pollution.

In this chapter we will focus on the dynamics of coherent vortex structures. On the larger (geophysical) scales the Earth's rotation is a key factor that tends to make the flow 2D. Other essential factors in environmental vortex motion are the ambient density stratification (by temperature and/or salinity gradients) and the shallowness of the flow domain. Some basic dynamical features of vortices in rotating and stratified fluids and in shallow fluid layers will be reviewed in Section 30.2.

Important insight in the dynamical and transport properties of environmental vortices has been gained through laboratory experiments, in which such vortex flows are simulated on a smaller scale under well-controlled conditions. In addition, the dynamics of vortices has been modeled analytically as well as by numerical flow simulations. Several modeling aspects will be discussed in Section 30.3.

Two concrete cases of environmental vortices will be considered in Section 30.4.

30.2 Principles

Large-scale vortex motion may to some extent be influenced by the Earth's rotation. Additionally, the ambient density stratification can also have a profound influence on the dynamics of environmental vortices. A third important factor is the shallowness of the flow domain, which tends to support large-scale horizontal flows and which is usually believed to suppress the vertical flow component. In this section we will review some essential characteristics of background rotation and stratification as well as the effects of shallowness on the flow evolution.

30.2.1 Effects of Background Rotation

Flows in a rotating fluid system can be described most conveniently relative to a co-rotating reference frame. Written in this frame of reference, the fluid acceleration now contains two additional terms in comparison with the acceleration in an inertial frame: the Coriolis acceleration $-2\vec{\Omega}\times\vec{v}$ and the centrifugal acceleration $\vec{\Omega}\times\vec{\Omega}\times\vec{r}$, with $\vec{\Omega}$ the system rotation vector, $\vec{v}$ the

Handbook of Environmental Fluid Dynamics, Volume One, edited by Harindra Joseph Shermal Fernando. © 2013 CRC Press/Taylor & Francis Group, LLC. ISBN: 978-1-4398-1669-1.

relative flow velocity, and $\vec{r}$ the position vector relative to the rotation axis. Since the centrifugal acceleration can be written as $-\nabla(\frac{1}{2}\Omega^2 r^2)$, with r the radius measured from the rotation axis, the (Navier-Stokes) equation of motion of the relative flow takes the following form:

$$\frac{\partial \vec{v}}{\partial t} + (\vec{v}\cdot\nabla)\vec{v} + 2\vec{\Omega}\times\vec{v} = -\frac{1}{\rho}\nabla P + \nu\nabla^2\vec{v} \tag{30.1}$$

with

$$P = p + \rho\phi_{gr} - \frac{1}{2}\rho\Omega^2 r^2 \tag{30.2}$$

the so-called reduced pressure, which contains contributions due to gravity ($\phi_{gr} = -gz$ is the gravitational potential, with g the magnitude of the gravitational acceleration and z the local vertical coordinate) and the centrifugal acceleration. Apparently, only the Coriolis acceleration in the equation of motion (30.1) plays an essential role, the centrifugal acceleration only contributing to the basic pressure distribution. Equation 30.1, together with the continuity equation $\nabla\cdot\vec{v} = 0$ for an incompressible fluid, forms the basis of a description of rotating fluid flow.

It is useful to nondimensionalize the equations by introducing characteristic length and velocity scales L and U, and using Ω^{-1} as a characteristic timescale; with the pressure nondimensionalized by the combination $\rho\Omega UL$, the nondimensional form of (30.1) becomes

$$\frac{\partial \vec{v}'}{\partial t'} + Ro(\vec{v}'\cdot\nabla')\vec{v}' + 2\vec{k}\times\vec{v}' = -\nabla' p' + E\nabla'^2\vec{v}' \tag{30.3}$$

with the primes indicating nondimensional quantities and $\vec{k} = \vec{\Omega}/\left|\vec{\Omega}\right|$ representing the unit vector in axial direction. Two nondimensional numbers appear in this equation: the Rossby number $Ro \equiv U/\Omega L$, which is a measure of the nonlinear acceleration $(\vec{v}\cdot\nabla)\vec{v}$ relative to the Coriolis acceleration $-2\vec{\Omega}\times\vec{v}$, and the Ekman number $E \equiv \nu/\Omega L^2$, which represents the ratio of the viscous term $\nu\nabla^2\vec{v}$ and the nonlinear acceleration.

30.2.1.1 Basic Balances

In many geophysical flows both the Rossby number and the Ekman number are small, that is, $Ro \ll 1$ and $E \ll 1$. For steady flow, (30.3) becomes (now again in dimensional notation)

$$2\vec{\Omega}\times\vec{v} = -\frac{1}{\rho}\nabla p \tag{30.4}$$

This equation describes flows that are in *geostrophic balance*, with the Coriolis force being balanced by the pressure gradient. It should be noted that the Coriolis force (per unit volume) acting on a fluid parcel moving with relative velocity $\vec{v}$ in a rotating frame is equal to $-2\rho\vec{\Omega}\times\vec{v}$, and thus acts perpendicular to $\vec{v}$, namely on the northern hemisphere to the right with respect to the parcel velocity $\vec{v}$ (and vice versa for the southern hemisphere). According to the balance (30.4), geostrophic flow is directed along isobars, as illustrated in Figure 30.1.

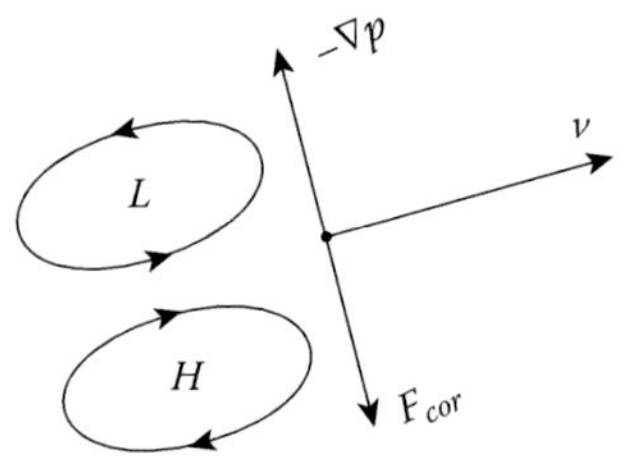

FIGURE 30.1 In geostrophically balanced flow the Coriolis force (acting perpendicular to the fluid velocity $\vec{v}$ is balanced by the pressure gradient force, so that the flow is directed along isobars.

An important property of this type of flows is derived by taking the curl of Equation 30.4:

$$(\vec{\Omega}\cdot\nabla)\vec{v} = 0 \rightarrow \frac{\partial \vec{v}}{\partial z} = 0 \tag{30.5}$$

with z the coordinate pointing in the direction of the rotation vector $\vec{\Omega}$. This result is the celebrated Taylor-Proudman theorem, which expresses that geostrophic flow is independent of the axial coordinate z. This theorem was verified experimentally by G.I. Taylor in 1923, by slowly towing a cylindrical obstacle through a fluid that was otherwise in solid-body rotation. Taylor's experiment revealed that a column of stagnant fluid was co-moving with the obstacle. This phenomenon is usually referred to as the "Taylor column." The Taylor-Proudman theorem (30.5) implies that small-Ro flow in a rotating fluid is organized in axially aligned columns, that is, such flows are uniform in the axial direction.

In the presence of background rotation, vortex flows may be governed by a number of different balances, depending on the value of the Rossby number Ro. Consider now a steady axisymmetric vortex motion in the horizontal plane (assuming that the vortex is columnar). For pure swirling flow, the radial and azimuthal velocity components are

$$v_r = 0, \quad v_\theta = V(r) \tag{30.6}$$

with r the radius measured from the vortex centre. When viscous effects are assumed negligibly small, the radial balance of forces acting on a fluid parcel in this vortex is governed by the radial component of (30.1):

$$\frac{v_\theta^2}{r} + 2\Omega v_\theta = -\frac{1}{\rho}\frac{\partial p}{\partial r} \tag{30.7}$$

This equation expresses a balance between the centrifugal force ($\rho v_\theta^2/r$; radially outward), the Coriolis force ($-2\Omega\rho v_\theta$; directed perpendicular to the fluid velocity v_θ, to the right on the NH), and the pressure gradient force ($-\partial p/\partial r$). In this particular flow, the Rossby number, defined earlier, represents the ratio of the centrifugal and the Coriolis forces. For small *Ro*-values ($Ro \ll 1$) the centrifugal term in (30.7) can be neglected, and the vortex motion is then governed by a balance between the Coriolis and pressure gradient forces: this is the *geostrophic* balance. The vortex motion can be either cyclonic (in the same direction as the planetary rotation) or anticyclonic, corresponding with a low pressure and a high pressure in the vortex centre, respectively. This type of vortex flow is encountered in the large-scale swirling flows in the atmosphere and oceans (e.g., atmospheric depressions, Gulf Stream rings).

For large *Ro*-values ($Ro \gg 1$), the Coriolis effect is negligible, and the resulting balance is between the centrifugal force and the radial pressure gradient force. In this case the motion can be in either direction (cyclonic or anticyclonic), in either case with a low pressure in the vortex centre. Flows that are governed by this so-called *cyclostrophic* balance are observed in many situations. The common bathtub vortex is a nice example (note that its rotation can be in either direction, irrespective of being on the northern or southern hemisphere).

On much larger scales, cyclostrophically balanced flows are also observed, for example, in the form of cyclones or hurricanes. These vortex flows are also characterized by large *Ro*-values, so that Coriolis effects are negligibly small. Note that their cyclonic rotation sense is determined by the dynamical balance at the initial stage of development, in which the *Ro*-value is still very small.

For intermediate *Ro*-values, that is, $Ro = O(1)$, the vortex motion is governed by the full equation (30.7), a balance which is commonly referred to as *gradient flow*. In this case, four different types of vortex balances are possible, although two of these are anomalous (for further details, see Holton, 1992).

30.2.1.2 Potential Vorticity

In the presence of background rotation, vortices are commonly columnar, even for larger *Ro*-values. When studying their behavior over bottom topography or while traveling over the spherical planet, it is convenient to use the concept of potential vorticity. In the so-called shallow-water approximation, that is, assuming columnar motion in a layer of inviscid fluid, it can be derived (see, e.g., Pedlosky, 1987) that for such a fluid column the potential vorticity

$$PV \equiv \frac{(f+\omega)}{H} \tag{30.8}$$

is conserved. Here ω is the (vertical component of the) relative vorticity of the flow, H is the column height, and f is the so-called Coriolis parameter: $f = 2\Omega \sin \phi$, with ϕ the geographical latitude. This parameter represents the component of the planetary vorticity in the local vertical direction, which obviously changes according to the latitudinal position: near the poles it takes the largest (absolute) values, while at the equator it is zero. Noting that the sum of the system vorticity and the relative vorticity is equal to the absolute vorticity, that is, $(f + \omega) = \omega_{abs}$, conservation of *PV* is essentially also formulated in Helmholtz's law, stating that ω_{abs}/H = constant for a segment of a vortex tube of length H in an inviscid barotropic flow. This conservation law implies that the absolute vorticity ω_{abs} increases when the segment of the vortex tube is stretched, while it decreases when the tube is squeezed.

This mechanism is also implied by conservation of the potential vorticity *PV* as defined by (30.8): the absolute vorticity $(f + \omega)$ increases or decreases when a vortex tube is stretched or squeezed, respectively. The presence of background rotation, denoted by the Coriolis parameter f, however, implies a symmetry-breaking: on the northern hemisphere ($f > 0$) a cyclonic vortex ($\omega > 0$) will become intenser when stretched, while an anticyclonic vortex ($\omega < 0$, with $|\omega| < f$) will become weaker when stretched. This has important consequences for vortex flow over bottom topography.

Conservation of *PV* also implies that a vortex moving in northern direction (increasing ϕ, so increasing f) undergoes the same change as when squeezed (decreasing H): an anticyclonic vortex ($\omega < 0$) intensifies, while a cyclonic vortex ($\omega < 0$) becomes less intensive. In fact, this dynamical equivalence forms the basis for the laboratory modeling of the latitudinal change in $f(\phi)$ by applying suitable bottom topography in a rotating fluid tank: a linear variation in $f(\phi)$–known as the "β-plane" approximation—can be simply modeled by a linearly sloping bottom (see, e.g., van Heijst, 1994).

The gradient in the planetary background vorticity (or: a gradient in the water depth) causes a vortex to drift, as can be understood from the principle of potential vorticity conservation. The fluid swirling around the vortex centre acquires additional relative vorticity when moving northward or southward; the northward moving fluid elements acquire additional negative relative vorticity, while those moving southward acquire positive relative vorticity. As a result, a dipolar vorticity perturbation is generated on top of the primary monopolar vorticity distribution of the vortex, which results in a drift of the structure. The axis of this dipolar structure is aligned to the NW for cyclonic and to the SW for anticyclonic vortices, respectively. As a result, these vortices will drift to the NW and to the SW, respectively. Likewise, a columnar vortex over a sloping bottom will show a similar drift, the compass direction $S \rightarrow N$ being equivalent to the upslope direction in the fluid tank.

30.2.1.3 Ekman Layers

The presence of a solid horizontal boundary imposing a no-slip condition implies the presence of a so-called Ekman layer. Likewise, such a layer is present at the free surface of the fluid layer, when imposed to wind stress. In such boundary layers the Coriolis force is of the same order as the viscous forces, that is, $\left|2\vec{\Omega}\times\vec{v}\right| \sim \left|\nu\nabla^2\vec{v}\right|$. Since $\nabla^2 \sim \partial^2/\partial z^2 \sim 1/\delta^2$, with δ the thickness of the Ekman layer, this balance of forces yields $\delta \sim (\nu/\Omega)^{1/2} = LE^{1/2}$. In most geophysical and environmental situations, and also in many laboratory tank experiments on rotating flows, the Ekman layer thickness is hence much smaller than the typical horizontal length scales: $\delta = LE^{1/2} \ll L$.

In addition to providing a matching to the no-slip horizontal boundary or to accommodating the wind stress exerted at the free surface of the fluid layer, the Ekman layer also induces a suction (or blowing) flow to the inviscid flow domain outside the boundary layer. For a geostrophically balanced flow with relative vorticity ω_I over a solid bottom boundary at $z = 0$, the Ekman layer at this bottom imposes a so-called *Ekman suction condition*, which is expressed as

$$\omega_I(z=0) = \frac{1}{2}\delta\omega_I \tag{30.9}$$

with $\delta = (\nu/\Omega)^{1/2}$ the Ekman-layer thickness. Likewise, the suction condition imposed by the Ekman layer accommodating the wind stress at the free surface is proportional to ω_I minus the vertical component of the curl of the wind stress $\vec{\tau}$ (i.e., $\vec{k}\cdot\nabla\times\vec{\tau}$, with $\vec{k}$ the unit vector in the vertical direction). Although the suction velocities induced by the Ekman layers are small (typically $O(E^{1/2}V)$, with V the horizontal velocity scale), they drive an internal circulation in the geostrophic fluid column that may result in a change in its relative vorticity ω_I. For example, in a columnar vortex in a fluid layer over a solid no-slip bottom the Ekman suction generally drives an internal circulation that gradually changes its radial vorticity distribution $\omega_I(r)$, eventually resulting in the complete spin-down of the vortex.

Also, the secondary circulation driven by the Ekman suction may gradually change the vorticity distribution associated with the primary vortex flow, which may subsequently become unstable. This is observed, for example, in the formation of a tripolar vortex arising as a result of an unstable monopolar vortex, see van Heijst and Kloosterziel (1989).

In most situations it suffices to use the linear Ekman condition (30.9). Zavala Sansón and van Heijst (2002) have extended the linear analysis by including the nonlinear terms associated with the secondary $O(E^{1/2})$ circulation driven by the Ekman layers.

30.2.2 Effects of Density Stratification

The dynamics of large- and meso-scale vortices is often to some extent influenced by the presence of density stratification, which may be either continuous or step-like. A stable density stratification has a tendency to suppress vertical motion, and hence to make a flow planar. On the other hand, stratification may support internal waves, as can be easily understood from the following "thought experiment" carried out in a linearly stratified fluid with a density distribution $\rho(z)$. Imagine a little fluid parcel that is displaced vertically upward over some distance—how will this parcel move when released? A displacement over a vertical distance ζ implies that the parcel is brought into an ambient with a smaller density, the difference being $\delta\rho = -\zeta d\rho/dz$. As a result, the parcel experiences a downward restoring gravity force, being equal to $g\zeta d\rho/dz$ (per unit volume). In absence of any other forces, this results in an acceleration $d^2\zeta/dt^2$ of the parcel. The equation of motion of the fluid parcel is then

$$\frac{d^2\zeta}{dt^2} + N^2\zeta = 0 \tag{30.10}$$

with

$$N^2 \equiv -\frac{g}{\rho}\frac{d\rho}{dz} \tag{30.11}$$

The quantity N is the so-called buoyancy frequency, which is the frequency of the oscillatory motion the displaced parcel will perform in the case of a stable stratification ($d\rho/dz < 0$). This type of motion corresponds with an internal wavelike motion in the stratified fluid with the natural frequency N. Of course, the action of viscosity will result in a gradual decay of the internal waves, until eventually a state of rest is reached.

In the case of unstable stratification ($d\rho/dz > 0$) the buoyancy frequency N is purely imaginary, and the equation of motion has exponential solutions of which one is unbounded for increasing t. This explosive behavior corresponds with strong overturning motions and hence mixing.

Obviously, stable density stratification supports internal waves and also provides a tendency for the flow to become planar. Vortices in continuously stratified fluids may have different appearances, for example, in the form of thin pancake-like vortex structures, and also in the form of blob-like volumes of well-mixed rotating fluid with a relatively sharp interface with the stratified ambient.

30.2.3 Effects of Shallowness

Fluid motion in a shallow layer is commonly assumed to be planar—and hence quasi-2D—because the horizontal length scales L are much larger than the vertical scale δ (the layer is assumed to be horizontal). According to a simple scaling argument applied to the continuity equation, vertical velocities would be typically of the order $O(V\delta/L)$, when horizontal velocities are $O(V)$. For example, the flow in a shallow fluid layer with a no-slip bottom and a stress-free upper surface would be planar, with a Poiseuille-like vertical structure. Vortex flows would then have a similar planar structure, with a Poiseuille-like profile in the vertical. It has been shown by Duran-Matute et al. (2010), however, that this scaling does not always apply, and that locally substantial vertical motions may occur. This feature will be discussed and illustrated in more detail in the next section.

30.2.4 Appearance of Vortices

Vortices observed in environmental situations, whether or not influenced by background rotation or stratification, or occurring in shallow-layer geometries or not, can have different appearances. The most common vortex is the single, *monopolar vortex*, characterized by a nested set of closed streamlines around the

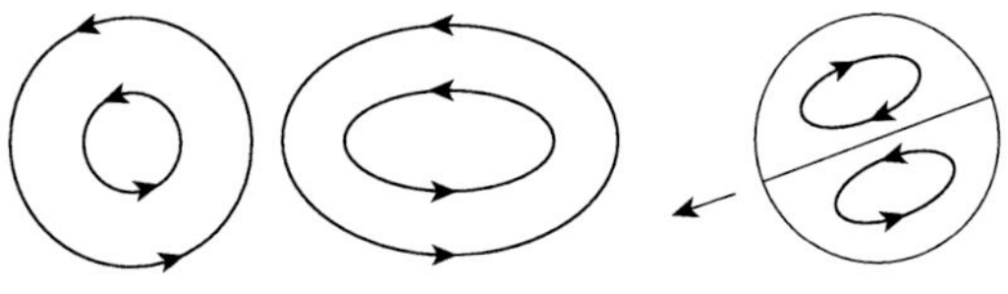

FIGURE 30.2 Schematic streamline patterns of a monopolar vortex with (a) circular streamlines or (b) elliptical streamlines, and of (c) a dipolar vortex.

vortex centre. In a quasi-steady state these streamlines are circular or elliptical. In terms of global quantities, such vortices contain a non-zero amount of angular momentum.

In some cases one may observe *dipolar vortices* consisting of a compact combination of two oppositely rotating vortices. This coherent structure shows a steady translation, along a straight line when symmetric, and moving along a circular trajectory when the dipole is not symmetric. The streamline pattern of a dipolar vortex (observed in a frame co-moving with the vortex) is characterized by two nested sets of closed streamlines, surrounded by a separatrix, with non-closed streamlines outside this separating streamline, see Figure 30.2.

In contrast to the single vortex, the dipolar structure contains net linear momentum (hence its propagation). The area enclosed by the separatrix is sometimes called the "atmosphere" of the dipole. When steady, this region remains closed, implying that material contained in this atmosphere remains trapped and is hence carried forward by the translating dipole. Because of this property, dipoles are believed to provide an important contribution to net transport of tracer material.

Although far less common, a *tripolar vortex* structure has also been observed to occur in natural situations (see, e.g., Pingree and Le Cann, 1992). This flow structure consists of an elliptical core vortex, flanked by two oppositely signed satellite vortices; the three vortices form a compact structure that rotates steadily as a whole, in the same direction as the rotation of the core vortex. It has been found in laboratory experiments and in numerical flow simulations that the tripolar vortex may emerge as the result of barotropic instability of a monopolar vortex (see, e.g., van Heijst et al., 1989, 1991). Such vortices may also be centrifugally unstable, usually giving rise to vigorous overturning motions, resulting in an explosive disintegration of the vortex. For a review of this type of behavior, see Hopfinger and van Heijst (1993).

Vortices in a density-stratified fluid may undergo instabilities in many different ways, see, for example, the review by Chomaz et al. (2010).

30.3 Methods and Analysis

Important insight in the basic dynamics and the transport properties of geophysical and environmental vortices has been gained by carefully designed laboratory experiments. In the first part of this section, we will discuss some techniques of generating vortices in the laboratory. Subsequently, some analytical modeling aspects will be reviewed.

30.3.1 Laboratory Modeling

In order to mimick environmental conditions, vortices have been studied experimentally in rotating and/or stratified fluids, and in shallow-layer configurations. The experimental techniques commonly applied in these experimental studies are dye visualization (providing qualitative information about the flow evolution) and quantitative measurement techniques like particle image velocimetry (PIV) and particle tracking velocimetry (PTV). The latter category of techniques provides quantitative information about the flow, for example, in the form of 3D velocity fields and vorticity distributions.

30.3.1.1 Vortices in a Rotating Fluid

Quasi-2D vortices can be generated most conveniently in a rotating fluid, since the background rotation tends to make the motion 2D, the vortices taking the appearance of columnar structures. One possible method is to place a thin-walled bottomless cylinder in the rotating fluid, and then stir the fluid inside this cylinder. After allowing the irregular small-scale motions to vanish and the vortex flow to get established, the vortex is released by quickly lifting the cylinder out of the fluid. The vortex structure thus generated is referred to as a "stirring vortex." When stirred in the anticyclonic direction, these vortices are usually centrifugally unstable, showing a rather explosive behavior. Somewhat depending on the vortex Rossby number, cyclonic stirring vortices usually do not show this centrifugal instability. Instead under certain conditions they may reveal a barotropic shear instability, leading to the formation of a tripolar vortex structure, as was first described by van Heijst and Kloosterziel (1989, 1991). Because these vortices are created within a solid cylinder with a no-slip wall, the total circulation—and hence the total vorticity—measured in the rotating frame is zero. In other words, stirring vortices are *isolated* vortex structures:

$$\Gamma = \oint_C \vec{v} \cdot d\vec{r} = \iint_A \omega_z dA = 0 \qquad (30.12)$$

implying that the vortex core is surrounded by a ring of oppositely signed vorticity.

An alternative way of producing vortices in a rotating fluid is by having the fluid level within the inner cylinder lower than outside: after lifting the cylinder, the resulting "gravitational collapse" implies a radial inward motion of the fluid, which by conservation of angular momentum results in a cyclonic swirling motion. After a short while, after allowing the small-scale motions created during the collapse, a columnar cyclonic vortex is formed. Alternatively, one may place a solid cylindrical object in the fluid and remove it vertically: the radially inward fluid motion will be deflected by the Coriolis force and a cyclonic swirling motion develops, as in the previous case.

Another vortex generation technique, also used by Kloosterziel and van Heijst (1991), is based on removing some volume of the rotating fluid by syphoning through a perforated tube placed vertically in the fluid. Again, the radial inward motion induced by the suction is deflected by the Coriolis force, soon resulting in a columnar cyclonic vortex. Careful flow measurements, based on tracking passive tracer particles floating at the free surface or floating in the water column when neutrally buoyant, have revealed that the radial distribution of the azimuthal velocity v_θ these "sink vortices" is well approximated by

$$v_{sink}(r) = \frac{\gamma}{2\pi r}\left[1 - \exp\left(-\frac{r^2}{L^2}\right)\right] \tag{30.13}$$

with γ the total circulation of the vortex and L a typical radial length scale.

In contrast, the stirring-induced vortices have a velocity distribution that is approximated by

$$v_{stir}(r) = \frac{1}{2}\omega_0 r \exp\left(-\frac{r^2}{L^2}\right) \tag{30.14}$$

with $\omega_0 = \omega_z(r = 0)$ the value of the vorticity in the vortex center.

In an alternative way, barotropic vortices—either cyclonic or anticyclonic—may be generated by rotating a solid flap around a vertical axis over a certain angle. A drawback of this generation technique is that, in addition to the starting vortex at the edge of the moving flap, a second vortex is produced when the flap decelerates and stops. The anticyclonic part of this vortex pair is usually observed to be centrifugally unstable, as also observed by Kloosterziel and van Heijst (1991).

Dipolar vortices have been generated in a rotating fluid by applying a pulsed planar jet through a vertical slit in a feeding cylinder. Deflection by the Coriolis force can be minimized by mounting a flat plate on the right-hand side of the split (seen in the downstream direction of the jet). In this way the pulsed jet gives rise to a more or less symmetric columnar vortex dipole. A different dipole generation method consists of moving a thin-walled open cylinder horizontally through the fluid, while gradually lifting it out of the fluid. When slowly moved along a straight line, after its removal the wake of this cylinder soon becomes organized in the form of a columnar dipolar structure.

Although viscous effects may be very small in the columnar vortex (which is in a geostrophic or gradient flow balance, depending on the Rossby number), the Ekman layer at the tank bottom produces a "suction" or "blowing" velocity that drives secondary circulation within the vortex column. Although this circulation is relatively weak (being $O(E^{1/2}V)$, with V the typical velocity of the primary vortex flow), it leads to the spin-down of the vortex. The e-folding time scale of this decay process is the so-called Ekman timescale $T_E = H/(\nu\Omega)^{1/2}$, with H the fluid depth.

The generation techniques of single and dipolar vortices in a rotating fluid have been applied by a number of researchers, for example, in studies of vortex spin-down, instability, interactions, and motion over various types of bottom typography. An overview of these applications, and also a more detailed description of the generation techniques, is given by Hopfinger and van Heijst (1993) and by van Heijst and Clercx (2009).

30.3.1.2 Vortices in a Stratified Fluid

In a continuously stratified fluid, single vortices have been created in a number of different ways, for example, by rotating an L-shaped rod or by a spinning sphere fixed to a vertical rod. In both cases the rotation of the spinning device adds angular momentum to the fluid, which is swept outward by centrifugal forces. After the forcing is stopped, the device is carefully lifted out of the fluid, and the flow becomes organized in a laminar horizontal vortex motion. Vortices generated in this way typically have a pancake-shape, with the vertical dimension of the swirling region being substantially smaller than its horizontal size.

Alternatively, a single vortex may be created by tangential injection of fluid in a thin-walled, bottomless cylinder that is immersed in the stratified fluid. By carefully lifting this injection device out of the fluid, the swirling fluid volume is released in the stratified ambient, thus resulting in a pancake-like vortex structure.

Also, translating bodies (spheres, cylinders, or "cigar-shaped bodies") produce wakes in which planar vortices are seen to emerge. Usually these vortex structures are very persistent, showing a very slow decay governed by vertical diffusion of vorticity. Van Heijst and Flór (1989) produced pancake-shaped dipolar vortices by a pulse horizontal injection of some fluid volume in a stratified ambient: after the pulsed injection was stopped, the mixed turbulent region showed a gravitational collapse, upon which the motion became planar and organized into a dipolar flow structure. When symmetric, such a dipolar pancake vortex was observed to translate along a straight trajectory. By using this dipole generation technique they could investigate the behavior of frontally colliding dipole vortices, both of equal and different strengths.

Columnar vortices in a stratified fluid have been created by applying one or more rotatable flaps. Such vortices may show a fascinating zig-zag-shaped instability behavior, as discussed by Chomaz et al. (2010).

30.3.1.3 Shallow-Layer Vortices

In many environmental situations the shallowness of the flow domain plays a crucial factor in the flow evolution. The dynamics of turbulent and nonturbulent flows in such a shallow-layer configuration has been studied in the laboratory quite extensively, see, for example, Jirka and Uijttewaal (2004). Vortex structures in shallow layers have been created experimentally in different ways, for example, by rotating flaps, by localized stirring, by pulsed horizontal jets, or by electromagnetic forcing. Lin et al. (2003) and Sous et al. (2004) have performed experiments on turbulent dipolar vortices generated by an impulsive jet. Depending on the forcing strength and the layer depth, a laminar vortex dipole emerges that slowly travels away from the forcing region. These experiments have revealed a remarkable feature: for larger Reynolds numbers a band of significant vertical motion exists in front of the translating dipole, in the form of a spanwise frontal circulation cell. Apparently, significant 3D motion may be observed locally in shallow flows. This feature was investigated in more detail by Akkermans et al. (2008), who generated dipolar flow structures by electromagnetic forcing. In their experiments, the flow was driven by applying an electrical current through a layer of salty fluid, with a disk-shaped magnet mounted underneath the bottom of the fluid tank. The combination of the magnetic field and the electric current implies a Lorentz force that sets the fluid in motion. By applying a single magnet, vortex dipoles can be generated in a well-controlled way. The Stereoscopic PIV measurements by Akkermans et al. (2008) have revealed the existence of substantial vertical (3D) motion in the shallow-layer vortex dipole, not only in the form of a spanwise circulation cell but also in the interior of the dipole and in its wake. The same electromagnetic generation technique has been applied, by using large arrays of magnets with alternating polarities, in an attempt to realize (quasi) 2D turbulent flows. High-accuracy 3D-PTV flow measurements by Cieślik et al. (2010) in such a shallow-layer configuration have demonstrated that these flows indeed exhibit significant 3D features. Obviously, the commonly adopted "classical" argument (see Section 30.2.3) that shallowness ensures such flows to be quasi-2D does not always apply. In this context, we also refer to the studies of Duran-Matute et al. 2010) and Kamp (2012), in which a more subtle scaling analysis is provided and in which it is shown that the vertical flow component is directly related to the strain and/or vorticity of the horizontal flow field.

30.3.2 Analytical Models

The most simple model of a single vortex is provided by the point vortex, whose azimuthal velocity is given by

$$v_\theta(r) = \frac{\gamma}{2\pi r} \tag{30.15}$$

with r the radius measured from the origin, and γ the strength of the point vortex. This strength γ is identical to the circulation of the point vortex, corresponding with its total vorticity—entirely concentrated singularly in its origin. The flow outside the singularity is irrotational. Multipolar vortices are nicely modeled by combinations of point vortices. For example, a dipolar vortex is conveniently described by a set of point vortices with strengths γ and $-\gamma$, at some distance d apart. Each point vortex is advected with the velocity locally induced by the other, in this case resulting in a translation speed $V = \gamma/2\pi d$. In the reference frame co-moving with the translating dipole, the streamline pattern contains a so-called separatrix, being a streamline that separates an inner region of nested sets of closed streamlines from the outer region in which the streamlines are not closed (see Figure 30.3). The region inside the separatrix is sometimes referred to as the "atmosphere" of the vortex dipole, since the material trapped within this closed streamline is not able to escape: it is carried with the translating point vortices. These characteristics of dipolar vortices (translation, transport of trapped tracers) are nicely captured by this point-vortex model, despite its simplicity.

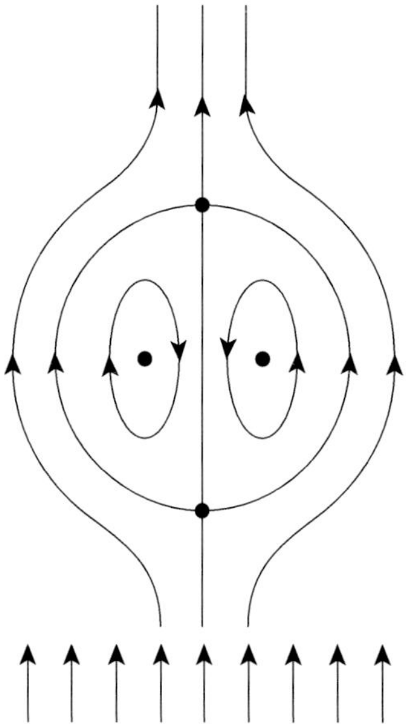

FIGURE 30.3 Streamline pattern associated with a dipolar vortex, as observed in a co-moving frame of reference.

Point vortices have also been used to model other multipolar vortex structure, like the tripolar vortex.

The singularity in the origin of the point vortex can be removed by replacing its central part by a core of solidly rotating fluid. In this so-called Rankine vortex, the azimuthal velocity v_θ is bounded, while the vorticity ω has a constant, finite value in the core region:

$$v_\theta = \Omega r, \quad \omega = 2\Omega \quad \text{for } 0 \le r \le a \tag{30.16a}$$

$$v_\theta = \frac{\Omega a^2}{r}, \quad \omega = 0 \quad \text{for } r > a \tag{30.16b}$$

Although the singularity in the origin has been removed, this vortex model is still not "ideal," because the jump in dv_θ/dr (and in ω) at $r = a$ implies a discontinuity in the shear stress, which is physically not realistic.

A smooth v_θ- and ω-distribution is provided by the Lamb-Oseen vortex solution:

$$v_\theta(r) = \frac{\gamma}{2\pi r}\left[1 - \exp\left(\frac{-r^2}{R^2}\right)\right] \tag{30.17a}$$

$$\omega(r) = \frac{\gamma}{\pi R^2}\exp\left(\frac{-r^2}{R^2}\right) \tag{30.17b}$$

with R a typical length scale. This vortex model can be considered as a "snapshot" of a viscously decaying point vortex of strength γ. This vortex contains a net amount of vorticity, as is easily verified by integrating $\omega(r)$ over a circular area with radius $r \to \infty$.

Isolated vortices $\left(\text{with zero net vorticity, i.e., } \int \omega dA = 0\right)$ are conveniently represented by the following relatively simple model:

$$v_\theta(r;\alpha) = \frac{1}{2} r \exp(-r^\alpha) \tag{30.18a}$$

$$\omega(r;\alpha) = \left[1 - \frac{1}{2}\alpha r^\alpha\right] \exp(-r^\alpha) \tag{30.18b}$$

with α a so-called steepness parameter that controls the shape of the profiles. This formulation (30.18) represents a wide class of swirl velocity and vorticity profiles, the latter characterized by a ring of oppositely signed vorticity around the vortex core region.

For the particular case $\alpha = 2$ one obtains

$$v_\theta(r) = \frac{1}{2} r \exp(-r^2) \tag{30.19a}$$

$$\omega(r) = (1 - r^2)\exp(-r^2) \tag{30.19b}$$

which is known as the "Gaussian vortex." This name is due to the shape of the streamline distribution $\psi(r)$, which is given by

$$\psi(r) = \frac{1}{2} \exp(-r^2) \tag{30.20}$$

with $v_\theta = -d\psi/dr$. Because of its relatively simple mathematical structure, the Gaussian vortex is often used to fit observational velocity data of, for example, oceanic, environmental, and laboratory vortices.

These vortex models concern 2D vortices, that is, they describe the radial distributions of v_θ and ω in a cross-sectional plane of the columnar vortex structure. The "Burgers vortex" is a 3D vortex model, essentially describing a radially diffusing line vortex (with circulation Γ) placed in an axial straining flow with velocity components

$$\vec{v} = (v_r, v_\theta, v_z) = \left(-\frac{1}{2}\alpha r, v_\theta(r), \alpha z\right) \tag{30.21}$$

with α now a constant defining the rate of strain of the flow in the r, z-plane, and $v_\theta(r)$ representing the swirl component. A steady solution of the following form exists:

$$v_\theta(r) = \frac{\Gamma}{2\pi r}\left[1 - \exp\left(\frac{-\alpha r^2}{4\nu}\right)\right] \tag{30.22a}$$

with the corresponding vorticity distribution

$$\omega(r) = \frac{\alpha\Gamma}{4\pi\nu} \exp\left(\frac{-\alpha r^2}{4\nu}\right) \tag{30.22b}$$

Obviously, the vorticity distribution of the Burgers vortex corresponds with that of the (2D) Lamb-Oseen vortex, and is concentrated within a core of radius $O(\nu/\alpha)^{1/2}$. The Burgers vortex is a remarkable example of a steady flow governed by an exact balance between advection, intensification (due to stretching), and diffusion of vorticity: the radially outward diffusion of vorticity is counteracted by the radially inward advection of vorticity in combination with the stretching-induced intensification associated with the axial strain flow. The Burgers vortex model may serve as a model (when only considering the half-planes $z > 0$ or $z < 0$) of an atmospheric tornado or of, for example, a bathtub vortex, respectively.

30.3.3 Numerical Modeling

The dynamics of vortex structures may be simulated numerically, by adopting a variety of approaches. A rather simple approach is based on the use of one or more point vortices, whose motion is simply governed by the velocity locally induced by the neighboring point vortices. Continuous vorticity distributions have been modeled by using a large number of interacting point vortices, usually of equal (in some cases oppositely signed) strength. The motion of the individual point vortices is conveniently calculated by using the so-called vortex-in-cell method, in which the advection of the point-vortices is computed from the vorticity and velocity distribution in the nodes of an interpolation grid. This method allows efficient, accurate calculations of the induced advection of the point vortices in the cloud. Remarkably, the collective behavior of the point vortices usually closely corresponds with the evolution of continuous vorticity distributions.

In order to model variations in the fluid column height H or in the background vorticity f, leading to changes in the relative vorticity ω as governed by conservation of the potential vorticity PV—see (30.8)—the strengths of the point vortices may be "modulated," see, for example, van Heijst (1994). Obviously the calculation of the evolution of systems of modulated point vortices can not be performed analytically, and has to be carried out numerically. A nice example of a study of a topography-modulated tripolar vortex based on three modulated point vortices is reported by Velasco Fuentes et al. (1996). The vortex evolution as well as the tracer transport according to this modulated model was found to agree remarkably well with the behavior observed in laboratory experiments.

Rather than using a cluster of point vortices to model a continuous vorticity distribution, a patch of uniform vorticity may be used. The evolution of such a vorticity patch may be conveniently studied by applying the so-called contour-dynamics method. This method is based on calculating the deformation of the contour of the vorticity patch by tracing the displacement of nodal points that are distributed over the contour. As the contour may become rather convoluted during the vortex evolution, additional nodal points are required in parts of the contour where substantial stretching or large curvature occurs. Whenever initially specified criteria of stretching or curvature are exceeded, the calculation is stopped, additional nodal points are added on the initial contour, and the calculation of the contour evolution is started again.

Since elongated, thin filaments of vorticity are dynamically less important, one often applies the technique of "contour surgery," which essentially cuts and removes parts of the contour that are filament-like.

Clearly, this contour-dynamics method allows modeling of continuous vorticity distributions by a nested set of patches of stepwise changing uniform vorticity, the boundary of each represented by a contour.

These vortex simulation methods are quite efficient for studying the evolution of inviscid vortices. When viscous effects are not negligibly small, one should apply full numerical simulations based on the Navier-Stokes equations, either in 2D or in 3D form.

30.4 Applications

In this section we will briefly discuss two specific cases of vortices in environmental situations: (1) the tidal flushing of an estuary by dipolar vortex structures and (2) the behavior of rip currents in the coastal zone. For both cases it is assumed that the size of the vortices is such that effects due to (gradients in) the Earth's rotation are negligibly small.

30.4.1 Tidal Flushing of an Estuary

Separation of tidal flow past a pronounced headland may result in the formation of a well-defined vortex structure. Likewise, the tidal flushing of an estuary through a narrow opening connecting it with the open sea may result in a pair of counter-rotating vortices. When closely enough, these vortices may form a dipolar structure that tends to propagate away from the tidal inlet (see, e.g., Fujiwara et al., 1994). When strong enough, such a dipolar vortex may propagate into the open sea during the ebb phase, while during the flood phase of the tidal cycle it may propagate in the opposite direction, that is, into the estuary. The flow on the high-water side of the inlet has an essentially different character: it is a convergent sink flow, directed more-or-less radially toward the inlet opening. This asymmetry of inflow and outflow results in a net exchange of tracer material between the estuary and the open sea, and is hence important for the transport of biological and chemical substances (see, e.g., Brown et al., 2001).

In an idealized model of the tidal channel between estuary and open sea, Wells and van Heijst (2003) have estimated the size and strength of the dipolar vortex in the outflow stage by an approximate analysis of the vorticity produced in the channel during one half of the tidal cycle. The dipolar vortex was modeled as a set of two oppositely signed point vortices, which allowed an easy calculation of its propagation speed. During the inflow phase the flow has the appearance of a potential sink flow, with the local fluid velocity in good approximation inversely proportional to the distance from the channel opening. By comparing the dipole propagation speed with the local velocity during the inlet stage, they derived a criterion for the dipole to escape and to propagate away from the returning flow.

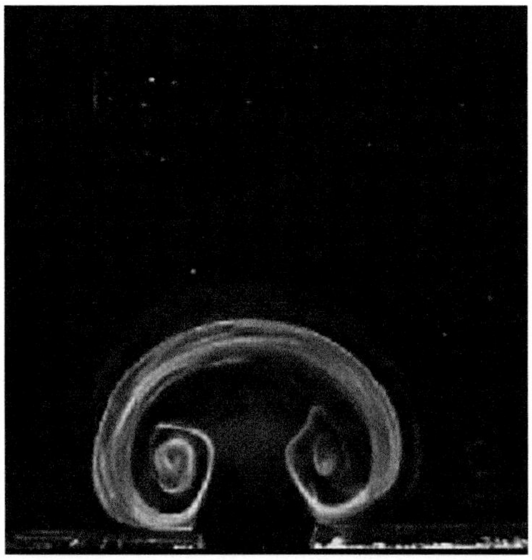

(a)

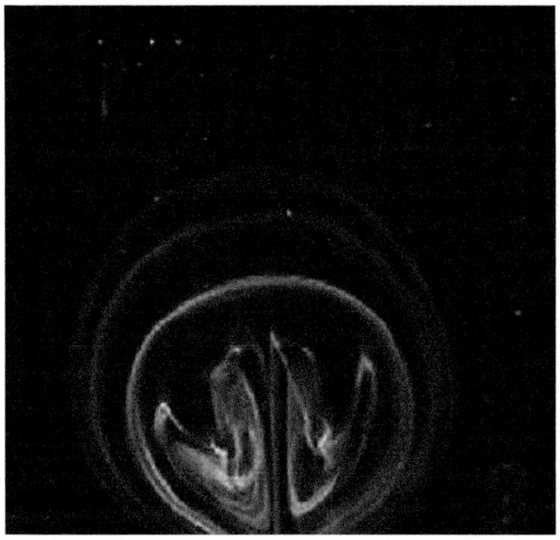

(b)

FIGURE 30.4 Laboratory visualization of a tidal exchange flow through an inlet for the case $W/UT = 0.21$, in which the dipolar vortex is not able to escape, but is instead sucked back into the return flow. Parameter values: inlet width $W = 30$ mm, maximum velocity $U = 12$ mm/s, tidal period $T = 12$ s. The photographs were taken at $t = 4$ s (a) and 10 s (b). (From Wells, M.G. and van Heijst, G.J.F., *Dyn. Atmos. Oceans*, 37, 223, 2003. With permission.)

The crucial parameter is the dimensionless ratio W/UT, with W the channel width, U the maximum velocity in the channel, and T the tidal period: for $W/UT < 0.13$ it was found that the dipoles can propagate away from the source region without being drawn back into the sink. When this parameter takes values greater than 0.13, the dipolar structure is unable to escape and is hence sucked back into the channel. Obviously, the net exchange during one tidal cycle is substantially reduced in the latter case. These model results were compared with laboratory observations, both for square and sinusoidal wave forcing, and the agreement was generally quite good (Wells and van Heijst, 2003). A laboratory visualization of a dipolar vortex produced by tidal exchange flow through an inlet is shown in Figure 30.4 for the case $W/UT > 0.13$: the dipole cannot escape and is sucked back by the return flow.

FIGURE 30.5 Rip currents observed at Rosarita Beach, Baja California, Mexico. (From Inman, Tait and Nordstrom, *J. Geophys. Res.*, 76, 1971. With permission.)

30.4.2 Rip Currents

Coastal regions show a plethora of interesting fluid-dynamical phenomena, including so-called rip currents. Incoming waves, incident from the ocean, carry energy, mass, and momentum, and dissipate by breaking in the surf zone. From mass and momentum balances, it may be understood that a wave-induced mean water level rise exists in the near-shore zone, which may drive large-scale flows. Because of the gentle slope of the beach, the predominant currents of the surf zone are horizontal, and can hence be considered as (quasi-)2D. In fact, the vertical flows in the surf zone are likely to take the character of 2D turbulence. As suggested by Peregrine (1998), it may be helpful to analyze the dynamics of surf zone flows on the basis of their vorticity, even though the measurement of large-scale vorticity in the surf zone is difficult.

An example of a flow structure commonly observed in the surf zone, and well-known for being hazardous for swimmers, is the "rip current." These currents may occur in a variety of forms of appearance, depending on the beach type and local bathymetry, see, e.g., the review by Dalrymple et al. (2011). A prominent feature is the rip current in the form of a narrow jet directed away from the coast, its head in the form of a dipolar vortex. Such currents may occur at regular intervals, as nicely illustrated in Figure 30.5.

Rip currents may thus provide an important mechanism for transport of material (pollutants, biochemical components) out of the surf zone, into the inner shelf region. Field observations have revealed that the rip currents indeed contain a relatively high concentration of material originating from the surf zone, although the flushing rate associated with their transport contribution appeared to be lower than previously estimated (see also Dalrymple et al., 2011).

When traveling away from the coastline, the current enters deeper water; and assuming a barotropic structure of the current, this implies a stretching of vortex columns. For the dipolar head structure the increasing depth would induce an intensification of the vortex dipole. At the same time, by conservation of volume the dipolar structure would become more compact. The combination of both effects tends to intensify the dipolar structure at the current head and to increase its propagation speed. This behavior is in remarkable contrast with a vortex dipole moving over a sloping bottom into shallower water: as a result of squeezing of vortex tubes and conservation of mass the vortex dipole widens and becomes less intensive. When getting closer to the coast, the dipole may even split up into two parts, with its original vortex cores initially moving in opposite directions parallel to the coastline. Additionally, secondary vorticity generated near the coastline causes the vortices to move back into the sea. This behavior has been observed in numerical simulations and in laboratory experiments with electromagnetically forced dipoles over a sloping bottom (unpublished results).

Although laboratory models of large-scale environmental flows have serious limitations (e.g., they fail to model turbulence in a proper way), these models can often be very instructive in providing important insight in the basic dynamics of such flow systems.

References

Akkermans, R.A.D., A.R. Cieslik, L.P.J. Kamp, R.R. Trieling, H.J.H. Clercx, and G.J.F. van Heijst. The three-dimensional structure of an electromagnetically generated dipolar vortex in a shallow fluid layer. *Phys. Fluids* **20**, 116601 (2008).

Brown, C.A., G.A. Jackson, and D.A. Brooks. Particle transport through a narrow tidal inlet due to tidal forcing and implications for larval transport. *J. Geophys. Res.* **105**, 24141–24156 (2001).

Chomaz, J.-M., S. Ortiz, F. Gallaire, and P. Billant. Stability of quasi two-dimensional vortices. In: *Fronts, Waves and Vortices in Geophysical Flows* (ed. J.B. Flór), *Lect. Notes Phys.* **805**, 35–59 (2010).

Cieślik, A.R., L.P.J. Kamp, H.J.H. Clercx, and G.J.F. van Heijst. Three-dimensional structures in a shallow flow. *J. Hydrol. Envir. Res.* **4**, 89–101 (2010).

Dalrymple, R.A., J.H. MacMahan, A.J.H.M. Reniers, and V. Nelko. Rip currents. *Annu. Rev. Fluid Mech.* **43**, 551–581 (2011).

Duran-Matute, M., L.P.J. Kamp, R.R. Trieling, and G.J.F. van Heijst. Scaling of shallow axisymmetric flows. *J. Fluid Mech.* **648**, 471–484 (2010).

Fujiwara, T., H. Nakata, and K. Nakatsuji. Tidal-jet and vortex-pair driving of the residual circulation in a tidal estuary. *Cont. Shelf Res.* **14**, 1025–1038 (1994).

van Heijst, G.J.F. Topography effects on vortices in a rotating fluid. *Meccanica* **29**, 431–451 (1994).

van Heijst, G.J.F. and H.J.H. Clercx. Laboratory modeling of geophysical vortices. *Annu. Rev. Fluid Mech.* **41**, 143–164 (2009).

van Heijst, G.J.F. and J.B. Flór. Dipole formation and collisions in a stratified fluid. *Nature* **340**, 212–215 (1989).

van Heijst, G.J.F. and R.C. Kloosterziel. Tripolar vortices in a rotating fluid. *Nature* **338**, 569–571 (1989).

van Heijst, G.J.F., R.C. Kloosterziel, and C.W.M. Williams. Laboratory experiments on the tripolar vortex in a rotating fluid. *J. Fluid Mech.* **225**, 301–331 (1991).

Holton, J.R. *An Introduction to Dynamic Meteorology*. Academic Press, New York (1979; 3rd edn. 1992).

Hopfinger, E.J. and G.J.F. van Heijst. Vortices in rotating fluids. *Annu. Rev. Fluid Mech.* **25**, 241–289 (1993).

Inman, D.L., R.J. Tait, and C.E. Nordstrom. Mixing in the surf zone. *J. Geophys. Res.* **76**, 3493–3514 (1971).

Jirka, G.H. and W.S.J. Uijttewaal (eds.). *Shallow Flows*. Balkema Publishers, Rotterdam, the Netherlands (2004).

Kamp, L.P.J. Strain-vorticity induced secondary motion in shallow flows. *Phys. Fluids* **24**, 023601 (2012).

Kloosterziel, R.C. and G.J.F. van Heijst. An experimental study of unstable barotropic vortices in a rotating fluid. *J. Fluid Mech.* **223**, 1–24 (1991).

Lin, J.C., M. Ozgoren, and D. Rockwell. Space-time development of the onset of a shallow-water vortex. *J. Fluid Mech.* **485**, 33–66 (2003).

Pedlosky, J. *Geophysical Fluid Dynamics*. Springer Verlag, New York (1st edn. 1979; 2nd edn. 1987).

Peregrine, D.H. Surf zone currents. *Theor. Comput. Fluid Dyn.* **10**, 295–309 (1998).

Pingree, R.D. and B. Le Cann. Anticyclonic Eddy X91 in the southern Bay of Biscay, May 1991 to February 1992. *J. Geophys. Res.* **97**, 14353–14367 (1992).

Sous, D., N. Bonneton, and J. Sommeria. Turbulent vortex dipoles in a shallow water layer. *Phys. Fluids* **16**, 2886–2898 (2004).

Velasco Fuentes, O.U., G.J.F. van Heijst, and N.P.M. van Lipzig. Unsteady behaviour of a topography-modulated tripolar vortex. *J. Fluid Mech.* **307**, 11–41 (1996).

Wells, M.G. and G.J.F. van Heijst. A model of tidal flushing of an estuary by dipole formation. *Dyn. Atmos. Oceans* **37**, 223–244 (2003).

Zavala Sansón, L. and G.J.F. van Heijst. Ekman effects in a rotating flow over bottom topography. *J. Fluid Mech.* **471**, 239–255 (2002).

31

Surface Waves in Coastal Waters

Chiang C. Mei
Massachusetts Institute of Technology

31.1 Introduction

Among different types of waves in essentially homogeneous seas, those with frequencies in the range between a few seconds to a few hours are considered as surface waves. Generated by wind, subsea earthquakes, or man-made disturbances, surface waves not only affect human activities and marine life directly but can also contribute to weather variation and the geological evolution of the coast. It is therefore natural that the phenomenon of surface waves has fascinated artists, poets, sailors, engineers, and scientists alike. While many aspects of the surface-wave physics and applications have been well expounded in textbooks (Dean and Dalrymple 1990, Mei 1989, or Mei et al. 2005a,b), more unknowns remain to be uncovered. A major source of scientific challenge is nonlinearity.

In the deep sea, the complete prediction of wave climate must take into account wind forcing, linear and nonlinear interaction of waves of various frequencies and dissipation by breaking. Due to the complex generation mechanisms and the inherent nonlinearity, waves in the open sea are in general highly irregular. Considerable progress has been made in the mathematical modeling of these aspects for the deep sea (Komen et al. 1994). In coastal waters friction losses on the seabed and radiation losses due to multiple scattering by depth irregularities are of additional importance, and are less studied.

For some irregularities, a convenient idealization is to assume randomness. In this chapter we limit our survey to some recent theories on the effects of random bathymetry and random waves when nonlinearity is taken into account. For simplicity the two kinds of randomness are treated separately. In Section 31.2 we first use the linearized theory to explain a method for analyzing the effects of multiple scattering by an irregular seabed. A nonlinear example of nonlinear harmonic generation is then briefly discussed. The horizontal length scale of bathymetry is assumed to be no greater than the typical wavelength. Otherwise, a deterministic and strictly computational model is more appropriate. In Section 31.3 attention is turned to the randomness in waves. The nonlinear excitation of long waves in a harbor by short random waves will be examined.

31.2 Irregular Bathymetry

Propagation of deterministic waves in disordered (random) media is an important topic in many branches of physics and engineering, and has been widely studied within the framework of linear theories (Chernov 1967, Ishihara 1997). For 1D monochromatic waves in a large region of random inhomogeneities, it is known that multiple scattering yields a small complex shift of the propagation constant with the real part corresponding to a change of wavenumber while the imaginary part to a gradual attenuation in space (Karal and Keller 1964, Keller 1964). Unlike waves through periodic inhomogeneities where strong (Bragg) scattering occurs only for certain discrete frequencies, effects of random scattering can be significant over a broad range of frequencies.

The irregular features of surface waves themselves have been treated by statistical theories with emphasis on wind input and nonlinear interaction among waves of different frequencies largely for deep seas. The literature of random waves over random seabed is scarce. Hasselman (1966) and Long (1973) employed diagramatic techniques to treat the spectral evolution of waves of intermediate length on a seabed of constant mean depth. Using the same technique, Elter and Molyneux (1972) studied tsunami propagation based on the linearized theory of

Handbook of Environmental Fluid Dynamics, Volume One, edited by Harindra Joseph Shermal Fernando. © 2013 CRC Press/Taylor & Francis Group, LLC.
ISBN: 978-1-4398-1669-1.

long waves in shallow water. Relevant to wind-induced waves, the behavior of the averaged wave height of infinitesimal waves in intermediate depth has been studied by Nachbin (1995, 1997) and Nachbin and Papanicolaou (1992). Nonlinear effects of long waves in shallow water have been studied by Rosales and Papanicolaou (1983), among others.

It is known for either an isolated obstacle or for many periodically spaced obstacles that scattering is the most significant when the wavelength is comparable to either the size of or the spacing between obstacles. In this review we shall focus on deterministic waves over a seabed with small and random depth fluctuations whose correlation length is comparable to the typical wavelength. Under these assumptions the disorder alters the leading waves only after a distance much longer than the typical wavelength by the cumulated effects of multiple scattering. For fluctuations with zero statistical mean, it is the mean square that matters, hence the length scale of slow evolution is on the order $1/\mu^2$ times that of the wave length. For this reason the asymptotic method of multiple scales will be used to examine the slow evolution of waves. The basic ideas will be demonstrated for the simplest example of linear nondispersive long waves in shallow water. Extensions to include nonlinearity and dispersion will be briefly discussed with less details.

31.2.1 Linearized Long Water Waves

Consider 1D long waves in shallow water with typical depth H and wavelength $2\pi/K$ where K is the characteristic wavenumber. For very long waves, $KH = \mu \ll 1$, the fluid velocity $u(x,t)$ is essentially horizontal and the pressure hydrostatic. The linearized conservation laws of mass and momentum are

$$\frac{\partial \eta^*}{\partial t^*} + \frac{\partial (h^* u^*)}{\partial x^*} = 0, \quad \text{and} \quad \frac{\partial u^*}{\partial t^*} + g\frac{\partial \eta^*}{\partial x^*} = 0 \tag{31.1}$$

where

$h^*(x)$ denotes the still water depth

$\eta^*(x, t)$ denotes the free surface displacement

Using the following normalization

$$x = Kx^*, \quad t = t^* K\sqrt{gH}, \quad \eta = \frac{\eta^*}{a}, \quad h = \frac{h^*}{H'} \quad u = \frac{u^*}{a\sqrt{g/H}} \tag{31.2}$$

where a is the characteristic wave amplitude, the conservation laws take the following form:

$$\frac{\partial \eta}{\partial t} + \frac{\partial (hu)}{\partial x} = 0, \quad \text{and} \quad \frac{\partial u}{\partial t} + \frac{\partial \eta}{\partial x} = 0 \tag{31.3}$$

Focusing on random fluctuates of depth only, we assume h deviates slightly from a constant mean value, that is, $h(x) = 1 - \mu b(x)$ where $b(x)$ is assumed to be a stationary random function of x with zero mean: $\langle b(x)\rangle = 0$. As long as $a/H \ll \mu^2$, nonlinear effects can be ignored. Combination of the two laws in (31.29) yields a single equation for the free surface displacement η.

$$\frac{\partial^2 \eta}{\partial t^2} - \frac{\partial^2 \eta}{\partial x^2} = -\mu\left(b\frac{\partial \eta}{\partial x}\right) \tag{31.4}$$

31.2.1.1 Asymptotic Analysis

Anticipating that the energy scattered within a wavelength is as small as the mean square of disorder, that is, $\mathcal{O}(\mu^2)$, the accumulated effect becomes significant only after a distance on the order $1/\mu^2$ times the typical wavelength. This suggests the introduction of fast and slow coordinates x and $X = \mu^2 x$ and expand η in ascending powers of μ:

$$\eta = \eta_0 + \mu\eta_1 + \mu^2\eta_2 + \cdots \tag{31.5}$$

where $\eta_j = \eta_j(x, X, t)$. At orders $\mathcal{O}(1)$, $\mathcal{O}(\mu)$, $\mathcal{O}(\mu^2)$, perturbation equations are found from (31.4):

$$\frac{\partial^2 \eta_o}{\partial t^2} - \frac{\partial^2 \eta_o}{\partial x^2} = 0 \tag{31.6}$$

$$\frac{\partial^2 \eta_1}{\partial t^2} - \frac{\partial^2 \eta_1}{\partial x^2} = -\frac{\partial}{\partial x}\left(b\frac{\partial \eta_o}{\partial x}\right) \tag{31.7}$$

$$\frac{\partial^2 \eta_2}{\partial t^2} - \frac{\partial^2 \eta_2}{\partial x^2} = -\frac{\partial}{\partial x}\left(b\frac{\partial \eta_1}{\partial x}\right) + 2\frac{\partial^2 \eta_o}{\partial X \partial x} \tag{31.8}$$

We first solve the perturbation problems at $\mathcal{O}(\mu^0)$. Consider the evolution of a train of progressive waves with amplitude A.*

$$\eta_o = \frac{1}{2}(A(X)e^{i\theta} + *), \quad \text{with} \quad \theta = kx - \omega t \tag{31.9}$$

The slow variation of amplitude $A(X)$ is yet unknown. At the order $O(\mu)$ we have

$$\frac{\partial^2 \eta_1}{\partial t^2} - \frac{\partial^2 \eta_1}{\partial x^2} = -\frac{\partial}{\partial x}\left(b\frac{\partial \eta_o}{\partial x}\right) = -Fe^{-i\omega t} + * \tag{31.10}$$

where the coefficient F

$$F = \frac{1}{2}ikA(X)\frac{d}{dx}\left[b(x)e^{ikx}\right] \tag{31.11}$$

* For ease of later extension, the frequency and wavenumber are denoted by ω and k respectively, although they can be taken as unity here.

is a random function of x. The solution can be found as

$$\eta_1 = \bar{\eta}_1(x)e^{-i\omega t} + *$$

$$= e^{-i\omega t}\int_{-\infty}^{\infty} G(|x - x'|)ik\frac{A(X)}{2}\frac{d}{dx'}\left[b(x')e^{ikx'}\right]dx' + c.c. \quad (31.12)$$

where

$$G(|x - x'|) = \frac{e^{ik|x-x'|}}{2ik} \quad (31.13)$$

is the Green function which behaves as outgoing waves at infinities. Clearly $\langle\eta_1\rangle = 0$.

Taking the ensemble average of Equation 31.8, we get

$$\left(\frac{\partial^2}{\partial t^2} - \frac{\partial^2}{\partial x^2}\right)\langle\eta_2\rangle = -\frac{\partial}{\partial x}\left\langle b\frac{\partial\eta_1}{\partial x}\right\rangle + 2\frac{\partial^2\eta_o}{\partial X\partial x} \quad (31.14)$$

Using the known solution for η_1, we can write

$$\left\langle b\frac{\partial\bar{\eta}_1}{\partial x}\right\rangle = \int_{-\infty}^{\infty} ik\,\mathrm{sgn}(x - x')G(|x - x'|)ik\frac{A}{2}\frac{d}{dx'}\left[\langle b(x)b(x')\rangle e^{ikx'}dx'\right]$$

$$= Ae^{ikx}\int_{-\infty}^{\infty}\mathrm{sgn}(x - x')\frac{1}{4}e^{ik(|x-x'|)}\frac{d}{dx'}\left[\Gamma(x - x')e^{ik(x'-x)}\right]dx'$$

$$= -Ae^{ikx}\int_{-\infty}^{\infty}\frac{1}{4}\mathrm{sgn}(\xi)e^{ik|\xi|}\frac{d}{d\xi}[\sigma^2\Gamma(\xi)e^{-ik\xi}]d\xi \quad (31.15)$$

where $\sigma^2\Gamma$ is the covariance of b, which is assumed to be a stationary random function of x on the fast scale

$$\langle b(x)b(\xi)\rangle = \sigma^2\Gamma(\xi) \quad (31.16)$$

Defining the complex coefficient β by

$$\beta = \frac{\sigma^2}{4}ik\int_{-\infty}^{\infty}\mathrm{sgn}(\xi)\left(\frac{d\Gamma}{d\xi} - ik\Gamma\right)e^{ik(|\xi|-\xi)}d\xi \quad (31.17)$$

(31.14) can be rewritten as

$$\left(\frac{\partial^2}{\partial t^2} - \frac{\partial^2}{\partial x^2}\right)\langle\eta_2\rangle = ikA\beta e^{i\theta} + ik\frac{dA}{dX}e^{i\theta} + c.c. \quad (31.18)$$

The forcing term on the right-hand side is proportional to the homogeneous solution and must be removed to avoid unbounded resonance; hence,

$$\frac{dA}{dX} + \beta A = 0 \quad (31.19)$$

The solution is

$$A(X) = A(0)e^{-\beta X} = A(0)e^{-X\Re\beta}e^{-iX\Im\beta} = A(0)e^{-x\mu^2\Re\beta}e^{-ix\mu^2\Im\beta} \quad (31.20)$$

Clearly the wave is attenuated (localized) with the decay distance $L_{loc} = 1/\mu^2\Re\beta$. The wavenumber is also modified by $\mu^2\Im\beta$. This result is a special case of a more general nonlinear theory for multiple harmonics to be discussed later.

For the special case of Gaussian correlation

$$\langle b(x)b(x')\rangle = \sigma^2\exp\left(-\frac{(x - x')^2}{2l^2}\right) = \sigma^2\exp\left(-\frac{\xi^2}{2l^2}\right), \quad \xi = x - x' \quad (31.21)$$

where

$\sigma(X)$ is the root-mean-square amplitude

l is the correlation distance, it has been shown by Grataloup and Mei (2003) that

$$\beta = \frac{\sigma^2}{8}k^2l\sqrt{2\pi}\left(1 + e^{-2k^2l^2}\right) - i\frac{\sigma^2}{2}k\left(1 - \frac{kl}{\sqrt{2}}e^{-2k^2l^2}\int_0^{\sqrt{2kl}} e^{u^2}du\right) \quad (31.22)$$

In Figure 31.1 $\Re\beta_m(k_m)$ and $\Im\beta_m(k_m)$ are plotted for all $m = 1, 2, 3, \ldots$ which are for the nonlinear theory to be discussed later. For the present problem we can take $k = k_m$ and $\beta = \beta_m$. Note that $\Re\beta > 0$ so that disorder leads to exponential attenuation, similar to the localization phenomenon in theoretical physics (Anderson 1958, Sheng 2006).

31.2.1.2 Energy Loss to Incoherent Scattering

Since no friction is included, the total energy should be conserved. Nevertheless, energy can be drained from the mean motion by random scattering. To see the physics better let us

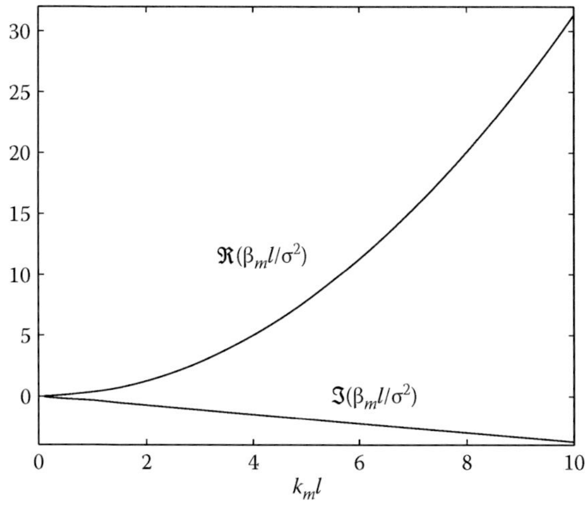

FIGURE 31.1 Real and imaginary parts of $\beta_m l/\sigma^2$.

rewrite the total free-surface amplitude $\bar{\eta}(x)$ without the time factor $e^{-i\omega t}$ as the sum of the ensemble mean $\langle\bar{\eta}\rangle$ and the random fluctuation $\bar{\eta}'$,

$$\bar{\eta} = \langle\bar{\eta}\rangle + \mu\bar{\eta}' \tag{31.23}$$

With an error of $\mathcal{O}(\mu^3)$ it is clear that $\langle\bar{\eta}\rangle = \bar{\eta}_0 + \mu^2\langle\bar{\eta}_2\rangle$ and $\bar{\eta}' = \bar{\eta}_1$. Without the artifice of fast and slow coordinates, we substitute (31.23) into (31.4) then take the ensemble average to get

$$\langle\bar{\eta}\rangle_{,xx} + k^2\langle\bar{\eta}\rangle = \mu^2\langle b\bar{\eta}_{1,x}\rangle_{,x} \tag{31.24}$$

By applying Green's formula to $\langle\bar{\eta}\rangle$ and its complex conjugate $\langle\bar{\eta}^*\rangle$ over a large region of disorder of $L = \mathcal{O}(1/\mu^2 k)$, we get

$$\int_0^L dx[\langle\bar{\eta}\rangle^*(\langle\bar{\eta}\rangle_{,xx} + k^2\langle\bar{\eta}\rangle) - \langle\bar{\eta}\rangle(\langle\bar{\eta}\rangle^*_{,xx} + k^2\langle\bar{\eta}\rangle^*]$$

$$= [\langle\bar{\eta}\rangle^*\langle\bar{\eta}_x\rangle - \langle\bar{\eta}\rangle\langle\bar{\eta}_x\rangle^*]_0^L \tag{31.25}$$

Let this equation be multiplied by $-i$. The right-hand side of (31.25) can be written as

$$2\Re[\langle\bar{\eta}\rangle^*(-i\langle\bar{\eta}\rangle_x)]_0^L \approx 2\Re[\bar{\eta}_0^*(-i\bar{\eta}_{0,x})]_0^L + O(\mu^2) \tag{31.26}$$

which is, except for the constant factor $\rho g^2/\omega$, the net power outflux of the leading-order coherent wave through the ends. By using (31.24), the left-hand side of (31.25) is

$$\mu^2\int_0^L dx[\langle\bar{\eta}\rangle^*(-i\langle b\bar{\eta}_{1,x}\rangle)_x + \langle\bar{\eta}\rangle(-i\langle b\eta_{1,x}\rangle)_x^*]$$

$$\approx 2\mu^2\Re\int_0^L dx\,\bar{\eta}_0^*(-i\langle b\bar{\eta}_{1,x}\rangle)_x$$

Since $\mu b(x)$ is the random fluctuation of depth and $-i\mu\bar{\eta}_{1,x}$ is the induced random fluctuation of the horizontal velocity, the product $-i\mu^2\langle b\bar{\eta}_{1,x}\rangle$ represents the averaged adjustment of volume flux. Thus the last integral given earlier is the accumulation of the rate of work done by the mean pressure through the fluctuating volume flux. Since the total extent of disorder is very large $L = O(1/k\mu^2)$, the integral given earlier is on the same order as (31.26). Thus

$$\Re[\bar{\eta}_0^*(-i\bar{\eta}_{0,x})]_0^L = \mu^2\Re\int_0^L dx\bar{\eta}_0^*(-i\langle b\bar{\eta}_{1,x}\rangle)_x \tag{31.27}$$

which means physically that the coherent motion loses energy by making random adjustment of velocity in response to fluctuations of the seabed. As an analogy, a jogger running on a rough road must get tired by having to make small and random adjustments of his/her speed. Equation 31.27 can be further confirmed by substituting the explicit result (31.20) on both sides.

We now turn to an example involving weak nonlinearity and dispersion in addition.

31.2.2 Harmonic Generation in Nonlinear Long Waves

When high-intensity light of ruby color shines through a slab of quartz crystal, it can turn to ultraviolet after transmission. This is the phenomenon of *second harmonic generation* (Armstrong et al. 1962). A similar phenomenon was first found experimentally for surface water waves over a horizontal bed by Goda (1967) and theoretically explained by Mei and Ünlüata (1972) borrowing the theory of nonlinear optics (See also Bryant 1976). As water waves climb a gently sloping beach, nonlinearity becomes more and more important so that the free surface tends to steepen. A train of long waves initially of single frequency can evolve into one with second, third, and higher harmonics in the course of propagation, leading to the broadening of the sea spectrum and ultimately to breaking. What if the seabed is also rough? How is harmonic generation affected by multiple scattering?

With weak nonlinearity and dispersion, long waves are best described by Boussinesq equations. In terms of the normalized variables defined in (31.2), the two conservation laws are

$$\frac{\partial\eta}{\partial t} + \frac{\partial}{\partial x}[(1 - \mu b + \varepsilon\eta)u] = 0, \text{ and } \frac{\partial u}{\partial t} + \varepsilon u\frac{\partial u}{\partial x} + \frac{\partial\eta}{\partial x} = \frac{\mu^2}{3}\frac{\partial^3 u}{\partial x^2\partial t} \tag{31.28}$$

where $\varepsilon = a/H$ signifies nonlinearity. These equations are valid when both nonlinearity $\varepsilon = a/H$ and dispersion $\mu^2 = K^2H^2$ are small and comparable, that is, $\varepsilon = \mathcal{O}(\mu^2) \ll 1$. The two conservation laws can be combined to yield

$$\frac{\partial^2\eta}{\partial t^2} - \frac{\partial^2\eta}{\partial x^2} = -\mu b\frac{\partial\eta}{\partial x} + \frac{\varepsilon}{2}\left(\frac{\partial^2 u^2}{\partial x^2} + \frac{\partial^2 u^2}{\partial t^2} + \frac{\partial^2\eta^2}{\partial t^2}\right) + \frac{\mu^2}{3}\frac{\partial^4\eta}{\partial x^4} \tag{31.29}$$

The two-scale analysis of the last section can be extended. Let

$$\eta = \eta_0 + \mu\eta_1 + \mu^2\eta_2 + \cdots, \quad u = u_0 + \mu u_1 + \mu^2 u_2 + \cdots \tag{31.30}$$

where $\eta_j = \eta_j(x, X, t)$ and $u_j = u_j(x, X, t)$. From (31.29) we find the perturbation equations for the first three orders,

$$\frac{\partial^2\eta_0}{\partial t^2} - \frac{\partial^2\eta_0}{\partial x^2} = 0 \tag{31.31}$$

$$\frac{\partial^2 \eta_1}{\partial t^2} - \frac{\partial^2 \eta_1}{\partial x^2} = -\frac{\partial}{\partial x}\left(b\frac{\partial \eta_0}{\partial x}\right) \tag{31.32}$$

$$\frac{\partial^2 \eta_2}{\partial t^2} - \frac{\partial^2 \eta_2}{\partial x^2} = -\frac{\partial}{\partial x}\left(b\frac{\partial \eta_1}{\partial x}\right) + 2\frac{\partial^2 \eta_0}{\partial x X \partial x} + \frac{3}{2}\frac{\partial^2 \eta_0^2}{\partial t^2} + \frac{v}{3}\frac{\partial^4 \eta_0}{\partial x^4} \tag{31.33}$$

The right-hand-side of (31.33) has been simplified by using the leading order approximation.

We now consider at the leading order a progressive wave train with many harmonics

$$\eta_0 = \frac{1}{2}\sum_{m=-\infty}^{\infty} A_m(X)e^{i\theta_m}, \quad u_0 = \frac{1}{2}\sum_{m=-\infty}^{\infty} B_m(X)e^{i\theta_m} \tag{31.34}$$

where

$$\theta_m = k_m x - \omega_m t, \quad k_m = mk, \quad \omega_m = m\omega \tag{31.35}$$

By similar reasoning as in the last subsection, solvability of $\langle \eta_2 \rangle$ gives rise to the following set of coupled equations (Grataloup and Mei 2003),

$$\frac{dA_m}{dX} + \beta_m A_m - i\frac{v}{6}k_m^3 A_m + \frac{3}{8}i\omega_m \times \left[\sum_{l=1}^{n-m} 2A_l^* A_{m+l} + \sum_{l=1}^{[\frac{m}{2}]} \alpha_l A_l A_{m-l}\right] = 0, \quad m = 1,2,\ldots,n \tag{31.36}$$

where

$[m/2]$ denotes the integer part of the fraction $m/2$

$\alpha_l = 1$ if $l = [m/2]$ and 2 otherwise

The coefficient $\beta_m(k_m)$ is of exactly the same form as $\beta(k)$ given by (31.17). For the Gaussian covariance (31.21), we have

$$\frac{\beta_m l}{\sigma^2} = k_m^2 l^2 \frac{\sqrt{2\pi}}{8}\left(1 + e^{-ek_m^2 l^2}\right) - \frac{i}{2}k_m l\left(1 - \frac{k_m l}{\sqrt{2}}e^{-2k_m^2 l^2}\int_0^{\sqrt{2}k_m l} e^{u^2}du\right) \tag{31.37}$$

Figure 31.1 shows the dependence of the real and imaginary parts of $\beta_m l/\sigma^2$ on $k_m l$.

Equation 31.36 has been solved by the numerical method of finite differences. In Figure 31.2 the variations of the first four harmonics are displayed for a semi-infinite ($X > 0$) region of disorder with $\sigma = 0.2$. Incident waves with only the first harmonic arrive from $X < 0$ where $\sigma = 0$. The input parameters are: $k_1 = 1$, $l = 1$, and $\nu = \mu^2/\varepsilon = 1$. At the start the second and higher harmonics grow at the expense of the first, because of nonlinearity.

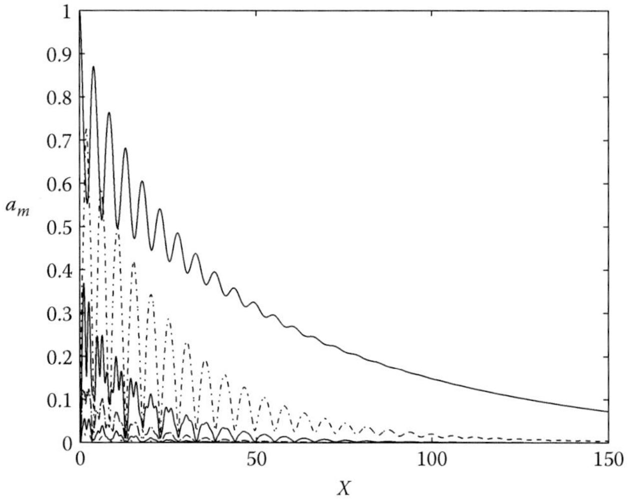

FIGURE 31.2 Evolution of the first four harmonics ($m = 1, 2, 3, 4$ from top down) over a semi-infinite region of disorder. $\sigma = 0.2$, $\nu = 1$, and $l = k_1 = 1$.

But they all attenuate with distance, with the higher harmonics dying out sooner. Once the second and higher harmonics are sufficiently diminished, energy exchange with the first harmonic becomes insignificant; the latter then attenuates monotonically with distance. Other examples can be found in Grataloup and Mei (2003).

31.2.2.1 Other Extensions

Similar treatment of transient solitary waves has been reported by Mei and Li (2004) who found the mean free-surface displacement to be governed by an integro-differential equation, which combines the features of Korteweg and de Vries and of Burgers. In addition to the usual nonlinear and dispersive terms, disorder adds diffusion and modifies the wave speed and dispersion. After entering a region of rough bed, a solution slows down gradually with the propagation distance. The tail of the wave profile flattens and lengthens. Extensions to interfacial waves can be found in Chen and Liu (1996) and Alam and Mei (2007). The theory has also been modified for weakly nonlinear Stokes waves over a rough bed with $KH = \emptyset(1)$ (Pihl et al. 2002, Mei and Hancock 2003). The wave envelope is found to be governed by a nonlinear Schrödinger equation with a new damping term. Random scattering is found to reduce side-band instability and alter the subsequent nonlinear evolution, for both 1D and 2D waves. Further generalization for the prediction of broad-banded random waves over rough and 2D bathymetry with variable mean depth is needed.

We now turn to random waves in a partially confined space.

31.3 Long-Period Oscillations in a Harbor by Short Wind Waves

Harbors are designed to protect ships from storm-induced sea waves whose typical periods are between a few to few tens of seconds. Yet oscillations with periods up to several minutes often occur inside some harbors such as Port of Long Beach, California, and can cause excessive sway of ships, damage to ship hulls and/or the mooring systems, or costly delay of loading and unloading operations. During Typhoon Long Wang on October 2, 2005, a 7000-ton cargo ship, originally moored near the northern end of the Hualien Harbor on the east coast of Taiwan (Figure 31.3), broke loose and drifted for 1 km southward, then ran aground outside the harbor and broke into halves.

The common engineering practice in coping with harbor oscillations is still guided by linear concepts suitable for the prediction of tsunami effects (Miles and Munk 1961). Specifically, in either numerical or experimental models, harbor oscillations are treated as the direct consequence of synchronous resonance by incident waves of similarly long periods. However, such an assumption overlooks the dominant presence of short waves in the incident sea. In the offshore industry there is a similar problem of slow-drift oscillations of semi-submersibles or tension-leg platforms in wind waves. Mathematical models needed for the design of dynamic positioning devices have been built on nonlinear deterministic theories for quite some time, mostly for deep waters.

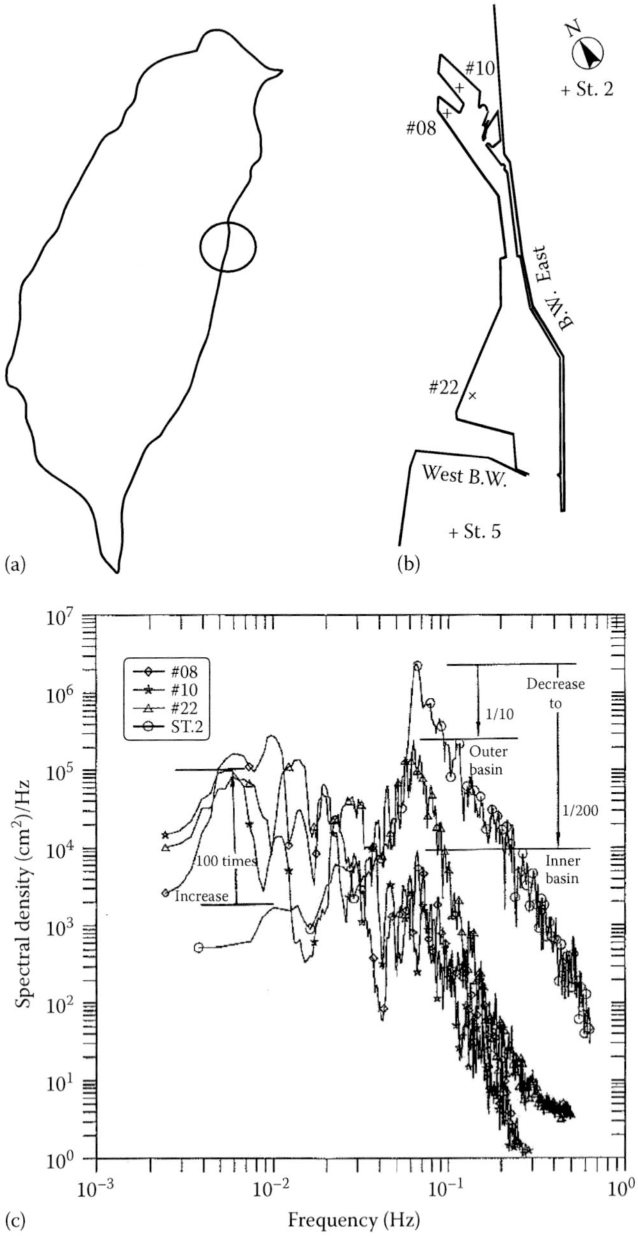

FIGURE 31.3 (a) Taiwan Island. Circle indicates Hualien Harbor facing Pacific Ocean. (b) Plan of Hualien Harbor with sites of wave-recording stations. Measuring Stations No. 8 and No. 10 are at the northern corner of the basin. Stations No. 2 and No. 5 are outside. (c) Wave spectra at various stations recorded during Typhoon Tim, 1994. Record for Station 5 is omitted.

With additional account of the finite water depth, nonlinear statistical theories are needed for long-period harbor oscillations due to short incident waves.

In an open sea of horizontal bottom, a long wave called set-down can be produced by two first-order progressive waves of much shorter and nearly equal periods, via second-order interactions. In a harbor or around a large structure, scattering and radiation at the first order further complicate the higher-order waves. For narrow-banded and deterministic short waves, Bowers (1977), Mei and Agnon (1989), and Wu and Liu (1990) have contributed perturbation theories to examine the resonance of second-order long waves. On the other hand, Sclavounos (1992) advanced a stochastic theory for random and broad-banded deep-water waves near a straight and vertical cliff. By assuming the incident waves to be a stationary and Gaussian process at the first order, he showed that the nonlinear correction to the spectrum starts at the fourth order in wave steepness. Calculation of the fourth-order correction for the entire spectral range requires the frequency responses at not only the first and second but also the third order. Although possible for the standing wave case treated by Sclavounos, the task for the more practical geometries appears daunting.

We review in the following the key ideas of a stochastic theory by Chen et al. (2006) for long-period oscillations in a harbor due to random incidence waves of broad band. Although numerical examples are so far limited to constant depth, with further development of numerical algorithms the theory is general to allow computations of scattering involving a complex structure, coastal line, and bathymetry, and modified for floating vessels or platforms.

31.3.1 Higher-Order Correlation and Spectrum

Let $\zeta(x, y, t)$ be the random free surface displacement. Denoting ensemble averages by angle brackets, we define the correlation function of the total free-surface height $\zeta(x, y, t)$ by

$$H(x,y,\tau) = \langle \zeta(x,y,t)\zeta^{*}(x,y,t+\tau) \rangle \qquad (31.38)$$

where ζ^* denotes the complex conjugate of ζ. Its Fourier transform gives the energy spectrum

$$S(x,y,\omega) = \frac{1}{2\pi}\int_{-\infty}^{\infty} d\tau e^{i\omega\tau} H(x,y,\tau,t) \qquad (31.39)$$

Assuming small steepness the free surface height can be expanded as $\zeta = \zeta_1 + \zeta_2 + \zeta_3 + \cdots$ where the subscripts also represent the orders, that is, $\zeta_j = \mathcal{O}(e^j)$, $j = 1, 2, 3, \ldots$. The total correlation function H is then the sum of products (ζ_1, ζ_1), (ζ_1, ζ_2), … $((\zeta_2, \zeta_2)$, $(\zeta_1, \zeta_3))$, … etc., which are on the orders $\mathcal{O}(\varepsilon^2)$, $\mathcal{O}(\varepsilon^3)$, $\mathcal{O}(\varepsilon^4)$, …, respectively.

Let us first recall certain known facts concerning the first-order incident wave. Let the free surface displacement be

$$\zeta_1^{(I)}(x,y,t) = \int_{-\infty}^{\infty} A(\omega)e^{i\mathbf{k}\cdot\mathbf{x}-\omega t}d\omega = \int_{-\infty}^{\infty} A(\omega)e^{ik(\omega)\tau\cos(\theta-\theta_I)-(i\omega t)}d\omega \tag{31.40}$$

where

A is the random amplitude spectrum

θ_I is the angle of incidence

For each frequency ω, the wave number $k(\omega)$ is given by the dispersion relation, $\omega^2 = gk \tanh kh$. The corresponding correlation function and spectrum are H_2^I and S_I, respectively, where

$$H_2^I(\mathbf{x},t,\tau) = \langle \zeta_1^{(I)}(\mathbf{x},t)\varsigma_1^{(1)^*}(\mathbf{x},t+\tau)\rangle = \int_{-\infty}^{\infty} S_I(\mathbf{x},t,\omega_2)e^{-i\omega(\omega_2)r\cos(\theta-\theta_I)+i\omega_2\tau}d\omega_2 \tag{31.41}$$

and

$$S_I(\mathbf{x},t,\omega_2) = \int_{-\infty}^{\infty} \langle A(\omega_1)A^*(\omega_2)\rangle e^{i[k(\omega_1)-k(\omega_2)]r\cos(\theta-\theta_I)-i(\omega_1-\omega_2)t}d\omega_1 \tag{31.42}$$

Both of these are on the order $\mathcal{O}(\varepsilon^2)$. Assuming stationarity, we must have

$$\langle A(\omega_1)A^*(\omega_2)\rangle = S_A(\omega_1)\,\delta(\omega_1-\omega_2) \tag{31.43}$$

It follows that $S_I(\mathbf{x}, t, \omega) = S_I(\omega) = S_A(\omega)$ and

$$H_2^I(\mathbf{x},t,\tau) = H_2^I(\mathbf{x},\tau) = \int_{-\infty}^{\infty} S_A(\omega_2)e^{-ik(\omega_2)r\cos(\theta-\theta_I)+i\omega_2\tau}d\omega_2 \tag{31.44}$$

Since H_2^I is real, S_A must be symmetric, that is, $S_A(\omega) = S_A(-\omega)$.

Because of the complex bathymetry and lateral boundaries, diffraction and refraction affect the free surface displacement near the scatterers. The total first-order displacement (incident plus scattered) is of the form

$$\varsigma_1(x,y,t) = \int_{-\infty}^{\infty} A(\omega)\Gamma_1(x,y,\omega)e^{-i\omega t}d\omega \tag{31.45}$$

where transfer function Γ_1 can be found from the linear scattering theory.

Now the nonlinear corrections ζ_2, ζ_3, … must involve double, threefold and fourfold Fourier transforms with integrands proportional to A, AA, AAA, … respectively, that is,

$$\zeta_2(x,y,t) = \iint_{-\infty}^{\infty} A(\omega_1)A(\omega_2)\Gamma_2(x,y,\omega_1,\omega_2)e^{-i(\omega_1+\omega_2)t}d\omega_1 d\omega_2 \tag{31.46}$$

and

$$\zeta_3(x,y,t) = \iiint_{-\infty}^{\infty} A(\omega_1)A(\omega_2)A(\omega_3)\Gamma_3 \times (x,y,\omega_1,\omega_2,\omega_3)e^{-i(\omega_1+\omega_2+\omega_3)t}d\omega_1 d\omega_2 d\omega_3 \tag{31.47}$$

respectively, where $\Gamma_2\,(x, y, \omega_1, \omega_2)$ and $\Gamma_3\,(x, y, \omega_1, \omega_2, \omega_3)$ are the transfer functions yet unknown.

Clearly, the right-hand side of (31.38) will involve ensemble averages of even and odd products of A. Assuming that wave motion is a Gaussian random process, all odd products give zero averages, while all even products can be reduced to averages of quadratic products. It follows that $H_3 = 0$. The leading order $\mathcal{O}(\varepsilon^2)$ and the $\mathcal{O}(\varepsilon^4)$ corrections are, respectively,

$$H_2(\tau) = \langle \zeta_1(t)\zeta_1^*(t+\tau)\rangle \quad \text{and} \quad H_4(\tau) = H_{22} + H_{13} + H_{31} \tag{31.48}$$

with

$$H_{22} = \langle \zeta_2(t)\zeta_2^*(t+\tau)\rangle, \quad H_{13} = \langle \zeta_1(t)\zeta_3^*(t+\tau)\rangle, \quad H_{31} = \langle \zeta_3(t)\zeta_1^*(t+\tau)\rangle \tag{31.49}$$

The frequency spectrum can be similarly expanded

$$S(\omega,t) = S_2(\omega,t) + S_4(\omega,t) + \cdots \tag{31.50}$$

where S_2 is the Fourier transform of H_2 as defined by (3.39), and is given by Wiener–Khintchine relation

$$S_2(x,y,\omega) = S_A(\omega)\,|\,\Gamma_1(x,y,\omega)\,|^2 \tag{31.51}$$

The nonlinear correction is

$$S_4(\omega,t) = S_{22}(\omega,t) + S_{13}(\omega,t) + S_{31}(\omega,t) \tag{31.52}$$

where S_{22}, S_{13}, and S_{31} are, respectively, the Fourier transforms of H_{22}, H_{13}, and H_{31}. As it now stands, computation of S_4 would require the full knowledge of ζ_1, ζ_2, and ζ_3, a formidable task indeed.

Making use of the assumption of Gaussianity and the properties of delta functions, Chen et al. (2006) have derived an expression for $H_{22}(\tau)$ and the corresponding energy spectrum

$$S_{22}(\omega) = \delta(\omega)\langle\zeta_2\rangle^2 + \mathcal{I}(\omega) \quad (31.53)$$

where $\mathcal{I}$ denotes the integral:

$$\mathcal{I}(\omega) = \int_{-\infty}^{\infty} S_A(\omega_1)\, S_A(\omega-\omega_1)\Gamma_2(\omega_1,\omega-\omega_1)\cdot [\Gamma_2^\star(\omega_1,\omega-\omega_1)+\Gamma_2^\star(\omega-\omega_1,\omega_1)]d\omega_1 \quad (31.54)$$

The first term in Equation 31.53 corresponds to the square of the mean sea-level; this quantity is usually not available from field data, as indicated by Figure 31.3, due likely to limits of instrumentation, and can be affected by tides. Focusing on low frequencies (small ω), the remaining integral $\mathcal{I}$ can be evaluated by first obtaining the transfer function $\Gamma_2\,(x, y, \omega_1, \omega_2)$ only in a narrow strip near the diagonal $\omega_1 + \omega_2 = \omega$ instead of the entire plane of ω_1,ω_2. This is a significant numerical advantage since it is not necessary to compute $\Gamma_2(\omega_1, \omega_2)$ in the entire plane of (ω_1, ω_2).

Chen et al. (2006) have also shown that

$$S_{13}(\omega) = S_A(\omega)\Gamma_1(x,y,\omega)\int_{-\infty}^{\infty} S_A(x,y,\omega_2)[\Gamma_3^\star(x,y,\omega,\omega_2,-\omega_2) + \Gamma_3^\star(x,y,\omega_2,\omega,-\omega_2)+\Gamma_3^\star(x,y,\omega_2,-\omega_2,\omega)]d\omega_2 \quad (31.55)$$

and that

$$S_{31}(\omega) = S_A(\omega)\Gamma_1^\star(\omega)\int_{-\infty}^{\infty} S_A(\omega_2)[\Gamma_3(\omega,\omega_2,-\omega_2) + \Gamma_3(\omega_2,\omega,-\omega_2)+\Gamma_3(\omega_2,-\omega_2,\omega)]d\omega_2 \quad (31.56)$$

Note the remarkable feature that S_{13} and S_{31} are both proportional to the incident spectrum $S_A(\omega)$.

In typical sea spectra such as JONSWAP, $S_A(\omega)$ is practically zero for small ω. It follows that S_2 is negligible in the low frequency range, where the total spectrum is dominated by the nonlinear correction S_4.

Referring to Figure 31.4, all computations can be confined in a narrow strip near the diagonal $\omega_1 + \omega_2 = 0$ of the (ω_1, ω_2) plane, as marked by the shaded portions of the diagonal strip.

In the range of high frequencies, contributions by S_{22}, S_{13}, and S_{31} are of $O(\varepsilon^4)$ importance, and are all much smaller than S_2. Nonlinear corrections are of minor significance for practical purposes.

The remaining tasks are to obtain the transfer functions $\Gamma_1(x, y; \omega)$ and $\Gamma(x, y; \omega_1, \omega_2)$. Assuming the lateral boundaries

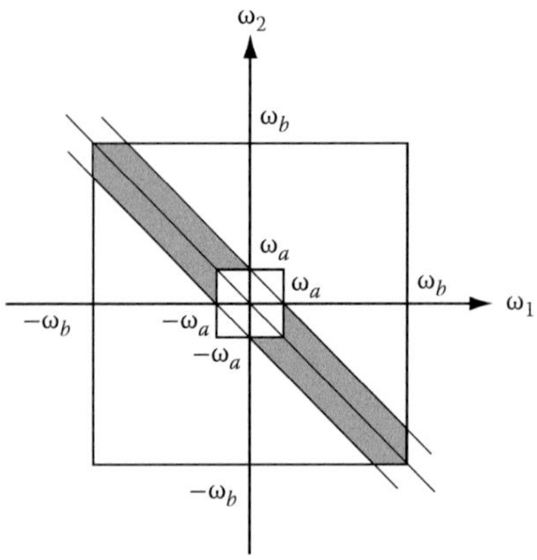

FIGURE 31.4 Plane of (ω_1, ω_2). Γ_2 is computed for frequency pairs inside the shaded strips. Frequencies ω_a and ω_b are the truncated limits of the incident sea spectrum. $S(\omega)$ is dominated by the nonlinear correction S_4. Now since S_{13} and S_{31} are in direct proportion to $S_A(\omega)$, they must likewise be negligible at low frequencies. Thus it is unnecessary to compute Γ_3 or ζ_3 to get S_4, for small ω. This is fortunate as it removes the need for finding Γ_3 in order to compute the integrals in S_{22}.

to be vertical and the seabed gently sloping, Γ_1 can be computed from the well-known mild slope equation of Berkhoff, as extended by Chamberlain and Porter (1995). For any harbor geometry, the associated elliptic boundary-value problem can be solved by the hybrid-element method of Chen and Mei (1974).

Chen and Mei (2006) and Chen et al. (2006) have further extended the mild-slope approximation for Γ_2, which is governed by an inhomogeneous elliptic equation and can also be solved by finite elements. Details are lengthy and omitted here.

31.3.2 Numerical Examples

For the one-sided spectrum $2S_A$ of the incident sea, we choose a modified JONSWAP spectrum for shallow seas (Bouws et al. 1985), as shown in Figure 31.5 by the dashed curve for wind speed $\bar{U}$ = 20 m/s and fetch distance $\bar{x}$ = 3000 m so that the peak frequency is ω_p = 0.767 rad/s. In numerical computations the spectrum is truncated so that it is non-zero only within the range $0.6 < \omega < 1.8$, shown by the solid curve.

We shall describe sample results for a small square harbor (300 m by 300 m) behind a straight coast in constant depth of 20 m. Only normal incidence ($\theta_I = 3\pi/2$) is considered.

It is well known in the linearized theory (Miles and Munk 1961) that, without accounting for frictional losses at the entrance, the resonance peaks in the plot of amplification versus frequency are taller and thinner for the Helmholtz and the lowest modes, if the entrance is narrower (the harbor paradox). To examine the possible nonlinear manifestations, we display the numerical results for two cases: (1) a wide entrance of width 60 m open to the sea, and (2) a narrow entrance of width 30 m protected by a detached breakwater as shown in Figure 31.5. All breakwaters are of caisson type and 5 m thickness. Figure 31.6 shows for the two harbors the basin-averaged response spectra $S(f) = S_2(f) \cup S_{22}(f)$ covering both the low- and high-frequency

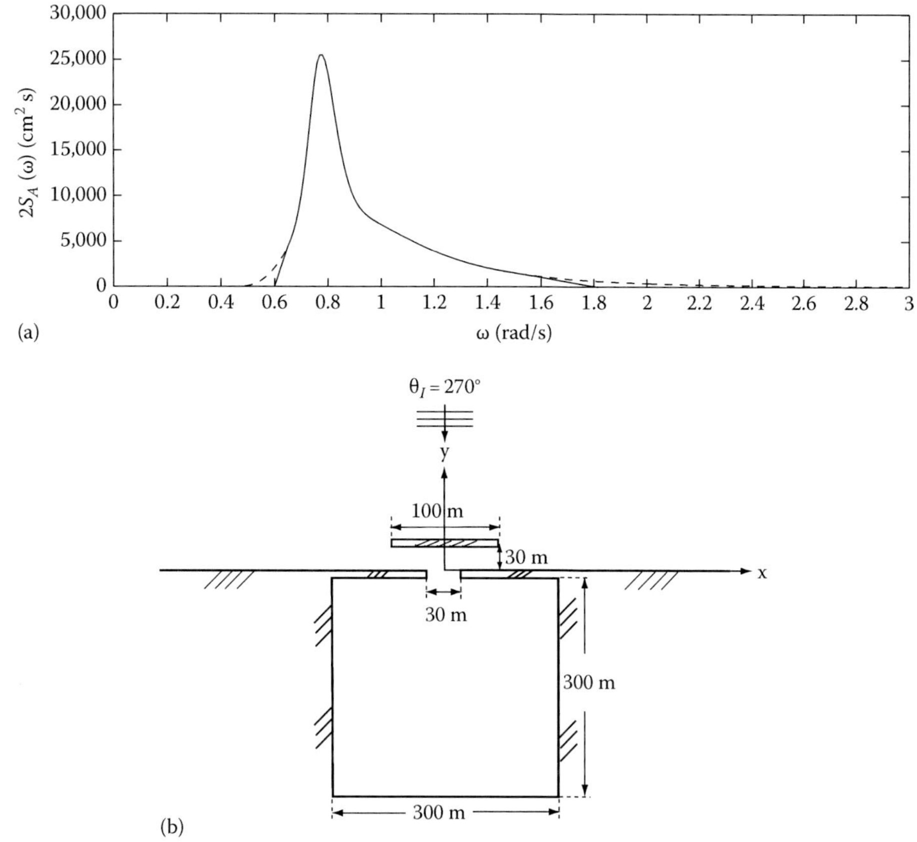

FIGURE 31.5 (a) TMA spectrum. Dashed curve: calculated with inputs $h = 20\,m$, $\bar{x} = 3000$ and $\bar{U} = 20$ m/s. Solid curve: truncated spectrum used in our computations as $2S_A$. (b) Plan view of a protected square basin (Case 2). For Case 1, the entrance width is 60 m without the detached breakwater.

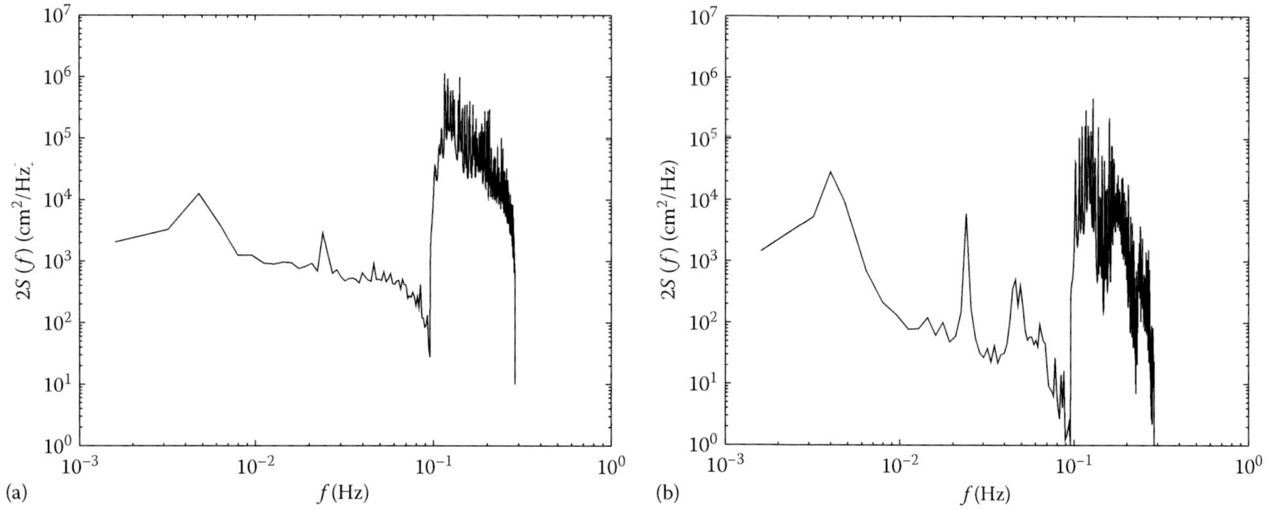

FIGURE 31.6 Spatial-averaged spectrum $2S(f)$ (Case 1). High-frequency range: $2S(f) = 2S_2(f)$, $f > 3/10\pi$ Hz. Low-frequency range: $2S(f) = 2S_{22}(f)$, $f < 3/10\pi$ Hz. (a) 60 m opening. (b) 30 m opening with breakwater.

ranges as a function of the circular frequency $f = \omega/2\pi$ Hz. In the high-frequency range ($3/10\pi < f < 9/10\pi$ Hz), S_2 is plotted without the correction for S_4. In the low-frequency range ($0 < f < 3/10\pi$ Hz), $S_{22} = S_4$ is shown instead. Use is made of the relation that $S_f(f) = 2\pi S(\omega)$. Comparing the spectra, the high-frequency peak is lower for Case 2 due to the better entrance protection. However, the low-frequency peak is higher, implying that harbor paradox is in effect.

Despite the differences in geometries, qualitative agreement between the theory for the small square harbor and the data from Hualien Harbor can be seen. In Figure 31.3, the high-frequency peaks at interior stations 8 and 10 are greatly reduced from those at the outside station 2 by a factor of 1/200, due to the effective protection by the breakwater, but the low-frequency peaks are enhanced by a factor of a hundred.

Application of the present theory to natural harbors of large area and variable depth requires not only further streamlining of the computing algorithms to allow bathymetric variations but also the additional account of directional distribution of the incident sea spectrum. Empirical corrections for losses due to breaking on a beach, separation at the entrance, and dissipation in rubble-mound breakwaters must be incorporated in the stochastic theory.

In principle, the present method can be modified for predicting slow-drift motions of moored ships or tethered offshore platforms in shallow seas. These problems are of increasing importance in the design of Liquefied Natural Gas terminals. Near the scatterers, 3D computations for floating and stationary structures are needed, and must be matched with the mild-slope theory away from the structures.

We remark in conclusion that many other important topics on the interaction of surface waves with the coastal environment are left out in this review. Interested readers can find them elsewhere in this Handbook or in the general literature.

Acknowledgment

This review is largely based on joint works with my past colleague, support financially by US Office of Naval Research and National Science Foundation over many years.

References

Alam, M. R. and C. C. Mei. 2007. Attenuation of long interfacial waves over a randomly rough seabed. *J. Fluid Mech.* **587**, 73–96.

Anderson, P. A. 1958. Absence of diffusion in certain random lattices. *Phys. Rev.* **109**, 1492–1505.

Armstrong, J. A., N. Bloembergen, J. Ducuing, and P. S. Pershan. 1962. Interactions between light waves in a nonlinear dielectric. *Phys. Rev.* **127**, 1918–1939.

Bouws, E. H. Gunthder, W. Rsenthal, and C. L. Vicent. 1985. Similarity of the wind wave spectrum in finite depth water. *J. Geophys. Res.* **90**, 975–986.

Bowers, E. C. 1977. Harbour resonance due to set-down beneath wave groups. *J. Fluid Mech.* **79**, 71–92.

Bryant, P. J. 1976. Periodic waves in shallow water. *J. Fluid Mech.* **59**, 625, 644.

Chamberlain, P. G. and D. Porter. 1995. The modified mild-slope equation. *J. Fluid Mech.* **291**, 393–407.

Chen, Y. and P. L.-F. Liu. 1996. On interfacial waves over random topography. *Wave Motion* **24**, 169–184.

Chen, H. S. and C. C. Mei. 1974. Oscillations and wave forces in an offshore harbor. Technical Report 190, Parsons Laboratory, Department of Civil and Environmental Engineering, MIT, Cambridge, MA.

Chen, M.-Y. and C. C. Mei. 2006. Second-order refraction and diffraction of surface water waves. *J. Fluid Mech.* **552**, 137–166.

Chen, M.-Y., C. C. Mei, and K. K. Chang. 2006. Low-frequency spectra in a harbour excited by short and random incident waves. *J. Fluid Mech.* **563**, 261–281.

Chernov, L. A. 1967. *Wave Propagation in a Random Medium.* New York: Dover.

Dean, R. G. and R. A. Dalrymple. 1990. *Water Wave Mechanics for Engineers and Scientists.* World Scientific: Singapore.

Elter, J. F. and J. E. Molyneux. 1972. The long-distance propagation of shallow water waves over an ocean of random depth. *J. Fluid Mech.* **53**, 1–15.

Goda, Y. 1967. Traveling secondary waves in channels. Report to Port and Harbor Research Institute, Ministry of Transport, Japan.

Grataloup, G. L. and C. C. Mei. 2003. Localization of harmonics generated in nonlinear shallow water waves. *Phys. Rev. E(3)*, **68**(2), 026314-1–026314-9.

Hasselman, K. 1966. Feynman diagrams and interaction rules of wave-wave scattering processes. *Rev. Geophys.* **4** (1), 1–32.

Ishihara, A. 1997, *Wave Propagation and Scattering in Random Media.* IEEE Press and Oxford University Press: New York.

Karal, F. C. and J. B. Keller. 1964. Elastic, electromagnetic and other waves in a random medium. *J. Math. Phys.* **5**(4), 537–547.

Keller, J. B. 1964. Stochastic equation and wave propagation in random media. *Proceedings of the 16th Symposium on Applied Mathematics*, pp. 145–170. American Mathematical Society, Providence, RI.

Komen, G. J., L. Cavaleri, M. Donelan, K. Hasselmann, S. Hasselmann, and P. A. E. M. Janssen. 1994. *Dynamics and Modelling of Ocean Waves.* Cambridge University Press: New York.

Long, R. B. 1973. Scattering of surface waves by an irregular bottom. *J. Geophys. Res.* **78**, 987–1004.

Mei, C. C. 1989. *Applied Dynamics of Ocean Surface Waves.* World Scientific: Singapore.

Mei, C. C. and Y. Agnon. 1989. Long-period oscillations in a harbour induced by incident short waves. *J. Fluid Mech.* **208**, 595–608.

Mei, C. C. and M. J. Hancock. 2003. Weakly nonlinear surface waves over a random seabed. *J. Fluid Mech.* **475**, 247–268.

Mei, C. C. and Y. L. Li. 2004. Evolution of solutions over a randomly rough seabed. *Phys. Rev. E(3)*, **70**(1), 016302-1–016302-11.

Mei, C. C., M. Stiassnie, and D. K.-P. Yue. 2005a. *Theory and Application of Ocean Surface Waves*, vol. 2, World Scientific: Singapore.

Mei, C. C., M. Stiassnie, and D. K.-P. Yue. 2005b. *Theory and Applications of Ocean Surface Waves, Part I.* World Scientific: Singapore.

Mei, C. C. and Ü. Ünluata. 1972. Harmonic generation in shallow water waves. In R. E. Meyer, ed., *Waves on Beaches.* Academic Press: New York, pp. 181–202.

Miles, J. W. and W. Munk. 1961. Harbor paradox. *J. Waterways and Harbor Div. ASCE* **87**, 111–130.

Nachbin, A. 1995. Localization length of randomly scattered water waves. *J. Fluid Mech.* **296**, 353–372.

Nachbin, A. 1997. The localization length of randomly scattered water waves. *J. Fluid Mech.* **296**, 353–372.

Nachbin, A. and G. C. Papanicolaou. 1992. Water waves in shallow channels of rapidly varying depth. *J. Fluid Mech.* **241**, 311–332.

Pihl, J. H., C. C. Mei, and M. J. Hancock. 2002. Surface gravity waves over a two-dimensional random seabed. *Phys. Rev. E.* **66**, 016611-1–016611-11.

Rosales, R. R. and G. C. Papanicolaou. 1983. Gravity waves in a channel with a rough bottom. *Stud. Appl. Math.* **68**, 89–102.

Sclavounos, P. D. 1992. On the quadratic effect of random gravity waves on a vertical boundary. *J. Fluid Mech.* **242**, 475–489.

Sheng, P. 2006. *Introduction to Wave Scattering, Localization and Macroscopic Phenomena*. Springer: New York.

Wu, J.-K. and P. L.-F. Liu. 1990. Harbour excitations by incident wave groups. *J. Fluid Mech.* **217**, 595–613.

32

Drops and Bubbles in the Environment

Lydia Bourouiba
Massachusetts Institute of Technology

John W.M. Bush
Massachusetts Institute of Technology

32.1 Introduction

Bubbles and drops are ubiquitous in nature and play critical roles in many important environmental processes. Most familiar is the life-sustaining role of rain. Less familiar are their roles in the thermal budget of the atmosphere, chemical and biological dispersion, volcanoes and exploding lakes, and disease transmission. The goal of this review is twofold. First, we review the fundamental physics of droplets and bubbles, thereby providing a framework for understanding their myriad roles in fluid transport and the sustenance of life within the aqueous and aerial environments. Second, we review the many environmental settings in which they arise, briefly reviewing well-studied problems while highlighting exciting new research directions.

Our review will highlight the processes that enable the creation of drops and bubbles. Bubbles may form in water through exsolution of dissolved gases, as may be prompted by change in chemistry or temperature, or through cavitation resulting from vigorous flow. Likewise, droplet formation in the atmosphere may arise through condensation of water vapor prompted by either cooling or dynamic low pressures. A solid phase can play a critical role in the birth of drops and bubbles: through reducing the energetic barrier to drop and bubble formation, rough solids often serve as nucleation sites. The lifetime of a water drop falling in the atmosphere is generally limited by evaporation and fracture by dynamic stresses, and may be extended through coalescence with other droplets and growth through condensation. Similarly, the lifetime of a bubble in a fluid, be it a lake, ocean, or magma chamber, will be limited by dissolution and flow-induced fracture, and may be extended via growth through coalescence with other bubbles and exsolution of dissolved gases. The fate of drops and bubbles is thus intimately connected to the chemical, thermal, and dynamic state of the ambient. The interplay between drops and bubbles will also be made clear: droplets impacting a free surface may generate underwater bubbles, while the bursting of bubbles may be a source of microdroplets.

Our review of the basic physics of drops and bubbles will indicate that stable water drops in air or bubbles in water are generally limited in size to the capillary length, a few millimeters. While this size limitation makes them less apparent on the macroscale, it increases their range in the presence of ambient flow, making them more efficient carriers of chemical and biological material. Indeed, our review will highlight the critical role of bubbles and drops in thermal, chemical, and biological transport between atmosphere and ocean, sea, and land. Since biomaterial is often surface active, it tends to adhere to interfaces; hence, biodispersion is greatly enhanced by the substantial range of drops and bubbles. Such dispersion can be beneficial, as in the case of the spread of life-promoting minerals or biomaterial in the surf zone, or deleterious, as in the case of pathogen transport in closed environments such as airplanes and hospitals. Finally, our review will highlight the myriad ways in which drops and bubbles are exploited by living creatures for drinking, feeding, and breathing. Just as the subtle water-proofing strategies of plants and insects have provided important guidance in the rapid advances in the development of superhydrophobic surfaces [14], we expect that such natural strategies may inform and inspire new technologies for microfluidic transport and liquid management.

Handbook of Environmental Fluid Dynamics, Volume One, edited by Harindra Joseph Shermal Fernando. © 2013 CRC Press/Taylor & Francis Group, LLC. ISBN: 978-1-4398-1669-1.

As an important caveat, we note that while this is a review article, the authors were limited to 50 references, making a complete scholarly account of this vast subject entirely impossible. Wherever possible, we have cited the most recent review on the various subjects described. It may thus appear that we are erroneously attributing the work of some researchers to writers of review articles; we thus note that most citations should be accompanied by "and references therein." We apologize preemptively to those authors whose work is not duly cited. Finally, while we have tried to be objective in terms of topic selection, our review is to an extent idiosyncratic, reflecting our own personal tastes as to what are interesting and important drop- and bubble-related problems. For example, we do not consider non-Newtonian effects as may influence bubbles and drops in complex fluids [9], but instead focus on air–water systems. Moreover, we chose not to consider the use of bubbles and drops in man-made devices with important biomedical and environmental applications, including drug transport, environmental sensing, and surgical ablation via acoustic cavitation.

In Section 32.2, we review the fundamental physics of bubbles and droplets, drawing largely upon the literature in surface physics and chemical engineering. The birth, life, and death of drops and bubbles will be discussed through elucidating the processes influencing their formation, growth, dynamics, and break-up. In Section 32.3, we highlight the roles played by drops and bubbles in the transport of heat, chemicals, and biomaterial in the aqueous and aerial environments, giving particular focus to drops in the atmosphere and bubbles in the hydrosphere. In Section 32.4, we examine some relatively novel problems involving drops and bubbles in the biosphere, including their roles in biodispersion, transmission of respiratory disease, and their myriad uses in the animal kingdom.

32.2 Principles

We begin by reviewing the fundamental physics of bubbles and droplets. The physical origins of surface tension and its theoretical characterization are reviewed in Section 32.2.1. In Section 32.2.2, we briefly review the influence of scale on both the dynamics and the shapes of drops and bubbles. The interactions of drops and bubbles with solids are reviewed in Section 32.2.3, while the role of surface tension gradients in their dynamics is considered in Section 32.2.4. The dynamics of bubble and droplet impact and coalescence on both fluid and solid interfaces are reviewed in Section 32.2.5. Bubble- and drop-induced mixing are discussed in Section 32.2.6.

32.2.1 Surface Tension

Drops and bubbles are, respectively, fluid or gaseous volumes bound by immiscible interfaces characterized by an interfacial tension. A fundamental treatment of drops and bubbles must begin with a discussion of the molecular origins of surface or interfacial tension [17].

Molecules in a fluid feel a mutual attraction. When this attractive force is overcome by thermal agitation, the molecules pass into a gaseous phase. Let us first consider the free surface between air and water. A water molecule in the fluid bulk is surrounded by attractive neighbors, while a molecule at the surface is attracted by a reduced number of neighbors and so in an energetically unfavorable state. The creation of new surface is thus energetically costly, and a fluid system will act to minimize surface areas. It is thus that small fluid bodies tend to evolve into spheres; for example, a thin fluid jet emerging from your kitchen sink will generally pinch off into spherical drops via the Rayleigh-Plateau instability. If U is the total cohesive energy per molecule, then a molecule at a flat surface will lose $U/2$. Surface tension is a direct measure of this energy loss per unit surface area. If the characteristic molecular dimension is R and its area thus $\sim R^2$, then the surface tension is $\sigma \sim U/(2R^2)$. Note that surface tension increases as the intermolecular attraction increases and the molecular size decreases. For most oils, $\sigma \sim 20$ dynes/cm, while for water, $\sigma \sim 70$ dynes/cm. The highest surface tensions are for liquid metals; for example, liquid mercury has $\sigma \sim 500$ dynes/cm.

The origins of interfacial tension are analogous. Interfacial tension is a material property of a fluid–fluid interface whose origins lie in the different attractive intermolecular forces that act in the two neighboring fluid phases. The result is an interfacial energy per area that acts to resist the creation of new interface. Fluids between which no interfacial tension arises are said to be miscible. For example, salt molecules will diffuse freely across a boundary between fresh and salt water; consequently, these fluids are miscible, and there is no interfacial tension between them. Our discussion of drops and bubbles will be confined to immiscible fluid–fluid interfaces (or fluid surfaces), at which an effective interfacial (or surface) tension acts.

Surface tension σ has the units of force per unit length or equivalently energy per unit area, and so may be thought of as a negative surface pressure. Pressure is an isotropic force per area that acts throughout the bulk of a fluid. Pressure gradients correspond to body forces within a fluid, and so appear explicitly in the Navier–Stokes equations. Conversely, the action of surface tension is confined to the free surface. Consequently, it does not appear in the Navier–Stokes equations, but rather enters through the boundary conditions. Its effects are twofold. First, there is a jump in normal stress at the interface equal to the curvature pressure $\sigma\nabla \cdot \mathbf{n}$, where $\mathbf{n}$ is the unit normal to the interface, and $\nabla \cdot \mathbf{n}$ the interfacial curvature. Second, the jump in tangential stress across an interface is balanced by the local surface tension gradient $\nabla\sigma$ as may arise due to gradients in temperature or chemistry along the interface.

In static fluid configurations, the normal stress balance takes the form $\Delta P = \sigma\nabla \cdot \mathbf{n}$: the pressure jump across the interface is balanced by the curvature pressure at the interface. We consider a spherical bubble of radius a submerged in a static fluid. Its curvature is simply $2/a$, so there is a pressure jump $2\sigma/a$ across the bubble surface. The pressure within the bubble is higher than that outside by an amount proportional to the surface tension, and inversely proportional to the bubble size. It is thus that small bubbles are louder than large ones when they burst at a free surface: champagne is louder than beer. We note that soap bubbles in air have two surfaces that define the inner and outer surfaces of the soap film; consequently, the pressure differential is twice that across a single interface.

Since the boundaries of drops and bubbles typically correspond to jumps in density, there is generally a net gravitational force acting on them: bubbles rise in water, while raindrops fall in air. In such dynamic situations, the jump in normal stress across the interface has a dynamic component and depends on position; consequently, moving drops and bubbles are not generally spherical. The hydrodynamic force resisting motion will in general have viscous, hydrostatic and dynamic pressure components. The relative magnitudes of these various stress components will determine both the shape and the speed of rising drops and bubbles.

32.2.2 Dynamics and Shapes of Drops and Bubbles

Bubbles may be formed in several ways. First, they may be generated by vigorous interfacial flow that mixes a gaseous phase into a liquid phase; for example, the breaking of a wave (Figure 32.1a). Second, they may arise from the exsolution of dissolved gases in a fluid, as arises when one opens a bottle of soda. Third, they may result from vigorous fluid motions that reduce the fluid pressure beneath the cavitation pressure: such cavitation bubbles arise, for example, in the wake of boat propellers. Similarly, fluid droplets may form in a gas phase by one of three principal routes. First, in response to vigorous flow, fluid volumes may fragment to sheets to filaments to drops [49] (Figure 32.1b). Second, drops may condense from moist air, as may arise due to the temperature drop accompanying an atmospheric updraft. Third, vigorous motions within the gaseous phase and associated Bernoulli low pressures can substantially reduce the pressure, thereby facilitating condensation. The generation of drops and bubbles via condensation and exsolution, respectively, is often facilitated by a solid phase providing nucleation sites. We thus present the theory of wetting in Section 32.2.3.

Consider a drop of density $\hat{\rho}$, viscosity $\hat{\mu} = \hat{\rho}\hat{\nu}$, undeformed radius a and surface tension σ falling at a characteristic speed U under the influence of gravity $\mathbf{g} = -g\hat{z}$ through a fluid of density ρ and viscosity $\mu = \rho\nu$. The eight physical variables (ρ, $\hat{\rho}$, ν, $\hat{\mu}$, σ, a, U, g) may be expressed in terms of three fundamental units (mass, length, and time), so Buckingham's Theorem indicates that the system may be uniquely prescribed in terms of five dimensionless groups. We choose $\rho/\hat{\rho}$, $\nu/\hat{\nu}$, the Reynolds number $R_e = Ua/\nu$, the Capillary number $C_a = \rho\nu U/\sigma$, the Weber number $W_e = \rho U^2 a/\sigma$, and the Bond number $B_o = \rho g a^2/\sigma$. The Reynolds number prescribes the relative magnitudes of inertial and viscous forces in the system, the Capillary number those of viscous and curvature forces, and the Weber number those of inertial and curvature forces. We note that we might also have chosen the Ohnesorge number, $O_h = (C_a/R_e)^{1/2} = \mu/(\rho a \sigma)^{1/2}$, which prescribes the relative importance of viscous stresses and curvature pressures. The Bond number indicates the relative importance of forces induced by gravity and surface tension. These forces are comparable when $B_o = 1$, that is, on a length scale corresponding to the capillary length: $\ell_c = (\sigma/(\rho g))^{1/2}$. For a water drop in air, for example, $\sigma \approx 70$ dynes/cm, $\rho = 1$ g/cc, and $g = 980\,\text{cm/s}^2$, so that $\ell_c \approx 2$ mm. Water drops are dominated by the influence of surface tension provided they are smaller than the capillary length: when placed on a surface, their tendency to flatten in response to gravity is then overcome by their tendency to remain spherical under the influence of surface tension. As a fluid system becomes progressively smaller, the relative importance of surface tension and gravity increases; it is thus that microfluidic systems are dominated by the influence of surface tension.

A great deal of work has been devoted to measuring and rationalizing the shape and speeds of drops falling and bubbles rising under the influence of gravity. The dependence of drop shape on B_o and R_e is detailed in Ref. [11]. The shape of small bubbles ($B_o \ll 1$) is typically dominated by the influence of surface tension, so they assume a spherical form. At low R_e, the Stokes drag on a bubble of radius a is given by $D = 4\pi\mu U a$. Balancing this drag with the buoyancy force yields the Stokes rise speed: $U = 1/3 a^2 g'/\nu$, where $g' = g\Delta\rho/\rho$ is the reduced gravity, and $\Delta\rho$ the density difference between bubble and ambient. As the R_e increases, both the shape

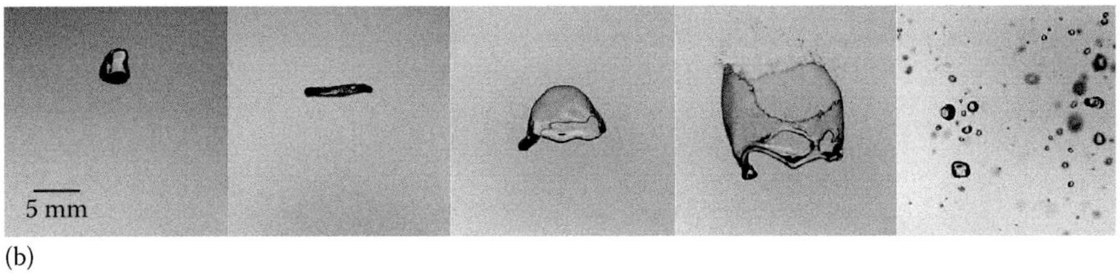

FIGURE 32.1 (a) The surf zone is an important site of drop- and bubble-induced thermal, chemical, and biological exchange between atmosphere and ocean, land and sea. (b) A water drop falls under the influence of gravity. (From Villermaux, E., *Annu. Rev. Fluid Mech.*, 39, 419, 2007. With permission.) The stagnation pressure at its nose causes it to expand into a parachute and then ultimately fracture into microdroplets.

of the bubble and the form of the flow change. The first effect of dynamic pressures on the shape of the bubble is to distort it into an oblate ellipsoid, an effect that can be understood in terms of the stagnation pressures at the nose and tail of the bubble. As R_e increases further, a stable wake adjoins the rising bubble, and the associated pressure difference breaks the bubble's fore-aft symmetry. As R_e is increased progressively, the wake becomes unstable, leading to vortex shedding and an associated path deviation: zigzag and helical paths have been reported and rationalized [28]. At high R_e, the characteristic drag is associated with the difference in dynamic pressures up- and downstream of the bubble and so takes the form $D \sim 1/2\rho U^2 a^2$. Balancing this with the buoyancy force on the drop yields a steady ascent speed $U_s \sim ((\Delta\rho/\rho)ga)^{1/2}$. Eventually, rise speeds will increase to the point that the associated dynamic pressures ρU^2 cause bubble cleavage. However, there is also a large bubble limit in which bubbles assume a spherical cap form, surface tension is entirely negligible, and there exists a balance between hydrostatic and dynamic pressures along the streamline corresponding to the bubble surface [11].

The progression with increasing size of the shape and dynamics of droplets falling through air is similar. The smallest drops ($B_o \ll 1$, $R_e \ll 1$) remain spherical and settle at their Stokes settling speed. As they get larger, they become distorted by aerodynamic forces, and, eventually, wake instability leads to their being deflected off a straight path, though these deflections are small relative to those of rising bubbles owing to the large mass of the drop relative to that of the shed vortices. At sufficient speed, aerodynamic stresses are sufficient to cleave a falling droplet. Given the critical life-sustaining role of rain drops, it is illustrative to review the considerations that determine their size. Raindrops have a diameter range of roughly 100 μm to 5 mm, and fall at high R_e, so the characteristic drag $D \sim 1/2\rho U^2 a^2$. Balancing this with the weight of the drop yields a steady descent speed $U_s \sim ((\Delta\rho/\rho)ga)^{1/2}$. The aerodynamic stresses experienced by the drop will cleave the drop unless they are balanced by curvature pressures. Specifically, the stagnation pressure on the nose of the drop will cause the drop to balloon out into a parachute before fracturing (Figure 32.1b). Drop integrity requires that $\rho U_s^2 a^2 < \sigma/a$. Substituting for U_s reveals that raindrops will be fractured by aerodynamic stresses unless they are smaller than the capillary length $\ell_c = (\sigma/(\rho g))^{1/2} \sim 2$ mm. This simple scaling relation has vast consequences for life: if surface tension was drastically higher, rainstorms would be perilous to humans and fatal for some of Earth's smaller denizens. As we shall see in what follows, however, the nucleation of raindrops would also be more energetically costly, so rain itself more of a rarity.

32.2.3 Wetting: Interaction with Solids

The interaction between fluid drops or bubbles with solids arises in a wide range of problems of environmental significance, including the nucleation of raindrops and the impact of rain on plants and soil. Wetting arises when a liquid–gas interface comes into contact with a solid [17]. The degree of wetting is in general determined by both the material properties of the solid and fluid phases and the topography of the solid surface. Just as a fluid–fluid interface has an associated energy per unit area, so do solid–fluid interfaces. For the case of a liquid–gas interface with interfacial tension σ in contact with a solid, the relevant surface energies are those of the wet and dry surfaces, respectively, γ_{SL} and γ_{SG} (Figure 32.2). The tendency of the liquid to wet the solid

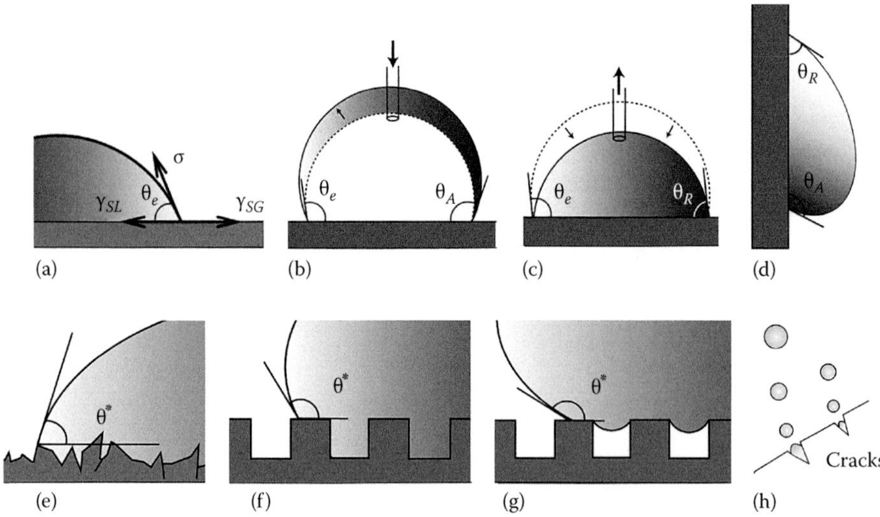

FIGURE 32.2 Wetting. (a) A droplet on a solid substrate makes contact over a region bounded by its contact line. The equilibrium contact angle θ_e is prescribed by the relative magnitudes of the surface energies of the three interfaces: σ, γ_{SL}, and γ_{SG}. (b–c) Static contact angles have a finite range, $\theta_A < \theta < \theta_R$; this so-called contact angle hysteresis results in drops sticking to rigid surfaces, for example, raindrops to window panes (d). (e) Through increasing the energetic cost of wetting, surface roughening generally acts to increase the apparent contact angle θ*. (f) If the fluid impregnates the roughness elements, a Wenzel state is achieved, and contact angle hysteresis enhanced. (g) If air is trapped within the roughness elements, a Cassie state obtains, the contact angle hysteresis is minimized, and water-repellency achieved. (h) Rough solids may reduce the energetic cost of forming bubbles and drops, and so serve as nucleation sites.

depends on the relative magnitudes of the surface energies σ, γ_{SL}, and γ_{SG} through the spreading parameter $S = \gamma_{SG} - (\gamma_{SL} + \sigma)$. If $S > 0$, the surface energy of the solid is lower wet than dry, so the liquid spreads completely into a thin film. Such is the case for oils spreading on most solids, including glass. Conversely, when $S < 0$, it is energetically favorable for the solid to stay dry, so the fluid remains in the form of a droplet with a finite equilibrium contact angle θ_e (Figure 32.2a). Given σ, γ_{SL}, and γ_{SG}, one can calculate the contact angle θ_e by considering the horizontal force balance at the contact line, $\sigma \cos \theta_e = \gamma_{SG} - \gamma_{SL}$, known as Young's relation. For an air–water interface in contact with a solid, complete wetting arises only if $\theta_e = 0$; otherwise, the normal situation of partial wetting is obtained. The solid is said to be hydrophilic if $\theta_e \leq 90°$, hydrophobic if $\theta_e \geq 90°$, and superhydrophobic if $\theta_e \geq 150°$. Water is totally wetting on perfectly clean glass, but partially wetting on most solids. Given the contact angle and drop volume, the static drop shape is uniquely prescribed by the normal stress boundary condition that applies at its surface: $p_0 + \rho g z = \sigma \nabla \cdot \mathbf{n}$.

While the definition of an equilibrium contact angle is useful conceptually, for a given solid–fluid combination, a range of static contact angles may be observed [17,37]. Consider a drop of fluid emplaced on a solid (Figure 32.2). If the drop is filled, it will grow, and its contact angle increase progressively until reaching a critical value, θ_A, at which the contact line begins to advance (Figure 32.2b). If conversely, fluid is withdrawn from the drop, its contact angle will decrease progressively until reaching a critical value, θ_R, at which the contact line begins to recede (Figure 32.2c). Observed static contact angles θ may thus lie anywhere within the range $\theta_A > \theta > \theta_R$, bounded above and below by the advancing and receding contact angles. This finite range of static values for a given three-phase system results from microscopic surface irregularities, and is referred to as contact angle hysteresis: $\Delta\theta = \theta_A - \theta_R$. An important consequence of this hysteresis is that there is a concomitant force of retention that causes droplets to adhere to surfaces: differences in the contact angle around the perimeter of a drop may result in a net contact force that resists its motion. For example, rain drops may stick to window panes because of the difference in the contact angles on their upper and lower edges (Figure 32.2d), provided the resulting retention force exceeds the drop's weight: $Mg \lesssim \pi R\sigma(\cos\theta_R - \cos\theta_A)$ where R is the drop's contact radius [17].

If a solid surface is not smooth, the microscopic contact angle θ_e will generally differ from that observed on a macroscopic scale, the apparent contact angle θ^* (Figure 32.2e). Moreover, the energetic cost of wetting will depend not only on the surface chemistry (which prescribes γ_{SG} and γ_{SL}) but also on the surface roughness. Consider the case of a water drop on a roughened surface. If the roughness is relatively mild, the water impregnates the roughness elements, and the drop will be in a so-called Wenzel state (Figure 32.2f). If the surface is sufficiently rough, air pockets are trapped beneath the drop, and a Cassie state is attained (Figure 32.2g). In both Wenzel and Cassie states, the apparent contact angle $\theta^* > \theta_e$; however, the contact angle hysteresis is, respectively, increased and decreased in the two states. The maintenance of a Cassie state is thus critical to reducing the force of adhesion of water droplets on rough solids, and so is considered the primary criterion for water-repellency [37]. It is noteworthy that roughness serves to amplify the intrinsic wettability of a solid: roughening a hydrophobic surface will tend to render it superhydrophobic, while doing likewise on a hydrophilic surface will make it more wettable. Water-repellency is generated in the same way in both plant and animal kingdoms (on plant leaves, animal fur, and bird feathers): roughness increases the energetic cost of wetting a hydrophobic (typically, waxy or oily) surface and so ensures the maintenance of a Cassie state [8].

The presence of rough solids may also facilitate the nucleation of gas bubbles from a saturated fluid (Figure 32.2h), or the condensation of water drops from moist air. It is readily observed that champagne bubbles tend to nucleate at specific points on the walls of a glass. It is less apparent that these points correspond to surface irregularities where either the glass has been scratched or a small piece of cloth has adhered. Such surface irregularities facilitate the degassing of champagne by reducing the surface energy of the incipient bubbles and so reducing the energetic barrier to bubble production. Likewise, adding a teaspoon of salt to a bottle of carbonated soda greatly facilitates exsolution, and can have explosive consequences. In an analogous fashion, particulate matter such as dust encourages the condensation of rain drops in clouds, a matter to be considered further in Sections 32.3 and 32.4.

32.2.4 Marangoni Flows

Marangoni flows are those driven by surface tension gradients. In general, surface tension σ depends on both temperature and chemical composition; consequently, Marangoni flows may be generated by gradients in either temperature or chemical concentration at an interface. Surfactants (or surface-active reagents) are molecules that have an affinity for interfaces; common examples include soap and oil. Owing to their molecular structure (often a hydrophilic head and hydrophobic tail), they find it energetically favorable to reside at the free surface. Their presence reduces the surface tension; consequently, gradients in surfactant concentration result in surface tension gradients and a special class of Marangoni flows. There are many different types of surfactants, some of which are insoluble (and so remain on the surface), others of which are soluble in the suspending fluid and so may diffuse into the bulk. For a wide range of common surfactants, surface tension is a monotonically decreasing function of surfactant concentration until a critical concentration is achieved, beyond which it remains constant. Most organic materials are surface active; therefore, surfactants are ubiquitous in the natural aqueous environment, and their transport via drops and bubbles plays an important role in the biosphere.

The presence of surfactants will generally serve not only to reduce σ but also to generate Marangoni stresses, that is, surface tension gradients. The principal dynamical influence of surfactants is thus to impart an effective elasticity to the interface through suppressing any fluid motions characterized by non-zero surface divergence. Consider a fluid motion characterized by a radially divergent surface motion. The presence of surfactants results in the redistribution of surfactants: surfactant concentration is reduced near the point of divergence. The resulting Marangoni stresses act to suppress the surface motion, resisting it through an effective surface elasticity. Similarly, if the flow is characterized by a radial convergence, the resulting accumulation of surfactant in the region of convergence will result in Marangoni stresses that serve to resist it. It is this effective elasticity that gives soap bubbles their longevity: the divergent motions that would cause a pure liquid bubble to rupture are suppressed by the surfactant layer on the soap film surface.

Theoretical predictions for the rise speed of very small drops or bubbles do not adequately describe observations. For example, in all but the most carefully cleaned fluids, air bubbles rise at low R_e at rates appropriate for rigid spheres with equivalent buoyancy. This discrepancy may be rationalized through consideration of the influence of surfactants on the surface dynamics. The flow generated by a clean spherical bubble of radius a rising at low R_e is a familiar one. The interior flow is toroidal, while the surface motion is characterized by divergence and convergence at, respectively, the leading and trailing surfaces. The presence of surface contamination changes the flow qualitatively. The effective surface elasticity imparted by the surfactants acts to suppress the surface motion. Surfactant is generally swept to the trailing edge of the bubble, where it accumulates, giving rise to a local decrease in surface tension. The resulting tail-to-nose surface tension gradient results in a Marangoni stress that resists surface motion, and so rigidifies the bubble surface. The air bubble thus moves as if its surface was stagnant, and it is thus that its rise speed is commensurate with that predicted for a rigid sphere: the no-slip boundary condition is more appropriate than the free-slip for small bubbles in contaminated fluids.

The ability of surfactant to suppress flows with non-zero surface divergence is evident throughout the natural world. It was remarked upon by Pliny the Elder, who noted the absence of capillary waves in the wake of ships. The ability of surfactants to suppress capillary waves was also widely known to spearfishermen, who poured oil on the water to increase their ability to see their prey, and by sailors, who would do similarly in an attempt to calm troubled seas [27]. The influence of natural surfactants is often evident as wind blows over puddles. While capillary waves are evident in the lee of the puddle, they are suppressed in the aft side by the surfactants swept downwind. The suppression of capillary waves by surfactant is also partially responsible for the "footprints of whales." In the wake of whales, even in turbulent roiling seas, one sees patches on the sea surface (of characteristic width 5–10 m) that are remarkably flat. These are generally thought to result in part from the whales sweeping biomaterial to the surface with their tails; this biomaterial serves as a surfactant that suppresses capillary waves. Finally, surfactants may have a dramatic effect on the wettability of rough solids such as the integument of plants, insects, and birds: by decreasing the equilibrium contact angle below 90°, they may in fact render an initially superhydrophobic surface superhydrophilic. It is thus that chemical contamination can be fatal to plants or sea-going creatures that rely on their water-repellency for survival.

32.2.5 Coalescence, Fracture, and Collapse

Droplets often arise in the form of suspensions, as they do in clouds and hurricanes. Likewise, bubbles often arise in swarms, as they do in magma chambers and volcanic vents. Given the dependence of the speed and transport properties of drops and bubbles on their size, the possibility of coalescence is an important one that has received considerable attention. The most striking coalescence event arises every time a raindrop hits a puddle. When a water droplet is placed on a quiescent water bath, it ultimately collapses into the bath due to gravity; however, this merger is generally delayed because the air layer between the droplet and the bath must first drain to a thickness at which Van der Waals forces between droplet and bath become important, approximately 100 nm. The resulting coalescence may take a number of distinct forms. Complete coalescence arises when the entirety of the drop merges with the underlying reservoir (Figure 32.3b). Partial coalescence arises when only some fraction of the drop coalesces, leaving behind a smaller daughter droplet that is ejected from the bath and bounces several times before itself undergoing a partial coalescence. This coalescence cascade continues until the daughter droplet becomes sufficiently small that viscosity comes into play, and complete coalescence occurs [5]. Droplet–droplet collision and coalescence have also received considerable attention owing to their importance in raindrop accretion in clouds [50]. While the geometry is slightly different, the physical picture is the same: drop merger is anticipated, provided the impact time is greater than the drainage time of the intervening air layer. The coalescence may be partial or complete. We note that when a drop or bubble coalesces at an interface, the impulsive release of curvature pressure generally produces a vigorous vortex ring (Figure 32.3b) that can transport their payload a considerable distance, and so enhance local mixing.

More energetic impacts of droplets on both fluid and solid surfaces have also been examined extensively. Indeed, progress in high-speed videography has allowed workers to go well beyond the original observations of Worthington to elucidate a host of subtle effects that continue to fuel supporting theoretical developments [47]. The normal impact of droplets generally gives rise to interfacial structure, including submerged bubbles and ejected jets, sheets and droplets, whose complexity increases with the energy of impact [36] (Figure 32.3). Jayaratne and Mason [24] considered the case of droplets striking a deep fluid layer obliquely, and experimentally deduced criteria for rebound, partial, and complete coalescence.

Richard et al. [38] have considered the impact of droplets on water-repellent surfaces, and deduced criteria for droplet rebound and fracture (Figure 32.4). In general, one expects

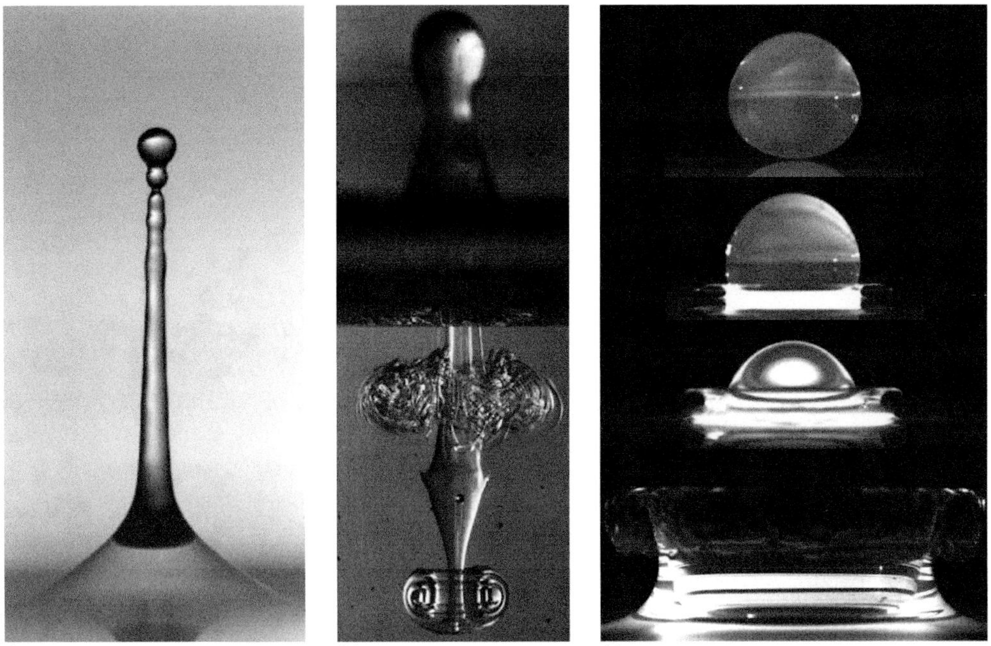

(a) (b) (c)

FIGURE 32.3 (a) The Worthington jet resulting from the collapse of the cavity generated by the impact of a 1 mm steel sphere on a water surface. Note the initiation of jet pinch-off via the Rayleigh-Plateau instability. (b) The coalescence of a droplet generates a subsurface vortex that transports the drop fluid downward. (c) The impact of a droplet onto a bath is accompanied by the ejection of a fluid sheet, the collapse of which may generate microbubbles and microdroplets. (a: Courtesy of J.M. Aristoff; b and c: Courtesy of S. Thoroddsen.)

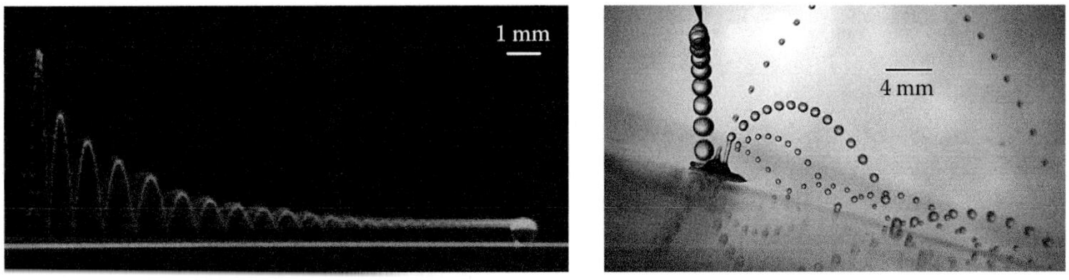

(a) (b)

FIGURE 32.4 (a) A water droplet bouncing off a slightly inclined superhydrophobic surface. (Courtesy of D. Quéré.) (b) A silicone oil droplet impacts a thin layer of the same fluid, fracturing then rebounding in the form of four microdroplets. Similar impacts may arise as pathogen-bearing droplets are transported along the respiratory tract. (Courtesy of T. Gilet.)

droplet fracture when the incident kinetic energy exceeds the surface energy of the drop, that is, when the droplet Weber number is greater than one. Moreover, one expects pure rebound only when the substrate is sufficiently rough to resist impregnation and so maintain a Cassie state. Deducing criteria for the rebound and adhesion of impacting droplets is a problem of considerable interest in the design of water-repellent materials; likewise, in understanding the dynamics of rain-splash and its roles in disease transmission between plants and the evolution of topsoil in arid regions (Section 32.4.1).

Bubble bursting may also be accompanied by remarkably subtle structure. Puncturing a bubble composed of a film of thickness h and density ρ causes the resulting hole to expand under the influence of the tensile force per unit length 2σ. If the O_h characterizing the rim is sufficiently low, the rim will retract at the constant Culick speed $(2\sigma/\rho h)^{1/2}$, engulfing the fluid film as it advances [42]. As the rim grows, it will eventually begin to pinch off via the Rayleigh–Plateau instability. As the resulting bulbous rim regions retract along the curved bubble surface, they may be flung outward and give rise to filaments that themselves pinch off via Rayleigh–Plateau, giving rise to a cloud of microdroplets [25]. This process is illustrated in Figure 32.5 and may play a significant role in the chemical and biological transport between atmosphere and ocean.

32.2.6 Mixing

Quite generally, drops and bubbles represent discrete density anomalies within their suspending fluid, and so are in a dynamic state, either falling or rising under the influence of gravity.

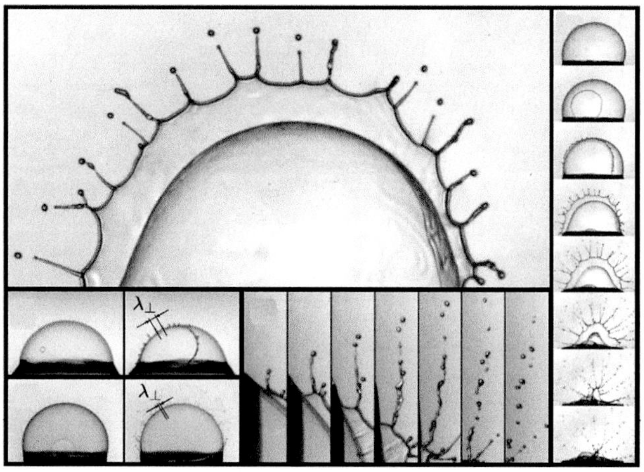

FIGURE 32.5 The fragmentation of a bursting bubble. Puncturing the bubble surface causes its rim to retract, and expel filaments that pinch off into microdroplets. Such a process accompanies bubble bursting in the surf zone, and influences the transport of chemicals and biomaterial between air, land, and sea. (From Lhuissier, H. and Villermaux, E., *Phys. Fluids*, 21, 091111, 2009. With permission.)

As such, they make important contributions to vertical transport and mixing within a two-phase system. Three principal modes of transport may arise. Their first contribution is the most obvious, the vertical transport of the fluid bound within their interface. Second, we have seen that surface-active materials tend to stick to the surface of drops and bubbles. Such material, which may be biological or chemical in nature, will be transported as part of their payload. As we shall see in Section 32.3, this scavenging of surface-active material by bubbles in the ocean and rain in the atmosphere plays a critical role in the biosphere.

The third mode of bubble- or drop-induced mixing results from the rearrangement of the ambient. For example, as a bubble rises at high R_e, it will typically transport ambient fluid in its wake, and also displace ambient fluid forward, the volume of which depends on the volume of the compound body that is the bubble plus wake. Potential flow theory indicates that the volume of fluid displaced vertically across an initially horizontal plane is equal to the Darwin drift volume, specifically, the added mass of the bubble plus wake [28]. This longitudinal dispersion may be accompanied by lateral dispersion in the case of vigorous shedding of vortices from the wake of the bubble. As we shall see in Section 32.3.2, bubble-induced mixing of stratified lakes has a number of important consequences.

It also bears mentioning that drops and bubbles arising in swarms will not necessarily settle or rise, respectively, at the rate expected for individuals. Consider, for example, a swarm of small bubbles that would rise at low R_e if in isolation. Gradients in bubble concentration will generally arise, prompting large-scale convective overturning. In such circumstances, bubble concentration can be considered to some extent as a passive scalar that affects the bulk fluid density as would, for example, temperature; consequently, all the rich behavior of thermal convection can be observed in two-phase fluids. More subtle, novel features may also arise in two-phase convection as a result of the discreteness of the density anomalies, including the formation of shock-like features. A familiar example arises in a glass of Guiness: the degassing of the beer is marked by horizontal stripes corresponding to regions of high bubble concentration that ultimately go unstable, giving rise to vertical bubble plumes. Similar dynamics may arise in magma chambers (Section 32.3.2).

32.3 Applications

32.3.1 Drops in the Atmosphere

Clouds play a critical role in the Earth's climate through their influence on the water cycle and radiative budget. Clouds are air masses with sufficient moisture to support either water droplets or ice crystals. Cloud droplets (ranging in size from μm to mm) form by condensation from chilled water vapor, while ice crystals form either by the direct deposition of water vapor (reverse sublimation) or by the freezing of liquid droplets [39]. The size, phase, and composition of the droplets determine the macroscopic properties of the clouds, including their spatial extent, lifetime, and radiative properties.

Condensation of pure water vapor into droplets requires higher vapor pressures than are typically found in clouds; however, condensation nuclei are ubiquitous in the atmosphere and facilitate the initiation of the condensation process in both warm (non ice-bearing) and mixed-phase (liquid- and ice-bearing) clouds. Droplets can remain suspended in clouds provided their settling speeds are small relative to the turbulent flow or updraft speed. Within clouds, updrafts of warm air expand and cool, leading to condensation and the release of latent heat; the hot air continues to rise due to its buoyancy. This process continues until some droplets reach a critical size at which their settling speeds exceed the updraft speeds. These large, falling droplets initiate the final phase of raindrop growth, which is dominated by collision and coalescence. Turbulence plays a key role in this last phase, by enhancing the collision and clustering process [45].

Tropical cyclone formation and intensification depend critically on the transfer of heat, moisture, and momentum at the sea–air interface. Moist air and sea spray droplets are the principal source of heat for these storms: the condensation of droplets leads to the release of latent heat at the sea surface and an associated buoyant updraft of warm, moist air. As in clouds, the upwelling of supersaturated air results in its expansion and cooling, leading to further condensation of droplets and the release of latent heat. Provided the droplets so formed fall back and coalesce with the sea before significantly evaporating, there will be a net transfer of heat and buoyancy to the air [2].

Condensation nuclei can be inorganic or organic, insoluble or soluble when wetted. Large condensation nuclei lead to larger droplets that are more likely to trigger precipitation, while a larger number of condensation nuclei lead to the formation of a large number of small droplets more likely to delay precipitation and increase albedo. Hence, anthropogenic and biological particles released into the atmosphere can have a drastic effect on the formation, evolution, and precipitation of clouds [34]. For the same ambient water content, larger concentrations of condensation nuclei typically lead to the formation of clouds with a larger number of relatively small cloud droplets. Typically, maritime cumulus clouds contain less condensation nuclei, resulting in low concentrations (~45 cm^{-3}) of large droplets (average diameter 30 μm); for the sake of comparison, continental cumulus clouds contain a relatively high concentration (~230 cm^{-3}) of small droplets (average diameter ~10 μm) [23]. The efficiency of coalescence upon collision increases with the size ratio between the colliding drops. All other things being equal (e.g., moisture and temperature), polluted clouds are less conducive to precipitation owing to the prevalence of more condensation nuclei resulting in small drops with a limited size range [23].

The incorporation of atmospheric aerosols and pollutants into cloud droplets is an important mechanism for atmospheric cleansing and planet-wide aerosol dispersion. Pollutants or suspended components can either play the role of condensation nuclei at the core of raindrops or remain in the cloud as free suspended aerosols that may be removed by adhesion onto cloud droplets. The aerosols, either scavenged during droplet nucleation or coalescence, may then be transported to the ground by precipitation. Indeed, precipitation removes 70%–80% of the mass of aerosols in the tropopause in temperate latitudes, the larger aerosols being removed more efficiently [23]. Some condensation nuclei such as organic compounds and microorganisms are surface active and can give rise to surface films on water droplets that may influence the drop dynamics and resulting cloud properties. Such films can inhibit droplet evaporation and decrease the diffusion of water-soluble compounds into the droplets, thus reducing the efficiency of droplet nuclei scavenging and cleansing. The radiative properties of the cloud may also be altered by the reduction in the UV reaching the droplet interior [19].

Acid rain is a dramatic illustration of the interplay between aerosols and droplets. When the pollution concentration is sufficiently high, the reduction of the rain's pH leads to damage of lake and river ecosystems, forest vegetation, as well as man-made materials and infrastructure. Acid clouds are the result of the scavenging (by nucleation and precipitation) of sulfate- and nitrate-bearing aerosols, such as SO_2 and NO_2, produced by the combustion of biomass and fossil fuels. Through a chain of chemical reactions, these compounds are converted into acidic sulfates and nitrates, which lower the pH below neutral, resulting in acid rain. Acid deposition poses a similar threat, and occurs when acidic drops come into direct contact with land; for example, when contaminated clouds collide with mountains. The damage resulting from such deposition can be relatively severe since suspended cloud droplets are generally more acidic than the acid rain droplets that strike the ground [43]. Owing to its enormous economic and ecological consequences, the factors influencing the production of acid rain and acid deposition continue to receive a great deal of attention [23,39].

Cirrus clouds cover up to 30% of the Earth's surface and have a major impact on the atmospheric radiative budget. Condensation trails, or contrails, form behind airplanes in the upper troposphere and are suspected to trigger the formation of cirrus-like clouds. Cirrus and cirrus contrails are both composed of ice particles; however, particles in contrails are smaller and more dense than those in natural cirrus clouds [54]. The cruise altitude of international flights ranges from 8 to 13 km. Two types of contrails are generated by such upper tropospheric flights: engine exhaust and aerodynamic contrails [18]. Exhaust contrails are the result of the mixing of the emitted hot moist gas and the cold, dry ambient. In order to form, the mixture of expelled gas and ambient air in the expanding plume must reach water saturation, in which case water vapor condenses onto the solid particles emitted by the exhaust that serve as droplet nuclei; these droplets subsequently freeze. Downstream, if conditions of ice supersaturation of the ambient are reached, the exhaust particles can also serve as ice nuclei and so directly trigger the formation of cirrus clouds. Aerodynamic contrails can form at higher temperatures than those required for exhaust contrails. They are the result of water condensation and subsequent possible freezing associated with the drop of pressure and temperature resulting from the air's acceleration over the wings [18,54].

32.3.2 Bubbles in the Lithosphere and Hydrosphere

Drops and bubbles generated at the sea surface play a critical role in coastal ecology. Most of the material ejected from the water surface into the atmosphere is due to droplet generation via wave breaking (Figure 32.1a), bubble bursting (Figure 32.5), and the impact of raindrops (Figure 32.3). When bubbles that have scavenged large concentrations of surface-active material (e.g., sea-salt and microorganisms) burst, they release their harvest into the atmosphere. The scavenging process is particularly effective for surface-active microorganisms (e.g., some bacteria) and can lead to the ejection of droplets with concentrations a thousand time higher than in the bulk fluid [4]. The minerals and biomatter produced by the sea and carried inland by droplets are typically advantageous to shoreline vegetation; however, this enrichment of droplet ejecta may have deleterious effects if the water is polluted. Aerosol production and biodispersion in the surf zone remains an active area of research [31].

Volcanic eruptions are characterized by the ejection of magma, a mixture of melt and bubbles; consequently, bubbles play a critical role in their dynamics. Various styles of eruption are observed, from relatively tranquil outpourings of lava to short-lived explosive eruptions, the most violent of which can release vast quantities of particulate matter high into the atmosphere. Indeed, it has been suggested that the biblical plagues of Egypt were a consequence of the particulate matter launched skyward by the explosion of Santorini. The rise of magma from the chambers underlying volcanoes is typically accompanied by decompression and the formation of bubbles of exsolved gas. As they form and expand with increasing height, the bubbles cause the acceleration of the rising melt [20]. The explosiveness of an eruption depends on the magma composition and viscosity, both of which are greatly affected by the mechanisms of bubble nucleation and growth.

In 1984, 37 people died from the degassing of CO_2 from Lake Monoun in Cameroon. In 1987, a similar eruption of Lake Nyos led to casualties as far as 26 km away. The increase of pressure with depth in the lakes supports the accumulation of dissolved CO_2 at depth, where it leaches in from the underlying volcanic crater. A stable stratification of the lake persists until the bottom layer becomes saturated in dissolved CO_2, at which point exsolved CO_2 bubbles (10–100 μm) can rise to the surface in the form of turbulent bubble plumes. The trigger for such an event may be any kind of disturbance causing substantial overturn, for example, landslides or vigorous wind-driven flows. As the bubbles rise, they can either reach the surface or be reabsorbed in the upper layer of the lake, where the water remains undersaturated. If they do reach the lake surface, they release a cloud of CO_2 more dense than air that spreads down the mountain as a gravity current and so may suffocate nearby residents. This form of natural bubble-induced disaster is only present in equatorial climates where seasonal mixing of the lakes is precluded, and is now mitigated by controlled degassing of such lakes via artificial bubble plumes [52].

Large quantities of methane hydrates are naturally stored in various geological formations and at the bottom of the ocean. Their stability at depth depends on factors including temperature and pressure [22], which may change in response to sea level changes or landslides, possibly leading to the release of gaseous methane into the atmosphere via bubble plumes. Indeed, adherents to the view of the Earth as a self-regulating organism [26] note that such deposits provide a natural check on runaway global cooling: planetary cooling is necessarily accompanied by the growth of polar ice caps, a concomitant diminution in sea level, and so a release of this greenhouse gas. The analogous capture and storage of anthropogenic carbon dioxide in deep-ocean reservoirs is currently being explored as a means of mitigating global warming. The sequestration of CO_2 might be accomplished by injecting it into either geological reservoirs (e.g., saline formations or depleted oil and gas reservoirs) or into the ocean. In the first approach, the CO_2 is injected at depth in either gaseous or liquid form in order to dissolve it into the ambient fluid or trigger its reaction with local minerals to form solid compounds. In the second, ocean-sequestration approach, liquid CO_2 is injected at depths where it is heavier than sea water, and so sinks to the ocean bottom to form CO_2 lakes [21,52].

32.4 Challenges: Drops and Bubbles in the Biosphere

32.4.1 Drops and the Spread of Life

Drops also play a key role in the spread of life through remote terrestrial regions. Bacteria, pollen, fungi, and marine algae serve as condensation nuclei, thereby enhancing their range of dispersal: clouds can transport and deposit such organisms even to the remotest areas [29,41]. Such bioaerosols serve as giant condensation nuclei and, when hydrophilic, are particularly favorable to initiating precipitation in warm clouds. A possible feedback loop between biological nucleation of droplets leading to precipitation and the natural emission of bioaerosols has recently been suggested [33].

As well as serving as droplet nuclei, some microbiological and organic material is known to multiply and grow in cloud water [1,41]. The resulting surfactant at the droplet surface can lead to major changes in the cloud microphysics. The increase of anthropogenic pollution, which is known to extend the life of clouds, could enhance this process and lead, in turn, to enhanced microbial activity within clouds [1]. Finally, the enrichment of surface-active organic compounds during the collision-coalescence phase of droplets and their ability to concentrate molecules has been suggested as a medium for key prebiotic reactions, and so may have played a role in the very conception of life [10,46].

Much of the early work on fluid–solid interactions was motivated by the desire to optimize insecticides designed to coat their insect target while leaving the plant unharmed. This task is made all the more challenging since the two coverings are often virtually identical due to their coevolution. Many plant leaves are known to be both water-repellent and self-cleaning owing to their complex surface structure. Their surface is characterized by roughness on two scales (the smallest being submicron) and a waxy coating, which together render it superhydrophobic [8]. When water drops strike a water-repellent plant, instead of sticking, they typically bounce or roll off, taking with them any residual material (e.g., dust, spores, pathogens) that might have collected on the plant. If the impact is sufficiently vigorous, it may lead to droplet fracture and the associated ejection of relatively small droplets [38]. The resulting rain-splash plays an important role in disease transfer between plants [15]. A similar process may arise when rain impacts sandy, water-repellent soils in arid regions, and the associated rain-induced stripping of top soil has been identified as an important contributor to the expansion of deserts [12].

32.4.2 Health Hazards

Bubble-bursting and atomization of jets in recreational and therapeutic aquatic facilities is known to generate pathogenic bioaerosols [4]. In therapy pools, the presence of detergents and elevated temperatures reduces surface tension and enhances aerosol release [3]. Moreover, the bacterial enrichment of the ejected droplets relative to their originating pool is thought to be facilitated by surface-active bacteria aggregating on gas bubble surfaces. Angenent et al. [3] examined the case of a particular therapeutic swimming pool and found greatly elevated levels of several pathogens such as those involved in sinusitis, ear infections, meningitis, and necrotizing fasciitis.

Droplets also play a critical role in the transmission of many respiratory diseases. Exhalation can transform an infectious

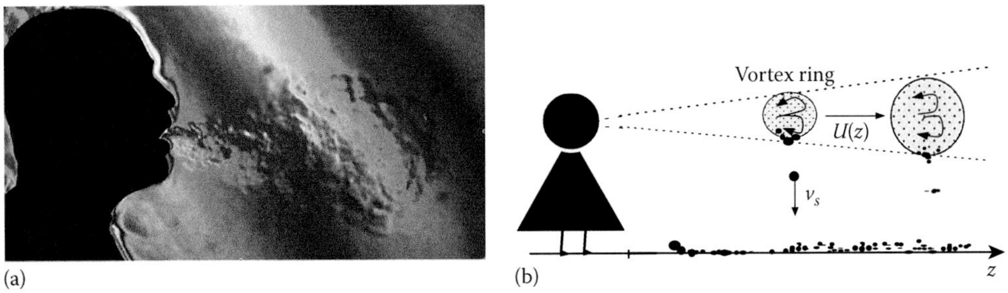

FIGURE 32.6 Coughing and sneezing may be accompanied by the ejection of pathogen-bearing droplets critical to the transmission of respiratory diseases. (a) The cough. (Courtesy of G.S. Settles.) (b) The cough releases a droplet-bearing vortex ring, the dynamics of which determine the range of the pathogens.

patient into a potent source of pathogen-bearing droplets [44] (Figure 32.6). Several routes of respiratory disease transmission have been examined [51]. Large droplet transmission may arise through the spraying of infected droplets via coughing or sneezing directly onto a susceptible host, while airborne transmission may arise through inhalation of relatively small (<5–10 μm) infected droplets or droplet nuclei, the pathogen-bearing residual solids that survive evaporation. The size of the droplets emitted by exhalation events (e.g., coughing, sneezing, speaking, and breathing) varies from a few μm to 1 mm, but the drop size distribution remains poorly understood [13,30]. As the fate of the emitted drops and their viral or bacterial payload is strongly dependent on their initial size, understanding the formation mechanisms and subsequent dynamics of the emitted droplets will be critical to mitigating airborne disease transmission. While the formation mechanisms of droplets in the respiratory tract have received some attention [6], their subsequent dynamics, including their impacts within the respiratory tract (Figure 32.4b) and their dispersal via vortex rings (Figure 32.6b), remain poorly understood. The role of droplets in the transmission of infectious diseases in confined environments such as hospitals and airplanes is an area of interdisciplinary research with enormous potential for human impact [13].

32.4.3 Biocapillarity

We have seen in Section 32.2.3 that surface tension dominates fluid systems on a scale small relative to the capillary length. It is thus that the lives of the world's smallest denizens are dominated by the influence of surface tension. Interfacial effects are used by the majority of water-walking creatures for propulsion [7], while bubbles and drops are used by insects and spiders in a variety of ingenious ways, for feeding, drinking, adhesion, and even respiration (Figure 32.7). A number of spiders can survive

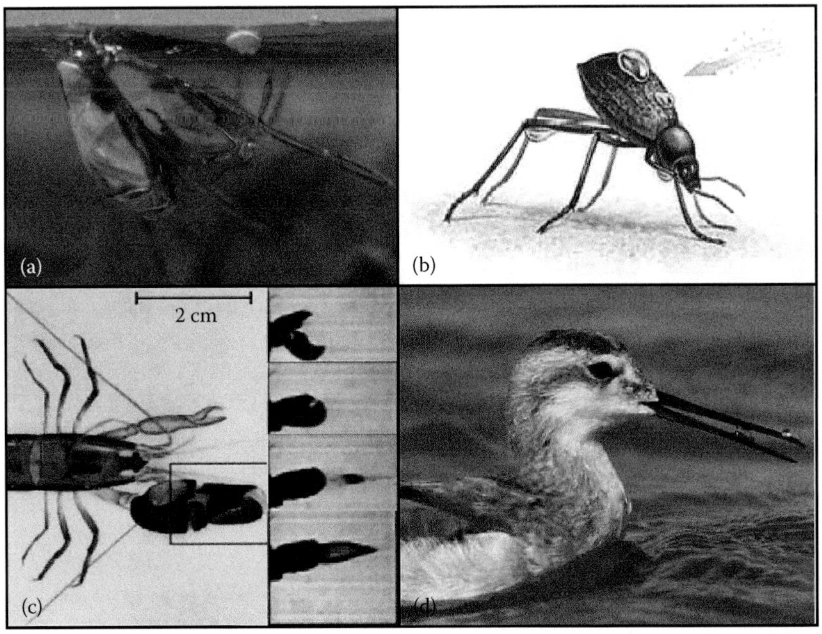

FIGURE 32.7 Bubbles and drops in the biological world. (a) The backswimmer breathes underwater by using its plastron, the thin bubble trapped on its water-repellent surface, as an external lung. (b) The Namib Desert beetle harvests drinking water by condensing fog droplets on its back. (Courtesy of Roberto Osti Illustrations.) (c) The trigger shrimp hunts by propelling a millimetric cavitation bubble toward its prey. (Courtesy of Detlef Lohse.) (d) The phalarope feeds by transporting prey-bearing droplets mouthward. (Courtesy of Robert Lewis.)

stints beneath the water surface using diving bubbles, while some insects employ a plastron bubble trapped on their water-repellent surfaces as an external lung that allows them to remain submerged indefinitely [16] (Figure 32.7a). Many insects, lizards, and birds employ curvature pressures associated with drops to drive flow, assisting them in either drinking or feeding. The interfacial mechanisms exploited by nature are as diverse as life itself. Given space limitations, we here simply highlight three of the more remarkable.

The Namib beetle resides in a desert where it rarely rains; nevertheless, it is able to condense water from micron-scale fog droplets that sweep in daily from the coast (Figure 32.7b). Their surface is composed of hydrophilic bumps on hydrophobic valleys. The fog droplets thus stick to the peaks, remaining pinned there by contact angle hysteresis, then grow through accretion until becoming large enough to be blown by the wind onto the hydrophobic valleys, across which they roll with little resistance. By guiding these droplets toward their mouths, the beetles may thus reap the rewards of the refrigeration-free condenser on their back [32]. This technique has inspired the creation of a new surface, "Super Plastic," whose analogous ability to condense water is being applied to water harvesting in arid regions of the developing world [53].

The trigger shrimp resides on the ocean floor, and uses bubbles for both communication and hunting [48]. By rapidly closing one of its claws, it ejects a water jet toward its prey at such high speed that the fluid pressure drops below the cavitation pressure, and a millimetric bubble of water vapor forms (Figure 32.7c). When it collapses, the bubble stuns the prey, which then makes for an easy target. The bubble collapse has also been known to emit light, an effect termed "shrimpoluminescence." This novel hunting strategy, which has been exploited by the trigger shrimp for millions of years, was thus a natural precursor to both depth charges (antisubmarine weaponry) and sonoluminescence.

Phalaropes are shorebirds that prey on small organisms such as miniature shrimp and phytoplankton. By swimming in a tight circle on the surface of shallow bodies of water, they generate a vortex that sweeps their prey upward, like tea leaves in a swirling cup [40]. By pecking the free surface, they capture a prey-bearing droplet in the tip of their beak. Then, by successively opening and closing their beaks in a tweezering motion, they draw the droplet mouthward (Figure 32.7d). This "capillary ratchet" mechanism relies critically on contact angle hysteresis, typically an impediment to drop motion [35].

This class of problems highlights the sensitivity of an enormous class of creatures to contamination at the free surface as may arise from pollution or chemical spills: the resulting alteration of their wetting properties may make it impossible for them to function. Moreover, they suggest novel mechanisms for discrete droplet transport in microfluidic systems. While microfluidics is a relatively new endeavor for mankind, it has long been explored by nature. As water management is a problem of increasing human concern, it would seem prudent to explore and familiarize ourselves with nature's own ingenious solutions.

References

1. P. Amato, F. Demeer, A. Melaouhi, S. Fontanella, A.-S. Martin-Biesse, M. Sancelme, P. Laj, and A.-M. Delort. A fate for organic acids, formaldehyde and methanol in cloud water: Their biotransformation by micro-organisms. *Atmos. Chem. Phys.*, 7:4159–4169, 2007.
2. E. L. Andreas and K. Emanuel. Effects of sea spray on tropical cyclone intensity. *J. Atmos. Sci.*, 58:3741–3751, 2001.
3. L. T. Angenent, S. T. Kelley, A. St. Amand, N. R. Pace, and M. T. Hernandez. Molecular identification of potential pathogens in water and air of a hospital therapy pool. *Proc. Natl Acad. Sci. U.S.A.*, 102:4860–4865, 2005.
4. D. C. Blanchard. The ejection of drops from the sea and their enrichment with bacteria and other materials: A review. *Eustuaries*, 12:127–137, 1989.
5. F. Blanchette and T. P. Bigioni. Partial coalescence of drops at liquid interfaces. *Nat. Phys.*, 2:254–257, 2006.
6. P. J. Brasser, T. A. McMahon, and P. Groffith. The mechanism of mucus clearance in cough. *J. Biomed. Eng.*, 111:288–297, 1989.
7. J. W. M. Bush and D. L. Hu. Walking on water: Biolocomotion at the interface. *Annu. Rev. Fluid Mech.*, 38:339–369, 2006.
8. J. W. M. Bush, D. L. Hu, and M. Prakash. The integument of water-walking arthropods: Form and function. *Adv. Insect Physiol.*, 34:117–192, 2008.
9. R. P. Chhabra. *Bubbles, Drops, and Particles in Non-Newtonian Fluids*. Taylor & Francis Group, Boca Raton, FL, 2007.
10. C. M. Dobson, E. Barney, A. F. Tuck, and V. Vaida. Atmospheric aerosols as prebiotic chemical reactors. *Proc. Natl Acad. Sci. U.S.A.*, 97:11864–11868, 2000.
11. R. Clift, J. R. Grace, and M. E. Weber. *Bubbles, Drops, and Particles*. Academic Press Inc, New York, 1978.
12. S. H. Doerr, R. A. Shakesby, and R. P. D. Walsh. Soil water repellency: its causes, characteristics and hydro-geomorphological significance. *Earth-Sci. Rev.*, 51:33–65, 2000.
13. I. Eames, J. W. Tang, Y. Li, and P. Wilson. Airborne transmission of disease in hospitals. *J. R. Soc. Interf.*, 6:S697–S702, 2009.
14. X.-Q. Feng and L. Jiang. Design and creation of superwetting/antiwetting surfaces. *Adv. Mater.*, 18:3063–3078, 2006.
15. B. D. L. Fitt, H. A. McCartney, and P. J. Walklate. The role of rain in dispersal of pathogen inoculum. *Annu. Rev. Phytopathol.*, 27:241–270, 1989.
16. M. R. Flynn and J. W. M. Bush. Underwater breathing. *J. Fluid Mech.*, 608:275–296, 2008.
17. P. G. de Gennes, F. Brochard-Wyart, and D. Quéré. *Capillarity and Wetting Phenomena: Drops, Bubbles, Pearls and Waves*. Springer-Verlag, Berlin, Germany, 2003.
18. K. Gierens, B. Karcher, H. Mannstein, and B. Mayer. Aerodynamic contrails: Phenomenology and flow physics. *J. Atmos. Sci.*, 66:217–226, 2009.
19. P. S. Gill, T. E. Graedel, and C. J. Weschler. Organic films on atmospheric aerosol particles, fog droplets, cloud droplets, raindrops, and snowflakes. *Rev. Geophys. Space Phys.*, 21:903–920, 1983.

20. H. M. Gonnermann and M. Manga. The fluid mechanics inside a volcano. *Annu. Rev. Fluid Mech.*, 39:321–356, 2007.
21. H. Herzog and D. Golomb. Carbon capture and storage from fossil fuel use. *Encyclopedia of Energy*, pp. 1–19, 2004. http://sequestration.mit.edu.
22. K. C. Hester, E.T. Peltzer, P. M. Walz, R. M. Dunk, E. D. Sloan, and P. G. Brewer. A natural hydrate dissolution experiment on complex multi-component hydrates on the sea floor. *Geochimica et Cosmochimica Acta*, 73:6747–6756, 2009.
23. P. V. Hobbs, ed. *Aerosol-Cloud-Climate Interactions*, International geophysics series. Academic Press, New York, 1993.
24. O. W. Jayaratne and B. J. Mason. The coalescence and bouncing of water drops at an air/water interface. *Proc. R. Soc. A*, 280:545–565, 1964.
25. H. Lhuissier and E. Villermaux. Bursting bubbles. *Phys. Fluids*, 21:091111, 2009.
26. J. Lovelock. *A New Look at Life on Earth*. Oxford University Press, Oxford, U.K., 1982.
27. E.H. Lucassen-Reynders and J. Lucassen. Properties of capillary waves. *Adv. Colloid Interf. Sci.*, 2:347–395, 1969.
28. J. Magnaudet and I. Eames. The motion of high-Reynolds-number bubbles in inhomogeneous flows. *Annu. Rev. Fluid Mech.*, 32:659–708, 2000.
29. O. Mohler, P. J. DeMott, G. Vali, and Z. Levin. Microbiology and atmospheric processes: The role of biological particles in cloud physics. *Biogeosciences*, 4:2559–2591, 2007.
30. L. Morawska, G. R. Johnson, Z. D. Ristovski, M. Hargreaves, K. Mengersen, S. Corbett, C. Y. H. Chao, Y. Li, and D. Katoshevski. Size distribution and sites of origin of droplets expelled from the human respiratory tract during expiratory activities. *J. Aerosol Sci.*, 40:256–269, 2009.
31. C. D. O'Dowd and G. de Leeuw. Marine aerosol production: A review of the current knowledge. *Philos. Trans. R. Soc. A*, 365:1753–1774, 2007.
32. A. R. Parker and C. R. Lawrence. Water capture by a desert beetle. *Nature*, 414:33–34, 2001.
33. V. T. J. Phillips, C. Andronache, B. Christner, C. E. Morris, D. C. Sands, A. Bansemer, A. Lauer, C. McNaughton, and C. Seman. Potential impacts from biological aerosols on ensembles of continental clouds simulated numerically. *Biogeosciences*, 6:987–1014, 2009.
34. M. Pósfai and P. R. Buseck. Nature and climate effects of individual tropospheric aerosol particles. *Annu. Rev. Earth Planet. Sci.*, 38:17–43, 2010.
35. M. Prakash, D. Quéré, and J. W. M. Bush. Surface tension transport of prey by feeding shorebirds: The capillary ratchet. *Science AAAS*, 320:931–934, 2008.
36. A. Prosperetti and H. N. Oguz. The impact of drops in liquid surfaces and the underwater noise of rain. *Annu. Rev. Fluid Mech.*, 25:577–602, 1993.
37. D. Quéré. Wetting and roughness. *Annu. Rev. Mater. Res.*, 38:71–99, 2008.
38. D. Richard, C. Clanet, and D. Quéré. Contact time of a bouncing drop. *Nature*, 417:811, 2002.
39. R. R. Rogers and M. K. Yau. *A Short Course in Cloud Physics*, 3rd edn. Pergamon Press, Oxford, U.K., 1989.
40. M. A. Rubega and B. S. Obst. Surface-tension feeding in phalaropes: Discovery of a novel feeding mechanism. *The Auk*, 110:169–178, 1993.
41. B. Sattler, H. Puxbaum, and R. Psenner. Bacterial growth in supercooled cloud droplets. *Geophys. Res. Lett.*, 28:239–242, 2001.
42. N. Savva and J. W. M. Bush. Viscous sheet retraction. *J. Fluid Mech.*, 626:211–240, 2009.
43. J. H. Seinfeld and S. N. Pandis. *Atmospheric Chemistry and Physics: From Air Pollution to Climate Change*. John Wiley & Sons Inc., New York, 2006.
44. G. S. Settles. Fluid mechanics and homeland security. *Annu. Rev. Fluid Mech.*, 38:87–110, 2006.
45. R. A. Shaw. Particle-turbulence interactions in atmospheric clouds. *Annu. Rev. Fluid Mech.*, 35:183–227, 2003.
46. I. Taraniuk, A. B. Kostinski, and Y. Rudich. Enrichment of surface-active compounds in coalescing cloud drops. *Geophys. Res. Lett.*, 35:L19810-1–4, 2008.
47. S. T. Thoroddsen, T. G. Etoh, and K. Takehara. High-speed imaging of drops and bubbles. *Annu. Rev. Fluid Mech.*, 40:257–285, 2008.
48. M. Versluis, B. Schmitz, A. von der Heydt, and D. Lohse. How snapping shrimp snap: Through cavitating bubbles. *Science*, 289:2114–2117, 2000.
49. E. Villermaux. Fragmentation. *Annu. Rev. Fluid Mech.*, 39:419–446, 2007.
50. E. Villermaux and B. Bossa. Single-drop fragmentation determines size distribution of raindrops. *Nat. Phys.*, 5:697–702, 2009.
51. W. F. Wells. *Airborne Contagion and Air Hygiene: An Ecological Study of Droplet Infection*. Harvard University Press, Cambridge, MA, 1955.
52. A. W. Woods. Turbulent plumes in nature. *Annu. Rev. Fluid Mech.*, 42:391–412, 2010.
53. L. Zhai, M. C. Berg, F. C. Cebeci, Y. Kim, J. M. Milwid, M. F. Rubner, and R. E. Cohen. Patterned superhydrophobic surfaces: Toward a synthetic mimic of the Namib desert beetle. *Nano Lett.*, 6:1213–1217, 2006.
54. M. A. Zondlo, P. K. Hudson, A. J. Prenni, and M. A. Tolbert. Chemistry and microphysics of polar stratospheric clouds and cirrus clouds. *Annu. Rev. Phys. Chem.*, 51:473–499, 2000.

33

Particle-Laden Flows

S. Balachandar
University of Florida

33.1 Introduction

Examples of particle-laden flows abound in environmental flows. Volcanic ash particles suspended in a Plinian eruption, powder snow avalanche down a mountain slope, turbidity current in a submarine canyon, wave-driven sedimentary transport in the coastal region, and dust storms are some of the classic examples of particle-laden flows that can be often observed in the environment.

The regime of particle-laden flows can be generally classified as "very dilute," "dilute," or "dense," and the dominant mechanisms that control the behavior of the flow substantially change in these three regimes. The fractional volume occupied by the particles (Φ_v) and the mass loading (Φ_m) are the two key parameters that determine the regime of the particle-laden flow. Here, mass loading is defined as the ratio of the mass of the particulates to the mass of the carrier phase, and, thus, the two parameters are related through the particle-to-fluid density ratio (ρ_p/ρ_f) as

$$\Phi_m = \frac{\Phi_v}{1-\Phi_v}\frac{\rho_p}{\rho_f} \tag{33.1}$$

Note that in a monodisperse system of spherical particles, the maximum possible volume fraction is $\pi/\sqrt{18} \approx 0.74$, and it corresponds to the case of a hexagonal close pack. But experiments and recent theoretical estimations show that typical random close pack volume fraction cannot exceed above 64% (Song et al. 2008). Under dilute conditions, we can approximate $\Phi_m \approx \Phi_v\rho_p/\rho_f$. Thus, in the environmentally relevant case of sand particles suspended in clear or salty water ($\rho_p/\rho_f \approx 2.65$), the mass loading is of the order of the volume fraction. Whereas, in the case of particles suspended in air (e.g., ash particles in air), mass loading will be typically three orders of magnitude larger than the corresponding volume fraction. In the other environmentally relevant limit of bubbly flows, mass loading (as defined earlier) will be three orders of magnitude smaller than the volume fraction. However, the problem of bubbly flows must be treated with care, since the added mass of the bubbles will make the effective mass loading to be of the order of volume fraction.

When both mass loading and volume fraction are very small (say $\Phi_m < 10^{-3}$, $\Phi_v < 10^{-2}$), the back effect of particles on the flow is typically unimportant, and only the effect of the carrier fluid on the dynamics of the particles and their dispersion is important. In this limit, the particle-laden flow is termed "one-way coupled." Here, the particles can be followed within the flow as a passive scalar marker. When mass loading becomes appreciable, the back effect of particles on the carrier fluid cannot be ignored. In case of bubbles even when mass loading is small, if volume fraction of bubbles is appreciable, the effect on the fluid cannot be ignored. The particle-laden flow is then termed "two-way coupled." In this regime, while the carrier fluid dictates the particle motion and dispersion, the particles exert a significant influence on the carrier phase flow.

The mean separation distance between the particles (l) can be estimated to go as $l/d \propto \Phi_v^{1/3}$, where d is the particle diameter. For the case of hexagonal packing

$$l/d = \left[\frac{\sqrt{18}\Phi_v}{\pi}\right]^{1/3} \approx 1.1\Phi_v^{1/3} \tag{33.2}$$

Handbook of Environmental Fluid Dynamics, Volume One, edited by Harindra Joseph Shermal Fernando. © 2013 CRC Press/Taylor & Francis Group, LLC. ISBN: 978-1-4398-1669-1.

Thus, for volume fractions greater than about 10^{-3}, the distance between the particles becomes of the order of particle diameter and interparticle interaction becomes important. Processes such as particle–particle collision, agglomeration, and breakup become important and the particle-laden flow is termed "four-way coupled." These classifications were originally introduced by Elghobashi (1991, 1994). In the case of very large volume fraction, approaching the close pack limit, the flow is often termed "granular flow." In this regime, the flow is dominated by interparticle interaction, and often the hydrodynamic force due to the surrounding carrier fluid can be ignored.

This chapter mainly addresses very dilute to dilute particle-laden flows. Issues pertaining to only one-way and two-way coupling are primarily discussed. Complications arising from four-way coupling and collision-dominated granular flows are not addressed in detail. Thus, our focus is on dispersed particle-laden flows (see Balachandar and Eaton 2010).

33.2 Range of Length and Time Scales

In the case of dilute particle-laden flows, it is useful to think of a globally undisturbed carrier flow that would exist in the absence of particles. The flow in this single-phase limit may be quiescent, laminar, or turbulent in nature. If the globally undisturbed carrier flow is turbulent, then it will be characterized by a range of length and time scales from the Kolmogorov scale to the integral scale. With the addition of particles, the perturbation flow around the particles must be included. Depending on particle Reynolds number (based on particle diameter and relative velocity), the wake flow around each particle may be laminar or turbulent. The overall perturbation flow due to the particulate phase will then be a superposition of contributions from the individual particles. Owing to the random distribution of particles, even when flow around each particle is laminar, the overall particle-induced perturbation to the carrier fluid will be spatially chaotic. The range of length scales associated with the perturbation flow will then be characterized by particle diameter (d) and interparticle separation (l).

Two different scenarios can be envisioned, and they are depicted in Figure 33.1 in terms of globally undisturbed energy spectra that would exist in the absence of the particles (solid line) and the energy spectra with the addition of particles (dashed line). In the scenario depicted in Figure 33.1a, the spectrum of scales generated around the particles and in their wakes can be considered microscale and well separated from the spectrum of the ambient carrier phase turbulence, which can be considered macroscale. If the particle-laden flow is *one-way coupled,* then the main part of the energy spectrum remains unchanged and the effect of the particles appears only at the microscale. If particle mass loading is significant, then particles can have a direct influence on the carrier phase even when the length scales associated with the particles are well separated from the scales of carrier phase turbulence. The *two-way coupled* energy spectrum is depicted as the dashed line.

As the particles approach the scales of the undisturbed flow, there will not be any scale separation between the macroscale turbulence and the microscale flow features around the particles. The dashed line in Figure 33.1b illustrates the energy spectrum. This situation significantly complicates the momentum exchange between the particles and the carrier phase that enters in the governing equation as interphase coupling force.

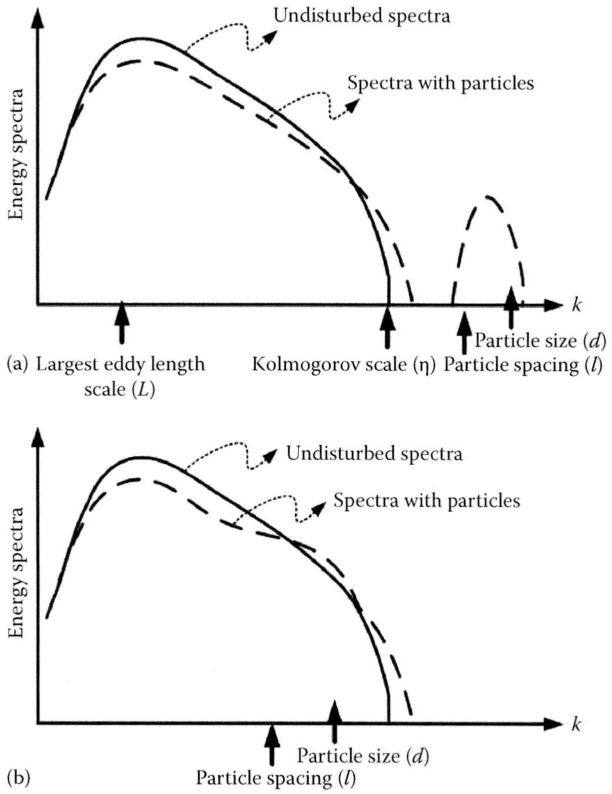

FIGURE 33.1 Schematic of the energy spectra of the globally undisturbed flow and that of the particle-laden flow. (a) Case when the length scales associated with the particles (d,l) are well separated from the carrier phase turbulence scales; (b) case when there is no scale separation.

33.3 Eulerian–Lagrangian and Eulerian–Eulerian Formulations

In the *one-way coupled* limit, the governing equations for the carrier fluid are the standard continuity and Navier–Stokes equations, with no back effect from the particles. The motion of each particle within the carrier fluid is then given by the Lagrangian equations for their position and velocity ($\mathbf{x}_p$ and $\mathbf{v}$)

$$\frac{d\mathbf{x}_p}{dt} = \mathbf{v} \tag{33.3}$$

$$m_p \frac{\mathrm{d}\mathbf{v}}{\mathrm{d}t} = \mathbf{F}_{hyd} + m_p\mathbf{g} = 3\pi\mu d(\mathbf{u}-\mathbf{v})\phi(\mathrm{Re}) + m_f \frac{D\mathbf{u}}{Dt} + m_f C_M\left(\frac{D\mathbf{u}}{Dt} - \frac{\mathrm{d}\mathbf{v}}{\mathrm{d}t}\right) + \frac{3}{2}d^2\sqrt{\pi\rho_f\mu} \times \int_{-\infty}^{t} K(t,\tau)\left.\frac{\mathrm{d}(\mathbf{u}-\mathbf{v})}{\mathrm{d}t}\right|_{t=\tau} d\tau + (m_p - m_f)\mathbf{g}, \tag{33.4}$$

where m_p and m_f are the mass of the particle and displaced fluid. The acceleration due to gravity is $\mathbf{g}$, the added mass coefficient (C_M) for a spherical particle can be taken to be ½, and μ is the dynamic viscosity of the fluid surrounding the particle. The terms on the right are the quasi-steady standard drag, pressure gradient, added-mass, Basset history [$K(t, \tau)$ is the kernel that weighs the past history of relative acceleration between the particle and the ambient fluid], and gravitational forces, respectively. Additional forces, such as the lift and electrostatic forces, can be included in the previous equation as necessary. We note that D/Dt is the total derivative following the fluid velocity and d/dt is the derivative following the particle velocity. In the previous expression for force, u and $D\mathbf{u}/Dt$ are defined as the undisturbed fluid velocity and total acceleration as seen by the particle at $\mathbf{x}_p$. Note that the force on the particle can be split into a hydrodynamic component ($\mathbf{F}_{hyd}$) and a gravitational body force.

This superposition of the hydrodynamic force into quasi-steady, added-mass, and other contributions can be rigorously obtained in the limit of small particle Reynolds number (Re) as shown by Maxey and Riley (1983) and Gatignol (1983). But at finite particle Reynolds number, Equation 33.4 is only empirical (see Magnaudet and Eames 2000; Bagchi and Balachandar 2002; Balachandar and Eaton 2010). For example, the finite Reynolds number effect in the quasi-steady drag can be accounted with the simple drag correlation $\phi(\text{Re}) = 1 + 0.15\,\text{Re}^{0.687}$.

The previous Lagrangian equations are appropriate to follow the motion of individual particles. When they are used along with the Navier–Stokes equations for the carrier fluid, we obtain the "Eulerian-Lagrangian approach" to the particle-laden flow. The alternative will be to use the "Eulerian-Eulerian approach," where the Lagrangian equations of motion for the particles are recast into mass and momentum equations for the particulate phase as:

$$\frac{\partial \Phi_v}{\partial t} + \nabla \cdot (\Phi_v \mathbf{v}) = 0 \tag{33.5}$$

$$m_p \left(\frac{\partial \mathbf{v}}{\partial t} + \mathbf{v}\nabla \cdot \mathbf{v} \right) = m_p \mathbf{g} + \mathbf{F}_{hyd} \tag{33.6}$$

Here, the particulate phase is treated as a continuum (as a second fluid) and particle velocity and volume fraction are given field representation and are denoted by $\mathbf{v}(\mathbf{x}, t)$ and $\Phi_v(\mathbf{x}, t)$.

Both Eulerian–Lagrangian and Eulerian–Eulerian approaches present distinct advantages and disadvantages. In the Eulerian–Eulerian approach by using field representations of particle properties, such as particle velocity, we implicitly assume the existence of unique values of these properties. In other words, all particles within a tiny volume of fluid at any given point in space and time are assumed to have the same velocity, etc. This requirement of uniqueness places restriction on the particle size and time scale. Ferry and Balachandar (2001) showed that provided the particle time scale is less than the inverse of the maximal compressional strain rate (this ratio can be defined as the particle Stokes number), uniqueness is guaranteed. If uniqueness is violated, then either a probabilistic framework must be adopted or the Eulerian fields of particle velocity, temperature, etc., must be thought of as ensemble averages. In the probabilistic framework, the phase space must be expanded to include velocity information as well, and this makes probabilistic framework impractical in most applications. If the Eulerian particle quantities represent ensemble averages, then the governing mass and momentum equations for the particles will require closure assumptions.

In case of polydisperse particle-laden flows, the standard Eulerian–Eulerian approach becomes considerably complex. For each particle size, a set of mass and momentum (partial differential) equations must be solved along with those of the carrier fluid. Thus, the Eulerian–Eulerian approach can be computationally expensive in systems where a wide range of particle sizes need to be considered. If the governing Eulerian equations are derived using the probability density function approach, one can include particle size as one of phase space variables (Pope 1985; Subramaniam 2001; Fox et al. 2008).

The significant advantage of the Lagrangian approach is that there is no fundamental limitation on the particle size or Stokes number, as there is no requirement of uniqueness. Furthermore, in the Lagrangian approach the size of each particle is independent, and, thus, polydisperse systems can be handled easily. On the other hand, coupling between the Lagrangian particles and the Eulerian carrier fluid poses interesting challenges. In the Lagrangian equations of particle motion, quantities such as the undisturbed fluid velocity and fluid acceleration (u and $D\mathbf{u}/Dt$) are needed at the particle location ($\mathbf{x}_p$). Particle locations do not, in general, coincide with the Eulerian grid, so interpolation is required. The interpolation is referred to as "forward" or Eulerian-to-Lagrangian coupling. At sufficiently high mass loadings, the particles begin to influence the fluid. This back effect represents Lagrangian-to-Eulerian or "backward" coupling.

In backward coupling, the forces from particles must be transferred to the grid points and applied as source/sink terms in the momentum equations of the carrier fluid. A smooth distribution of the particle number density, that is, the number of particles per unit volume, on the grid is essential for an accurate accounting of the back effect of particles on the carrier fluid. Large grid-scale fluctuation in the particle number density will introduce spurious small-scale oscillations in the carrier phase through back coupling (Ling et al. 2009). These oscillations, in turn, can then be fed back to the particle motion and distribution through forward coupling and lead to accumulation of errors. In this sense, it is easier to implement two-way coupling in the Eulerian–Eulerian formulation.

If we consider the class of physical problems where the particles are small and numerous, we can envision a conceptually smooth representation of the particle number density and the back coupling. Because the cost of Eulerian–Lagrangian simulations is proportional to the number of particles, it is common to limit the average number of particles per cell and define so-called *computational particles* that represent a large number (or cloud) of real particles. There needs to be sufficient number of particles within each cell in order to have a smooth Eulerian representation of the feedback force from the particles. With less number of particles per cell, ad hoc smoothing of the feedback forcing is needed. Furthermore, with less number of particles per cell, the algorithm by which the feedback is apportioned to the neighboring cells is necessarily ad hoc and approximate. Thus, in the Eulerian–Lagrangian approach the computational cost increases, since a large number of particles must be tracked in order to contain the level of statistical fluctuation. Even when the computation employs on average a large number of particles, due to preferential accumulation, there are local regions of low particle number density adversely affecting accurate representation of local back coupling to the carrier fluid. One other disadvantage of the Lagrangian approach is that it becomes exceedingly more expensive, due to restriction in time step size, as the particle size decreases, for example, in reacting flow applications with evaporating droplets.

33.4 Two-Way Coupling Mechanisms

The back effect of particles on the carrier fluid manifests in several ways. These effects must be accurately accounted in the mathematical model of the carrier phase, while still striving for simplicity in the formulation. In the one-way coupled limit, the back effect of particles can be ignored altogether, and as a result the single-phase governing equations are sufficient. One approach to accurate incorporation of the back effect of particles is to fully resolve the flow around each and every particle by imposing the correct translational and rotational motion of the particles as time-dependent boundary conditions in solving the carrier fluid. Such fully resolved simulations of particle-laden flows have typically accounted for only $O(1000)$ particles or less (Kajishima et al. 2001; Ten cate et al. 2004; Lu and Tryggvason 2006; Uhlmann 2008). The fully resolved formulation is fundamental (does not involve simplifying assumptions) but is computationally impractical.

Continuum equations for the conservation of mass, momentum, energy, etc., of the carrier phase that accounts for the back effect of particles are obtained through an appropriate averaging process. An averaging is required in order to account for the random location of the particles that vary from realization to realization. A variety of averaging techniques—time, space, and ensemble averages—have been employed in the past (Joseph and Lundgren 1990; Zhang and Prosperetti 1994, 1997; Jackson 1997). Here, we seek appropriate governing equations that are suitable for direct and large eddy simulations of the particle-laden flows at the macroscale. This requires that the averaging process leave the range of scales associated with the macroscale dynamics unaffected and only average out the randomness associated with the random microscale distribution of particles. Such selective averaging can be achieved when the flow scales are well separated from those associated with the particles (see Figure 33.1a). In this limit, we carefully interpret the averaging process (such as those employed by Joseph and Lundgren 1990; Zhang and Prosperetti 1994, 1997) to be only over an ensemble of small-scale particle arrangements.

When particle volume fraction is sufficiently small, it is reasonable to ignore its variation in the continuity equation of the carrier fluid. If the carrier fluid is incompressible, this simplifies the formulation to the standard incompressibility condition. However, as the particle volume fraction increases, the effect of spatial and temporal variation in fluid volume fraction becomes important and an averaging process yields

$$\nabla \cdot \mathbf{u} = \frac{\partial \Phi_v}{\partial t} + \nabla \cdot (\Phi_v \mathbf{u}) \tag{33.7}$$

The momentum equation for the carrier phase can be similarly obtained from rigorous averaging and after some simplification can be expressed as

$$\rho_f(1-\Phi_v)\left(\frac{\partial \mathbf{u}}{\partial t} + \mathbf{u}\nabla \cdot \mathbf{u}\right) = -\nabla p - \rho_f \Phi_v \mathbf{g} + \nabla \cdot \left(\mu_{eff}\left[\nabla \mathbf{u} + \nabla \mathbf{u}^T\right]\right) - \frac{\Phi_v}{V}\mathbf{F}_{hyd} \tag{33.8}$$

where V is the volume of a particle and thus (Φ_v/V) corresponds to local particle number density. These mass and momentum balance equations of the carrier fluid can be used to highlight the important mechanisms of two-way coupling:

- The *volumetric effect* of the space occupied by the particles introduces divergence in the macroscale flow and makes the carrier phase to be effectively compressible. This effect depends on the local variation in particle volume fraction as given in Equation 33.7.
- The *inertial effect* of the particle-laden flow is captured by the coefficient $\rho_f(1 - \Phi_v)$ on the left-hand side of the momentum equation. Since fluid occupies only part of the volume, the particles contribute $\rho_p\Phi_v$ to the local mass. In simple terms, as mass loading increases, the inertia of the mixture increases, and a higher force will be required to impart the same acceleration.
- The presence of particles contributes to *enhanced effective local viscosity*. This is purely a consequence of enhanced stress that arises due to the particles for a specified macroscale strain-rate field. This enhancement will be present even when the particles are perfectly following the local fluid motion. When particles are randomly distributed, without any preferred small-scale structure or arrangement, the stress–strain relation remains isotropic, and the net effect of particles is to increase the effective viscosity of the mixture above that of the pure fluid. The effective viscosity can be expressed in terms of the Einstein's correction factor as $\mu_{eff} = \mu(1 + 5\Phi_v/2)$ (Batchelor 1970).
- The drag force experienced by the particles will contribute to *enhanced dissipation*. The relative velocity between the particle and the surrounding fluid $(\mathbf{v} - \mathbf{u}(\mathbf{x}_p))$ contributes

to the quasi-steady drag force (first term on the right-hand side of Equation 33.4, which we will call $\mathbf{F}_{qs}$). The resulting particle wake and the small-scale fluid motion around the particle lead to added dissipation. Unlike enhanced viscosity discussed in the previous bullet, enhanced dissipation arises from relative motion between the particle and the fluid. The increase in kinetic energy of the fluid at the macroscale, per unit volume of the mixture, due to the quasi-steady drag force is $-(\Phi_v/V)\mathbf{F}_{qs} \cdot \mathbf{u}$, while the increase in kinetic energy of particles within the unit volume of the mixture is $(\Phi_v/V)\mathbf{F}_{qs} \cdot \mathbf{v}$. Thus, the kinetic energy lost/gained by the fluid at the macroscale does not entirely comes/goes to the particles. The balance between the two goes as $\propto|\mathbf{u} - \mathbf{v}|^2$. This nonnegative flow of energy contributes to the kinetic energy of the small-scale fluid motion and is dissipative as far as the macroscale motion is concerned. Similarly, the Bassett history force in Equation 33.4 also contributes to dissipation. The other force contributions in Equation 33.4 are inviscid in origin and therefore do not contribute to dissipation.

- Spatial variation in particle concentration in combination with gravity (second term on the right-hand side of Equation 33.8) contributes to the *buoyancy effect*. Many geophysical and environmental flows, such as turbidity currents, are driven by this buoyancy effect.

33.5 Approximations and Simplifications

The momentum equation given in Equation 33.8 involves several approximations. In the macroscale momentum equation, the averaging process leads to pressure effect expressed in terms of the mixture pressure as ∇p_m. In Equation 33.8, we have employed an approximation and expressed the pressure effect in terms of fluid pressure as ∇p. The relation between the two forms has been explored in detail by Marchioro et al. (1999). The difference between the two is asymptotically small for the case of dilute dispersion of small particles.

In the momentum equation, the strain rate in the viscous term must be defined in terms of volume-averaged mixture velocity, $\mathbf{u}_v = (1 - \Phi_v)\mathbf{u} + \Phi_v\mathbf{v}$, and the third term on the right-hand side of Equation 33.8 will take the following more complex form:

$$\nabla \cdot \left(\mu_{eff}\left[\nabla \mathbf{u}_v + \nabla \mathbf{u}_v^T\right]\right) \tag{33.9}$$

The expression for mixture stress, as given in Marchioro et al. (1999), includes additional terms. The velocity fluctuation at the microscale, upon ensemble average over all microscale arrangements, gives rise to the kinematic Reynolds stress, defined as $-\rho_f\langle\tilde{\mathbf{u}}\tilde{\mathbf{u}}\rangle_{ss}$, where $\tilde{\mathbf{u}}$ is the local small-scale perturbation velocity field away from the macroscale flow due to the presence of particles and the angle brackets represent average over all subscale particle distributions. Here, such additional terms that arise from rigorous small-scale ensemble averaging procedure are ignored, and they can be shown to be of smaller order (Cantero et al. 2008a).

In problems where volumetric, inertial, enhanced viscosity, enhanced dissipation, and buoyancy effects are important, Equations 33.7 and 33.8 provide very good approximation to the mass and momentum balance. In what follows we will employ additional approximations to obtain even simpler governing equations. In many applications of geophysical interest, since particle volume fraction is small, the volumetric effect can be ignored and the continuity equation can be simplified. Furthermore, the enhanced viscosity effect is typically small and the viscous term in the momentum equation can be simplified as $\mu\nabla^2\mathbf{u}$.

More importantly, in the limit of dilute distribution of small particles, it can be assumed that the particles are sufficiently small that the equation of motion (Equation 33.4) can be explicitly solved independent of the initial condition. In other words, the particles are taken to be in equilibrium with the local carrier phase and their initial conditions can be taken to be forgotten. As shown in Ferry and Balachandar (2001) and Ferry et al. (2003), under equilibrium assumption particle velocity can be explicitly expressed as an expansion in terms of local fluid velocity and its gradients with the particle Stokes number (ratio of particle time scale to Kolmogorov scale) as the small parameter. In dimensional terms, the equilibrium particle velocity field can be expressed as

$$\mathbf{v} = \mathbf{u} + \mathbf{w} - \frac{(\rho_p - \rho_f)d^2}{18\mu\,\phi(\mathrm{Re})}\frac{D\mathbf{u}}{Dt} \tag{33.10}$$

where $\mathbf{w}$ is the setting velocity of an isolated particle in still fluid. This equation can be interpreted in the following way. Deviation in particle velocity ($\mathbf{v}$) from fluid velocity ($\mathbf{u}$) arises due to two different mechanisms: first, due to particle settling ($\mathbf{w}$) and, second, due to particles of finite inertia not responding to ambient acceleration the same way as a fluid element. For Stokes number less than unity, even the first-order correction to local fluid velocity provides accurate representation of the slip velocity between the particle and the surrounding fluid.

Equation 33.10 is an explicit expression for the particle velocity field, and it can be readily evaluated with knowledge of the carrier phase velocity field. The biggest advantage is that we do not need to solve either a set of ordinary differential equations in the Lagrangian approach (Equation 33.4) or a set of partial differential equations in the Eulerian approach (Equation 33.6) for the particle velocity. The previous approach has been termed the "equilibrium Eulerian approach" and has been shown to accurately capture important phenomena, such as preferential particle accumulation (Squires and Eaton 1991) and turbophoresis (Reeks 1983) (we will discuss these phenomena in later sections). The accuracy of Equation 33.10 has been tested in a variety of homogeneous and wall-bounded turbulent flows (Rani and Balachandar 2003; Shotorban and Balachandar 2006, 2007,

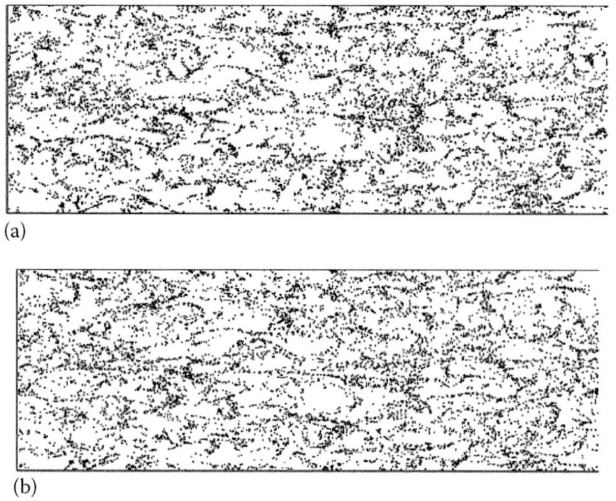

FIGURE 33.2 Comparison of particle distribution (St = 3) in a turbulent channel flow of Re_τ = 180 around a horizontal plane of y_+ = 30 parallel to the channel wall. (a) Particle velocity is computed with the Lagrangian equation of motion. (b) Particle velocity computed with equilibrium Eulerian velocity (Equation 33.10). (From Ferry, J. et al., *Int. J. Multiphase Flow*, 29, 869, 2003. With permission.)

2009). For example, in Figure 33.2 we present distribution of Lagrangian particles of Stokes number (St = 3) around a horizontal plan located at y_+ = 30 in a DNS turbulent channel flow. The Lagrangian particles shown in the top frame were advected using velocity computed from the equation of motion (simplified Equation 33.4), while the particles in the lower frame were computed using the equilibrium particle velocity (Equation 33.10). Despite the large Stokes number, the comparison is quite good, and for St < 1, accuracy of Equation 33.10 has been shown to be excellent (Ferry and Balachandar 2003).

In the explicit expression for particle velocity (Equation 33.10), forces such as the Bassett history and lift have been ignored. Consistent with this level of approximation, the previous velocity can be substituted into the right-hand side of Equation 33.4 to obtain

$$\mathbf{F}_{hyd} = -m_p\mathbf{g} + m_p\frac{D\mathbf{u}}{Dt} \quad (33.11)$$

This expression for the hydrodynamic force can then be substituted into Equation 33.8, and the resulting simplified continuity and momentum equations are written as

$$\nabla \cdot \mathbf{u} = 0 \quad (33.12)$$

$$\rho_f\left(\frac{\partial \mathbf{u}}{\partial t} + \mathbf{u}\nabla \cdot \mathbf{u}\right) = -\nabla p - (\rho_f(1-\Phi_v) + \rho_p\Phi_v)\mathbf{g} + \mu\nabla^2\mathbf{u} \quad (33.13)$$

These simplified equations, along with Equation 33.7 for particle concentration and Equation 33.10 for particle velocity, have often been used to characterize particle-laden flows. In fact, Equation 33.10 is frequently further simplified to $\mathbf{v} = \mathbf{u} + \mathbf{w}$. For example, such simplified models have been used in the simulation of turbidity currents, powder snow avalanches, etc. (Necker et al. 2005; Cantero et al. 2009a,b). The effect of the additional inertial term has also been studied by Cantero et al. (2008a,b).

33.6 Turbulence Modulation

In the context of turbulent particle-laden flows, the back effect of particles is primarily discussed in terms of turbulence modulation. Before we address this, let us first discuss the effect of particles on the mean motion of the carrier fluid. There are particle-laden flows where the mean flow is significantly altered due to the particles. The experiments by Tsuji et al. (1984) provide a good example that illustrates the dramatic effect the dispersed phase can have on the flow. Two of their figures are reproduced here (Figure 33.3), which show the mean velocity profile in the

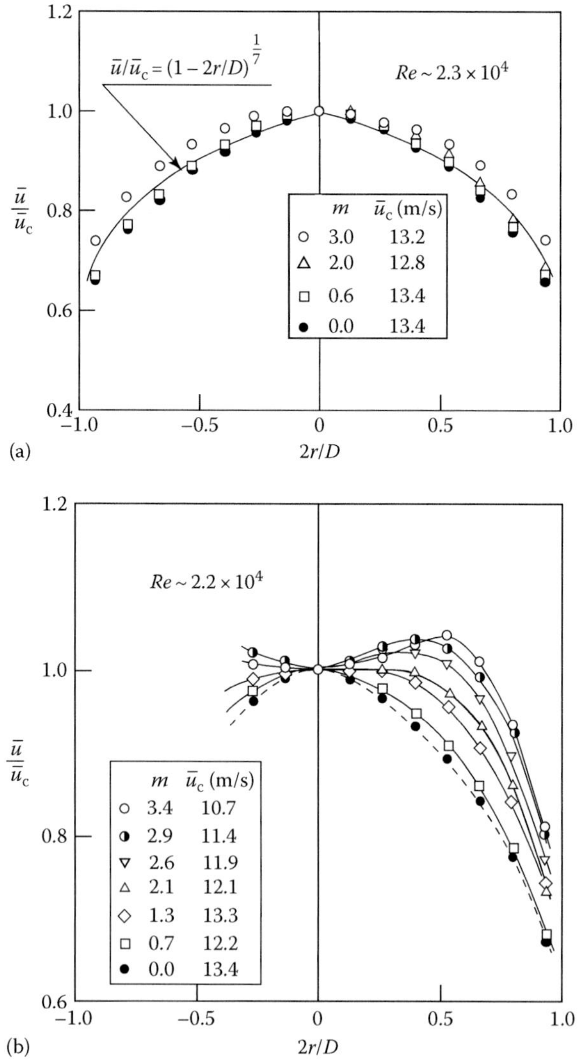

FIGURE 33.3 The results of Tsuji et al. (1984) on the normalized turbulent mean velocity, measured in a pipe flow of Re ~ 2.2 × 10^4 for particle mass loading ranging from 0.0 to 3.4 for two different sized particles: (a) 1 mm and (b) 500 μm. Also plotted in frame (a) as the solid line is the expected single-phase turbulent mean flow velocity. The dramatic effect of the inclusion of the particles for the case of 500 μm particles can be clearly seen.

presence of 500 μm and 1 mm sized particles for varying mass loading ranging from 0 to 3.5. The larger particles have only a small effect on the mean velocity profile, but for the same mass loading, the 500 μm particles dramatically alter the mean velocity distribution—the peak velocity is not at the pipe center but has moved halfway between the center and the pipe wall. Interestingly, particles smaller than 500 μm do not influence the flow as much. These results clearly illustrate the importance of the back effect of particles on the flow and the need to consider simultaneously the mechanism of preferential accumulation (see Section 33.7). Otherwise, the critical dependence on particle size cannot be explained.

Particles can have an even more dramatic effect on the fluctuating turbulent quantities. If the mean velocity is substantially altered due to the presence of the particles, turbulence production that depends on mean velocity gradients will be altered as well. In this case, large differences in the rms fluctuation and higher-order turbulence statistics can be expected. There are particle-laden flows where the mean flow may not be significantly altered by particles; however, turbulent fluctuation is substantially influenced. The interaction between particles and suspending fluid remains complex and poorly understood. Conditions under which turbulence will be enhanced or suppressed remain to be fully demarcated. Based on a collection of experimental measurements, Crowe et al. (1996) suggested that turbulence modulation is dictated by the ratio of particle diameter to the characteristic size of the energy-containing eddies. If this ratio is greater than 0.1, turbulence is augmented and otherwise suppressed. On the other hand, Elghobashi and Truesdell (1993) observed turbulence enhancement even for particles of diameter comparable to the Kolmogorov scale of turbulence. Hetsroni (1989) argued that the vortex shedding process is responsible for turbulence augmentation and, therefore, suggested Re > 400 to be the criterion for turbulence enhancement. Bagchi and Balachandar (2004), however, have observed the shedding process to initiate at a much lower Re, triggered by the free-stream turbulence. Clearly, the back effect of particles on the flow is perhaps the most challenging of all the mechanisms. It is of fundamental importance, without which we cannot answer the basic question of whether the carrier phase turbulence is augmented or suppressed by the inclusion of the dispersed phase. For further discussion on turbulence modulation in dilute particle-laden flows, see Balachandar and Eaton (2010).

33.7 Preferential Particle Concentration

Distribution of particles in particle-laden flows presents interesting surprises. In a turbulent flow, it appears well justified to assume that an initial uniform distribution of particles will remain uniformly distributed at all later times. In fact, it is tempting to even suppose that an initially nonuniform distribution of particles will be made more uniform due to the action of turbulent eddies. After all, carrier phase turbulence can be argued to introduce an effective diffusion of particles and smooth concentration peaks and valleys. However, there is now ample evidence that particle concentration can be highly nonuniform, even in isotropic turbulence. Two primary mechanisms contribute to this nonuniform distribution of particles.

One of the key features of particle dispersion in turbulent flows is the phenomenon of *preferential accumulation*. Heavier-than-fluid particles tend to accumulate in regions of high strain rate and avoid regions of intense vorticity. In contrast, lighter-than-fluid particles (or bubbles) tend to congregate in vortical regions. This behavior of preferential particle concentration is due to particle inertia. A heavier-than-fluid particle does not follow the curved streamlines of a vortex the same way a fluid parcel does and as a result spins out of the vortex. Such particles thus tend to collect in the high strain regions between vortices. In contrast, bubbles are drawn toward vortex centers.

Using asymptotic analysis and kinematic simulations, Maxey (1987) showed the effect of preferential accumulation on enhanced settling velocity of particles. Squires and Eaton (1991) performed direct numerical simulations of forced homogeneous turbulence over a range of particle Stokes number and demonstrated more than an order of magnitude increase in local concentration for Stokes numbers of $O(1)$. The mechanism of preferential concentration can be easily illustrated with the equation of equilibrium particle velocity (Equation 33.10). Taking the divergence of this equation and non-dimensionalizing with the Kolmogorov length and time scales (η and τ_k), we obtain (tilde denotes nondimensional quantities)

$$\nabla \cdot \tilde{\mathbf{v}} = -St(1-\beta)\left(\|\tilde{\mathbf{S}}\| - \|\tilde{\Omega}\|\right) \qquad (33.14)$$

where particle Stokes number is defined as the ratio of particle time scale $\tau_p = (2\rho + 1)d^2/(36\nu)$. Here, ρ is particle-to-fluid density ratio, ν is kinematic viscosity of the fluid to the Kolmogorov time scale, and β is density parameter defined as $\beta = 3/(2\rho + 1)$. In Equation 33.14, $\|\tilde{\mathbf{S}}\|$ and $\|\tilde{\Omega}\|$ are the norms of the local strain-rate and rotation-rate tensors. Thus, heavier-than-fluid particles ($0 < \beta < 1$) will accumulate in regions where strain-rate dominates over vorticity ($\|\tilde{\mathbf{S}}\| > \|\tilde{\Omega}\|$), while lighter-than-fluid particles and bubbles ($1 < \beta < 3$) will collect in regions where $\|\tilde{\mathbf{S}}\| < \|\tilde{\Omega}\|$. According to this expression, preferential concentration increases linearly with St, but this dependence applies only for small St (since the expansion given in Equation 33.10 is valid only for small St). Large Stokes number particles are very sluggish, and they only weakly respond to the turbulent eddies during the eddy's lifetime. In this case, turbulence can be considered to provide only random uncorrelated perturbations to the large particle settling velocity. Thus, both for small and large Stokes numbers, the mechanism of preferential particle concentration can be expected to be weak. At intermediate St, turbulent eddies induce significant coherent motion of the particles, so preferential concentration is most important when $St \approx 1$.

There has been substantial subsequent work on preferential concentration and its effect in particle-laden flows. Preferential concentration has been experimentally confirmed by Lazaro and Lasheras (1989), Fessler et al. (1994), among others.

Sundaram and Collins (1997) and Wang et al. (2000) studied particle collisions for a range of particle Stokes numbers. They found strong preferential concentration for *St* near one which resulted in substantially different collision behavior than for smaller or larger particles. Sundaram and Collins (1997) used the radial distribution function (RDF) as a quantitative measure of preferential concentration, where RDF is the probability of finding a second particle at a given radial distance from any particle. Preferential concentration plays an important role in the natural world. For example, preferential accumulation of water droplets in clouds plays an important role in precipitation and initiation of rain. Shaw (2003) discussed the potential impact of preferential concentration on the formation and evolution of clouds. Preferential concentration may affect both the spatial distribution of water saturation and droplet coalescence rates.

Another mechanism for nonuniform particle distribution in turbulent flows is *turbophoresis* (Reeks 1983). In inhomogeneous turbulence, particles tend to migrate on average between regions of higher turbulent fluctuation velocities and regions of lower fluctuation. This phenomenon (coined turbophoresis) plays a critical role in generating a nonuniform distribution of particles. This behavior again can be explored using Equation 33.10. Consider inhomogeneous wall turbulence close to a flat boundary. Averaging Equation 33.10 along homogeneous directions parallel to the wall and over time will result in (Ferry and Balachandar 2001)

$$\langle \tilde{v}_2 \rangle = -St(1-\beta)\frac{d\langle \tilde{u}_2^2 \rangle}{d\tilde{y}} \tag{33.15}$$

where angle brackets indicate average over horizontal plane and time. This equation for the average wall normal particle velocity ($\langle \tilde{v}_2 \rangle$) shows that in the limit of very small particles ($St \to 0$), there is no net migration of particles close to or away from the wall. But for finite Stokes number particles, wall normal ($\tilde{y}$) gradient of the wall normal fluid velocity fluctuation will drive heavier-than-fluid particles ($0 < \beta < 1$) toward the wall and lighter-than-fluid particles ($1 < \beta < 3$) away from the wall. The previous expression for turbophoretic migration obtained from Equation 33.10 is the same as that obtained by Reeks (1983). Furthermore, in wall turbulence once close to the wall heavier-than-fluid particles tend to accumulate in regions of low speed streaks (Ferry and Balachandar 2002).

33.8 Particle Settling

The settling of heavier-than-fluid particles and droplets and the rising of bubbles through turbulent flows are common phenomena of significance in many environmental and geophysical applications. In the atmosphere, the settling of particles naturally suspended by dust storms or volcanic eruption and man-made pollutants, such as smoke and ash particles, is of importance to air quality. In cloud physics, the settling rate of water droplets play a critical role in their growth process and in the eventual precipitating as rain (Shaw 2003). The erosion and settling of mud and sand particles in rivers, beaches, and submarine turbidity currents play an important role in their dynamics (McCaffrey et al. 2001). In the oceanic context, the settling of biological particles from the mixed top layer to the deep ocean is an important mechanism for vertical transport of carbon (Ittekkot et al. 1996). In all these examples (and many others), the turbulent nature of the ambient flow has been known to significantly alter the mean settling velocity of the particles from the still fluid settling velocity.

If the carrier phase has a mean nonzero vertical velocity, clearly it will add to the still fluid settling velocity and contribute to a decrease or increase in the net particle settling velocity. More interestingly, even in the absence of a mean carrier phase velocity, turbulence has been established to substantially influence the mean particle settling velocity through several mechanisms. First, in the case of sufficiently large particles, whose Reynolds number based on relative velocity is larger than unity, the effect of nonlinear relation between drag coefficient and particle Reynolds number is to reduce the mean settling velocity. Tunstall and Houghton (1968) demonstrated this nonlinear drag effect for the case of particles settling in a spatially uniform flow that oscillated over time about a zero mean. Their results were confirmed in later experiments (Schoneborn 1975). Mei (1994) considered the nonlinear drag effect both theoretically and computationally using isotropic Gaussian pseudo turbulence, and using the empirical nonlinear standard drag relation he related the mean settling velocity of a particle (V_T) in a turbulent flow to the still fluid settling velocity (V_{Ts}) by the following relation

$$V_T\left\{1+0.15\langle \mathrm{Re}\rangle^{0.687}\left(1+0.58\frac{\langle |\mathbf{u}_r'|^2\rangle}{V_T^2}\right)\right\} = V_{Ts} \tag{33.16}$$

where Re is particle Reynolds number based on relative velocity and the angle bracket represents an ensemble average. By definition, $V_T = |\langle \mathbf{v} - \mathbf{u}\rangle|$, and the perturbation relative velocity is defined as $\mathbf{u}_r' = \mathbf{v} - \mathbf{u} - \langle \mathbf{v} - \mathbf{u}\rangle$. Since the correction term on the left is positive, it can be seen that the effect of nonlinear drag relation is to yield $V_T \le V_{Ts}$. Only in the limit of Stokes drag, $V_T \to V_{Ts}$.

Even in the linear drag regime, the mean settling velocity can be influenced by the loitering effect. Consider a particle settling through a cellular flow where part of the time when the particle is in the down-flow region settling is enhanced and part of the time when the particle is in the up-flow region settling is opposed. If we assume the particle trajectory to equi-partition and pass through up and down flow regions with equal probability, it can be argued that the particle will spend more time in the up-flow region, since the settling velocity is lowered here, than in the down-flow region, where settling is enhanced. This temporal bias toward up-flow region is termed the loitering effect (Nielsen 1993).

Trapping of particles in regions of strong updraft is another mechanism that can result in substantial reduction in the mean settling velocity. Such trapping can be considered as an extreme case of the loitering effect. The possibility of permanent and temporary trapping of particles in regions where the downward buoyancy force is balanced by the upward drag force has been well recognized for over a century (Stommel 1949; Nielsen 1984), and the resulting closed particle trajectories are often referred to as the Stommel retention zone. These trapped particles, when taken into account in the calculation of the mean settling velocity, clearly contribute to a reduction in the mean settling velocity. While particles can be permanently trapped in steady cellular flows of sufficient strength, in a time-dependent cellular flow particles can be temporarily trapped in the cells in a chaotic manner (Fung 1997). Recent efforts have shown the possibility of trapping of heavier-than-fluid particles even in open flows (Vilela and Motter 2007). Furthermore, inertial particles are not always spun out of closed streamlines. In regions where streamline curvature changes, permanent trapping of inertial particles can occur (Pasquero et al. 2003). These features clearly make the question of particle trapping in turbulent flows quite interesting.

Maxey (1987) demonstrated an important mechanism of interaction between the particles and the turbulent eddies which can result in substantial increase in particle settling velocity. Two physical processes were observed to drive this mechanism: First, due to preferential accumulation, heavier-than-fluid particles tend to spin out of vortices and concentrate in the high strain-rate regions on the periphery of the vortices. Second, due to the combined effect of particle inertia and local flow field, particles tend to prefer down-flow regions than up-flow regions (Wang and Maxey 1993). The net effect is that the fluid velocity averaged over the particle trajectory is not the same as the volume (or ensemble) average. Since the particles preferentially sample down moving fluid, the net settling velocity of particles is higher than the still fluid settling velocity.

This mechanism is referred to as "trajectory bias" or "fast-tracking" (Nielsen 1993) and is most active when particle Stokes number is of order one (i.e., when $St \sim O(1)$). Such particles respond best to the turbulent eddies and are able to follow the trajectory bias. Much smaller particles ($St \ll 1$) better follow the fluid and behave closer to tracer particles and as a result do not fast-track. On the other hand, bigger particles of much larger Stokes number fail to respond to the turbulent eddies and simply follow a near vertical path. The importance of Kolmogorov scale on trajectory bias has been debated in the literature (Yang and Lei 1998). Yang and Shy (2005) argued that while Kolmogorov time scale is appropriate in the evaluation of particle Stokes number, the appropriate velocity scale for estimating the turbulence effect is the rms velocity and not the Kolmogorov velocity scale.

Recent experiments of particle settling in isotropic turbulence (Aliseda et al. 2002; Yang and Shy 2003, 2005) have measured increase in settling velocity due to trajectory bias and confirmed theoretical prediction that the mechanism is most effective for particles of $O(1)$ Stokes number. The results of Aliseda et al. (2002) have further shown the effect of trajectory bias to increase with increasing particle concentration. The mechanistic reason of this particle concentration effect is simple. When a region of higher particle concentration is surrounded by lower particle concentration, the density variation of the mixture (carrier phase plus particles) begins to play a role. In other words, the collective effect of the excess particles in the high concentration region is to induce an additional vertical motion. Bosse et al. (2006), in their recent simulations, explored this two-way coupling effect on particle settling and observed it to be effective for mean particle volume fractions in excess of 10^{-5}. It is to be noted that this two-way coupling effect is primarily through the role of gravity. In the absence of gravity effect, simulations of particle-laden isotropic turbulence at such low concentration levels have shown two-way coupling effect to be negligible. Furthermore, the two-way coupling effect on particle settling requires spatial variation in particle concentration for it to be active. Thus, increase in settling velocity with particle concentration was observed to be most effective only for $St \sim O(1)$. For otherwise, particles remain well mixed and settling velocity is not influenced by two-way coupling, at least at low concentration levels. It is thus intriguing that even at very low concentration, particles of $O(1)$ Stokes number can substantially influence turbulence and enhance the settling velocity. Of course at larger particle concentration, back coupling of particles on turbulence can be even more substantial, and the resulting influence on mean settling velocity can be quite complex and remains unexplored.

Of the five mechanisms discussed, namely, nonlinear effect, loitering effect, particle trapping, trajectory bias, and two-way coupling, the first three contribute to a reduction in settling velocity, while the latter two result in an increase in settling velocity. The three primary parameters that influence these effects are the particle Stokes number, the particle Reynolds number, and the particle volume fraction. The relative importance of the aforementioned mechanisms depend on the value of the three parameters, and as a result past experiments and computations have shown conflicting influence on settling velocity, ranging from substantial increase (Maxey 1987) to modest increase or even reduction in settling velocity (Fung 1993). Recently, Davila and Hunt (2001) have systematically considered different (Eulerian average, Lagrangian average, and bulk average) definitions of settling velocity and their interrelation to illustrate how such differences in the definition of mean settling velocity could partially account for the observed and computed differences in settling velocity.

33.9 Major Challenges

There still remain several significant gaps in our understanding of particle-laden flows, and there are many major experimental/computational challenges to be overcome. Here, we will list a few:

1. Detailed experimental measurements in particle-laden flows have been limited. Experimental techniques that can yield well-resolved simultaneous measurements of both the particles and the carrier phase are needed.
2. Fully resolved simulations can yield deeper insight into complex interactions between particles and turbulence. Further progress in this direction, where tens of thousands of particles can be simulated using petaflop computations must

be pursued. However, it must be recognized that fully resolved simulations cannot provide answers to all our questions in the foreseeable future. Eulerian–Lagrangian and Eulerian–Eulerian techniques must be advanced further. They are on solid theoretical footing when particles are smaller than the flow scales. Unanswered questions remain as the particle size approaches the scales of resolved turbulence.

3. Our current understanding of preferential accumulation generally applies only to particles of size much smaller than the fluid scales. In this limit, particles simply respond to the local flow structure, such as vortices and shear layers. As the particles become larger, their presence will itself influence the local flow structure, and the question of preferential concentration must be carefully addressed.
4. In the context of turbulence modulation, despite several important contributions, even basic questions of critical parameters that influence turbulence modification are not fully settled. Mechanisms of turbulence modulation and the role of key parameters such as volume fraction, mass loading, and density ratio are issues ripe for further investigation.
5. It is important to address the influence of turbulence on particle motion and dispersion, in the context when particle–particle hydrodynamic interaction begins to play a role. Current implementations of Lagrangian models for particle motion are generally based on results of a single particle of small size. It is important to reevaluate these drag laws by investigating the screening effect arising from interactions between the particles and their wake. Such interactions can be more important in a turbulent flow due to local preferential accumulation of particles. Thus, the investigation of particle–turbulence interaction must be extended from a single particle to a collection of interacting particles in a turbulent flow.
6. Furthermore, our understanding of the effect of turbulence on particles in phenomena such as preferential concentration and the back effect of particles in turbulence modulation is primarily limited to particles of size smaller than the Kolmogorov length scale. As the particles become larger and overlap with the scales of ambient turbulence, it is unclear how the finite size of the particle will influence the forces on the particle, dispersion and preferential concentration of particles, and the back effect on carrier fluid turbulence. Future research must address these questions.

References

Aliseda A, Cartellier A, Hainaux F, Lasheras JC. 2002. Effect of preferential concentration on the settling velocity of heavy particles in homogeneous turbulence. *J. Fluid Mech.* 468:77–105.

Bagchi P, Balachandar S. 2002. Steady planar straining flow past a rigid sphere at moderate Reynolds number. *J. Fluid Mech.* 466:365–407.

Bagchi P, Balachandar S. 2004. Response of the wake of an isolated particle to an isotropic turbulent flow. *J. Fluid Mech.* 518:95–123.

Balachandar S, Eaton JK. 2010. Turbulent dispersed multiphase flow. *Annu. Rev. Fluid Mech.* 42:111–133.

Batchelor GK. 1970. Stress system in a suspension of force-free particles. *J. Fluid Mech.* 41:545–570.

Bosse T, Kleiser L, Meiburg E. 2006. Small particles in homogeneous turbulence: Settling velocity enhancement by two-way coupling. *Phys. Fluids* 18:027102.

Cantero MI, Balachandar S, Garcia MH. 2008a. An Eulerian–Eulerian model for gravity currents driven by inertial particles. *Int. J. Multiphase Flow* 34:484–501.

Cantero MI, Balachandar S, Garcia MH. 2008b. Effect of particle inertia on the dynamics of depositional particulate density currents. *Comput. Geosci.* 34:1307–1318.

Cantero MI, Balachandar S, Cantelli A, Pitmez C, Parker G. 2009a. Turbidity current with a roof: Direct numerical simulation of self-stratified turbulent channel flow driven by suspended sediment. *J. Geophys. Res.* 114:C03008.

Cantero MI, Balachandar S, Parker G. 2009b. Direct numerical simulation of stratification effects in a sediment-laden turbulent channel flow. *J. Turbul.* 10:1–28.

Crowe CT, Troutt TR, Chung JN. 1996. Numerical models for two-phase turbulent flows. *Annu. Rev. Fluid Mech.* 28:11–43.

Davila J, Hunt JCR. 2001. Settling of small particles near vortices and in turbulence. *J. Fluid Mech.* 440:117–145.

Elghobashi S. 1991. Particle-laden turbulent flows: Direct simulation and closure models. *Appl. Sci. Res.* 48:301–314.

Elghobashi S. 1994. On predicting particle-laden turbulent flows. *Appl. Sci. Res.* 52:309–329.

Elghobashi S, Truesdell GC. 1993. On the 2-way interaction between homogeneous turbulence and dispersed solid particles. 1: Turbulence modification. *Phys. Fluids* 5:1790–1801.

Ferry J, Balachandar S. 2001. A fast Eulerian method for two-phase flow. *Int. J. Multiphase Flow* 27:1199–1226.

Ferry J, Balachandar S. 2002. Equilibrium expansion for the Eulerian velocity of small particles. *Powder Technol.* 125:131–139.

Ferry J, Rani SL, Balachandar S. 2003. A locally implicit improvement of the equilibrium Eulerian method. *Int. J. Multiphase Flow* 29:869–891.

Fessler JR, Kulick JD, Eaton JK. 1994. Preferential concentration of heavy particles in a turbulent channel flow. *Phys. Fluids A* 6:3742–3749.

Fox RO, Laurent F, Massot F. 2008. Numerical simulation of spray coalescence in an Eulerian framework: Direct quadrature method of moments and multi-fluid method. *J. Comput. Phys.* 277:3058–3088.

Fung JCH. 1993. Gravitational settling of particles and bubbles in homogeneous turbulence. *J. Geophys. Res.* 98:20,287–20,297.

Fung JCH. 1997. Gravitational settling of small spherical particles in unsteady cellular flow fields. *J. Aerosol. Sci.* 28:753–787.

Gatignol R. 1983. The Faxen formulae for a rigid particle in an unsteady non-uniform Stokes flow. *J. Mech. Theor. Appl.* 1:143–160.

Hetsroni G. 1989. Particle-turbulence interaction. *Int. J. Multiphase Flow* 15:735.

Ittekkot V, Schafer P, Honjo S, Depetris PJ. 1996. *Particle Flux in the Ocean*. John Wiley & Sons, Chichester, U.K.

Jackson R. 1997. Locally averaged equations of motion for mixture of identical spherical particles and a Newtonian fluid. *Chem. Eng. Sci.* 52:2457–2469.

Joseph DD, Lundgren TS. 1990. Ensemble averaged and mixture theory equations for incompressible fluid-particle suspensions. *Int. J. Multiphase Flow* 16:35–42.

Kajishima T, Takiguchi S, Hamasaki H, Miyake Y. 2001. Turbulence structure of particle-laden flow in a vertical plane channel due to vortex shedding. *JSME Int. J. Ser. B* 44:526–535.

Lazaro BJ, Lasheras JC. 1989. Particle dispersion in a turbulent plane, free shear layer. *Phys. Fluids A* 1:1035–1044.

Ling Y, Haselbacher A, Balachandar S. 2009. A numerical source of small-scale number-density fluctuations in Eulerian-Lagrangian simulations of multiphase flows. *J. Comput. Phys.* 229:1828–1851.

Lu JC, Tryggvason G. 2006. Numerical study of turbulent bubbly downflows in a vertical channel. *Phys. Fluids* 18:103302.

Magnaudet J, Eames I. 2000. The motion of high Reynolds number bubbles in inhomogeneous flows. *Annu. Rev. Fluid Mech.* 32:659–708.

Marchioro M, Tanksley M, Prosperetti A. 1999. Mixture pressure and stress in disperse two-phase flow. *Int. J. Multiphase Flow* 25:1395–1429.

Maxey MR. 1987. The gravitational settling of aerosol particles in homogeneous turbulence and random flow fields. *J. Fluid Mech.* 174:441–465.

Maxey MR, Riley JJ. 1983. Equation of motion for a small sphere in a nonuniform flow. *Phys. Fluids* 26:883–889.

McCaffrey B, Kneller B, Peakall J. 2001. *Particulate Gravity Currents*. Blackwell Publishing, Oxford, U.K.

Mei R. 1994. Effect of turbulence on the particle settling velocity in the nonlinear drag range. *Int. J. Multiphase Flow* 20:273–284.

Necker F, Hartel C, Kleiser L, Meiburg E. 2005. Mixing and dissipation in particle-driven gravity currents. *J. Fluid Mech.* 545:339–372.

Nielsen P. 1984. On the motion of suspended sand particles. *J. Geophys. Res.* 89:616–626.

Nielsen P. 1993. Turbulence effects on the settling of suspended particles. *J. Sedimen. Petrol.* 63:835–838.

Pasquero C, Provenzale A, Spiegel EA. 2003. Suspension and fall of heavy particles in random two-dimensional flows. *Phys. Rev. Lett.* 91:054502.

Pope SB. 1985. PDF methods for turbulent reactive flows. *Prog. Energ. Combust. Sci.* 11:119–192.

Rani SL, Balachandar S. 2003. Evaluation of the equilibrium Eulerian approach for the evolution of particle concentration in isotropic turbulence. *Int. J. Multiphase Flow* 29:1793–1816.

Reeks MW. 1983. The transport of discrete particles in inhomogeneous turbulence. *J. Aerosol Sci.* 14:729–739.

Schoneborn PR. 1975. The interaction between a single particle and oscillating fluid. *Int. J. Multiphase Flow* 2:307–317.

Shaw RA. 2003. Particle-turbulence interactions in atmospheric clouds. *Annu. Rev. Fluid Mech.* 35:183–227.

Shotorban B, Balachandar S. 2006. Particle concentration in homogeneous shear turbulence simulated via Lagrangian and equilibrium Eulerian approaches. *Phys. Fluids* 18:065105.

Shotorban B, Balachandar S. 2007. A Eulerian model for large-eddy simulation of concentration of particles with small Stokes numbers. *Phys. Fluids* 19:118107.

Shotorban B, Balachandar S. 2009. Two-fluid approach for direct numerical simulation of particle laden turbulent flows at small Stokes numbers. *Phys. Rev. E.* 79:056703.

Song C, Wang P, Makse HA. 2008. A phase diagram for jammed matter. *Nature* 453:629–632.

Squires KD, Eaton JK. 1991. Preferential concentration of particles by turbulence. *Phys. Fluids A* 3:1169–1179.

Stommel H. 1949. Trajectories of small bodies sinking slowly through convection cells. *J. Mar. Res.* 8–24.

Subramaniam S. 2001. Statistical modeling of sprays using the droplet distribution function. *Phys. Fluids* 13:624–642.

Sundaram S, Collins LR. 1997. Collision statistics in an isotropic particle-laden turbulent suspension. Part 1. Direct numerical simulation. *J. Fluid Mech.* 335:75–109.

Ten Cate A, Derksen JJ, Portela L, van den Akker HEA. 2004. Fully resolved simulations of colliding monodisperse spheres in forced isotropic turbulence. *J. Fluid Mech.* 519:233–271.

Tsuji Y, Morikawa Y, Shiomi H. 1984. LDV measurements of an air-solid two-phase flow in a vertical pipe. *J. Fluid Mech.* 139:417–434.

Tunstall EB, Houghton G. 1968. Retardation of falling spheres by hydraulic oscillations. *Chem. Eng. Sci.* 23:1067–1081.

Uhlmann M. 2008. Interface-resolved direct numerical simulation of vertical particulate channel flow in the turbulent regime. *Phys. Fluids* 20:053305.

Vilela RD, Motter AE. 2007. Can aerosols be trapped in open flows? *Phys. Rev. Lett.* 99:264101.

Wang LP, Maxey MR. 1993. Settling velocity and concentration distribution of heavy particles in homogeneous isotropic turbulence. *J. Fluid Mech.* 256:27–68.

Wang L-P, Wexler AS, Zhou Y. 2000. Statistical mechanical description and modeling of turbulent collision of inertial particles. *J. Fluid Mech.* 415:117–153.

Yang CY, Lei U. 1998. The role of the turbulent scales in the settling velocity of heavy particles in homogeneous isotropic turbulence. *J. Fluid Mech.* 371:179–205.

Yang TS, Shy SS. 2003. The settling velocity of heavy particles in an aqueous near-isotropic turbulence. *Phys. Fluids* 15:868.

Yang TS, Shy SS. 2005. Two-way interaction between solid particles and homogeneous air turbulence: Particle settling and turbulence modification measurements. *J. Fluid Mech.* 526:171–216.

Zhang DZ, Prosperetti A. 1994. Averaged equations for inviscid disperse two-phase flow. *J. Fluid Mech.* 267:185–219.

Zhang DZ, Prosperetti A. 1997. Momentum and energy equations for disperse two-phase flows and their closure for dilute suspensions. *Int. J. Multiphase Flow* 23:425–453.

34

Sediment Transport

Jørgen Fredsøe
Technical University of Denmark

34.1 Introduction

Transport of solid particles by flowing water comprises a wide range of transport phenomena from coal transport in pipes (slurry flow) to transport of mud, clay, sand, or big blocks in the fluvial and marine environment.

The gradients in the sediment transport govern the morphological changes (erosion and deposition) in rivers, estuaries, and coastlines.

In the *fluvial environment*, sediment transport is important for the sedimentation in reservoirs behind dams, for evolution of plan form movement of rivers caused by bank erosion, for changes in the bed levels of rivers, for scour development around river structures, and it is also important for the habitat environment in natural rivers.

In the *estuarine and coastal environment*, the sediment transport pattern becomes more complex than in the fluvial environment due to the presence of tides and waves. Determination of these patterns is necessary to evaluate coastal behavior like coastal erosion and the functioning of coastal structures. Also an evaluation of the need for dredging of trapped sediment in navigation channels requires a complete picture of the sediment transport pattern.

Sediment can also be transported by blowing wind, Aeolian transport. This is of relevance regarding desertification, but wind-blown transport is also important in other contexts like the further transport of sand from a beach to the hinterland.

34.2 Basic Principles of Sediment Transport

34.2.1 Sediment Properties

Transport of sediment depends on the properties of the sediment as well as the property of the transporting fluid. This chapter mainly treats the case where the transporting medium is water.

34.2.1.1 Cohesive versus Noncohesive

The most important feature regarding sediment transport description is whether the sediment is cohesive or noncohesive. Cohesive sediment is mainly in the clay-silt domain, where physicochemical forces become just as important as gravity forces for these very fine particles with a large specific area. For noncohesive sediment such as sand these forces are negligible, and in this case it is possible to describe the sediment transport applying mechanical principles. This is not the case for the cohesive sediment, where the motion therefore can be described less accurate than for the noncohesive sediment. The main part of this chapter focuses on noncohesive materials, simply because the description of this is far advanced to that

Handbook of Environmental Fluid Dynamics, Volume One, edited by Harindra Joseph Shermal Fernando. © 2013 CRC Press/Taylor & Francis Group, LLC.
ISBN: 978-1-4398-1669-1.

of the cohesive environment. At the end of the next section, the cohesive transport is briefly touched.

34.2.1.2 Grain Sizes

The grain size can be described by its median diameter d, corresponding to 50% of the material by weight being finer than *d*. The gradation of the sediment can be described by the geometric standard deviation $\sigma = \sqrt{d_{84}/d_{16}}$, where d_{16} corresponds to 16% are finer than that value. Clay is usually assumed to have medium diameters smaller than 0.002–0.004 mm (2–4 μm), while silt is in the fraction between 4–60 μm. Above this value we have sand (60–200 μm, covering fine, medium, and coarse), gravel (2–64 mm), and cobble (64–256 mm).

In a sieve analysis, the shape of sediment is not taken into account. Sometimes you therefore introduce a sedimentation diameter d_f for a particle, defined as the diameter of a sphere having the same settling velocity w_s in water at 24°C.

34.2.2 Transport Modes of Noncohesive Sediment

The transport of sediment takes place in different ways, depending on the flow velocity and the property of the sediment. For noncohesive sediment, we usually have the following modes:

- Bed load (in air this is called creeping flow), saltation
- Suspended load
- Wash load

34.2.2.1 Bed Load

In flowing water, bed load is that part of the load which is sliding, rolling, or jumping past the bed, so sometimes it is also called "contact load." A special case of this kind of transportation is the so-called sheet flow, where several layers of heavy grains are moving along the bed at high flow velocities, see also the photo Figure 34.4.

34.2.2.1.1 Saltation

Occasionally, you can observe a mode of transport in which a grain is hurled up from the bed, temporarily supported by the fluid above, and then crashes back to bed again, Sleath (1984) and Figure 34.1c. Regarding Aeolian transport, this process is quite important because when the crash-back occurs, this causes new grains to be hurled up to repeat the process. In the momentum-rich water this process is considered to be less important. Some modelers describe the bed load transport as a saltation process: Wiberg and Smith (1985) considered the bed load layer to consist of saltating grains, and described the particle path of the individual grains by considering lift and drag forces from the flow. They considered how much momentum the saltating grain extracts from the fluid, thereby decreasing the shear in the fluid. By considerations analogue to those described in Section 34.2.3, they also argued that the fluid shear should be reduced to the critical, so no more net erosion of sediment will occur.

34.2.2.1.2 Sheet Flow

With increasing flow velocities—or strictly speaking Shields parameters—the sediment is not any more only moving in a single layer. Depending on the sediment properties, some of the sediment will either go into suspension, saltate, or move as contact load "travelling in a dense layer immediately above the bed supported by intergranular collisions rather than by fluid turbulence" (Wilson 1966). This flow is called "sheet flow," and in a fluid the particles in the sheet flow layer are carried by intergranular collisions as well as by fluid turbulence in a complicated manner. This is in contrast to airborne transport, where the intergranular collisions dominate.

34.2.2.2 Suspended Load

Suspended sediment is that part of the total load, which is supported by the fluid turbulence rather than intergranular collisions. As seen on the pictures in Figure 34.4, there is a smooth transition from the one transport mode to the other.

34.2.2.3 Wash Load

Flows usually also transport significant quantities of very fine sediment, which is not found in the bed. This is called wash load. The quantity of this load is determined by conditions occurring well upstream like caving of river banks, etc. (Sleath 1984). It is not possible to estimate the amount of wash load with the aid of formulas based on local conditions, since wash load is not supplied from the bed.

34.2.2.4 Total Load

The term total load is used for the sum of bed load and the suspended load, but without wash load.

34.2.3 Principles of Modeling Noncohesive Bed Load Transport

In this section we outline some of the principles behind the development of a bed load transport formulation. The purpose is to describe some of the features behind bed load transport, and to obtain the most important dimensionless parameters.

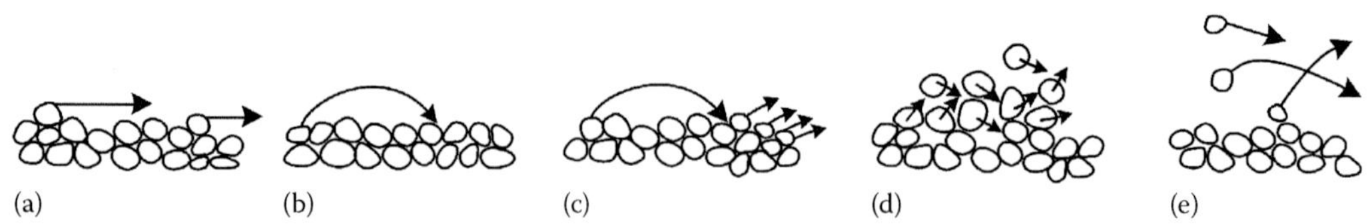

FIGURE 34.1 Sketches of transport modes. (a) Bed load at small shear stresses. (b) Saltation in water. (c) Saltation in air (creeping flow). (d) Sheet flow. (e) Suspended load.

Actually, there are many different views on how to define bed load, and many formulas for the load are total-load formula, which overcome this problem. However, in some cases it is of importance to distinguish in between the bed load and the suspended load. For instance, in morphological modeling, the impact of bed slope on the sediment transport is sometimes crucial, and only contact load can feel this slope, sees Engelund and Fredsoe (1982b). Another important thing is that bed load responds immediately on spatial and temporal changes, while suspended sediment responds with a lag (Engelund and Fredsoe 1982a).

34.2.3.1 Modeling of Bed Load Transport at Small Shear Stresses

In this and the next section some of the basic features of sediment transport are illustrated by use of very simple time-averaged models. First we consider the transport at moderate transport rates, and we develop a semi-deterministic approach. At small shear stresses, the sediment is assumed to move in only one single layer, and not all the particles are moving at all times.

34.2.3.1.1 Nonmoving Grain: Forces and Critical Shields Parameter

If we consider a grain located in the bottom in between other grains, the flow above will provide a lift and a drag on the grain. The mobility of a sand grain depends on the ratio in between these agitating forces, and the stabilizing forces originating from gravity.

For *nonmoving grains*, the *lift* is caused by two contributions, both due to the curvilinear character of the flow just above the grains. Firstly, the curvature of streamlines just above the grain will be upward convex as sketched in Figure 34.2. To make this curved flow possible requires a negative pressure $-\Delta p_1$ at the center of the curvature, that is, at the upper surface of the grains.

Secondly, you have a positive pressure Δp_2 in between two grains, because the flow here is downward convex. This excess pressure is transferred to the lower part of the grain because the flow is in rest between the grains due to flow separation as sketched in Figure 34.2a, which is a simplified model of spheres illustrating the grains densely packed in a hexagonal pattern.

The lift from this upward pressure lift is given by $\Delta p_1 d^2$ and $\Delta p_2 d^2$, respectively, and can be estimated from the required centrifugal forces for the curvilinear flow. This centrifugal force is

$$M\frac{U^2}{r} \sim \rho d^3 \frac{U_f^2}{d} = \rho d^2 U_f^2 \tag{34.1}$$

Here the typical near bed flow velocity is taken to be the friction velocity multiplied by a certain constant factor α of order 10, one to two grain diameters above the bed. M is the mass of fluid above the grains which is part of the curvilinear flow, and ρ is the fluid density. Thus we have

$$\Delta p_1 d^2 \sim \Delta p_2 d^2 \sim \rho d^2 U_f^2 \tag{34.2}$$

From this it is seen that the excess pressure are proportional to the bottom shear stress $\tau_0 = \rho U_f^2$.

The horizontal *drag* consists of skin friction acting on the surface of the grain and a form drag due to flow separation, which creates a pressure difference up- and downstream the grain. The drag is given by

$$F_D = c_D \frac{1}{2}\rho\left(\alpha U_f^2\right)d^2 \tag{34.3}$$

Here c_D is the drag coefficient, which depends on the grain Reynolds number $R = Ud/\nu$, where ν is the kinematic viscosity of the fluid. By comparing Equations 34.2 and 34.3 it is seen that the drag and the lift force both varies proportional with the bed shear stress τ_0. The total agitating forces acting on a grain is therefore proportional to $\tau_0 d^2$, and depends on the grain Reynolds number and the shape of the grains.

The *stabilizing forces* can—very simplified—be modeled as a friction term against movement, and therefore proportional to a friction coefficient μ multiplied by the submerged weight $W \sim (\rho_s - \rho)gd^3$, where $\rho_s = \rho s$ is the specific sediment gravity and s the relative density of the sediment. For a resting grain, μ should be taken as the static friction coefficient μ_s, while for moving grains it should be the dynamic μ_d.

The mobility of a sand grain depends on the relative strength of the agitating forces drag and lift and the stabilizing force proportional to W. The most important parameter in

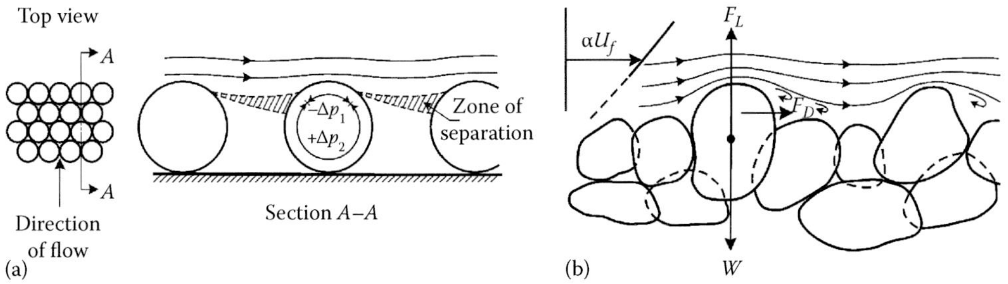

FIGURE 34.2 Forces acting on a grain resting in the bed: (a) curvilinear effects and (b) real grains.

noncohesive sediment transport is this ratio, which is called the Shields parameter θ, given by

$$\theta = \frac{\tau_0 d^2}{\rho(s-1)gd^3} = \frac{U_f^2}{(s-1)gd} \quad (34.4)$$

θ is seen to be a dimensionless form of the bed shear stress τ_0. At small values of θ the agitating force is smaller than the stabilizing, and even an exposed grain will not move. At larger values—above a critical value θ_c—the grains begin to move. θ_c depends slightly on the Reynolds number as described earlier, but is in the range 0.04–0.06.

34.2.3.1.2 Moving Grains

The principles outlined in the previous section can easily be extended to moving grains. When moving, the grain is assumed to move with a constant speed U_B, so the *relative* speed between grain and fluid become $(\alpha U_f - U_B)$. By expressing equilibrium between agitating and stabilizing forces, you get

$$\frac{1}{2}\rho c_D \frac{\pi}{4} d^2(\alpha U_f - U_B)^2 = \frac{\pi}{6}\rho g(s-1)d^3\mu_d \quad (34.5)$$

From which you get

$$\frac{U_B}{U_f} = \alpha\left[1 - \sqrt{\frac{\theta_c}{\theta}}\right] \quad (34.6)$$

in which α is equal to 10 as explained earlier. From the knowledge to the mean velocity of a moving grain, you can get the bed load transport q_B as volume of one grain multiplied by the numbers of moving grains N per unit area multiplied by velocity of the moving grains:

$$q_B = \frac{\pi}{6} d^3 \frac{p}{d^2} U_B \quad (34.7)$$

Here p is the probability for a grain in a single layer to move (so $N \simeq p/d^2$).

Much effort has been done to estimate p, which requires detailed knowledge on the flow structure very close to the bottom. However, this problem can be overcome by formulating how the fluid shear stress close to the bed is transferred immobile bed as done by Fernandez Luque and van Beek (1976). They suggested that of the total shear τ only the critical shear stress τ_c was transferred directly to the immobile bed, while the residual part $\tau - \tau_c$ is carried as drag on the moving bed particles and indirectly transferred to the bed by occasional encounters:

$$\tau = \tau_c + NF_D \quad (34.8)$$

From this N is easily obtained, and (34.7) gives

$$\Phi_B = \frac{q_B}{\sqrt{(s-1)gd^3}} = \frac{30}{\pi\mu_d}(\theta - \theta_c)\left(\sqrt{\theta} - \sqrt{\theta_c}\right) \quad (34.9)$$

where Φ_B is the dimensionless sediment transport (sometimes called the Einstein number). For $\mu_d = 1.0$, Equation 34.9 is very close to the most commonly used bed load equation, the Meyer–Peter formula

$$\Phi_B = 8(\theta - 0.047)^{3/2} \quad (34.10)$$

μ_d should actually rather be around 0.65 than unity, and in this case, the factor 8 in Equation 34.10 should rather be 12 than 8 to fit with Equation 34.9.

34.2.3.2 Modeling the Sheet Flow Layer

In the following a simple model of the sheet flow originally made by Engelund (1981) is given, in which only the intergranular collisions are included, that is, turbulence is neglected.

In a dense particle-laden flow, the collisions between particles create a stress field introducing normal and tangential forces due to transfer of momentum by intergranular collisions, see Figure 34.4. Bagnold (1954) did some very famous experiments in a small concentric cylinder rheometer to measure the dispersive stresses, and attained the following experimental expression for the dispersive normal stress σ

$$\frac{\sigma}{\rho} = 0.013s\left(\lambda d \frac{dU}{dy}\right)^2 \quad (34.11)$$

where

- y is distance from bed
- λ is the linear concentration related to the volumetric concentration by $c = 0.65/(1 + 1/\lambda)^3$

The thickness of the sheet flow layer is called δ_s, and in this layer the concentration is assumed to be constant equal c_0 (around 0.33 for moving grains). Outside this layer c rapidly falls to a small value, so that the dispersive stress drops to nearly zero, see Figures 34.3 and 34.4. That the particles in the layer are supported by the dispersive stress rather than fluid turbulence is expressed by

$$(\rho_s - \rho_f)cg + \frac{d\sigma}{dy} = 0 \quad (34.12)$$

which by integration gives

$$\frac{\sigma_0}{\rho} = c_0(s-1)g\delta_s \quad (34.13)$$

The dispersive pressure σ_0 can be evaluated from Bagnold's expression Equation 34.11, by evaluating the flow velocity gradient in the sheet flow layer to be $dU/dy = U_f/\kappa y$ for $y \simeq d$ as for a normal logarithmic boundary layer, κ being the von Karman's

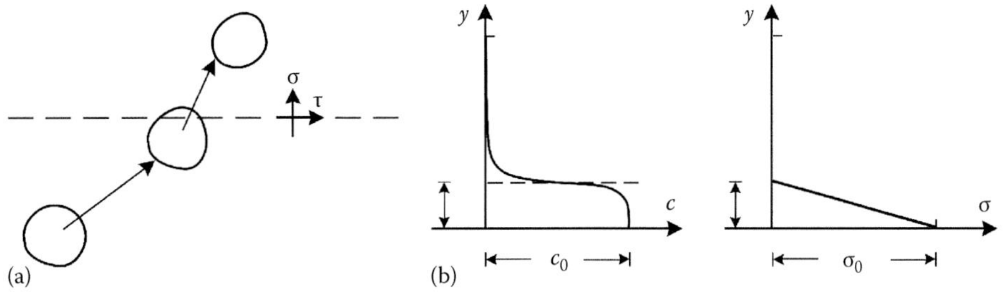

FIGURE 34.3 (a) Sketch of intergranular collisions, which are one of the causes for the dispersive stresses. (b) Idealized distribution of sediment concentration and dispersive stress σ in the sheet flow layer applied in the calculation example in this section.

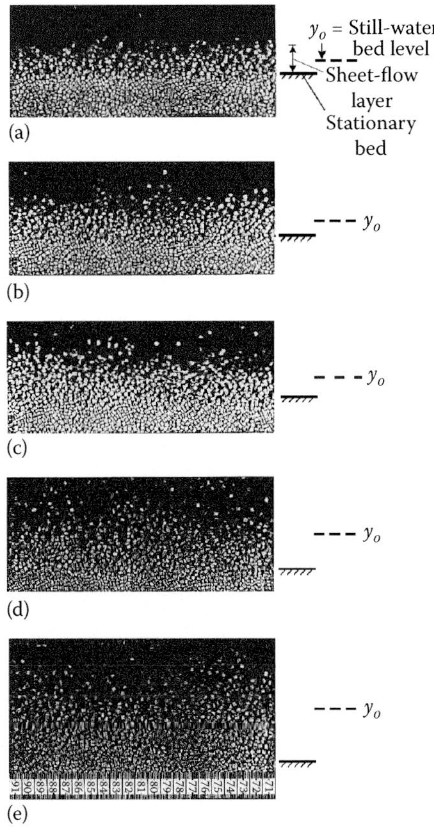

FIGURE 34.4 Sequence illustrating transition from "no-suspension" sheet flow regime to "suspension" sheet flow regime. (a) No suspension ($W/U_f = 1.1$) $\theta = 1.2$. (b) ($W/U_f = 1.0$) $\theta = 1.6$. (c) ($W/U_f = 0.85$) $\theta = 2.0$. (d) ($W/U_f = 0.70$) $\theta = 3.0$. (e) ($W/U_f = 0.61$) $\theta = 4.0$. (Reproduced from Sumer, B.M. et al., *J. Hydraul. Eng.-ASCE*, 122(10), 549, 1996. With permission from ASCE.)

constant (~0.4). This is certainly a crude approximation, but at least it gives the order of magnitude. From this we get

$$\frac{\sigma_0}{\rho} = 0.013s\left(\lambda_0 d\frac{U_f}{\kappa d}\right)^2 \approx 1.3sU_f^2 \qquad (34.14)$$

Equations 34.13 and 34.14 combined give the thickness of the sheet flow layer to be

$$\frac{\delta_s}{d} = 1.3s\theta \qquad (34.15)$$

From this it is possible to make an estimate the transport rate q_B by $q_B \simeq c_0\delta_s U_B$, which gives

$$\Phi_b = 4.5s\theta^{3/2} \qquad (34.16)$$

applying Equation 34.6 and $c_0 = 0.33$. It is seen that the transport in this case depends on the relative grain density *s*, which were not the case for the bed load. This finding is supported by Wilson's (1966) measurements.

34.2.3.3 Modeling the Suspended Sediment

34.2.3.3.1 Vertical Distribution of Suspended Sediment

Consider a steady flow over a plane bed. In this case, the time-averaged flow velocity is parallel with the bed, and the sediment is kept in suspension by turbulent fluctuations. The classical approach to obtain the vertical distribution of suspended sediment is to consider it as a pure diffusion process, so settling of suspended particles is counteracted by a vertical diffusion term. Usually the settling is given by *cw*, where *c* is the temporal mean value of the concentration and *w* is the fall velocity of the grains. The diffusion is given by $-\varepsilon_s(dc/dy)$, where ε_s is the diffusion coefficient, which usually is equal or proportional to the eddy viscosity of the flow, so that

$$\varepsilon_s \sim \varepsilon = \kappa U_f y(1 - y/h)\sigma_s \qquad (34.17)$$

where

- *h* is the flow depth
- σ_s is a factor of proportionality (the Smith number) reflecting that the exchange of sediment does not need to be the same as that for the fluid

The equilibrium distribution is given by

$$-\varepsilon_s\frac{dc}{dy} = wc \qquad (34.18)$$

Inserting Equation 34.17 into Equation 34.18 gives

$$c = c_b\left[\frac{D-y}{y}\frac{b}{D-b}\right]^Z \qquad (34.19)$$

In which Z is the so-called Rouse number

$$Z = \frac{w}{\sigma_s \kappa U_f} \tag{34.20}$$

Equation 34.19 is called the Vanoni distribution, and it is seen that it contain a reference concentration c_b at a distance $y = b$ above the bed.

Stratification effects: As seen from Equation 34.19, the sediment concentration decreases away from the bed, so the flow actually is stratified. This might lead to dampening of the turbulence, and many researchers modify the Vanoni distribution by including stratification effects, for instance, by introducing a Flux Richardson number in the eddy viscosity description Equation 34.17 for a continuously sediment-stratified flow (Smith and Mclean 1977).

34.2.3.3.2 *Reference Concentration*

There is still much dispute about the bed or reference concentration, and at which distance b above the bed it shall be taken. Some scientists take the distance b to be only a few grain sizes above the bed (Engelund and Fredsoe 1976, Zyserman and Fredsoe 1994). One way to obtain the reference concentration experimentally is to get it indirectly from measurements of the total lad q_T, which is the sum of the bed load q_B and the suspended load q_S, the latter being given by

$$q_S = \int_b^h cU dy \tag{34.21}$$

Applying a logarithmic velocity profile for the flow and inserting Equation 34.19 into Equation 34.21 putting $b = 2d$, and estimating the bed load from a formula like Equation 34.10, the only unknown property is c_b. Zyserman and Fredsoe did this exercise, and got consistent variation in c_b with θ.

Other researchers like Van Rijn (1984) and Smith and McLean (1977) assume the reference concentration to be taken further away from the bottom, at 1% of the water depth and at least 5 cm away from the immobile bed. This makes sense from a practical engineering point of view, since field measurements cannot handle to measure only one grain diameter away from the immobile bed. However, as seen from Equation 34.19, the vertical gradients in the concentration of suspended sediment are very large near the bed. Therefore, if you prefer to measure the reference concentration away from the bed, other parameters next to θ like the Rouse number and water depth play a role to in order to describe the reference concentration.

34.2.3.3.3 *Pickup and Deposition*

The near-bed concentration c_b is the result of the combined effect of sediment entrained (picked up) from the bed and deposition of grains from above. There is some debate whether these two processes can be treated independently of each other or not. In steady uniform flow this does not matter, while it will have some impact on other kinds of flow. As an example, when the flow passes a pit or trench, some of the suspended sediment may settle due to reduced value of the Shields parameter in the pit. This problem is usually modeled mathematically by expressing the distribution of suspended sediment as a diffusion process like

$$\frac{dc}{dt} = \frac{\partial c}{\partial t} + U\frac{\partial c}{\partial x} + v\frac{\partial c}{\partial y} = w\frac{\partial c}{\partial y} + \frac{\partial}{\partial y}\left(\varepsilon_s \frac{\partial c}{\partial y}\right) + \frac{\partial}{\partial x}\left(\varepsilon_s \frac{\partial c}{\partial x}\right) \tag{34.22}$$

which expresses that the change in concentration is caused by advection, settling, and diffusion, U and v being the horizontal and vertical flow velocity. To solve Equation 34.22 you need a bed boundary condition, which either can be that the near bed concentration is determined by the local value of the bed shear stress which is the bed boundary concept. If settling and erosion, on the other hand, is independent, then the erosion is still determined by the local bed shear stress

$$E = -\varepsilon_s \frac{\partial c}{\partial y} = E(\tau_0, d) \tag{34.23}$$

while deposition, settling, is given by

$$D = wc_b \tag{34.24}$$

In steady uniform flow, the erosion and deposition must equal each other, leading to

$$E = wc_b \tag{34.25}$$

so the bed concentration and the pickup function are related through Equation 34.25. Garcia and Parker (1991) reviewed several formulations for the pickup function based on the formulation Equation 34.25, and recommended the pickup rates suggested by Van Rijn (1984) and Smith and Mclean (1977) to be the best.

34.2.3.4 Graded Sediment

Natural sediment is not uniform but graded, which makes the transport description much more complicated. If we consider the bed load, it is easier to move a finer particle than a coarse at small shear stresses. Hence you get a sorting of sediment, which, for instance, at small Shields numbers leads to a coarsening of the surface bed material (Parker and Klingeman 1982). The mathematical description can be done by splitting the sediment up into fractions, and apply a bed load transport formula like Equation 34.9 on each individual fractions, which, for instance, explains selective sorting by differential transport. However, a number of effects like hiding factors for the fine sediment and presence of bed forms (see the following) complicate the description. For suspended sediment, a graded sediment distribution implies that the mean grain diameter of the sand into suspension becomes finer than the mean diameter of the total load. As described earlier, the sheet flow is the transport mode, where the bed shear stress is large, but the sediment is too heavy to enter the suspension regime. The suspended sediment need to be transported away from the bed by vertical turbulent fluctuations. One way to describe this is by assuming that the vertical velocity fluctuations must be larger than the settling velocity of the grains. Since the vertical velocity fluctuations can be scaled by the friction velocity, this gives a relation like $w < U_f$ as can be observed in Figure 34.4.

34.2.4 Cohesive Sediment

The transport of fine-grained particles like clay differs significantly from the description earlier, because of the presence of physicochemical interparticle forces, which becomes just as important as gravity forces for very fine particles with a large specific area. Clay deposited in the bed shows different kinds of plasticity depending on the content of water, and content of other material like organic matter do change the properties of clay as well. Muddy sediments (primary composed of silt and clay) occur commonly in estuaries and coastal embayment and in areas of the continental shelf where the bed shear stresses from current and waves are weak (Whitehouse et al. 2000). Many of these environments are ecologically active zones, so the natural sediment also comprises biological components like biofilm, which also affects the mobility of the sediment (biostabilization). Biological activity from, for instance, worms can have the opposite impact by loosening the bottom. Also the chemical environment of the bottom sediments can affect their properties. Whereas in sandy sediment it is the particle size of the bed sediment which controls the mobility and transport of sediment as described earlier, it is the bulk properties of the mixture of cohesive and noncohesive sediment that determine the behavior of sediment like natural mud. Sediment containing more than 10% by mass of sediment finer than 60 μm usually exhibits cohesive properties.

34.2.4.1 Flocculation

In the marine environment, the clay particles form flocs, which are loose aggregates formed because of saline water, and these flocs have much higher fall velocities compared to the individual clay particles. Above a certain concentration, however, the aggregates start to hinder each other to settle (hindered settling) and the settling velocity decreases rapidly.

34.2.4.2 Consolidation

When flocs reach the bed by settling, they form a dense fluid mud layer with strongly non-Newtonian rheological properties, which with time will be compacted (consolidated) as the water is expelled by the weight of the deposited mud above (self-weight consolidation). Since the behavior of cohesive sediments depends on the physicochemical properties of the fluid and the nature of the sediment itself, the description of cohesive sediment transport is usually empirical and site specific.

34.2.4.3 Stratification

Stratification can be very important in a mud layer. Often you have a sharp interface between the mud layer and the water outside. This is called the lutocline, and this often dramatically suppresses the development of turbulence and suspension outside the boundary layer.

34.2.4.4 Deposition and Erosion

The mathematical modeling of cohesive sediment transport follows most often the principles outlined earlier for noncohesive sediments regarding pickup and deposition. If τ_d is a critical shear stress above which no deposition occurs, deposition is given by

$$D = \left(1 - \frac{\tau_b}{\tau_d}\right) w_s C \quad (34.26)$$

The erosion is similarly described by

$$E = M_e(\tau_b - \tau_e) \quad (34.27)$$

Equations 34.26 and 34.27 are just two simple suggestions out of a wider range. The unknown coefficients in the two equations are usually obtained from either field or laboratory experiments, and the values depend among other things on where you are in the consolidation phase. Whitehouse et al. (2000) provides a number of numerical examples on how to estimate the coefficients given in Equations 34.26 and 34.27.

34.2.5 Impact of Bed Configurations on Sediment Transport

Morphology and sediment transport are interlinked. The interaction between the flow and the sediment transport molds the geometry of the channel and hence determines the hydraulic roughness. The sediment transport, on the other hand, depends to a large extent on the hydraulic resistance developed by the bed configurations.

As seen from the previous section you need information on bed shear stress and near bed distribution of turbulence to be able to predict the rate of sediment transport. In a river with uniform steady flow and an energy slope I, the bed shear stress $\tau_0 = \rho g D I$ can be found from the Colebrook–White formula

$$\sqrt{\frac{2}{f}} = \frac{V}{\sqrt{gDI}} = 6 + 2.5\ln\left(\frac{D}{k}\right) \quad (34.28)$$

Here f is the Colebrook–White friction factor and I is the surface slope of the flow. In Equation 34.28 the bed is assumed to be plane and hydraulic rough. The roughness k is usually taken not to be equal d but rather $2.5d$ for loose deposited sand, due to their irregular position.

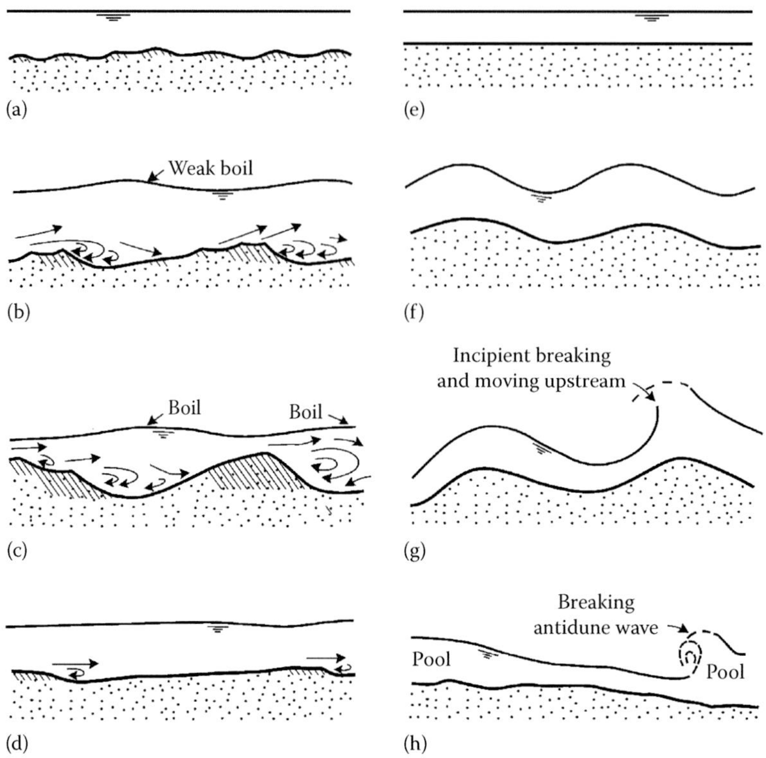

FIGURE 34.5 Typical bed forms in a stream in the order of increased stream power. (a) Typical ripple pattern, (b) dunes with ripples superposed, (c) dunes, (d) washed-out dunes or transition, (e) plane bed, (f) standing waves, (g) antidunes, (h) antidunes.

In a river the bed is usually not plane, but covered with different kind of bed forms, see Figure 34.5. A review of the occurrence of these bed forms can be found in Engelund and Fredsoe (1982a).

Figure 34.6 gives a general review of what happens in a flume covered with loose sand, in which the velocity is gradually increased. The bed shear stress τ_0 is plotted against the mean current velocity V, and the dashed line shows the usually expected parabolic variation in flow over a rigid bed. The full drawn line is the actually observed, and it is seen that the (schematized) variation is quite complicated, due to presence of ripples (short crested triangular shaped sand waves), dunes (much longer bed forms), etc. The flow over a large sand wave is sketched in Figure 34.7. The longitudinal profile of a dune is often nearly triangular, having a slightly curved upstream surface and a downstream slope close to the angle of repose of the bed material.

At the lee side of a dune, a bottom roller is formed, and above this a zone of free turbulence is formed in the immediate continuation of the crest. A certain part of the bed shear stress τ_b is actually carried as a drag force stemming from the expansion

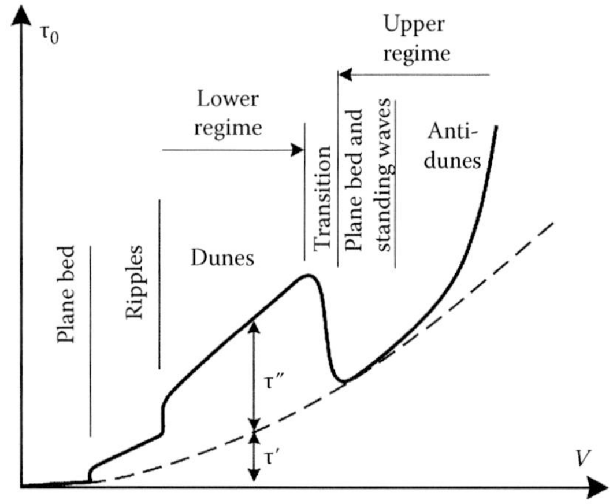

FIGURE 34.6 Variation in bed shear stress against flow velocity in a flume, where the velocity gradually is increased.

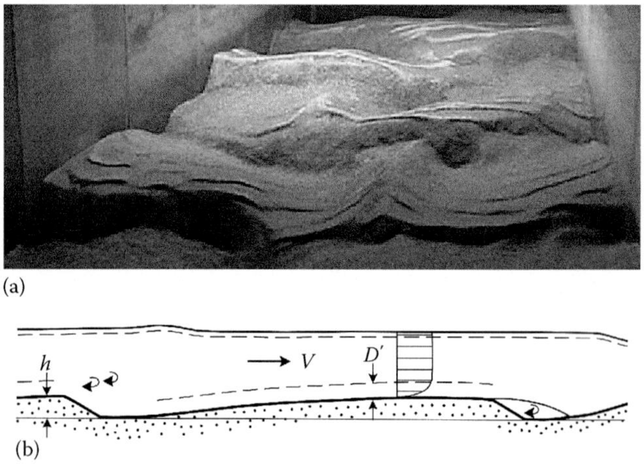

FIGURE 34.7 (a) Dunes superposed by smaller ripples. (b) Schematic representation of the flow over dunes. (Courtesy of Stephen E. Coleman, Auckland University, New Zealand.)

loss behind the dunes, due to the fact that the water pressure is larger at the gentle upstream slope side than at the lee face slope. This is named τ'. The other part τ' is due to the skin friction. This value can be calculated as a boundary layer of thickness D' along the gentle side of the sand wave to be given by

$$\sqrt{\frac{2}{f'}} = \frac{V}{\sqrt{gD'I}} = 6 + 2.5\ln\left(\frac{D'}{k}\right) \tag{34.29}$$

as already suggested by Einstein (1950). The presence of the bed forms complicates the sediment transport calculation significantly, since the sediment transport rate now becomes nonuniform. This is easily realized by considering a train of sand waves, which are assumed to move with a constant velocity a and without any change in shape $h = h(x - at)$, where t is time and h is the local height of the sand wave above the x-coordinate placed through the troughs, see Figure 34.7. The sediment transport continuity equation (the Exner equation) reads

$$\frac{\partial q}{\partial x} = -(1-n)\frac{\partial h}{\partial t} = a(1-n)\frac{\partial h}{\partial x} \tag{34.30}$$

where the last expression is obtained by expressing that $\partial h/\partial t = -a\partial h/\partial x$ since $h = h(x - at)$. In Equation 34.30 n is the porosity of the bed material, and the factor $1 - n$ must be included to account for the fact that all the sediment transport equations derived earlier only gives the transport of solid matter as seen from Equation 34.7. Equation 34.30 can easily be integrated to

$$q = q_0 + a(1-n)h \tag{34.31}$$

The quantity q_0 is a constant, which is equal q in the trough, where $h = 0$. In the trough the bed load vanishes, so q_0 can be interpreted as the suspended load. The last term in Equation 34.31 represents the bed load, which shows that the local intensity of bed load transport is proportional to the local height h of the bed. Hence, also the shear stress τ' varies from zero in the trough to a maximum near the crest. From this description it is seen that the hydraulic resistance is determined by two types of roughness elements, the grains and the sand waves.

In the ripple case this is not that complicated, because the ripple geometry mainly is scaled by the grain length scale, the ripple length being around 1000 times the grain diameter, so the combined roughness originating from grains and ripples is nearly constant. Regarding the larger sand waves called dunes, these change their properties with water depth and flow velocity in a complicated manner, so the combined apparent bed roughness varies with the flow.

Based on similarity principles, Engelund (1966) suggested a unique relation between the skin friction and the total friction like

$$\theta' = f(\theta) \tag{34.32}$$

An example on Equation 34.32 is

$$\theta' = 0.06 + 0.3\theta^{3/2} \tag{34.33}$$

see Engelund and Fredsoe (1982a). More detailed calculations of relations like Equation 34.32 require knowledge to the shape of the bed forms and the spatial variation in bottom pressure and bed shear stress, see for instance Tjerry and Fredsoe (2005). A relation like Equation 34.33 only holds for steady flow. In unsteady flow a lag is introduced between bed form dimensions and flow velocity, because it takes time for the flow to modify the bed form dimensions, see Fredsoe (1979).

If the distribution of the total shear in between skin friction and form drag is known, it is still not easy directly to relate these findings to the sediment transport. Usually you relate the bed load to the skin friction θ', while it is more questionable regarding the suspended load: the reference concentration can also be related to the skin friction, but the vertical distribution of sediment is probably more related to the total shear stress since the turbulence formed in the shear layer downstream the crest also contribute to the turbulence, see Zyserman and Fredsoe (1994).

34.3 Practical Formulae for Sediment Transport in the Fluvial Environment

There exists an enormous amount of different sediment transport formulae for transport of noncohesive sediment transport in streams, at least more than 200 formulas are published. As described earlier, the sediment transport depends on a number of different quantities like discharge Q, slope I, fall velocity w, bed shear stress, turbulence intensity, grain size and shape, temperature and bed configurations. There exists roughly speaking two different major schools to get a universal formulation of how to predict the sediment transport in a fluvial system, namely, energetic approaches and process-based formulations.

34.3.1 Energetic Approach

A large numbers of researchers have tried to correlate the transport of sediment to energetic considerations, namely, the so-called stream power, defined as the loss of potential energy of the fluid per unit time. Bagnold (1966) introduced the *stream power theory,* in which he related the rate of sediment transport to the rate of energy dissipation. Bagnold's dissipation rate is given by τV, and deals with power per unit area. He related the *bed load* to the stream power by

$$(s-1)q_b\mu = \tau_0 V e_b \tag{34.34}$$

And the suspended load similarly by

$$(s-1)q_s = \left(\frac{V}{w}\tau_0 V\right)e_s \tag{34.35}$$

The *e*-factors appearing in Equations 34.34 and 34.35 are efficiency factors, which Bagnold took to be a constant. Regarding the suspended sediment, *e* can be interpreted as the ratio in between keeping the suspended sediment into suspension and the total energy dissipation. Bagnold took this value to be 0.01. Wiuff (1985) reanalyzed a number of experiments and obtained an efficiency function *e*, which mainly depended on the Shields parameter, and he obtained efficiency factors as high as 6% at large θ-values.

Anyway, it is not easy fully to understand the physics behind the energetic approaches. In a flow without sediments, the entire turbulent energy will be transformed into heat. With sediment present, a small fraction of the turbulent energy is used to keep the suspension steady. Another small fraction is applied to move the bed load. One way to explain it could be that if you have a large dissipation, then you also need a large production of turbulence, and that large turbulence will cause a large transport of solid material.

Ackers and White (1973) used Bagnold's stream power concept, in combination with a huge number of laboratory and field data, and also introduced a grain particle Reynolds number together with a sediment mobility number to get better agreement with data.

Yang (1972) elaborated further on the stream power theory. While Bagnold's approach deals with power per unit area, Yang modified this by introducing *the unit stream power concept*, which deals with power per unit weight of fluid: it is defined as the rate in loss of potential energy per unit weight or *VI*.

Yang (1972) suggested the following relation

$$\ln(C) = 6.3 + 1.35\ln(VI - VI_c) \quad (34.36)$$

where

C is the sediment concentration

VI_c reflects the critical shear stress

Yang put this last term to be 0.00002 based on laboratory data and has other constants than those appearing in Equation 34.36 for field data.

34.3.2 Process-Based Formulations

The examples given in the Section 34.2 illustrate the process-based formulation, where a relative simple formulation provides the essential dimensionless parameters. Einstein (1950) was one of the first to construct a bed load formula by describing the movement of the individual grains, and Engelund (1966) did one of the earliest approaches (but still quite successful) to include the presence of bed forms in the sediment transport formulation as described in Section 34.2.5. Next he combined this with an energetic approach for the sediment transport, where he assumed that the work done on a unit area done by drag forces (skin friction) was used to gain potential energy to elevate the sediment discharge a height of order equal to the sand wave height. Hereby he got the following transport formula for the total load

$$f\Phi = 0.1\theta^{5/2} \quad (34.37)$$

Van Rijn (1984) elaborated on earlier process-based models, and got formulas on bed load as well as total load with larger accuracy than before by incorporating additional dimensionless parameters, not that different from those adapted by Ackers and White (1973).

34.4 Sediment Transport in the Marine Environment

In the fluvial environment the flow is generally not steady or uniform, but as long as the temporal or spatial variations are slow and gradual, the sediment transport is usually described applying the formulations given earlier for the local and instantaneous parameters. In the marine environment, however, short period waves change the flow pattern drastically, and you need another hydrodynamic description to account for the changes in the flow.

34.4.1 Wave Boundary Layers

In a wave motion, you have a thin—typically 1–10 cm—turbulent wave boundary layer just above the bed. Within this layer, the turbulence level changes significantly in time with outer orbital flow velocity as shown in Figure 34.8. The bed shear stress varies accordingly. Because the boundary layer is thin, the bed shear stresses usually becomes much larger from the oscillatory motion than from a co-existent current motion, implying that the waves are responsible for the pickup of the sediment (the agitating mechanism), while the current is the transporting mechanism. Much of the current is introduced by the waves itself: streaming, undertow, long shore current, while other contributions stem from tide and wind.

34.4.2 Bed Load

In this unsteady environment, the *bed load transportation* may differ from that in steady flow due to the inertia of the sediment grains. However, it can be proved that the inertia effect is small, and therefore it is common still to apply a bed load formula as developed in Section 34.1, but now bases on the instantaneous value of the bed shear stress. However, at least two things may change the bed load transport pattern. One is related to the pore water flow induced in the bed by the waves, where you have an upward directed flow under the wave through (low bottom pressure) and a similar downward directed under the wave crest. In extreme cases the upward directed flow can be so strong, so the effective stresses in between the grains disappear, and you get *temporary liquefaction*. This leads to the so-called plug flow of sediment as described by Sleath (1999). Another mechanism

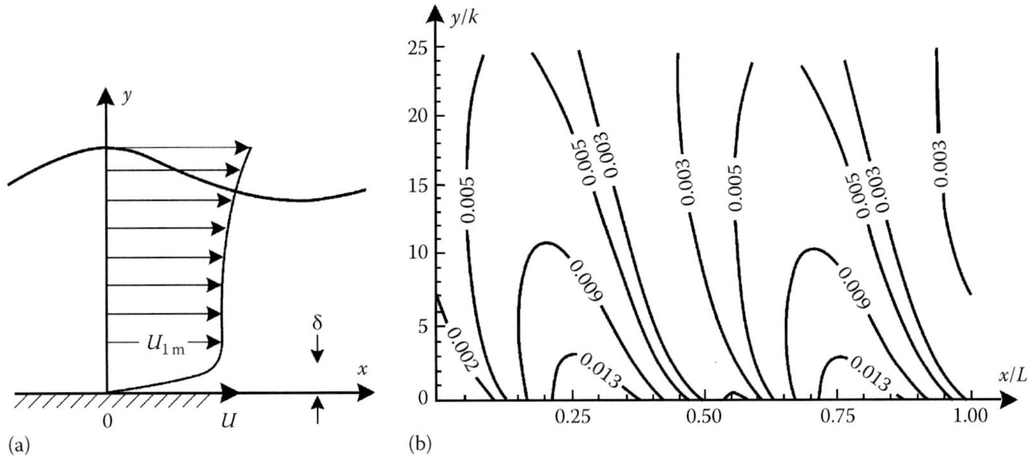

FIGURE 34.8 (a) The wave boundary layer and (b) contour plots of turbulent kinetic energy in this layer.

is that the spatial variation in pressure along the bed in the unsteady case becomes very large, and contributes significantly to the driving forces acting on a sediment grain (Figure 34.1) as discussed by Foster et al. (2006).

34.4.3 Suspended Sediment

Some of the earlier concept from steady flow breaks down in oscillatory flow: you still have a reference concentration for suspended sediment, based on the Shields parameter, but the vertical distribution becomes quite different because of the different vertical distribution of turbulence under waves than in steady flow. In *nonbreaking* waves, the additional turbulence stems from the wave boundary layer as seen in Figure 34.8, in which additional turbulence next to that from the mean current is created. In *breaking/broken* waves also surface-generated turbulence contributes significantly to increase the turbulence intensity, and thus increases the amount of sediment in suspension.

34.4.4 Stratification

The effect of stratification under waves becomes complicated because of the unsteadiness character of the flow. Conley et al. (2008) demonstrated that the sediment suspension affects even those turbulence properties with timescales less than the wave period. The stratification is also important for further transport of riverine sediment offshore on the Continental Shelf. This transport process ranges from transport by waves and current to density-driven "hyperpycnal" flows of highly concentrated sediment suspension as observed by Traykovski et al. (2000).

34.4.5 Bed Forms

Like in steady flow the bed will not remain plane in the oscillatory case for small values of the Shields parameter. For values of θ less than 0.8–1.0, the bed is covered by wave ripples, which in length is scaled to the orbital amplitude. This creates additional coherent flow structures over the bottom, which increases the amount of suspended sediment significantly, Figure 34.9.

34.4.6 Sediment Transport Modeling in the Coastal Environment

Along a coast you usually divide the transport up into two components, the alongshore component and the cross-shore component.

34.4.6.1 Long Shore Sediment Transport

This is driven by a current in the combination with waves. In the surf zone, the long shore current is dominated by the wave-induced current, which is due to the shear component of the radiation stress (Longuet-Higgins 1970). Like in the fluvial

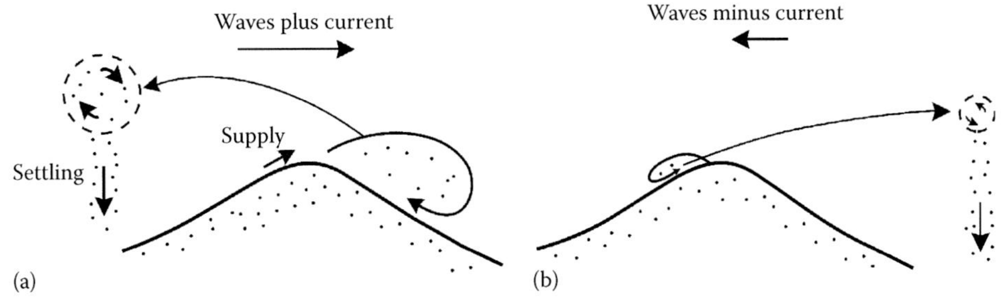

FIGURE 34.9 Flow over wave-generated ripples: the vortex (a) formed during forward motion in combination with a current is larger than the vortex (b) formed in other part of the wave period, where the wave is against the current. Therefore, the net transport can sometimes be opposite of the average current direction.

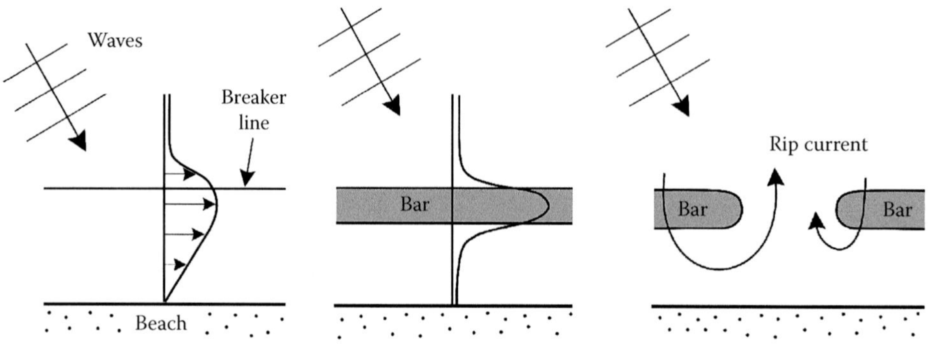

FIGURE 34.10 The wave-generated long shore current is often the main transporting medium for the long shore sediment transport. The strength and distribution depend significantly on the bottom morphology as sketched earlier, so like in the fluvial environment there is a strong interaction between sediment transport and morphology.

environment, there are two approaches to calculate the long shore sediment transport. The most widely used (probably because of its simplicity) is the energetic approach. Here the transport is related to the "long shore energy flux factor" P_{ls} defined by

$$P_{ls} = E_{fb}\cos(\alpha_b)\sin(\alpha_b) \tag{34.38}$$

where

α_b is the angle between the waves and the coast at the point of breaking

E_{fb} is the wave energy flux at the point of wave breaking, see Figure 34.10

Wave energy flux is the transport of wave energy (potential plus kinetic energy multiplied by the group velocity). Komar and Inman (1970) suggested proportionality between P_{ls} and the long shore transport of sand, given by the submerged weight of the transported sediment. This is later known as the CERC-formula.

A process-based description of the long shore sediment transport requires detailed knowledge to the current and wave properties in the area under investigation. The current can either be wave driven or external driven. Next step requires a detailed analysis of the net transport of sand in combined waves and current. Here the level of turbulence depends on whether the waves are nonbreaking or breaking/broken, so additional turbulence is added from the wave-breaking process.

This might increase the transport capacity of especially fine sand. An example on such a process-based description is given by Deigaard et al. (1986).

34.4.6.2 Cross-Shore Sediment Transport

While the long shore transport is relatively easy to calculate using deterministic tools, the understanding and modeling of the cross-shore transport is one of the big challenges for the future. Since erosion of a coast is a combination of variations in long shore transport and the cross-shore transport, the cross-shore component will be an important issue in the coming years, among other things in order to understand whether the coastal profile can keep in pace with the sea level rise.

Since the wave orbital motion back and forth results in a very small net transport either in the direction of wave propagation or opposite, you need to investigate a number of mechanisms like wave asymmetry, wave skewness, which both will push the sediment onshore, streaming (caused by spatial changes in wave boundary layer thickness) and wave drift which also will contribute to onshore movements, undertow (caused by broken waves) which will transport the sediment offshore, and wave ripples which in the case of nonbreaking waves can lead to transport against the direction of wave propagation (see Figure 34.9). To complicate the picture even more, the transport of sediment in oscillatory flow superposed by a weak current like the undertow, the resulting sediment flux can differ significantly from the flux based on time-averaged quantities of concentration and velocity because of the phase lag between bed shear stress and concentration.

The different cross-shore mechanisms lead to the formation of breaker bars, because the sediment transport is onshore outside breaking and offshore inside. On larger scales, large 3D circulation pattern formed by rip holes in breaker bars also modify cross-shore (and long shore) sediment pattern, Figure 34.10. Each of these contributions is so delicate that a sufficient accurate description of the cross-shore profile still lacks. For a review see Fredsoe and Deigaard (1992).

34.5 Major Challenges

There will always be some uncertainties connected to the description and prediction of sediment transport. Regarding noncohesive sediment transport, this can today be predicted within a factor of two in the fluvial environment, and the progress in the marine sediment transport description is so fast today, so a similar accuracy might be obtained during the next decade. However, the output from a mathematical/numerical model is never better than the input, and especially information of sediment properties will always be related with a certain degree of inaccuracy (fall velocity, gradation, content of clay, etc.). So it can be questioned how detailed the turbulence modeling shall be, when the sediment properties and met-ocean information always will be related with some uncertainty.

From a theoretical point of view, some major problems still exists, of which a few are listed in the following.

34.5.1 Sheet Flow Modeling

There has been put some doubt in Bagnold's expression for the dispersive stresses, see Hunt et al. (2002), who demonstrated that the top and bottom plates in Bagnold's experiments generated secondary flow, which especially disturbed the experimental findings regarding shear component of the dispersive stress, while it is more difficult to evaluate whether errors also are related to the normal stress given by Equation 34.11. Advanced modeling on the grain collisions are done by Hsu et al. (2004), who used kinetic theory for rapid grain flows together with a turbulent flow model to describe essential features of the sheet flow.

34.5.2 Coherent Structures

The near bed turbulence are often more regular than expected, and the so-called Bursting process, where water is ejected in large-scale vortices away from the bed (Grass 1971), shows its clear signature in near bed temporal variations in concentration of sediment even in the sheet flow regime as reported by Sumer et al. (1996). These coherent structures make the semi-detailed modeling of particle motion slightly problematic, because the trajectory of the particle rather is determined by the appearance of the fluid structure, than anything else. Vittori and Verzicco (1998) included the bursting phenomenon in a detailed DNS-modeling of the wave boundary layer, but their findings were limited to moderate Reynolds numbers, and grain–grain interaction was not included. However, with the increased computational capacity, a more realistic description of the near wall turbulence will be one of the major tasks in improving the sediment transport description including sheet flow modeling and suspended sediment.

34.5.3 Morphology

As explained earlier, the interaction between the bed morphology and the sediment transport plays a dominant role for the sediment transport. The description of the dynamic behavior of bed forms and their interaction between the different scales (like ripples and dunes) and its 3D features is still in its initial phase.

References

Ackers, P. and White, W.R. 1973. Sediment transport—New approaches and analysis. *J. Hydraul. Div.—ASCE* 99(11): 2041–2060.

Bagnold, R.A. 1954. Experiments on a gravity-free dispersion of large solid spheres in a Newtonian fluid under shear. *Proc. R. Soc. Lond. A* 225: 49–63.

Bagnold, R.A. 1966. An approach to the sediment transport problem from general physics. USGS Professional Paper 422-I.

Conley, D.C., Falchetti, S., Lohmann, I.P., and Brocchini, M. 2008. The effects of flow stratification by non-cohesive sediment on transport in high-energy wave-driven flows. *J. Fluid Mech.* 610: 43–67.

Deigaard, R., Fredsoe, J., and Hedegaard, I.B. 1986. Mathematical model for littoral drift. *J. Waterw. Port Coastal Ocean Eng—ASCE* 112(3): 351–369.

Einstein, H.A. 1950. The bed-load function for sediment transportation in open channel flows. Technical Bulletin No.1026, U.S. Department of Agriculture, Washington, DC.

Engelund, F. 1966. Hydraulic resistance of alluvial streams. *J. Hydraul. Div.—ASCE* 92(2): 315–326 plus closure 93(4): 287–296.

Engelund, F. 1981. Transport of bed load at high shear stresses. *Prog. Rep.* 53: 31–35, ISVA, DTU, Denmark.

Engelund, F. and Fredsoe, J. 1976. A sediment transport model for straight alluvial channels. *Nord. Hydrol.* 7: 293–306.

Engelund, F. and Fredsoe, J. 1982a. Sediment ripples and dunes. *Annu. Rev. Fluid Mech.* 14: 13–37.

Engelund, F. and Fredsoe, J. 1982b. Hydraulic theory of alluvial rivers. *Adv. Hydrosci.* 13: 187–215.

Fernandez Luque, R. and van Beek, R. 1976. Erosion and transport of Bed-load sediment. *J. Hydraul. Res.* 14(2): 127–144.

Foster, D.L., Bowen, A.J., Holman, R.A., and Natoo, P. 2006. Field evidence of pressure gradient induced incipient motion. *J. Geophys. Res.* 111(C5): C05004.

Fredsoe, J. 1979. Unsteady flow in straight alluvial streams—Modification of individual dunes. *J. Fluid Mech.* 91: 497–512.

Fredsoe, J. and Deigaard, R. 1992. *Mechanics of Coastal Sediment Transport*. World Scientific.

Garcia, M. and Parker, G. 1991. Entrainment of bed sediment into suspension. *J. Hydraul. Eng.—ASCE* 117(4): 414–435.

Grass, A.J. 1971. Structural features of turbulent flow over smooth and rough boundaries. *J. Fluid Mech.* 50: 233–255.

Hsu, T.J., Jenkins, J.T., and Liu, P.L.F. 2004. On two-phase sediment transport: Sheet flow of massive particles. *Proc. R. Soc. Lond. A* 460(2048): 2223–2250.

Hunt, M.L., Zenit, R., Campbell, C.S., and Brennen, C.E. 2002. Revisiting the 1954 suspension experiments of R.A. Bagnold. *J. Fluid Mech.* 452: 1–24.

Komar, P.D. and Inman, D.L. 1970. Long shore sand transport on beaches. *J. Geophys. Res.* 75(30): 5914–5927.

Longuet-Higgins, M.S. 1970. Long shore currents generated by obliquely incident sea waves. Part 1. *J. Geophys. Res.* 75(33): 6778–6789.

Parker, G. and Klingeman, P.C. 1982. On why gravel bed streams are paved. *Water Resour. Res.* 18(5): 1409–1423.

Sleath, J.F.A. 1984. *Sea Bed Mechanics*. John Wiley & Sons, New York.

Sleath, J.F.A. 1999. Conditions for plug formation in oscillatory flow. *Cont. Shelf Res.* 19(3): 1643–1664.

Smith, J.D. and McLean, S.R. 1977. Spatially averaged flow over a wavy surface. *J. Geophys. Res.—Oceans* 82(12): 1735–1746.

Sumer, B.M., Kozakiewicz, A., Fredsoe, J., and Deigaard, R. 1996. Velocity and concentration profiles in sheet-flow layer of movable bed. *J. Hydraul. Eng.—ASCE* 122(10): 549–558.

Tjerry, S. and Fredsoe, J. 2005. Calculation of dune morphology. *J. Geophys. Res.*, 110: F04013, doi:10.1029/2004JF000171.

Traykovski, P., Geyer, W.R., Irish, J.D., and Lynch, J.F. 2000. The role of wave-induced density-driven fluid mud flows for cross-shelf transport on the Eel River continental shelf. *Cont. Shelf Res.* 20(16): 2113–2140.

Van Rijn, L.C. 1984. Sediment pick-up functions. *J. Hydraul. Eng.* 110(10): 1494–1502.

Vittori, G. and Verzicco, R. 1998. Direct simulation of transition in an oscillatory boundary layer. *J. Fluid Mech.* 371: 207–232.

Whitehouse, R., Soulsby, R., Roberts, W., and Mitchener, H. 2000. *The Dynamics of Estuarine Muds*. Telford, London, U.K.

Wiberg, P.L. and Smith, J.D. 1985. A theoretical model for saltating grains in water. *J. Geophys. Res.—Ocean* 90(NC4): 7341–7354.

Wilson, K.C. 1966. Bed-load transport at high shear stress. *Proc. ASCE*, 92 (HY6): 49–60.

Wiuff, R 1985. Transport of suspended material in open and submerged streams. *J. Hydraul. Eng.—ASCE* 111(5): 774–792.

Yang, C.T. 1972 Unit stream power and sediment transport. *J. Hydraul Div.—ASCE* 98(19): 1805–1826.

Zyserman, J.A. and Fredsoe, J. 1994. Data-analysis of bed concentration of suspended sediment. *J. Hydraul. Eng.—ASCE* 120(9): 1021–1042.

35

Flows Involving Phase Change

Herbert E. Huppert
University of Cambridge

M. Grae Worster
University of Cambridge

35.1 Introduction

Flows involving phase change are ubiquitous in geophysical and industrial settings and are vital for life on Earth. For example, the solid iron inner core of the Earth solidifies from the molten outer core and releases buoyant constituents that drive compositional convection and power the geo-dynamo that generates our magnetic field. Melting and re-solidification of silicates in the mantle and crust generate new rocks and segregate economically important minerals into mineable seams. Groundwater is converted to steam in geothermal reservoirs that can be tapped for energy production. The oceans freeze and thaw in annual cycles, moderating global temperatures and driving ocean circulations. Surface waters evaporate, rise, and re-condense to form clouds and life-giving rain. And from the Bronze Age onward, melting and solidification have been central to many technological developments.

In this chapter, we introduce the key thermodynamic and fluid-mechanical principles that govern phase changes, explaining how flow in fluid melts can both be driven by and influence those changes. Our discussions are illustrated mostly by geophysical examples, particularly phenomena occurring in our oceans and atmosphere but also some that occur in water-saturated soils, in magmatic systems, and in industrial processes.

35.2 Principles

Changes of phase, from liquid water to vapor or to solid ice, for example, are thermodynamic processes requiring the transport of heat and sometimes, as in the case of freezing salt water, of a chemical component. Therefore, flows involving phase change are inevitably associated with temperature gradients and possibly also compositional gradients. Both can cause density gradients and associated buoyancy forces. The importance of phase change to a given flow can be assessed by estimating certain key dimensionless parameters.

35.2.1 Thermodynamic Considerations

We can distinguish two important situations: those in which phase change occurs at the boundary of a fluid domain or at the interface between fluid domains and those in which phase change occurs in the interior of a fluid domain. Examples of the former are the melting of an iceberg and the evaporation of a pool of water. Examples of the latter are condensation in the interiors of clouds and dissolution in the interior of sea ice.

When phase change occurs at a boundary, a key dimensionless parameter is the Stefan number

$$\mathcal{S} = \frac{L}{c_p \Delta T},$$

where

L is the latent heat per unit mass associated with change of phase at constant temperature

c_p is the specific heat capacity of the medium through which the latent heat is principally transported in order to effect the phase change

ΔT is the characteristic temperature difference driving that heat transport

Handbook of Environmental Fluid Dynamics, Volume One, edited by Harindra Joseph Shermal Fernando. © 2013 CRC Press/Taylor & Francis Group, LLC.
ISBN: 978-1-4398-1669-1.

When the Stefan number is large, the motion of the phase boundary is slow compared to the rate of thermal processes in the media on either side, whether or not the media are flowing. To a good approximation, therefore, the heat transfer rates and any fluid flow can be calculated assuming that the phase boundary is stationary, and the evolution of the phase boundary can be determined subsequently.

In the transition from ice to water, L/c_p is about 80°C in the water and about 160°C in the ice, while in the transition from liquid water to water vapor, L/c_p is about 600°C in the water and about 2500°C in the vapor. In all these cases and for geophysically relevant flows, for which ΔT is typically only a few degrees Celsius, the Stefan number is therefore indeed large and the quasi-stationary approximation can be used with confidence.

Regions of mixed phase can exist when there are variations of two (or more) thermodynamic variables that determine the state of the matter. A well known and geophysically important example occurs during the freezing of salt water. The equilibrium (liquidus) temperature T_L between ice and salt water decreases with the salinity S of the latter. In many circumstances, this relationship can be approximated linearly by

$$T_L(S) = T_0 - m(S - S_0),$$

where

S_0 is a reference value of the salinity at which $T_L = T_0$
m is the slope of the liquidus

For seawater $m \approx 0.05$°C psu^{-1}, where the practical salinity unit (psu), defined in terms of conductivity, corresponds closely to parts per thousand of sodium chloride.

Another important thermodynamic variable in geophysical contexts is pressure, which influences the equilibrium freezing temperature $T_e(p)$ according to the Clausius–Clapeyron relationship

$$L\frac{T_m - T_e(p)}{T_m} = (p - p_m)\left(\frac{1}{\rho_s} - \frac{1}{\rho_l}\right),$$

where

T_m is the freezing temperature at pressure p_m
ρ_s and ρ_l are the densities of the solid and liquid phases, respectively

Therefore, the inner core of the Earth, which is virtually pure iron with $\rho_s > \rho_l$, is solid owing to the high pressures there, even though it is hotter than the outer core of molten iron. On the other hand, because the density of ice is less than that of water, marine ice sheets in Antarctica melt at their basal contact with the deep ocean, even though the temperature there is about −2°C.

Pressure is continuous across a planar interface between bulk phases in equilibrium. But if there is a pressure difference between a solid and its liquid melt, either because surface tension acts across a curved interface between them or because disjoining forces act in proximity to a substrate, then the equilibrium freezing temperature T_e is depressed according to

$$\rho_s L\frac{T_m - T_e}{T_m} = p_s - p_l.$$

In the case of a curved interface with surface tension γ_{sl} and principal radius of curvature R, this gives the Gibbs–Thomson relationship

$$\rho_s L\frac{T_m - T_e}{T_m} = \frac{\gamma_{sl}}{R},$$

which stabilizes morphological instabilities of phase boundaries (see later) and hence influences the scale and patterns of snowflakes, for example.

When a mixture, like salt water, begins to freeze, one of the components is preferentially incorporated into the solid phase, while the remaining liquid is enriched in the other. At low salinities, typical of the ocean, the solid phase is almost pure ice, and salt is rejected as sea ice forms. The rate of phase change at a planar interface is governed by the rate of removal of excess salt, rather than by the rate of removal of latent heat. The important dimensionless parameter is the concentration ratio

$$\mathcal{C} = \frac{(S_0 - S_s)}{\Delta S},$$

where

S_s is the salinity of the solid phase (equal to zero in the case of ice formed from salt water)
ΔS is the characteristic salinity difference driving salt transport in the liquid phase

Usually the solid–liquid interface is close to equilibrium and

$$\Delta S = \frac{(T_0 - T_i)}{m},$$

where T_i is the temperature of the interface.

In consequence of the fact that salt diffuses slower than heat, a phenomenon called constitutional supercooling usually occurs in the liquid adjacent to the phase boundary during solidification of a mixture. This is the phenomenon of having liquid at a temperature below its freezing (liquidus) temperature in consequence of locally altered salinity. It is relieved in most natural and many industrial settings by a morphological instability of the phase boundary that results in the formation of a mushy layer: a region of mixed phase comprising small-scale, often dendritic, solid crystals bathed in residual melt. Mushy layers are reactive porous media (Worster 2000) whose porosity and permeability can change in response to changes in temperature and salinity caused by advection and diffusion.

Latent heat is released in the interior of a mushy layer during solidification as the volume fraction of solid crystals increases. The effective specific heat capacity (the heat energy required to raise or lower the temperature by 1°) of a mushy layer is thereby augmented by a factor characterized by the dimensionless group

$$\Omega = 1 + \frac{\mathcal{S}}{\mathcal{C}} = 1 + \frac{L}{c_p m(S_0 - S_s)}.$$

If there is a larger salinity difference between solid and liquid phases, then the change in solid fraction associated with a given change in temperature is smaller, and hence the internal release of latent heat is inversely related to $\mathcal{C}$. A similar effect occurs in moist convection, where the effective heat capacity is equal to the heat capacity of dry air augmented by a factor approximately equal to

$$\Omega = 1 + \frac{L^2 q}{c_p \mathcal{R} T^2}$$

(Stevens 2005), where

q is the mass fraction of water vapor
$\mathcal{R}$ is the gas constant

The analogy is clear once we associate q with S, the saturation of the condensed phase $q_s = 1 \gg q$ with S_s and the liquidus slope m with $-\partial T/\partial q \approx -\mathcal{R}T^2/Lq$.

35.2.2 Buoyancy-Driven Convection

These sorts of considerations have a profound influence on the buoyancy forcing of the oceans and clouds. The density of salt water, for example, varies approximately linearly as

$$\rho = \rho_0\left[1 - \alpha(T - T_0) + \beta(S - S_0)\right],$$

for small temperature and salinity variations, where ρ_0 is the reference density when $(T, S) = (T_0, S_0)$ and α and β are the coefficients of thermal expansion and solutal density variation, respectively. Pure water attains a maximum density at 4°C, at which temperature $\alpha = 0$. The temperature at which salt water attains its maximum density decreases as salinity increases until, for S greater than about 14 psu, there is no density maximum and α is positive for all temperatures above the liquidus. However, α remains very sensitive to temperature at temperatures close to the liquidus, and seawater, whose salinity is around 30–35 psu, has a value of $\alpha \approx 10^{-5}$ near its liquidus temperature of about −2°C (typical of polar conditions), compared with a value of $\alpha \approx 10^{-4}$ at 20°C.

In a *polynya*, growing ice crystals are blown down wind and the open ocean is continuously exposed to the cold atmosphere. A heat flux F from ocean to atmosphere creates ice crystals, which increases the salinity of the remaining liquid and causes a buoyancy flux

$$F_{BS} = \rho_0 g\beta \frac{\partial S}{\partial t} = \rho_0 g\beta S_0 \frac{\partial \phi}{\partial t} = \rho_0 g\beta S_0 \frac{F}{\rho_s L},$$

where

S and ϕ are the salinity and solid fraction in a parcel of freezing ocean of unit volume
g is the acceleration due to gravity

This can be contrasted with the buoyancy flux

$$F_{BT} = -\rho_0 g\alpha \frac{\partial T}{\partial t} = \frac{g\alpha F}{c_p}$$

that results from cooling the ocean without freezing. The ratio of these buoyancy fluxes

$$\frac{F_{BS}}{F_{BT}} = \frac{\rho_0}{\rho_s}\frac{\beta S_0}{\alpha}\frac{c_p}{L}$$

is about 20 in the polar oceans, which is why the densest abyssal waters are generated around Antarctica, where polynyas are commonly formed by strong katabatic winds off the continent, and, to a lesser extent, in the Greenland Sea. In contrast, where sea ice becomes consolidated, typical of the Arctic Ocean, it insulates the ocean from atmospheric cooling, and brine production is much weaker.

A similar expression relates the buoyancy fluxes associated with salinity increase and cooling resulting from evaporation in warmer waters, such as the Mediterranean. However, because α is significantly larger at those temperatures and salinities, and the latent heat of vaporization is much larger than that of freezing, the balance is reversed: cooling is then more potent than salinity increase at producing buoyancy fluxes by a factor of about 4 (Gill 1982).

Buoyancy forces can cause convection in a fluid (Turner 1979), which can have profound effects on phase changes in the fluid. Whether or not convection occurs and the strength and character of any convection when it does occur are determined by the magnitude of the Rayleigh number. This dimensionless number characterizes the potential energy of the fluid due to buoyancy relative to the dissipation of buoyancy by diffusion and of energy by viscosity as the system attempts to convert its potential energy into the kinetic energy of fluid motion.

In a fluid whose density is affected only by temperature, the Rayleigh number is

$$Ra_T = \frac{\alpha \Delta T g h^3}{\kappa \nu},$$

where

h is a characteristic vertical length scale for the fluid system
κ and ν are the thermal diffusivity and kinematic viscosity of the fluid, respectively

This reflects the fact that the buoyancy is due to temperature gradients and is dissipated by thermal diffusion. By direct analogy, if the density is affected only by salinity, then the appropriate Rayleigh number is

$$Ra_S = \frac{\beta \Delta S g h^3}{D\nu},$$

where D is the diffusivity of salt in water. In fluids whose density is simultaneously affected by thermal and salinity variations, both Rayleigh numbers are important. Indeed, fascinating forms of convection, so-called double-diffusive convection, can occur even when the overall density of fluid decreases with height, so that the system is apparently stably stratified, provided that there is potential energy associated with either the temperature or the salinity field (Huppert and Turner 1981). Such flows owe their existence to the fact that heat and salt diffuse at different rates, which allows the potential energy of the unstably stratified component to be exploited. A ubiquitous feature of double-diffusive convection is the spontaneous development of discrete layers of fluid motion, illustrated in Figure 35.6. It has been suggested that such fluid-mechanical layering during solidification can result in layered structures within igneous intrusions, for example.

There are many interesting and important geophysical flows involving phase change that take place in porous media. Examples include natural hydrothermal flow of groundwater and the flow of supercritical carbon dioxide in aquifers during attempts at sequestering unwanted CO_2 from gas and oil fields and other industrial sources. In a porous medium, the relevant thermal Rayleigh number is

$$Ra_p = \frac{\alpha \Delta T g h \Pi}{\kappa\nu},$$

where Π is the permeability of the medium. This reflects the fact that viscous dissipation occurs on the scale of the pores of the medium, which is proportional to $\sqrt{\Pi}$.

A fascinating hybrid of these results occurs in a mushy layer. The temperature and interstitial salinity of a mushy layer are coupled by the liquidus relationship, so the density variation is given by

$$\rho = \rho_0 \left[1 + \left(\frac{(\beta - \alpha)}{m}\right)(S - S_0)\right]$$

and there is no propensity for double-diffusive convection. Typically $\beta \gg \alpha/m$, and the characteristic Rayleigh number is

$$Ra_m = \frac{\beta \Delta S g h \Pi}{\kappa\nu}.$$

This reflects the facts that a mushy layer acts as a porous medium, that the buoyancy is dominated by salinity variations, but that the dissipation of buoyancy results from internal phase change, mediated by the thermal field altering the interstitial salinity (Worster 2000). As described earlier, the internal release of latent heat augments the effective heat capacity of the mushy layer, so the effective thermal diffusivity is κ/Ω and, as we shall see later, the critical parameter for convection in a mushy layer is therefore ΩRa_m.

In this section, we have introduced many of the dimensionless parameters governing flows involving phase change. They will appear and be illustrated further as we explore specific examples later.

35.3 Methods of Analysis

Our understanding of flows involving phase change has benefited enormously from studies that have combined laboratory experimentation with theoretical and computational developments. Here we outline some of the most common approaches, starting again with single-component systems.

35.3.1 Scaling and Similarity Solutions

Significant physical insight and understanding can be gained from scaling analyses, which in some important circumstances can lead to similarity solutions. This is the case in what has become known as Neumann's solution to the Stefan problem, though Neumann's analysis seems to predate Stefan's own investigation! In the simplest version, depicted in Figure 35.1 and described in Carslaw and Jaeger (1959) for example, a deep layer of liquid (ocean) at its freezing temperature T_m has its fixed boundary maintained at a constant lower temperature T_0. Solid (ice) forms a layer of thickness $h(t)$ adjacent to the cooled boundary. The temperature $T(z, t)$ in the solid satisfies the diffusion equation

$$T_t = kT_{zz} \quad (0 < z < h(t))$$

and conservation of heat at the solid–liquid interface is expressed by the Stefan condition

$$\rho L\dot{h} = \kappa T_z \,|_{z=h},$$

where $k = \rho c_p \kappa$ is the thermal conductivity. Since there is no externally imposed nor any intrinsic length scale in the problem, there is a similarity solution with

$$T(z,t) = T_0 + (T_m - T_0)\,\mathrm{erf}\left(\frac{z}{2\sqrt{\kappa t}}\right) \quad \text{and} \quad h(t) = 2\lambda\sqrt{\kappa t}.$$

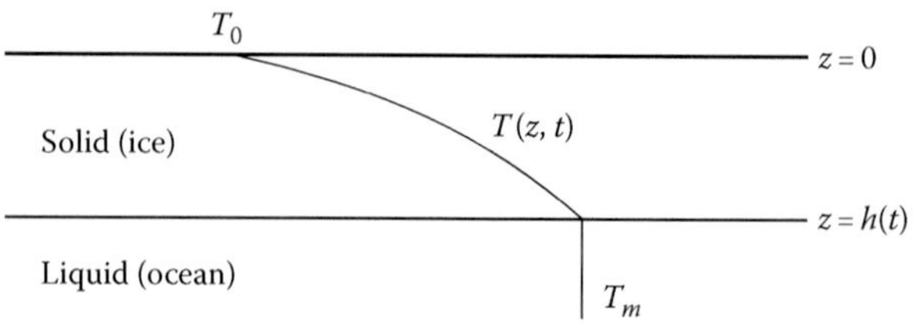

FIGURE 35.1 A simple Stefan problem, in which solid forms at the cooled boundary of a deep layer of melt at its freezing temperature.

The growth parameter λ is determined by the transcendental equation

$$\sqrt{\pi}\lambda e^{\lambda^2}\,\mathrm{erf}\lambda = \mathcal{S}^{-1},$$

where $\mathcal{S} = L/c_p(T_m - T_0)$.

A useful and instructive approximation can be found when $\mathcal{S}$ is large, in which case the temperature field is quasi-steady and therefore linear. The Stefan condition then has the simple form

$$\rho L\dot{h} = \frac{k(T_m - T_0)}{h},$$

which is readily integrated to give

$$h = \sqrt{\frac{2\kappa t}{\mathcal{S}}}.$$

A comparison between this approximate result and the exact solution is shown in Figure 35.2a.

Many important analytical and computational methods have been developed in terms of the enthalpy of the system. A new and revealing example of such an approach is provided by the case of a solid at temperature T_m growing into a supercooled liquid of temperature $T_\infty < T_m$, as depicted in Figure 35.3. The total enthalpy of this system relative to the far field (at constant pressure) is

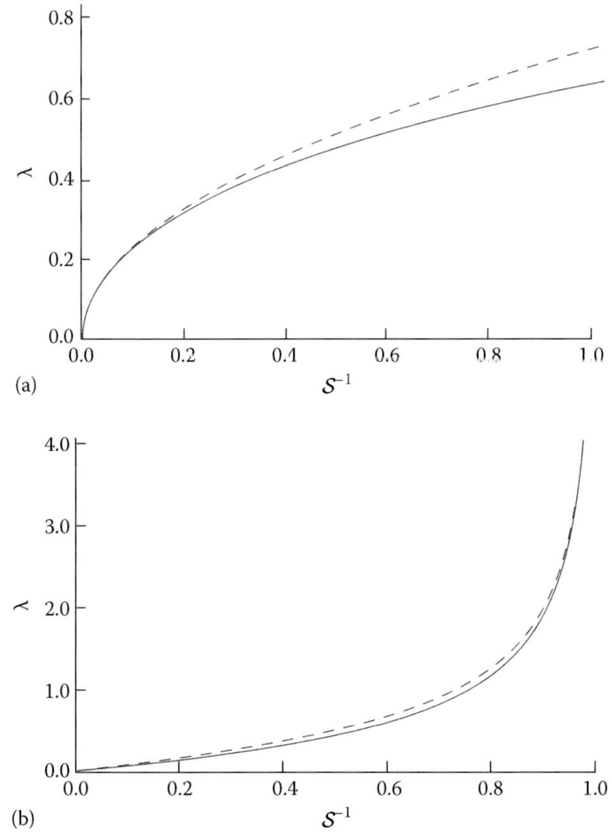

FIGURE 35.2 The growth parameter λ as a function of the inverse Stefan number $\mathcal{S}^{-1}$ for solidification at a cooled boundary (a) and into a supercooled melt (b). In the latter case, there is no solution for $S < 1$. The solid curves show the similarity solutions of the full diffusion equation. The dashed curves show the approximate solutions described in the text.

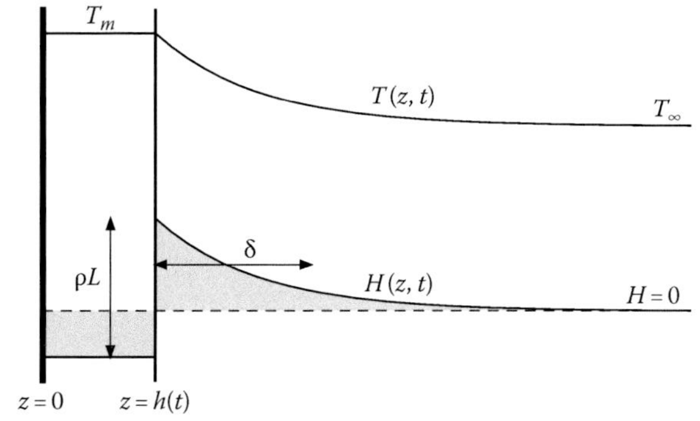

FIGURE 35.3 The temperature T and enthalpy H for a solid growing into a supercooled melt. Conservation of enthalpy means that the two shaded regions must have equal area, which cannot be achieved if the latent heat L is too small.

$$H = \left[-\rho L + \rho c_p(T_m - T_\infty)\right]h + \int_h^\infty \rho c_p(T - T_\infty)dz.$$

We can approximate the integral by $\rho c_p(T_m - T_\infty)\delta$, where δ is a scale length (boundary-layer thickness) for the temperature field, and use conservation of enthalpy, given the initial condition that $H = 0$, to deduce that $\delta = (\mathcal{S} - 1)h$, where $\mathcal{S} = L/c_p(T_m - T_\infty)$. In this case, the Stefan condition can be approximated by

$$\rho L\dot{h} = \frac{k(T_m - T_\infty)}{\delta},$$

whence

$$h = \sqrt{\frac{2\kappa t}{\mathcal{S}(\mathcal{S}-1)}}.$$

This is compared with the exact similarity solution $h = 2\lambda\sqrt{\kappa t}$, with $\sqrt{\pi}\lambda e^{\lambda^2}\,\mathrm{erfc}\lambda = \mathcal{S}^{-1}$ in Figure 35.2b. The quantitative effectiveness of the approximation is of secondary importance to the insight that can be gained from the structure revealed by it. We see immediately that the predicted rate of solidification tends to infinity as $\mathcal{S} \to 1+$, and that there is no solution when $\mathcal{S} < 1$. The enthalpy formulation reveals the fact that the solid is predicted to have positive enthalpy when $\mathcal{S} < 1$, and that the total enthalpy cannot therefore be conserved. In this case, one must take account of the kinetics of molecular attachment and recognize on the continuum level that the solid–liquid interface is below the equilibrium freezing temperature T_m during solidification (Worster 2000).

The two previous approaches can be combined or the method of Neumann extended straightforwardly to consider the case of a melt initially at a temperature greater than T_m brought into contact with a boundary held at a temperature less than T_m. The additional diffusive heat transport from the liquid typically plays a minor role in comparison with the latent heat that must be removed through the solid phase to the boundary.

35.3.2 Perturbation Analysis

An important phenomenon that occurs when the liquid is supercooled is morphological instability: small corrugations of the solid–liquid interface grow into the branched patterns of snowflakes or Jack Frost, for example. The same phenomenon is common during solidification of a mixture, as depicted in Figure 35.4. In this illustrative example, we employ the frozen temperature approximation (Davis 2001) that the temperature field remains unperturbed with a uniform gradient G. In the liquid near the interface, the salt rejected during solidification creates a salinity gradient G_S. When the initially planar interface at $z = 0$ is perturbed to position

$$z = \eta(x,t) = \hat{\eta} A e^{i\alpha x + \sigma t},$$

there is a corresponding perturbation to the salinity field, which to good approximation satisfies Laplace's equation so that

$$S(x,z,t) = S_0 - G_S z + \Theta e^{-\alpha z} e^{i\alpha x + \sigma t}$$

where $\hat{\eta}$ and Θ are constants, while

$$T(x,z,t) = T_0 + Gz,$$

where we have chosen $(S, T) = (S_0, T_0)$ to correspond to the salinity and the corresponding liquidus temperature at the unperturbed interface.

The temperature of the curved interface

$$T(x,\eta,t) = -mS(x,\eta,t) - \Gamma(-\eta_{xx}) \quad (z = \eta)$$

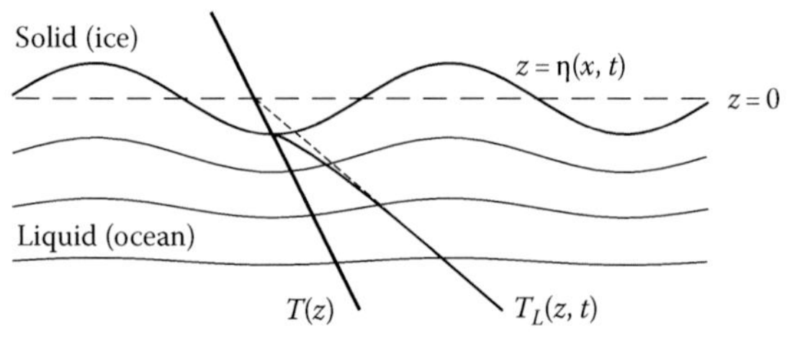

FIGURE 35.4 The thick sinusoidal curve shows the perturbed position $z = \eta(x, t)$ of the solid–liquid interface, originally at $z = 0$. The thinner sinusoidal curves show perturbed contours of concentration in the liquid: they are closer together near protrusions of the solid into the liquid, which enhances diffusion of solute away from the solidification front and promotes instability. The thick solid line shows the (frozen) temperature field. The thinner curve and short-dashed curve show the perturbed and original vertical salinity profiles, expressed as the local liquidus temperature.

is equal to the liquidus temperature depressed by an amount proportional to its curvature, with constant of proportionality Γ, according to the Gibbs–Thomson effect. Conservation of salt is expressed by

$$S\eta_t = -DS_x \quad (z = \eta),$$

which is similar in form to the Stefan condition.

By substituting the aforementioned expressions for the temperature and salinity fields into the two interfacial conditions and linearizing in the small perturbation quantities, we find the dispersion relation

$$\sigma = \frac{D}{mS_0}\alpha\left[(mG_S - G) - \Gamma\alpha^2\right]$$

for the growth rate σ of disturbances of wavenumber α.

In this simple, idealized model, we see that there are positive growth rates and disturbances will grow if $G < mG_S$, which corresponds to there being constitutional supercooling (temperatures less than the liquidus) of the unperturbed liquid.

35.3.3 Continuum Modeling of Mushy Layers

Once significant morphological instability has occurred and a mushy layer has formed, it is no longer practical to follow the evolution of the solid–liquid interface, and it is usual to adopt an averaged, continuum description of the resulting reactive, two-phase medium, called a mushy layer. Dependent variables of a mushy layer are the mean temperature, the mean interstitial salinity, and the mean volume fraction of the solid phase, with averages taken over a scale larger than the pore scale. Because the salt rejected by the solid phase can be accommodated within the interstices of a mushy layer, transport of salt is no longer a rate-limiting factor and the macroscopic envelopes of mushy layers are determined principally by heat transfer (Huppert and Worster 1985).

However, gradients of the interstitial salinity can drive convection in the interior of a mushy layer, and this is the dominant mechanism by which salt is delivered to the ocean during formation of sea ice: the salt flux from sea ice to ocean would be negligible in the absence of buoyancy forces, a fact that is insufficiently recognized in the parameterizations used in most current climate models.

The key ideas can be gleaned from the following simplified model. The equations governing an ideal mushy layer (Worster 1997) can be written as

$$\rho c_p\left(\frac{\partial T}{\partial t} + \mathbf{u}\cdot\nabla T\right) = k\nabla^2 T + \rho L\frac{\partial\phi}{\partial t},$$

$$(1-\phi)\frac{\partial S}{\partial t} + \mathbf{u}\cdot\nabla S = S\frac{\partial\phi}{\partial t},$$

$$\mu\mathbf{u} = \Pi(-\nabla p + \rho g),$$

together with the liquidus constraint and the continuity equation $\nabla \cdot \mathbf{u} = 0$. An approximate dimensionless form, when $\mathcal{S} \gg 1$, $\mathcal{C} \gg 1$ with $\mathcal{S}/\mathcal{C} = O(1)$, can be written as

$$\Omega \frac{D\theta}{Dt} = \nabla^2\theta,$$

$$\mathbf{u} = R_m(-\nabla p + \theta k),$$

$$\frac{\partial \phi}{\partial t} = -\mathcal{C}^{-1}\frac{D\theta}{Dt},$$

where $\theta = [T - T_B]/[T_L(S_0) - T_B] = [S - S_L(T_B)]/[S_0 - S_L(T_B)]$ and $D/Dt \equiv \partial/\partial t + \mathbf{u} \cdot \nabla$ is the material derivative following the Darcy flow. Note that the first two equations are de-coupled from the third, which determines the evolution of the solid fraction in terms of the redistribution of solute. In particular, the third equation shows that there is dissolution ($\partial\phi/\partial t < 0$) whenever there is flow from cooler to warmer regions within a mushy layer. This is the cause of the brine channels that form in sea ice, for example.

The first two equations admit a steady solution with a linear vertical temperature gradient and no flow in an infinite horizontal layer, as depicted in Figure 35.5a. Perturbations to this solution can be analyzed by considering the normal modes

$$\theta = 1 - z + \hat{\theta}(z)e^{i\alpha x + \sigma t},$$

$$\omega = \hat{\omega}(z)e^{i\alpha x + \sigma t},$$

where ω is the vertical component of the velocity field. If these expressions are substituted into the governing equations and the resulting equations are linearized in the perturbation quantities, we obtain

$$-\Omega\hat{\omega} = (D^2 - \alpha^2)\hat{\theta},$$

$$(D^2 - \alpha^2)\hat{\omega} = \alpha^2 Ra_m \hat{\theta},$$

which can be combined to give

$$(D^2 - \alpha^2)^2\hat{\omega} = -\alpha^2\Omega Ra_m\hat{\omega}.$$

For illustration, we solve this perturbation equation subject to boundary conditions that the upper surface is impermeable and at constant temperature, so that

$$\hat{\omega} = 0, \quad \hat{\theta} = 0 \Rightarrow D^2\hat{\omega} = 0 \quad (z = 0)$$

and that the lower boundary is held at constant pressure and heat flux, as is appropriate at a mush–liquid interface, so that

$$D\hat{\omega} = 0, \quad D\hat{\theta} = 0 \Rightarrow D^3\hat{\omega} = 0 \quad (z = 1).$$

The eigenfunctions of this system are

$$\hat{\omega}_n = \sin\left[\left(n + \frac{1}{2}\right)\pi(1 - z)\right],$$

with corresponding eigenvalues

$$\Omega Ra_m = \frac{\left[\left(n + \frac{1}{2}\right)^2 \pi^2 + \alpha^2\right]^2}{\alpha^2}.$$

The most unstable mode has $n = 0$ and

$$\Omega Ra_m = \frac{(\pi^2/4 + \alpha^2)^2}{\alpha^2}.$$

This neutral curve has a minimum value of $\Omega Ra_m = \pi^2$ at $\alpha = \pi/2$. As a rule of thumb, a mushy layer growing from a region of melt will begin to convect when its modified Rayleigh number ΩRa_m exceeds a value of about 10 (Worster 1992).

35.3.4 Numerical Analysis

Numerical studies of phase change have employed a range of algorithms and different approaches. When phase boundaries are involved, a choice has to be made whether to "fit" the boundaries or to "capture" them. In the former case, separate computational domains are established for the solid and the liquid or the mush and the liquid, for example, and conditions are employed to relate variables across the interface between the domains as

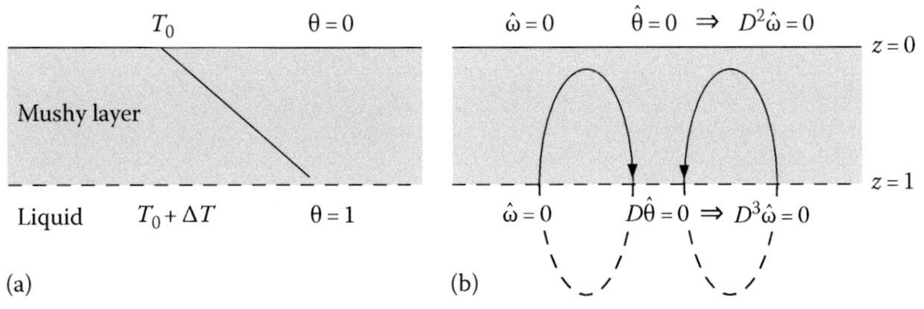

FIGURE 35.5 (a) A quiescent mushy layer above a deep liquid region with a linear temperature gradient imposed across it. (b) Small perturbations can lead to convective rolls, shown by the stream lines. Boundary conditions for the perturbed fields are shown.

well as to evolve the location of the interface. For solidification of a pure melt, for example, the diffusion equation would be applied to the solid domain, the Navier–Stokes equations and the advection–diffusion equation would be applied to the liquid domain, and an equilibrium constraint $T = T_m$ plus the Stefan condition could be applied at the solid–liquid interface. This mimics the approach usually taken in analytical studies, such as those described earlier.

Alternatively, equations are sought that apply, at least approximately, throughout the system, and the interface is subsequently located as a contour of one of the dependent variables. In the case of a pure melt of temperature $T_\infty > T_m$ brought into contact with a boundary of temperature $T_0 < T_m$, for example, the *enthalpy method* would start with uniform enthalpy $H(x, t = 0) = c_{ps}(T_\infty - T_m)$ throughout a fixed domain then, at each point, evolve the enthalpy according to

$$\frac{\partial H}{\partial t} = \frac{\partial}{\partial x}\left[k(\varphi)\frac{\partial T}{\partial x}\right],$$

where $k(\phi)$ is the conductivity, with the temperature T and the phase fraction ϕ determined by inverting

$$\begin{aligned} H &= c_{ps}(T - T_m), & \phi &= 0 \quad \text{if} & H &> 0, \\ H &= -\rho L\phi, & T &= T_m \quad \text{if} & -\rho L < H &< 0, \\ H &= c_{pl}(T - T_m) - \rho L, & \phi &= 1 \quad \text{if} & H &< -\rho L. \end{aligned}$$

Note that the phase fraction ϕ is an artifact of the method, related to the fact that the phase boundary would like to be between the fixed grid points of the numerical scheme and is not the same as the solid fraction ϕ of a mushy layer. Such methods can smear the solid–liquid interface over several grid points. The location of the phase boundary is usually identified with the contour $\phi = 0.5$. There are some drawbacks to such methods, but they can be outweighed by the convenience of computing on a single domain, particularly in higher dimensions if the phase boundary becomes convoluted.

Another single-domain approach is the *phase-field method*, which starts with a weak form of the governing equations and includes a gradient energy proportional to $|\nabla\phi|^2$, which leads to behavior that mimics the curvature undercooling associated with the Gibbs–Thomson effect. This method has been used extensively to study branched or dendritic growth of single crystals, including cases in which fluid flow is important (Boettinger et al. 2002).

An additional complication in using a single domain to compute solidification of systems involving mushy regions is the need for a single momentum equation that reduces to the Navier–Stokes equation in fully liquid regions and reduces to (or at least approximates) Darcy's equation in the interior of mushy regions. The most common choice is to use the Brinkmann equation, which adds Darcy friction to the Navier–Stokes equations with a permeability that tends to infinity as the solid fraction tends to zero. The two-domain approach for such problems has its own difficulties. Principally, because of the hyperbolic character of the differential operator acting on the solid fraction in the mushy layer equations, the treatment of mush–liquid interfaces depends crucially on whether the material flow is from liquid to mush or vice versa, which may not be known in advance of a computational time step.

As with all computational endeavors, there are horses for courses, and it pays to consider carefully the likely behavior of a system before choosing an appropriate algorithm.

35.3.5 Experimental Studies

Many solidifying systems of interest are opaque and have melting temperatures that are difficult or inconvenient for laboratory study. There have been significant advances in our understanding of the solidification of mixtures, in particular, made by studying the crystallization of aqueous salt solutions (Huppert 1990). Many of these are transparent, which has made possible visual observations of crystal habit and of convective flows in the melt. Flow observations have been made quite simply using dye streaks, shadowgraph images, or schlieren. These have revealed many fascinating phenomena, including double-diffusive convection of the melt (Figure 35.6) and plumes of residual fluid emanating from chimneys (called "brine channels" in the context of sea ice) formed by convection in mushy layers (Figure 35.7).

Temperature and composition (salinity) in solidifying systems can be measured using standard techniques: thermocouples or thermistors for temperature and conductivity probes or refractometers for concentration. More difficult to measure is the solid-fraction distribution in mushy regions. Probably the first direct measurement of solid fraction was made using X-ray tomography in a medical facility after completion of an experiment (Chen 1995). More recently, magnetic resonance imaging (MRI)

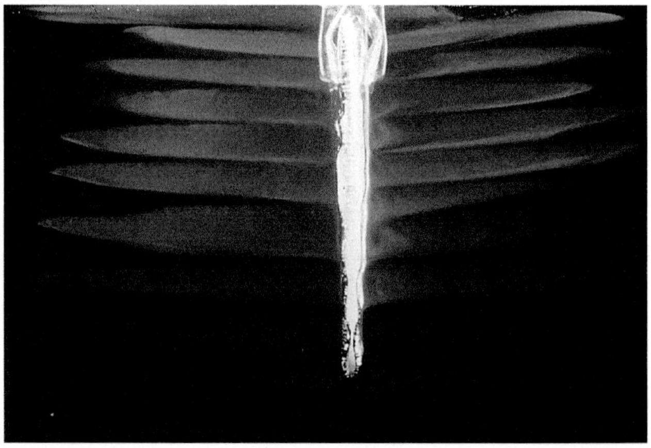

FIGURE 35.6 **(See color insert.)** A vertical block of ice melting into a warm, stratified salt solution. Double-diffusive convection results in the formation of nearly horizontal, nearly uniform layers of water that are colder and fresher than the environment.

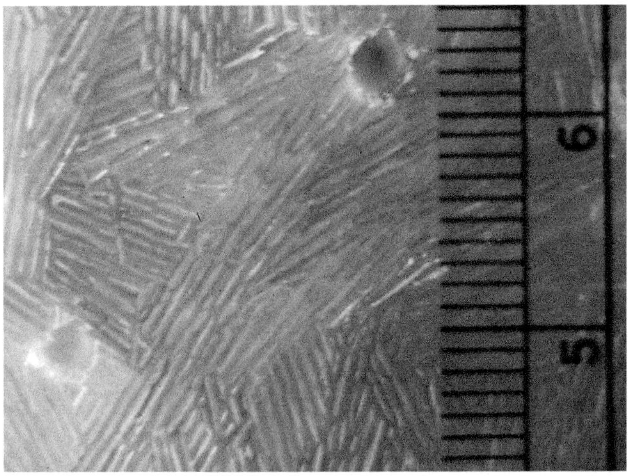

(a)

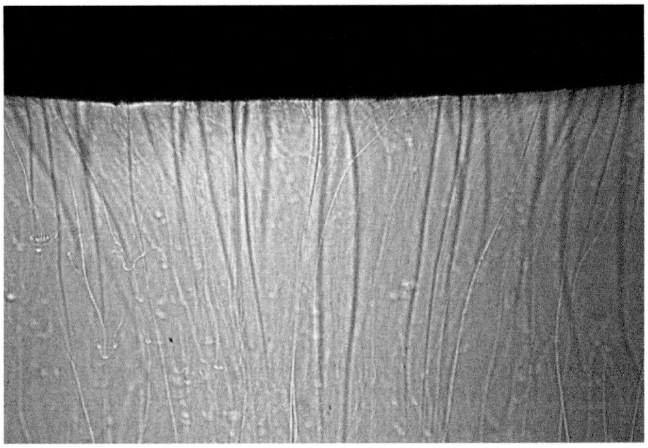

(b)

FIGURE 35.7 **(See color insert.)** (a) The underside of columnar sea ice, showing ice platelets spaced slightly less than a millimeter apart and two brine channels of approximately 3 mm diameter. (b) A shadow graph image of the side view of growing sea ice (the black region at the top), showing the plumes of dense brine that emanate from brine channels and deliver salt to the underlying ocean. The image is of a region approximately 20 cm wide.

has been used under laboratory conditions to measure the evolution of solid fraction as well as the internal structure of chimneys (Aussillous et al. 2006). MRI has also been used in field experiments, exploiting the Earth's magnetic field to measure the brine diffusivity and volume fraction of ice in thick multi-year sea ice with a vertical resolution of about 20 cm (Callaghan et al. 1999). The evolution of solid fraction in growing sea ice has been measured in situ with a vertical resolution of 1 cm by measuring the electrical impedance between pairs of horizontal wires around which the ice grew (Notz et al. 2005). Such non-destructive techniques as MRI or electrical impedance methods are important to develop if accurate remote monitoring of sea ice is to be achieved.

To this end, there is considerable effort in developing ways to sense the state of sea ice from orbiting satellites. Some variables, such as ice extent and ice concentration (area fraction covered by ice), are relatively straightforward to detect by electromagnetic methods. On the other hand, ice thickness is very difficult to determine, in part because of snow and in part because the reflectivity to the microwaves used depends on the electrical properties of sea ice, which are sensitive to its brine content. Currently, more accurate measurements of ice thickness are made using upward-looking sonar from submarines, but such data is sparse compared to the coverage that could in principle be obtained by satellite (Kwok and Rothrock 2009).

35.4 Applications

As indicated in the Introduction, there is a vast range of applications involving fluid flow coupled with phase change. Here we sketch a few interesting examples.

The inner core of the Earth, of current radius 1221 km, representing 0.5% by volume and 2% by mass of the Earth, grows slowly by deposition of almost pure iron from the fluid outer core of molten iron and roughly 10% of impurities such as oxygen, sulfur, and nickel. The deposition rate is 2×10^6 kg s^{-1}, approximately two orders of magnitude larger than the total iron and steel production on the surface of the Earth, yet the radius of the inner core currently increases at only about 1 mm per year. The compositional convection driven by the relatively light fluid (enriched in impurities) released during solidification through an electrically conducting medium maintains the magnetic field of the Earth. Without this field, life would be impossible because the field deflects the solar wind that would otherwise blow away the atmosphere. The inner core was initiated approximately 2×10^9 years ago and will become completely solidified after another 2×10^{10} years (by which time the sun will have burned itself out). The growth of the inner core is currently controlled by a balance between the decrease in heat content of the cooling outer core and the heat flux out at the top because the effective Stefan number is relatively small.

Solidification plays an essential role in igneous petrology—the study of rocks derived from a liquid melt of volcanic origin. The rich multicomponent melts solidify in magma chambers (large storage reservoirs of partially molten rock that power volcanic eruptions), in volcanic conduits, and in lava flows. An important component, even though present in small amounts (of order 1% by weight) is dissolved gas (such as H_2O and CO_2) which can exsolve due to pressure release and temperature change and alter the form of the resulting solid as well as drive vigorous motion. This is an area of extreme laboratory investigations, using very high temperatures and pressures. An example of considerable financial importance is the rather exotic magma formed up to 450 km beneath the surface of the Earth that rises in so-called kimberlite pipes and is the major source of the world's diamonds (Sparks et al. 2009).

A vertical surface of ice surrounded by water stratified by salt melts to produce a layered structure, as shown in Figure 35.6. While the relatively fresh meltwater rises in a thin, turbulent boundary layer adjacent to the ice, cooling extends further into the liquid region and drives a downward flow that detrains into the stratified exterior when it reaches a level of neutral buoyancy. Mixing between these counter-flowing layers increases the salinity of the rising meltwater and takes some of it into the horizontal layers, which thus ruins the potential to harvest freshwater simply from icebergs. The thickness of the intruding layers can be determined quantitatively; and the theory has been compared successfully against numerous laboratory experiments (Huppert and Turner 1981) and the layers formed adjacent to the Erebus Glacier Tongue in the Antarctic.

The solidification of multicomponent melts in molds is an essential and very successful aspect of the metallurgical industry (Hurle 1993). Amorphous (noncrystalline) alloys called glasses can be produced by spin casting at rapid cooling rates of millions of degrees per second. Glasses also occur as a result of rapid volcanic eruptions, both within volcanic conduits and in the resultant hot plume cooled by contact with the atmosphere. By contrast, epitaxy is a process of slow deposition of a thin film onto a monocrystalline substrate from either gaseous or liquid precursors. The substrate acts as a seed crystal and the epitaxial film adopts its crystalline structure and orientation. Epitaxy is widely used to manufacture silicon-based semiconductors such as bipolar junction transistors within integrated-circuit microcontrollers and microprocessors. At a more prosaic level, concrete solidifies on mixing and reacting with water to be the world's most used synthetic material, approximately 8 km^3 being manufactured each year (more than 1 m^3 per person per year) in a multibillion-dollar industry that contributes approximately 5% of the world's anthropogenic emission of CO_2.

Clathrate hydrates, first documented by Sir Humphrey Davy in 1810, consist of a molecular cage of solid ice enclosing gases such as water vapor, methane, carbon dioxide, and hydrogen sulfide (Buffett 2000). They form at high pressures, and on Earth they exist within sea-floor sediments along deep continental margins. In the Martian ice caps, they sequester large amounts of water vapor and carbon dioxide from the atmosphere. It has been estimated that on Earth about 10^{12} kg of carbon is trapped in oceanic sediments as methane hydrates with a smaller amount in permafrost regions, a total that represents the largest hydrocarbon source on Earth. Because methane is such a strong greenhouse gas, the release of large quantities of methane from these hydrates could have serious consequences for global climate.

Intermolecular disjoining forces exist between ice and many other materials when they are brought into contact. These cause the ice to melt (*pre-melt* because this phenomenon occurs at temperatures below the bulk melting temperature) in a thin surface film (of the order of tens of nanometers) and generate high pressures between the ice and its substrate that cause water to be sucked toward colder regions of partially frozen, saturated soils (Wettlaufer and Worster 2006). This is the underlying cause of frost heave, which results in potholes in roads and is a major geomorphological agent, fracturing rocks and forming various types of patterned ground.

Significant attention is currently being paid to the fate of sea ice, which may soon be absent from the Arctic Ocean during summer months, though the ocean is likely to remain ice-covered in winter for the foreseeable future. The seasonal waxing and waning of the sea-ice cover drives a high-latitude distillation process in the oceans: salt is rejected by the growing sea ice and is carried buoyantly into the mixed layer or deeper into the abyss; melting sea ice forms a relatively fresh cap on the oceans that resists convective overturning. After the initial buoyancy-driven instability described earlier, brine convecting in the interstices of sea ice dissolves the ice matrix to form channels that provide the principal conduit for salty plumes to be injected into the ocean (Figure 35.7, Wettlaufer et al. 1997).

On the surface of very young sea ice, when it is only a few centimeters thick, beautiful fern-like clusters of ice crystals called frost flowers can form. Concentrated brine is wicked up from the interstices of the underlying sea ice to form a liquid surface layer on frost flowers in which bromides from seawater are converted to bromine monoxide, which is carried into the atmosphere and contributes to the depletion of ozone. High in the atmosphere, the pre-melted surfaces of ice crystals enhance a wide range of reactions, making ice a key component of atmospheric chemistry.

35.5 Major Challenges

Interactions between phase change and fluid flow involve physics on a vast range of scales, from the nanoscale of pre-melted liquid films through the microscale of dendritic snowflakes and the interstices of mushy layers, to the macroscales of brine channels and the megascales of ocean circulations. The major challenges facing this branch of science relate to understanding the physics of these multiscale processes sufficiently to codify them efficiently within predictive mathematical models. There are important questions relating to the evolution of sea ice and the associated brine fluxes that contribute significantly to the thermohaline circulation driving, together with wind stresses, the Gulf Stream, and other major oceanic features that affect regional climate.

The mathematics of phase change is relevant to other systems that do not involve thermodynamic phase change. Examples include predicting the shape and evolution of alluvial fans, where the advancing phase is the sedimentary bed; calculating the heat and salt fluxes across double-diffusive interfaces, where the phases are turbulent and quiescent fluid regions; and establishing the location and stability of marine-ice-sheet grounding lines, separating grounded ice sheets from floating ice shelves. In all these examples, the major challenge is to understand the physical processes sufficiently to derive tractable mathematical models capable of making robust predictions.

It is clear that flows involving phase change are an integral part of many phenomena and processes that affect our environment. Their understanding requires collaborations across many scientific disciplines, and it is essential that the mathematical and numerical approaches that are increasingly used to make predictions of our future environment are tested against well controlled and characterized experiments.

Acknowledgments

We are grateful to Jerome Neufeld, Dominic Vella, and John Wettlaufer for their helpful comments on a draft of the manuscript. Herbert E. Huppert is supported by a Royal Society Wolfson Merit Award, for which he is grateful.

References

Aussillous, P., Sederman, A.J., Gladden, L.F., Huppert, H.E., and Worster, M.G. 2006. Magnetic resonance imaging of structure and convection in solidifying mushy layers. *J. Fluid Mech.* **552**, 99–125.

Boettinger, W.J., Warren, J.A., Beckermann, C., and Karma, A. 2002. Phase-field simulation of solidification. *Annu. Rev. Mater. Res.* **32**, 163–194.

Buffett, B.A. 2000. Clathrate hydrates. *Annu. Rev. Earth Planet. Sci.* **28**, 477507.

Callaghan, P.T., Dykstraa, R., Ecclesa, C.D., Haskellb, T.G., and Seymour, J.D. 1999. A nuclear magnetic resonance study of Antarctic sea ice brine diffusivity. *Cold Reg. Sci. Technol.* **29**(2), 53–171.

Carslaw, H.S. and Jaeger, J.C. 1959. *Conduction of Heat in Solids*. Oxford University Press, Oxford, U.K.

Chen, C.F. 1995. Experimental study of convection in a mushy layer during directional solidification. *J. Fluid Mech.* **293**, 81–98.

Davis, S.H. 2001. *Theory of Solidification*. Cambridge University Press, Cambridge, U.K.

Gill, A.E. 1982. *Atmosphere–Ocean Dynamics*. International Geophysics Series, Vol. 30. Academic Press, New York.

Huppert, H.E. 1990. The fluid mechanics of solidification. *J. Fluid Mech.* **212**, 209–240.

Huppert, H.E. and Turner, J.S.T. 1981. Double-diffusive convection. *J. Fluid Mech.* **106**, 299–329.

Huppert, H.E. and Worster, M.G. 1985. Dynamic solidification of a binary melt. *Nature* **314**, 703–707.

Hurle, D.T.J. (ed.) 1993. *Handbook of Crystal Growth*. Elsevier, Amsterdam, the Netherlands.

Kwok, R. and Rothrock, D.A. 2009. Decline in Arctic sea ice thickness from submarine and ICESat records: 19582008. *Geophys. Res. Lett.* **36**, L15501.

Notz, D., Wettlaufer, J.S., and Worster, M.G. 2005. A non-destructive method for measuring salinity and solid fraction of growing sea ice in situ. *J. Glac.* **51**(172), 159–166.

Sparks, R.S.J., Brooker, R.A., Field, M., Kavanagh, J., Schumacher, J.C., and Walter, M.J. 2009. The nature of erupting kimberlite melts. *Lithos* **112**, 429–438.

Stevens, B. 2005. Atmospheric moist convection. *Annu. Rev. Earth Planet. Sci.* **33**, 605–643.

Turner, J.S. 1979. *Buoyancy Effects in Fluids*. Cambridge University Press, Cambridge, U.K.

Wettlaufer, J.S. and Worster, M.G. 2006. Premelting dynamics. *Annu. Rev. Fluid Mech.* **38**, 427–452.

Wettlaufer, J.S., Worster, M.G., and Huppert, H.E. 1997. Natural convection during solidification of an alloy from above with application to the evolution of sea ice *J. Fluid Mech.* **344**, 291–316.

Worster, M.G. 1992. Instabilities of the liquid and mushy regions during solidification of alloys. *J. Fluid Mech.* **237**, 649–669.

Worster, M.G. 1997. Convection in mushy layers. *Annu. Rev. Fluid Mech.* **29**, 91–122.

Worster, M.G. 2000. Solidification of fluids. In *Perspectives in Fluid Dynamics*, eds. G.K. Batchelor, H.K. Moffatt, and M.G. Worster, pp. 393–496. Cambridge University Press, Cambridge, U.K.

36

Turbulent Gas Transfer across Air–Water Interfaces

Satoru Komori
Kyoto University

36.1 Introduction

Turbulent gas transfer across air–water interfaces is used as a separation technology in many industrial processes, including gas absorption, evaporation, and condensation, and it often occurs in geophysical flows such as in oceans, lakes, and rivers. It is, therefore, of great practical interest to investigate the gas transfer mechanism across air–water interfaces both in designing and controlling industrial gas–liquid contact-type equipment and in discussing geophysical and environmental problems such as air–sea gas exchange and eutrophication of lake and river waters. Recently the accurate estimation of the gas transfer rate has attracted special interest in the global climate change problem related to the exchange of carbon dioxide (CO_2) between atmosphere and oceans.

For low-solubility gases such as CO_2 and O_2 (but not for extremely soluble gases such as NH_3), the resistance on the water side usually dominates gas transfer across air–water interfaces and therefore the gas transfer mechanism has to be clarified in relation to the turbulence structure on the water side. In a turbulent flow with an unsheared flat air–water interface, turbulence is mainly generated in the bulk water flow away from the interface, as shown in Figure 36.1a. Komori et al. (1982) first measured turbulence quantities near the free surface in a fully developed open-channel flow by means of a laser-Doppler velocimeter (LDV), and suggested that energy-containing eddies are extremely elongated longitudinally and laterally by the presence of the free surface and intermittently renew the free surface. Furthermore, Komori et al. (1989) clarified that the surface-renewal eddies originate in bursting motions generated in the bottom wall region, that is, energy-containing turbulent eddies ejected by bursting phenomena rise from the buffer region toward the interfacial (outer) region and arrive at and renew the free surface. They also showed that the surface-renewal eddies control the CO_2 transfer across the free surface on the liquid side and the gas transfer coefficient on the water side k_L is proportional to the square root of the surface-renewal frequency.

On the other hand, for wind-driven turbulence with a sheared interface (see Figure 36.1b), turbulence structure on the water side should be discussed together with turbulent motions on the air side which induce turbulence on the water side. Most geophysical studies have accumulated measurements of the gas transfer coefficient on the water side k_L through laboratory and field experiments (e.g., Liss and Merlivat, 1986; Wanninkhof, 1992), but from the fluid-mechanical point of view they have not clarified the gas transfer mechanism across wavy-sheared air–water interfaces. In the engineering field, a number of studies have measured k_L in thin-film flows and have also discussed the effect of the wind shear on the gas transfer only in terms of the correlation between k_L and friction velocity at the bottom wall (e.g., McCready and Hanratty, 1985). However, the relation between turbulence structure and gas transfer has not been fully clarified since very complicated turbulence is generated by the shear in both wall and free-surface regions in thin-film flows. On the other hand, Komori et al. (1993a, 1999, 2010) clarified that organized motions are intermittently generated on the air side in front of the wave crest in a wind-wave tank and there the motions induce surface-renewal motions in the water flow through high wind shear on the interface, which controls gas transfer across a wavy-sheared air–water interface. Furthermore, they showed that k_L increases with wind speed and it has a kink in the middle

Handbook of Environmental Fluid Dynamics, Volume One, edited by Harindra Joseph Shermal Fernando. © 2013 CRC Press/Taylor & Francis Group, LLC. ISBN: 978-1-4398-1669-1.

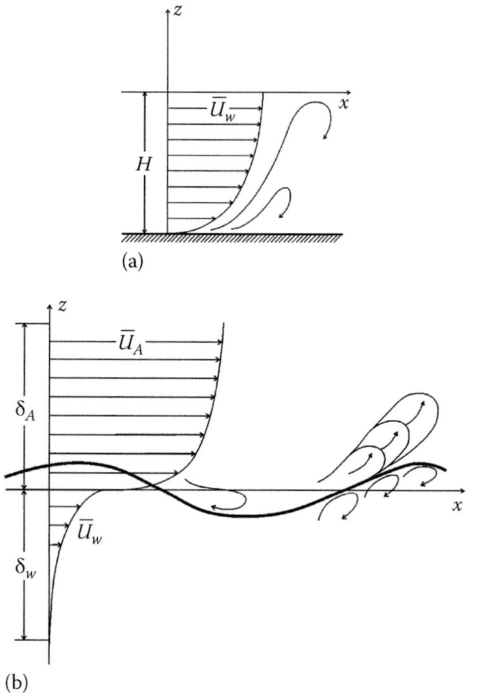

FIGURE 36.1 Sketch of turbulent motions with (a) unsheared and (b) sheared interfaces.

wind-speed region where the turbulence structure changes from a streaky structure, due to longitudinal bursting like vortices in low wind-speed region, to a patchy spanwise structure due to breaking waves with ripples in the high wind-speed region.

In this chapter, the gas transfer mechanism at these two sheared and unsheared interfaces will be discussed together with the fundamental turbulence structure near the interface.

36.2 Principles (Gas Transfer Coefficient)

The nondimensional governing equations for an incompressible Newtonian fluid flow with air–water gas transfer are the equation of continuity, the Navier–Stokes (N–S) equation, and conservation equation of passive mass, using the Einstein summation convection:

$$\frac{\partial U_i}{\partial x_i} = 0, \tag{36.1}$$

$$\frac{\partial U_i}{\partial t} + U_j \frac{\partial U_i}{\partial x_j} = -\frac{\partial p}{\partial x_i} + \frac{1}{Re}\frac{\partial^2 U_i}{\partial x_j \partial x_j} + \frac{1}{Fr}\delta_{i3}, \tag{36.2}$$

$$\frac{\partial C}{\partial t} + U_j \frac{\partial C}{\partial x_j} = \frac{1}{Re \cdot Sc}\frac{\partial^2 C}{\partial x_j \partial x_j}, \tag{36.3}$$

where

- U_i is the ith component of the velocity vector (i = 1, 2, and 3 denote the streamwise, spanwise, and vertical directions, respectively)
- p is the pressure
- δ_{ij} is the Kronecker's delta
- C is the *concentration* of passive mass

The nondimensional parameters, Re, Sc, and Fr based on the reference length L_0 and velocity U_0, are defined as

$$Re = \frac{U_0 L_0}{\nu}, \quad Sc = \frac{\nu}{D}, \quad Fr = \frac{U_0^2}{g L_0}, \tag{36.4}$$

where

- ν is the kinematic viscosity
- D is the molecular diffusivity of mass
- g is the acceleration of gravity

On the air–water interface, two boundary conditions should be satisfied as shown in Figure 36.2. One is the kinematic boundary condition that describes the Lagrangian behavior of a fluid particle on the free surface:

$$\frac{\partial G}{\partial t} + U_i \frac{\partial G}{\partial x_i} = 0, \tag{36.5}$$

where $G(x_1, x_2, x_3, t)$ is the function representing the deformation of the interface. When the free surface is not overturning or breaking due to strong wind shear, that is, for wind speeds less than several meters per second above the air–water interface, the kinematic boundary condition of Equation 36.5 at the free surface can be rewritten using a single-valued function f of x_1 and x_2:

$$\frac{\partial f}{\partial t} + U_1 \frac{\partial f}{\partial x_1} + U_2 \frac{\partial f}{\partial x_2} = U_3, \tag{36.6}$$

where f is defined by

$$G(x_1, x_2, x_3, t) = f(x_1, x_2, t) - x_3 = 0. \tag{36.7}$$

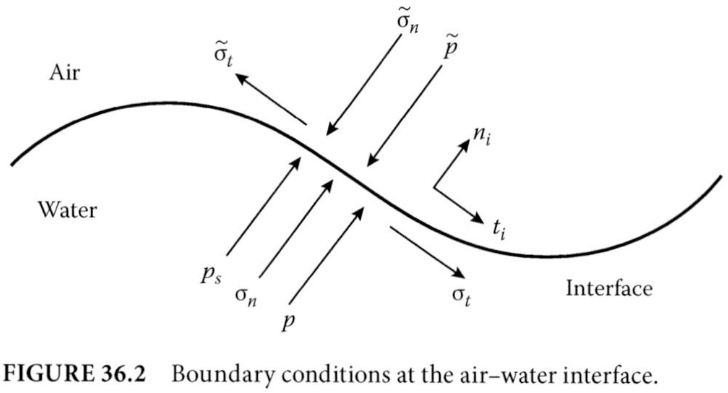

FIGURE 36.2 Boundary conditions at the air–water interface.

The other is the dynamic boundary condition which is determined from the balance of stresses acting on the interface in the normal and tangential directions:

$$p+\sigma_n+p_s=\tilde{p}+\tilde{\sigma}_n,\quad \sigma_n=\frac{1}{Re}e_{ij}n_jn_i,\quad p_s=\gamma k_m, \tag{36.8}$$

$$\sigma_t=\tilde{\sigma}_t,\quad \sigma_t=\frac{1}{Re}e_{ij}n_jt_i, \tag{36.9}$$

where σ_n, σ_t, p_s, e_{ij}, γ, k_m, n_i, and t_i are the normal components of the viscous force vector, the tangential component of the viscous force vector, the pressure variation due to the surface tension, the deformation rate tensor, the surface tension coefficient, the mean curvature of the free surface, the wall normal unit vector, and the wall tangential unit vector, respectively. The parameters with and without the tilde "~" denote the values on the air and water sides, respectively. The passive mass conservation equation (36.3) is needed only on the water side, since a gas such as CO_2 is generally homogeneously mixed on the air side. Details of the previous equations are described in Komori et al. (1993b).

The gas flux at the air–water interface per unit area is given by

$$F=\frac{1}{A}\int_{interface}D\frac{\partial C}{\partial n}dS\approx-\overline{wc}\big|_{suf}, \tag{36.10}$$

where

n is the normal direction with respect to the air–water interface

A is the surface area of the interface

$\overline{wc}\big|_{suf}$ is the vertical turbulent gas flux in the immediate vicinity of the free surface

On the other hand, F is given using the gas transfer coefficient on the water side k_L:

$$F=k_L\Delta C=k_LS\Delta P, \tag{36.11}$$

where

ΔC and ΔP are the differences of concentration and partial pressure of gas between the interface and bulk water, respectively

S is the solubility of gas into water

From (36.10) and (36.11), k_L is obtained by

$$k_L=\frac{1}{A\Delta C}\int_{interface}D\frac{\partial C}{\partial n}dS\approx-\frac{\overline{wc}\big|_{suf}}{\Delta C}. \tag{36.12}$$

As the concentration C is determined by (36.3), all the effects of turbulence are summarized in k_L when we discuss gas transfer across air–water interfaces using (36.11).

36.3 Methods of Analysis (Simulation and Experiments)

Gas transfer across air–water interfaces can be perfectly described by solving the governing equations (36.1 through 36.3) with boundary conditions (36.5), (36.8), and (36.9). A method for directly solving the governing equations without using any turbulence model is called "direct numerical simulation (DNS)," which has been recently applied to interfacial turbulent flows with not only an unsheared interface (Komori et al., 1993b) but also a wavy interface sheared by wind speeds less than several meters per second (Fulgosi et al., 2003; Lin et al., 2008; Komori et al., 2009, 2010). If the concentration field can be precisely described by the DNS, the gas flux F and gas transfer coefficient k_L can be estimated by (36.10) and (36.12). However, it is very difficult to apply the DNS to practical interfacial flows with gas transfer since the Batchelor scale (i.e., the smallest scale of concentration field) is too much smaller than the Kolmogorov scale to resolve the concentration field of high Schmidt number gases such as CO_2. Even for the velocity field, the DNS is inadequate to high Reynolds number flows and intensive breaking waves with ripples in high wind-speed conditions. Therefore, the DNS should be applied only to investigate the fundamental turbulence structure and gas transfer mechanism in low Reynolds and Schmidt number flows and to explain the measurements supplementarily. In order to apply any numerical simulation to interfacial flows with high Re and Sc numbers, a more sophisticated method should be developed in future study.

The best way to discuss gas transfer across air–water interfaces is to perform laboratory or field experiments and to measure the gas transfer coefficient k_L together with turbulent quantities such as turbulence intensities. In oceans, lakes, and rivers, we can directly measure the vertical turbulent gas flux $\overline{wc}\big|_{suf}$ in the vicinity of the free surface by means of the combination of a CO_2 analyzer based on infrared spectroscopy with a sonic anemometer. This measuring technique is called the "eddy-correlation technique." In open-channel or wind-wave tank experiments, we cannot directly measure $\overline{wc}\big|_{suf}$ because the spatial resolution of a CO_2 analyzer cannot resolve energy-containing eddies in laboratory channels and tanks; therefore, we are forced to measure the gas flux by taking the mass balance in a gas flow or liquid flow (see Komori et al., 1989, 1993a, 1999). In contrast, most previous laboratory experiments estimated the gas flux by taking the time derivative of concentration in a channel or tank under the assumption of homogeneity of the concentration field in the vertical direction. However, such a method cannot accurately measure the gas flux at a local station (fetch) in the streamwise direction, especially in the low wind-speed region.

36.4 Examples (Gas Transfer and Turbulence Structure)

36.4.1 Unsheared Air–Water Interface Case

36.4.1.1 Turbulence Structure

A typical turbulent flow with an unsheared gas–liquid interface is an open-channel flow. In an open-channel flow, turbulent organized motions are generated by bursting phenomena in the buffer region above the bottom wall, and most of the energy-containing eddies reach the free surface as shown in Figure 36.1a (Komori et al., 1989). The eddies promote gas transfer at the free surface as surface-renewal eddies. A fundamental difference between a rigid wall and a free surface is that the free surface satisfies the slip boundary condition in (36.8) and (36.9). That is, the vertical velocity component is damped to almost zero at the air–water interface as well as the rigid wall, whereas the streamwise velocity component is not damped by the interface. Figure 36.3 shows the vertical distributions of turbulence intensities and the Reynolds stress measured by LDV near the free surface in an open-channel flow with water depth H. Here, the turbulence intensities are normalized by the friction velocity at the bottom wall. In fact, both the vertical velocity fluctuation and the Reynolds stress are rapidly damped by the free surface, whereas streamwise and spanwise velocity fluctuations are promoted in the vicinity of the interface. The trends of the turbulence intensities suggest that in the free surface region the energy of the vertical velocity fluctuation is redistributed into the energy of the streamwise and spanwise velocity fluctuations through the mean pressure–strain terms $(2/\rho)\overline{p(\partial u/\partial x)}$, $(2/\rho)\overline{p(\partial v/\partial y)}$, and $(2/\rho)\overline{p(\partial w/\partial z)}$ in the following transport equations for mean-squared values of three velocity component fluctuations:

$$\frac{D\overline{u^2}}{Dt} = -2\overline{uv}\frac{\partial U}{\partial y} - 2\nu\overline{\frac{\partial u}{\partial x_k}\frac{\partial u}{\partial x_k}} + 2\overline{\frac{p}{\rho}\frac{\partial u}{\partial x}} - \frac{\partial}{\partial x_k}\left[\overline{u^2 u_k} - \nu\frac{\partial \overline{u^2}}{\partial x_k} + 2\delta_{1k}\frac{\overline{up}}{\rho}\right] \quad (36.13)$$

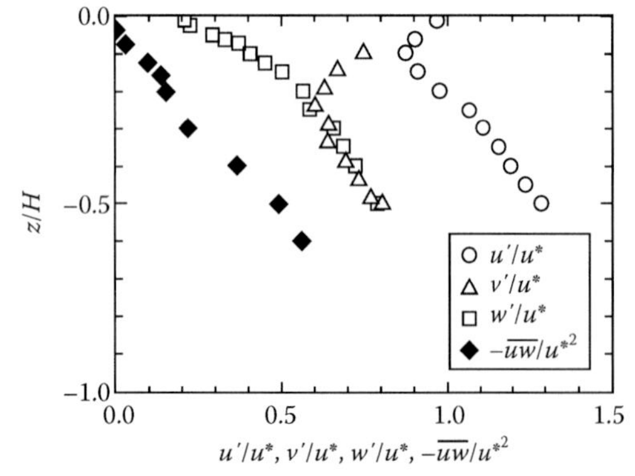

FIGURE 36.3 Vertical distributions of the turbulence intensities and the Reynolds stress below the unsheared interface in an open-channel flow. (From Komori, S. et al., *Int. J. Heat Mass Transfer*, 25, 513, 1982. With permission.)

$$\frac{D\overline{v^2}}{Dt} = -2\nu\overline{\frac{\partial v}{\partial x_k}\frac{\partial v}{\partial x_k}} + 2\overline{\frac{p}{\rho}\frac{\partial v}{\partial y}} - \frac{\partial}{\partial x_k}\left[\overline{v^2 u_k} - \nu\frac{\partial \overline{v^2}}{\partial x_k} + 2\delta_{2k}\frac{\overline{vp}}{\rho}\right] \quad (36.14)$$

$$\frac{D\overline{w^2}}{Dt} = -2\nu\overline{\frac{\partial w}{\partial x_k}\frac{\partial w}{\partial x_k}} + 2\overline{\frac{p}{\rho}\frac{\partial w}{\partial z}} - \frac{\partial}{\partial x_k}\left[\overline{w^2 u_k} - \nu\frac{\partial \overline{w^2}}{\partial x_k} + 2\delta_{3k}\frac{\overline{wp}}{\rho}\right] \quad (36.15)$$

Of course, the summation of the three pressure–strain terms is equal to zero from the equation of continuity. Although the redistribution of energy through the pressure–strain terms is not verified by experiments because of the difficulty of instantaneous pressure measurement, the DNSs of open-channel flows with almost flat interfaces by Lombardi et al. (1996) and Komori et al. (1993b) verified the energy transfer through pressure–strain terms. In addition to the DNS, closure models such as $k - \varepsilon$ or Reynolds stress closure have been applied to open-channel flows with a flat free surface and have shown good agreement with measurements (see Gibson and Rodi, 1989; Nakayama et al., 2003). Recently, a large-eddy simulation (LES) has been successfully applied to an open-channel flow with higher Reynolds number (Calmet and Magnaudet, 2003).

36.4.1.2 Gas Transfer

The gas transfer mechanism at a flat interface has been mainly studied through measurements of gas flux, since it is difficult to estimate the gas flux by DNS because of high *Sc* numbers. Figure 36.4 shows the gas transfer coefficient k_L for an almost flat interface obtained by CO_2 absorption experiments of Komori et al. (1989) in an open-channel flow together with k_L for a sheared interface in a wind-wave tank. The measured values of k_L in an open channel are well correlated by the frequency of

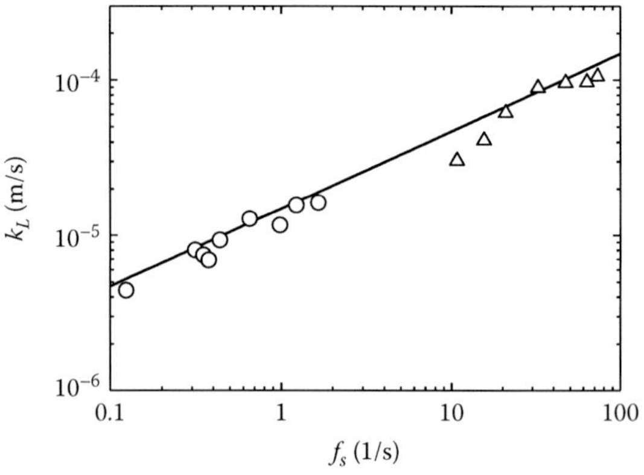

FIGURE 36.4 The gas transfer coefficient k_L obtained by CO_2 transfer experiments against the frequency of appearance of energy-containing surface-renewal eddies f_S. ○, in an open-channel flow with unsheared interface; △, in a wind-driven turbulent flow with sheared interface. (From Komori, S. et al., *J. Fluid Mech.*, 203, 103, 1989; Komori, S. et al., *J. Fluid Mech.*, 249, 161, 1993a. With permission.)

appearance of surface-renewal eddies originating in bursting events in the bottom wall region. The correlation is given by

$$k_L = 0.34\sqrt{D_L f_S}, \qquad (36.16)$$

where

f_S is the surface-renewal frequency

D_L is the *molecular* diffusivity of CO_2 in the water

Here f_S was determined using a conditionally averaging method called the "variable interval time-averaging (VITA) technique." The constant 0.34 in (36.16) depends on the cleanness of the interface and corresponds to the value for tap water. The value 1.3 is regarded as the value for a perfectly clean interface (Komori et al., 1999). A series of laboratory experiments by Komori et al. (1982, 1989, 1990) showed that energy-containing (large-scale) eddies control the gas transfer, and these gas transfer mechanisms have also been confirmed by a sophisticated LES that simulates an open-channel flow with high *Sc* number (Wang et al., 2005). On the other hand, Moog and Jirka (1999) indicated that small-scale eddies play a larger rule in gas transfer at high Reynolds numbers. Whether the gas transfer is controlled by large-scale eddies or small-scale eddies in environmental flows with high Reynolds numbers is still controversial. However, Moog and Jirka (1999) did not directly measure the local gas flux; they estimated the gas flux from the time derivative of concentration under the rough assumption of a homogeneous concentration profile in the vertical direction. Furthermore, k_L in various flows, including a river current, is well correlated with the mean surface water velocity independent of the Reynolds number (Komori et al., 1982). The result suggests that the surface-renewal frequency f_S is determined only by an outer parameter such as the surface velocity, and the frequency of bursting motions in the bottom wall region which are sources of surface-renewal motions may be scaled with the outer parameters such as the free-stream velocity at the outer edge of the turbulent boundary layer.

36.4.2 Sheared Wavy Gas–Liquid Interface Case

36.4.2.1 Turbulence Structure

A typical flow with a sheared interface is an oceanic flow with wind waves. In the free surface region below the wavy air–water interface, turbulence is driven by wind shear acting on the interface, as shown in Figure 36.1b. Such wind-driven turbulence can be simulated in a wind-wave tank. Therefore, many field and laboratory studies have been conducted in both oceans and laboratory wind-wave tanks in order to clarify the structure of wind-driven turbulence. The turbulence structure depends on the fetch (the distance from where the wind begins to contact the interface); fully developed conditions with constant wave height and length cannot be attained unless the fetch reaches several kilometers even for wind speeds of several meters per second. In addition, a credible parameter for correlating turbulence quantities has not yet been proposed. Therefore, we mainly discuss the turbulence structure and gas transfer at short fetches of several meters observed in wind-wave tanks.

When the free-stream wind velocity U_∞ is lower than several meters per second and the wave height is small, turbulence structures like wall turbulence on a flat rigid wall are observed in both air and water flows. For this low wind-speed region, the DNS based on the arbitrary Lagrangian-Eulerian formulation (ALE) method by (36.6) and (36.7) in Section 36.2 is applicable. It shows a realistic configuration of the wavy interface with micro-breaking waves, that is, small ripples, as shown in Figure 36.5 (Komori et al., 2009, 2010). Figures 36.6a and 36.7 show the instantaneous surface temperature at U_∞ = 3.1 m/s measured in a wind-wave tank and the instantaneous virtual gas flux with Sc = 1.0 at U_∞ = 5.2 m/s predicted by the DNS, respectively (Komori et al., 2009). Both distributions suggest the existence of streaky structures in the streamwise direction. In fact, the prediction of the instantaneous second invariant Q on the water side in Figure 36.8 shows the presence of typical horseshoe vortices similar to bursting motions over the flat rigid wall. Here the second invariant Q is defined as

$$Q = \frac{1}{2}(r_{ij}r_{ij} - s_{ij}s_{ij}), \qquad (36.17)$$

where r_{ij} and s_{ij} are given by

$$r_{ij} = \frac{1}{2}\left(\frac{\partial u_i}{\partial x_j} - \frac{\partial u_j}{\partial x_i}\right)$$
$$s_{ij} = \frac{1}{2}\left(\frac{\partial u_i}{\partial x_j} + \frac{\partial u_j}{\partial x_i}\right). \qquad (36.18)$$

The positive Q ($Q > 0$) means that vortices are present; higher Q corresponds to stronger vortices.

When the free-stream wind velocity U_∞ increases beyond several meters per second, the turbulence structure begins to

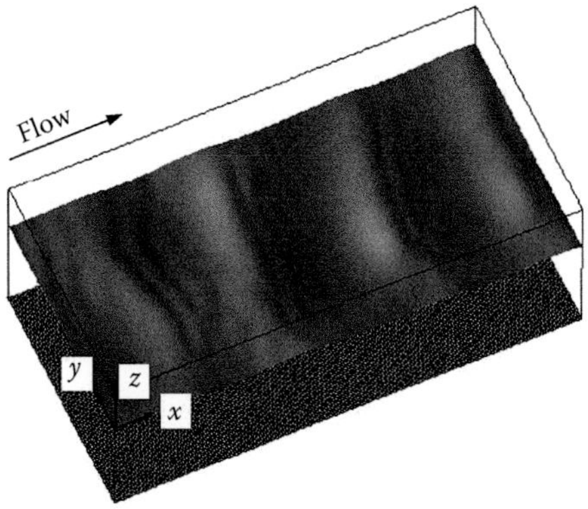

FIGURE 36.5 Instantaneous configuration of wind waves predicted by the DNS at U_∞ = 5.2 m/s. (From Komori, S. et al., *J. Turbulence*, 11, 1, 2010. With permission.)

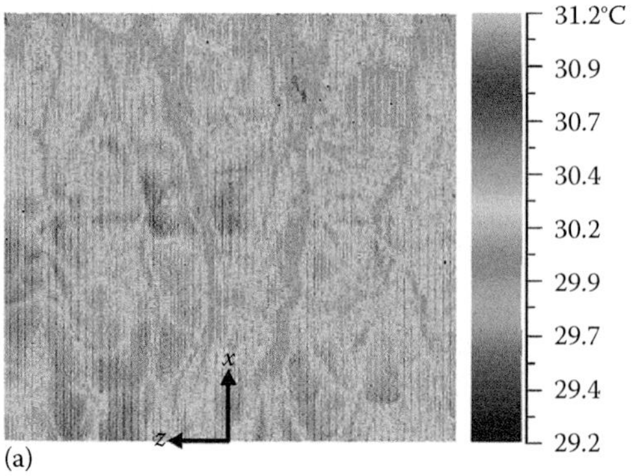

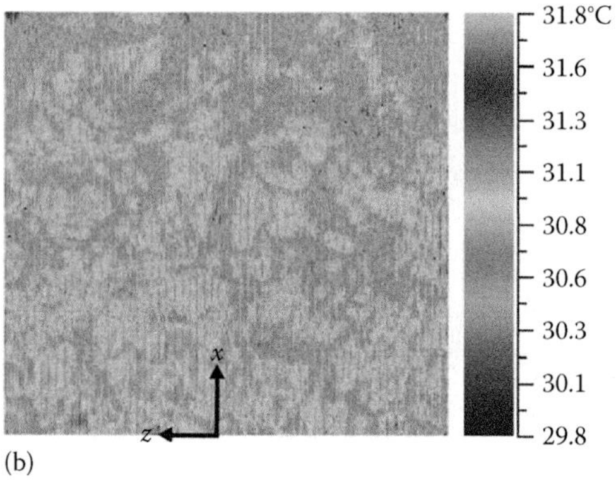

FIGURE 36.6 (See color insert.) Instantaneous surface temperature at (a) U_∞ = 3.1 m/s and (b) U_∞ = 13.2 m/s measured by a high-speed scanning infrared thermometer in a wind-wave tank. (From Komori, S. et al., Heat and mass transfer across the wavy sheared interface in wind-driven turbulence, In *Proceedings of 6th International Symposium on Turbulent Shear Flow and Phenomena*, Eds. Kasagi, N., Eaton, J.K., Friedrich, R., Humphrey, J.A.C., and Sung, H.J., Vol. II, pp. 399–407, Seoul National University, Seoul, Korea, 2009. With permission.)

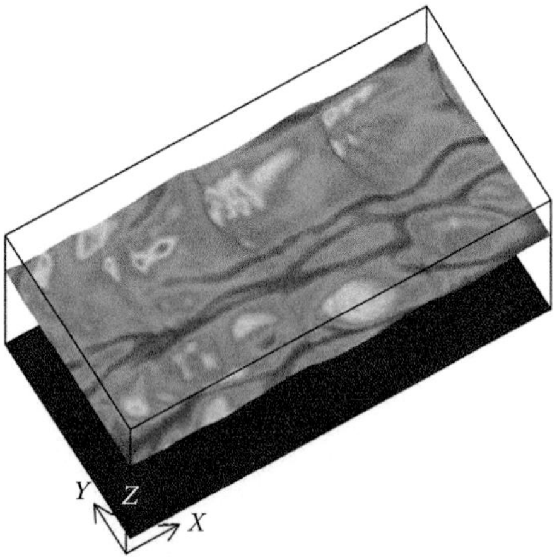

FIGURE 36.7 (See color insert.) Instantaneous surface flux of virtual gas at U_∞ = 5.2 m/s predicted by the DNS. (From Komori, S. et al., Heat and mass transfer across the wavy sheared interface in wind-driven turbulence, In *Proceedings of 6th International Symposium on Turbulent Shear Flow and Phenomena*, Eds. Kasagi, N., Eaton, J.K., Friedrich, R., Humphrey, J.A.C., and Sung, H.J., Vol. II, pp. 399–407, Seoul National University, Seoul, Korea, 2009. With permission.)

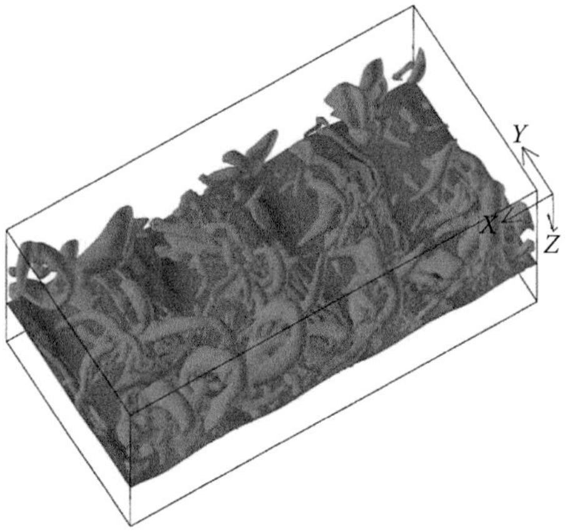

FIGURE 36.8 (See color insert.) Instantaneous second invariant on the water side at U_∞ = 5.2 m/s predicted by the DNS. (From Komori, S. et al., Heat and mass transfer across the wavy sheared interface in wind-driven turbulence, In *Proceedings of 6th International Symposium on Turbulent Shear Flow and Phenomena*, Eds. Kasagi, N., Eaton, J.K., Friedrich, R., Humphrey, J.A.C., and Sung, H.J., Vol. II, pp. 399–407, Seoul National University, Seoul, Korea, 2009. With permission.)

change from a longitudinal streaky structure to a patchy structure. In such a middle wind-speed region, a recirculating flow is intermittently generated by flow separation behind the wave crest on the air side. Strong shear stress acts on the interface in front of the wave crest (see Komori et al., 1993a). By the strong shear, organized motions are generated in front of the wave crest on both the air and water sides. For higher values of U_∞ than about 10 m/s, the wind waves are intensively broken by strong shear, and many strong ripples are generated there. As shown in Figure 36.6b, the surface temperature shows a complicated patchy structure mixed with a longitudinal streaky structure and spanwise breaking-wave structure. In extremely high wind-speed conditions, many bubbles and droplets are entrained into the water side and dispersed into the air side, respectively, and turbulence is intensively promoted by the wave breaking.

Near the interface the mean velocities normalized by friction velocities on the air and water sides showed logarithmic distributions on both the air and water sides. Figures 36.9 and 36.10 show the vertical distributions of turbulence intensities and the Reynolds stress normalized by the friction velocity on the air and water sides, respectively (Kawamura and Toba, 1988; Magnaudet and Thais, 1995). Since turbulence is generated in the interfacial region by wind shear, the turbulence intensities on both the air and water sides increase approaching the free surface. Compared to the turbulence intensities for the unsheared interface, the difference in intensity among the three directions for the sheared wavy interface is not big and the intensities have 1.5 ~ 3 times larger values than the friction velocities on both the air and water sides. The Reynolds stress also has a peak value near the interface. For the flat sheared interface, the shear stress acting on the interface should be equal to the Reynolds stress, but for

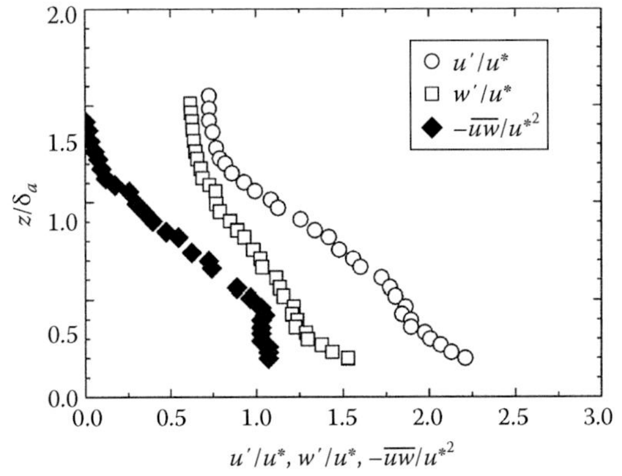

FIGURE 36.9 Vertical distributions of turbulence intensities and the Reynolds stress on the air side at $U_\infty = 5.75$ m/s in a wind-wave tank. (From Kawamura, H. and Toba, Y., *J. Fluid Mech.*, 197, 105, 1988. With permission.)

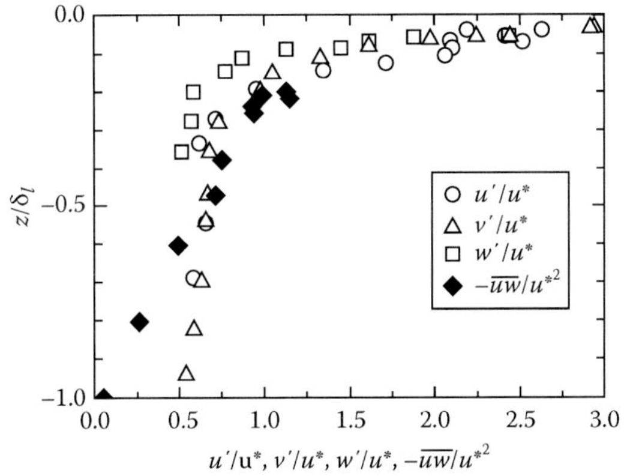

FIGURE 36.10 Vertical distributions of turbulence intensities and the Reynolds stress on the water side at $U_\infty = 4.5$ m/s in a wind-wave tank. (From Magnaudet, J. and Thais, L., *J. Geophys. Res.*, 100, 757, 1995. With permission.)

the wavy sheared interface the ratio of the shear stress to the Reynolds stress is reduced by the increase of pressure drag due to developing waves.

As shown in Figures 36.5, 36.7, and 36.8, the DNS based on the arbitrary Lagrangian-Eulerian formulation (ALE) method of (36.6) and (36.7) is useful for investigating wind-driven turbulence with low or middle wind speeds (Fulgosi et al., 2003; Komori et al., 2009, 2010), but it cannot be applied to wind-driven turbulence with overturning breaking waves for higher wind speeds than several meters per second. In order to describe the overturning breaking waves, we have to introduce a more sophisticated method such as a level set or a volume of fluid (VOF) technique, but such a method will take huge computation time to accurately describe the breaking waves even if the fastest parallel supercomputers are available. On the other hand, LES may be useful for simulating wind-driven turbulence with higher wind speeds but it is not so easy to develop sub-grid scale models suitable for interfacial flows with breaking waves. Therefore, a useful simulation technique for the wind-driven turbulence with overturning breaking waves has not yet been developed.

36.4.2.2 Gas Transfer

In a number of previous studies on gas transfer across sheared air–water interfaces, the gas transfer coefficient k_L has been measured in oceans or wind-wave tanks in order to estimate the CO_2 transfer rate across the air–sea interface. As shown in Figure 36.4, k_L measured in a wind-wave tank is about one order of magnitude larger than k_L for an unsheared interface in an open channel, since turbulence is generated by wind shear in the interfacial region. The values of k_L roughly follow the correlation (36.16) with the surface-renewal frequency f_S in the low and middle wind-speed regions with a free-stream wind speed U_∞ less than 10 m/s. This suggests that the organized motions like bursting motions control the gas transfer across the sheared interface as shown in Figures 36.5, 36.7, and 36.8. Of course, the relation between k_L and f_S will be collapsed in the extremely high wind-speed region associated with many entrained bubbles and dispersing droplets.

In practical oceanic flows, it is not easy to measure f_S, and therefore k_L has been conventionally correlated with the wind-speed, especially the wind-speed U_{10} estimated at elevation 10 m above the air–water interface using the vertical logarithmic distribution of mean air velocity and friction velocity on the air side. However, when k_L measured in a wind-wave tank is plotted against U_{10}, together with scattered data measured by the eddy-correlation technique in oceans as shown in Figure 36.11, it is obvious that the conventional correlation with U_{10} is suspicious and the effects of fetch, the Reynolds number and other factors cannot be represented by U_{10}. On the other hand, Komori et al. (2009) found that k_L is well correlated by the free-stream wind speed U_∞ independently of fetch as shown in Figure 36.12. They also showed that k_L linearly increases with U_∞ in the low

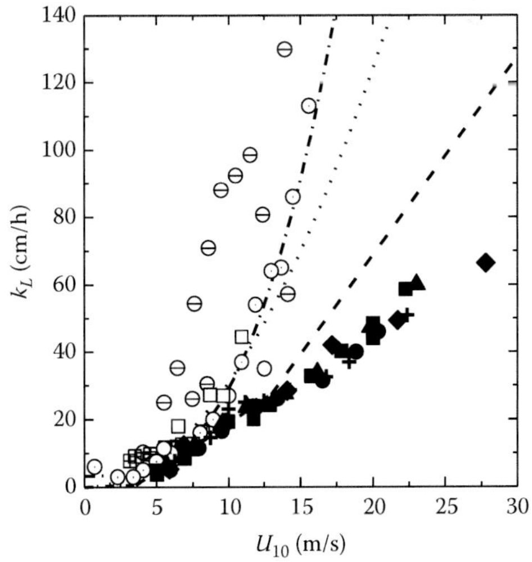

FIGURE 36.11 Comparison of k_L between laboratory and field measurements against U_{10}. Closed and open symbols are the laboratory measurements of Komori et al. (2009) and field measurements of McGillis et al. (2001, 2004), Jacobs et al. (2002), and Fairall et al. (2000), respectively. Lines show the conventional k_L correlations proposed by; - - - - Liss and Merlivat (1986); ------ Wanninkhof (1992); -- - -- McGillis et al. (2001).

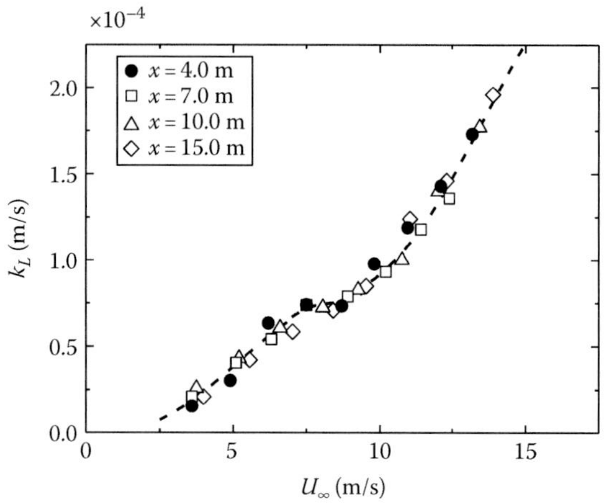

FIGURE 36.12 Gas transfer coefficient on the water side k_L against U_∞ at the fetch of $x = 4, 7, 10$, and 15 m in a wind-wave tank. (From Komori, S. et al., Heat and mass transfer across the wavy sheared interface in wind-driven turbulence, In *Proceedings of 6th International Symposium on Turbulent Shear Flow and Phenomena*, Eds. Kasagi, N., Eaton, J.K., Friedrich, R., Humphrey, J.A.C., and Sung, H.J., Vol. II, pp. 399–407, Seoul National University, Seoul, Korea, 2009. With permission.)

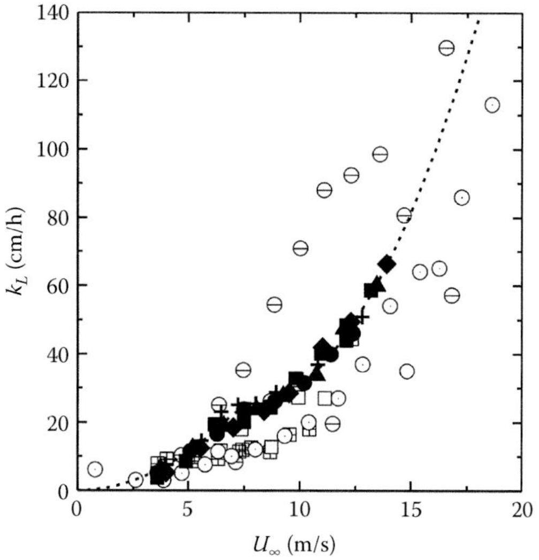

FIGURE 36.13 Comparison of k_L with field measurements against U_∞. Symbols as in Figure 36.9.

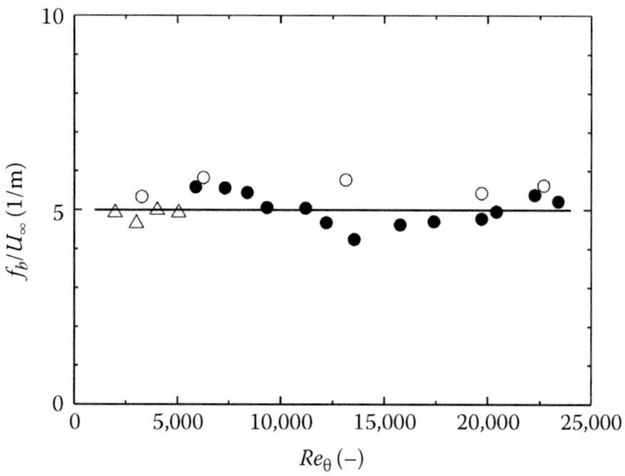

FIGURE 36.14 Comparison of the ratio of bursting frequency f_b to U_∞ in rough boundary layers (○, by Antonia and Krogstad, 1993; △, by Mochizuki and Osaka, 1992) as functions of the ratio of f_S to U_∞ in wind-driven turbulence (● by Komori et al., 1993a).

wind-speed region where longitudinal organized motions are dominant, and k_L has a small kink in the middle wind-speed region around $U_\infty = 8$ m/s. In the middle wind region, the turbulence structure changes from a longitudinal streaky structure to a patchy structure as shown in Figure 36.6. In the transient region, the shear stress may be maintained at a constant level by growing waves associated with flow separation behind the wave crest in spite of the increase of U_∞, whereas the pressure drag may increase with U_∞. In the wind-speed region higher than $U_\infty = 10$ m/s, k_L rapidly increases with increasing U_∞, since the turbulence is intensively promoted by breaking waves associated with strong ripple-like motions. In extremely high wind-speed conditions, the effect of entrained bubbles will become significant.

Figure 36.13 compares k_L between laboratory and field measurements by using the free stream wind-speed U_∞. The correlation between k_L and U_∞ is better than that between k_L and U_{10} in Figure 36.11. Here the free stream wind speed at the outer edge of the atmospheric boundary layer above the air–sea interface was approximately given by the wind speed estimated at the elevation of 65 m above the air–sea interface (Anderson et al., 2005). The field measurements are so scattered that it is difficult to show explicitly the adequateness of U_∞ as a dominant parameter. However, the agreement with the field data suggests that the outer variable U_∞ may be a more suitable parameter for correlating k_L than inner variables such as U_{10}.

When the wavy interface is considered as a kind of roughness boundary, the reason why U_∞ becomes a predominant parameter can be explained. According to the measurements of bursting frequency f_S in the turbulent rough wall boundary layers by Antonia and Krogstad (1993) and Mochizuki and Osaka (1992), the ratio of bursting frequency f_b to U_∞ has an almost constant value of 5.0, irrespective of the Reynolds number Re_θ based on the momentum thickness as shown in Figure 36.14. When the surface-renewal frequency f_s measured on the water side in a wind-wave tank is plotted against Re_θ on the air side, the ratio f_S/U_∞ surprisingly shows a value close to 5.0. This fact together with the surface-renewal concept (36.16) suggests that the free-stream wind speed U_∞ is a more suitable parameter for correlating the gas transfer coefficient k_L than U_{10}.

36.5 Major Challenges

One of the major challenges in the study of air–water gas transfer is to improve a climate-change model by accurately estimating the air–sea CO_2 flux. In order to accomplish this purpose,

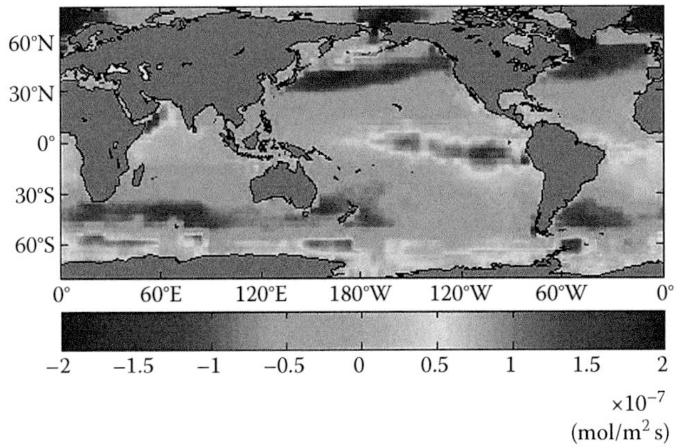

FIGURE 36.15 Distribution of mean annual air-sea CO_2 flux for year 2001.

we have to clarify the similarity of gas transfer between laboratory and field scales. If the assumption that k_L for the air–sea interface is given by the dashed line in Figure 36.13 is acceptable, we can estimate the global carbon exchange rate using the database of both partial pressure difference in CO_2 between atmosphere and ocean, ΔP_{CO_2} (Takahashi et al., 2002) and wind speed from NCEP/NCAR (National Center for Environmental Prediction/National Center for Atmospheric Research) Reanalysis. Figure 36.15 shows the distribution of the mean annual air–sea CO_2 flux for the year 2001 estimated by using the empirical correlation of Komori et al. (2009) in Figure 36.13. It is found that uptake and discharge of CO_2 into and from oceans are seen in high- and low-latitude regions, respectively. Table 36.1 also shows the global CO_2 exchange rate for the year 2001 with the predictions by the correlations of Liss and Merlivat (1986), Wanninkhof (1992), and McGillis et al. (2001) for U_{10} (see Figure 36.11). The empirical correlation of k_L with U_∞ proposed by Komori et al. (2009) shows that the net uptake of carbon into the oceans is about 2.2 PgC/year and this value is larger than other predictions. In order to confirm whether the empirical correlation in Figure 36.13 is really right, k_L should be more accurately measured in the oceans together with a more detailed turbulence structure. Especially, the surface-renewal motions below the air–sea interface should be investigated.

TABLE 36.1 Global CO_2 Exchange Rate for the Year 2001 Estimated by the Correlation between k_L and U_∞ or U_{10}

Authors	Net (PgC/Year)	Sea to Air (PgC/Year)	Air to Sea (PgC/Year)
Komori et al. (2009) for k_L versus U_∞	−2.19	2.5	−4.69
McGillis et al. (2001) for k_L versus U_{10}	−1.59	1.52	−3.11
Wanninkhof (1992) for k_L versus U_{10}	−1.39	1.55	−2.94
Liss and Merlivat (1986) for k_L versus U_{10}	−0.80	0.93	−1.73

The unit of the exchange rate is PgC/year.

The other challenge is to develop a sophisticated numerical simulation technique for accurately describing both wind-driven turbulence with intensive breaking waves, ripples, entrained bubbles, and dispersing droplets in high wind-speed conditions, and a gas transfer mechanism for high Schmidt numbers. Especially, the development of a LES applicable for both high Reynolds and Schmidt numbers will be desired.

References

Anderson, PS, Ladkin, RS, Renfrew, IA. 2005. An autonomous Doppler sodar wind profiling system. *J. Atmos. Oceanic Technol.* 22: 1309–1325.

Antonia, RA, Krogstad, PA. 1993. Scaling of the bursting period in turbulent rough wall boundary layers. *Exp. Fluids.* 15: 82–84.

Calmet, I, Magnaudet, J. 2003. Statistical structure of high-Reynolds-number turbulence close to the free surface of an open-channel flow. *J. Fluid Mech.* 474: 355–378.

Fairall, CW, Hare, JE, Edson, JB, McGillis, W. 2000. Parameterization and micrometeorological measurement of air-sea gas transfer. *Boundary Layer Meteorol.* 96: 63–105.

Fulgosi, M, Lakehal, D, Banerjee, S, De Angelis, V. 2003. Direct numerical simulation of turbulence in a sheared air-water flow with a deformable interface. *J. Fluid Mech.* 482: 319–345.

Gibson, MM, Rodi, W. 1989. Simulation of free surface effects on turbulence with a Reynolds stress model. *J. Hydraul. Res.* 27: 233–244.

Jacobs, C, Kjeld, F, Nightingale, P, Upstill-Goddard, R, Larsen, S, Oost, W. 2002. Possible errors in CO_2 air-sea transfer velocity from deliberate tracer releases and eddy covariance measurements due to near-surface concentration gradients. *J. Geophys. Res.* 107: doi/10.1029/2001JC000983.

Kawamura, H, Toba, Y. 1988. Ordered motion in the turbulent boundary layer over wind waves. *J. Fluid Mech.* 197: 105–138.

Komori, S, Kurose, R, Ohstubo, S, Tanno, K, Suzuki, N. 2009. Heat and mass transfer across the wavy sheared interface in wind-driven turbulence. In *Proceedings of 6th International Symposium on Turbulent Shear Flow and Phenomena*, Eds. N Kasagi, JK Eaton, R Friedrich, JAC Humphrey HJ Sung, Vol. II, pp. 399–407. Seoul National University, Seoul, Korea.

Komori, S, Murakami, Y, Ueda, H. 1989. The relationship between surface-renewal and bursting motions in an open-channel flow. *J. Fluid Mech.* 203: 103–123.

Komori, S, Nagaosa, R, Murakami, Y. 1990. Mass transfer into a turbulent liquid across the zero-shear gas-liquid interface. *AIChE J.* 36: 957–960.

Komori, S, Nagaosa, R, Murakami, Y. 1993a. Turbulence structure and mass transfer across a sheared air-water interface in wind-driven turbulence. *J. Fluid Mech.* 249: 161–183.

Komori, S, Nagaosa, R, Murakami, Y, Chiba, S, Ishii, K, Kuwahara, K. 1993b. Direct numerical simulation of three-dimensional open-channel flow with zero-shear gas-liquid interface. *Phys. Fluids* A5: 115–125.

Komori, S, Shimada, T, Misumi, R. 1999. Turbulence structure and mass transfer at a wind-driven air-water interface. In *Wind over Wave Couplings: Perspectives and Prospects*, Eds. SG Sajjadi, NH Thomas, JCR Hunt, pp. 273–285. Oxford University Press, Oxford, U.K.

Komori, S, Kurose, R, Iwano, K, Ukai, T, Suzuki, N. 2010. Direct numerical simulation of wind-driven turbulence and scalar transfer at sheared gas-liquid interfaces. *J. Turbulence.* 11: 1–20.

Komori, S, Ueda, H, Ogino, F, Mizushina, T. 1982. Turbulent structure and transport mechanism at the free surface in an open-channel flow. *Int. J. Heat Mass Transfer* 25: 513–521.

Lin, MY, Moeng, CH, Tsai, WT, Sullivan, PP, Belcher, SE. 2008. Direct numerical simulation of wind-wave generation processes. *J. Fluid Mech.* 616: 1–30.

Liss, PS, Merlivat, L. 1986. Air-sea gas exchange rates: Introduction and synthesis. In *The Role of Air-Sea Exchange in Geochemical Cycling*, Ed. P Menard, pp. 113–127. Reidel, Dordrecht, the Netherlands.

Lombardi, P, De Angelis, V, Banerjee, S. 1996. Direct numerical simulation of near-interface turbulence in coupled gas-liquid flow. *Phys. Fluids* 8: 1643–1665.

Magnaudet, J, Thais, L. 1995. Orbital rotational motion and turbulence below laboratory wind water waves. *J. Geophys. Res.* 100: 757–771.

McCready, MJ, Hanratty, TJ. 1985 Effect of air shear on gas absorption by a liquid film. *AIChE J.* 31: 2066–2074.

McGillis, WR, Edson, JB, Hare, JE, Fairall, CW. 2001. Direct covariance air-water CO_2 fluxes. *J. Geophys. Res.* 106: 16729–16745.

McGillis, W, Edson, JB, Zappa, CJ, Ware, JD, McKenna, SP, Terray, EA., Hare, JE, Fairall, CW, Dorennan, W, Donelan, M, DeGrandpre, MD, Wanninkhof, R, Feely, RA. 2004. Air-sea CO_2 exchange in the equatorial Pacific. *J. Geophys. Res.* 109: C08S02, doi:10.1029/2003JC002256.

Mochizuki, S, Osaka, H. 1992. Coherent structure of a d-type rough wall boundary layer in a transitionally rough and a fully rough regime. *Trans. JSME (B).* 58: 1392–1399.

Moog, DB, Jirka, GH. 1999. Air–water gas transfer in uniform channel flow. *J. Hydrol. Eng.* 125: 3–10.

Nakayama, A, Yokojima, S. 2003. Free-surface fluctuation effects for calculation of turbulent open-channel flows. *Environmental Fluid Mech.* 3: 1–21.

Takahashi, T, Sutherland, SC, Sweeney, C, Poisson, A, Metzl, N, Tilbrook, B, Bates, N, Wanninkhof, R, Feely, RA, Sabine, C, Olafsson, F, Nojiri, Y. 2002. Global sea-air CO_2 flux based on climatological surface ocean pCO_2, and seasonal biological and temperature effects. *Deep-Sea Res. II.* 49: 1601–1622.

Wang, L, Dong, YHD, Lu, XY. 2005. An investigation of turbulent open channel flow with heat transfer by large eddy simulation. *Comput. Fluids.* 34: 23–47.

Wanninkhof, RH. 1992. Relationship between wind speed and gas exchange over the ocean. *J. Geophys. Res.* 97: 7373–7382.

37

Convection

Jaywant H. Arakeri
Indian Institute of Science

37.1 Introduction

Convection traditionally refers to heat transfer associated with a fluid which is in motion. When the fluid motion is caused by an external agency, for example, a fan or a "wind," the convection is termed as forced convection. In free convection, in contrast, fluid motion is predominantly due to buoyancy forces caused by density differences in the fluid; the density difference is generally due to variations of temperature or variations of concentration of species or both.

In this chapter, we consider mainly free convection flows over surfaces. Many of the free convection flows we observe around us are turbulent. Of practical relevance is the amount of heat transferred by convection due to a certain temperature difference, the nature of the flow, and its velocity and temperature fields.

Figure 37.1 shows a few common free convection flows. One basic mechanism consists of fluid next to a hot (cold) surface becoming warmer (cooler) and rising (falling) due to gravity. Convection also occurs within the interior of a fluid due to any horizontal or vertical gradients of density. Vertically oriented flows (Figure 37.1a and b), such as next to a vertical heated wall or the plume from a hot wire, are easy to imagine. An unstable temperature difference across the two ends of a tall vertical tube leads to an "exchange" convection with simultaneous rising and falling fluid (Figure 37.1c). Convective flows from horizontal surfaces are more complicated, such as in the gap between a hot bottom plate and a top cold plate (Figure 37.1e). In shallow horizontal layers, stratification plays an important part in determining the type of convection. Stable stratification (density decreasing with height) generally suppresses turbulence and convection; unstable stratification enhances the turbulence and the convection. The flow over a large heated horizontal plate (Figure 37.1d) displays both types of motion: Near the plate surface, the convection is akin to that on a horizontal heated surface, whereas away from the plate we obtain a plume with flow predominantly in the vertical direction.

37.2 Principles

The governing equations are those of conservation of mass, momentum, and energy. Using the usual tensor notation, equations of incompressible fluid motion under the Boussinesq approximation (see Turner, 1973) for mass and momentum conservation are

$$-\frac{1}{\rho}\frac{D\rho}{Dt} = \frac{\partial u_i}{\partial x_i} = 0 \tag{37.1}$$

$$\frac{Du_i}{Dt} = \frac{\partial u_i}{\partial t} + u_j\frac{\partial u_i}{\partial x_j} = -\frac{1}{\rho}\frac{\partial P}{\partial x_i} - \frac{\delta\rho}{\rho}g\delta_{ik} + \nu\frac{\partial^2 u_i}{\partial x_j \partial x_j} \tag{37.2}$$

where

u_i is velocity component in direction of coordinate x_i
ν is kinematic viscosity
g is the acceleration due to gravity
(D/Dt) represents time derivative following a fluid element

The first equation states that as a fluid element convects, its volume and density remain constant. In the momentum equation,

Handbook of Environmental Fluid Dynamics, Volume One, edited by Harindra Joseph Shermal Fernando. © 2013 CRC Press/Taylor & Francis Group, LLC. ISBN: 978-1-4398-1669-1.

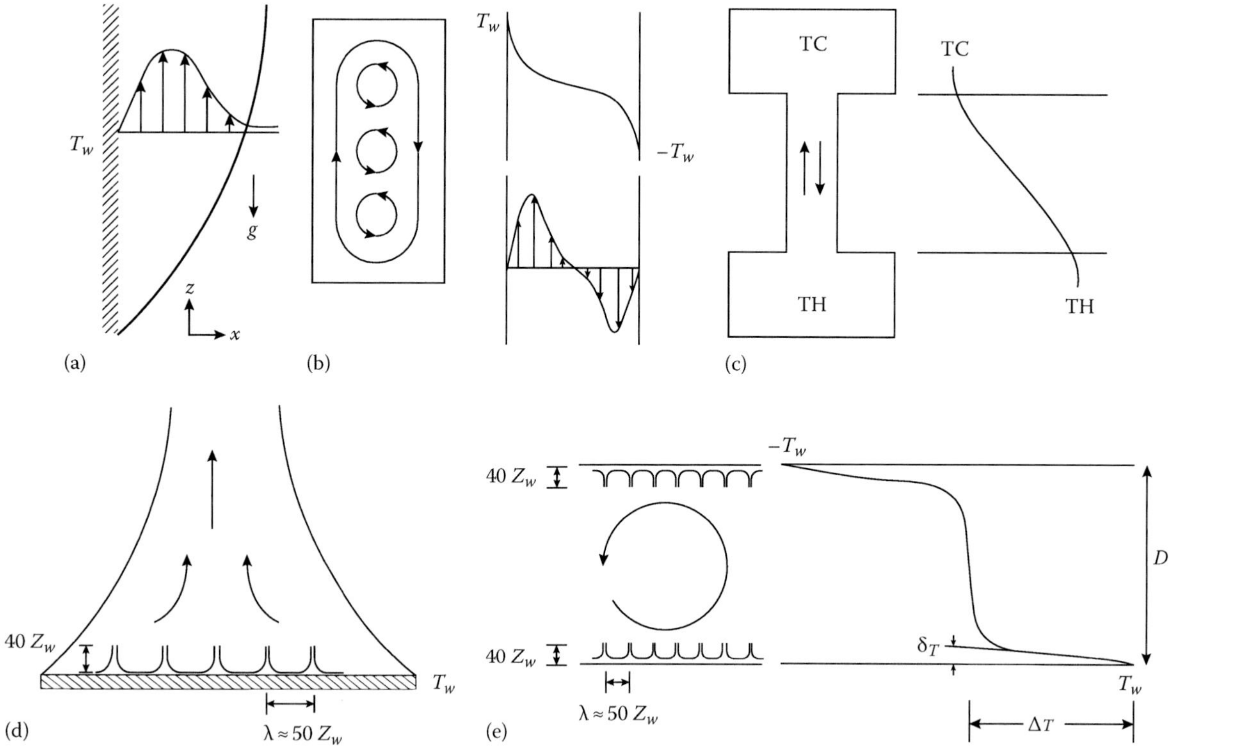

FIGURE 37.1 (a) Convection next to heated vertical plate. (b) Convection in a vertical slot with left side wall temperature (T_w) greater than the right side ($-T_w$). (c) Exchange flow in a vertical pipe due to an unstable density difference maintained across the two ends; TH > TC. (d) Convection over a horizontal hot plate. Small-scale wall plumes agglomerate to form a single large plume. (e) Rayleigh–Benard (R-B) convection. When the convection is turbulent, near-wall plumes and large-scale rolls (~D) are observed.

the density difference appears only in the body force term ($\delta\rho g$), which is a buoyancy force on the fluid element due to its density being different from the density (ρ) in a static fluid by an amount $\delta\rho$. Here, P is pressure that is subtracted from the hydrostatic value. If the density difference is caused by temperature difference δT, we may write

$$\delta\rho = -\beta\rho\delta T$$

where β is the temperature expansion coefficient of the fluid. A similar expression is obtained for density variation caused by composition.

Neglecting viscous dissipation of kinetic energy into heat and assuming no internal heat generation, the equation for thermal energy is

$$\frac{\partial T}{\partial t} + u_j \frac{\partial T}{\partial x_j} = \alpha \frac{\partial^2 T}{\partial x_j \partial x_j} \qquad (37.3)$$

where α is thermal diffusivity.

In this chapter, we are mainly concerned with turbulent convection, and use the Reynolds averaged equations. First, consider a 2-d boundary layer with the mean flow in a horizontal (say x) direction with vertically upward coordinate represented by z (z, coordinate being antiparallel to gravity is the convention followed in the rest of the chapter); the Reynolds averaged momentum equation in the x direction is

$$\bar{u}\frac{\partial \bar{u}}{\partial x} + \bar{w}\frac{\partial \bar{u}}{\partial z} = -\frac{1}{\rho}\frac{\partial \bar{P}}{\partial x} - \frac{\partial\left(\overline{u'w'}\right)}{\partial z} + \nu\frac{\partial^2 \bar{u}}{\partial z^2} \qquad (37.4)$$

Over bar represents the mean and prime the fluctuating part of a variable; for example,

$$u = \bar{u} + u'$$

where $\bar{u}$ is the mean velocity and u' is the fluctuating velocity; depending on the context the mean could be a *temporal* average or a *spatial* average. The second term on the right-hand side is the gradient of the Reynolds or turbulent shear stress, $-\overline{u'w'}$. Note that buoyancy force does not enter the momentum equation directly; but density field can induce horizontal motion through a horizontal pressure gradient, and in addition buoyancy can influence through its production of turbulence kinetic energy. In contrast, for a vertical boundary layer flow (for example, on a heated vertical wall or in a plume), the buoyancy force directly affects the mean flow,

$$\bar{u}\frac{\partial \bar{w}}{\partial x} + \bar{w}\frac{\partial \bar{w}}{\partial z} = -\frac{1}{\rho}\frac{\partial \bar{P}}{\partial z} - \frac{\partial\left(\overline{u'w'}\right)}{\partial x} + g\beta\delta T \qquad (37.5)$$

The mechanisms by which turbulent kinetic energy (k), is produced, diffused, or distributed, and dissipated are of interest. The equation for turbulent kinetic energy (k) is

$$\frac{D}{Dt}(k) = -\frac{\partial}{\partial x_j}\left(\frac{\overline{P'u_j'}}{\rho} + \frac{\overline{(u_i'^2 u_j')}}{2} - 2\nu\overline{u_i'e_{ij}'}\right) - \left(\overline{u_i' u_j'}\frac{\partial \overline{u_i}}{\partial x_j}\right) + \left(g\beta\overline{w'T'}\right) - (\epsilon) \quad (37.6)$$

where k is

$$k = \frac{\overline{u'^2} + \overline{v'^2} + \overline{w'^2}}{2}$$

The terms within the first brackets on the right-hand side represents diffusion of k by pressure fluctuations, velocity fluctuations, and by viscosity; the term neither produces nor destroys k but just redistributes it in space. The last term represents viscous dissipation of k. In convective flows, turbulent kinetic energy may be produced by shear (term 2) and by buoyancy (term 3).

$$\text{Buoyant production} = g\beta\overline{w'T'} \quad (37.7)$$

is associated with buoyant (heavier) fluid elements gaining velocity while rising (falling) due to gravity; in stably stratified flows, this term may be negative, buoyancy attenuating the turbulence. In a horizontal boundary-layer flow, for example,

$$\text{shear production} = -\overline{u'w'}\frac{\partial \overline{u}}{\partial z} \quad (37.8)$$

Shear production is due to the mean stress doing work on the mean flow.

37.2.1 Nondimensional Numbers

Rayleigh number (Ra), a measure of the ratio of buoyancy effects to diffusive (momentum and thermal) effects, is the key nondimensional parameter in free convection flows, much like the Reynolds number in forced flows.

$$Ra = \frac{g\left(\Delta\rho_{ref}/\rho\right)L_{ref}^3}{\nu\alpha}$$

where

$\Delta\rho_{ref}$ is a reference density difference
L_{ref} is a reference length scale

Prandtl number, $Pr = (\nu/\alpha)$, is the ratio of momentum diffusivity to thermal diffusivity. Sometimes it is more appropriate to use Grashof number = (Ra/Pr) instead of Ra. Ra and Pr determine the convection, in particular the heat transfer, velocity and temperature fields. Ra may be considered to be the forcing that creates the convection.

A nondimensional heat transfer rate usually Nusselt number (Nu) and a suitably defined Reynolds number (Re) based on the convective velocity can be expressed in terms of Ra, Pr, and geometrical parameters like an aspect ratio (AR).

$$Nu = f_1(Ra, Pr, \text{ geom. par.}), Re = f_2(Ra, Pr, \text{ geom. par.})$$

In horizontal shear layers, ratios relating inertia or shear to buoyancy occur in the form of Richardson numbers (R_f, R_i) or Froude numbers. In this chapter, the flux Richardson number (R_f) enters in the context of transition from free or buoyancy-driven convection to forced convection.

37.3 Examples

37.3.1 Rayleigh–Benard Convection

We will consider Rayleigh–Benard (R-B) convection [the free convection in a fluid placed between two horizontal plates (distance D apart) with the bottom plate temperature = T_w, and the top plate temperature = $-T_w$ (Figure 37.1e)] in detail and relate it to other free convection flows, in particular, free and mixed convection on horizontal surfaces. Rayleigh–Benard convection has been extensively studied not only because of its application to engineering and in the atmospheric and oceanic sciences, but because it is a "simple" buoyancy driven flow that exhibits a rich variety of flow phenomena. Importantly, it contains many of the features obtained in convection in horizontal shallow layers in general. Despite the extensive study, a number of unresolved issues remain, mainly relating to turbulent convection; (see reviews by Adrian et al., 1986; Siggia, 1994; Ahlers et al., 2009). The unstable density differential—heavier fluid at the top and lighter fluid at the bottom—due to the temperature difference is the cause of the convection.

The three nondimensional parameters that determine the flow are Ra, Pr, and aspect ratio (AR) of the convection cell, defined as width/height. Generally Ra is based on the temperature difference between the plates ($2T_w$) and distance between the plates (D),

$$Ra = \frac{g\beta(2T_w)D^3}{\nu\alpha} \quad (37.9)$$

Linear stability analysis predicts that independent of the value of Pr, for $Ra < 1708$, no convection is obtained, and all the heat transfer is by conduction. Just above the critical Ra, convection is steady and may be in the form of rolls, or hexagonal or square cells extending across the convection cell height; the original experiments of Benard showed hexagonal cells, though this convection, we now know, was caused by temperature-induced surface-tension gradients. With increasing Ra, the rolls or cells become three dimensional but are still steady, then they become unsteady and at some large Ra > transition Rayleigh number (Ra_{tr}), flow is turbulent (see Krishnamurthy, 1970; Turner, 1973). These later transitions depend on Pr. For $Pr \sim 1$, the $Ra_{tr} \sim 10^5$, though it is not possible to give a precise value for Ra_{tr}. Ra_{tr} increases with Pr. As we shall see later, a rich variety of phenomena is obtained in turbulent convection, many not understood including the type of flow to be expected at $Ra \to \infty$.

In R-B convection (laminar or turbulent), taking the mean over a plane, z = constant, there is no mean flow, $\bar{u}$, $\bar{w}$ = 0. (Note that, time mean will be nonzero in flows with "steady" convection rolls.) The Reynolds averaged energy equation for R-B convection is

$$\overline{w'T'} - \alpha\frac{d\overline{T}}{dz} = \frac{q}{\rho C_p} = Q \quad (37.10)$$

where

q is the heat flux (in W/m²)
Q ($=q/\rho C_p$) is the kinematic heat flux
C_p is the heat capacity

Applying the equation at the walls, where w' and T' = 0,

$$q = -K\left(\frac{dT}{dz}\right)_{wall}$$

where $K = \rho C_p \alpha$ is the thermal conductivity of the fluid. The heat is carried at a constant rate across the cell purely by conduction $\left(-\alpha\frac{d\overline{T}}{dz}\right)$ at the walls, and a combination of conduction and the turbulence ($\overline{w'T'}$) away from the walls. In turbulent convection, except in thin regions, boundary layers, near the walls, the gradient of the mean temperature ($d\overline{T}/dz$) is negligible; almost all the heat is carried by the turbulence in the interior of the convection cell. $\overline{w'T'}$ is the turbulent heat flux and note it also enters directly in buoyant production of turbulent kinetic energy (Equation 37.7).

Figures 37.2 and 37.3 show for air and water, respectively, the heat flux as a function of plate spacing (D) for several values of temperature differences, ΔT ($=T_w$). The heat flux was calculated using correlations given in a handbook (Rohsenowe et al., 1985). Lines of constant Ra demarcating different regimes of flow (conduction, laminar convection, etc.) are also shown. It is clear that for the two common fluids, air and water, and for the temperature differences (~10 K) and length scales (~1 m) we normally encounter, the Rayleigh numbers are very large and turbulent convection is the norm rather than the exception. Also, in the turbulent regime, there is very little change in heat transfer rate with the distance between the plates; for example, in air for ΔT = 20 K, the heat transfer flux is nearly constant = 100 W/m², as D changes from 10 cm to 2 m.

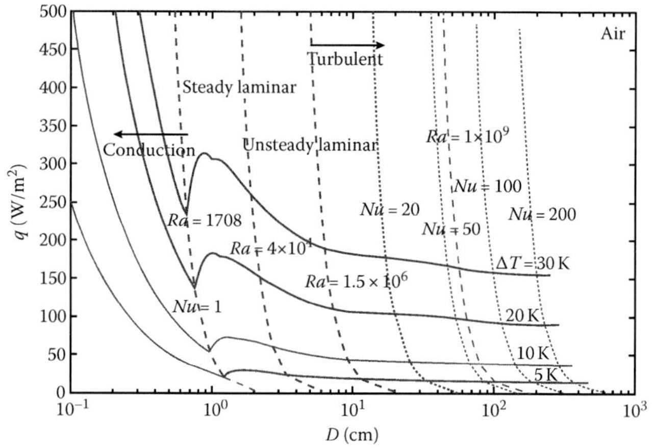

FIGURE 37.2 Calculated heat flux for air ($Pr \sim 0.7$) using correlations in R-B convection. Lines of constant Ra and Nu are shown. Boundaries for transition to unsteady and turbulent convection cannot be given precisely.

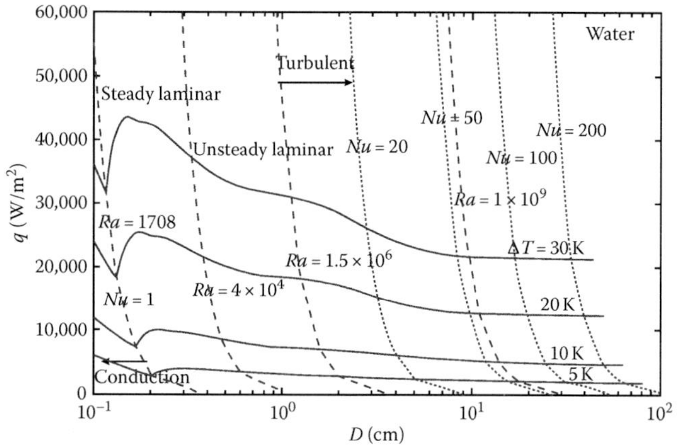

FIGURE 37.3 Calculated heat flux for water ($Pr \sim 6$) using correlations in R-B convection. Lines of constant Ra and Nu are shown. Boundaries for transition to unsteady and turbulent convection cannot be given precisely.

37.3.1.1 Flow Structure and Scales in R-B Convection

It is generally agreed that a turbulent flow, even though apparently random, consists of distinctive coherent structures which determine the important dynamics of the flow; the turbulent flow can be thought to consist of an ensemble of such structures. The transverse vortices in a mixing layer and the sublayer vortices in a turbulent boundary layer are examples of coherent structures. In turbulent R-B convection, it is possible to identify two types of coherent structures, one relating to the *near-wall* or *inner flow* and the other corresponding to the convection away from the wall, the *outer flow*. These are schematically shown in Figure 37.1e as near-wall plumes and circulations or rolls scaling with the height of the convection, D.

Corresponding to these motions, two scales may be defined—an inner or wall scale, where diffusive effects of viscosity and thermal conduction are important, and an outer scale that does not explicitly contain viscosity and thermal diffusivity. (A similar type of twin scaling is obtained in turbulent boundary layers.)

The inner scales may be derived on the basis of temperature difference between the wall and the core (ΔT) or on the basis of the wall heat flux (Townsend, 1976). Townsend's scales, based on the heat flux at the wall Q_o ($=q/\rho C_p$), for velocity (W_o), temperature (θ_o), and length (Z_o) are

$$W_o = (g\beta Q_o\alpha)^{(1/4)}, \quad \theta_o = \frac{Q_o}{W_o}, \quad Z_o = \frac{\alpha}{W_o} \tag{37.11}$$

Alternatively with ΔT as the independent variable instead of Q_o, we get the following inner scales for velocity (U_w), temperature (T_w), and length (Z_w); (Theerthan and Arakeri, 1998),

$$U_w = (g\beta T_w)^{(1/3)}(\vartheta\alpha)^{(1/6)}, \quad T_w = \Delta T, \quad Z_w = \frac{(\alpha\nu)^{1/2}}{U_w} \tag{37.12}$$

Another convenient inner or wall scale is the conduction layer thickness, (δ_T), defined as (see Figure 37.1e)

$$\delta_T = \frac{(\Delta T)}{(\partial T/\partial z)_{wall}} \tag{37.13}$$

Since δ_T contains both heat flux and ΔT, it is a mixed scale, and may be written in terms of Z_w and Z_o.

In the laboratory scale R–B convection experiments, the mean temperature profile shows sharp changes in temperature near the two walls and a nearly isothermal core (Figure 37.1e), indicating a thorough mixing due to the outer flow eddies. We will see later that the dynamics close to the wall are crucial in determining the heat transfer. At very large scales of convection, which, for example, would be obtained in the atmosphere, the mean temperature would however be expected to be nonuniform in the outer region as well, though with temperature gradients much more lower than at the wall.

In the near-wall region, two types of structures have been observed. One is periodic eruption of hot blobs of fluid termed thermals from the bottom wall (cold thermals would be obtained from the top wall). The second type is line plumes, where hot/cold fluid rises/falls from lines. In R-B convection, the predominant structure seems to be the second type: line plumes, which move about randomly on the surface. Figure 37.4 shows visualizations of the planform structure close to the wall from several studies.

In Figure 37.4a through c, the flow was visualized using an electrochemical technique using pH indicator thymol blue. The "dye" which is produced uniformly on the surface collects along lines. Each line indicates the base of a line or sheet plume from which hot fluid rises. In Figure 37.4b through d, alignment of plumes is caused by large-scale flow.

The line plumes are randomly oriented, have some average spacing, and randomly move about on the horizontal surface.

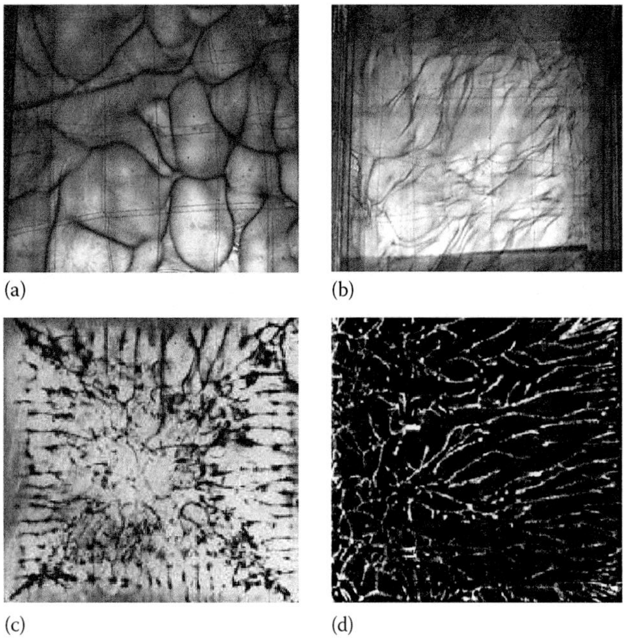

FIGURE 37.4 (a) R–B convection, $AR = 6$, $Ra = 4.63 \times 10^5$, $q = 65\,W/m^2$, $\Delta T = 0.13\,K$. (b) R–B convection, $AR = 1.5$, $Ra = 1.1 \times 10^9$ $q = 4150\,W/m^2$, $\Delta T = 4.87\,K$. (c) Convection over 8.9 cm × 8.9 cm, heated horizontal plate, $\Delta T = 27.5\,K$, $Ra = 5.5 \times 10^8$. (From Husar, R.B. and Sparrow, E.M., *Int. J. Heat Mass Transfer*, 11, 1206, 1968. With permission.) (d) High *Pr* convection, $Ra = 2.03 \times 10^{11}$, $AR = 0.435$. (From Puthenveettil, B.A. and Arakeri, J.H., *J. Fluid Mech.*, 542, 217, 2005. With permission.)

Depending on the Rayleigh number and aspect ratio (AR) of the flow cell, Theerthan and Arakeri (2000) have observed different arrangements of the line plumes—cell type, broken lines, and aligned lines. The cell type structure (Figure 37.4a) is seen at lower *Ra*, when a strong outer circulation is not set up. Increasing *Ra* results in breaking up the cells and alignment of the plumes.

The outer flow scales or the Deardorff scales are derived using the basic parameters, β, g, Q, and Z^*, and neglecting the diffusivities, α and ν,

$$W^* = (\beta g Q Z^*)^{1/3}, \quad \theta^* = \frac{Q}{W^*} \tag{37.14}$$

Z^* can be taken to be the height of the convection layer; for R-B convection $Z^* \sim D$. The Deardorff velocity scale may also be determined as a free-fall velocity of a fluid element with temperature difference θ^* rising over a distance Z^*. W^* represents a velocity scale for the large-scale circulation. Relations between the various scales are given in Theerthan and Aakeri (1998). Grossmann and Lohse (2000) derive scaling laws from a different viewpoint.

An interesting and relevant question is the interaction of the outer flow and the inner flow, both in terms of the structures of the two flows and the effect on heat transfer rate. One effect of the outer flow on the plumes is clear: The plumes are aligned in the direction of the outer flow (Figure 37.4).

37.3.1.2 Nondimensional Heat Transfer

The usual nondimensional representation of heat transfer flux is by the Nusselt number (*Nu*), which can be thought of as the ratio of convective heat flux to the heat flux if the heat transfer is by *conduction* over the *length D*:

$$Nu = \frac{q}{q_{conduction}} = \frac{-K(\partial T/\partial z)}{-K(2T_w/D)} = \frac{qD}{k(2T_w)} \tag{37.15}$$

Thus $Nu = (D/2\delta t)$ gives a *ratio of the outer to the inner scale.*

In turbulent free convection, the usual form of the correlation is $Nu = CRa^n Pr^p$, where C is a constant, and $n \approx (1/3)$, and p is small, indicating a weak Pr dependence of Nu. The value of n close to 1/3 has the important implication that heat flux becomes nearly independent of length scale: $q \sim D^{(n-(1/3))}$. For R-B convection in air and water, Figures 37.2 and 37.3 show that the heat flux becomes independent of D beyond about $D = 10$ cm for the temperature differences considered.

There are two important facts about turbulent free convection. One is the near independence of heat flux with length scale; and the second, as we shall see later, for a given temperature difference and fluid, heat flux is nearly same for *different types* of turbulent free convection. Whereas Nu is a useful indicator of the ratio of outer to inner length scales, it does not give a direct measure of the heat flux.

There is a need to represent the nondimensional heat transfer in an alternate way that makes apparent the two features of turbulent free convection mentioned above. A skin-friction coefficient (C_f) type representation for nondimensionalizing heat flux removes the inadequacies of Nu in representing free convection turbulent heat transfer.

The near independence of heat flux from the length scales suggests that the heat transfer rate depends on ΔT and the fluid properties. From dimensional analysis, taking Q to depend on g, β, ΔT, ν, and α, and independent of D, we may write a nondimensional heat transfer (Theerthan and Arakeri, 2000),

$$(Ra_{\delta_T})^{-1/3} = \frac{q}{K(g\beta/\nu\alpha)^{(1/3)}\Delta T^{(4/3)}} \tag{37.16}$$

where

$$Ra_{\delta_T} = \frac{g\beta\Delta T\delta_T^3}{\nu\alpha}$$

is the Rayleigh number based on the conduction layer thickness, δ_T. The heat flux in terms of Ra_{δ_T} is

$$q = K\left(\frac{g\beta}{\nu\alpha}\right)^{(1/3)} \Delta T^{4/3}(Ra_{\delta_T})^{-1/3} \tag{37.17}$$

If only the above parameters are relevant, then Ra_{δ_T} will be a function of Pr alone. Clearly, however, the heat transfer rate will depend on other factors as well: temperature fluctuations, velocity fluctuations, etc. (which, in turn, will depend on the configuration and Ra). $Ra_{\delta_T}^{(-1/3)}$ is a direct measure of the heat transfer rate and is the alternate nondimensional representation of heat flux. Townsend (1970) used these arguments to represent turbulent convective heat transfer, but did not bring in the Ra_{δ_T} interpretation.

Comparison of the heat transfer rates in the different flows and under different conditions is now possible by just looking at the values of $Ra_{\delta_T}^{(-1/3)}$. We may write correlations for $Ra_{\delta_T}^{(-1/3)}$ in the form

$$(Ra_{\delta_T})^{-1/3} = ARa^m$$

where A and m are related to C and n in the Nu-Ra correlation; $m = n - (1/3)$.

Thus, since $n \cong (1/3)$ in turbulent convection, we may expect $Ra_{\delta_T}^{(-1/3)}$ unlike Nu, not to vary much. For example, the correlation, $Nu = 0.183\ Ra^{0.278}$, of Chu and Goldstein (1973) for Rayleigh–Benard convection gives $Ra_{\delta_T}^{(-1/3)} = ARa^{-0.055}$, a much more weakly varying function of Ra than Nu.

Figures 37.5 and 37.6 show plots $Ra_{\delta_T}^{(-1/3)}$ versus Ra for Rayleigh–Benard convection for $Pr \sim 0.7$ and 6.0, respectively. Correlations (and not the actual data points) proposed in the literature have been used and they are restricted to the Ra ranges over which the experiments were performed. The $n = 1/3$ correlations of Globe and Dropkin (1959) gives $Ra_{\delta_T}^{(-1/3)}$ = constant for each Pr. The remarkable result is that, for the range of Ra covered, the $Ra_{\delta_T}^{(-1/3)}$ value has a variation of only between about 0.1 and 0.3; in the same range of Ra there would be more than thousand fold change in Nu. For very high Pr values (=2750), Goldstein et al. (See Puthenveettil and Arakeri, 2005) have proposed the correlation $Nu = 0.066\ Ra^{(1/3)}$ in the Ra range 10^9 – $5*10^{12}$, which gives $Ra_{\delta_T}^{(-1/3)} = 0.166$.

Table 37.1 gives the $Ra_{\delta_T}^{(-1/3)}$ values for other configurations. Here, the advantage of the alternate representation over Ra-Nu representation is readily apparent since different length scales are involved. It would be impossible to compare heat transfer rates from Nu for different configurations.

For the finite horizontal plate case, the $Ra_{\delta_T}^{(-1/3)}$ values are similar to the values obtained in R-B convection. In the case of heat transfer by evaporation, the configuration is similar to R-B except that the boundary condition is not one of no slip but closer to zero shear stress. It is remarkable even for mixed convection the heat transfer rates are not very different and only slightly higher than that for convection over a horizontal plate. Finally, it is intriguing that the heat transfer rates are similar in the vertical and horizontal geometries even though the turbulence production mechanisms are different. The fact that the turbulent heat transfer rates for different configurations do not vary much had been noticed by earlier investigators. The value of $Ra_{\delta_T}^{-(1/3)} = 0.1$ gives $Ra_{\delta_T} = 1000$, indicating that the heat transfer and the conduction layer thickness is probably due to instability.

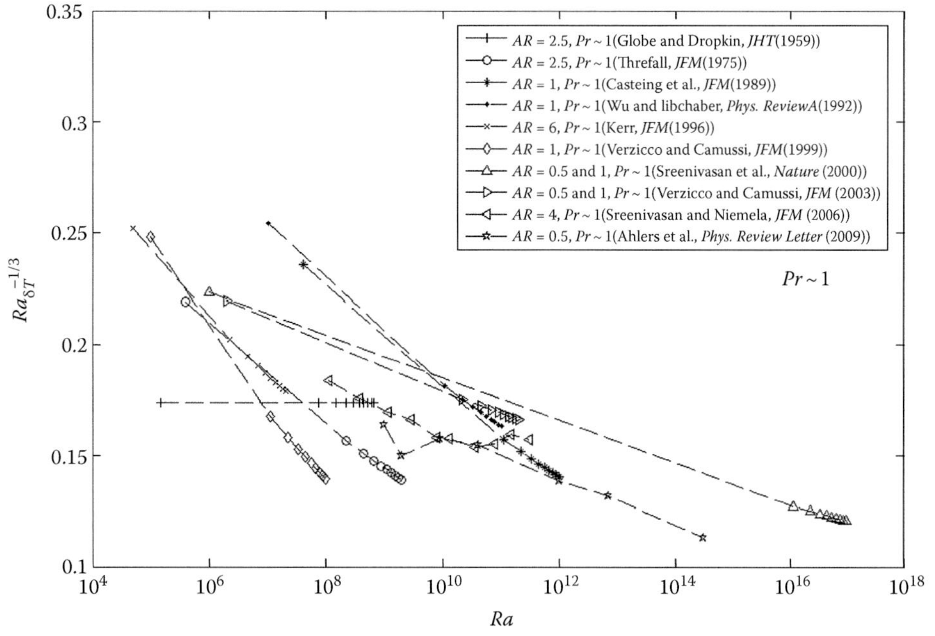

FIGURE 37.5 Nondimensional heat flux calculated from correlations for $Pr \sim 1$.

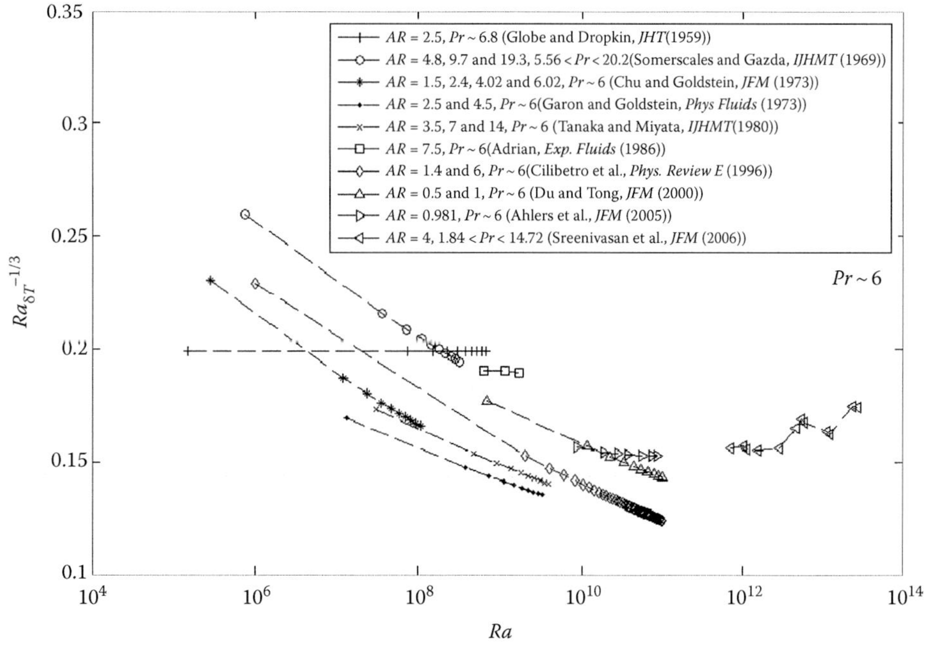

FIGURE 37.6 Nondimensional heat flux calculated from correlations for $Pr \sim 6$.

37.3.1.2.1 Temperature and Velocity Fields

Figure 37.7 shows the profiles of mean temperature (Figure 37.7a), and the profiles of the rms of the fluctuations of temperature (Figure 37.7b) and fluctuation of the two components of velocity (Figure 37.7c and d), scaled using the near-wall scales (T_w, Z_w, and U_w). As discussed above, the mean temperature profile shows a rapid variation near the wall and near isothermal core (Figure 37.7a). The temperature fluctuations profile shows a characteristic peak at about the conduction layer thickness level, which decays with increasing z; the peak appears due to fluid motion created by the near-wall plumes. The amplitude of the vertical velocity fluctuations (Figure 37.7d) increases monotonically with z. Note that in the outer region the flux $\overline{w'T'} = Q_0$ is constant, so we may expect $w' \sim 1/T'$.

TABLE 37.1 $Ra_{\delta_T}^{(-1/3)}$ Values Obtained from *Nu* to *Ra* Correlations for Different Types of Configurations Other than Rayleigh–Bernard Convection

Configuration	$Ra_{\delta T}^{-1/3}$	Conditions of Validity	Reference
Finite horizontal plate	0.12	$Ra_x > 8 \times 10^8$, $Pr = 6.0$	Kitamura and Kimura (*IJHMT*, 38, 1995)
Finite horizontal plate	0.13	$Ra > 5 \times 10^8$, $Pr = 6.0$	Fuji and Imura (*Intl. JHMT*, 15, 1972)
Mixed convection	0.158	$Ra_x > 39\ Re_x^{(5/3)}$, $Pr = 0.7$	Wang (*IJHT*, 104, 1982)
Convection driven by evaporative cooling	0.156	$3 \times 10^8 < Ra < 4 \times 10^9$, $Pr = 6.0$	Katsaros et al. (*JFM*, 83, 1977)
Vertical plate	0.1	$Ra > 10^9$, $Pr = 0.7$	Warner and Arpaci (*IJHMT*, 11, 1968)
Vertical rectangular cavity	0.12	$10^5 < Ra < 10^8$, $1 < Pr < 20$	McGregor and Emery (*IJHT*, 91, 1969)

The horizontal velocity fluctuations ($\overline{u'^2}$) will have contributions from the large-scale circulations and as well as from the wall plumes. The contribution from the outer flow is expected to scale with the Deardorff scale (W^*). For convection in $AR = 1$ cells, a single roll spanning the height has been observed, in which case a clear boundary layer due to this roll will be observed, and a time average will give a nonzero value; Figure 37.8a shows a time mean velocity profile and Figure 37.8b the horizontal component fluctuations both scaled by W^*; z is scaled by the wall scale. Data is from experiments in the 6.3 m high and 7.15 m diameter "Barrel of Ilmenau" (Puits et al., 2009). The mean wind is quite strong with magnitude going to about $2W^*$; at the highest $Ra = 9.77{*}10^{11}$, this corresponds to $\bar{u}_{max} \sim 0.6$ m/s. The $(\overline{u'^2})^{(1/2)}$ values are comparable to those obtained in lower *Ra* experiments (e.g., Deardroff and Willis, 1967; Fernando and Smith, 2001; Sun et al., 2008). The boundary layer thicknesses are very small, $\sim 20Z_w$ (16 mm for $Ra = 9.77{*}10^{11}$), comparable with the conduction layer thickness. The Deardroff scale seems to scale well both the horizontal mean velocity and velocity fluctuations for $AR = 1$ cells, at least for this limited set of data. For *large* AR cells, it may be expected that the large-scale circulations themselves may be moving about, in which case a fixed probe will measure the contributions from the circulations as fluctuations (in time). Whether to treat the large-scale circulation as a mean flow or a fluctuating component depends on what aspect is being studied. As noted above, however, the outer flow at least for the *Ra* values that have been studied has a negligible effect on the heat transfer rate; heat transfer is nearly independent of the nature and velocity value of the large-scale circulation.

Figure 37.9 shows the profiles of the various terms in the equation for the turbulence kinetic energy. The production of k, which is only due to buoyancy in R-B convection, is large and constant away from the wall and is balanced mainly by the viscous dissipation and to lesser extent by turbulent diffusion. The wall plumes probably contribute to the large positive value of the turbulent diffusion term obtained near the wall.

37.3.1.3 Models for the Plumes

For turbulent free convection over hot horizontal surfaces, heated fluid is released in narrow regions, either in the form of thermals or in the form of line plumes (Figure 37.4). Similarly, cold thermals or plumes are found below a cold horizontal surface.

Howard (see Turner, 1973) proposed a, now very widely known, model for *thermals*. It states that the near-wall convection is one of cyclical growth and eruption of a thermal conduction layer. A conduction layer grows by diffusion, becomes unstable, and erupts giving rise to a thermal. Then the cycle repeats. Thermals have been observed under special circumstances, but based on

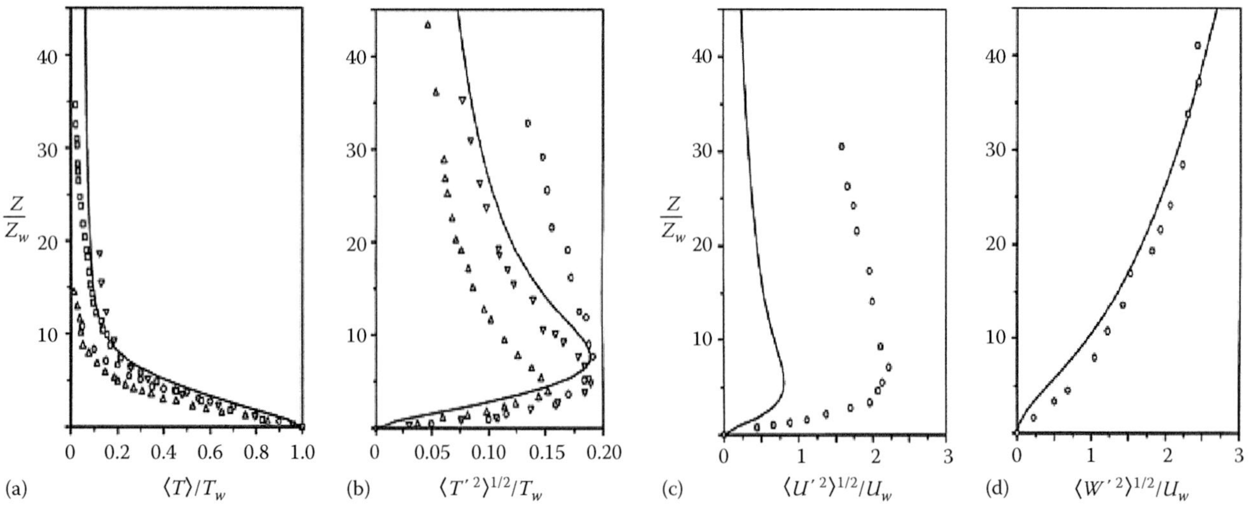

FIGURE 37.7 Near-wall distribution of various quantities calculated for air ($Pr \sim 6$) using the model (line). Symbols indicate data from various experiments done in air. Here < > indicates mean. (a) Mean temperature, (b) RMS of temperature fluctuations, (c) RMS of horizontal velocity fluctuations, (d) RMS of vertical velocity fluctuations. (From Theerthan, S.A. and Arakeri, J.H., *J. Fluid Mech.*, 373, 221, 1998. With permission.)

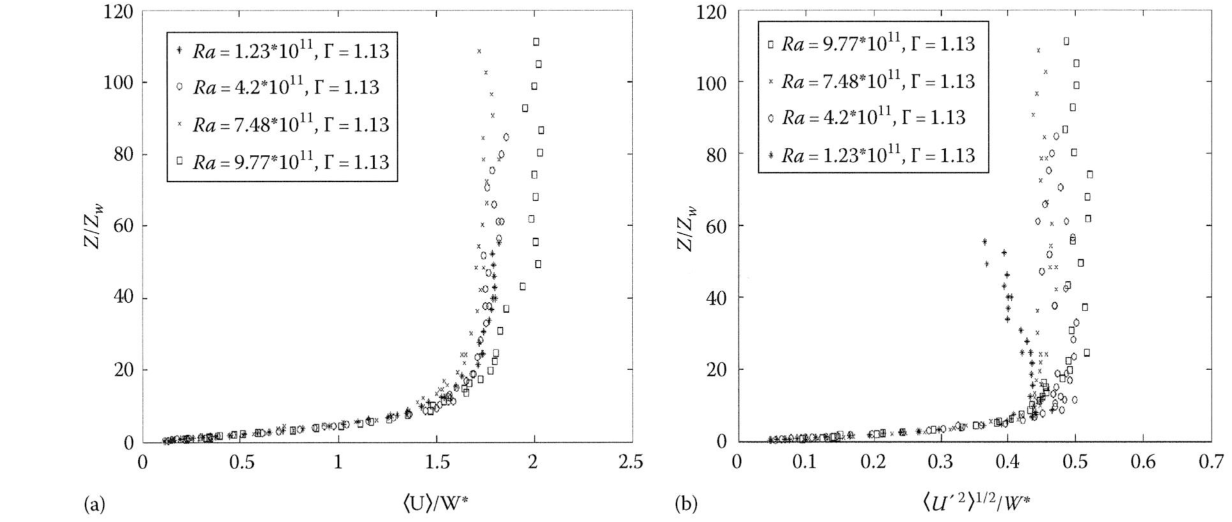

FIGURE 37.8 Profiles at different Ra and AR (Γ) = 1.13, for R-B convection in 6.3 m high cell. (a) The mean horizontal velocity and (b) the rms velocity fluctuation. Here < > indicates mean. (Data from du Puits, R., Resagk, C., and Thess, A. Structure of viscous boundary layers in turbulent Rayleigh-Benard convection. *Phy. Rev. E.*, 80, 036318, 2009, rescaled and personal communication.)

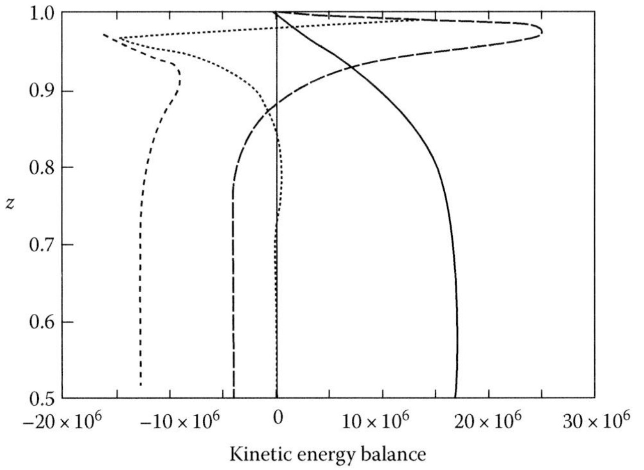

FIGURE 37.9 Vertical profiles of terms in kinetic energy equation at Ra = 2.5 × 10^6 near the top wall. –, production; ---, dissipation; ·····, molecular transfer; -- -- -, diffusion. (From Deardroff, J.W. and Willis, G.E., *J. Fluid Mech.*, 28, 675, 1967. With permission.)

the experimental observation that the near-wall flow consists of line plumes, Theerthan and Arakeri (1998) proposed a model consisting of a periodic array of viscous line plumes with a spacing λ (Figure 37.10). Clearly, this array of steady plumes is only a model of the real flow in which the plume spacing is not fixed and the plumes are randomly oriented and moving about.

Boundary layers on either side feed a plume with hot fluid; a plume results when the boundary layer becomes unstable. This model may be considered as a spatial equivalent of Howard's model. The heat is carried into the interior by the upward moving plume fluid. Figure 37.11 shows the flow associated with this model plume. The idea is that heat is carried by conduction near the walls, then by the plumes a little distance away from the walls, and then in the interior by the large-scale turbulence. Figure 37.12 shows the proportions of conductive heat transfer and convective heat transfer (i.e., by advection of fluid), and shows that conduction is negligible beyond Z_w > 15. Sub-layer vortices in turbulent wall-bounded flows play a role similar to the plumes;

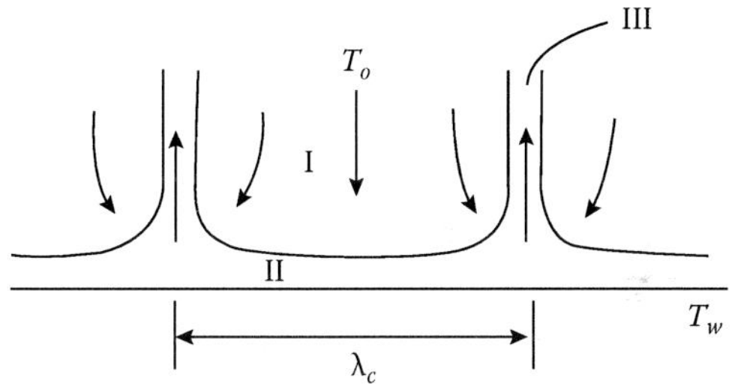

FIGURE 37.10 A periodic array of 2D plumes as a model for near-wall flow in turbulent convection.

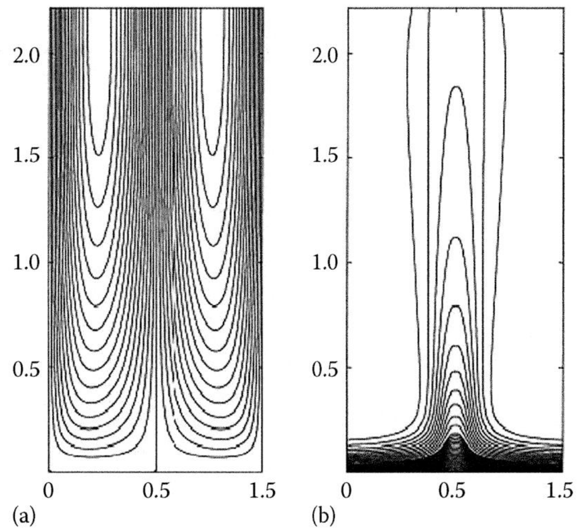

FIGURE 37.11 Streamlines (a) and isotherms (b) associated with a plume.

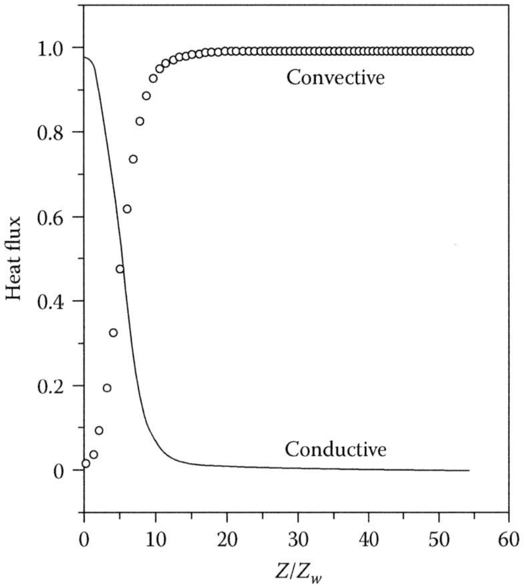

FIGURE 37.12 Proportions of conduction and convective heat transfer near a wall. At (Z/Z_w) >15, heat transfer is entirely by convection. (From Theerthan, S.A. and Arakeri, J.H., *J. Fluid Mech.*, 373, 221, 1998. With permission.)

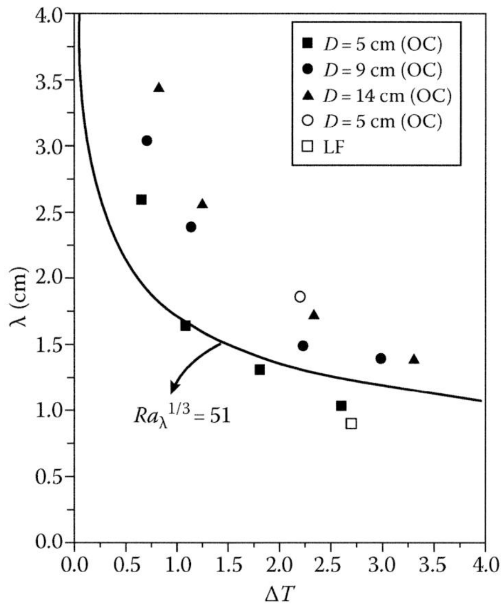

FIGURE 37.13 Comparisons of measured (symbols) and predicted (line) plume spacing. (From Theerthan, S.A. and Arakeri, J.H., *Phys. Fluids*, 12(4), 884, 2000. With permission.)

both represent a transition from molecular diffusion at the wall to turbulent transport far away from the wall. The plumes represent the mixing region proposed by Castaing et al. (1989).

One main result of the model is the prediction of the average spacing between adjacent plumes.

$$Ra_{\lambda}^{1/3} = \left(\frac{g\beta\Delta T\lambda^3}{\nu\alpha}\right)^{1/3} = 52Pr^{-0.012} \tag{37.18}$$

$$Ra_{\lambda_f}^{1/4} = \left(\frac{g\beta q\lambda^4}{k\nu\alpha}\right)^{1/4} = 33.5Pr^{0.0065} \tag{37.19}$$

These relations state that the Rayleigh number (Ra_{λ}) based on plume spacing and ΔT or the flux Rayleigh number $(Ra_{\lambda f})$ based on plume spacing and heat flux is nearly a constant with a weak dependence on Prandtl number (for $Pr \sim 1$). The model extended to large Pr (Puthenveettil and Arakeri, 2005) gives nondimensional spacing of about 90. One can predict, knowing either the temperature difference or the bottom heat flux, the average plume spacing in turbulent free convection over a horizontal surface. For a given fluid,

$$\lambda \sim \frac{1}{(\Delta T)^{1/3}} \quad \text{or} \quad \lambda \sim \frac{1}{(q)^{1/4}}$$

Higher temperature differences or larger fluxes leads to closer spaced and more number of plumes per unit area, which is evident in Figure 37.4a through c. Assuming the number of plumes, $N_p \sim (AR)D/\lambda$, we can show $N_p \sim AR(Ra)^{(1/3)}$, a scaling obtained in (Zhou and Xia, 2010). Some experimental verification of the prediction for plume spacing is shown in Figure 37.13.

The model predicts profiles of mean temperature, temperature fluctuations, and of velocity fluctuations (Figure 37.7). Except for the horizontal component velocity fluctuations, all the quantities agree reasonably well with the experimental data. The disagreement between model prediction and experimental data in the case of horizontal velocity component fluctuations is because the model does not take into account the horizontal movement of the plumes and the large-scale circulation, which as discussed above is an outer scale phenomenon (Deardroff). These results can be used to generate wall functions in k–ε type turbulence modeling, and large eddy simulation (LES) of turbulent free convection flows over horizontal surfaces.

37.3.2 Turbulent Mixed Convection over a Horizontal Surface

We will now consider turbulent mixed convection over a horizontal surface. In free convection the imposed flow is zero; in forced convection the imposed flow velocity is so large that the flow is dominated by shear, buoyancy is unimportant. Between these two extremes *mixed convection flow* is obtained where forces due to both buoyancy and shear are of the same order. A number of experiments have studied the flow structure and measured the heat transfer in mixed convection flows.

Mixed convection flows are frequently obtained when a "light wind" blows over a horizontal heated surface. Even in R-B convection, the large-scale circulation may be considered as an imposed flow on the top cold plate and the bottom heated plate. One certain fact is that as the Rayleigh number increases, the outer flow characterized by the Deardorff scale becomes stronger in relation to the inner flow, characterized by the plumes; it can be shown that $W^*/W_o \sim Ra^{(1/9)}$. So at high enough Rayleigh numbers we may expect the outer flow to start affecting the wall flow and in particular the heat transfer rate. We may think of the outer flow as an eddy scaling with the height D of the fluid layer. To the wall region this outer flow is like a wind (Figure 37.1). It appears that the wind at least for the *Ras* achieved in the laboratory experiments has a small and unclear effect on the heat flux, as evidenced by the fact that the exponent "*n*" in the Nusselt number correlation is close to 1/3. The main effect of the large-scale circulation is that it aligns the plumes in the flow direction.

The alignment of plumes due to an outer flow mentioned above occurs in other related situations—free convection on a hot surface that is inclined to the horizontal (see, Theerthan and Arakeri, 1998) and in mixed convection on a horizontal surface. In the first case the inclination causes buoyancy-driven mean flow along the surface. In the second case, which is of interest to us, a small flow is imposed along the surface.

The mixed-convection turbulent flow is best understood by considering the equation for the turbulent kinetic energy (k). The flux Richardson number (Turner, 1973), defined as the ratio of negative of buoyant production to shear production of k,

$$R_f = \frac{-g\beta\overline{w'T'}}{-\overline{u'w'}\dfrac{\partial\bar{u}}{\partial z}}$$

$R_f > 0$ is obtained in a stable environment, where buoyancy attenuates the turbulence. A combination of turbulence destruction by the stable stratification and by viscous dissipation gives a critical value of $R_f = 0.25$, above which turbulence ceases to be self supporting. In an unstable environment, R_f is negative and both buoyancy and shear produce turbulent kinetic energy.

Richardson number may be defined based on the more easily measured gradients of the mean temperature and mean velocity,

$$R_i = \frac{g\beta\left(\partial\bar{T}/\partial z\right)}{\left(\partial\bar{u}/\partial z\right)^2}$$

37.3.2.1 Monin–Obukhov Theory

Monin–Obukhov (M-O) theory (Monin and Yaglom, 1971; Kader and Yaglom, 1990) is the standard theory that describes turbulent mixed convection over a horizontal surface. The theory states that there are two regions. At and near the wall, the shear is high enough that forced convection conditions are obtained and turbulence production is mainly due to shear, $-\overline{u'w'}\dfrac{\partial\bar{u}}{\partial z}$. Away from the wall, where the shear is small, free convection dominates and turbulence production is by buoyancy, $\beta g\overline{w'T'}$. In a turbulent boundary layer, in the constant shear stress region,

$$-\overline{u'w'} = u_*^2,\quad \frac{\partial\bar{u}}{\partial z} = \frac{u_*}{\kappa z}$$

here u_* is the friction velocity given by

$$u_* = \sqrt[2]{\frac{\tau_w}{\rho}}$$

τ_w is the wall shear stress and κ the Karman constant. In this region, thus, shear production of turbulence is $(u_*^3)/\kappa z$. Equating the shear and buoyant productions, we obtain the Monin–Obhukov length scale:

$$L_m = -\frac{(u_*^3/\kappa)}{\beta g\overline{w'T'}}$$

Using the shear production relation in the constant stress region, we get $R_f = z/L_m$. (L_m is negative in unstably stratified flows, but we will ignore the negative sign in the following discussion).

Thus three layers are obtained (Kader and Yaglom, 1990): for $z \gg L_m$, free convection; for $z \ll L_m$, forced convection; and for $z \sim L_m$ mixed convection. This regime is schematically shown in Figure 37.14b. M-O theory implicitly makes the assumption that $Re_L = (u_* L_m/\nu) \rightarrow \infty$, and the flow near the wall corresponds to an equilibrium turbulent boundary layer. In particular, the velocity fluctuations, $u', w' \sim u_*$. Importantly, as in forced convection flows, the wall shear stress, scaling as ρu_∞^2 determines the heat flux. We may write the mean velocity gradients $\dfrac{d\bar{u}}{dz}, \dfrac{d\bar{T}}{dz}$ as functions of z/L_m, with the forced convection limits obtained as $z/L_m \rightarrow 0$.

Now consider the case of *very low-wind speeds,* that is the limit of $L_m \rightarrow 0$, in particular $Re_L \rightarrow 0$. It has been found in experiments, under these conditions, when the free stream velocity is very low that the heat flux is independent of the free stream velocity and is close to the value obtained in the free convection limit (Wang, 1982). The flow near the wall is dominated by the wall plumes such as those seen in Figure 37.4.

Under these very low velocity conditions the Monin–Obukhov theory is inapplicable. In fact, turbulent mixed convection may be demarcated into two regimes: high-speed mixed convection (HSM) where the M-O theory is applicable and low-speed mixed convection (LSM) where an alternate theory is needed. In LSM, the *whole boundary layer is dominated by free convection*; buoyancy dominates across the height of the boundary layer. Figure 37.14 shows schematically the differences between LSM and HSM. In forced convection and in HSM, the shear determines the heat flux; in LSM, the buoyancy, in particular, the plumes and the vertical mixing they cause, determine the Reynolds stress and the wall friction. Now instead of friction velocity (u_*) and friction length (ν/u_*), the near-wall scales of free convection (U_w, Z_w, T_w) become the relevant scales. Extending the plume model (Figure 37.15), the near-wall distributions for mean velocity and Reynolds shear stress may be computed, and the scaling for the different quantities are obtained (Ramesh Chandra, 2000): streamwise velocity fluctuation $\sim U_o$, vertical and lateral

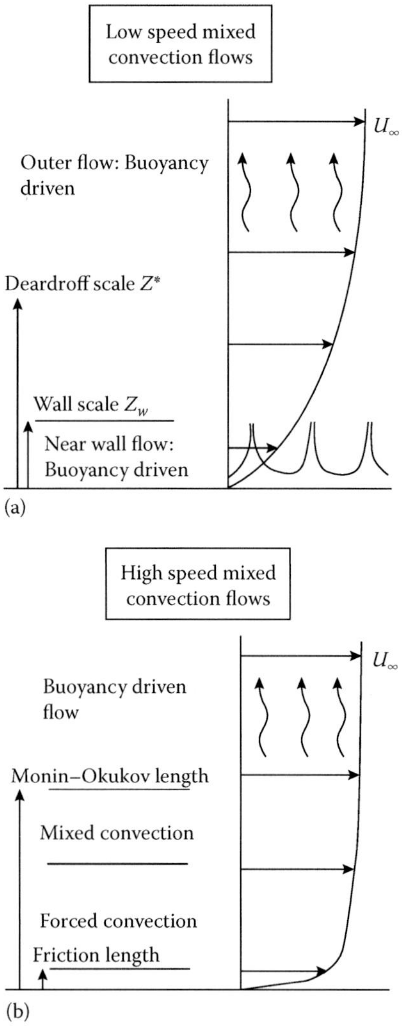

FIGURE 37.14 (a) Low-speed mixed convection. (b) High-speed mixed convection.

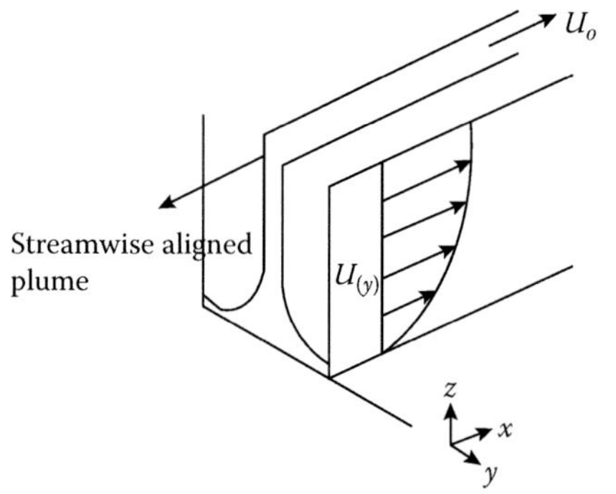

FIGURE 37.15 Model for low-speed turbulent convection. Mixed convection.

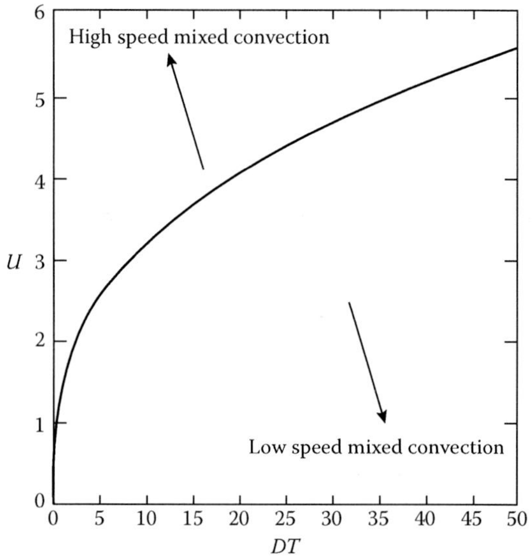

FIGURE 37.16 Demarcation of low-speed and high-speed mixed convection regimes for air. Here $U = U_\infty$ and $DT = \Delta T$.

velocity fluctuations $\sim U_w$, Reynolds shear stress $\sim U_o U_w$. Here, U_o can be taken as U_∞.

The boundary between HSM and LSM may be obtained by assuming the velocity scale of fluctuation, U_w, is equal to the velocity scale of fluctuations in a forced turbulent flow, u_*. This is equivalent to assuming $L_m \sim$ sublayer thickness, $15\nu/u_*$. Assuming the skin friction coefficient, $C_f = 0.004 = 2u_*^2/U\infty^2$, we get $U_\infty/U_w = 22$ as the boundary between LSM and HSM, which is plotted for air in Figure 37.16. Whether the flow is HSM or LSM is determined by the ratio of the buoyancy velocity U_w to the free stream velocity U_∞. It is interesting that in air for a temperature difference of about 10 K, free stream velocities as high as 3 m/s would still correspond to LSM. No skin friction and velocity measurements in LSM have been reported to test the predictions of the model. In the atmosphere, convection under low wind conditions is often obtained, especially in the tropics. Similar scaling as above has been used to collapse drag data under low wind conditions in the atmosphere (Rao and Narasimha, 2006, and references there in).

In the context of R-B convection, the large-scale circulation is a wind for convection over the hot and cold plates. At the *Ra*s that have been studied so far, the wind is not strong enough, the regime appears to be LSM and heat transfer rate is essentially unaffected by the wind. Kraichnan (see Ahlers et al., 2009) predicted at high enough *Ra* ($\sim 10^{17}$), the wind will be strong enough, and the $Nu \sim Ra^{(1/3)}$ free convection scaling will change to $Nu \sim Ra^{(1/2)}$ forced convection scaling. This regime is commonly referred to as the "ultimate regime" of R-B convection.

37.3.3 Convection in Long Vertical Tubes: Ultimate Regime

An interesting and a different type of convection is obtained in a long vertical tube due to an unstable density difference (Figures 37.1c and 37.17) (Cholemari and Arakeri, 2009; Schmidt et al., 2011).

This convection may be considered to be low AR R-B convection but with the bottom and top plates removed. Removal of the plates, it turns out, changes totally the convection and scalings

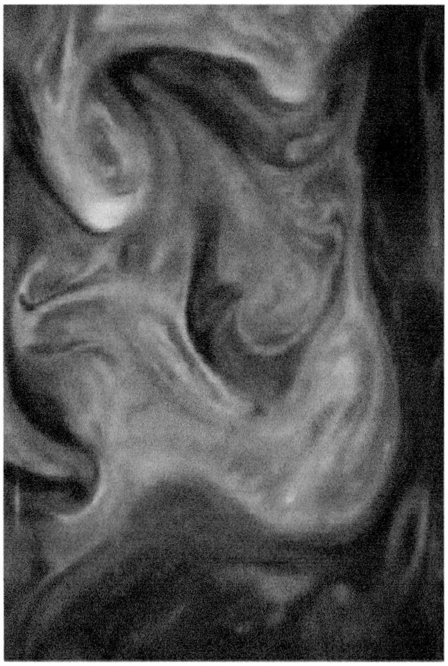

FIGURE 37.17 Laser-induced fluorescence visualization of turbulent convection in the midsection of a vertical pipe.

for flux and velocity. The flow has several unique features: (1) For sufficiently tall tubes ($L/d \geq 10$), the convection is axially homogeneous, driven by a linear density gradient. (2) The mean velocity and mean Reynolds shear stress at any point are zero; thus, all turbulence production is by buoyancy and none by shear. (3) At sufficiently high Rayleigh numbers, the Nusselt number and Reynolds number scalings are given by

$$Nu = C_1 Ra^{1/2} Pr^{1/2}, \quad Re = C_2 Ra^{1/2} Pr^{-1/2}$$

implying heat flux independent of thermal diffusivity and viscosity. Thus the $Nu \sim Ra^{1/2}$ "ultimate regime" scaling expected in R-B convection at very high Ra is achieved relatively easily in the convection with "open ends." Moreover the fluxes and Nusselt numbers are orders of magnitude larger than in R-B convection. Though this type of convection is not directly obtained in the atmosphere, it provides a prototype for fundamental studies of buoyancy turbulence interaction especially at high Taylor microscale Reynolds numbers, which are much more easily obtained than in R-B convection. Also such type of convective flows may be obtained within buildings, for example, in the stairway between rooms with bottom room being warmer, say due to a fire.

37.4 Applications

Examples of convection abound in environmental flows. In the atmosphere, mesoscale shallow convection with vertical scales ~1–2 km and horizontal scales of tens of kilometers display many of the features found in laboratory scale experiments described above: narrow plumes along which heated fluid rises accompanied by slow sinking motion of the colder fluid between the plumes; an aspect ratio of plume spacing to height of convection of 5–20; alignment of the plumes by weak winds along the wind direction. A major difference is these plumes would be turbulent unlike the viscous plumes found in the laboratory scales. Nearer the ground, locally, these large-scale convective motions would cause, depending on the velocity magnitude, forced or free convection. It is likely that a cascade of similar plume like structures make up the total convection.

In the atmosphere, often moisture and heat transfer transport simultaneously occur with both composition and temperature contributing to buoyancy force, and the relations have to be suitably modified. Associated with water bodies (oceans, lakes, tanks), convection may be driven in the air above by evaporation and temperature difference, and below the water surface, convection would result from cooling (Spangenberg and Rowland, 1961) and salt concentration increase (in the case of oceans) due to evaporation and from any temperature difference. At large scales, the earth's rotation will also start affecting the convection (see Fernando and Smith, 2001).

At smaller scales in natural and urban environments, convection is observed over heated surfaces (or colder surfaces), such as roof tops and side walls of buildings, large grounds, and roads. Exchange convection of the type obtained in long vertical pipes would be obtained in tall enclosed spaces between or within buildings. Judicious application of the relations given above may be used for estimating heat transfer rates ($Ra_{\delta_T}^{-1/3} \sim 0.15$) and velocities.

37.5 Major Challenges

It appears that for a variety of free convection flows on solid surfaces, the heat transfer rate is reasonably well represented by $Ra_{\delta_T}^{-1/3} \sim 0.15$. Most of the detailed measurements have been done at laboratory scales under controlled conditions. High Rayleigh numbers have been achieved in $AR \sim 1$ cells using special fluids (Niemela et al., 2000). To achieve high Rayleigh numbers and large aspect ratios of relevance to flows encountered in the environment would require large experimental setups similar to the 7 m high Barrel of Ilmenau (du Puits et al., 2007). One question is at what Ra would the $Nu \sim Ra^{(1/3)}$ scaling transition to the so-called ultimate regime scaling of $Nu \sim Ra^{(1/2)}$ predicted by Kraichnan, and whether this transition Ra is relevant to environmental flows. It is quite certain that the single coherent roll observed in $AR \sim 1$ cells would change to many, probably chaotic, rolls at large ARs, although all indications are that heat transfer rates will be minimally affected.

Further work needs to be done to study the effect of roughness on free convection flows, in particular the heat transfer rates. Table 37.2 lists calculated values of several lengths (δ_T, plume spacing λ) and near-wall velocity scales for convection over horizontal surfaces in water and air. Roughness would start affecting the heat transfer when its height is larger than the conduction layer thickness, δ_T, which is about 5 mm (Table 37.2) for $\Delta T \sim 15$ K; such roughness heights are commonly observed in atmospheric flows. The limited amount data seems to suggest an

TABLE 37.2 Various Parameters for Turbulent Convection in Water and Air

Water					
ΔT(K)	5	10	20	30	50
δ_T (mm)	1.1	0.9	0.7	0.6	0.5
λ (mm)	11.4	9.1	7.2	6.3	5.3
q (w/m^2)	2,743	6,913	17,419	29,909	59,102
U_w (mm/s)	1.6	2.1	2.6	3.0	3.5
W_o (mm/s)	0.7	0.9	1.1	1.3	1.5
Air					
ΔT (k)	5	10	20	30	50
δ_T (mm)	7.1	5.6	4.4	3.9	3.3
λ (mm)	66	52	42	36	31
q (w/m^2)	18.06	45.5	113	197	389
U_w (mm/s)	14.4	18.2	22.9	26.2	31.1
W_o (mm/s)	10.2	12.8	16.1	18.5	21.9

increase of 70% in heat transfer rates due to roughness. And a second important question would be whether the $Nu \sim Ra^{(1/3)}$ is maintained on rough surfaces. Similarly, systematic studies of effect of wall conductivity are missing (Hunt et al., 2003).

Large-scale controlled experiments on mixed convection flows, both high-speed mixed convection flows (Figure 37.14), where Monin–Obhukov theory would be applicable, and low-speed mixed convection flows would be useful. Detailed measurements of mean and turbulent temperature and velocity fields would help in development of turbulence models for simulating environmental flows. The coexistence and interaction of both free and forced convection scales make such flows interesting and complicated.

Acknowledgments

We thank Dr. du Puits for providing data related to Figure 37.8, and Aloka for help in the preparation of this chapter.

References

Adrian, R. J., Ferreira, R. T. D. S., and Boberg, T. Turbulent thermal convection in wide horizontal fluid layers. *Exp. Fluids*, 4, 121–141, 1986.

Ahlers, G., Grossmann, S., and Lohse, D. Heat transfer and large scale dynamics in turbulent Rayleigh-Bénard convection. *Rev. Mod. Phys*., 81, 503–538, 2009.

Castaing, B., Gunaratne, G., Heslot, F., Kadanoff, L., Libchaber, A., Thomae, S., Wu, X., Zaleski, S., and Zanetti, G. Scaling of hard thermal turbulence in Rayleigh-Bénard convection. *J. Fluid Mech*., 204, 1–30, 1989.

Cholemari, M. R. and Arakeri, J. H. Axially homogeneous, zero mean flow buoyancy-driven turbulence in a vertical pipe. *J. Fluid Mech*., 621, 69–102, 2009.

Chu, T. Y. and Golddtein, R. J. Turbulent convection in a horizontal layer of water. *J. Fluid Mech*., 60, 141–159, 1973.

Deardroff, J. W. and Willis, G. E. Investigation of turbulent thermal convection between horizontal plates. *J. Fluid Mech*., 28, 675–704, 1967.

Fernando, H. J. S. and Smith IV, D. C. Vortex structures in geophysical convection. *Eur. J. Mech. B—Fluids*, 20, 437–470, 2001.

Fuji, T., and Imura, H. Natural convection heat transfer from a plate with arbitrary inclination. *Int. J. Heat Mass Transfer.*, Vol. 15, 755–767, 1972.

Globe, S. and Dropkin, D. Natural convection heat transfer in liquids confined by two horizontal plates and heated from below. *J. Heat Transfer*, 81, 24–28, 1959.

Grossmann, S. and Lohse, D. Scaling in thermal convection: A unifying theory. *J. Fluid Mech*., 407, 27, 2000.

Hunt, J. C. R., Vrieling, A. J., Nieuwstadt, F. T. M., and Fernando, H. J. S. The influence of the thermal diffusivity of the lower boundary on eddy motion in convection. *J. Fluid Mech*., 491, 183–205, 2003.

Husar, R. B. and Sparrow, E. M., Patterns of free convection flow adjacent to horizontal heated surfaces. *Int. J. Heat Mass Transfer*, 11, 1206–1208, 1968.

Kader, B. A. and Yaglom, A. M. Mean fields and fluctuation moments in unstably stratified turbulent boundary layers. *J. Fluid Mech*., 212, 637–662, 1990.

Katsaros, K. B., Liu, W. T., Businger, J. A., and Tillman, J. E. Heat transport and thermal structure in the interfacial boundary layer measured in an open tank of water in turbulent free convection. *J. Fluid Mech*., 83, 311–335, 1977.

Kitamura, K. and Kimura, F. Heat transfer and fluid flow of natural convection adjacent to upward facing horizontal plates. *Int. J. Heat Mass Transfer*, 38, 3149–3159, 1995.

Krishnamurthy, R. On the transition to turbulent convection part I. *J. Fluid Mech*., 42, 295–307, 1970.

Macgregor, R. K., and Emery, A. F. Free convection through vertical plane layers - moderate and high Prandtl number fluids. *J. Heat Transfer*., Vol. 91, 391–403, 1969.

Monin, A. S. and Yaglom, A. M. *Statistical Fluid Mechanics-Mechanics of Turbulence*, Vol. 1, Dover Publications, 2007.

Niemela, J. J., Skrbek, L., Sreenivasan, K. R., and Donnely, R. J. Turbulent convection at very high Rayleigh number. *Nature*, 404, 837–840, 2000.

du Puits, R., Resagk, C., and Thess, A. Structure of viscous boundary layers in turbulent Rayleigh-Benard convection. *Phy. Rev. E*., 80, 036318, 2009.

du Puits, R., Resagk, C., and Tilgner, A. Structure of thermal boundary layers in Rayleigh-Bernard convection. *J. Fluid Mech*., 572, 231–254, 2007.

Puthenveettil, B. A. and Arakeri, J. H. Plume structure in high Rayleigh number convection. *J. Fluid Mech*., 542, 217–249, 2005.

Ramesh Chandra, D. S. Turbulent mixed convection. MSc (Engineering) thesis, Department of Mechanical Engineering, Indian Institute of Science, Bangalore, India, 2000.

Rao, K. G. and Narasimha, R. Heat-flux scaling for weakly forced turbulent convection in the atmosphere. *J. Fluid Mech*., 547, 115–135, 2006.

Rohsenow, W. M., Hartnett, J. P., and Ganić, E. N. *Handbook of Heat Transfer Fundamentals*. McGraw-Hill, New York, 1985.

Schmidt, L. E., Calzavarini, E., Lohse, D., Toschi, F., and Verzicco, R. Axially-homogeneous Rayleigh-Benard convection in a cylindrical cell. *J. Fluid. Mech.*, online, 1–17, 2011.

Siggia, E. D. High Rayleigh number convection. *Annu. Rev. Fluid Mech.*, 26, 137–168, 1994.

Spangenberg, W. G. and Rowland, W. G. Convective circulation in water induced by evaporative cooling. *Phys. Fluids*, 4, 743–750, 1961.

Sun, C., Cheung, Y., and Xia, K. Experimental studies of the viscous boundary layer properties in turbulent Rayleigh–Benard convection. *J. Fluid Mech.*, 605, 79–113, 2008.

Theerthan, S. A. and Arakeri, J. H. A model for near-wall dynamics in turbulent Rayleigh-Bénard convection. *J. Fluid Mech.*, 373, 221–254, 1998.

Theerthan, S. A. and Arakeri, J. H. Planform structure and heat transfer in turbulent free convection over horizontal surfaces. *Phys. Fluids*, 12(4), 884–894, 2000.

Townsend, A. A. *The Structure of Turbulent Flows*. Cambridge University Press, London, UK, 1976.

Turner, J. S. *Buoyancy Effects in Fluids*. Cambridge University Press, London, U.K., 1973.

Wang, X. A. An experimental study of mixed, forced and free convection heat transfer from a horizontal flat plate exposed to air. *J. Heat Transfer*, 104, 139–144, 1982.

Zhou, Q. and Xia, K. Physical and geometrical properties of thermal plumes in turbulent Rayleigh–Bénard convection. *New J. Phys.*, 12, 075006, 2010.

38

Convection (Rotating Fluids)

Robert E. Ecke
Los Alamos National Laboratory

38.1 Introduction

The combination of thermal buoyancy and rotation plays a crucial role in the formation of structures and the transport of heat in planetary and stellar atmospheres and in oceans. On Earth, rotation is not very important on the scale of convective motions in the atmosphere (Julien et al. 1996) but contributes in both oceanic flows and in the motion in the Earth's core (King et al. 2009). In particular, rotation influences geophysical flows by inducing vertical vorticity in lateral velocity fields through the action of the Coriolis force. There have been a number of reviews of rotating convection from the perspectives of geophysical systems (Boubnov and Golitsyn 1995) and of the formation of vortical structures (Fernando and Smith 2001). Here we focus on the relationships among heat transfer, vortical structures, and fluctuations of temperature, velocity, and vertical vorticity for the well-posed Rayleigh–Bénard (RB) convection conditions with rotation about a vertical axis. For RB convection, as shown in Figure 38.1, the system is bounded above and below by conducting boundaries; there is a temperature difference ΔT across the layer of height d, and the axis of rotation and gravity are aligned along the z-axis, perpendicular to the plane of the top and bottom boundaries. Although such a geometry does not have the full complexity of geophysical situations, it is a realization that includes the essential ingredients of rotation and buoyancy and is amenable to precise measures of steady-state heat transport.

The phase diagram for rotating RB convection is shown in Figure 38.2 in the space of Ra number, the ratio of buoyancy to viscous/thermal dissipation and proportional to ΔT, and the Ta number, a dimensionless measure of rotation proportional to the square of the angular rotation rate Ω, that is, $Ta \sim \Omega^2$. A similar phase diagram (Fernando et al. 1991) can be constructed for systems with an open top boundary by plotting the flux Rayleigh number $Ra_f = Nu\ Ra$ where Nu is the molecular-conduction-normalized heat transport. For small Ra, the system is stable, there is no fluid motion, and heat is transported by molecular diffusion. For a laterally infinite system (Chandrasekhar 1961), convection begins when $Ra > Ra_c$. As indicated in Figure 38.2, convection is suppressed by rotation so that Ra_c increases with Ta; at large Ta, one has $Ra_c \sim Ta^{2/3}$. In the presence of lateral (insulating) walls, traveling modes localized near the boundaries form for $Ra < Ra_c$ at large enough Ta (Zhong et al. 1993, Liu and Ecke 1999) as indicated in Figure 38.2. For finite amplitude convection, heat transport increases approximately linearly with increasing $Ra > Ra_c$ (e.g., Zhong et al. 1993) before transitioning to a much more gradual increase for Ra above about 5 Ra_c in the turbulent regime. In Figure 38.3, we show representative data of heat flux Nu (normalized by heat flux from molecular diffusion) for modest $Ta < 10^7$ where these basic trends can be seen. For turbulent convection, one finds buoyancy-dominated flow at small Ta and rotation-dominated flow at large Ta with an approximate separation for $Ra = Pr\ Ta$ where Pr is the ratio of viscous and thermal diffusion.

Although we will be mostly interested in turbulent conditions, it is instructive to consider weaker solutions near the onset of convection. Without rotation, convection occurs in the form of roll pairs with up-flow and down-flow regions alternating periodically in space. With rotation, bulk convection is suppressed, and the form of the convection is substantially modified into vortical structures such as illustrated in Figure 38.1, where black shading indicates cyclonic (with the rotation direction) circulation and gray shading indicates anticyclonic (counter to the rotation direction) circulation. The sense of the rotation of the flow is determined by the action of the Coriolis force on flow that is either converging or diverging

Handbook of Environmental Fluid Dynamics, Volume One, edited by Harindra Joseph Shermal Fernando.
ISBN: 978-1-4398-1669-1.

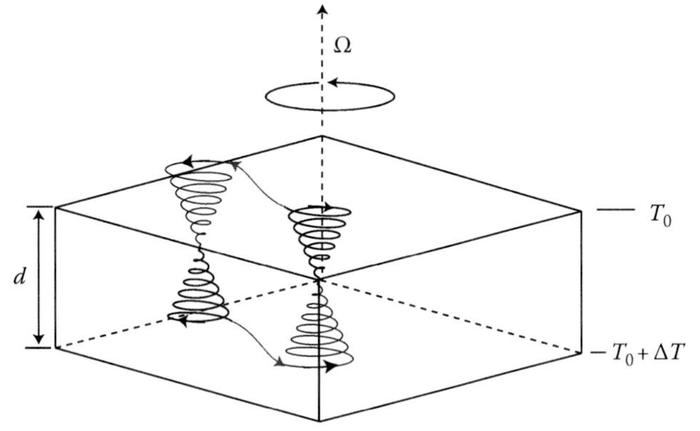

FIGURE 38.1 Schematic illustration of rotating thermal convection (following Veronis 1959) in which a temperature difference ΔT is maintained between no-slip top and bottom boundaries that are separated by a distance d. The convection cell is rotated about a vertical axis with angular rotation rate Ω. The black (gray) shading of the flow paths indicate cyclonic (anticyclonic) portions of the flow. Note the change from cyclonic to anticyclonic circulation at the mid-plane.

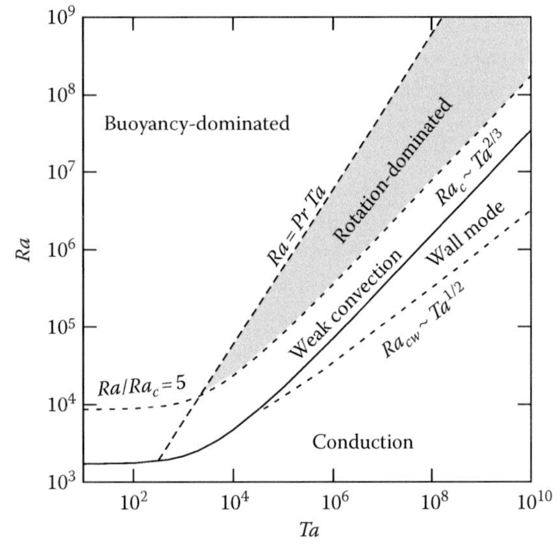

FIGURE 38.2 Phase diagram of rotating convection in space of Rayleigh number Ra and Taylor number Ta. Bulk convection is suppressed by rotation so that at large Ta the bulk critical Rayleigh number $Ra_c \sim Ta^{2/3}$ (Chandrasekhar 1961). In the presence of lateral boundaries, a traveling wall mode is unstable for Ra below the bulk onset (Zhong et al. 1993) with $Ra_{cw} \sim Ta^{1/2}$. For Ra/Ra_c of order 5 or less, convection is relatively weak. The line labeled $Ra = Pr\ Ta$ approximately divides buoyancy-dominated flow from rotation-dominated convection. The region of interest here is turbulent rotating convection that is observed for parameter values indicated in the shaded region.

near boundaries with a change in the sense of rotation about the mid-plane (Veronis 1959). For weak convection, there remains an approximate symmetry of cyclonic and anticylonic vortices, a condition roughly shown in the top view of Figure 38.4 (Sakai 1997), which was visualized in water ($Pr \approx 6$) rather close to the onset of bulk convection, that is, $Ra/Ra_c = 8$ with $Ra = 10^7$, using thermochromic liquid crystal particles. Note

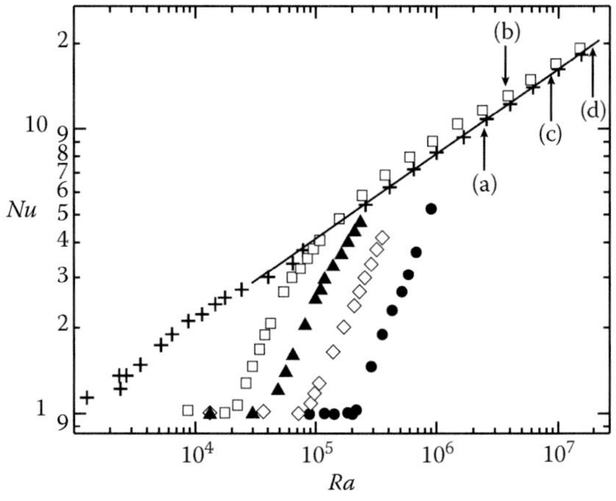

FIGURE 38.3 Normalized heat transport Nu as a function of Ra in a cylindrical convection cell (Zhong et al. 1993) for different Ta: plus (0), open square (3.5×10^5), solid triangle (1.3×10^6), open diamond (5.3×10^6), and solid circle (1.8×10^7). Solid line is $Nu \sim Ra^{0.29}$. Labels a–d correspond to images in Figure 22 of Zhong et al., 1993. (Reproduced with permission from Zhong, F. et al., *J. Fluid Mech.*, 249, 135, 1993. Copyright 2006 Cambridge Journals.)

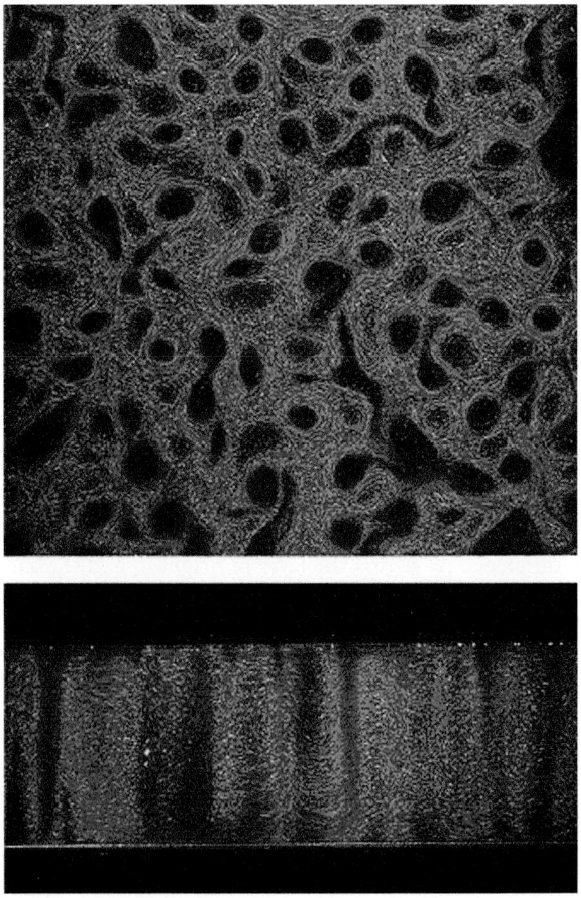

FIGURE 38.4 **(See color insert.)** Thermochromic liquid crystal flow visualization (upper–top view, lower–side view) of rotating convection in water with $Pr = 6$, $Ra = 8 \times 10^6$, $Ta = 10^8$, $Ro = 0.1$, $Ra/Ra_c = 8$. (Reproduced with permission from Sakai, S., *J. Fluid Mech.*, 333, 85, 1997. Copyright 2006 Cambridge Journals.)

also from the side view in Figure 38.4 that the vortices extend vertically over most of the cell height, a structure known as a Taylor column that arises from the Taylor–Proudman theorem discussed in detail later.

38.2 Principles

The equations of rotating RB convection are those representing conservation of mass, momentum, and energy. We assume that thermal expansion is relatively weak so that the Boussinesq approximation is valid and that centrifugal effects can be collected in the pressure. In that case, the equations of motion are

$$\frac{\partial \vec{u}}{\partial t} + \vec{u}\cdot\nabla\vec{u} = -\frac{\nabla P}{\rho} + \nu\nabla^2\vec{u} + g\alpha\theta\hat{z} - 2\Omega\hat{z}\times\vec{u},$$
$$\frac{\partial \theta}{\partial t} + \vec{u}\cdot\nabla\theta = \kappa\nabla^2\theta, \quad (38.1)$$
$$\nabla\cdot\vec{u} = 0,$$

where
- $\vec{u}(\vec{r},t)$ is the full 3D velocity field
- $\theta(\vec{r},t) = [T(\vec{r},t) - T_0]/\Delta T$ is the normalized temperature field
- $P(\vec{r}, t)$ is the pressure field
- ρ is the mean fluid density
- g is the acceleration of gravity
- α is the thermal expansion coefficient
- ν is the kinematic viscosity
- Ω is the angular rotation rate
- κ is the thermal diffusivity

The space vector $\vec{r}$ has component unit vectors $\{\hat{x}, \hat{y}, \hat{z}\}$ where $\hat{z}$ is the vertical direction aligned with gravity, rotation, and the applied temperature gradient. The boundary conditions for temperature are that the upper boundary located at $z = d$ is at temperature T_0 whereas the bottom boundary located at $z = 0$ has a fixed temperature $T_0 + \Delta T$. Using scales of time, length, and temperature as d^2/ν, d, and ΔT, respectively, we have $\theta(0) = 1$ and $\theta(1) = 0$. For no-slip upper and lower boundaries, we have $u_z = \partial u_x/\partial z = \partial u_y/\partial z = 0$ at both $z = 0$ and $z = 1$. The resulting nondimensional parameters representing rotating convection, obtained by scaling Equations 38.1 using d^2/ν, d, and ΔT, are the Rayleigh number Ra, the Taylor number Ta, and the Prandtl number Pr, respectively:

$$Ra = \frac{g\alpha\Delta T d^3}{\nu\kappa},$$
$$Ta = \left(\frac{2\Omega d^2}{\nu}\right)^2, \quad (38.2)$$
$$Pr = \frac{\nu}{\kappa}.$$

There are several other dimensionless numbers that are commonly used in descriptions of rotating convection. A different dimensionless measure of rotation rate is the Ekman number $Ek = \nu/(2\Omega d^2) = Ta^{-1/2}$ with a related Ekman length $\delta_E = d(Ek)^{1/2}$. The dimensionless ratio of inertia to viscous dissipation is the Reynolds number $Re = Ud/\nu$, where U is a characteristic velocity of convection. A useful combination of Ra, Ta, and Pr is the convective Rossby number Ro (Julien et al. 1996) that measures the ratio of a rotational time $\tau_R \sim \Omega^{-1}$ to a buoyancy rise time $\tau_B \sim [d\,(g\alpha\Delta T)]^{1/2}$ so $Ro = \tau_R/\tau_B = \sqrt{Ra/(Pr\,Ta)}$. As one varies Ra or Ta at fixed Ro, the relative dominance of buoyancy or rotation stays roughly constant: $Ro \gg 1$ implies buoyancy dominates whereas $Ro \ll 1$ indicates the dominance of rotation over buoyancy. Here we will be interested mostly in the latter condition of rotation-dominated convection.

Rotating flows have a tendency to be uniform along an axis perpendicular to the rotation axis as a result of the Taylor–Proudman theorem: A steady and slow flow in an inviscid rotating fluid is 2D in the plane perpendicular to the vector of angular velocity. The proof of this theorem is straightforward. We first obtain an equation in terms of the fluid vorticity $\vec{\omega} = \nabla \times \hat{u}$ for momentum conservation by taking the curl of the velocity equation in (38.1):

$$\frac{\partial \vec{\omega}}{\partial t} + \vec{u}\cdot\nabla\vec{\omega} - \vec{\omega}\,\nabla\vec{u} = \nu\nabla^2\vec{\omega} + g\alpha\nabla\theta\times\hat{z} - 2\Omega\nabla\times\hat{z}\times\vec{u}. \quad (38.3)$$

The terms on the left are small for slow flow, and the dissipation term vanishes for inviscid flow, that is, $\nu = 0$. This leaves us with $g\alpha\nabla\theta \times \hat{z} = 2\Omega\nabla \times \hat{z} \times \vec{u}$. Because $\nabla\theta$ is along the z-axis (for convection near onset), the left-hand side is zero. Rewriting the right-hand side using vector identities and assuming spatially uniform rotation, that is, $\nabla\cdot\vec{\Omega} = 0$, one obtains the result $(\Omega\hat{z}\cdot\nabla)\vec{u} = 0$. Since $\Omega \neq 0$, we have that any component of $\vec{u}$ must be uniform, that is, zero gradient, in the z direction. The Taylor–Proudman constraint manifests itself most strongly in suppressing the onset of convection so that for large rotation one has that the onset of bulk convection increases with rotation rate as $Ra_c = 8.7\ Ta^{2/3}$ (Chandrasekhar 1961). For strong convection with $Ra/Ra_c \gg 1$, the Taylor–Proudman theorem is not valid because the flow is both time-dependent and strongly nonlinear. Nevertheless, some tendency toward extended vertical structures remains provided rotation dominates buoyancy, that is, $Ro \ll 1$.

Thermal convection acts to transport heat more effectively than heat can be carried by molecular diffusion alone over an extended range in Ra, Ta, and Pr. One has Fourier's heat transport relationship that the heat transported by molecular diffusion is $\dot{Q}_d = (kA/d)\Delta T$, where k is the thermal conductivity, A is the cross sectional area, d is the length over which heat is transported, and ΔT is the applied temperature difference. For RB convection, d is the cell height and ΔT is the temperature difference between the upper and lower boundaries. With the advection of heat by fluid motion, the heat transport is enhanced by an amount usually called the Nusselt number Nu, defined as the ratio of total heat transported normalized by $\dot{Q}_d$. In turbulent convection, thin boundary layers form near solid surfaces whereas the interior is fairly isothermal

(Ahlers et al. 2009). In that case, one has that heat transport is dominated by the approximately linear variation of temperature over a thermal boundary layer distance δ_T so that $Nu = (k\Delta T/(2\delta_T))/(k\Delta T/d) = d/(2\delta_T)$. Experiments can directly determine Nu by measuring the power needed to maintain ΔT in the presence of convection although care needs to be taken to reduce parasitic heat leaks and to account for heat transport through, for example, the sidewalls (Zhong et al. 1993, Funfschilling et al. 2005, Liu and Ecke 2009). Such experiments demonstrate, perhaps unexpectedly, that heat transport is enhanced by rotation provided the flow is turbulent and Pr is sufficiently large for a given Ro. This enhancement of heat transport, first measured by (Rossby 1969), must arise from rotation effects on the turbulent boundary layer, which leads us to consider local structure.

The Coriolis force spins up lateral flow into vortices in rotating convection. For turbulent flow with relatively thin thermal and viscous boundary layers, for example, $\delta_T \ll d$, thermal disturbances are formed in the boundary layer and penetrate into the bulk of the flow. These structures stir up the bulk fluid through thermal buoyancy and nonlinear inertial forcing. In addition to the thermal and kinetic boundary layers (and associated instabilities of these layers) that dominate nonrotating RB convection (Ahlers et al. 2009), there is an additional effect for rotating flows near boundaries, namely, the formation of Ekman layers (e.g., Vallis 2006, see Julien et al. 1996 for a detailed discussion of Ekman layers in the presence of convection). For a thermal instability of the boundary layer, fluid is drawn in laterally to supply the vertical velocity induced by thermal buoyancy. In nonrotating convection, these thermal instabilities tend to form thermal plumes as shown in Figure 38.5a for experimental data of the temperature field near the top (colder) surface of a convection cell (Vorobieff and Ecke 2002). The lateral (dark) lines in the image are horizontal cross sections of colder, downwardly buoyant thermal plumes. In rotating convection, on the other hand, the Coriolis force acts on the convergent flow to produce cyclonic vorticity centered around the thermal instability as illustrated in Figure 38.5b where localized dark regions in the temperature field have cyclonic vorticity (as deduced from data animations). In Figure 38.5c, dark and light regions are in closer balance as the flow becomes only weakly turbulent, approaching the conditions in Figure 38.4. The cyclonic vortices lead to the formation of an Ekman layer in which the bulk geostrophic balance between the Coriolis force and radial pressure gradients (in the linear hydrostatic theory of classical Ekman analysis) is affected by frictional surface stresses near the boundary. As a result, there are unbalanced radial pressure gradients that accelerate the convergent flow and, by conservation of mass, actively pump fluid in the vertical direction away from the boundary layer into the bulk, so-called Ekman pumping. The vertical distance over which an Ekman layer forms is estimated as the Ekman length δ_E. The actual nonlinear process governing the interactive thermal instability of the boundary layer and the formation of cyclonic vorticity is complex (Julien et al. 1996), leading to the indication that the thermal vortices in rotating convection may have a more complicated form than a classical Ekman vortex with converging lateral flow producing a central vertical velocity at the core. For low Prandtl number fluids, $Pr < 1$ (Julien et al. 1996), there is positive horizontal divergence in the core implying vertical suction of bulk fluid into the boundary layer that opposes the inward lateral flow from the far field. For larger Prandtl number, this mechanism may not operate and the Ekman pumping may be simpler. The resulting vortex pumps fluid into the bulk near its core with possible inflow of cooler fluid around the edge as indicated schematically in Figure 38.6.

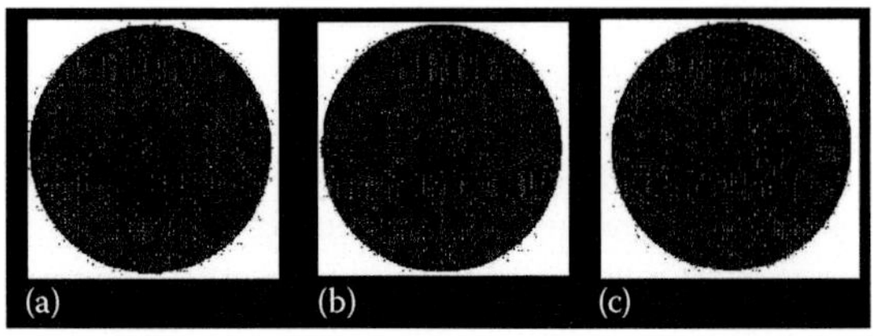

FIGURE 38.5 Thermochromic liquid crystal temperature field near the upper surface for $Ra = 3.2 \times 10^8$. (a) $Ra/Ra_c = 1500$, $Ro = 3.9$, (b) $Ra/Ra_c = 26$, $Ro = 0.19$, and (c) $Ra/Ra_c = 8$, $Ro = 0.08$. (From Vorobieff, P. and Ecke, R.E., *J. Fluid Mech.*, 458, 191, 2002.)

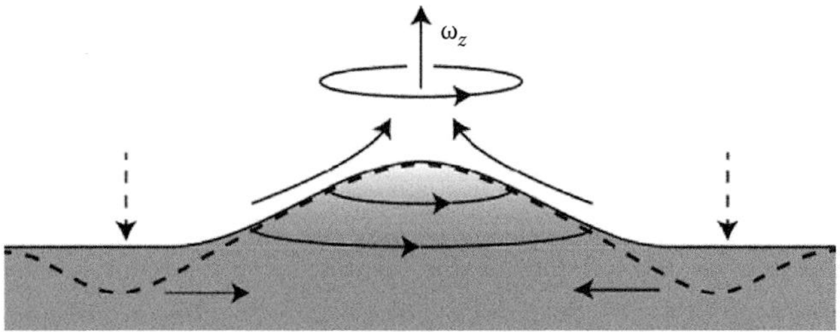

FIGURE 38.6 Schematic illustration of Ekman pumping/suction for thermal boundary-layer instability in rotating convection. A planar representation of the temperature field (light color) showing the growing unstable buoyant boundary layer where the arrows indicate flow directions. Lateral flow is spun up by the Coriolis force to form a complex thermal vortex with cyclonic vertical vorticity ω_z. The dashed lines with arrows show possible inflow of cooler fluid owing to the rapidly converging flow in the vortex core and the dashed line indicates the possible suppression of temperature contours by that inflow.

38.3 Examples

Here we focus on precise experimental measurements of heat transport in fluids with $Pr \approx 5$, for example, water. One finds that, contrary to ones expectations that heat transport would be suppressed in rotating convection owing to the tendency of the flow toward the vertical uniformity suggested by the Taylor–Proudman constraint, heat transport is enhanced by rotation. In Figure 38.7, we show data obtained in two different experiments (Liu and Ecke 1997, 2009, King et al. 2009) using water as the working fluid. Similar to Figure 38.3 but at considerably higher Ra, there is a region of rapid increase in $Nu \sim Ra$ (a slight steeper dependence $Nu \sim Ra^{6/5}$) has been proposed recently (King et al. 2009) followed by a more gradual increase of Nu for higher Ra that follows an approximate power-law $Nu \sim Ra^{0.29}$.

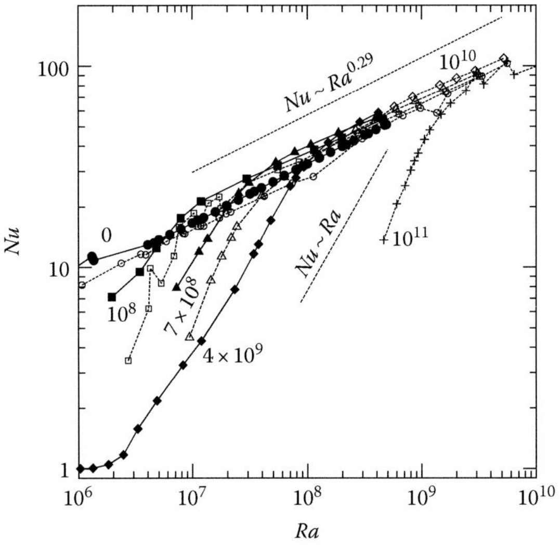

FIGURE 38.7 Experimental normalized heat transport *Nu* as a function of *Ra* for different values of *Ta* indicated on the plot. Solid lines are data from Liu and Ecke 2009 and dashed lines are from King et al. (2009). Labeled dashed lines show empirical $Nu \sim Ra^{0.29}$ turbulent scaling and linear $Nu \sim Ra$ in the nonturbulent regime.

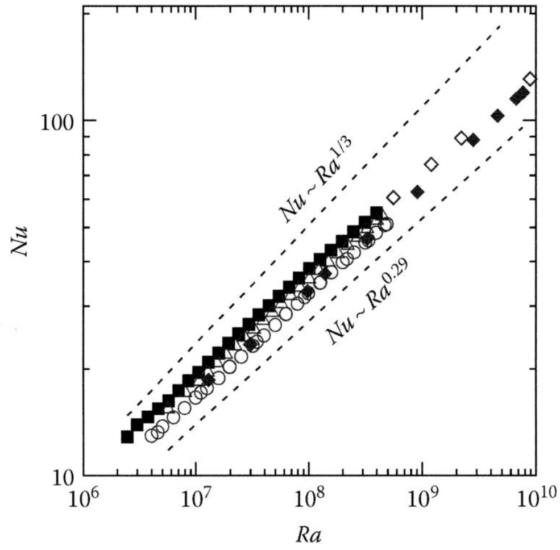

FIGURE 38.8 *Nu* versus *Ra* at constant *Ro*. Open circles, $Ro = \infty$; open triangles, $Ro = 0.75$; solid squares, $Ro = 0.30$ (Liu and Ecke 2009); solid diamonds, $Ro = \infty$ (Funfschilling et al. 2005); open diamonds, $Ro = 0.30$ (Zhong et al. 2009). Labeled dashed lines show $Nu \sim Ra^{1/3}$ and $Nu \sim Ra^{0.29}$ turbulent scaling.

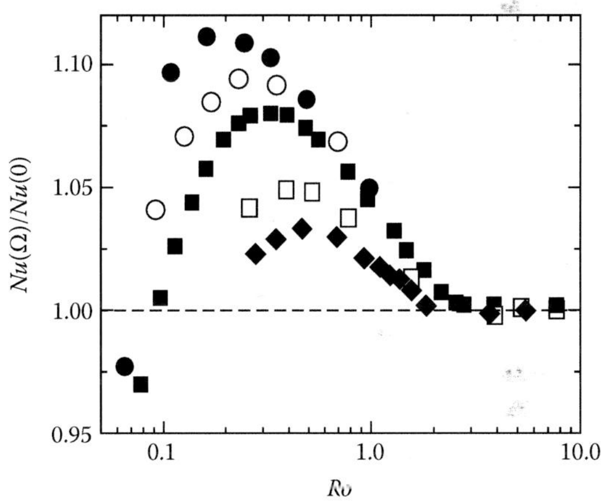

FIGURE 38.9 Ratio of *Nu* (Ω) for rotation rate Ω and *Nu* (Ω = 0) versus Rossby number *Ro* on a logarithmic scale. Solid circles, $Ra = 5.6 \times 10^8$; open circles, $Ra = 1.2 \times 10^9$; solid squares, $Ra = 2.2 \times 10^9$, open squares, $Ra = 8.9 \times 10^9$; solid diamonds: $Ra = 1.8 \times 10^{10}$. (With permission from Zhong, J.-Q., Stevens, R.J.A.M., Clercx, H.J.H., Verzicco, R., Lohse, D., and Ahlers, G., *Phys. Rev. Lett.*, 102, 044502. Copyright 2009 by the American Physical Society.)

Looking more carefully at the data in both Figures 38.3 and 38.7, one sees that *Nu* at a particular *Ta* rises toward the *Nu* (*Ta* = 0) value, surpasses it over some interval in *Ra* before appearing to asymptote to the *Nu* (*Ta* = 0) scaling at large *Ra*. As indicated in the phase diagram in Figure 38.2, a path at constant *Ta* passes through a region of rotation dominated convection but enters buoyancy-dominated convection at higher *Ra*. To sort out the conflicting influence of buoyancy and rotation, it is useful to plot *Nu* as a function of *Ra* at fixed convective Rossby number *Ro* (Julien et al. 1996, Liu and Ecke 1997) as in Figure 38.8 where experimental data from three sources (Liu and Ecke 1997, 2009, Funfschilling et al. 2005, Zhong et al. 2009) are presented. There is excellent correspondence between the data sets in overlapping ranges of *Ra* and with *Ro* values of ∞ (no rotation), 0.75 and 0.30. Overall the data scale approximately as $Nu \sim Ra^{0.29}$, which is noticeably less steep than classical scaling $Nu \sim Ra^{1/3}$ (Ahlers et al. 2009). Over the range $10^6 < Ra < 10^{10}$, there is enhancement of heat transport for smaller *Ro* but the enhancement is less at the very highest values of *Ra*. To unravel further the dependence of heat transport on *Ra* and *Ta*, we show in Figure 38.9 for different *Ra* (Zhong et al. 2009) the relative enhancement of heat transport for rotating convection normalized by its value at zero rotation, that is, $Nu(\Omega)/Nu(\Omega = 0)$. For $Ro > 3$, $Nu(\Omega)/Nu(0) = 1$ to within experimental uncertainty but the ratio increases to a peak with decreasing *Ro* before falling at still lower *Ro* (presumably because of the decreasing Ra/Ra_c). The magnitude of the enhancement is modest, of order 10% or less, and is reduced for increasing *Ra*. The value of *Ro* at the peak maximum also shifts toward higher values as *Ra* decreases. The amount of enhancement of heat transport depends on *Pr* such that there may be little or no enhancement for small *Pr* (Zhong et al. 2009). Finally, as indicated in all the data presented here, there is no evidence for a scaling of heat transport of the form $Nu \sim Ra^3 Ta^{-2}$ that has been proposed for the weak convection regime near onset based on dimensional arguments (Boubnov and Golitsyn 1990, 1995).

In order to understand the enhancement of heat transport in rotating convection, one needs to consider the small-scale features of the flow. In particular, the competing interactions of thermal instability of the top and bottom boundary layers, kinetic boundary layers arising from shear on the boundary layer, and the effects of rotation-induced Ekman layers need to be addressed. In the next section, we consider the character of the boundary layer and how it affects heat transport.

38.4 Methods of Analysis

In order to understand heat transport, it is essential to probe the internal structure of the convecting fluid. In particular, one would like to obtain the full 3D velocity and temperature fields of the flow. This task is possible using direct numerical simulations (Julien et al. 1996, Kunnen et al. 2009, Zhong et al. 2009) for modest value of Ra but experimental measurements are typically limited to a single point as in a local probe of temperature or to a plane as in velocity field measurements using the particle-image velocimetry (PIV) technique (e.g., see Vorobieff and Ecke 2002). Here we consider both experimental and numerical experiments to provide insight into the flow structure in the vicinity of the boundary layer.

As mentioned earlier, laterally converging fluid is drawn in by thermal instability of the boundary layers and is spun up by the Coriolis force, leading to localized strong cyclonic vortices. One can measure the vertical vorticity near the boundary layer to determine if this picture is verified for the physical system. We plot the probability distribution functions (PDF) of vertical vorticity ω_z (normalized to its standard deviation σ_ω) for experimental data with $Pr = 6$ and $Ra = 3.2 \times 10^8$ (Vorobieff and Ecke 2002) in Figure 38.10a and b and from direct numerical simulation with $Pr = 1$ and $Ra = 1.1 \times 10^8$ (Julien et al. 1996) in Figure 38.10c. The symmetry of the distribution in Figure 38.10c indicates no preference for cyclonic vorticity production for nonrotating convection. On the other hand, the experimental data for $Ro = 0.19$ and the numerical data for $Ro = 0.75$ in Figures 38.10b and c, respectively, both demonstrate enhanced probability for strong cyclonic vorticity, consistent with the picture presented earlier. Numerous other data, both experimental and numerical, support the predominance of strong cyclonic vorticity in rotating convection.

In Figure 38.6, we sketched out a form for these cyclonic vortices and described how the action of Ekman layers could actively pump fluid from the boundary layer into the bulk. The more detailed picture suggested by numerical simulation for $Pr = 1$ (Julien et al. 1996) or the possibility of suction near the edge of the vortex owing to the inability of the boundary layer to supply the unstable vortex with hot fluid added the possibility of a reverse suction of bulk fluid into the boundary layer. Such active exchange would tend to enhance heat transport compared to the more diffusively controlled thermal instability of the boundary layer in the absence of rotation. There is modest evidence for the picture outlined earlier. In Figure 38.11, we show results for the temperature, horizontal divergence, and vertical vorticity from direct numerical simulation (Julien et al. 1996) for no-slip boundary conditions $Pr = 1$ and $Ra = 1.1 \times 10^8$. A single localized cyclonic vortex is shown with a ring-like shape. Notice in particular the positive horizontal divergence indicating convergent flow (bright) around a core region of diverging flow and the weaker temperature and vertical vorticity

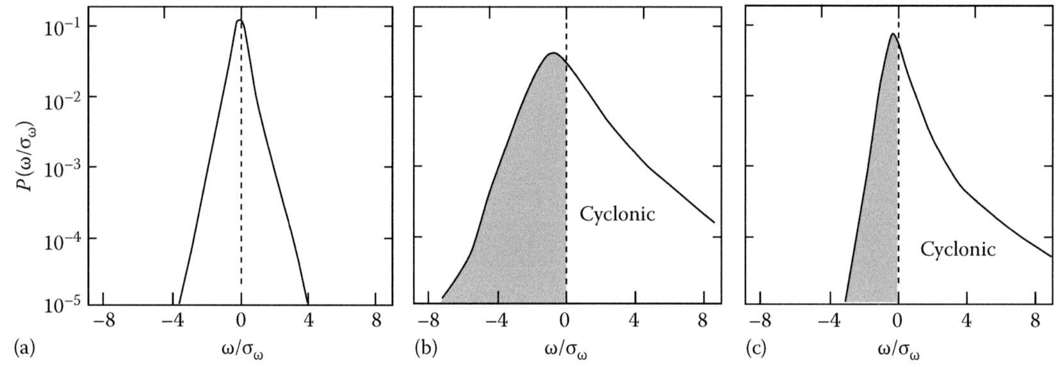

FIGURE 38.10 Probability distribution of vertical vorticity ω near the upper boundary of rotating convection for experimental data with $Pr = 5.8$, $Ra = 3.2 \times 10^8$, and (a) $Ro = \infty$ and (b) $Ro = 0.19$ (Vorobieff and Ecke 2002). Data in (c) are from a numerical simulation with $Pr = 1$, $Ra = 1.1 \times 10^8$, and $Ro = 0.75$ (Julien et al. 1996). The data for rotating convection (b, c) are skewed toward more cyclonic (positive) vorticity whereas the PDF for nonrotating convection is quite symmetric; the shaded regions in (b, c) indicate anticyclonic (negative) vorticity.

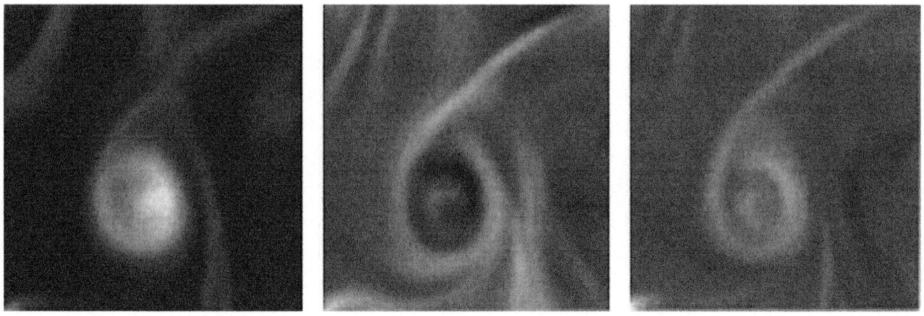

FIGURE 38.11 Temperature field, horizontal divergence, and vertical vorticity for direct numerical simulation with $Pr = 1$, $Ra = 1.1 \times 10^8$, $Ro = 0.75$ for no-slip boundary conditions. (Reproduced with permission from Julien, K. et al., *J. Fluid Mech.*, 322, 243, 1996. Copyright 2006 Cambridge Journals.)

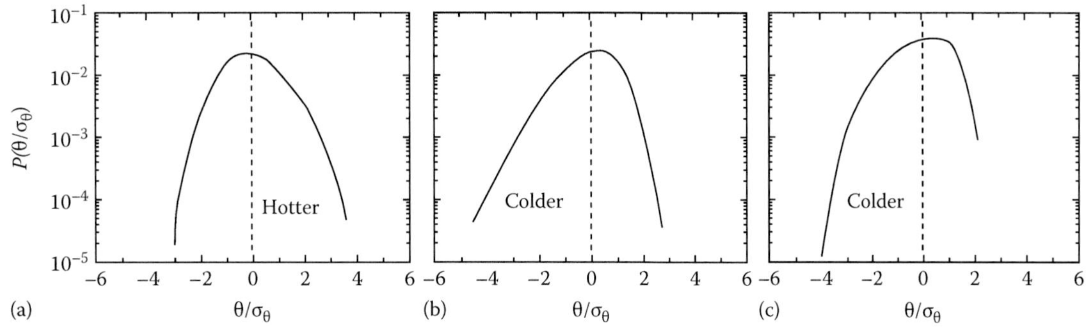

FIGURE 38.12 Probability distribution of temperature q near the bottom boundary of rotating convection for $Pr = 5$, $Ra = 4 \times 10^8$, (a) $Ro = \infty$ and (b) $Ro = 0.37$ (Liu and Ecke, 2011). Data in (c) are from a numerical simulation (no-slip top/bottom boundaries) with $Pr = 1$, $Ra = 1.1 \times 10^8$, and $Ro = 0.75$ (Julien et al. 1996). Whereas there are more frequent hotter fluctuations for nonrotating convection, in rotating convection fluctuations are skewed in the colder direction inside the boundary layer.

in the vortex core. Measurements at higher Prandtl $Pr = 6$ and $Ra = 3.2 \times 10^8$ (Vorobieff and Ecke 2002) seemed to support this picture but recent higher resolution measurements (Kunnen et al. 2010) have demonstrated that the previous result seems to be an artifact of low spatial resolution. Nevertheless, some active inflow appears to be needed to explain other local temperature data. The PDFs of temperature, slightly within or at the edge of the boundary layer, are shown in Figure 38.12. The PDFs for experimental data taken with a local temperature probe near the bottom hot boundary for $Pr = 5$ and $Ra = 4 \times 10^8$ are shown in Figure 38.12a and b, and numerical data for $Pr = 1$, $Ra = 1.1 \times 10^8$, and $Ro = 0.75$ are shown in Figure 38.12c. For the nonrotating case, Figure 38.12a, the PDF is skewed toward more probable hot fluctuations as upwelling thermals carry hotter fluid from below but colder bulk fluid only weakly penetrates the boundary layer. In contrast, for rotating convection there is a tendency toward colder fluctuations. This is consistent with the active suction of colder fluid by the thermal cyclonic vortices, either at the core or near the vortex edge, combined with the greater thermal contrast of the colder bulk fluid compared with hotter fluid. To drive home the connection between Ekman layer effects and localized vortex structures, we show in Figure 38.13 the typical size of a lateral cyclonic vortex as a function of Ro. The size decreases with decreasing Ro consistent with the Ro variation of a multiple (factor of about 8) of the Ekman length δ_E/d that can be written as $\delta_E/d = (Pr\,Ta)^{-1/4}\,Ro^{1/2}$.

There are many more properties of rotating convection that might be presented. One of particular interest is the formation of mean temperature gradients along the vertical direction (Boubnov and Golitsyn 1990, Julien et al. 1996, Hart and Ohlsen 1999). In nonrotating convection thermal plumes induce vigorous vertical mixing, leading to very little variation of the mean temperature in the bulk of the fluid, that is, $d\theta/dz = 0$. As rotation increases $d\theta/dz$ becomes more negative indicating a less vertically mixed interior. Figure 38.14 shows this effect dramatically

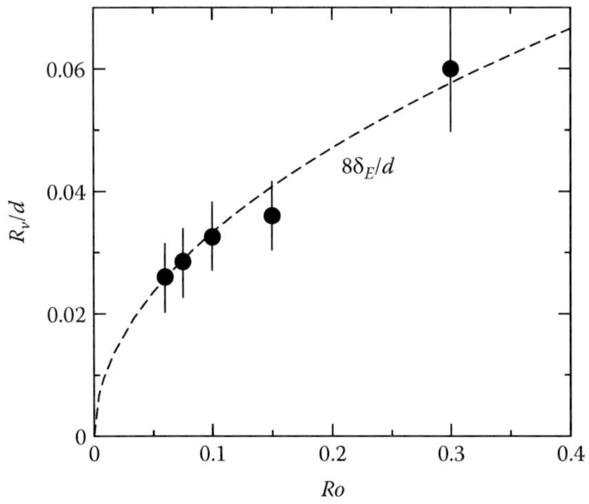

FIGURE 38.13 Typical size (radius) of a thermal vortex normalized by cell depth d, R_v/d, versus Rossby number (Vorobieff and Ecke 1998). Dashed line is $8\delta_E/d$ where δ_E is the Ekman length. Vertical lines indicate standard deviation of distribution of R_v/d.

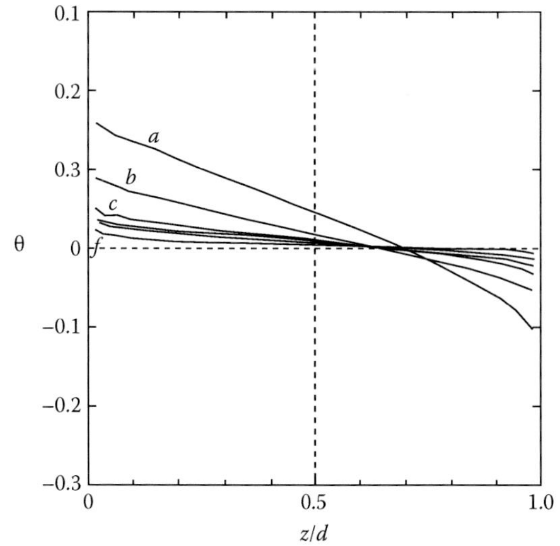

FIGURE 38.14 Mean temperature profiles θ versus vertical height z/d for turbulent convection experiments in silicone oil. $Ra = 1.9 \times 10^{11}$, $Pr = 8.2$, and from a to f, respectively: $Ta = [35, 4.9, 0.62, 0.31, 0.15, 0.078] \times 10^{11}$ corresponding, respectively, to values of $Ro = [0.08, 0.22, 0.61, 0.86, 1.2, 1.7]$. (Reprinted with permission from Hart, J.E. and Ohlsen, D.R., *Phys. Fluids*, 11, 2101–2107. Copyright 1999, American Institute of Physics.)

for rotating convection in a fluid with $Pr = 8$ and $Ra = 1.9 \times 10^{11}$ (Hart and Ohlsen 1999). One explanation of the finite vertical temperature gradient in rotating convection (Julien et al. 1996) is that the production of vortices enhances horizontal mixing through vortex merger at the expense of vertical mixing. Another aspect of the data in Figure 38.14 is that there is a thermal offset at the cell mid-plane, $z/d = 0.5$, consistent with a model of a large-scale Ekman circulation (Hart and Ohlsen 1999).

38.5 Major Challenges

The basic characteristics of rotating RB convection including heat transport enhancement over nonrotating convection, the formation of cyclonic vortices near top and bottom boundaries, and the role of Ekman pumping in heat transport enhancement are well supported by existing experimental and numerical data. To better understand the delicate coupling of thermal, kinetic, and Ekman boundary layers and their role in heat transport, new and more precise measurements at higher *Ra* and *Ta* will be required. Recent work in this direction (King et al. 2009, Kunnen et al. 2009, Zhong et al. 2009) appears poised to contribute further insights into the mechanisms of heat transport in rotating convection.

The importance of free-slip boundaries that often arise in geophysical, planetary, or stellar situations make the interpretation of these results challenging as does the very different range of parameters for typical planetary scale convection phenomena. In that case, one uses the flux Rayleigh number $Ra_f = NuRa$ to describe the convecting state. Nevertheless, the competition between buoyancy and rotation can lead to circumstances very similar to those described here. For example, the core of the earth that supports a geo-dynamo that generates the Earth's magnetic field is estimated (King et al. 2009) to have $Ra \sim 10^{25}$ and $Ta \sim 10^{30}$, close to the conditions for the crossover from weak convection to turbulent flow. In any case, the set of precise heat transport and local probe measurements made over the last two decades puts important constraints on the physical phenomenon of rotating thermal convection that will serve as the foundation for extensions to more complex and geophysically relevant applications.

References

Ahlers, G, Grossman, S, Lohse, D. 2009. Heat transfer and large scale dynamics in turbulent Rayleigh-Bénard convection. *Rev. Mod. Phys.* 81:503–537.

Boubnov, BM, Golitsyn, GS. 1990. Temperature and velocity field regimes of convective motions in a rotating plane fluid layer. *J. Fluid Mech.* 219:215–239.

Boubnov, BM, Golitsyn, GS. 1995. *Convection in Rotating Fluids.* Dordrecht, the Netherlands: Kluwer Academic.

Chandrasekhar, S. 1961. *Hydrodynamic and Hydromagnetic Stability.* Oxford, U.K.: Oxford University Press.

Fernando, HJS, Chen, RR, Boyer, D. 1991. Effects of rotation on convective turbulence. *J. Fluid Mech.* 228:513–547.

Fernando, HJS, Smith, DC. 2001. Vortex structures in geophysical convection. *Euro. J. Mech. B-Fluids* 20:437–470.

Funfshilling, D, Brown, E, Nikolaenko, A, Ahlers, G. 2005. Heat transport by turbulent Rayleigh-Bénard convection in cylindrical samples with aspect ratio one and larger. *J. Fluid Mech.* 536:145–154.

Hart, JE, Ohlsen, DR. 1999. On the thermal offset in rotating thermal convection. *Phys. Fluids* 11:2101–2107.

Julien, K, Legg, S, McWilliams, J, Werne, J. 1996. Rapidly rotating turbulent Rayleigh-Bénard convection. *J. Fluid Mech.* 322:243–273.

King, EM, Stellmach, S, Noir, J, Hansen, U, Aurnou, JM. 2009. Boundary layer control of rotating convection systems. *Nature* 457:301–304.

Kunnen, RPJ, Clercx, HJH, Geurts, BJ. 2010. Vortex statistics in turbulent rotating convection. *Phys. Rev. E* 82:036306.

Kunnen, RPJ, Geurts, BJ, Clercx, HJH. 2009. Turbulent statistics and energy budget in rotating Rayleigh-Bénard convection. *Euro. J. Mech. B/Fluids* 28:578–589.

Liu, Y, Ecke, RE. 1997. Heat transport scaling in turbulent Rayleigh-Bénard convection: Effects of rotation and Prandtl number. *Phys. Rev. Lett.* 79:2257–2260.

Liu, Y, Ecke, RE. 1999. Nonlinear traveling waves in rotating Rayleigh-Bénard convection: Stability boundaries and phase diffusion. *Phys. Rev. E* 59:4091–4105.

Liu, Y, Ecke, RE. 2009. Heat transport measurements in turbulent rotating Rayleigh-Bénard convection. *Phys. Rev. E* 80: 036314.

Liu, Y, Ecke, RE. 2011. Local temperature measurements in turbulent rotating Rayleigh-Bénard convection. *Phys. Rev. E* 84: 016311.

Rossby, HT. 1969. A study of Bénard convection with and without rotation. *J. Fluid Mech.* 36:309–335.

Sakai, S. 1997. The horizontal scale of rotating convection in the geostrophic regime. *J. Fluid Mech.* 333:85–95.

Vallis, GK. 2006. *Atmospheric and Oceanic Fluid Dynamics.* Cambridge, U.K.: Cambridge University Press.

Veronis, G. 1959. Cellular convection in a rotating fluid. *J. Fluid Mech.* 5(3):401–435.

Vorobieff, P, Ecke, RE. 1998. Vortex structures in rotating. Rayleigh-Bénard convection. *Physica D* 123: 153–160.

Vorobieff, P, Ecke, RE. 2002. Turbulent rotating convection: An experimental study. *J. Fluid Mech.* 458:191–218.

Zhong, F, Steinberg, V, Ecke, RE. 1993. Rotating Rayleigh-Bénard convection: Asymmetric modes and vortex states. *J. Fluid Mech.* 249:135–159.

Zhong, J-Q, Stevens, RJAM, Clercx, HJH, Verzicco, R, Lohse, D, Ahlers, G. 2009. Prandtl-, Rayleigh-, and Rossby-number dependence of heat transport in turbulent rotating Rayleigh-Bénard convection. *Phys. Rev. Lett.* 102:044502.

39

Double-Diffusive Instabilities

Ruby Krishnamurti
Florida State University

39.1 Introduction

As discussed in the editor's Introduction, double-diffusive flows in the ocean would be classified as fine scale (1–100 m) to microscale or Kolmogorov scale (1 mm) in that they rely specifically on the process of molecular diffusion. Yet the mixing that they achieve, particularly when they occur far from other sources of turbulent mixing, may be having profound effects on the global thermohaline or meridional overturn circulation (the MOC). The following is presented from this viewpoint.

Unlike the faster wind-driven ocean circulation with flows in horizontal gyres and particle orbit times of a few years, the MOC is vertical overturning with sinking in very narrow polar regions and slow rising over much of the rest of the ocean. Particle orbit times may be of the order of 10^2–10^3 years. Antarctic bottom water is produced by winter cooling primarily around the Weddell Sea; the resulting water is the coldest and densest ocean water and is found in the abyss. North Atlantic deep water is produced by winter cooling of the saline North Atlantic waters in polar or subpolar regions. (Deep water is not produced in the North Pacific where salinities are significantly smaller than those of the North Atlantic.) A problem lies in understanding how this deep dense water is converted to lower density surface water, which must happen for a sustained circulation. In their review article Wunsch and Ferrari (2004) point out that when a stratified fluid is homogenized by mixing, the center of mass has been raised from below mid-depth up to mid-depth. Expressing this mixing as an effective diffusivity κ_e requires for the MOC a value of $\kappa_e \simeq 1.0\,\text{cm}^2\,\text{s}^{-1}$. This may be compared to molecular diffusivities $\kappa_T = 10^{-3}\,\text{cm}^2\,\text{s}^{-1}$ for heat and $\kappa_S = 10^{-5}\,\text{cm}^2\,\text{s}^{-1}$ for salt. They point out further that there is sufficient energy in winds and tides to accomplish this. On the other hand studies like those of St. Laurent and Schmitt (1999) show that turbulent mixing occurs from double-diffusive instabilities as well as from turbulence created by tidal flows against bottom topography. Their microstructure measurements when sorted by Richardson's number *Ri* and density ratio R_ρ (defined later) shows a clear bimodal structure to the dissipation rates of turbulent kinetic energy. This they interpret as resulting from salt-fingering double-diffusive instability when *Ri* is large (and internal wave turbulence less likely) and R_ρ near unity. Certainly salt fingers were observed at the site in these cases. St Laurent and Schmitt determine $\kappa_{es} \simeq 0.13\,\text{cm}^2\,\text{s}^{-1}$ and $\kappa_{eT} \simeq 0.08\,\text{cm}^2\,\text{s}^{-1}$ in these conditions.

With this in mind, we will see in Section 39.3.3 that salt fingering produces a larger downward transport of buoyancy due to salt than to heat, with the result that there is a net *lowering* of the center of mass of the fluid by this kind of instability. In Section 39.1.2 we will describe an oceanic double-diffusive flow observed to be a lateral intrusive or interleaving flow which can arise from horizontal property gradients and which can surprisingly accomplish a mixing of deep water with upper layer water. Numerical models of double diffusive processes have not been included in this review.

39.1.1 Vertical Forcing in the Ocean

There are maps of the world oceans (e.g., Kluikov and Karlin 1995) indicating locations susceptible to double-diffusive instability. From these we see that, for depths 200–250 m, most of the

Handbook of Environmental Fluid Dynamics, Volume One, edited by Harindra Joseph Shermal Fernando.

ISBN: 978-1-4398-1669-1.
This is a contribution no. 467 of the Geophysical Fluid Dynamics Institute, Florida State University.

tropical and subtropical oceans are prone to a double-diffusive instability called salt finger instability, which can occur when warm salty water lies above cold freshwater. (In all cases we consider double-diffusive instability occurring in a fluid whose density increases downward with depth, i.e., in situations immune to convective overturning.) Of course, not all parts of this salt-fingering domain have equally vigorous finger flow. In the western North Atlantic east of Barbados, and in the eastern Mediterranean vertical profiles of temperature T and salinity S show striking displays of thermohaline staircases, which may be regarded as a further instability of salt fingers. For example, from the Tyrrhenean Sea, Molcard and Williams (1975) show that when salinity, or temperature, is plotted against depth, the profile is stairs-shaped; the "risers" are tens of m deep, the "treads" one or two *m* deep. The latter have large drops of T and S across them and are the salt finger interfaces, the former are well-mixed convecting layers.

From Kluikov and Karlin's (1995) map we see further that most of the remainder of the oceans at that depth (most of the Arctic and some of the Southern Ocean) are susceptible to another form of double-diffusive instability, sometimes called "diffusive" convection. It can occur when warm saline but more dense water lies below cold fresh less dense water. It onsets as an oscillatory flow but further instabilities lead again to a thermohaline staircase (Neshyba et al. 1971).

Double-diffusive stable regions are definitely a rarity, according to this map. Furthermore such a map shows double-diffusive instabilities resulting from vertical property gradients only. There is another family of double-diffusive instabilities resulting from horizontal property gradients which results in lateral interleaving.

39.1.2 Lateral Forcing in the Ocean

Vertical soundings with multiple temperature inversions were observed as early as 1967 by Stommel and Federov (1967) in the East Timor Sea and by Gregg (1980) in the North Pacific. These are known to be accompanied by matching salinity inversions, such that the density could even be monotonically increasing with depth. They are often seen across frontal zones where one water mass (of temperature T_1, salinity S_1, density ρ_1) is adjacent to another (of temperature T_2, salinity S_2, density ρ_2). If $\rho_1 = \rho_2$ there might be no other instability possible than double diffusive. The vertical scale of these inversions or reversals in property anomalies is on the order tens of meters (see Figure 39.1a), but their horizontal coherence distance is large: 400 km across the equator (Richards and Pollard 1991) more than 1000 km in the Arctic, but less in more energetic confluence zones like the Brazil–Malvinas Current Confluence. They have been described as interleaving or lateral intrusive flows. When the vertical sounding data is plotted on a T–S plot as is commonly done with ocean sounding data, they have a most distinctive saw-toothed appearance. One example from the equatorial Pacific is shown in Figure 39.1.

Although we lack a world map of lateral intrusive instability, we do have a general understanding of confluence zones and fronts. Broadly speaking, surface winds can cause Ekman pumping or suction. Anticyclonic wind stress curl pumps warm saline surface water downward, as in the warm saline lenses associated with the subtropical gyres, while cyclonic wind stress curl sucks up deep water toward the surface as in the subpolar gyres. Confluence of currents from these separate gyres (such as Gulf Stream/Labrador Current and Kuroshio/Oyashio) means originally warm saline

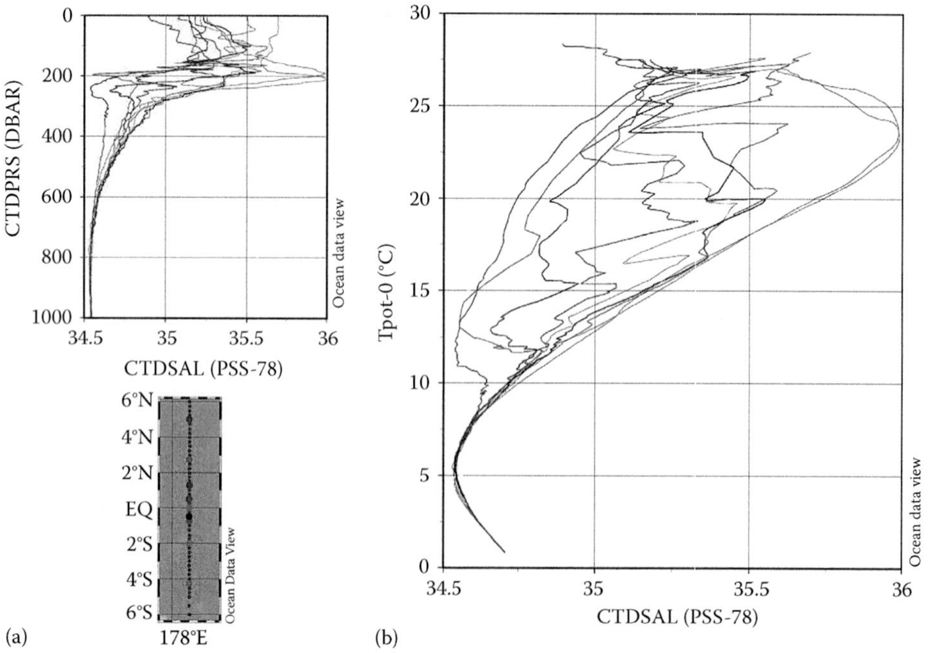

FIGURE 39.1 (See color insert.) (a) Salinity (CTDSAL) versus depth (CTDPRS). (b) Potential temperature versus salinity, showing "sawtooth" characteristic of intrusions.

surface waters are placed adjacent to originally cold fresh deep water. Any instability that mixes these laterally, as with double-diffusive interleaving, in effect mixes deep water with surface water, hence entering the mixing "equation" of the MOC.

Likewise with equatorial interleaving, because of the sign change of Coriolis parameter across the equator, near-equatorial easterly winds produce surface divergence, leading to upwelling of cold freshwater from depths. These now lie adjacent to the down-welled pools of warm saline surface waters of the subtropics. A clear picture can be seen in the WOCE section P14 which shows the field of salinity versus depth along the date line (180 E) from New Zealand to the Bering Sea. The two warm saline pools on either side of the Equator are separated by approximately 500 km, which might well have been spanned by the 400 km interleaving observed by Richards and Pollard. The T–S plot of Figure 39.1 is from WOCE data on section P14.

The manner in which double-diffusive interleaving results in horizontal mixing has itself a surprising dynamics. Many details of these phenomena such as salt-fingering or lateral interleaving will be clarified by the stability theories and laboratory observations of relevant simple models.

39.2 Principles

When the density ρ of a fluid is determined by the concentrations of two components C_1 and C_2 which diffuse at different rates, then an initially balanced state in a gravitational field may lose that balance simply because one component has diffused faster than the other.

For simplicity in demonstrating the essentials of double-diffusive flow, we may take Fickian diffusion of each component:

$$\frac{dC_1}{dt} = \kappa_1 \nabla^2 C_1 \tag{39.1}$$

$$\frac{dC_2}{dt} = \kappa_2 \nabla^2 C_2 \tag{39.2}$$

with $\kappa_1 \neq \kappa_2$ (cross-diffusion is being neglected). Suppose further that the equation of state takes the simplest linear form:

$$\rho = \rho_0 \left[1 + \alpha_1 C_1 + \alpha_2 C_2\right] \tag{39.3}$$

where α_1 and α_2 are expansion (a contraction) coefficients

$$\alpha_i = \frac{1}{\rho_0}\frac{\partial \rho}{\partial C_i} \quad i = 1,2 \tag{39.4}$$

Then, for example, at a given depth z in the fluid, the density ρ may be initially independent of horizontal position x:

$$\alpha_1 \frac{\partial C_1(x)}{\partial x} = -\alpha_2 \frac{\partial C_2(x)}{\partial x} \tag{39.5}$$

(C_1 and C_2 may be called "compensating") and no gravity currents are expected. However, with unequal diffusivities, say $\kappa_1 \gg \kappa_2$ then $C_1(x)$, by diffusing down its gradient, may change relatively quickly, while $C_2(x)$ has hardly changed; density compensation is lost and gravity currents may result. This is the nature of double-diffusive instability and onset of flow. Sustaining this flow will be discussed later.

Equations 39.1 through 39.3 are coupled to an equation of motion through the velocity field included in the substantial time derivatives, and through the buoyancy force in the equation of motion. The Navier–Stokes equation in the Boussinesq approximation is written

$$\frac{du_i}{dt} = -\frac{1}{\rho_0}\frac{\partial p}{\partial x_i} + \nu \nabla^2 u_i - g(1 + \alpha_1 C_1 + \alpha_2 C_2)\lambda_i \tag{39.6}$$

$$i = 1,2,3; \quad \lambda_i = (0,0,1)$$

where

ν is the kinematic viscosity
g is the acceleration of gravity
p is the pressure

With linear stability theories d/dt is usually approximated by $\partial/\partial t$ except for imposed vertical stratifications of T_b or S_b in which case $w\partial T_b/\partial z$ and $w\partial S_b/\partial z$ must be retained. If T_b and S_b are independent of time then the equations governing the dependent variables u_i, c_1, c_2, p have constant coefficients and separation of variables allows us to write $\partial/\partial t = \sigma$ the growth rate of the disturbance.

A problem with many diffusing components, each having diffusivity κ_i and each satisfying a reaction-diffusion equation was solved earlier by A.M. Turing (1952). He posed the problem as one of symmetry breaking for a spherically symmetrical model of a blastula to be able to differentiate to form ultimately appendages and organs. He writes of applications to dappling of deer and spotting on salamanders. In our double-diffusive context, Griffiths (1979) has shown the modifications to salt fingers achieved when a third component is considered.

39.3 Methods of Analysis

39.3.1 Stability Theory for Vertical Forcing

Although originally formulated by Stern (references in Turner (1985) and Schmitt (1994)) a most convenient one-figure stability diagram showing both salt finger and diffusive regimes would that be due to Baines and Gill (1969) shown in Figure 39.2a. Their theory is for a model which is of finite depth d, verti-

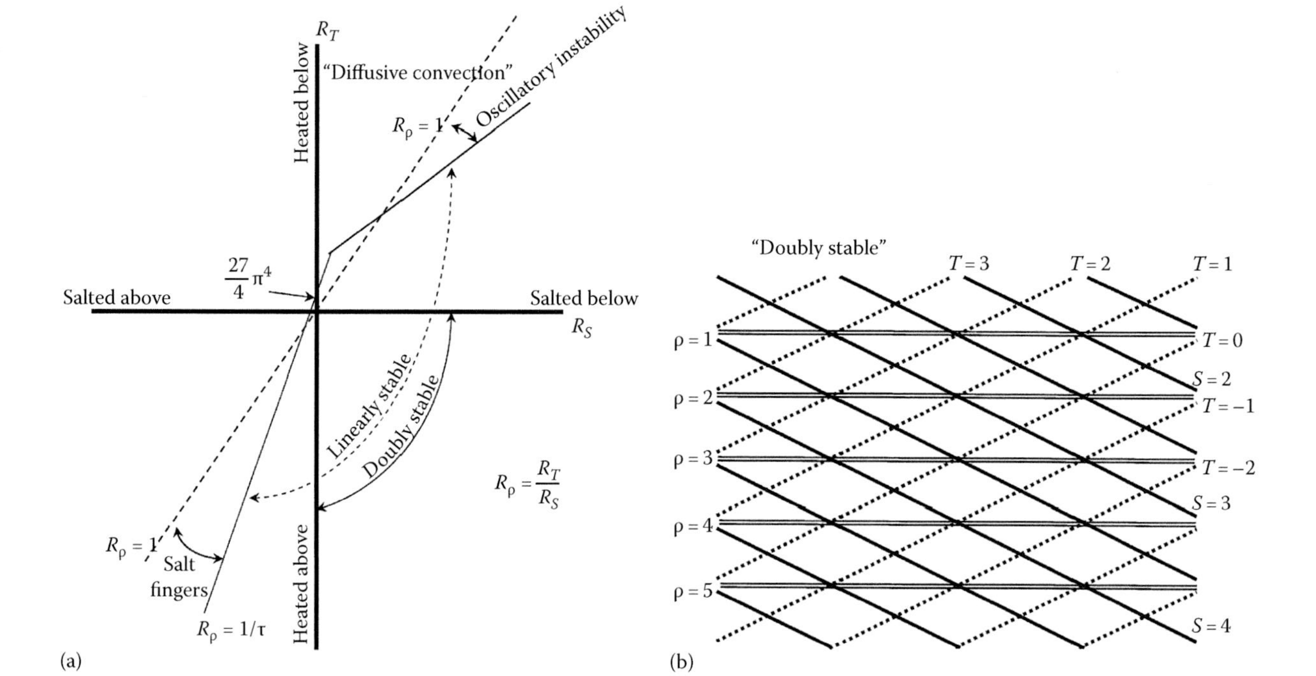

FIGURE 39.2 (a) Stability diagram, vertical forcing only. (After Baines, P.G. and Gill, A.E., *J. Fluid Mech.*, 37, 289, 1969. With permission.) (b) Isopycnals, isotherms, and isohalines in compensated lateral forcing. (After Krishnamurti, R., *J. Fluid Mech.*, 558, 113, 2006. With permission.)

cally stratified with temperature T and salinity S, where $T = C_1$ and $S = C_2$.

The boundary conditions are $T = T_0$, $S = S_0$ (T_0 and S_0 are constants) at $z = 0$, and $T = T_0 + \Delta T$, $S = S_0 + \Delta S$ at $z = d$. Positive ΔT, ΔS can lead to salt fingers, negative values to diffusive convection. This allows definition of the dimensionless parameters:

$$R_T = g\alpha\Delta T d^3/\kappa_T \nu \quad \text{the thermal Rayleigh number}$$

$$R_S = g\beta\Delta S d^3/\kappa_T \nu \quad \text{the salt Rayleigh number}$$

$$R_\rho = R_T/R_S \quad \text{the "stability ratio"}$$

where

- α is the thermal expansion coefficient
- β is the salt contraction coefficient
- κ_T is the diffusivity of heat in water

$$\alpha = \frac{-1}{\rho_0}\frac{\partial \rho}{\partial T}, \quad \beta = \frac{1}{\rho_0}\frac{\partial \rho}{\partial S}$$

In the first quadrant of the stability diagram, a Hopf bifurcation occurs when crossing from below, the line labeled "oscillatory instability." The slope of this line is $Pr/1+Pr$ for negligible τ where $Pr = \nu/\kappa_T$, $\tau = \kappa_S/\kappa_T$. The oscillatory convection that results has often been called "diffusive" convection. The physical nature of this flow is explained in Section 39.3.3.

Salt finger instability occurs in the third quadrant between the lines $R_\rho = 1$ and $R_\rho = 1/\tau$. The onset of instability is direct, i.e., one with monotonic growth in time.

For either case, the further instabilities that lead to staircases still need to be mapped onto such a stability diagram.

39.3.2 Stability Theory for Lateral Forcing

Again originally formulated by Stern, the early models had a parameterized vertical salt finger flux throughout the fluid. Horizontal gradients were imposed upon such a state. In fact we see in experiments that fingers occur only in layers that alternate with convecting layers. Holyer (1983) considered the fingers as a secondary instability resulting after the lateral intrusive flow. She shows that the critical horizontal Rayleigh number R_H is zero where, for salt,

$$R_{H,S} = \frac{g\beta L^4}{\kappa_T \nu}\frac{\partial S}{\partial x}$$

For compensating horizontal gradients of T and S, she computed the vertical wave numbers and the slope of the intrusions at onset.

39.3.3 Laboratory Experiments with Vertical Forcing

Many of the early pioneering experiments conducted by Turner have yielded much of the understanding that we have today of double-diffusive flows. These were performed as two-layer run-down experiments in which hot, salty, lower density water was carefully placed over higher density cold freshwater. As the heat diffused more quickly than salt across the interface, originally hot salty water is now cool salty water just above the interface, so becomes unstable under gravity. After some time a salt-fingering layer forms across the interface (see Turner 1985) in which columns of salty water flow downward alternating with columns of freshwater flowing upward. The width of these columns is several mm to 1 cm depending upon parameters of the experiment. Their height d is several cm but their growth in height appears to be limited and generally does not extend top to bottom of the tank. The original upper layer acts as reservoir from which the down-fingers extract salt and deliver it to the original lower layer which acts as fresh reservoir for the up-fingers. This is a one-step staircase with a shallow finger layer between two deep mixed convecting layers. The convecting layers have small property gradients, the large jumps are across the finger layer. Flux laws are then derived from measured changes in reservoir temperature and salinity with time. The salt flux F_s data is usually fitted to a $(\Delta S)^{4/3}$ power law, where ΔS is the current drop in salinity across the finger layer. The implications of this power law is that the flux F_s is independent of total depth d, for then, the Nusselt number N_s which is defined as the ratio of total flux F_s to the diffusive flux $\kappa_s \Delta S/d$, would be proportional to d, i.e., to salt Rayleigh number R_s to the 1/3 power. $N_s \sim R_s^{1/3}$ is the nondimensional equivalent of $F_s \sim (\Delta S)^{4/3}$. References can be found in the review articles by Schmitt (1994) and by Turner (1985).

A small modification to Turner's original setup turns out to allow reaching a wider range of parameters and flow types. This arrangement is based on the stability theory model in which the fluid has bottom boundary at $z = 0$ maintained at temperature T_o, salinity S_o and top boundary at $z = d$ maintained at $T_o + \Delta T$, $S_o + \Delta S$. Note that d is now a free parameter to be chosen by the experimenter. Initially the fluid between these boundaries is given a double linear profile of temperature and salinity.

A schematic diagram of the apparatus to produce this state is shown in Figure 39.3. But first to avoid heat exchange between working fluid and laboratory, sugar is used as proxy for the slower diffusing ocean salt, and salt is used as proxy for the faster diffusing ocean cold. The diffusivity ratio is $\tau = 0.3$ for sugar and salt. It is 10^{-2} for salt and heat.

The apparatus is built in three segments: the bottom reservoir, the test section, and the top reservoir. The fluids in the three segments are separated from each other by tautly stretched porous membranes which allow sugar and salt to diffuse through it. The bottom reservoir is filled with "cold freshwater," i.e., water with no sugar but with salt concentration ΔT. The top reservoir is filled with "hot salty water," i.e., no salt but with sugar concentration ΔS. The test section is prepared using the two-bucket method to have a double linear gradient,

$$\left.\begin{aligned} T &= \Delta T(1 - z/d), \\ S &= \Delta S(z/d), \end{aligned}\right\} \tag{39.7}$$

where d is the total height of the test section (d = 2, 1, or 0.3 m). Each of the segments had horizontal dimensions 30 cm × 10 cm for these three depths. A fourth tank had the important distinction that it had horizontal dimensions 150 cm × 15 cm and d = 0.3 m. The large aspect ratio allowed a new development.

The reservoirs are constantly stirred and flushed at controlled rates by the apparatus shown in Figure 39.3. While the upper reservoir is being flushed with sugar water the lines to the lower reservoir are closed by the solenoid valves. Likewise when the lower reservoir is being flushed with saltwater the upper reservoir lines are closed. This is to prevent forcing fluid through the membranes. This alternation can be set for desired duration, such as 15 min, then runs at this set protocol indefinitely (for days or months as needed). The outflow from each reservoir is collected and the salinity and density as well as the volume flux are measured. Thus, the outflow from the top reservoir contains some salt because of the transport by the fingers. From these measurements the salt flux and sugar flux are determined.

Some of the nature of salt fingers can be seen through shadowgraphs. The apparatus shown in Figure 39.3 is kept in a photographically dark room. Parallel light shone through the working fluid is refracted by the small density (hence refractive index) variations and may be focused at some distance from the fluid, where photograph paper is placed and exposed. (For the 2 m tank the paper is 2 m long.) At the lowest values of ΔS and ΔT, these shadowgraphs show straight salt fingers (Figure 39.4a) which extend from the bottom reservoir to the top reservoir for all three test sections. At higher ΔS and ΔT, the fingers are lumpy (Figure 39.4b) but still extend from the bottom reservoir to the top reservoir. At still higher ΔS and ΔT the flow becomes layered, Figure 39.4c and d (having a staircase profile in T and S) with finger layers and convecting layers alternating in the vertical. We still do not have an adequate mapping in parameter space of these various flow forms.

The most surprising result is that, not only is flux depth dependent, but for larger depths, the flux is *smaller.* With the same ΔS, ΔT imposed at the reservoirs the 0.3 m tank (with the smallest d) had the largest flux. This was finally unraveled by observing the shadowgraphs from the three tanks during a run-down experiment. The 2 m tank had seven layers (four finger layers alternating with three convecting layers each approximately 30 cm tall) while the 30 cm tank had only one finger layer 30 cm tall. With time, ΔS and ΔT decrease slowly in a time of several days. The convecting layers become less vigorous and the finger layers grow taller. Occasionally pairs of finger layers merge as the convecting layer vanishes. Since merging in pairs is not always simultaneous, the appearance is one of an irregular staircase. As each convecting layer vanished, the flux increased markedly, almost jumped. Finally at the lowest ΔS, ΔT and only one finger layer and no convecting layers, the flux was at its largest. While ΔS, ΔT had changed by a few percent, the flux had increased by a factor of 4. This is shown in Figure 39.5. Figure 39.6 shows a simple model of this process. In the left panel salt fingers operate between two reservoirs of salinity ΔS at the top, 0 at the bottom. Down-going fingers carry salinity ΔS, from the top to bottom reservoir, up-going fingers carry zero salinity. In the right panel the finger layer has been stirred at mid-depth, producing a mixed layer of salinity approximately $\Delta S/2$. The down-giving fingers leaving the

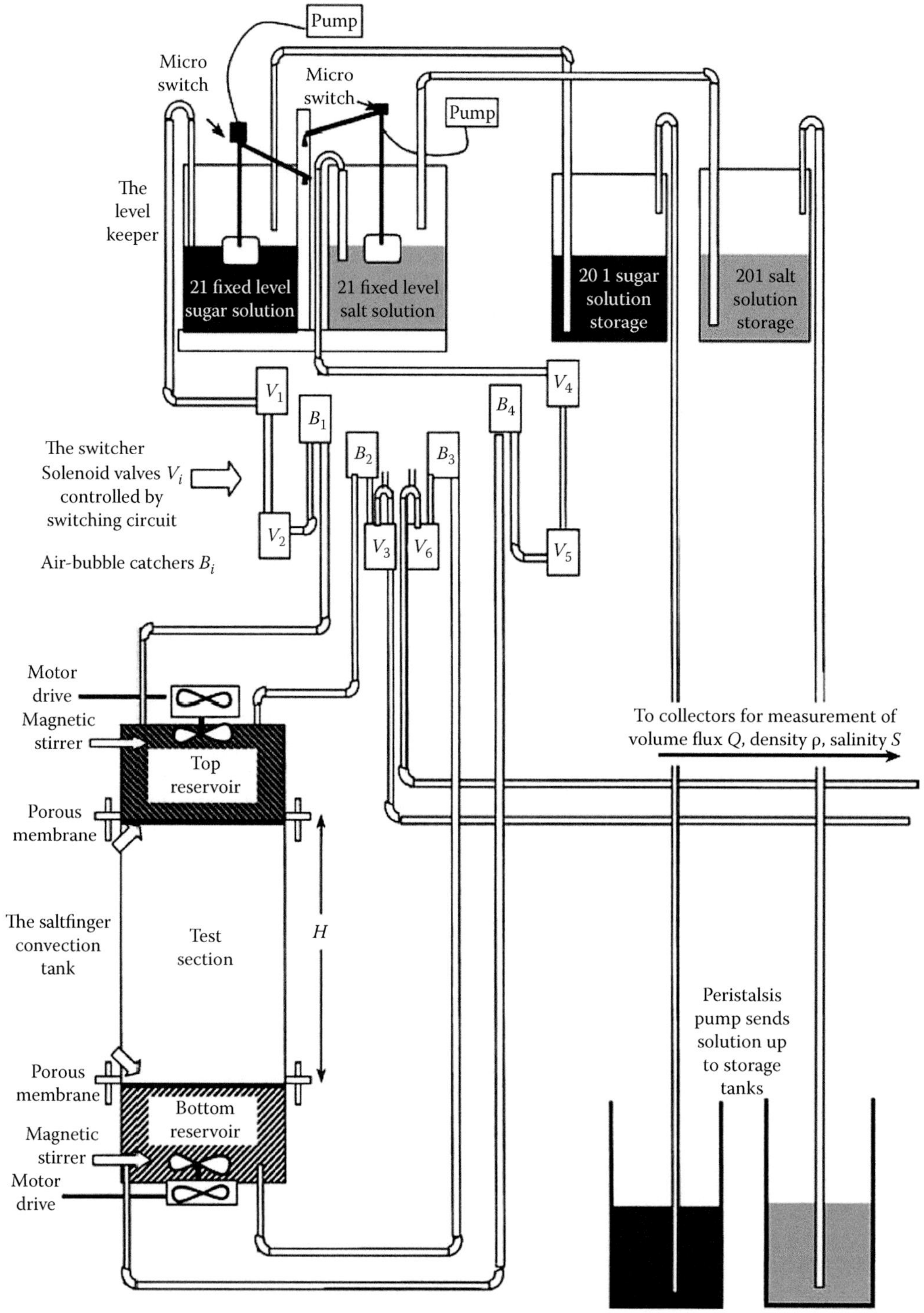

FIGURE 39.3 Schematic diagram of apparatus. (After Krishnamurti, R., *J. Fluid Mech.*, 483, 287, 2003. With permission.)

top reservoir still carry salinity ΔS and the up-going fingers leaving the bottom reservoir still carry $S = 0$. However the up-going fingers leaving the mixed layer now carry salinity $\Delta S/2$. Thus half of the export from the top reservoir is returned. The net flux has been decreased to ½. Now the two states (left and right panels) may not occur at the same ΔS, ΔT, but they do occur at ΔS, ΔT that differ by only a few percent, while the flux has been changed by a factor of 2. It is an easy algebraic exercise to show that with $n + 1$ finger layers, the flux is reduced by a factor $1/(n + 1)$.

To summarize, the salt flux is not determined by ΔS and ΔT alone; it is depth dependent in that it depends upon how many layers will fit into the given depth. Each additional convecting layer

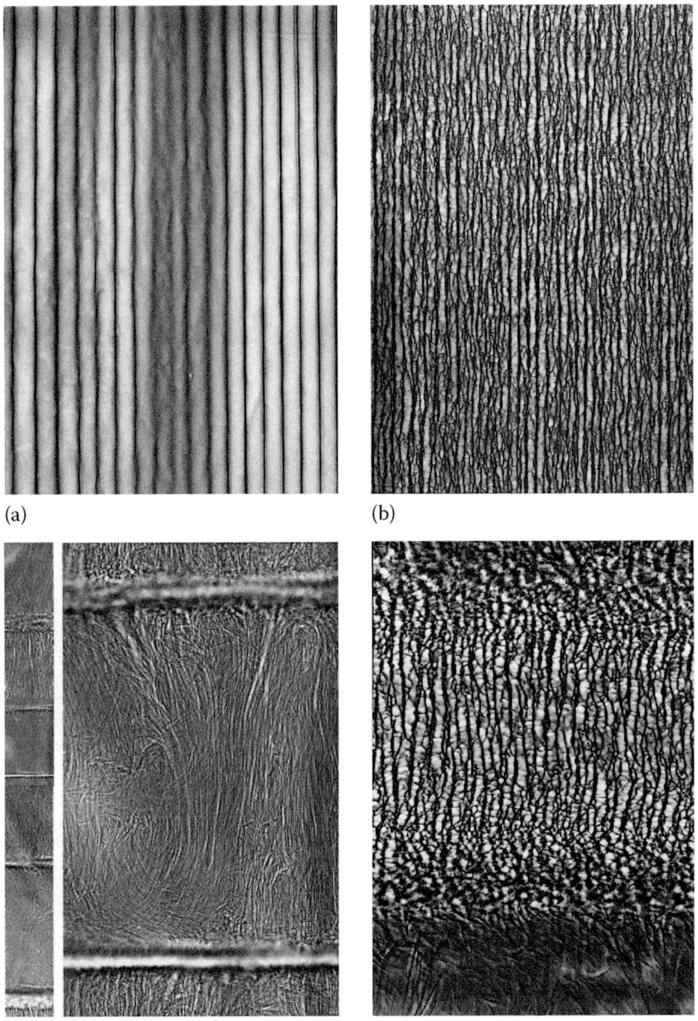

(a) (b) (c) (d)

FIGURE 39.4 Shadowgraphs of salt fingers (a) straight fingers (photo excerpted from 2 m tall shadowgraph) $R_S = 7.12 \times 10^{15}$, $R_\rho = 1.25$. (b) As in (a) except lumpy fingers (excerpted from 2 m tall shadowgraph) $R_S = 7.60 \times 10^{15}$, $R_\rho = 1.25$. (c) Portion of thermohaline staircase $R_S = 10.0 \times 10^{15}$, $R_\rho = 1.25$. (d) Portion of staircase showing finger zone $R_S = 9.42 \times 10^{15}$, $R_\rho = 1.25$. (After Krishnamurti, R., *J. Fluid Mech.*, 483, 287, 2003. With permission.)

decreases the flux. Thus the determination of ultimate heights of finger layer and convecting layer is crucial to determining fluxes.

In oceanic staircases, the vertical profile of salinity shows thin (1–2 m depth) finger interfaces with large salinity drops, while the well-mixed convecting layers are tens of meters in depth and of nearly constant salinity. They are very long-lasting (Molcard and Williams 1975). Laboratory thermohaline staircases have this appearance initially, but as the finger layers lengthen, the "treads" become "slides." It is the decline of the convecting layer with time that made study of layer depths dynamics difficult.

It was recognized that the convecting layers, confined by the side walls, often have aspect ratio A of order unity. Here A is the ratio of available horizontal to vertical dimension, L_x/L_z. Three tanks have $L_x = 30$ cm, in which the convecting layers often had $L_z = 30$ cm or greater. The observed flow (Figure 39.7a) shows unorganized plumes of small scale (mm in width). It should be pointed out that simply making A large may not be sufficient; if L_z is too small the effective Rayleigh number $Ra \sim L_z^3$ may not be large enough to support large-scale flow.

However with the fourth tank $L_x = 150$ cm, and the convecting layer occupied depth $L_z = 10$ cm. (The remaining 20 cm of the total tank depth $d = 30$ cm was occupied by a finger layer, making a one-step staircase.) Now the convecting layers, with aspect ratio $A = 15$, displays an organized large-scale flow with horizontal flow in one direction near the top and in the opposite direction near the bottom. It is in effect a uni-cell 150 cm wide, 10 cm deep. The vertical turn-around of the flow occupies approximately 10 cm at each end wall. All details of the flow cannot be seen in a single form of visualization. In shadowgraph the finger layer shows well (Figure 39.7b) but the convecting layer shows almost no detail. With particle image velocimetry (PIV), the particles in the finger layer appear stationary since speeds in the finger layer are often two orders of magnitude slower than

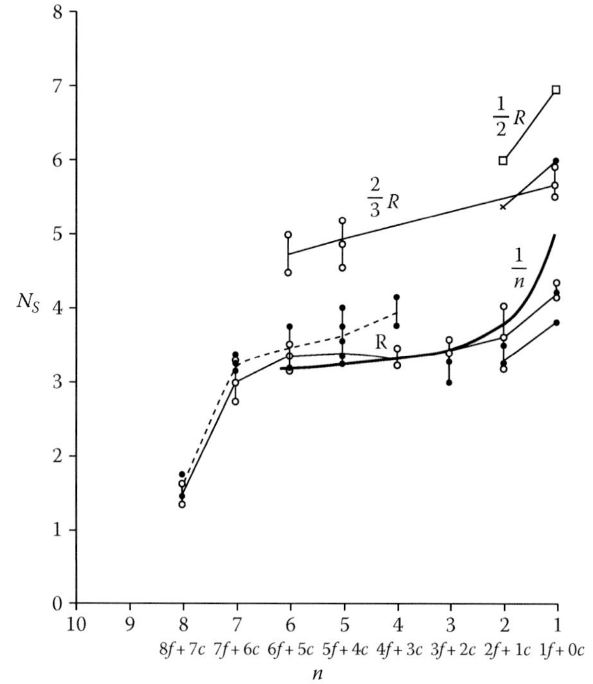

FIGURE 39.5 The variation of the overall Nusselt number N_S with the number n of finger layers in the 3 m tall tank. The labels R, $2/3R$, and $1/2R$ refer to relative initial concentration gradients of 1, 2/3, and 1/2 for S and T. $R\rho = 1.17$ for all these experiments. The filled circles and the broken lines are for the sugar Nusselt number N_S; the open symbols are for the salt Nusselt number N_T. For both Nusselt numbers $C = 0.304$. The heavy solid curve represents a function $N \sim 1/n$. Along the abscissa, under $n = 8$ for example, the notation $8f + 7c$ indicates eight finger layers and seven convection layers. (After Krishnamurti, R., *J. Fluid Mech.*, 483, 287, 2003. With permission.)

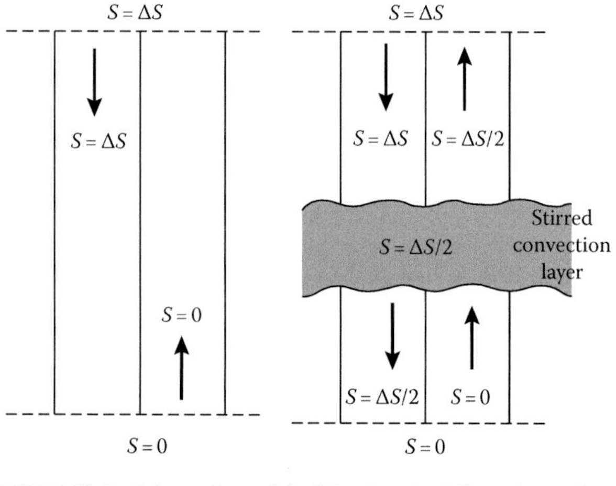

FIGURE 39.6 Schematic model of the drop in S-flux when coherent fingers are stirred by convection layers. (After Krishnamurti, R., *J. Fluid Mech.*, 483, 287, 2003. With permission.)

flows in the convecting layer. To see PIV in the finger layer, the convecting layer appears blurred. One of the best views, shown in Figure 39.8, was made as follows.

In preparation for PIV, nearly neutrally buoyant tracer particles approximately 30 μ diameter, suspended in 5 mL of sugar solution of appropriate density, was introduced at the top left end of the 70 L test section. This disrupted local salt fingers which have healed when the photo in Figure 39.8 was taken 5 h later. By this time, tracers have traveled down the down-going fingers and have entered the convecting layer below it. Here the large-scale flow, from left to right, carry the particles to the right. At the bottom of the convecting layer, the large-scale flow is from right to left, but the tracer particles are still absent in that part of the tank. Thus, when a sheet of green laser light illuminates the x, z plane at mid-distance in y, the tracer particles shine green, the particle-free fluid is black. The lower part of Figure 39.8 shows a transient tilted plume. PIV of this flow

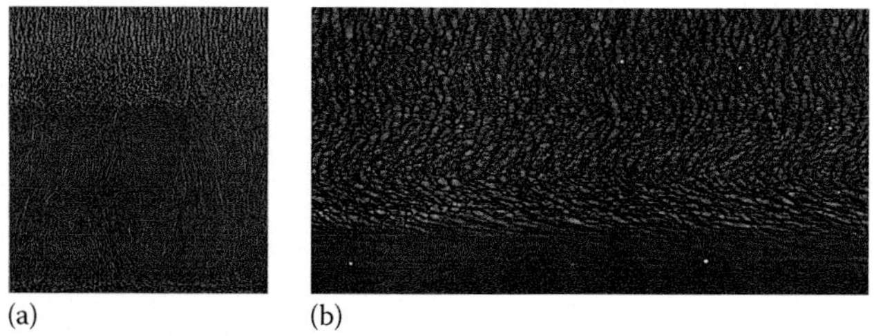

FIGURE 39.7 (a) An unorganized convection layer in the 30 cm wide tank. (b) Shadowgraph of salt fingers above a convecting layer, with large-scale flow from left to right directly under the fingers. The image represents 9.2 cm in the x direction, 5.2 cm in the z direction, excerpted from a photograph 150 cm wide by 30 cm tall, which was uniformly of this character at all x. The fingers occupied 20 cm, the convection occupied 10 cm in depth. Note the tilt of the fingers that end in the convecting flow. (From Krishnamurti, R., *J. Fluid Mech.*, 638, 491, 2009. With permission.)

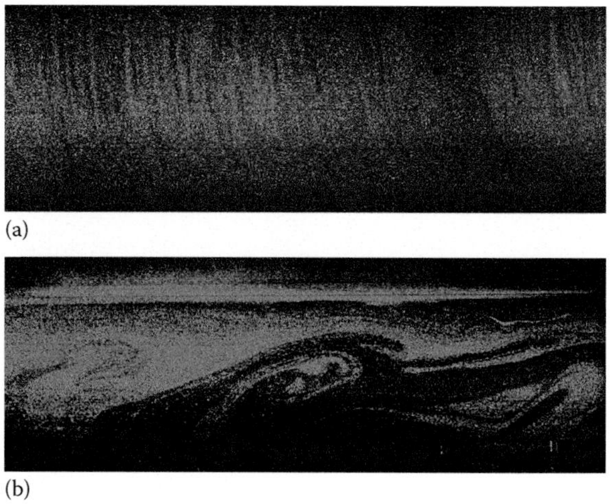

FIGURE 39.8 Tracer particles in a vertical sheet (x, z) at mid-depth in y, showing (a) fingers in the upper part of the tank and (b) tilted plumes in a large-scale shearing convection in the lower part of the tank. The tracers were introduced from the upper left, 5 h prior to this imaging. They traveled downward in the down-going fingers and entered the convecting fluid. Meanwhile fluid free of tracers came from the far right along the lower part of the convecting layer, leading to this visualization. Both (a) and (b) represent approximately 10 cm in z and 22 cm in x out of the total horizontal dimension of 150 cm. (From Krishnamurti, R., *J. Fluid Mech.*, 638, 491, 2009. With permission.)

is shown in Figure 39.9a. The *x*-averaged *x*-component <*u*> is shown in Figure 39.9b, and the *x*-averaged Reynolds stress <*uw*> is shown in Figure 39.9c. Between depth $z = 60$ and $z = 90$, <*uw*> is positive where ∂<*u*>/∂*z* is also positive, indicating that the kinetic energy of the mean flow increases at the expense of the kinetic energy of the fluctuating flow. Also analysis shows that the ratio of *ν*∂<*u*>/∂*z* to <*uw*> is very nearly equal to unity. It appears that the large-scale flow is maintained against viscous dissipation by the Reynolds stress divergence. As the photograph implies, particles in the tilted plume are traveling upward and rightward carrying rightward horizontal momentum to a region where the mean flow is already rightward.

Shadowgraphs of the bottom of the finger layer/top of the convecting layer show finger tips all being uniformly dragged to the right (Figure 39.7b) in agreement with the PIV results.

One additional piece of information is obtained from *y*-averaged sugar concentration measurements using optical rotation of polarized light. By scanning the fluid in the convecting layer just below the dragged finger tips, it is found that the sugar concentration progressively increases from left to right. It appears that the large-scale flow collects the small quantities of sugar delivered by each finger tip. This small quantity of sugar could probably produce a low Rayleigh number shallow convection (since the overall stratification is stable) but apparently is not sustainable. Instead, the large-scale flow collects the small deliveries from hundreds of fingers to make the resulting density sufficient to sink to 10 cm depth in spite of the mean stable stratification. It is reminiscent of fixed flux Rayleigh–Benard convection where at onset, the critical Rayleigh number $Ra_c = 5!$ and the critical wave number $a_c = 0$. The flow in this experiment has apparently sufficiently large *Ra* to support the large-scale flow via the Reynolds stresses.

For the diffusive regime, the oscillatory nature of the flow is easily visualized in the "Bouncing bottle" demonstration

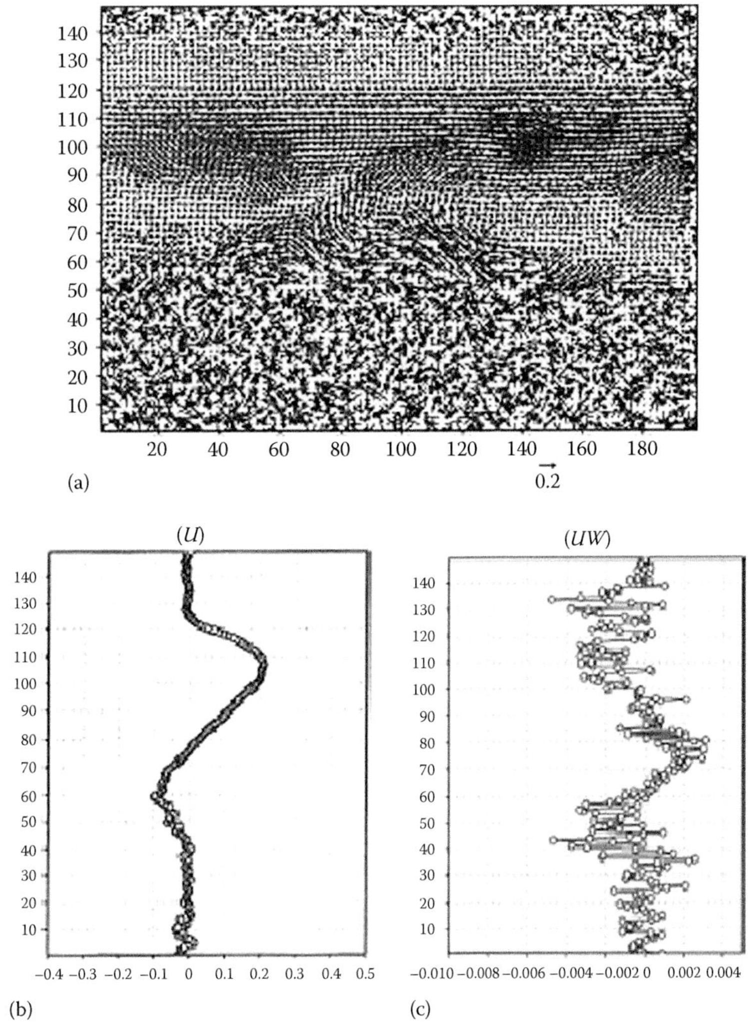

FIGURE 39.9 (a) A portion of the PIV-derived flow vectors on the vertical (*x*, *z*) sheet, (*b*, *c*) Analysis of the flow vectors of (a). The first graph (b) shows the horizontally averaged horizontal component of velocity <*u*> as it varies with vertical coordinate *z*. The second (c) shows the horizontally averaged Reynolds stress <*uw*> as it varies with *z*. Positive <*uw*> where ∂<*u*>/∂*z* is also positive (between $z = 60$ and $z = 90$) indicates that the kinetic energy of the mean flow <*u*> increases at the expense of the kinetic energy of the fluctuating flow. (From Krishnamurti, R., *J. Fluid Mech.*, 638, 491, 2009. With permission.)

experiment. A beaker of water with salt crystals in the bottom is heated from below. An inverted bottle of air bounces up and down in this beaker, extracting heat, but no salt, as it sits near the bottom of the beaker, causing the air to expand. The increased buoyancy force makes the bottle rise where it cools, air contracts, buoyancy force decreases so the bottle sinks and the process repeats.

39.3.4 Laboratory Experiments with Lateral Forcing

The original lateral interleaving experiment performed by Ruddick and Turner (1979) consisted of a two-compartment tank, a left and a right compartment separated by a removable partition. The left compartment was stratified with salt so that its density $\rho_L(z) = \alpha T(z)$ while the right was stratified with sugar such that its density $\rho_R(z) = \beta S(z)$. The fluid was prepared so that at each z, $\rho_L(z) = \rho_R(z)$. When the partition was removed there would be no gravity current, but salt would diffuse more quickly into the right than the sugar diffuse into the left compartment. The density imbalance would create a flow. However before diffusion could have progressed even 1 mm, mechanical disturbances from removing the partition had initiated interleaving flows. Whatever the initiation, once started, the dynamics that drives the intrusion could be studied. They showed that the "hot salty" intrusion (i.e., salt-free sugar intrusion) lying over "cooler fresher" water would drop its "salt" in the form of salt fingers, thereby producing a horizontal pressure gradient that would propel it leftward, as well as to make it buoyant to rise upward.

These are intrusive flows into horizontally unstratified regions. On the other hand the mathematical models (Holyer 1983) treat instability-generated flows in regions with horizontal property gradients, namely, with density profile dependent linearly on horizontal coordinate x and vertical coordinate z;

$$\rho(x,z) = \rho_0\left[1 - \alpha T(x,z) + \beta S(x,z)\right] \tag{39.8a}$$

with

$$\alpha T(x,z) = ax + bz \tag{39.8b}$$

$$\beta S(x,z) = ax + cz \tag{39.8c}$$

where a, b, c are constants to be chosen for reasonable ease in handling.

$$\frac{\alpha \partial T}{\partial x} = \frac{\beta \partial S}{\partial x} \tag{39.9}$$

Then T and S are density compensating. To produce this fluid we proceeded as follows. As an approximation to the field in Equation 39.8a we consider the fluid as discretized into bins, with each bin labeled for its location in x, z by indices i, j. Fluid for each bin is made with the appropriate T and S concentration, as dictated by (39.8b and 39.8c), usually in 5 mL volumes per bin, and each stored in one of several hundred syringes which are carefully labeled by its i, j index. For each value of j, i.e., for each depth, syringes are lined up and connected to the correct input lines for delivery to each i. These syringes are slowly and simultaneously plunged. Simultaneity is crucial since, if any one syringe delivers before others, its contents would flood into neighboring bins. The process is repeated level by level. The technique described in Krishnamurti (2006) works well according to all tests.

Many experiments (in parameter space R_S,R_T where this refers only to the vertical stratification) were run from finger-favorable to diffusive-convection-favorable. The most interesting were those cases stable to either fingers or diffusive convection, called "doubly stable". Even in the doubly stable regime fingers and convection become possible as a result of the lateral intrusive motion which carries hot salty water, say, laterally into fluid of different T, S properties where differential diffusion then leads to instability (see Figure 39.2b).

Typical sample shadowgraphs and associated PIV are shown in Figure 39.10a and b. The results may be summarized as follows. Lateral motion leads to localized salt fingering. The potential energy released thereby leads to convection in the layer below it. The convective plumes, under the influence of the shear from the alternating intrusive flows, tilt and the measured Reynolds stress is found to be re-amplifying the shear flow (Figure 39.10c). It is in the shear flow of the convecting layers that "hot salty" and "cold fresh" are separated rightward and leftward.

The theoretically predicted tilts of the intrusive flows were not observed except perhaps as an initial transient as the tank was slowly filled. After the filling process, which usually took at least 2 h to dispense the hundreds of syringes, the layers are seen to be horizontal within experimental error. Previous experiments into regions with no horizontal gradients do show a tilt upward toward hot salty regions. Hence we tried the following experiment: discretized and binned fluid samples were introduced into the tank as described earlier, but the choice of $T(x, z)$, $S(x, z)$ was such that only the middle third of the tank had $\alpha\partial T/\partial x = \beta\partial S/\partial x$ with T and S varying linearly in x. The left 1/3 and right 1/3 were prepared to have no horizontal gradients of properties. The shadowgraph of the resulting flow is shown in Figure 39.11. It shows that the nose of the intrusion does tilt upward to the "hot salty" corner, only when the intrusion is into horizontally unstratified regions. The tilt upward might also be interpreted as a wearing away of the lower part of the nose as the salt fingers are shed off. Likewise it tilts downward in the right 1/3. Otherwise the Reynolds-stress-driven part of the flow (middle 1/3) is horizontal.

Obviously from these results it is clear that the "salt" and "heat" transports are complicated. There is not a straight lateral transport of salt, for example, since some is shed vertically while some is carried laterally. In this last geometry described earlier (middle thirds), it will be possible to measure the vertically averaged sugar concentration as it changes from zero in the right 1/3, and progressively increasing with time as the intrusions come in, using optical rotation. This would be far easier than measuring sugar concentration pixel by pixel, where the velocities are obtained from PIV.

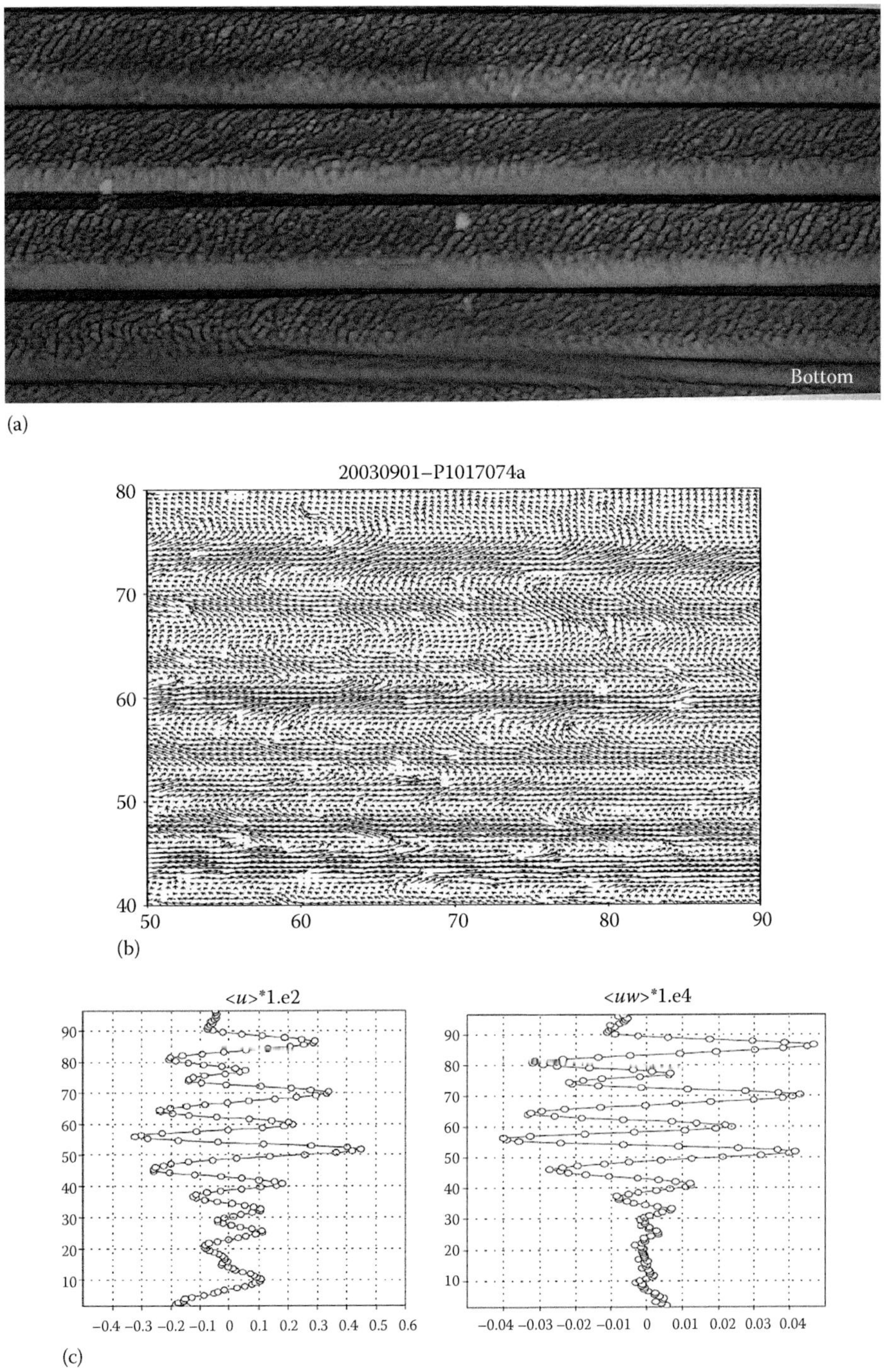

FIGURE 39.10 (a) Shadowgraphs showing salt-finger favorable interleaving. (b) Corresponding PIV. (c) Mean flow and Reynolds stress derived from PIV. (After Krishnamurti, R., *J. Fluid Mech.*, 558, 113, 2006. With permission.)

39.4 Applications

It has been suggested that double-diffusive flows must occur in magma chambers and on binary stars where eruptions from one of the star pairs falls upon the surface of the other thereby supplying the "salt" to the surface.

Closer to home we see large "boils" of freshwater erupting through the seafloor and into the warm salty waters of the Gulf of Mexico. This is part of the groundwater flow where large rainfalls over Florida and Georgia enter the Floridan Aquifer and through the Karst environment travels underground, at places erupting at famous springs, but ultimately, through the sands into the seawater. The west coast of Florida has a very wide shelf; the 50 m deep isobaths is on average 200 km from the coast. Thus isolating part of this profile would suggest a saturated porous medium, fresh and cold at the coastline, warm and salty offshore

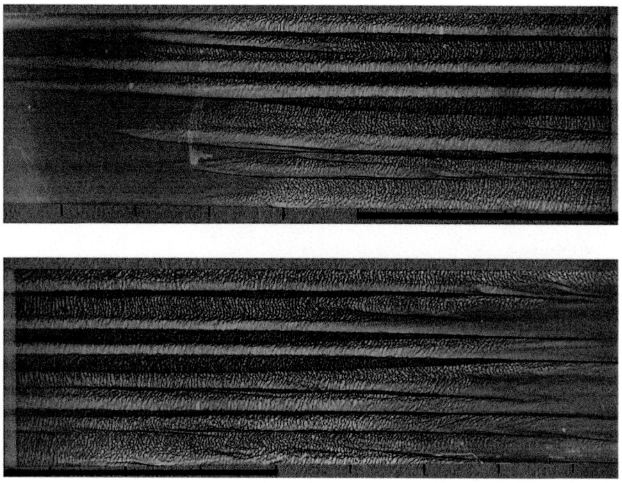

FIGURE 39.11 Shadowgraph showing interleaving on a continuous gradient, intruding into regions of no horizontal gradient. The distance between lines on the scale is 2.54 cm. The heavy black line along the bottom indicates the "middle one third" where continuous property gradients were initially imposed. (After Krishnamurti, R., *J. Fluid Mech.*, 558, 113, 2006. With permission.)

(in deep water) implying lateral intrusive porous medium flow. Isolating an even smaller part of this profile would suggest a warm salty seawater reservoir above cold, fresh Karst water filling the porous matrix.

Figure 39.12 shows a photograph of salt fingers (the "hot salty" fingers shown as dyed fluid from the reservoir on top). The right half of the tank is filled with glass beads of median diameter 0.3 mm. Although sieved, there is some distribution of sizes around this value. The tank was filled layer by layer; a few mm depth of glass beads saturated with "cold fresh" water, the next few mm of beads placed, then saturated with slighter "warmer saltier" water from a two-bucket procedure until the top was saturated with "hot salty" water. The left half of the tank separated from the right half by an impermeable vertical wall was filled in the same way, but at certain depths, a layer of coarser beads (5 mm diameter which is still smaller than salt-finger widths) was laid down. Salt fingers, which start simultaneously at all depths, start to flow from the fine layer. Upon reaching the coarse layer, each finger flow is obstructed and dispersed laterally, effectively mixing the down-going salty with the up-going fresh fingers. In approximation this is like the mixing by the convective layer in a staircase. The up-going fingers now operate not from the bottom reservoir but from this new mixed layer in the coarse beads. The salt flux determined from top reservoir salinity measurements is considerably smaller for the left half tank with the layered medium than the right half. Once more, the flux might not be determined easily by ΔS, ΔT between seawater and freshwater at depths, but is dependent upon the number of disrupting layers.

Sandy beds of seafloors must have many nonuniformities since sands are laid or move episodically. Nature is more complicated than simple models.

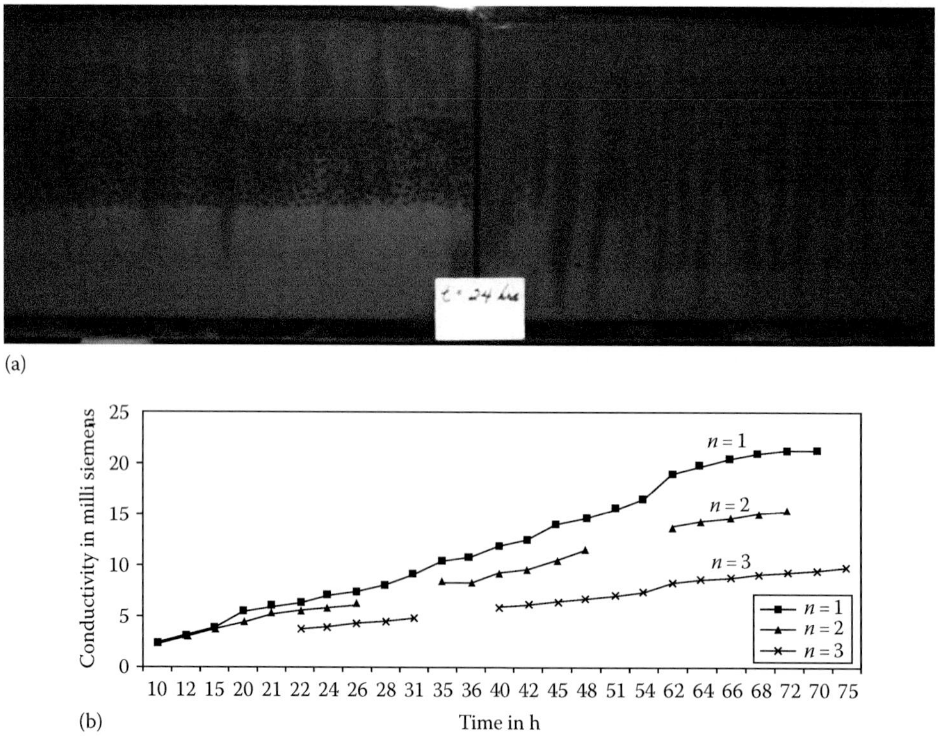

FIGURE 39.12 (a) Salt fingers in uniform porous medium (right), disrupted by dispersing layer of beads (left); (b) Slopes show larger salt flux through uniform (right) than through disrupted (left). (n is the number of uniform porous layers.)

39.5 Major Challenges

We have seen that when double-diffusive staircases set in, the fluxes of T and S become highly depth dependent. For ΔS, ΔT being the salinity and temperature drop (from top to bottom) across such a staircase, the flux drops off as $1/(n + 1)$ where n is the number of convecting layers in the staircase. Thus it is crucial to know the thickness of each layer, convecting and fingering, in order to know how many steps fit into the space.

Recent attempts have been made by Radko (2005) for the salt finger case. It is an ad hoc model in which he matches parameterized salt finger flux by a $(\Delta S)^{4/3}$ form and a convective flux by $Ra^{1/4}$. This is a reasonable first attempt, but will no doubt be improved. The convective flux, for example, is from a Rayleigh–Benard convection experiment with rigid heat-conducting fixed temperature boundaries. Krishnamurti (2009) shows that the convecting layer in a staircase develops a large-scale flow which collects the salt delivered from above by the many salt fingers until it has collected "enough" to convect to depth, in spite of the overall stable stratification. "Enough" appears to dictate the horizontal scale of the large-scale flow. However this depth must be large enough so that the Rayleigh number is large enough to sustain the large-scale flow via the Reynolds stress. Otherwise the local trickle of salt from salt fingers could sustain only a shallow convection which has small transport capability and large viscous dissipation. The staircase would then vanish and pure finger flow result.

For ocean application to mixing and the MOC, the magnitude of the fluxes by lateral intrusive flows needs to be measured and made parameterizable.

Our understanding of physical processes has been enhanced by laboratory experiments with sugar/salt experiment with $\tau = 0.3$. However, for quantitative flux laws for ocean application, we need to learn how to transfer laws for $\tau = 0.3$ to $\tau - 10^{-2}$.

References

Baines, P. G. and A. E. Gill, 1969, On thermohaline convection with linear gradients, *J. Fluid Mech.*, **37**, 289–306, 253, 256.

Gregg, M. C., 1980, The three-dimensional mapping of a small thermohaline intrusion, *J. Phys. Oceanogr.*, **10**, 1468–1492.

Griffiths, R. W., 1979, The influence of a third diffusing component upon the onset of convection, *J. Fluid Mech.*, **92**, 659–670.

Holyer, J. Y., 1983, Double-diffusive interleaving due to horizontal gradients, *J. Fluid Mech.*, **137**, 347–362.

Kluikov, Ye. Yu. and L. N. Karlin, 1995, A model of the ocean thermocline stepwise stratification cause by double diffusion. In *Double Diffusive Convection*, A. Brandt and H. J. S. Fernando, eds. Geophysical monograph 94, AGU, Washington, DC.

Krishnamurti, R., 2003, Double-diffusive transport in laboratory thermohaline staircases, *J. Fluid Mech.*, **483**, 287–314.

Krishnamurti, R., 2006, Double-diffusive interleaving on horizontal gradients, *J. Fluid Mech.*, **558**, 113–131.

Krishnamurti, R., 2009, Heat, salt and momentum transport in a laboratory thermohaline staircase, *J. Fluid Mech.*, **638**, 491–506.

Molcard, R. and A. J. Williams, 1975, Deep stepped structure in the Tyrrhenian Sea. *Mem. Soc. R. Sci. Liege,* 6, *Perie* **VII**, 191–210.

Neshyba, S., V. T. Neal, and W. Denner, 1971, Temperature and conductivity measurements under Ice Island T-3, *J. Geophys. Res.*, **76**, 8107–8120.

Radko, T., 2005, What determines the thickness of layers in a thermohaline staircase? *J. Fluid Mech.*, **523**, 79–98.

Richards, K. J. and R. T. Pollard, 1991, Fronts, layers and mixing in the western equatorial Pacific, *Nature*, **350**, 48–50.

Ruddick, B. R. and J. S. Turner, 1979, The vertical length scale of double-diffusive intrusions, *Deep-Sea Res.*, **26A**, 903–913.

Schlitzer, R., 2000, Electronic atlas of WOCE. *EOS Trans.* AGU, **81**(5), 45.

Schmitt, R. W., 1994, Double diffusion in oceanography, *Annu. Rev. Fluid Mech.*, **26**, 255–285.

St. Laurent, L. and R. Schmitt, 1999, The contribution of salt fingers to vertical mixing in the North Atlantic Tracer release experiment, *J. Phys. Oceanogr.*, **29**, 1404–1424.

Stommel, H. and K. N. Fedorov, 1967, Small scale structure in temperature and salinity near Timor and Mindinao, *Tellus*, **19**, 306–325.

Turing, A. M., 1952, The chemical basis for morphogenesis, *Philos. Trans. R. Soc. Lon.*, **237**, 37–72.

Turner, J. S., 1985, Multicomponent convection, *Annu. Rev. Fluid Mech.*, **17**, 11–44.

Wunsch, C. and R. Ferrari, 2004, Vertical mixing, energy and the general circulation of the oceans, *Annu. Rev. Fluid Mech.*, **36**, 281–314.

40

Shallow Shear Flows in Surface Water

M.S. Ghidaoui
Hong Kong University of Science and Technology

M.Y. Lam
Hong Kong University of Science and Technology

40.1 Introduction

Shallow surface water flows refer to problems where the horizontal dimension of the dominant flow process far exceeds the water depth. When waves are the dominant flow process, such flows are often referred to as shallow water waves and are characterized by wavelengths significantly larger than the mean water depth (see Figure 40.1a). Examples of shallow water waves include tsunamis, tidal and flood waves, where the wavelength is often 100–1000 times larger than the water depth. For example, a typical length scale of a tsunami wave with a period of 20 min in an ocean with a depth of the order of 4 km is about 200 km. When turbulence is the dominant process of interest, such flows are often called shallow turbulent flows and are characterized by vortical structures aligned with the gravitational axis and whose diameter is far greater than the water depth (see Figure 40.1b). For example, the horizontal scale of the eddies in the wake of Rattray Island in northeast Australia is of the order of 1000 m, while the water depth is of the order of 20 m (Wolanski et al. 1984). In addition, the horizontal scale of the eddies in the wake of islands in Rupert Bay in northern Canada ranges from 100 to 3000 m, while water depth ranges from 1 to 4 m (Ingram and Chu 1987). Moreover, the horizontal scale of the eddies in the mixing layer downstream of a junction of the Mekong river is of the order of a 1000 m, while the water depth ranges from 5 to 10 m (Jirka and Seol 2010).

This chapter introduces the properties of turbulent shallow shear flows in a homogeneous fluid and the methods used to examine them. Such flows are omnipresent in lowland rivers, estuaries and coastal waters. For example, notice the large-scale eddies generated by a channel that discharges into an oxbow lake connected to the Lower Mississippi River near Baton Rouge, United States (see Figure 40.2). Shallow turbulent flows also occur in other geophysical and engineering applications. For example, shallowness can be the result of stratification, where the fluid is arranged in thin layers overlaying one another in pancake-like structures such that the vertical scale of the motion in each layer is limited by the buoyancy forces in comparison to the horizontal scale. Examples of such flows occur in stratified reservoirs, lakes, oceans, and atmospheric boundary layers (Fernando and Voropayev 2010). Other shallow flows include large-scale synoptic metrological flows and large-scale atmospheric island wakes.*

The turbulent shallow shear flows that concern this chapter consist of a layer of water bounded by the fixed bed and a free surface with strong three-dimensional turbulence in which the length scale of the turbulent eddies are of the order of the water depth or less and a strong quasi-two-dimensional turbulence in which the length scale of the eddies are much larger than the water depth. An example of such configuration is shown in

* An example of atmospheric turbulent shear flows in the wake of Guadalupe Island can be found in http://eosweb.larc.nasa.gov/HPDOCS/misr/misr_html/von_karman_vortex.html.

Handbook of Environmental Fluid Dynamics, Volume One, edited by Harindra Joseph Shermal Fernando. © 2013 CRC Press/Taylor & Francis Group, LLC.
ISBN: 978-1-4398-1669-1.

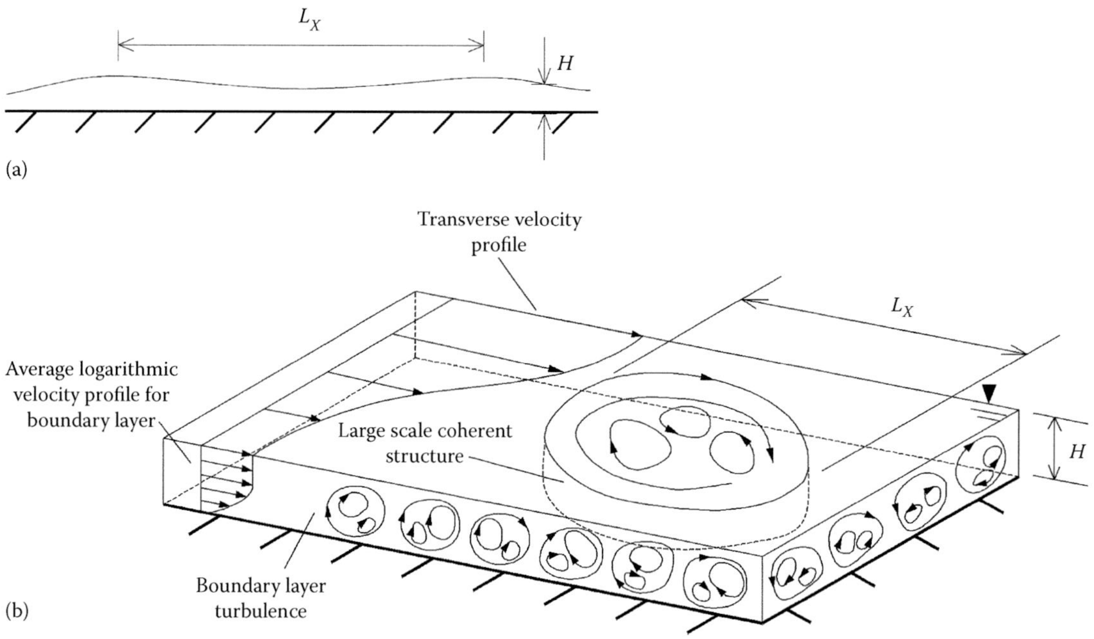

FIGURE 40.1 (a) Shallow water waves (i.e., gravity waves with horizontal length scale far greater than the water depth). (b) Shallow shear flows (i.e., large-scale turbulent eddies with horizontal length scale far greater than the water depth).

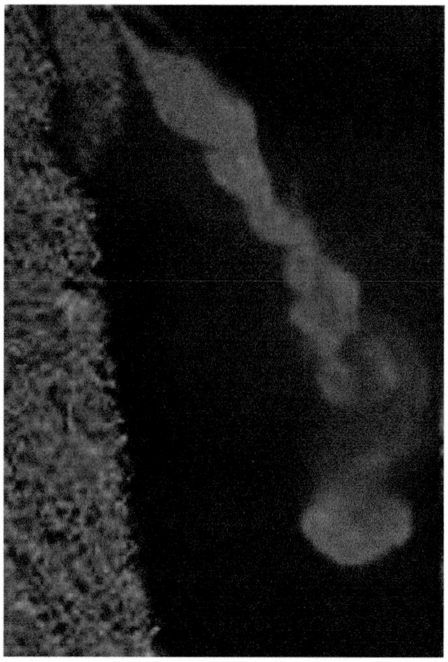

FIGURE 40.2 Aerial photograph of a river mouth, in which a tie channel discharges into an oxbow lake connected to the Lower Mississippi River near Baton Rouge, LA. (Courtesy of 1998 U.S. Geological Survey Digital Ortho-Quarter Quadrangle.)

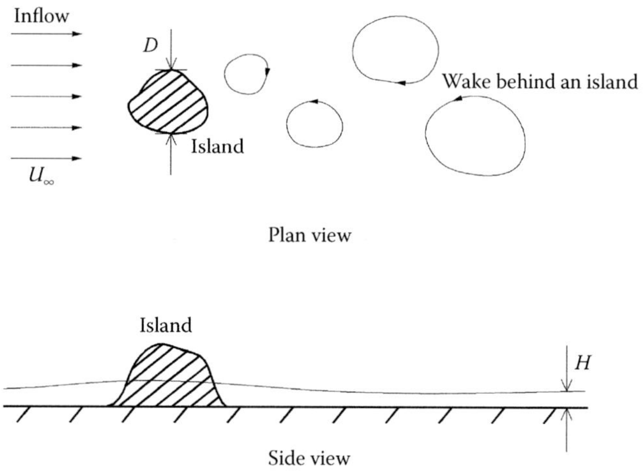

FIGURE 40.3 Schematic diagram of a shallow wake, with an island diameter greater than the water depth.

Figure 40.3, where the large-scale turbulent structure is generated by an island. The three-dimensional turbulence is generated at the bed and possibly at the free surface.

The large-scale quasi-two-dimensional turbulence is generated by a wide range of mechanisms, which Jirka (2001) subdivided into three types, namely, A, B, and C. Type A refers to large-scale turbulence generated by topographic forcing such as islands, headlands, jetties, and groynes. Type B refers to large-scale turbulence induced by an excess or a deficit of horizontal momentum (e.g., jets or mixing layers) or by a lateral variation in bed friction, bed elevation, or both (e.g., compound and composite channels). Type C refers to large-scale turbulence induced by local spatial and/or temporal acceleration or deceleration of the flow or by the vertical tilting and lift up of the bed-generated three-dimensional turbulent vortices. Boils are typical example of large-scale eddies produced by the tilting and lift-up mechanism (Nezu and Nakagawa 1993).

The large-scale turbulent structures in shallow flows are associated with significant horizontal mixing of mass, momentum, and energy. Therefore, an understanding of these flows is important for a wide range of engineering problems. For example, the study of flow patterns and mixing processes in the wake of natural protrusions such as mountains, islands, and headlands is critical for the prediction, assessment, analysis, and mitigation of pollution problems in air and water. Consider Hong Kong as an example. It comprises about 235 islands with sizes ranging from dozens of meters (i.e., large rocks) to 25 km (e.g., Lantau Island) in length. The flow pattern in the wake of these islands plays a decisive role in locating outfall discharges, mud disposal, cooling intakes, reclaimed land, marine parks, and marine reserves, to name a few only. In addition, the large-scale coherent structures that often occur at confluences of rivers, in rivers with lateral variation in roughness and/or bed elevation, in rivers with groynes, etc., are critical for estimating the horizontal spread of mass, momentum, and energy and for determining flood discharge and stage.

40.2 Principles

The mathematical modeling of shallow water flows is based on the principles of mass and momentum. Before introducing these principles, it is instructive to understand the main flow parameters that govern such flows. In general, free surface flows are driven by gravity, shear, Coriolis, and buoyancy forces. The relative contributions of the various driving forces to the inertial force are measured by a set of dimensionless parameters. For example, the ratio of inertial to viscous forces is the Reynolds number (*Re*), the ratio of inertial to gravity forces is the Froude number (*Fr*), the ratio of inertial to Coriolis forces is the Rossby number (*Ro*), and the ratio of inertial to buoyancy forces is the Richardson number (*Ri*). In this chapter, we limit ourselves to flows with negligible Coriolis and buoyancy forces so that the focus is on large-scale turbulent shallow shear flows in free surface where *Re* and *Fr* may be important, but not *Ro* and *Ri*. The implications are that the size of the large structures is much larger than the water depth, but much smaller than the Rossby radius and the fluid is either of uniform density or well mixed.

The Reynolds number for the large-scale turbulent structures is too large for viscous forces to play a major role in the dynamics of shallow flows. Wolanski et al. (1984) and Ingram and Chu (1987) found that the Reynolds number cannot explain the flow behavior in the wake of Rattray Island in northeast Australia as well as in the wake of islands in Rupert Bay in northern Canada since the observed flow features are reminiscent of a flow around a flat plate with a *Re* of order 10–100 when in actual fact the island diameter-based *Re* in the field is of the order 10^7–10^9. The linear analyses of Chen and Jirka (1997, 1998) and Ghidaoui et al. (2006) show that the stability of shallow shear flows is insensitive to the effective (i.e., eddy viscosity-based) Reynolds number once its value is of the order of 1000 or more—a condition that is met in the majority of environmental and hydraulic applications. It must be stressed that the Reynolds number still plays a central role in the small-scale three-dimensional turbulence generated at the bed of the flow and at the sides of islands and headlands. In fact, the bed (vertical) shear stress plays a critical role in the dynamics of large-scale turbulence—a fact that will become more clear in the rest of the chapter—and the horizontal shear stresses are required for resolving flow separation at islands, headlands and other type A generators. In addition, the horizontal shear stresses are essential for removing the mathematical singularity associated with the existence of a point of zero-velocity in separated flows (Ghidaoui et al. 2006).

The role of Froude number on the large-scale turbulent structures has not been fully investigated and is generally not explicitly considered in experimental design and also ignored in works geared at deriving analytical solutions for shallow shear flows. The theoretical studies of Falques and Iranzo (1994), Camassa et al. (1997), and Ghidaoui and Kolyshkin (1999) show that the stability results of shallow shear flows with and without gravity waves are within about 10% of each other when the Froude number is less than about 0.6. That is, Froude dependence is small when *Fr* is smaller than 0.6. On the other hand, the linear analysis of Kolyshkin and Ghidaoui (2002) and the nonlinear analysis of Liu et al. (2010) indicate that the influence of gravity waves on the stability, spread, and dynamics shallow of shear flows becomes more pronounced as the Froude number increases beyond 0.6. For *Fr* values greater than 0.6, gravity waves cause part of the energy of the flow to radiate away from the shear layer and part to dissipate at chocks (hydraulic jumps); thus, reducing the amount of energy available for the growth of the shear layer. In addition, unpublished preliminary research by the writers indicate that the shear layer is further stabilized due to the increased water depth (i.e., pressure) at the convergence points in the braid regions which acts in the opposite direction to the sense of rotation of the turbulent eddies and pushes different eddies apart in a way that tends to suppress pairing. It is stressed that a large majority of flows in rivers, lakes, and well mixed coastal waters are of low enough *Fr* that gravity wave effects are negligible. However, gravity waves effects on shallow shear flows need to be considered in problems involving shallow mixed layers overlaying a denser water, where the densimetric Froude number is often quite large (Chu 2010). In addition, Froude number effects for peak flows in natural or artificial compound and composite channels are generally moderate to high (Kolyshkin and Ghidaoui 2002).

If the scale and density are such *Ro* and *Ri* are not important and if *Re* and *Fr* cannot explain large-scale features in shallow shear flows, then what is the other parameter that can explain *Re* and *Fr*? Field, experimental, and theoretical studies indicate that the onset, growth, and dynamics of large-scale turbulence in shallow flows require a parameter that involves the bed friction force and the transverse large-scale horizontal shear. For example, Wolanski et al. (1984) carried out flow visualization and flow measurements of turbulent shallow wakes at Rattray Island in northeast Australia and found that, unlike deep wake flows which are governed by the Reynolds number, shallow wakes are governed by a different mechanism. The proposed mechanism involves the generation of vorticity by the island and the dissipation of vorticity by the bottom friction in the lee of the island (Wolanski et al. 1984). In particular, the island injects vorticity into the wake and promotes the formation of large-scale eddies while the bed friction impedes the eddy rotation and suppresses the formation of the large-scale structures. As a result, Wolanski et al. (1984) developed the following dimensionless parameter, P, on the basis of the ratio of the vorticity injection to the vorticity cancellation in order to understand and characterize shallow wake behavior:

$$P = \frac{U_\infty H^2}{\upsilon_{H_T} D} \tag{40.1}$$

where
- U_∞ is the ambient flow velocity
- H is the water depth
- υ_{H_T} is the vertical eddy diffusion coefficient
- D is the island width

Wolanski et al. (1984) suggested friction dominates and quasi-potential flow occurs if $P \ll 1$; a stable wake can be observed in Rattray Island if $P \approx 1$; wake instabilities if $P > 1$; and a 2D vortex street flow, similar in appearance to flows behind obstacles in deep flows where Reynolds number is of order 50 and where bed friction effects are negligible, develops.

The eddy viscosity in free surface flows, υ_{H_T} scales as $HU_\star$, where $U_\star$ is the shear velocity, such that $\upsilon_{H_T} = HU_\star(1/Re_{H_T})$ and $1/Re_{H_T}$ is the dimensionless eddy viscosity which can range from about 0.07 to 0.3 (Vionnet et al. 2004). The shear velocity is related to the bottom shear stress, τ_b, through the Darcy–Weisback friction factor, f, as follows: $\tau_b = \rho f U^2/8 = \rho U_\star^2$ (Pope 2000; Schlichting and Gersten 2000). As a result, Equation 40.1 can be rewritten as:

$$P = \frac{U_\infty H^2}{\upsilon_{H_T} D} = \frac{Re_{H_T} H}{\sqrt{f/8} D} = \frac{Re_{H_T} H}{\sqrt{c_f/2} D} \tag{40.2}$$

where $c_f = f/4$ is the bottom friction coefficient and can be approximated using relations such the Colebrook–White formula and its variants.

Another example that highlights the need for a new parameter, in addition to Re and Fr, to classify and understand shallow shear flows is the field study in Rupert Bay in Northern Canada conducted by Ingram and Chu (1987). They observed a turbulent wake bubble trailed by a Karman vortex street. Two turbulent motion scales were identified: (1) the large-scale vortex motions generated by the transverse shear and (2) the small-scale motions created by the bottom shear and limited by the shallow water depth. Ingram and Chu (1987) classified wakes on the basis of the bottom friction stability parameter S, introduced by Chu et al. (1983) for turbulent shear flows in shallow environments such as shallow wakes, shallow jets, and shallow mixing layers. The S parameter is defined on the basis of the ratio of the rate of dissipation of energy from the large-scale eddies by bottom friction to the rate of energy input into these eddies by the mean shear. Mathematically, its form is:

$$S_{IP} = \frac{c_f U_{IP}}{2H(U_y)_{IP}} \tag{40.3}$$

where
- y is the coordinate in the transverse direction
- U_y is the transverse gradient of the longitudinal velocity
- the subscript "IP" indicates that the function is evaluated at the inflection point so that U_{IP} and $(U_y)_{IP}$ are the velocity and velocity gradient at the inflexion point, respectively

For the case of island wake flows, the velocity at the inflexion point scales as $U_\infty/2$, the shear layer width scales as D and the velocity difference across the shear layer scales as U_∞ implying that the velocity gradient at the inflexion point scales as U_∞/D. Thus, Equation 40.3 for shallow island wakes is:

$$S_{IP} = \frac{c_f(U_\infty/4)}{H(U_\infty/D)} = \frac{1}{4}\frac{c_f D}{H} \tag{40.4}$$

where S is the island wake parameters first proposed by Ingram and Chu (1987) and later used in experimental and theoretical studies of shallow wakes (e.g., Chen and Jirka 1995, 1997; Kolyshkin and Ghidaoui 2003; Uijttewaal and Jirka 2003; Ghidaoui et al. 2006). Most researchers drop the ¼ in Equation 40.4 and use $S = c_f D/H$. This form is adopted in this chapter. The field data from Rupert Bay in Northern Canada conducted by Ingram and Chu (1987) indicate the presence of vortex street wakes for $0.19 < S < 0.52$ and stable wake with low horizontal turbulence for $S > 0.52$. The few anomalies that did not fall in these ranges were explained by the authors to be due to the stabilization effect of stratification—a process not accounted for in the S parameter. It is noted that the island parameters (i.e., $S = c_f D/H$ and P) are related through the relation $P = [Re_{H_T}(2c_f)^{1/2}]S^{-1}$. Since c_f and Re_{H_T} do not change much within a particular flow, the two parameters, P and S, are approximately the same and can be used interchangeably in the analysis of wake flows.

The bed friction parameter given by Equation 40.3 is robust and applicable to shallow shear flows in general including mixing layers and jets. Its form for the mixing layer is:

$$S_{IP} = \frac{c_f\left(U_\infty^1 + U_\infty^2\right)/(2H)}{2\left(U_\infty^1 - U_\infty^2\right)/\Delta_0} \tag{40.5}$$

where
- U_∞^1 and U_∞^2 are the ambient velocities of the fast and slow streams, respectively
- Δ_0 is the width of the mixing layer so that $(U_y)_{IP}$ can be approximated by $\left(U_\infty^1 - U_\infty^2\right)/\Delta_0$ and U_{IP} by $\left(U_\infty^1 - U_\infty^2\right)/2$

The form of Equation 40.3 for jets in still environments is (Chen and Jirka 1998):

$$S = \frac{c_f B}{H} \tag{40.6}$$

where B is the half-width. Note that (40.6) can be recovered from (40.4) by setting $D = B$ and $U_\infty^2 = 0$ so that $\left(U_\infty^1 - U_\infty^2\right) = U_\infty^1$, $\left(U_\infty^1 + U_\infty^2\right)/2 = U_\infty^1/2$, and dropping the 1/4 as before.

40.2.1 Governing Equations

Consider the free surface flow given in Figure 40.4. Let x_α^* be the horizontal coordinates with $\alpha = 1,2$ so that $x_\alpha^* = (x_1^*, x_2^*)$; the index 1 designates stream-wise direction and 2 designates the cross stream (transverse) direction; z^* the vertical coordinate, t^* the time coordinate, $(u_\alpha^*, w^*) = (u_1^*, u_2^*, w^*)$ are the velocities along (x_1^*, x_2^*, z^*), respectively; P^* is the pressure; υ is the kinematic viscosity; ρ the fluid density; υ the fluid kinematic viscosity; and g is the gravitational acceleration. The superscript "*" denotes dimensional quantities. The governing equations of this free surface flow are the Navier–Stokes equations:

$$\frac{\partial u_\alpha^*}{\partial x_\alpha^*} + \frac{\partial w^*}{\partial z^*} = 0 \tag{40.7}$$

$$\frac{\partial u_\alpha^*}{\partial t^*} + \frac{\partial u_\kappa^* u_\alpha^*}{\partial x_\kappa^*} + \frac{\partial w^* u_\alpha^*}{\partial z^*} = -\frac{1}{\rho}\frac{\partial p^*}{\partial x_\alpha^*} + \upsilon\frac{\partial^2 u_\alpha^*}{\partial x_\kappa^* \partial x_\kappa^*} + \upsilon\frac{\partial^2 u_\alpha^*}{\partial z^{*2}} \tag{40.8}$$

$$\frac{\partial w^*}{\partial t^*} + \frac{\partial u_\kappa^* w^*}{\partial x_\kappa^*} + \frac{\partial w^{*2}}{\partial z^*} = -g - \frac{1}{\rho}\frac{\partial p^*}{\partial z^*} + \upsilon\frac{\partial^2 w^*}{\partial x_\kappa^* \partial x_\kappa^*} + \upsilon\frac{\partial^2 w^*}{\partial z^{*2}} \tag{40.9}$$

along with initial condition, the no slip and no flux conditions at the bed and the kinematic condition at the free surface. Like α, κ takes on values 1 and 2 (i.e., $\kappa = 1,2$). As stated earlier, this chapter focuses on hydraulic and environmental fluid mechanics problem where (1) the fluid is either of uniform density or can be approximated as being of uniform density (e.g., a well-mixed layer) and (2) the horizontal scale much larger than the water depth, but much smaller than the Rossby radius.

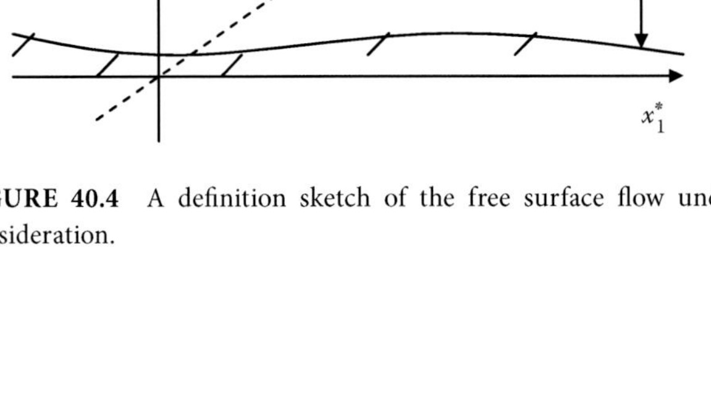

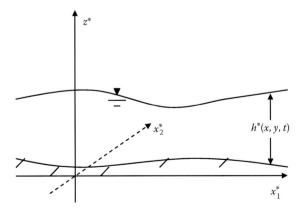

FIGURE 40.4 A definition sketch of the free surface flow under consideration.

Let the horizontal and vertical length and velocity scales be (L_X, U) and (H, W), respectively, and the pressure scale be ρU^2 and time scale be $T = L_X/U$ then system (40.7) through (40.9) becomes:

$$\frac{\partial u_\alpha}{\partial x_\alpha} + \frac{WL_X}{UH}\frac{\partial w}{\partial z} = 0 \tag{40.10}$$

$$\frac{\partial u_\alpha}{\partial t} + \frac{\partial u_\kappa u_\alpha}{\partial x_\kappa} + \frac{WL_X}{UH}\frac{\partial w u_\alpha}{\partial z} = -\frac{\partial p}{\partial x_\alpha} + \frac{1}{Re_{L_X}}\frac{\partial^2 u_\alpha}{\partial x_\kappa \partial x_\kappa} + \frac{L_X^2}{H^2}\frac{1}{Re_{L_X}}\frac{\partial^2 u_\alpha}{\partial z^2} \tag{40.11}$$

$$\frac{WH}{UL_X}\left(\frac{\partial w}{\partial t} + \frac{\partial u_\kappa w}{\partial x_\kappa}\right) + \frac{W^2}{U^2}\frac{\partial w^2}{\partial z} = -\frac{gH}{U^2} - \frac{\partial p}{\partial z} + \frac{WH}{UL_X}\frac{1}{Re_{L_X}}\frac{\partial^2 w}{\partial x_\kappa \partial x_\kappa} + \frac{WL_X}{UH}\frac{1}{\mathrm{Re}_{L_X}}\frac{\partial^2 w}{\partial z^2} \tag{40.12}$$

where

the lower case represents dimensionless variables

$Re_{L_X} = UL_X/\upsilon$ is the Reynolds number based on the velocity and length scales (L_X, U)

Mass balance requires that all three terms in (40.10) are of the same order of magnitude so that the vertical transport of mass balances the horizontal mass transport. Therefore, WL_X/HU is of order 1 implying the vertical and horizontal advective timescales are comparable and W/U is of order H/L_X. Since the focus is on shallow flows where $H/L_X \ll 1$, then $W/U \ll 1$ (i.e., the vertical motion is far limited compared to the horizontal motion). With $H/L_X \ll 1$, $W/U \ll 1$ and $1/Re_{L_X} \ll 1$, the vertical momentum Equation 40.12 simplifies to simple balance between the vertical pressure gradient and gravity, leading to the hydrostatic pressure distribution and system (40.10) through (40.12) becomes:

$$\frac{\partial u_\alpha}{\partial x_\alpha} + \frac{\partial w}{\partial z} = 0 \tag{40.13}$$

$$\frac{\partial u_\alpha}{\partial t} + \frac{\partial u_\kappa u_\alpha}{\partial x_\kappa} + \frac{\partial w u_\alpha}{\partial z} = -\frac{\partial p}{\partial x_\alpha} + \frac{1}{Re_{L_X}}\frac{\partial^2 u_\alpha}{\partial x_\kappa \partial x_\kappa} + \frac{L_X^2}{H^2}\frac{1}{Re_{L_X}}\frac{\partial^2 u_\alpha}{\partial z^2} \tag{40.14}$$

$$0 = -\frac{gH}{U^2} - \frac{\partial p}{\partial z} \tag{40.15}$$

While it is tempting to neglect the second term on the right-hand side of (40.14), these horizontal viscous terms are essential for the development of horizontal boundary layers past islands and other obstacles. Without these terms, flow separation and vorticity generation due to islands, headlands, groyne fields, and other type A shallow flow generators would be impossible. Models that produce flow separations without the inclusion of such horizontal viscous terms must be producing them from numerical and not physical viscosity. It is important to note that while the vertical motion is weak in comparison to the horizontal motion, the flow is not two-dimensional. In fact, since WL_X/HU is of order 1, then L_X/U is of the order of H/W (i.e., the vertical and horizontal advective timescales are of the same order) and the fine vertical turbulence structure and the bed friction it generates is important for the dynamics of the large-scale turbulent structures as implied by the S parameter defined in the previous section.

Integrating the continuity equation (40.13) and the momentum equations (40.14) and (40.15) with respect to depth and imposing the kinematic boundary condition at free surface and the no slip condition and zero flux at the bed yields the well-known shallow water equations

$$\frac{\partial h}{\partial t}+\frac{\partial h\bar{u}_\alpha}{\partial x_\alpha}=0 \tag{40.16}$$

$$\frac{\partial h\bar{u}_\alpha}{\partial t}+\frac{\partial \beta_{\kappa\alpha}h\bar{u}_\kappa\bar{u}_\alpha+\delta_{\kappa\alpha}(gHh^2/2U^2)}{\partial x_\kappa}=\frac{gH}{U^2}hS_{0\alpha}+\frac{1}{Re_{L_X}}\frac{\partial}{\partial x_\kappa}\left(h\frac{\partial \bar{u}_\alpha}{\partial x_\kappa}\right)-\frac{L_X{}^2}{H^2}\frac{1}{Re_{L_X}}\left.\frac{\partial u_\alpha}{\partial z}\right|_{z=b} \tag{40.17}$$

where

$\delta_{\kappa\alpha}=1$ when $\kappa=\alpha$ and $\delta_{\kappa\alpha}=0$ otherwise
$S_{0\alpha}$ is the bed slope
$\partial u/\partial z|_{z=b}$ is the vertical velocity gradient at the bed
$\beta_{\kappa\alpha}$ are the momentum correction factors
$h=h^*/H$ is the dimensionless distance from the bed to water surface
H is the un-scaled (dimensional) distance from the bed to water surface
the over-bar denotes the depth-averaged velocity

The last term in (40.17) represents the along-stream and cross-stream bed shear stresses, which is represented by τ_α/ρ in Equation 40.19.

$$\frac{\partial h}{\partial t}+\frac{\partial h\bar{u}_\alpha}{\partial x_\alpha}=0 \tag{40.18}$$

$$\frac{\partial h\bar{u}_\alpha}{\partial t}+\frac{\partial \beta_{\kappa\alpha}h\bar{u}_\kappa\bar{u}_\alpha+\delta_{\kappa\alpha}(1/Fr^2)h^2/2}{\partial x_\kappa}=\frac{1}{Fr^2}hS_{0\alpha}+\frac{1}{Re_{L_X}}\frac{\partial}{\partial x_\kappa}\left(h\frac{\partial \bar{u}_\alpha}{\partial x_\kappa}\right)-\frac{\tau_\alpha}{\rho} \tag{40.19}$$

in which $1/Fr^2 = gH/U^2$ is introduced. In integrating system (40.13) through (40.15) to arrive at system (40.16) through (40.17), motion with scales smaller than the water depth are filtered out. Therefore, in system (40.18) through (40.19), the effects of the depth-filtered scales on the residual scales are represented by the bottom friction as well as by $\partial(\beta_{\kappa\alpha}-1)h\bar{u}_\kappa\bar{u}_\alpha/\partial x_\kappa$. This dispersion term is often neglected by setting $\beta_{\kappa\alpha}=1$. The assumption $\beta_{\kappa\alpha}=1$ is strictly valid only if the velocity profile is exactly or nearly uniform with respect to depth. Such a condition only occurs in steady, uniform, and turbulent channel flows and its adoption for shallow shear flows is dictated by expediency and not necessarily by accuracy. An excellent exposition of the momentum correction coefficient can be found in Yen (1973). Preliminary work by Kolyshkin et al. (1999) indicates that shallow flows may be influenced by $\beta_{\kappa\alpha}$, but no definite conclusion has yet been reached.

Applying the Favre filtering to system (40.18) through (40.19) gives:

$$\frac{\partial \hat{h}}{\partial t}+\frac{\partial \hat{h}\tilde{\bar{u}}_\alpha}{\partial x_\alpha}=0 \tag{40.20}$$

$$\frac{\partial \hat{h}\tilde{\bar{u}}_\alpha}{\partial t}+\frac{\partial}{\partial x_\kappa}\left(\hat{h}\tilde{\bar{u}}_\kappa\tilde{\bar{u}}_\alpha+\delta_{\kappa\alpha}\frac{\hat{h}^2}{2Fr^2}\right)=\frac{\hat{h}}{Fr^2}S_{0\alpha}-\frac{\hat{\tau}_\alpha}{\rho}+\frac{1}{Re_{L_X}}\frac{\partial}{\partial x_\kappa}\left(\hat{h}\frac{\partial \tilde{\bar{u}}_\alpha}{\partial x_\kappa}\right)+\frac{\partial\left(\hat{h}\tilde{\bar{u}}_\kappa\tilde{\bar{u}}_\alpha-\widetilde{h\bar{u}_\kappa\bar{u}_\alpha}\right)}{\partial x_\kappa} \tag{40.21}$$

where

the hat denotes ordinary filtered quantities
the tilde denotes water height-weighted filtered quantities (i.e., $\tilde{f}=\widehat{hf}/\hat{h}$)

The bottom shear stresses $\hat{\tau}_\alpha$ are commonly modeled by the quadratic friction law (Pope 2000; Schlichting and Gersten 2000). The last term in the right-hand side of (40.21) requires a turbulence closure model (see Ghidaoui et al. 2006; Hinterberger et al. 2007). An eddy viscosity-type model leads to:

$$\frac{\partial \hat{h}}{\partial t}+\frac{\partial \hat{h}\tilde{\bar{u}}_\alpha}{\partial x_\alpha}=0 \tag{40.22}$$

$$\frac{\partial \hat{h}\tilde{\bar{u}}_\alpha}{\partial t}+\frac{\partial}{\partial x_\kappa}\left(\hat{h}\tilde{\bar{u}}_\kappa\tilde{\bar{u}}_\alpha+\delta_{\kappa\alpha}\frac{\hat{h}^2}{2Fr^2}\right)=\frac{\hat{h}}{Fr^2}S_{0\alpha}-c_f\frac{\tilde{\bar{u}}_\alpha\sqrt{\tilde{\bar{u}}_\kappa\tilde{\bar{u}}_\kappa}}{2}+\frac{\partial}{\partial x_\kappa}\left[\left(\frac{1}{Re_{L_X}}+\frac{1}{Re_{L_{XT}}}\right)\hat{h}\frac{\partial \tilde{\bar{u}}_\alpha}{\partial x_\kappa}\right] \tag{40.23}$$

in which the used total turbulent viscosity is replaced by the sum of the turbulent viscosity, $\upsilon_{L_{XT}}$, and fluid viscosity, υ, so that $Re_{L_{XT}}=UL_X/\upsilon_{L_{XT}}$ and $Re_{L_X}=UL_X/\upsilon$.

It is important to note that, unless otherwise stated, the hat and tilde will be dropped from system (40.22) and (40.23) in the remainder of this chapter. That is, h is used instead of $\hat{h}$ and u_α is used instead of $\tilde{\tilde{u}}_\alpha$.

40.3 Examples

Typical shallow flow patterns that can be obtained by the numerical solution of system (40.22) and (40.23), together with a turbulence model to estimate the eddy viscosity and appropriate initial and boundary conditions, are given in Figures 40.5 and 40.6.

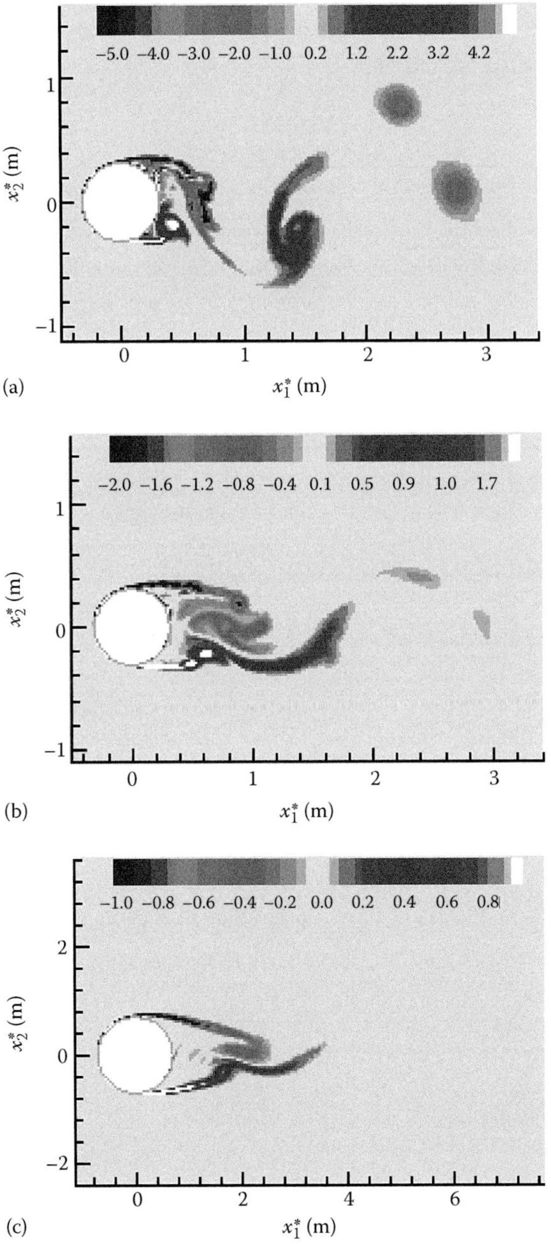

FIGURE 40.5 Vorticity in a shallow wake flow: (a) a vortex-street pattern (S = 0.17); (b) an unsteady bubble pattern (S = 0.24); (c) a steady bubble pattern (S = 0.66).

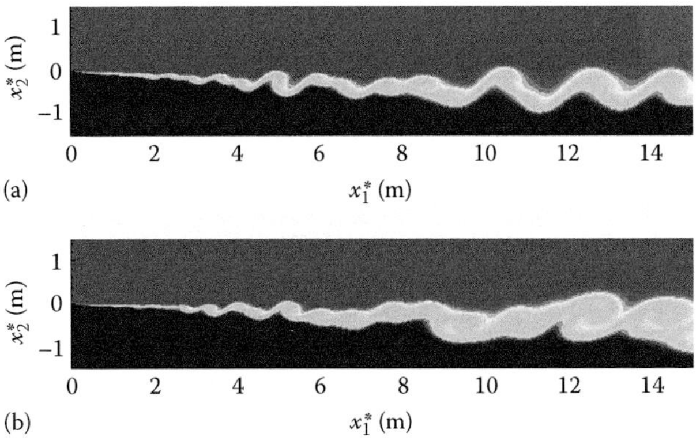

FIGURE 40.6 Scalar contours for a mixing layer corresponding to a laboratory test in Van Prooijen and Uijttewaal (2002): (a) water depth of 42 mm and (b) case with water depth of 67 mm. Note the faster spread for case (b) compared to case (a). This is due to the bed friction S (given by Equation 40.5) being larger for the shallower case.

The details of the numerical code with which these figures were obtained can be found in Ghidaoui et al. 2006 and references given in it. The figures display the results of the numerical model when applied to the laboratory tests conducted in Chen and Jirka (1995) and van Prooijen and Uijttewaal (2002), respectively, and illustrate the development of large-scale coherent structures and show the dependence of the flow pattern on the bed friction parameter. In particular, the coherent structures grow in scale and oscillate more vigorously as the bed friction parameter becomes smaller. Various studies have concluded that models based on system (40.22) and (40.23) can reproduce the large-scale features with reasonable accuracy that is generally sufficient for practical purposes. Detailed assessment of how well depth-averaged models represent shallow flows can be found in Lloyd and Stansby (1997a,b), Ghidaoui et al. (2006), Hinterberger et al. (2007), Ghidaoui and Liang (2008), and others.

Examples of analytical solutions of system (40.22) and (40.23), which give expressions for velocity, spread, and momentum deficit in shallow flows have been derived in the literature (Ingram and Chu 1987; van Prooijen and Uijttewaal 2002; Negretti et al. 2006), are illustrated below. These solutions highlight the role of bed friction and vertical confinement on the properties of shallow flows, provide base flow states for stability analysis, and serve as test cases for numerical models. It turns out that such solutions can only be derived either (1) by neglecting the hydrostatic pressure gradient in the stream-wise momentum equation so that the flow is driven by the bed slope (Negretti et al. 2006) or (2) by adopting the rigid-lid assumption, where the free surface is replaced by an inviscid lid so that water depth is fixed (i.e., its derivatives are zero) and the flow is either driven by gravity or pressure gradient (Ingram and Chu 1987; van Prooijen and Uijttewaal 2002). Both approaches are similar and lead to similar analytical solutions. Here, the rigid-lid-based solutions are illustrated.

System (40.22) and (40.23), together with the rigid-lid assumption, gives:

$$\frac{\partial u_\alpha}{\partial x_\alpha}=0 \tag{40.24}$$

$$\frac{\partial u_\alpha}{\partial t}+\frac{\partial}{\partial x_\kappa}(u_\kappa u_\alpha+\delta_{\kappa\alpha}p)$$

$$=S_{0\alpha}-c_f\frac{u_\alpha\sqrt{u_\kappa u_\kappa}}{2h}+\frac{\partial}{\partial x_\kappa}\left[\left(\frac{1}{Re_{L_X}}+\frac{1}{Re_{L_{XT}}}\right)\frac{\partial u_\alpha}{\partial x_\kappa}\right] \tag{40.25}$$

Although already stated, it is important to note that h is used instead of $\hat{h}$ and u_α is used instead of $\bar{\tilde{u}}_\alpha$. In (40.25), p is the dimensionless pressure used to replace the hydrostatic pressure. Therefore, the pressure force that drives the flow with a rigid-lid has the same magnitude and direction as the hydrostatic pressure force in the free surface flow (i.e., the gradient of p in Equation 40.25 plays the same role as the gradient of the hydrostatic pressure in momentum equation (40.23)). The rigid-lid assumption is investigated in details in Ghidaoui and Kolyshkin (1999) and found to be sufficiently accurate when $Fr < 0.6$.

40.3.1 Analytical Solution for a Shallow Wake

Here, the analytical solution of Ingram and Chu (1987) is summarized, but the reader should consult Negretti et al. (2006) for an alternate approach. The appropriate length scale is the island diameter (i.e., $L_X = D$) and the appropriate velocity scale is the free stream velocity (i.e., $U = U_\infty$). Therefore, $Re_{L_{XT}} = U_\infty D/\upsilon_{L_{XT}}$ and $h = H/D$. As a result, $c_f/h = c_fD/H = S$, which shows that bed friction parameter naturally emerges.

Let $u_1^s = u_1^s(x_1,x_2)$ be the steady state stream-wise wake velocity, $u_1^s(x_1,x_2\to\pm\infty)=1$ be the free stream velocity (i.e., velocity outside wake), $u_d = 1-u_1^s$ be the velocity deficit, and H be the vertical distance between the two parallel planes made up of the flow bed and the fictitious rigid-lid (i.e., water depth). These definitions, together with system (40.24) and (40.25) and the far wake approximation (i.e., $u_d \ll 1$ and $\partial/\partial x_1 \ll \partial/\partial x_2$), leads to the following stream-wise component of the momentum within the wake, which is:

$$-\frac{\partial u_d}{\partial x_1}=f_1-\frac{c_f}{2h}(1-2u_d)-\frac{\partial}{\partial x_2}\left[\frac{1}{Re_{L_{XT}}}\frac{\partial u_d}{\partial x_2}\right] \tag{40.26}$$

where $f_1 = S_{01} - \partial\hat{p}/\partial x_1$ is the sum of pressure gradient and bed slope. Therefore, although Ingram and Chu (1987) assume that the bed is horizontal, using $f_1 = S_{01} - \partial\hat{p}/\partial x_1$ generalizes the solution to flows driven by gravity, pressure gradient, or both.

Outside the wake, the velocity deficit is zero (i.e., $u_d \to 0$ as $x_2 \to \infty$). As a result, (40.26) becomes

$$0=f_1-\frac{S}{2} \tag{40.27}$$

Subtracting (40.26) from (40.27), gives,

$$\frac{\partial u_d}{\partial x_1}=-Su_d+\frac{\partial}{\partial x_2}\left[\frac{1}{Re_{L_{XT}}}\frac{\partial u_d}{\partial x_2}\right] \tag{40.28}$$

Integration of (40.28) across the wake gives:

$$\frac{d}{dx_1}\left[\int_{-\infty}^{+\infty}u_d dx_2\right]=-S\left[\int_{-\infty}^{+\infty}u_d dx_2\right]\to\frac{dm}{dx_1}=-Sm \tag{40.29}$$

where $\int_{-\infty}^{+\infty}u_d dx_2 = m$ is the momentum deficit due to the presence of the obstacle. Integrating (40.29) gives:

$$m=m_0\exp(-Sx_1) \tag{40.30}$$

where m_0 is the momentum deficit at the origin (i.e., x_1 = 0). The magnitude of m_0 due to the island can be estimated using $m_0 = C_D/2$ in which C_D is the drag coefficient of the island. Equation 40.30 shows that the momentum deficit decays exponentially with distance from the island and the flow tends to return to its original state far downstream. In addition, the momentum deficit decays faster for flows with larger S. That is, the downstream distance over which the island has marked influence on the flow field gets smaller as the flow becomes shallower (i.e., S becomes larger). In fact, the downstream longitudinal distance over which the island influences the flow tends to zero when the flow is infinitely shallow (i.e., when $S \to \infty$), but tends to infinity when the flow is deep (i.e., when $S \to 0$). Therefore, the influence of a vortex street wake, being characterized by S values of order 0.2 or less, extends further downstream than, say, a recirculating bubble wake, being characterized by S values larger than 0.2 (Negretti et al. 2006). In fact, it is clear from Equation 40.30 that the length of influence of the island is of order $1/S$. Consider an island whose cross flow width is about 2 km. A vortex street within the wake of this island, for say S = 0.1, would extend over a dimensionless distance of 1/0.1 = 10, which is equivalent to a dimensional distance of 10*2 = 20 km. On the other hand, a recirculating pair of eddies, for say S = 0.7, would extend over a dimensionless distance of 1/0.7 which is equivalent to a dimensional distance of (1/0.7)*2 = 2.8 km. Note that $m = m_0$ if $S \to 0$. That is, the deep wake momentum conservation relation is recovered in the limit $S \to 0$.

Now, a solution for Equation 40.28 is sought. Ingram and Chu (1987) found that a similarity solution in the form of $u_d(x_1, x_2)/u_m(x_1) = g(\eta)$, where $\eta = x_2/\delta(x_1)$ is the similarity variable, $\delta(x_1)$ is the dimensionless wake half-width, and g is the similarity function, can be derived provided that the eddy viscosity has the following form:

$$\upsilon_{L_{XT}} = U_\infty D \frac{\delta}{2\ln 2}\frac{d\delta}{dx_1} \rightarrow \frac{\upsilon_{L_{XT}}}{DU_\infty} = \frac{1}{Re_{L_{XT}}} = \frac{\delta}{2\ln 2}\frac{d\delta}{dx_1} \quad (40.31)$$

The form of (40.31) reflects (1) the assumption that the mixing length for the large-scale structures is of the order of the wake width and (2) the fact that far downstream of the bluff body the large-scale turbulence gets annihilated by the bed friction. In fact, as will be shown at end of this derivation, $d\delta(x_1)/dx_1 = 0$ as $x_1 \rightarrow \infty$, implying from (40.30) that $\upsilon_{L_{XT}} = 0$ (i.e., complete annihilation of turbulence far downstream).

Solving (40.28) and (40.31) gives:

$$g(\eta) = \frac{u_d(x_1,x_2)}{u_m(x_1,x_2=0)} = e^{-\ln 2\left(\frac{x_2}{\delta(x_1)}\right)^2} = e^{-\ln 2(\eta)^2} \quad (40.32)$$

in which $u_m(x_1,x_2 = 0)$ is the maximum velocity deficit at the wake centerline (i.e., $x_2 = 0$), $\eta = x_2/\delta(x_1)$ is the similarity variable and $\delta(x_1)$ is the wake half-width, which is the transverse distance from the wake centerline to the point where $u_d/u_m = 1/2$.

Up to now, the shape of the velocity profile has been derived. In the remainder of this section, the growth of the wake half-width $\delta(x_1)$ will be derived. In arriving to the similarity solution, it requires that:

$$\frac{d\delta}{dx_1} = a_\star u_m \quad (40.33)$$

in which $a_\star$ is a constant,

By using (40.30), (40.32), and (40.33), we can obtain

$$\frac{d(\delta^2)}{dx_1} = 2a_\star u_m\delta = \frac{a_\star}{2.13}C_D \exp\left[-Sx_1\right] \quad (40.34)$$

Integrating Equation 40.34 with respect to x_1 gives:

$$\delta^2 = \int_0^{x_1} aC_D \exp\left[-Sx_1\right]dx_1 + \delta_0^2 \quad (40.35)$$

in which $a = a_\star/2.13$ and δ_0 is the initial wake half-width.

When the flow approaches deep water ($S \rightarrow 0$), the entrainment coefficient is then a constant, and Equation 40.35 becomes:

$$\delta^2 = aC_D x_1 + 1 \quad (40.36)$$

since $\delta_0 \approx D = 1$. From the experiments of Townsend (1956), $aC_D = 0.09$. If $C_D = 1$, $a = 0.09$.

For shallow conditions, the experimental results in Babarutsi and Chu (1985) suggest the following:

$$a = a_0[1-(S_{IP}/S_{IPc})], \quad S_{IP} \le S_{IPc}$$

$$a = 0 \quad S_{IP} > S_{IPc} \quad (40.37)$$

Ingram and Chu (1987) calculate $S_c \approx 0.16 - 0.17$ with stability analysis. For the current self-similar shape (40.32), one can obtain $U_{IP} = 1 - 0.61u_m$ $(U_y)_{IP} = 0.71\, u_m/\delta$. Thus

$$S_{IP} = 0.7\left[S\delta\left(\frac{1}{u_m}\right) - 0.6065\left(S\delta\right)\right] \quad (40.38)$$

in which

$$\frac{1}{u_m} = \frac{2(2.13)}{C_D}\delta \exp\left(Sx_1\right) \quad (40.39)$$

Equations 40.35, 40.37 through 40.39 give a complete set of relationships which can be used to compute $\delta(x_1)$ numerically. Experiments in Ingram and Chu (1987) have shown that the computed $\delta(x_1)$ is in reasonable agreement with data for the vortex-street wake, but not for the unsteady bubble wake.

40.3.2 Analytical Solution for a Shallow Mixing Layer

The analytical solution given here is due to van Prooijen and Uijttewaal (2002). The appropriate length scale is the initial width of the mixing layer Δ_0, such that $L_X = \Delta_0 = \delta_0 = 1$ is the dimensionless initial width of the mixing layer. The appropriate velocity scale is $(U_\infty^1 + U_\infty^2)/2$.

As in the case of shallow wakes, the rigid-lid assumption is used so that the flow is governed by system (40.24) and (40.25). The analytical solutions for the velocity and spread are:

$$u_1^s(x_1,x_2) = \bar{u}(x_1) + \frac{\Delta u_0 \exp(-(c_f\Delta_0/h)x_1)}{2}\tanh\left(\frac{x_2 - l(x_1)}{0.5\delta(x_1)}\right) \quad (40.40)$$

$$\delta(x_1) = a\frac{\Delta u_0}{\bar{u}}\frac{h}{c_f}\left(1 - \exp\left(-\frac{c_f\Delta_0}{h}x_1\right)\right) + 1 \quad (40.41)$$

where

$\Delta u_0 = u_1^s(x_1 = 0, x_2 \rightarrow \infty) - u_1^s(x_1 = 0, x_2 \rightarrow -\infty)$, $u_{1\infty}^1 = u_1^s(x_1 = 0, x_2 \rightarrow \infty)$ is the free stream velocity on the fast side of the mixing layer at the inflow boundary

$u_{1\infty}^2 = u_1^s(x_1 = 0, x_2 \rightarrow -\infty)$ is the free stream velocity on the slow side of the mixing layer at the inflow boundary

c_f^1 is the bed friction on the side of the fast stream

c_f^2 is the bed friction on the side of the slow stream

$c_f = (c_f^1 + c_f^2)/2$ is the average bed friction

a is the entrainment coefficient

$l(x_1)$ is transverse distance from $x_2 = 0$ to the center of the mixing layer (i.e., where $\bar{u}(x_1) = (u_1^s(x_1, x_2 \rightarrow \infty) + u_1^s(x_1, x_2 \rightarrow -\infty))/2$

Experiments by van Prooijen and Uijttewaal (2002) suggest a constant value of $a \approx 0.085$. In the course of arriving at the analytical solution, van Prooijen and Uijttewaal (2002) assumed the bed is horizontal, the transverse gradients are much larger than longitudinal gradients (i.e., $\partial/\partial x_1 << \partial/\partial x_2$), and the flow is self-similar so that the velocity profile is given by a hyperbolic tangent function. In terms of the initial bed friction number, the above analytical solution becomes:

$$u_1^s(x_1,x_2) = \bar{u}(x_1) + \frac{\Delta u_0 \exp\left(-4S\dfrac{U_\infty^1 - U_\infty^2}{U_\infty^1 + U_\infty^2}x_1\right)}{2}\tanh\left(\frac{x_2 - l(x_1)}{0.5\delta(x_1)}\right) \tag{40.42}$$

$$\delta(x_1) = \frac{a\Delta_0}{2S}\left(1 - \exp\left(-4S\frac{U_\infty^1 - U_\infty^2}{U_\infty^1 + U_\infty^2}x_1\right)\right) + 1 \tag{40.43}$$

where S is defined by Equation 40.5. Note that $\delta \to 1 \Rightarrow \Delta \to \Delta_0$ as S becomes large. That is, the spread in a mixing layer in a highly shallow environment and/or excessively large bed friction experience is very small. As expected, the solution shows that the mixing layer spreads less as the bed friction parameter increases. The analytical solution for the spread and velocity are in reasonable agreement with laboratory data (van Prooijen and Uijttewaal 2002) and numerical solution (Ghidaoui and Liang 2008).

40.4 Methods of Analysis

40.4.1 Laboratory Experiments

Laboratory experimentation is an invaluable tool in understanding shallow shear flows and allows the observation and understanding of such flows under controlled environments. The data generated can be used for model verification and calibration and the understanding gained can be used for guiding model improvements. A major challenge in laboratory experiments of shallow flows is to develop a long and wide enough shallow table in order (1) to study the dynamics of such flows under realistic ranges of S, Fr, and Re and (2) to investigate the full range of spatial development of the coherent structures including birth, growth, and then decay. Examples of typical shallow water tables are the basin in the hydromechanics laboratory at the University of Karlsruhe which is 13.5 m long and 5.48 m wide and the wide channel in Delft University of Technology which is 20 m long and 3 m wide.

Experiments on shallow wakes, mixing layers, and jets resulted in the following consistent picture. The water depth constraint suppresses the mechanisms of bending and stretching in such a way that the large-scale vortices are not susceptible to 3D instabilities and do not breakdown into 3D turbulence. Instead, once formed, vertical vortices extending from the bed to the water surface go through a stage of growth pairing and diffusion followed by a stage of decay due to bed friction leading to the loss of coherence and eventually disappearance at large enough distances downstream. These stages reflect the lack of spatial equilibrium between the source and sink of large-scale turbulent energy—the source is larger than the sink in the upstream part of the domain, but smaller in the downstream part provided that the domain is long enough. In the growth stage (see Uijttewaal and Jirka [2003] for wakes; van Prooijen and Uijttewaal [2002] for mixing layers; Biggs et al. [2010] and Dracos et al. [1992] for jets), the turbulent energy spectra develops a peak around the most unstable mode, announcing the birth of the vortex. The mean transverse shear takes energy from the mean flow and passes it to the large-scale turbulent structures. The magnitude of this energy peak and the scale at which it occurs grows with distance downstream signaling vortex pairing and diffusion. The large-scale structures tend to cause the flow trajectory to oscillate laterally. The decay stage is less understood as it requires very long shallow tables to observe it. The shallow water table used by van Prooijen and Uijttewaal (2002) is 20 m long and allowed the observation of the decay stage. They found that the decay stage is not uniform and needs to be subdivided into two substages. In the first substage, the dominant mode decays, but some other modes are still growing. In the second substage, all modes decay signaling the beginning of the end of the life of the large-scale structures. Another experiment that is relevant to the decay stage of the large-scale structures is conducted by Uijttewaal and Jirka (2003). In this experiment, the turbulence is said to be decaying in the sense that there is a sink for the large-scale turbulent energy in the form of bed friction, but there is no source of energy since there is no mean lateral shear. It is found that there is a peak in the energy spectra and that this peak decays and shifts toward larger scales. Such growth in scale signals vortex pairing. The slope of the energy spectra is close to −3, supporting the hypothesis that large-scale turbulence is akin to the ideal 2D turbulence.

Shallow turbulent flows are subjected to instabilities with widely separated scales. For example, the scale of the bed-generated instability, responsible for the small-scale three-dimensional turbulence, is smaller than the water depth. On the other hand, the scale of the instability due to type A, B, or C mechanisms, responsible for the large-scale quasi-two-dimensional turbulence, is much larger than the water depth. Therefore, it is plausible to expect the energy spectra to contain two peaks—one peak associated with the large-scale instability and another associated with small-scale instability. To test the scale separation hypothesis, laboratory experiments where (1) a large-scale instability is generated by a local sudden narrowing of the width of an otherwise wide basin by installing a horizontal bar and (2) a small-scale instability is generated by local scour hole-like in an otherwise flat bed are conducted by Carrasco and Vionnet (2004). They found evidence of double peak turbulent energy spectra with two-dimensional-like turbulence spectra at the large scales, associated with the discontinuity in basin width, and three-dimensional-like turbulence spectra at the small scales, associated with the scour hole. Further experiments are required to understand and characterize this interesting and important problem. It is noted that the scale separation hypothesis has been used in model development for shallow flows (Nadaoka and Yagi 1998) and, if confirmed, would place the modeling of shallow shear flows by two-dimensional shallow water models on stronger footing.

Recognizing that large-scale vortices are the fundamental building blocks of shallow shear flows prompted experimental studies of the dynamics of elementary vortices under shallow conditions (Akkermans et al. 2008; Jirka and Seol 2010). Methods used to generate elementary vortex structures include electromagnetic forcing, flaps, rotating a hollow cylinder, and fluid injection. Jirka and Seol (2010) used flaps to generate dipolar vortices and a rotating hollow cylinder to generate a single forced vortex, which quickly evolved into a tri-polar vortex system. It is found that the spin down of the vortices is governed by bottom friction and that large Reynolds and bed friction numbers are required to generate fully turbulent shallow vortex structures. Akkermans et al. (2008) used electromagnetic forcing to generate a dipolar vortex in a shallow fluid layer and showed that the intrinsic three dimensionality of electromagnetic forcing produces strong three-dimensional behavior with significant circulation, causing the spin down of the vortices to be independent of bottom friction. This casts doubt on the ability of electromagnetic forcing to study shallow shear flows.

40.4.2 Stability Theory

In principle, system (40.22) and (40.23) admits steady solutions if the boundary conditions and the external forces are not time dependent. In fact, examples of steady solutions are analytically derived in the previous section. However, it turns out that such steady solutions, also called base flows or base states, only exist when the stability parameter, S given by Equation 40.3, exceeds a critical value S_c. Conversely, if $S < S_c$, a steady solution of system (40.22) and (40.23) is not stable and cannot exist in practice. For example, the field study by Ingram and Chu (1987), laboratory experiments by Chen and Jirka (1995), and nonlinear analysis by Ghidaoui et al. (2006) show that there is a critical bed friction number, S_c, which lies in the range 0.5–0.6 such that the wake flow is steady if $S = c_f D/H > S_c$ and unsteady if $S = c_f D/H < S_c$. Figure 40.7, shows the solution for a shallow wake, where (1) the boundary conditions of the flow and all other input parameters are time-independent and (2) $S = 0.16$ (i.e., $S < S_c \approx 0.6$). It is clear that, despite the fact that the conditions are steady, the resulting flow is unsteady as described by the variation of the streamlines with time. That is, a steady flow is unstable and cannot exist in this case.

For shallow mixing layers, the laboratory experiments by van Prooijen and Uijttewaal (2002) and nonlinear analysis by Ghidaoui and Liang (2008) show that the critical bed friction number, S_c, is in the range 0.06–0.12 such that the flow is steady if $S = c_f(U_\infty^1 + U_\infty^2)\Delta_0/[4H(U_\infty^1 - U_\infty^2)] > S_c$. Otherwise, the flow is unsteady and large-scale horizontal vortices are formed. Chen and Jirka (1998) found that shallow jets are steady if $S = c_f B/H > S_c \approx 0.7$; otherwise, they are unsteady and meander.

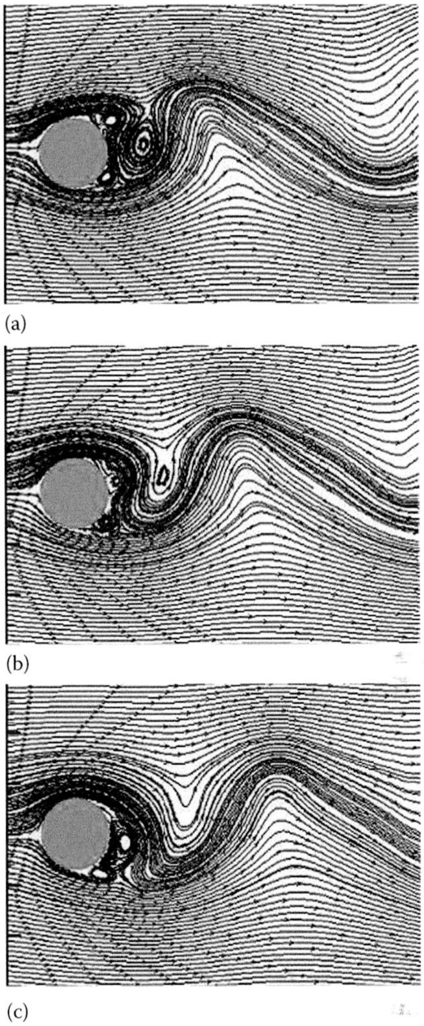

FIGURE 40.7 Shallow wake streamlines at different time instants. In fact, the flow is oscillatory with period T. (a) $i^* = T/8 = 12.19$ s. (b) $i^* = T/4 = 24.38$ s. (c) $i^* = 3T/8 = 36.75$ s.

The fact that steady solutions to system (40.22) and (40.23) are only realizable when $S > S_c$ implies that such solutions are stable when $S > S_c$ and unstable otherwise. That is, if a shallow steady flow that obeys system (40.22) and (40.23) were to be slightly perturbed, the perturbations would grow and lead to new and unsteady state flow if $S < S_c$. Conversely, the perturbations would die out and lead to the flow returning to its original steady state when $S > S_c$.

Clearly, S varies from one flow to another. It must be stressed, however, that the local value of S also varies with downstream distance within a particular flow field. In a given flow setting, S increases with downstream distance in proportion to the width of the shear layer. Essentially, the transverse shear, being the generation mechanism of coherent structures, decays with distance downstream and results in more stable conditions

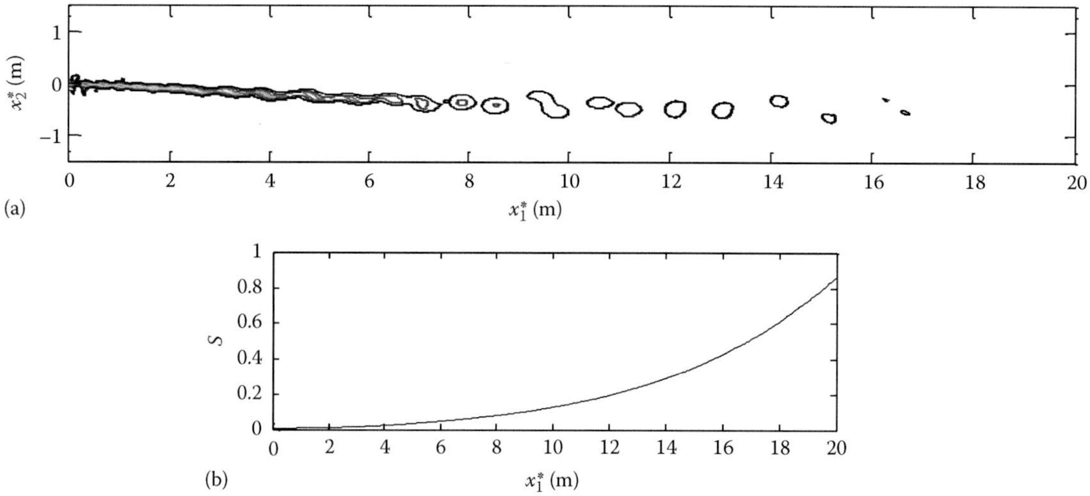

FIGURE 40.8 Spatial development of a mixing layer. (a) Vorticity contour and (b) stability parameter. The dimensional inflow velocity is $u_1^{s*}(0,x_2^*) = \bar{u}^*(0) + \frac{\Delta u_0^* \exp\left(-\frac{c_f \Delta_0^*}{h^*}(0)\right)}{2} \tanh\left(\frac{x_2^* - l^*(0)}{0.5\delta^*(0)}\right)$, where $\bar{u}^*(0) = 0.1925(m/s)$, $\Delta u_0^* = 0.14\ (m/s)$, $\delta^*(0) = 0.05(m)$ and $l^*(0) = 0(m)$ and the perturbation is the form of a white noise perturbation.

further down the flow domain. Therefore, the flow is generally unstable in the upstream region of the flow domain and stable further downstream. In fact, this situation is illustrated in Figure 40.8. It is clear from Figure 40.8a that the eddies develop and then go through a stage of by pairing and diffusion which extends up to $x_1^* \approx 10$m, followed by a stage of decay which can lead to eventual complete annihilation if the domain is long enough. The growth stage corresponds to $S < S_c \approx 0.12$ and the decay stage corresponds to $S > S_c \approx 0.12$ (Figure 40.8b).

As an important aside, note that the flow domain needs to be sufficiently long in order to observe both the growth and decay stages. This implies that the full dynamics of shallow flows can only be studied experimentally in shallow tables that are quite long. van Prooijen and Uijttewaal (2002) found that a shallow water table 12 m long allowed them to study the growth, transition, and decay stages of the eddies. On the other hand, the decay stage for shallow jets requires a much longer shallow table and, to our knowledge, has not been yet achieved in laboratory settings. The reason is that the critical bed friction number for jets is 0.7 which is about six times larger than that of a mixing layer. For shallow jets, even the width requirements poses space problem. Consider a typical laboratory setting where the water depth H is about 5 cm and c_f is about 0.005, then the jet would become steady when $c_f B/H > 0.7$ implying a half-width of $B > 0.7(5/0.007) = 700$ cm, which requires a water table of more than 14 m wide—i.e., far larger than the width of usual water tables used in laboratories. In fact, the recent shallow jet experiments by Biggs et al. (2010) in a facility that is 4.3 m long covered part of the growth region only.

The growth stage involves the onset of Kelvin–Helmholtz (KH) instability and the growth of this instability with distance downstream. These features are well seen in Figure 40.6. Take Figure 40.6a, for example, the KH instability is perceptible at about 2.0 m, where the mixing layer begins to oscillate transversally with a small amplitude. This region of early development of the instability is the onset region and can be described well by the linear stability theory that is discussed in the next section. The roll up mechanism, where the oscillation gets folded into eddies (KH vortices), occurs in the region between $x_1^* = 2.5$–3.5 m. The latter part of the onset region and the early part of the roll up region is what weakly nonlinear stability theory is for. Beyond about 6.0–7.0 m, vortex pairing is obvious. Such pairing belongs to the domain of nonlinear analysis and requires the solution of system (40.22) and (40.23), together with a turbulence model to estimate the eddy viscosity and appropriate initial and boundary conditions.

40.4.2.1 Linear Stability: Onset of Large-Scale Turbulent Structures

The mathematical problem of determining the onset of the instability and how small amplitude disturbances evolve in a certain base flow belongs to the domain of linear stability theory and originates from several seminal theoretical and experimental analysis by Helmholtz, Kelvin, Rayleigh, and Reynolds (for a historical development of the subject, see Drazin and Reid 2004). In short, linear stability analysis entails the following steps: (1) small perturbations are introduced into a given base flow; (2) the governing equations are linearized in the neighborhood of the base flow; (3) the linearized equations are solved analytically or numerically to determine the fate of small perturbations. To elucidate the mathematical formulation of the linear stability problem, consider a steady solution of system (40.22) and (40.23) of the form $u_\alpha^s = u_\alpha^s(x_1, x_2)$ and

$h^s = h^s(x_1,x_2)$, where the superscript "s" emphasizes that the basic state is steady. Superposing a small unsteady perturbation field $u'_\alpha(x_1,x_2,t)$ and $h'(x_1,x_2,t)$ onto the basic state such that the total field $u_\alpha = u^s_\alpha(x_1,x_2)+u'_\alpha(x_1,x_2,t)$ and $h = h^s(x_1,x_2) + h'(x_1,x_2,t)$ satisfies system (40.22) and (40.23), invoking the smallness of the perturbations so that terms involving their products are negligible and using steady base field and the equations that govern it leads to:

$$\frac{\partial h'}{\partial t}+\frac{\partial h'u^s_\alpha + h^s u'_\alpha}{\partial x_a}=0 \qquad (40.44)$$

$$\frac{\partial h'u^s_\alpha + h^s u'_\alpha}{\partial t}+\frac{\partial h'u^s_\alpha u^s_\kappa + h^s u'_\alpha u^s_\kappa + h^s u^s_\alpha u'_\kappa + \delta_{\kappa\alpha}(1/Fr^2)h^s h'}{\partial x_\kappa}$$
$$=\frac{1}{Fr^2}h'S_{0_\alpha}+\frac{1}{Re_{L_{XT}}}\frac{\partial^2 h'u^s_\alpha + h^s u'_\alpha}{\partial x_\kappa \partial x_\kappa}$$
$$-c_f\frac{u'_\alpha\sqrt{u^s_\kappa u^s_\kappa}}{2}+c_f\frac{u^s_\kappa u^s_\alpha u'_\kappa}{2\sqrt{u^s_\kappa u^s_\kappa}} \qquad (40.45)$$

This is a linear homogeneous system with time-independent coefficient. As a result, its general solution depends on time in an exponential manner (i.e., $u'_\alpha(x_1,x_2,t)\sim\exp(-\lambda t)$ and $h'(x_1,x_2,t)\sim\exp(-\lambda t)$, where $\lambda = \sigma + i\gamma$ is the complex eigenvalue). Clearly, the modulus of the perturbations vary with time as $\exp(-\sigma t)$. As a result, these perturbations grow with time (i.e., unstable) if $\sigma < 0$, decay (i.e., stable) if $\sigma > 0$ and neither grow nor decay (i.e., neutrally stable) if $\sigma = 0$. A further approximation that is made in the stability of shallow flows is that the flow is quasi-parallel. This approximation is rooted in the assumption that the variation in the base flow field along the transverse direction, x_2, is much more significant than the variation along the streamwise direction, x_1. As a result, the longitudinal distance x_1 appearing in the coefficients of system (40.44) and (40.45) can be treated in a parametric manner. That is, one takes a particular location x_1 and evaluates the coefficients of the system (40.44) and (40.45) at this location. The result is a linear homogeneous system of equations where the perturbations are a function of x_1, but not the coefficients. The general solution of this depends on x_1 in an exponential manner (i.e., $u'_\alpha(x_1,x_2,t)\sim\exp(ikx_1)$ and $h'(x_1,x_2,t)\sim\exp(ikx_1)$, where k is a real wave number. This process is repeated for different positions x_2. Therefore, the general solution of system (40.44) and (40.45) varies with time as $\exp(-\lambda t)$ and with x_1 as $\exp(ikx_1)$ so that

$$u'_\alpha = (u'_1(x_1,x_2,t),u'_2(x_1,x_2,t))$$
$$=(u'(x_2),v'(x_2))e^{ikx_1}e^{-\lambda t} \quad \text{and} \quad h' = h'(x_2)e^{ikx_1}e^{-\lambda t} \qquad (40.46)$$

Plugging (40.46) into system (40.44) and (40.45) gives the following:

$$ikh^s u' + h^s\frac{dv'}{dx_2}+v'\frac{dh^s}{dx_2}+\left(-\lambda + iku^s_1\right)h' = 0 \qquad (40.47)$$

$$\left(-\lambda + iku^s_1 + \frac{c_f u^{s2}_1}{h^s}+\frac{\left(k^2-\frac{1}{h^s}\frac{d^2h^s}{dx_2^2}\right)}{Re_{L_{XT}}}\right)$$
$$u' - 2\frac{1}{Re_{L_{XT}}}\frac{1}{h^s}\frac{dh^s}{dx_2}\frac{du'}{dx_2}-\frac{1}{Re_{L_{XT}}}\frac{d^2u'}{dx_2^2}+\frac{du^s_1}{dx_2}v'$$
$$+\left(\frac{ik}{Fr^2}-\frac{1}{Fr^2}\frac{c_f(u^s_1)^2}{2(h^s)^2}+\frac{\left(\frac{k^2u^s_1}{h^s}-\frac{1}{h^s}\frac{d^2u^s_1}{dx_2^2}\right)}{Re_{L_{XT}}}\right)$$
$$h' - 2\frac{1}{Re_{L_{XT}}}\frac{1}{h^s}\frac{du^s_1}{dx_2}\frac{dh'}{dx_2}-\frac{1}{Re_{L_{XT}}}\frac{u^s_1}{h^s}\frac{d^2h'}{dx_2^2}=0 \qquad (40.48)$$

$$\left(-\lambda + \frac{c_f u^s_1}{2h^s}+iku^s_1+\frac{\left(k^2-\frac{1}{h^s}\frac{d^2h^s}{dx_2^2}\right)}{Re_{L_{XT}}}\right)$$
$$v' - 2\frac{1}{Re_{L_{XT}}}\frac{1}{h^s}\frac{dh^s}{dx_2}\frac{dv'}{dx_2}-\frac{1}{Re_{L_{XT}}}\frac{d^2v'}{dx_2^2}+\frac{1}{Fr^2}\frac{dh'}{dx_2}=0 \qquad (40.49)$$

Recall that the bed friction coefficient, c_f, is connected to the bed friction parameter S, through Equation 40.3. Therefore, as expected from the discussion in Section 40.2, the stability problem, described by the above linear system of linear equations, depends on S, Fr, and Re. Ghidaoui and Kolyshkin (1999) Kolyshkin and Ghidaoui et al. (2002) and Ghidaoui et al. (2006) show that the solution of the linear stability problem is weakly dependent on the Froude number Fr if $Fr < 0.6$. As a result, the rigid-lid assumption used in most researches (Chu et al. 1983, 1991; Chen and Jirka 1997; van Prooijen and Uijttewaal 2002) is valid when $Fr < 0.6$. In addition, Chen and Jirka (1997, 1998) and Ghidaoui et al. (2006) found the solution of the linear stability problem to be weakly dependent on Reynolds number when the eddy viscosity-based $Re > 1000$.

In general, the linear stability problem (40.47) and (40.49) cannot be solved exactly and a numerical scheme is needed. A robust scheme for such problems is the pseudo-spectral collocation method based on Chebyshev polynomials. The details of this method of solution are described in Ghidaoui and Kolyshkin (1999). The resulting discrete form of the generalized eigenvalue problem is solved by the IMSL routine GVLCG Fortran Subroutines for Mathematical Applications 1994, IMSL Math/Library, Visual Numerics. It must be stressed that other methods such as finite difference can be also used. For a given base flow, Fr and Re, the aim of the solution is to find pairs of wave number k and bed friction parameter S for which the flow is neutrally stable (i.e., if $\sigma = 0$). The curve made up by these pairs constitutes what is the neutral stability curve, which divides the stable and unstable regions. Note that different neutral stability curves would result for different Fr and $Re_{L_{XT}}$ (i.e., $S_c = S_c(Fr, Re_{L_{XT}})$). Clearly, regions where the distance between these curves is small indicate insensitivity to Fr and $Re_{L_{XT}}$. Stability analyses that investigate the sensitivity of neutral curves to Fr and to $Re_{L_{XT}}$ can be found in Ghidaoui and Kolyshkin (1999) and Chen and Jirka (1998), respectively.

For completeness, it is important to emphasize that the stability approach presented above is referred to as the temporal modal analysis to differentiate it from the spatial modal analysis and the transient stability analysis. The transient analysis solves the linear stability of shallow flows as an initial-value problem. This approach is called for when a complete perturbation dynamics (including short-time transients and long-time asymptotics) is desired and/or when the base flow is unsteady such that the timescale of unsteadiness is more rapid than the timescale of growth of perturbations. This approach is described in detail in Schmid and Henningson (2001). On the other hand, if the major purpose of the study is the long-term behavior of perturbations, then the method of normal modes is used. In this case the perturbations are assumed to be proportional to $\exp[i\alpha x_1 - \lambda t]$, where both parameters α and λ may be complex (i.e., $\alpha = k + i\beta$ and $\lambda = \sigma + i\gamma$). Two particular cases are usually considered: temporal stability analysis and spatial stability analysis. In the temporal problem used in this chapter and the majority of stability analyses of shallow flows, the wave number is assumed to be real (i.e., $\beta = 0$ and $\alpha = k$) while the frequency of oscillations, λ, is complex. From the point of view of temporal analysis, the base flow is said to be stable if $\sigma > 0$ for all modes and unstable if $\sigma < 0$ for at least one mode. Spatial stability analysis assumes $\lambda = i\gamma$ and $\alpha = k + i\beta$. The flow is said to be stable if $\beta > 0$ for all modes and unstable if $\beta < 0$ for at least one mode. From a computational point of view temporal analysis is simpler since the corresponding eigenvalue problem is linear with respect to $\lambda = \sigma + i\gamma$. In this case the discretized stability equations are transformed to a generalized matrix eigenvalue problem which can be solved using standard IMSL routines. This is the reason why temporal analysis is so popular among researchers. On the other hand, spatial stability problem is not linear with respect to $\alpha = k + i\beta$ since diffusion terms, which involve second-order derivative, result in terms involving α^2 and complicates the computational process.

40.4.2.2 Weakly Nonlinear* and Nonlinear Stability: Growth and Dynamics of the Large-Scale Turbulent Structures

In linear stability theory, discussed above, the amplitude of perturbations in shallow flows is $A(t) = \exp(-\sigma t)\exp(-i\gamma t)$ where $\sigma = \sigma(S, Fr, Re_{L_{XT}})$. The solution of the linear problem confirms that there exists a critical bed friction number $S_c = S_c(Fr, Re_{L_{XT}})$ such $\sigma(S) < 0$ when $S < S_c$ (i.e., unstable); $\sigma(S) > 0$ when $S > S_c$ (i.e., stable) and $\sigma(S) = 0$ when $S = S_c$ (i.e., neutrally stable). Consider the unsteady case given by $\sigma(S) < 0$ when $S < S_c$. The amplitude of perturbations is

$$A(t) = e^{-\lambda t} \rightarrow \frac{dA}{dt} = -\lambda e^{-\lambda t} = -\lambda A \rightarrow \frac{dA}{dt} = -\lambda A \tag{40.50}$$

where its modulus is $|A|^2 = |e^{-\sigma t}\, e^{i\gamma t}|^2 = e^{-2\sigma t}$ and implies that the time growth of the amplitude is unlimited. Such growth is mathematically and physically incorrect, implying that (40.50) is only valid for small time. Mathematically, an unlimited growth violates a key premise used in the derivation of system (40.44) and (40.45), namely, that amplitude of the perturbations are so small that their products are negligible. Physically, an unlimited growth is unrealistic. Initially the oscillations grow rapidly, but once oscillations reach to a certain magnitude, vortices develop (see Figures 40.2 and 40.5) and the subsequent growth is far lower than exponential. Further downstream, the growth ceases and gives way to a decay stage.

Linear theory and the amplitude relation derived from it (Equation 40.50), are valid when time is small enough so that velocity and water depth perturbations are negligible compared to the mean flow velocity and water depth. For larger time, linear theory gives way to weakly nonlinear theory. It is shown in Kolyshkin and Ghidaoui (2003) that the amplitude equation near S_c, (i.e., $S - S_c$ is small), is the complex Ginzburg–Landau equation

$$\frac{\partial A}{\partial t} = -\lambda A + \zeta \frac{\partial^2 A}{\partial x_1^2} - \mu |A|^2 A \tag{40.51}$$

where λ, ζ, and μ are complex coefficients. They found that (40.51) describes the nonlinear characteristics of shallow wake flows in the unstable regime reasonably well, confirms the existence of a finite equilibrium amplitude, where the unstable mode evolves with time into a finite amplitude oscillation, and produces results that are consistent with flow patterns observed in experiments.

It is important to note that (40.51) is an extension of (40.50). In fact, when the amplitude is small and using the quasi-parallel assumption—as required in the linear stability section above—Equations 40.51 becomes identical to (40.50). On the other hand, when the amplitude is large enough so that the last term in (40.51) is not negligible, but the quasi-parallel assumption is valid, (40.51) gives the well-known Landau equation:

$$\frac{dA}{dt} = -\lambda A - \mu |A|^2 A \rightarrow \frac{d|A|^2}{dt} = -2\sigma |A|^2 - 2\theta |A|^4 \rightarrow \frac{1}{|A|^2} = -\frac{\theta}{\sigma} + Ce^{-2\sigma t} \tag{40.52}$$

* An excellent introduction to weak nonlinear analysis can be found in Landau and Lifshitz (2002, Chapter III). The discussion below on weak nonlinear analysis of shallow flows is based on their presentation.

where

θ is the real part of μ and equals half the Landau constant

C is a constant of integration

The limit of (40.52) for large time is:

$$\lim_{t\to\infty}\frac{1}{|A|^2}=-\frac{\theta}{\sigma}\rightarrow |A|_{\max}=\sqrt{-\frac{\sigma}{\theta}} \quad (40.53)$$

showing that, unlike linear theory which predicts infinite amplitude for large time, weakly nonlinear analysis gives a finite amplitude. To judge the accuracy of (40.53), it is instructive to expand $\sigma = \sigma(S)$ in a Taylor series around S_c

$$\sigma=\sigma(S_c)+(S-S_c)\left.\frac{d\sigma}{ds}\right|_{S_c}+O(S-S_c)^2\rightarrow$$

$$\sigma=(S-S_c)\left.\frac{d\sigma}{ds}\right|_{S_c}+O(S-S_c)^2 \quad (40.54)$$

where O is the order function. Recall that $S_c(Fr, Re_{L_{XT}})$, the focus is on $S < S_c$, but with $S - S_c$ small, and combine (40.54) with (40.53) gives:

$$|A|^2_{\max}=-\frac{\sigma}{\theta}\cong-\left.\frac{d\sigma}{dS}\right|_{S_c}\frac{1}{\theta}\left(S-S_c\left(Fr,Re_{L_{XT}}\right)\right) \quad (40.55)$$

Figure 40.9 shows a good fit of expression (40.55) to the numerical data for a shallow wake obtained by solving nonlinear system (40.22) and (40.23) for fixed Re and variable Fr. The dependence on Froude number is also clear and shows that the flow becomes more stable and the maximum amplitude reduces as Froude number increases. In addition, (40.55) provides simple, convenient, and direct link between the stability parameter and the amplitude of oscillations.

While important, weakly nonlinear analysis cannot provide detailed velocity and pressure fields and is generally limited to small values of $S - S_c$; thus, the need for nonlinear theory. Such a theory is founded either on three-dimensional Navier–Stokes-based models [i.e., system (40.7) and (40.9)] or two-dimensional shallow-water based models [i.e., system (40.22) and (40.23)] along with appropriate turbulence models. Both three and two-dimensional models, generally, need to be solved numerically and require much computer resources for real-life problems.

Examples of two-dimensional nonlinear studies of shallow flows can be found in Lloyd and Stansby (1997a,b), Nadaoka and Yagi (1998), Ghidaoui et al. (2006), Hinterberger et al. (2007), Ghidaoui and Liang (2008), Chu (2010) and others. These models provide reasonable predictions of the large-scale instability features such flow pattern, amplitude and frequency of oscillations for shallow flows including wakes and mixing layers. However, these models cannot represent large-scale structures, which originate due to the lift up and tilting of bed-generated three-dimensional turbulent vortices, such as boils. In fact, it is important to note that large-scale unsteady vertical structures can be observed in free surface flows even in the absence of horizontal mean velocity gradients (Nezu and Nakagawa, 1993). Such flows require either three-dimensional models or two-dimensional shallow water models with a forcing function that mimics the momentum transfer from the three-dimensional small scales to the large scales. Examples of forcing functions are developed and applied in van Prooijen (2004) and Hinterberger et al. (2007).

Three-dimensional models received far less attention, partly, due to their excessive computational requirements and partly due to the belief and satisfaction that large-scale coherent structures can be well represented by two-dimensional models. Examples of three-dimensional models are developed and tested by Stansby (2003, 2006) and Hinterberger et al. (2007). Stansby (2006) found that the transitional bed friction number between unstable and stable wakes as well as the size of stable wakes are

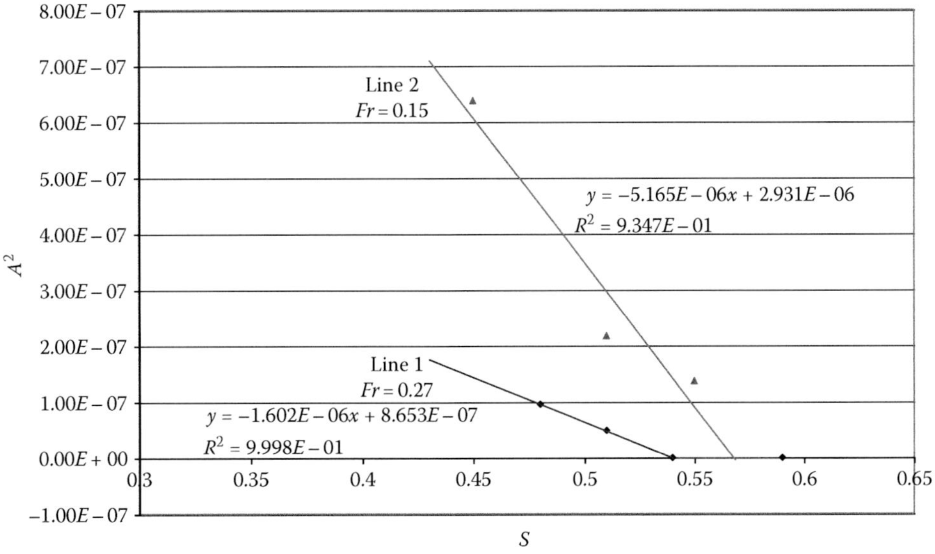

FIGURE 40.9 Linear relationship between A^2 and S *for different Froude numbers.*

poorly defined by two-dimensional models and require three-dimensional models. This is likely to due to the inability of quadratic friction laws, such as the one incorporated in Equation 40.23, to represent the complicated three-dimensional motion that are known to occur in the vicinity of type A generator such as in the wake of islands and headlands. Hinterberger et al. (2007) reported that their three-dimensional model reproduces experimental results fairly well, but requires much computer time. On the other hand, the two-dimensional shallow water model is not as accurate as the three-dimensional models especially for structures whose size is less than about five water depth, but still accurate enough for the large-scale features which are of concern in practical applications.

40.5 Major Challenges

To the extent possible and with the view to highlight possible research directions, the assumptions and limitations of the various methods and approaches as well as gaps in our knowledge have been throughout the chapter and will not be repeated here. In summary, major challenges, research challenges include the following.*

1. How well and to what extent does stability theories based on shallow water equations represent shallow shear flows? It is important to recall that such theories involve a host of gross assumptions such as quasi-parallel and quasi-steady assumptions, use base flows that are assumed rather than derived and frictional laws that are based on log-law profiles. It is important to note that the combination of strong vertical motion with the bed boundary layer can produce strong circulation in the vertical plane with upwelling at the core and inflow towards the core near the bed. As a result, the validity of log-law-based bed friction laws needs to be assessed.
2. What shallow flow problems can be modeled by two-dimensional shallow water models and, by implication, which require three-dimensional models?
3. The coherent quasi-two-dimensional turbulence with one set of scales develops on a pre-existing three-dimensional turbulence with another set of scales. What is the influence of these distinct and different scales on mass, momentum and energy transport and how to develop appropriate closure models for it? Under what conditions are these scales spectrally segregated? What turbulence models can faithfully reproduce the reverse energy cascade at large scales and forward energy cascade at small scale? How does the large-scale turbulence evolve with distance downstream and interact with small-scale background three-dimensional turbulence? How does this interaction changes from deep to intermediate to shallow flows? Note that with few exceptions, the tendency is to lump the interaction between large-scale turbulence and the background turbulence on which it develops by a simplistic "catch all" friction parameter (see system (40.22) and (40.23)).
4. Under what conditions does quasi-two dimensional regime prevail? In this regard, Uijttewaal and Jirka (2003) concluded by stating "Future experiments should be aimed at finding the conditions for the quasi-two dimensional regime. At large water depths, three-dimensional instabilities will affect the large-scale dynamics, but at small water depths, where bottom friction dominates, three-dimensional effects play a role in prohibiting the development of large structures or even the shedding of vortices."
5. How to experimentally study elementary vortex structures, such as dipoles, under realistic parameter ranges and without inducing unwanted effects such as strong vertical motion? If performed well, such prototype experiments are essential for developing fundamental results for shallow turbulent flows.
6. How to develop ingenious ways to get around the scale requirement to study shallow jets and wakes under realistic ranges of S, Fr, and Re and where the full range of spatial development of the coherent structures including birth, growth, and then decay can be investigated? A noteworthy development in this area is the novel experimental setup by Briggs et al. (2010) which removes the need for excessively wide basins in the study of shallow jets by using variable flow lateral boundaries within the pond.

* The challenges listed here are biased by the authors' background, interest, views, and gaps in their knowledge. This invariably leads to other major challenges being overlooked and/or the inclusion of challenges that others may deem not major.

Acknowledgment

This work is supported by the Hong Kong Grant Council (project 613407). Some of the figures are generated by Ms. C. F. Chan and Mr. J.H. Liang.

References

Akkermans, R. A. D., Cieslik, A. R., Kamp, L. P. J., Trieling, R. R., Clercx, H. J. H., and van Heijst, G. J. F. 2008. The three-dimensional structure of an electromagnetically generated dipolar vortex in a shallow fluid layer. *Phys. Fluids* 20(11): 116601–116615.

Babarutsi, S. and Chu, V. H. 1985. Experimental study of turbulent mixing layers in shallow open channel flows. Technical Report 85-1, Fluid Mechanics Laboratory, Department of Civil Engineering and Applied Mechanics, McGill University, Montreal, Quebec, Canada.

Biggs, C., Nokes, R., and Vennell, R. 2010. The dynamics of a steady shallow turbulent jet. *Proceedings of the 17th IAHR-APD Congress*, Auckland, New Zealand.

Camassa, R., Holm, D. D., and Levermore, C. D. 1997. Long-time shallow-water equations with a varying bottom. *J. Fluid Mech.* 349: 173–189.

Carrasco, A. and Vionnet, C. A. 2004. Separation of scales on a broad shallow turbulent flow. *J. Hydraul. Res.* 42(6): 630–638.

Chen, D. and Jirka, G. H. 1995. Experimental study of plane turbulent wake in a shallow water layer. *Fluid Dyn. Res.* 16(1): 11–41.

Chen, D. and Jirka, G. H. 1997. Absolute and convective instabilities of plane turbulent wakes in a shallow water layer. *J. Fluid Mech.* 338: 157–172.

Chen, D. and Jirka, G. H. 1998. Linear stability analysis of turbulent mixing layers and jets in shallow water layers. *J. Hydraul. Res.* 36(5): 815–830.

Chu, V. H., Wu, J. H., and Khayat, R. E. 1983. Stability of turbulent shear flows in shallow channel. *Proceedings of the 20th Congress of the International Association for Hydraulic Research*, Moscow, Russia, Vol. 3, pp. 128–133.

Chu, V. H., Wu, J. H., and Khayat, R. E. 1991. Stability of transverse shear flows in shallow open channels. *J. Hydraul. Eng. ASCE* 117(10): 1370–1388.

Chu, V. H. 2010. Shear instability, wave and turbulence simulations using the shallow-water equations. *J. Hydro.-Environ. Res.* 3(4): 173–178. Special issue I on Shallow Flows, Eds. M.S. Ghidaoui and Y.K. Tung.

Dracos, T., Giger, M., and Jirka, G. H. 1992. Plane turbulent jets in a bounded fluid layer. *J. Fluid Mech.* 241: 587–614.

Drazin, P. G. and Reid, W. H. 2004. *Hydrodynamic Stability*. Cambridge University Press, Cambridge, U.K.

Falques, A. and Iranzo, V. 1994. Numerical simulation of vorticity waves in the nearshore. *J. Geophys. Res.* 99(C1): 825–841.

Fernando, H. J. S. and Voropayev, S. I. 2010. Shallow flows in the atmosphere and oceans: Geophysical and engineering applications. *J. Hydro.-Environ. Res.* 4(2): 83–87.

Ghidaoui, M. S. and Kolyshkin, A. A. 1999. Linear stability analysis of lateral motions in compound channels with free surface. *J. Hydraul. Eng. ASCE* 125(8): 871–880.

Ghidaoui, M. S., Kolyshkin, A. A., Chan, F. C., Liang, J. H., and Xu, K. 2006. Linear and nonlinear analysis of shallow wakes. *J. Fluid Mech.* 548: 309–340.

Ghidaoui, M. S. and Liang, J. H. 2008. Investigation of shallow mixing layers by BGK Finite volume model. *Int. J. Comput. Fluid Dyn.* 22(7): 523–537.

Hinterberger, C., Fröhlich, J., and Rodi, W. 2007. Three-dimensional and depth-averaged large-eddy simulations of some shallow water flows. *J. Hydraul. Eng. ASCE* 133(8): 857–872.

Ingram, R. G. and Chu, V. H. 1987. Flow around islands in Rupert Bay: An investigation of the bottom friction effect. *J. Geophys. Res.* 92(C13): 14521–14533.

Jirka, G. H. 2001. Large scale flow structures and mixing processes in shallow flows. *J. Hydraul. Res.* 39(6): 567–573.

Jirka, G. H. and Seol, D. G. 2010. Dynamics of isolated vortices in shallow flows. *J. Hydro.-Environ. Res.* 4(2): 65–73.

Kolyshkin, A. A. and Ghidaoui, M. S. 2002. Gravitational and shear instabilities in compound and composite channels. *J. Hydraul. Eng.* 128(12): 1076–1086.

Kolyshkin, A. A. and Ghidaoui, M. S. 2003. Stability analysis of shallow wake flows. *J. Fluid Mech.* 494: 355–377.

Kolyshkin, A. A., Ghidaoui, M. S., and Yen, B. 1999. Influence of momentum correction coefficients on linear stability of open channels. *XXVIII IAHR Congress*, Graz, Austria, Theme E5, CD-ROM.

Landau, L. D. and Lifshitz, E. M. 2002. *Fluid Mechanics*. 2nd edn., London, Pergamon.

Liu, H., Lam, M. Y., and Ghidaoui, M. S. 2010. Numerical study of temporal shallow mixing layers using BGK-based schemes. *Comput. Math. Appl.* 59: 2393–2402.

Lloyd, P. M. and Stansby, P. K. 1997a. Shallow-water flow around model conical islands of small side slope. I: Surface piercing. *J. Hydraul. Eng. ASCE* 123(12): 1057–1067.

Lloyd, P. M. and Stansby, P. K. 1997b. Shallow-water flow around model conical islands of small side slope. II: Submerged. *J. Hydraul. Eng. ASCE* 123(12): 1068–1077.

Nadaoka, K. and Yagi, H. 1998. Shallow-water turbulence modeling and horizontal large-eddy computation of river flow. *J. Hydraul. Eng.* 124(5): 493–500.

Negretti, M. E., Vignoli, G., Tubino, M., and Brocchini, M. 2006. On shallow-water wakes: An analytical study. *J. Fluid Mech.* 567: 457–475.

Nezu, I. and Nakagawa, H. 1993. *Turbulence in Open-Channel Flows*. Balkema, Rotterdam, the Netherlands.

Pope, S. B. 2000. *Turbulent Flows*. Cambridge University Press, New York.

van Prooijen, B. C. 2004. Shallow mixing layers. PhD thesis, Delft University of Technology, Delft, the Netherlands.

van Prooijen, B. C. and Uijttewaal, W. S. J. 2002. A linear approach for the evolution of coherent structures in shallow mixing layers. *Phys. Fluid* 14(12): 4105–4114.

Schlichting, H. and Gersten, K. 2000. *Boundary-Layer Theory*. Springer, Hong Kong.

Schmid, P. J. and Henningson, D. S. 2001. *Stability and Transition in Shear Flows*. Springer, New York.

Stansby, P. K. 2003. A mixing-length model for shallow turbulent wakes. *J. Fluid Mech.* 495: 369–384.

Stansby, P. K. 2006. Limitations of depth-averaged modelling for shallow wakes. *J. Hydraul. Eng.* 132(7): 737–740.

Townsend, A. A. 1956. *The Structure of Turbulent Shear Flows*. Cambridge University Press, New York.

Uijttewaal, W. S. J. and Jirka, G. H. 2003. Grid turbulence in shallow flows. *J. Fluid Mech.* 489: 325–344.

Vionnet, C. A., Tassi, P. A., and Vide, J. P. M. 2004. Estimates of flow resistance and eddy viscosity coefficients for 2D modelling on vegetated floodplains. *Hydrol. Processes* 18: 2907–2926.

Wolanski, E., Imberger, J., and Heron, M. L. 1984. Island wakes in shallow coastal water. *J. Geophys. Res.* 89(C6): 10533–10569.

Yen, B. C. 1973. Open-channel flow equations revisited. *J. Eng. Mech. Div. ASCE* 99(5): 979–1009.

41

Shallow Granular Flows

O. Bokhove
University of Leeds and University of Twente

A.R. Thornton
University of Twente

41.1 Introduction

Straightforward shallow granular flow consists of the movement of dry particulate material traveling in relatively thin layers over (inclined) topography. Often the fluidized material involved is more complex, consisting of a multiphase mixture of predominantly solid matter with an interstitial fluid filling the pore space. Small-scale laboratory experiments of glass beads or sand flowing down inclined chutes are an example of the dry case, these are performed to study idealized granular systems. In nature, the situation is often complicated by the presence of a fluid; examples include the collapse of water-saturated slopes into debris flows or the flow of pyroclastic material into a river forming a lahar.

The fascination of shallow granular flows lies in their multifaceted behavior. Particles are strongly forced down-slope by gravity, delivering energy to the system and allowing the particles to flow. This energy is balanced by losses through inelastic particle–particle collisions and frictional interaction with the substrate over which the particles are moving. After a short initialization process, these two effects are approximately in balance and there is no net flow of energy into the system. We will see that, to leading order, a separate conservation of energy consideration will not be required to model these shallow flows. From a macro-scale viewpoint, the dynamics of fast granular flows are then akin to the fluid mechanics of shallow layers of water with a free surface. A notable difference, in the granular case, is that the constitutive or closure laws for the bottom stress relations are generally unknown, except perhaps in very special cases. The separation between the macroscopic granular flow scales of interest and the microscopic scales, as given by the mean free path of granular particles, is much smaller than in molecular fluids; however, the continuum formulation for granular flows is still valid, even though it is less distinct. Depending on the particle size, the free surface in these fast shallow granular flows is clearly seen by eye to be grainy, on the macroscopic scale, and only exists in a statistical sense using space and/or time averages.

When gravitational forcing is small, the collective particle behavior is (quasi) static and concerns the realm of solid mechanics involving larger particle deformations, multiple simultaneous interparticle contacts and force chains. Mixed continuum behavior arises in transient granular flows with static and dynamic zones. In all these cases, micro-scale discrete particle methods (DPMs) form a useful tool. In DPMs the individual forces on all the particles are calculated and their positions updated via the integration of Newton's equations of motion. In DPMs, the deformation of the particles during collisions is simplified using a contact model to enhance computational speed (e.g., Van der Hoef et al. 2006). We will limit ourselves to fluidized shallow granular flows in that the flow scales and speeds are small in one direction normal to the topography relative to the other directions. Hence, "depth-averaging" will be applied in the direction approximately normal to the terrain, to obtain leading-order models. For highly varying terrain, curvilinear coordinates will be required, but we will restrict attention to mildly undulating topography with one clearly defined average slope angle.

A typical industrial application of granular flows concerns the inflow of sinter, pellets, and cokes, via a rotating inclined

Handbook of Environmental Fluid Dynamics, Volume One, edited by Harindra Joseph Shermal Fernando. © 2013 CRC Press/Taylor & Francis Group, LLC. ISBN: 978-1-4398-1669-1.

channel, into the blast furnace for iron-ore melting, at Corus Hoogovens, IJmuiden, the Netherlands. These particles have different sizes and densities, and are also irregular in shape. In this situation particle segregation occurs, which is stimulated by the channel's rotation. Both the dynamics of the segregation and the motion of irregularly shaped particles are poorly understood problems. Comprehension of such problems is required to control these industrial flows, such that costly bottlenecks in the production process can be avoided. Study of these flows often provides insights in similar granular flows in the natural environment. A typical geological application concerns the 2005 landslide and debris flow after the collapse of a water-saturated slope of pyroclastic material high on Tolimán Volcano in Guatemala. This debris flow devastated the community of Panabaj and was caused by heavy rainfall. During flow, segregation occurred with the coarse-grained fraction of sand, gravel, and rocks coming to rest partly down slope, while the fine-grained portion of water and fine sediment (40% of the volume) continued to move in a less turbulent manner, see Figure 41.1.

Current research on granular flows ranges from idealized laboratory setups, in which rather uniform dry granular matter is considered, to water-saturated polydispersed debris flows over complex terrain. Monodispersed irregular and spherical particles have been used by Denlinger and Iverson (2001) (dry sand), Gray et al. (2003) (non-pareil sugar grains or "sprinkles") Hákonardóttir and Hogg (2005) (almost spherical glass ballotini), Börzsönyi and Ecke (2007) (sand, glass, and copper particles), and Vreman et al. (2007) (glass beads and poppy seeds) to study granular flows, whereas bidispersed (in size) spherical particles were used to study segregation by Thornton et al. (2006). Denlinger and Iverson (2001) also investigated water-saturated sand and gravel, bringing the problem under consideration closer to real natural debris flows. We will distinguish broadly two modeling philosophies in the research on shallow granular flows: detailed modeling using idealized particle types and flow geometries and more realistic modeling with complex particle types over complex terrain.

The first modeling philosophy aims to verify and validate the shallow granular flow models accurately against experimental flow measurements for uniform particle sizes, shapes, and densities in simple geometries. Often solutions from hydraulic theory are used, and these carry over to granular flows: hydraulic jumps/bores for shallow water flows become granular jumps and bores. A jump or bore is a discontinuity in the depth and velocity of the flow that propagates at a well-defined velocity. In granular flows this bore is localized over 5–10 particle diameters, in sharp contrast to water flows where a longer turbulent and bubbly region around the broken wave is observed. Figure 41.2 compares a hydraulic and granular bore and illustrates the sharper transition in the granular case. For dilute granular flows, kinetic theory has validity (e.g., Jenkins 2006) and has been used to determine stress and constitutive laws for both mono- and bidispersed spherical particles: again strengthening granular flow research in idealized circumstances. In the second modeling philosophy, experiments use more realistic fluid-saturated particles with wide size distributions (Denlinger and Iverson 2001, 2004). Then, an averaged constitutive law is considered, and validation is done against shallow granular flow over complex terrain, such that the interactions between flow and terrain are the discriminating factors. While the second philosophy is more relevant for environmental applications, the first one allows closer comparisons between experimental data, theory, and simulations. Of course, models used in the second philosophy should

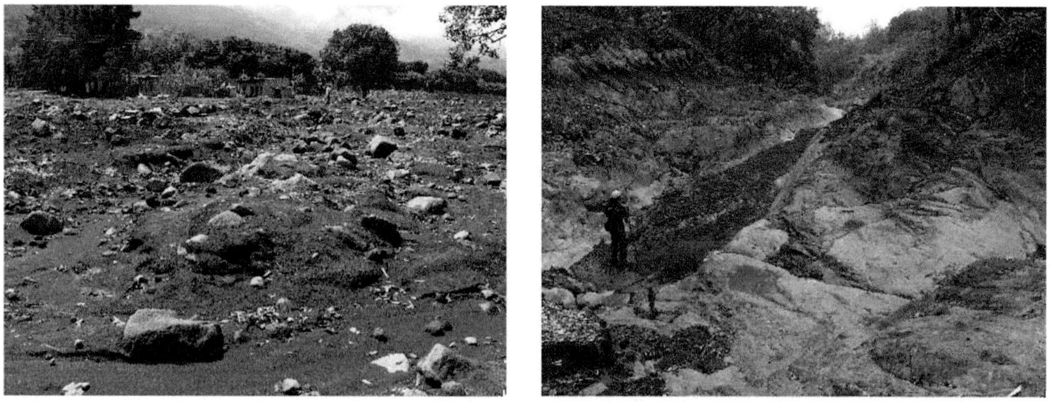

(a) (b)

FIGURE 41.1 The Panabaj debris flow displayed strong segregation into coarse- and fine-grained material such as sand, gravel, and rock as seen from the remnants at higher elevations (coarser debris (a)); and fine-grained water and fine-sediment constituents in the channel lower down the volcano (finer debris (b)). (Courtesy of Connor, C.N., Connor, L., Sheridan, M., Assessment of Oct. 2005 debris flows at Panabaj, Guatemala, and recommendations for hazard mitigation. Report for Oxfam G.B., 2006.)

FIGURE 41.2 (a) A hydraulic jump in a horizontal channel has steadied against the incoming fast flow from the left. (Photo courtesy of Benjamin Akers, the University of Illinois, Chicago, Illinois.) (b) A granular bore on an inclined channel is moving upstream due to the contraction in the channel; the incoming fast flow of poppy seeds slows down after the jump. Bore fronts are indicated and in the hydraulic case also water heights. Flows are from left to right.

be reducible to ones used in the first philosophy. We will limit ourselves mainly to the first philosophy with its idealizations.*

In the remainder of this chapter, we will start by deriving shallow depth-averaged granular flow equations in the absence of particle segregation, and the associated bore relations. Then we consider the validity of these equations for several basal stress relations for both smooth and rough topography (Section 41.2). Our second goal is to analyze a model of particle segregation for bidispersed flows with an interstitial fluid and compare its predictions to chute flow experiments (Section 41.3). In both cases, we discuss limitations and possible extensions to these models. Finally, we combine the two goals and outline future challenges in Section 41.4.

41.2 Depth-Averaged Shallow Granular Flow Models

41.2.1 Cauchy Balance Equations

In this section a model for shallow granular flow will be derived. The starting point will be Cauchy's equations for mass and momentum

$$\frac{D\rho}{Dt}+\rho(\nabla\cdot\underline{u})=0 \quad \text{and} \quad \rho\frac{D\underline{u}}{Dt}=\nabla\cdot\underline{\underline{\sigma}}+\rho\underline{B} \tag{41.1}$$

These conservation laws must be satisfied by all solids, liquids, and gases. In these equations, the material density is ρ, velocity $\underline{u} = (u, v, w)^T$, the stress tensor is $\underline{\underline{\sigma}}$, and the external body force is $\underline{B}$. The material derivative $D/Dt = \partial/\partial t + \underline{u} \cdot \nabla$, with gradient ∇ and time t, is the rate of change in time taken along a path moving with a material element. To solve these equations a model for the stress tensor and the body forces is required. From conservation of angular momentum, it can be shown that the stress tensor must be symmetric and this means it has six independent components in three dimensions and only three in two dimensions.

* See the chapter 43 on debris flows by R.M. Iverson.

41.2.2 Shallow Granular Flows

We will proceed by making a series of approximations to obtain a simpler set of equations for shallow granular flows. This was first done by Savage and Hutter (1989)†; however, here we follow the derivation by Gray et al. (2003). The fluidized granular material is assumed to be incompressible and homogeneous with constant density ρ_0. It thus includes certain granular mixtures with particles of different densities, shapes, and sizes, in the absence of segregation. The only external body force acting is gravity, and attention will be restricted to 2D flows, as illustrated in Figure 41.3. Hence, system (41.1) reduces to the following system of partial differential equations for u, w, and stress components σ_{xx}, σ_{xz}, and σ_{zz}:

$$\frac{\partial u}{\partial x}+\frac{\partial w}{\partial z}=0$$

$$\rho\left(\frac{\partial u}{\partial t}+\frac{\partial u^2}{\partial x}+\frac{\partial}{\partial z}(uw)\right)=\rho g\sin\theta+\frac{\partial\sigma_{xx}}{\partial x}+\frac{\partial\sigma_{xz}}{\partial z} \tag{41.2}$$

$$\rho\left(\frac{\partial w}{\partial t}+\frac{\partial}{\partial x}(uw)+\frac{\partial w^2}{\partial z}\right)=-\rho g\cos\theta+\frac{\partial\sigma_{xz}}{\partial x}+\frac{\partial\sigma_{zz}}{\partial z}$$

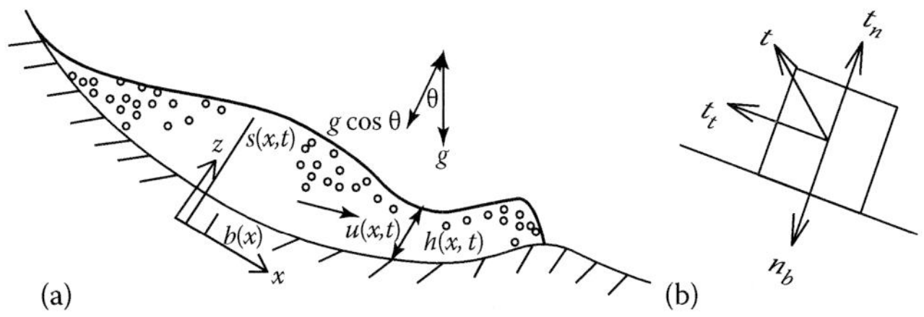

FIGURE 41.3 (a) Sketches are given of the domain with coordinate z normal to and coordinate x along the inclined base with layer thickness $h(x, t) = s(x, t) - b(x)$, and (b) the contact forces involved at the base.

† It should be noted that similar models appeared in the Russian literature in the 1960/1970s, but English translations are hard to obtain.

with g the acceleration of gravity and θ the inclination of the coordinate system (Figure 41.3).

41.2.3 Boundary Conditions

As follows from Figure 41.3, the free and basal surfaces are described by $z - s(x, t) = 0$ and $b(x, t) - z$, respectively, with coordinate z lying normal to an average slope. The *kinematic* boundary conditions must hold there, such that

$$\frac{D(s-z)}{Dt} = \frac{\partial s}{\partial t} + u^s \frac{\partial s}{\partial x} - w^s = 0 \quad \text{and}$$

$$\frac{D(b-z)}{Dt} = \frac{\partial b}{\partial t} + u^b \frac{\partial b}{\partial x} - w^b = 0 \tag{41.3}$$

Superscripts here denote evaluation at free and basal surfaces. The boundaries are impermeable, which for fixed boundaries reduces to the more familiar condition $\underline{u} \cdot \underline{n} = 0$ with outward normal $\underline{n}$ at the fixed boundary. We limit ourselves to one coordinate frame, and refer to the literature for extensions of the shallow layer equations in curvilinear coordinates (e.g., Denlinger and Iverson 2004). The latter is required for shallow flows over highly variable terrain.

The force per unit area applied to each of the boundaries (the traction) must be specified. Since the free surface is assumed to be open, it is traction-free (i.e., no force is applied), which means $\underline{\underline{\sigma}}^s \underline{n}^s = \underline{0}$. On the base a Coulomb friction model (for flowing material) will be implemented $\underline{t}_t^b = -\mu \left|\underline{t}_n^b\right| \underline{u}^b \big/ \left|\underline{u}^b\right|$, stating that the tangential component of the traction is proportional to the normal component of the traction and acts in a direction opposing the motion. The proportionality factor μ can be a function of both the local depth and the velocity of the flow. It should, however, be noted that this friction model is a limiting state, valid for flowing material, and care needs to be taken when considering material that is arresting or starts to flow. The traction on the base is given by $\underline{t}^b = \underline{\underline{\sigma}}^b \underline{n}^b$, which can be decomposed into normal and tangential components as follows: $\underline{t}^b = \underline{t}_t^b + \underline{t}_n^b = \underline{t}_t^b - \underline{n}^b \left|\underline{t}_n^b\right|$ (the minus sign indicates that the normal forces act into the domain, see Figure 41.3). Now clearly $\left|\underline{t}_n^b\right| = -\underline{n}^b \cdot \underline{t}^b = -\underline{n}^b \cdot \left(\underline{\underline{\sigma}}^b \underline{n}^b\right)$, which implies

$$\underline{\underline{\sigma}}^b \underline{n}^b = \mu\left(\underline{n}^b \cdot \underline{\underline{\sigma}}^b \underline{n}^b\right) \frac{\underline{u}^b}{\left|\underline{u}^b\right|} + \left(\underline{n}^b \cdot \underline{\underline{\sigma}}^b \underline{n}^b\right) \underline{n}^b \tag{41.4}$$

Finally, to apply these boundary conditions, the surface normal vectors $\underline{n}^b$ and $\underline{n}^s$ are expressed in terms of the unit normal vectors $\underline{e}_x$ and $\underline{e}_z$. From the basic geometry, see Figure 41.3, it is clear that $\Delta^s \underline{n}^s = -(\partial s/\partial x)\underline{e}_x + \underline{e}_z$, and $\Delta^b \underline{n}^b = -(\partial b/\partial x)\underline{e}_x - \underline{e}_z$ where the normalization factors Δ^v are given by $\Delta^v = \sqrt{(\partial v/\partial x)^2 + 1}$ for $v = s, b$.

41.2.4 Dimensionless Form

We proceed by writing the equations in dimensionless form. This results in the appearance of a dimensionless aspect ratio ε and its smallness will be employed later to simplify the equations.
Let L be a typical flow length, H a typical flow depth with g the magnitude of the gravitational acceleration. Then the following dimensionless variables are introduced:

$$x = L\tilde{x}, \quad z = H\tilde{z}, \quad t = \sqrt{L/g}\,\tilde{t}, \quad u = \sqrt{Lg}\,\tilde{u}, \quad w = \varepsilon\sqrt{Lg}\,\tilde{w}$$

$$\sigma_{xx} = \rho_0 g H \tilde{\sigma}_{xx}, \quad \sigma_{zz} = \rho_0 g H \tilde{\sigma}_{zz} \quad \text{and} \quad \sigma_{xz} = \varepsilon^\gamma \rho_0 g H \tilde{\sigma}_{xz}$$

where $\varepsilon = H/L$ and $0 < \gamma < 1$. Applying this scaling to the governing equations (41.2) yields

$$\frac{\partial \tilde{u}}{\partial \tilde{x}} + \frac{\partial \tilde{w}}{\partial \tilde{z}} = 0 \tag{41.5}$$

$$\frac{\partial \tilde{u}}{\partial \tilde{t}} + \frac{\partial \tilde{u}^2}{\partial \tilde{x}} + \frac{\partial}{\partial \tilde{z}}(\tilde{u}\tilde{w}) = \sin\theta + \varepsilon \frac{\partial \tilde{\sigma}_{xx}}{\partial \tilde{x}} + \varepsilon^\gamma \frac{\partial(\tilde{\sigma}_{xz})}{\partial \tilde{z}} \tag{41.6}$$

$$\varepsilon\left(\frac{\partial \tilde{w}}{\partial \tilde{t}} + \frac{\partial}{\partial \tilde{x}}(\tilde{u}\tilde{w}) + \frac{\partial \tilde{w}^2}{\partial \tilde{z}}\right) = -\cos\theta + \varepsilon^{1+\gamma} \frac{\partial(\tilde{\sigma}_{xz})}{\partial \tilde{x}} + \frac{\partial \tilde{\sigma}_{zz}}{\partial \tilde{z}} \tag{41.7}$$

Similarly, applying the scaling to boundary equations (41.3) and (41.4) leads to the following conditions at the free surface

$$\frac{\partial \tilde{s}}{\partial \tilde{t}} + \tilde{u}^s \frac{\partial \tilde{s}}{\partial \tilde{x}} - \tilde{w}^s = 0, \quad -\varepsilon \tilde{\sigma}_{xx}^s \frac{\partial \tilde{s}}{\partial \tilde{x}} + \varepsilon^\gamma \tilde{\sigma}_{xz}^s = 0 \quad \text{and}$$

$$-\varepsilon^{1+\gamma} \tilde{\sigma}_{xz}^s \frac{\partial \tilde{s}}{\partial \tilde{x}} + \tilde{\sigma}_{zz}^s = 0 \tag{41.8}$$

and at the basal surface

$$\frac{\partial \tilde{b}}{\partial \tilde{t}} + \tilde{u}^b \frac{\partial \tilde{b}}{\partial \tilde{x}} - \tilde{w}^b = 0,$$

$$\varepsilon \tilde{\sigma}_{xx}^b \frac{\partial \tilde{b}}{\partial \tilde{x}} - \varepsilon^\gamma \tilde{\sigma}_{xz}^b = \left(\underline{\tilde{n}}^b \cdot \underline{\underline{\tilde{\sigma}}}^b \underline{\tilde{n}}^b\right)\left(\tilde{\Delta}^b \left(\tilde{u}^b / \left|\underline{\tilde{u}}^b\right|\right)\mu + \varepsilon \frac{\partial \tilde{b}}{\partial \tilde{x}}\right) \tag{41.9}$$

$$\text{and} \quad \varepsilon^{\gamma+1} \tilde{\sigma}_{xz}^b \frac{\partial \tilde{b}}{\partial \tilde{x}} - \tilde{\sigma}_{zz}^b = \left(\underline{\tilde{n}}^b \cdot \underline{\underline{\tilde{\sigma}}}^b \underline{\tilde{n}}^b\right)\left(\tilde{\Delta}^b \left(\varepsilon \tilde{w} / \left|\underline{\tilde{u}}\right|\right)\mu - 1\right)$$

41.2.5 Depth-Averaged Modeling

The main idea of depth-averaging is to integrate out the z-dependence (recall that z is the spatial coordinate normal to the base). This process results in a model of lower dimension, yet it is not equivalent to setting the z-derivative to zero since velocity component w remains dependent on z. The height or depth of the flow in the z-direction is defined as $h = s - b$. The depth-averaged value $\bar{f}$ of a variable f is given by $\bar{f} = (1/h)\int_b^s f dz$. Integration of (41.5) over depth while using kinematic conditions (41.8) and (41.9) yields, after dropping the tildes,

$$\frac{\partial h}{\partial t} + \frac{\partial}{\partial x}(h\bar{u}) = 0 \tag{41.10}$$

Depth-averaging the x-momentum equation (41.6) in the same manner yields

$$\frac{\partial}{\partial t}(h\bar{u}) + \frac{\partial h\overline{u^2}}{\partial x} - \left[u\left(\frac{\partial z}{\partial t} + u\frac{\partial z}{\partial x} - w\right)\right]_b^s$$

$$= h\sin\theta + \varepsilon\frac{\partial}{\partial x}(h\bar{\sigma}_{xx}) - \left[\varepsilon\sigma_{xx}\frac{\partial z}{\partial x} - \varepsilon^\gamma\sigma_{xz}\right]_b^s \tag{41.11}$$

Using the kinematic and traction boundary conditions (41.8) and (41.9), the terms in the square brackets of (41.11) can be simplified to give

$$\frac{\partial}{\partial t}(h\bar{u}) + \frac{\partial h\overline{u^2}}{\partial x} = h\sin\theta + \varepsilon\frac{\partial}{\partial x}(h\bar{\sigma}_{xx})$$

$$+\left(\mu\Delta^b\frac{\bar{u}^b}{|\underline{\bar{u}}^b|} + \varepsilon\frac{\partial b}{\partial x}\right)\left(\underline{n}^b\cdot\underline{\underline{\sigma}}^b\underline{n}^b\right) \tag{41.12}$$

Exercise 41.1 Verify derivations (41.10) and (41.12). Employ Leibniz' rule and use $\bar{u} = u^b + O(\varepsilon)$.

41.2.6 Order of Magnitude Estimates

As discussed in detail in previous sections, granular flows are often long and shallow and so far we have not exploited this fact; mathematically this implies that $\varepsilon = H/L \ll 1$. In the following treatment, terms of order greater than ε will be neglected and the order of the tangential stress must be discussed. If it is assumed that $0 < \gamma < 1$, then only leading order terms in the scalar factor $(\underline{n}^b\cdot\underline{\underline{\sigma}}^b\underline{n}^b)$ need to be retained. Expansion gives $(\underline{n}^b\cdot\underline{\underline{\sigma}}^b\underline{n}^b) = \sigma_{zz}|_b + O(\varepsilon)$. From (41.7), the leading order behavior of σ_{zz} is governed by the solution of $\partial\sigma_{zz}/\partial z = \cos\theta$, whose integration (using (41.8) at leading order such that $\sigma_{zz}^s = 0$) gives

$$\sigma_{zz} = (z - s)\cos\theta + O(\varepsilon) \tag{41.13}$$

Hence, $\sigma_{zz}|_b = -h\cos\theta + O(\varepsilon)$ and by retaining terms to $O(\varepsilon)$, Equation (41.12) is rewritten as

$$\frac{\partial}{\partial t}(h\bar{u}) + \frac{\partial}{\partial x}(h\overline{u^2}) = h\left(\sin\theta - \mu\cos\theta\frac{\bar{u}}{|\bar{u}|}\right)$$

$$+\varepsilon\frac{\partial}{\partial x}(h\bar{\sigma}_{xx}) - \varepsilon h\cos\theta\frac{\partial b}{\partial x} + O(\varepsilon^{1+\gamma}) \tag{41.14}$$

since in one dimension $|\bar{u}| = \sqrt{\overline{u^2} + \varepsilon^2\overline{w^2}} \approx |\bar{u}|$ and $\Delta^b \approx 1$. To close the model a constitutive relation for $\bar{\sigma}_{xx}$, a model for μ, and the depth-averaged square velocity $\overline{u^2}$ need to be specified. These closures are still open research questions and will be discussed in more detail in the following.

41.2.7 Constitutive Model for $\bar{\sigma}_{xx}$ and Depth-Integrated Square Velocity

In the original paper of Savage and Hutter (1989), they assumed the granular material behaved as a Mohr–Coulomb material in yield, which gives $\sigma_{xx} = K\sigma_{zz}$. The material constant K is called the Earth pressure coefficient and is given by

$$K = 2\sec^2\phi_i(1 \mp (1 - \cos^2\phi_i\sec^2\phi)^{1/2}) - 1 \tag{41.15}$$

where ϕ_i is the internal angle of friction and ϕ is the basal friction angle. The minus sign is taken for material in an active state, that is, $\partial\bar{u}/\partial x > 0$, and the positive sign for material in a passive state, that is, $\partial\bar{u}/\partial x < 0$ (see also Iverson and Denlinger 2001). A simpler model is to assume a flowing granular material acts like an inviscid fluid, implying $K = 1$. We will retain K and leave its choice to the reader. Finally, from $\sigma_{xx} = K\sigma_{zz}$ and (41.13) it follows that

$$\bar{\sigma}_{xx} = -Kh\cos\theta/2 \tag{41.16}$$

Using $h\int_b^s u^2 dz = \alpha_1\left(\int_b^s u dz\right)^2$, the depth-average square velocity can be expressed in terms of the squared depth-averaged velocity as follows:

$$\overline{u^2} = \alpha_1\bar{u}^2 \tag{41.17}$$

The shape-factor α_1 gives information about the vertical velocity profile. For a parabolic velocity profile (with zero basal velocity) $\alpha_1 = 6/5$, and for realistic nearly uniform velocity profiles $\alpha_1 \approx 1$ (e.g., GDR MiDi 2004). The latter is assumed here.

41.2.8 Shallow Layer Equations and Granular Jumps or Shocks

Substitution of (41.16) and (41.17) into (41.14) leads together with (41.10) to shallow layer equations

$$\frac{\partial h}{\partial t} + \frac{\partial}{\partial x}(hu) = 0 \tag{41.18a}$$

$$\frac{\partial}{\partial t}(hu) + \frac{\partial}{\partial x}(hu^2) + \varepsilon K\cos\theta\frac{\partial}{\partial x}\left(\frac{h^2}{2}\right)$$

$$= h(\sin\theta - \mu\frac{u}{|u|}\cos\theta) - \varepsilon h\cos\theta\frac{\partial b}{\partial x} \tag{41.18b}$$

where basal friction $\mu = \tan\phi$ with basal friction angle ϕ. Bars and higher-order terms in ε have been dropped. The basal friction μ and hence ϕ, in general, depend on the flow variables h and u, as will be discussed in detail later. The system of Equations (41.18) is hyperbolic and hence discontinuous (shock) solutions are possible. In the hydraulic literature, these discontinuous solutions are generally called hydraulic jumps or bores instead of shocks as is usual in gas dynamics. The differential (strong) form implies implicitly that both h and u are continuous; therefore, to obtain shock relations we must use the integral (weak) form. Here, the weak form and associated shock relation will be derived for the mass balance equation, whereas similar results for the momentum balance will be left as an exercise.

First, we integrate the depth-average mass balance equations (41.18a) from $X(t) - \delta$ to $X(t) + \delta$ and take the limit $\delta \to 0$, with both h and u discontinuous at $x = X(t)$. Defining X^- as the limit position on the left side of the jump and X^+ the limit on the right side, and bore speed $S = dX/dt$, this integral equation becomes $\frac{d}{dt}\int_{X^-(t)}^{X^+(t)} h dx - S[h]_{X^-}^{X^+} + [hu]_{X^-}^{X^+} = 0$, after applying Leibniz's rule. Of course, the term $\int_{X^-(t)}^{X^+(t)} h dx = 0$. Altogether, we obtain the jump relation for the continuity equations and similarly the one for the momentum equation:

$$[h(u - S)]_-^+ = 0 \quad \text{and} \quad [hu(u - S)] = -[\varepsilon K h^2 \cos\theta/2]_-^+ \tag{41.19}$$

where the +/– subscripts indicate the states on the right/left of the shock.

Exercise 41.2 Derive the preceding bore/shock relation (41.19) for the momentum equation.

By analogy with (41.18), the 2D shallow layer equations for the simplified case (for K = 1) are

$$\frac{\partial h}{\partial t} + \frac{\partial}{\partial x}(hu) + \frac{\partial}{\partial y}(hv) = 0$$

$$\frac{\partial}{\partial t}(hu) + \frac{\partial}{\partial x}\left(hu^2 + \varepsilon K \cos\theta \frac{h^2}{2}\right) + \frac{\partial}{\partial y}(huv)$$

$$= h(\sin\theta - \mu \frac{u}{|\underline{u}|}\cos\theta) - \varepsilon h \cos\theta \frac{\partial b}{\partial x} \tag{41.20}$$

$$\frac{\partial}{\partial t}(hv) + \frac{\partial}{\partial x}(huv) + \frac{\partial}{\partial y}\left(hv^2 + \varepsilon K \cos\theta \frac{h^2}{2}\right)$$

$$= -\mu \frac{hv}{|\underline{u}|}\cos\theta - \varepsilon h \cos\theta \frac{\partial b}{\partial y}$$

with velocity v in the y-direction. The following shock relations arise from (41.20):

$$[h(\underline{u} \cdot \underline{n} - S)]_-^+ = 0 \quad \text{and}$$

$$[h\underline{u}(\underline{u} \cdot \underline{n} - S)]_-^+ + [\varepsilon K h^2 \cos\theta/2]_-^+ \underline{n} = 0 \tag{41.21}$$

where we defined the unit vector $\underline{n}$ normal to the bore in the direction of the drop in depth; S is the bore speed in that direction.

Exercise 41.3 Derive the shallow granular layer equations (41.20) by extending the derivation given for the 1D case to the 2D case, cf. Gray et al. (2003), for the simplified case with $K = 1$ such that $\sigma_{xx} = \sigma_{yy} = \sigma_{zz}$.

Exercise 41.4 Similarly derive the bore relations (41.21) corresponding to the 2D shallow layer equations (41.20).

41.2.9 Applications

Several studies have compared solutions of the shallow layer equations in 1D and 2D over inclined topography, or extensions thereof, against data of laboratory or field experiments. The particulate matter used in these studies involved a range of materials, such as relatively uniform sugar grains, sand, or copper particles; nearly spherical glass beads in small diameter ranges; and, poppy and mustard seeds, among others. The surface of the topography varied from smooth (wood or aluminum) to rough (sand paper, velvet, or particles glued onto the surface). We will consider several idealized results with uniform particles and specific geometries, and the corresponding choices of K and μ.

Starting with studies on smooth chutes, Gray et al. used the two dimension version of (2.18), (see Exercise 41.2), by using $K = 1$ with laboratory experiments on non-accelerating slopes. Forcing and basal friction were therefore in balance with $\mu = \tan\phi = \tan\theta$. Nearly, 1D granular bores were generated by entirely blocking the chute. We follow Vreman et al. (2007), who also included the solid fraction $\alpha_\pm$ before and after the bore. The bore speed in this 1D case becomes

$$S = -\sqrt{\frac{1}{2}\varepsilon K \cos\theta(\alpha_- h_-/\alpha_+ h_+)(\alpha_+ h_+^2 - \alpha_- h_-^2)/(\alpha_+ h_+ - \alpha_- h_-)} \tag{41.22}$$

with constant inflow speed u_-, depth h_- and solid fraction α_-. For dilute inflows, Vreman et al. (2007) showed that including the difference in solid fraction, for example, $\alpha_- = 0.36 \pm 0.06$ and $\alpha_+ \approx 0.64$, before and after the bore passes, yields a more accurate bore speed. The comparisons between numerical simulations of their shallow flow model in 2D and experiments of granular flow around a pyramid obstacle placed in the middle of the flow in Gray et al. (2003) were quite successful. These predictions included transient flow with particle-free regions and dead zones, where the particles are motionless.

Exercise 41.5 Derive the bore speed (41.22) for the case with the solid's volume fraction α constant, and roughly equal to the random packing fraction. To compute the bore speed what information must be the known or measured, also for the case with variable α?

Hákonardóttir and Hogg (2005) considered oblique shocks in rapid granular flows more extensively than Gray et al. (2003). They also took $K = 1$ and $\mu = \tan\phi = \tan\theta$ and a relatively smooth, wooden inclined chute was used. A key parameter in these flows is the Froude number $F = u/\sqrt{\varepsilon \cos\theta K h}$, that is, the ratio of fluid velocity u over the speed $\sqrt{\varepsilon K h \cos\theta}$ of granular free-surface gravity waves. In dimensional form this speed is $\sqrt{g\cos\theta Kh}$ and $F = u/\sqrt{g\cos\theta Kh}$.

Flows with Froude number greater than one are called supercritical, whereas other flows are termed subcritical. This is akin to the demarcation of compressible flows into sub- and supersonic flows based on the Mach number, the ratio of the flow speed over the speed of sound. Hákonardóttir and Hogg (2005) investigate supercritical flows with inflow numbers $F > 4$ against a straight barrier with a deflection angle θ_c with respect to a half-channel wall. For certain sufficiently supercritical flows, steady oblique hydraulic jumps with shock speed $S = 0$ emerge with an angle $\theta_s > \theta_c$ to the wall. The expressions follow from (41.21) as

$$2F_-^2\sin^2\theta_s = \frac{1}{h_-}\frac{\alpha_+ h_+}{\alpha_- h_-}\frac{(\alpha_- h_-^2 - \alpha_+ h_+^2)}{(\alpha_- h_- - \alpha_+ h_+)} \quad \text{and}$$

$$\frac{\alpha_+ h_+}{\alpha_- h_-} = \frac{\tan\theta_s}{\tan(\theta_s - \theta_c)} \tag{41.23}$$

where F_- is the Froude number of the incoming flow. Again, we have included the solid's volume fraction α. Very good agreement was found between theory and experiment for constant α, while for dilute inflows the inclusion of different solid fractions remains important.

Exercise 41.6 (harder) Derive (41.23) from (41.21) by considering the momentum component along and orthogonal to the bore "propagation" direction. First draw a defining sketch of the situation, including the unit vectors normal and tangential to the granular jump.

Inspired by lake formation in the Rhine River due to ash flows after the Laacher See volcanic eruption 12.900aBP and research in hydraulics, Vreman et al. (2007) considered supercritical granular flow on a smooth aluminum incline through a linear and localized contraction, see Figure 41.4 for a series of snapshots. The linear contraction consisted of two equal triangular aluminum blocks snug tightly to the channel walls. Glass beads in several diameter ranges and poppy seeds were used. While the incoming flows were always supercritical, the contraction allowed the emergence of several steady states including ones with a transition to subcritical flows. Two parameters govern the flow: upstream Froude number $F = u/\sqrt{g\cos\theta Kh}$ and scaled minimum contraction width $0 \leq B_c = b_c/b_- \leq 1$ with constant upstream channel width b_-. Without the contraction $B_c = 1$, the

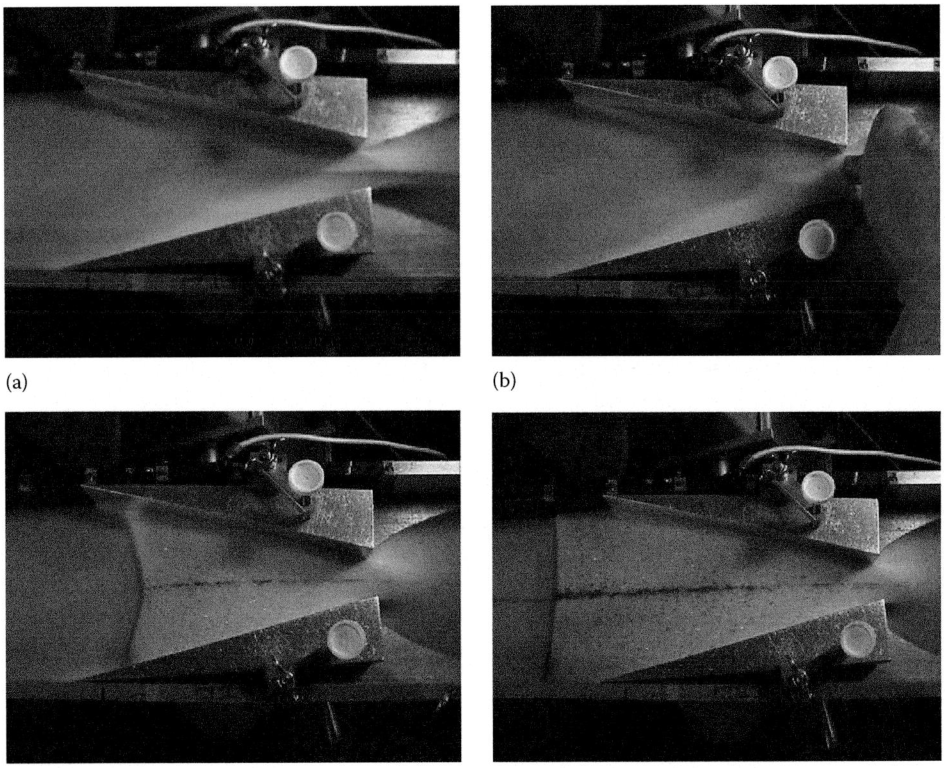

(a) (b) (c) (d)

FIGURE 41.4 We view the contraction normal to the inclined chute. Panel (a) shows the smooth granular flow of glass beads with two weak oblique jumps. When we manually disturb this flow by partially blocking the contraction exit for a short time, panel (b), the smooth flow state I in (a) changes to the reservoir state III with a clear granular jump in (c) and upstream bore state II in (d). This transition is displayed in the plates going from top left to bottom right. The top left plate manifests flow with smooth oblique jumps and the bottom right plate flow with an upstream halted jump and lake. All cases clearly show a jet behind the contraction. Inflow conditions at the left, upstream side are constant. Stray poppy seeds play the role of markers. (From Vreman, A. W. et al., *J. Fluid Mech.*, 578, 233, 2007. © J. Fluid Mech.)

chute inclination θ was adjusted such that the flow was steady and uniform (far enough away from the upstream inflow sluice gate). In this case we can take $\mu = \tan \phi = \tan \theta$ and also $K = 1$. In the presence of the contraction, several steady sometimes coexisting states can emerge. When the contraction width is relatively large, including $B_c \sim 1$, two oblique jumps emerge from the contraction walls and potentially collide in the middle of the contraction to form two new oblique jumps, etc. This state I is relatively smooth because the jumps are weak, see Figure 41.4a. After the contraction, a free jet flows downstream. When the contraction is sufficiently closed, including $B_c \sim 0$, then it acts as a block and an upstream moving granular bore emerges traveling toward the upstream sluice gate. This is state II, as seen in Figure 41.4d. For intermediate values of the contraction width, regions emerge where the various states coexist, including a reservoir state III with a complex pattern of jumps in the contraction, see Figure 41.4c. Note that the Froude number becomes subcritical behind the hydraulic jump for the state with the upstream moving bore and the reservoir state. While for the supercritical inflows, and also for the state with the weak oblique jumps, the Froude number stays nearly constant such that the forcing due to gravity and the friction due to interparticle and particle–wall collisions are (nearly) in balance, with $\mu(F > 1) = \tan \phi \approx \tan \theta$, the Coulomb friction factor is seen to decrease for the subcritical case such that $\mu(F < 1) = \tan \phi < \tan \theta$. From (41.18) or (41.20), we thus see that the flow effectively accelerates in subcritical regions, relative to supercritical flow regions. In Figure 41.5, a comparison between the laboratory experiments and 1D hydraulic model is made using adjustable parameters Z_1 and Z_2, which are related to $\mu(F > 1)$ and $\mu(F < 1)$. When forcing and friction are in balance, parameter Z_1 is zero and $\mu = \tan \phi = \tan \theta$. The curves shown emerge from analytical expressions, which are extensions of classical hydraulic theory for a width average of 2D equations (41.20). These width-averaged equations are simplified in a similar manner as the depth-averaged case. Unfortunately, Vreman et al. (2007) could not further constrain the dependence of μ on, for example, F and h. This result is important because it reveals that, although fixing μ may be valid for certain supercritical granular flows, it is not for subcritical flows. Further comparison (Vreman et al. 2007) between DPMs, in which all particles were simulated microscopically and stress relations were analyzed, revealed the shortcomings of several shallow layer equations used in the literature. This comparison was an a priori analysis because shallow layer models were not analyzed and validated, but only their specific stress relations.

Kerswell (2005) investigated granular dam break or slumping problems with Coulomb friction on a horizontal surface. One-dimensional equations (41.18) were used with $b = 0$, $\theta = 0$, and constant $\mu = \tan \phi$. Corresponding analytical solutions were compared with a series of experimental data from the literature on both smooth and rough surfaces. Essentially, K, g, and $\mu = \tan \phi$ could be combined into one fitting parameter after suitable rescaling. Such a recombination was argued to lead to a convenient (simplest) model to study granular slumping.

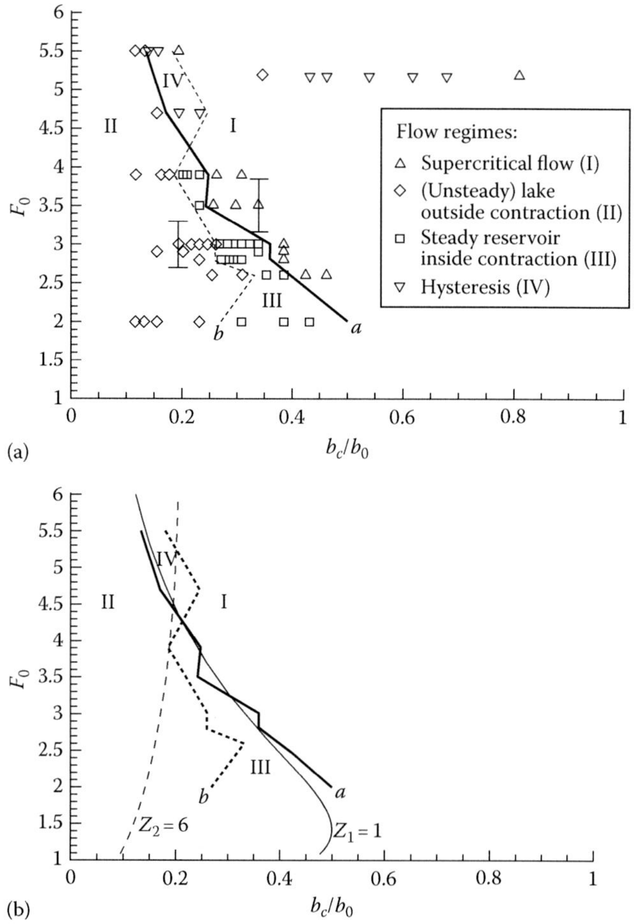

FIGURE 41.5 (a) Experimental granular results collected in a phase diagram spanned by Froude number versus nozzle width. Flow regimes observed are (I) smooth supercritical flows, (II) steady lakes halting outside the contraction or upstream moving bore, (III) steady reservoirs with a granular or hydraulic jump inside the contraction, and (IV) a hysteretic region with multiple flow states (Figure 41.4). Drawn dashed and solid lines demarcate transitions between regimes IV–I and II–III, and II–IV and III–I. Representative error bars are shown. Tiny symbols (upper right corner) denote poppy seed experiments falling outside our classification. (b) Comparison between analytical and experimental demarcation lines. (From Vreman, A. W. et al., *J. Fluid Mech.*, 578, 233, 2007. © J. Fluid Mech.)

In contrast, steady flows on rough inclined chutes have been studied extensively (Pouliquen and Forterre 2002; GDR MiDi 2004; Börzsönyi and Ecke 2007) and reveal that $\mu = \tan \phi$ depends on the (dimensional) Froude number $F = u/\sqrt{gh\cos\theta}$ and the depth h, as well as material fitting parameters. For relatively uniform sand grains, copper particles, and glass beads, the constitutive model determining μ was shown to follow the Pouliquen–Jenkins law. This was theoretically formulated by Jenkins (2006) using a phenomenological modification of granular kinetic theory to account for enduring particle contacts, revealing a slight modification to the law previously experimentally obtained by Pouliquen. The idea is that enduring contacts between grains, forced by the shearing (gradient of the particle velocity), reduce the collision rate of dissipation. Therefore, a modification to the dissipation is introduced, which does not affect the stress. This (dimensional) Pouliquen–Jenkins law for flowing material reads

$$F_c = \frac{u}{\sqrt{gh}} = \beta\left(\frac{h}{h_s}\right)\left(\frac{\tan^2\phi}{\tan^2\phi_1}\right) \tag{41.24}$$

with slope β as material parameter and the thickness

$$h_s(\phi) = Ad\left(\frac{\tan\phi_2 - \tan\phi}{\tan\phi\ - \tan\phi_1}\right) \quad \text{with} \quad \phi_1 < \phi < \phi_2 \tag{41.25}$$

of the layer when the flow subsides, with particle diameter d and coefficient A. Hence, we must have $h \geq h_s$, which places a restriction on (41.24). The angles ϕ_1 and ϕ_2 are the angles where h_s diverges and approaches zero, respectively. Börzsönyi and Ecke (2007) confirmed this law against data from their laboratory experiments measured in the middle of the channel where wall effects are minimal. For steady uniform flow, 1D balance follows from (41.18) as $\mu = \tan\phi = \tan\theta$ with a corresponding depth and velocity. Substitution of $\mu = \tan\phi$ and $\vartheta = uAd/(\beta h\sqrt{gh})$ in the two expressions (41.24) and (41.25) and elimination of h_s yield the cubic equation

$$\mu^3 - \mu^2\tan\phi_1 + \mu\vartheta\tan^2\phi_1 - \vartheta\tan^2\phi_1\tan\phi_2 = 0 \tag{41.26}$$

Through division by $\tan^2\phi_1$ and use of $\tilde{\mu} = \mu/\tan\phi_1$ and $\eta = \tan\phi_2/\tan\phi_1$, we rewrite this cubic equation as $\tilde{\mu}^3 - \tilde{\mu}^2 + \tilde{\mu}\vartheta - \vartheta\eta = 0$. It has only one real solution since $\eta > 1$ and the discriminant of this equation is negative. Given material parameters and the flow state combined into ϑ, this variable friction $\mu = \tan\phi$ can be determined and used in the equations of motion.

This first version of the friction law (41.24) was obtained from experiments by allowing a flowing layer to become stationary and measuring $h_s(\phi)$. For most granular materials a greater angle is required to start stationary than to retard flowing material. By starting with a stationary layer and measuring the angle required to start motion, Pouliquen and Forterre (2002) measured $h_{start}(\phi)$. They extended the friction law (41.24) to include this effect, allowing the study of problems with emerging and disappearing dead zones. We have plotted μ versus the Froude number in Figure 41.6, and it shows that the basal friction decreases with decreasing Froude number, similar to the Froude number dependence in the granular flow experiments in Vreman et al. (2007) on a smooth chute. How relevant these detailed studies of granular flows under idealized conditions are for (dry) granular particle flows in the field, such as landslides and pyroclastic flows, is an important question, which will be discussed in Section 41.4.

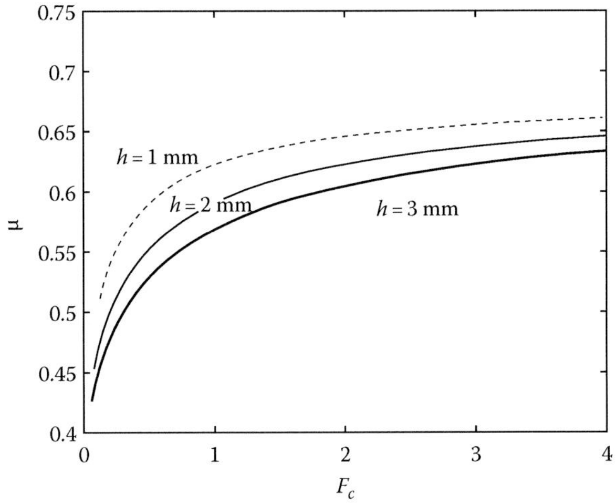

FIGURE 41.6 Basal friction versus Froude number $F_c = F\sqrt{\cos\theta} = u/\sqrt{gh}$ for glass beads: $A = 0.95$, $d = 0.72$ mm, $\phi_1 = 20.8°$, $\phi_2 = 34.2°$, and $\beta = 0.06$. (Based on measurements in Börzsönyi, T. and Ecke, R.E., *Phys. Rev. E*, 76, 031301, 2007.)

41.3 Particle Segregation

The shallow granular model presented in Section 41.2 assumes that all the flow either is comprised of identical monodispersed particles or does not segregate for polydispersed ones, but this is far from reality. As discussed in the introduction, real geological flows are polydispersed and contain particles of different shape, roughness, size, and density. In the high solid fraction regions of these flows the large particles commonly segregate to the surface, where they are transported to the margins to form bouldery flow fronts (see Figure 41.1 and the associated discussion in the introduction). In many natural flows these bouldery margins experience a much greater frictional force, leading to frontal instabilities (Iverson 1997). These instabilities create levees that channel the flow, thus vastly increasing the run-out distance. Gaining in-depth understanding of this effect is one of the major challenges in improving the predictive power of shallow granular models.

A similar effect can be observed in dry granular experiments with a combination of small smooth and large rough particles. When this mixture is poured down an inclined plane, particle size segregation causes large particles to accumulate near the margins. Being rougher, the large particles experience a greater frictional force and this configuration (rougher material in front of smoother) can be unstable (see Figure 41.7). This instability causes the uniform flow front to break up into a series of fingers (Pouliquen and Vallance 1999).

There are many mechanisms of segregation of dissimilar grains in granular materials (Drahun and Bridgewater 1983), but kinetic sieving is the mechanism that normally dominates in the dense granular flow regime where shallow granular avalanche models are applicable. The basic idea is that as grains avalanche down the slope, the local void ratio fluctuates, and the small particles fall into gaps that open up beneath them because they are

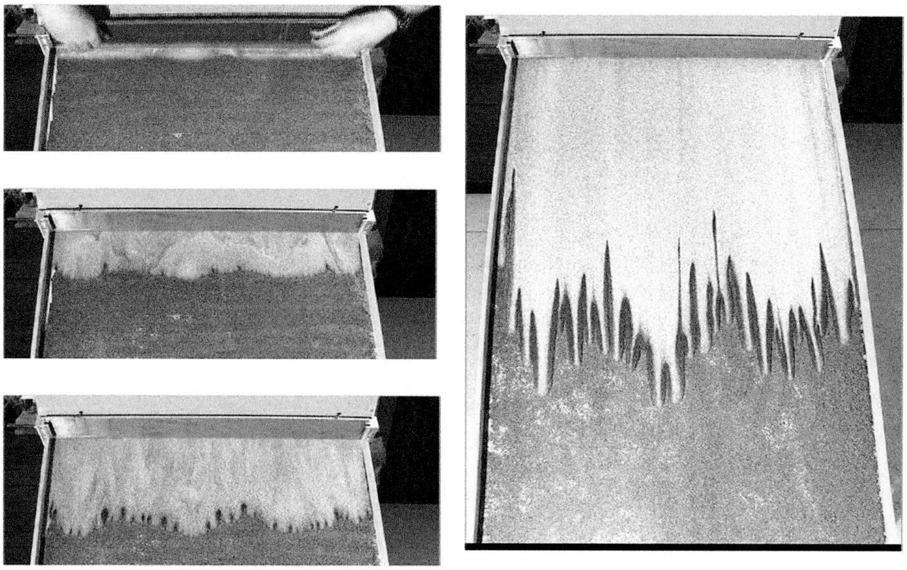

FIGURE 41.7 Four images show the fingering instability created by particle segregation at t = 0, 1.5, 3, 14 s. The chute measures 1 × 2 m^2. The experiment is performed with a mixture of 17% black rough carborundum (315–355 μm) and 83% glass ballotini (75–150 μm). (Courtesy of Nico Gray, University of Manchester, U.K. and Pete Kokelaar, University of Liverpool, U.K.)

simply more likely to fit into the available space than the large ones. The small particles, therefore, migrate toward the bottom of the flow and lever the large particles upward by force imbalances.

The first model of kinetic sieving was developed by Savage and Lun (1988), using a statistical argument about the distribution of void space. This model was able to predict steady-state size distributions for simple shear flows with bidisperse granular materials. Recently, Thornton et al. (2006) developed similar results from the framework of mixture theory. The two assumptions of this model are as follows: (i) firstly, as the different particles percolate past each other, there is a Darcy-style drag between the different constituents (i.e., the small and large particles) and (ii) secondly, particles falling into void spaces do not support any of the bed weight. Since the number of voids available for small particles to fall into is greater than for large particles, it follows that a higher percentage of the small particles will be falling and, hence, not supporting any of the bed load.

Mixture theory assumes that all constituents, ν, simultaneously occupy every point in the material, with a local volume fraction Φ^ν. Then overlapping partial densities, ρ^ν, partial velocities, $\underline{u}^\nu$, and partial pressures, p^ν, can be defined for each constituent. These partial quantities can be related to their intrinsic (laboratory measured) value via the following relationships, $\rho^\nu = \Phi^\nu \rho^{\nu*}$, $p^\nu = f^\nu p$, and $\underline{u}^\nu = \underline{u}^{\nu*}$, where functions f^ν are to be determined and superscripts "*" indicate intrinsic variables. This formulation is very powerful as the following simple relationships hold for the bulk density, $\rho = \sum_\nu \rho^\nu$ and bulk pressure $p = \sum_\nu p^\nu$. Each of these constituents must satisfy individual mass and momentum balances

$$\frac{\partial \rho^\nu}{\partial t} + \nabla\cdot(\rho^\nu \underline{u}^\nu) = 0,\ \rho^\nu \frac{D^\nu \underline{u}^\nu}{D^\nu t} = -\nabla p^\nu + \rho^\nu \underline{g} + \underline{\beta}^\nu \qquad (41.27)$$

where $\underline{\beta}^\nu$ is the interaction force exerted on phases ν by the other constituents. By definition the sum across all constituents of the $\underline{\beta}^\nu$'s must be zero. For simplicity, we will consider a two-constituent mixture theory (large, l, and small, s, particles), in which the interstitial pore space is incorporated into the volume of the grain, that is, $\Phi^l + \Phi^s = 1$.

Exercise 41.7 Assume that large and small particles have the same intrinsic density $\rho^{s^*} = \rho^{l^*} = \rho_0$ and that Dw/Dt is small, and then show from vertical momentum balance that the bulk pressure is $p = \rho_0 g(h - z)\cos\theta$, with h the depth of the flow and θ the angle the flow makes to gravity.

To close this model we must specify the interaction force between large and small particles, and the pressure functions f^ν. Experiments show that the kinetic sieving process is similar to the percolation of fluids through a porous solid, and we therefore use a Darcy-style law

$$\underline{\beta}^\nu = p\nabla\cdot f^\nu - \rho^\nu c(\underline{u}^\nu - \underline{u}), \quad \nu = l, s \qquad (41.28)$$

The material constant c is the interparticle drag coefficient and will in general depend on both species' surface properties, whereas $\underline{u} = (\rho^l \underline{u}^l + \rho^s \underline{u}^s)/\rho$ is the bulk velocity.

Exercise 41.8 Assuming that vertical accelerations are small, show that drag relation (41.26) implies

$$\Phi^\nu w^\nu = \Phi^\nu w + (f^\nu - \Phi^\nu)(g/c)\cos\theta, \quad \nu = l, s \qquad (41.29)$$

All that remains to close the model is to specify the f^ν functions that divide the bed weight between the large and small particles. Mathematically, we have the constraint $f^s + f^l = 1\ \forall\Phi^s$ and physically it follows that $f^v(\Phi^v = 0) = 0$ and $f^v(\Phi^v = 1) = 1$ must also hold.

Exercise 41.9 By considering $f^l = a_0 + a_1\Phi^l + a_2(\Phi^l)^2$ and $f^s = b_0 + b_1\Phi^s + b_2(\Phi^s)^2$, show that these constraints imply that $f^s = \Phi^s + B\Phi^s(\Phi^s - 1)$ where $B = b_2$. Note that the additional physical argument that the small particles are more likely to be falling implies $B > 0$.

Exercise 41.10 Using the mass balance equations (41.25) and the relationships for f^v obtained earlier, derive equations for the evolution of the small particle volume fraction Φ^s. Assume small chute inclination such that $\sin\theta \approx 0$.

Using (41.26) with (41.25) in dimensionless variables leads to a scalar conservation law for the small particle concentration $\Phi^s = \Phi(x, y, z, t) \in [0, 1]$, (see Exercise 41.10 for details)

$$\frac{\partial \Phi}{\partial t} + \frac{\partial}{\partial x}(\Phi u) + \frac{\partial}{\partial y}(\Phi v) + \frac{\partial}{\partial z}(\Phi w) - S_r(\Phi(1-\Phi)) = 0 \quad (41.30)$$

where S_r is the nondimensional segregation number. For the more general fluid saturated case (Thornton et al. 2006) $S_r = LB\hat{\rho}g\cos\theta/cHU$, where (as in Section 41.2) L, H, U are typical flow length, height, and down slope velocity scales, and, θ is the chute angle. The relative density difference $\hat{\rho}$ is a dimensionless measure of the density contrast between the granular phase and the fluid phase (often air) that fills the interstitial pore space between the particles. The main weakness of this theory is that no model is provided for the overburden pressure on each constituent, but only a series of physical and mathematical constraints are derived. Thornton et al. (2006) then obtain the simplest function satisfying all constraints with B a free dimensionless constant, which appears in this function (Exercise 41.9). Physically, B is a measure of the increased bed weight (overburden pressure) carried by the large particles compared to the small. Measuring pressure functions f^v directly in experiments is not easy, but these functions could be determined using particle-based simulations. We have carried out these simulations.*

Taken the simple form for f^v, see Exercise 41.9, the aforementioned segregation model gives very good agreement (using S_r as a fitting parameter) to both dry experiments performed by Savage and Lun (1988) and later liquid saturated experiments of Vallance and Savage (2000). The model predicts the correct dependence of the segregation length (distance before the large and small particles are fully segregated) on the background fluid density, and that no segregation takes place if the particles and fluid are matched in density. The question of how to obtain the velocity field in the segregation model (41.30) is a challenge discussed in Section 41.4.

* Thornton, A.R., Weinhart, T., Luding, S., Bokhove, O. 2012: Modelling of particle size segregation: calibration using DPM. *Int. J. Mod. Phys. C*.

41.4 Major Challenges

In this section, we will pose several challenges concerning the relevance of idealized studies of shallow granular flow and depth-averaged modeling, and the scope of particle segregation models and multiscale approaches to efficient and accurate modeling of shallow granular flows.

Firstly, the question is how relevant the detailed studies of granular flows under idealized conditions, as reviewed in Section 41.2, are to (dry) granular particle flows in the environment. In contrast to the cases reviewed, dry particulate flows in nature consist of non-uniform and polydispersed particles over terrain of varying roughness. Furthermore, an interstitial fluid may be involved, for example, when the soil of a hill or the rocky debris of a volcanic slope is saturated with water. The papers of Iverson and Denlinger (2001) and Denlinger and Iverson (2001, 2004) focus more strongly on debris flows with an interstitial fluid. In their depth-averaged model, fluid and solid stress components are taken into account, but in the dry limit with only a solid fraction their model is closely related to (41.18) and (41.20). Their model contains an extra stress component in the x–y-direction, which concerns the internal friction angle and also uses the more complex Earth pressure coefficient K, that is, (41.15). Additionally, extra nonlinear effects, due to the curvature of the bed, are taken into account but the basal friction coefficient is fixed to the tangent of the bed slope. One problem is, of course, that it is unclear what the dependence of the basal friction on the flow state is, for example, on the instantaneous Froude number and the layer depth. The research reviewed indicates that basal friction must be variable as the flow changes from supercritical to subcritical. It remains a somewhat open question which effects are most important to forecast environmental particle flows. Dalbey et al. (2008) mention that granular flows are relatively insensitive to the internal friction angle, and the key parameters are the bed friction angle and the mass. They explored the input uncertainty with statistical methods and using a depth-averaged model similar to the one presented in Section 41.2 produced hazard maps indicating danger regions for geophysical granular mass flows.

Secondly, measurements indicate that the velocity of shallow granular flows is depth-dependent (GDR MiDi 2004). Depth-averaged models keep only one degree of freedom in the vertical. When this restriction is too strong relative to the other approximations, it is useful to include some more degrees of freedom, as is done in Boussinesq water wave models. In addition, the solid's volume fraction is variable, especially in the horizontal direction when flow changes from super- to subcritical or from fast thin to slow thick flows. The extension of shallow granular models to include horizontal variation of the solid's fraction while still keeping a limited number of degrees of freedom in the vertical is of particular interest.

Thirdly, particle size segregation can have a major effect on the dynamics of shallow granular flows. Understanding this process is going to be a key step in improving the predictive power of existing models. There are two possible ways to couple the depth-averaged and segregation models. Fully 3D velocity fields could be reconstructed from a depth-integrated avalanche theory and solved in conjunction with 3D segregation equations. When one depth-averages the equations of motion, a series of assumptions about the vertical dependence of flow variables are made. These assumptions can be used to reconstruct fully 3D information from 2D depth-averaged theory and then employed in 3D segregation theory. A second approach is to depth-average 3D segregation Equation (41.30) and then direct coupling with the bulk flow theory (41.18) is possible. The coupling of a segregation model with shallow-water theories is thus an open research question, which will lead to further investigation in the next few years.

Finally, efficient DPMs would in principle eliminate the need to formulate continuum models altogether, but in practice this is impossible. In particular, the application of DPMs to flows with dead (static) and dynamic zones is of interest, in combination with formulating associated continuum models for shallow granular flow. Adequate determination of the normal and tangential coefficients of restitution is then required, yet efficient calculation of flows with non-uniform and polydispersed particles using DPMs is a challenge. New alleys of research are multi-scale modeling approaches (e.g., Van der Hoef et al. 2006). In these approaches macro-scale continuum models are coupled with local DPMs, to save computational time.*

References

Börzsönyi, T., Ecke, R.E. 2007. Flow rule of dense granular flows down a rough incline. *Phys. Rev. E* 76:031301.

Dalbey, K., Patra, A.K., Pitman, E.B., Bursik, M.I., Sheridan, M.F. 2008. Input uncertainty Input uncertainty propagation methods and hazard mapping of geophysical mass flows. *J. Geophys. Res.* 113:05203.

Denlinger, R.P., Iverson, R.M. 2001. Flow of variably fluidized granular masses across three-dimensional terrain 2. Numerical predictions and experimental tests. *J. Geophys. Res.* 106:553–566.

Denlinger, R.P., Iverson, R.M. 2004. Granular avalanches across irregular three-dimensional terrain: 1. Theory and computation. *J. Geophys. Res.* 109:01014.

Drahun, J.A., Bridgewater, J. 1983. The mechanisms of free surface segregations. *Powder Tech.* 36:39–53.

GDR MiDi. 2004. On dense granular flows. *Eur. Phys. J. E* 14:341–365.

Gray, J.M.N.T., Tai, Y.-C., Noelle, S. 2003. Shock waves, dead zones and particle-free regions in rapid granular free-surface flows. *J. Fluid Mech.* 491:161–181.

Hákonardóttir, K.M., Hogg, A.J. 2005. Oblique shocks in rapid granular flows. *Phys. Fluids* 17:077101.

van der Hoef, M.A., Ye, M., van Sint Annaland, M., Andrews IV, A.T., Sundaresan, S., Kuipers, J.A.M. 2006. Multi-scale modeling of gas-fluidized beds. *Adv. Chem. Eng.* 31:65.

Iverson, R.M. 1997. The physics of debris flows. *Rev. Geophys.* 35:245–296.

Iverson, R.M., Denlinger, R.P. 2001. Flow of variably fluidized granular masses across three-dimensional terrain 1. Coulomb mixture theory. *J. Geophys. Res.* 106:537–552.

Jenkins, J.T. 2006. Dense shearing flows of inelastic disks. *Phys. Fluids* 18:103307.

Kerswell, R.R. 2005. Dam break with Coulomb friction: A model for granular slumping? *Phys. Fluids* 17:057101.

Pouliquen, O., Forterre, Y. 2002. Friction law for dense granular flows: application to the motion of a mass down a rough inclined plane. *J. Fluid Mech.* 453:133–151.

Pouliquen, O., Vallance, J.W. 1999. Segregation induced instabilities of granular fronts. *Chaos* 9:621–630.

Savage, S.B., Hutter, K.J.F.M. 1989. On the motion of a finite granular mass of material down a rough incline. *J. Fluid Mech.* 199:177–215.

Savage, S.B., Lun, C.K.K. 1988. Particle size segregation in inclined chute flow of dry cohesionless granular material. *J. Fluid Mech.* 189:311–335.

Thornton, A.R., Gray, J.M.N.T., Hogg, A. 2006. A three-phase model of segregation in granular avalanche flows. *J. Fluid Mech.* 550:1–25.

Vallance, J.W., Savage, S.B. 2000. Particle segregation in granular flows down chutes. *IUTAM Symposium on Segregation in Granular Materials* (Eds. A.D. Rosato and D.L. Blackmore), pp. 31–51. Kluwer, Dordrecht, the Netherlands.

Vreman, A.W., Al-Tarazi, M., Kuipers, J.A.M., van Sint Annaland, M., Bokhove, O. 2007. Supercritical shallow granular flow through a contraction: Experiment, theory and simulation. *J. Fluid Mech.* 578:233–269.

* We thank L. Courtland, V. Orgarko, S. Rhebergen, and T. Weinhart for careful proofreading.

42

Turbidity Currents and Powder Snow Avalanches

Eckart Meiburg
University of California, Santa Barbara

Jim McElwaine
University of Cambridge

Ben Kneller
University of Aberdeen

42.1 Introduction

Powder snow avalanches and turbidity currents are particle-laden gravity currents (Simpson 1997) in which the particles are largely or wholly suspended by fluid turbulence. They are closely related to pyroclastic flows from volcanoes and dust storms in the desert. The turbulence is typically generated by the forward motion of the current interacting along the lower boundary of the domain, at the upper boundary with the ambient fluid, and within the shear layer above the velocity maximum (Kneller et al. 1999; Felix et al. 2005). The forward motion being in turn driven by the action of gravity on the density difference between the particle–fluid mixture and the ambient fluid. The ambient fluid is generally of similar composition to (and miscible with) the interstitial fluid, and consists of water for turbidity currents and air for avalanches. These flows are nonconservative in that they may exchange particles at the lower boundary by deposition or suspension, and may exchange fluid with the ambient by entrainment or detrainment (Figure 42.1). Such flows dissipate when the turbulence can no longer hold the particles in suspension and they are deposited on the lower boundary. When the turbulence is strong enough to maintain the particles in suspension, it is said to be *auto-suspending*; and when it is stronger yet and capable of entraining new material from the bed or the underlying dense flow, the current is said to be *igniting* (Bagnold 1962; Pantin 1979). Particle concentrations in the suspension cloud are usually sufficiently low (0.1%–7% by volume) that particle–particle interactions play a small or negligible role in maintaining the suspension (Bagnold 1954). However, this upper limit is contentious and many authors would put it higher (Lowe 1982). At these concentrations in turbidity currents the extra mass of the suspended particles is small relative to that of water, so the Boussinesq approximation, where density differences are considered negligible in inertia terms, is reasonable. In contrast, in powder snow avalanches the density difference between the suspended particles and the interstitial air is high and the particles may carry most of the flow's momentum. Nonetheless, due to the extreme difficulty in estimating particle concentrations in natural flows, there remains considerable uncertainty—and debate—concerning the particle loading in large submarine turbidity currents and the validity of the Boussinesq approximation.

42.1.1 Turbidity Currents

Although several nineteenth century investigations of the ocean (such as the Challenger expedition of 1872–1876) suggested that deep sea sediments consisted almost exclusively of pelagic sediments, sands had been known to occur on the ocean floor since the Gazelle expedition of 1874–1876 (Andrée 1920), and the occurrence of submarine cable breaks along a canyon axis

Handbook of Environmental Fluid Dynamics, Volume One, edited by Harindra Joseph Shermal Fernando. © 2013 CRC Press/Taylor & Francis Group, LLC. ISBN: 978-1-4398-1669-1.

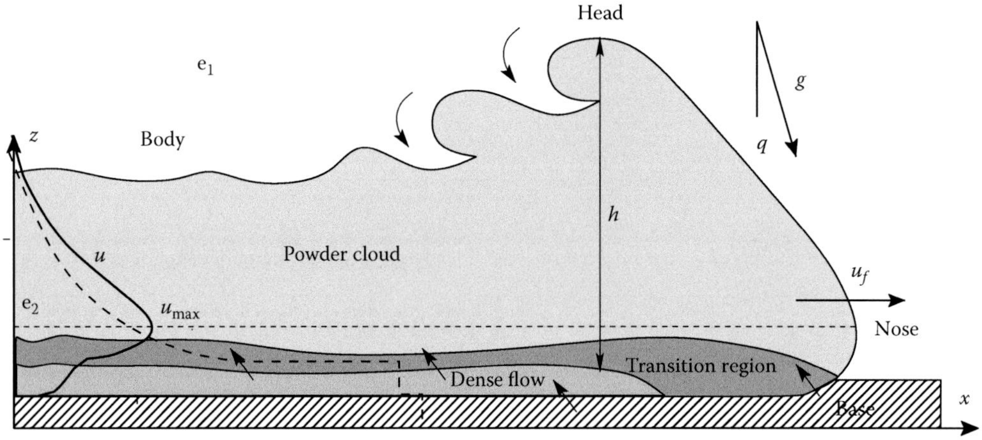

FIGURE 42.1 Schematic of a current showing generalized velocity and density profiles. Entrainment of ambient fluid is occurring on the upper surface. Either erosion or deposition can occur between the base region, dense flow, transition region, and powder cloud. Note that not all layers and processes will occur in all currents.

(Milne 1897) indicated energetic submarine flows. The existence of dense, sediment-driven undercurrents was first postulated by Forel (1885) who proposed that a sub-aqueous canyon in Lake Geneva had been created by such flows generated by the Rhone River. A similar mechanism was suggested by Daly (1936) for the formation of submarine canyons; the name turbidity current was coined by Johnson (1939). In a geophysical context, turbidity currents are important as agents of sediment transport in deep sub-aqueous environments such as deep seas, lakes, and oceans (Figure 42.2). The recognition of their potential importance as agents of sediment transport to the deep sea is due to Kuenen (1938, 1951) and Kuenen and Migliorini (1950) who conducted the first experiments on turbidity currents. Indeed, turbidity currents, along with submarine landslides, are the principal means by which sediment is transported into deep water. In natural sub-aqueous environments, the particles consist of sediment that is moved downslope by turbidity currents to be redeposited in deeper water. The sediment consists of fragments of rock or minerals delivered to shallow marine environments by rivers or calcium carbonate particles (mainly fragments of invertebrate shells), formed in shallow marine environments. Transport distances range from a few hundreds of meters down the fronts of deltas to thousands of kilometers within the ocean (e.g., North Atlantic Mid-Ocean Channel [Klaucke et al. 1998]).

Turbidites (the geological term for the deposits of turbidity currents) make up a large proportion of the sediments in the deep sea, and also in deep lakes such as Lake Baikal (Nelson et al. 1995), often accumulating thickness of hundreds to thousands

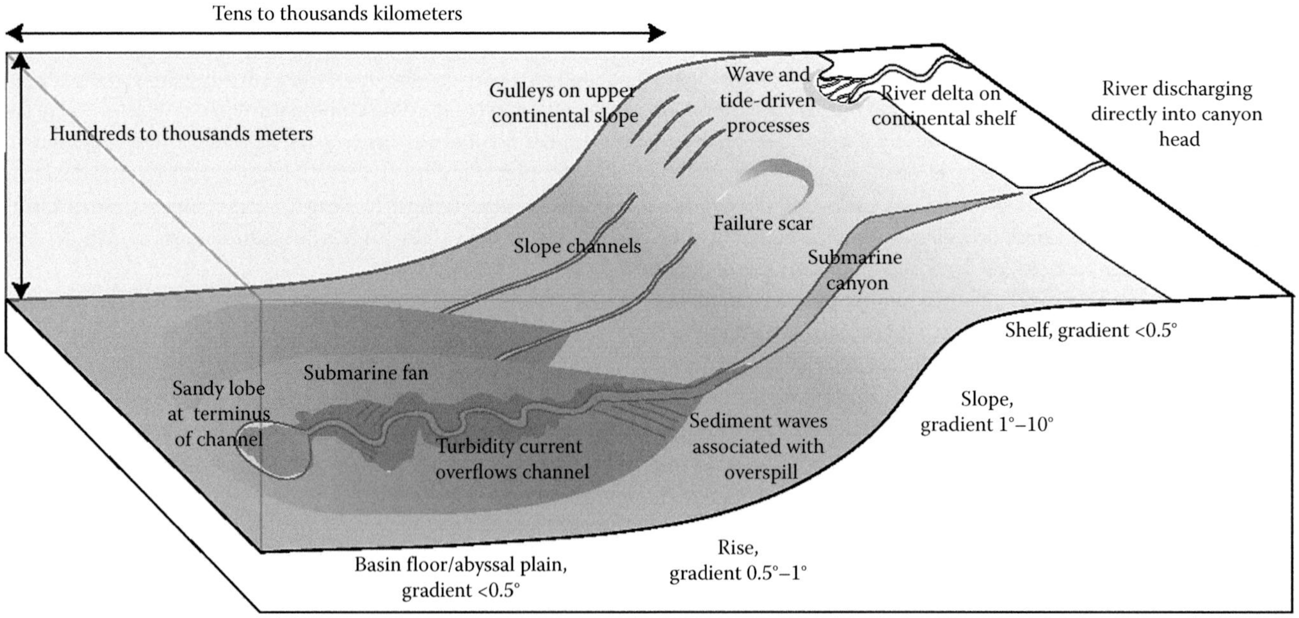

FIGURE 42.2 Context of turbidity currents on the margins of continents and intra-continental basins, including deep lakes.

of meters by deposition from large numbers of individual turbidity current events. Such events are typically infrequent (at intervals of years to millenia) and sporadic, but may generate accumulations with volumes up to millions of km^3 over periods of the order of 10^4–10^6 years [e.g., Bengal Fan; Curray et al. (2003)]. Many oil and gas reservoirs consist of ancient deposits of turbidite sand in the subsurface, and certain gold deposits are turbidite-hosted (Keppie et al. 1987). Sedimentation in water reservoirs, with consequent loss of storage capacity, is largely due to turbidity currents (Fan 1986; De Cesare et al. 2001). In the ocean turbidity currents not infrequently sever submarine cables (Milne 1897; Heezen and Ewing 1952; Dengler et al. 1984), and sea-floor equipment and instrumentation may be damaged or destroyed even by fairly small turbidity currents (Inman et al. 1976; Prior et al. 1987; Khripounoff et al. 2003), which—along with their inaccessibility and unpredictability—accounts for our relative ignorance of the properties of such currents in *Nature*.

42.1.2 Powder Snow Avalanches

The effects of avalanches have been documented for millenia, especially in the heavily populated European Alps. The Roman poet Silius Italicus recounts how Hannibal's army lost 18,000 men, 2,000 horses, and several elephants crossing the Alps in 218 BC. In 1618 the Swiss village of Plurs was completely buried by an avalanche killing 2427. During World War I more than 60,000 Italian and Austrian troops were killed by avalanches often triggered by artillery fire. The worst documented accident occurred in Peru in 1970. An earthquake triggered an avalanche on Nevado de Huascarán which flowed into a lake causing the moraine dam to fail. The resulting debris flow traveled down the valley completely burying the town of Yungay killing 20,000 people. For these reasons great efforts have been devoted to understanding when avalanches will occur and what their consequences are. One aim is the construction of avalanche hazard maps, which regulate land use, for example, forbidding the construction of dwellings in areas of high risk. These hazard maps are mostly based on statistics of past events, but changing land use and changing climate make past statistics unreliable and in many parts of the world such statistics are unavailable (Höller 2007). Another way to reduce risk is the construction of protection structures which either act to prevent avalanches starting or control their path. The design of such structures and the manufacture of avalanche hazard maps for areas with non-existent or unreliable statistics is a major motivation for research into avalanche dynamics, the ultimate objective of which is the development and validation of accurate models that can predict the motion of avalanche over complex topography and their interaction with structures such as pylons, defense dams, and buildings (Jóhannesson and Tracy 2001). In Europe and North America, avalanche hazard zoning and the construction of avalanche defenses have made deaths in buildings much less common. The most frequent causalities now are skiers, snow boarders, and snow mobilers, but in other countries such as Iran, Turkey, Pakistan, and Afghanistan large disasters killing hundreds of people still occur (Naaim and Gurer 1998).

Avalanches can be classified in different ways according to the type of snow, the form of initiation, and the nature of the flow (UNESCO 1981; Hopfinger 1983). In general, an avalanche consists of two layers: a dense layer where interaction with the air is unimportant and the stresses are generated by contacts between the grains, and a powder cloud where the volume fraction of snow is small and grain–grain interactions can be ignored. Typical densities in the dense layer are 200–500 kg m^{-3} (Louge et al. 1997) and <20 kg m^{-3} in the powder cloud. Avalanches with wetter snow or those moving more slowly may have no powder cloud, and in other cases all of the snow in the dense layer may be entrained into the powder cloud resulting in the absence of a dense layer. It is also possible for topographic effects to cause a complete separation of the layers with the dense layer following the contours of a gully bottom, whereas the powder cloud may jump into another drainage.

Damage in most cases is caused by the dense part of an avalanche due to its high densities, but in the largest avalanches the speeds are sufficiently high that the powder cloud is very destructive (Gruber and Margreth 2001). Snow avalanches are easy to document compared to turbidity currents, but they occur in inaccessible and dangerous locations, often at times of bad weather. Observation instruments frequently malfunction in the harsh conditions or are destroyed. Measurements of powder snow avalanches are particularly difficult, as these occur less frequently and are usually very large.

42.1.3 Overview

This chapter will cover the structure of turbidity currents and powder snow avalanches as deduced from natural flows and experiments, theoretical approaches to modeling, and some current controversies. It will not cover other types of particulate gravity currents such as pyroclastic flows, neither will it cover the dense flows that often underlie gravity currents which more closely resemble debris flows, rock avalanches, or dense snow avalanches. Henceforth, we refer solely to turbidity currents and powder snow avalanches when mentioning gravity currents. Various topics in gravity and turbidity current research have previously been reviewed. First and foremost, the book by Simpson (1997) offers a beautiful and accessible introduction to the field. The chapter by Rottman and Linden (2001) reviews the basic scaling laws and force balances for idealized compositional gravity currents. Several articles by Huppert review various aspects of gravity and turbidity currents and Huppert (2006) also includes dense granular flows. Middleton (1993) gives an elegant review of the literature on turbidity currents and their deposits, including experimental results and field data up to that time. Parsons et al. (2007) summarizes work in the ocean, and Meiburg and Kneller (2010) is a recent review of turbidity current ranging from field data to simulations. Hopfinger (1983) summarizes avalanche research up to the time, and Harbitz et al. (1999) review avalanche models. Höller (2007) reviews practical mitigation measures including defenses and hazard zoning.

42.2 Field Observations

42.2.1 Powder Snow Avalanches

Over the previous 30 years, measurement campaigns have been carried out in Norway, Japan, France, Austria, and Switzerland to observe snow avalanches (Issler 2003; Barbolini and Issler 2006). These can either be artificially initiated by explosive charges or the instruments can be designed to record naturally occurring avalanches, usually triggered by seismic sensors (Figure 42.3). There are several different measurement techniques useful to investigate the powder cloud. The most frequently used is videogrammetry. This can range in quality from a single hand-held camera without calibration to several synchronized cameras filming from precisely determined locations with known ground control points in the field of view and an accurate digital terrain model. By tracking points in stereo on the surface of the powder cloud, it is then possible to calculate the avalanche's height, width, volume, and speed down the avalanche track. Vallet et al. (2004) report for one small avalanche the speed rising from 10 to 50 m s^{-1} and then decreasing to 10 m s^{-1}, while the height near the front increases linearly to 20 m and the volume grows cubically to 7×10^6 m^3. Nishimura et al. (1995) demonstrate an ingenious method for calculating the height from only one camera if there is a clear shadow and the exact time of the avalanche is known.

Once a powder snow avalanche is developed, its dynamics are primarily influenced by its interaction with the ambient air. The interaction with the basal surface is mainly through deposition and entrainment of snow. This suggests that measurements of the air flow, outside and inside an avalanche, would be particularly useful for understanding their dynamics. Conventional sensors for measuring air velocities are hot-wire probes, ultrasonic anemometers, and Pitot tubes. Hot-wire probes are very fragile and not suitable for use outdoors. Ultrasonic anemometers have been used successfully in front of an avalanche (Nishimura et al. 1989), but they are unreliable inside due to the snow particles, as they are easily destroyed and their frequency response is low. For these reasons subsequent work on air flow measurements has focused on Pitot type sensors (Nishimura and Ito 1997). However, the data from these sensors are difficult to interpret, because of the complex dependence of the measured pressure on the local pressure and velocity, and the resonance effects of the instruments. Measurements have also been made at the tube-bridge site in Bschlabs, Austria (Rammer et al. 1998), but there were difficulties with low sampling frequency and with the sensor becoming blocked with snow. McElwaine and Turnbull (2005) developed a robust Pitot probe with a steady reference pressure, high frequency response, and simple geometry that was deployed in Vallée de la Sionne (Figure 42.3) and reported speeds for five avalanches from 20 to 70 m s^{-1}.

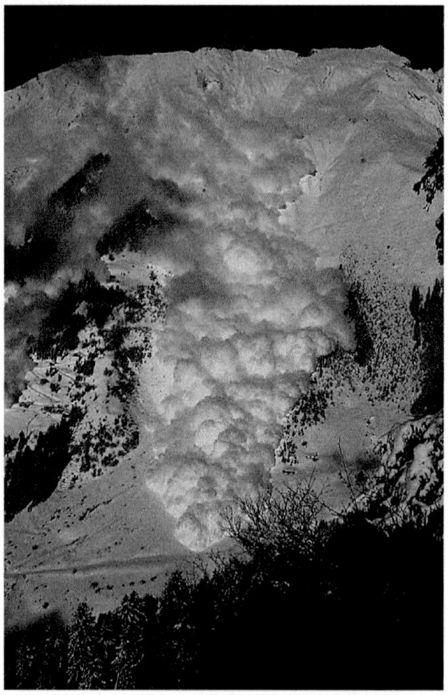

FIGURE 42.3 Typical dry-mixed avalanche artificially released at the Vall'ee da la Sionne test site in the west of Switzerland.

Radar systems for studying avalanches can be divided into two types: (1) those that are fixed in the avalanche slope, look upward, and reflect off layers in the snow pack; and (2) those that are mounted, often on a counter-slope, so as to see a large fraction of the avalanche track. Gauer et al. (2007) discusses the history of the second type of radar systems and recent observations from Himmelegg Austria, Ryggfonn Norway and Vallée de la Sionne, Switzerland. The earliest radars were continuous wave and used only for avalanche detection. The next generation introduced continuous wave Doppler systems that give a velocity distribution averaged over the entire region in view of the radar. The third generation were pulsed Doppler radar systems that split the return signal from a pulse into short intervals of time corresponding to *range-gates* along the avalanche track 25–150 m in size. This gives information on the velocity distribution averaged over each individual range-gate. Ash et al. (2010) discusses the most recently deployed system involving a phased-array frequency modulated Doppler system. This uses upward and downward chirps to provide a 1 m resolution along the slope, and the array of receiving elements gives a horizontal resolution of order 10 m. These radar systems usually operate at different frequencies depending on the target. 5.8 GHz corresponds to a wavelength of ≈50 mm and therefore penetrates the powder cloud and reflects of the transition zone or top of the dense layer (Figure 42.1). 35.8 GHz corresponds to a wavelength of ≈8 mm and reflects off the larger particles in the powder cloud. A difficulty with the interpretation of these signals is that the amplitude of the return signal depends in detail on the micro-structure of the snow and therefore only the velocity of maximum amplitude and the range of velocities can be determined. Rammer et al. (2007) compares radar measurements of velocity with other instruments in the powder cloud and in the dense layer and shows that the velocities are in good agreement and lie in the range of 20–70 m s^{-1}.

Gubler and Hiller (1984) developed upward pointing FMCW (frequency-modulated continuous wave) radars, which have been deployed in pairs at the Ryggfonn and Vallée de la Sionne test sites. They present the time evolution of the snow depth, showing the initial erosion and final deposition as an avalanche passes, and also provide information about the density in the dense layer, the transition region, and bottom of the powder cloud. Gauer et al. (2009) use FMCW radar data to show that the transition region and powder cloud may move at distances 200–300 m in front of the dense flowing snow and may directly entrain particles from the snow-pack. Correlating the signals between closely spaced FMCW sensors also allows the calculation of the local speed.

Kern et al. (2009) and Sovilla et al. (2008) discuss direct measurements of impact pressures using load cells, and velocity profiles using correlation between opto-electronic sensors. Though they reveal some information about the intermittent nature of the transition layer, densities in the powder layer are too low for useful signals. Louge et al. (1997) has pioneered measurements of density in various media using capacitance probes. Though they should work in theory in the powder cloud, problems of instability have so far prevented this and there are unfortunately few direct measurements of density.

Vallet et al. (2001) use helicopter-based photogrammetry to map snow surface height before and after avalanches at Vallée de la Sionne. Photogrammetry is expensive and limited by the poor contrast of the snow surface and the frequent destruction of the necessary ground control points. More recently, Sovilla et al. (2010) have used laser scanning, which is much more accurate (<100 mm in three dimensions) and easily achieves good coverage. These measurements can provide valuable information about erosion and deposition by the dense avalanche core, but deposits from the powder cloud are usually of the order of 10 mm and too small to be measured with laser scanning. Manual methods can be used, but they are difficult to apply in hazardous areas, have poor spatial resolution, and are extremely time-consuming. Nevertheless, observations after avalanches can distinguish areas where the powder cloud has moved separately from the dense core either by the nature of the surface deposit or by damage to vegetation (Gauer et al. 2009). Such separation occurs on the side of steep valleys or when a valley makes an abrupt turn. The dense core follows the topography whereas the powder cloud continues and may even jump to a different drainage. Such behavior makes the effectiveness of protection dams against powder snow avalanches questionable.

42.2.2 Turbidity Currents

Many of the features seen on the sea floor have been produced by turbidity currents. Erosional features range from small gullies, a few tens of meters deep and hundreds of meters wide (Hall et al. 2008, and references therein) to submarine canyons several kilometers wide and hundreds of meters deep (Inman et al. 1976; Babonneau et al. 2002). Depositional features include laterally extensive, sheet-like deposits of the abyssal plains, and submarine fans, which range in size from a few kilometers across to the Bengal Fan which is over 3500 km end to end. Channels within these systems similarly range up to thousands of kilometers in length. These channels often have levees analogous to those of rivers, formed by overspill from the channel; deposition within the channel and on the levees often results in elevation of the channel-levee system above the surrounding fan surface (Normark et al. 1997). Various bed-forms may be produced by turbidity currents, especially within channels, that are similar to those produced by unidirectional flow in shallow water. Larger-scale sediment waves may also be generated, especially where turbidity currents pass over topographic inflections such as the crests of submarine levees, or the base of the continental slope (Lee et al. 2002), generating fields of sediment waves with heights of tens of meters, wavelengths typically around 1 km and crests oriented perpendicular to flow that migrate upstream (Meiburg and Kneller 2010).

Our understanding of such deposits on the sea floor is based on remote observation techniques including side-scan sonar images, multibeam bathymetry surveys, shallow coring, and industrial seismic surveys, such as those used to assess seafloor hazards to drilling. Outcrops of ancient turbidite deposits have generated numerous qualitative models, and considerable conjecture about the nature of the flows responsible for their deposition (Mulder and Alexander 2001). In a fluid mechanical context, ancient turbidite deposits may provide benchmarking data for future numerical simulations.

Evidence of the nature of turbidity currents in the ocean has until very recently been largely indirect, often as result of the breakage of submarine cables. A well-known example is the event of November 18, 1929, off the Grand Banks of Newfoundland (Piper et al. 1999). This followed a magnitude 7.2 earthquake with an epicenter beneath the upper continental slope at around 600 m water depth, which liquefied large areas of the sea floor. The total volume of material removed by the resulting failure was perhaps 100 km^3. Some of this material formed a turbidity current, by entraining water, and flowed down the slope valleys, eroding 50–100 km^3 of sand en route. Increasing wavelengths of bed-forms down the first 100 km of the valley indicate an accelerating current, despite a decrease in gradient from 8° to 1°. Below about 4700 m water depth the flow became depositional, probably triggered by radial expansion as it began to exit the valley, leading to a decrease in turbulence to levels below that necessary to maintain all the material in suspension. The resulting deposit has an estimated volume of 150–175 km^3, and extends for over 1400 km from the presumed source. The current broke a number of cables in sequence down the continental slope over the 13 h following the earthquake; the timing of these breaks yields estimates of the maximum front velocity of around 19 m s^{-1}. Indirect estimates of maximum flow thickness range from 300 to 400 m. This suggests overall Reynolds numbers of O (10^9–10^{10}) on the slope.

A smaller event (>8 × 10^6 m^3) occurred near the mouth of the Var River SE, France, in 1979 (Dan et al. 2007). The subsequent current was around O (10^2) m thick and the average speed (again estimated from cable breaks) was 7.4 m s^{-1} over the first 95 km and 1.7 m s^{-1} over the next 22 km on a slope of only 0.15° (Piper and Savoye 1993).

The most complete picture of any marine currents to date comes from currents in the Monterey and Hueneme Canyons, California, using acoustic Doppler current profilers moored 70 m above the sea bed, yielding vertical profiles of downstream velocity (Xu et al. 2004; Xu 2010). None of the flows was seismically triggered. Peak velocities up to 2.8 m s^{-1} were measured. Flow thickness increased down-canyon whereas the height of the velocity maximum decreased. In some cases the peak velocity appeared to increase down-canyon, suggesting a self-acceleration due to erosion of bed material. Flows persisted for several hours each, but the duration of peak flow decreased down-canyon and became more surge-like.

Measurements in lakes and reservoirs have been made by Best et al. (2005), Chikita (1989), Gould (1951), Normark (1989), amongst others. Other turbidity current data comes from the Congo-Zaire submarine fan (Khripounoff et al. 2003), and Bute Inlet, a British Columbia fjord (Prior et al. 1987), and gives typical speeds of 1–4 m s^{-1} and flow thicknesses ranging from 30 m to at least 150 m on slopes generally less than 1°.

42.3 Experiments and Computer Simulations

42.3.1 Similarity Criteria

To better understand turbidity currents and powder snow avalanches, experiments and computer simulations are invaluable and there is a long history of such work. As in all modeling it is essential to understand the relevant similarity criteria. Table 42.1 summarizes a range of experiments and simulations on suspension currents in terms of the relevant dimensionless numbers that characterize them and gives representative values for natural flows.

The nondimensional density difference can be defined as $\Delta = (\rho_2 - \rho_1)/\rho_1$, where ρ_1 is the density of the ambient fluid and ρ_2 is a representative density of the suspension. The particle concentration by volume is low in both fully developed powder snow avalanches and turbidity currents, because the density of air is a thousand times less than that of water, Δ is much larger in avalanches. When Δ is small, the extra inertia of the suspended particles can be ignored at lowest order and the Boussinesq approximation is valid. In addition, the mass-averaged velocity field can be treated as divergence free which is not the case for large Δ (Birman et al. 2005; Étienne et al. 2005). Both natural powder snow avalanches and the snow–air experimental currents are non-Boussinesq, and density differences are important in both dynamic and buoyancy terms (Boussinesq 1903). This contrasts with experiments where a dense ambient fluid is used, and the currents are in the Boussinesq regime. In a snow–air suspension current, the relatively heavy snow particles carry a significant proportion of the current's momentum. For suspension currents with a relatively low density ratio, for example, saw dust-water currents (Bozhinskiy and Sukhanov 1998b), the inertia of the particles is small compared with the inertia of the ambient fluid. Hampton (1972) and Ancey (2004a) achieved density differences that approach the non-Boussinesq case by using high concentration suspensions in water.

The Reynolds number of a suspension current can be defined as

$$\mathrm{Re} = \frac{u_f h \rho_1}{\mu} \tag{42.1}$$

where

μ is the dynamic viscosity of the ambient fluid
u_f is the front velocity
h is the head height (Figure 42.1)

TABLE 42.1 Magnitudes of Five Dimensionless Numbers That Characterize Suspension Current Experiments

Experiment	Materials	Re_p	Ri	St	Δ	Re
Powder snow avalanches	Snow–air	3000	1	0.02	10	10^9
Turbidity currents	Sediment–water	150	1	10^{-4}	0.01	10^{10}
Bozhinskiy and Sukhanov (1998a)	Sawdust and aluminum mixture–air	0.1	20	0.03	1	10^3
Hampton (1972)	Kaolinite/water	—	<0.5	—	0.1	—
Turnbull and McElwaine (2008)	Snow–air	150	1	10	10	10^5
Turnbull and McElwaine (2010)	Polystyrene–air	150	2	1	5	10^4
McElwaine and Nishimura (2001)	Ping-pong balls	20,000	2	10	50	10^7
Tochon-Danguy and Hopfinger (1975)	Salt–water	0	2	0	0.01	10^3
Beghin and Olagne (1991)	Salt–water	0	5	0	0.02	10^4
Beghin et al. (1981a)	Sand–water	150	5	10^{-4}	0.01	10^3
Hermann et al. (1987)	Polystyrene powder-water	1.5	0.1	10^{-4}	0.002	10^4
Ancey (2004a)	Saw dust suspension–water	50	1.7	0.006	0.05	10^4
Gröbelbauer et al. (1993)	Gas–gas	0	0.1	0	20	10^4
Étienne et al. (2005)	2D DNS	0	0.1	0	20	10^5
Necker et al. (2005)	3D DNS	0	1	0.02	0	10^3

Re_p, particle Reynolds number; Ri, Richardson number; St, Stokes number; Δ, density ratio; and Re, Reynolds number.

The similarity criteria for previous experiments are compared with those for powder snow avalanches and with those for the snow currents and the polystyrene currents in the present study.

The Reynolds number measures the ratio of inertial forces to viscous forces and determines the degree of turbulence in the flow. For all of the experiments and simulations listed in Table 42.1, the Reynolds number of the flow is significantly lower than the Reynolds number for a natural flow (the ping-pong ball avalanches of McElwaine and Nishimura [2001] provide the closest similitude). The Reynolds numbers are high enough, however, to ensure that the flows are within a turbulent regime and that viscous drag from the basal surface can be ignored as a first approximation. Thus it is reasonable to expect rough agreement of, appropriately scaled, front speeds, for example, but other features such as the nose height will not match up as this is thought to decay as a power of Re.

The bulk Richardson number (hereafter referred to as the Richardson number, Ri) determines whether a sheared interface between two fluids is stable or unstable (Ellison and Turner 1959). It can be thought of as the ratio of potential energy to kinetic energy of a parcel of the dense, particle-laden fluid at the interface if it moves up in to the ambient fluid. The bulk Richardson number is defined as

$$\mathrm{Ri} = \frac{g' h \cos\theta}{u_f^2} \tag{42.2}$$

The reduced gravity is $g' = g\Delta$. A dense flow will entrain air on the upper surface and become suspended if the Richardson number is low enough, and subsequently the suspension current will maintain the particles in suspension and further entrain air if the Richardson number stays low enough. Thus the value of Ri provides an indication of the stability of the flow. A local gradient Ri can also be defined in a continuous stratification and linear stability analysis of a stratified shear flow shows that the flow is unstable to 2D disturbances if the gradient Ri < 1/4 (Turner 1973).

Some experiments modeling powder clouds (Beghin et al. 1981a; Beghin and Olagne 1991) are characterized by values of Ri of order 5 (Table 42.1) and the interface is relatively stable. Over most of its lifetime a powder snow avalanche typically has values of Ri less than 1, where the interface will be locally unstable (Simpson 1997), and consequently for the natural flows there is a higher entrainment rate of ambient fluid compared with Beghin's work. In the early stages of powder snow avalanche motion Ri is greater than 1, and Beghin's experiments provide a closer match. However, these elevated values of Ri in a developing powder snow avalanche are due to the high density ratios and low velocities, whereas the Beghin et al. (1981a), Beghin and Olagne (1991) experiments were carried out in water resulting in very low density ratios and thus elevated Ri. Turnbull and McElwaine (2008) performed experiments of snow particles in air on steep slopes (between 40° and 90°) to keep a low Ri while preserving the particle-ambient density ratio ($\Delta \approx 200$ initially).

There are two possible drag regimes for a particle of diameter d, mass m moving with speed v relative to the interstitial fluid. In a viscous regime the drag varies linearly with velocity and has an associated time scale

$$\tau_1 = \frac{m}{d\mu} \tag{42.3}$$

(Batchelor 1967). At higher speeds, the drag is dominated by pressure forces resulting in form drag and varies quadratically with velocity. In this case the associated time scale is

$$\tau_2 = \frac{m}{\rho_2 d^2 u} \tag{42.4}$$

The ratio of the viscous and form drag time scales is

$$\frac{\tau_1}{\tau_2} = \frac{\rho_2 u d}{\mu} \tag{42.5}$$

which is the particle Reynolds number, Re_p, of the flow. The particle Reynolds number determines whether the drag is dominated by viscous or pressure forces. For experiments such as Bozhinskiy and Sukhanov (1998a) and Hermann et al. (1987), the particle Reynolds number is very low ($\ll 10$) as a result of using very fine particles. With very small particles and a low Re_p, the drag force exerted on the particles from the ambient fluid is almost entirely through viscous forces. For a turbidity current, Re_p is an order of magnitude larger and for a powder snow avalanche Re_p is even larger and therefore viscous drag forces play a relatively minor role compared to the form drag of the particle. The ping-pong ball experiments (Nishimura et al. 1998) modeled the interaction between snow particles and ambient air relatively well with a very high particle Reynolds number $\mathrm{Re}_p \approx 2 \times 10^4$ (compared with 3000 for powder snow avalanches). For values $500 < \mathrm{Re}_p < 10^5$, the regime for powder snow avalanches, the drag coefficient for a spherical particle is essentially independent of Re_p (Batchelor 1967).

The Stokes number is used as a measure of the interplay between the particles and the fluid and the degree to which the suspension behaves as a two-phase or a single-phase flow (Ancey 2007). This depends on the drag regime and is defined in terms of the particle drag time scale τ and time scale of the flow h/u

$$\mathrm{St} = \frac{\tau u}{h} \tag{42.6}$$

Thus for the viscous drag regime (where $\tau = \tau_1$, Equation 42.3) and form drag regime (where $\tau = \tau_2$, Equation 42.4), we have Stokes numbers

$$\mathrm{St}_1 = \frac{\rho_2 u}{\mu}\frac{d^2}{h}, \quad \mathrm{St}_2 = \frac{\rho_2 d}{\rho_1 h} \tag{42.7}$$

respectively, where $m = \rho_2 d^3$. With high values of Stokes number, the drag is small and the particle is unaffected by fast variations in the fluid velocity. Thus, the particle and fluid have a weak interplay and the suspension has a strong multiphase character. For low values of Stokes number, the motion of particles and fluid is strongly linked and the suspension behaves as a single phase (Batchelor 1989; Ancey 2007). Experiments in air typically have a higher Stokes number compared to either avalanches or turbidity currents, but this is usually not a problem for experiments in water, because of the much higher viscosity. For experiments with dissolved salts and in some simulations the Stokes number is zero.

One further consideration is the sedimentation velocity of a particle normal to the slope, v_t. In order to maintain a suspension, the vertical fluctuations in the flow velocity must be the same order of magnitude as the settling velocity. For turbulent fluctuations of 10% of the characteristic large-scale flow velocity u, suspension will be maintained if $u > 10v_t$. Thus the requirement for the flow to be well suspended is that the nondimensional group v_t/u is small, which is $\mathrm{St}_1/\mathrm{Fr}^2$ for viscous drag and $\sqrt{\mathrm{St}_2}/\mathrm{Fr}$ for form drag; thus, it is satisfied if the Stokes number is small since $\mathrm{Fr} = O(1)$ for a flow at equilibrium.

42.3.2 Velocity and Turbulence

A key result of most experiments is that the front velocity is largely independent of slope angle. This is because the increased driving force of gravity on steeper slopes is countered by entrainment of the ambient fluid. The vertical structure of density currents is analyzed by Stacey and Bowen (1988). The mean velocity structure of turbidity currents consists of an inner (near-wall) region with a positive velocity gradient, similar to a conventional turbulent boundary layer, and an outer region (shear layer), generally many times thicker than the inner region, with a negative velocity gradient and shear stress of opposite sign (Figure 42.1). The fractional height of the velocity maximum appears to be lower for Froude-supercritical flows than for subcritical flows (Sequeiros et al. 2009; Xu 2010). The velocity structure has been compared to that of plane turbulent wall jets (Parker et al. 1987; Kneller and Buckee 2000; Gray et al. 2005; Leeder et al. 2005). Launder and Rodi (1983) advocated the use of $y^{1/2}$ (the height at which the downstream velocity falls to half its maximum) as a characteristic length scale for wall jets. However, this yields a rather unsatisfactory similarity collapse of velocity profiles derived from laboratory experiments (Kneller and Buckee 2000; Gray et al. 2005), suggesting that the shear layer does not possess a Gaussian profile. In fact, for some currents the shear layer profile is almost linear (Ellison and Turner 1959). Xu (2010), combining data from full-scale currents in the ocean and model currents in the laboratory, obtained a good similarity collapse of the velocity profiles when the streamwise velocity was normalized by the depth-averaged velocity and the elevation was normalized by the turbidity current thickness; this can be generalized to an empirical function $Y = \exp(-\alpha X^\beta)$ for the jet region above the velocity maximum (Xu 2010).

Turbulent kinetic energy profiles in turbidity currents are similar to those of saline gravity currents, being close to zero at the height of the downstream velocity maximum (Kneller et al. 1999; Kneller and Buckee 2000; Gray et al. 2005; Leeder et al. 2005), reflecting the dominance of turbulence production by shear related to the mean streamwise velocity profile (Felix et al. 2005). There is little data available, however, for powder snow avalanches, and it may be that their strongly non-Boussinesq character significantly changes the turbulence structure.

42.3.3 Density

The sediment concentration usually decreases monotonically with height from the bed. Commonly, the gradients of concentration also decrease with height, though there may be a more step-like change from a dense basal flow into a more dilute upper region (Figure 42.1; see review in Kneller and Buckee [2000]). These results are for experiments in water, but there are little data for experiments in air or for natural flows. The vertical distribution of suspended sediment shows a much weaker dependence on the Rouse number (a function of the ratio of the shear velocity to the sediment fall velocity) than is the case in open-channel suspensions (Parker et al. 1987). Ultimately, the vertical mass flux of sediment is controlled by the balance between the turbulent kinetic energy and the potential energy of the suspended sediment. Leeder et al. (2005) proposed a criterion for the maintenance of suspension based on the ratio of maximum vertical turbulent stress to immersed weight of suspended load over unit bed area. Suspended sediment distributions may be highly unsteady, however, and controlled by the ratio of particle settling velocity to the local upward vertical turbulent velocity associated with coherent flow structures (Baas et al. 2005).

42.4 Theoretical Models

42.4.1 Dimensional Analysis

Much of the elegant experimental and theoretical research carried out over the last three decades demonstrates the ability of dimensional analysis to provide fundamental insight into the dynamics of gravity and turbidity currents (Rottman and Linden 2001). First and foremost, the celebrated result that the front velocity u_f of a lock-exchange gravity current is proportional to the square root of the reduced gravity and the front height

$$u_f \propto (\tilde{g}h)^{1/2} \tag{42.8}$$

follows from dimensional considerations of the balance between inertial and buoyancy forces. Gröbelbauer et al. (1993) performed experiments on gasses with density ratios up to 20 on flat slopes and proposed $g' = g\dfrac{\rho_2 - \rho_1}{\rho_2 + \rho_1}$ which is correct in the limit of $\rho_1 = \rho_2$ and $\rho_1 = 0$. When a finite volume V of material is released, the length scale of the flow will scale as $V^{1/3}$ ($V^{1/2}$ in a 2d geometry), and one therefore expects

$$u_f \propto \tilde{g}^{1/2} V^{1/6} \tag{42.9}$$

This scaling has been tested with field observations in Turnbull and McElwaine (2007), laboratory experiments in Turnbull and McElwaine (2010), and even ping-pong ball avalanches in McElwaine and Nishimura (2001). The classical analysis by Benjamin (1968) finds that the proportionality factor, commonly referred to as the Froude number Fr of the current, has a value of $\sqrt{2}$ for inviscid flows in deep ambients, and this result was extended by McElwaine (2005) and shown to hold on slopes of any angle even with internal motions, that is, the hydrostatic assumption of Benjamin is not needed. The remarkable similarity between the heads of experiments, simulations, and field observations is shown in Figure 42.4.

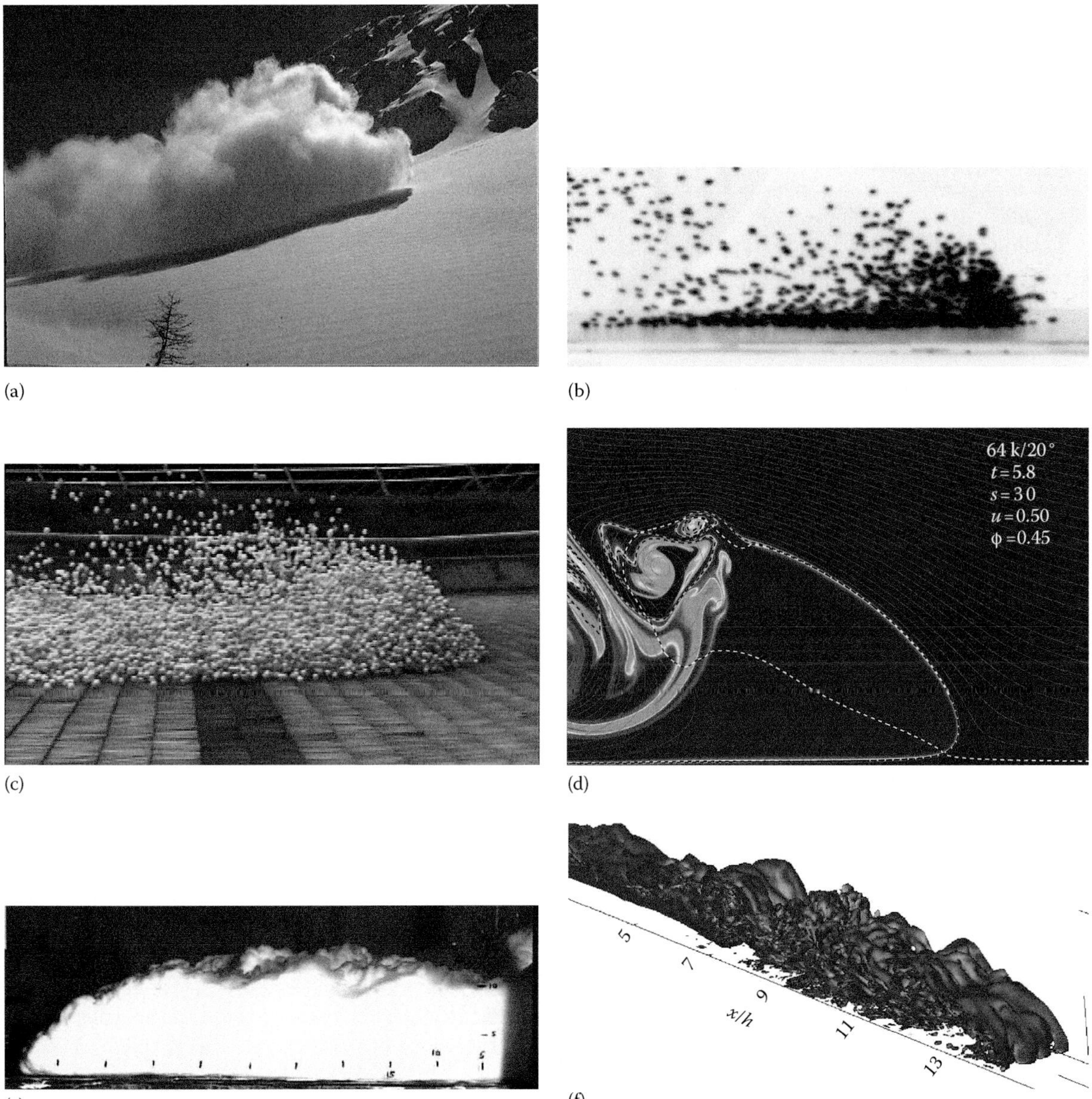

FIGURE 42.4 (See color insert.) (a) Head of a powder snow avalanche (T. Castelle, Cemagref). (b) Head of a 1 L polystyrene ball avalanche 0.05 m high on 70° slope. (c) Head of a 550,000 ball ping pong ball avalanche approximately 0.5 m high. (d) Head of a 2D DNS on 20° slope at Re = 64,000. The color represents flow density and the streamlines in the frame of reference of the current are plotted. (e) Head of a turbidity current experiment. (From Baas, J.H. et al., *J. Geophys. Res.*, 110, C11015, 2005.) (f) Head of a 3D DNS showing iso-surfaces of concentration. (From Necker, F., et al., *Int. J. Multiphase Flow*, 28, 279, 2002.)

Another theoretical treatment by Shin et al. (2004) yields Fr = 1 and for environments of finite depth H Nokes et al. (2008) determine the dependence of Fr on the ratio h/H. Since the flow will slow down when $\text{Ri} \ll 1$, because of entrainment of the ambient fluid, and it may accelerate when $\text{Ri} \gg 1n$, since there is no entrainment, the flow will tend to adjust its speed until $\text{Ri} \approx 1$. With our definitions $\text{Fr} = 1/\sqrt{\text{Ri}}$ so that $\text{Fr} \approx 1$, and this then explains why currents flow at roughly the same speed regardless of the slope angle.

42.4.2 Integral Models

For finite volume releases integral models are widely used. At their simplest the flow is modeled as a box with constant volume and changing aspect ratio, and despite its simplicity even this model can reproduce several aspects of experimentally observed flows (Huppert and Simpson 1980). The Kulikovskiy-Sveshnikova–Beghin (KSB) model extends this approach to incorporate slope angle effects and entrainment of the ambient fluid. The KSB model originated from the work of Kulikovskiy and Sveshnikova (1977), and was developed by Beghin (1979) who introduced a slope angle-dependence to supplement the density ratio-dependence of ambient entrainment. This was then developed by Ancey (2004b) by comparing Beghin's slope angle-dependent ambient entrainment assumption to a growth rate governed by overall Richardson number (Turner 1973), consistent with the entrainment assumption proposed for inclined plumes by Ellison and Turner (1959). Turnbull et al. (2007) further extended this approach to include the volume of the entrained snow and tested this against field observations of powder snow avalanches. Dai and García (2010) argue that detrainment of particles and the ambient fluid is important and show a better agreement with Beghin's data than his original model. A general model of this form consists of three ODEs describing the evolution of current mass, volume, and momentum. Highly elongated currents can also be modeled as plumes (Beghin et al. 1981b; Turnbull and McElwaine 2007).

42.4.3 Shallow Water Models

At the next level of complexity one finds so-called depth-averaged or shallow water models, first introduced for compositional gravity currents by Rottman and Simpson (1983), and later extended to turbidity currents by Bonnecaze et al. (1993) and powder snow avalanches by Fukushima and Parker (1990). These models are reviewed in detail by Huppert (1998, 2006) and Parsons et al. (2007). The shallow water approach expands the underlying equations in terms of a shallowness parameter $\epsilon = h/L$, where L is a horizontal length scale over which quantities vary slowly and h denotes the current height. For high Re gravity currents, viscous forces can be neglected and to lowest order in ϵ the pressure field is hydrostatic. A self-consistent model can be formed with only one PDE, since the velocity is slaved to the height, but most basic models consist of two PDEs representing conservation of momentum and particle mass. This approach can easily be extended to two dimensions in space by including a second momentum conservation equation. In models for turbidity currents, it is usually assumed that the ambient fluid is neither entrained nor detrained; but for powder snow avalanches entrainment of the air is always included and is critical, resulting in an extra equation for flow height (or equivalently volume fraction). These models have not been widely used for powder snow avalanches, however, and their validity is unclear partly because the flows are not very shallow. Because of entrainment the avalanche is often higher than it is wide so that rather than ϵ being small it may be greater than one. The more sophisticated model of Parker et al. (1986) also includes an equation for turbulent kinetic energy, which gives more realistic solutions in some circumstances. Furthermore, the suspended phase is considered to be well-mixed across the height of the current. This assumption may hold for very fine sediment, but it is questionable for coarser particles or during the late stages of the flow when the decaying turbulence may no longer be fully able to distribute the particles across the entire current height. For the case of a deep ambient, the motion of the overlying fluid can be neglected, and the so-called single-layer shallow water equations hold. For shallow ambients, on the other hand, it is necessary to extend this approach by formulating a two-layer system that also accounts for the dynamics of the overlying fluid layer (Baines 1995). We note that the equations have to be closed by prescribing the front velocity, which is commonly accomplished based on the Froude number relations discussed earlier.

42.4.3.1 Entrainment

Turbulent eddies on the upper surface of a gravity current can entrain the ambient fluid. This entrained fluid must be accelerated by the flow, and this inertial contribution is often the dominant drag mechanism in powder snow avalanches. Modeling this entrainment is therefore key to accurately understanding the flow speed. The initiation of a turbulent suspension current from a dense granular flow can also be thought of as entrainment of the ambient fluid, but this time by a dense granular flow. Entrainment into the body is usually modeled as a function of the overall Richardson number, (Ellison and Turner 1959) and is developed from the classical theory of plumes. Modeling is based on an entrainment coefficient e such that the flow of ambient into the current is ue at all points on the surface, where both can be functions of space and time depending on the model. Parker et al. (1987), based on experiments with turbidity currents, propose an empirical relation for the entrainment coefficient $e = 0.075/(1 + 718\text{Ri}^{2.4})^{0.5}$. (N.B. Ri has a different definition than in this chapter). Another relation, first proposed by Ancey (2004a) from analysis of unpublished data of Beghin, is

$$e = \begin{cases} \exp(-1.6\text{Ri}^2), & \text{Ri} \le 1, \\ \exp(-1.6)/\text{Ri}, & \text{Ri} > 1. \end{cases} \tag{42.10}$$

This has similar functional behavior as the relation of Parker but is 100 times larger for large Ri. This difference is only partly due to the differing definitions of Ri but also indicates the difficulties of performing such measurements and the lack of any predictive theory.

High Richardson numbers for turbidity currents flowing over low gradients are indicated by very low entrainment rates and extremely long run-out distances of channelized flows on the ocean floor (Srivatsan et al. 2004; Birman et al. 2009). This argues for gradient Richardson numbers, Ri_g, sufficiently above the critical value of 0.25 to suppress mixing (i.e., stable density stratification in the shear layer).

$$\mathrm{Ri}_g = \frac{g}{\rho}\frac{(\partial\rho/\partial z)}{(\partial u/\partial z)^2} \tag{42.11}$$

where z is the vertical coordinate. In contrast, those on continental slopes may be supercritical and strongly entraining (Parsons et al. 2007). Powder snow avalanches are similarly strongly entraining as can clearly be seen in Figure 42.3. Indeed without the supply of fresh snow they will rapidly entrain sufficient air that they come almost to a halt. The differences are presumably due to the greater slope angles usually experienced by avalanches and also the greater entrainment expected from non-Boussinesq flows.

42.5 Outlook and Future Directions

Data from direct observations of snow avalanches have greatly increased in recent years but are still limited and very difficult to obtain. Data from turbidity currents are much rarer and will remain so for the foreseeable future. Experimental investigations are therefore crucial despite the impossibility of simultaneously satisfying all similarity criteria. Typically one is restricted to considering limited aspects of whole flow behavior, where the relevant parameters can be maintained within a critical range [e.g., above the threshold values of the Reynolds number where similarity applies; Parsons and Garcia (1998)] and others are relaxed. Nevertheless, experiments have enabled the validation and development of theories; and by checking the theories in different regions at the edges of the parameter space, one can hope that they are also accurate in the experimentally inaccessible regions that correspond to the natural flows. Experiments are also very important for the verification of numerical simulations, especially in areas such as sediment erosion mechanisms.

On the modeling side, there are several substantial challenges waiting to be addressed. Understanding of the initiation process by which a dense flow transitions into a suspension current is still lacking, though Blanchette et al. (2005) have performed simulations of the related problem of resuspension. Performing experiments covering this transition is difficult. Here the main difficulties lie in understanding the mechanisms that govern the initial fluidization of the sediment bed, a process that involves the interaction of densely packed particles that may or may not be cohesive with the interstitial fluid. Toward this end, it may be helpful to incorporate recent advances from the field of granular flows (Forterre and Pouliquen 2008) or debris flows (Iverson 1997 is a comprehensive review). During the later stages of the flow, the transition layer involves dense particle concentrations, so that particle/particle interactions cannot be neglected. One approach is to have a different model for the three regions (or more if the ambient fluid and static bed are included), with suitable closures chosen for the mass and momentum fluxes between them. Alternatively one model can be used throughout whose character changes strongly with solid fraction. For example, to first order the effects may be captured by allowing both the viscosity of the suspension and the particle settling velocity to depend on the local volume fraction of the particles. However, the true dynamics frequently will be substantially more complex, involving the interaction of suspended load and bed-load, and the exchange of particles between the current and the bed, and possibly non-Boussinesq flow effects. Specifically, the development of advanced erosion and deposition models will be highly beneficial for improving the fidelity of numerical simulations (Birman et al. 2007).

Once the particles are in suspension, their interaction with the fluid turbulence still involves open questions. While both experimental (Aliseda et al. 2002, and references therein) and computational (Bosse et al. 2006, and references therein) investigations of particles in homogeneous turbulence suggest that particle settling should be enhanced by two-way coupling effects, experiments and field observations generally indicate that particles are kept in suspension by the turbulence for long times, which allows turbidity currents especially to travel over very long distances. The back reaction of the particles on the turbulence is also poorly understood. A strong stratification will suppress the turbulence into layers as is seen with salinity and temperature gradients in the ocean. However, particle sedimentation acts to maintain the stratification and entrainment of new material at the base and ambient fluid on the upper surface also increase sedimentation. The interplay between these effects is important, but turbulence models have not been well tested in strongly stratified flows, and even less is known when non-Boussinesq effects are important.

Given that many hydrocarbon reservoirs consist of turbidity current deposits, it will be attractive to couple the flow simulation to a realistic substrate model for the sediment bed that accounts for spatially varying distributions of particle sizes, porosity and permeability. These properties could then feed into a reservoir model that, in turn, would form the basis of subsequent porous media flow simulations. The focus on powder snow avalanche modeling is understanding their interaction with structures whether these are pylons, buildings, or defensive dams. Ultimately, this will require models that are simple enough to be computationally tractable over complex terrain. Addressing the aforementioned goals will require the integration of complimentary research approaches from the fields of geology and fluid mechanics, involving field observations and measurements, laboratory experiments, development of fundamental models, high-resolution simulations, and linear stability analysis.

42.5.1 Future Directions

1. Improved models for the mass and momentum fluxes between the different layers need to be developed that can be employed in numerical simulations.
2. Turbulence closures suitable for strongly stratified, non-Boussinesq flows near a boundary need to be developed.
3. The coupling between the evolution of the current and that of the underlying substrate will have to be explored.

Acknowledgments

EM's work has been supported by the National Science Foundation, most recently under grant CBET-0854338. JNM is supported by the U.K. Engineering and Physical Sciences Research Council.

References

Aliseda A, Cartellier A, Hainaux F, Lasheras JC. 2002. Effect of preferential concentration on the settling velocity of heavy particles in homogenous isotropic turbulence. *J. Fluid Mech.* 468:77–105.

Ancey C. 2004a. Powder snow avalanches: Approximation as non-boussinesq clouds with a Richardson number dependent entrainment function. *J. Geophys. Res.* 109:F01005:F01005.

Ancey C. 2004b. Powder snow avalanches: Approximation as non-boussinesq clouds with a Richardson number dependent entrainment function. *J. Geophys. Res.* 109:F01005

Ancey C. 2007. Plasticity and geophysical flows: A review. *J. Non-Newtonian Fluid Mech.* 142:4–35.

Andrée K. 1920. *Geologie des Meeresbodens Bornträger*. Leipzig, Germany.

Ash M, Brennan PV, Chetty K, Mcelwaine J, Keylock C. 2010. FMCW radar imaging of avalanche-like snow movements. In *Proceedings of the 2010 IEEE International Radar Conference "Global Innovation in Radar."* Washington, DC.

Baas JH, McCaffrey WD, Haughton PDW, Choux C. 2005. Coupling between suspended sediment distribution and turbulence structure in a laboratory turbidity current. *J. Geophys. Res.* 110:C11015, doi:10.1029/2004JC002668.

Babonneau N, Savoye B, Cremer M, Klein B. 2002. Morphology and architecture of the present canyon and channel system of the zaire deep sea fan. *Mar. Pet. Geol.* 19:445–467.

Bagnold RA. 1954. Experiments on gravity free dispersion of large solid spheres in a Newtonian fluid under stress. *Proc. R. Soc. Lond. Sem A: Math Phys.* 225:49–63.

Bagnold RA. 1962. Auto-suspension of transported sediment: Turbidity currents. *Proc. R. Soc. Lond. A* 265:315–319.

Baines PG. 1995. *Topographic Effects in Stratified Flows*. Cambridge University Press, London, U.K.

Barbolini M, Issler D. 2006. Avalanche test sites and research equipment in Europe: An updated overview. Technical report, Final Report Deliverable D9, SATSIE Avalanche Studies and Model Validation in Europe.

Batchelor GK. 1967. *An Introduction to Fluid Dynamics*. Cambridge University Press, Cambridge, MA.

Batchelor G. 1989. A brief guide to two-phase flow. In *Theoretical and Applied Mechanics*, eds. P. Germain, J. Piau, and D. Caillerie. Elsevier Science Publishers, Amsterdam, the Netherlands, pp. 27–41.

Beghin P. 1979. *Etude des Bouffées Bidimensionnelles de Densité en écoulement sur Pente avec Application aux Avalanches de Neige Poudreuse*. PhD thesis, Institut National Polytechnique de Grenoble, Grenoble, France.

Beghin P, Hopfinger EJ, Britter RE. 1981a. Gravitational convection from instantaneous sources on inclined boundaries. In Beghin et al. (1981b), 407–422:407–422

Beghin P, Hopfinger EJ, Britter RE. 1981b. Gravitational convection from instantaneous sources on inclined boundaries. *J. Fluid Mech.* 107:407–422.

Beghin P, Olagne X. 1991. Experimental and theoretical study of the dynamics of powder snow avalanches. *Cold Reg. Sci. Tech.* 19:317–326.

Benjamin TB. 1968. Gravity currents and related phenomena. *J. Fluid Mech.* 31:209–248.

Best JL, Kostaschuk RA, Peakall J, Villard PV, Franklin M. 2005. Whole flow field dynamics and velocity pulsing within natural sediment-laden underflows. *Geology* 33:765–768.

Birman VK, Battandier BA, Meiburg E, Linden PF. 2007. Lock exchange flows in sloping channels. *J. Fluid Mech.* 577:53.

Birman VK, Martin JE, Meiburg E. 2005. The non-Boussinesq lock-exchange problem. Part 2. High resolution simulations. *J. Fluid Mech.* 537:125.

Birman VK, Meiburg E, Kneller B. 2009. The shape of submarine levees: Exponential or power law? *J. Fluid Mech.* 619:367–376.

Blanchette F, Strauss M, Meiburg E, Kneller B, Glinsky ME. 2005. High-resolution numerical simulations of resuspending gravity currents: Conditions for self-sustainment. *J. Geophys. Res.* 110:C12022.

Bonnecaze RT, Huppert HE, Lister JR. 1993. Particle-driven gravity currents. *J. Fluid Mech.* 250:339–369.

Bosse T, Kleiser L, Meiburg E. 2006. Small particles in homogenous turbulence: Settling velocity enhancement by two-way coupling. *Phys. Fluids* 18:027102.

Boussinesq J. 1903. *Théorie Analytique de la Chaleur*, Vol. 2. Gathier-Villars, Paris, France (in French).

Bozhinskiy A, Sukhanov L. 1998a. Physical modelling of avalanches using an aerosol cloud of powder material. In Bozhinskiy and Sukhanov (1998b), 242–394:242–394.

Bozhinskiy A, Sukhanov L. 1998b. Physical modelling of avalanches using an aerosol cloud of powder material. *Ann. Glaciol.* 26:242–394.

Chikita K. 1989. A field study on turbidity currents initiated from spring runoffs. *Water Resour. Res.* 25:257–271.

Curray JR, Emmel FJ, Moore DG. 2003. The Bengal Fan: Morphology, geometry, stratigraphy, history and processes. *Mar. Pet. Geol.* 19:1191–1223.

Dai A, Garc'ıa MH. 2010. Gravity currents down a slope in deceleration phase. *Dyn. Atmos. Oceans* 49:75–82.

Daly RA. 1936. Origin of submarine canyons. *Am. J. Sci. Ser.* 31:401–420.

Dan G, Sultan N, Savoye B. 2007. The 1979 nice harbour catastrophe revisited: Trigger mechanism inferred from geotechnical measurements and numerical modeling. *Mar. Geol.* 245:40–64.

De Cesare G, Schleiss A, Hermann F. 2001. Impact of turbidity currents on reservoir sedimentation. *J. Hydraul. Eng.* 127:6–16.

Dengler AT, Wilde P, Noda EK, Normark WR. 1984. Turbidity currents generated by hurricane Iwa. *Geo-Mar. Lett.* 4:5–11.

Ellison T, Turner J. 1959. Turbulent entrainment in stratified flow. *J. Fluid Mech.* 6:423–448.

Étienne J, Hopfinger EJ, Saramito P. 2005. Numerical simulations of high density ratio lock-exchange flows. *Phys. Fluids* 17:036601.

Fan J. 1986. Turbid density currents in reservoirs. *Water Int.* 11:107–116.

Felix M, Sturton S, Peakall J. 2005. Combined measurements of velocity and concentration in experimental turbidity currents. *Sed. Geol.* 179:31–47.

Forel FA. 1885. Les ravins sous-lacustres des fleuves glaciaires. *C. R. Acad. Sci. Paris* 101:725–728.

Forterre Y, Pouliquen O. 2008. Flows of dense granular media. *Ann. Rev. Fluid Mech.* 65:144–164.

Fukushima Y, Parker G. 1990. Numerical simulation of powder-snow avalanches. *J. Glaciol.* 36:229–237.

Gauer P, Issler D, Lied K, Kristensen K, Sandersen F. 2009. On snow avalanche flow regimes: Inferences from observations and measurements. In *International Snow Science Workshop, 21–27 September 2008, Whistler Canada*. Canadian Avalanche Association, Vancouver, British Columbia, Canada.

Gauer P, Kern M, Kristensen K, Lied K, Rammer L, Schreiber H. 2007. On pulsed doppler radar measurements of avalanches and their implication to avalanche dynamics. *Cold Reg. Sci. Tech.* 50:55–71.

Gould HR. 1951. Some quantitative aspects of Lake Mead turbidity currents. In *Society of Economic Paleontologists and Mineralogists Special Publication*, Vol. 2, pp. 34–52.

Gray TE, Alexander J, Leeder MR. 2005. Quantifying velocity and turbulence structure in depositing sustained turbidity currents across breaks in slope. *Sedimentology* 52:467–488.

Gröbelbauer HP, Fanneløp TK, Britter RE. 1993. The propagation of intrusion fronts of high density ratio. *J. Fluid Mech.* 250:669.

Gruber U, Margreth S. 2001. Winter 1999: A valuable test of the avalanche-hazard mapping procedure in Switzerland. *Ann. Glaciol.* 32:328–332.

Gubler H, Hiller M. 1984. The use of microwave FMCW radar in snow and avalanche research. *Cold Reg. Sci. Technol.* 9:109–119.

Hall B, Meiburg E, Kneller B. 2008. Channel formation by turbidity currents: Navier-Stokes based linear stability analysis. *J. Fluid Mech.* 615:185–210.

Hampton MA. 1972. The role of subaqueous debris flow in generating turbidity currents. *J. Sed. Petrol.* 42:775–793.

Harbitz C, Issler D, Keylock CJ. 1998. Conclusions from a recent survey of avalanche computational models. In *25 years of Snow Avalanche Research*, Hestnes, E. (ed.) Norwegian Geotechnical Institute Publication, Voss, Norway, 203:128–139.

Heezen BC, Ewing WM. 1952. Turbidity currents and submarine slumps and the 1929 Grand Banks (Newfoundland) earthquake. *Am. J. Sci. Ser.* 250:849–873.

Hermann F, Hermann J, Hutter K. 1987. Laboratory experiments on the dynamics of powder snow avalanches. In *International Symposium on Avalanche Formation, Movement and Effects, Proceedings of the Davos Symposium*, September 1986, Vol. IAHS-Publ. No. 162, Davos, Switzerland.

Höller P. 2007. Avalanche hazards and mitigation in Austria: A review. *JNH* 43:81–101.

Hopfinger EJ. 1983. Snow avalanches motion and related phenomena. *Annu. Rev. Fluid Mech.* 15:47–96.

Huppert HE. 1998. Quantitative modeling of granular suspension flows. *Trans. R. Soc. Lond.* 356:2471.

Huppert HE. 2006. Gravity currents: A personal perspective. *J. Fluid Mech.* 554:299–322.

Huppert HE, Simpson JE. 1980. The slumping of gravity currents. *J. Fluid Mech.* 99:785.

Inman DL, Nordstrom CE, Flick RE. 1976. Currents in submarine canyons: An air-sea-land interaction. *Ann. Rev. Fluid Mech.* 8:275–310.

Issler D. 2003. Experimental information on the dynamics of dry-snow avalanches. In *Dynamic Response of Granular and Porous Materials under Large and Catastrophic Deformations*, eds. K. Hutter and N. Kirchner, No. 11 in Lecture notes in applied and Computational mechanics. Springer, Berlin, Germany, pp. 109–160.

Iverson R. 1997. The physics of debris flows. *Rev. Geophys.* 35:245–296.

Jóhannesson T, Tracy L. 2001. Results of the 2d avalanche model SAMOS. Technical report, 01011,01019,02008,02013,02018, 03012,04008, Icel. Met. Office.

Johnson DW. 1939. *The Origin of Submarine Canyons*. Columbia University Press, New York.

Keppie SJD, Boyle RW, Haynes SJ. 1987. Turbidite-hosted gold deposits. *Geol. Assoc. Canada Spec. Paper* 32:186.

Kern MA, Bartelt P, Sovilla B, Buser O. 2009. Measured shear rates in large dry and wet snow avalanches. *J. Glaciol.* 55:327–338.

Khripounoff A, Vangriesheim A, Babonneau N, Crassous P, Dennielou B, Savoye B. 2003. Direct observation of intense turbidity current activity in the zair submarine valley at 4000 m water depth. *Mar. Geol.* 194:151–158.

Klaucke I, Hesse R, Ryan BF. 1998. Seismic stratigraphy of the northwest Atlantic mid-ocean channel: Growth pattern of a mid-ocean channel-levee complex. *Mar. Pet. Geol.* 15:575–585.

Kneller B, Bennett SJ, McCaffrey WD. 1999. Velocity structure, turbulence and fluid stresses in experimental gravity currents. *J. Geophys. Res. Oceans* 104:5281–5291.

Kneller B, Buckee C. 2000. The structure and fluid mechanics of turbidity currents: A review of some recent studies and their geological implications. *Sedimentology* 47:62–94.

Kuenen PH. 1938. Density currents in connection with the problem of submarine canyons. *Geol. Mag.* 75:241–249.

Kuenen PH. 1951. Properties of turbidity currents of high density. *Society of Economic Paleontologists and Mineralogists Special Publication,* Vo. 2, pp. 14–33.

Kuenen PH, Migliorini CI. 1950. Turbidity currents as a cause of graded bedding. *J. Geol.* 58:91–127.

Kulikovskiy A, Sveshnikova E. 1977. Model dlja rascheta dvizhija pilevoi snezhnoi lavini [A model for computing powdered snow avalanche motion]. *Materiali Glatsiologicheskih Isseledovanii [Data of Glaciological Studies]* 31:74–80 (in Russian).

Launder BE, Rodi W. 1983. The turbulent wall jet—Measurements and modeling. *Ann. Rev. Fluid Mech.* 15:429–459.

Lee SE, Talling PJ, Ernst GCJ, Hogg AJ. 2002. Occurence and origin of submarine plunge pools at the base of the US continental slope. *Mar. Geol.* 185:363–377.

Leeder MR, Gray TE, Alexander J. 2005. Sediment suspension dynamics and a new criterion for the maintainance of turbulent suspensions. *Sedimentology* 52:683–691.

Louge MY, Steiner R, Keast S, Decker R, Dent J, Schneebeli M. 1997. Application of capacitance instrumentation to the measurement of density and velocity of flowing snow. *Cold Reg. Sci. Technol.* 25:47–63.

Lowe DR. 1982. Sediment gravity flows: Ii. Depositional models with special reference to the deposits of high-density turbidity currents. *J. Sed. Pet.* 52:279–297.

McElwaine JN. 2005. Rotational flow in gravity current heads. *Phil. Trans. R. Soc. Lond.* 363:1603–1623.

McElwaine JN, Nishimura K. 2001. Ping-pong ball avalanche experiments. In *Particulate Gravity Currents*, eds. W.D. McCaffrey, B.C. Kneller, and J. Peakall, No. 31 in Special Publication of the International Association of Sedimentologists. Blackwell Science, Oxford, U.K., pp. 135–148.

McElwaine JN, Turnbull B. 2005. Air pressure data from the Vallée de la Sionne avalanches of 2004. *J. Geophys. Res.* 110:F03010.

Meiburg E, Kneller B. 2010. Turbidity currents and their deposits. *Annu. Rev. Fluid Mech.* 42:135–156.

Middleton GV. 1993. Sediment deposition from turbidity currents. *Ann. Rev. Earth Planet. Sci.* 21:89–114.

Milne J. 1897. Sub-oceanic changes. *Geogr. J.* 10:259–285.

Mulder T, Alexander J. 2001. The physical character of subaqueous sedimentary density flows and their deposits. *Sedimentology* 48:269–299.

Naaim M, Gurer I. 1998. Two-phase numerical model of powder avalanche theory and application. *Nat. Haz.* 17:129–145.

Necker F, Härtel C, Kleiser L, Meiburg E. 2002. High-resolution simulations of particle-driven gravity currents. *Int. J. Multiphase Flow* 28:279–300.

Necker F, Härtel C, Kleiser L, Meiburg E. 2005. Mixing and dissipation in particle-laden gravity currents. *J. Fluid Mech.* 545:339–372.

Nelson CH, Karabanov EB, Coleman SM. 1995. Atls of deep water environments: Architectural style in turbidite systems. *Late Quaternary Turbidite Systems in Lake Baikal, Russia*, ed. Smith. Chapman and Hall, London, U.K., pp. 29–33.

Nishimura K, Ito Y. 1997. Velocity distribution in snow avalanches. *J. Geophys. Res. B* 102:27297–27303.

Nishimura K, Keller S, McElwaine JN, Nohguchi Y. 1998. Ping-pong ball avalanche at a ski jump. *Gran. Matt.* 1:51–56.

Nishimura K, Narita H, Maeno N. 1989. The internal structure of powder-snow avalanches. *Ann. Glaciol.* 13:207–210.

Nishimura K, Sandersen F, Kristensen K, Lied K. 1995. Measurements of powder snow avalanche—nature—. *Surv. Geophys.* 16:649–660.

Nokes R, Davidson M, Stepien C, Veale W, Oliver R. 2008. The front condition for intrusive gravity currents. *J. Hydraul. Res.* 46:788–801.

Normark WR. 1989. Observed parameters for turbidity-current flow in channels, reserve fan, Lake Superior. *J. Sed. Pet.* 59:423–431.

Normark WR, Damuth DE, Wickens HDV. 1997. Sedimentary facies and associated depositional elements of the Amazon fan. In *Proceedings of the Ocean Drilling Program, Scientific Results, Leg 155*, Flood, R.D., Piper, D.J.W., Klaus, A., and Peterson, L.C., (eds.), College Station, TX (Ocean Drilling Program), pp. 611–651.

Pantin HM. 1979. Interaction between velocity and effective density in turbidity flow: Phase-plane analysis, with criteria for autosuspension. *Mar. Geol.* 31:59–99.

Parker G, Fukushima Y, Pantin HM. 1986. Self-accelerating turbidity currents. *J. Fluid Mech.* 171:145–181.

Parker G, Garcia M, Fukushima Y, Yu W. 1987. Experiments on turbidity currents over an erodible bed. *J. Hydraul. Eng.* 52:123–147.

Parsons JD, Friedrichs CT, Traykovski PA, Mohrig D, Imran J et al. 2007. The mechanics of marine sediment gravity flows. In *Continental Margin Sedimentation: From Sediment Transport to Sequence Stratigraphy*, eds. C.A. Nittrouer, J.A. Austin, M.E. Field, J. Syvitski, and P.L. Wiberg. Blackwell Publishing, Oxford, U.K.

Parsons JD, Garcia M. 1998. Similarity of gravity current fronts. *Phys. Fluids* 10:3209–3213.

Piper DJW, Cochonat P, Morrison ML. 1999. The sequence of events around the epicentre of the 11929 grand banks earthquake: Initiation of debris flows and turbidity currents inferred from sidescan sonar. *Sedimentology* 46:79–97.

Piper DJW, Savoye B. 1993. Processes of late quaternary turbidity current flow and deposition on the Var deep-sea fan, northwest Mediterranean Sea. *Sedimentology* 40:3528–3542.

Prior DB, Bornhold BD, Wiseman WJ, Lowe DR. 1987. Turbidity current activity in a British Columbia fjord. *Science* 237:1330–1333.

Rammer L, Kern MA, Gruber U, Tiefenbacher F. 2007. Comparison of avalanche velocity measurements by means of pulsed doppler radar, continuous wave radar and optical methods. *Cold Reg. Sci. Tech.* 50:35–54.

Rammer L, Schaffhauser H, Sampl P. 1998. Computed powder avalanche impact pressures on a tunnel-bridge in Ausserfern-Tirol. In *Environmental Forest Science, Proceedings of the IUFRO Division 8th Conference*, ed. K. Sassa. Kluwer, Dordrecht, the Netherlands.

Rottman JW, Linden PF. 2001. Gravity currents. In *Environmental Stratified Flows*, ed. R. Grimshaw. Kluwer Academic Publishers, Dordrecht, the Netherlands.

Rottman JW, Simpson JE. 1983. Gravity currents produced by instantaneous releases of heavy fluids in a rectangular channel. *J. Fluid Mech.* 135:95–110.

Sequeiros OE, Naruse H, Endo N, Garcia MH, Parker G. 2009. Experimental study on self-accelerating turbidity currents. *J. Geophys. Res.* 114:C05025.

Shin JO, Dalziel SB, Linden PF. 2004. Gravity currents produced by lock exchange. *J. Fluid Mech.* 521:1–34.

Simpson JE. 1997a. *Gravity Currents in the Environment and the Laboratory*. Cambridge University Press, Cambridge, U.K.

Sovilla B, McElwaine JN, Schaer M, Vallet J. 2010. Variation of deposition depth with slope angle in snow avalanches: Measurements from Vallée de la sionne. *J. Geophys. Res.* 115:F02016.

Sovilla B, Schaer M, Kern M, Bartelt P. 2008. Impact pressures and flow regimes in dense snow avalanches observed at the Vallée de la Sionne test site. *J. Geophys. Res.* 113:F01010, doi:10.1029/2006JF000688.

Srivatsan L, Lake LW, Bonnecaze RT. 2004. Scaling analysis of deposition from turbidity currents. *Geo-Mar. Lett.* 24:63–74.

Stacey MW, Bowen AJ. 1988. Vertical structure of density and turbidity currents: Theory and observations. *J. Geophys. Res.* 93:3528–3542.

Tochon-Danguy JC, Hopfinger EJ. 1975. Simulations of the dynamics of powder snow avalanches. In *Proceeding of the Grindelwald Symposium April 1974—Snow Mechanics*. IAHS, Wallingford, CT.

Turnbull B, McElwaine JN. 2007. A comparison of powder snow avalanches at Vallée de la Sionne with plume theories. *J. Glaciol.* 53:30–40.

Turnbull B, McElwaine JN. 2008. Experiments on the non-boussinesq flow of self-igniting suspension currents on a steep open slope. *J. Geophys. Res.* 113:F01003.

Turnbull B, McElwaine JN. 2010. Potential flow models of suspension current air pressure. *Ann. Glaciol.* A51:113–122.

Turnbull B, McElwaine JN, Ancey CJ. 2007. The Kulikovskiy-Sveshnikova-Beghin model of powder snow avalanches: Development and application. *J. Geophys. Res.* 112:F0100.

Turner JS. 1973. *Buoyancy Effects in Fluids*. Cambridge University Press, Cambridge, U.K.

UNESCO. 1981. Avalanche atlas; illustrated international avalanche classification. Technical report 2, UNESCO.

Vallet J, Gruber U, Dufour F. 2001. Photogrammetric avalanche volume measurements at Vallée de la Sionne, Switzerland. *Ann. Glaciol.* 32:141–146.

Vallet J, Turnbull B, Joly S, Dufour F. 2004. Observations on powder snow avalanches using videogrammetry. *Cold Reg. Sci. Technol.* 39:153–159.

Xu JP. 2010. Normalized velocity profiles of field-measured turbidity currents. *Geology* 38:563–566.

Xu JP, Noble M, Rosenfeld LK. 2004. In-situ measurements of velocity structure within turbidity currents. *Geophys. Res. Lett.* 31:L09311, doi:10.1029/2004GL019718.

43

Mechanics of Debris Flows and Rock Avalanches

Richard M. Iverson
United States Geological Survey

43.1 Introduction

Debris flows are geophysical phenomena intermediate in character between rock avalanches and flash floods. They commonly originate as water-laden landslides on steep slopes and transform into liquefied masses of fragmented rock, muddy water, and entrained organic matter that disgorge from canyons onto valley floors. Typically including 50%–70% solid grains by volume, attaining speeds >10 m/s, and ranging in size up to $\sim10^9$ m^3, debris flows can denude mountainsides, inundate floodplains, and devastate people and property (Figure 43.1). Notable recent debris-flow disasters resulted in more than 20,000 fatalities in Armero, Colombia, in 1985 and in Vargas state, Venezuela, in 1999.

Alternative terms such as mudflow, mudslide, debris torrent, and lahar are sometimes used to describe debris flows, but the terms "debris" and "flow" have precise geological meanings. "Debris" implies that grains with greatly differing sizes are present. The largest grains can exceed 10 m in diameter, but the presence of at least a few weight percent of mud-sized grains (<62 μm) is more critical because persistent hydrodynamic suspension of these small grains effectively increases the viscosity of the muddy water that fills pore spaces between the larger grains. This enhanced viscosity promotes development of high pore-fluid pressures that facilitate debris-flow motion by exerting lubrication forces at grain contacts. The term "flow" implies that slip at grain contacts is pervasive, and granular debris that is liquefied by high pore pressures can appear to flow almost as fluidly as water.

This chapter emphasizes the physical basis and mathematical structure of models that analyze two-phase debris-flow behavior by considering gravity-driven motion of granular solids that transport pore fluid with evolving pressure. In this modeling framework, granular rock avalanches represent a limiting class of flows in which effects of pore-fluid pressure are negligible. Differences between rock avalanches and debris flows can be gradational, however: relatively dry rock avalanches can sometimes engulf enough water to gradually transform to debris flows, for example. A further complication is that debris flows can occur in submarine environments, where water not only fills intergranular pores but also surrounds the flows and thereby exerts buoyancy and inertia forces. This chapter focuses exclusively on terrestrial flows in which the effects of the surrounding air are negligible.

43.2 Observations and Data

Although debris flows can be difficult and dangerous to observe directly, the chief qualitative features of debris-flow behavior can readily be observed in extensive video documentation obtained

Handbook of Environmental Fluid Dynamics, Volume One, edited by Harindra Joseph Shermal Fernando. © 2013 CRC Press/Taylor & Francis Group, LLC. ISBN: 978-1-4398-1669-1.

FIGURE 43.1 Oblique aerial photograph of a lethal debris flow that began as a rainfall-triggered landslide, Minamata, Japan, July 20, 2003. (After Sidle, R.C. and Chigira, M., *EOS*, 85, 145, 2004.) (Photo by courtesy of R.C. Sidle, Reproduced by permission of the American Geophysical Union.)

in the field and laboratory (e.g., Costa and Williams 1984; Logan and Iverson 2007):

1. Debris flows exhibit unsteady, nonuniform motion, with distinct starting and ending points in space and time. Debris-flow models must, therefore, include explicit time dependence.
2. Debris flows typically originate from discrete or distributed source areas that have slopes >30° mantled with soil and fragmented rock. This debris becomes thoroughly wet through introduction of surface water or groundwater, commonly as a result of intense rain or snowmelt. The water-laden debris starts to move downslope when frictional forces no longer can resist driving forces, and it then liquefies and begins to flow.
3. Many debris flows entrain additional sediment and water as they descend steep slopes and channels. Entrainment can occur by scour of bed material or collapse of stream banks, and it can cause the mass of a debris flow to increase 10-fold or more before deposition begins on flatter terrain downstream.
4. Abrupt, steep surge fronts generally form at the heads of moving debris flows. Large grains accumulate at surge fronts as a result of grain-size segregation and migration within the debris, but large grains can also be scoured from the bed and retained at surge fronts.
5. Water-saturated debris that trails surge fronts commonly resembles watery, flowing concrete or roiling quicksand. Thus, a debris-flow surge front commonly behaves as a "bouldery dam... pushed along by the finer, more fluid debris impounded behind..." (Sharp and Nobles 1953).
6. Lateral levees form where liquefied debris shoulders aside high-friction debris at surge fronts, most commonly where debris flows escape lateral confinement by overtopping stream banks or discharging onto broad alluvial fans or plains.
7. Depositional lobes form where the frictional resistance imposed by coarse-grained flow fronts and margins is sufficient to halt motion of the trailing, liquefied debris. Bodies of fresh debris-flow deposits are generally too weak for humans to traverse on foot, although the coarse-grained lateral levees and distal margins of fresh deposits commonly afford more secure footing.
8. Following emplacement, bodies of debris-flow deposits gradually dewater and consolidate to a degree that allows secure passage on foot. As desiccation proceeds, deposits become nearly rigid, but this process commonly requires several days to weeks.

Data that quantify many of the phenomena described above come from nearly field-scale experiments conducted in the 95 m long, 2 m wide USGS debris-flow flume (Iverson et al. 2010). A set of eight of these experiments in which the debris consisted of water-saturated sand and gravel containing 7% mud-sized grains (<62 μm) yielded results that contrast with those from a set of nine experiments that were identical in every respect, except that the mud content was <1%. In each of the 17 experiments, flow was initiated by suddenly releasing 10 m^3 of thoroughly mixed, loosely packed debris from behind a vertical headgate. The ensuing debris flows descended the 31° flume and swept past two instrumented cross sections (located at x = 32 and 66 m downslope from the headgate) before discharging onto a nearly flat runout surface and forming deposits at a third instrumented cross section (located at x = 90 m) (Figure 43.2). Each instrumented cross section was equipped with an overhead laser that measured the flow thickness and with bed sensors that measured basal normal stress and pore-fluid pressure at frequencies of 500 Hz.

Ensemble averages of the time-series data recorded at the instrumented cross sections provide comprehensive summaries of the behavior of the experimental debris flows. Each panel of Figure 43.3 depicts simultaneous evolution of the ensemble-averaged flow thickness, h, total basal normal stress, σ_{bed}, and basal pore-fluid pressure, p_{bed}, for a particular instrumented cross section and experiment set (identified as "SGM" for experiments with 7% mud content and as "SG" for experiments with <1% mud content). Vertical axes in Figure 43.3 are scaled so that the three time series in each panel exhibit perfect superposition if a liquefied state exists in which $p_{bed} = \sigma_{bed} = \rho g h$ (where $g \approx$ 9.8 m/s^2 is the magnitude of gravitational acceleration, and $\rho \approx$ 2040 kg/m^3 is a bulk density typical of water-saturated debris).

The most obvious implication of the data summarized in Figure 43.3 is that the fronts of SGM debris flows move downslope more rapidly than those of SG flows. The difference in

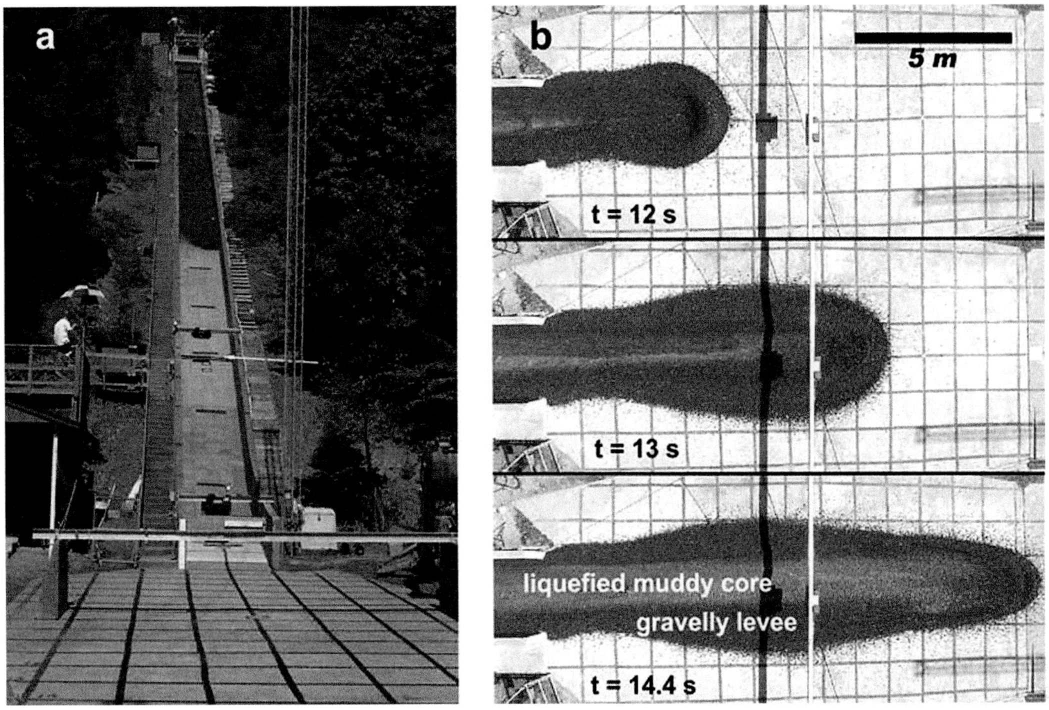

FIGURE 43.2 Photographs of a 10 m^3 experimental debris flow in the USGS flume. (a) View of flow descending the flume. (b) Sequential aerial views of flow crossing runout surface and forming levees at base of flume. "t" denotes time elapsed since opening of headgate. Shadow is cast by crossbeam suspending laser at x = 90 m. (Reproduced from Iverson, R.M. et al., *J. Geophys. Res.*, 115, 2010, doi:10.1029/2009JF001514. With permission of the American Geophysical Union.)

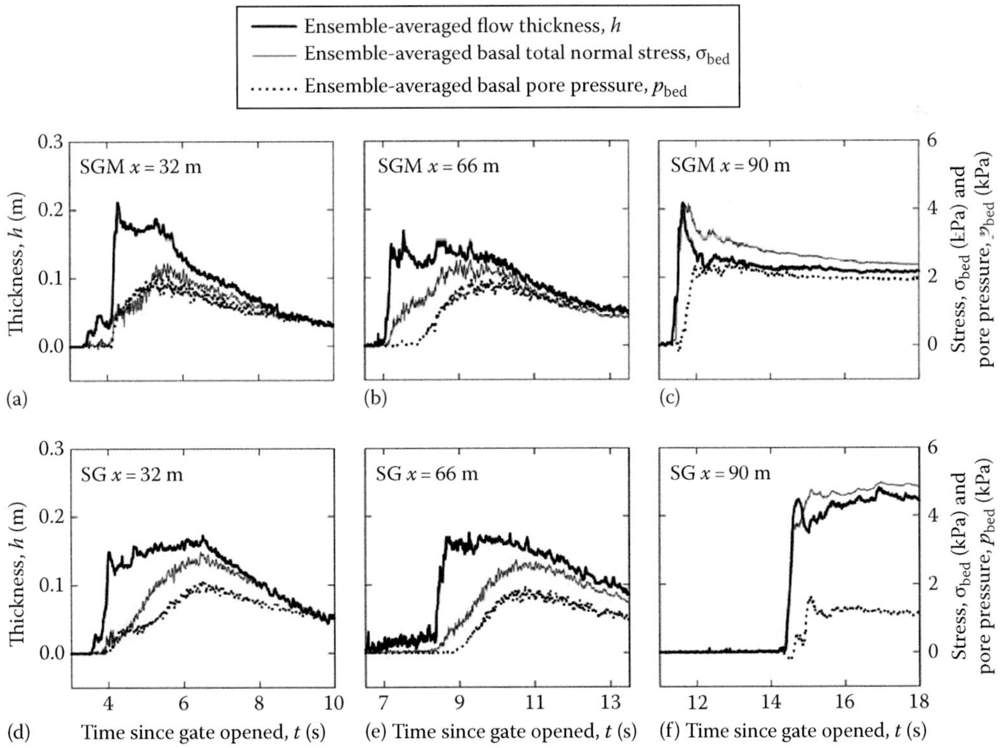

FIGURE 43.3 **(See color insert.)** Ensemble averages of flow thicknesses, basal total normal stresses, and basal pore-fluid pressures measured at three distances from the headgate (x = 32, 66, and 90 m) in two sets of experimental debris flows. (a–c) Data from eight SGM flows containing 7% mud. (d–f) Data from nine SG flows containing <1% mud. (Reproduced from Iverson, R.M. et al., *J. Geophys. Res.*, 115, 2010, doi:10.1029/2009JF001514. With permission of the American Geophysical Union.)

speeds takes time to develop; however, the fronts of both types of flows arrive at $x = 32$ m about $t = 4$ s after their release (cf. Figure 43.3a and d). This arrival time implies that all flows initially attain speeds only slightly less than that of a frictionless body, which theoretically reaches $x = 32$ m at $t = 3.56$ s when released on a 31° slope. The large initial flow-front speeds result not from near-zero friction, but instead from a strong downslope thrust (roughly proportional to $-\partial h/\partial x$) that is produced during collapse of the debris as the headgate opens. As this thrust diminishes, the effects of friction become more apparent: after $t = 4$ s, the SG flows begin to decelerate (arriving at $x = 66$ m at $t \approx 8.5$ s), but the SGM flows continue to accelerate (arriving at $x = 66$ m at $t \approx 7$ s). The disparity in SG and SGM arrival times grows to about 3 s when the flows reach $x = 90$ m (cf. Figure 43.3c and f). Furthermore, after the debris flows issue from the mouth of the flume, the SGM flows run out about twice as far as the SG flows (Iverson et al. 2010).

The high mobility of the SGM flows may seem counterintuitive, given that mud increases the viscosity of the fluid phase of the SGM, but relationships between the three time series in each of the upper panels of Figure 43.3 reveal the cause. The data in Figure 43.3a show that, after passage of a dilated, gravel-rich flow front from $t \approx 4$ to 6 s (wherein h averages roughly twice the magnitude of $\sigma_{bed}/\rho g$ and $p_{bed}/\rho g$), the approximation $p_{bed} = \sigma_{bed} = \rho gh$ holds reasonably well in SGM flows. This result implies that the debris trailing the dilated front at $x = 32$ m is almost completely liquefied by high pore-fluid pressure that reduces grain-contact stresses. Figure 43.3b and c show that this liquefaction persists at $x = 66$ and 90 m, although at these locations the rise in p_{bed} lags behind the rise in σ_{bed} because a more mature, coarse-grained flow front is present. Particularly noteworthy is the nearly perfect data superposition of time series in the final ~6 s shown in Figure 43.3c, which indicates that deposits behind the flow front are fully liquefied. Probing of the deposits demonstrates that the liquefied debris spreads into a thin puddle if it is not impounded by gravel-rich lateral levees like those pictured in Figure 43.2b.

Pore-pressure behavior in the SG flows differs markedly from that in the SGM flows. The differences first appear at $x = 32$ m, where p_{bed} remains significantly less than σ_{bed} during passage of most of the flow (Figure 43.3d). The discrepancy between p_{bed} and σ_{bed} grows more persistent at $x = 66$ m, and becomes most pronounced in deposits formed at $x = 90$ m (Figure 43.3e and f). Moreover, the SG deposits have $p_{bed} < 0.5\ \sigma_{bed}$, implying that the debris has not only lost some pore pressure but has also begun to dewater. Lack of sustained high pore pressure in the SG debris accounts for its relatively low mobility. Indeed, it may be appropriate to regard the behavior of the SG flows as intermediate between that of rock avalanches and true debris flows.

Irrespective of their differences, the SGM and SG flows both exhibit rapid formation and persistence of gravel-rich snouts, which are products of grain-size segregation. Videotapes of the experiments reveal that within the first 4 s of downslope travel, gravel migrates to the surface of the flows and then advances to the snouts, where much of it is retained (Logan and Iverson 2007). As the flows travel further downslope, the maturing gravelly snouts develop the pore-pressure deficits that are evident in Figure 43.3b through f. The pore-pressure deficits arise because of the high hydraulic permeability of the gravel, which makes it incapable of maintaining much pore pressure. As a result, gravelly snouts exert more frictional resistance than the liquefied, finer-grained debris that pushes them from behind.

43.3 Physical Principles Used in Modeling

The data summarized above illustrate the crucial role of evolving pore-fluid pressure in debris flows, and recent models have emphasized that debris consists of a distinct solid phase and fluid phase in which such pressure can exist (Iverson 1997; Iverson and Denlinger 2001; Savage and Iverson 2003; Pitman and Le 2005; Takahashi 2007; Kowalski 2008). The physical principles used to construct two-phase models of debris-flow dynamics involve concepts from continuum mixture theory, soil mechanics, and fluid mechanics, which are described below. Also described is the rationale for approximating two-phase debris flows as one-phase granular flows with evolving porosities and pore-fluid pressures.

43.3.1 Continuum Conservation Laws

For each phase of a debris-flow mixture individually, as well as for the mixture as a whole, the principle of mass conservation is expressed by

$$\frac{\partial \rho}{\partial t} + \nabla \cdot \rho \vec{v} = 0, \tag{43.1}$$

where

- ρ is the mass density
- $\vec{v}$ is the velocity vector
- $\nabla \cdot \rho\vec{v}$ is the divergence of the linear momentum vector, $\rho\vec{v}$.

Conservation of linear momentum is expressed by

$$\frac{\partial \rho \vec{v}}{\partial t} + \nabla \cdot \rho \vec{v}\vec{v} = \rho \vec{g} - \nabla \cdot T, \tag{43.2}$$

where

- $\vec{g}$ is the acceleration due to gravity
- T is the stress (defined using a soil-mechanics convention in which compression is positive, because granular debris can sustain little or no tension)

Note that T and $\vec{v}\vec{v}$ are each 3 × 3 tensors, implying that $\nabla \cdot T$ and $\nabla \cdot \rho\vec{v}\vec{v}$ in (43.2) represent vectors rather than scalars obtained from vector divergences (e.g., Gidaspow 1994).

43.3.2 Density and Velocity Definitions

When mixture models are applied to debris flows, the solid and fluid constituents are generally assumed to have fixed mass densities, ρ_s and ρ_f, respectively. The definition of the mixture bulk density

$$\rho = \rho_s(1-n)+\rho_f n, \tag{43.3}$$

consequently shows that variation of the mixture porosity n wholly determines variation of ρ, provided that the debris remains fully saturated with pore fluid. Although n ranges from only about 0.3 in the densest debris flows to 0.5 in the most dilute, evolution of n plays a crucial role in debris-flow mechanics.

Like the mixture density, the mixture momentum is weighted by the mass of solid grains $\rho_s(1 - n)$ and mass of pore fluid $\rho_f n$ per unit volume, but it also depends on the velocities of each phase. Thus, the linear momentum of the mixture is defined as $\rho\vec{v} = \vec{v}_s\rho_s(1 - n) + \vec{v}_f\rho_f n$, where $\vec{v}_s$ is the velocity of the solid grains and $\vec{v}_f$ is the velocity of the pore fluid. This definition of mixture momentum implies that the mixture velocity is defined as

$$\vec{v} = \frac{\vec{v}_s\rho_s(1-n)+\vec{v}_f\rho_f n}{\rho}. \tag{43.4}$$

If separate momentum-conservation equations are written for the solid and fluid phases, these equations sum to yield the momentum-conservation equation for the mixture as a whole only if the mixture velocity is defined as in (43.4).

Another velocity that plays a key role in debris-flow mechanics is the pore-fluid velocity relative to the solid velocity, $\vec{v}_f - \vec{v}_s$. To an observer moving with the local solid velocity $\vec{v}_s$, the apparent fluid velocity is the volumetric flux of pore fluid per unit area of mixture, $\vec{q}$:

$$\vec{q} = n(\vec{v}_f - \vec{v}_s). \tag{43.5}$$

In porous media theory, $\vec{q}$ is known as the specific discharge (Bear 1972).

The definition of $\vec{q}$ can be combined algebraically with Equations 43.3 and 43.4 to obtain the important relationship

$$\frac{\vec{v}}{\vec{v}_s} = \frac{\rho_f}{\rho}\frac{\vec{q}}{\vec{v}_s}+1. \tag{43.6}$$

For debris flows this relationship commonly reduces to $\vec{v}/\vec{v}_s \approx 1$, because values $\rho_f/\rho \approx 1/2$, $\vec{v}_s > 0.1$ m/s, and $\vec{q} \ll 0.1$ m/s are typical. The approximation $\vec{v}/\vec{v}_s \approx 1$ allows the mixture momentum to be approximated as $\rho\vec{v}_s$, thereby reducing the two-phase flow problem to an equivalent one-phase problem in which $\vec{v}_s$ (or $\vec{v}$) is influenced by the solid–fluid interaction stress associated with $\vec{q}$.

43.3.3 Effective Stress and Pore-Fluid Pressure

The use of an equivalent one-phase formulation implies that the stress T in (43.2) must account for all solid–fluid interaction stresses, including those due to $\vec{q}$ (Iverson 1997). A simple but nevertheless useful approach to this problem employs a key concept from soil mechanics: the effective stress principle. This principle states that the total stress tensor T can be decomposed into components of stress borne by the solid and fluid phases, such that

$$T = T_e + Ip + nT_{vis}, \tag{43.7}$$

where

- T_e is the effective stress borne by the solid grains
- p is the pressure borne by the fluid
- I is the identity tensor
- T_{vis} is the deviatoric fluid stress that results from macroscopic viscous shearing

The stresses T_e and p are treated as if they act throughout the entire mixture, whereas T_{vis} acts only within the fluid volume fraction n. Separation of the fluid stresses in (43.7) into an isotropic component p and deviatoric component T_{vis} is similar to the convention used in fluid mechanics, but it differs owing to the definition of p as a mixture-spanning quantity.

The most important ramification of (43.7) is that increases in the pore-fluid pressure p imply attendant reductions in the mean effective normal stress borne by the solid grains, σ_e (where σ_e is a scalar equaling the mean of the diagonal components of the tensor T_e). This normal-stress reduction reduces intergranular Coulomb friction and thereby facilitates debris-flow motion.

43.3.4 Coulomb Friction

Coulomb friction generates most of the shear stress in debris flows and in other dense granular flows (Iverson 1997). In its simplest form the Coulomb friction rule states that the maximum shear resistance attainable before intergranular slip occurs is equal to the product of a constant friction coefficient and the normal stress at grain contacts. Furthermore, as slip occurs, the shear stress retains this limiting equilibrium value.

Two important modifications of the Coulomb friction rule pertain to debris flows. The first involves application of the effective-stress principle (43.7), which states that T_e is the relevant normal stress at grain contacts. The second modification accounts for the observation that the intergranular friction coefficient $\tan\phi$ can evolve as the shear rate and effective normal stress evolve. (This chapter expresses friction coefficients by using the tangent of the friction angle, ϕ, which is similar to the steepest angle of repose attainable by a static, tabular layer of grains.) These modifications result in a Coulomb friction rule expressed in 1D form as

$$\tau_s = -\mathrm{sgn}(\vec{v})[\sigma_s - p_s]\tan\phi(S), \tag{43.8}$$

where

- τ_s is the shear stress on a plane of slippage
- $-\mathrm{sgn}(\vec{v})$ denotes that this shear stress always resists motion
- σ_s and p_s are the total normal stress and pore-fluid pressure acting on the same plane as τ_s
- $\phi(S)$ denotes dependence of ϕ on the state parameter S

Multidimensional versions of (43.8) can be formulated, but they require considerable tensor algebra, a topic best reserved for detailed treatises (e.g., Desai and Siriwardane 1984).

The importance of the dimensionless parameter S in (43.8) has recently been emphasized by Forterre and Pouliquen (2008), among others; and for application to debris flows, S can be defined as

$$S = \frac{\rho_s \dot{\gamma}^2 \delta^2}{\sigma_e}, \tag{43.9}$$

where

$\dot{\gamma}$ is the local shear rate (which has dimensions of t^{-1})
δ is the characteristic diameter of grains involved in the shearing
σ_e is the mean effective normal stress defined above (Iverson et al. 2010)

Physically, S expresses the ratio of grain-scale inertial stresses (caused by dynamic grain interactions during shearing) to bulk-scale quasi-static stresses (caused by gravitational forces and reduced by pore-fluid pressure). Experiments and simulations indicate that $\tan\phi$ increases smoothly as a function of S, although variation of $\tan\phi$ is probably less than twofold over the entire domain $S = 0$ to $S \to \infty$. Remarkably, even in the limit $S \to \infty$, which indicates a liquefied state in which granular momentum exchange occurs by brief collisions rather than enduring friction at grain contacts, τ_s obeys an equation analogous to (43.8), as first demonstrated long ago by Bagnold (Hunt et al. 2002).

The Coulomb friction rule (43.8) applies to flow boundaries as well as flow interiors. Values of $\tan\phi$ along boundaries can differ from those in interiors, however, and these differences can be crucial because boundary slip can be responsible for a large fraction of the total frictional energy dissipation in debris flows (Iverson et al. 2010).

43.3.5 Dilatancy and Porosity Change

Granular materials like those in debris flows can exhibit porosity change for several reasons. Mathematically, the rate of porosity change is related to the dilation rate (i.e., the divergence of the solid grain velocity, $\nabla \cdot \vec{v}_s$) by

$$\nabla \cdot \vec{v}_s = \frac{1}{1-n}\frac{dn}{dt}, \tag{43.10}$$

where $d/dt = \partial/\partial t + \vec{v}_s \cdot \nabla$ is a total time derivative in a frame of reference that moves with the granular velocity $\vec{v}_s$ (Bear 1972). If porosity change occurs in response to shearing, the phenomenon is known as dilatancy. In densely packed states, rotund grains must move apart (exhibiting positive dilatancy) to attain sufficient space to shear past one another, whereas in loosely packed states they contract (exhibiting negative dilatancy) as shearing occurs.

In debris flows the net dilation rate depends on two interacting effects: dilatancy associated with shearing and compression caused by increases in the mean effective normal stress, σ_e. Iverson (2009) proposed that the net dilation rate can be expressed by a linear sum of these effects, yielding

$$\nabla \cdot \vec{v}_s = \dot{\gamma}\psi - \alpha \frac{d\sigma_e}{dt}, \tag{43.11}$$

where ψ is the debris' shear-induced dilatancy (a dimensionless quantity commonly expressed as an angle, $-\pi/2 \le \psi \le \pi/2$), and α is the debris' compressibility (the reciprocal of a bulk modulus). Note that if shearing of debris were to occur in a closed container that imposes the condition $\nabla \cdot \vec{v}_s = 0$, then (43.11) reduces to $d\sigma_e/dt = \dot{\gamma}\psi/\alpha$, which implies that σ_e increases with time if shearing proceeds at a constant rate $\dot{\gamma}$ with constant $\psi > 0$. This specious prediction demonstrates that ψ cannot be a material constant. Rather, it must evolve and ultimately become zero during steady-state shearing with no volume change. In soil mechanics this type of steady state is known as a critical state.

Experiments show that the dilatancy ψ evolves in a manner that depends on the current value of n relative to a value n_{eq} that is in equilibrium with the ambient state of stress and shear rate, and that this dependence roughly obeys a linear relation, $\psi = -C_1(n - n_{eq})$, where C_1 is a positive constant of order 1 (Pailha and Pouliquen 2009). Other experiments demonstrate that the dependence of n_{eq} on the ambient state of stress and shear rate can be summarized as a dependence on S (Forterre and Pouliquen 2008). Although the exact form of this dependence has not been determined for debris-flow materials, the relation $\psi = -C_1[n - n_{eq}(S)]$ can nevertheless be combined with (43.10) and (43.11) to infer that a differential equation describing evolution of n is (cf. Iverson 2009; Pailha and Pouliquen 2009)

$$\frac{dn}{dt} = C_1\dot{\gamma}(n-1)[n - n_{eq}(S)] - \alpha(1-n)\frac{d\sigma_e}{dt}. \tag{43.12}$$

The implications of (43.12) can be complicated, but some insight can be gained by assuming that $d\sigma_e/dt = 0$ and that $\dot{\gamma}$ and S are constants, such that n_{eq} is also a constant and evolution of n is decoupled from evolving debris-flow dynamics. Then an exact solution of (43.12) demonstrates that n relaxes exponentially toward its equilibrium value, n_{eq}, with a relaxation time $1/(1 - n_{eq})C_1\dot{\gamma}$. This result implies that the speed of porosity relaxation is directly proportional to the shear rate $\dot{\gamma}$—but only if changes in σ_e and S do not intercede.

43.3.6 Excess Pore-Fluid Pressure and Darcy Drag

Equation 43.12 indicates that porosity evolution depends not only on shear-induced dilatancy but also on evolution of the mean effective normal stress, σ_e (and, thus, on evolution of pore-fluid pressure). Indeed, evolution of porosity implies that relative motion of the solid and fluid phases must occur (i.e., $\vec{v}_f - \vec{v}_s \neq 0$), and this relative motion necessarily results in momentum exchange that modifies pore-fluid pressure. In debris-flow mechanics it is conventional to approximate the effects of solid–fluid momentum exchange by using a simple linear drag rule (Darcy's law), which can be expressed as

$$\vec{q} = n(\vec{v}_f - \vec{v}_s) = -\frac{k}{\mu}\nabla p_e, \tag{43.13}$$

where

- $\vec{q}$ is the specific discharge defined in (43.5)
- k is the hydraulic permeability of the granular assemblage (a quantity with dimensions of length squared)
- μ is the viscosity of the pore fluid

The quantity p_e is the "excess" pore-fluid pressure defined as $p_e = p - \rho_f g(h - z)\cos\theta$, where p is the total pore-fluid pressure, and $\rho_f g(h - z)\cos\theta$ is the hydrostatic equilibrium pressure at a slope-normal height z in a debris flow of thickness h on a slope with angle θ. Values of k for typical debris-flow materials range from about 10^{-13} to $10^{-8}\,m^2$, and values of μ for muddy pore fluid range from about 10^{-2} to 10 Pa-s, so that k/μ ranges from about 10^{-14} to $10^{-6}\,m^2/Pa\,s$. As a consequence, (43.13) implies that very significant excess pore-pressure gradients ($\nabla p_e \sim \rho_f g \sim 10^4$ Pa/m) will develop in reaction to solid–fluid momentum exchange if the magnitude of $\vec{q}$ exceeds 10^{-2} m/s for debris flows with the largest k/μ values and if it exceeds 10^{-10} m/s for debris flows with the smallest k/μ values. The large excess pore-pressure gradients associated with these $\vec{q}$ magnitudes tend to inhibit development of greater $\vec{q}$ magnitudes, owing to the effects of pore-pressure diffusion.

43.3.7 Pore-Pressure Diffusion

The relationship between the porosity change described by (43.10) and the excess pore-pressure gradient described by (43.13) implies that pore-pressure evolution is mathematically analogous to forced diffusion. This analogy is revealed by first observing that, if the mixture remains saturated, mass conservation dictates that the divergence of the solid grain velocity $\nabla \cdot \vec{v}_s$ in (43.10) must be balanced by a counter-flow of pore fluid, such that

$$\nabla \cdot \vec{v}_s = -\nabla \cdot \vec{q}. \tag{43.14}$$

Substitution of (43.13) and (43.14) into (43.10) then yields a fundamental equation that shows how porosity change is related to the divergence of ∇p_e:

$$\frac{1}{1-n}\frac{dn}{dt} = \nabla \cdot \frac{k}{\mu}\nabla p_e. \tag{43.15}$$

The porosity n can be eliminated from this equation by combining (43.15) with (43.10) and (43.11) to obtain an equation with $d\sigma_e/dt$ on the left-hand side. For the case in which k/μ is constant, this equation reduces to

$$\frac{d\sigma_e}{dt} = -\frac{k}{\alpha\mu}\nabla^2 p_e + \frac{\dot{\gamma}\psi}{\alpha}, \tag{43.16}$$

where $k/\alpha\mu$ plays the role of a pore-pressure diffusivity (which has dimensions of length squared per unit time). Finally, the definition of effective stress (43.7) can be used to infer that $d\sigma_e/dt = d\sigma/dt - dp_e/dt - d[\rho_f g(h - z)\cos\theta]/dt$, where σ is the mean total stress, and substitution of this relationship in (43.16) enables the equation to be recast as a forced, advection-diffusion equation for p_e,

$$\frac{dp_e}{dt} - \frac{k}{\alpha\mu}\nabla^2 p_e = \frac{d}{dt}[\sigma - \rho_f g(h-z)\cos\theta] - \frac{\dot{\gamma}\psi}{\alpha}. \tag{43.17}$$

The total time derivative $dp_e/dt = \partial p_e/\partial t + \vec{v}_s \cdot \nabla p_e$ on the left-hand side of (43.17) includes the effects of advection, and the forcing terms on the right-hand side of (43.17) express the evolving effects of the shear-induced dilation rate $\dot{\gamma}\psi$, the mean total stress σ, and the hydrostatic pore-pressure component $\rho_f g(h - z)\cos\theta$. Note that if all of the time derivatives in (43.17) are zero and $\dot{\gamma}\psi$ is constant, the equation reduces to the steady-state balance $(k/\mu)\nabla^2 p_e = \dot{\gamma}\psi$, which can alternatively be written as $-\nabla \cdot \vec{q} = \dot{\gamma}\psi$. This result shows that porosity creation during steady dilation is balanced by a steady influx of fluid that fills the enlarging pores.

The forcing effects described by the right-hand side of (43.17) can drive pore-pressure change, but p_e can evolve even in the absence of forcing owing to diffusion described by the left-hand side of (43.17). Normalization of the left-hand side of (43.17) shows that excess pore pressure relaxes diffusively with a characteristic time $h^2\alpha\mu/k$. This relaxation time includes not only the pore-pressure diffusivity $k/\alpha\mu$ but also the square of the debris-flow thickness h, which is the length scale over which pore-pressure diffusion typically occurs. Owing to this dependence on h^2, pore-pressure relaxation proceeds more slowly in large debris flows than in small ones. Thus, once excess pore pressure develops, large debris flows can maintain lower Coulomb friction and exhibit greater mobility than can small flows.

43.3.8 Disparate Relaxation Times and Limits on Feedback

Equations 43.12, 43.15, and 43.17 indicate a strong interdependence between evolution of porosity, dilatancy, excess pore-fluid pressure, and effective stress. They also show that a large disparity exists between the characteristic timescales for porosity relaxation in the absence of pore pressure ($1/(1 - n_{eq})C_1\dot{\gamma}$ commonly ~ 1 s) and for dissipation of excess pore-fluid pressure that occurs in response to porosity change ($h^2\alpha\mu/k$ commonly > 10^3 s). Such disparate values imply that the inherently fast process of shear-induced porosity change can rapidly generate pore-pressure changes that inhibit further porosity change. Thus, in the absence of changes in forcing (such as changes in bed slope that drive changes in $\dot{\gamma}$), diffusive pore-pressure responses tend to stabilize debris-flow motion by regulating pore-pressure feedback that influences frictional resistance to flow.

Pore-pressure feedback may also be subject to lower and upper bounds due to phenomena not explicitly represented in (43.12), (43.15), and (43.17). For example, the effective lower limit on pore-fluid pressure in debris flows is probably zero (i.e., the atmospheric reference pressure). Negative pore pressures might occur in dewatering debris, but they would result from surface tension at air–water interfaces (i.e., meniscuses) in partly filled pore spaces, and such delicate features seem unlikely to have significant effects on agitated, coarse-grained debris. To date, debris-flow models have ignored them. A practical upper bound on pore-fluid pressure in debris flows is probably the liquefaction pressure, $p = \sigma$, which produces $\sigma_e = 0$. Higher pore pressures theoretically could be produced by forced debris contraction, but the propensity for contraction largely vanishes as $\sigma_e \to 0$.

43.3.9 Effects of Grain-Size Segregation

The development and dissipation of excess pore-fluid pressure described by (43.17) underscores the significance of grain-size segregation in debris flows, because the value of k in (43.17) depends strongly on the local grain grain-size distribution. As discussed in Section 43.2, grain-size segregation leads to a characteristic debris-flow architecture in which coarse-grained, high-friction snouts that lack much pore pressure impede the motion of trailing, liquefied, fine-grained debris. At present no satisfactory model exists for predicting grain-size segregation in debris flows, although recent advances in granular mechanics indicate that progress may be forthcoming (Gray and Kokelaar 2010). A stop-gap approach mimics the effect of grain-size segregation by specifying a heterogeneous k distribution (Savage and Iverson 2003).

43.3.10 Boundary Erosion and Mass Change

As noted in Section 43.2, debris flows commonly gain mass as they descend steep, erodible slopes and channels, and they begin to lose mass and form deposits when they reach flatter terrain. No precise criteria exist for determining where this transition occurs, however, and no widely accepted formula exists for predicting the rate of mass change. Many debris-flow models, such as the one described in the following, account for the *effects* of mass change in conservation laws, but they take no account of the forces necessary to *cause* the mass change. Better understanding of the mechanics of mass change awaits further research.

43.4 Model Formulation and Analysis

Key steps in the development of most mathematical models of debris-flow motion include depth integration of the governing equations and shallow-flow scaling that justifies neglect of some terms. Depth integration removes the explicit appearance of one velocity component (here denoted by v_z) and thereby reduces the number of dependent variables. It also embeds, within the conservation equations, kinematic boundary conditions that describe the position of the free upper surface and basal flow boundary, thereby eliminating the need to track motion of these boundaries separately. Finally, it readily incorporates mass-change terms that describe fluxes of debris through the upper and basal boundaries.

Because of the need to specify the direction of depth integration (the z direction) *a priori*, the choice of a coordinate system is crucial. Some models use an Earth-centered, orthogonal Cartesian coordinate system with z vertical, which has the advantage of being universal and independent of terrain geometry. Such a system leads to complicated mechanical considerations when computing motion across steep, irregular slopes, however (Denlinger and Iverson 2004). Other models, including the one presented here, use a z coordinate normal to the local ground surface, such that the x coordinate is directed downslope and the y coordinate cross-slope (Figure 43.4). This approach simplifies the mechanics, but it requires use of curvilinear coordinate systems to adapt it to natural terrain. This chapter omits consideration of the complex mathematics associated with curvilinear coordinates, and instead focuses on the mechanical implications of depth-integrated conservation laws.

43.4.1 Depth-Integrated Conservation Laws with Mass and Bulk-Density Change

Consider motion of a debris flow of variable bulk density ρ moving down a planar slope inclined at the angle θ (Figure 43.4). (Recall from (43.3) that a simple relation exists between variations in debris-flow bulk density and porosity, provided that the

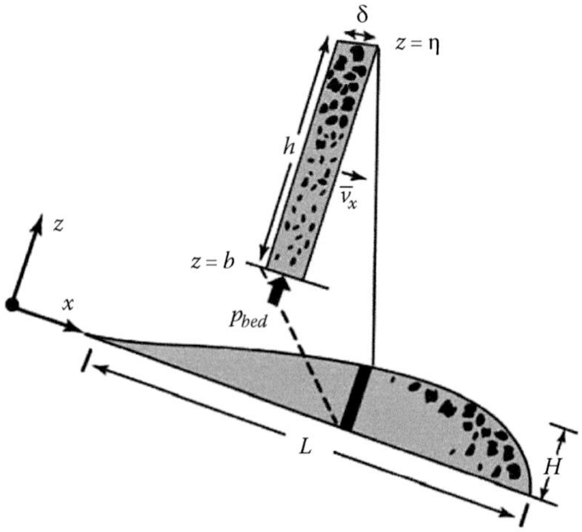

FIGURE 43.4 Schematic vertical cross section of a debris flow descending a uniform slope inclined at the angle θ. The x–z coordinate system and flow length scales H and L are defined. Magnified slice illustrates the dependent variables, $\bar{v}_x$, h, and p_{bed} as well as a local grain length scale, δ. (Reproduced from Iverson, R.M. et al., *J. Geophys. Res.*, 115, 2010, doi:10.1029/2009JF001514. With permission of the American Geophysical Union.)

flow remains saturated with pore fluid. Thus, a variable-bulk-density model can be used in place of a variable-porosity model, and it can be extended to include cases with variable saturation.) The vector conservation laws (43.1) and (43.2) imply that the scalar equations describing conservation of mass and the downslope (x) component of linear momentum are

$$\frac{\partial \rho}{\partial t}+\frac{\partial(\rho v_x)}{\partial x}+\frac{\partial(\rho v_y)}{\partial y}+\frac{\partial(\rho v_z)}{\partial z}=0, \tag{43.18}$$

$$\frac{\partial(\rho v_x)}{\partial t}+\frac{\partial(\rho v_x^2)}{\partial x}+\frac{\partial(\rho v_x v_y)}{\partial y}+\frac{\partial(\rho v_x v_z)}{\partial z}$$
$$=\rho g_x-\frac{\partial \tau_{xx}}{\partial x}-\frac{\partial \tau_{yx}}{\partial y}-\frac{\partial \tau_{zx}}{\partial z}=\Sigma F_x, \tag{43.19}$$

where v_x, v_y, and v_z, are, respectively, the x, y, and z components of the debris velocity (either $\vec{v}$ or $\vec{v}_s$, as described in Section 43.3); $\rho g_x = \rho g \sin\theta$ is the x component of the debris weight per unit volume; τ_{xx}, τ_{yx} and τ_{zx} are the components of the stress tensor acting in the x direction; and ΣF_x is shorthand notation for the sum of the weight and stress terms. Equations analogous to (43.19) describe conservation of the y and z momentum components.

Depth averages of the dependent variables in (43.18) and (43.19) are defined as

$$\bar{\rho}(x,y,t)=\frac{1}{h}\int_b^\eta \rho\,dz \quad \bar{v}_x(x,y,t)=\frac{1}{h}\int_b^\eta v_x\,dz$$
$$\bar{v}_y(x,y,t)=\frac{1}{h}\int_b^\eta v_y\,dz \quad \bar{v}_z(x,y,t)=\frac{1}{h}\int_b^\eta v_z\,dz, \tag{43.20}$$

where $z = b(x, y, t)$ is the position of the debris-flow base, $z = \eta(x,y,t)$ is the position of the free upper surface, and $h(x,y,t) = \eta - b$ is the flow thickness (Figure 43.4). Subsequent equations are simplified by assuming that ρ varies only as a function of x, y, and t, such that $\rho = \bar{\rho}$. Similar assumptions are necessary in any depth-integrated model.

Integration of (43.18) and (43.19) through the debris-flow thickness from its base at $z = b$ to its upper surface at $z = \eta$ employs kinematic boundary conditions that relate $v_z(b)$ and $v_z(\eta)$ to the other velocity components at $z = b$ and $z = \eta$ and to variations in the boundary positions:

$$v_z(b)=\frac{\partial b}{\partial t}+v_x(b)\frac{\partial b}{\partial x}+v_y(b)\frac{\partial b}{\partial y}-B(x,y,t), \tag{43.21}$$

$$v_z(\eta)=\frac{\partial \eta}{\partial t}+v_x(\eta)\frac{\partial \eta}{\partial x}+v_y(\eta)\frac{\partial \eta}{\partial y}-A(x,y,t). \tag{43.22}$$

Here A and B are the boundary-migration velocities (positive upward) caused by the possible entry of debris at the free upper surface or basal surface of the flow, respectively. For example, if a static heap of debris with $v_x = v_y = v_z = 0$ is subject to upper-surface accretion at a rate $A > 0$, (43.22) shows that the heap's height η increases at the rate $\partial\eta/\partial t = A$. The situation is more complicated at the base of a moving debris flow, where either erosion or sedimentation can occur and all of the terms in (43.21) can evolve simultaneously, but $B > 0$ always characterizes upward bed migration due to sedimentation, and $B < 0$ characterizes bed lowering due to erosion. Subsequent equations assume that the bulk density of bed and bank material potentially incorporated in the debris flow locally equals $\bar{\rho}$ of the flow itself.

For the case with $\rho = \bar{\rho}$, depth integration of the mass-conservation Equation 43.18 yields

$$\int_b^\eta\left[\frac{\partial \bar{\rho}}{\partial t}+\frac{\partial(\bar{\rho} v_x)}{\partial x}+\frac{\partial(\bar{\rho} v_y)}{\partial y}+\frac{\partial(\bar{\rho} v_z)}{\partial z}\right]dz$$
$$=h\frac{\partial \bar{\rho}}{\partial t}+h\bar{v}_x\frac{\partial \bar{\rho}}{\partial x}+h\bar{v}_y\frac{\partial \bar{\rho}}{\partial y}+\bar{\rho}\left[v_z(\eta)-v_z(b)\right]$$
$$+\bar{\rho}\left[\frac{\partial}{\partial x}\int_b^\eta v_x dz-v_x(\eta)\frac{\partial \eta}{\partial x}+v_x(b)\frac{\partial b}{\partial x}\right.$$
$$\left.+\frac{\partial}{\partial y}\int_b^\eta v_y dz-v_y(\eta)\frac{\partial \eta}{\partial y}+v_y(b)\frac{\partial b}{\partial y}\right]$$
$$=h\frac{\partial \bar{\rho}}{\partial t}+\frac{\partial(\bar{\rho}h\bar{v}_x)}{\partial x}+\frac{\partial(\bar{\rho}h\bar{v}_y)}{\partial y}-\bar{\rho}\left[v_x(\eta)\frac{\partial \eta}{\partial x}+v_y(\eta)\frac{\partial \eta}{\partial y}-v_z(\eta)\right]$$
$$+\bar{\rho}\left[v_x(b)\frac{\partial b}{\partial x}+v_y(b)\frac{\partial b}{\partial y}-v_z(b)\right]$$
$$=\frac{\partial(\bar{\rho}h)}{\partial t}+\frac{\partial(\bar{\rho}h\bar{v}_x)}{\partial x}+\frac{\partial(\bar{\rho}h\bar{v}_y)}{\partial y}-\bar{\rho}A+\bar{\rho}B=0. \tag{43.23}$$

The third and fourth lines of (43.23) illustrate the result of using Leibniz rule for interchanging the order of integration and differentiation during evaluation of $\int_b^\eta[\partial v_x/\partial x]dz$ and $\int_b^\eta[\partial v_y/\partial y]dz$. The last line of (43.23) results from substituting the kinematic boundary conditions (43.21) and (43.22) into the fifth and sixth lines of (43.23) in place of $v_z(\eta)$ and $v_z(b)$, and then cancelling terms that sum to zero and making the identification $\partial(\eta - b)/\partial t = \partial h/\partial t$. If $A = B = 0$ and $\bar{\rho}$ is constant, then Equation 43.23 reduces to the form of the mass-conservation equation used in standard shallow-water flow theory, $\partial h/\partial t + \partial(h\bar{v}_x)/\partial x + \partial(h\bar{v}_y)/\partial y = 0$.

Depth integration of the x-component momentum-conservation Equation (43.19) employs Leibniz rule and uses the kinematic boundary conditions (43.21) and (43.22) to cancel many terms. For the case with $\rho = \bar{\rho}$ the integration yields

$$\int_b^\eta \left[\frac{\partial(\bar{\rho} v_x)}{\partial t} + \frac{\partial(\bar{\rho} v_x^2)}{\partial x} + \frac{\partial(\bar{\rho} v_x v_y)}{\partial y} + \frac{\partial(\bar{\rho} v_x v_z)}{\partial z} \right] dz$$

$$= \frac{\partial}{\partial t} \bar{\rho} \int_b^\eta v_x dz - \bar{\rho} v_x(\eta) \frac{\partial \eta}{\partial t} + \bar{\rho} v_x(b) \frac{\partial b}{\partial t}$$

$$+ \frac{\partial}{\partial x} \bar{\rho} \int_b^\eta v_x^2 dz - \bar{\rho} v_x^2(\eta) \frac{\partial \eta}{\partial x} + \bar{\rho} v_x^2(b) \frac{\partial b}{\partial x}$$

$$+ \frac{\partial}{\partial y} \bar{\rho} \int_b^\eta v_x v_y dz - \bar{\rho} v_x(\eta) v_y(\eta) \frac{\partial \eta}{\partial y} + \bar{\rho} v_x(b) v_y(b) \frac{\partial b}{\partial y}$$

$$+ \bar{\rho} v_x(\eta) v_z(\eta) - \bar{\rho} v_x(b) v_z(b)$$

$$= \frac{\partial(\bar{\rho} h \bar{v}_x)}{\partial t} + \frac{\partial(\bar{\rho} h \bar{v}_x^2)}{\partial x} + \frac{\partial(\bar{\rho} h \bar{v}_x \bar{v}_y)}{\partial y}$$

$$+ \frac{\partial}{\partial x} \bar{\rho} \int_b^\eta (v_x - \bar{v}_x)^2 dz + \frac{\partial}{\partial y} \bar{\rho} \int_b^\eta (v_x - \bar{v}_x)(v_y - \bar{v}_y) dz - \bar{\rho} v_x(\eta) A$$

$$+ \bar{\rho} v_x(b) B$$

$$= \int_b^\eta \Sigma F_x dz. \tag{43.24}$$

The integrals in the seventh line of (43.24) arise from the use of the identities

$$\int_b^\eta v_x^2 \, dz = h\bar{v}_x^2 + \int_b^\eta (v_x - \bar{v}_x)^2 dz$$

$$\int_b^\eta v_x v_y dz = h\bar{v}_x \bar{v}_y + \int_b^\eta (v_x - \bar{v}_x)(v_y - \bar{v}_y) dz, \tag{43.25}$$

which show how the depth integrals of products are related to the products of depth integrals. Physically, the integrands $(v_x - \bar{v}_x)^2$ and $(v_x - \bar{v}_x)(v_y - \bar{v}_y)$ in (43.24) and (43.25) describe the effects of differential advection of momentum due to variations of v_x and v_y with depth z. Most debris-flow models neglect these effects and assume that $v_x = \bar{v}_x$ and $v_y = \bar{v}_y$, and the same approach is used here. Differential advection of momentum might play a particularly important role, however, if erosion or deposition occurs and velocity gradients near the bed are significant.

With neglect of the differential advection terms, (43.24) reduces to

$$\frac{\partial(\bar{\rho} h \bar{v}_x)}{\partial t} + \frac{\partial(\bar{\rho} h \bar{v}_x^2)}{\partial x} + \frac{\partial(\bar{\rho} h \bar{v}_x \bar{v}_y)}{\partial y} = \int_b^\eta \Sigma F_x dz + \bar{\rho} v_x(\eta) A - \bar{\rho} v_x(b) B. \tag{43.26}$$

This equation can be manipulated into useful alternative forms by expanding the derivatives on the left-hand side to obtain

$$\bar{v}_x \bar{\rho} \left[\frac{\partial h}{\partial t} + \frac{\partial(h\bar{v}_x)}{\partial x} + \frac{\partial(h\bar{v}_y)}{\partial y} \right] + \bar{v}_x h \left[\frac{\partial \bar{\rho}}{\partial t} + \bar{v}_x \frac{\partial \bar{\rho}}{\partial x} + \bar{v}_y \frac{\partial \bar{\rho}}{\partial y} \right]$$

$$+ \bar{\rho} h \left[\frac{\partial \bar{v}_x}{\partial t} + \bar{v}_x \frac{\partial \bar{v}_x}{\partial x} + \bar{v}_y \frac{\partial \bar{v}_x}{\partial y} \right] = \int_b^\eta \Sigma F_x dz + \bar{\rho} v_x(\eta) A - \bar{\rho} v_x(b) B. \tag{43.27}$$

The second term in brackets in (43.27) can be replaced with the depth-averaged total time derivative $\bar{d}\bar{\rho}/\bar{d}t$, where $\bar{d}/\bar{d}t = \partial/\partial t + \bar{v}_x \partial/\partial x + \bar{v}_y \partial/\partial y$, and the first and third terms in brackets can be recombined to reduce (43.27) to

$$\bar{\rho} \left[\frac{\partial(h\bar{v}_x)}{\partial t} + \frac{\partial(h\bar{v}_x^2)}{\partial x} + \frac{\partial(h\bar{v}_x \bar{v}_y)}{\partial y} \right] + \bar{v}_x h \frac{\bar{d}\bar{\rho}}{\bar{d}t}$$

$$= \int_b^\eta \Sigma F_x dz + \bar{\rho} v_x(\eta) A - \bar{\rho} v_x(b) B. \tag{43.28}$$

This form of the x-momentum equation is called "conservative" because the left-hand side represents evolution of a conserved variable, x-momentum per unit area $dxdy$. The terms $\bar{\rho} v_x(\eta) A$ and $\bar{\rho} v_x(b) B$ on the right-hand side of (43.28) account for x-momentum carried into or out of the flow by material with $v_x \neq 0$ passing through its upper or lower boundaries.

An alternative, "primitive" form of the x-momentum equation is obtained by using (43.23) and $\bar{d}/\bar{d}t$ as defined earlier to replace the first term in brackets in (43.27) with $-(h/\bar{\rho})(\bar{d}\bar{\rho}/\bar{d}t) + A - B$, and then replace the second and third terms in brackets with $\bar{d}\bar{\rho}/\bar{d}t$ and $\bar{d}\bar{v}_x/\bar{d}t$, respectively. These substitutions result in cancelations that reduce (43.27) to

$$\bar{\rho} h \frac{\bar{d}\bar{v}_x}{\bar{d}t} = \int_b^\eta \Sigma F_x dz + \bar{\rho} v_x(\eta) A - \bar{\rho} v_x(b) B - \bar{\rho} \bar{v}_x A + \bar{\rho} \bar{v}_x B, \tag{43.29}$$

Although (43.29) is correct mathematically, it describes evolution of $\bar{v}_x$, a variable that is not physically conserved. As a result, the right-hand side of (43.29) contains the added terms $-\bar{\rho}\bar{v}_x A$ and $+\bar{\rho}\bar{v}_x B$, which misleadingly appear to represent momentum sources or sinks not present in (43.26), (43.27), or (43.28). In fact, $-\bar{\rho}\bar{v}_x A$ and $+\bar{\rho}\bar{v}_x B$ merely account for terms cancelled from the left-hand side of (43.29), and this fact must be borne in mind if Lagrangian numerical methods that employ (43.29) are used to compute solutions.

Evaluation of the forcing term $\int_b^\eta \Sigma F_x \, dz$ on the right-hand sides of (43.24) and (43.26) through (43.29) employs Leibniz rule during depth integration of the stress-gradient components, which appear explicitly in (43.19). Three terms generated in this integration vanish because they involve $\tau_{xx}(\eta)$, $\tau_{yx}(\eta)$ or $\tau_{zx}(\eta)$, which all equal zero owing to the stress-free condition of the upper flow boundary. The remaining terms yield

$$\int_b^\eta \Sigma F_x \, dz = \bar{\rho} g_x h - \int_b^\eta \left[\frac{\partial \tau_{xx}}{\partial x} + \frac{\partial \tau_{yx}}{\partial y} + \frac{\partial \tau_{zx}}{\partial z} \right] dz$$

$$= \bar{\rho} g h \sin\theta - \frac{\partial(\bar{\tau}_{xx} h)}{\partial x} - \tau_{xx}(b)\frac{\partial b}{\partial x} - \frac{\partial(\bar{\tau}_{yx} h)}{\partial y} - \tau_{yx}(b)\frac{\partial b}{\partial y} + \tau_{zx}(b). \tag{43.30}$$

Note that according to the sign convention used here, $\tau_{zx}(b) < 0$ when $\tau_{zx}(b)$ resists distortion associated with $[\partial v_x/\partial z]_{z=b} > 0$. Thus, in (43.30) and succeeding equations, the basal shear stress $\tau_{zx}(b)$ helps resist the gravitational driving term $\bar{\rho} g h \sin\theta$.

Simplification of (43.30) is possible if the bed surface remains parallel to the x–y plane during erosion or sedimentation, such that $\partial b/\partial x = 0$ and $\partial b/\partial y = 0$. Then (43.30) reduces to

$$\int_b^\eta \Sigma F_x \, dz = \bar{\rho} g h \sin\theta - \frac{\partial(\bar{\tau}_{xx} h)}{\partial x} - \frac{\partial(\bar{\tau}_{yx} h)}{\partial y} + \tau_{zx}(b). \tag{43.31}$$

Substitution of (43.31) in (43.28) enables the x-momentum equation to be written as

$$\bar{\rho}\left[\frac{\partial(h\bar{v}_x)}{\partial t} + \frac{\partial(h\bar{v}_x^2)}{\partial x} + \frac{\partial(h\bar{v}_x\bar{v}_y)}{\partial y} \right] + \bar{v}_x h \frac{\bar{d}\bar{\rho}}{\bar{d}t}$$

$$= \bar{\rho} g h \sin\theta - \frac{\partial(\bar{\tau}_{xx} h)}{\partial x} - \frac{\partial(\bar{\tau}_{yx} h)}{\partial y} + \tau_{zx}(b)$$

$$+ \bar{\rho} v_x(\eta) A - \bar{\rho} v_x(b) B, \tag{43.32}$$

and the y-component momentum equation has a form exactly analogous to that of (43.32). The z-component momentum equation is also closely analogous, but it is useful to write it explicitly because of its critical role in scaling and stress evaluation:

$$\bar{\rho}\left[\frac{\partial(h\bar{v}_z)}{\partial t} + \frac{\partial(h\bar{v}_z\bar{v}_x)}{\partial x} + \frac{\partial(h\bar{v}_z\bar{v}_y)}{\partial y} \right] + \bar{v}_z h \frac{\bar{d}\bar{\rho}}{\bar{d}t}$$

$$= -\bar{\rho} g h \cos\theta - \frac{\partial(\bar{\tau}_{xz} h)}{\partial x} - \frac{\partial(\bar{\tau}_{yz} h)}{\partial y}$$

$$+ \tau_{zz}(b) + \bar{\rho} v_z(\eta) A - \bar{\rho} v_z(b) B. \tag{43.33}$$

43.4.2 Scaling and the Shallow-Flow Approximation

Further simplification of the depth-integrated momentum-conservation equations relies on the identification of characteristic scales for all variables they contain. Scaling, in turn, leads to identification of small terms that can be neglected. As illustrated in Figure 43.4, the characteristic length of a debris flow can be defined as L, and the characteristic thickness can be defined as H, so that the length scale for the x and y coordinates is L and the length scale for the z coordinate is H. Similarly, the scale for b, η, and h (all measured in the z direction) is H. Because debris-flow motion is driven by gravitational potential, the scale for the velocity components in the x and y directions is $(gL)^{1/2}$, and the scale for the z-direction velocity component is $(gH)^{1/2}$. A z-direction velocity scale $\beta(gH)^{1/2}$, which is adjusted by the arbitrary factor β, applies to the erosion and sedimentation rates A and B, because the magnitudes of these quantities are poorly constrained. Values $\beta \ll 1$ seem probable in most circumstances, however. The timescale for debris-flow motion $(L/g)^{1/2}$ is the downslope length scale L divided by the downslope velocity scale $(gL)^{1/2}$. The scale for $\bar{\rho}$ is an equilibrium value, such as the initial static value, $\bar{\rho}_0$. Finally, the scale for all stress components (τ_{xx}, τ_{yy}, τ_{zz}, τ_{yx}, τ_{zx}, τ_{yz}) is the equilibrium lithostatic stress, $\bar{\rho}_0 gH$. The use of these scales enables definition of the following dimensionless variables, denoted by asterisks:

$$x^* = x/L \quad y^* = y/L \quad z^* = z/H \quad t^* = t/(L/g)^{1/2}$$

$$v_x^* = v_x/(Lg)^{1/2} \quad v_y^* = v_y/(Lg)^{1/2} \quad v_z^* = v_z/(Hg)^{1/2}$$

$$h^* = h/H \quad \bar{\rho}^* = \bar{\rho}/\bar{\rho}_0$$

$$\eta^* = \eta/H \quad b^* = b/H \quad A^* = A/\beta(Hg)^{1/2} \quad B^* = B/\beta(Hg)^{1/2}$$

$$(\tau_{xx}^*, \tau_{yy}^*, \tau_{zz}^*, \tau_{yx}^*, \tau_{zx}^*, \tau_{yz}^*) = (\tau_{xx}, \tau_{yy}, \tau_{zz}, \tau_{yx}, \tau_{zx}, \tau_{yz})/\bar{\rho}_0 gH \tag{43.34}$$

Substitution of (43.34) into (43.32) and (43.33) results in scaled forms of the equations

$$\bar{\rho}^*\left[\frac{\partial(h^*\bar{v}_x^*)}{\partial t^*} + \frac{\partial(h^*\bar{v}_x^{*2})}{\partial x^*} + \frac{\partial(h\bar{v}_x^*\bar{v}_y^*)}{\partial y^*} \right] + \bar{v}_x^* h^* \frac{\bar{d}\bar{\rho}^*}{\bar{d}t^*}$$

$$= \bar{\rho}^* h^* \sin\theta - \varepsilon\left[\frac{\partial(\bar{\tau}_{xx}^* h^*)}{\partial x^*} + \frac{\partial(\bar{\tau}_{yx}^* h^*)}{\partial y^*} \right]$$

$$+ \tau_{zx}^*(b^*) + \varepsilon^{-1/2}\beta\bar{\rho}^*\left[v_x^*(\eta^*)A^* - v_x^*(b^*)B^* \right] \tag{43.35}$$

and

$$\varepsilon^{1/2}\bar{\rho}^*\left[\frac{\partial(h^*\bar{v}_z^*)}{\partial t^*} + \frac{\partial(h^*\bar{v}_z^*\bar{v}_x^*)}{\partial x^*} + \frac{\partial(h^*\bar{v}_z^*\bar{v}_y^*)}{\partial y^*} \right]$$

$$+ \varepsilon^{1/2}\bar{v}_z^* h^* \frac{\bar{d}\bar{\rho}^*}{\bar{d}t^*}$$

$$= -\bar{\rho}^* h^* \cos\theta - \varepsilon\left[\frac{\partial(\bar{\tau}_{xz}^* h^*)}{\partial x^*} + \frac{\partial(\bar{\tau}_{yz}^* h^*)}{\partial y^*} \right]$$

$$+ \tau_{zz}^*(b^*) + \beta\bar{\rho}^*\left[v_z^*(\eta^*)A^* - v_z^*(b^*)B^* \right], \tag{43.36}$$

where

$$\varepsilon = H/L \tag{43.37}$$

is a fundamental length-scale ratio. The condition $\varepsilon \ll 1$ commonly applies in debris flows, which generally have thicknesses much smaller than their lengths and widths. Neglect of all terms containing ε or $\varepsilon^{1/2}$ in (43.35) and (43.36) therefore constitutes a rigorous *shallow-flow approximation*. The factor ε does not appear in the mass-conservation Equation 43.23 when it is scaled using (43.34), implying that no terms in (43.23) can be neglected.

43.4.3 Stress Estimation

Several physical implications of the shallow-flow approximation are noteworthy. Most significantly, if $\beta \ll 1$ and the erosion and sedimentation terms involving A^* and B^* are neglected, the condition $\varepsilon \ll 1$ reduces (43.36) to a static balance between the basal normal stress and the slope-normal component of the debris weight,

$$\tau_{zz}^*(b^*) = \bar{\rho}^* h^* \cos\theta \quad \text{(or, dimensionally, } \tau_{zz}(b) = \bar{\rho} g h \cos\theta). \tag{43.38}$$

This approximation is employed in most shallow-flow theories, and as shown in the following, it can be used as a basis for estimating the stress components in (43.35). If $\beta \neq 0$, however, (43.38) is modified by the flux of z-direction momentum due to debris entering or leaving the flow. If, for example, basal sedimentation occurs at a rate $B^* > 0$, accompanied by a downward basal velocity $v_z^*(b^*) < 0$, the term $\beta\bar{\rho}^* B^* v_z^*(b^*)$ reduces the basal normal stress $\tau_{zz}^*(b^*)$.

Estimation of the stress components in (43.32) or (43.35) hinges on the validity of (43.38) or some other approximation of the z-momentum balance, and on the use of a constitutive model such as the Coulomb equation (43.8). Substitution of (43.38) in (43.8), and inclusion of the viscous stresses specified in (43.7), enables the basal shear stress in (43.32) to be estimated as

$$\tau_{xz}(b) = -[\bar{\rho} g h \cos\theta - p_{bed}]\tan\phi(S) - 2\bar{n}\mu\left(\frac{\bar{v}_x}{h}\right), \tag{43.39}$$

where

- $\bar{\rho} g h \cos\theta$ $(=\sigma_{bed})$ is the total basal normal stress
- p_{bed} is the basal pore-fluid pressure
- $\bar{n}$ is the depth-averaged porosity
- $-2\bar{n}\mu(\bar{v}_x/h)$ is the pore-fluid stress associated with viscous shearing at an estimated depth-averaged rate $2\bar{v}_x/h$

Importantly, σ_{bed}, p_{bed}, h, and $\bar{v}_x$ are readily measured quantities, as illustrated in Figure 43.3.

Although the lateral stress-gradient terms $\partial(\bar{\tau}_{xx}h)/\partial x$ and $\partial(\bar{\tau}_{yx}h)/\partial y$ in (43.32) are typically small, as indicated by the factor ε that precedes them in (43.35), these terms are generally included in depth-averaged models because their neglect would leave only rigid-body forcing effects on the right-hand side of (43.32). The terms can be approximated by inferring that $\bar{\tau}_{zz} \approx \tau_{zz}(b)/2$ and $\bar{p} \approx p_{bed}/2$, and postulating that $\bar{\tau}_{xx} = \kappa_1\bar{\tau}_{zz} + \bar{p}$ and $\bar{\tau}_{yx} = \kappa_2\bar{\tau}_{zz} + \bar{n}\mu(\partial\bar{v}_x/\partial y)$, where κ_1 and κ_2 are proportionality coefficients (of order 1) that describe the magnitude of lateral stress transfer by solid grains, and $\bar{p}$ and $\bar{n}\mu(\partial\bar{v}_x/\partial y)$ are depth-averaged fluid stresses due to pressure and viscous shearing, respectively (Iverson and Denlinger 2001). The use of these expressions in conjunction with (43.38) yields

$$\frac{\partial(h\bar{\tau}_{xx})}{\partial x} = \frac{1}{2}\frac{\partial}{\partial x}\kappa_1[\bar{\rho} g h^2\cos\theta - h p_{bed}] + \frac{1}{2}\frac{\partial(h p_{bed})}{\partial x} \tag{43.40}$$

$$\frac{\partial(h\bar{\tau}_{yx})}{\partial y} = \frac{1}{2}\frac{\partial}{\partial y}\kappa_2[\bar{\rho} g h^2\cos\theta - h p_{bed}] + \bar{n}\mu\left[h\frac{\partial^2\bar{v}_x}{\partial y^2} + \frac{\partial h}{\partial y}\frac{\partial\bar{v}_x}{\partial y}\right] \tag{43.41}$$

Note that (43.40) reduces to the analogous expression used in conventional shallow-water theory, $\partial(h\bar{\tau}_{xx})/\partial x = \bar{\rho} g h\cos\theta(\partial h/\partial x)$, if $\bar{\rho}$ is constant and either $\kappa_1 = 1$ (implying hydrostatic intergranular stress), or $p_{bed} = \bar{\rho} g h\cos\theta$ (implying complete mixture liquefaction).

Equations 43.39 through 43.41 and analogous y-component equations provide mathematical closure of the depth-averaged conservation laws describing evolution of h, $\bar{v}_x$, and $\bar{v}_y$, but only if equations governing simultaneous evolution of p_{bed}, $\bar{n}$, and $\bar{\rho}$ are also specified—and only if erosion and sedimentation are negligible ($\beta \to 0$). Analyses of cases with $\beta \neq 0$ are in their earliest stages and are not presented here, but derivations of the depth-integrated evolution equations for p_{bed}, $\bar{n}$, and $\bar{\rho}$ are relatively straightforward.

43.4.4 Depth-Integrated Pore-Pressure Evolution

Depth integration of the pore-pressure evolution equation (43.17) relies on some simplifying approximations to obtain an equation that contains p_{bed} rather than p_e. Preliminary steps involve recasting (43.17) in terms of the total pore-fluid pressure, $p = p_e + \rho_f g(h - z)\cos\theta$, and invoking shallow-flow scaling that applies if $\varepsilon \ll 1$. This scaling indicates that $\partial^2 p/\partial z^2$ is much greater than $\partial^2 p/\partial x^2$ and $\partial^2 p/\partial y^2$ because $\partial^2/\partial z^2$ scales with $1/H^2$, whereas $\partial^2/\partial x^2$ and $\partial^2/\partial y^2$ scale with $1/L^2$. Consequent neglect of $\partial^2 p/\partial x^2$ and $\partial^2 p/\partial y^2$ reduces (43.17) to

$$\frac{dp}{dt} - \frac{k}{\alpha\mu}\frac{\partial^2 p}{\partial z^2} = \frac{d\sigma}{dt} - \frac{\dot{\gamma}\psi}{\alpha}. \tag{43.42}$$

Another step involves the use of the approximations $v_z = (z/h)\,dh/dt$, $v_x = \bar{v}_x$, and $v_y = \bar{v}_y$ to recast the total time derivatives in (43.42) as $d/dt = \bar{d}/\bar{d}t + (z/h)(\bar{d}h/\bar{d}t)\partial/\partial z$. Then (43.42) can be rewritten as

$$\frac{\bar{d}p}{\bar{d}t} - \frac{k}{\alpha\mu}\frac{\partial^2 p}{\partial z^2} = \frac{\bar{d}\sigma}{\bar{d}t} + \frac{z}{h}\frac{\bar{d}h}{\bar{d}t}\frac{\partial(\sigma - p)}{\partial z} - \frac{\dot{\gamma}\psi}{\alpha}. \tag{43.43}$$

Depth integration of (43.43) is accomplished term-by-term using Leibniz rule and applying the stress-free surface boundary conditions $p(\eta) = \sigma(\eta) = 0$, yielding

$$\frac{\bar{d}(\bar{p}h)}{\bar{d}t} - \frac{k}{\alpha\mu}\left[\left.\frac{\partial p}{\partial z}\right|_{z=\eta} + \rho_f g\cos\theta\right]$$
$$= \frac{\bar{d}(\bar{\sigma}h)}{\bar{d}t} + (\bar{\sigma} - \bar{p})\frac{\bar{d}h}{\bar{d}t} - \sqrt{\bar{v}_x^2 + \bar{v}_y^2}\,\frac{\psi}{2\alpha}, \qquad (43.44)$$

where overbars denote depth-averaged variables and $\sqrt{\bar{v}_x^2 + \bar{v}_y^2}\,/2h$ is used to approximate the depth-averaged shear rate, $(1/h)\int_b^\eta \dot{\gamma}dz$.

The term $\rho_f g \cos\theta$ arises in (43.44) from depth-integration of the pore-pressure diffusion term in (43.43) and application of a zero-flux basal boundary condition that requires the pore-pressure gradient at the bed to remain hydrostatic: $[\partial p/\partial z]_{z=b} = -\rho_f g \cos\theta$. The term $(\bar{\sigma} - \bar{p})\bar{d}h/\bar{d}t$ arises from depth-integrating the term that includes $\partial(\sigma - p)/\partial z$ in (43.43) by parts. This term cancels some other terms and thereby reduces (43.44) to

$$\frac{\bar{d}\bar{p}}{\bar{d}t} - \frac{k}{\alpha\mu h}\left[\left.\frac{\partial p}{\partial z}\right|_{z=\eta} + \rho_f g\cos\theta\right] = \frac{\bar{d}\bar{\sigma}}{\bar{d}t} - \sqrt{\bar{v}_x^2 + \bar{v}_y^2}\,\frac{\psi}{2h\alpha}, \qquad (43.45)$$

where

$$\bar{\sigma} = \frac{\bar{\tau}_{zz} + \bar{\tau}_{xx} + \bar{\tau}_{yy}}{3} = \frac{(1+2\kappa_1)}{6}\bar{\rho}gh\cos\theta \qquad (43.46)$$

and κ_1 is the lateral stress coefficient introduced in (43.40). Equation 43.45 retains two pore-pressure variables, p and $\bar{p}$, however, rather than the single desired variable, p_{bed}.

To express (43.45) in terms of p_{bed}, approximations of $\bar{p}$ and $[\partial p/\partial z]_{z=}\eta$ are necessary. First-order approximations assume that p varies linearly with depth, ranging from $p = p_{bed}$ at $z = b$ to $p = 0$ at $z = \eta$. This linear distribution of p implies that

$$\bar{p} = \frac{1}{2}p_{bed} \qquad \left.\frac{\partial p}{\partial z}\right|_{z=\eta} = -\frac{p_{bed}}{h}. \qquad (43.47)$$

Higher-order (in z/h) approximations allow for nonlinearity of the pore-pressure profile, particularly near the bed (Savage and Iverson 2003), but such details complicate the results without revealing effects of fundamental importance, and they are omitted here for the sake of brevity. Substitution of (43.46) and (43.47) into (43.45) then yields

$$\frac{\bar{d}p_{bed}}{\bar{d}t} = \frac{-2k}{\alpha\mu h^2}\left[p_{bed} - \rho_f gh\cos\theta\right] + \frac{(1+2\kappa_1)g\cos\theta}{3}\left[\frac{\bar{d}(\bar{\rho}h)}{\bar{d}t}\right]$$
$$- \sqrt{\bar{v}_x^2 + \bar{v}_y^2}\,\frac{\psi}{h\alpha}. \qquad (43.48)$$

The derivative $\bar{d}(\bar{\rho}h)/\bar{d}t$ can be eliminated from the right-hand side of (43.48) by using the mass-conservation equation (43.23) to find that $\bar{d}(\bar{\rho}h)/\bar{d}t = -\bar{\rho}[h(\partial\bar{v}_x/\partial x + \partial\bar{v}_y/\partial y) - A + B]$. Making this substitution in (43.48) yields the final form of the evolution equation for p_{bed}:

$$\frac{\bar{d}p_{bed}}{\bar{d}t} = \frac{-2k}{\alpha\mu h^2}\left[p_{bed} - \rho_f gh\cos\theta\right]$$
$$- \bar{\rho}gh\cos\theta\frac{1+2\kappa_1}{3}\left[\frac{\partial\bar{v}_x}{\partial x} + \frac{\partial\bar{v}_y}{\partial y} - \frac{A-B}{h}\right] - \sqrt{\bar{v}_x^2 + \bar{v}_y^2}\,\frac{\psi}{h\alpha}. \qquad (43.49)$$

43.4.5 Depth-Integrated Porosity and Bulk Density Evolution

The derivation of the depth-integrated evolution equation for n utilizes the scaling inference $\partial^2 p/\partial z^2 \gg \partial^2 p/\partial x^2$ and $\partial^2 p/\partial z^2 \gg \partial^2 p/\partial y^2$ described above, and also utilizes the assumption $\rho = \bar{\rho}$, which implies that $n = \bar{n}$. Under these conditions, depth integration of Equation 43.15 yields

$$\frac{h}{1-\bar{n}}\frac{\bar{d}\bar{n}}{\bar{d}t} = \frac{k}{\mu}\left.\frac{\partial p_e}{\partial z}\right|_{z=b}^{z=\eta}. \qquad (43.50)$$

Like the preceding equations involving $\bar{d}/\bar{d}t$, this equation assumes that $v_x = \bar{v}_x$ and $v_y = \bar{v}_y$. The linear pore-pressure distribution defined in conjunction with (43.47) indicates that $\partial p_e/\partial z = -(p_{bed} - \rho_f gh\cos\theta)/h$, and use of this expression in (43.50) yields the evolution equation for $\bar{n}$:

$$\frac{\bar{d}\bar{n}}{\bar{d}t} = (1-\bar{n})\frac{-k}{h^2\mu}(p_{bed} - \rho_f gh\cos\theta). \qquad (43.51)$$

This result also provides an evolution equation for $\bar{\rho}$ if the debris remains saturated with pore fluid, because (43.3) implies that

$$\bar{\rho} = \rho_s(1-\bar{n}) + \rho_f\bar{n}. \qquad (43.52)$$

The similarity of terms on the right-hand side of (43.51) and some of those on the right-hand side of (43.49) is significant. Indeed, the unforced version of (43.49) can be substituted into (43.51) to reduce it to

$$\frac{1}{1-\bar{n}}\frac{\bar{d}\bar{n}}{\bar{d}t} = \frac{\alpha}{2}\frac{\bar{d}p_{bed}}{\bar{d}t} \qquad (43.53)$$

This equation implies that, in the absence of external forcing, $\bar{n}$ declines logarithmically as p_{bed} declines linearly. Such nonlinear behavior is typical of quasi-static, water-saturated granular debris as it consolidates during dissipation of pore-fluid pressure (Iverson 1997).

43.4.6 Model Summary

The system of depth-integrated equations governing simultaneous evolution of h, $h\bar{v}_x$, p_{bed}, $\bar{n}$, and $\bar{\rho}$ is (43.23), (43.32), (43.49), (43.51), and (43.52). In addition, (43.39) through (43.41) specify the stress terms that appear in (43.32). A system of equations exactly analogous to (43.32) and (43.39) through (43.41), but with x and y transposed, describes evolution of $h\bar{v}_y$. All stress calculations are predicated on (43.38), which approximates the z-momentum equation and constitutes the central postulate of the depth-integrated model.

43.5 Solution Techniques and Model Predictions

Numerical solution of the full set of model equations described above is the object of ongoing research. Two broad classes of techniques have proven useful for solving similar systems of conservation equations, such as the shallow-water equations and Savage-Hutter granular avalanche equations. One type of technique employs Lagrangian numerical methods in which the computational mesh translates with the local flow velocity. This approach has the advantage of replacing partial derivatives of the nonlinear terms on the left-hand side of (43.28) with total time derivatives such as that on the left-hand side of (43.29). Furthermore, these methods are relatively easy to implement if the flow path is simple (such as a uniformly inclined plane). Classical Lagrangian techniques have limited potential for computing motion across complex, 3D terrain, however, because deformation of the computational mesh can become exceedingly complicated. Meshless Lagrangian techniques such as those used in smooth-particle hydrodynamics may have promise, but their structure makes it difficult to determine if conservation equations are rigorously satisfied (McDougall and Hungr 2004).

A more rigorous approach utilizes a fixed, Eulerian computational mesh, but also requires the use of shock-capturing numerical methods to accurately account for the potentially severe effects of nonlinearities, which can give rise to discontinuous solutions. (Conventional finite-difference and finite-element methods tend to smear out these effects, leading to inaccurate solutions.) Finite-volume methods enable shock capturing by reframing the numerical problem as a series of elementary Riemann problems that describe fluxes of conserved variables between adjacent computational cells (e.g., Denlinger and Iverson 2001, 2004; Pitman and Le 2005). Such methods also lend themselves to adaptive mesh refinement (AMR), a sophisticated technique that can greatly accelerate computation speeds by automatically implementing mesh refinement only where high resolution is needed (George and LeVeque 2008).

Despite significant advances, application of computational models to forecasting behavior of debris flows and rock avalanches remains in its earliest stages, largely because most model predictions have not been subject to rigorous, controlled tests. Instead, models have generally been calibrated to fit field observations (mostly by tuning resistive stress terms), and such models cannot be regarded as truly predictive. A basis for more stringent model testing is provided by recently acquired experimental data such as those summarized in Figure 43.3. The availability of such data, along with increasingly sophisticated numerical methods, makes prospects for better understanding and modeling of debris flows appear promising.

References

Bear, J. 1972. *Dynamics of Fluids in Porous Media*. New York: Dover.

Costa, J.E. and G.P. Williams. 1984. Debris-flow dynamics (videotape). U.S. Geological Survey Open-file Report 84-606. Videotape.

Denlinger, R.P. and R.M. Iverson. 2001. Flow of variably fluidized granular masses across three-dimensional terrain: 2. Numerical predictions and experimental tests. *Journal Geophysical Research* 106: 553–566.

Denlinger, R.P. and R.M. Iverson. 2004. Granular avalanches across irregular three-dimensional terrain: 1. Theory and computation. *Journal Geophysical Research* 109: F01014, doi:10.1029/2003JF000085.

Desai, C.S. and H.J. Siriwardane. 1984. *Constitutive Laws for Engineering Materials, with Emphasis on Geological Materials*. Englewood Cliffs, NJ: Prentice-Hall.

Forterre, Y. and O. Pouliquen. 2008. Flows of dense granular media. *Annual Review of Fluid Mechanics* 40: 1–24.

George, D.L. and R.J. LeVeque. 2008. High-resolution methods and adaptive refinement for tsunami propagation and inundation. In *Hyperbolic Problems: Theory, Numerics, Application*. eds. S. Benzoni-Gavage and D. Serre, pp. 541–549. Berlin, Germany: Springer.

Gidaspow, D. 1994. *Multiphase Flow and Fluidization*. Boston, MA: Academic Press.

Gray, J.M.N.T. and B.P. Kokelaar. 2010. Large particle segregation, transport and accumulation in granular free-surface flows. *Journal Fluid Mechanics* 652: 105–137.

Hunt, M.L., R. Zenit, C.S. Campbell, and C.E. Brennen. 2002. Revisiting the 1954 suspension experiments of R.A. Bagnold. *Journal Fluid Mechanics* 452: 1–24.

Iverson, R.M. 1997. The physics of debris flows. *Reviews of Geophysics* 35: 245–296.

Iverson, R.M. 2009. Elements of an improved model of debris-flow motion. In *Powders and Grains 2009*, ed. M. Nakagawa and S. Luding, pp. 9–16. Melville, NY: American Institute of Physics.

Iverson, R.M. and R.P. Denlinger. 2001. Flow of variably fluidized granular masses across three-dimensional terrain: 1. Coulomb mixture theory. *Journal Geophysical Research* 106: 537–552.

Iverson, R.M., M. Logan, R.G. LaHusen, and M. Berti. 2010. The perfect debris flow: Aggregated results from 28 large-scale experiments. *Journal Geophysical Research* 115: doi:10.1029/2009JF001514.

Kowalski, J. 2008. Two-phase modeling of debris flows. Dr. Sci. dissertation, Zurich, Switzerland: ETH Zurich.

Logan, M. and R.M. Iverson. 2007. Video documentation of experiments at the USGS debris-flow flume, 1992–2006. U.S. Geological Survey Open-file Report 2007-1315. http://pubs.usgs.gov/of/2007/1315/

McDougall, S. and O. Hungr. 2004. A model for the analysis of rapid landslide motion across three-dimensional terrain. *Canadian Geotechnical Journal* 41: 1084–1097.

Pailha, M. and O. Pouliquen. 2009. A two-phase flow description of the initiation of underwater granular avalanches. *Journal of Fluid Mechanics* 633: 115–135.

Pitman, E.B. and L. Le. 2005. A two-fluid model for avalanche and debris flow. *Philosophical Transactions Royal Society London* 363: 1573–1601.

Savage, S.B. and R.M. Iverson. 2003. Surge dynamics coupled to pore-pressure evolution in debris flows. In *Debris-Flow Hazards Mitigation: Mechanics, Prediction, and Assessment*, ed. D. Rickenmann and C.-L. Chen, Vol. 1, pp. 503–514. Rotterdam, the Netherlands: Millpress.

Sharp, R.P. and L.H. Nobles. 1953. Mudflow of 1941 at Wrightwood, southern California. *Geological Society America Bulletin* 64: 547–560.

Sidle, R.C. and M. Chigira. 2004. Landslides and debris flows strike Kyushu, Japan. *EOS* 85: 145–156.

Takahashi, T. 2007. *Debris Flow Mechanics, Prediction and Countermeasures*. London, U.K.: Taylor & Francis.

Index

A

B

C

D

E

M

N

R

S